# Electricity and Magnetism

For 50 years, Edward M. Purcell's classic textbook has introduced students to the world of electricity and magnetism. This third edition has been brought up to date and is now in SI units. It features hundreds of new examples, problems, and figures, and contains discussions of real-life applications.

The textbook covers all the standard introductory topics, such as electrostatics, magnetism, circuits, electromagnetic waves, and electric and magnetic fields in matter. Taking a nontraditional approach, magnetism is derived as a relativistic effect. Mathematical concepts are introduced in parallel with the physical topics at hand, making the motivations clear. Macroscopic phenomena are derived rigorously from the underlying microscopic physics.

With worked examples, hundreds of illustrations, and nearly 600 end-of-chapter problems and exercises, this textbook is ideal for electricity and magnetism courses. Solutions to the exercises are available for instructors at www.cambridge.org/Purcell-Morin.

EDWARD M. PURCELL (1912–1997) was the recipient of many awards for his scientific, educational, and civic work. In 1952 he shared the Nobel Prize for Physics for the discovery of nuclear magnetic resonance in liquids and solids, an elegant and precise method of determining the chemical structure of materials that serves as the basis for numerous applications, including magnetic resonance imaging (MRI). During his career he served as science adviser to Presidents Dwight D. Eisenhower, John F. Kennedy, and Lyndon B. Johnson.

DAVID J. MORIN is a Lecturer and the Associate Director of Undergraduate Studies in the Department of Physics, Harvard University. He is the author of the textbook *Introduction to Classical Mechanics* (Cambridge University Press, 2008).

**THIRD EDITION**

# ELECTRICITY AND MAGNETISM

**EDWARD M. PURCELL**

**DAVID J. MORIN**

Harvard University, Massachusetts

CAMBRIDGE
UNIVERSITY PRESS

**CAMBRIDGE**
UNIVERSITY PRESS

Shaftesbury Road, Cambridge CB2 8EA, United Kingdom

One Liberty Plaza, 20th Floor, New York, NY 10006, USA

477 Williamstown Road, Port Melbourne, VIC 3207, Australia

314–321, 3rd Floor, Plot 3, Splendor Forum, Jasola District Centre, New Delhi – 110025, India

103 Penang Road, #05–06/07, Visioncrest Commercial, Singapore 238467

Cambridge University Press is part of Cambridge University Press & Assessment,
a department of the University of Cambridge.

We share the University's mission to contribute to society through the pursuit of
education, learning and research at the highest international levels of excellence.

www.cambridge.org
Information on this title: www.cambridge.org/Purcell-Morin

Previously published by Mc-Graw Hill, Inc., 1985
First edition published by Education Development Center, Inc., 1963, 1964, 1965
First published by Cambridge University Press & Assessment 2013 (version 14, April 2024)

Printed in the United Kingdom by TJ Books Limited, Padstow Cornwall, April 2024

*A catalogue record for this publication is available from the British Library*

*Library of Congress Cataloging-in-Publication data*
Purcell, Edward M.
Electricity and magnetism / Edward M. Purcell, David J. Morin, Harvard University,
Massachusetts. – Third edition.
pages cm
ISBN 978-1-107-01402-2 (Hardback)
1. Electricity.   2. Magnetism.   I. Title.
QC522.P85  2012
537–dc23

2012034622

ISBN  978-1-107-01402-2  Hardback

Additional resources for this publication at www.cambridge.org/Purcell-Morin

Cambridge University Press & Assessment has no responsibility for the persistence
or accuracy of URLs for external or third-party internet websites referred to in this
publication and does not guarantee that any content on such websites is, or will
remain, accurate or appropriate.

# CONTENTS

# Preface to the third edition of Volume 2

For 50 years, physics students have enjoyed learning about electricity and magnetism through the first two editions of this book. The purpose of the present edition is to bring certain things up to date and to add new material, in the hopes that the trend will continue. The main changes from the second edition are (1) the conversion from Gaussian units to SI units, and (2) the addition of many solved problems and examples.

The first of these changes is due to the fact that the vast majority of courses on electricity and magnetism are now taught in SI units. The second edition fell out of print at one point, and it was hard to watch such a wonderful book fade away because it wasn't compatible with the way the subject is presently taught. Of course, there are differing opinions as to which system of units is "better" for an introductory course. But this issue is moot, given the reality of these courses.

For students interested in working with Gaussian units, or for instructors who want their students to gain exposure to both systems, I have created a number of appendices that should be helpful. Appendix A discusses the differences between the SI and Gaussian systems. Appendix C derives the conversion factors between the corresponding units in the two systems. Appendix D explains how to convert formulas from SI to Gaussian; it then lists, side by side, the SI and Gaussian expressions for every important result in the book. A little time spent looking at this appendix will make it clear how to convert formulas from one system to the other.

The second main change in the book is the addition of many solved problems, and also many new examples in the text. Each chapter ends with "problems" and "exercises." The solutions to the "problems" are located in Chapter 12. The only official difference between the problems and exercises is that the problems have solutions included, whereas the exercises do not. (A separate solutions manual for the exercises is available to instructors.) In practice, however, one difference is that some of the more theorem-ish results are presented in the problems, so that students can use these results in other problems/exercises.

Some advice on using the solutions to the problems: problems (and exercises) are given a (very subjective) difficulty rating from 1 star to 4 stars. If you are having trouble solving a problem, it is critical that you don't look at the solution too soon. Brood over it for a while. If you do finally look at the solution, don't just read it through. Instead, cover it up with a piece of paper and read one line at a time until you reach a hint to get you started. Then set the book aside and work things out for real. That's the only way it will sink in. It's quite astonishing how unhelpful it is simply to read a solution. You'd *think* it would do some good, but in fact it is completely ineffective in raising your understanding to the next level. Of course, a careful reading of the text, including perhaps a few problem solutions, is necessary to get the basics down. But if Level 1 is understanding the basic concepts, and Level 2 is being able to *apply* those concepts, then you can read and read until the cows come home, and you'll never get past Level 1.

The overall structure of the text is essentially the same as in the second edition, although a few new sections have been added. Section 2.7 introduces dipoles. The more formal treatment of dipoles, along with their applications, remains in place in Chapter 10. But because the fundamentals of dipoles can be understood using only the concepts developed in Chapters 1 and 2, it seems appropriate to cover this subject earlier in the book. Section 8.3 introduces the important technique of solving differential equations by forming complex solutions and then taking the real part. Section 9.6.2 deals with the Poynting vector, which opens up the door to some very cool problems.

Each chapter concludes with a list of "everyday" applications of electricity and magnetism. The discussions are brief. The main purpose of these sections is to present a list of fun topics that deserve further investigation. You can carry onward with some combination of books/internet/people/pondering. There is effectively an infinite amount of information out there (see the references at the beginning of Section 1.16 for some starting points), so my goal in these sections is simply to provide a springboard for further study.

The intertwined nature of electricity, magnetism, and relativity is discussed in detail in Chapter 5. Many students find this material highly illuminating, although some find it a bit difficult. (However, these two groups are by no means mutually exclusive!) For instructors who wish to take a less theoretical route, it is possible to skip directly from Chapter 4 to Chapter 6, with only a brief mention of the main result from Chapter 5, namely the magnetic field due to a straight current-carrying wire.

The use of non-Cartesian coordinates (cylindrical, spherical) is more prominent in the present edition. For setups possessing certain symmetries, a wisely chosen system of coordinates can greatly simplify the calculations. Appendix F gives a review of the various vector operators in the different systems.

Compared with the second edition, the level of difficulty of the present edition is slightly higher, due to a number of hefty problems that have been added. If you are looking for an extra challenge, these problems should keep you on your toes. However, if these are ignored (which they certainly can be, in any standard course using this book), then the level of difficulty is roughly the same.

I am grateful to all the students who used a draft version of this book and provided feedback. Their input has been invaluable. I would also like to thank Jacob Barandes for many illuminating discussions of the more subtle topics in the book. Paul Horowitz helped get the project off the ground and has been an endless supplier of cool facts. It was a pleasure brainstorming with Andrew Milewski, who offered many ideas for clever new problems. Howard Georgi and Wolfgang Rueckner provided much-appreciated sounding boards and sanity checks. Takuya Kitagawa carefully read through a draft version and offered many helpful suggestions. I thank Ali Woollatt and Irene Pizzie for their professional work in producing the layout of the book and copy editing the manuscript. Other friends and colleagues whose input I am grateful for are: Lindsay Barnes, Simon Capelin, Allen Crockett, David Derbes, John Doyle, Gary Feldman, Melissa Franklin, Jerome Fung, Jene Golovchenko, Doug Goodale, Robert Hart, Tom Hayes, Peter Hedman, Jennifer Hoffman, Charlie Holbrow, Gareth Kafka, Alan Levine, Aneesh Manohar, Kirk McDonald, Masahiro Morii, Lev Okun, Joon Pahk, Dave Patterson, Mara Prentiss, Dennis Purcell, Frank Purcell, Daniel Rosenberg, Emily Russell, Roy Schwitters, Nils Sorensen, Charlotte Thomas, Josh Winn, and Amir Yacoby.

Despite careful editing, there is zero probability that this book is error free. A great deal of new material has been added, and errors have undoubtedly crept in. If anything looks amiss, please check the webpage www.cambridge.org/Purcell-Morin for a list of typos, updates, etc. And please let me know if you discover something that isn't already posted. Suggestions are always welcome.

**David Morin**

# Preface to the second edition of Volume 2

This revision of "Electricity and Magnetism," Volume 2 of the Berkeley Physics Course, has been made with three broad aims in mind. First, I have tried to make the text clearer at many points. In years of use teachers and students have found innumerable places where a simplification or reorganization of an explanation could make it easier to follow. Doubtless some opportunities for such improvements have still been missed; not too many, I hope.

A second aim was to make the book practically independent of its companion volumes in the Berkeley Physics Course. As originally conceived it was bracketed between Volume I, which provided the needed special relativity, and Volume 3, "Waves and Oscillations," to which was allocated the topic of electromagnetic waves. As it has turned out, Volume 2 has been rather widely used alone. In recognition of that I have made certain changes and additions. A concise review of the relations of special relativity is included as Appendix A. Some previous introduction to relativity is still assumed. The review provides a handy reference and summary for the ideas and formulas we need to understand the fields of moving charges and their transformation from one frame to another. The development of Maxwell's equations for the vacuum has been transferred from the heavily loaded Chapter 7 (on induction) to a new Chapter 9, where it leads naturally into an elementary treatment of plane electromagnetic waves, both running and standing. The propagation of a wave in a dielectric medium can then be treated in Chapter 10 on Electric Fields in Matter.

A third need, to modernize the treatment of certain topics, was most urgent in the chapter on electrical conduction. A substantially rewritten

Chapter 4 now includes a section on the physics of homogeneous semiconductors, including doped semiconductors. Devices are not included, not even a rectifying junction, but what is said about bands, and donors and acceptors, could serve as starting point for development of such topics by the instructor. Thanks to solid-state electronics the physics of the voltaic cell has become even more relevant to daily life as the number of batteries in use approaches in order of magnitude the world's population. In the first edition of this book I unwisely chose as the example of an electrolytic cell the one cell—the Weston standard cell—which advances in physics were soon to render utterly obsolete. That section has been replaced by an analysis, with new diagrams, of the lead-acid storage battery—ancient, ubiquitous, and far from obsolete.

One would hardly have expected that, in the revision of an elementary text in classical electromagnetism, attention would have to be paid to new developments in particle physics. But that is the case for two questions that were discussed in the first edition, the significance of charge quantization, and the apparent absence of magnetic monopoles. Observation of proton decay would profoundly affect our view of the first question. Assiduous searches for that, and also for magnetic monopoles, have at this writing yielded no confirmed events, but the possibility of such fundamental discoveries remains open.

Three special topics, optional extensions of the text, are introduced in short appendixes: Appendix B: Radiation by an Accelerated Charge; Appendix C: Superconductivity; and Appendix D: Magnetic Resonance.

Our primary system of units remains the Gaussian CGS system. The SI units, ampere, coulomb, volt, ohm, and tesla are also introduced in the text and used in many of the problems. Major formulas are repeated in their SI formulation with explicit directions about units and conversion factors. The charts inside the back cover summarize the basic relations in both systems of units. A special chart in Chapter 11 reviews, in both systems, the relations involving magnetic polarization. The student is not expected, or encouraged, to memorize conversion factors, though some may become more or less familiar through use, but to look them up whenever needed. There is no objection to a "mixed" unit like the ohm-cm, still often used for resistivity, providing its meaning is perfectly clear.

The definition of the meter in terms of an assigned value for the speed of light, which has just become official, simplifies the exact relations among the units, as briefly explained in Appendix E.

There are some 300 problems, more than half of them new.

It is not possible to thank individually all the teachers and students who have made good suggestions for changes and corrections. I fear that some will be disappointed to find that their suggestions have not been followed quite as they intended. That the net result is a substantial improvement I hope most readers familiar with the first edition will agree.

Mistakes both old and new will surely be found. Communications pointing them out will be gratefully received.

It is a pleasure to thank Olive S. Rand for her patient and skillful assistance in the production of the manuscript.

**Edward M. Purcell**

Preface to the first edition of Volume 2

# Preface to the first edition of Volume 2

The subject of this volume of the Berkeley Physics Course is electricity and magnetism. The sequence of topics, in rough outline, is not unusual: electrostatics; steady currents; magnetic field; electromagnetic induction; electric and magnetic polarization in matter. However, our approach is different from the traditional one. The difference is most conspicuous in Chaps. 5 and 6 where, building on the work of Vol. I, we treat the electric and magnetic fields of moving charges as manifestations of relativity and the invariance of electric charge. This approach focuses attention on some fundamental questions, such as: charge conservation, charge invariance, the meaning of field. The only formal apparatus of special relativity that is really necessary is the Lorentz transformation of coordinates and the velocity-addition formula. It is essential, though, that the student bring to this part of the course some of the ideas and attitudes Vol. I sought to develop—among them a readiness to look at things from different frames of reference, an appreciation of invariance, and a respect for symmetry arguments. We make much use also, in Vol. II, of arguments based on superposition.

Our approach to electric and magnetic phenomena in matter is primarily "microscopic," with emphasis on the nature of atomic and molecular dipoles, both electric and magnetic. Electric conduction, also, is described microscopically in the terms of a Drude-Lorentz model. Naturally some questions have to be left open until the student takes up quantum physics in Vol. IV. But we freely talk in a matter-of-fact way about molecules and atoms as electrical structures with size, shape, and stiffness, about electron orbits, and spin. We try to treat carefully a question that is sometimes avoided and sometimes beclouded in introductory texts, the meaning of the macroscopic fields $\mathbf{E}$ and $\mathbf{B}$ inside a material.

In Vol. II, the student's mathematical equipment is extended by adding some tools of the vector calculus—gradient, divergence, curl, and the Laplacian. These concepts are developed as needed in the early chapters.

In its preliminary versions, Vol. II has been used in several classes at the University of California. It has benefited from criticism by many people connected with the Berkeley Course, especially from contributions by E. D. Commins and F. S. Crawford, Jr., who taught the first classes to use the text. They and their students discovered numerous places where clarification, or something more drastic, was needed; many of the revisions were based on their suggestions. Students' criticisms of the last preliminary version were collected by Robert Goren, who also helped to organize the problems. Valuable criticism has come also from J. D. Gavenda, who used the preliminary version at the University of Texas, and from E. F. Taylor, of Wesleyan University. Ideas were contributed by Allan Kaufman at an early stage of the writing. A. Felzer worked through most of the first draft as our first "test student."

The development of this approach to electricity and magnetism was encouraged, not only by our original Course Committee, but by colleagues active in a rather parallel development of new course material at the Massachusetts Institute of Technology. Among the latter, J. R. Tessman, of the MIT Science Teaching Center and Tufts University, was especially helpful and influential in the early formulation of the strategy. He has used the preliminary version in class, at MIT, and his critical reading of the entire text has resulted in many further changes and corrections.

Publication of the preliminary version, with its successive revisions, was supervised by Mrs. Mary R. Maloney. Mrs. Lila Lowell typed most of the manuscript. The illustrations were put into final form by Felix Cooper.

The author of this volume remains deeply grateful to his friends in Berkeley, and most of all to Charles Kittel, for the stimulation and constant encouragement that have made the long task enjoyable.

**Edward M. Purcell**

**1**

**Overview** The existence of this book is owed (both figuratively and literally) to the fact that the building blocks of matter possess a quality called *charge*. Two important aspects of charge are *conservation* and *quantization*. The electric force between two charges is given by *Coulomb's law.* Like the gravitational force, the electric force falls off like $1/r^2$. It is *conservative,* so we can talk about the potential energy of a system of charges (the work done in assembling them). A very useful concept is the *electric field,* which is defined as the force per unit charge. Every point in space has a unique electric field associated with it. We can define the *flux* of the electric field through a given surface. This leads us to *Gauss's law,* which is an alternative way of stating Coulomb's law. In cases involving sufficient symmetry, it is much quicker to calculate the electric field via Gauss's law than via Coulomb's law and direct integration. Finally, we discuss the *energy density* in the electric field, which provides another way of calculating the potential energy of a system.

# Electrostatics: charges and fields

## 1.1 Electric charge

Electricity appeared to its early investigators as an extraordinary phenomenon. To draw from bodies the "subtle fire," as it was sometimes called, to bring an object into a highly electrified state, to produce a steady flow of current, called for skillful contrivance. Except for the spectacle of lightning, the ordinary manifestations of nature, from the freezing of water to the growth of a tree, seemed to have no relation to the curious behavior of electrified objects. We know now that electrical forces largely determine the physical and chemical properties of matter over the whole range from atom to living cell. For this understanding we have to thank the scientists of the nineteenth century, Ampère, Faraday, Maxwell, and many others, who discovered the nature of electromagnetism, as well as the physicists and chemists of the twentieth century who unraveled the atomic structure of matter.

Classical electromagnetism deals with electric charges and currents and their interactions as if all the quantities involved could be measured independently, with unlimited precision. Here *classical* means simply "nonquantum." The quantum law with its constant $h$ is ignored in the classical theory of electromagnetism, just as it is in ordinary mechanics. Indeed, the classical theory was brought very nearly to its present state of completion before Planck's discovery of quantum effects in 1900. It has survived remarkably well. Neither the revolution of quantum physics nor the development of special relativity dimmed the luster of the electromagnetic field equations Maxwell wrote down 150 years ago.

Of course the theory was solidly based on experiment, and because of that was fairly secure within its original range of application – to coils, capacitors, oscillating currents, and eventually radio waves and light waves. But even so great a success does not guarantee validity in another domain, for instance, the inside of a molecule.

Two facts help to explain the continuing importance in modern physics of the classical description of electromagnetism. First, special relativity required no revision of classical electromagnetism. Historically speaking, special relativity *grew out* of classical electromagnetic theory and experiments inspired by it. Maxwell's field equations, developed long before the work of Lorentz and Einstein, proved to be entirely compatible with relativity. Second, quantum modifications of the electromagnetic forces have turned out to be unimportant down to distances less than $10^{-12}$ meters, 100 times smaller than the atom. We can describe the repulsion and attraction of particles in the atom using the same laws that apply to the leaves of an electroscope, although we need quantum mechanics to predict how the particles will behave under those forces. For still smaller distances, a fusion of electromagnetic theory and quantum theory, called *quantum electrodynamics,* has been remarkably successful. Its predictions are confirmed by experiment down to the smallest distances yet explored.

It is assumed that the reader has some acquaintance with the elementary facts of electricity. We are not going to review all the experiments by which the existence of electric charge was demonstrated, nor shall we review all the evidence for the electrical constitution of matter. On the other hand, we do want to look carefully at the experimental foundations of the basic laws on which all else depends. In this chapter we shall study the physics of stationary electric charges – *electrostatics*.

Certainly one fundamental property of electric charge is its existence in the two varieties that were long ago named *positive* and *negative*.

The observed fact is that all charged particles can be divided into two classes such that all members of one class repel each other, while attracting members of the other class. If two small electrically charged bodies *A* and *B*, some distance apart, attract one another, and if *A* attracts some third electrified body *C*, then we always find that *B* repels *C*. Contrast this with gravitation: there is only one kind of gravitational mass, and every mass attracts every other mass.

One may regard the two kinds of charge, positive and negative, as opposite manifestations of one quality, much as *right* and *left* are the two kinds of handedness. Indeed, in the physics of elementary particles, questions involving the sign of the charge are sometimes linked to a question of handedness, and to another basic symmetry, the relation of a sequence of events, *a*, then *b*, then *c*, to the temporally reversed sequence *c*, then *b*, then *a*. It is only the duality of electric charge that concerns us here. For every kind of particle in nature, as far as we know, there can exist an *antiparticle,* a sort of electrical "mirror image." The antiparticle carries charge of the opposite sign. If any other intrinsic quality of the particle has an opposite, the antiparticle has that too, whereas in a property that admits no opposite, such as mass, the antiparticle and particle are exactly alike.

The electron's charge is negative; its antiparticle, called a *positron,* has a positive charge, but its mass is precisely the same as that of the electron. The proton's antiparticle is called simply an *antiproton*; its electric charge is negative. An electron and a proton combine to make an ordinary hydrogen atom. A positron and an antiproton could combine in the same way to make an atom of antihydrogen. Given the building blocks, positrons, antiprotons, and antineutrons,[1] there could be built up the whole range of antimatter, from antihydrogen to antigalaxies. There is a practical difficulty, of course. Should a positron meet an electron or an antiproton meet a proton, that pair of particles will quickly vanish in a burst of radiation. It is therefore not surprising that even positrons and antiprotons, not to speak of antiatoms, are exceedingly rare and short-lived in our world. Perhaps the universe contains, somewhere, a vast concentration of antimatter. If so, its whereabouts is a cosmological mystery.

The universe around us consists overwhelmingly of matter, not antimatter. That is to say, the abundant carriers of negative charge are electrons, and the abundant carriers of positive charge are protons. The proton is nearly 2000 times heavier than the electron, and very different, too, in some other respects. Thus matter at the atomic level incorporates negative and positive electricity in quite different ways. The positive charge is all in the atomic nucleus, bound within a massive structure no more than $10^{-14}$ m in size, while the negative charge is spread, in

---

[1] Although the electric charge of each is zero, the neutron and its antiparticle are not interchangeable. In certain properties that do not concern us here, they are opposite.

effect, through a region about $10^4$ times larger in dimensions. It is hard to imagine what atoms and molecules – and all of chemistry – would be like, if not for this fundamental electrical asymmetry of matter.

What we call negative charge, by the way, could just as well have been called positive. The name was a historical accident. There is nothing essentially negative about the charge of an electron. It is not like a negative integer. A negative integer, once multiplication has been defined, differs essentially from a positive integer in that its square is an integer of opposite sign. But the product of two charges is not a charge; there is no comparison.

Two other properties of electric charge are essential in the electrical structure of matter: charge is *conserved,* and charge is *quantized.* These properties involve *quantity* of charge and thus imply a measurement of charge. Presently we shall state precisely how charge can be measured in terms of the force between charges a certain distance apart, and so on. But let us take this for granted for the time being, so that we may talk freely about these fundamental facts.

## 1.2 Conservation of charge

The total charge in an isolated system never changes. By *isolated* we mean that no matter is allowed to cross the boundary of the system. We could let light pass into or out of the system, since the "particles" of light, called *photons,* carry no charge at all. Within the system charged particles may vanish or reappear, but they always do so in pairs of equal and opposite charge. For instance, a thin-walled box in a vacuum exposed to gamma rays might become the scene of a "pair-creation" event in which a high-energy photon ends its existence with the creation of an electron and a positron (Fig. 1.1). Two electrically charged particles have been newly created, but the net change in total charge, in and on the box, is zero. An event that *would* violate the law we have just stated would be the creation of a positively charged particle *without* the simultaneous creation of a negatively charged particle. Such an occurrence has never been observed.

Of course, if the electric charges of an electron and a positron were not precisely equal in magnitude, pair creation would still violate the strict law of charge conservation. That equality is a manifestation of the particle–antiparticle duality already mentioned, a universal symmetry of nature.

One thing will become clear in the course of our study of electromagnetism: nonconservation of charge would be quite incompatible with the structure of our present electromagnetic theory. We may therefore state, either as a postulate of the theory or as an empirical law supported without exception by all observations so far, the charge conservation law:

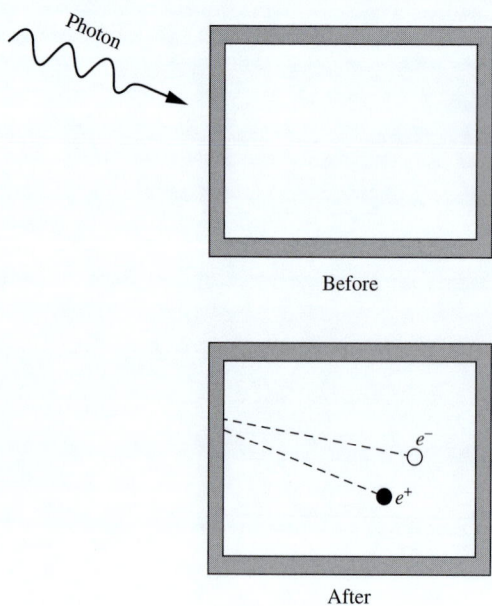

**Figure 1.1.**
Charged particles are created in pairs with equal and opposite charge.

The total electric charge in an isolated system, that is, the algebraic sum of the positive and negative charge present at any time, never changes.

Sooner or later we must ask whether this law meets the test of relativistic invariance. We shall postpone until Chapter 5 a thorough discussion of this important question. But the answer is that it does, and not merely in the sense that the statement above holds in any given inertial frame, but in the stronger sense that observers in different frames, measuring the charge, obtain the same number. In other words, the total electric charge of an isolated system is a relativistically invariant number.

## 1.3 Quantization of charge

The electric charges we find in nature come in units of one magnitude only, equal to the amount of charge carried by a single electron. We denote the magnitude of that charge by $e$. (When we are paying attention to sign, we write $-e$ for the charge on the electron itself.) We have already noted that the positron carries precisely that amount of charge, as it must if charge is to be conserved when an electron and a positron annihilate, leaving nothing but light. What seems more remarkable is the apparently exact equality of the charges carried by all other charged particles – the equality, for instance, of the positive charge on the proton and the negative charge on the electron.

That particular equality is easy to test experimentally. We can see whether the net electric charge carried by a hydrogen molecule, which consists of two protons and two electrons, is zero. In an experiment carried out by J. G. King,[2] hydrogen gas was compressed into a tank that was electrically insulated from its surroundings. The tank contained about $5 \cdot 10^{24}$ molecules (approximately 17 grams) of hydrogen. The gas was then allowed to escape by means that prevented the escape of any ion – a molecule with an electron missing or an extra electron attached. If the charge on the proton differed from that on the electron by, say, one part in a billion, then each hydrogen molecule would carry a charge of $2 \cdot 10^{-9}e$, and the departure of the whole mass of hydrogen would alter the charge of the tank by $10^{16}e$, a gigantic effect. In fact, the experiment could have revealed a residual molecular charge as small as $2 \cdot 10^{-20}e$, and none was observed. This proved that the proton and the electron do not differ in magnitude of charge by more than 1 part in $10^{20}$.

Perhaps the equality is really *exact* for some reason we don't yet understand. It may be connected with the possibility, suggested by certain

---

[2] See King (1960). References to previous tests of charge equality will be found in this article and in the chapter by V. W. Hughes in Hughes (1964).

theories, that a proton can, *very* rarely, decay into a positron and some uncharged particles. If that were to occur, even the slightest discrepancy between proton charge and positron charge would violate charge conservation. Several experiments designed to detect the decay of a proton have not yet, as of this writing, registered with certainty a single decay. If and when such an event is observed, it will show that exact equality of the magnitude of the charge of the proton and the charge of the electron (the positron's antiparticle) can be regarded as a corollary of the more general law of charge conservation.

That notwithstanding, we now know that the *internal* structure of all the strongly interacting particles called *hadrons* – a class that includes the proton and the neutron – involves basic units called *quarks,* whose electric charges come in multiples of $e/3$. The proton, for example, is made with three quarks, two with charge $2e/3$ and one with charge $-e/3$. The neutron contains one quark with charge $2e/3$ and two quarks with charge $-e/3$.

Several experimenters have searched for single quarks, either free or attached to ordinary matter. The fractional charge of such a quark, since it cannot be neutralized by any number of electrons or protons, should betray the quark's presence. So far no fractionally charged particle has been conclusively identified. The present theory of the strong interactions, called *quantum chromodynamics,* explains why the liberation of a quark from a hadron is most likely impossible.

The fact of charge quantization lies outside the scope of classical electromagnetism, of course. We shall usually ignore it and act as if our point charges $q$ could have any strength whatsoever. This will not get us into trouble. Still, it is worth remembering that classical theory cannot be expected to explain the structure of the elementary particles. (It is not certain that present quantum theory can either!) What holds the electron together is as mysterious as what fixes the precise value of its charge. Something more than electrical forces must be involved, for the electrostatic forces between different parts of the electron would be repulsive.

In our study of electricity and magnetism we shall treat the charged particles simply as carriers of charge, with dimensions so small that their extension and structure is, for most purposes, quite insignificant. In the case of the proton, for example, we know from high-energy scattering experiments that the electric charge does not extend appreciably beyond a radius of $10^{-15}$ m. We recall that Rutherford's analysis of the scattering of alpha particles showed that even heavy nuclei have their electric charge distributed over a region smaller than $10^{-13}$ m. For the physicist of the nineteenth century a "point charge" remained an abstract notion. Today we are on familiar terms with the atomic particles. The graininess of electricity is so conspicuous in our modern description of nature that we find a point charge less of an artificial idealization than a smoothly varying distribution of charge density. When we postulate such smooth charge distributions, we may think of them as averages over very large numbers of elementary charges, in the same way that we can define the macroscopic density of a liquid, its lumpiness on a molecular scale notwithstanding.

## 1.4 Coulomb's law

As you probably already know, the interaction between electric charges at rest is described by Coulomb's law: two stationary electric charges repel or attract one another with a force proportional to the product of the magnitude of the charges and inversely proportional to the square of the distance between them.

We can state this compactly in vector form:

$$\mathbf{F}_2 = k\frac{q_1 q_2 \hat{\mathbf{r}}_{21}}{r_{21}^2}. \tag{1.1}$$

Here $q_1$ and $q_2$ are numbers (scalars) giving the magnitude and sign of the respective charges, $\hat{\mathbf{r}}_{21}$ is the unit vector in the direction[3] from charge 1 to charge 2, and $\mathbf{F}_2$ is the force acting on charge 2. Thus Eq. (1.1) expresses, among other things, the fact that like charges repel and unlike charges attract. Also, the force obeys Newton's third law; that is, $\mathbf{F}_2 = -\mathbf{F}_1$.

The unit vector $\hat{\mathbf{r}}_{21}$ shows that the force is parallel to the line joining the charges. It could not be otherwise unless space itself has some built-in directional property, for with two point charges alone in empty and isotropic space, no other direction could be singled out.

If the point charge itself had some internal structure, with an axis defining a direction, then it would have to be described by more than the mere scalar quantity $q$. It is true that some elementary particles, including the electron, do have another property, called *spin*. This gives rise to a magnetic force between two electrons in addition to their electrostatic repulsion. This magnetic force does not, in general, act in the direction of the line joining the two particles. It decreases with the inverse fourth power of the distance, and at atomic distances of $10^{-10}$ m the Coulomb force is already about $10^4$ times stronger than the magnetic interaction of the spins. Another magnetic force appears if our charges are moving – hence the restriction to stationary charges in our statement of Coulomb's law. We shall return to these magnetic phenomena in later chapters.

Of course we must assume, in writing Eq. (1.1), that both charges are well localized, each occupying a region small compared with $r_{21}$. Otherwise we could not even define the distance $r_{21}$ precisely.

The value of the constant $k$ in Eq. (1.1) depends on the units in which $r$, $\mathbf{F}$, and $q$ are to be expressed. In this book we will use the International System of Units, or "SI" units for short. This system is based on the

---

[3] The convention we adopt here may not seem the natural choice, but it is more consistent with the usage in some other parts of physics and we shall try to follow it throughout this book.

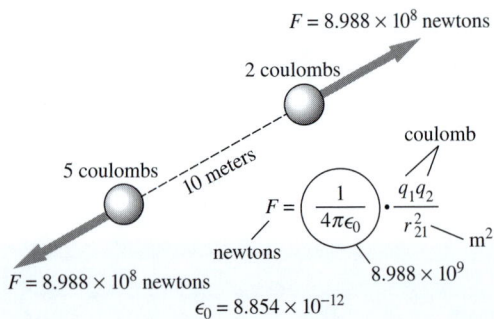

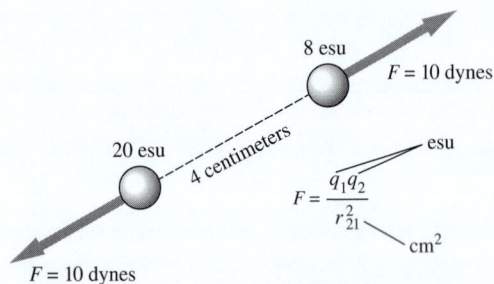

meter, kilogram, and second as units of length, mass, and time. The SI unit of charge is the *coulomb* (C). Some other SI electrical units that we will eventually become familiar with are the volt, ohm, ampere, and tesla. The official definition of the coulomb involves the magnetic force, which we will discuss in Chapter 6. For present purposes, we can define the coulomb as follows. Two like charges, each of 1 coulomb, repel one another with a force of $8.988 \cdot 10^9$ newtons when they are 1 meter apart. In other words, the $k$ in Eq. (1.1) is given by

$$k = 8.988 \cdot 10^9 \, \frac{\text{N m}^2}{\text{C}^2}. \qquad (1.2)$$

In Chapter 6 we will learn where this seemingly arbitrary value of $k$ comes from. In general, approximating $k$ by $9 \cdot 10^9 \, \text{N m}^2/\text{C}^2$ is quite sufficient. The magnitude of $e$, the fundamental quantum of electric charge, happens to be about $1.602 \cdot 10^{-19}$ C. So if you wish, you may think of a coulomb as defined to be the magnitude of the charge contained in $6.242 \cdot 10^{18}$ electrons.

Instead of $k$, it is customary (for historical reasons) to introduce a constant $\epsilon_0$ which is defined by

$$k \equiv \frac{1}{4\pi\epsilon_0} \implies \boxed{\epsilon_0 \equiv \frac{1}{4\pi k} = 8.854 \cdot 10^{-12} \, \frac{\text{C}^2}{\text{N m}^2}} \left(\text{or} \, \frac{\text{C}^2 \text{s}^2}{\text{kg m}^3}\right).$$

$$\qquad (1.3)$$

In terms of $\epsilon_0$, Coulomb's law in Eq. (1.1) takes the form

$$\boxed{\mathbf{F} = \frac{1}{4\pi\epsilon_0} \frac{q_1 q_2 \hat{\mathbf{r}}_{21}}{r_{21}^2}} \qquad (1.4)$$

The constant $\epsilon_0$ will appear in many expressions that we will meet in the course of our study. The $4\pi$ is included in the definition of $\epsilon_0$ so that certain formulas (such as Gauss's law in Sections 1.10 and 2.9) take on simple forms. Additional details and technicalities concerning $\epsilon_0$ can be found in Appendix E.

Another system of units that comes up occasionally is the *Gaussian* system, which is one of several types of cgs systems, short for centimeter–gram–second. (In contrast, the SI system is an mks system, short for meter–kilogram–second.) The Gaussian unit of charge is the "electrostatic unit," or esu. The esu is defined so that the constant $k$ in Eq. (1.1) *exactly* equals 1 (and this is simply the number 1, with no units) when $r_{21}$ is measured in cm, **F** in dynes, and the $q$ values in esu. Figure 1.2 gives some examples using the SI and Gaussian systems of units. Further discussion of the SI and Gaussian systems can be found in Appendix A.

**Figure 1.2.**
Coulomb's law expressed in SI units (top) and in Gaussian electrostatic units (bottom). The constant $\epsilon_0$ and the factor relating coulombs to esu are connected, as we shall learn later, with the speed of light. We have rounded off the constants in the figure to four-digit accuracy. The precise values are given in Appendix E.

**Example (Relation between 1 coulomb and 1 esu)** Show that 1 coulomb equals $2.998 \cdot 10^9$ esu (which generally can be approximated by $3 \cdot 10^9$ esu).

**Solution** From Eqs. (1.1) and (1.2), two charges of 1 coulomb separated by a distance of 1 m exert a (large!) force of $8.988 \cdot 10^9$ N $\approx 9 \cdot 10^9$ N on each other. We can convert this to the Gaussian unit of force via

$$1\,\text{N} = 1\,\frac{\text{kg m}}{\text{s}^2} = \frac{(1000\,\text{g})(100\,\text{cm})}{\text{s}^2} = 10^5\,\frac{\text{g cm}}{\text{s}^2} = 10^5\,\text{dynes}. \qquad (1.5)$$

The two 1 C charges therefore exert a force of $9 \cdot 10^{14}$ dynes on each other. How would someone working in Gaussian units describe this situation? In Gaussian units, Coulomb's law gives the force simply as $q^2/r^2$. The separation is 100 cm, so if 1 coulomb equals $N$ esu (with $N$ to be determined), the $9 \cdot 10^{14}$ dyne force between the charges can be expressed as

$$9 \cdot 10^{14}\,\text{dyne} = \frac{(N\,\text{esu})^2}{(100\,\text{cm})^2} \implies N^2 = 9 \cdot 10^{18} \implies N = 3 \cdot 10^9. \qquad (1.6)$$

Hence,[4]

$$1\,\text{C} = 3 \cdot 10^9\,\text{esu}. \qquad (1.7)$$

The magnitude of the electron charge is then given approximately by $e = 1.6 \cdot 10^{-19}$ C $\approx 4.8 \cdot 10^{-10}$ esu.

   If we had used the more exact value of $k$ in Eq. (1.2), the "3" in our result would have been replaced by $\sqrt{8.988} = 2.998$. This looks suspiciously similar to the "2.998" in the speed of light, $c = 2.998 \cdot 10^8$ m/s. This is no coincidence. We will see in Section 6.1 that Eq. (1.7) can actually be written as $1\,\text{C} = (10\{c\})$ esu, where we have put the $c$ in brackets to signify that it is just the number $2.998 \cdot 10^8$ without the units of m/s.

   On an everyday scale, a coulomb is an extremely large amount of charge, as evidenced by the fact that if you have two such charges separated by 1 m (never mind how you would keep each charge from flying apart due to the self repulsion!), the above force of $9 \cdot 10^9$ N between them is about one million tons. The esu is a much more reasonable unit to use for everyday charges. For example, the static charge on a balloon that sticks to your hair is on the order of 10 or 100 esu.

---

   The only way we have of detecting and measuring electric charges is by observing the interaction of charged bodies. One might wonder, then, how much of the apparent content of Coulomb's law is really only definition. As it stands, the significant physical content is the statement of inverse-square dependence and the implication that electric charge

---

[4] We technically shouldn't be using an "=" sign here, because it suggests that the units of a coulomb are the same as those of an esu. This is not the case; they are units in different systems and cannot be expressed in terms of each other. The proper way to express Eq. (1.7) is to say, "1 C is equivalent to $3 \cdot 10^9$ esu." But we'll usually just use the "=" sign, and you'll know what we mean. See Appendix A for further discussion of this.

(a)

(b)

(c)

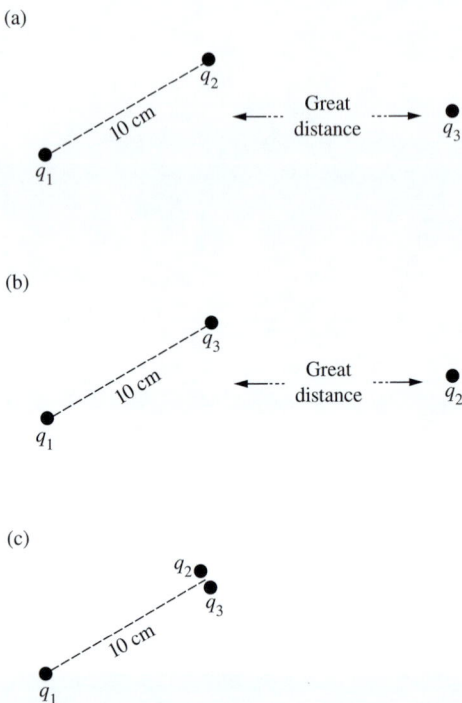

**Figure 1.3.**
The force on $q_1$ in (c) is the sum of the forces on $q_1$ in (a) and (b).

is *additive* in its effect. To bring out the latter point, we have to consider *more* than two charges. After all, if we had only two charges in the world to experiment with, $q_1$ and $q_2$, we could never measure them separately. We could verify only that $F$ is proportional to $1/r_{21}^2$. Suppose we have *three* bodies carrying charges $q_1$, $q_2$, and $q_3$. We can measure the force on $q_1$ when $q_2$ is 10 cm away from $q_1$, with $q_3$ very far away, as in Fig. 1.3(a). Then we can take $q_2$ away, bring $q_3$ into $q_2$'s former position, and again measure the force on $q_1$. Finally, we can bring $q_2$ and $q_3$ very close together and locate the combination 10 cm from $q_1$. We find by measurement that the force on $q_1$ is equal to the sum of the forces previously measured. This is a significant result that could *not* have been predicted by logical arguments from symmetry like the one we used above to show that the force between two point charges *had* to be along the line joining them. *The force with which two charges interact is not changed by the presence of a third charge.*

No matter how many charges we have in our system, Coulomb's law in Eq. (1.4) can be used to calculate the interaction of every pair. This is the basis of the principle of *superposition,* which we shall invoke again and again in our study of electromagnetism. Superposition means combining two sets of sources into one system by adding the second system "on top of" the first without altering the configuration of either one. Our principle ensures that the force on a charge placed at any point in the combined system will be the vector sum of the forces that each set of sources, acting alone, causes to act on a charge at that point. This principle must not be taken lightly for granted. There may well be a domain of phenomena, involving very small distances or very intense forces, where superposition *no longer holds.* Indeed, we know of quantum phenomena in the electromagnetic field that do represent a failure of superposition, seen from the viewpoint of the classical theory.

Thus the physics of electrical interactions comes into full view only when we have *more* than two charges. We can go beyond the explicit statement of Eq. (1.1) and assert that, with the three charges in Fig. 1.3 occupying any positions whatsoever, the force on any one of them, such as $q_3$, is correctly given by the following equation:

$$\mathbf{F} = \frac{1}{4\pi\epsilon_0}\frac{q_3 q_1 \hat{\mathbf{r}}_{31}}{r_{31}^2} + \frac{1}{4\pi\epsilon_0}\frac{q_3 q_2 \hat{\mathbf{r}}_{32}}{r_{32}^2}. \tag{1.8}$$

The experimental verification of the inverse-square law of electrical attraction and repulsion has a curious history. Coulomb himself announced the law in 1786 after measuring with a torsion balance the force between small charged spheres. But 20 years earlier Joseph Priestly, carrying out an experiment suggested to him by Benjamin Franklin, had noticed the absence of electrical influence within a hollow charged container and made an inspired conjecture: "May we not infer from this experiment that the attraction of electricity is subject to the same laws with that of gravitation and is therefore according to the square of the distances; since it is easily demonstrated that were the earth in the form of a shell, a body in the inside of it would not be attracted to one side more than the other." (Priestly, 1767).

The same idea was the basis of an elegant experiment in 1772 by Henry Cavendish. Cavendish charged a spherical conducting shell that contained within it, and temporarily connected to it, a smaller sphere. The outer shell was then separated into two halves and carefully removed, the inner sphere having been first disconnected. This sphere was tested for charge, the absence of which would confirm the inverse-square law. (See Problem 2.8 for the theory behind this.) Assuming that a deviation from the inverse-square law could be expressed as a difference in the exponent, $2 + \delta$, say, instead of 2, Cavendish concluded that $\delta$ must be less than 0.03. This experiment of Cavendish remained largely unknown until Maxwell discovered and published Cavendish's notes a century later (1876). At that time also, Maxwell repeated the experiment with improved apparatus, pushing the limit down to $\delta < 10^{-6}$. The present limit on $\delta$ is a fantastically small number – about one part in $10^{16}$; see Crandall (1983) and Williams *et al.* (1971).

Two hundred years after Cavendish's experiment, however, the question of interest changed somewhat. Never mind how perfectly Coulomb's law works for charged objects in the laboratory – is there a range of distances where it completely breaks down? There are two domains in either of which a breakdown is conceivable. The first is the domain of very small distances, distances less than $10^{-16}$ m, where electromagnetic theory as we know it may not work at all. As for very large distances, from the geographical, say, to the astronomical, a test of Coulomb's law by the method of Cavendish is obviously not feasible. Nevertheless we do observe certain large-scale electromagnetic phenomena that prove that the laws of classical electromagnetism work over very long distances. One of the most stringent tests is provided by planetary magnetic fields, in particular the magnetic field of the giant planet Jupiter, which was surveyed in the mission of Pioneer 10. The spatial variation of this field was carefully analyzed[5] and found to be entirely consistent with classical theory out to a distance of at least $10^5$ km from the planet. This is tantamount to a test, albeit indirect, of Coulomb's law over that distance.

To summarize, we have every reason for confidence in Coulomb's law over the stupendous range of 24 decades in distance, from $10^{-16}$ to $10^8$ m, if not farther, and we take it as the foundation of our description of electromagnetism.

## 1.5 Energy of a system of charges

In principle, Coulomb's law is all there is to electrostatics. Given the charges and their locations, we can find all the electrical forces. Or, given

[5] See Davis *et al.* (1975). For a review of the history of the exploration of the outer limit of classical electromagnetism, see Goldhaber and Nieto (1971).

(a)

Great
distance

$q_1$

$q_2$

(b)

$r_{12}$

$q_1$

$q_2$

(c)

$r_{21}$

$r_{32}$

$q_1$

$r_{31}$

$q_2$

Final
position
of $q_3$

$q_3$ in
transit

$ds$

$\mathbf{F}_{32}$      $\mathbf{F}_{31}$

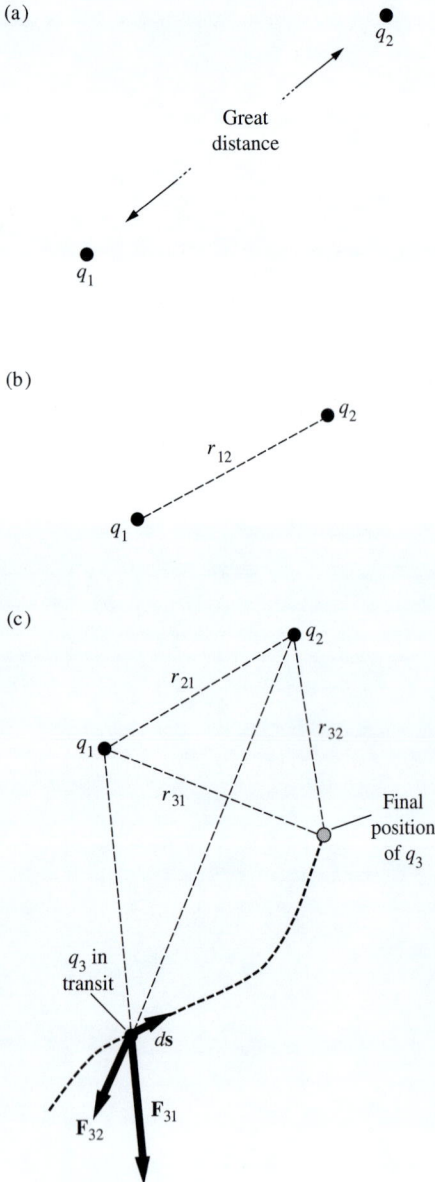

**Figure 1.4.**
Three charges are brought near one another.
First $q_2$ is brought in; then, with $q_1$ and $q_2$ fixed,
$q_3$ is brought in.

that the charges are free to move under the influence of other kinds of forces as well, we can find the equilibrium arrangement in which the charge distribution will remain stationary. In the same sense, Newton's laws of motion are all there is to mechanics. But in both mechanics and electromagnetism we gain power and insight by introducing other concepts, most notably that of energy.

Energy is a useful concept here because electrical forces are *conservative*. When you push charges around in electric fields, no energy is irrecoverably lost. Everything is perfectly reversible. Consider first the work that must be done *on* the system to bring some charged bodies into a particular arrangement. Let us start with two charged bodies or particles very far apart from one another, as indicated in Fig. 1.4(a), carrying charges $q_1$ and $q_2$. Whatever energy may have been needed to create these two concentrations of charge originally we shall leave entirely out of account. How much work does it take to bring the particles slowly together until the distance between them is $r_{12}$?

It makes no difference whether we bring $q_1$ toward $q_2$ or the other way around. In either case the work done is the integral of the product: force times displacement, where these are signed quantities. The force that has to be applied to move one charge toward the other is equal and opposite to the Coulomb force. Therefore,

$$W = \int (\text{applied force}) \cdot (\text{displacement})$$

$$= \int_{r=\infty}^{r_{12}} \left( -\frac{1}{4\pi\epsilon_0} \frac{q_1 q_2}{r^2} \right) dr = \frac{1}{4\pi\epsilon_0} \frac{q_1 q_2}{r_{12}}. \qquad (1.9)$$

Note that because $r$ is changing from $\infty$ to $r_{12}$, the differential $dr$ is negative. We know that the overall sign of the result is correct, because the work done on the system must be positive for charges of like sign; they have to be pushed together (consistent with the minus sign in the applied force). Both the displacement and the applied force are negative in this case, resulting in positive work being done on the system. With $q_1$ and $q_2$ in coulombs, and $r_{12}$ in meters, Eq. (1.9) gives the work in joules.

This work is the same whatever the path of approach. Let's review the argument as it applies to the two charges $q_1$ and $q_2$ in Fig. 1.5. There we have kept $q_1$ fixed, and we show $q_2$ moved to the same final position along two different paths. Every spherical shell, such as the one indicated between $r$ and $r + dr$, must be crossed by both paths. The increment of work involved, $-\mathbf{F} \cdot d\mathbf{s}$ in this bit of path (where $\mathbf{F}$ is the Coulomb force), is the same for the two paths.[6] The reason is that $\mathbf{F}$ has the same magnitude at both places and is directed radially from $q_1$, while

---

[6] Here we use for the first time the scalar product, or "dot product," of two vectors. A reminder: the scalar product of two vectors $\mathbf{A}$ and $\mathbf{B}$, written $\mathbf{A} \cdot \mathbf{B}$, is the number $AB\cos\theta$, where $A$ and $B$ are the magnitudes of the vectors $\mathbf{A}$ and $\mathbf{B}$, and $\theta$ is the angle between them. Expressed in terms of Cartesian components of the two vectors, $\mathbf{A} \cdot \mathbf{B} = A_x B_x + A_y B_y + A_z B_z$.

$ds = dr/\cos\theta$; hence $\mathbf{F}\cdot d\mathbf{s} = F\,dr$. Each increment of work along one path is matched by a corresponding increment on the other, so the sums must be equal. Our conclusion holds even for paths that loop in and out, like the dotted path in Fig. 1.5. (Why?)

Returning now to the two charges as we left them in Fig. 1.4(b), let us bring in from some remote place a third charge $q_3$ and move it to a point $P_3$ whose distance from charge 1 is $r_{31}$, and from charge 2, $r_{32}$. The work required to effect this will be

$$W_3 = -\int_\infty^{P_3} \mathbf{F}_3 \cdot d\mathbf{s}. \qquad (1.10)$$

Thanks to the additivity of electrical interactions, which we have already emphasized,

$$-\int \mathbf{F}_3 \cdot d\mathbf{s} = -\int (\mathbf{F}_{31} + \mathbf{F}_{32}) \cdot d\mathbf{s}$$

$$= -\int \mathbf{F}_{31} \cdot d\mathbf{s} - \int \mathbf{F}_{32} \cdot d\mathbf{s}. \qquad (1.11)$$

That is, the work required to bring $q_3$ to $P_3$ is the sum of the work needed when $q_1$ is present alone and that needed when $q_2$ is present alone:

$$W_3 = \frac{1}{4\pi\epsilon_0} \frac{q_1 q_3}{r_{31}} + \frac{1}{4\pi\epsilon_0} \frac{q_2 q_3}{r_{32}}. \qquad (1.12)$$

The total work done in assembling this arrangement of three charges, which we shall call $U$, is therefore

$$U = \frac{1}{4\pi\epsilon_0} \left( \frac{q_1 q_2}{r_{12}} + \frac{q_1 q_3}{r_{13}} + \frac{q_2 q_3}{r_{23}} \right). \qquad (1.13)$$

We note that $q_1$, $q_2$, and $q_3$ appear symmetrically in the expression above, in spite of the fact that $q_3$ was brought in last. We would have reached the same result if $q_3$ had been brought in first. (Try it.) Thus $U$ is independent of the *order* in which the charges were assembled. Since it is independent also of the route by which each charge was brought in, $U$ must be a unique property of the final arrangement of charges. We may call it the *electrical potential energy* of this particular system. There is a certain arbitrariness, as always, in the definition of a potential energy. In this case we have chosen the zero of potential energy to correspond to the situation with the three charges already in existence but infinitely far apart from one another. The potential energy *belongs to the configuration as a whole*. There is no meaningful way of assigning a certain fraction of it to one of the charges.

It is obvious how this very simple result can be generalized to apply to any number of charges. If we have $N$ different charges, in any arrangement in space, the potential energy of the system is calculated by summing over all pairs, just as in Eq. (1.13). The zero of potential energy, as in that case, corresponds to all charges far apart.

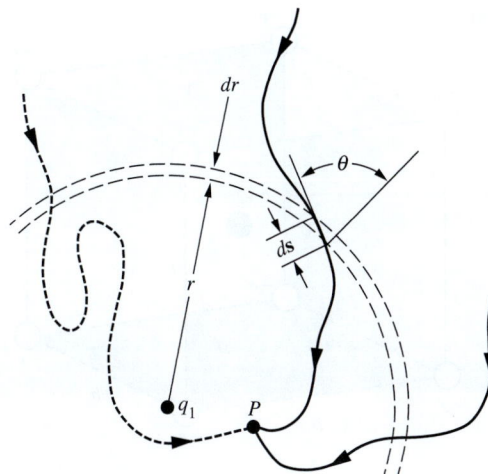

**Figure 1.5.**
Because the force is central, the sections of different paths between $r + dr$ and $r$ require the same amount of work.

(a)

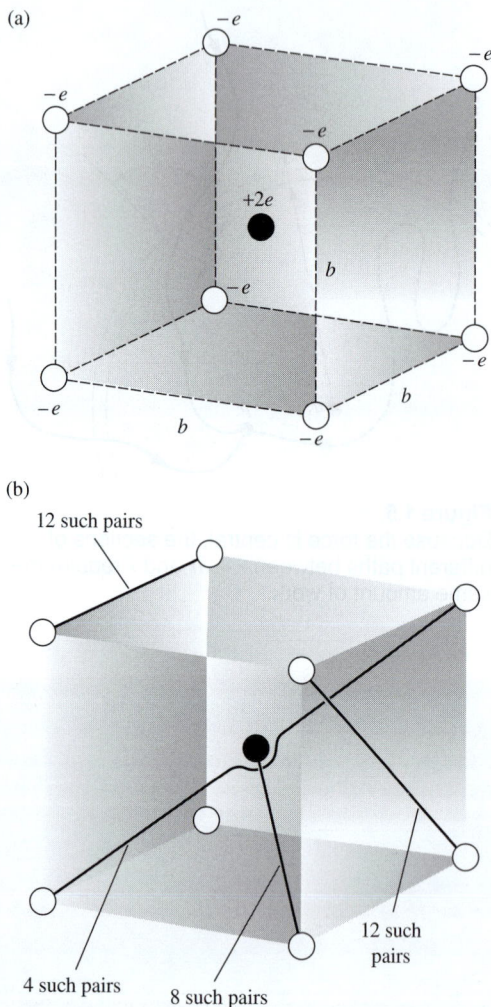

(b)

12 such pairs

12 such pairs

4 such pairs          8 such pairs

**Figure 1.6.**
(a) The potential energy of this arrangement of
nine point charges is given by Eq. (1.14).
(b) Four types of pairs are involved in the sum.

**Example (Charges in a cube)**   What is the potential energy of an arrangement of eight negative charges on the corners of a cube of side $b$, with a positive charge in the center of the cube, as in Fig. 1.6(a)? Suppose each negative charge is an electron with charge $-e$, while the central particle carries a double positive charge, $2e$.

**Solution**   Figure 1.6(b) shows that there are four different types of pairs. One type involves the center charge, while the other three involve the various edges and diagonals of the cube. Summing over all pairs yields

$$U = \frac{1}{4\pi\epsilon_0}\left(8 \cdot \frac{(-2e^2)}{(\sqrt{3}/2)b} + 12 \cdot \frac{e^2}{b} + 12 \cdot \frac{e^2}{\sqrt{2}b} + 4 \cdot \frac{e^2}{\sqrt{3}b}\right) \approx \frac{1}{4\pi\epsilon_0}\frac{4.32e^2}{b}.$$
(1.14)

The energy is positive, indicating that work had to be done on the system to assemble it. That work could, of course, be recovered if we let the charges move apart, exerting forces on some external body or bodies. Or if the electrons were simply to fly apart from this configuration, the *total kinetic energy* of all the particles would become equal to $U$. This would be true whether they came apart simultaneously and symmetrically, or were released one at a time in any order. Here we see the power of this simple notion of the total potential energy of the system. Think what the problem would be like if we had to compute the resultant vector force on every particle at every stage of assembly of the configuration! In this example, to be sure, the geometrical symmetry would simplify that task; even so, it would be more complicated than the simple calculation above.

One way of writing the instruction for the sum over pairs is this:

$$U = \frac{1}{2}\sum_{j=1}^{N}\sum_{k\neq j}\frac{1}{4\pi\epsilon_0}\frac{q_j q_k}{r_{jk}}.$$
(1.15)

The double-sum notation, $\sum_{j=1}^{N}\sum_{k\neq j}$, says: take $j=1$ and sum over $k=2,3,4,\ldots,N$; then take $j=2$ and sum over $k=1,3,4,\ldots,N$; and so on, through $j=N$. Clearly this includes every pair *twice*, and to correct for that we put in front the factor $1/2$.

## 1.6 Electrical energy in a crystal lattice

These ideas have an important application in the physics of crystals. We know that an ionic crystal like sodium chloride can be described, to a very good approximation, as an arrangement of positive ions ($Na^+$) and negative ions ($Cl^-$) alternating in a regular three-dimensional array or lattice. In sodium chloride the arrangement is that shown in Fig. 1.7(a). Of course the ions are not point charges, but they are nearly spherical distributions of charge and therefore (as we shall prove in Section 1.11) the electrical forces they exert on one another are the same as if each ion were replaced by an equivalent point charge at its center. We show this electrically equivalent system in Fig. 1.7(b). The electrostatic potential energy of the lattice of charges plays an important role in the explanation of the stability and cohesion of the ionic crystal. Let us see if we can estimate its magnitude.

We seem to be faced at once with a sum that is enormous, if not doubly infinite; any macroscopic crystal contains $10^{20}$ atoms at least. Will the sum converge? Now what we hope to find is the potential energy per unit volume or mass of crystal. We confidently expect this to be independent of the size of the crystal, based on the general argument that one end of a macroscopic crystal can have little influence on the other. Two grams of sodium chloride ought to have twice the potential energy of one gram, and the shape should not be important so long as the surface atoms are a small fraction of the total number of atoms. We would be *wrong* in this expectation if the crystal were made out of ions of one sign only. Then, 1 g of crystal would carry an enormous electric charge, and putting two such crystals together to make a 2 g crystal would take a fantastic amount of energy. (You might estimate how much!) The situation is saved by the fact that the crystal structure is an alternation of equal and opposite charges, so that any macroscopic bit of crystal is very nearly neutral.

To evaluate the potential energy we first observe that every positive ion is in a position equivalent to that of every other positive ion. Furthermore, although it is perhaps not immediately obvious from Fig. 1.7, the arrangement of positive ions around a negative ion is exactly the same as the arrangement of negative ions around a positive ion, and so on. Hence we may take one ion as a center, it matters not which kind, sum over *its* interactions with all the others, and simply multiply by the total number of ions of both kinds. This reduces the double sum in Eq. (1.15) to a single sum and a factor $N$; we must still apply the factor $1/2$ to compensate for including each pair twice. That is, the energy of a sodium chloride lattice composed of a total of $N$ ions is

$$U = \frac{1}{2}N\sum_{k=2}^{N}\frac{1}{4\pi\epsilon_0}\frac{q_1 q_k}{r_{1k}}. \tag{1.16}$$

Taking the positive ion at the center as in Fig. 1.7(b), our sum runs over all its neighbors near and far. The leading terms start out as follows:

$$U = \frac{1}{2}N\frac{1}{4\pi\epsilon_0}\left(-\frac{6e^2}{a} + \frac{12e^2}{\sqrt{2}\,a} - \frac{8e^2}{\sqrt{3}\,a} + \cdots\right). \tag{1.17}$$

The first term comes from the 6 nearest chlorine ions, at distance $a$, the second from the 12 sodium ions on the cube edges, and so on. It is clear, incidentally, that this series does not converge *absolutely*; if we were so

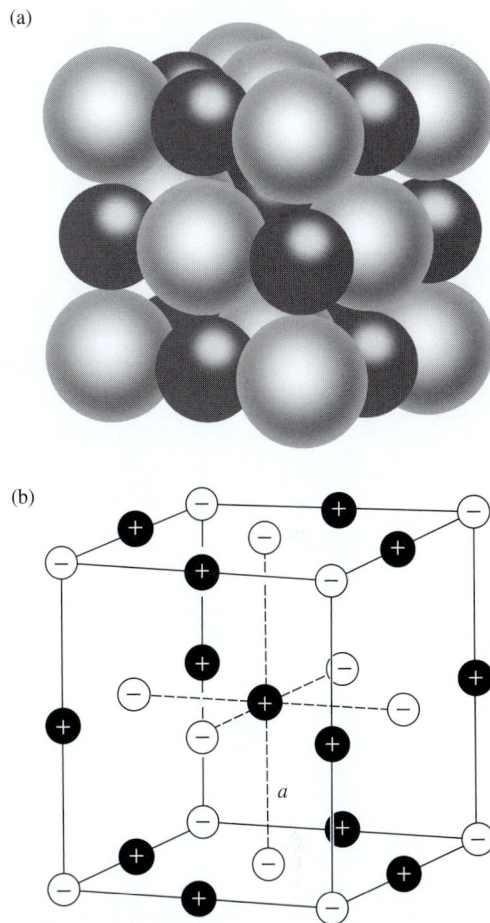

**Figure 1.7.**
A portion of a sodium chloride crystal, with the ions Na$^+$ and Cl$^-$ shown in about the right relative proportions (a), and replaced by equivalent point charges (b).

foolish as to try to sum all the positive terms first, that sum would diverge. To evaluate such a sum, we should arrange it so that as we proceed outward, including ever more distant ions, we include them in groups that represent nearly neutral shells of material. Then if the sum is broken off, the more remote ions that have been neglected will be such an even mixture of positive and negative charges that we can be confident their contribution would have been small. This is a crude way to describe what is actually a somewhat more delicate computational problem. The numerical evaluation of such a series is easily accomplished with a computer. The answer in this example happens to be

$$U = \frac{-0.8738 N e^2}{4 \pi \epsilon_0 a}. \tag{1.18}$$

Here $N$, the number of ions, is twice the number of NaCl molecules.

The negative sign shows that work would have to be *done* to take the crystal apart into ions. In other words, the electrical energy helps to explain the cohesion of the crystal. If this were the whole story, however, the crystal would collapse, for the potential energy of the charge distribution is obviously *lowered* by shrinking all the distances. We meet here again the familiar dilemma of classical – that is, nonquantum – physics. No system of stationary particles can be in stable equilibrium, according to classical laws, under the action of electrical forces alone; we will give a proof of this fact in Section 2.12. Does this make our analysis useless? Not at all. Remarkably, and happily, in the quantum physics of crystals the electrical potential energy can still be given meaning, and can be computed very much in the way we have learned here.

## 1.7 The electric field

Suppose we have some arrangement of charges, $q_1, q_2, \ldots, q_N$, fixed in space, and we are interested not in the forces they exert on one another, but only in their effect on some other charge $q_0$ that might be brought into their vicinity. We know how to calculate the resultant force on this charge, given its position which we may specify by the coordinates $x, y, z$. The force on the charge $q_0$ is

$$\mathbf{F} = \frac{1}{4 \pi \epsilon_0} \sum_{j=1}^{N} \frac{q_0 q_j \hat{\mathbf{r}}_{0j}}{r_{0j}^2}, \tag{1.19}$$

where $\mathbf{r}_{0j}$ is the vector from the $j$th charge in the system to the point $(x, y, z)$. The force is proportional to $q_0$, so if we divide out $q_0$ we obtain a vector quantity that depends only on the structure of our original system of charges, $q_1, \ldots, q_N$, and on the position of the point $(x, y, z)$. We call this vector function of $x, y, z$ the *electric field* arising from the $q_1, \ldots, q_N$

and use the symbol **E** for it. The charges $q_1, \ldots, q_N$ we call *sources* of the field. We may take as the *definition* of the electric field **E** of a charge distribution, at the point $(x, y, z)$,

$$\mathbf{E}(x, y, z) = \frac{1}{4\pi\epsilon_0} \sum_{j=1}^{N} \frac{q_j \hat{\mathbf{r}}_{0j}}{r_{0j}^2}. \qquad (1.20)$$

The force on some other charge $q$ at $(x, y, z)$ is then

$$\boxed{\mathbf{F} = q\mathbf{E}} \qquad (1.21)$$

Figure 1.8 illustrates the vector addition of the field of a point charge of 2 C to the field of a point charge of $-1$ C, at a particular point in space. In the SI system of units, electric field strength is expressed in newtons per unit charge, that is, newtons/coulomb. In Gaussian units, with the esu as the unit of charge and the dyne as the unit of force, the electric field strength is expressed in dynes/esu.

After the introduction of the electric potential in Chapter 2, we shall have another, and completely equivalent, way of expressing the unit of electric field strength; namely, volts/meter in SI units and statvolts/ centimeter in Gaussian units.

So far we have nothing really new. The electric field is merely another way of describing the system of charges; it does so by giving the force per unit charge, in magnitude and direction, that an exploring charge $q_0$ would experience at any point. We have to be a little careful with that interpretation. Unless the source charges are really immovable, the introduction of some finite charge $q_0$ may cause the source charges to shift their positions, so that the field itself, as defined by Eq. (1.20), is different. That is why we assumed fixed charges to begin our discussion. People sometimes define the field by requiring $q_0$ to be an "infinitesimal" test charge, letting **E** be the limit of $\mathbf{F}/q_0$ as $q_0 \to 0$. Any flavor of rigor this may impart is illusory. Remember that in the real world we have never observed a charge smaller than $e$! Actually, if we take Eq. (1.20) as our *definition* of **E**, without reference to a test charge, no problem arises and the sources need not be fixed. If the introduction of a new charge causes a shift in the source charges, then it has indeed brought about a change in the electric field, and if we want to predict the force on the new charge, we must use the new electric field in computing it.

Perhaps you still want to ask, what *is* an electric field? Is it something real, or is it merely a name for a factor in an equation that has to be multiplied by something else to give the numerical value of the force we measure in an experiment? Two observations may be useful here. First, since it works, it doesn't make any difference. That is not a frivolous answer, but a serious one. Second, the fact that the electric field vector

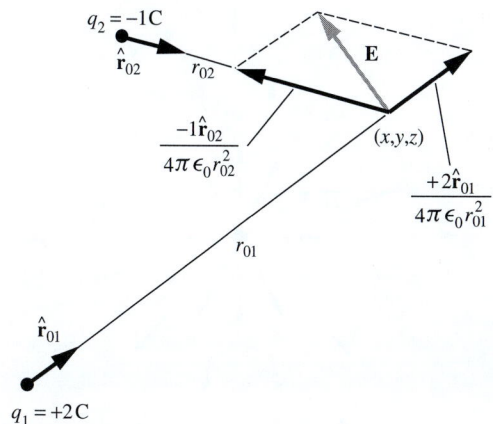

**Figure 1.8.**
The field at a point is the vector sum of the fields of each of the charges in the system.

(a)

(b)

● Charge +3
○ Charge −1

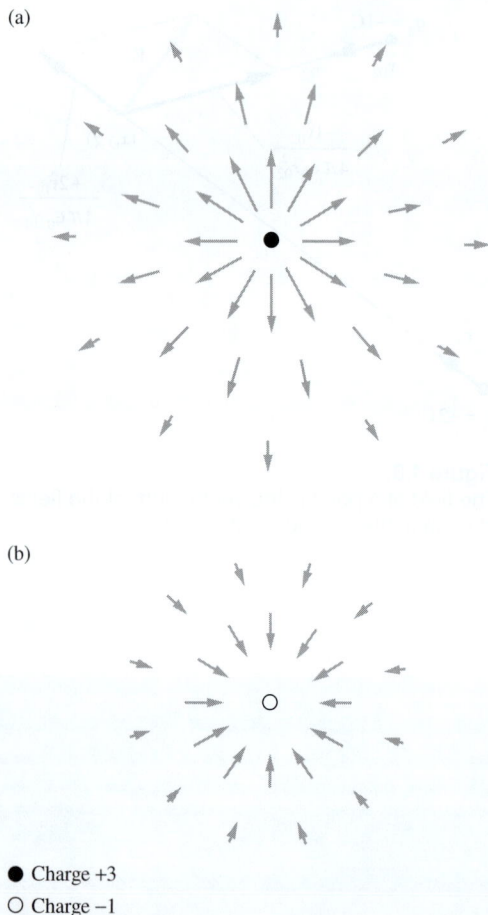

**Figure 1.9.**
(a) Field of a charge $q_1 = 3$. (b) Field of a charge $q_2 = -1$. Both representations are necessarily crude and only roughly quantitative.

at a point in space is all we need know to predict the force that will act on *any* charge at that point is by no means trivial. It might have been otherwise! If no experiments had ever been done, we could imagine that, in two different situations in which unit charges experience equal force, test charges of strength 2 units might experience unequal forces, depending on the nature of the other charges in the system. If that were true, the field description wouldn't work. The electric field attaches to every point in a system a *local property,* in this sense: if we know **E** in some small neighborhood, we know, *without further inquiry,* what will happen to any charges in that neighborhood. We do not need to ask what produced the field.

To visualize an electric field, you need to associate a vector, that is, a magnitude and direction, with every point in space. We shall use various schemes in this book, none of them wholly satisfactory, to depict vector fields.

It is hard to draw in two dimensions a picture of a vector function in three-dimensional space. We can indicate the magnitude and direction of **E** at various points by drawing little arrows near those points, making the arrows longer where $E$ is larger.[7] Using this scheme, we show in Fig. 1.9(a) the field of an isolated point charge of 3 units and in Fig. 1.9(b) the field of a point charge of $-1$ unit. These pictures admittedly add nothing whatsoever to our understanding of the field of an isolated charge; anyone can imagine a simple radial inverse-square field without the help of a picture. We show them in order to combine (side by side) the two fields in Fig. 1.10, which indicates in the same manner the field of two such charges separated by a distance $a$. All that Fig. 1.10 can show is the field in a plane containing the charges. To get a full three-dimensional representation, one must imagine the figure rotated around the symmetry axis. In Fig. 1.10 there is one point in space where **E** is zero. As an exercise, you can quickly figure out where this point lies. Notice also that toward the edge of the picture the field points more or less radially outward all around. One can see that at a very large distance from the charges the field will look very much like the field from a positive point charge. This is to be expected because the separation of the charges cannot make very much difference for points far away, and a point charge of 2 units is just what we would have left if we superimposed our two sources at one spot.

Another way to depict a vector field is to draw *field lines.* These are simply curves whose tangent, at any point, lies in the direction of the field at that point. Such curves will be smooth and continuous except at singularities such as point charges, or points like the one in the example of Fig. 1.10 where the field is zero. A field line plot does not directly give

---

[7] Such a representation is rather clumsy at best. It is hard to indicate the point in space to which a particular vector applies, and the range of magnitudes of $E$ is usually so large that it is impracticable to make the lengths of the arrows proportional to $E$.

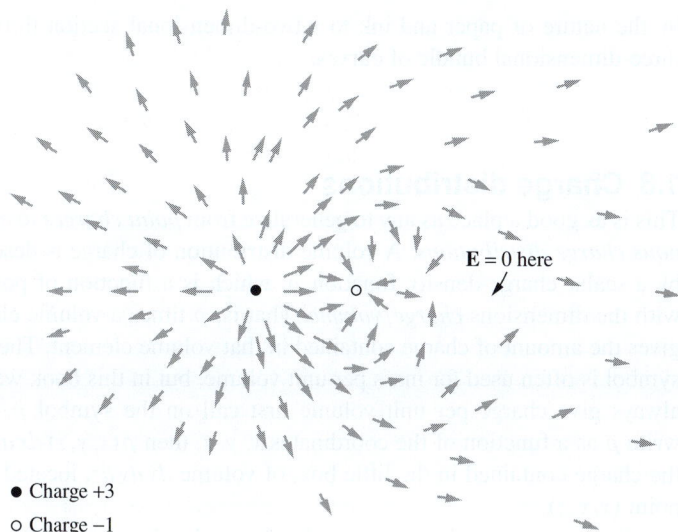

**Figure 1.10.**
The field in the vicinity of two charges, $q_1 = +3$, $q_2 = -1$, is the superposition of the fields in Figs. 1.9(a) and (b).

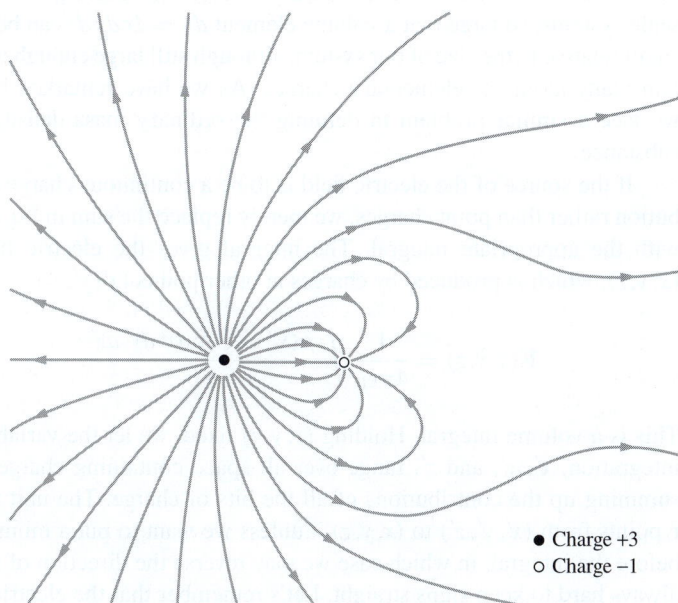

**Figure 1.11.**
Some field lines in the electric field around two charges, $q_1 = +3$, $q_2 = -1$.

the magnitude of the field, although we shall see that, in a general way, the field lines converge as we approach a region of strong field and spread apart as we approach a region of weak field. In Fig. 1.11 are drawn some field lines for the same arrangement of charges as in Fig. 1.10, a positive charge of 3 units and a negative charge of 1 unit. Again, we are restricted by the nature of paper and ink to a two-dimensional section through a three-dimensional bundle of curves.

## 1.8 Charge distributions

This is as good a place as any to generalize from *point charges* to *continuous charge distributions*. A volume distribution of charge is described by a scalar charge-density function $\rho$, which is a function of position, with the dimensions *charge/volume*. That is, $\rho$ times a volume element gives the amount of charge contained in that volume element. The same symbol is often used for mass per unit volume, but in this book we shall always give charge per unit volume first call on the symbol $\rho$. If we write $\rho$ as a function of the coordinates $x$, $y$, $z$, then $\rho(x, y, z)\, dx\, dy\, dz$ is the charge contained in the little box, of volume $dx\, dy\, dz$, located at the point $(x, y, z)$.

On an atomic scale, of course, the charge density varies enormously from point to point; even so, it proves to be a useful concept in that domain. However, we shall use it mainly when we are dealing with large-scale systems, so large that a volume element $dv = dx\, dy\, dz$ can be quite small relative to the size of our system, although still large enough to contain many atoms or elementary charges. As we have remarked before, we face a similar problem in defining the ordinary mass density of a substance.

If the source of the electric field is to be a continuous charge distribution rather than point charges, we merely replace the sum in Eq. (1.20) with the appropriate integral. The integral gives the electric field at $(x, y, z)$, which is produced by charges at other points $(x', y', z')$:

$$\mathbf{E}(x, y, z) = \frac{1}{4\pi\epsilon_0} \int \frac{\rho(x', y', z')\hat{\mathbf{r}}\, dx'\, dy'\, dz'}{r^2}. \tag{1.22}$$

This is a volume integral. Holding $(x, y, z)$ fixed, we let the variables of integration, $x'$, $y'$, and $z'$, range over all space containing charge, thus summing up the contributions of all the bits of charge. The unit vector $\hat{\mathbf{r}}$ points from $(x', y', z')$ to $(x, y, z)$ – unless we want to put a minus sign before the integral, in which case we may reverse the direction of $\hat{\mathbf{r}}$. It is always hard to keep signs straight. Let's remember that the electric field points *away* from a positive source (Fig. 1.12).

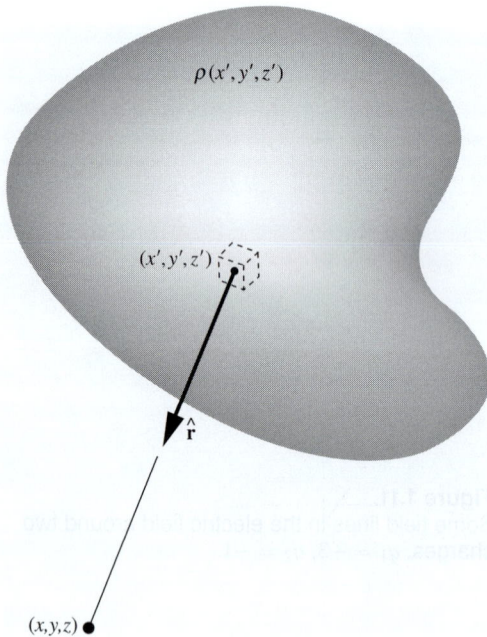

**Figure 1.12.**
Each element of the charge distribution $\rho(x', y', z')$ makes a contribution to the electric field $\mathbf{E}$ at the point $(x, y, z)$. The total field at this point is the sum of all such contributions; see Eq. (1.22).

---

**Example (Field due to a hemisphere)**    A solid hemisphere has radius $R$ and uniform charge density $\rho$. Find the electric field at the center.

Solution    Our strategy will be to slice the hemisphere into rings around the symmetry axis. We will find the electric field due to each ring, and then integrate over the rings to obtain the field due to the entire hemisphere. We will work with polar coordinates (or, equivalently, spherical coordinates), which are much more suitable than Cartesian coordinates in this setup.

The cross section of a ring is (essentially) a little rectangle with side lengths $dr$ and $r\,d\theta$, as shown in Fig. 1.13. The cross-sectional area is thus $r\,dr\,d\theta$. The radius of the ring is $r\sin\theta$, so the volume is $(r\,dr\,d\theta)(2\pi r\sin\theta)$. The charge in the ring is therefore $\rho(2\pi r^2\sin\theta\,dr\,d\theta)$. Equivalently, we can obtain this result by using the standard spherical-coordinate volume element, $r^2\sin\theta\,dr\,d\theta\,d\phi$, and then integrating over $\phi$ to obtain the factor of $2\pi$.

Consider a tiny piece of the ring, with charge $dq$. This piece creates an electric field at the center of the hemisphere that points diagonally upward (if $\rho$ is positive) with magnitude $dq/4\pi\epsilon_0 r^2$. However, only the vertical component survives, because the horizontal component cancels with the horizontal component from the diametrically opposite charge $dq$ on the ring. The vertical component involves a factor of $\cos\theta$. When we integrate over the whole ring, the $dq$ simply integrates to the total charge we found above. The (vertical) electric field due to a given ring is therefore

$$dE_y = \frac{\rho(2\pi r^2\sin\theta\,dr\,d\theta)}{4\pi\epsilon_0 r^2}\cos\theta = \frac{\rho\sin\theta\cos\theta\,dr\,d\theta}{2\epsilon_0}. \qquad (1.23)$$

Integrating over $r$ and $\theta$ to obtain the field due to the entire hemisphere gives

$$E_y = \int_0^R\int_0^{\pi/2}\frac{\rho\sin\theta\cos\theta\,dr\,d\theta}{2\epsilon_0} = \frac{\rho}{2\epsilon_0}\left(\int_0^R dr\right)\left(\int_0^{\pi/2}\sin\theta\cos\theta\,d\theta\right)$$

$$= \frac{\rho}{2\epsilon_0}\cdot R\cdot\frac{\sin^2\theta}{2}\Big|_0^{\pi/2} = \frac{\rho R}{4\epsilon_0}. \qquad (1.24)$$

Note that the radius $r$ canceled in Eq. (1.23). For given values of $\theta$, $d\theta$, and $dr$, the volume of a ring grows like $r^2$, and this exactly cancels the $r^2$ in the denominator in Coulomb's law.

REMARK   As explained above, the electric field due to the hemisphere is vertical. This fact also follows from considerations of symmetry. We will make many symmetry arguments throughout this book, so let us be explicit here about how the reasoning proceeds. Assume (in search of a contradiction) that the electric field due to the hemisphere is *not* vertical. It must then point off at some angle, as shown in Fig. 1.14(a). Let's say that the **E** vector lies above a given dashed line painted on the hemisphere. If we rotate the system by, say, 180° around the symmetry axis, the field now points in the direction shown in Fig. 1.14(b), because it must still pass over the dashed line. But we have *exactly the same hemisphere* after the rotation, so the field must still point upward to the right. We conclude that the field due to the hemisphere points both upward to the left and upward to the right. This is a contradiction. The only way to avoid this contradiction is for the field to point along the symmetry axis (possibly in the negative direction), because in that case it doesn't change under the rotation.

In the neighborhood of a true point charge the electric field grows infinite like $1/r^2$ as we approach the point. It makes no sense to talk about the field *at* the point charge. As our ultimate physical sources of field are

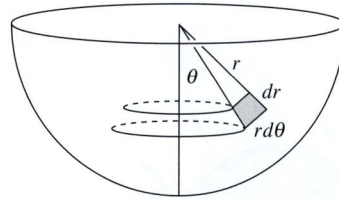

**Figure 1.13.**
Cross section of a thin ring. The hemisphere may be considered to be built up from rings.

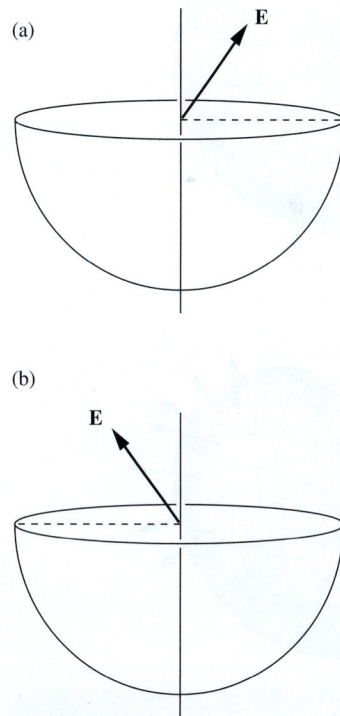

(a)

(b)

**Figure 1.14.**
The symmetry argument that explains why **E** must be vertical.

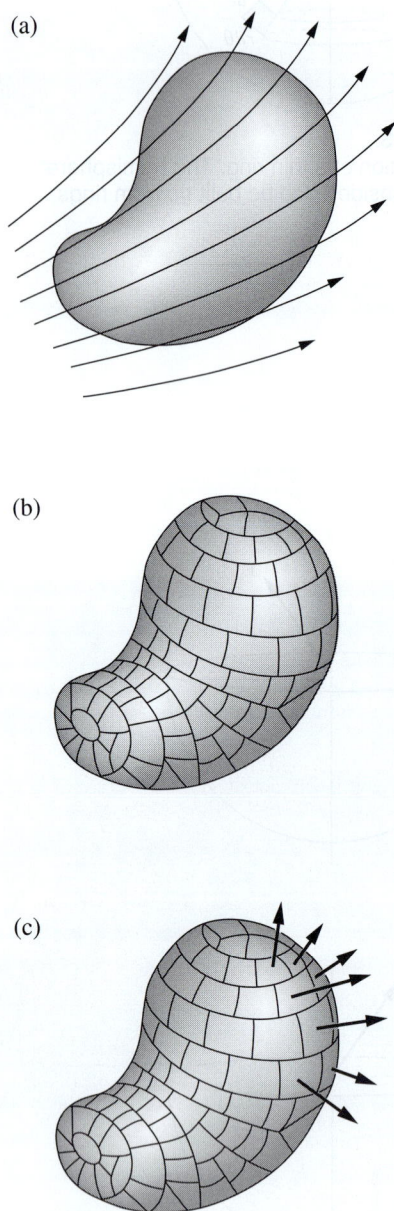

**Figure 1.15.**
(a) A closed surface in a vector field is divided
(b) into small elements of area. (c) Each
element of area is represented by an outward
vector.

not, we believe, infinite concentrations of charge in zero volume, but instead finite structures, we simply ignore the mathematical singularities implied by our point-charge language and rule out of bounds the interior of our elementary sources. A continuous charge distribution $\rho(x', y', z')$ that is nowhere infinite gives no trouble at all. Equation (1.22) can be used to find the field at any point within the distribution. The integrand doesn't blow up at $r = 0$ because the volume element in the numerator equals $r^2 \sin \phi \, d\phi \, d\theta \, dr$ in spherical coordinates, and the $r^2$ here cancels the $r^2$ in the denominator in Eq. (1.22). That is to say, so long as $\rho$ remains finite, the field will remain finite everywhere, even in the interior or on the boundary of a charge distribution.

## 1.9 Flux

The relation between the electric field and its sources can be expressed in a remarkably simple way, one that we shall find very useful. For this we need to define a quantity called *flux*.

Consider some electric field in space and in this space some arbitrary closed surface, like a balloon of any shape. Figure 1.15 shows such a surface, the field being suggested by a few field lines. Now divide the whole surface into little patches that are so small that over any one patch the surface is practically flat and the vector field does not change appreciably from one part of a patch to another. In other words, don't let the balloon be too crinkly, and don't let its surface pass right through a singularity[8] of the field such as a point charge. The area of a patch has a certain magnitude in square meters, and a patch defines a unique direction – the outward-pointing normal to its surface. (Since the surface is closed, you can tell its inside from its outside; there is no ambiguity.) Let this magnitude and direction be represented by a vector. Then for every patch into which the surface has been divided, such as patch number $j$, we have a vector $\mathbf{a}_j$ giving its area and orientation. The steps we have just taken are pictured in Figs. 1.15(b) and (c). Note that the vector $\mathbf{a}_j$ does not depend at all on the shape of the patch; it doesn't matter how we have divided up the surface, as long as the patches are small enough.

Let $\mathbf{E}_j$ be the electric field vector at the location of patch number $j$. The scalar product $\mathbf{E}_j \cdot \mathbf{a}_j$ is a number. We call this number the *flux* through that bit of surface. To understand the origin of the name, imagine a vector function that represents the velocity of motion in a fluid – say in a river, where the velocity varies from one place to another but is constant in time at any one position. Denote this vector field by $\mathbf{v}$, measured in

---

[8] By a singularity of the field we would ordinarily mean not only a point source where the field approaches infinity, but also any place where the field changes magnitude or direction discontinuously, such as an infinitesimally thin layer of concentrated charge. Actually this latter, milder, kind of singularity would cause no difficulty here unless our balloon's surface were to coincide with the surface of discontinuity over some finite area.

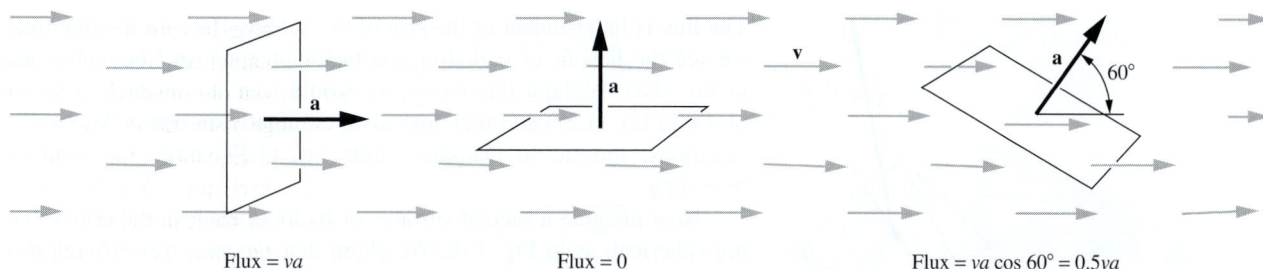

Flux = $va$          Flux = 0          Flux = $va \cos 60° = 0.5va$

**Figure 1.16.**
The flux through the frame of area **a** is **v** · **a**, where **v** is the velocity of the fluid. The flux is the volume of fluid passing through the frame, per unit time.

meters/second. Then, if **a** is the oriented area in square meters of a frame lowered into the water, **v** · **a** is the *rate of flow* of water through the frame in cubic meters per second (Fig. 1.16). The $\cos\theta$ factor in the standard expression for the dot product correctly picks out the component of **v** along the direction of **a**, or equivalently the component of **a** along the direction of **v**. We must emphasize that our definition of flux is applicable to any vector function, whatever physical variable it may represent.

Now let us add up the flux through all the patches to get the flux through the entire surface, a scalar quantity which we shall denote by $\Phi$:

$$\Phi = \sum_{\text{all } j} \mathbf{E}_j \cdot \mathbf{a}_j. \tag{1.25}$$

Letting the patches become smaller and more numerous without limit, we pass from the sum in Eq. (1.25) to a surface integral:

$$\Phi = \int_{\substack{\text{entire} \\ \text{surface}}} \mathbf{E} \cdot d\mathbf{a}. \tag{1.26}$$

A surface integral of any vector function **F**, over a surface $S$, means just this: divide $S$ into small patches, each represented by a vector outward, of magnitude equal to the patch area; at every patch, take the scalar product of the patch area vector and the local **F**; sum all these products, and the limit of this sum, as the patches shrink, is the surface integral. Do not be alarmed by the prospect of having to perform such a calculation for an awkwardly shaped surface like the one in Fig. 1.15. The surprising property we are about to demonstrate makes that unnecessary!

## 1.10 Gauss's law

Take the simplest case imaginable; suppose the field is that of a single isolated positive point charge $q$, and the surface is a sphere of radius $r$ centered on the point charge (Fig. 1.17). What is the flux $\Phi$ through this surface? The answer is easy because the magnitude of **E** at every point on the surface is $q/4\pi\epsilon_0 r^2$ and its direction is the same as that of the outward normal at that point. So we have

$$\Phi = E \cdot (\text{total area}) = \frac{q}{4\pi\epsilon_0 r^2} \cdot 4\pi r^2 = \frac{q}{\epsilon_0}. \tag{1.27}$$

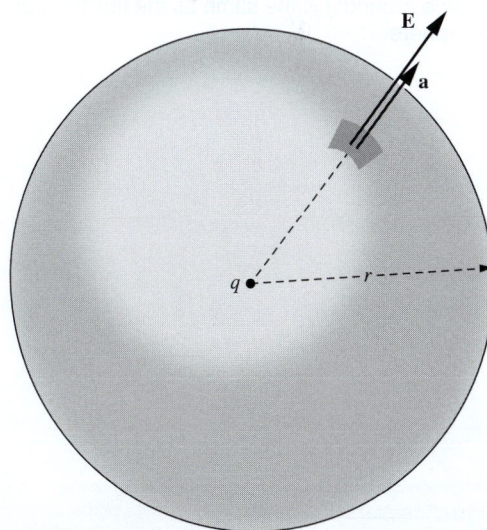

**Figure 1.17.**
In the field **E** of a point charge $q$, what is the outward flux over a sphere surrounding $q$?

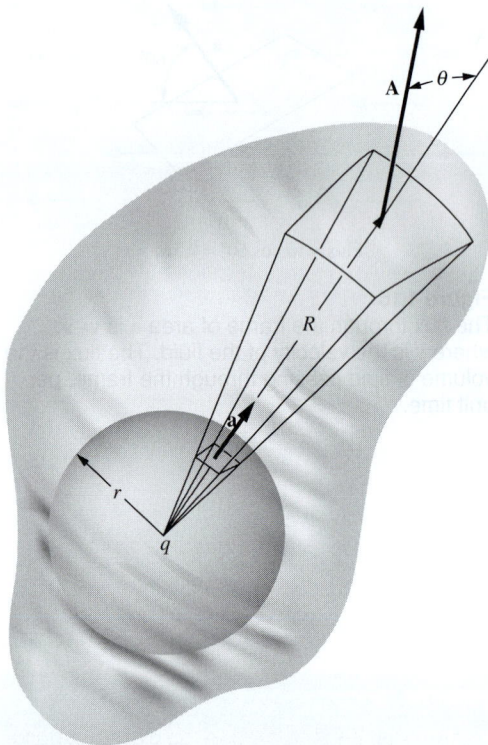

**Figure 1.18.**
Showing that the flux through any closed surface around $q$ is the same as the flux through the sphere.

The flux is independent of the size of the sphere. Here for the first time we see the benefit of including the factor of $1/4\pi$ in Coulomb's law in Eq. (1.4). Without this factor, we would have an uncanceled factor of $4\pi$ in Eq. (1.27) and therefore also, eventually, in one of Maxwell's equations. Indeed, in Gaussian units Eq. (1.27) takes the form of $\Phi = 4\pi q$.

Now imagine a second surface, or balloon, enclosing the first, but *not* spherical, as in Fig. 1.18. We claim that *the total flux through this surface is the same as that through the sphere.* To see this, look at a cone, radiating from $q$, that cuts a small patch **a** out of the sphere and continues on to the outer surface, where it cuts out a patch **A** at a distance $R$ from the point charge. The area of the patch **A** is larger than that of the patch **a** by two factors: first, by the ratio of the distance squared $(R/r)^2$; and second, owing to its inclination, by the factor $1/\cos\theta$. The angle $\theta$ is the angle between the outward normal and the radial direction (see Fig. 1.18). The electric field in that neighborhood is reduced from its magnitude on the sphere by the factor $(r/R)^2$ and is still radially directed. Letting $\mathbf{E}_{(R)}$ be the field at the outer patch and $\mathbf{E}_{(r)}$ be the field at the sphere, we have

$$\text{flux through outer patch} = \mathbf{E}_{(R)} \cdot \mathbf{A} = E_{(R)}A\cos\theta,$$
$$\text{flux through inner patch} = \mathbf{E}_{(r)} \cdot \mathbf{a} = E_{(r)}a. \qquad (1.28)$$

Using the above facts concerning the magnitude of $\mathbf{E}_{(R)}$ and the area of **A**, the flux through the outer patch can be written as

$$E_{(R)}A\cos\theta = \left[E_{(r)}\left(\frac{r}{R}\right)^2\right]\left[a\left(\frac{R}{r}\right)^2\frac{1}{\cos\theta}\right]\cos\theta = E_{(r)}a, \qquad (1.29)$$

which equals the flux through the inner patch.

Now every patch on the outer surface can in this way be put into correspondence with part of the spherical surface, so the total flux must be the same through the two surfaces. That is, the flux through the new surface must be just $q/\epsilon_0$. But this was a surface of *arbitrary* shape and size.[9] We conclude: the flux of the electric field through *any* surface enclosing a point charge $q$ is $q/\epsilon_0$. As a corollary we can say that the total flux through a closed surface is *zero* if the charge lies *outside* the surface. We leave the proof of this to the reader, along with Fig. 1.19 as a hint of one possible line of argument.

There is a way of looking at all this that makes the result seem obvious. Imagine at $q$ a source that emits particles – such as bullets or photons – in all directions at a steady rate. Clearly the flux of particles through a window of unit area will fall off with the inverse square of the window's distance from $q$. Hence we can draw an analogy between the electric field strength $E$ and the intensity of particle flow in bullets per unit area per

[9] To be sure, we had the second surface enclosing the sphere, but it didn't have to, really. Besides, the sphere can be taken as small as we please.

unit time. It is pretty obvious that the flux of bullets through any surface completely surrounding $q$ is independent of the size and shape of that surface, for it is just the total number emitted per unit time. Correspondingly, the flux of $E$ through the closed surface must be independent of size and shape. The common feature responsible for this is the inverse-square behavior of the intensity.

The situation is now ripe for superposition! Any electric field is the sum of the fields of its individual sources. This property was expressed in our statement, Eq. (1.19), of Coulomb's law. Clearly flux is an additive quantity in the same sense, for if we have a number of sources, $q_1, q_2, \ldots, q_N$, the fields of which, if each were present alone, would be $\mathbf{E}_1, \mathbf{E}_2, \ldots, \mathbf{E}_N$, then the flux $\Phi$ through some surface $S$ in the actual field can be written

(a)

$$\Phi = \int_S \mathbf{E} \cdot d\mathbf{a} = \int_S (\mathbf{E}_1 + \mathbf{E}_2 + \cdots + \mathbf{E}_N) \cdot d\mathbf{a}. \qquad (1.30)$$

We have just learned that $\int_S \mathbf{E}_i \cdot d\mathbf{a}$ equals $q_i/\epsilon_0$ if the charge $q_i$ is inside $S$ and equals zero otherwise. So every charge $q$ inside the surface contributes exactly $q/\epsilon_0$ to the surface integral of Eq. (1.30) and all charges outside contribute nothing. We have arrived at Gauss's law.

(b)

---

The flux of the electric field $\mathbf{E}$ through any closed surface, that is, the integral $\int \mathbf{E} \cdot d\mathbf{a}$ over the surface, equals $1/\epsilon_0$ times the total charge enclosed by the surface:

$$\int \mathbf{E} \cdot d\mathbf{a} = \frac{1}{\epsilon_0} \sum_i q_i = \frac{1}{\epsilon_0} \int \rho \, dv \qquad \text{(Gauss's law)} \qquad (1.31)$$

---

**Figure 1.19.**
To show that the flux through the closed surface in (a) is zero, you can make use of (b).

We call the statement in the box a *law* because it is equivalent to Coulomb's law and it could serve equally well as the basic law of electrostatic interactions, after charge and field have been defined. Gauss's law and Coulomb's law are not two independent physical laws, but the same law expressed in different ways.[10] In Gaussian units, the $1/\epsilon_0$ in Gauss's law is replaced with $4\pi$.

Looking back over our proof, we see that it hinged on the inverse-square nature of the interaction and of course on the additivity of interactions, or superposition. Thus the theorem is applicable to any inverse-square field in physics, for instance to the gravitational field.

---

[10] There is one difference, inconsequential here, but relevant to our later study of the fields of moving charges. Gauss's law is obeyed by a wider class of fields than those represented by the electrostatic field. In particular, a field that is inverse-square in $r$ but not spherically symmetrical can satisfy Gauss's law. In other words, Gauss's law alone does not imply the symmetry of the field of a point source which is implicit in Coulomb's law.

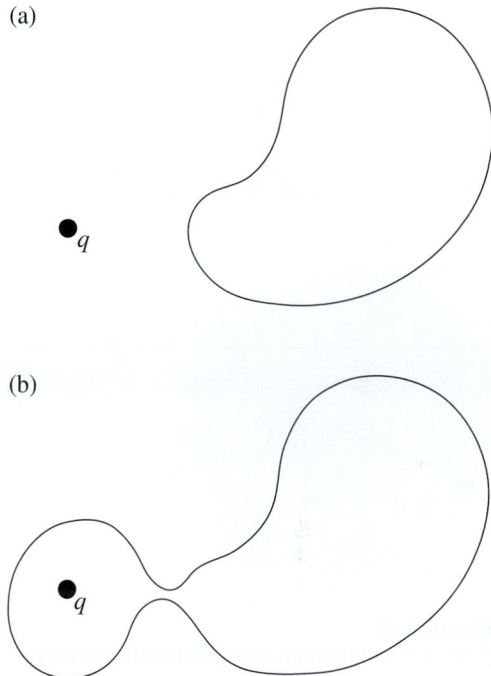

It is easy to see that Gauss's law would *not* hold if the law of force were, say, inverse-cube. For in that case the flux of electric field from a point charge $q$ through a sphere of radius $R$ centered on the charge would be

$$\Phi = \int \mathbf{E} \cdot d\mathbf{a} = \frac{q}{4\pi\epsilon_0 R^3} \cdot 4\pi R^2 = \frac{q}{\epsilon_0 R}. \tag{1.32}$$

By making the sphere large enough we could make the flux through it as small as we pleased, while the total charge inside remained constant.

This remarkable theorem extends our knowledge in two ways. First, it reveals a connection between the field and its sources that is the converse of Coulomb's law. Coulomb's law tells us how to derive the electric field if the charges are given; with Gauss's law we can determine how much charge is in any region if the field is known. Second, the mathematical relation here demonstrated is a powerful analytic tool; it can make complicated problems easy, as we shall see in the following examples. In Sections 1.11–1.13 we use Gauss's law to calculate the electric field due to various nicely shaped objects. In all of these examples the symmetry of the object will play a critical role.

## 1.11 Field of a spherical charge distribution

We can use Gauss's law to find the electric field of a spherically symmetrical distribution of charge, that is, a distribution in which the charge density $\rho$ depends only on the radius from a central point. Figure 1.20 depicts a cross section through some such distribution. Here the charge density is high at the center, and is zero beyond $r_0$. What is the electric field at some point such as $P_1$ outside the distribution, or $P_2$ inside it (Fig. 1.21)? If we could proceed only from Coulomb's law, we should have to carry out an integration that would sum the electric field vectors at $P_1$ arising from each elementary volume in the charge distribution. Let's try a different approach that exploits both the symmetry of the system and Gauss's law.

Because of the spherical symmetry, the electric field at any point must be radially directed – no other direction is unique. Likewise, the field magnitude $E$ must be the same at all points on a spherical surface $S_1$ of radius $r_1$, for all such points are equivalent. Call this field magnitude $E_1$. The flux through this surface $S_1$ is therefore simply $4\pi r_1^2 E_1$, and by Gauss's law this must be equal to $1/\epsilon_0$ times the charge enclosed by the surface. That is, $4\pi r_1^2 E_1 = (1/\epsilon_0) \cdot$ (charge inside $S_1$) or

$$E_1 = \frac{\text{charge inside } S_1}{4\pi\epsilon_0 r_1^2}. \tag{1.33}$$

Comparing this with the field of a point charge, we see that *the field at all points on $S_1$ is the same as if all the charge within $S_1$ were concentrated at the center*. The same statement applies to a sphere drawn

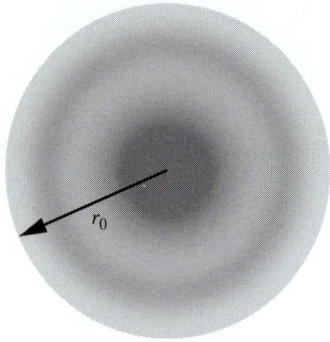

**Figure 1.20.**
A charge distribution with spherical symmetry.

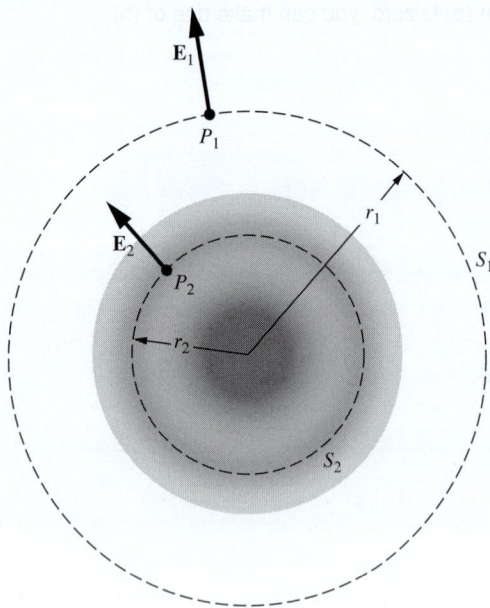

**Figure 1.21.**
The electric field of a spherical charge distribution.

*inside* the charge distribution. The field at any point on $S_2$ is the same as if all charge within $S_2$ were at the center, and all charge *outside* $S_2$ absent. Evidently the field inside a "hollow" spherical charge distribution is zero (Fig. 1.22). Problem 1.17 gives an alternative derivation of this fact.

**Example (Field inside and outside a uniform sphere)** A spherical charge distribution has a density $\rho$ that is constant from $r=0$ out to $r=R$ and is zero beyond. What is the electric field for all values of $r$, both less than and greater than $R$?

**Solution** For $r \geq R$, the field is the same as if all of the charge were concentrated at the center of the sphere. Since the volume of the sphere is $4\pi R^3/3$, the field is therefore radial and has magnitude

$$E(r) = \frac{(4\pi R^3/3)\rho}{4\pi \epsilon_0 r^2} = \frac{\rho R^3}{3\epsilon_0 r^2} \qquad (r \geq R). \tag{1.34}$$

For $r \leq R$, the charge outside radius $r$ effectively contributes nothing to the field, while the charge inside radius $r$ acts as if it were concentrated at the center. The volume inside radius $r$ is $4\pi r^3/3$, so the field inside the given sphere is radial and has magnitude

$$E(r) = \frac{(4\pi r^3/3)\rho}{4\pi \epsilon_0 r^2} = \frac{\rho r}{3\epsilon_0} \qquad (r \leq R). \tag{1.35}$$

In terms of the total charge $Q = (4\pi R^3/3)\rho$, this can be written as $Qr/4\pi \epsilon_0 R^3$. The field increases linearly with $r$ inside the sphere; the $r^3$ growth of the effective charge outweighs the $1/r^2$ effect from the increasing distance. And the field decreases like $1/r^2$ outside the sphere. A plot of $E(r)$ is shown in Fig. 1.23. Note that $E(r)$ is continuous at $r = R$, where it takes on the value $\rho R/3\epsilon_0$. As we will see in Section 1.13, field discontinuities are created by surface charge densities, and there are no surface charges in this system. The field goes to zero at the center, so it is continuous there also. How should the density vary with $r$ so that the magnitude $E(r)$ is uniform inside the sphere? That is the subject of Exercise 1.68.

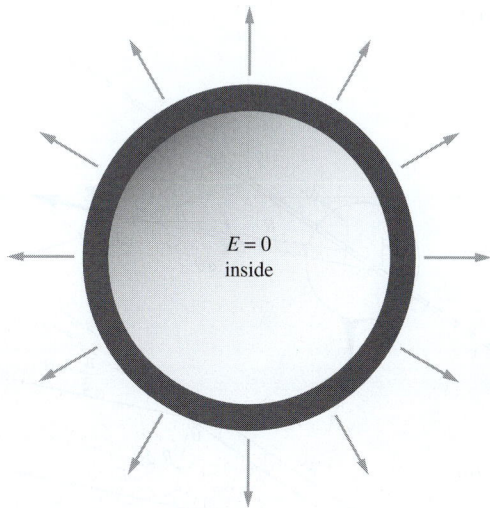

**Figure 1.22.**
The field is zero inside a spherical shell of charge.

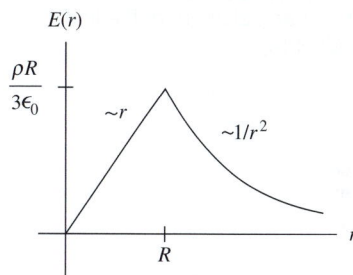

**Figure 1.23.**
The electric field due to a uniform sphere of charge.

The same argument applied to the gravitational field would tell us that the earth, assuming it is spherically symmetrical in its mass distribution, attracts outside bodies as if its mass were concentrated at the center. That is a rather familiar statement. Anyone who is inclined to think the principle expresses an obvious property of the center of mass must be reminded that the theorem is not even true, in general, for other shapes. A perfect cube of uniform density does *not* attract external bodies as if its mass were concentrated at its geometrical center.

Newton didn't consider the theorem obvious. He needed it as the keystone of his demonstration that the moon in its orbit around the earth and a falling body on the earth are responding to similar forces. The delay of nearly 20 years in the publication of Newton's theory of gravitation

(a)

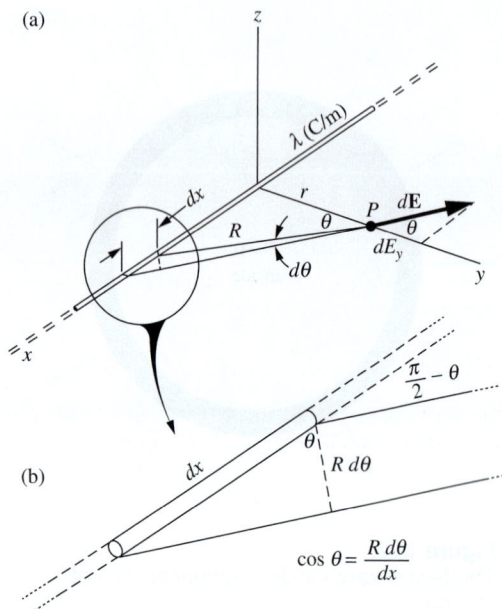

(b)

$$\cos \theta = \frac{R\,d\theta}{dx}$$

**Figure 1.24.**
(a) The field at $P$ is the vector sum of contributions from each element of the line charge. (b) Detail of (a).

was apparently due, in part at least, to the trouble he had in proving this theorem to his satisfaction. The proof he eventually devised and published in the *Principia* in 1686 (Book I, Section XII, Theorem XXXI) is a marvel of ingenuity in which, roughly speaking, a tricky volume integration is effected without the aid of the integral calculus as we know it. The proof is a good bit longer than our whole preceding discussion of Gauss's law, and more intricately reasoned. You see, with all his mathematical resourcefulness and originality, Newton lacked Gauss's law – a relation that, once it has been shown to us, seems so obvious as to be almost trivial.

## 1.12  Field of a line charge

A long, straight, charged wire, if we neglect its thickness, can be characterized by the amount of charge it carries per unit length. Let $\lambda$, measured in coulombs/meter, denote this *linear charge density*. What is the electric field of such a line charge, assumed infinitely long and with constant linear charge density $\lambda$? We'll do the problem in two ways, first by an integration starting from Coulomb's law, and then by using Gauss's law.

To evaluate the field at the point $P$, shown in Fig. 1.24, we must add up the contributions from all segments of the line charge, one of which is indicated as a segment of length $dx$. The charge $dq$ on this element is given by $dq = \lambda\,dx$. Having oriented our $x$ axis along the line charge, we may as well let the $y$ axis pass through $P$, which is a distance $r$ from the nearest point on the line. It is a good idea to take advantage of symmetry at the outset. Obviously the electric field at $P$ must point in the $y$ direction, so that $E_x$ and $E_z$ are both zero. The contribution of the charge $dq$ to the $y$ component of the electric field at $P$ is

$$dE_y = \frac{dq}{4\pi\epsilon_0 R^2}\cos\theta = \frac{\lambda\,dx}{4\pi\epsilon_0 R^2}\cos\theta, \qquad (1.36)$$

where $\theta$ is the angle the electric field of $dq$ makes with the $y$ direction. The total $y$ component is then

$$E_y = \int dE_y = \int_{-\infty}^{\infty} \frac{\lambda\cos\theta}{4\pi\epsilon_0 R^2}\,dx. \qquad (1.37)$$

It is convenient to use $\theta$ as the variable of integration. Since Figs. 1.24(a) and (b) tell us that $R = r/\cos\theta$ and $dx = R\,d\theta/\cos\theta$, we have $dx = r\,d\theta/\cos^2\theta$. (This expression for $dx$ comes up often. It also follows from $x = r\tan\theta \implies dx = r\,d(\tan\theta) = r\,d\theta/\cos^2\theta$.) Eliminating $dx$ and $R$ from the integral in Eq. (1.37), in favor of $\theta$, we obtain

$$E_y = \int_{-\pi/2}^{\pi/2} \frac{\lambda\cos\theta\,d\theta}{4\pi\epsilon_0 r} = \frac{\lambda}{4\pi\epsilon_0 r}\int_{-\pi/2}^{\pi/2}\cos\theta\,d\theta = \frac{\lambda}{2\pi\epsilon_0 r}. \qquad (1.38)$$

We see that the field of an infinitely long, uniformly dense line charge is proportional to the reciprocal of the distance from the line. Its direction is of course radially outward if the line carries a positive charge, inward if negative.

Gauss's law leads directly to the same result. Surround a segment of the line charge with a closed circular cylinder of length $L$ and radius $r$, as in Fig. 1.25, and consider the flux through this surface. As we have already noted, symmetry guarantees that the field is radial, so the flux through the ends of the "tin can" is zero. The flux through the cylindrical surface is simply the area, $2\pi rL$, times $E_r$, the field at the surface. On the other hand, the charge enclosed by the surface is just $\lambda L$, so Gauss's law gives us $(2\pi rL)E_r = \lambda L/\epsilon_0$ or

$$E_r = \frac{\lambda}{2\pi \epsilon_0 r}, \qquad (1.39)$$

in agreement with Eq. (1.38).

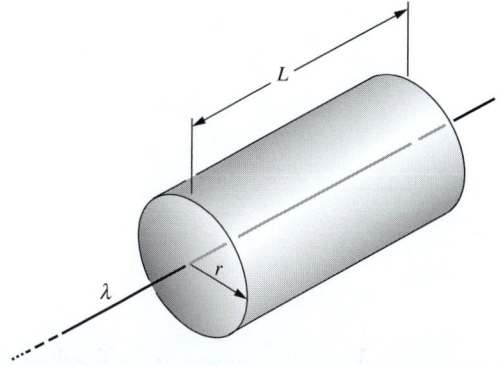

**Figure 1.25.**
Using Gauss's law to find the field of a line charge.

## 1.13 Field of an infinite flat sheet of charge

Electric charge distributed smoothly in a thin sheet is called a *surface charge distribution*. Consider a flat sheet, infinite in extent, with the constant surface charge density $\sigma$. The electric field on either side of the sheet, whatever its magnitude may turn out to be, must surely point perpendicular to the plane of the sheet; there is no other unique direction in the system. Also, because of symmetry, the field must have the same magnitude and the opposite direction at two points $P$ and $P'$ equidistant from the sheet on opposite sides. With these facts established, Gauss's law gives us at once the field intensity, as follows: draw a cylinder, as in Fig. 1.26 (actually, any shape with uniform cross section will work fine), with $P$ on one side and $P'$ on the other, of cross-sectional area $A$. The outward flux is found only at the ends, so that if $E_P$ denotes the magnitude of the field at $P$, and $E_{P'}$ the magnitude at $P'$, the outward flux is $AE_P + AE_{P'} = 2AE_P$. The charge enclosed is $\sigma A$, so Gauss's law gives $2AE_P = \sigma A/\epsilon_0$, or

$$E_P = \frac{\sigma}{2\epsilon_0}. \qquad (1.40)$$

We see that the field strength is independent of $r$, the distance from the sheet. Equation (1.40) could have been derived more laboriously by calculating the vector sum of the contributions to the field at $P$ from all the little elements of charge in the sheet.

In the more general case where there are other charges in the vicinity, the field need not be perpendicular to the sheet, or symmetric on either side of it. Consider a very squat Gaussian surface, with $P$ and $P'$ infinitesimally close to the sheet, instead of the elongated surface in Fig. 1.26. We can then ignore the negligible flux through the cylindrical "side" of the pillbox, so the above reasoning gives $E_{\perp,P} + E_{\perp,P'} = \sigma/\epsilon_0$, where the "$\perp$" denotes the component perpendicular to the sheet. If you want

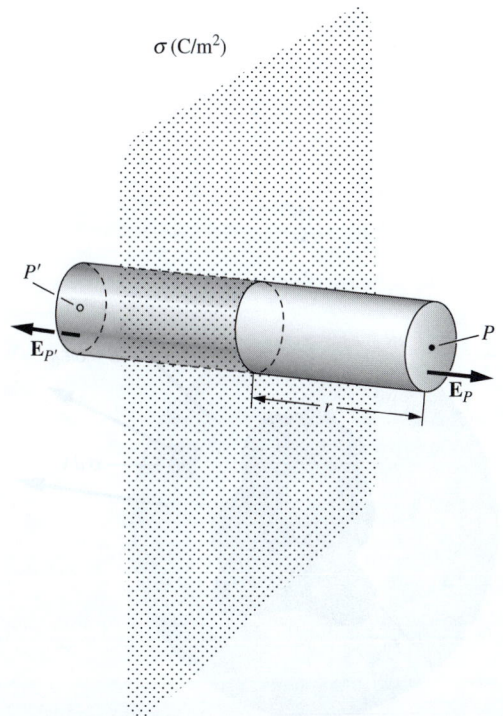

**Figure 1.26.**
Using Gauss's law to find the field of an infinite flat sheet of charge.

to write this in terms of vectors, it becomes $\mathbf{E}_{\perp,P} - \mathbf{E}_{\perp,P'} = (\sigma/\epsilon_0)\hat{\mathbf{n}}$, where $\hat{\mathbf{n}}$ is the unit vector perpendicular to the sheet, in the direction of $P$. In other words, the discontinuity in $\mathbf{E}_\perp$ across the sheet is given by

$$\Delta\mathbf{E}_\perp = \frac{\sigma}{\epsilon_0}\hat{\mathbf{n}}. \tag{1.41}$$

Only the normal component is discontinuous; the parallel component is continuous across the sheet. So we can just as well replace the $\Delta\mathbf{E}_\perp$ in Eq. (1.41) with $\Delta\mathbf{E}$. This result is also valid for any finite-sized sheet, because from up close the sheet looks essentially like an infinite plane, at least as far as the normal component is concerned.

The field of an infinitely long line charge, we found, varies inversely as the distance from the line, while the field of an infinite sheet has the same strength at all distances. These are simple consequences of the fact that the field of a point charge varies as the inverse square of the distance. If that doesn't yet seem compellingly obvious, look at it this way: roughly speaking, the part of the line charge that is mainly responsible for the field at $P$ in Fig. 1.24 is the near part – the charge within a distance of order of magnitude $r$. If we lump all this together and forget the rest, we have a concentrated charge of magnitude $q \approx \lambda r$, which ought to produce a field proportional to $q/r^2$, or $\lambda/r$. In the case of the sheet, the amount of charge that is "effective," in this sense, increases proportionally to $r^2$ as we go out from the sheet, which just offsets the $1/r^2$ decrease in the field from any given element of charge.

## 1.14 The force on a layer of charge

The sphere in Fig. 1.27 has a charge distributed over its surface with the uniform density $\sigma$, in C/m$^2$. Inside the sphere, as we have already learned, the electric field of such a charge distribution is zero. Outside the sphere the field is $Q/4\pi\epsilon_0 r^2$, where $Q$ is the total charge on the sphere, equal to $4\pi r_0^2\sigma$. So just outside the surface of the sphere the field strength is

$$E_{\text{just outside}} = \frac{\sigma}{\epsilon_0}. \tag{1.42}$$

Compare this with Eq. (1.40) and Fig. 1.26. In both cases Gauss's law is obeyed: the *change* in the normal component of $\mathbf{E}$, from one side of the layer to the other, is equal to $\sigma/\epsilon_0$, in accordance with Eq. (1.41).

What is the electrical force experienced by the charges that make up this distribution? The question may seem puzzling at first because the field $\mathbf{E}$ arises from these very charges. What we must think about is the force on some small element of charge $dq$, such as a small patch of area $dA$ with charge $dq = \sigma\, dA$. Consider, separately, the force on $dq$ due to all the other charges in the distribution, and the force on the patch due to the charges within the patch itself. This latter force is surely zero. Coulomb repulsion between charges within the patch is just another example of

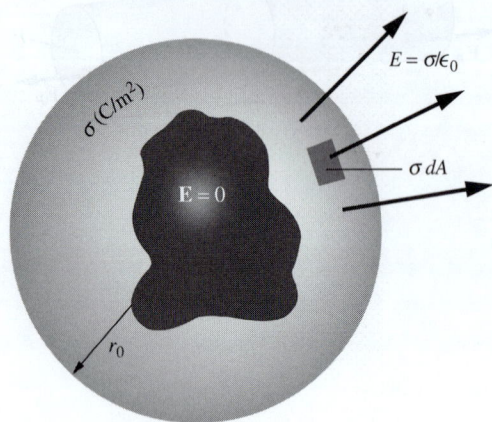

**Figure 1.27.**
A spherical surface with uniform charge density $\sigma$.

Newton's third law; the patch as a whole cannot push on itself. That simplifies our problem, for it allows us to use the entire electric field **E**, *including* the field due to all charges in the patch, in calculating the force d**F** on the patch of charge $dq$:

$$d\mathbf{F} = \mathbf{E}\,dq = \mathbf{E}\sigma\,dA. \tag{1.43}$$

But what $E$ shall we use, the field $E = \sigma/\epsilon_0$ outside the sphere or the field $E = 0$ inside? The correct answer, as we shall prove in a moment, is the *average* of the two fields that is,

$$dF = \frac{1}{2}(\sigma/\epsilon_0 + 0)\sigma\,dA = \frac{\sigma^2\,dA}{2\epsilon_0}. \tag{1.44}$$

To justify this we shall consider a more general case, and one that will introduce a more realistic picture of a layer of surface charge. Real charge layers do not have zero thickness. Figure 1.28 shows some ways in which charge might be distributed through the thickness of a layer. In each example, the value of $\sigma$, the total charge per unit area of layer, is the same. These might be cross sections through a small portion of the spherical surface in Fig. 1.27 on a scale such that the curvature is not noticeable. To make it more general, however, we can let the field on the left be $E_1$ (rather than 0, as it was inside the sphere), with $E_2$ the field on the right. The condition imposed by Gauss's law, for given $\sigma$, is, in each case,

$$E_2 - E_1 = \frac{\sigma}{\epsilon_0}. \tag{1.45}$$

Now let us look carefully within the layer where the field is changing continuously from $E_1$ to $E_2$ and there is a volume charge density $\rho(x)$ extending from $x = 0$ to $x = x_0$, the thickness of the layer (Fig. 1.29). Consider a much thinner slab, of thickness $dx \ll x_0$, which contains per unit area an amount of charge $\rho\,dx$. If the area of this thin slab is $A$, the force on it is

$$dF = E\rho\,dx \cdot A. \tag{1.46}$$

Thus the total force per unit area of our original charge layer is

$$\frac{F}{A} = \int \frac{dF}{A} = \int_0^{x_0} E\rho\,dx. \tag{1.47}$$

But Gauss's law tells us via Eq. (1.45) that $dE$, the change in $E$ through the thin slab, is just $\rho\,dx/\epsilon_0$. Hence $\rho\,dx$ in Eq. (1.47) can be replaced by $\epsilon_0\,dE$, and the integral becomes

$$\frac{F}{A} = \int_{E_1}^{E_2} \epsilon_0 E\,dE = \frac{\epsilon_0}{2}\left(E_2^2 - E_1^2\right). \tag{1.48}$$

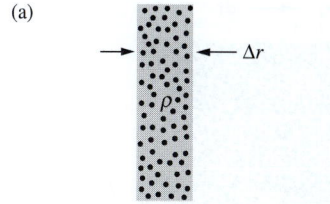

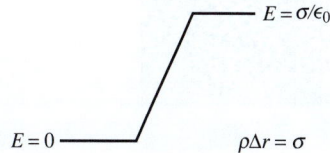

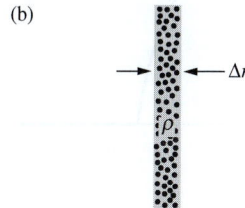

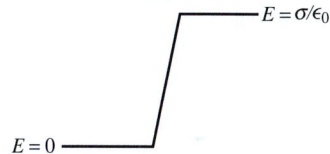

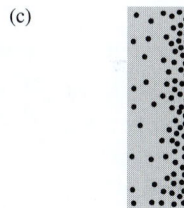

(a)

$E = \sigma/\epsilon_0$

$E = 0$

$\rho\,\Delta r = \sigma$

(b)

$E = \sigma/\epsilon_0$

$E = 0$

(c)

$E = \sigma/\epsilon_0$

$E = 0$

**Figure 1.28.**
The net change in field at a charge layer depends only on the total charge per unit area.

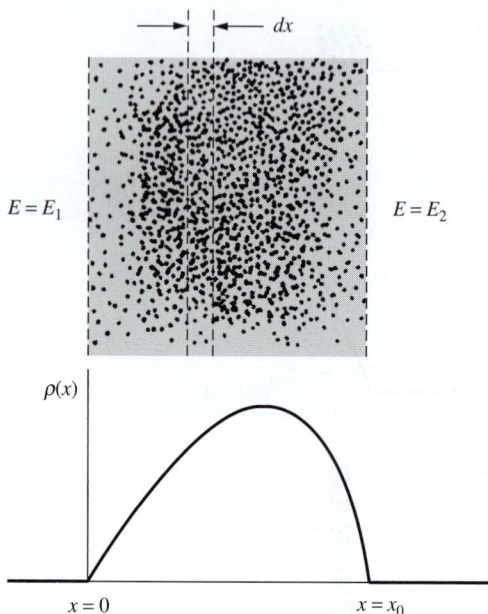

**Figure 1.29.**
Within the charge layer of density $\rho(x)$,
$E(x + dx) - E(x) = \rho \, dx/\epsilon_0$.

Since $E_2 - E_1 = \sigma/\epsilon_0$, the force per unit area in Eq. (1.48), after being factored, can be expressed as

$$\frac{F}{A} = \frac{1}{2}(E_1 + E_2)\sigma \qquad (1.49)$$

We have shown, as promised, that for given $\sigma$ the force per unit area on a charge layer is determined by the mean of the external field on one side and that on the other.[11] This is independent of the thickness of the layer, as long as it is small compared with the total area, and of the variation $\rho(x)$ in charge density within the layer. See Problem 1.30 for an alternative derivation of Eq. (1.49).

The direction of the electrical force on an element of the charge on the sphere is, of course, outward whether the surface charge is positive or negative. If the charges do not fly off the sphere, that outward force must be balanced by some inward force, not included in our equations, that can hold the charge carriers in place. To call such a force "nonelectrical" would be misleading, for electrical attractions and repulsions are the dominant forces in the structure of atoms and in the cohesion of matter generally. The difference is that these forces are effective only at short distances, from atom to atom, or from electron to electron. Physics on that scale is a story of individual particles. Think of a charged rubber balloon, say 0.1 m in radius, with $10^{-8}$ C of negative charge spread as uniformly as possible on its outer surface. It forms a surface charge of density $\sigma = (10^{-8} \text{ C})/4\pi(0.1 \text{ m})^2 = 8 \cdot 10^{-8} \text{ C/m}^2$. The resulting outward force, per area of surface charge, is given by Eq. (1.44) as

$$\frac{dF}{dA} = \frac{\sigma^2}{2\epsilon_0} = \frac{(8 \cdot 10^{-8} \text{ C/m}^2)^2}{2(8.85 \cdot 10^{-12} \text{ C}^2/(\text{N m}^2))} = 3.6 \cdot 10^{-4} \text{ N/m}^2. \quad (1.50)$$

In fact, our charge consists of about $6 \cdot 10^{10}$ electrons attached to the rubber film, which corresponds to about 50 million extra electrons per square centimeter. So the "graininess" in the charge distribution is hardly apparent. However, if we could look at one of these extra electrons, we would find it roughly $10^{-4}$ cm – an enormous distance on an atomic scale – from its nearest neighbor. This electron would be stuck, electrically stuck, to a local molecule of rubber. The rubber molecule would be attached to adjacent rubber molecules, and so on. If you pull on the electron, the force is transmitted in this way to the whole piece of rubber. Unless, of course, you pull hard enough to tear the electron loose from the molecule to which it is attached. That would take an electric field many thousands of times stronger than the field in our example.

---

[11] Note that this is *not* necessarily the same as the average field within the layer, a quantity of no special interest or significance.

## 1.15 Energy associated with the electric field

Suppose our spherical shell of charge is compressed slightly, from an initial radius of $r_0$ to a smaller radius, as in Fig. 1.30. This requires that work be done against the repulsive force, which we found above to be $\sigma^2/2\epsilon_0$ newtons for each square meter of surface. The displacement being $dr$, the total work done is $(4\pi r_0^2)(\sigma^2/2\epsilon_0)\,dr$, or $(2\pi r_0^2\sigma^2/\epsilon_0)\,dr$. This represents an *increase* in the energy required to assemble the system of charges, the energy $U$ we talked about in Section 1.5:

$$dU = \frac{2\pi r_0^2\sigma^2}{\epsilon_0}\,dr. \tag{1.51}$$

**Figure 1.30.**
Shrinking a spherical shell or charged balloon.

Notice how the electric field $E$ has been changed. Within the shell of thickness $dr$, the field was zero and is now $\sigma/\epsilon_0$. Beyond $r_0$ the field is unchanged. In effect we have created a field of strength $E = \sigma/\epsilon_0$ filling a region of volume $4\pi r_0^2\,dr$. We have done so by investing an amount of energy given by Eq. (1.51) which, if we substitute $\epsilon_0 E$ for $\sigma$, can be written like this:

$$dU = \frac{\epsilon_0 E^2}{2}\,4\pi r_0^2\,dr. \tag{1.52}$$

This is an instance of a general theorem which we shall not prove now (but see Problem 1.33): *the potential energy U of a system of charges, which is the total work required to assemble the system, can be calculated from the electric field itself simply by assigning an amount of energy $(\epsilon_0 E^2/2)$ dv to every volume element dv and integrating over all space where there is electric field:*

$$U = \frac{\epsilon_0}{2}\int_{\substack{\text{entire}\\\text{field}}} E^2\,dv \tag{1.53}$$

$E^2$ is a scalar quantity, of course: $E^2 \equiv \mathbf{E}\cdot\mathbf{E}$.

One may think of this energy as "stored" in the field. The system being conservative, that amount of energy can of course be recovered by allowing the charges to go apart; so it is nice to think of the energy as "being somewhere" meanwhile. Our accounting comes out right if we think of it as stored in space with a density of $\epsilon_0 E^2/2$, in joules/m$^3$. There is no harm in this, but in fact we have no way of identifying, quite independently of anything else, the energy stored in a particular cubic meter of space. Only the total energy is physically measurable, that is, the work required to bring the charge into some configuration, starting from some other configuration. Just as the concept of electric field serves in place of Coulomb's law to explain the behavior of electric charges, so when we use Eq. (1.53) rather than Eq. (1.15) to express the total potential energy of an electrostatic system, we are merely using a different kind of bookkeeping. Sometimes a change in viewpoint, even if it is at

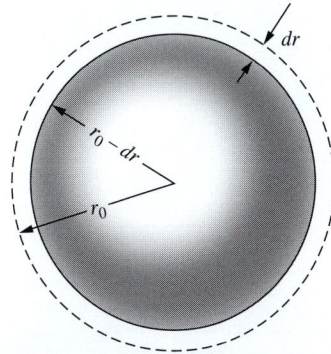

first only a change in bookkeeping, can stimulate new ideas and deeper understanding. The notion of the electric field as an independent entity will take form when we study the dynamical behavior of charged matter and electromagnetic radiation.

---

**Example (Potential energy of a uniform sphere)**    What is the energy stored in a sphere of radius $R$ with charge $Q$ uniformly distributed throughout the interior?

**Solution**    The electric field is nonzero both inside and outside the sphere, so Eq. (1.53) involves two different integrals. Outside the sphere, the field at radius $r$ is simply $Q/4\pi\epsilon_0 r^2$, so the energy stored in the external field is

$$U_{\text{ext}} = \frac{\epsilon_0}{2}\int_R^\infty \left(\frac{Q}{4\pi\epsilon_0 r^2}\right)^2 4\pi r^2\, dr = \frac{Q^2}{8\pi\epsilon_0}\int_R^\infty \frac{dr}{r^2} = \frac{Q^2}{8\pi\epsilon_0 R}. \quad (1.54)$$

The example in Section 1.11 gives the field at radius $r$ inside the sphere as $E_r = \rho r/3\epsilon_0$. But the density equals $\rho = Q/(4\pi R^3/3)$, so the field is $E_r = (3Q/4\pi R^3)r/3\epsilon_0 = Qr/4\pi\epsilon_0 R^3$. The energy stored in the internal field is therefore

$$U_{\text{int}} = \frac{\epsilon_0}{2}\int_0^R \left(\frac{Qr}{4\pi\epsilon_0 R^3}\right)^2 4\pi r^2\, dr = \frac{Q^2}{8\pi\epsilon_0 R^6}\int_0^R r^4\, dr = \frac{Q^2}{8\pi\epsilon_0 R}\cdot\frac{1}{5}. \quad (1.55)$$

This is one-fifth of the energy stored in the external field. The total energy is the sum of $U_{\text{ext}}$ and $U_{\text{int}}$, which we can write as $(3/5)Q^2/4\pi\epsilon_0 R$. We see that it takes three-fifths as much energy to assemble the sphere as it does to bring in two point charges $Q$ to a separation of $R$. Exercise 1.61 presents an alternative method of calculating the potential energy of a uniformly charged sphere, by imagining building it up layer by layer.

---

We run into trouble if we try to apply Eq. (1.53) to a system that contains a point charge, that is, a finite charge $q$ of zero size. Locate $q$ at the origin of the coordinates. Close to the origin, $E^2$ will approach $q^2/(4\pi\epsilon_0)^2 r^4$. With $dv = 4\pi r^2\, dr$, the integrand $E^2\, dv$ will behave like $dr/r^2$, and our integral will blow up at the limit $r = 0$. That simply tells us that it would take infinite energy to pack finite charge into zero volume – which is true but not helpful. In the real world we deal with particles like electrons and protons. They are so small that for most purposes we can ignore their dimensions and think of them as point charges when we consider their electrical interaction with one another. How much energy it took to make such a particle is a question that goes beyond the range of classical electromagnetism. We have to regard the particles as supplied to us ready-made. The energy we are concerned with is the work done in moving them around.

The distinction is usually clear. Consider two charged particles, a proton and a negative pion, for instance. Let $\mathbf{E}_p$ be the electric field of the proton, $\mathbf{E}_\pi$ that of the pion. The total field is $\mathbf{E} = \mathbf{E}_p + \mathbf{E}_\pi$, and $\mathbf{E}\cdot\mathbf{E}$

equals $E_\mathrm{p}^2 + E_\pi^2 + 2\mathbf{E}_\mathrm{p} \cdot \mathbf{E}_\pi$. According to Eq. (1.53) the total energy in the electric field of this two-particle system is

$$U = \frac{\epsilon_0}{2} \int E^2 \, dv$$

$$= \frac{\epsilon_0}{2} \int E_\mathrm{p}^2 \, dv + \frac{\epsilon_0}{2} \int E_\pi^2 \, dv + \epsilon_0 \int \mathbf{E}_\mathrm{p} \cdot \mathbf{E}_\pi \, dv. \qquad (1.56)$$

The value of the first integral is a property of any isolated proton. It is a constant of nature which is not changed by moving the proton around. The same goes for the second integral, involving the pion's electric field alone. It is the third integral that directly concerns us, for it expresses the energy required to assemble the system *given* a proton and a pion as constituents.

The distinction could break down if the two particles interact so strongly that the electrical structure of one is distorted by the presence of the other. Knowing that both particles are in a sense composite (the proton consisting of three quarks, the pion of two), we might expect that to happen during a close approach. In fact, nothing much happens down to a distance of $10^{-15}$ m. At shorter distances, for strongly interacting particles like the proton and the pion, nonelectrical forces dominate the scene anyway.

That explains why we do not need to include "self-energy" terms like the first two integrals in Eq. (1.56) in our energy accounts for a system of elementary charged particles. Indeed, we want to omit them. We are doing just that, in effect, when we replace the actual distribution of discrete elementary charges (the electrons on the rubber balloon) by a perfectly continuous charge distribution.

## 1.16 Applications

Each chapter of this book concludes with a list of "everyday" applications of the topics covered in the chapter. The discussions are brief. It would take many pages to explain each item in detail; real-life physics tends to involve countless variations, complications, and subtleties. The main purpose here is just to say a few words to convince you that the applications are interesting and worthy of further study. You can carry onward with some combination of books/internet/people/pondering. There is effectively an infinite amount of information out there, so you should take advantage of it! Two books packed full of real-life applications are:

- *The Flying Circus of Physics* (Walker, 2007);
- *How Things Work* (Bloomfield, 2010).

And some very informative websites are:

- *The Flying Circus of Physics* website: www.flyingcircusofphysics.com;
- *How Stuff Works*: www.howstuffworks.com;

- *Explain That Stuff*: www.explainthatstuff.com;
- and *Wikipedia*, of course: www.wikipedia.org.

These websites can point you to more technical sources if you want to pursue things at a more advanced level.

With the exception of the gravitational force keeping us on the earth, and ignoring magnets for the time being, essentially all "everyday" forces are *electrostatic* in origin (with some quantum mechanics mixed in, to make things stable; see Earnshaw's theorem in Section 2.12). Friction, tension, normal force, etc., all boil down to the electric forces between the electrons in the various atoms and molecules. You can open a door by pushing on it because the forces between neighboring molecules in the door, and also in your hand, are sufficiently strong. We can ignore the gravitational force between everyday-sized objects because the gravitational force is so much weaker than the electric force (see Problem 1.1). Only if one of the objects is the earth does the gravitational force matter. And even in that case, it is quite remarkable that the electric forces between the molecules in, say, a wooden board that you might be standing on can completely balance the gravitational force on you due to the entire earth. However, this wouldn't be the case if you attempt to stand on a lake (unless it's frozen!).

If you want to give an object a net charge, a possible way is via the *triboelectric effect*. If certain materials are rubbed against each other, they can become charged. For example, rubbing wool and Teflon together causes the wool to become positively charged and the Teflon negatively charged. The mechanism is simple: the Teflon simply grabs electrons from the wool. The determination of which material ends up with extra electrons depends on the electronic structure of the molecules in the materials. It turns out that actual rubbing isn't necessary. Simply touching and separating the materials can produce an imbalance of charge. Triboelectric effects are mitigated by humid air, because the water molecules in the air are inclined to give or receive electrons, depending on which of these actions neutralizes the object. This is due to the fact that water molecules are *polar*, that is, they are electrically lopsided. (Polar molecules will be discussed in Chapter 10.)

The *electrical breakdown* of air occurs when the electric field reaches a strength of about $3 \cdot 10^6$ V/m. In fields this strong, electrons are ripped from molecules in the air. They are then accelerated by the field and collide with other molecules, knocking electrons out of these molecules, and so on, in a cascading process. The result is a *spark*, because eventually the electrons will combine in a more friendly manner with molecules and drop down to a lower energy level, emitting the light that you see. If you shuffle your feet on a carpet and then bring your finger close to a grounded object, you will see a spark.

The electric field near the surface of the earth is about 100 V/m, pointing downward. You can show that this implies a charge of $-5 \cdot 10^5$ C on the earth. The atmosphere contains roughly the opposite charge, so that the earth-plus-atmosphere system is essentially neutral, as it must be. (Why?) If there were no regenerative process, charge would leak between the ground and the atmosphere, and they would neutralize each other in about an hour. But there *is* a regenerative process: *lightning*. This is a spectacular example of electrical breakdown. There are millions of lightning strikes per day over the surface of the earth, the vast majority of which transfer negative charge to the earth. A lightning strike is the result of the strong electric field that is produced by the buildup (or rather, the separation) of charge in a cloud. This separation arises from the charge carried on moving raindrops, although the exact process is rather complicated (see the interesting discussion in Chapter 9 of Feynman *et al.* (1977)). "Lightning" can also arise from the charge carried on dust particles in coal mines, flour mills, grain storage facilities, etc. The result can be a deadly explosion.

A more gentle form of electrical breakdown is *corona discharge*. Near the tip of a charged pointy object, such as a needle, the field is large but then falls off rapidly. (You can model the needle roughly as having a tiny charged sphere on its end.) Electrons are ripped off the needle (or off the air molecules) very close to the needle, but the field farther away isn't large enough to sustain the breakdown. So there is a slow leakage instead of an abrupt spark. This leakage can sometimes be seen as a faint glow. Examples are *St. Elmo's fire* at the tips of ship masts, and a glow at the tips of airplane wings.

*Electrostatic paint sprayers* can produce very even coats of paint. As the paint leaves the sprayer, an electrode gives it a charge. This causes the droplets in the paint mist to repel each other, helping to create a uniform mist with no clumping. If the object being painted is grounded (or given the opposite charge), the paint will be attracted to it, leading to less wasted paint, less mess, and less inhalation of paint mist. When painting a metal pipe, for example, the mist will wrap around and partially coat the back side, instead of just sailing off into the air.

*Photocopiers* work by giving the toner powder a charge, and giving certain locations on a drum or belt the opposite charge. These locations on the drum can be made to correspond to the locations of ink on the original paper. This is accomplished by coating the drum with a *photoconductive* material, that is, one that becomes conductive when exposed to light. The entire surface of the drum is given an initial charge and then exposed to light at locations corresponding to the white areas on the original page (accomplished by reflecting light off the page). The charge can be made to flow off these newly conductive locations on the drum, leaving charge only at the locations corresponding to the ink. When the oppositely charged toner is brought nearby, it is attracted to these locations on the drum. The toner is then transferred to a piece of paper, producing the desired copy.

*Electronic paper*, used in many eBook readers, works by using electric fields to rotate or translate small black and white objects.

One technique uses tiny spheres (about $10^{-4}$ m in diameter) that are black on one side and white on the other, with the sides being oppositely charged. Another technique uses similarly tiny spheres that are filled with many even tinier charged white particles along with a dark dye. In both cases, a narrow gap between sheets of electrodes (with one sheet being the transparent sheet that you look through) is filled with the spheres. By depositing a specific pattern of charge on the sheets, the color of the objects facing your eye can be controlled. In the first system, the black and white spheres rotate accordingly. In the second system, the tiny white particles pile up on one side of the sphere. In contrast with a standard LCD computer screen, electronic paper acts like normal paper, in that it doesn't produce its own light; an outside light source is needed to view the page. An important advantage of electronic paper is that it uses a very small amount of power. A battery is needed only when the page is refreshed, whereas an LCD screen requires continual refreshing.

## CHAPTER SUMMARY

- Electric charge, which can be positive or negative, is both *conserved* and *quantized*. The force between two charges is given by *Coulomb's law*:

$$\mathbf{F} = \frac{1}{4\pi\epsilon_0} \frac{q_1 q_2 \hat{\mathbf{r}}_{21}}{r_{21}^2}. \tag{1.57}$$

Integrating this force, we find that the *potential energy* of a system of charges (the work necessary to bring them in from infinity) equals

$$U = \frac{1}{2} \sum_{j=1}^{N} \sum_{k \neq j} \frac{1}{4\pi\epsilon_0} \frac{q_j q_k}{r_{jk}}. \tag{1.58}$$

- The *electric field* due to a charge distribution is (depending on whether the distribution is continuous or discrete)

$$\mathbf{E} = \frac{1}{4\pi\epsilon_0} \int \frac{\rho(x',y',z')\hat{\mathbf{r}}\,dx'\,dy'\,dz'}{r^2} \quad \text{or} \quad \frac{1}{4\pi\epsilon_0} \sum_{j=1}^{N} \frac{q_j \hat{\mathbf{r}}_j}{r_j^2}. \tag{1.59}$$

The force on a test charge $q$ due to the field is $\mathbf{F} = q\mathbf{E}$.

- The *flux* of an electric field through a surface $S$ is

$$\Phi = \int_S \mathbf{E} \cdot d\mathbf{a}. \tag{1.60}$$

*Gauss's law* states that the flux of the electric field $\mathbf{E}$ through any closed surface equals $1/\epsilon_0$ times the total charge enclosed by the surface. That is (depending on whether the distribution is continuous or discrete),

$$\int \mathbf{E} \cdot d\mathbf{a} = \frac{1}{\epsilon_0} \int \rho \, dv = \frac{1}{\epsilon_0} \sum_i q_i. \qquad (1.61)$$

Gauss's law gives the fields for a sphere, line, and sheet of charge as

$$E_{\text{sphere}} = \frac{Q}{4\pi\epsilon_0 r^2}, \qquad E_{\text{line}} = \frac{\lambda}{2\pi\epsilon_0 r}, \qquad E_{\text{sheet}} = \frac{\sigma}{2\epsilon_0}. \qquad (1.62)$$

More generally, the discontinuity in the normal component of $\mathbf{E}$ across a sheet is $\Delta E_\perp = \sigma/\epsilon_0$. Gauss's law is always valid, although it is useful for calculating the electric field only in cases where there is sufficient symmetry.

- The force per unit area on a layer of charge equals the density times the average of the fields on either side:

$$\frac{F}{A} = \frac{1}{2}(E_1 + E_2)\sigma. \qquad (1.63)$$

- The *energy density* of an electric field is $\epsilon_0 E^2/2$, so the total energy in a system equals

$$U = \frac{\epsilon_0}{2} \int E^2 \, dv. \qquad (1.64)$$

## Problems

1.1  *Gravity vs. electricity*  ∗

(a) In the domain of elementary particles, a natural unit of mass is the mass of a *nucleon,* that is, a proton or a neutron, the basic massive building blocks of ordinary matter. Given the nucleon mass as $1.67 \cdot 10^{-27}$ kg and the gravitational constant $G$ as $6.67 \cdot 10^{-11}$ m$^3$/(kg s$^2$), compare the gravitational attraction of two protons with their electrostatic repulsion. This shows why we call gravitation a very *weak* force.

(b) The distance between the two protons in the helium nucleus could be at one instant as much as $10^{-15}$ m. How large is the force of electrical repulsion between two protons at that distance? Express it in newtons, and in pounds. Even stronger is the *nuclear* force that acts between any pair of hadrons (including neutrons and protons) when they are that close together.

1.2  *Zero force from a triangle*  ∗∗

Two positive ions and one negative ion are fixed at the vertices of an equilateral triangle. Where can a fourth ion be placed, along the symmetry axis of the setup, so that the force on it will be zero? Is there more than one such place? You will need to solve something numerically.

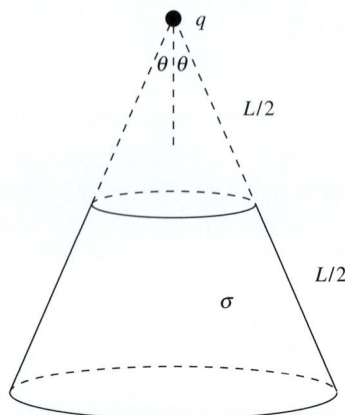

**Figure 1.31.**

**Figure 1.32.**

1.3    *Force from a cone* **

(a) A charge $q$ is located at the tip of a hollow cone (such as an ice cream cone without the ice cream) with surface charge density $\sigma$. The slant height of the cone is $L$, and the half-angle at the vertex is $\theta$. What can you say about the force on the charge $q$ due to the cone?

(b) If the top half of the cone is removed and thrown away (see Fig. 1.31), what is the force on the charge $q$ due to the remaining part of the cone? For what angle $\theta$ is this force maximum?

1.4    *Work for a rectangle* **

Two protons and two electrons are located at the corners of a rectangle with side lengths $a$ and $b$. There are two essentially different arrangements. Consider the work required to assemble the system, starting with the particles very far apart. Is it possible for the work to be positive for either of the arrangements? If so, how must $a$ and $b$ be related? You will need to solve something numerically.

1.5    *Stable or unstable?* **

In the setup in Exercise 1.37, is the charge $-Q$ at the center of the square in stable or unstable equilibrium? You can answer this by working with either forces or energies. The latter has the advantage of not involving components, although things can still get quite messy. However, the math is simple if you use a computer. Imagine moving the $-Q$ charge infinitesimally to the point $(x, y)$, and use, for example, the Series operation in *Mathematica* to calculate the new energy of the charge, to lowest nontrivial order in $x$ and $y$. If the energy decreases for at least one direction of displacement, then the equilibrium is unstable. (The equilibrium is certainly stable with respect to displacements perpendicular to the plane of the square, because the attractive force from the other charges is directed back toward the plane. The question is, what happens in the plane of the square?)

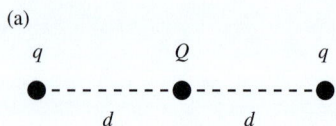

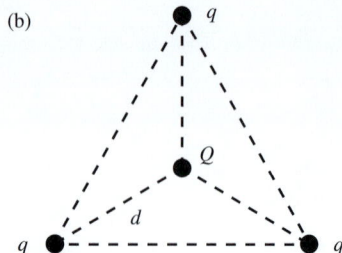

1.6    *Zero potential energy for equilibrium* **

(a) Two charges $q$ are each located a distance $d$ from a charge $Q$, as shown in Fig. 1.32(a). What should the charge $Q$ be so that the system is in equilibrium; that is, so that the force on each charge is zero? (The equilibrium is an unstable one, which can be seen by looking at longitudinal displacements of the (negative) charge $Q$. This is consistent with a general result that we will derive Section 2.12.)

(b) Same question, but now with the setup in Fig. 1.32(b). The three charges $q$ are located at the vertices of an equilateral triangle.

(c) Show that the total potential energy in each of the above systems is zero.

(d) In view of the previous result, we might make the following conjecture: "The total potential energy of any system of charges in equilibrium is zero." Prove that this conjecture is indeed true. *Hint:* The goal is to show that zero work is required to move the charges out to infinity. Since the electrostatic force is conservative, you need only show that the work is zero for one particular set of paths of the charges. And there is indeed a particular set of paths that makes the result clear.

1.7 *Potential energy in a two-dimensional crystal* ✶✶
Use a computer to calculate numerically the potential energy, per ion, for an infinite two-dimensional square ionic crystal with separation $a$; that is, a plane of equally spaced charges of magnitude $e$ and alternating sign (as with a checkerboard).

1.8 *Oscillating in a ring* ✶✶✶
A ring with radius $R$ has uniform positive charge density $\lambda$. A particle with positive charge $q$ and mass $m$ is initially located at the center of the ring and is then given a tiny kick. If it is constrained to move in the plane of the ring, show that it undergoes simple harmonic motion (for small oscillations), and find the frequency. *Hint:* Find the potential energy of the particle when it is at a (small) radius, $r$, by integrating over the ring, and then take the negative derivative to find the force. You will need to use the law of cosines and also the Taylor series $1/\sqrt{1+\epsilon} \approx 1 - \epsilon/2 + 3\epsilon^2/8$.

1.9 *Field from two charges* ✶✶
A charge $2q$ is at the origin, and a charge $-q$ is at $x = a$ on the $x$ axis.
   (a) Find the point on the $x$ axis where the electric field is zero.
   (b) Consider the vertical line passing through the charge $-q$, that is, the line given by $x = a$. Locate, at least approximately, a point on this line where the electric field is parallel to the $x$ axis.

1.10 *45-degree field line* ✶✶
A half-infinite line has linear charge density $\lambda$. Find the electric field at a point that is "even" with the end, a distance $\ell$ from it, as shown in Fig. 1.33. You should find that the field always points up at a 45° angle, independent of $\ell$.

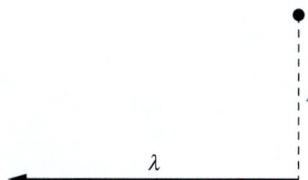

**Figure 1.33.**

1.11 *Field at the end of a cylinder* ✶✶
   (a) Consider a half-infinite hollow cylindrical shell (that is, one that extends to infinity in one direction) with radius $R$ and uniform surface charge density $\sigma$. What is the electric field at the midpoint of the end face?
   (b) Use your result to determine the field at the midpoint of a half-infinite *solid* cylinder with radius $R$ and uniform volume

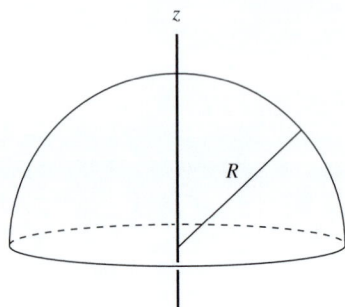

**Figure 1.34.**

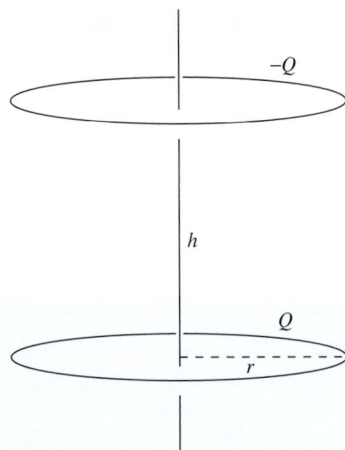

**Figure 1.35.**

charge density $\rho$, which can be considered to be built up from many cylindrical shells.

**1.12** *Field from a hemispherical shell* ***

A hemispherical shell has radius $R$ and uniform surface charge density $\sigma$ (see Fig. 1.34). Find the electric field at a point on the symmetry axis, at position $z$ relative to the center, for any $z$ value from $-\infty$ to $\infty$.

**1.13** *A very uniform field* ***

(a) Two rings with radius $r$ have charge $Q$ and $-Q$ uniformly distributed around them. The rings are parallel and located a distance $h$ apart, as shown in Fig. 1.35. Let $z$ be the vertical coordinate, with $z = 0$ taken to be at the center of the lower ring. As a function of $z$, what is the electric field at points on the axis of the rings?

(b) You should find that the electric field is an even function with respect to the $z = h/2$ point midway between the rings. This implies that, at this point, the field has a local extremum as a function of $z$. The field is therefore fairly uniform there; there are no variations to first order in the distance along the axis from the midpoint. What should $r$ be in terms of $h$ so that the field is *very* uniform?

By "very" uniform we mean that additionally there aren't any variations to second order in $z$. That is, the second derivative vanishes. This then implies that the leading-order change is *fourth* order in $z$ (because there are no variations at any odd order, since the field is an even function around the midpoint). Feel free to calculate the derivatives with a computer.

**1.14** *Hole in a plane* **

(a) A hole of radius $R$ is cut out from a very large flat sheet with uniform charge density $\sigma$. Let $L$ be the line perpendicular to the sheet, passing through the center of the hole. What is the electric field at a point on $L$, a distance $z$ from the center of the hole? *Hint:* Consider the plane to consist of many concentric rings.

(b) If a charge $-q$ with mass $m$ is released from rest on $L$, very close to the center of the hole, show that it undergoes oscillatory motion, and find the frequency $\omega$ of these oscillations. What is $\omega$ if $m = 1$ g, $-q = -10^{-8}$ C, $\sigma = 10^{-6}$ C/m$^2$, and $R = 0.1$ m?

(c) If a charge $-q$ with mass $m$ is released from rest on $L$, a distance $z$ from the sheet, what is its speed when it passes through the center of the hole? What does your answer reduce to for large $z$ (or, equivalently, small $R$)?

1.15  *Flux through a circle* **

A point charge $q$ is located at the origin. Consider the electric field flux through a circle a distance $\ell$ from $q$, subtending an angle $2\theta$, as shown in Fig. 1.36. Since there are no charges except at the origin, any surface that is bounded by the circle and that stays to the right of the origin must contain the same flux. (Why?) Calculate this flux by taking the surface to be:

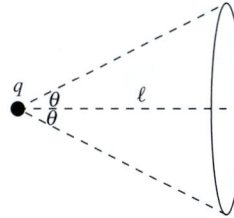

**Figure 1.36.**

(a)  the flat disk bounded by the circle;

(b)  the spherical cap (with the sphere centered at the origin) bounded by the circle.

1.16  *Gauss's law and two point charges* **

(a)  Two point charges $q$ are located at positions $x = \pm\ell$. At points close to the origin on the $x$ axis, find $E_x$. At points close to the origin on the $y$ axis, find $E_y$. Make suitable approximations with $x \ll \ell$ and $y \ll \ell$.

(b)  Consider a small cylinder centered at the origin, with its axis along the $x$ axis. The radius is $r_0$ and the length is $2x_0$. Using your results from part (a), verify that there is zero flux through the cylinder, as required by Gauss's law.

1.17  *Zero field inside a spherical shell* **

Consider a hollow spherical shell with uniform surface charge density. By considering the two small patches at the ends of the thin cones in Fig. 1.37, show that the electric field at any point $P$ in the interior of the shell is zero. This then implies that the electric potential (defined in Chapter 2) is constant throughout the interior.

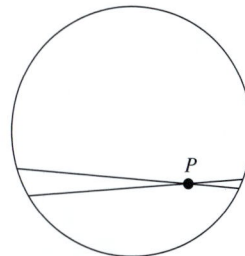

**Figure 1.37.**

1.18  *Fields at the surfaces* **

Consider the electric field at a point on the surface of (a) a sphere with radius $R$, (b) a cylinder with radius $R$ whose length is infinite, and (c) a slab with thickness $2R$ whose other two dimensions are infinite. All of the objects have the same volume charge density $\rho$. Compare the fields in the three cases, and explain physically why the sizes take the order they do.

1.19  *Sheet on a sphere* **

Consider a large flat horizontal sheet with thickness $x$ and volume charge density $\rho$. This sheet is tangent to a sphere with radius $R$ and volume charge density $\rho_0$, as shown in Fig. 1.38. Let $A$ be the point of tangency, and let $B$ be the point opposite to $A$ on the top side of the sheet. Show that the net upward electric field (from the sphere plus the sheet) at $B$ is larger than at $A$ if $\rho > (2/3)\rho_0$. (Assume $x \ll R$.)

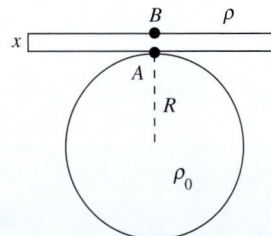

**Figure 1.38.**

1.20 *Thundercloud* **

You observe that the passage of a particular thundercloud over-head causes the vertical electric field strength in the atmosphere, measured at the ground, to rise to 3000 N/C (or V/m).

(a) How much charge does the thundercloud contain, in coulombs per square meter of horizontal area? Assume that the width of the cloud is large compared with the height above the ground.

(b) Suppose there is enough water in the thundercloud in the form of 1 mm diameter drops to make 0.25 cm of rainfall, and that it is those drops that carry the charge. How large is the electric field strength at the surface of one of the drops?

1.21 *Field in the end face* *

Consider a half-infinite hollow cylindrical shell (that is, one that extends to infinity in one direction) with uniform surface charge density. Show that at all points in the circular end face, the electric field is parallel to the cylinder's axis. *Hint:* Use superposition, along with what you know about the field from an infinite (in both directions) hollow cylinder.

1.22 *Field from a spherical shell, right and wrong* **

The electric field outside and an infinitesimal distance away from a uniformly charged spherical shell, with radius $R$ and surface charge density $\sigma$, is given by Eq. (1.42) as $\sigma/\epsilon_0$. Derive this in the following way.

(a) Slice the shell into rings (symmetrically located with respect to the point in question), and then integrate the field contributions from all the rings. You should obtain the incorrect result of $\sigma/2\epsilon_0$.

(b) Why isn't the result correct? Explain how to modify it to obtain the correct result of $\sigma/\epsilon_0$. *Hint:* You could very well have performed the above integral in an effort to obtain the electric field an infinitesimal distance *inside* the shell, where we know the field is zero. Does the above integration provide a good description of what's going on for points on the shell that are very close to the point in question?

1.23 *Field near a stick* **

A stick with length $2\ell$ has uniform linear charge density $\lambda$. Consider a point $P$, a distance $\eta\ell$ from the center (where $0 \leq \eta < 1$), and an infinitesimal distance away from the stick. Up close, the stick looks infinitely long, as far as the **E** component perpendicular to the stick is concerned. So we have $E_\perp = \lambda/2\pi\epsilon_0 r$. Find the **E** component parallel to the stick, $E_\parallel$. Does it approach infinity, or does it remain finite at the end of the stick?

1.24 *Potential energy of a cylinder* ✷✷✷

A cylindrical volume of radius $a$ is filled with charge of uniform density $\rho$. We want to know the potential energy per unit length of this cylinder of charge, that is, the work done per unit length in assembling it. Calculate this by building up the cylinder layer by layer, making use of the fact that the field outside a cylindrical distribution of charge is the same as if all the charge were located on the axis. You will find that the energy per unit length is infinite if the charges are brought in from infinity, so instead assume that they are initially distributed uniformly over a hollow cylinder with large radius $R$. Write your answer in terms of the charge per unit length of the cylinder, which is $\lambda = \rho\pi a^2$. (See Exercise 1.83 for a different method of solving this problem.)

1.25 *Two equal fields* ✷✷

The result of Exercise 1.78 is that the electric field at the center of a small hole in a spherical shell equals $\sigma/2\epsilon_0$. This happens to be the same as the field due to an infinite flat sheet with the same density $\sigma$. That is, at the center of the hole at the top of the spherical shell in Fig. 1.39, the field from the shell equals the field from the infinite horizontal sheet shown. (This sheet could actually be located at any height.) Demonstrate this equality by explaining why the rings on the shell and sheet that are associated with the angle $\theta$ and angular width $d\theta$ yield the same field at the top of the shell.

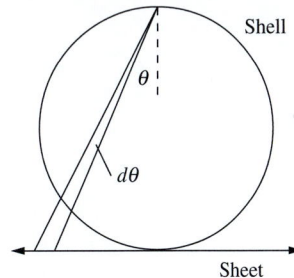

**Figure 1.39.**

1.26 *Stable equilibrium in electron jelly* ✷✷

The task of Exercise 1.77 is to find the equilibrium positions of two protons located inside a sphere of electron jelly with total charge $-2e$. Show that the equilibria are *stable*. That is, show that a displacement in any direction will result in a force directed back toward the equilibrium position. (There is no need to know the exact locations of the equilibria, so you can solve this problem without solving Exercise 1.77 first.)

1.27 *Uniform field in a cavity* ✷✷

A sphere has radius $R_1$ and uniform volume charge density $\rho$. A spherical cavity with radius $R_2$ is carved out at an arbitrary location inside the larger sphere. Show that the electric field inside the cavity is uniform (in both magnitude and direction). *Hint:* Find a vector expression for the field in the interior of a charged sphere, and then use superposition.

What are the analogous statements for the lower-dimensional analogs with cylinders and slabs? Are the statements still true?

1.28 *Average field on/in a sphere* ✷✷

(a) A point charge $q$ is located at an arbitrary position *inside* a sphere (just an imaginary sphere in space) with radius $R$. Show that the average electric field over the *surface* of the sphere is zero. *Hint:* Use an argument involving Newton's third law, along with what you know about spherical shells.

(b) If the point charge $q$ is instead located *outside* the sphere, a distance $r$ from the center, show that the average electric field over the *surface* of the sphere has magnitude $q/4\pi\epsilon_0 r$.

(c) Return to the case where the point charge $q$ is located *inside* the sphere of radius $R$. Let the distance from the center be $r$. Use the above results to show that the average electric field over the entire *volume* of the sphere of radius $R$ has magnitude $qr/4\pi\epsilon_0 R^3$ and points toward the center (if $q$ is positive).

**1.29** *Pulling two sheets apart* **

Two parallel sheets each have large area $A$ and are separated by a small distance $\ell$. The surface charge densities are $\sigma$ and $-\sigma$. You wish to pull one of the sheets away from the other, by a small distance $x$. How much work does this require? Calculate this by:

(a) using the relation $W = $ (force) $\times$ (distance);

(b) calculating the increase in energy stored in the electric field. Show that these two methods give the same result.

**1.30** *Force on a patch* **

Consider a small patch of charge that is part of a larger surface. The surface charge density is $\sigma$. If $\mathbf{E}_1$ and $\mathbf{E}_2$ are the electric fields on either side of the patch, show that the force per unit area on the patch equals $\sigma(\mathbf{E}_1 + \mathbf{E}_2)/2$. This is the result we derived in Section 1.14, for the case where the field is perpendicular to the surface. Derive it here by using the fact that the force on the patch is due to the field $\mathbf{E}^{\text{other}}$ from all the *other* charges in the system (excluding the patch), and then finding an expression for $\mathbf{E}^{\text{other}}$ in terms of $\mathbf{E}_1$ and $\mathbf{E}_2$.

**1.31** *Decreasing energy?* *

A hollow spherical shell with radius $R$ has charge $Q$ uniformly distributed over it. The task of Problem 1.32 is to show that the energy stored in this system is $Q^2/8\pi\epsilon_0 R$. (You can derive this here if you want, or you can just accept it for the purposes of this problem.) Now imagine taking all of the charge and concentrating it in two point charges $Q/2$ located at diametrically opposite positions on the shell. The energy of this new system is $(Q/2)^2/4\pi\epsilon_0(2R) = Q^2/32\pi\epsilon_0 R$, which is less than the energy of the uniform spherical shell. Does this make sense? If not, where is the error in this reasoning?

1.32 *Energy of a shell* **

A hollow spherical shell with radius $R$ has charge $Q$ uniformly distributed over it. Show that the energy stored in this system is $Q^2/8\pi\epsilon_0 R$. Do this in two ways as follows.

(a) Use Eq. (1.53) to find the energy stored in the electric field.

(b) Imagine building up the shell by successively adding on infinitesimally thin shells with charge $dq$. Find the energy needed to add on a shell when the charge already there is $q$, and then integrate over $q$.

1.33 *Deriving the energy density* ***

Consider the electric field of two protons a distance $b$ apart. According to Eq. (1.53) (which we stated but did not prove), the potential energy of the system ought to be given by

$$U = \frac{\epsilon_0}{2} \int \mathbf{E}^2 \, dv = \frac{\epsilon_0}{2} \int (\mathbf{E}_1 + \mathbf{E}_2)^2 \, dv$$
$$= \frac{\epsilon_0}{2} \int \mathbf{E}_1^2 \, dv + \frac{\epsilon_0}{2} \int \mathbf{E}_2^2 \, dv + \epsilon_0 \int \mathbf{E}_1 \cdot \mathbf{E}_2 \, dv, \qquad (1.65)$$

where $\mathbf{E}_1$ is the field of one particle alone and $\mathbf{E}_2$ that of the other. The first of the three integrals on the right might be called the "electrical self-energy" of one proton; an intrinsic property of the particle, it depends on the proton's size and structure. We have always disregarded it in reckoning the potential energy of a system of charges, on the assumption that it remains constant; the same goes for the second integral. The third integral involves the distance between the charges. Evaluate this integral. This is most easily done if you set it up in spherical polar coordinates with one of the protons at the origin and the other on the polar axis, and perform the integration over $r$ before the integration over $\theta$. Thus, by direct calculation, you can show that the third integral has the value $e^2/4\pi\epsilon_0 b$, which we already know to be the work required to bring the two protons in from an infinite distance to positions a distance $b$ apart. So you will have proved the correctness of Eq. (1.53) for this case, and by invoking superposition you can argue that Eq. (1.53) must then give the energy required to assemble any system of charges.

## Exercises

1.34 *Aircraft carriers and specks of gold* *

Imagine (quite unrealistically) removing one electron from every atom in a tiny cube of gold 1 mm on a side. (Never mind how you would hold the resulting positively charged cube together.) Do the same thing with another such cube a meter away. What is the repulsive force between the two cubes? How many aircraft carriers

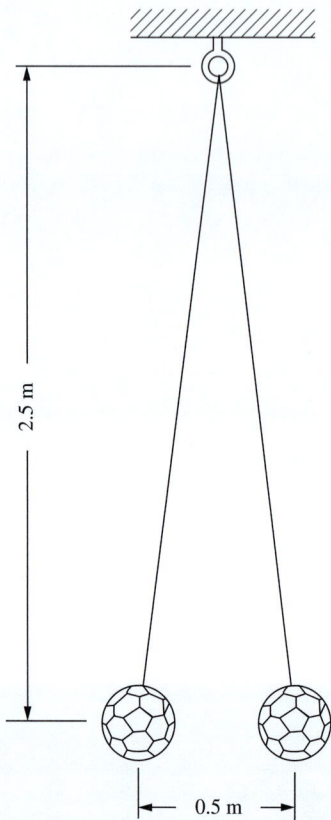

2.5 m

0.5 m

**Figure 1.40.**

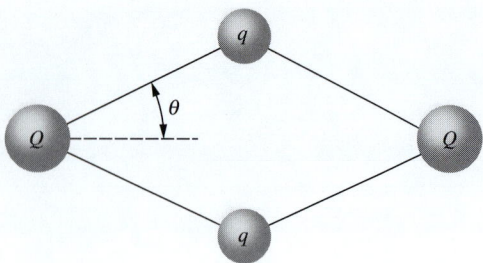

$q$

$\theta$

$Q$                    $Q$

$q$

**Figure 1.41.**

would you need in order to have their total weight equal this force? Some data: The density of gold is $19.3 \, g/cm^3$, and its molecular weight is 197; that is, 1 mole $(6.02 \cdot 10^{23})$ of gold atoms has a mass of 197 grams. The mass of an aircraft carrier is around 100 million kilograms.

1.35 *Balancing the weight* ∗
On the utterly unrealistic assumption that there are no other charged particles in the vicinity, at what distance below a proton would the upward force on an electron equal the electron's weight? The mass of an electron is about $9 \cdot 10^{-31}$ kg.

1.36 *Repelling volley balls* ∗
Two volley balls, mass 0.3 kg each, tethered by nylon strings and charged with an electrostatic generator, hang as shown in Fig. 1.40. What is the charge on each, assuming the charges are equal?

1.37 *Zero force at the corners* ∗∗
(a) At each corner of a square is a particle with charge $q$. Fixed at the center of the square is a point charge of opposite sign, of magnitude $Q$. What value must $Q$ have to make the total force on each of the four particles zero?
(b) With $Q$ taking on the value you just found, show that the potential energy of the system is zero, consistent with the result from Problem 1.6.

1.38 *Oscillating on a line* ∗∗
Two positive point charges $Q$ are located at points $(\pm \ell, 0)$. A particle with positive charge $q$ and mass $m$ is initially located midway between them and is then given a tiny kick. If it is constrained to move along the line joining the two charges $Q$, show that it undergoes simple harmonic motion (for small oscillations), and find the frequency.

1.39 *Rhombus of charges* ∗∗
Four positively charged bodies, two with charge $Q$ and two with charge $q$, are connected by four unstretchable strings of equal length. In the absence of external forces they assume the equilibrium configuration shown in Fig. 1.41. Show that $\tan^3 \theta = q^2/Q^2$. This can be done in two ways. You could show that this relation must hold if the total force on each body, the vector sum of string tension and electrical repulsion, is zero. Or you could write out the expression for the energy $U$ of the assembly (like Eq. (1.13) but for four charges instead of three) and minimize it.

1.40 *Zero potential energy* ∗∗
Find a geometrical arrangement of one proton and two electrons such that the potential energy of the system is exactly zero. How

many such arrangements are there with the three particles on the same straight line? You should find that the ratio of two of the distances involved is the golden ratio.

**1.41** *Work for an octahedron* **

Three protons and three electrons are to be placed at the vertices of a regular octahedron of edge length $a$. We want to find the energy of the system, that is, the work required to assemble it starting with the particles very far apart. There are two essentially different arrangements. What is the energy of each?

**1.42** *Potential energy in a one-dimensional crystal* **

Calculate the potential energy, per ion, for an infinite 1D ionic crystal with separation $a$; that is, a row of equally spaced charges of magnitude $e$ and alternating sign. *Hint:* The power-series expansion of $\ln(1 + x)$ may be of use.

**1.43** *Potential energy in a three-dimensional crystal* **

In the spirit of Problem 1.7, use a computer to calculate numerically the potential energy, per ion, for an infinite 3D cubic ionic crystal with separation $a$. In other words, derive Eq. (1.18).

**1.44** *Chessboard* **

An infinite chessboard with squares of side $s$ has a charge $e$ at the center of every white square and a charge $-e$ at the center of every black square. We are interested in the work $W$ required to transport one charge from its position on the board to an infinite distance from the board. Given that $W$ is finite (which is plausible but not so easy to prove), do you think it is positive or negative? Calculate an approximate value for $W$ by removing the charge from the central square of a $7 \times 7$ board. (Only nine different terms are involved in that sum.) For larger arrays you can write a program to compute the work numerically. This will give you some idea of the rate of convergence toward the value for the infinite array; see Problem 1.7.

**1.45** *Zero field?* **

Four charges, $q$, $-q$, $q$, and $-q$, are located at equally spaced intervals on the $x$ axis. Their $x$ values are $-3a$, $-a$, $a$, and $3a$, respectively. Does there exist a point on the $y$ axis for which the electric field is zero? If so, find the $y$ value.

**1.46** *Charges on a circular track* **

Suppose three positively charged particles are constrained to move on a fixed circular track. If the charges were all equal, an equilibrium arrangement would obviously be a symmetrical one with the particles spaced 120° apart around the circle. Suppose that two of the charges are equal and the equilibrium arrangement is such that these two charges are 90° apart rather than 120°. What is the relative magnitude of the third charge?

1.47  *Field from a semicircle* ∗

A thin plastic rod bent into a semicircle of radius $R$ has a charge $Q$ distributed uniformly over its length. Find the electric field at the center of the semicircle.

1.48  *Maximum field from a ring* ∗∗

A charge $Q$ is distributed uniformly around a thin ring of radius $b$ that lies in the $xy$ plane with its center at the origin. Locate the point on the positive $z$ axis where the electric field is strongest.

1.49  *Maximum field from a blob* ∗∗

(a) A point charge is placed somewhere on the curve shown in Fig. 1.42. This point charge creates an electric field at the origin. Let $E_y$ be the vertical component of this field. What shape (up to a scaling factor) should the curve take so that $E_y$ is independent of the position of the point charge on the curve?

(b) You have a moldable material with uniform volume charge density. What shape should the material take if you want to create the largest possible electric field at a given point in space? Be sure to explain your reasoning clearly.

1.50  *Field from a hemisphere* ∗∗

(a) What is the electric field at the center of a hollow hemispherical shell with radius $R$ and uniform surface charge density $\sigma$? (This is a special case of Problem 1.12, but you can solve the present exercise much more easily from scratch, without going through all the messy integrals of Problem 1.12.)

(b) Use your result to show that the electric field at the center of a *solid* hemisphere with radius $R$ and uniform volume charge density $\rho$ equals $\rho R/4\epsilon_0$.

1.51  *N charges on a circle* ∗∗∗

$N$ point charges, each with charge $Q/N$, are evenly distributed around a circle of radius $R$. What is the electric field at the location of one of the charges, due to all the others? (You can leave your answer in the form of a sum.) In the $N \to \infty$ limit, is the field infinite or finite? In the $N \to \infty$ limit, is the force on one of the charges infinite or finite?

1.52  *An equilateral triangle* ∗

Three positive charges, $A$, $B$, and $C$, of $3 \cdot 10^{-6}$, $2 \cdot 10^{-6}$, and $2 \cdot 10^{-6}$ coulombs, respectively, are located at the corners of an equilateral triangle of side 0.2 m.

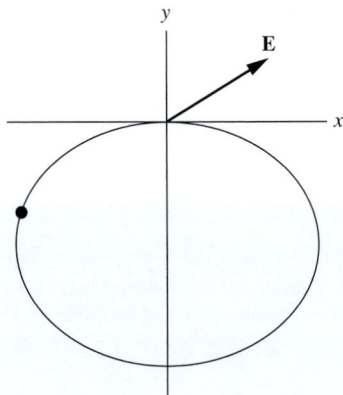

**Figure 1.42.**

(a) Find the magnitude in newtons of the force on each charge.

(b) Find the magnitude in newtons/coulomb of the electric field at the center of the triangle.

**1.53** *Concurrent field lines* **∗∗**

A semicircular wire with radius $R$ has uniform charge density $-\lambda$. Show that at all points along the "axis" of the semicircle (the line through the center, perpendicular to the plane of the semicircle, as shown in Fig. 1.43), the vectors of the electric field all point toward a common point in the plane of the semicircle. Where is this point?

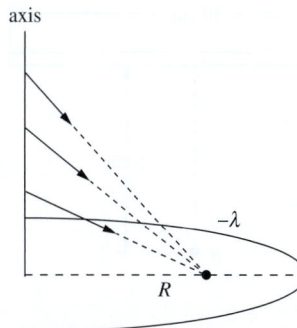

**Figure 1.43.**

**1.54** *Semicircle and wires* **∗∗**

(a) Two long, thin parallel rods, a distance $2b$ apart, are joined by a semicircular piece of radius $b$, as shown in Fig. 1.44. Charge of uniform linear density $\lambda$ is deposited along the whole filament. Show that the field **E** of this charge distribution vanishes at the point $C$. Do this by comparing the contribution of the element at $A$ to that of the element at $B$ which is defined by the same values of $\theta$ and $d\theta$.

(b) Consider the analogous two-dimensional setup involving a cylinder and a hemispherical end cap, with uniform surface charge density $\sigma$. Using the result from part (a), do you think that the field at the analogous point $C$ is directed upward, downward, or is zero? (No calculations needed!)

**1.55** *Field from a finite rod* **∗∗**

A thin rod 10 cm long carries a total charge of 24 esu $= 8 \cdot 10^{-9}$ C uniformly distributed along its length. Find the strength of the electric field at each of the two points $A$ and $B$ located as shown in Fig. 1.45.

**1.56** *Flux through a cube* **∗**

(a) A point charge $q$ is located at the center of a cube of edge $d$. What is the value of $\int \mathbf{E} \cdot d\mathbf{a}$ over one face of the cube?

(b) The charge $q$ is moved to one corner of the cube. Now what is the value of the flux of **E** through each of the faces of the cube? (To make things well defined, treat the charge like a tiny sphere.)

**1.57** *Escaping field lines* **∗∗**

Charges $2q$ and $-q$ are located on the $x$ axis at $x = 0$ and $x = a$, respectively.

(a) Find the point on the $x$ axis where the electric field is zero, and make a rough sketch of some field lines.

(b) You should find that some of the field lines that start on the $2q$ charge end up on the $-q$ charge, while others head off to infinity. Consider the field lines that form the cutoff between these two cases. At what angle (with respect to the $x$ axis) do

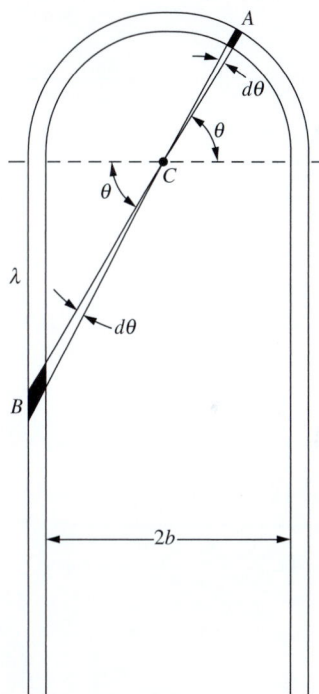

**Figure 1.44.**

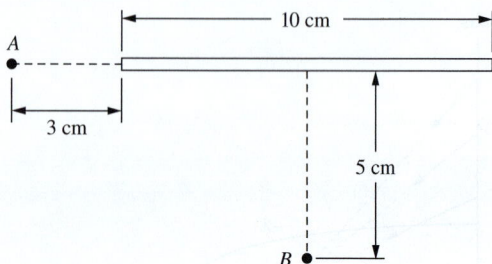

**Figure 1.45.**

these lines leave the $2q$ charge? *Hint:* Draw a wisely chosen Gaussian surface that mainly follows these lines.

1.58 *Gauss's law at the center of a ring*  **
   (a) A ring with radius $R$ has total charge $Q$ uniformly distributed around it. To leading order, find the electric field at a point along the axis of the ring, a very small distance $z$ from the center.
   (b) Consider a small cylinder centered at the center of the ring, with small radius $r_0$ and small height $2z_0$, with $z_0$ lying on either side of the plane of the ring. There is no charge in this cylinder, so the net flux through it must be zero. Using a result given in the solution to Problem 1.8, verify that this is indeed the case (to leading order in the small distances involved).

1.59 *Zero field inside a cylindrical shell*  *
   Consider a distribution of charge in the form of a hollow circular cylinder, like a long charged pipe. In the spirit of Problem 1.17, show that the electric field inside the pipe is zero.

1.60 *Field from a hollow cylinder*  *
   Consider the hollow cylinder from Exercise 1.59. Use Gauss's law to show that the field inside the pipe is zero. Also show that the field outside is the same as if the charge were all on the axis. Is either statement true for a pipe of square cross section on which the charge is distributed with uniform surface density?

1.61 *Potential energy of a sphere*  **
   A spherical volume of radius $R$ is filled with charge of uniform density $\rho$. We want to know the potential energy $U$ of this sphere of charge, that is, the work done in assembling it. In the example in Section 1.15, we calculated $U$ by integrating the energy density of the electric field; the result was $U = (3/5)Q^2/4\pi\epsilon_0 R$. Derive $U$ here by building up the sphere layer by layer, making use of the fact that the field outside a spherical distribution of charge is the same as if all the charge were at the center.

1.62 *Electron self-energy*  *
   At the beginning of the twentieth century the idea that the rest mass of the electron might have a purely electrical origin was very attractive, especially when the equivalence of energy and mass was revealed by special relativity. Imagine the electron as a ball of charge, of constant volume density out to some maximum radius $r_0$. Using the result of Exercise 1.61, set the potential energy of this system equal to $mc^2$ and see what you get for $r_0$. One defect of the model is rather obvious: nothing is provided to hold the charge together!

1.63  *Sphere and cones* **

(a) Consider a fixed hollow spherical shell with radius $R$ and surface charge density $\sigma$. A particle with mass $m$ and charge $-q$ that is initially at rest falls in from infinity. What is its speed when it reaches the center of the shell? (Assume that a tiny hole has been cut in the shell, to let the charge through.)

(b) Consider two fixed hollow conical shells (that is, ice cream cones without the ice cream) with base radius $R$, slant height $L$, and surface charge density $\sigma$, arranged as shown in Fig. 1.46. A particle with mass $m$ and charge $-q$ that is initially at rest falls in from infinity, along the perpendicular bisector line, as shown. What is its speed when it reaches the tip of the cones? You should find that your answer relates very nicely to your answer for part (a).

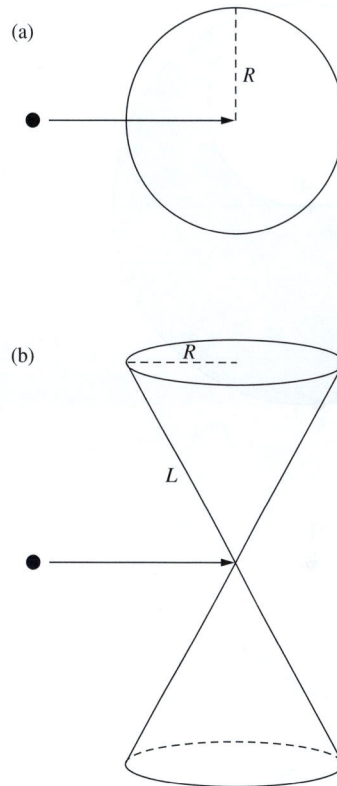

(a)

(b)

**Figure 1.46.**

1.64  *Field between two wires* *

Consider a high-voltage direct current power line that consists of two parallel conductors suspended 3 meters apart. The lines are oppositely charged. If the electric field strength halfway between them is 15,000 N/C, how much excess positive charge resides on a 1 km length of the positive conductor?

1.65  *Building a sheet from rods* **

An infinite uniform sheet of charge can be thought of as consisting of an infinite number of adjacent uniformly charged rods. Using the fact that the electric field from an infinite rod is $\lambda/2\pi\epsilon_0 r$, integrate over these rods to show that the field from an infinite sheet with charge density $\sigma$ is $\sigma/2\epsilon_0$.

1.66  *Force between two strips* **

(a) The two strips of charge shown in Fig. 1.47 have width $b$, infinite height, and negligible thickness (in the direction perpendicular to the page). Their charge densities per unit area are $\pm\sigma$. Find the magnitude of the electric field due to one of the strips, a distance $x$ away from it (in the plane of the page).

(b) Show that the force (per unit height) between the two strips equals $\sigma^2 b(\ln 2)/\pi\epsilon_0$. Note that this result is finite, even though you will find that the field due to a strip diverges as you get close to it.

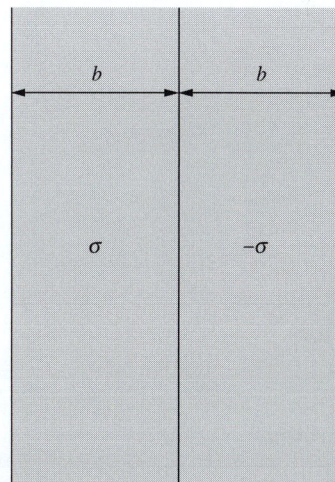

1.67  *Field from a cylindrical shell, right and wrong* **

Find the electric field outside a uniformly charged hollow cylindrical shell with radius $R$ and charge density $\sigma$, an infinitesimal distance away from it. Do this in the following way.

(a) Slice the shell into parallel infinite rods, and integrate the field contributions from all the rods. You should obtain the incorrect result of $\sigma/2\epsilon_0$.

**Figure 1.47.**

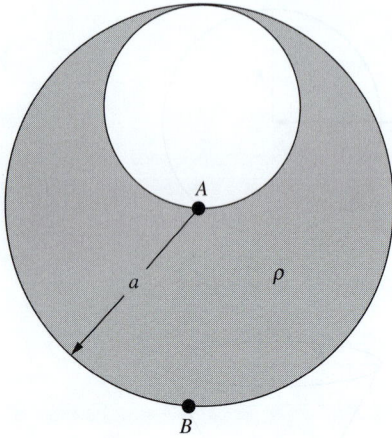

**Figure 1.48.**

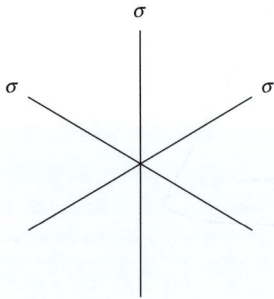

**Figure 1.49.**

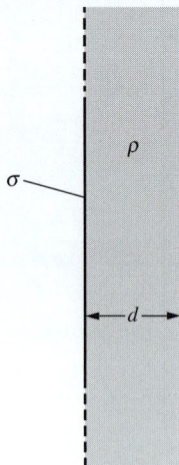

**Figure 1.50.**

(b) Why isn't the result correct? Explain how to modify it to obtain the correct result of $\sigma/\epsilon_0$. *Hint:* You could very well have performed the above integral in an effort to obtain the electric field an infinitesimal distance *inside* the cylinder, where we know the field is zero. Does the above integration provide a good description of what's going on for points on the shell that are very close to the point in question?

**1.68** *Uniform field strength* ∗

We know from the example in Section 1.11 that the electric field inside a solid sphere with uniform charge density is proportional to $r$. Assume instead that the charge density is not uniform, but depends only on $r$. What should this dependence be so that the magnitude of the field at points inside the sphere is independent of $r$ (except right at the center, where it isn't well defined)? What should the dependence be in the analogous case where we have a cylinder instead of a sphere?

**1.69** *Carved-out sphere* ∗∗

A sphere of radius $a$ is filled with positive charge with uniform density $\rho$. Then a smaller sphere of radius $a/2$ is carved out, as shown in Fig. 1.48, and left empty. What are the direction and magnitude of the electric field at $A$? At $B$?

**1.70** *Field from two sheets* ∗

Two infinite plane sheets of surface charge, with densities $3\sigma_0$ and $-2\sigma_0$, are located a distance $\ell$ apart, parallel to one another. Discuss the electric field of this system. Now suppose the two planes, instead of being parallel, intersect at right angles. Show what the field is like in each of the four regions into which space is thereby divided.

**1.71** *Intersecting sheets* ∗∗

(a) Figure 1.49 shows the cross section of three infinite sheets intersecting at equal angles. The sheets all have surface charge density $\sigma$. By adding up the fields from the sheets, find the electric field at all points in space.

(b) Find the field instead by using Gauss's law. You should explain clearly why Gauss's law is in fact useful in this setup.

(c) What is the field in the analogous setup where there are $N$ sheets instead of three? What is your answer in the $N \to \infty$ limit? This limit is related to the cylinder in Exercise 1.68.

**1.72** *A plane and a slab* ∗∗

An infinite plane has uniform surface charge density $\sigma$. Adjacent to it is an infinite parallel layer of charge of thickness $d$ and uniform volume charge density $\rho$, as shown in Fig. 1.50. All charges are fixed. Find **E** everywhere.

1.73 *Sphere in a cylinder* ✶✶

An infinite cylinder with uniform volume charge density $\rho$ has its axis lying along the $z$ axis. A sphere is carved out of the cylinder and then filled up with a material with uniform density $-\rho/2$. Assume that the center of the sphere is located on the $x$ axis at position $x = a$. Show that inside the sphere the component of the field in the $xy$ plane is uniform, and find its value. *Hint:* The technique used in Problem 1.27 will be helpful.

1.74 *Zero field in a sphere* ✶✶

In Fig. 1.51 a sphere with radius $R$ is centered at the origin, an infinite cylinder with radius $R$ has its axis along the $z$ axis, and an infinite slab with thickness $2R$ lies between the planes $z = -R$ and $z = R$. The uniform volume densities of these objects are $\rho_1$, $\rho_2$, and $\rho_3$, respectively. The objects are superposed on top of each other; the densities add where the objects overlap. How should the three densities be related so that the electric field is zero everywhere throughout the volume of the sphere? *Hint:* Find a vector expression for the field inside each object, and then use superposition.

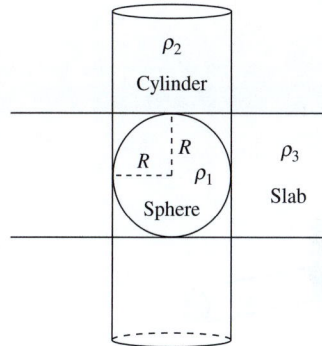

**Figure 1.51.**

1.75 *Ball in a sphere* ✶✶

We know that if a point charge $q$ is located at radius $a$ in the interior of a sphere with radius $R$ and uniform volume charge density $\rho$, then the force on the point charge is effectively due only to the charge that is located inside radius $a$.

(a) Consider instead a uniform ball of charge located entirely inside a larger sphere of radius $R$. Let the ball's radius be $b$, and let its center be located at radius $a$ in the larger sphere. Its volume charge density is such that its total charge is $q$. Assume that the ball is superposed on top of the sphere, so that all of the sphere's charge is still present. Can the force on the ball be obtained by treating it like a point charge and considering only the charge in the larger sphere that is inside radius $a$?

(b) Would the force change if we instead remove the charge in the larger sphere where the ball is? So now we are looking at the force on the ball due to the sphere with a cavity carved out, which is a more realistic scenario.

1.76 *Hydrogen atom* ✶✶

The neutral hydrogen atom in its normal state behaves, in some respects, like an electric charge distribution that consists of a point charge of magnitude $e$ surrounded by a distribution of negative charge whose density is given by $\rho(r) = -Ce^{-2r/a_0}$. Here $a_0$ is the *Bohr radius*, $0.53 \cdot 10^{-10}$ m, and $C$ is a constant with the value required to make the total amount of negative charge exactly $e$.

What is the net electric charge inside a sphere of radius $a_0$? What is the electric field strength at this distance from the nucleus?

1.77  *Electron jelly* **
Imagine a sphere of radius $a$ filled with negative charge of uniform density, the total charge being equivalent to that of two electrons. Imbed in this jelly of negative charge two protons, and assume that, in spite of their presence, the negative charge distribution remains uniform. Where must the protons be located so that the force on each of them is zero? (This is a surprisingly realistic caricature of a hydrogen molecule; the magic that keeps the electron cloud in the molecule from collapsing around the protons is explained by quantum mechanics!)

1.78  *Hole in a shell* **
Figure 1.52 shows a spherical shell of charge, of radius $a$ and surface density $\sigma$, from which a small circular piece of radius $b \ll a$ has been removed. What is the direction and magnitude of the field at the midpoint of the aperture? There are two ways to get the answer. You can integrate over the remaining charge distribution to sum the contributions of all elements to the field at the point in question. Or, remembering the superposition principle, you can think about the effect of replacing the piece removed, which itself is practically a little disk. Note the connection of this result with our discussion of the force on a surface charge – perhaps that is a third way in which you might arrive at the answer.

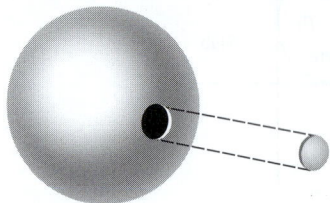

**Figure 1.52.**

1.79  *Forces on three sheets* **
Consider three charged sheets, $A$, $B$, and $C$. The sheets are parallel with $A$ above $B$ above $C$. On each sheet there is surface charge of uniform density: $-4 \cdot 10^{-5}$ C/m$^2$ on $A$, $7 \cdot 10^{-5}$ C/m$^2$ on $B$, and $-3 \cdot 10^{-5}$ C/m$^2$ on $C$. (The density given includes charge on both sides of the sheet.) What is the magnitude of the electrical force per unit area on each sheet? Check to see that the total force per unit area on the three sheets is zero.

1.80  *Force in a soap bubble* **
Like the charged rubber balloon described at the end of Section 1.14, a charged soap bubble experiences an outward electrical force on every bit of its surface. Given the total charge $Q$ on a bubble of radius $R$, what is the magnitude of the resultant force tending to pull any hemispherical half of the bubble away from the other half? (Should this force divided by $2\pi R$ exceed the surface tension of the soap film, interesting behavior might be expected!)

1.81  *Energy around a sphere* *
A sphere of radius $R$ has a charge $Q$ distributed uniformly over its surface. How large a sphere contains 90 percent of the energy stored in the electrostatic field of this charge distribution?

1.82 *Energy of concentric shells* ∗

(a) Concentric spherical shells of radius $a$ and $b$, with $a < b$, carry charge $Q$ and $-Q$, respectively, each charge uniformly distributed. Find the energy stored in the electric field of this system.

(b) Calculate the stored energy in a second way: start with two neutral shells, and then gradually transfer positive charge from the outer shell to the inner shell in a spherically symmetric manner. At an intermediate stage when there is charge $q$ on the inner shell, find the work required to transfer an additional charge $dq$. And then integrate over $q$.

1.83 *Potential energy of a cylinder* ∗∗

Problem 1.24 gives one way of calculating the energy per unit length stored in a solid cylinder with radius $a$ and uniform volume charge density $\rho$. Calculate the energy here by using Eq. (1.53) to find the total energy per unit length stored in the electric field. Don't forget to include the field inside the cylinder.

You will find that the energy is infinite, so instead calculate the energy relative to the configuration where all the charge is initially distributed uniformly over a hollow cylinder with large radius $R$. (The field outside radius $R$ is the same in both configurations, so it can be ignored when calculating the relative energy.) In terms of the total charge $\lambda$ per unit length in the final cylinder, show that the energy per unit length can be written as $(\lambda^2/4\pi\epsilon_0)\big(1/4 + \ln(R/a)\big)$.

# 2

# The electric potential

**Overview** The first half of this chapter deals mainly with the *potential* associated with an electric field. The second half covers a number of mathematical topics that will be critical in our treatment of electromagnetism. The potential difference between two points is defined to be the negative *line integral* of the electric field. Equivalently, the electric field equals the negative *gradient* of the potential. Just as the electric field is the force per unit charge, the potential is the potential energy per unit charge. We give a number of examples involving the calculation of the potential due to a given charge distribution. One important example is the *dipole*, which consists of two equal and opposite charges. We will have much more to say about the applications of dipoles in Chapter 10.

Turning to mathematics, we introduce the *divergence*, which gives a measure of the flux of a vector field out of a small volume. We prove *Gauss's theorem* (or the *divergence theorem*) and then use it to write Gauss's law in differential form. The result is the first of the four equations known as *Maxwell's equations* (the subject of Chapter 9). We explicitly calculate the divergence in Cartesian coordinates. The divergence of the gradient is known as the *Laplacian* operator. Functions whose Laplacian equals zero have many important properties, one of which leads to *Earnshaw's theorem*, which states that it is impossible to construct a stable electrostatic equilibrium in empty space. We introduce the *curl*, which gives a measure of the line integral of a vector field around a small closed curve. We prove *Stokes' theorem* and explicitly calculate the curl in Cartesian coordinates. The conservative nature of a static electric field implies that its curl is zero. See Appendix F for a discussion of the various vector operators in different coordinate systems.

## 2.1  Line integral of the electric field

Suppose that $\mathbf{E}$ is the field of some stationary distribution of electric charges. Let $P_1$ and $P_2$ denote two points anywhere in the field. The line integral of $E$ between the two points is $\int_{P_1}^{P_2} \mathbf{E} \cdot d\mathbf{s}$, taken along some path that runs from $P_1$ to $P_2$, as shown in Fig. 2.1. This means: divide the chosen path into short segments, each segment being represented by a vector connecting its ends; take the scalar product of the path-segment vector with the field $\mathbf{E}$ at that place; add these products up for the whole path. The integral as usual is to be regarded as the limit of this sum as the segments are made shorter and more numerous without limit.

Let's consider the field of a point charge $q$ and some paths running from point $P_1$ to point $P_2$ in that field. Two different paths are shown in Fig. 2.2. It is easy to compute the line integral of $\mathbf{E}$ along path $A$, which is made up of a radial segment running outward from $P_1$ and an arc of

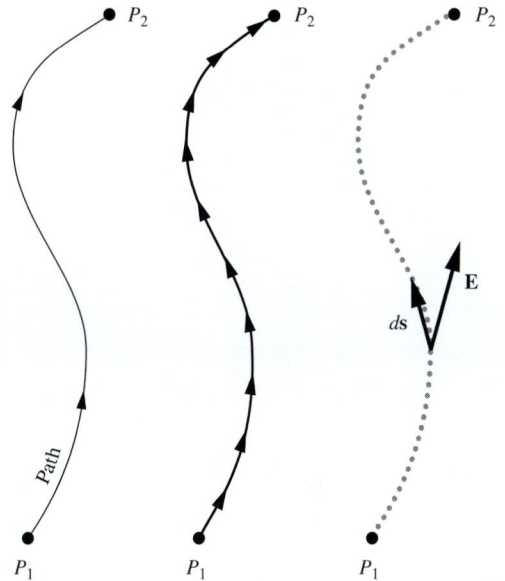

**Figure 2.1.**
Showing the division of the path into path elements $d\mathbf{s}$.

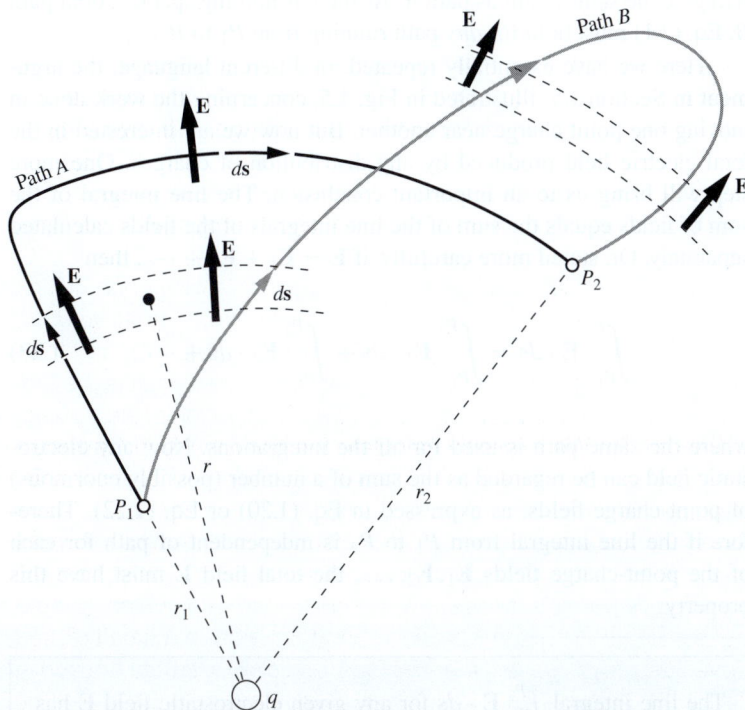

**Figure 2.2.**
The electric field $\mathbf{E}$ is that of a positive point charge $q$. The line integral of $\mathbf{E}$ from $P_1$ to $P_2$ along path $A$ has the value $(q/4\pi\epsilon_0)(1/r_1 - 1/r_2)$. It will have exactly the same value if calculated for path $B$, or for any other path from $P_1$ to $P_2$.

radius $r_2$. Along the radial segment of path $A$, $\mathbf{E}$ and $d\mathbf{s}$ are parallel, the magnitude of $\mathbf{E}$ is $q/4\pi\epsilon_0 r^2$, and $\mathbf{E} \cdot d\mathbf{s}$ is simply $(q/4\pi\epsilon_0 r^2)\,ds$. Thus the line integral on that segment is

$$\int_{r_1}^{r_2} \frac{q\,dr}{4\pi\epsilon_0 r^2} = \frac{q}{4\pi\epsilon_0}\left(\frac{1}{r_1} - \frac{1}{r_2}\right). \tag{2.1}$$

The second leg of path $A$, the circular segment, gives zero because $\mathbf{E}$ is perpendicular to $d\mathbf{s}$ everywhere on that arc. The entire line integral is therefore

$$\int_{P_1}^{P_2} \mathbf{E} \cdot d\mathbf{s} = \frac{q}{4\pi\epsilon_0}\left(\frac{1}{r_1} - \frac{1}{r_2}\right). \tag{2.2}$$

Now look at path $B$. Because $\mathbf{E}$ is radial with magnitude $q/4\pi\epsilon_0 r^2$, $\mathbf{E} \cdot d\mathbf{s} = (q/4\pi\epsilon_0 r^2)\,dr$ even when $d\mathbf{s}$ is not radially oriented. The corresponding pieces of path $A$ and path $B$ indicated in the diagram make identical contributions to the integral. The part of path $B$ that loops beyond $r_2$ makes a net contribution of zero; contributions from corresponding outgoing and incoming parts cancel. For the entire line integral, path $B$ will give the same result as path $A$. As there is nothing special about path $B$, Eq. (2.1) must hold for *any* path running from $P_1$ to $P_2$.

Here we have essentially repeated, in different language, the argument in Section 1.5, illustrated in Fig. 1.5, concerning the work done in moving one point charge near another. But now we are interested in the total electric field produced by any distribution of charges. One more step will bring us to an important conclusion. The line integral of the sum of fields equals the sum of the line integrals of the fields calculated separately. Or, stated more carefully, if $\mathbf{E} = \mathbf{E}_1 + \mathbf{E}_2 + \cdots$, then

$$\int_{P_1}^{P_2} \mathbf{E} \cdot d\mathbf{s} = \int_{P_1}^{P_2} \mathbf{E}_1 \cdot d\mathbf{s} + \int_{P_1}^{P_2} \mathbf{E}_2 \cdot d\mathbf{s} + \cdots, \tag{2.3}$$

where the same path is used for all the integrations. Now any electrostatic field can be regarded as the sum of a number (possibly enormous) of point-charge fields, as expressed in Eq. (1.20) or Eq. (1.22). Therefore if the line integral from $P_1$ to $P_2$ is independent of path for each of the point-charge fields $\mathbf{E}_1, \mathbf{E}_2, \ldots$, the total field $\mathbf{E}$ must have this property:

The line integral $\int_{P_1}^{P_2} \mathbf{E} \cdot d\mathbf{s}$ for any given electrostatic field $\mathbf{E}$ has the same value for all paths from $P_1$ to $P_2$.

The points $P_2$ and $P_1$ may coincide. In that case the paths are all closed curves, among them paths of vanishing length. This leads to the following corollary:

> The line integral $\int \mathbf{E} \cdot d\mathbf{s}$ around any closed path in an electrostatic field is zero.

By *electrostatic field* we mean, strictly speaking, the electric field of stationary charges. Later on, we shall encounter electric fields in which the line integral is *not* path-independent. Those fields will usually be associated with rapidly moving charges. For our present purposes we can say that, if the source charges are moving slowly enough, the field $\mathbf{E}$ will be such that $\int \mathbf{E} \cdot d\mathbf{s}$ is practically path-independent. Of course, if $\mathbf{E}$ itself is varying in time, the $\mathbf{E}$ in $\int \mathbf{E} \cdot d\mathbf{s}$ must be understood as the field that exists over the whole path at a given instant of time. With that understanding we can talk meaningfully about the line integral in a changing electrostatic field.

## 2.2 Potential difference and the potential function

Because the line integral in the electrostatic field is path-independent, we can use it to define a scalar quantity $\phi_{21}$, without specifying any particular path:

$$\phi_{21} = - \int_{P_1}^{P_2} \mathbf{E} \cdot d\mathbf{s}. \qquad (2.4)$$

With the minus sign included here, $\phi_{21}$ is the work per unit charge done by an external agency in moving a positive charge from $P_1$ to $P_2$ in the field $\mathbf{E}$. (The external agency must supply a force $\mathbf{F}_{\text{ext}} = -q\mathbf{E}$ to balance the electrical force $\mathbf{F}_{\text{elec}} = q\mathbf{E}$; hence the minus sign.) Thus $\phi_{21}$ is a single-valued scalar function of the two positions $P_1$ and $P_2$. We call it the *electric potential difference* between the two points.

In our SI system of units, potential difference is measured in joule/coulomb. This unit has a name of its own, the *volt:*

$$1 \text{ volt} = 1 \frac{\text{joule}}{\text{coulomb}}. \qquad (2.5)$$

One joule of work is required to move a charge of one coulomb through a potential difference of one volt. In the Gaussian system of units, potential difference is measured in erg/esu. This unit also has a name of its own, the statvolt ("stat" comes from "electrostatic"). As an exercise, you can use the $1 \text{ C} \approx 3 \cdot 10^9$ esu relation from Section 1.4 to show that one volt is equivalent to approximately $1/300$ statvolt. These two relations are accurate to better than 0.1 percent, thanks to the accident that $c$ is that close to $3 \cdot 10^8$ m/s. Appendix C derives the conversion factors between all of the corresponding units in the SI and Gaussian systems. Further discussion of the exact relations between SI and Gaussian electrical units is given in Appendix E, which takes into account the definition of the meter in terms of the speed of light.

Suppose we hold $P_1$ fixed at some reference position. Then $\phi_{21}$ becomes a function of $P_2$ only, that is, a function of the spatial coordinates $x$, $y$, $z$. We can write it simply $\phi(x, y, z)$, without the subscript, if we remember that its definition still involves agreement on a reference point $P_1$. We can say that $\phi$ is the potential associated with the vector field $\mathbf{E}$. It is a scalar function of position, or a scalar field (they mean the same thing). Its value at a point is simply a number (in units of work per unit charge) and has no direction associated with it. Once the vector field $\mathbf{E}$ is given, the potential function $\phi$ is determined, except for an arbitrary additive constant allowed by the arbitrariness in our choice of $P_1$.

**Example** Find the potential associated with the electric field described in Fig. 2.3, the components of which are $E_x = Ky$, $E_y = Kx$, $E_z = 0$, with $K$ a constant. This is a possible electrostatic field; we will see why in Section 2.17. Some field lines are shown.

**Solution** Since $E_z = 0$, the potential will be independent of $z$ and we need consider only the $xy$ plane. Let $x_1$, $y_1$ be the coordinates of $P_1$, and $x_2$, $y_2$ the coordinates of $P_2$. It is convenient to locate $P_1$ at the origin: $x_1 = 0$, $y_1 = 0$. To evaluate $-\int \mathbf{E} \cdot d\mathbf{s}$ from this reference point to a general point $(x_2, y_2)$ it is easiest to use a path like the dashed path $ABC$ in Fig. 2.3:

$$\phi(x_2, y_2) = -\int_{(0,0)}^{(x_2, y_2)} \mathbf{E} \cdot d\mathbf{s} = -\int_{(0,0)}^{(x_2, 0)} E_x \, dx - \int_{(x_2, 0)}^{(x_2, y_2)} E_y \, dy. \quad (2.6)$$

The first of the two integrals on the right is zero because $E_x$ is zero along the $x$ axis. The second integration is carried out at constant $x$, with $E_y = Kx_2$:

$$-\int_{(x_2, 0)}^{(x_2, y_2)} E_y \, dy = -\int_0^{y_2} Kx_2 \, dy = -Kx_2 y_2. \quad (2.7)$$

There was nothing special about the point $(x_2, y_2)$ so we can drop the subscripts:

$$\phi(x, y) = -Kxy \quad (2.8)$$

for any point $(x, y)$ in this field, with zero potential at the origin. Any constant could be added to this. That would only mean that the reference point to which zero potential is assigned had been located somewhere else.

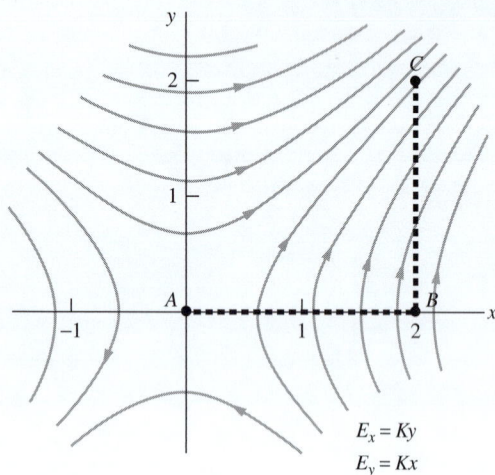

$E_x = Ky$
$E_y = Kx$

**Figure 2.3.**
A particular path, $ABC$, in the electric field $E_x = Ky$, $E_y = Kx$. Some field lines are shown.

**Example (Potential due to a uniform sphere)** A sphere has radius $R$ and uniform volume charge density $\rho$. Use the results from the example in Section 1.11 to find the potential for all values of $r$, both inside and outside the sphere. Take the reference point $P_1$ to be infinitely far away.

Solution   From the example in Section 1.11, the magnitude of the (radial) electric field inside the sphere is $E(r) = \rho r/3\epsilon_0$, and the magnitude outside is $E(r) = \rho R^3/3\epsilon_0 r^2$. Equation (2.4) tells us that the potential equals the negative of the line integral of the field, from $P_1$ (which we are taking to be at infinity) down to a given radius $r$. The potential outside the sphere is therefore

$$\phi_{\text{out}}(r) = -\int_\infty^r E(r')\,dr' = -\int_\infty^r \frac{\rho R^3}{3\epsilon_0 r'^2}\,dr' = \frac{\rho R^3}{3\epsilon_0 r}. \tag{2.9}$$

In terms of the total charge in the sphere, $Q = (4\pi R^3/3)\rho$, this potential is simply $\phi_{\text{out}}(r) = Q/4\pi\epsilon_0 r$. This is as expected, because we already knew that the potential *energy* of a charge $q$ due to the sphere is $qQ/4\pi\epsilon_0 r$. And the potential $\phi$ equals the potential energy per unit charge.

To find the potential inside the sphere, we must break the integral into two pieces:

$$\phi_{\text{in}}(r) = -\int_\infty^R E(r')\,dr' - \int_R^r E(r')\,dr' = -\int_\infty^R \frac{\rho R^3}{3\epsilon_0 r'^2}\,dr' - \int_R^r \frac{\rho r'}{3\epsilon_0}\,dr'$$

$$= \frac{\rho R^3}{3\epsilon_0 R} - \frac{\rho}{6\epsilon_0}(r^2 - R^2) = \frac{\rho R^2}{2\epsilon_0} - \frac{\rho r^2}{6\epsilon_0}. \tag{2.10}$$

Note that Eqs. (2.9) and (2.10) yield the same value of $\phi$ at the surface of the sphere, namely $\phi(R) = \rho R^2/3\epsilon_0$. So $\phi$ is continuous across the surface, as it should be. (The field is everywhere finite, so the line integral over an infinitesimal interval must yield an infinitesimal result.) The slope of $\phi$ is also continuous, because $E(r)$ (which is the negative derivative of $\phi$, because $\phi$ is the negative integral of $E$) is continuous. A plot of $\phi(r)$ is shown in Fig. 2.4.

The potential at the center of the sphere is $\phi(0) = \rho R^2/2\epsilon_0$, which is $3/2$ times the value at the surface. So if you bring a charge in from infinity, it takes $2/3$ of your work to reach the surface, and then $1/3$ to go the extra distance of $R$ to the center.

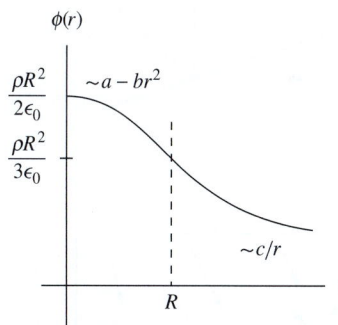

**Figure 2.4.**
The potential due to a uniform sphere of charge.

We must be careful not to confuse the potential $\phi$ associated with a given field $\mathbf{E}$ with the potential energy of a system of charges. The potential energy of a system of charges is the total work required to assemble it, starting with all the charges far apart. In Eq. (1.14), for example, we expressed $U$, the potential energy of the charge system in Fig. 1.6. The electric potential $\phi(x, y, z)$ associated with the field in Fig. 1.6 would be the work per unit charge required to move a unit positive test charge from some chosen reference point to the point $(x, y, z)$ in the field of that structure of nine charges.

## 2.3 Gradient of a scalar function

Given the electric field, we can find the electric potential function. But we can also proceed in the other direction; from the potential we can derive the field. It appears from Eq. (2.4) that the field is in some sense the *derivative* of the potential function. To make this idea precise we introduce the *gradient* of a scalar function of position. Let $f(x, y, z)$ be

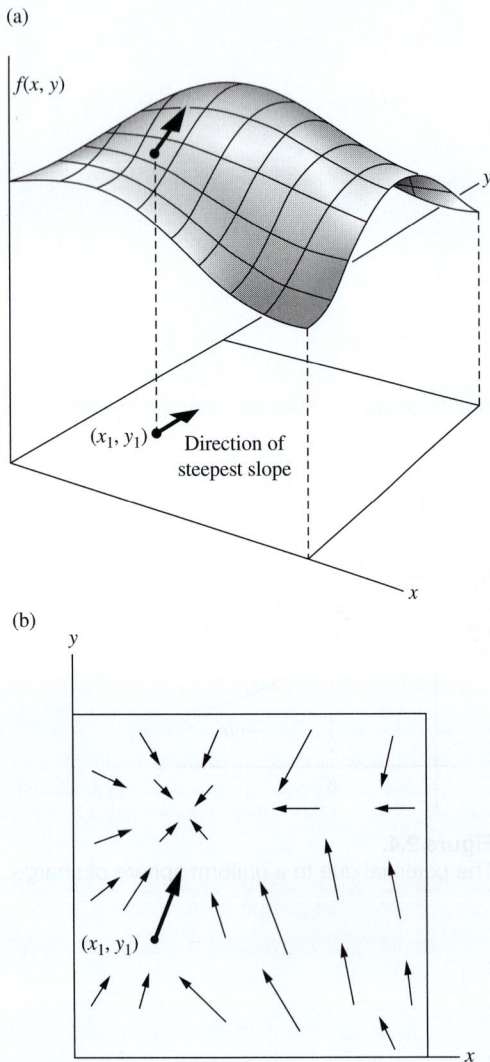

(a)

(b)

**Figure 2.5.**
The scalar function $f(x, y)$ is represented by the surface in (a). The arrows in (b) represent the vector function, grad $f$.

some continuous, differentiable function of the coordinates. With its partial derivatives $\partial f / \partial x$, $\partial f / \partial y$, and $\partial f / \partial z$ we can construct at every point in space a vector, the vector whose $x$, $y$, $z$ components are equal to the respective partial derivatives.[1] This vector we call the *gradient* of $f$, written "grad $f$," or $\nabla f$:

$$\nabla f \equiv \hat{\mathbf{x}} \frac{\partial f}{\partial x} + \hat{\mathbf{y}} \frac{\partial f}{\partial y} + \hat{\mathbf{z}} \frac{\partial f}{\partial z}. \qquad (2.13)$$

$\nabla f$ is a vector that tells how the function $f$ varies in the neighborhood of a point. Its $x$ component is the partial derivative of $f$ with respect to $x$, a measure of the rate of change of $f$ as we move in the $x$ direction. The direction of the vector $\nabla f$ at any point is the direction in which one must move from that point to find the most rapid increase in the function $f$. Suppose we were dealing with a function of two variables only, $x$ and $y$, so that the function could be represented by a surface in three dimensions. Standing on that surface at some point, we see the surface rising in some direction, sloping downward in the opposite direction. There is a direction in which a short step will take us higher than a step of the same length in any other direction. The gradient of the function is a vector in that direction of steepest ascent, and its magnitude is the slope measured in that direction.

Figure 2.5 may help you to visualize this. Suppose some particular function of two coordinates $x$ and $y$ is represented by the surface $f(x, y)$ sketched in Fig. 2.5(a). At the location $(x_1, y_1)$ the surface rises most steeply in a direction that makes an angle of about 80° with the positive $x$ direction. The gradient of $f(x, y)$, $\nabla f$, is a vector function of $x$ and $y$. Its character is suggested in Fig. 2.5(b) by a number of vectors at various points in the two-dimensional space, including the point $(x_1, y_1)$. The vector function $\nabla f$ defined in Eq. (2.13) is simply an extension of this idea to three-dimensional space. (Be careful not to confuse Fig. 2.5(a) with real three-dimensional $xyz$ space; the third coordinate there is the value of the function $f(x, y)$.)

As one example of a function in three-dimensional space, suppose $f$ is a function of $r$ only, where $r$ is the distance from some fixed point $O$. On a sphere of radius $r_0$ centered about $O$, $f = f(r_0)$ is constant. On a slightly larger sphere of radius $r_0 + dr$ it is also constant, with the value $f = f(r_0 + dr)$. If we want to make the change from $f(r_0)$ to $f(r_0 + dr)$,

---

[1] We remind the reader that a partial derivative with respect to $x$, of a function of $x, y, z$, written simply $\partial f / \partial x$, means the rate of change of the function with respect to $x$ with the other variables $y$ and $z$ held constant. More precisely,

$$\frac{\partial f}{\partial x} = \lim_{\Delta x \to 0} \frac{f(x + \Delta x, y, z) - f(x, y, z)}{\Delta x}. \qquad (2.11)$$

As an example, if $f = x^2 y z^3$,

$$\frac{\partial f}{\partial x} = 2xyz^3, \quad \frac{\partial f}{\partial y} = x^2 z^3, \quad \frac{\partial f}{\partial z} = 3x^2 y z^2. \qquad (2.12)$$

the shortest step we can make is to go radially (as from $A$ to $B$) rather than from $A$ to $C$, in Fig. 2.6. The "slope" of $f$ is thus greatest in the radial direction, so $\nabla f$ at any point is a radially pointing vector. In fact $\nabla f = \hat{\mathbf{r}}(df/dr)$ in this case, $\hat{\mathbf{r}}$ denoting, for any point, a unit vector in the radial direction. See Section F.2 in Appendix F for further discussion of the gradient.

## 2.4 Derivation of the field from the potential

It is now easy to see that the relation of the scalar function $f$ to the vector function $\nabla f$ is the same, except for a minus sign, as the relation of the potential $\phi$ to the field $\mathbf{E}$. Consider the value of $\phi$ at two nearby points, $(x, y, z)$ and $(x + dx, y + dy, z + dz)$. The change in $\phi$, going from the first point to the second, is, in first-order approximation,

$$d\phi = \frac{\partial \phi}{\partial x}\, dx + \frac{\partial \phi}{\partial y}\, dy + \frac{\partial \phi}{\partial z}\, dz. \qquad (2.14)$$

On the other hand, from the definition of $\phi$ in Eq. (2.4), the change can also be expressed as

$$d\phi = -\mathbf{E} \cdot d\mathbf{s}. \qquad (2.15)$$

The infinitesimal vector displacement $d\mathbf{s}$ is just $\hat{\mathbf{x}}\, dx + \hat{\mathbf{y}}\, dy + \hat{\mathbf{z}}\, dz$. Thus if we identify $\mathbf{E}$ with $-\nabla \phi$, where $\nabla \phi$ is defined via Eq. (2.13), then Eqs. (2.14) and (2.15) become identical. So the electric field is the negative of the gradient of the potential:

$$\boxed{\mathbf{E} = -\nabla \phi} \qquad (2.16)$$

The minus sign came in because the electric field points from a region of greater potential toward a region of lesser potential, whereas the vector $\nabla \phi$ is defined so that it points in the direction of increasing $\phi$.

To show how this works, we go back to the example of the field in Fig. 2.3. From the potential given by Eq. (2.8), $\phi = -Kxy$, we can recover the electric field we started with:

$$\mathbf{E} = -\nabla(-Kxy) = -\left(\hat{\mathbf{x}}\frac{\partial}{\partial x} + \hat{\mathbf{y}}\frac{\partial}{\partial y}\right)(-Kxy) = K(\hat{\mathbf{x}}y + \hat{\mathbf{y}}x). \quad (2.17)$$

## 2.5 Potential of a charge distribution

We already know the potential that goes with a single point charge, because we calculated the work required to bring one charge into the neighborhood of another in Eq. (1.9). The potential at any point, in the field of an isolated point charge $q$, is just $q/4\pi\epsilon_0 r$, where $r$ is the distance

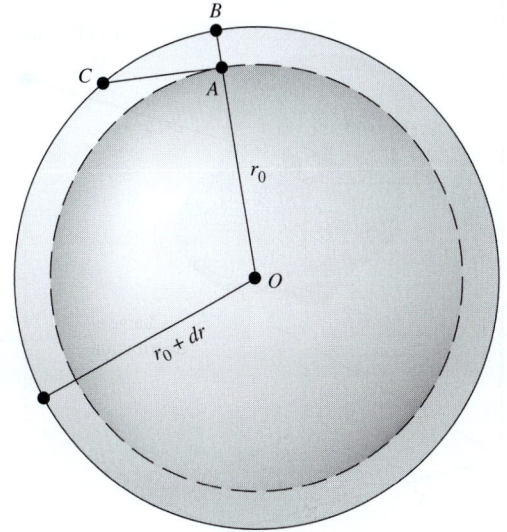

**Figure 2.6.**
The shortest step for a given change in $f$ is the radial step $AB$, if $f$ is a function of $r$ only.

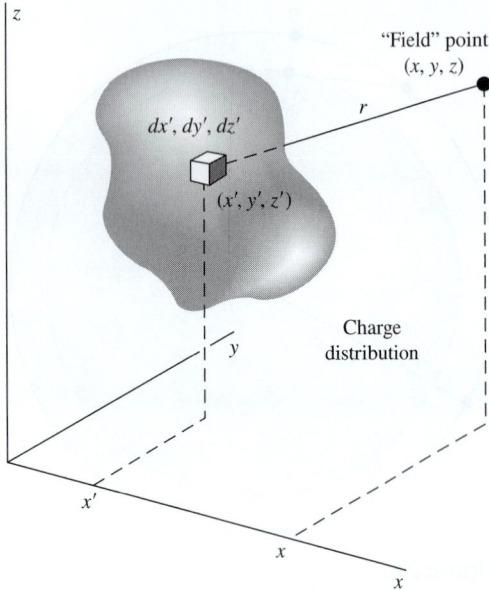

**Figure 2.7.**
Each element of the charge distribution
$\rho(x', y', z')$ contributes to the potential $\phi$ at the
point $(x, y, z)$. The potential at this point is the
sum of all such contributions; see Eq. (2.18).

from the point in question to the source $q$, and where we have assigned
zero potential to points infinitely far from the source.

Superposition must work for potentials as well as fields. If we have
several sources, the potential function is simply the sum of the potential functions that we would have for each of the sources present alone –
*providing* we make a consistent assignment of the zero of potential in
each case. If all the sources are contained in some finite region, it is
always possible, and usually the simplest choice, to put zero potential at
infinite distance. If we adopt this rule, the potential of any charge distribution can be specified by the integral

$$\phi(x, y, z) = \int_{\substack{\text{all} \\ \text{sources}}} \frac{\rho(x', y', z') \, dx' \, dy' \, dz'}{4\pi\epsilon_0 r}, \qquad (2.18)$$

where $r$ is the distance from the volume element $dx' \, dy' \, dz'$ to the point
$(x, y, z)$ at which the potential is being evaluated (Fig. 2.7). That is, $r =
[(x - x')^2 + (y - y')^2 + (z - z')^2]^{1/2}$. Notice the difference between
this and the integral giving the electric field of a charge distribution; see
Eq. (1.22). Here we have $r$ in the denominator, not $r^2$, and the integral
is a scalar not a vector. From the scalar potential function $\phi(x, y, z)$ we
can always find the electric field by taking the negative gradient of $\phi$,
according to Eq. (2.16).

In the case of a discrete distribution of source charges, the above
integral is replaced by a sum over all the charges, indexed by $i$:

$$\phi(x, y, z) = \sum_{\text{all sources}} \frac{q_i}{4\pi\epsilon_0 r}, \qquad (2.19)$$

where $r$ is the distance from the charge $q_i$ to the point $(x, y, z)$.

**Example (Potential of two point charges)** Consider a very simple example, the potential of the two point charges shown in Fig. 2.8. A positive charge of
$12\,\mu\text{C}$ is located 3 m away from a negative charge, $-6\,\mu\text{C}$. (The "$\mu$" prefix stands
for "micro," or $10^{-6}$.) The potential at any point in space is the sum of the potentials due to each charge alone. The potentials for some selected points in space
are given in the diagram. No vector addition is involved here, only the algebraic
addition of scalar quantities. For instance, at the point on the far right, which is
6 m from the positive charge and 5 m from the negative charge, the potential has
the value

$$\frac{1}{4\pi\epsilon_0}\left(\frac{12 \cdot 10^{-6}\,\text{C}}{6\,\text{m}} + \frac{-6 \cdot 10^{-6}\,\text{C}}{5\,\text{m}}\right) = \frac{0.8 \cdot 10^{-6}\,\text{C/m}}{4\pi\epsilon_0}$$

$$= 7.2 \cdot 10^3\,\text{J/C} = 7.2 \cdot 10^3\,\text{V}, \quad (2.20)$$

where we have used $1/4\pi\epsilon_0 \approx 9 \cdot 10^9\,\text{N}\,\text{m}^2/\text{C}^2$ (and also $1\,\text{N}\,\text{m} = 1\,\text{J}$). The
potential approaches zero at infinite distance. It would take $7.2 \cdot 10^3$ J of work

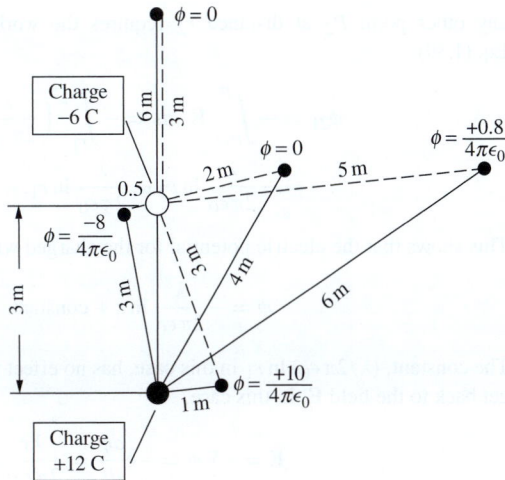

**Figure 2.8.**
The electric potential $\phi$ at various points in a system of two point charges. $\phi$ goes to zero at infinite distance and is given in units of volts, or joules per coulomb.

to bring a unit positive charge in from infinity to a point where $\phi = 7.2 \cdot 10^3$ V. Note that two of the points shown on the diagram have $\phi = 0$. The net work done in bringing in any charge to one of these points would be zero. You can see that there must be an infinite number of such points, forming a surface in space surrounding the negative charge. In fact, the locus of points with any particular value of $\phi$ is a surface – an *equipotential surface* – which would show on our two-dimensional diagram as a curve.

There is one restriction on the use of Eq. (2.18): it may not work unless all sources are confined to some finite region of space. A simple example of the difficulty that arises with charges distributed out to infinite distance is found in the long charged wire whose field **E** we studied in Section 1.12. If we attempt to carry out the integration over the charge distribution indicated in Eq. (2.18), we find that the integral diverges – we get an infinite result. No such difficulty arose in finding the electric *field* of the infinitely long wire, because the contributions of elements of the line charge to the field decrease so rapidly with distance. Evidently we had better locate the zero of potential somewhere close to home, in a system that has charges distributed out to infinity. Then it is simply a matter of calculating the difference in potential $\phi_{21}$, between the general point $(x, y, z)$ and the selected reference point, using the fundamental relation, Eq. (2.4).

**Example (Potential of a long charged wire)**   To see how this goes in the case of the infinitely long charged wire, let us arbitrarily locate the reference point $P_1$ at a distance $r_1$ from the wire. Then to carry a charge from $P_1$ to any other point $P_2$ at distance $r_2$ requires the work per unit charge, using Eq. (1.39):

$$\phi_{21} = -\int_{P_1}^{P_2} \mathbf{E} \cdot d\mathbf{s} = -\int_{r_1}^{r_2} \left(\frac{\lambda}{2\pi\epsilon_0 r}\right) dr$$

$$= -\frac{\lambda}{2\pi\epsilon_0}\ln r_2 + \frac{\lambda}{2\pi\epsilon_0}\ln r_1. \qquad (2.21)$$

This shows that the electric potential for the charged wire can be taken as

$$\phi = -\frac{\lambda}{2\pi\epsilon_0}\ln r + \text{constant.} \qquad (2.22)$$

The constant, $(\lambda/2\pi\epsilon_0)\ln r_1$ in this case, has no effect when we take $-\text{grad}\,\phi$ to get back to the field $\mathbf{E}$. In this case,

$$\mathbf{E} = -\nabla\phi = -\hat{\mathbf{r}}\frac{d\phi}{dr} = \frac{\lambda\hat{\mathbf{r}}}{2\pi\epsilon_0 r}. \qquad (2.23)$$

## 2.6 Uniformly charged disk

Let us now study the electric potential and field around a uniformly charged disk. This is a charge distribution like that discussed in Section 1.13, except that it has a limited extent. The flat disk of radius $a$ in Fig. 2.9 carries a positive charge spread over its surface with the constant density $\sigma$, in $C/m^2$. (This is a single sheet of charge of infinitesimal thickness, not two layers of charge, one on each side. That is, the total charge in the system is $\pi a^2\sigma$.) We shall often meet surface charge distributions in the future, especially on metallic conductors. However, the object just described is *not* a conductor; if it were, as we shall soon see, the charge could not remain uniformly distributed but would redistribute itself, crowding more toward the rim of the disk. What we have is an insulating disk, like a sheet of plastic, upon which charge has been "sprayed" so that every square meter of the disk has received, and holds fixed, the same amount of charge.

---

**Example (Potential on the axis)**   Let us find the potential due to our uniformly charged disk, at some point $P_1$ on the axis of symmetry, which we have made the $y$ axis. All charge elements in a thin, ring-shaped segment of the disk lie at the same distance from $P_1$. If $s$ denotes the radius of such an annular segment and $ds$ is its width, its area is $2\pi s\,ds$. The amount of charge it contains, $dq$, is therefore $dq = \sigma\,2\pi s\,ds$. Since all parts of this ring are the same distance away from $P_1$, namely, $r = \sqrt{y^2 + s^2}$, the contribution of the ring to the potential at $P_1$ is $dq/4\pi\epsilon_0 r = \sigma s\,ds/\left(2\epsilon_0\sqrt{y^2 + s^2}\right)$. To get the potential due to the whole disk, we have to integrate over all such rings:

$$\phi(0, y, 0) = \int \frac{dq}{4\pi\epsilon_0 r} = \int_0^a \frac{\sigma s\,ds}{2\epsilon_0\sqrt{y^2 + s^2}} = \frac{\sigma}{2\epsilon_0}\sqrt{y^2 + s^2}\,\Big|_0^a . \qquad (2.24)$$

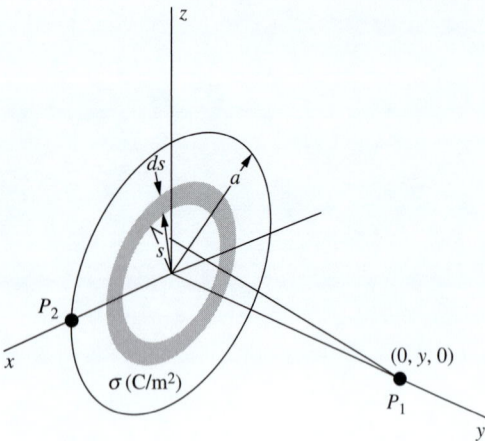

**Figure 2.9.**
Finding the potential at a point $P_1$ on the axis of a uniformly charged disk.

Putting in the limits, we obtain

$$\phi(0, y, 0) = \frac{\sigma}{2\epsilon_0}\left(\sqrt{y^2 + a^2} - y\right) \qquad \text{for } y > 0. \qquad (2.25)$$

A minor point deserves a comment. The result we have written down in Eq. (2.25) holds for all points on the *positive* y axis. It is obvious from the physical symmetry of the system (there is no difference between one face of the disk and the other) that the potential must have the same value for negative and positive y, and this is reflected in Eq. (2.24), where only $y^2$ appears. But in writing Eq. (2.25) we made a choice of sign in taking the square root of $y^2$, with the consequence that it holds only for positive y. The correct expression for $y < 0$ is obtained by the other choice of root and is given by

$$\phi(0, y, 0) = \frac{\sigma}{2\epsilon_0}\left(\sqrt{y^2 + a^2} + y\right) \qquad \text{for } y < 0. \qquad (2.26)$$

In view of this, we should not be surprised to find a kink in the plot of $\phi(0, y, 0)$ at $y = 0$. Indeed, the function has an abrupt change of slope there, as we see in Fig. 2.10, where we have plotted as a function of y the potential on the axis. The potential at the center of the disk is

$$\phi(0, 0, 0) = \frac{\sigma a}{2\epsilon_0}. \qquad (2.27)$$

This much work would be required to bring a unit positive charge in from infinity, by any route, and leave it sitting at the center of the disk.

The behavior of $\phi(0, y, 0)$ for very large y is interesting. For $y \gg a$ we can approximate Eq. (2.25) as follows:

$$\sqrt{y^2 + a^2} - y = y\left[\left(1 + \frac{a^2}{y^2}\right)^{1/2} - 1\right] = y\left[1 + \frac{1}{2}\left(\frac{a^2}{y^2}\right) + \cdots - 1\right] \approx \frac{a^2}{2y}. \qquad (2.28)$$

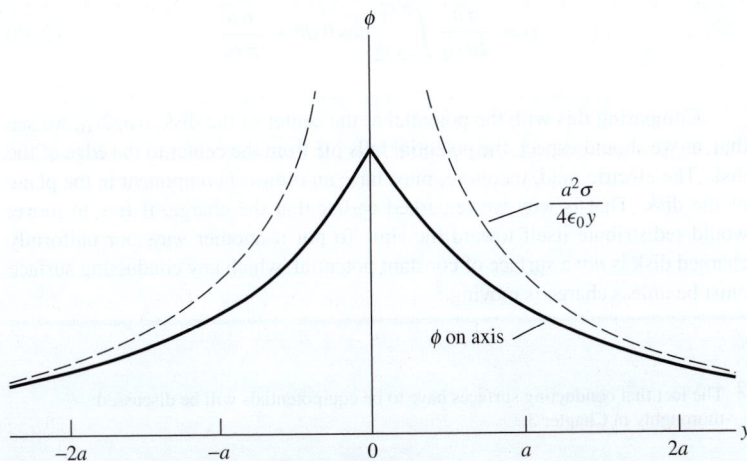

Figure 2.10.
A graph of the potential on the axis. The dashed curve is the potential of a point charge $q = \pi a^2 \sigma$.

Hence

$$\phi(0, y, 0) \approx \frac{a^2 \sigma}{4\epsilon_0 y} \qquad \text{for } y \gg a. \tag{2.29}$$

Now $\pi a^2 \sigma$ is the total charge $q$ on the disk, and Eq. (2.29), which can be written as $\pi a^2 \sigma / 4\pi \epsilon_0 y$, is just the expression for the potential due to a point charge of this magnitude. As we should expect, at a considerable distance from the disk (relative to its diameter), it doesn't matter much how the charge is shaped; only the total charge matters, in first approximation. In Fig. 2.10 we have drawn, as a dashed curve, the function $a^2 \sigma / 4\epsilon_0 y$. You can see that the axial potential function approaches its asymptotic form pretty quickly.

It is not quite so easy to derive the potential for general points away from the axis of symmetry, because the definite integral isn't so simple. It proves to be something called an *elliptic integral*. These functions are well known and tabulated, but there is no point in pursuing here mathematical details peculiar to a special problem. However, one further calculation, which is easy enough, may be instructive.

**Example (Potential on the rim)**   We can find the potential at a point on the very edge of the disk, such as $P_2$ in Fig. 2.11. To calculate the potential at $P_2$ we can consider first the thin wedge of length $R$ and angular width $d\theta$, as shown. An element of the wedge, the black patch at distance $r$ from $P_2$, contains an amount of charge $dq = \sigma r \, d\theta \, dr$. Its contribution to the potential at $P_2$ is therefore $dq/4\pi \epsilon_0 r = \sigma \, d\theta \, dr/4\pi \epsilon_0$. The contribution of the entire wedge is then $(\sigma \, d\theta / 4\pi \epsilon_0) \int_0^R dr = (\sigma R/4\pi \epsilon_0) \, d\theta$. Now $R$ is $2a \cos \theta$, from the geometry of the right triangle, and the whole disk is swept out as $\theta$ ranges from $-\pi/2$ to $\pi/2$. Thus we find the potential at $P_2$:

$$\phi = \frac{\sigma a}{2\pi \epsilon_0} \int_{-\pi/2}^{\pi/2} \cos \theta \, d\theta = \frac{\sigma a}{\pi \epsilon_0}. \tag{2.30}$$

Comparing this with the potential at the center of the disk, $\sigma a/2\epsilon_0$, we see that, as we should expect, the potential falls off from the center to the edge of the disk. The electric field, therefore, must have an *outward* component in the plane of the disk. That is why we remarked earlier that the charge, if free to move, would redistribute itself toward the rim. To put it another way, our uniformly charged disk is *not* a surface of constant potential, which any conducting surface must be unless charge is moving.[2]

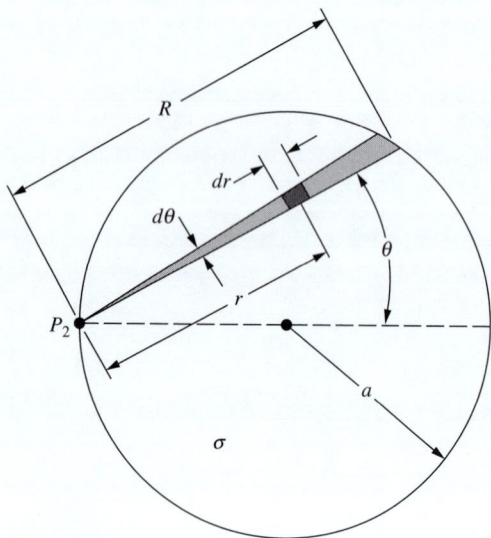

**Figure 2.11.**
Finding the potential at a point $P_2$ on the rim of a uniformly charged disk.

---

[2]   The fact that conducting surfaces have to be equipotentials will be discussed thoroughly in Chapter 3.

Let us now examine the electric field due to the disk. For $y > 0$, the field on the symmetry axis can be computed directly from the potential function given in Eq. (2.25):

$$E_y = -\frac{\partial \phi}{\partial y} = -\frac{d}{dy} \frac{\sigma}{2\epsilon_0} \left( \sqrt{y^2 + a^2} - y \right)$$

$$= \frac{\sigma}{2\epsilon_0} \left[ 1 - \frac{y}{\sqrt{y^2 + a^2}} \right] \qquad y > 0. \qquad (2.31)$$

To be sure, it is not hard to compute $E_y$ directly from the charge distribution, for points on the axis. We can again slice the disk into concentric rings, as we did prior to Eq. (2.24). But we must remember that $\mathbf{E}$ is a vector and that only the $y$ component survives in the present setup, whereas we did not need to worry about components when calculating the scalar function $\phi$ above.

As $y$ approaches zero from the positive side, $E_y$ approaches $\sigma/2\epsilon_0$. On the negative $y$ side of the disk, which we shall call the back, $\mathbf{E}$ points in the other direction and its $y$ component $E_y$ is $-\sigma/2\epsilon_0$. This is the same as the field of an infinite sheet of charge of density $\sigma$, derived in Section 1.13. It ought to be, for at points close to the center of the disk, the presence or absence of charge out beyond the rim can't make much difference. In other words, any sheet looks infinite if viewed from close up. Indeed, $E_y$ has the value $\sigma/2\epsilon_0$ not only at the center, but also all over the disk.

For large $y$, we can find an approximate expression for $E_y$ by using a Taylor series approximation as we did in Eq. (2.28). You can show that $E_y$ approaches $a^2\sigma/4\epsilon_0 y^2$, which can be written as $\pi a^2\sigma/4\pi\epsilon_0 y^2$. This is correctly the field due to a point charge with magnitude $\pi a^2\sigma$.

In Fig. 2.12 we show some field lines for this system and also, plotted as dashed curves, the intersections on the $yz$ plane of the surfaces of constant potential. Near the center of the disk these are lens-like surfaces, while at distances much greater than $a$ they approach the spherical form of equipotential surfaces around a point charge.

Figure 2.12 illustrates a general property of field lines and equipotential surfaces. A field line through any point and the equipotential surface through that point *are perpendicular to one another*, just as, on a contour map of hilly terrain, the slope is steepest at right angles to a contour of constant elevation. This must be so, because if the field at any point had a component parallel to the equipotential surface through that point, it would require work to move a test charge along a constant-potential surface.

The energy associated with this electric field could be expressed as the integral over all space of $(\epsilon_0/2)E^2 \, dv$. It is equal to the work done in assembling this distribution, starting with infinitesimal charges far apart. In this particular example, as Exercise 2.56 will demonstrate, that work

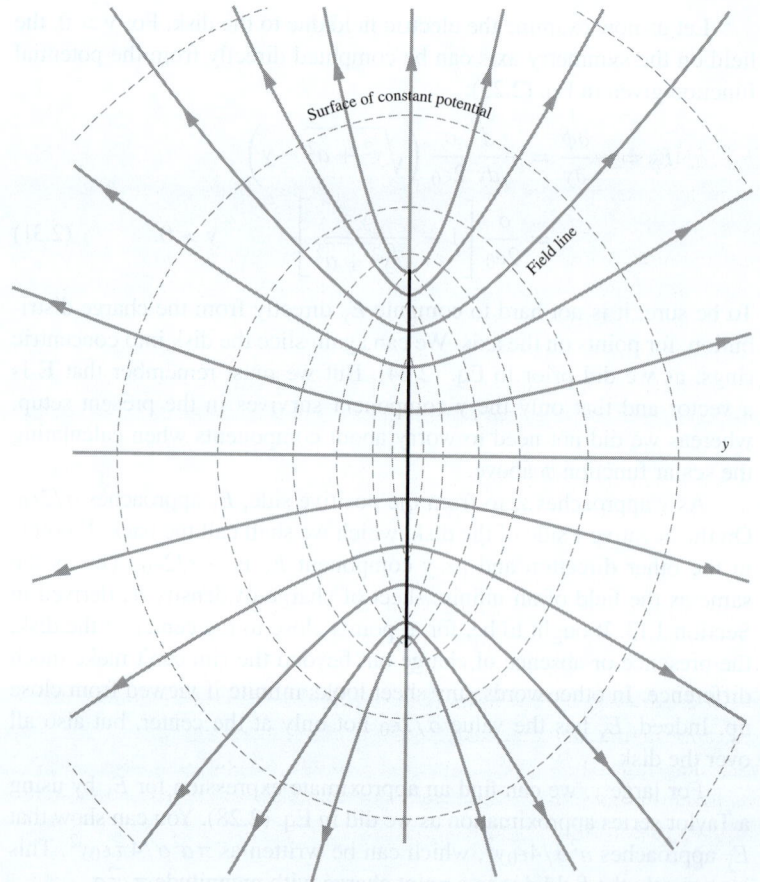

**Figure 2.12.**
The electric field of the uniformly charged disk.
Solid curves are field lines. Dashed curves are
intersections, with the plane of the figure, of
surfaces of constant potential.

is not hard to calculate directly if we know the potential at the rim of a
uniformly charged disk.

There is a general relation between the work $U$ required to assemble a charge distribution $\rho(x, y, z)$ and the potential $\phi(x, y, z)$ of that
distribution:

$$U = \frac{1}{2} \int \rho \phi \, dv \tag{2.32}$$

Equation (1.15), which gives the energy of a system of discrete point
charges, could have been written in this way:

$$U = \frac{1}{2} \sum_{j=1}^{N} q_j \sum_{k \neq j} \frac{1}{4\pi \epsilon_0} \frac{q_k}{r_{jk}}. \tag{2.33}$$

The second sum is the potential at the location of the $j$th charge, due to all
the other charges. To adapt this to a continuous distribution we merely

replace $q_j$ with $\rho \, dv$ and the sum over $j$ by an integral, thus obtaining Eq. (2.32).

## 2.7 Dipoles

Consider a setup with two equal and opposite charges $\pm q$ located at positions $\pm \ell/2$ on the $y$ axis, as shown in Fig. 2.13. This configuration is called a *dipole*. The purpose of this section is to introduce the basics of dipoles. We save further discussion for Chapter 10, where we define the word "dipole" more precisely, derive things in more generality, and discuss examples of dipoles in actual matter. For now we just concentrate on determining the electric field and potential of a dipole. We have all of the necessary machinery at our disposal, so let's see what we can find.

We will restrict the treatment to points far away from the dipole (that is, points with $r \gg \ell$). Although it is easy enough to write down an exact expression for the potential $\phi$ (and hence the field $\mathbf{E} = -\nabla \phi$) at any position, the result isn't very enlightening. But when we work in the approximation of large distances, we obtain a result that, although isn't exactly correct, is in fact quite enlightening. That's how approximations work – you trade a little bit of precision for a large amount of clarity.

Our strategy will be to find the potential $\phi$ in polar (actually spherical) coordinates, and then take the gradient to find the electric field $\mathbf{E}$. We then determine the shape of the field-line and constant-potential curves. To make things look a little cleaner in the calculations below, we write $1/4\pi\epsilon_0$ as $k$ in some intermediate steps.

### 2.7.1 Calculation of $\phi$ and E

First note that, since the dipole setup is rotationally symmetric around the line containing the two charges, it suffices to find the potential in an arbitrary plane containing this line. We will use spherical coordinates, which reduce to polar coordinates in a plane because the angle $\phi$ doesn't come into play (but note that $\theta$ is measured down from the vertical axis). Consider a point $P$ with coordinates $(r, \theta)$, as shown in Fig. 2.14. Let $r_1$ and $r_2$ be the distances from $P$ to the two charges. Then the exact expression for the potential at $P$ is (with $k \equiv 1/4\pi\epsilon_0$)

$$\phi_P = \frac{kq}{r_1} - \frac{kq}{r_2}. \tag{2.34}$$

If desired, the law of cosines can be used to write $r_1$ and $r_2$ in terms of $r$, $\theta$, and $\ell$.

Let us now derive an approximate form of this result, valid in the $r \gg \ell$ limit. One way to do this is to use the law-of-cosines expressions for $r_1$ and $r_2$; this is the route we will take in Chapter 10. But for the present purposes a simpler method suffices. In the $r \gg \ell$ limit, a closeup view of the dipole is shown in Fig. 2.15. The two lines from the charges to $P$ are essentially parallel, so we see from the figure that the lengths of

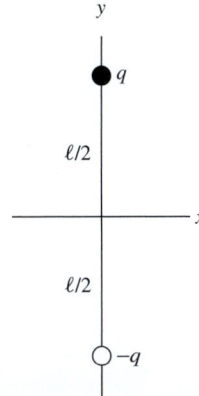

**Figure 2.13.**
Two equal and opposite charges form a dipole.

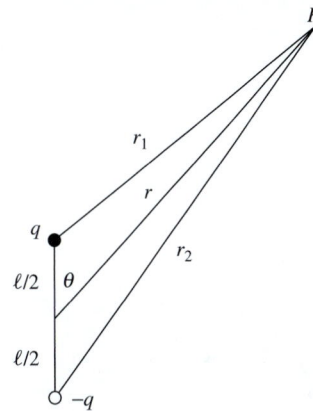

**Figure 2.14.**
Finding the potential $\phi$ at point $P$.

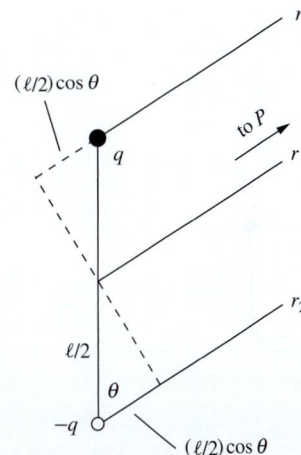

**Figure 2.15.**
Closeup view of Fig. 2.14.

these lines are essentially $r_1 = r - (\ell/2)\cos\theta$ and $r_2 = r + (\ell/2)\cos\theta$. Using the approximation $1/(1 \pm \epsilon) \approx 1 \mp \epsilon$, Eq. (2.34) becomes

$$
\phi(r,\theta) = \frac{kq}{r - \dfrac{\ell\cos\theta}{2}} - \frac{kq}{r + \dfrac{\ell\cos\theta}{2}} = \frac{kq}{r}\left[\frac{1}{1 - \dfrac{\ell\cos\theta}{2r}} - \frac{1}{1 + \dfrac{\ell\cos\theta}{2r}}\right]
$$

$$
\approx \frac{kq}{r}\left[\left(1 + \frac{\ell\cos\theta}{2r}\right) - \left(1 - \frac{\ell\cos\theta}{2r}\right)\right]
$$

$$
= \frac{kq\ell\cos\theta}{r^2} \equiv \boxed{\frac{q\ell\cos\theta}{4\pi\epsilon_0 r^2}} \equiv \frac{p\cos\theta}{4\pi\epsilon_0 r^2}, \tag{2.35}
$$

where $p \equiv q\ell$ is called the *dipole moment*.

There are three important things to note about this result. First, $\phi(r,\theta)$ depends on $q$ and $\ell$ only through their product, $p \equiv q\ell$. This means that if we make $q$ ten times larger and $\ell$ ten times smaller, the potential at a given point $P$ stays the same (at least in the $r \gg \ell$ approximation). An *idealized dipole* or *point dipole* is one where $\ell \to 0$ and $q \to \infty$, with the product $p = q\ell$ taking on a particular finite value. In the other extreme, if we make $q$ smaller and $\ell$ proportionally larger, the potential at $P$ again stays the same. Of course, if we make $\ell$ too large, our $r \gg \ell$ assumption eventually breaks down.

Second, $\phi(r,\theta)$ is proportional to $1/r^2$, in contrast with the $1/r$ dependence for a point-charge potential. We will see below that the present $1/r^2$ dependence in $\phi(r,\theta)$ leads to an **E** field that falls off like $1/r^3$, in contrast with the $1/r^2$ dependence for a point-charge field. It makes sense that the potential (and field) falls off faster for a dipole, because the potentials from the two opposite point charges nearly cancel. The dipole potential is somewhat like the derivative of the point-charge potential, in that we are taking the difference of two nearby values.

Third, there is angular dependence in $\phi(r,\theta)$, in contrast with the point-charge potential. This is expected, in view of the fact that the dipole has a preferred direction along the line joining the charges, whereas a point charge has no preferred direction.

We will see in Chapter 10 that the $q/r$ point-charge (or *monopole*) potential and the $q\ell/r^2$ dipole potential (just looking at the $r$ dependence) are the first two pieces of what is called the *multipole expansion*. A general charge distribution also has a *quadrupole* term in the potential that goes like $q\ell^2/r^3$ (where $\ell$ is some length scale of the system), and an *octupole* term that goes like $q\ell^3/r^4$, and so on. These pieces have more complicated angular dependences. Two examples of quadrupole arrangements are shown in Fig. 2.16. A quadrupole is formed by placing two oppositely charged dipoles near each other, just as a dipole is formed by placing two oppositely charged monopoles near each other. The various terms in the expansion are called the *moments* of the distribution.

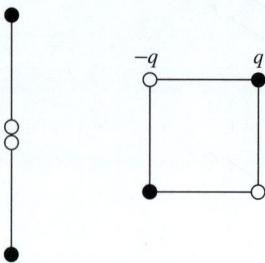

**Figure 2.16.**
Two possible kinds of quadrupoles.

Even the simple system of the dipole shown in Fig. 2.13 has higher terms in its multipole expansion. If you keep additional terms in the $1/(1 \pm \epsilon)$ Taylor series in Eq. (2.35), you will find that the quadrupole term is zero, but the octupole term is nonzero. It is easy to see that the terms with even powers of $r$ are nonzero. However, in the limit of an idealized dipole ($\ell \to 0$ and $q \to \infty$, with $q\ell$ fixed), only the dipole potential survives, because the higher-order terms are suppressed by additional powers of $\ell/r$.

Along the same lines, we can back up a step in the expansion and consider the monopole term. If an object has a nonzero net charge (note that our dipole does not), then the monopole potential, $q/r$, dominates, and all higher-order terms are comparatively negligible in the $r \gg \ell$ limit. The distribution of charge in an object determines which of the terms in the expansion is the first nonzero one, and it is this term that determines the potential (and hence field) at large distances. We label the object according to the first nonzero term; see Fig. 2.17.

Let's now find the electric field, $\mathbf{E} = -\nabla\phi$, associated with the dipole potential in Eq. (2.35). In spherical coordinates (which reduce to polar coordinates in this discussion) the gradient of $\phi$ is $\nabla\phi = \hat{\mathbf{r}}(\partial\phi/\partial r) + \hat{\boldsymbol{\theta}}(1/r)(\partial\phi/\partial\theta)$; see Appendix F. So the electric field is

$$\mathbf{E}(r,\theta) = -\hat{\mathbf{r}}\frac{\partial}{\partial r}\left(\frac{kq\ell\cos\theta}{r^2}\right) - \hat{\boldsymbol{\theta}}\frac{1}{r}\frac{\partial}{\partial\theta}\left(\frac{kq\ell\cos\theta}{r^2}\right)$$

$$= \frac{kq\ell}{r^3}\left(2\cos\theta\,\hat{\mathbf{r}} + \sin\theta\,\hat{\boldsymbol{\theta}}\right)$$

$$\boxed{\equiv \frac{q\ell}{4\pi\epsilon_0 r^3}\left(2\cos\theta\,\hat{\mathbf{r}} + \sin\theta\,\hat{\boldsymbol{\theta}}\right)}$$

$$\equiv \frac{p}{4\pi\epsilon_0 r^3}\left(2\cos\theta\,\hat{\mathbf{r}} + \sin\theta\,\hat{\boldsymbol{\theta}}\right). \tag{2.36}$$

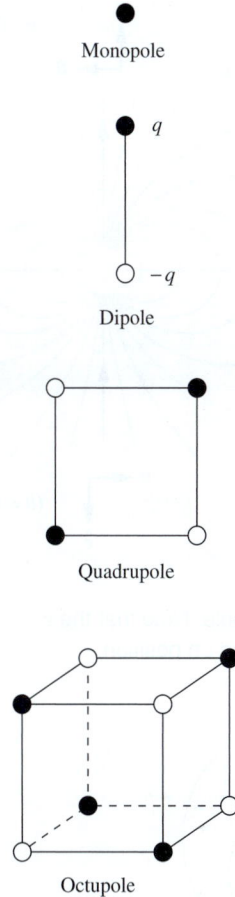

**Figure 2.17.**
Examples of different objects in the multipole expansion.

A few field lines are shown in Fig. 2.18. Let's look at some special cases for $\theta$. Equation (2.36) says that $\mathbf{E}$ points in the positive radial direction for $\theta = 0$ and the negative radial direction for $\theta = \pi$. These facts imply that $\mathbf{E}$ points *upward* everywhere on the $y$ axis. Equation (2.36) also says that $\mathbf{E}$ points in the positive tangential direction for $\theta = \pi/2$ and the negative tangential direction for $\theta = 3\pi/2$. In view of the local $\hat{\mathbf{r}}$ and $\hat{\boldsymbol{\theta}}$ basis vectors shown in Fig. 2.18 (which vary with position, unlike the Cartesian $\hat{\mathbf{x}}$ and $\hat{\mathbf{y}}$ basis vectors), this means that $\mathbf{E}$ points *downward* everywhere on the $x$ axis. We haven't drawn the lines for small $r$, to emphasize the fact that our results are valid only in the limit $r \gg \ell$. There *is* a field for small $r$, of course (and it diverges near each charge); it's just that it doesn't take the form given in Eq. (2.36).

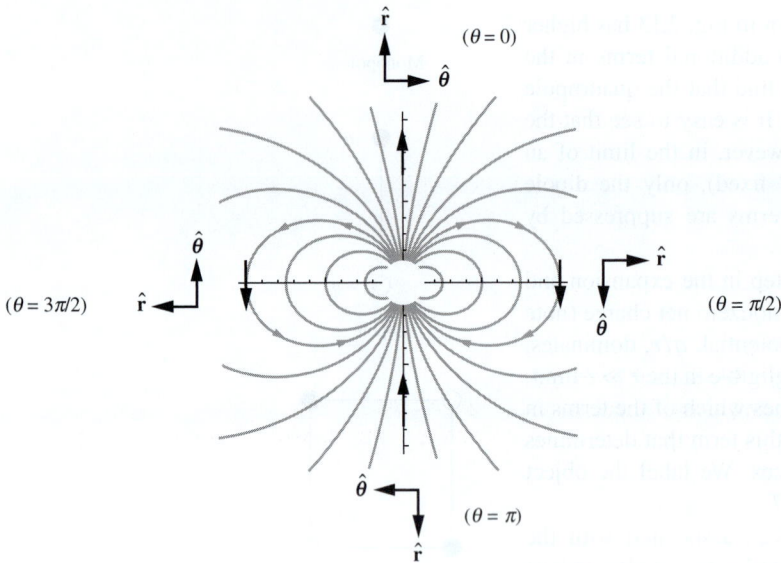

**Figure 2.18.**
Electric field lines for a dipole. Note that the $\hat{\mathbf{r}}$ and $\hat{\boldsymbol{\theta}}$ basis vectors depend on position.

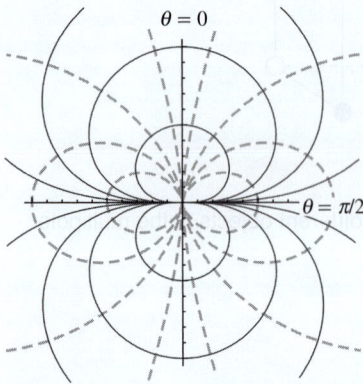

**Figure 2.19.**
Field lines and constant-potential curves for a dipole. The two sets of curves are orthogonal at all intersections. The solid lines show constant-$\phi$ curves ($r = r_0\sqrt{\cos\theta}$), and the dashed lines show $\mathbf{E}$ field lines ($r = r_0\sin^2\theta$).

## 2.7.2 The shapes of the curves

Let us now be quantitative about the shape of the $\mathbf{E}$ and $\phi$ curves. More precisely, let us determine the equations that describe the field-line curves and the constant-potential curves. In the process we will also determine the slopes of the tangents to these curves. We know that the two classes of curves are orthogonal wherever they meet, because $\mathbf{E}$ is the (negative) gradient of $\phi$, and because the gradient of a function is always perpendicular to the level-surface curves. This orthogonality is evident in Fig. 2.19. Our task now is to derive the two expressions for $r$ given in this figure.

Let's look at $\phi$ first. We will find the equation for the constant-potential curves and then use this to find the slope of the tangent at any point. The equation for the curves is immediately obtained from Eq. (2.35). The set of points for which $\phi$ takes on the constant value $\phi_0$ is given by

$$\frac{kq\ell\cos\theta}{r^2} = \phi_0 \implies r^2 = \left(\frac{kq\ell}{\phi_0}\right)\cos\theta \implies \boxed{r = r_0\sqrt{\cos\theta}}$$

(2.37)

where $r_0 \equiv \sqrt{kq\ell/\phi_0}$ is the radius associated with the angle $\theta = 0$. This result is valid in the upper half-plane where $-\pi/2 < \theta < \pi/2$. In the lower half-plane, both $\phi_0$ and $\cos\theta$ are negative, so we need to add in some absolute-value signs. That is, $r = r_0\sqrt{|\cos\theta|}$, where $r_0 \equiv \sqrt{kq\ell/|\phi_0|}$. The constant-potential curves in Fig. 2.19 are the intersections of the constant-potential *surfaces* with the plane of the paper. These surfaces are generated by rotating the curves around the vertical axis. The curves are stretched horizontally compared with the circles described by the relation $r = r_0\cos\theta$ (which you can verify is indeed a circle).

The slope of a given curve at a given point, relative to the local $\hat{\mathbf{r}}$ and $\hat{\boldsymbol{\theta}}$ basis vectors at that point, is $dr/r\,d\theta$. The $r$ is needed in the denominator because $r\,d\theta$ is the actual distance associated with the angular span $d\theta$ in the $r$-$\theta$ plane; see Fig. 2.20. So the slope of the $r = r_0\sqrt{\cos\theta}$ curve is

$$\frac{1}{r}\frac{dr}{d\theta} = \frac{1}{r_0\sqrt{\cos\theta}}\frac{d\left(r_0\sqrt{\cos\theta}\right)}{d\theta} = \frac{1}{r_0\sqrt{\cos\theta}}\frac{-r_0\sin\theta}{2\sqrt{\cos\theta}} = -\frac{\sin\theta}{2\cos\theta}.$$

$$(2.38)$$

Remember that this is the slope with respect to the local $\hat{\mathbf{r}}$, $\hat{\boldsymbol{\theta}}$ basis (which varies with position), and *not the fixed* $\hat{\mathbf{x}}$, $\hat{\mathbf{y}}$ *basis*. For $\theta = 0$ or $\pi$, the slope is 0, which means that the tangent is parallel to the $\hat{\boldsymbol{\theta}}$ direction. This is horizontal in Fig. 2.19; the constant-$\phi$ curves all intersect the $y$ axis horizontally. For $\theta = \pm\pi/2$ the slope is $\pm\infty$, which means that the tangent is parallel to the $\hat{\mathbf{r}}$ direction. Due to the orientation of the local $\hat{\mathbf{r}}$, $\hat{\boldsymbol{\theta}}$ basis vectors (see Fig. 2.18), this is also horizontal in Fig. 2.19; the curves all feed into the origin directly along the $x$ axis.

Now consider the **E** field. We will do things in reverse order here, first finding the slope of the tangent, and then using that to find the equation of the field-line curves. The slope of the tangent is immediately obtained from the $E_r$ and $E_\theta$ components given in Eq. (2.36). We have

$$\frac{E_r}{E_\theta} = \frac{2\cos\theta}{\sin\theta}.$$

$$(2.39)$$

This slope is the negative reciprocal of the slope of the tangent to the constant-$\phi$ curves, given in Eq. (2.38). This means that the two classes of curves are orthogonal at all intersections, as we know is the case.

To find the equation for the field-line curves, we can use the fact that the slope in Eq. (2.39) must be equal to $dr/r\,d\theta$. We can then separate variables and integrate to obtain

$$\frac{1}{r}\frac{dr}{d\theta} = \frac{2\cos\theta}{\sin\theta} \quad\Longrightarrow\quad \int\frac{dr}{r} = \int\frac{2\cos\theta\,d\theta}{\sin\theta}$$

$$\Longrightarrow\quad \ln r = 2\ln\sin\theta + C.$$

$$(2.40)$$

Exponentiating both sides gives

$$\boxed{r = r_0\sin^2\theta}$$

$$(2.41)$$

where $r_0 \equiv e^C$ is the radius associated with the angle $\theta = \pi/2$. The curves are squashed vertically compared with the circles described by the relation $r = r_0\sin\theta$. If you want to get some practice with the concepts in this section, the task of Exercise 2.63 is to repeat everything we've done here, but for the case of a dipole in two dimensions.

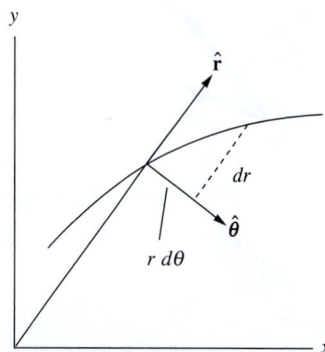

**Figure 2.20.**
The slope with respect to the $\hat{\mathbf{r}}$, $\hat{\boldsymbol{\theta}}$ basis equals $dr/(r\,d\theta)$.

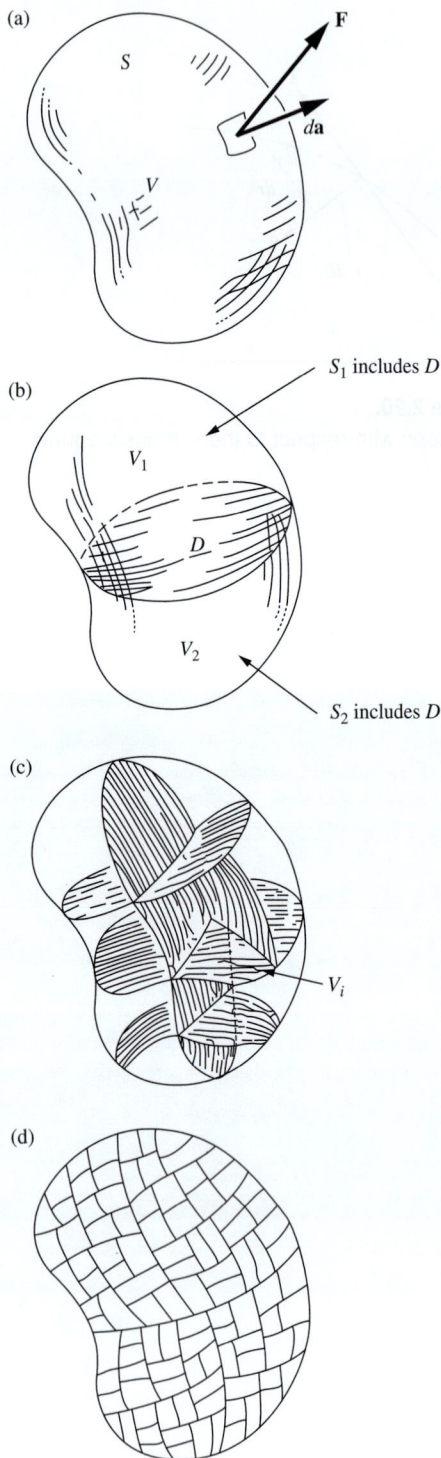

(a)

(b)

$S_1$ includes $D$

$S_2$ includes $D$

(c)

(d)

## 2.8 Divergence of a vector function

The electric field has a definite direction and magnitude at every point. It is a vector function of the coordinates, which we have often indicated by writing $\mathbf{E}(x, y, z)$. What we are about to say can apply to any vector function, not just to the electric field; we shall use another symbol, $\mathbf{F}(x, y, z)$, as a reminder of that. In other words, we shall talk mathematics rather than physics for a while and call $\mathbf{F}$ simply a general vector function. We shall keep to three dimensions, however.

Consider a finite volume $V$ of some shape, the surface of which we shall denote by $S$. We are already familiar with the notion of the total flux $\Phi$ emerging from $S$. It is the value of the surface integral of $\mathbf{F}$ extended over the whole of $S$:

$$\Phi = \int_S \mathbf{F} \cdot d\mathbf{a}. \tag{2.42}$$

In the integrand $d\mathbf{a}$ is the infinitesimal vector whose magnitude is the area of a small element of $S$ and whose direction is the outward-pointing normal to that little patch of surface, indicated in Fig. 2.21(a).

Now imagine dividing $V$ into two parts by a surface, or a diaphragm, $D$ that cuts through the "balloon" $S$, as in Fig. 2.21(b). Denote the two parts of $V$ by $V_1$ and $V_2$ and, treating them as distinct volumes, compute the surface integral over each separately. The boundary surface $S_1$ of $V_1$ includes $D$, and so does $S_2$. It is pretty obvious that the sum of the two surface integrals

$$\int_{S_1} \mathbf{F} \cdot d\mathbf{a}_1 + \int_{S_2} \mathbf{F} \cdot d\mathbf{a}_2 \tag{2.43}$$

will equal the original integral over the whole surface expressed in Eq. (2.42). The reason is that any given patch on $D$ contributes with one sign to the first integral and the same amount with opposite sign to the second, the "outward" direction in one case being the "inward" direction in the other. In other words, any flux *out* of $V_1$, through this surface $D$, is flux *into* $V_2$. The rest of the surface involved is identical to that of the original entire volume.

We can keep on subdividing until our internal partitions have divided $V$ into a large number of parts, $V_1, \ldots, V_i, \ldots, V_N$, with surfaces $S_1, \ldots, S_i, \ldots, S_N$. No matter how far this is carried, we can still be sure that

$$\sum_{i=1}^{N} \int_{S_i} \mathbf{F} \cdot d\mathbf{a}_i = \int_S \mathbf{F} \cdot d\mathbf{a} = \Phi. \tag{2.44}$$

**Figure 2.21.**
(a) A volume $V$ enclosed by a surface $S$ is divided (b) into two pieces enclosed by $S_1$ and $S_2$. No matter how far this is carried, as in (c) and (d), the sum of the surface integrals over all the pieces equals the original surface integral over $S$, for any vector function $\mathbf{F}$.

What we are after is this: in the limit as $N$ becomes enormous we want to identify something which is characteristic of a particular small region – and, ultimately, of the neighborhood of a point. Now the surface integral

$$\int_{S_i} \mathbf{F} \cdot d\mathbf{a}_i \qquad (2.45)$$

over one of the small regions is *not* such a quantity, for if we divide everything again, so that $N$ becomes $2N$, this integral divides into two terms, each smaller than before since their sum is constant. In other words, as we consider smaller and smaller volumes in the same locality, the surface integral over one such volume gets steadily smaller. But we notice that, when we divide, the volume is also divided into two parts that sum to the original volume. This suggests that we look at the ratio of surface integral to volume for an element in the subdivided space:

$$\frac{\int_{S_i} \mathbf{F} \cdot d\mathbf{a}_i}{V_i}. \qquad (2.46)$$

It seems plausible that for $N$ large enough, that is, for sufficiently fine-grained subdivision, we can halve the volume every time we halve the surface integral, so we find that, with continuing subdivision of any particular region, this ratio approaches a limit. If so, this limit is a property characteristic of the vector function $\mathbf{F}$ in that neighborhood. We call it the *divergence* of $\mathbf{F}$, written div $\mathbf{F}$. That is, the value of div $\mathbf{F}$ at any point is defined as

$$\boxed{\operatorname{div} \mathbf{F} \equiv \lim_{V_i \to 0} \frac{1}{V_i} \int_{S_i} \mathbf{F} \cdot d\mathbf{a}_i} \qquad (2.47)$$

where $V_i$ is a volume including the point in question, and $S_i$, over which the surface integral is taken, is the surface of $V_i$. We must include the proviso that the limit exists and is independent of our method of subdivision. For the present we shall assume that this is true.

The meaning of div $\mathbf{F}$ can be expressed in this way: div $\mathbf{F}$ is the flux out of $V_i$, per unit of volume, in the limit of infinitesimal $V_i$. It is a scalar quantity, obviously. It may vary from place to place, its value at any particular location $(x, y, z)$ being the limit of the ratio in Eq. (2.47) as $V_i$ is chopped smaller and smaller while always enclosing the point $(x, y, z)$. So div $\mathbf{F}$ is simply a scalar function of the coordinates.

## 2.9 Gauss's theorem and the differential form of Gauss's law

If we know this scalar function of position, div $\mathbf{F}$, we can work our way right back to the surface integral over a large volume. We first write

Eq. (2.44) in this way:

$$\int_S \mathbf{F} \cdot d\mathbf{a} = \sum_{i=1}^{N} \int_{S_i} \mathbf{F} \cdot d\mathbf{a}_i = \sum_{i=1}^{N} V_i \left[ \frac{\int_{S_i} \mathbf{F} \cdot d\mathbf{a}_i}{V_i} \right]. \tag{2.48}$$

In the limit $N \to \infty$, $V_i \to 0$, the term in brackets becomes the divergence of $\mathbf{F}$, and the sum goes into a volume integral:

$$\boxed{\int_S \mathbf{F} \cdot d\mathbf{a} = \int_V \text{div } \mathbf{F} \, dv} \qquad \text{(Gauss's theorem).} \tag{2.49}$$

This result is called *Gauss's theorem*, or the *divergence theorem*. It holds for any vector field for which the limit involved in Eq. (2.47) exists. Note that the entire content of the theorem is contained in Eq. (2.44), which itself is simply the statement that the fluxes cancel in pairs over the interior boundaries of all the little regions. The other steps in the proof were the multiplication by 1 in the form of $V_i/V_i$, the use of the definition in Eq. (2.47), and the conversion of an infinite sum to an integral. None of these steps contains much content.

Let us see what Eq. (2.49) implies for the electric field $\mathbf{E}$. We have Gauss's law, Eq. (1.31), which assures us that

$$\int_S \mathbf{E} \cdot d\mathbf{a} = \frac{1}{\epsilon_0} \int_V \rho \, dv. \tag{2.50}$$

If the divergence theorem holds for any vector field, it certainly holds for $\mathbf{E}$:

$$\int_S \mathbf{E} \cdot d\mathbf{a} = \int_V \text{div } \mathbf{E} \, dv. \tag{2.51}$$

Equations (2.50) and (2.51) hold for *any* volume we care to choose – of any shape, size, or location. Comparing them, we see that this can only be true if, at every point,

$$\boxed{\text{div } \mathbf{E} = \frac{\rho}{\epsilon_0}} \tag{2.52}$$

If we adopt the divergence theorem as part of our regular mathematical equipment from now on, we can regard Eq. (2.52) simply as an alternative statement of Gauss's law. It is Gauss's law in differential form, that is, stated in terms of a local relation between charge density and electric field.

**Example (Field and density in a sphere)**   Let's use the result from the example in Section 1.11 to verify that Eq. (2.52) holds both inside and outside a sphere with radius $R$ and uniform density $\rho$. Spherical coordinates are of course the most convenient ones to use here, given that we are dealing with a sphere. For the purposes of this example we will simply accept the expression given in

Eq. (F.3) in Appendix F for the divergence (also written as $\nabla \cdot \mathbf{E}$) in spherical coordinates. This appendix explains how to derive the various vector operators, including the divergence, in the common systems of coordinates (Cartesian, cylindrical, spherical). You are encouraged to read it in parallel with this chapter. In Section 2.10 we give a detailed derivation of the form of the divergence in Cartesian coordinates.

Since the electric field due to the sphere has only an $r$ component, Eq. (F.3) tells us that the divergence of $\mathbf{E}$ is div $\mathbf{E} = (1/r^2)\partial(r^2 E_r)/\partial r$. Inside the sphere, we have $E_r = \rho r/3\epsilon_0$ from Eq. (1.35), so

$$\text{div } \mathbf{E}_{\text{in}} = \frac{1}{r^2}\frac{\partial}{\partial r}\left(r^2\frac{\rho r}{3\epsilon_0}\right) = \frac{1}{r^2}\frac{\rho r^2}{\epsilon_0} = \frac{\rho}{\epsilon_0}, \tag{2.53}$$

as desired. Outside the sphere, the field is $E_r = \rho R^3/3\epsilon_0 r^2$ from Eq. (1.34), which equals the standard $Q/4\pi\epsilon_0 r^2$ result when written in terms of the total charge $Q$. However, the exact form doesn't matter here. All that matters is that $E_r$ is proportional to $1/r^2$, because then

$$\text{div } \mathbf{E}_{\text{out}} \propto \frac{1}{r^2}\frac{\partial}{\partial r}\left(r^2\frac{1}{r^2}\right) = 0. \tag{2.54}$$

This agrees with Eq. (2.52) because $\rho = 0$ outside the sphere. Of course, it is no surprise that these relations worked out – we originally derived $E_r$ from Gauss's law, and Eq. (2.52) is simply the differential form of Gauss's law.

Although we used spherical coordinates in this example, Eq. (2.52) must still be true for any choice of coordinates. The task of Exercise 2.68 is to redo this example in Cartesian coordinates. If you are uneasy about invoking the above form of the divergence in spherical coordinates, you should solve Exercise 2.68 after reading the following section.

## 2.10 The divergence in Cartesian coordinates

While Eq. (2.47) is the fundamental definition of *divergence*, independent of any system of coordinates, it is useful to know how to calculate the divergence of a vector function when we are given its explicit form. Suppose a vector function $\mathbf{F}$ is expressed as a function of Cartesian coordinates $x$, $y$, and $z$. That means that we have three scalar functions, $F_x(x, y, z)$, $F_y(x, y, z)$, and $F_z(x, y, z)$. We'll take the region $V_i$ in the shape of a little rectangular box, with one corner at the point $(x, y, z)$ and sides $\Delta x$, $\Delta y$, and $\Delta z$, as in Fig. 2.22(a). Whether some other shape will yield the same limit is a question we must face later.

Consider two opposite faces of the box, the top and bottom for instance, which would be represented by the $d\mathbf{a}$ vectors $\hat{\mathbf{z}}\,\Delta x\,\Delta y$ and $-\hat{\mathbf{z}}\,\Delta x\,\Delta y$. The flux through these faces involves only the $z$ component of $\mathbf{F}$, and the net contribution depends on the *difference* between $F_z$ at the top and $F_z$ at the bottom or, more precisely, on the difference between the average of $F_z$ over the top face and the average of $F_z$ over the bottom face of the box. To the first order in small quantities this difference is $(\partial F_z/\partial z)\,\Delta z$. Figure 2.22(b) will help to explain this. The average value of $F_z$ on the bottom surface of the box, if we consider only first-order

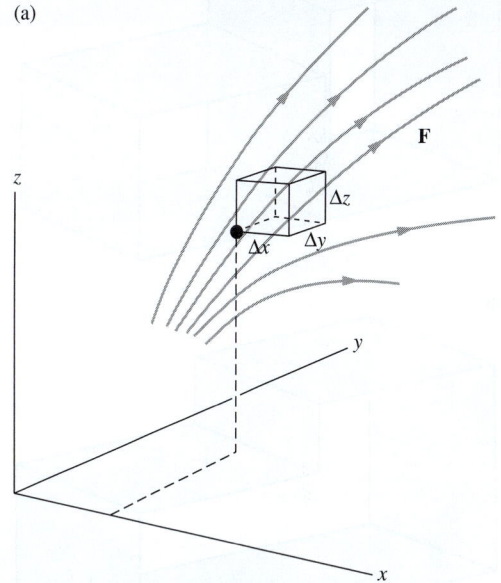

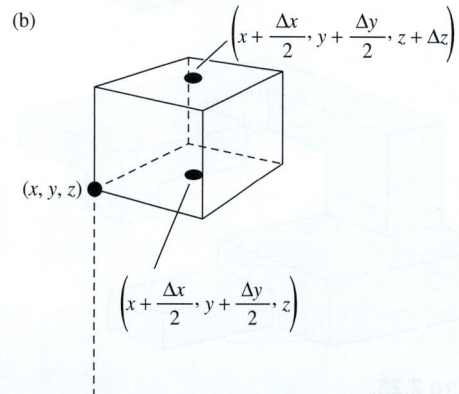

(a)

(b)

**Figure 2.22.**
Calculation of flux from the box of volume $\Delta x\,\Delta y\,\Delta z$.

variations in $F_z$ over this small rectangle, is its value at the center of the rectangle. That value is, to first order[3] in $\Delta x$ and $\Delta y$,

$$F_z(x, y, z) + \frac{\Delta x}{2} \frac{\partial F_z}{\partial x} + \frac{\Delta y}{2} \frac{\partial F_z}{\partial y}. \tag{2.55}$$

For the average of $F_z$ over the top face we take the value at the center of the top face, which to first order in the small displacements is

$$F_z(x, y, z) + \frac{\Delta x}{2} \frac{\partial F_z}{\partial x} + \frac{\Delta y}{2} \frac{\partial F_z}{\partial y} + \Delta z \frac{\partial F_z}{\partial z}. \tag{2.56}$$

The net flux out of the box through these two faces, each of which has the area of $\Delta x \, \Delta y$, is therefore

$$\underbrace{\Delta x \, \Delta y \left[ F_z(x, y, z) + \frac{\Delta x}{2} \frac{\partial F_z}{\partial x} + \frac{\Delta y}{2} \frac{\partial F_z}{\partial y} + \Delta z \frac{\partial F_z}{\partial z} \right]}_{\text{(flux out of box at top)}}$$

$$\underbrace{- \Delta x \, \Delta y \left[ F_z(x, y, z) + \frac{\Delta x}{2} \frac{\partial F_z}{\partial x} + \frac{\Delta y}{2} \frac{\partial F_z}{\partial y} \right]}_{\text{(flux into box at bottom)}}, \tag{2.57}$$

which reduces to $\Delta x \, \Delta y \, \Delta z \, (\partial F_z / \partial z)$. Obviously, similar statements must apply to the other pairs of sides. That is, the net flux out of the box is $\Delta x \, \Delta z \, \Delta y \, (\partial F_y / \partial y)$ through the sides parallel to the $xz$ plane and $\Delta y \, \Delta z \, \Delta x \, (\partial F_x / \partial x)$ through the sides parallel to the $yz$ plane. Note that the product $\Delta x \, \Delta y \, \Delta z$ occurs in all of these expressions. Thus the total flux out of the little box is

$$\Phi = \Delta x \, \Delta y \, \Delta z \left( \frac{\partial F_x}{\partial x} + \frac{\partial F_y}{\partial y} + \frac{\partial F_z}{\partial z} \right). \tag{2.58}$$

The volume of the box is $\Delta x \, \Delta y \, \Delta z$, so the ratio of flux to volume is $\partial F_x / \partial x + \partial F_y / \partial y + \partial F_z / \partial z$, and as this expression does not contain the dimensions of the box at all, it remains as the limit when we let the box shrink. (Had we retained terms proportional to $(\Delta x)^2$, $(\Delta x \, \Delta y)$, etc., in the calculation of the flux, they would of course vanish on going to the limit.)

Now we can begin to see why this limit is going to be independent of the shape of the box. Obviously it is independent of the proportions of the rectangular box, but that isn't saying much. It is easy to see that it will be the same for any volume that we can make by sticking together little rectangular boxes of any size and shape. Consider the two boxes in Fig. 2.23. The sum of the flux $\Phi_1$ out of box 1 and $\Phi_2$ out of box 2 is not

(a)

(b)

(c)

**Figure 2.23.**
The limit of the flux/volume ratio is independent of the shape of the box.

---

[3] This is simply the beginning of a Taylor expansion of the scalar function $F_z$, in the neighborhood of $(x, y, z)$. That is, $F_z(x + a, y + b, z + c) = F_z(x, y, z) + \left( a \frac{\partial}{\partial x} + b \frac{\partial}{\partial y} + c \frac{\partial}{\partial z} \right) F_z + \cdots + \frac{1}{n!} \left( a \frac{\partial}{\partial x} + b \frac{\partial}{\partial y} + c \frac{\partial}{\partial z} \right)^n F_z + \cdots$ . The derivatives are all to be evaluated at $(x, y, z)$. In our case $a = \Delta x / 2$, $b = \Delta y / 2$, $c = 0$, and we drop the higher-order terms in the expansion.

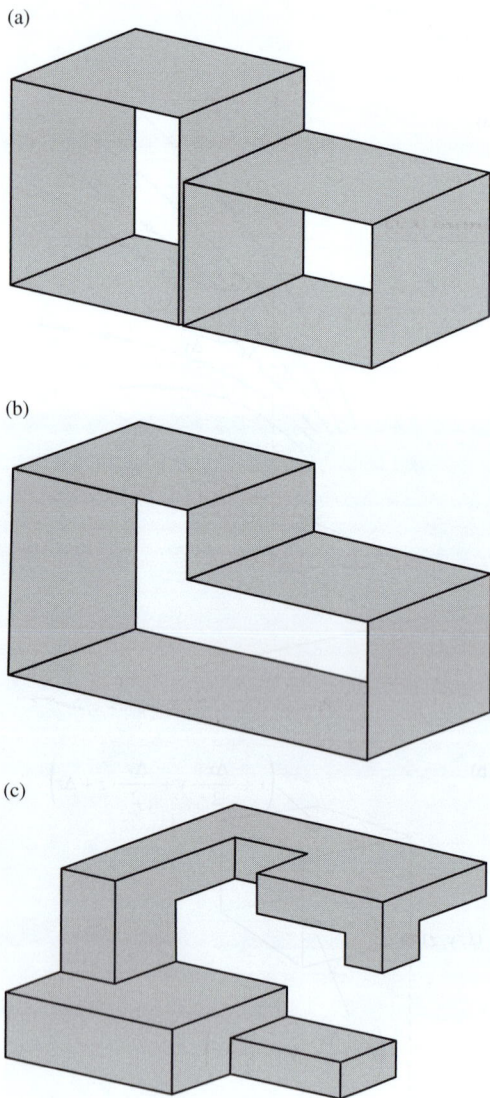

changed by removing the adjoining walls to make one box, for whatever flux went through that plane was negative flux for one and positive for the other. So we could have a bizarre shape like Fig. 2.23(c) without affecting the result. We leave it to the reader to generalize further. Tilted surfaces can be taken care of if you first prove that the vector sum of the four surface areas of the tetrahedron in Fig. 2.24 is zero.

We conclude that, assuming only that the functions $F_x$, $F_y$, and $F_z$ are differentiable, the limit does exist and is given by

$$\text{div } \mathbf{F} = \frac{\partial F_x}{\partial x} + \frac{\partial F_y}{\partial y} + \frac{\partial F_z}{\partial z} \qquad (2.59)$$

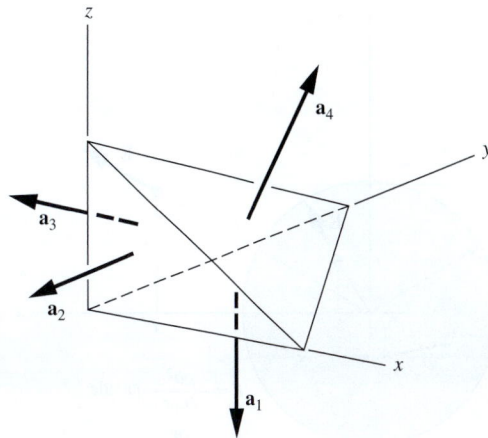

**Figure 2.24.**
You can prove that $\mathbf{a}_1 + \mathbf{a}_2 + \mathbf{a}_3 + \mathbf{a}_4 = 0$.

We can also write the divergence in a very compact form using the "$\nabla$" symbol. From Eq. (2.13) we see that the gradient operator (symbolized by $\nabla$ and often called "del") can be treated in Cartesian coordinates as a vector consisting of derivatives:

$$\nabla = \hat{\mathbf{x}}\frac{\partial}{\partial x} + \hat{\mathbf{y}}\frac{\partial}{\partial y} + \hat{\mathbf{z}}\frac{\partial}{\partial z}. \qquad (2.60)$$

In terms of this vector operator, we can write the divergence in the simple form, as you can quickly verify,

$$\text{div } \mathbf{F} = \nabla \cdot \mathbf{F}. \qquad (2.61)$$

If $\text{div } \mathbf{F}$ has a positive value at some point, we find – thinking of $\mathbf{F}$ as a velocity field – a net "outflow" in that neighborhood. For instance, if all three partial derivatives in Eq. (2.59) are positive at a point $P$, we might have a vector field in that neighborhood something like that suggested in Fig. 2.25. But the field could look quite different and still have positive divergence, for any vector function $\mathbf{G}$ such that $\text{div } \mathbf{G} = 0$ could be superimposed. Thus one or two of the three partial derivatives could be negative, and we might still have $\text{div } \mathbf{F} > 0$. The divergence is a quantity that expresses only one aspect of the spatial variation of a vector field.

---

**Example (Field due to a cylinder)** Let's find the divergence of an electric field that is rather easy to visualize. An infinitely long circular cylinder of radius $a$ is filled with a distribution of positive charge of density $\rho$. We know from Gauss's law that outside the cylinder the electric field is the same as that of a line charge on the axis. It is a radial field with magnitude proportional to $1/r$, given by Eq. (1.39) with $\lambda = \rho(\pi a^2)$. The field inside is found by applying Gauss's law to a cylinder of radius $r < a$. You can do this as an easy problem

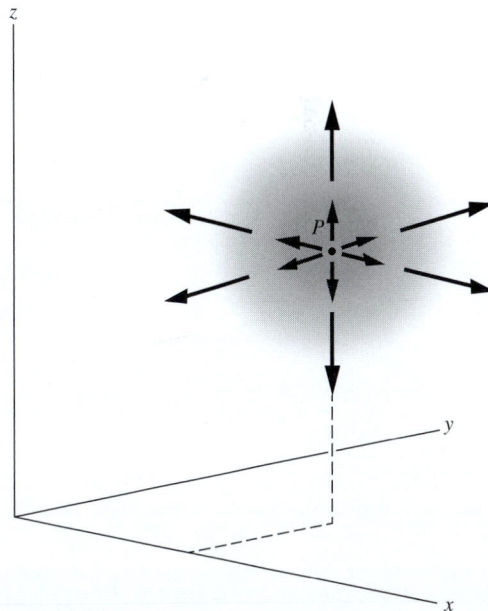

**Figure 2.25.**
Showing a field that in the neighborhood of point $P$ has a nonzero divergence.

# The electric potential

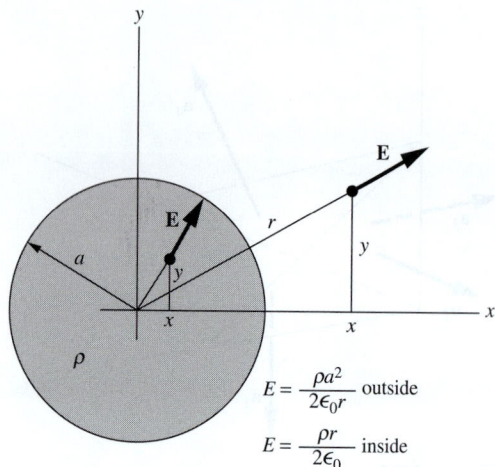

$$E = \frac{\rho a^2}{2\epsilon_0 r} \quad \text{outside}$$

$$E = \frac{\rho r}{2\epsilon_0} \quad \text{inside}$$

**Figure 2.26.**
The field inside and outside a uniform cylindrical distribution of charge.

(see Exercise 2.42). You will find that the field inside is directly proportional to $r$, and of course it is radial also. The exact values are:

$$E^{\text{out}} = \frac{\rho a^2}{2\epsilon_0 r} \quad \text{for } r > a,$$

$$E^{\text{in}} = \frac{\rho r}{2\epsilon_0} \quad \text{for } r < a. \tag{2.62}$$

Figure 2.26 is a section perpendicular to the axis of the cylinder. Rectangular coordinates aren't the most natural choice here, but we'll use them anyway to get some practice with Eq. (2.59). With $r = \sqrt{x^2 + y^2}$, the field components are expressed as follows:

$$E_x^{\text{out}} = \left(\frac{x}{r}\right) E^{\text{out}} = \frac{\rho a^2 x}{2\epsilon_0 (x^2 + y^2)} \quad \text{for } r > a,$$

$$E_y^{\text{out}} = \left(\frac{y}{r}\right) E^{\text{out}} = \frac{\rho a^2 y}{2\epsilon_0 (x^2 + y^2)} \quad \text{for } r > a,$$

$$E_x^{\text{in}} = \left(\frac{x}{r}\right) E^{\text{in}} = \frac{\rho x}{2\epsilon_0} \quad \text{for } r < a,$$

$$E_y^{\text{in}} = \left(\frac{y}{r}\right) E^{\text{in}} = \frac{\rho y}{2\epsilon_0} \quad \text{for } r < a. \tag{2.63}$$

And $E_z$ is zero everywhere, of course.

Outside the cylinder of charge, div $\mathbf{E}$ has the value given by

$$\frac{\partial E_x^{\text{out}}}{\partial x} + \frac{\partial E_y^{\text{out}}}{\partial y} = \frac{\rho a^2}{2\epsilon_0} \left[ \frac{1}{x^2 + y^2} - \frac{2x^2}{(x^2 + y^2)^2} + \frac{1}{x^2 + y^2} - \frac{2y^2}{(x^2 + y^2)^2} \right] = 0. \tag{2.64}$$

Inside the cylinder, div $\mathbf{E}$ is

$$\frac{\partial E_x^{\text{in}}}{\partial x} + \frac{\partial E_y^{\text{in}}}{\partial y} = \frac{\rho}{2\epsilon_0}(1 + 1) = \frac{\rho}{\epsilon_0}. \tag{2.65}$$

We expected both results. Outside the cylinder, where there is no charge, the net flux emerging from any volume – large or small – is zero, so the limit of the ratio *flux/volume* is certainly zero. Inside the cylinder we get the result required by the fundamental relation Eq. (2.52).

Having gotten some practice with Cartesian coordinates, let's redo this example in a much quicker manner by using cylindrical coordinates. Since $\mathbf{E}$ has only a radial component, Eq. (F.2) in Appendix F gives the divergence in cylindrical coordinates as div $\mathbf{E} = (1/r) \, \partial(r E_r)/\partial r$ (see Section F.3 for the derivation). Inside the cylinder, the field is $E_r = \rho r/2\epsilon_0$, so we quickly find div $\mathbf{E} = \rho/\epsilon_0$, as above. Outside the cylinder, the field is $E_r = \rho a^2/2\epsilon_0 r$, so we immediately find div $\mathbf{E} = 0$, which is again correct. All that matters in this latter case is that the field is proportional to $1/r$. Any such field will have div $\mathbf{E} = 0$.

## 2.11 The Laplacian

We have now met two scalar functions related to the electric field, the potential function $\phi$ (see Eq. (2.16)) and the divergence, div $\mathbf{E}$. In Cartesian coordinates the relationships are expressed as follows:

$$\mathbf{E} = -\text{grad}\,\phi = -\left(\hat{\mathbf{x}}\frac{\partial\phi}{\partial x} + \hat{\mathbf{y}}\frac{\partial\phi}{\partial y} + \hat{\mathbf{z}}\frac{\partial\phi}{\partial z}\right), \tag{2.66}$$

$$\text{div}\,\mathbf{E} = \frac{\partial E_x}{\partial x} + \frac{\partial E_y}{\partial y} + \frac{\partial E_z}{\partial z}. \tag{2.67}$$

Equation (2.66) shows that the $x$ component of $\mathbf{E}$ is $E_x = -\partial\phi/\partial x$. Substituting this and the corresponding expressions for $E_y$ and $E_z$ into Eq. (2.67), we get a relation between div $\mathbf{E}$ and $\phi$:

$$\text{div}\,\mathbf{E} = -\text{div grad}\,\phi = -\left(\frac{\partial^2\phi}{\partial x^2} + \frac{\partial^2\phi}{\partial y^2} + \frac{\partial^2\phi}{\partial z^2}\right). \tag{2.68}$$

The operation on $\phi$ that is indicated by Eq. (2.68), except for the minus sign, we could call "div grad," or "taking the divergence of the gradient of ...." The symbol used to represent this operation is $\nabla^2$, called *the Laplacian operator*, or just *the Laplacian*. The expression

$$\frac{\partial^2}{\partial x^2} + \frac{\partial^2}{\partial y^2} + \frac{\partial^2}{\partial z^2} \tag{2.69}$$

is the prescription for the Laplacian in Cartesian coordinates. So we have

$$\boxed{\text{div}\,\mathbf{E} = -\nabla^2\phi} \tag{2.70}$$

The notation $\nabla^2$ is explained as follows. With the vector operator $\nabla$ given in Eq. (2.60), its square equals

$$\nabla \cdot \nabla = \frac{\partial^2}{\partial x^2} + \frac{\partial^2}{\partial y^2} + \frac{\partial^2}{\partial z^2}, \tag{2.71}$$

the same as the Laplacian in Cartesian coordinates. So the Laplacian is often called "del squared," and we say "del squared $\phi$," meaning "div grad $\phi$." *Warning:* In other coordinate systems, spherical coordinates, for instance, the explicit forms of the gradient operator and the Laplacian operator are not so simply related. This is evident in the list of formulas at the beginning of Appendix F. It is well to remember that the fundamental definition of the Laplacian operation is "divergence of the gradient of."

We can now express directly a *local* relation between the charge density at some point and the potential function in that immediate neighborhood. Combining Eq. (2.70) with Gauss's law in differential form, div $\mathbf{E} = \rho/\epsilon_0$, we have

$$\nabla^2 \phi = -\frac{\rho}{\epsilon_0} \tag{2.72}$$

Equation (2.72), sometimes called *Poisson's equation*, relates the charge density to the second derivatives of the potential. Written out in Cartesian coordinates it is

$$\frac{\partial^2 \phi}{\partial x^2} + \frac{\partial^2 \phi}{\partial y^2} + \frac{\partial^2 \phi}{\partial z^2} = -\frac{\rho}{\epsilon_0}. \tag{2.73}$$

One may regard this as the differential expression of the relationship expressed by the integral in Eq. (2.18), which tells us how to find the potential at a point by summing the contributions of all sources near and far.[4]

---

**Example (Poisson's equation for a sphere)**   Let's verify that Eq. (2.72) holds for the potential due to a sphere with radius $R$ and uniform charge density $\rho$. This potential was derived in the second example in Section 2.2. Spherical coordinates are the best choice here, so we will invoke the expression for the Laplacian in spherical coordinates, given in Eq. (F.3) in Appendix F. Since the potential depends only on $r$, we have $\nabla^2 \phi = (1/r^2)\partial(r^2 \, \partial\phi/\partial r)/\partial r$.

The potential outside the sphere is $\phi = \rho R^3/3\epsilon_0 r$. All that matters here is the fact that $\phi$ is proportional to $1/r$, because this makes $\partial\phi/\partial r$ proportional to $1/r^2$, from which we immediately see that $\nabla^2 \phi = 0$. This agrees with Eq. (2.72), because $\rho = 0$ outside the sphere.

Inside the sphere, we have $\phi = \rho R^2/2\epsilon_0 - \rho r^2/6\epsilon_0$. The constant term vanishes when we take the derivative, so we have

$$\nabla^2 \phi = \frac{1}{r^2}\frac{\partial}{\partial r}\left(r^2 \frac{\partial \phi}{\partial r}\right) = \frac{1}{r^2}\frac{\partial}{\partial r}\left(r^2 \cdot \frac{-\rho r}{3\epsilon_0}\right) = -\frac{1}{r^2}\frac{\rho r^2}{\epsilon_0} = -\frac{\rho}{\epsilon_0}, \tag{2.74}$$

as desired.

---

## 2.12 Laplace's equation

Wherever $\rho = 0$, that is, in all parts of space containing no electric charge, the electric potential $\phi$ has to satisfy the equation

$$\nabla^2 \phi = 0. \tag{2.75}$$

This is called *Laplace's equation*. We run into it in many branches of physics. Indeed one might say that from a mathematical point of view the

---

[4]   In fact, it can be shown that Eq. (2.73) is the *mathematical* equivalent of Eq. (2.18). This means, if you apply the Laplacian operator to the integral in Eq. (2.18), you will come out with $-\rho/\epsilon_0$. We shall not stop to show how this is done; you'll have to take our word for it or figure out how to do it in Problem 2.27.

theory of classical fields is mostly a study of the solutions of this equation. The class of functions that satisfy Laplace's equation are called *harmonic functions*. They have some remarkable properties, one of which is the following.

**Theorem 2.1** *If $\phi(x, y, z)$ satisfies Laplace's equation, then the average value of $\phi$ over the surface of any sphere (not necessarily a small sphere) is equal to the value of $\phi$ at the center of the sphere.*

*Proof* We can easily prove that this must be true of the electric potential $\phi$ in regions containing no charge. (See Section F.5 in Appendix F for a more general proof.) Consider a point charge $q$ and a spherical surface $S$ over which a charge $q'$ is uniformly distributed. Let the charge $q$ be brought in from infinity to a distance $R$ from the center of the charged sphere, as in Fig. 2.27. The electric field of the sphere being the same as if its total charge $q'$ were concentrated at its center, the work required is $qq'/4\pi\epsilon_0 R$.

Now suppose, instead, that the point charge $q$ was there first and the charged sphere was later brought in from infinity. The work required for that is the product of $q'$ and the average over the surface $S$ of the potential due to the point charge $q$. Now the work is surely the same in the second case, namely $qq'/4\pi\epsilon_0 R$, so the average over the sphere of the potential due to $q$ must be $q/4\pi\epsilon_0 R$. That is indeed the potential at the center of the sphere due to the external point charge $q$. That proves the assertion for any single point charge outside the sphere. But the potential of many charges is just the sum of the potentials due to the individual charges, and the average of a sum is the sum of the averages. It follows that the assertion must be true for *any* system of sources lying wholly outside the sphere. □

This property of the potential, that its average over an empty sphere is equal to its value at the center, is closely related to the following fact that you may find disappointing.

**Theorem 2.2** *(Earnshaw's theorem) It is impossible to construct an electrostatic field that will hold a charged particle in stable equilibrium in empty space.*

This particular "impossibility theorem," like others in physics, is useful in saving fruitless speculation and effort. We can prove it in two closely related ways, first by looking at the field **E** and using Gauss's law, and second by looking at the potential $\phi$ and using the above fact concerning the average of $\phi$ over the surface of a sphere.

*Proof* First, suppose we have an electric field in which, contrary to the theorem, there *is* a point $P$ at which a positively charged particle would be in stable equilibrium. That means that *any* small displacement of the particle from $P$ must bring it to a place where an electric field acts to push it back toward $P$. But that means that a little sphere around $P$ must have **E**

**Figure 2.27.**
The work required to bring in $q'$ and distribute it over the sphere is $q'$ times the *average*, over the sphere, of the potential $\phi$ due to $q$.

pointing inward *everywhere* on its surface, which in turn means that there is a net inward flux through the sphere. This contradicts Gauss's law, for there is no negative source charge within the region. (Our charged test particle doesn't count; besides, it's positive.) In other words, you can't have an empty region where the electric field points all inward or all outward, and that's what you would need for *stable* equilibrium. Note that since this proof involved only Gauss's law, we could have presented this theorem back in Chapter 1.

A second proof, using Theorem 2.1, proceeds as follows. A stable position for a charged particle must be one where the potential $\phi$ is either lower than that at all neighboring points (if the particle is positively charged) or higher than that at all neighboring points (if the particle is negatively charged). Clearly neither is possible for a function whose average value over a sphere is always equal to its value at the center.  □

Of course, one can have a charged particle in *equilibrium* in an electrostatic field, in the sense that the force on it is zero. The point where $\mathbf{E} = 0$ in Fig. 1.10 is such a location. The position midway between two equal positive charges is an equilibrium position for a third charge, either positive or negative. But the equilibrium is not stable. (Think what happens when the third charge is slightly displaced, either transversely or longitudinally, from its equilibrium position.) It *is* possible, by the way, to trap and hold stably an electrically charged particle by electric fields that vary in *time*. And it is certainly possible to hold stably a charged particle within a nonzero charge distribution. For example, a positive charge located at the center of a solid sphere of uniform negative charge is in stable equilibrium.

## 2.13 Distinguishing the physics from the mathematics

In the preceding sections we have been concerned with mathematical relations and new ways of expressing familiar facts. It may help to sort out physics from mathematics, and law from definition, if we try to imagine how things would be if the electric force were *not* a pure inverse-square force but instead a force with a finite range, for instance, a force varying like[5]

$$F(r) = \frac{e^{-\lambda r}}{r^2}. \tag{2.76}$$

Then Gauss's law in the integral form expressed in Eq. (2.50) would surely fail, for, by taking a very large surface enclosing some sources, we would find a vanishingly small field on this surface. The flux would go to zero as the surface expanded, rather than remain constant. However, we

---

[5]  This force technically has an infinite range, but the exponential decay causes it to become essentially zero far away. So the range is finite, for all practical purposes.

could still define a field at every point in space. We could calculate the divergence of that field, and Eq. (2.51), which describes a mathematical property of *any* vector field, would still be true. Is there a contradiction here? No, because Eq. (2.52) would also fail. The divergence of the field would no longer be the same as the source density. We can understand this by noting that a small volume empty of sources could still have a net flux through it owing to the effect of a source *outside* the volume, if the field has finite range. As suggested in Fig. 2.28, more flux would enter the side near the source than would leave the volume.

Thus we may say that Eqs. (2.50) and (2.52) express the same *physical law*, the inverse-square law that Coulomb established by direct measurement of the forces between charged bodies, while Eq. (2.51) is an expression of a *mathematical theorem* that enables us to translate our statement of this law from differential to integral form or the reverse. The relations that connect **E**, $\rho$, and $\phi$ are gathered together in Fig. 2.29(a). The analogous expressions in Gaussian units are shown in Fig. 2.29(b).

How can we justify these differential relations between source and field in a world where electric charge is really not a smooth jelly but is concentrated on particles whose interior we know very little about? Actually, a statement like Eq. (2.72), Poisson's equation, is meaningful on a macroscopic scale only. The charge density $\rho$ is to be interpreted as an average over some small but finite region containing many particles. Thus the function $\rho$ cannot be continuous in the way a mathematician might prefer. When we let our region $V_i$ shrink down in the course of demonstrating the differential form of Gauss's law, we know as physicists that we musn't let it shrink too far. That is awkard perhaps, but the fact is that we make out very well with the continuum model in large-scale

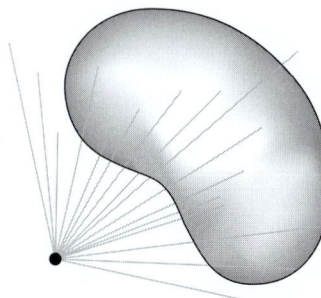

**Figure 2.28.**
In a non-inverse-square field, the flux through a closed surface is not zero.

**Figure 2.29.**
(a) How electric charge density, electric potential, and electric field are related. The integral relations involve the line integral and the volume integral. The differential relations involve the gradient, the divergence, and div · grad (equivalently $\nabla^2$), the Laplacian operator. The charge density $\rho$ is in coulomb/meter$^3$, the potential $\phi$ is in volts, the field **E** is in volt/meter, and all lengths are in meters. (b) The same relations in Gaussian units. The charge density $\rho$ is in esu/cm$^3$, the potential $\phi$ is in statvolts, the field **E** is in statvolt/meter, and all lengths are in centimeters.

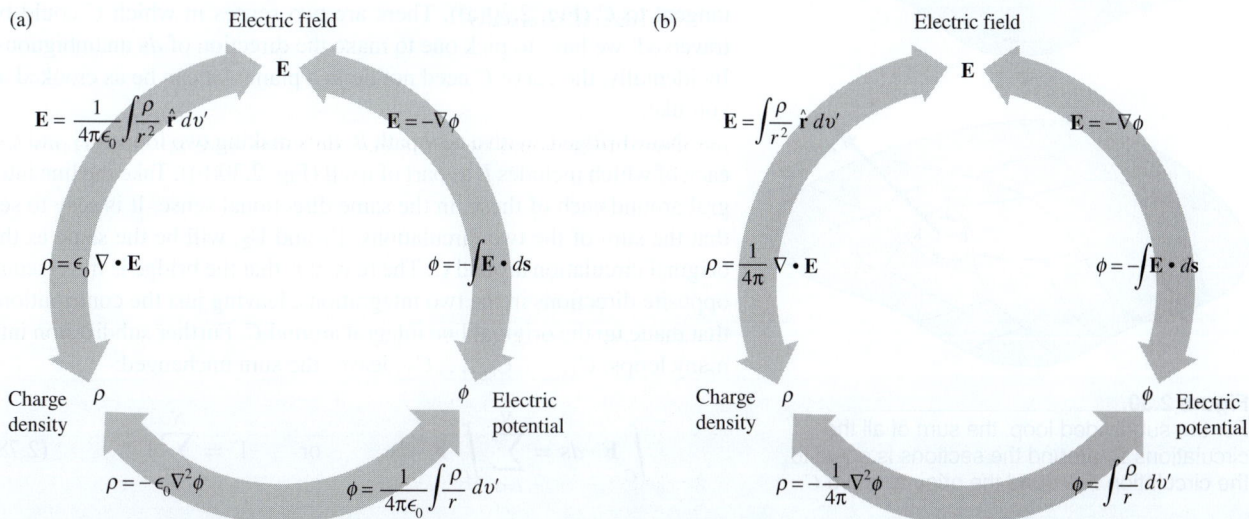

(a)

Electric field

**E**

$$\mathbf{E} = \frac{1}{4\pi\epsilon_0} \int \frac{\rho}{r^2} \hat{\mathbf{r}} \, dv'$$

$$\mathbf{E} = -\nabla\phi$$

$$\rho = \epsilon_0 \nabla \cdot \mathbf{E}$$

$$\phi = -\int \mathbf{E} \cdot d\mathbf{s}$$

Charge density $\rho$

$\phi$ Electric potential

$$\rho = -\epsilon_0 \nabla^2 \phi$$

$$\phi = \frac{1}{4\pi\epsilon_0} \int \frac{\rho}{r} \, dv'$$

(b)

Electric field

**E**

$$\mathbf{E} = \int \frac{\rho}{r^2} \hat{\mathbf{r}} \, dv'$$

$$\mathbf{E} = -\nabla\phi$$

$$\rho = \frac{1}{4\pi} \nabla \cdot \mathbf{E}$$

$$\phi = -\int \mathbf{E} \cdot d\mathbf{s}$$

Charge density $\rho$

$\phi$ Electric potential

$$\rho = -\frac{1}{4\pi} \nabla^2 \phi$$

$$\phi = \int \frac{\rho}{r} \, dv'$$

electrical systems. In the atomic world we have the elementary particles, and vacuum. Inside the particles, even if Coulomb's law turns out to have some kind of meaning, much else is going on. The vacuum, so far as electrostatics is concerned, is ruled by Laplace's equation. Still, we cannot be sure that, even in the vacuum, passage to a limit of zero size has *physical* meaning.

## 2.14 The curl of a vector function

*Note: Study of this section and the remainder of Chapter 2 can be postponed until Chapter 6 is reached. Until then our only application of the curl will be the demonstration that an electrostatic field is characterized by* curl $\mathbf{E} = 0$*, as explained in Section 2.17. The reason we are introducing the curl now is that the derivation so closely parallels the above derivation of the divergence.*

We developed the concept of divergence, a local property of a vector field, by starting from the surface integral over a large closed surface. In the same spirit, let us consider the line integral of some vector field $\mathbf{F}(x, y, z)$, taken around a closed path, some curve $C$ that comes back to join itself. The curve $C$ can be visualized as the boundary of some surface $S$ that spans it. A good name for the magnitude of such a closed-path line integral is *circulation*; we shall use $\Gamma$ (capital gamma) as its symbol:

$$\Gamma = \int_C \mathbf{F} \cdot d\mathbf{s}. \tag{2.77}$$

In the integrand, $d\mathbf{s}$ is the element of path, an infinitesimal vector locally tangent to $C$ (Fig. 2.30(a)). There are two senses in which $C$ could be traversed; we have to pick one to make the direction of $d\mathbf{s}$ unambiguous. Incidentally, the curve $C$ need not lie in a plane – it can be as crooked as you like.

Now bridge $C$ with a new path $B$, thus making two loops, $C_1$ and $C_2$, each of which includes $B$ as part of itself (Fig. 2.30(b)). Take the line integral around each of these, in the same directional sense. It is easy to see that the sum of the two circulations, $\Gamma_1$ and $\Gamma_2$, will be the same as the original circulation around $C$. The reason is that the bridge is traversed in opposite directions in the two integrations, leaving just the contributions that made up the original line integral around $C$. Further subdivision into many loops, $C_1, \ldots, C_i, \ldots, C_N$, leaves the sum unchanged:

$$\int_C \mathbf{F} \cdot d\mathbf{s} = \sum_{i=1}^N \int_{C_i} \mathbf{F} \cdot d\mathbf{s}_i, \qquad \text{or} \qquad \Gamma = \sum_{i=1}^N \Gamma_i. \tag{2.78}$$

(a)
(b)
(c)

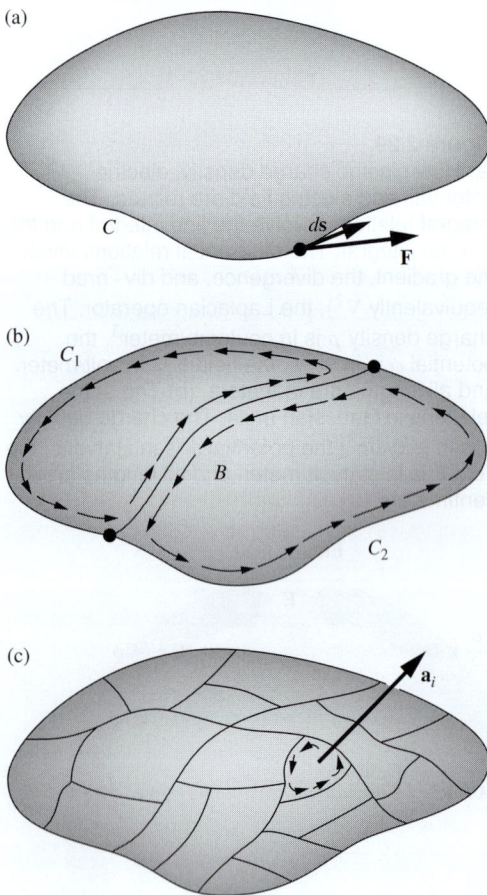

**Figure 2.30.**
For the subdivided loop, the sum of all the circulations $\Gamma_i$ around the sections is equal to the circulation $\Gamma$ around the original curve $C$.

In the same manner as in our discussion of divergence in Section 2.8, we can continue indefinitely to subdivide, now by adding new bridges instead of new surfaces, seeking in the limit to arrive at a quantity characteristic of the field **F** in a local neighborhood. When we subdivide the loops, we make loops with smaller circulation, but also with smaller area. So it is natural to consider the ratio of *loop circulation* to *loop area*, just as we considered in Section 2.8 the ratio of *flux* to *volume*. However, things are a little different here, because the area $a_i$ of the bit of surface that spans a small loop $C_i$ is really a vector (Fig. 2.30(c)), in contrast with the scalar volume $V_i$ in Section 2.8. A surface has an orientation in space, whereas a volume does not. In fact, as we make smaller and smaller loops in some neighborhood, we can arrange to have a loop oriented in any direction we choose. (Remember, we are not committed to any particular surface over the whole curve $C$.) Thus we can pass to the limit in essentially different ways, and we must expect the result to reflect this.

Let us choose some particular orientation for the patch as it goes through the last stages of subdivision. The unit vector $\hat{\mathbf{n}}$ will denote the normal to the patch, which is to remain fixed in direction as the patch surrounding a particular point $P$ shrinks down toward zero size. The limit of the ratio of *circulation* to *patch area* will be written this way:

$$\lim_{a_i \to 0} \frac{\Gamma_i}{a_i} \quad \text{or} \quad \lim_{a_i \to 0} \frac{\int_{C_i} \mathbf{F} \cdot d\mathbf{s}}{a_i}. \tag{2.79}$$

The rule for sign is that the direction of $\hat{\mathbf{n}}$ and the sense in which $C_i$ is traversed in the line integral shall be related by a right-hand-screw rule, as in Fig. 2.31. The limit we obtain by this procedure is a scalar quantity that is associated with the point $P$ in the vector field **F**, and with the direction $\hat{\mathbf{n}}$. We could pick three directions, such as $\hat{\mathbf{x}}$, $\hat{\mathbf{y}}$, and $\hat{\mathbf{z}}$, and get three different numbers. It turns out that these numbers can be considered components of a vector. We call the vector "curl **F**." That is to say, the number we get for the limit with $\hat{\mathbf{n}}$ in a particular direction is the component, in that direction, of the vector curl **F**. To state this in an equation,

$$(\text{curl } \mathbf{F}) \cdot \hat{\mathbf{n}} = \lim_{a_i \to 0} \frac{\int_{C_i} \mathbf{F} \cdot d\mathbf{s}}{a_i} \tag{2.80}$$

where $\hat{\mathbf{n}}$ is the unit vector normal to the curve $C_i$.

For instance, the $x$ component of curl **F** is obtained by choosing $\hat{\mathbf{n}} = \hat{\mathbf{x}}$, as in Fig. 2.32. As the loop shrinks down around the point $P$, we keep

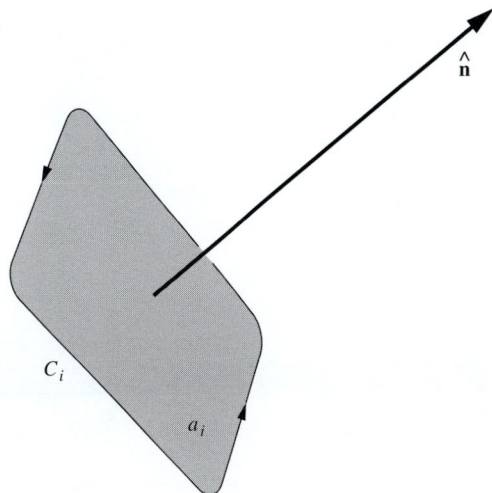

**Figure 2.31.**
Right-hand-screw relation between the surface normal and the direction in which the circulation line integral is taken.

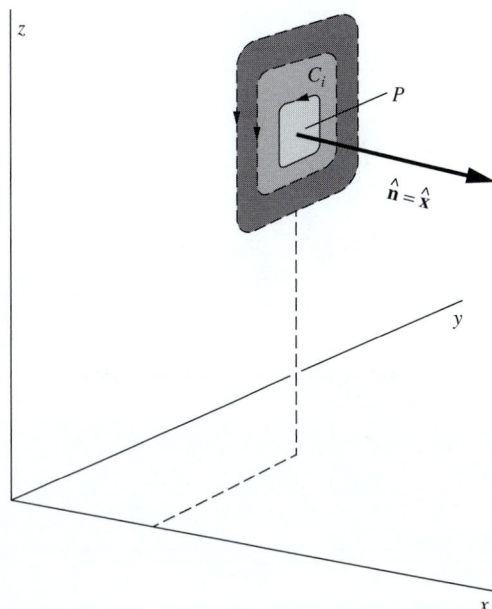

**Figure 2.32.**
The patch shrinks around $P$, keeping its normal pointing in the $x$ direction.

it in a plane perpendicular to the $x$ axis. In general, the vector curl $\mathbf{F}$ will vary from place to place. If we let the patch shrink down around some other point, the ratio of circulation to area may have a different value, depending on the nature of the vector function $\mathbf{F}$. That is, curl $\mathbf{F}$ is itself a vector function of the coordinates. Its direction at each point in space is normal to the plane through this point in which the circulation is a maximum. Its magnitude is the limiting value of circulation per unit area, in this plane, around the point in question.

The last two sentences might be taken as a definition of curl $\mathbf{F}$. Like Eq. (2.80) they make no reference to a coordinate frame. We have not proved that the object so named and defined is a vector; we have only asserted it. Possession of direction and magnitude is not enough to make something a vector. The components as defined must behave like vector components. Suppose we have determined certain values for the $x$, $y$, and $z$ components of curl $\mathbf{F}$ by applying Eq. (2.80) with $\hat{\mathbf{n}}$ chosen, successively, as $\hat{\mathbf{x}}$, $\hat{\mathbf{y}}$, and $\hat{\mathbf{z}}$. If curl $\mathbf{F}$ is a vector, it is uniquely determined by these three components. If some fourth direction is now chosen for $\hat{\mathbf{n}}$, the left side of Eq. (2.80) is fixed and the quantity on the right, the circulation in the plane perpendicular to the new $\hat{\mathbf{n}}$, had better agree with it! Indeed, until one is sure that curl $\mathbf{F}$ is a vector, it is not even obvious that there can be at most one direction for which the circulation per unit area at $P$ is maximum – as was tacitly assumed in the latter definition. In fact, Eq. (2.80) does define a vector, but we shall not give a proof of that.

## 2.15 Stokes' theorem

From the circulation around an infinitesimal patch of surface we can now work back to the circulation around the original large loop $C$:

$$\Gamma = \int_C \mathbf{F} \cdot d\mathbf{s} = \sum_{i=1}^{N} \Gamma_i = \sum_{i=1}^{N} a_i \left( \frac{\Gamma_i}{a_i} \right). \qquad (2.81)$$

In the last step we merely multiplied and divided by $a_i$. Now observe what happens to the right-hand side as $N$ is made enormous and all the $a_i$ areas shrink. From Eq. (2.80), the quantity in parentheses becomes (curl $\mathbf{F}$) $\cdot \hat{\mathbf{n}}_i$, where $\hat{\mathbf{n}}_i$ is the unit vector normal to the $i$th patch. So we have on the right the sum, over all patches that make up the entire surface $S$ spanning $C$, of the product "patch area times normal component of (curl $\mathbf{F}$)." This is simply the *surface integral*, over $S$, of the vector curl $\mathbf{F}$:

$$\sum_{i=1}^{N} a_i \left( \frac{\Gamma_i}{a_i} \right) = \sum_{i=1}^{N} a_i (\text{curl } \mathbf{F}) \cdot \hat{\mathbf{n}}_i \longrightarrow \int_S \text{curl } \mathbf{F} \cdot d\mathbf{a}, \qquad (2.82)$$

because $d\mathbf{a} = a_i\hat{\mathbf{n}}_i$, by definition. We thus find that

$$\boxed{\int_C \mathbf{F} \cdot d\mathbf{s} = \int_S \operatorname{curl} \mathbf{F} \cdot d\mathbf{a}} \qquad \text{(Stokes' theorem)}. \qquad (2.83)$$

The relation expressed by Eq. (2.83) is a mathematical theorem called *Stokes' theorem*. Note how it resembles Gauss's theorem, the divergence theorem, in structure. Stokes' theorem relates the line integral of a vector to the surface integral of the curl of the vector. Gauss's theorem, Eq. (2.49), relates the surface integral of a vector to the volume integral of the divergence of the vector. Stokes' theorem involves a surface and the curve that bounds it. Gauss's theorem involves a volume and the surface that encloses it.

## 2.16 The curl in Cartesian coordinates

Equation (2.80) is the fundamental definition of curl $\mathbf{F}$, stated without reference to any particular coordinate system. In this respect it is like our fundamental definition of divergence, Eq. (2.47). As in that case, we should like to know how to calculate curl $\mathbf{F}$ when the vector function $\mathbf{F}(x, y, z)$ is explicitly given. To find the rule, we carry out the integration called for in Eq. (2.80), but we do it over a path of very simple shape, one that encloses a rectangular patch of surface parallel to the $xy$ plane (Fig. 2.33). That is, we are taking $\hat{\mathbf{n}} = \hat{\mathbf{z}}$. In agreement with our rule about sign, the direction of integration around the rim must be clockwise as seen by someone looking up in the direction of $\hat{\mathbf{n}}$. In Fig. 2.34 we look down onto the rectangle from above.

The line integral of $\mathbf{A}$ around such a path depends on the variation of $A_x$ with $y$ and the variation of $A_y$ with $x$. For if $A_x$ had the same average value along the top of the frame, in Fig. 2.34, as along the bottom of the frame, the contribution of these two pieces of the whole line integral would obviously cancel. A similar remark applies to the side members. To the first order in the small quantities $\Delta x$ and $\Delta y$, the difference between the average of $A_x$ over the top segment of path at $y + \Delta y$ and its average over the bottom segment at $y$ is

$$\left( \frac{\partial A_x}{\partial y} \right) \Delta y. \qquad (2.84)$$

This follows from an argument similar to the one we used with Fig. 2.22(b):

$$A_x = A_x(x, y) + \frac{\Delta x}{2} \frac{\partial A_x}{\partial x} \qquad \left( \begin{array}{l} \text{at midpoint of} \\ \text{bottom of frame} \end{array} \right),$$

$$A_x = A_x(x, y) + \frac{\Delta x}{2} \frac{\partial A_x}{\partial x} + \Delta y \frac{\partial A_x}{\partial y} \qquad \left( \begin{array}{l} \text{at midpoint of} \\ \text{top of frame} \end{array} \right). \qquad (2.85)$$

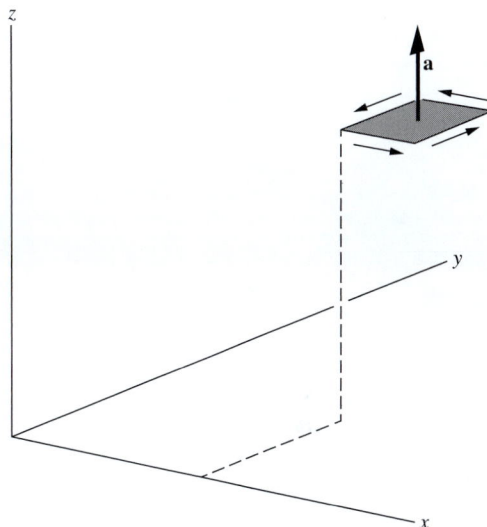

**Figure 2.33.**
Circulation around a rectangular patch with $\hat{\mathbf{n}} = \hat{\mathbf{z}}$.

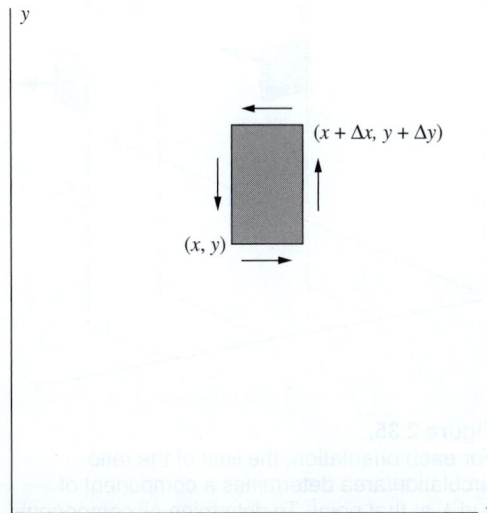

**Figure 2.34.**
Looking down on the patch in Fig. 2.33.

These are the average values referred to, to first order in the Taylor expansion. It is their difference, times the length of the path segment $\Delta x$, that determines their net contribution to the circulation. This contribution is $-\Delta x \, \Delta y \, (\partial A_x / \partial y)$. The minus sign comes in because we are integrating toward the left at the top, so that if $A_x$ is more positive at the top, it results in a negative contribution to the circulation. The contribution from the sides is $\Delta y \, \Delta x \, (\partial A_y / \partial x)$, and here the sign is positive, because if $A_y$ is more positive on the right, the result is a positive contribution to the circulation.

Thus, neglecting any higher powers of $\Delta x$ and $\Delta y$, the line integral around the whole rectangle is

$$\int \mathbf{A} \cdot d\mathbf{s} = -\Delta x \cdot \left( \frac{\partial A_x}{\partial y} \right) \Delta y + \Delta y \cdot \left( \frac{\partial A_y}{\partial x} \right) \Delta x$$

$$= \Delta x \, \Delta y \left( \frac{\partial A_y}{\partial x} - \frac{\partial A_x}{\partial y} \right). \tag{2.86}$$

Now $\Delta x \, \Delta y$ is the magnitude of the area of the enclosed rectangle, which we have represented by a vector in the $z$ direction. Evidently the quantity

$$\frac{\partial A_y}{\partial x} - \frac{\partial A_x}{\partial y} \tag{2.87}$$

is the limit of the ratio

$$\frac{\text{line integral around patch}}{\text{area of patch}} \tag{2.88}$$

as the patch shrinks to zero size. If the rectangular frame had been oriented with its normal in the positive $y$ direction, like the left frame in Fig. 2.35, we would have found the expression

$$\frac{\partial A_x}{\partial z} - \frac{\partial A_z}{\partial x} \tag{2.89}$$

for the limit of the corresponding ratio. And if the frame had been oriented with its normal in the positive $x$ direction, like the right frame in Fig. 2.35, we would have obtained

$$\frac{\partial A_z}{\partial y} - \frac{\partial A_y}{\partial z}. \tag{2.90}$$

Although we have considered rectangles only, our result is actually independent of the shape of the little patch and its frame, for reasons much the same as in the case of the integrals involved in the divergence theorem. For instance, it is clear that we can freely join different

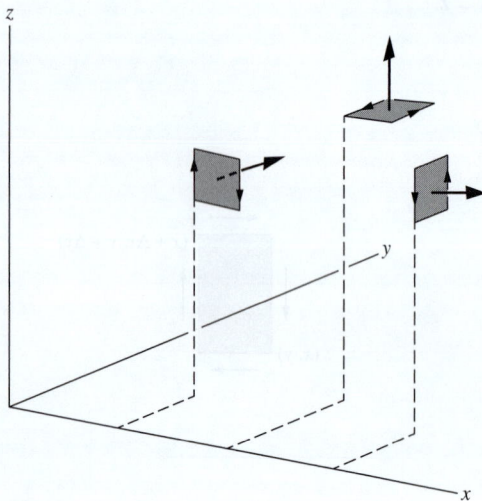

**Figure 2.35.**
For each orientation, the limit of the ratio circulation/area determines a component of curl $\mathbf{A}$ at that point. To determine all components of the vector curl $\mathbf{A}$ at any point, the patches should all cluster around that point; here they are separated for clarity.

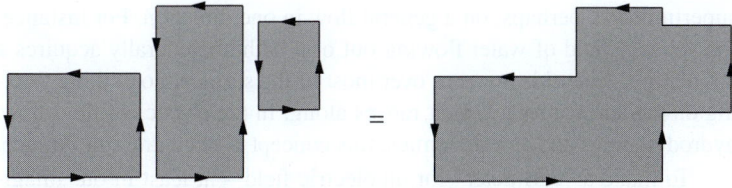

rectangles to form other figures, because the line integrals along the merging sections of boundary cancel one another exactly (Fig. 2.36).

We conclude that, for any of these orientations, the limit of the ratio of circulation to area is independent of the shape of the patch we choose. Thus we obtain as a general formula for the components of the vector curl $\mathbf{F}$, when $\mathbf{F}$ is given as a function of $x$, $y$, and $z$:

$$\text{curl } \mathbf{F} = \hat{\mathbf{x}} \left( \frac{\partial F_z}{\partial y} - \frac{\partial F_y}{\partial z} \right) + \hat{\mathbf{y}} \left( \frac{\partial F_x}{\partial z} - \frac{\partial F_z}{\partial x} \right) + \hat{\mathbf{z}} \left( \frac{\partial F_y}{\partial x} - \frac{\partial F_x}{\partial y} \right). \tag{2.91}$$

You may find the following rule easier to remember than the formula itself. Make up a determinant like this:

$$\begin{vmatrix} \hat{\mathbf{x}} & \hat{\mathbf{y}} & \hat{\mathbf{z}} \\ \partial/\partial x & \partial/\partial y & \partial/\partial z \\ F_x & F_y & F_z \end{vmatrix}. \tag{2.92}$$

Expand it according to the rule for determinants, and you will get curl $\mathbf{F}$ as given by Eq. (2.91). Note that the $x$ component of curl $\mathbf{F}$ depends on the rate of change of $F_z$ in the $y$ direction and the negative of the rate of change of $F_y$ in the $z$ direction, and so on.

The symbol $\nabla \times$, read as "del cross," where $\nabla$ is interpreted as the "vector"

$$\nabla = \hat{\mathbf{x}} \frac{\partial}{\partial x} + \hat{\mathbf{y}} \frac{\partial}{\partial y} + \hat{\mathbf{z}} \frac{\partial}{\partial z}, \tag{2.93}$$

is often used in place of the name *curl*. If we write $\nabla \times \mathbf{F}$ and follow the rules for forming the components of a vector cross product, we get automatically the vector curl $\mathbf{F}$. So curl $\mathbf{F}$ and $\nabla \times \mathbf{F}$ mean the same thing.

## 2.17 The physical meaning of the curl

The name *curl* reminds us that a vector field with a nonzero curl has circulation, or vorticity. Maxwell used the name *rotation*, and in German a similar name is still used, abbreviated rot. Imagine a velocity vector field $\mathbf{G}$, and suppose that curl $\mathbf{G}$ is not zero. Then the velocities in this field have something of this character:

$$\underset{\rightarrow}{\overset{\leftarrow}{\downarrow\uparrow}} \quad \text{or} \quad \underset{\leftarrow}{\overset{\rightarrow}{\uparrow\downarrow}}$$

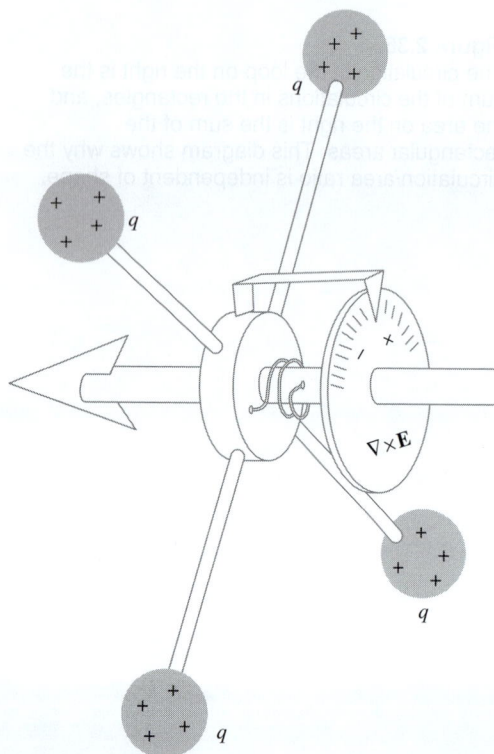

**Figure 2.37.**
The curlmeter.

superimposed, perhaps, on a general flow in one direction. For instance, the velocity field of water flowing out of a bathtub generally acquires a circulation. Its curl is not zero over most of the surface. Something floating on the surface rotates as it moves along. In the physics of fluid flow, hydrodynamics and aerodynamics, this concept is of central importance.

To make a "curlmeter" for an electric field – at least in our imagination – we could fasten positive charges to a hub by insulating spokes, as in Fig. 2.37. Exploring an electric field with this device, we would find, wherever curl **E** is not zero, a tendency for the wheel to turn around the shaft. With a spring to restrain rotation, the amount of twist could be used to indicate the torque, which would be proportional to the component of the vector curl **E** in the direction of the shaft. If we can find the direction of the shaft for which the torque is maximum and clockwise, that is the direction of the vector curl **E**. (Of course, we cannot trust the curlmeter in a field that varies greatly within the dimensions of the wheel itself.)

What can we say, in the light of all this, about the *electrostatic* field **E**? The conclusion we can draw is a simple one: the curlmeter will always read zero! That follows from a fact we have already learned; namely, in the electrostatic field the line integral of **E** around *any* closed path is zero. Just to recall why this is so, remember from Section 2.1 that the line integral of **E** between any two points such as $P_1$ and $P_2$ in Fig. 2.38 is independent of the path. (This then implies that **E** can be written as the negative gradient of the well-defined potential function given by Eq. (2.4).) As we bring the two points $P_1$ and $P_2$ close together, the line integral over the shorter path in the figure obviously vanishes – unless the final location is at a singularity such as a point charge, a case we can rule out. So the line integral must be zero over the closed loop in Fig. 2.38(d). But now, if the circulation is zero around *any* closed path, it follows from Stokes' theorem that the surface integral of curl **E** is zero over a patch of any size, shape, or location. But then curl **E** must be zero *everywhere*, for if it were not zero somewhere we could devise a patch in that neighborhood to violate the conclusion. We can sum all of this up by saying that if **E** equals the negative gradient of a potential function $\phi$ (which is the case for any electrostatic field **E**), then

$$\text{curl } \mathbf{E} = 0 \qquad \text{(everywhere)}. \qquad (2.94)$$

The converse is also true. If curl **E** is known to be zero everywhere, then **E** must be describable as the gradient of some potential function $\phi$. This follows from the fact that zero curl implies that the line integral of **E** is path-independent (by reversing the above reasoning), which in turn implies that $\phi$ can be defined in an unambiguous manner as the negative line integral of the field. If curl **E** = 0, then **E** could be an electrostatic field.

**Example**   This test is easy to apply. When the vector function in Fig. 2.3 was first introduced, it was said to represent a possible electrostatic field. The components were specified by $E_x = Ky$ and $E_y = Kx$, to which we should add $E_z = 0$ to complete the description of a field in three-dimensional space. Calculating curl **E** we find

$$(\text{curl } \mathbf{E})_x = \frac{\partial E_z}{\partial y} - \frac{\partial E_y}{\partial z} = 0,$$

$$(\text{curl } \mathbf{E})_y = \frac{\partial E_x}{\partial z} - \frac{\partial E_z}{\partial x} = 0,$$

$$(\text{curl } \mathbf{E})_z = \frac{\partial E_y}{\partial x} - \frac{\partial E_x}{\partial y} = K - K = 0. \tag{2.95}$$

This tells us that **E** is the (negative) gradient of some scalar potential, which we know from Eq. (2.8), and which we verified in Eq. (2.17), is $\phi = -Kxy$. Incidentally, this particular field **E** happens to have zero divergence also:

$$\frac{\partial E_x}{\partial x} + \frac{\partial E_y}{\partial y} + \frac{\partial E_z}{\partial z} = 0. \tag{2.96}$$

It therefore represents an electrostatic field in a *charge-free* region.

On the other hand, the equally simple vector function defined by $F_x = Ky$; $F_y = -Kx$; $F_z = 0$, does not have zero curl. Instead,

$$(\text{curl } \mathbf{F})_z = -2K. \tag{2.97}$$

Hence no electrostatic field could have this form. If you sketch roughly the form of this field, you will see at once that it has circulation.

**Example (Field from a sphere)**   We can also verify that the electric field due to a sphere with radius $R$ and uniform charge density $\rho$ has zero curl. From the example in Section 1.11, the fields inside and outside the sphere are, respectively,

$$E_r^{\text{in}} = \frac{\rho r}{3\epsilon_0} \quad \text{and} \quad E_r^{\text{out}} = \frac{\rho R^3}{3\epsilon_0 r^2}. \tag{2.98}$$

As usual, we will work with spherical coordinates when dealing with a sphere. The expression for the curl in spherical coordinates, given in Eq. (F.3) in Appendix F, is unfortunately the most formidable one in the list. However, the above electric field has only a radial component, so only two of the six terms in the lengthy expression for the curl have a chance of being nonzero. Furthermore, the radial component depends only on $r$, being proportional to either $r$ or $1/r^2$. So the two possibly nonzero terms, which involve the derivatives $\partial E_r / \partial \phi$ and $\partial E_r / \partial \theta$, are both zero ($\phi$ here is an angle, not the potential!). The curl is therefore zero. This result holds for *any* radial field that depends only on $r$. The particular $r$ and $1/r^2$ forms of our field are irrelevant.

You can develop some feeling for these aspects of vector functions by studying the two-dimensional fields pictured in Fig. 2.39. In four of

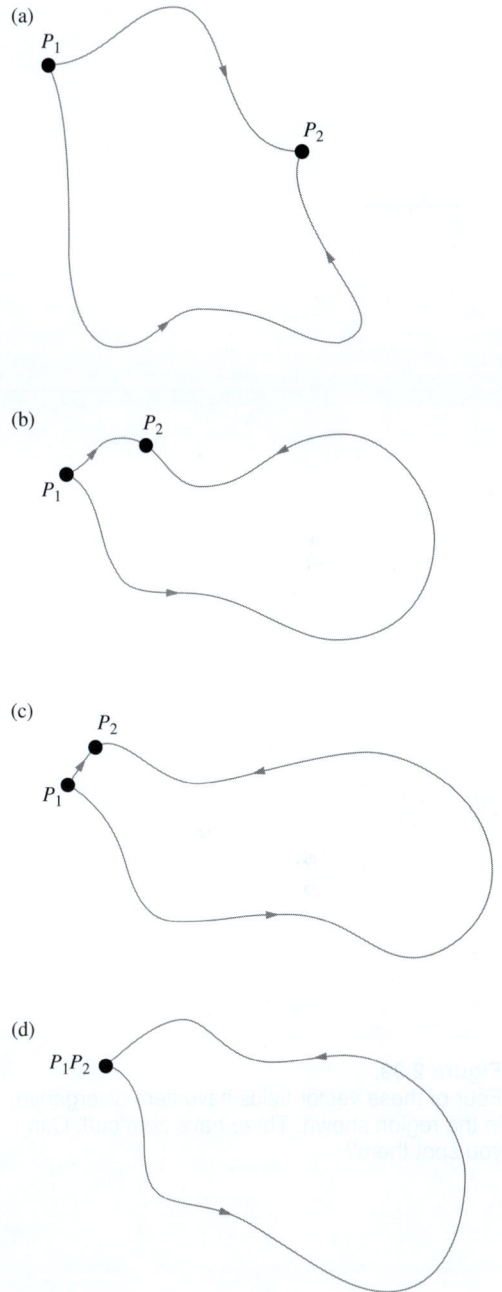

(a)

$P_1$

$P_2$

(b)

$P_2$

$P_1$

(c)

$P_2$

$P_1$

(d)

$P_1 P_2$

**Figure 2.38.**
If the line integral between $P_1$ and $P_2$ is independent of path, the line integral around a closed loop must be zero.

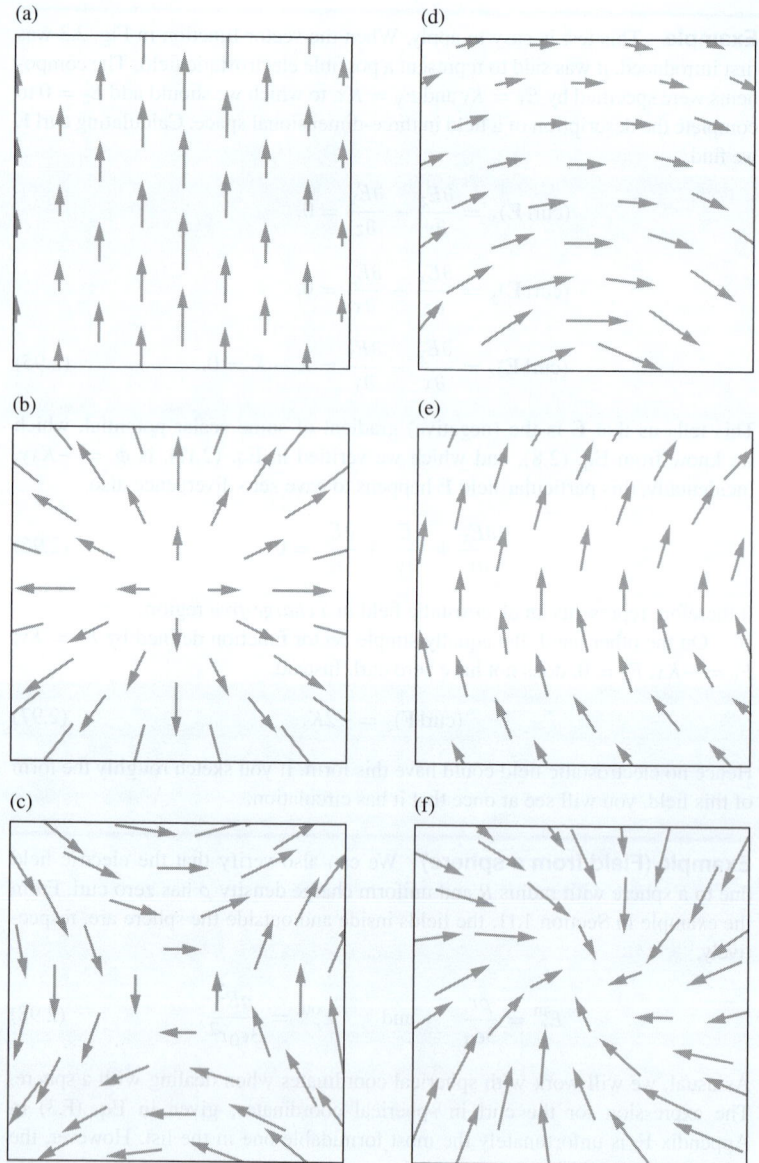

(a)

(d)

(b)

(e)

(c)

(f)

**Figure 2.39.**
Four of these vector fields have zero divergence in the region shown. Three have zero curl. Can you spot them?

these fields the divergence of the vector function is zero throughout the region shown. Try to identify the four. Divergence implies a net flux into, or out of, a neighborhood. It is easy to spot in certain patterns. In others you may be able to see at once that the divergence is zero. In three of the fields the curl of the vector function is zero throughout the region shown. Try to identify the three by deciding whether a line

(a)

Note that the vector remains constant as you advance in the direction in which it points. That is, $\partial F_y/\partial y = 0$, with $F_x = 0$. Hence div $\mathbf{F} = 0$. Note that the line integral around the dashed path is not zero.

div $\mathbf{F} = 0$     curl $\mathbf{F} \neq 0$

(b)

This is a central field. That is, $\mathbf{F}$ is radial and, for given $r$, its magnitude is constant. Any central field has zero curl; the circulation is zero around the dashed path, and any other path. But the divergence is obviously not zero.

div $\mathbf{F} \neq 0$     curl $\mathbf{F} = 0$

(c)

The circulation evidently *could* be zero around the paths shown. Actually, this is the same field as that in Fig. 2.3 and is a possible electrostatic field.

It is not obvious that div $\mathbf{F} = 0$ from this picture alone, but you can see that it too *could* be zero.

div $\mathbf{F} = 0$     curl $\mathbf{F} = 0$

(d)

Note that there is no change in the magnitude of $\mathbf{F}$, to first order, as you advance in the direction $\mathbf{F}$ points. That is enough to ensure zero divergence.     It appears that the     circulation *could* be     zero around the path     shown, for $F$ is weaker     on the long leg than on the short leg. Actually, this is a possible electrostatic field, with $F$ proportional to $1/r$, where $r$ is the distance to a point outside the picture.

div $\mathbf{F} = 0$     curl $\mathbf{F} = 0$

(e)

For the same reason as in (d), we deduce that div $\mathbf{F}$ is zero. Here the magnitude of $F$ is the same everywhere, so the line integral over the long     leg of the path shown is     not canceled by the integral     over the short leg, and the     circulation is not zero.

div $\mathbf{F} = 0$     curl $\mathbf{F} \neq 0$

(f)

Clearly the circulation around the dashed path is not zero. There appears also to be a nonzero divergence, since we see vectors converging toward the center from all directions.

div $\mathbf{F} \neq 0$     curl $\mathbf{F} \neq 0$

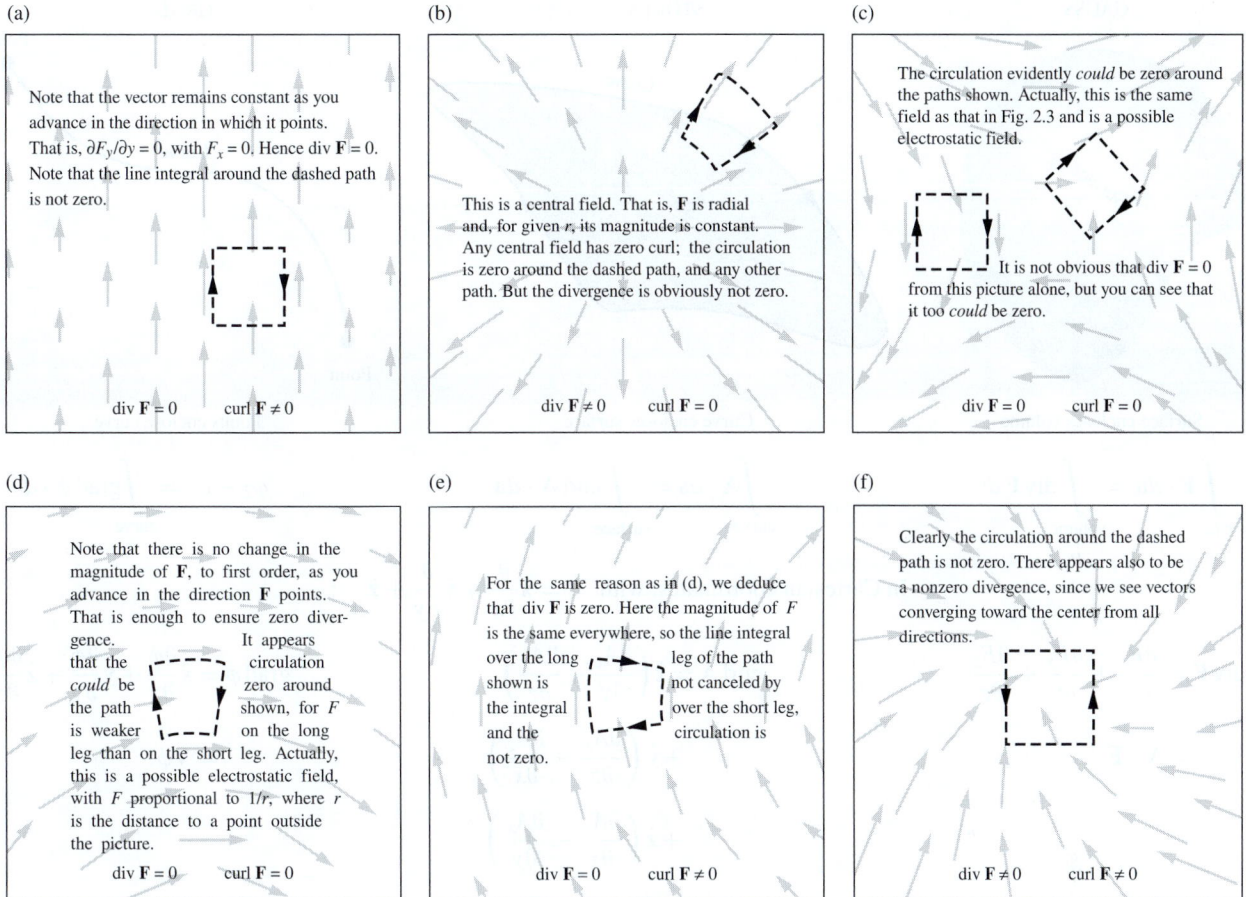

**Figure 2.40.**
Discussion of Fig. 2.39.

integral around any loop would or would not be zero in each picture. That is the essence of *curl*. After you have studied the pictures, think about these questions before you compare your reasoning and your conclusions with the explanation given in Fig. 2.40.

The curl of a vector field will prove to be a valuable tool later on when we deal with electric and magnetic fields whose curl is *not* zero. We have developed it at this point because the ideas involved are so close to those involved in the divergence. We may say that we have met two kinds of derivatives of a vector field. One kind, the divergence, involves the rate of change of a vector component in its own direction, $\partial F_x/\partial x$, and so on. The other kind, the curl, is a sort of "sideways derivative," involving the rate of change of $F_x$ as we move in the $y$ or $z$ direction.

GAUSS

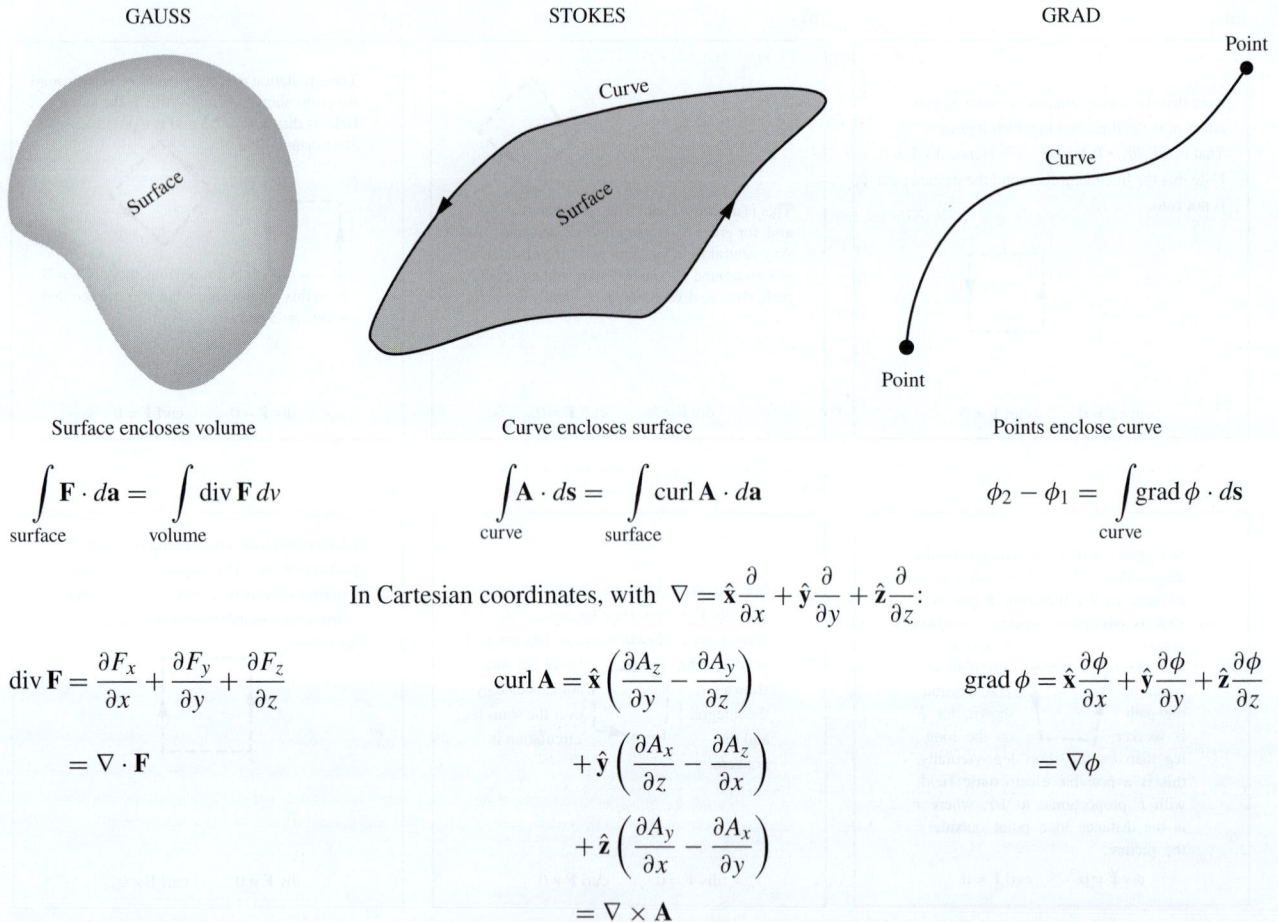

Surface

Surface encloses volume

$$\int_{\text{surface}} \mathbf{F} \cdot d\mathbf{a} = \int_{\text{volume}} \text{div } \mathbf{F} \, dv$$

STOKES

Curve

Surface

Curve encloses surface

$$\int_{\text{curve}} \mathbf{A} \cdot d\mathbf{s} = \int_{\text{surface}} \text{curl } \mathbf{A} \cdot d\mathbf{a}$$

GRAD

Point

Curve

Point

Points enclose curve

$$\phi_2 - \phi_1 = \int_{\text{curve}} \text{grad } \phi \cdot d\mathbf{s}$$

In Cartesian coordinates, with $\nabla = \hat{\mathbf{x}} \dfrac{\partial}{\partial x} + \hat{\mathbf{y}} \dfrac{\partial}{\partial y} + \hat{\mathbf{z}} \dfrac{\partial}{\partial z}$:

$$\text{div } \mathbf{F} = \frac{\partial F_x}{\partial x} + \frac{\partial F_y}{\partial y} + \frac{\partial F_z}{\partial z}$$

$$= \nabla \cdot \mathbf{F}$$

$$\text{curl } \mathbf{A} = \hat{\mathbf{x}} \left( \frac{\partial A_z}{\partial y} - \frac{\partial A_y}{\partial z} \right)$$

$$+ \hat{\mathbf{y}} \left( \frac{\partial A_x}{\partial z} - \frac{\partial A_z}{\partial x} \right)$$

$$+ \hat{\mathbf{z}} \left( \frac{\partial A_y}{\partial x} - \frac{\partial A_x}{\partial y} \right)$$

$$= \nabla \times \mathbf{A}$$

$$\text{grad } \phi = \hat{\mathbf{x}} \frac{\partial \phi}{\partial x} + \hat{\mathbf{y}} \frac{\partial \phi}{\partial y} + \hat{\mathbf{z}} \frac{\partial \phi}{\partial z}$$

$$= \nabla \phi$$

**Figure 2.41.**
Some vector relations summarized.

The relations called Gauss's theorem and Stokes' theorem are summarized in Fig. 2.41. The connection between the scalar potential function and the line integral of its gradient can also be looked on as a member of this family of theorems and is included in the third column. In all three of these theorems, the right-hand side of the equation involves an integral over an $N$-dimensional space, while the left-hand side involves an integral over the $(N - 1)$-dimensional boundary of the space. In the "grad" theorem, this latter integral is simply the discrete sum over two points.

## 2.18 Applications

As mentioned in Section 1.16, the electrical breakdown of air occurs at a field of about $3 \cdot 10^6$ V/m. So if you shuffle your feet on a carpet and then generate a spark with a grounded object, and if the spark is 1 mm long, then (assuming that the field is roughly constant over this 1 mm distance) your potential was given by $\phi = Ed \implies \phi = (3 \cdot 10^6 \, \text{V/m})(10^{-3} \, \text{m}) = 3000 \, \text{V}$. Conversely, if you rub a balloon on your hair and it achieves a potential of 3000 V, and if the radius is $r = 0.1$ m, then the charge on it is given by $\phi = kq/r \implies q = (3000)(0.1)/(9 \cdot 10^9) \approx 3 \cdot 10^{-8} \, \text{C}$, or about 100 esu.[6] Although 3000 V might sound like a dangerous voltage, it's quite safe in this case, because one of the factors that determines the severity of a shock is the amount of charge involved. And there simply isn't enough charge residing on the balloon to cause much pain when it flows off; 3000 V and an unlimited supply of charge would be a different story!

However, the spark from an object like a balloon *can* be dangerous if it occurs in the presence of a flammable gas, such as the gasoline vapors present at a gas station. If a voltage difference is somehow generated between different objects, a dangerous electrical breakdown can occur. One way of generating a voltage difference is to get back in your car while pumping the gas. Triboelectric effects between you and the car seat can produce charge transfer, causing you and the car to end up with different potentials, assuming that you don't touch any of its metal as you get back out. So don't get back in your car. Or if you must, be sure to touch a grounded object far from the hose nozzle after you get out. There are also triboelectric effects from the gasoline moving through the hose. The relative motion between the fast-moving gasoline in the middle of the hose and the stationary boundary layer near the wall causes a transfer of electrons. The gasoline flowing into the tank is therefore charged. This is the reason why it is dangerous to fill a gas container that isn't grounded, for example one that is sitting in the bed of a pickup truck. A nongrounded container can achieve a significant potential.

In the event that there is an unlimited supply of charge, it isn't the particular voltage of a given object that makes things dangerous, but rather the voltage *difference* between two objects. (In the end, it's the current that you need to worry about, but a larger voltage difference means a larger current, all other things being equal.) If you stand on a wooden stool with your hands on a small Van de Graaff generator held at 100 kV, then you are also at this high potential. But there are no ill effects (assuming you don't mind your hair sticking up, and barring pacemakers, etc.), because all parts of your body are at the same potential, so there is no current flowing anywhere. Likewise, when flying birds take a break and sit on power lines, their potential is very high, but it is essentially the same throughout their bodies; the potential difference between their

---

[6] This charge can be obtained in a quicker way, using the following rule of thumb, which you can verify after you learn about capacitance in Chapter 3: the capacitance of a sphere with a radius of $N$ centimeters is approximately $N$ picofarads, that is, $N \cdot 10^{-12}$ farads. Multiplying this capacitance by $\phi$ gives $q$.

feet isn't large enough to push a noticeable current through their bodies. Also, of course, each bird is touching only one wire. This should be contrasted with the case of flying deer, whose bodies are large enough to touch two of the power lines (which have different potentials) simultaneously. In the summer of 2011, a flying deer caused a power outage in a Montana town when it landed on power lines. Well, it was either a flying deer ... or a fawn dropped by an eagle.[7]

Although the potential difference between a bird's feet on a wire is insignificant, the potential difference between your feet on the ground can be quite significant if there is a lightning strike nearby. The main difference between these two cases is that the resistivity (discussed in Chapter 4) of the ground is much larger than that of the metal in a wire. The voltage difference between your feet can be thousands of volts, so current will travel up one leg and down the other. And there is more than enough charge in a lightning strike to do damage. So if there is a threat of lightning, then, after taking all other proper precautions, you should stand with your feet close together. Livestock are at a bit of a disadvantage in this regard, since they don't stand that way.

When the tires of a car roll along the ground, there is a triboelectric effect, so the car acquires a charge, and hence also a voltage difference with the ground. When the car stops, this charge leaks off, but it takes a few seconds. This means that if there is no line at a toll booth, so that you encounter the toll collector right after stopping, and if you happen to touch their hand, you may receive (and give) a shock. Toll collectors are therefore probably happy to see at least a short line of cars.

Helicopters can build up significant charge due to triboelectric effects between the blades and the air; the potential can reach 100 kV. And because a helicopter is much larger than the balloon we discussed earlier, it *can* hold enough charge to generate a serious (and perhaps lethal) shock. If a cable is lowered from a hovering helicopter to a person being rescued, it is critical that the cable touch the ground (or water) before it touches the person, so that the discharge doesn't occur through the person. This is often accomplished by attaching to the bottom of the main cable a *static discharge cable,* which has a breakaway safety mechanism in case it gets snagged.

The signals that travel along *neurons* take the form of modifications to the potential difference between the inside and outside of the axon (the long slender part) of the neuron. A certain enzyme pumps $Na^+$ (sodium) ions out of the axon, and $K^+$ (potassium) ions into it. But it pumps more of the former, so the exterior has a positive charge relative to the interior; its potential is about 70 mV higher. This is the resting potential. The nerve signal consists of a wave of depolarization, where $Na^+$ ions

---

[7] From photographs, it looks like the deer actually didn't touch two wires. So perhaps it instead tripped some sort of safety mechanism when it struck the wire, given that it was probably traveling at a high speed. In any event, a hypothetical flying deer could hypothetically touch two wires.

rush into the axon and then $K^+$ ions rush out. This modification to the potential is called the *action potential*. At a given location, it takes a few milliseconds for the signal to pass, and the potential then returns to its original value. The speed of the depolarization pulse along the neuron is roughly 100 m/s. Although this nerve signal is in some sense an electrical signal, it isn't an actual current. There is no net flow of charge along the neuron; there is only the transverse motion of the $Na^+$ and $K^+$ ions. The propagation of the signal depends on the opening and closing of various enzyme channels, so its speed isn't remotely close to the speed of an electrical signal in a metal wire, which is on the order of the speed of light.

## CHAPTER SUMMARY

- For an electrostatic field, the line integral $\int_{P_1}^{P_2} \mathbf{E} \cdot d\mathbf{s}$ is independent of the path from $P_1$ to $P_2$. This allows us to define uniquely the *electric potential difference*:

$$\phi_{21} = -\int_{P_1}^{P_2} \mathbf{E} \cdot d\mathbf{s}. \tag{2.99}$$

Relative to infinity, the potential due to a charge distribution is (depending on whether the distribution is continuous or discrete)

$$\phi(x, y, z) = \int \frac{\rho(x', y', z')\, dx'\, dy'\, dz'}{4\pi\epsilon_0 r} \qquad \text{or} \qquad \sum \frac{q_i}{4\pi\epsilon_0 r}. \tag{2.100}$$

- In Cartesian coordinates, the *gradient* of a scalar function (written as $\mathrm{grad}\,f$ or $\nabla f$) is

$$\nabla f \equiv \hat{\mathbf{x}}\frac{\partial f}{\partial x} + \hat{\mathbf{y}}\frac{\partial f}{\partial y} + \hat{\mathbf{z}}\frac{\partial f}{\partial z}. \tag{2.101}$$

The gradient gives the direction in which $f$ has the largest rate of increase. In terms of the gradient, the differential form of Eq. (2.99) is

$$\mathbf{E} = -\nabla \phi. \tag{2.102}$$

This relation implies that the lines of the electric field are perpendicular to the surfaces of constant potential.

- The electrostatic potential *energy* difference of a charge $q$ between points $P_1$ and $P_2$ equals $q\phi_{21}$. The energy required to assemble a group of charges from infinity is (depending on whether the distribution is continuous or discrete)

$$U = \frac{1}{2}\int \rho\phi\, dv \qquad \text{or} \qquad \frac{1}{2}\sum_{j=1}^{N} q_j \sum_{k \neq j} \frac{1}{4\pi\epsilon_0} \frac{q_k}{r_{jk}}. \tag{2.103}$$

The factor of 1/2 addresses the double counting of each pair of charges.

- A *dipole* consists of two charges $\pm q$ located a distance $\ell$ apart. The dipole moment is $p \equiv q\ell$. At large distances, the potential and field due to a dipole are

$$\phi(r,\theta) = \frac{p\cos\theta}{4\pi\epsilon_0 r^2},$$

$$\mathbf{E}(r,\theta) = \frac{p}{4\pi\epsilon_0 r^3}\left(2\cos\theta\,\hat{\mathbf{r}} + \sin\theta\,\hat{\boldsymbol{\theta}}\right). \tag{2.104}$$

We verified that the electric field is everywhere perpendicular to the surfaces of constant potential.

- In Cartesian coordinates, the *divergence* of a vector function (written as div $\mathbf{F}$ or $\nabla \cdot \mathbf{F}$) is

$$\operatorname{div}\mathbf{F} = \frac{\partial F_x}{\partial x} + \frac{\partial F_y}{\partial y} + \frac{\partial F_z}{\partial z}. \tag{2.105}$$

The divergence appears in *Gauss's theorem*, or the *divergence theorem*,

$$\int_S \mathbf{F}\cdot d\mathbf{a} = \int_V \operatorname{div}\mathbf{F}\,dv. \tag{2.106}$$

Physically, the divergence equals the flux of $\mathbf{F}$ out of a volume, divided by the volume, in the limit where the volume becomes infinitesimal. Combining Gauss's theorem with Gauss's law, Eq. (1.31), gives

$$\operatorname{div}\mathbf{E} = \frac{\rho}{\epsilon_0}. \tag{2.107}$$

This is the first of Maxwell's equations.

- In Cartesian coordinates, the *Laplacian* of a scalar function (written as div grad $f$, or $\nabla \cdot \nabla f$, or $\nabla^2 f$) is

$$\nabla^2 f = \frac{\partial^2 f}{\partial x^2} + \frac{\partial^2 f}{\partial y^2} + \frac{\partial^2 f}{\partial z^2}. \tag{2.108}$$

In terms of the Laplacian, Eqs. (2.102) and (2.107) can be combined to give

$$\nabla^2\phi = -\frac{\rho}{\epsilon_0}. \tag{2.109}$$

This is called *Poisson's equation*. It is the final link in Fig. 2.29, which shows all the relations among the electrostatic quantities $\mathbf{E}$, $\phi$, and $\rho$. A special case of Eq. (2.109) is *Laplace's equation*:

$$\nabla^2\phi = 0. \tag{2.110}$$

If $\phi$ satisfies this equation, then the average value of $\phi$ over the surface of any sphere equals the value of $\phi$ at the center of the sphere. This fact (or alternatively Gauss's law) implies that it is impossible to construct an electrostatic field that will hold a charged particle in *stable* equilibrium in empty space.

- In Cartesian coordinates, the *curl* of a vector function (written as curl $\mathbf{F}$ or $\nabla \times \mathbf{F}$) is

$$\text{curl } \mathbf{F} = \begin{vmatrix} \hat{\mathbf{x}} & \hat{\mathbf{y}} & \hat{\mathbf{z}} \\ \partial/\partial x & \partial/\partial y & \partial/\partial z \\ F_x & F_y & F_z \end{vmatrix}. \quad (2.111)$$

The curl appears in *Stokes' theorem*,

$$\int_C \mathbf{F} \cdot d\mathbf{s} = \int_S \text{curl } \mathbf{F} \cdot d\mathbf{a}. \quad (2.112)$$

Physically, the curl equals the line integral of $\mathbf{F}$ around an area, divided by the area, in the limit where the area becomes infinitesimal. Since the line integral of an electrostatic field $\mathbf{E}$ around any closed path is zero, Stokes' theorem implies that curl $\mathbf{E} = 0$. See Appendix F for a discussion of the various vector operators in different coordinate systems.

## Problems

2.1   *Equivalent statements*  *
It is arbitrary which of the two boxed statements in Section 2.1 we regard as the corollary of the other. Show that, if the line integral $\int \mathbf{E} \cdot d\mathbf{s}$ is zero around any closed path, it follows that the line integral between two different points is path-independent.

2.2   *Combining two shells*  *
We know from Problem 1.32 that the self-energy of a spherical shell of radius $R$ with charge $Q$ uniformly distributed over it is $Q^2/8\pi\epsilon_0 R$. What if we put two such shells right on top of each other, to make a shell with charge $2Q$? Since we now just have two copies of the original system, it seems like the energy should be twice as large, or $Q^2/4\pi\epsilon_0 R$. However, the above formula gives an energy of $(2Q)^2/8\pi\epsilon_0 R = Q^2/2\pi\epsilon_0 R$. Which answer is correct, and what is wrong with the reasoning for the wrong answer?

2.3   *Equipotentials from four charges*  *
Two point charges of strength $2q$ each and two point charges of strength $-q$ each are symmetrically located in the $xy$ plane as follows. The two positive charges are at $(0, 2\ell)$ and $(0, -2\ell)$, the two negative charges are at $(\ell, 0)$ and $(-\ell, 0)$. Some of the equipotentials in the $xy$ plane have been plotted in Fig. 2.42. (Of course these curves are really the intersection of some three-dimensional

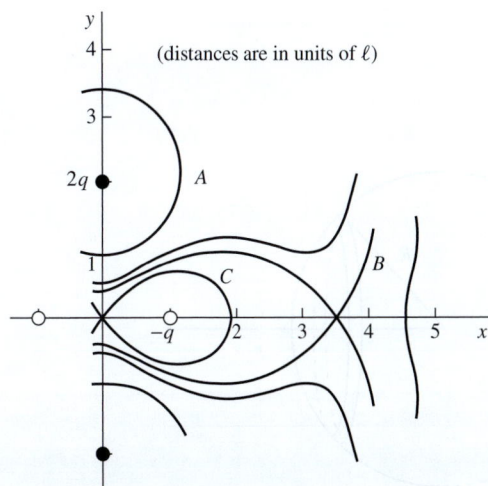

**Figure 2.42.**

equipotential surfaces with the $xy$ plane.) Study this figure until you understand its general appearance. Now find the value of the potential $\phi$ on each of the curves $A$, $B$, and $C$, as usual taking $\phi = 0$ at infinite distance. Curve $A$ has been arbitrarily chosen to cross the $y$ axis at $y = \ell$, and it can be shown (how?) that curve $B$ crosses the $x$ axis at $x \approx (3.44)\ell$. Roughly sketch some intermediate equipotentials.

2.4　*Center vs. corner of a cube* **

Consider a charge distribution that has the constant density $\rho$ everywhere inside a cube of edge $b$ and is zero everywhere outside that cube. Letting the electric potential $\phi$ be zero at infinite distance from the cube of charge, denote by $\phi_0$ the potential at the center of the cube and by $\phi_1$ the potential at a corner of the cube. Determine the ratio $\phi_0/\phi_1$. The answer can be found with very little calculation by combining a dimensional argument with superposition. (Think about the potential at the center of a cube with the same charge density and with twice the edge length.)

2.5　*Escaping a cube* **

Suppose eight protons are permanently fixed at the corners of a cube. A ninth proton floats freely near the center of the cube. There are no other charges around, and no gravity. Is the ninth proton trapped? Can it find an escape route that is all downhill in potential energy? Feel free to analyze this numerically/graphically.

2.6　*Electrons on a basketball* *

A sphere the size of a basketball is charged to a potential of $-1000$ volts. About how many extra electrons are on it, per square centimeter of surface?

2.7　*Shell field via direct integration* **

Consider the electric field $E$ due to a spherical shell of radius $R$ with charge $Q$ uniformly distributed over its surface. In Section 1.11 we found $E$ by using Gauss's law. Find $E$ here (both inside and outside the shell) by directly calculating the potential at a given value of $r$ by integrating the contributions from the different parts of the shell, and then using $E_r = -d\phi/dr$. The simplest strategy is to slice the shell into rings, as shown in Fig. 2.43. You will need to use the law of cosines.

2.8　*Verifying the inverse square law* ****

As mentioned in Section 1.4, Cavendish and Maxwell conducted experiments to test the inverse-square nature of Coulomb's law. This problem gives the theory behind their experiments.

(a) Assume that Coulomb's law takes the form of $kq_1q_2/r^{2+\delta}$. Given a hollow spherical shell with radius $R$ and uniformly

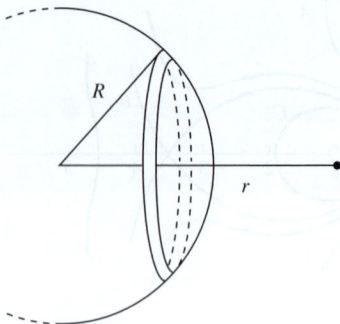

**Figure 2.43.**

distributed charge $Q$, show that the potential at radius $r$ is (with $f(x) = x^{1-\delta}$ and $k \equiv 1/4\pi\epsilon_0$)

$$\phi(r) = \frac{kQ}{2(1-\delta^2)rR}\left[f(R+r)-f(R-r)\right] \qquad \text{(for } r < R\text{)},$$

$$\phi(r) = \frac{kQ}{2(1-\delta^2)rR}\left[f(R+r)-f(r-R)\right] \qquad \text{(for } r > R\text{)}.$$

$$(2.113)$$

The calculation requires only a slight modification of the analogous direct calculation (that is, one that doesn't use the Gauss's-law shortcut) of the potential in the case of the standard Coulomb $1/r^2$ law; see Problem 2.7.

*Note:* We are usually concerned with $\delta \ll 1$, in which case the $(1 - \delta^2)$ factor in the denominators in Eq. (2.113) can be reasonably approximated by 1. We will ignore it for the remainder of this problem.

(b) Consider two concentric shells with radii $a$ and $b$ (with $a > b$) and uniformly distributed charges $Q_a$ and $Q_b$. Show that the potentials on the shells are given by

$$\phi_a = \frac{kQ_a}{2a^2}f(2a) + \frac{kQ_b}{2ab}\left[f(a+b)-f(a-b)\right],$$

$$\phi_b = \frac{kQ_b}{2b^2}f(2b) + \frac{kQ_a}{2ab}\left[f(a+b)-f(a-b)\right]. \qquad (2.114)$$

(c) Show that if the shells are connected, so that they are at the same potential $\phi$, then the charge on the inner shell is

$$Q_b = \frac{2b\phi}{k} \cdot \frac{bf(2a) - a\left[f(a+b)-f(a-b)\right]}{f(2a)f(2b) - \left[f(a+b)-f(a-b)\right]^2}. \qquad (2.115)$$

If $\delta = 0$ so that $f(x) = x$, then $Q_b$ equals zero, as it should. So if $Q_b$ is measured to be nonzero, then $\delta$ must be nonzero.

For small $\delta$ it is possible to expand $Q_b$ to first order in $\delta$ by using the approximation $f(x) = xe^{-\delta \ln x} \approx x(1 - \delta \ln x)$, but this gets very messy. You are encouraged instead to use a computer to calculate and plot $Q_b$ for various values of $a$, $b$, and $\delta$. You can also trivially expand $Q_b$ to first order in $\delta$ by using the Series operation in *Mathematica*.

**2.9** $\phi$ *from integration* **

(a) A solid sphere has radius $R$ and uniform volume charge density $\rho$. Find the potential at the center by evaluating the integral in Eq. (2.18).

(b) A spherical shell has radius $R$ and uniform surface charge density $\sigma$. Find the potential at a point on the surface by evaluating the integral in Eq. (2.18).

(c) When written in terms of the total charge involved, how do the above two results compare?

2.10 *A thick shell* ∗∗
(a) A spherical shell with charge $Q$ uniformly distributed through-out its volume has inner radius $R_1$ and outer radius $R_2$. Calculate (and make a rough plot of) the electric field as a function of $r$, for $0 \leq r \leq \infty$.
(b) What is the potential at the center of the shell? You can let $R_2 = 2R_1$ in this part of the problem, to keep things from getting too messy. Give your answer in terms of $R \equiv R_1$.

2.11 *E for a line, from a cutoff potential* ∗∗
Consider the electric field $E$ due to an infinite straight wire with uniform linear charge density $\lambda$. In Section 1.12 we found $E$ by direct integration of Coulomb's law, and again by using Gauss's law. Find $E$ here by calculating the potential and then taking the derivative.

You will find that the potential (relative to infinity) due to an infinite wire diverges. But you can get around this difficulty by instead finding the potential due to a very long but finite wire of length $2L$, at a point lying on its perpendicular bisecting plane. Use a Taylor series to simplify your result, and then take the derivative to find $E$. Explain why this procedure is valid, even though it cuts off an infinite amount from the potential.

2.12 *E and $\phi$ from a ring* ∗∗
(a) Consider a ring with charge $Q$ and radius $R$. Let point $P$ be a distance $x$ from the plane of the ring, along the axis through its center. By adding up the contributions from all the pieces of the ring, find the electric field $E(x)$ at point $P$.
(b) In the same manner, find the potential $\phi(x)$ at point $P$.
(c) Show that $E = -d\phi/dx$.
(d) If a charge $-q$ with mass $m$ is released from rest far away along the axis, what is its speed when it passes through the center of the ring? Assume that the ring is fixed in place.

2.13 *$\phi$ at the center of an N-gon* ∗∗
Use the technique from the second example in Section 2.6 to calculate the potential at the center of a sheet in the shape of a regular $N$-gon with surface charge density $\sigma$. Let the distance from the center to the midpoint of a side be $a$. Show that your answer reduces to the result in Eq. (2.27) in the $N \to \infty$ limit.

2.14 *Energy of a sphere* ∗∗
A spherical volume of radius $R$ is filled with charge of uniform density $\rho$. Exercise 1.61 and the example in Section 1.15 presented two methods for calculating the energy stored in the system. Calculate the energy in a third way, by using Eq. (2.32).

2.15 *Crossed dipoles* *

Two dipoles, each with dipole moment $p$, are oriented perpendicularly as shown in Fig. 2.44. What is the dipole moment of the system?

2.16 *Disks and dipoles* **

Two parallel disks each have radius $R$ and are separated by a distance $\ell$. The surface charge densities are $\sigma$ and $-\sigma$. What is the electric field at a large distance $r$ along the axis of the disks? Solve this in two ways.

(a) Treat the disks like a collection of a large number of dipoles standing next to each other.

(b) Explain why the parts of the two disks that are contained within the cone in Fig. 2.45 produce canceling fields at point $P$, and then find the field due to the uncanceled part of the top disk.

2.17 *Linear quadrupole* **

Consider a "linear quadrupole" consisting of two adjacent dipoles oriented oppositely and placed end to end; see the left quadrupole in Fig. 2.16. There is effectively a point charge $-2q$ at the center. By adding up the electric fields from the charges, find the electric field at a distant point (a) along the axis and (b) along the perpendicular bisector.

2.18 *Field lines near the origin* **

(a) Two equal positive charges $q$ are located at the points $(\pm a, 0, 0)$. Write down the potential $\phi(x, y)$ for points in the $xy$ plane, and then use a Taylor expansion to find an approximate expression for $\phi$ near the origin. (You can set $a = 1$ to make things simpler.)

(b) Find the electric field at points near the origin. Then find the equations for the field lines near the origin by demanding that the slope $dy/dx$ of a curve at a given point equals the slope $E_y/E_x$ of the field at that point.

2.19 *Equipotentials for a ring* ***

(a) A ring with radius $R$ has charge $Q$ uniformly distributed on it. It lies in the $xy$ plane, with its center at the origin. Find the electric field at all points on the $z$ axis. For what value of $z$ is the field maximum?

(b) Make a rough sketch of the equipotential curves everywhere in space (or rather, everywhere in a plane containing the $z$ axis; you can represent the ring by two dots where it intersects the plane). Be sure to indicate what the curves look like very close to and very far from the ring, and how the transition from close to far occurs.

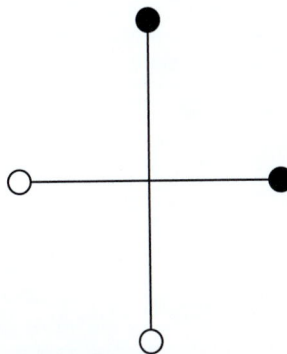

**Figure 2.44.**

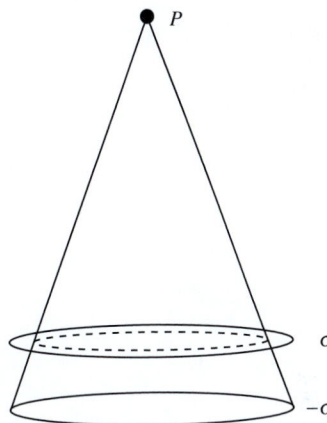

**Figure 2.45.**

(c) There is a particular $z$ value (along with its negative) at which the equipotentials make the transition from concave up to concave down. Explain why this $z$ value equals the $z$ value you found in part (a). *Hint:* The divergence of **E** is zero.

**2.20** *A one-dimensional charge distribution* **\*\***
Find (and make rough plots of) the electric field and charge distribution associated with the following potential:

$$\phi(x) = \begin{cases} 0 & \text{(for } x < 0) \\ \rho_0 x^2 / 2\epsilon_0 & \text{(for } 0 < x < \ell) \\ \rho_0 \ell^2 / 2\epsilon_0 & \text{(for } \ell < x). \end{cases} \qquad (2.116)$$

**2.21** *A cylindrical charge distribution* **\*\***
A distribution of charge has cylindrical symmetry. As a function of the distance $r$ from the symmetry axis, the electric potential is

$$\phi(r) = \begin{cases} \dfrac{3\rho_0 R^2}{4\epsilon_0} & \text{(for } r \leq R) \\ \dfrac{\rho_0}{4\epsilon_0}(4R^2 - r^2) & \text{(for } R < r < 2R) \\ 0 & \text{(for } 2R \leq r), \end{cases} \qquad (2.117)$$

where $\rho_0$ is a quantity with the dimensions of volume charge density.

(a) Find (and make rough plots of) the electric field and charge distribution, for all values of $r$. The derivative operators in cylindrical coordinates are listed in Appendix F.
(b) From your charge distribution, calculate the total charge per unit length along the cylinder.

**2.22** *Discontinuous E and $\phi$* **\*\***
(a) What kind of charge distribution yields a discontinuous electric field?
(b) Can you think of a charge distribution (or perhaps the limit of a charge distribution) that yields a discontinuous potential?

**2.23** *Field due to a distribution* **\*\***
Each of the objects described below has uniform volume charge density $\rho$. There are no other charges present in addition to the given object. In each case use $\nabla \cdot \mathbf{E} = \rho/\epsilon_0$ to show that the electric field takes the stated form.

(a) A rectangular slab has thickness $\ell$ in the $x$ direction and infinite extent in the $y$ and $z$ directions. Show that $E_x = \rho x/\epsilon_0$ inside the slab, where $x$ is measured from the midplane of the slab.
(b) An infinitely long cylinder has radius $R$. Show that $E_r = \rho r/2\epsilon_0$ inside the cylinder.

(c) A sphere has radius $R$. Show that $E_r = \rho r/3\epsilon_0$ inside the sphere.

(d) Given that the above three setups all involve the same charge density and the same relation $\nabla \cdot \mathbf{E} = \rho/\epsilon_0$, why do they give different results for the electric field?

2.24 *Two expressions for the energy* **

(a) Prove the identity

$$\nabla \cdot (\phi \mathbf{E}) = (\nabla \phi) \cdot \mathbf{E} + \phi \, \nabla \cdot \mathbf{E} \qquad (2.118)$$

by explicitly calculating the various derivatives in Cartesian coordinates.

(b) This identity holds for any scalar function $\phi$ and any vector function $\mathbf{E}$. In particular, it holds for the electric potential and field. Use this fact to show that Eqs. (1.53) and (2.32) are equivalent expressions for the energy stored in a charge distribution of finite extent. You will want to apply the divergence theorem with a wisely chosen volume.

2.25 *Never trapped* **

A number of positive point charges, with various magnitudes, are located at fixed positions in space. Show that no matter where an additional positive charge $q$ is placed, there exists an escape route to infinity that is all downhill in potential energy. The "impossibility theorem" discussed in Section 2.12 will be helpful, but that deals only with small displacements, so you will need to extend the argument.

2.26 *The delta function* **

In spherical coordinates, consider the Laplacian of the function $f(r) = 1/r$, that is, $\nabla^2(1/r)$. From Appendix F we have $\nabla^2 f = (1/r^2)(\partial/\partial r)(r^2 \, \partial f/\partial r)$ for a function that depends only on $r$. Since $r^2 \, \partial(1/r)/\partial r$ takes on the constant value of $-1$, we see that $\nabla^2(1/r)$ equals zero. Well, almost. It is certainly zero for $r \neq 0$, but we must be careful at the origin, due to the infinite $1/r^2$ factor out front.

   Show that $\nabla^2(1/r)$ is *not* equal to zero at $r = 0$. Do this by showing that it is large enough (or more precisely, infinite enough) to make the volume integral $\int \nabla^2(1/r) \, dv$ equal to $-4\pi$, provided that the volume contains the origin. The divergence theorem will be helpful.

2.27 *Relations between $\phi$ and $\rho$* ***

Figure 2.29 gives two relations between $\phi$ and $\rho$, namely $\phi = (1/4\pi\epsilon_0) \int (\rho/r) \, dv'$ and $\nabla^2 \phi = -\rho/\epsilon_0$. Show that these relations are consistent, by operating on the first one with the Laplacian $\nabla^2$

operator. Be careful that there are two types of coordinates in the equation, primed and unprimed; it can be written more precisely as

$$\phi(\mathbf{r}) = \frac{1}{4\pi\epsilon_0} \int \frac{\rho(\mathbf{r}')\,dv'}{|\mathbf{r}' - \mathbf{r}|}. \tag{2.119}$$

You will want to solve Problem 2.26 first.

2.28  *Zero curl* ∗

Consider the electric field, $\mathbf{E} = (2xy^2 + z^3, 2x^2y, 3xz^2)$. We have ignored a multiplicative factor with units of $V/m^4$ necessary to make the units correct. Show that curl $\mathbf{E} = 0$, and then find the associated potential function $\phi(x, y, z)$.

2.29  *Ends of the lines* ∗

Explain why electrostatic field lines can't form closed loops, and why their ends must be located either at charges or at infinity.

2.30  *Curl of a gradient* ∗∗

The electric field equals the negative gradient of the potential, that is, $\mathbf{E} = -\nabla\phi$. Show that this implies that the curl of $\mathbf{E}$, which we can write as $\nabla \times \mathbf{E}$, is identically zero. Do this by:

(a)  calculating $\nabla \times \nabla\phi$ in Cartesian coordinates;
(b)  making judicious use of Stokes' theorem.

## Exercises

2.31  *Finding the potential* ∗

The following vector function represents a possible electrostatic field:

$$E_x = 6xy, \qquad E_y = 3x^2 - 3y^2, \qquad E_z = 0. \tag{2.120}$$

(We have ignored a multiplicative factor with units of $V/m^3$ necessary to make the units correct.) Calculate the line integral of $\mathbf{E}$ from the point $(0, 0, 0)$ to the point $(x_1, y_1, 0)$ along the path that runs straight from $(0, 0, 0)$ to $(x_1, 0, 0)$ and thence to $(x_1, y_1, 0)$. Make a similar calculation for the path that runs along the other two sides of the rectangle, via the point $(0, y_1, 0)$. You ought to get the same answer if the assertion above is true. Now you have the potential function $\phi(x, y, z)$. Take the gradient of this function and see that you get back the components of the given field.

2.32  *Line integral the easy way* ∗

Designate the corners of a square, $\ell$ on a side, in clockwise order, $A, B, C, D$. Put charges $2q$ at $A$ and $-3q$ at $B$. Determine the value of the line integral of $\mathbf{E}$, from point $C$ to point $D$. (No actual integration needed!) What is the numerical answer if $q = 10^{-9}$ C and $\ell = 5$ cm?

2.33 *Plot the potential* *
Consider the system of two charges shown in Fig. 2.8. Let $z$ be the coordinate along the line on which the two charges lie, with $z = 0$ at the location of the positive charge. Make a plot of the potential $\phi$ (or rather $4\pi\epsilon_0\phi$, for simplicity) along this line, from $z = -5$ m to $z = 15$ m.

2.34 *Extremum of $\phi$* *
A charge of 2 C is located at the origin. Two charges of $-1$ C each are located at the points $(1, 1, 0)$ and $(-1, 1, 0)$. If the potential $\phi$ is taken to be zero at infinity (as usual), then it is easy to see that $\phi$ is also zero at the point $(0, 1, 0)$. It follows that somewhere on the $y$ axis beyond $(0, 1, 0)$ the function $\phi(0, y, 0)$ must have a minimum or a maximum. At that point the electric field $\mathbf{E}$ must be zero. Why? Locate the point, at least approximately.

2.35 *Center vs. corner of a square* **
A square sheet has uniform surface charge density $\sigma$. Letting the electric potential $\phi$ be zero at infinite distance from the square, denote by $\phi_0$ the potential at the center of the square and by $\phi_1$ the potential at a corner. Determine the ratio $\phi_0/\phi_1$. The answer can be found with very little calculation by combining a dimensional argument with superposition. (Think about the potential at the center of a square with the same charge density and with twice the edge length.)

2.36 *Escaping a cube, toward an edge* **
Consider the setup in Problem 2.5. Will the proton escape if it moves from the center directly toward the midpoint of an edge? Feel free to analyze this numerically/graphically.

2.37 *Field on the earth* *
A sphere the size of the earth has 1 C of charge distributed evenly over its surface. What is the electric field strength just outside the surface? What is the potential of the sphere, with zero potential at infinity?

2.38 *Interstellar dust* *
An interstellar dust grain, roughly spherical with a radius of $3 \cdot 10^{-7}$ m, has acquired a negative charge such that its potential is $-0.15$ volt. How many extra electrons has it picked up? What is the strength of the electric field at its surface?

2.39 *Closest approach* **
By means of a Van de Graaff generator, protons are accelerated through a potential difference of $5 \cdot 10^6$ volts. The proton beam then passes through a thin silver foil. The atomic number of silver is 47, and you may assume that a silver nucleus is so massive compared with the proton that its motion may be neglected. What is

the closest possible distance of approach, of any proton, to a silver nucleus? What will be the strength of the electric field acting on the proton at that position? What will be the proton's acceleration?

**2.40** *Gold potential* **

As a distribution of electric charge, the gold nucleus can be described as a sphere of radius $6 \cdot 10^{-15}$ m with a charge $Q = 79e$ distributed fairly uniformly through its interior. What is the potential $\phi_0$ at the center of the nucleus, expressed in megavolts? (First derive a general formula for $\phi_0$ for a sphere of charge $Q$ and radius $a$. Do this by using Gauss's law to find the internal and external electric field and then integrating to find the potential. You should redo this here, even though it was done in an example in the text.)

**2.41** *A sphere between planes* **

A spherical shell with radius $R$ and surface charge density $\sigma$ is sandwiched between two infinite sheets with surface charge densities $-\sigma$ and $\sigma$, as shown in Fig. 2.46. If the potential far to the right at $x = +\infty$ is taken to be zero, what is the potential at the center of the sphere? At $x = -\infty$?

**2.42** *E and $\phi$ for a cylinder* **

For the cylinder of uniform charge density in Fig. 2.26:

(a) show that the expression there given for the field inside the cylinder follows from Gauss's law;

(b) find the potential $\phi$ as a function of $r$, both inside and outside the cylinder, taking $\phi = 0$ at $r = 0$.

**2.43** *Potential from a rod* **

A thin rod extends along the $z$ axis from $z = -d$ to $z = d$. The rod carries a charge uniformly distributed along its length with linear charge density $\lambda$. By integrating over this charge distribution, calculate the potential at a point $P_1$ on the $z$ axis with coordinates $(0, 0, 2d)$. By another integration find the potential at a general point $P_2$ on the $x$ axis and locate this point to make the potential equal to the potential at $P_1$.

**2.44** *Ellipse potentials* ***

The points $P_1$ and $P_2$ in Exercise 2.43 happen to lie on an ellipse that has the ends of the rod as its foci, as you can readily verify by comparing the sums of the distances from $P_1$ and from $P_2$ to the ends of the rod. This suggests that the whole ellipse might be an equipotential. Test that conjecture by calculating the potential at the point $(3d/2, 0, d)$, which lies on the same ellipse. Indeed it is true, though there is no obvious reason why it should be, that the equipotential surfaces of this system are a family of confocal prolate spheroids. See if you can prove that. You will have to derive

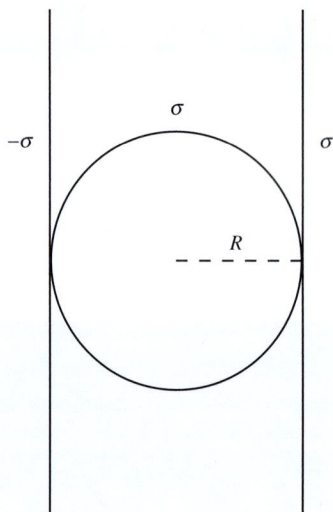

$\sigma$

$-\sigma$

$\sigma$

$R$

**Figure 2.46.**

a formula for the potential at a general point $(x, 0, z)$ in the $xz$ plane. Then show that, if $x$ and $z$ are related by the equation $x^2/(a^2-d^2)+z^2/a^2 = 1$, which is the equation for an ellipse with foci at $z = \pm d$, the potential will depend only on the parameter $a$ (in addition to $d$), not on $x$ or $z$.

**2.45 A stick and a point charge** **

A stick with length $\ell$ has charge $Q$ uniformly distributed on it. It lies along the $x$ axis between the points $x = -\ell$ and $x = 0$. A point charge also with charge $Q$ lies on the $x$ axis at the point $x = \ell$; see Fig. 2.47.

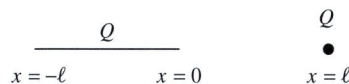

**Figure 2.47.**

(a) Let $x = a$ be the point on the $x$ axis between the two objects where the electric field is zero. Find $a$.

(b) There happens to be another point where the electric field is zero (it's inside the stick). In addition to this one, are there any other points in space where the electric field is zero? Why or why not?

(c) Make a rough sketch of the field lines and equipotential curves everywhere in the plane of the paper. Be sure to indicate how the lines and curves make the transition from their shapes close to the objects to their shapes far from them. (Don't worry about what's going on extremely close to the stick.) What do things look like near the point you found in part (a)?

**2.46 Right triangle $\phi$** **

The right triangle shown in Fig. 2.48 with vertex $P$ at the origin has base $b$, altitude $a$, and uniform density of surface charge $\sigma$. Determine the potential at the vertex $P$. First find the contribution of the vertical strip of width $dx$ at $x$. Show that the potential at $P$ can be written as $\phi_P = (\sigma b/4\pi\epsilon_0) \ln[(1 + \sin\theta)/\cos\theta]$.

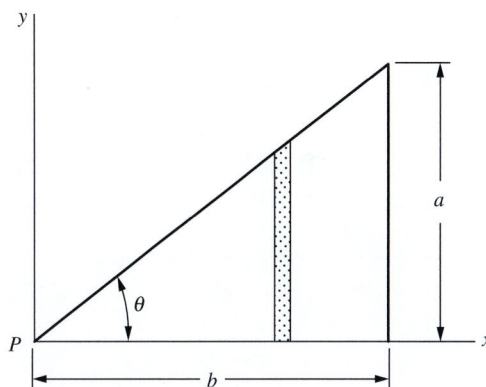

**2.47 A square and a disk** **

Use the result from Exercise 2.46 to answer the following question. If a square with surface charge density $\sigma$ and side $s$ has the same potential at its center as a disk with the same surface charge density and diameter $d$, what must be the ratio $s/d$? Is your answer reasonable?

**Figure 2.48.**

**2.48 Field from a hemisphere** **

Following the strategy in Problem 2.7, find the electric field at the center of a hemispherical shell with radius $R$ and uniform surface charge density $\sigma$. That is, find $\phi$ as a function of $r$ and then take the derivative. You might find it easier to Taylor-expand $\phi$ before differentiating. (You already found this electric field in a simpler manner if you solved Exercise 1.50. The present method is more involved because we need to do more than calculate $\phi$ at just one point; we need to know $\phi$ as a function of $r$ so that we can take its derivative.)

**2.49** *E for a sheet, from a cutoff potential* **

Consider the electric field $E$ due to an infinite sheet with uniform surface charge density $\sigma$. In Section 1.13 we found $E$ by using Gauss's law. Find $E$ here by calculating the potential and then taking the derivative.

You will find that the potential (relative to infinity) due to an infinite sheet diverges. But in the spirit of Problem 2.11 you can get around this difficulty by instead finding the potential due to a very large but finite disk with radius $R$, at a point lying on the perpendicular line through the center. Use a Taylor series to simplify the potential, and then take the derivative to find $E$. Explain why this procedure is valid, even though it cuts off an infinite amount from the potential.

**2.50** *Dividing the charge* **

We have two metal spheres, of radii $R_1$ and $R_2$, quite far apart from one another compared with these radii. Given a total amount of charge $Q$ which we have to divide between the spheres, how should it be divided so as to make the potential energy of the resulting charge distribution as small as possible? To answer this, first calculate the potential energy of the system for an arbitrary division of the charge, $q$ on one sphere and $Q - q$ on the other. Then minimize the energy as a function of $q$. You may assume that any charge put on one of these spheres distributes itself uniformly over the surface of the sphere, the other sphere being far enough away so that its influence can be neglected. When you have found the optimum division of the charge, show that with that division the potential difference between the two spheres is zero. (Hence they could be connected by a wire, and there would still be no redistribution. This is a special example of a very general principle we shall meet in Chapter 3: on a conductor, charge distributes itself so as to minimize the total potential energy of the system.)

**2.51** *Potentials on the axis* **

A hollow circular cylinder, of radius $a$ and length $b$, with open ends, has a total charge $Q$ uniformly distributed over its surface. What is the difference in potential between a point on the axis at one end and the midpoint of the axis? Show by sketching some field lines how you think the field of this thing ought to look.

**2.52** *Spherical cavity in a slab* **

Figure 2.49 shows a cross section of a slab with uniform volume charge density $\rho$. It has thickness $2R$ in one dimension and is infinite in the other two dimensions. A spherical cavity with radius

$R$ is hollowed out. A few equipotential curves are drawn in the figure.

(a) Show that an equipotential curve that starts at the center of the cavity (curve $A$ shown) ends up meeting the surface of the slab at infinity. *Hint:* Superpose two oppositely charged objects.

(b) Show that curve $A$ is a straight line inside the cavity, and find its slope.

(c) Show that a curve that is tangent to the sphere (curve $B$ shown) ends up a distance $R/3$ outside the slab at infinity.

**2.53** *Field from two shells* **

One of two nonconducting spherical shells of radius $a$ carries a charge $Q$ uniformly distributed over its surface, the other carries a charge $-Q$, also uniformly distributed. The spheres are brought together until they touch. What does the electric field look like, both outside and inside the shells? How much work is needed to move them far apart?

**2.54** *An equipotential for a disk* *

For the system in Fig. 2.11 sketch the equipotential surface that touches the rim of the disk. Find the point where it intersects the symmetry axis.

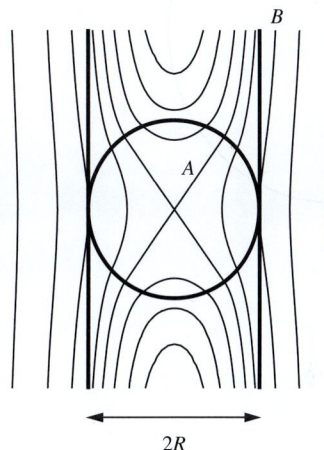

**Figure 2.49.**

**2.55** *Hole in a disk* **

A thin disk, radius 3 cm, has a circular hole of radius 1 cm in the middle. There is a uniform surface charge of $-10^{-5}\,\mathrm{C/m^2}$ on the disk.

(a) What is the potential at the center of the hole? (Assume zero potential at infinite distance.)

(b) An electron, starting from rest at the center of the hole, moves out along the axis, experiencing no forces except repulsion by the charges on the disk. What velocity does it ultimately attain? (Electron mass $= 9.1 \cdot 10^{-31}$ kg.)

**2.56** *Energy of a disk* **

Use the result stated in Eq. (2.30) to show that the energy stored in the electric field of the charged disk described in Section 2.6 equals $(2/3\pi^2\epsilon_0)(Q^2/a)$. (*Hint:* Consider the work done in building the disk of charge out from zero radius to radius $a$ by adding successive rings of width $dr$.) Compare this with the energy required to build up a hollow spherical shell with radius $a$ and uniform charge $Q$.

**2.57** *Field near a disk* ****

(a) A disk with radius $R$ has uniform surface charge density $\sigma$. Consider a point $P$ a distance $\eta R$ from the center of the disk (where $0 \le \eta < 1$) and an infinitesimal distance away from the plane of the disk. Very close to the disk, the disk looks

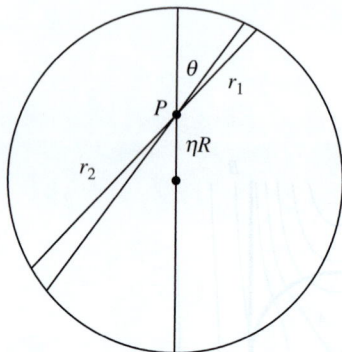

**Figure 2.50.**

essentially like an infinite plane, as far as the **E** component perpendicular to the disk is concerned. So we have $E_\perp = \sigma/2\epsilon_0$. Show that the **E** component parallel to the disk equals

$$E_\parallel = \frac{\sigma}{2\pi\epsilon_0} \int_0^{\pi/2} \ln\left(\frac{\sqrt{1 - \eta^2 \sin^2\theta} + \eta\cos\theta}{\sqrt{1 - \eta^2 \sin^2\theta} - \eta\cos\theta}\right) \cos\theta \, d\theta.$$

(2.121)

*Hint:* Use the technique from the second example in Section 2.6 (but now with $E$ instead of $\phi$) to find the sum of the fields from the two wedges shown in Fig. 2.50 (two diverging terms will cancel). The distances $r_1$ and $r_2$ shown can be obtained with the help of the law of cosines.

(b) Given $\eta$, the above integral can be evaluated numerically. However, if $\eta$ is very small or very close to 1, it is possible to make some analytic progress. Show that the leading-order dependence on $\eta$ in the limit $\eta \to 0$ is $E_\parallel = \sigma\eta/4\epsilon_0$. And show that the leading-order dependence on $\epsilon$ (where $\epsilon \equiv 1 - \eta$) in the limit $\epsilon \to 0$ is $E_\parallel = -(\sigma/2\pi\epsilon_0) \ln\epsilon$. You can verify these results numerically.

2.58  *Energy of a shell* *

A hollow spherical shell with radius $R$ has charge $Q$ uniformly distributed on it. Problem 1.32 presented two methods for calculating the potential energy of this system. Calculate the energy in a third way, by using Eq. (2.32).

2.59  *Energy of a cylinder* ***

Problem 1.24 and Exercise 1.83 presented two methods for calculating the energy per unit length stored in a cylinder with radius $a$ and uniform charge density $\rho$. Calculate the energy in a third way, by using Eq. (2.32). If you take the $\phi = 0$ point to be at infinity, you will obtain an infinite result. So instead take it to be at a given radius $R$ outside the cylinder. You will then be calculating the energy relative to the configuration where the charge is distributed over a cylinder with radius $R$. In terms of the total charge $\lambda$ per unit length in the final cylinder, show that the energy per unit length can be written as $(\lambda^2/4\pi\epsilon_0)(1/4 + \ln(R/a))$.

2.60  *Horizontal field lines* **

Using the $r = r_0 \sin^2\theta$ expression for the dipole **E** field lines in Fig. 2.19, find the locations where the curves are horizontal.

2.61  *Dipole field on the axes* **

A dipole is centered at the origin and has charges $q$ and $-q$ located at $z = \ell/2$ and $z = -\ell/2$, respectively. Find the electric field at position $r$ along the $z$ axis, and also at position $r$ along the $x$ axis (or anywhere at radius $r$ in the $xy$ plane). Do this by writing down the fields from the two charges and then adding them, making suitable approximations in the $r \gg \ell$ limit. Check that your answers agree with Eq. (2.36) when $\theta = 0$ and $\theta = \pi/2$.

2.62 *Square quadrupole* **

Consider a square quadrupole consisting of two adjacent dipoles oppositely oriented and placed side by side to form a square, as shown in Fig. 2.16. If the side length is $\ell$, find the electric field at a large distance $r$ along the diagonal containing the two positive charges. Be careful to take into account *all* quantities that are second order in $\ell/r$.

2.63 *Two-dimensional dipole* ***

Two parallel wires with uniform linear charge densities $\lambda$ and $-\lambda$ are separated by a distance $\ell$. Consider the electric field at points in a given plane perpendicular to the wires. The fields from the wires fall off like $1/r$. So at points far from the wires, we effectively have a 2D version of a dipole, where the wires act like point charges with $1/r$ fields instead of the usual $1/r^2$ Coulomb fields.

Repeat the process in Section 2.7 for this 2D dipole. That is, find $\phi(r, \theta)$ and $\mathbf{E}(r, \theta)$, and also find the shapes of the field-line and constant-potential curves. (The individual potentials relative to infinity diverge, so you will want to pick a local point as the reference point. Any choice will work, but you may as well pick the point midway between the wires.)

2.64 *Field lines near the equilibrium point* **

(a) Charges $4q$ and $-q$ are located at the points $(-2a, 0, 0)$ and $(-a, 0, 0)$, respectively. Write down the potential $\phi(x, y)$ for points in the $xy$ plane, and then use a Taylor expansion to find an approximate expression for $\phi$ near the origin, which you can quickly show is the equilibrium point. (You can set $a = 1$ to make things simpler.)

(b) Find the electric field at points near the origin. Then find the equations for the field lines near the origin by demanding that the slope $dy/dx$ of a curve at a given point equals the slope $E_y/E_x$ of the field at that point.

2.65 *A theorem on field lines* **

If you solved Exercise 2.64, you probably noted that the final result is the same as for Problem 2.18. This suggests a theorem. Consider two point charges with arbitrary values $q_1$ and $q_2$.

(a) First explain why there is exactly one point where $E = 0$ on the line containing the charges, except in a couple of special cases. (What are they?)

(b) Show that at the point where $E = 0$, the closeup views of the equipotentials and field lines always look like Figs. 12.39 and 12.40, independent of the values of $q_1$ and $q_2$. In other words, the constant-$\phi$ lines passing through the equilibrium point always have slope $\pm\sqrt{2}$. (You will want to look at the solution to Problem 2.18.)

2.66 *Equipotentials for two point charges* **

(a) Two point charges $Q$ are located at $(\pm R, 0, 0)$. Find the electric field at all points on the $z$ axis. For what value of $z$ is the field maximum?

(b) Make a rough sketch of the equipotential curves everywhere in space (or rather, everywhere in the $xz$ plane). Be sure to indicate what the curves look like very close to and very far from the charges, and how the transition from close to far occurs.

(c) There is a particular point on the $z$ axis (along with its negative) at which the equipotentials make the transition from concave up to concave down. In Problem 2.19 we saw that in the analogous setup with a ring, this point coincided with the point on the $z$ axis where the field was maximum. Does the same result hold here? Explain why the reasoning we used in Problem 2.19 (involving the divergence of $\mathbf{E}$) is still valid, or why it is now invalid.

2.67 *Product of $\rho$ and $\phi$* **

Consider a charge distribution $\rho_1(\mathbf{r})$ and the potential $\phi_1(\mathbf{r})$ due to it. Consider another charge distribution $\rho_2(\mathbf{r})$ and the potential $\phi_2(\mathbf{r})$ due to it. Both distributions have finite extent, but are otherwise arbitrary and need not have anything to do with each other. Show that $\int \rho_1\phi_2\, dv = \int \rho_2\phi_1\, dv$, where the integrals are taken over all space. Solve this in two different ways as follows.

(a) Consider the two collections of charges to be rigid objects that can be moved around. Start with them initially located very far apart, and then bring them together. How much work does this require? Imagine moving collection 1 toward collection 2, and then the other way around.

(b) Consider the integral $\int \mathbf{E}_1\mathbf{E}_2\, dv$ over all space, where $\mathbf{E}_1$ and $\mathbf{E}_2$ are the electric fields due to the two distributions. By using the vector identity $\nabla \cdot (\mathbf{E}_1\phi_2) = (\nabla \cdot \mathbf{E}_1)\phi_2 + \mathbf{E}_1 \cdot \nabla\phi_2$ (and similarly with the 1s and 2s switched), rewrite the integral $\int \mathbf{E}_1\mathbf{E}_2\, dv$ in two different ways.

2.68 *E and $\rho$ for a sphere* **

In the example in Section 2.9, we used spherical coordinates to verify the relation $\operatorname{div}\mathbf{E} = \rho/\epsilon_0$ for a sphere with radius $R$ and uniform density $\rho$. Verify this relation (both inside and outside the sphere) by working in Cartesian coordinates. You will first need to write out the Cartesian components of **E**.

2.69 *E and $\phi$ for a slab* **

A rectangular slab with uniform volume charge density $\rho$ has thickness $2\ell$ in the $x$ direction and infinite extent in the $y$ and $z$ directions. Let the $x$ coordinate be measured relative to the center plane of the slab. For values of $x$ both inside and outside the slab:

(a) find the electric field $E(x)$ (you can do this by considering the amount of charge on either side of $x$, or by using Gauss's law);

(b) find the potential $\phi(x)$, with $\phi$ taken to be zero at $x = 0$;

(c) verify that $\rho(x) = \epsilon_0 \nabla \cdot \mathbf{E}(x)$ and $\rho(x) = -\epsilon_0 \nabla^2 \phi(x)$.

2.70 *Triangular E* **

Find the charge density $\rho$ and potential $\phi$ associated with the electric field shown in Fig. 2.51. $E$ is independent of $y$ and $z$. Assume that $\phi = 0$ at $x = 0$.

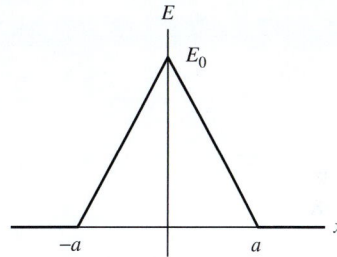

**Figure 2.51.**

2.71 *A one-dimensional charge distribution* **

Find (and make rough plots of) the electric field and charge distribution that go with the following potential: $\phi(x) = B(\ell^2 - x^2)$ for $|x| \le \ell$, and $\phi(x) = 0$ for $|x| > \ell$.

2.72 *A spherical charge distribution* ***

Find (and make rough plots of) the electric field and charge distribution that go with the following potential:

$$\phi = \begin{cases} \dfrac{\rho_0}{4\pi\epsilon_0}(x^2 + y^2 + z^2) & (\text{for } x^2 + y^2 + z^2 < a^2) \\[2ex] \dfrac{\rho_0}{4\pi\epsilon_0}\left(-a^2 + \dfrac{2a^3}{(x^2 + y^2 + z^2)^{1/2}}\right) & (\text{for } x^2 + y^2 + z^2 > a^2), \end{cases}$$

$$(2.122)$$

where $\rho_0$ is a quantity with the dimensions of volume charge density. Note that we are not assuming that $\phi = 0$ at infinity.

2.73 *Satisfying Laplace* *

Does the function $f(x, y) = x^2 + y^2$ satisfy the two-dimensional Laplace's equation? Does the function $g(x, y) = x^2 - y^2$? Sketch the latter function, calculate the gradient at the points $(x, y) = (0, 1)$, $(1, 0)$, $(0, -1)$, and $(-1, 0)$, and indicate by little arrows the directions in which these gradient vectors point.

2.74 *Oscillating exponential $\phi$* ***

A flat nonconducting sheet lies in the $xy$ plane. The only charges in the system are on this sheet. In the half-space above the sheet,

$z > 0$, the potential is $\phi = \phi_0 e^{-kz} \cos kx$, where $\phi_0$ and $k$ are constants.

(a) Verify that $\phi$ satisfies Laplace's equation in the space above the sheet.

(b) What do the electric field lines look like?

(c) Describe the charge distribution on the sheet.

**2.75** *Curls and divergences* *

Calculate the curl and the divergence of each of the following vector fields. If the curl turns out to be zero, try to discover a scalar function $\phi$ of which the vector field is the gradient.

(a) $\mathbf{F} = (x + y, -x + y, -2z)$;

(b) $\mathbf{G} = (2y, 2x + 3z, 3y)$;

(c) $\mathbf{H} = (x^2 - z^2, 2, 2xz)$.

**2.76** *Zero curl* *

By explicitly calculating the components of $\nabla \times \mathbf{E}$, show that the vector function specified in Exercise 2.31 is a possible electrostatic field. (Of course, if you worked that exercise, you have already proved it in another way by finding a scalar function of which it is the gradient.) Evaluate the divergence of this field.

**2.77** *Zero dipole curl* *

Verify that the curl of the dipole field in Eq. (2.36) is zero. We know that it must be zero, of course, because the field is the sum of the fields from two point charges, but demonstrate this here by explicitly calculating the curl, using the expression given in Eq. (F.3) in Appendix F.

**2.78** *Divergence of the curl* **

If $\mathbf{A}$ is any vector field with continuous derivatives, div (curl $\mathbf{A}$) = 0 or, using the "del" notation, $\nabla \cdot (\nabla \times \mathbf{A}) = 0$. We shall need this theorem later. The problem now is to prove it. Here are two different ways in which that can be done.

(a) (Uninspired straightforward calculation in a particular coordinate system.) Using the formula for $\nabla$ in Cartesian coordinates, work out the string of second partial derivatives that $\nabla \cdot (\nabla \times \mathbf{A})$ implies.

(b) (With the divergence theorem and Stokes' theorem, no coordinates are needed.) Consider the surface $S$ in Fig. 2.52, a balloon almost cut in two which is bounded by the closed curve $C$. Think about the line integral, over a curve like $C$, of any vector field. Then invoke Stokes and Gauss with suitable arguments. (The reasoning also works if the curve $C$ is a very tiny loop on the surface.)

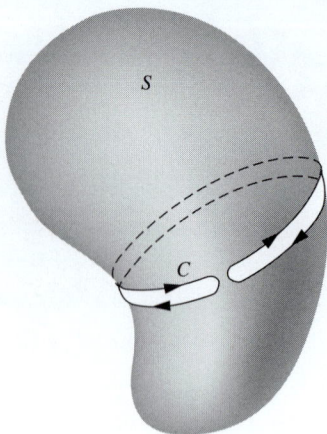

**Figure 2.52.**

2.79 *Vectors and squrl*  ∗

To show that it takes more than direction and magnitude to make a vector, let's try to define a vector, which we'll name squrl **F**, by a relation like Eq. (2.80) but with the right-hand side squared:

$$(\text{squrl } \mathbf{F}) \cdot \hat{\mathbf{n}} = \left[ \lim_{a_i \to 0} \frac{\int_{C_i} \mathbf{F} \cdot d\mathbf{s}}{a_i} \right]^2 . \qquad (2.123)$$

Prove that this does *not* define a vector. (*Hint:* Consider reversing the direction of $\hat{\mathbf{n}}$.)

# 3

# Electric fields around conductors

**Overview** In the first two chapters, we were concerned with the electric field and potential due to charges whose positions were fixed and known. We will now study the field and potential due to charges on *conductors,* where the charges are free to move around. This is a more difficult task, because on one hand we need to know the field to determine the positions of the charges, but on the other hand we need to know the positions of the charges to determine the field. Fortunately, there are some facts and theorems that make this tractable, and indeed in some cases trivial. The most important fact is that in an electrostatic setup, the electric field inside the material of a conductor is zero. Equivalently, all points in a given conductor have the same potential. This leads to the somewhat surprising effect called electrical *shielding;* the electric field inside an empty conducting shell is zero, independent of whatever arbitrary charge distribution exists outside. We prove the very helpful *uniqueness theorem,* which states that, given the values of the potential $\phi$ on the surfaces of a set of conductors, the solution for $\phi$ throughout space is unique. This theorem often makes things so easy that you may wonder if you're actually cheating. A byproduct of the theorem is the topic of *image charges,* which allow us to construct the electric field near conductors in certain cases. We define the *capacitance coefficient(s)* of a set of conductors; these tell us how much charge resides on a conductor at a given potential. *Capacitors* are a fundamental circuit element, as we will see in Chapter 8. Finally, we discuss the energy stored in a capacitor.

## 3.1 Conductors and insulators

The earliest experimenters with electricity observed that substances differed in their power to hold the "Electrick Vertue." Some materials could be easily electrified by friction and maintained in an electrified state; others, it seemed, could not be electrified that way, or did not hold the Vertue if they acquired it. Experimenters of the early eighteenth century compiled lists in which substances were classified as "electricks" or "nonelectricks." Around 1730, the important experiments of Stephen Gray in England showed that the Electrick Vertue could be conducted from one body to another by horizontal string, over distances of several hundred feet, provided that the string was itself supported from above by silk threads.[1] Once this distinction between conduction and nonconduction had been grasped, the electricians of the day found that even a nonelectrick could be highly electrified if it were supported on glass or suspended by silk threads. A spectacular conclusion of one of the popular electric exhibitions of the time was likely to be the electrification of a boy suspended by many silk threads from the rafters; his hair stood on end and sparks could be drawn from the tip of his nose.

After the work of Gray and his contemporaries, the elaborate lists of electricks and nonelectricks were seen to be, on the whole, a division of materials into electrical *insulators* and electrical *conductors*. This distinction is still one of the most striking and extreme contrasts that nature exhibits. Common good conductors like ordinary metals differ in their electrical conductivity from common insulators like glass and plastics by factors on the order of $10^{20}$. To express it in a way the eighteenth-century experimenters like Gray or Benjamin Franklin would have understood, a metal globe on a metal post can lose its electrification in a millionth of a second; a metal globe on a glass post can hold its Vertue for many years. (To make good on the last assertion we would need to take some precautions beyond the capability of an eighteenth-century laboratory. Can you suggest some of them?)

The electrical difference between a good conductor and a good insulator is as vast as the mechanical difference between a liquid and a solid. That is not entirely accidental. Both properties depend on the *mobility* of atomic particles: in the electrical case, the mobility of the carriers of charge, electrons or ions; in the case of the mechanical properties, the mobility of the atoms or molecules that make up the structure of the material. To carry the analogy a bit further, we know of substances whose fluidity is intermediate between that of a solid and that of a liquid – substances such as tar or ice cream. Indeed some substances – glass is a good example – change gradually and continuously from a mobile

---

[1] The "pack-thread" Gray used for his string was doubtless a rather poor conductor compared with metal wire, but good enough for transferring charge in electrostatic experiments. Gray found, too, that fine copper wire was a conductor, but mostly he used the pack-thread for the longer distances.

liquid to a very permanent and rigid solid with a few hundred degrees' lowering of the temperature. In electrical conductivity, too, we find examples over the whole wide range from good conductor to good insulator, and some substances that can change conductivity over nearly as wide a range, depending on conditions such as their temperature. A fascinating and useful class of materials called semiconductors, which we shall meet in Chapter 4, have this property.

Whether we call a material solid or liquid sometimes depends on the time scale, and perhaps also on the scale of distances involved. Natural asphalt seems solid enough if you hold a chunk in your hand. Viewed geologically, it is a liquid, welling up from underground deposits and even forming lakes. We may expect that, for somewhat similar reasons, whether a material is to be regarded as an electrical insulator or a conductor will depend on the time scale of the phenomenon we are interested in.

## 3.2 Conductors in the electrostatic field

We shall look first at electrostatic systems involving conductors. That is, we shall be interested in the *stationary* state of charge and electric field that prevails after all redistributions of charge have taken place in the conductors. Any insulators present are assumed to be perfect insulators. As we have already mentioned, quite ordinary insulators come remarkably close to this idealization, so the systems we shall discuss are not too artificial. In fact, the air around us is an extremely good insulator. The systems we have in mind might be typified by some such example as this: bring in two charged metal spheres, insulated from one another and from everything else. Fix them in positions relatively near one another. What is the resulting electric field in the whole space surrounding and between the spheres, and how is the charge that is on each sphere distributed? We begin with a more general question: after the charges have become stationary, what can we say about the electric field inside conducting matter?

In the static situation there is no further motion of charge. You might be tempted to say that the electric field must then be zero within conducting material. You might argue that, if the field were *not* zero, the mobile charge carriers would experience a force and would be thereby set in motion, and thus we would not have a static situation after all. Such an argument overlooks the possibility of *other* forces that may be acting on the charge carriers, and that would have to be counterbalanced by an electric force to bring about a stationary state. To remind ourselves that it is physically possible to have other than electrical forces acting on the charge carriers, we need only think of gravity. A positive ion has weight; it experiences a steady force in a gravitational field, and so does an electron; also, the forces they experience are not equal. This is a rather absurd example. We know that gravitational forces are utterly negligible on an atomic scale.

There are other forces at work, however, which we may very loosely call "chemical." In a battery and in many, many other theaters of chemical reaction, including the living cell, charge carriers sometimes move *against* the general electric field; they do so because a reaction may thereby take place that yields more energy than it costs to buck the field. One hesitates to call these forces nonelectrical, knowing as we do that the structure of atoms and molecules and the forces between them can be explained in terms of Coulomb's law and quantum mechanics. Still, from the viewpoint of our *classical* theory of electricity, they must be treated as quite extraneous. Certainly they behave very differently from the inverse-square force upon which our theory is based. The general necessity for forces that are in this sense nonelectrical was already foreshadowed by our discovery in Chapter 2 that inverse-square forces alone cannot make a stable, static structure (see Earnshaw's theorem in Section 2.12).

The point is simply this: we must be prepared to find, in some cases, unbalanced, non-Coulomb forces acting on charge carriers inside a conducting medium. When that happens, the electrostatic situation is attained when there *is* a finite electric field in the conductor that just offsets the influence of the other forces, whatever they may be.

Having issued this warning, however, we turn at once to the very familiar and important case in which there is no such force to worry about, the case of a homogeneous, isotropic conducting material. In the interior of such a conductor, in the static case, we can state confidently that the electric field must be zero.[2] If it weren't, charges would have to move. It follows that all regions inside the conductor, including all points just below its surface, must be at the same potential. Outside the conductor, the electric field is not zero. The surface of the conductor must be an equipotential surface of this field.

The vanishing of the electric field in the interior of a conductor implies that the volume charge density $\rho$ also vanishes in the interior. This follows from Gauss's law, $\nabla \cdot \mathbf{E} = \rho/\epsilon_0$. Since the field is identically zero inside the conductor, its divergence, and hence $\rho$, are also identically zero. Of course, as with the field, this holds only in an average sense. The charge density at the location of, say, a proton is most certainly not zero.

Imagine that we could change a material from insulator to conductor at will. (It's not impossible – glass becomes conducting when heated; any gas can be ionized by x-rays.) Figure 3.1(a) shows an uncharged nonconductor in the electric field produced by two fixed layers of charge.

---

[2] In speaking of the electric field inside matter, we mean an average field, averaged over a region large compared with the details of the atomic structure. We know, of course, that very strong fields exist in all matter, including the good conductors, if we search on a small scale near an atomic nucleus. The nuclear electric field does not contribute to the average field in matter, ordinarily, because it points in one direction on one side of a nucleus and in the opposite direction on the other side. Just how this average field ought to be defined, and how it could be measured, are questions we consider in Chapter 10.

(a)

(b)

(c)

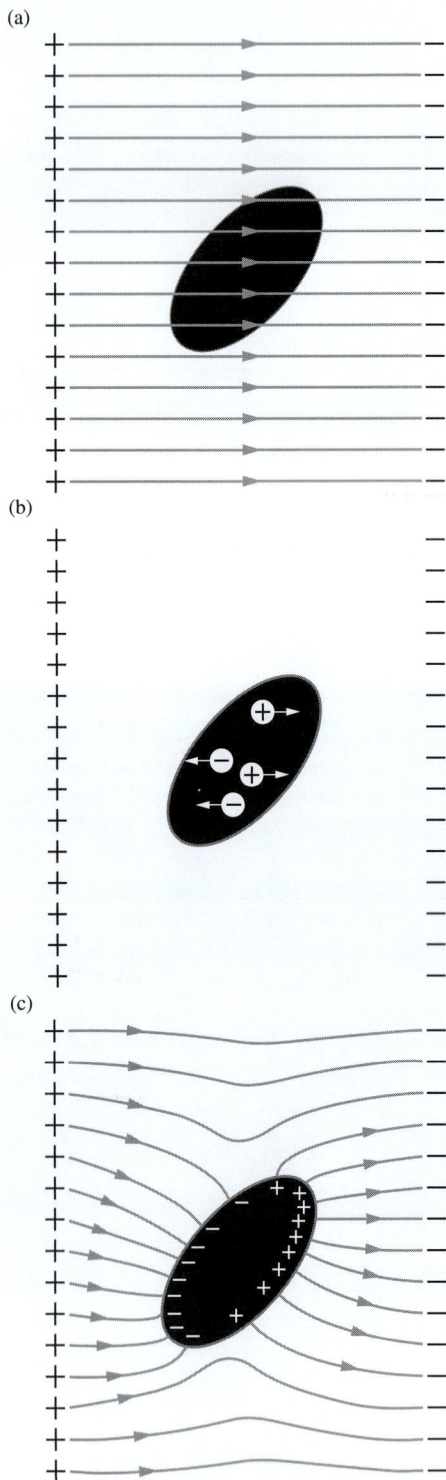

The electric field is the same inside the body as outside. (A dense body such as glass would actually distort the field, an effect we will study in Chapter 10, but that is not important here.) Now, in one way or another, let mobile charges (or *ions*) be created, making the body a conductor. Positive ions are drawn in one direction by the field, negative ions in the opposite direction, as indicated in Fig. 3.1(b). They can go no farther than the surface of the conductor. Piling up there, they begin themselves to create an electric field inside the body which tends to *cancel* the original field. And in fact the movement goes on until that original field is *precisely* canceled. The final distribution of charge at the surface, shown in Fig. 3.1(c), is such that its field and the field of the fixed external sources combine to give *zero* electric field in the interior of the conductor. Because this "automatically" happens in every conductor, it is really only the surface of a conductor that we need to consider when we are concerned with the external fields.

With this in mind, let us see what can be said about a system of conductors, variously charged, in otherwise empty space. In Fig. 3.2 we see some objects. Think of them, if you like, as solid pieces of metal. They are prevented from moving by invisible insulators – perhaps by Stephen Gray's silk threads. The total charge of each object, by which we mean the net excess of positive over negative charge, is fixed because there is no way for charge to leak on or off. We denote it by $Q_k$, for the $k$th conductor. Each object can also be characterized by a particular value $\phi_k$ of the electric potential function $\phi$. We say that conductor 2 is "at the potential $\phi_2$." With a system like the one shown, where no physical objects stretch out to infinity, it is usually convenient to assign the potential zero to points infinitely far away. In that case $\phi_2$ is the work per unit charge required to bring an infinitesimal test charge in from infinity and put it anywhere on conductor 2. (Note, by the way, that this is just the kind of system in which the test charge needs to be kept small, a point raised in Section 1.7.)

Because the surface of a conductor in Fig. 3.2 is necessarily a surface of constant potential, the electric field, which is $-\text{grad}\,\phi$, must be *perpendicular* to the surface at every point on the surface. Proceeding from the interior of the conductor outward, we find at the surface an abrupt change in the electric field; **E** is not zero outside the surface, and it is zero inside. The discontinuity in **E** is accounted for by the presence of a surface charge, of density $\sigma$, which we can relate directly to **E** by Gauss's law. We can use a flat box enclosing a patch of surface (Fig. 3.3), similar to the cylinder we used when considering the infinite

**Figure 3.1.**
The object in (a) is a neutral nonconductor. The charges in it, both positive and negative, are immobile. In (b) the charges have been released and begin to move. They will move until the final condition, shown in (c), is attained.

flat sheet in Section 1.13. However, here there is *no* flux through the "bottom" of the box, which lies inside the conductor, so we conclude that $E_n = \sigma/\epsilon_0$ (instead of the $\sigma/2\epsilon_0$ we found in Eq. (1.40)), where $E_n$ is the component of electric field normal to the surface. As we have already seen, there *is* no other component in this case, the field being always perpendicular to the surface. The surface charge must account for the whole charge $Q_k$. That is, the surface integral of $\sigma$ over the whole conductor must equal $Q_k$. In summary, we can make the following statements about *any* such system of conductors, whatever their shape and arrangement:

(1) $\mathbf{E} = 0$ inside the material of a conductor;

(2) $\rho = 0$ inside the material of a conductor;

(3) $\phi = \phi_k$ at all points inside the material and on the surface of the $k$th conductor;

(4) At any point just outside the conductor, $\mathbf{E}$ is perpendicular to the surface, and $E = \sigma/\epsilon_0$, where $\sigma$ is the local density of surface charge;

(5) $Q_k = \int_{S_k} \sigma\, da = \epsilon_0 \int_{S_k} \mathbf{E} \cdot d\mathbf{a}$.

$\mathbf{E}$ is the total field arising from *all* the charges in the system, near and far, of which the surface charge is only a part. The surface charge on a conductor is obliged to "readjust itself" until relation (4) is fulfilled. That the conductor presents a special case, in contrast to other surface charge distributions, is brought out by the comparison in Fig. 3.4.

---

**Example (A spherically symmetric field)**   A point charge $q$ is located at an arbitrary position inside a neutral conducting spherical shell. Explain why the electric field outside the shell is the same as the spherically symmetric field due to a charge $q$ located at the *center* of the shell (with the shell removed, although the point is that this doesn't matter).

Solution   The spherical shell has an inner surface and an outer surface. Between these surfaces (inside the material of the conductor) we know that the electric field is zero. So if we draw a Gaussian surface that lies entirely inside the material, signified by the dashed line in Fig. 3.5, there is zero flux through it, so it must enclose zero charge. The charge on the inner surface of the shell is therefore $-q$. This leaves $+q$ for the outer surface. The charge $-q$ on the inner surface won't be uniformly distributed unless the point charge $q$ is located at the center, but that doesn't concern us.

The only question is how the $+q$ charge is distributed over the outer surface. Imagine that we have removed this $+q$ charge, so that we have only the point charge $q$ and the inner-surface charge $-q$. The combination of these charges produces zero field in the material of the conductor. It also produces zero field outside the conductor. This is true because field lines must have at least one end on a charge (the other end may be at infinity); they can't form closed loops because the electric field has zero curl. However, in the present setup, external field lines have no possibility of touching any of the charges on the inside, because the lines can't pass through the material of the conductor to reach them, since the field is zero there. Therefore there can be no field lines outside the conductor.

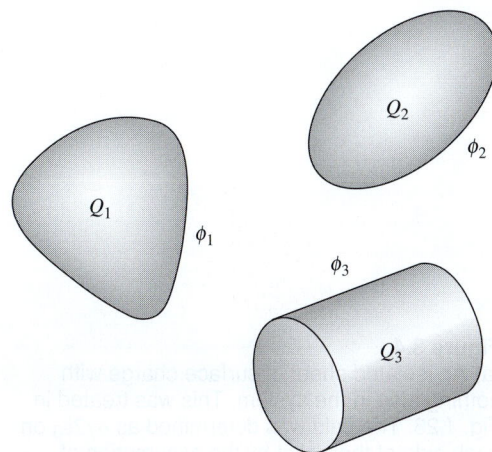

**Figure 3.2.**
A system of three conductors: $Q_1$ is the charge on conductor 1, $\phi_1$ is its potential, etc.

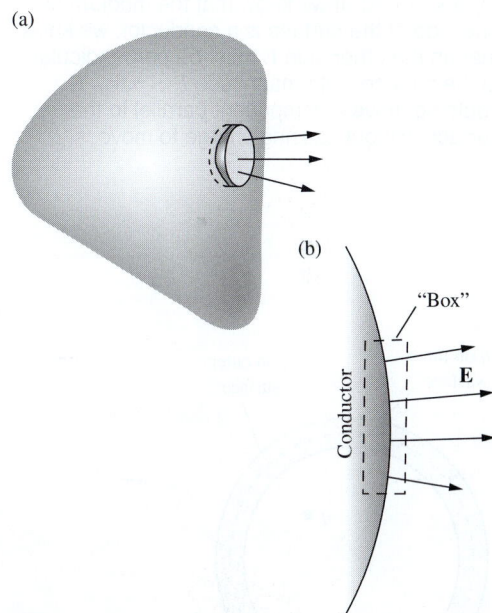

**Figure 3.3.**
(a) Gauss's law relates the electric field strength at the surface of a conductor to the density of surface charge; $E = \sigma/\epsilon_0$. (b) Cross section through surface of conductor and box.

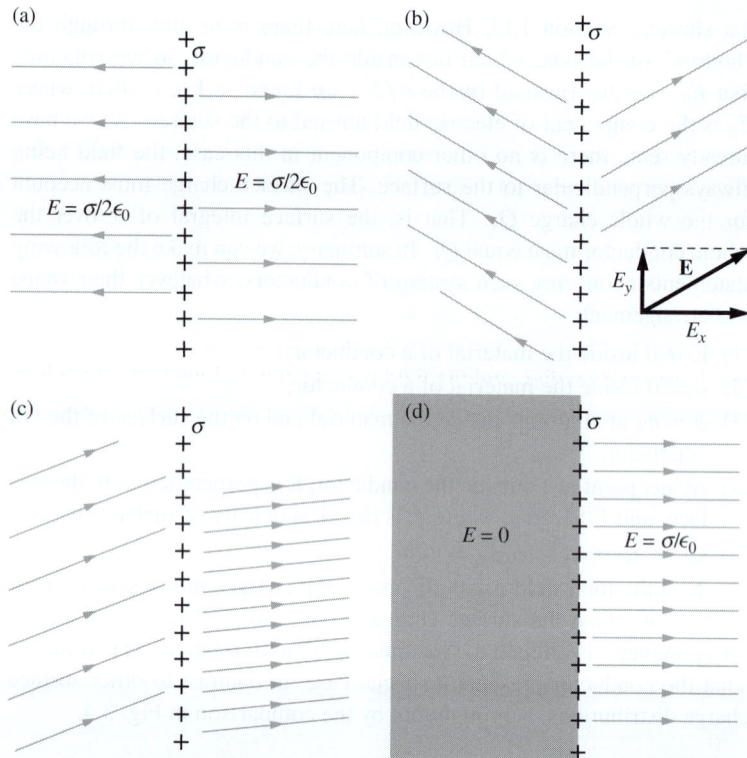

(a)

$E = \sigma/2\epsilon_0$

$E = \sigma/2\epsilon_0$

(b)

$E_y$    **E**

$E_x$

**Figure 3.4.**
(a) An isolated sheet of surface charge with nothing else in the system. This was treated in Fig. 1.26. The field was determined as $\sigma/2\epsilon_0$ on each side of the sheet by the assumption of symmetry. (b) If there are other charges in the system, we can say only that the change in $E_x$ at the surface must be $\sigma/\epsilon_0$, with zero change in $E_y$. Many fields other than the field of (a) above could have this property. Two such are shown in (b) and (c). (d) If we know that the medium on one side of the surface is a conductor, we know that on the other side **E** must be perpendicular to the surface, with magnitude $E = \sigma/\epsilon_0$. **E** could not have a component parallel to the surface without causing charge to move.

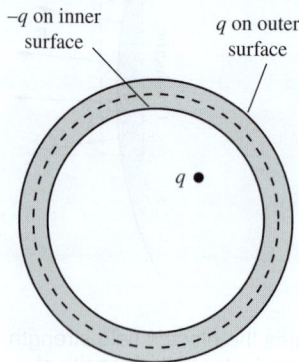

(c)

(d)

$E = 0$

$E = \sigma/\epsilon_0$

$-q$ on inner surface

$q$ on outer surface

$q$

**Figure 3.5.**
A Gaussian surface (dashed line) inside the material of a conducting spherical shell.

If we gradually add back on the outer-surface charge $+q$, it will distribute itself in a spherically symmetric manner because it feels no field from the other charges. Furthermore, due to this spherical symmetry, the outer-surface charge will produce no field at the other charges (because a uniform shell produces zero field in its interior), so we don't have to worry about any shifting of these charges.

Since the combination of the point charge and the inner-surface charge produces no field outside the shell, the external field is due only to the spherically symmetric outer-surface charge. By Gauss's law, the external field is therefore radial (with respect to the center of the shell and *not* the point charge $q$) and has magnitude $q/4\pi\epsilon_0 r^2$. Note that the shape of the inner surface was irrelevant in the above reasoning. If we have the setup shown in Fig. 3.6, the external field is still spherically symmetric with magnitude $q/4\pi\epsilon_0 r^2$.

More generally, if the neutral conducting shell takes an odd nonspherical shape, we can't say that the external field is spherically symmetric. But we *can* say that the external field, whatever it may be, is *independent of the location* of the point charge $q$ inside. Whatever the location, the external field equals the field in a system where the point charge $q$ is absent and where we instead dump a total charge $q$ on the shell (which will distribute itself in a particular manner).[3]

---

[3] There is a slight subtlety that arises in this case, namely the effect of the outer-surface charge on the inner-surface charge. It turns out that, as with the sphere, there is no effect. We'll see why in Section 3.3.

Figure 3.7 shows the field and charge distribution for a simple system like the one mentioned at the beginning of this section. There are two conducting spheres, a sphere of unit radius carrying a total charge of +1 unit, the other a somewhat larger sphere with total charge zero. Observe that the surface charge density is not uniform over either of the conductors. The sphere on the right, with total charge zero, has a negative surface charge density in the region that faces the other sphere, and a positive surface charge on the rearward portion of its surface. The dashed curves in Fig. 3.7 indicate the equipotential surfaces or, rather, their intersection with the plane of the figure. If we were to go a long way out, we would find the equipotential surfaces becoming nearly spherical and the field lines nearly radial, and the field would begin to look very much like that of a point charge of magnitude 1 and positive, which is the net charge on the entire system.

Figure 3.7 illustrates, at least qualitatively, all the features we anticipated, but we have an additional reason for showing it. Simple as the system is, the exact mathematical solution for this case cannot be obtained

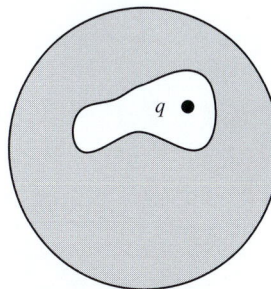

**Figure 3.6.**
The external field is radial even if the cavity takes an odd shape.

**Figure 3.7.**
The electric field around two spherical conductors, one with total charge +1, and one with total charge zero. Dashed curves are intersections of equipotential surfaces with the plane of the figure. Zero potential is at infinity.

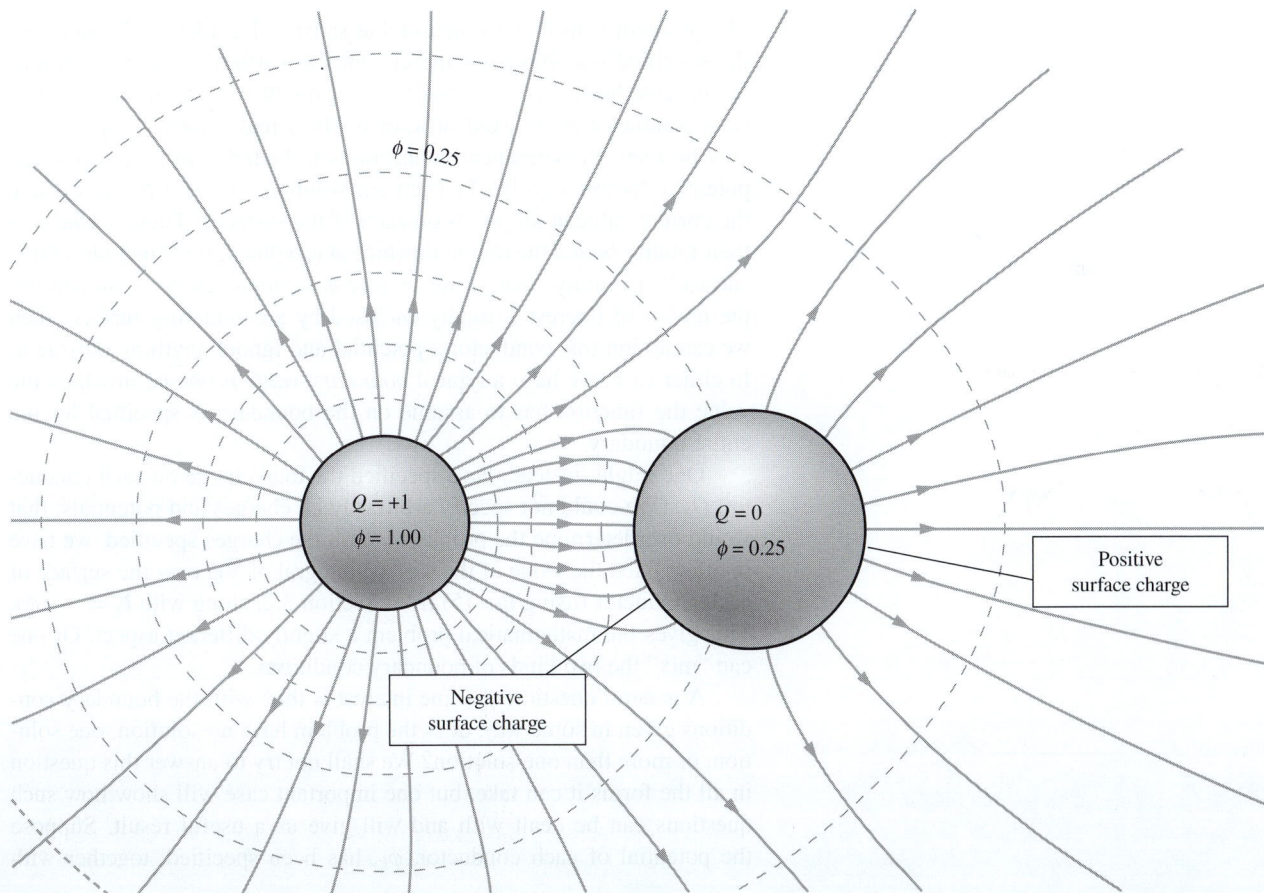

$\phi = 0.25$

$Q = +1$
$\phi = 1.00$

$Q = 0$
$\phi = 0.25$

Positive surface charge

Negative surface charge

in a straightforward way. Our Fig. 3.7 was constructed from an approximate solution. In fact, the number of three-dimensional geometrical
arrangements of conductors that permit a mathematical solution in closed
form is lamentably small. One does not learn much physics by concentrating on the solution of the few neatly soluble examples. Let us instead
try to understand the general nature of the mathematical problem such
a system presents.

## 3.3 The general electrostatic problem and the uniqueness theorem

We can state the problem in terms of the potential function $\phi$, for if
$\phi$ can be found, we can at once get $\mathbf{E}$ from it. Everywhere outside the
conductors, $\phi$ has to satisfy the partial differential equation we met in
Section 2.12, Laplace's equation: $\nabla^2 \phi = 0$. Written out in Cartesian
coordinates, Laplace's equation reads

$$\frac{\partial^2 \phi}{\partial x^2} + \frac{\partial^2 \phi}{\partial y^2} + \frac{\partial^2 \phi}{\partial z^2} = 0. \tag{3.1}$$

The problem is to find a function that satisfies Eq. (3.1) and also meets
the specified conditions on the conducting surfaces. These conditions
might have been set in various ways. It might be that the potential of
each conductor $\phi_k$ is fixed or known. (In a real system the potentials
may be fixed by permanent connections to batteries or other constant-
potential "power supplies.") Then our solution $\phi(x, y, z)$ has to assume
the correct value at all points on each of the surfaces. These surfaces in
their totality *bound* the region in which $\phi$ is defined, if we include a large
surface "at infinity," where we require $\phi$ to approach zero. Sometimes
the region of interest is totally enclosed by a conducting surface; then
we can assign this conductor a potential and ignore anything outside it.
In either case, we have a typical *boundary-value problem,* in which the
value the function has to assume on the boundary is specified for the
entire boundary.

One might, instead, have specified the total charge on each conductor, $Q_k$. (We could not specify arbitrarily all charges and potentials; that
would overdetermine the problem.) With the charges specified, we have
in effect fixed the value of the surface integral of $\nabla \phi$ over the surface of
each conductor (using fact (5) from Section 3.2, along with $\mathbf{E} = -\nabla \phi$).
This gives the mathematical problem a slightly different aspect. Or one
can "mix" the two kinds of boundary conditions.

A general question of some interest is this: with the boundary conditions given in some way, does the problem have no solution, one solution, or more than one solution? We shall not try to answer this question
in all the forms it can take, but one important case will show how such
questions can be dealt with and will give us a useful result. Suppose
the potential of each conductor, $\phi_k$, has been specified, together with

the requirement that $\phi$ approach zero at infinite distance, or on a conductor that encloses the system. We shall prove that this boundary-value problem has no more than one solution. It seems obvious, as a matter of physics, that it has *a* solution, for if we should actually arrange the conductors in the prescribed manner, connecting them by infinitesimal wires to the proper potentials, the system would have to settle down in *some* state. However, it is quite a different matter to prove mathematically that a solution always exists, and we shall not attempt it. Instead, we shall prove the following theorem.

**Theorem 3.1** *(Uniqueness theorem) Assuming that there <u>is</u> a solution $\phi(x, y, z)$ for a given set of conductors with potentials $\phi_k$, this solution must be unique.*

*Proof* The argument, which is typical of proofs of this sort, runs as follows. Assume there is another function $\psi(x, y, z)$ that is also a solution meeting the same boundary conditions. Now Laplace's equation is *linear*. That is, if $\phi$ and $\psi$ satisfy Eq. (3.1), then so does $\phi + \psi$ or any linear combination such as $c_1\phi + c_2\psi$, where $c_1$ and $c_2$ are constants. In particular, the difference between our two solutions, $\phi - \psi$, must satisfy Eq. (3.1). Call this function $W$:

$$W(x, y, z) \equiv \phi(x, y, z) - \psi(x, y, z). \tag{3.2}$$

Of course, $W$ does *not* satisfy the boundary conditions. In fact, at the surface of every conductor $W$ is zero, because $\phi$ and $\psi$ take on the same value, $\phi_k$, at the surface of a conductor $k$. Thus $W$ is a solution of *another* electrostatic problem, one with the same conductors but with all conductors held at zero potential.

We can now assert that if $W$ is zero on all the conductors, then $W$ must be zero at all points in space. For if it is not, it must have either a maximum or a minimum somewhere – remember that $W$ is zero at infinity as well as on all the conducting boundaries. If $W$ has an extremum at some point $P$, consider a sphere centered on that point. As we saw in Section 2.12, the average over a sphere of a function that satisfies Laplace's equation is equal to its value at the center. This could not be true if the center is a maximum or minimum. Thus $W$ cannot have a maximum or minimum;[4] it must therefore be zero everywhere. It follows that $\psi = \phi$ everywhere, that is, there can be only *one* solution of Eq. (3.1) that satisfies the prescribed boundary conditions. $\square$

In proving this theorem, we assumed that $\phi$ and $\psi$ satisfied Laplace's equation. That is, we assumed that the region outside the conductors was empty of charge. However, the uniqueness theorem actually holds even if

---

[4] If you want to demonstrate this without invoking the "average over a sphere" fact, you can use the related reasoning involving Gauss's law: if the potential at $P$ is a maximum (or minimum), then **E** must point outward (or inward) everywhere around $P$. This implies a net flux through a small sphere surrounding $P$, contradicting the fact that there are no charges enclosed.

there are charges present, provided that these charges are fixed in place. These charges could come in the form of point charges or a continuous distribution. The proof for this more general case is essentially the same. In the above reasoning, you will note that we never used the fact that $\phi$ and $\psi$ satisfied Laplace's equation, but rather only that their *difference* W did. So if we instead start with the more general Poisson's equations, $\nabla^2 \phi = -\rho/\epsilon_0$ and $\nabla^2 \psi = -\rho/\epsilon_0$, where the *same* $\rho$ appears in both of these equations, then we can take their difference to obtain $\nabla^2 W = 0$. That is, W satisfies Laplace's equation. The proof therefore proceeds exactly as above, and we again obtain $\phi = \psi$.

As a quick corollary to the uniqueness theorem, we can demonstrate a remarkable fact as follows.

**Corollary 3.2** *In the space inside a hollow conductor of any shape whatsoever, if that space itself is empty of charge, the electric field is zero.*

*Proof*  The potential function inside the conductor, $\phi(x, y, z)$, must satisfy Laplace's equation. The entire boundary of this region, namely the conductor, is an equipotential, so we have $\phi = \phi_0$, a constant everywhere on the boundary. One solution is obviously $\phi = \phi_0$ throughout the volume. But there can be only one solution, according to the above uniqueness theorem, so this is it. And then "$\phi = $ constant" implies $\mathbf{E} = 0$, because $\mathbf{E} = -\nabla\phi$. $\qquad\square$

This corollary is true whatever the field may be outside the conductor. We are already familiar with the fact that the field is zero inside an isolated uniform spherical shell of charge, just as the gravitational field inside the shell of a hollow spherical mass is zero. The corollary we just proved is, in a way, more surprising. Consider the closed metal box shown partly cut away in Fig. 3.8. There are charges in the neighborhood of the box, and the external field is approximately as depicted. There is a highly nonuniform distribution of charge over the surface of the box. Now the field everywhere in space, *including the interior of the box,* is the sum of the field of this charge distribution and the fields of the external sources. It seems hardly credible that the surface charge has so cleverly arranged itself on the box that its field precisely *cancels* the field of the external sources at every point inside the box. Yet this must indeed be what has happened, in view of the above proof.

As surprising as this may seem for a hollow conductor, it is really no more surprising than the fact that the charges on the surface of a *solid* conductor arrange themselves so that the electric field is zero inside the material of the conductor (which we know is the case, otherwise charges in the interior would move). These two setups are related because the interior of the solid conductor is neutral (since $\nabla \cdot \mathbf{E} = \rho/\epsilon_0$, and $\mathbf{E}$ is identically zero). So if we remove this neutral material from the solid conductor (a process that can't change the electric field anywhere,

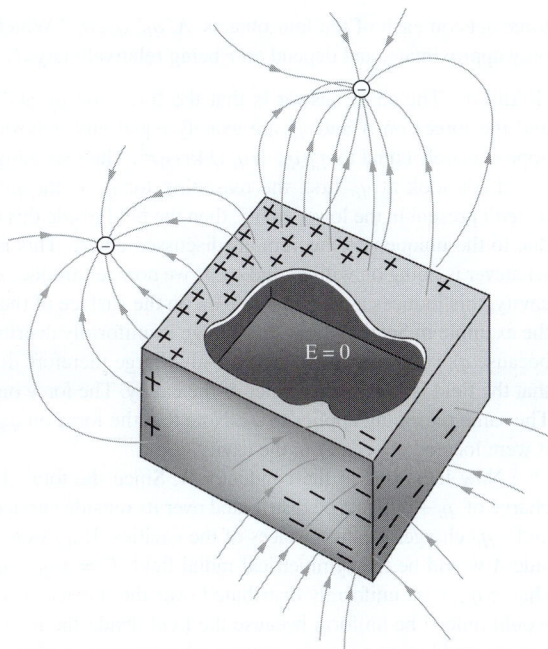

**Figure 3.8.**
The field is zero everywhere inside a closed conducting box.

because we aren't moving any particles with net charge), then we end up with a hollow conductor with zero field inside.

The corollary is also consistent with what we know about field lines. If there were field lines inside the shell, they would have to start at one point on the shell and end at another (there can't be any closed loops because curl $\mathbf{E} = 0$). But this would imply a nonzero potential difference between these two points on the shell, contradicting the fact that all points on the shell have the same potential. Therefore there can be no field lines inside the shell.

The absence of electric field inside a conducting enclosure is useful, as well as theoretically interesting. It is the basis for electrical shielding. For most practical purposes the enclosure does not need to be completely tight. If the walls are perforated with small holes, or made of metallic screen, the field inside will be extremely weak except in the immediate vicinity of a hole. A metal pipe with open ends, if it is a few diameters long, very effectively shields the space inside that is not close to either end. We are considering only static fields of course, but for slowly varying electric fields these remarks still hold. (A rapidly varying field can become a wave that travels through the pipe. *Rapidly* means here "in less time than light takes to travel a pipe diameter.")

**Example (Charges in cavities)** A spherical conductor $A$ contains two spherical cavities. The total charge on the conductor itself is zero. However, there is a point charge $q_b$ at the center of one cavity and $q_c$ at the center of the other, as shown in Fig. 3.9. A considerable distance $r$ away is another charge $q_d$. What

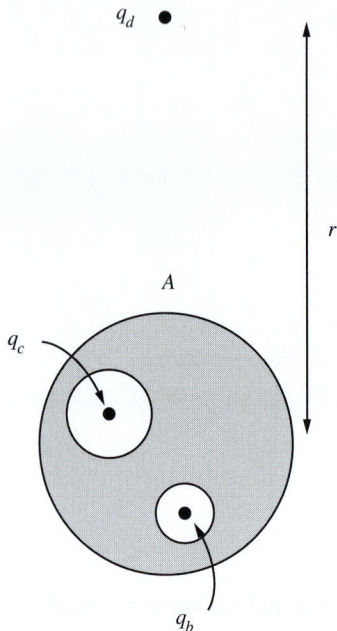

**Figure 3.9.**
Point charges are located at the centers of spherical cavities inside a neutral spherical conductor. Another point charge is located far away.

force acts on each of the four objects, $A$, $q_b$, $q_c$, $q_d$? Which answers, if any, are only approximate, and depend on $r$ being relatively large?

**Solution**   The short answer is that the forces on $q_b$ and $q_c$ are exactly zero, and the forces on $A$ and $q_d$ are exactly equal and opposite, with a magnitude approximately equal to $q_d(q_b + q_c)/4\pi\epsilon_0 r^2$. The reasoning is as follows.

Let's look at $q_b$ first; the reasoning for $q_c$ is the same. If the charge $q_b$ *weren't* present in the lower cavity, then the field inside this cavity would be zero, due to the uniqueness theorem, as discussed above. This fact is independent of whatever is going on with $q_c$ and $q_d$. If we now reintroduce $q_b$ at the center of the cavity, this induces a total charge $-q_b$ on the surface of the cavity (as we saw in the example in Section 3.2). This charge is uniformly distributed over the surface because $q_b$ is located at the center. This charge therefore doesn't change the fact that the field is zero at the center of the cavity. The force on $q_b$ is therefore zero. The same reasoning applies to $q_c$. Note that the force on $q_b$ would *not* be zero if it were located off-center in the cavity.

Now let's look at the conductor $A$. Since the total charge on $A$ is zero, a charge of $q_b + q_c$ must be distributed over its outside surface, to balance the $-q_b$ and $-q_c$ charges on the surfaces of the cavities. If $q_d$ were absent, the field outside $A$ would be the symmetrical radial field, $E = (q_b + q_c)/4\pi\epsilon_0 r^2$, with the charge $q_b + q_c$ uniformly distributed over the outside surface. The distribution would indeed be uniform because the field inside the material of the conductor is zero, and because we are assuming that there is no charge external to the conductor. The setup is therefore spherically symmetric, as far as the outside surface of the conductor is concerned. (Any effect of the interior charges on the outside surface charge can be felt only through the field. And since the field is zero just inside the outside surface, there is therefore no effect.)

If we now reintroduce the charge $q_d$, its influence will slightly alter the distribution of charge on the outside surface of $A$, but without affecting the total amount. If $q_d$ is positive, then negative charge will be drawn toward the near side of $A$, or equivalently positive charge will be pushed to the far side. Hence for large $r$, the force on $q_d$ will be approximately equal to $q_d(q_b + q_c)/4\pi\epsilon_0 r^2$, but it will be slightly more attractive than this; you can check that this is true for either sign of $q_d(q_b + q_c)$. The force on $A$ must be exactly equal and opposite to the force on $q_d$.

The *exact* value of the force on $q_d$ is the sum of the force just given, $q_d(q_b + q_c)/4\pi\epsilon_0 r^2$, and the force that would act on $q_d$ if the total charge *on and within* $A$ were zero (it is $q_b + q_c$ here). This latter force (which is always attractive) can be determined by applying the "image charge" technique that we will learn about in the following section; see Problem 3.13.

## 3.4 Image charges

About the simplest system in which the mobility of the charges in the conductor makes itself evident is the point charge near a conducting plane. Suppose the $xy$ plane is the surface of a conductor extending out to infinity. Let's assign this plane the potential zero. Now bring in a positive charge $Q$ and locate it $h$ above the plane on the $z$ axis, as in Fig. 3.10(a). What sort of field and charge distribution can we expect? We expect the positive charge $Q$ to attract negative charge, but we hardly expect the negative charge to pile up in an infinitely dense concentration at the foot of the perpendicular from $Q$. (Why not?) Also, we remember that the electric field is always perpendicular to the surface of a conductor, at the conductor's surface. Very near the point charge $Q$, on the other hand, the presence of the conducting plane can make little difference; the field lines must *start out* from $Q$ as if they were leaving a point charge radially. So we might expect something qualitatively like Fig. 3.10(b), with some of the details still a bit uncertain. Of course the whole thing is bound to be quite symmetrical about the $z$ axis.

But how do we really solve the problem? The answer is, by a trick, but a trick that is both instructive and frequently useful. We find an easily soluble problem whose solution, or a piece of it, can be made to fit the problem at hand. Here the easy problem is that of two equal and opposite point charges, $Q$ and $-Q$. On the plane that bisects the line joining the two charges, the plane indicated in cross section by the line $AA$ in Fig. 3.10(c), the electric field is everywhere perpendicular to the plane. If we make the distance of $Q$ from the plane agree with the distance $h$ in our original problem, the upper half of the field in Fig. 3.10(c) meets all our requirements: the field is perpendicular to the plane of the conductor, and in the neighborhood of $Q$ it approaches the field of a point charge.

The boundary conditions here are not quite those that figured in our uniqueness theorem in Section 3.3. The potential of the conductor is fixed, but we have in the system a point charge at which the potential approaches infinity. We can regard the point charge as the limiting case of a small, spherical conductor on which the total charge $Q$ is fixed. For this mixed boundary condition – potentials given on some surfaces, total charge on others – a uniqueness theorem also holds. If our "borrowed" solution fits the boundary conditions, it must be *the* solution.

Figure 3.11 shows the final solution for the field above the plane, with the density of the surface charge suggested. We can calculate the field strength and direction at any point by going back to the two-charge problem, Fig. 3.10(c), and using Coulomb's law. Consider a point on the surface, a distance $r$ from the origin. The square of its distance from $Q$ is $r^2 + h^2$, and the $z$ component of the field of $Q$, at this point, is $-Q \cos\theta / 4\pi\epsilon_0(r^2 + h^2)$. The "image charge," $-Q$, below the plane contributes an equal $z$ component. Thus the electric field here is given by

$$E_z = \frac{-2Q}{4\pi\epsilon_0(r^2 + h^2)} \cos\theta = \frac{-2Q}{4\pi\epsilon_0(r^2 + h^2)} \cdot \frac{h}{(r^2 + h^2)^{1/2}}$$

$$= \frac{-Qh}{2\pi\epsilon_0(r^2 + h^2)^{3/2}}. \tag{3.3}$$

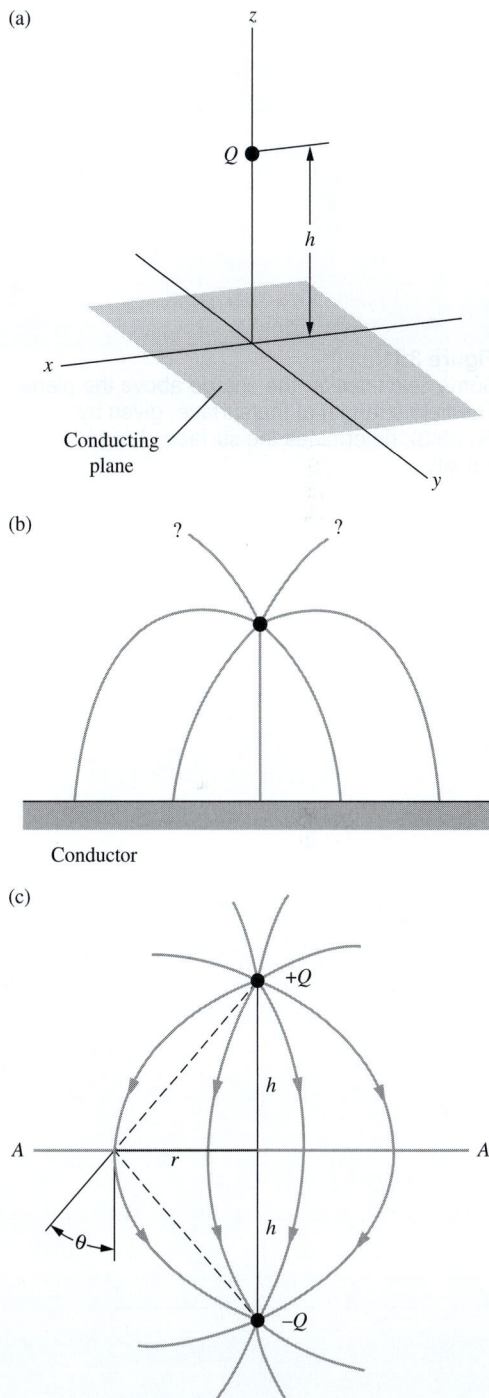

**Figure 3.10.**
(a) A point charge $Q$ above an infinite plane conductor. (b) The field must look something like this. (c) The field of a pair of opposite charges.

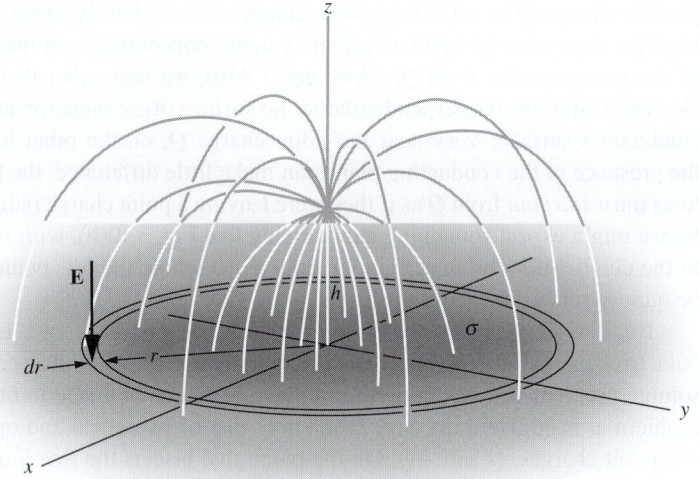

**Figure 3.11.**
Some field lines for the charge above the plane.
The field strength at the surface, given by
Eq. (3.3), determines the surface charge
density $\sigma$.

Returning to the actual setup with the conducting plane, we know that in terms of the surface charge density $\sigma$, the electric field just above the plane is $E_z = \sigma/\epsilon_0$. There is no factor of 2 in the denominator here, because when using Gauss's law with a small pillbox, there is zero field below the conducting plane, so there is zero flux out the bottom of the box. The field is indeed zero below the plane because we can consider the conducting plane to be the top of a very large conducting sphere, and we know that the field inside a conductor is zero. Using $E_z = \sigma/\epsilon_0$, the density $\sigma$ is given by

$$\sigma = \epsilon_0 E_z = \frac{-Qh}{2\pi(r^2 + h^2)^{3/2}}. \tag{3.4}$$

Let us calculate the total amount of charge on the surface by integrating over the distribution:

$$\int_0^\infty \sigma \cdot 2\pi r \, dr = -Qh \int_0^\infty \frac{r \, dr}{(r^2 + h^2)^{3/2}} = \left. \frac{Qh}{(r^2 + h^2)^{1/2}} \right|_0^\infty = -Q. \tag{3.5}$$

This result was to be expected. It means that all the flux leaving the charge $Q$ ends on the conducting plane.

There is one puzzling point. We never said what the charge on the conducting plane was, but what if we had chosen it to be zero before the charge $Q$ was put in place above it? (You might have just assumed this was the case anyway.) How can the conductor now exhibit a net charge $-Q$? The answer is that a compensating positive charge, $+Q$ in amount, must be distributed over the whole plane. The combination of the given point charge $Q$ and the surface density $\sigma$ in Eq. (3.4) produces the $E_z$

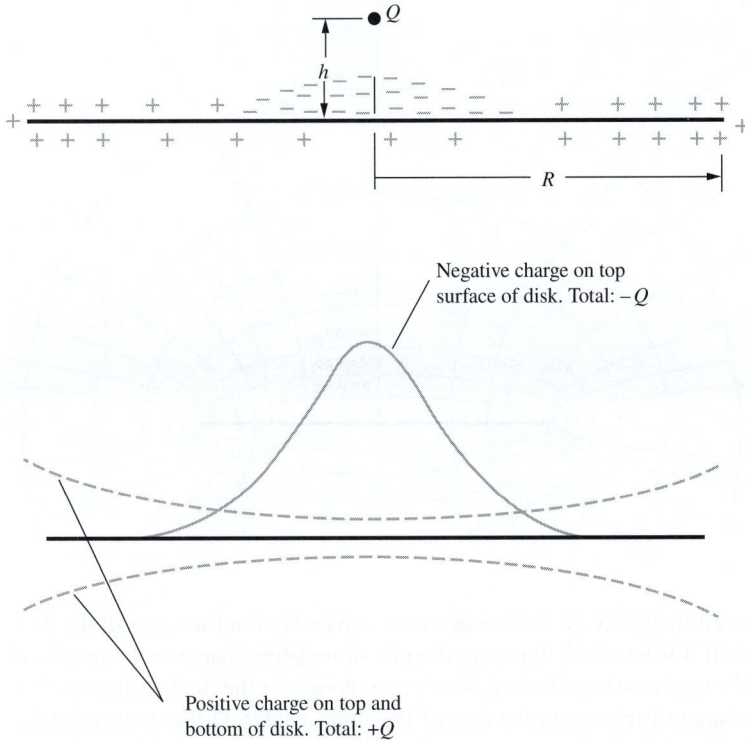

Negative charge on top
surface of disk. Total: $-Q$

Positive charge on top and
bottom of disk. Total: $+Q$

**Figure 3.12.**
The distribution of charge on a conducting disk
with total charge zero, in the presence of a
positive point charge $Q$ at height $h$ above the
center of the disk. The actual surface charge
density at any point is of course the algebraic
sum of the positive and negative densities
shown.

field in Eq. (3.3), but nothing precludes us from superposing additional
charge on the conducting plane which will produce an additional field.

To see what is going on here, imagine that the conducting plane is
actually a metal disk, not infinite but finite and with a radius $R \gg h$. If a
charge $+Q$ were to be spread uniformly over this disk, on *both* sides (so
$Q/2$ is on each side), the resulting surface density on each side would
be $Q/2\pi R^2$, which would cause an electric field of strength $Q/2\pi\epsilon_0 R^2$
normal to the plane of the disk. Since our disk is a conductor, on which
charge can move, the charge density and the resulting field strength will
be even *less* than $Q/2\pi\epsilon_0 R^2$ near the center of the disk because of the
tendency of the charge to spread out toward the rim. In any case the field
of this distribution is smaller in order of magnitude by a factor $h^2/R^2$
than the field described by Eq. (3.3), because the latter field behaves like
$1/h^2$ in the vicinity of $r = 0$. As long as $R \gg h$ we were justified in
ignoring the former field, and of course it vanishes completely for an
unbounded conducting plane with $R = \infty$.

Figure 3.12 shows in separate plots the surface charge density $\sigma$,
given by Eq. (3.4), and the distribution of the compensating charge $Q$
on the upper and lower surfaces of the disk. Here we have made $R$ not
very much larger than $h$, in order to show both distributions clearly in the
same diagram. Note that the compensating positive charge has arranged

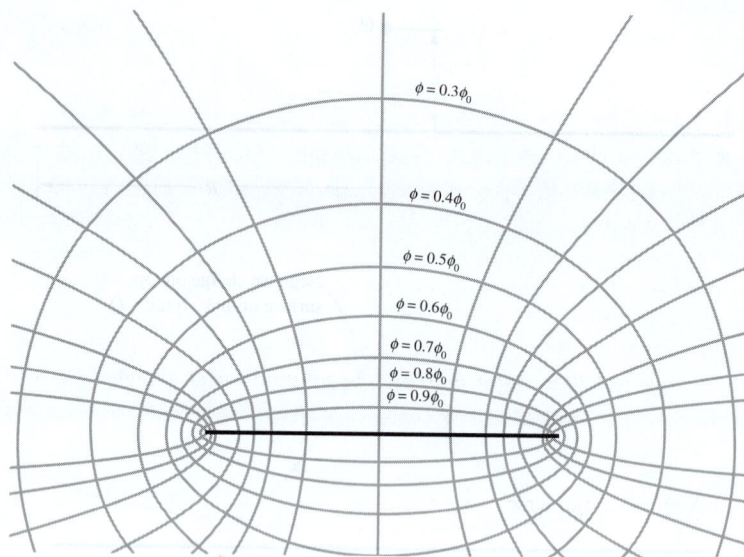

**Figure 3.13.**
Equipotentials and field lines for a charged
conducting disk.

itself in exactly the same way on the top and bottom surfaces of the disk,
as if it were utterly ignoring the pile of negative charge in the middle of
the upper surface! Indeed, it is free to do so, for the field of that negative
charge distribution *plus* that of the point charge $Q$ that induced it has
horizontal component zero at the surface of the disk, and hence has no
influence whatsoever on the distribution of the compensating positive
charge.

The isolated conducting disk mentioned above belongs to another
class of soluble problems, a class that includes any isolated conductor in
the shape of a spheroid, an ellipsoid of revolution. Without going into the
mathematics[5] we show in Fig. 3.13 some electric field lines and equipotential surfaces around the conducting disk. The field lines are hyperbo-
las. The equipotentials are oblate ellipsoids of revolution enclosing the
disk. The potential $\phi$ of the disk itself, relative to infinity, turns out to be

$$\phi_0 = \frac{(\pi/2)Q}{4\pi\epsilon_0 a}, \tag{3.6}$$

where $Q$ is the total charge of the disk and $a$ is its radius. (Written this
way, we see that $\phi_0$ is larger than the potential of a sphere of charge
$Q$ and radius $a$, by a factor $\pi/2$.) Compare this picture with Fig. 2.12,
the field of a *uniformly* charged *non*conducting disk. In that case the
electric field at the surface was not normal to the surface; it had a radial
component outward. If you could make that disk in Fig. 2.12 a conductor,
the charge would flow outward until the field in Fig. 3.13 was established.

---

[5] Mathematically speaking, this class of problems is soluble because a spheroidal
coordinate system happens to be one of those systems in which Laplace's equation
takes on a particularly simple form.

According to the mathematical solution on which Fig. 3.13 is based, the charge density at the center of the disk would then be just half as great as it was at the center of the uniformly charged disk. This fact also follows as a corollary to Problem 3.4.

Figure 3.13 shows us the field not only of the conducting disk, but also of any isolated oblate spheroidal conductor. To see that, choose one of the equipotential surfaces of revolution – say the one whose trace in the diagram is the ellipse marked $\phi = 0.6\,\phi_0$. Imagine that we could plate this spheroid with copper and deposit charge $Q$ on it. Then the field shown outside it already satisfies the boundary conditions: electric field normal to surface; total flux $Q/\epsilon_0$. It is *a* solution, and in view of the uniqueness theorem it must be *the* solution for an isolated charged conductor of that particular shape. All we need to do is erase the field lines *inside* the conductor. We can also imagine copperplating two of the spheroidal surfaces, putting charge $Q$ on the inner surface, $-Q$ on the outer. The section of Fig. 3.13 between these two equipotentials shows us the field between two such concentric spheroidal conductors. The field is zero elsewhere.

This suggests a general strategy. Given the solution for any electrostatic problem with the equipotentials located, we can extract from it the solution for any other system made from the first by copperplating one or more equipotential surfaces. Perhaps we should call the method "a solution in search of a problem." The situation was well described by Maxwell:

> "It appears, therefore, that what we should naturally call the inverse problem of determining the forms of the conductors when the expression for the potential is given is more manageable than the direct problem of determining the potential when the form of the conductors is given."[6]

If you worked Exercise 2.44, you already possess the raw material for an important example. You found that a uniform line charge of finite length has equipotential surfaces in the shape of prolate ellipsoids of revolution. This solves the problem of the potential and field of any isolated charged conductor of prolate spheroidal shape, reducing it to the relatively easy calculation of the potential due to a line charge. You can try it in Exercise 3.62.

## 3.5 Capacitance and capacitors

An isolated conductor carrying a charge $Q$ has a certain potential $\phi_0$, with zero potential at infinity; $Q$ is proportional to $\phi_0$. The constant of proportionality depends only on the size and shape of the conductor.

---

[6] See Maxwell (1891). Every student of physics ought sometime to look into Maxwell's book. Chapter VII is a good place to dip in while we are on the present subject. At the end of Volume I you will find some beautiful diagrams of electric fields, and shortly beyond the quotation we have just given, Maxwell's reason for presenting these figures. One may suspect that he also took delight in their construction and their elegance.

We call this factor the *capacitance* of that conductor and denote it by $C$:

$$\boxed{Q = C\phi_0} \tag{3.7}$$

Obviously the units for $C$ depend on the units in which $Q$ and $\phi_0$ are expressed. In our usual SI units, charge is measured in coulombs and potential in volts, so the capacitance $C$ is measured in coulombs/volt. This combination of units is given its own name, the *farad:*

$$1 \text{ farad} = 1 \frac{\text{coulomb}}{\text{volt}}. \tag{3.8}$$

Since one volt equals one joule per coulomb, a farad can be expressed in terms of other units as[7]

$$1 \text{ farad} = 1 \frac{\text{C}^2 \, \text{s}^2}{\text{kg m}^2}. \tag{3.9}$$

For an isolated spherical conductor of radius $a$ we know that $\phi_0 = Q/4\pi\epsilon_0 a$. Hence the capacitance of the sphere, defined by Eq. (3.7), must be

$$C = \frac{Q}{\phi_0} = 4\pi\epsilon_0 a. \tag{3.10}$$

For an isolated conducting disk of radius $a$, according to Eq. (3.6), $Q = 8\epsilon_0 a\phi_0$, so the capacitance of such a conductor is $C = 8\epsilon_0 a$. It is somewhat smaller than the capacitance of a sphere of the same radius. In other words, the disk requires a smaller amount of charge to attain a given potential than does the sphere. This seems reasonable.

The farad happens to be a gigantic unit; the capacitance of an isolated sphere the size of the earth is only

$$C_e = 4\pi\epsilon_0 a = 4\pi \left( 8.85 \cdot 10^{-12} \frac{\text{C}^2 \, \text{s}^2}{\text{kg m}^3} \right) (6.4 \cdot 10^6 \text{ m})$$

$$\approx 7 \cdot 10^{-4} \frac{\text{C}^2 \, \text{s}^2}{\text{kg m}^2} = 7 \cdot 10^{-4} \text{ farad}. \tag{3.11}$$

But this causes no trouble. We deal on more familiar terms with the *microfarad* ($\mu$F), $10^{-6}$ farad, and the *picofarad* (pF), $10^{-12}$ farad. Note that the units of the constant $\epsilon_0$ can be conveniently expressed as farads/meter. The capacitance will always involve one factor of $\epsilon_0$ and one net power of length, so for conductors of a given shape, capacitance scales as a linear dimension of the object.

That applies to single, isolated conductors. The concept of capacitance is also useful whenever we are concerned with charges on and potentials of a general number of conductors. By far the most common

---

[7] In Gaussian units, $Q$ is measured in esu and $\phi_0$ in statvolts, so $C$ is measured in esu/statvolt. Since in Gaussian units the esu can be written in terms of other fundamental units, you can show that the unit of capacitance is simply the centimeter, so it needs no other name.

case of interest is that of two conductors oppositely charged, with $Q$ and $-Q$, respectively. Here the capacitance is defined as the ratio of the charge $Q$ to the potential difference between the two conductors. The object itself, comprising the two conductors, insulating material to hold the conductors apart, and perhaps electrical terminals or leads, is called a *capacitor*. Most electronic circuits contain numerous capacitors. The parallel-plate capacitor is the simplest example.

Two similar flat conducting plates are arranged parallel to one another, separated by a distance $s$, as in Fig. 3.14(a). Let the area of each plate be $A$ and suppose that there is a charge $Q$ on one plate and $-Q$ on the other; $\phi_1$ and $\phi_2$ are the values of the potential at each of the plates. Figure 3.14(b) shows in cross section the field lines in this system. Away from the edge, the field is very nearly uniform in the region between the plates. When it is treated as uniform, its magnitude must be $(\phi_1 - \phi_2)/s$. The corresponding density of the surface charge on the inner surface of one of the plates is

$$\sigma = \epsilon_0 E = \frac{\epsilon_0(\phi_1 - \phi_2)}{s}. \tag{3.12}$$

If we may neglect the actual variation of $E$, and therefore of $\sigma$, which occurs principally near the edge of the plates, we can write a simple expression for the total charge, $Q = A\sigma$, on one plate:

$$Q = A\frac{\epsilon_0(\phi_1 - \phi_2)}{s} \qquad \text{(neglecting edge effects).} \tag{3.13}$$

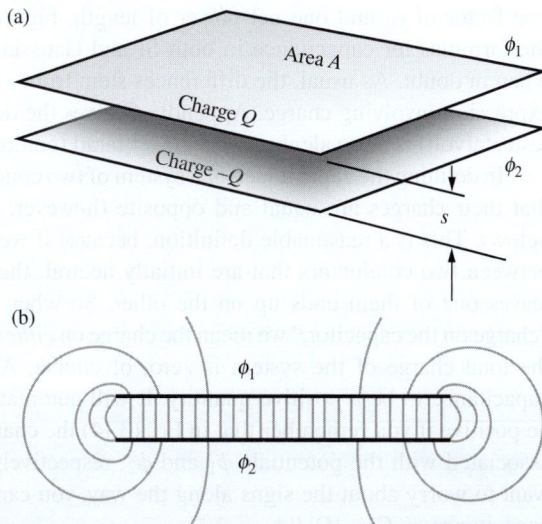

**Figure 3.14.**
(a) Parallel-plate capacitor. (b) Cross section of (a) showing field lines. The electric field is essentially uniform inside the capacitor.

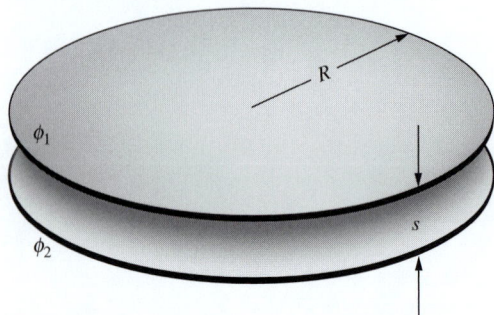

**Figure 3.15.**
The true capacitance of parallel circular plates, compared with the prediction of Eq. (3.13), for various ratios of separation to plate radius. The effect of the edge correction can be represented by writing the charge $Q$ as

$$Q = \frac{\epsilon_0 A (\phi_1 - \phi_2)}{s} f.$$

For circular plates, the factor $f$ depends on $s/R$ as follows:

| $s/R$ | $f$ |
|-------|-------|
| 0.2 | 1.286 |
| 0.1 | 1.167 |
| 0.05 | 1.094 |
| 0.02 | 1.042 |
| 0.01 | 1.023 |

We should expect Eq. (3.13) to be more nearly accurate the smaller the ratio of the plate separation $s$ to the lateral dimension of the plates. Of course, if we were to solve exactly the electrostatic problem, edge and all, for a particular shape of plate, we could replace Eq. (3.13) by an exact formula. To show how good an approximation Eq. (3.13) is, there are listed in Fig. 3.15 values of the correction factor $f$ by which the charge $Q$ given in Eq. (3.13) differs from the exact result, in the case of two conducting disks at various separations. The total charge is always a bit greater than Eq. (3.13) would predict. That seems reasonable as we look at Fig. 3.14(b), for there is evidently an extra concentration of charge at the edge, and even some charge on the outer surfaces near the edge.

We are not concerned now with the details of such corrections but with the general properties of a two-conductor system, the *capacitor*. We are interested in the relation between the charge $Q$ on one of the plates and the potential difference between the two plates. For the particular system to which Eq. (3.13) applies, the quotient $Q/(\phi_1 - \phi_2)$ is $\epsilon_0 A/s$. Even if this is only approximate, it is clear that the exact formula will depend only on the size and geometrical arrangement of the plates. That is, for a fixed pair of conductors, the ratio of charge to potential difference will be a constant. We call this constant the *capacitance* of the capacitor and denote it usually by $C$.

$$Q = C(\phi_1 - \phi_2). \tag{3.14}$$

Thus the capacitance of the parallel-plate capacitor, with edge fields neglected, is given by

$$\boxed{C = \frac{\epsilon_0 A}{s}} \tag{3.15}$$

As with the above cases of the sphere and disk, this capacitance contains one factor of $\epsilon_0$ and one net power of length. Figure 3.16 summarizes the formulas for capacitance in both SI and Gaussian units. Refer to it when in doubt. As usual, the differences stem from a factor $4\pi\epsilon_0$ in any expression involving charge. Appendix C gives the derivation that 1 cm (esu/statvolt) is equivalent to $1.11 \cdot 10^{-12}$ farad (coulomb/volt).

In defining the capacitance of a system of two conductors, we assume that their charges are equal and opposite (however, see the discussion below). This is a reasonable definition, because if we hook up a battery between two conductors that are initially neutral, then whatever charge leaves one of them ends up on the other. So when we talk about the "charge on the capacitor," we mean the charge on *either* of the conductors; the total charge of the system is zero, of course. Also, we define the capacitance to be a positive quantity. It will automatically come out to be positive if you remember that in Eq. (3.14) the charges $Q$ and $-Q$ are associated with the potentials $\phi_1$ and $\phi_2$, respectively. But if you don't want to worry about the signs along the way, you can simply define the capacitance as $C = |Q|/|\phi_1 - \phi_2|$.

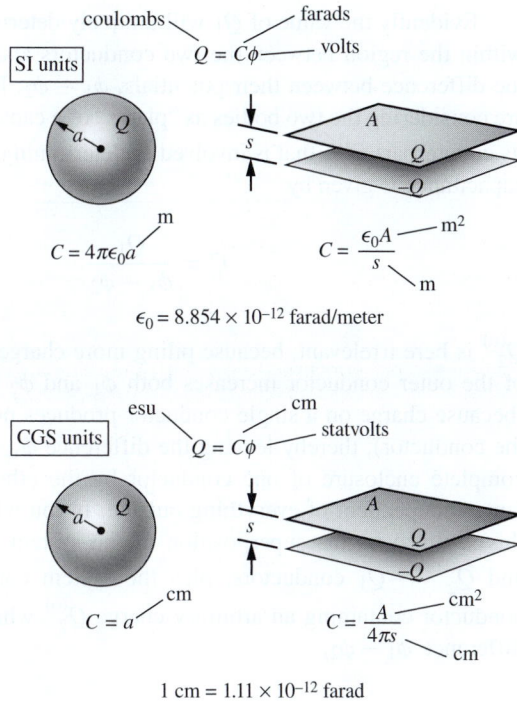

**Figure 3.16.**
Summary of units associated with capacitance.

Any pair of conductors, regardless of shape or arrangement, can be considered a capacitor. It just happens that the parallel-plate capacitor is a common arrangement and one for which an approximate calculation of the capacitance is very easy. Figure 3.17 shows two conductors, one inside the other. We can call this arrangement a capacitor too. As a practical matter, some mechanical support for the inner conductor would be needed, but that does not concern us. Also, to convey electric charge to or from the conductors we would need leads, which are themselves conducting bodies. Since a wire leading out from the inner body, numbered 1, necessarily crosses the space between the conductors, it is bound to cause some perturbation of the electric field in that space. To minimize this we may suppose the lead wires to be extremely thin, so that any charge residing on them is negligible. Or we might imagine the leads removed before the potentials are determined.

In this system we can distinguish three charges: $Q_1$, the total charge on the inner conductor; $Q_2^{(i)}$, the amount of charge on the inner surface of the outer conductor; $Q_2^{(o)}$, the charge on the outer surface of the outer conductor. Observe first that $Q_2^{(i)}$ must equal $-Q_1$. As we have seen in earlier examples, we know this because a surface such as $S$ in Fig. 3.17 encloses both these charges and no others, and the flux through this surface is zero. The flux is zero because on the surface $S$, lying, as it does, in the interior of a conductor, the electric field is zero.

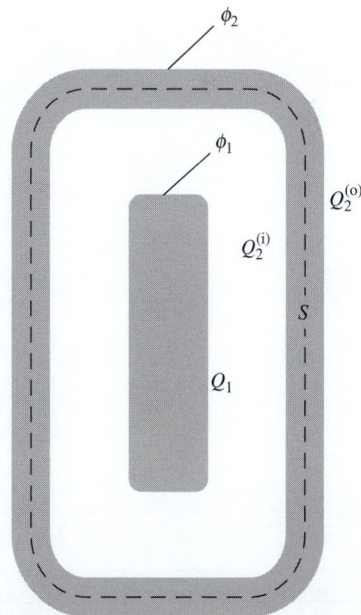

**Figure 3.17.**
A capacitor in which one conductor is enclosed by the other.

Evidently the value of $Q_1$ will uniquely determine the electric field within the region between the two conductors and thus will determine the difference between their potentials, $\phi_1 - \phi_2$. For that reason, if we are considering the two bodies as "plates" of a capacitor, it is only $Q_1$, or its counterpart $Q_2^{(i)}$, that is involved in determining the capacitance. The capacitance is given by

$$C = \frac{Q_1}{\phi_1 - \phi_2}. \tag{3.16}$$

$Q_2^{(o)}$ is here irrelevant, because piling more charge on the outer surface of the outer conductor increases both $\phi_1$ and $\phi_2$ by the *same* amount (because charge on a single conductor produces no electric field inside the conductor), thereby leaving the difference $\phi_1 - \phi_2$ unchanged. The complete enclosure of one conductor by the other makes the capacitance independent of everything outside. If you wish, you can consider this setup to be the superposition of the system consisting of the $Q_1$ and $Q_2^{(i)} = -Q_1$ conductors, plus the system consisting of the outer conductor containing an arbitrary charge $Q_2^{(o)}$ which doesn't affect the difference $\phi_1 - \phi_2$.

---

**Example (Capacitance of two spherical shells)**   What is the capacitance of a capacitor that consists of two concentric spherical metal shells? The inner radius of the outer shell is $a$; the outer radius of the inner shell is $b$.

**Solution**   Let there be charge $Q$ on the inner shell and charge $-Q$ on the outer shell. As mentioned above, any additional charge on the outside surface of the outer shell doesn't affect the potential difference. The field between the shells is due only to the inner shell, so it equals $Q/4\pi\epsilon_0 r^2$. The magnitude of the potential difference is therefore

$$\Delta\phi = \int_b^a E\, dr = \int_b^a \frac{Q\, dr}{4\pi\epsilon_0 r^2} = \frac{Q}{4\pi\epsilon_0}\left(\frac{1}{b} - \frac{1}{a}\right). \tag{3.17}$$

The capacitance is then

$$C = \frac{Q}{\Delta\phi} = \frac{4\pi\epsilon_0}{\dfrac{1}{b} - \dfrac{1}{a}} = \frac{4\pi\epsilon_0 ab}{a - b}. \tag{3.18}$$

We can check this result by considering the limiting case where the gap between the conductors, $a - b$, is much smaller than $b$. In this limit the capacitor should be essentially the same as a flat-plate capacitor with separation $s = a - b$ and area $A = 4\pi r^2$, where $r \approx a \approx b$. And indeed, in this limit Eq. (3.18) gives $C \approx 4\pi\epsilon_0 r^2/s = \epsilon_0 A/s$, in agreement with Eq. (3.15). If we let $r$ be the geometric mean of $a$ and $b$, then the equivalence is exact, because the product $ab$ in the numerator of $C$ exactly equals $r^2$.

Also, in the $a \gg b$ limit, Eq. (3.18) gives $C = 4\pi\epsilon_0 b$, which is the correct result for the capacitance of an isolated sphere with radius $b$, with its counterpart at infinity; see Eq. (3.10).

## 3.6 Potentials and charges on several conductors

We have been skirting the edge of a more general problem, the relations among the charges and potentials of any number of conductors of some given configuration. The two-conductor capacitor is just a special case. It may surprise you that anything useful can be said about the general case. In tackling it, about all we can use is the uniqueness theorem and the superposition principle. To have something definite in mind, consider three separate conductors, all enclosed by a conducting shell, as in Fig. 3.18. The potential of this shell we may choose to be zero; with respect to this reference the potentials of the three conductors, for some particular state of the system, are $\phi_1$, $\phi_2$, and $\phi_3$. The uniqueness theorem guarantees that, with $\phi_1$, $\phi_2$, and $\phi_3$ given, the electric field is determined throughout the system. It follows that the charges $Q_1$, $Q_2$, and $Q_3$ on the individual conductors are likewise uniquely determined.

We need not keep account of the charge on the inner surface of the surrounding shell, since it will always be $-(Q_1 + Q_2 + Q_3)$. If you prefer, you can let "infinity" take over the role of this shell, imagining the shell to expand outward without limit. We have kept it in the picture because it makes the process of charge transfer easier to follow, for some people, if we have something to connect to.

Among the possible states of this system are ones with $\phi_2$ and $\phi_3$ both zero. We could enforce this condition by connecting conductors 2 and 3 to the zero-potential shell, as indicated in Fig. 3.18(a). As before, we may suppose the connecting wires are so thin that any charge residing on them is negligible. Of course, we really do not care how the specified condition is brought about. In such a state, which we shall call state I, the electric field in the whole system and the charge on every conductor is determined uniquely by the value of $\phi_1$. Moreover, if $\phi_1$ were doubled, that would imply a doubling of the field strength everywhere, and hence a doubling of each of the charges $Q_1$, $Q_2$, and $Q_3$. That is, with $\phi_2 = \phi_3 = 0$, each of the three charges must be proportional to $\phi_1$. Stated mathematically:

- State I ($\phi_2 = \phi_3 = 0$):

$$Q_1 = C_{11}\phi_1; \quad Q_2 = C_{21}\phi_1; \quad Q_3 = C_{31}\phi_1. \quad (3.19)$$

The three constants, $C_{11}$, $C_{21}$, and $C_{31}$, can depend only on the shape and arrangement of the conducting bodies.

In just the same way we could analyze states in which $\phi_1$ and $\phi_3$ are zero, calling such a condition state II (Fig. 3.18(b)). Again, we find

(a) State I

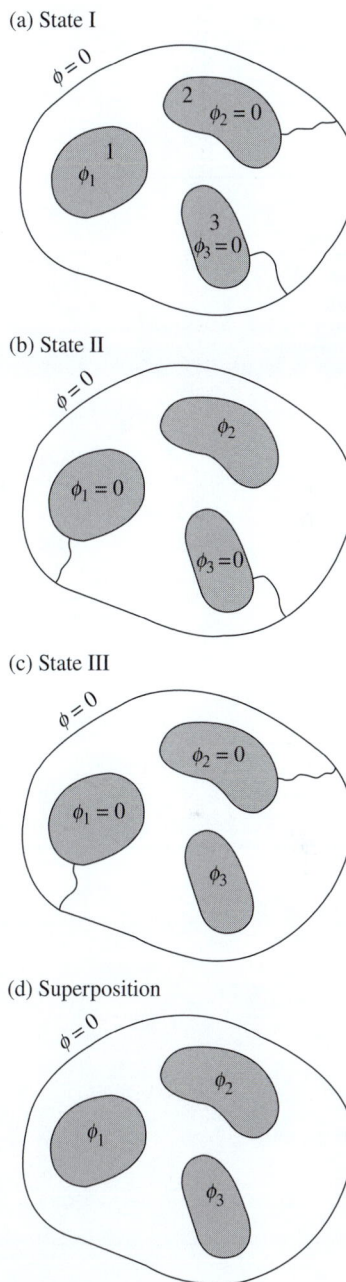

(b) State II

(c) State III

(d) Superposition

**Figure 3.18.**
A general state of this system can be analyzed as the superposition (d) of three states (a)–(c) in each of which all conductors but one are at zero potential.

a linear relation between the only nonzero potential, $\phi_2$ in this case, and the various charges:

- State II ($\phi_1 = \phi_3 = 0$):

$$Q_1 = C_{12}\phi_2; \quad Q_2 = C_{22}\phi_2; \quad Q_3 = C_{32}\phi_2. \tag{3.20}$$

Finally, when $\phi_1$ and $\phi_2$ are held at zero, the field and the charges are proportional to $\phi_3$:

- State III ($\phi_1 = \phi_2 = 0$):

$$Q_1 = C_{13}\phi_3; \quad Q_2 = C_{23}\phi_3; \quad Q_3 = C_{33}\phi_3. \tag{3.21}$$

Now the superposition of three states like I, II, and III is also a possible state. The electric field at any point is the vector sum of the electric fields at that point in the three cases, while the charge on a conductor is the sum of the charges it carried in the three cases. In this new state the potentials are $\phi_1$, $\phi_2$, and $\phi_3$, none of them necessarily zero. In short, we have a completely general state. The relation connecting charges and potentials is obtained simply by adding Eqs. (3.19) through (3.21):

$$Q_1 = C_{11}\phi_1 + C_{12}\phi_2 + C_{13}\phi_3,$$
$$Q_2 = C_{21}\phi_1 + C_{22}\phi_2 + C_{23}\phi_3,$$
$$Q_3 = C_{31}\phi_1 + C_{32}\phi_2 + C_{33}\phi_3. \tag{3.22}$$

It appears that the electrical behavior of this system is characterized by the nine constants $C_{11}$, $C_{12}$, ..., $C_{33}$. In fact, only six constants are necessary, for it can be proved that in *any* system $C_{12} = C_{21}$, $C_{13} = C_{31}$, and $C_{23} = C_{32}$. Why this should be so is not obvious. Exercise 3.64 will suggest a proof based on conservation of energy, but for that purpose you will need an idea developed in Section 3.7. The $C$'s in Eq. (3.22) are called the *coefficients of capacitance*. It is clear that our argument would extend to any number of conductors.

A set of equations like Eq. (3.22) can be solved for the $\phi$'s in terms of the $Q$'s. That is, there is an equivalent set of linear relations of the form

$$\phi_1 = P_{11}Q_1 + P_{12}Q_2 + P_{13}Q_3,$$
$$\phi_2 = P_{21}Q_1 + P_{22}Q_2 + P_{23}Q_3,$$
$$\phi_3 = P_{31}Q_1 + P_{32}Q_2 + P_{33}Q_3. \tag{3.23}$$

The $P$'s are called the *potential coefficients*; they could be computed from the $C$'s, or vice versa.

We have here a simple example of the kind of relation we can expect to govern any *linear* physical system. Such relations turn up in the study of mechanical structures (connecting the strains with the loads), in the analysis of electrical circuits (connecting voltages and currents), and, generally speaking, wherever the superposition principle can be applied.

---

**Example (Capacitance coefficients for two plates)**   Figure 3.19 shows in cross section a flat metal box in which there are two flat plates, 1 and 2, each of area $A$. The potential of the box is chosen to be zero. The various distances separating the plates from each other and from the top and bottom of the box, labeled $r$, $s$, and $t$ in the figure, are to be assumed small compared with the width and length of the plates, so that it will be a good approximation to neglect the edge fields in estimating the charges on the plates. In this approximation, work out the capacitance coefficients, $C_{11}$, $C_{22}$, $C_{12}$, and $C_{21}$. Check that $C_{12} = C_{21}$.

**Solution**   With the potential of the box chosen to be zero, we can write, in general,

$$Q_1 = C_{11}\phi_1 + C_{12}\phi_2,$$
$$Q_2 = C_{21}\phi_1 + C_{22}\phi_2. \qquad (3.24)$$

Consider the case where $\phi_2$ is made equal to zero by connecting plate 2 to the box. Then (see Fig. 3.20) the fields in the three regions are $E_r = \phi_1/r$, $E_s = \phi_1/s$, and $E_t = 0$. Gauss's law with a thin box completely surrounding plate 1 tells us that $Q_1 = \epsilon_0(AE_r + AE_s)$. Eliminating the $E$'s in favor of the $\phi$'s gives

$$Q_1 = \epsilon_0 A \phi_1 \left(\frac{1}{r} + \frac{1}{s}\right) \implies C_{11} = \epsilon_0 A \left(\frac{1}{r} + \frac{1}{s}\right). \qquad (3.25)$$

Also, Gauss's law with a box around plate 2 tells us that $Q_2 = -\epsilon_0(AE_s + 0)$. Hence,

$$Q_2 = -\frac{\epsilon_0 A \phi_1}{s} \implies C_{21} = -\frac{\epsilon_0 A}{s}. \qquad (3.26)$$

We can repeat the above arguments, but now with $\phi_1 = 0$ instead of $\phi_2 = 0$. This basically just involves switching the 1's and 2's, and letting $r \to t$ (but $s$ remains $s$). We quickly find

$$C_{22} = \epsilon_0 A \left(\frac{1}{t} + \frac{1}{s}\right) \quad \text{and} \quad C_{12} = -\frac{\epsilon_0 A}{s}. \qquad (3.27)$$

As expected, $C_{12} = C_{21}$. How do these four coefficients reduce to the $C = \epsilon_0 A/s$ capacitance we found for a parallel-plate capacitor in Eq. (3.15)? That is the subject of Problem 3.23.

---

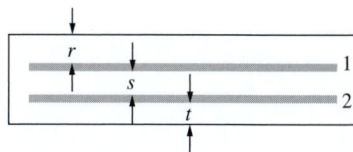

**Figure 3.19.**
Two capacitor plates inside a conducting box.

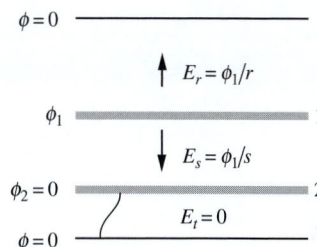

**Figure 3.20.**
The situation with the bottom plate grounded to the box.

# 3.7 Energy stored in a capacitor

Consider a capacitor of capacitance $C$, with a potential difference $\phi$ between the plates. The charge $Q$ is equal to $C\phi$. There is a charge $Q$ on one plate and $-Q$ on the other. Suppose we *increase* the charge from $Q$ to $Q+dQ$ by transporting a positive charge $dQ$ from the negative to the positive plate, working against the potential difference $\phi$. The work that has to be done is $dW = \phi\,dQ = Q\,dQ/C$. Therefore to charge the capacitor starting from the uncharged state to some final charge $Q_f$ requires the work

$$W = \frac{1}{C}\int_0^{Q_f} Q\,dQ = \frac{Q_f^2}{2C}. \tag{3.28}$$

This is the energy $U$ that is "stored" in the capacitor. Since $Q_f = C\phi$, it can also be expressed by

$$\boxed{U = \frac{1}{2}C\phi^2} \tag{3.29}$$

where $\phi$ is the final potential difference between the plates. Using $Q = C\phi$ again, we can also write the energy as $U = Q\phi/2$. This result is consistent with the energy we would obtain from Eq. (2.32); see Exercise 3.65.

For the parallel-plate capacitor with plate area $A$ and separation $s$, we found the capacitance $C = \epsilon_0 A/s$ and the electric field $E = \phi/s$. Hence Eq. (3.29) is also equivalent to

$$U = \frac{1}{2}\left(\frac{\epsilon_0 A}{s}\right)(Es)^2 = \frac{\epsilon_0 E^2}{2}\cdot As = \frac{\epsilon_0 E^2}{2}\cdot(\text{volume}). \tag{3.30}$$

This agrees with our general formula, Eq. (1.53), for the energy stored in an electric field.[8]

Equation (3.29) applies as well to the isolated charged conductor, which can be thought of as the inner plate of a capacitor, enclosed by an outer conductor of infinite size and potential zero. For the isolated sphere of radius $a$, we found $C = 4\pi\epsilon_0 a$, so that $U = (1/2)C\phi^2 = (1/2)(4\pi\epsilon_0 a)\phi^2$ or, equivalently, $U = (1/2)Q^2/C = (1/2)Q^2/4\pi\epsilon_0 a$, agreeing with the calculation in Problem 1.32 for the energy stored in the electric field of the charged sphere.

The oppositely charged plates of a capacitor will attract one another; some mechanical force will be required to hold them apart. This is obvious in the case of the parallel-plate capacitor, for which we could easily calculate the force on the surface charge. But we can make a more general statement based on Eq. (3.28), which relates stored energy to charge $Q$ and capacitance $C$. Suppose that $C$ depends in some manner on a linear coordinate $x$ that measures the displacement of one "plate" of a capacitor, which might be a conductor of any shape, with respect to the other. Let $F$ be the magnitude of the force that must be applied to each plate to overcome their attraction and keep $x$ constant. Now imagine the distance $x$ is increased by an increment $\Delta x$ with $Q$ remaining constant and one

---

[8] All this applies to the *vacuum capacitor* consisting of conductors with empty space in between. As you may know from the laboratory, most capacitors used in electric circuits are filled with an insulator or "dielectric." We are going to study the effect of that in Chapter 10.

plate fixed. The external force $F$ on the other plate does work $F \, \Delta x$ and, if energy is to be conserved, this must appear as an increase in the stored energy $Q^2/2C$. That increase at constant $Q$ is

$$\Delta U = \frac{dU}{dx} \Delta x = \frac{Q^2}{2} \frac{d}{dx} \left( \frac{1}{C} \right) \Delta x. \qquad (3.31)$$

Equating this to the work $F \, \Delta x$ we find

$$F = \frac{Q^2}{2} \frac{d}{dx} \left( \frac{1}{C} \right). \qquad (3.32)$$

**Example (Parallel-plate capacitor)**   Let's verify that Eq. (3.32) yields the correct force on a plate in a parallel-plate capacitor. If the plate separation is $x$, Eq. (3.15) gives the capacitance as $C = \epsilon_0 A/x$. So Eq. (3.32) gives the (attractive) force as

$$F = \frac{Q^2}{2} \frac{d}{dx} \left( \frac{x}{\epsilon_0 A} \right) = \frac{Q^2}{2\epsilon_0 A}. \qquad (3.33)$$

Is this correct? We know from Eq. (1.49) that the force (per unit area) on a sheet of charge equals the density $\sigma$ times the average of the fields on either side. The total force on the entire plate of area $A$ is then the total charge $Q = \sigma A$ times the average of the fields. The field is zero outside the capacitor, and it is $\sigma/\epsilon_0$ inside. So the average of the two fields is $\sigma/2\epsilon_0$. (This is correctly the field due to the other plate, which is the field that the given plate feels.) The force on the plate is therefore

$$F = Q \frac{\sigma}{2\epsilon_0} = Q \frac{Q/A}{2\epsilon_0} = \frac{Q^2}{2\epsilon_0 A}, \qquad (3.34)$$

as desired.

## 3.8 Other views of the boundary-value problem

It would be wrong to leave the impression that there are no general methods for dealing with the Laplacian boundary-value problem. Although we cannot pursue this question much further, we shall mention some useful and interesting approaches that you are likely to meet in future study of physics or applied mathematics.

First, an elegant method of analysis, called conformal mapping, is based on the theory of functions of a complex variable. Unfortunately it applies only to two-dimensional systems. These are systems in which $\phi$ depends only on $x$ and $y$, for example, all conducting boundaries being cylinders (in the general sense) with elements running parallel to $z$. Laplace's equation then reduces to

$$\frac{\partial^2 \phi}{\partial x^2} + \frac{\partial^2 \phi}{\partial y^2} = 0, \qquad (3.35)$$

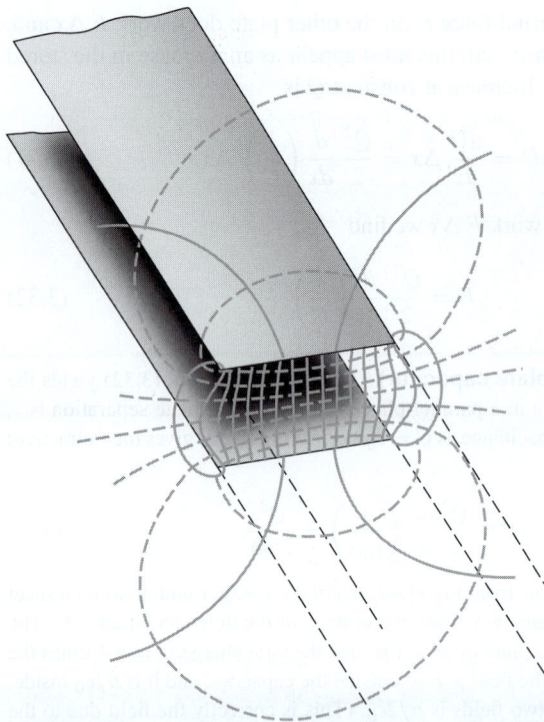

**Figure 3.21.**
Field lines and equipotentials for two infinitely
long conducting strips.

with boundary values specified on some lines or curves in the $xy$ plane. Many systems of practical interest are like this or sufficiently like this to make the method useful, quite apart from its intrinsic mathematical interest. For instance, the exact solution for the potential around two long parallel strips is easily obtained by the method of conformal mapping. The field lines and equipotentials are shown in a cross-sectional plane in Fig. 3.21. This provides us with the edge field for any parallel-plate capacitor in which the edge is long compared with the gap. The field shown in Fig. 3.14(b) was copied from such a solution. You will be able to apply this method after you have studied functions of a complex variable in more advanced mathematics courses.

Second, we mention a numerical method for finding approximate solutions of the electrostatic potential with given boundary values. Surprisingly simple and almost universally applicable, this method is based on that special property of harmonic functions with which we are already familiar: the value of the function at a point is equal to its average over the neighborhood of the point. In this method the potential function $\phi$ is represented by values at an array of discrete points only, including discrete points on the boundaries. The values at nonboundary points are then adjusted until each value is equal to the average of the neighboring values. In principle one could do this by solving a large number of simultaneous linear equations – as many as there are interior points.

But an approximate solution can be obtained by the following procedure, called a *relaxation method*. Start with the boundary points of the array, or grid, set at the values prescribed. Assign starting values arbitrarily to the interior points. Now visit, in some order, all the interior points. At each point reset its value to the average of the values at the four (for a square grid) adjacent grid points. Repeat again and again, until all the changes made in the course of one sweep over the network of interior points are acceptably small. If you want to see how this method works, Exercises 3.76 and 3.77 will provide an introduction. Whether convergence of the relaxation process can be ensured, or even hastened, and whether a relaxation method or direct solution of the simultaneous equations is the better strategy for a given problem, are questions in applied mathematics that we cannot go into here. It is the high-speed computer, of course, that makes both methods feasible.

## 3.9 Applications

The purpose of a *lightning rod* on a building is to provide an alternative path for the lightning's current on its way to ground, that is, a path that travels along a metal rod as opposed to through the building itself. Should the tip of the rod be pointed or rounded? The larger the field generated by the tip, the better the chance that a conductive path for the lightning is formed, meaning that the lightning is more likely to hit the rod than some other point on the building. On one hand, a pointed tip generates a large electric field very close to the tip, but on the other hand the field falls off more quickly than the field due to a more rounded tip (you can model the tip roughly as a small sphere). It isn't obvious which of these effects wins, but experiments suggest that a somewhat rounded tip has a better chance of being struck.

*Capacitors* have many uses; we will look at a few here. Capacitors can be used to store energy, for either slow discharge or fast discharge. In the slow case, the capacitor acts effectively like a battery. Examples include shake flashlights and power adapters. For the fast case, capacitors also have the ability to release their energy very quickly (unlike a normal battery). Examples include flashbulbs, stun guns, defibrillators, and the National Ignition Facility (NIF), whose goal is to create sustained fusion. The capacitor for a flashbulb might store 10 J of energy, while the huge capacitor bank at the NIF can store $4 \cdot 10^8$ J.

In many electronic devices, capacitors are used to smooth out fluctuations in the voltage in a DC circuit. If a capacitor is placed in parallel with the load, it acts like a reserve battery. If the voltage from the power supply dips, the capacitor will (temporarily) continue to push current through the load.

The *dynamic random access memory (DRAM)* in your computer works by storing charge on billions of tiny capacitors. Each capacitor represents a *bit* of information; uncharged is 0, charged is 1. However, the capacitors are leaky, so their charges must be refreshed many times each second (64 ms is a common refresh time); hence the adjective "dynamic." The memory is lost when the power is shut off. The permanent memory on the hard disk must therefore use a different method – the orientation of tiny magnetic domains, as we will see in Chapter 11.

Capacitors are also used for tuning electronic circuits. We will see in Chapter 8 that the *resonant frequency* of a circuit containing a resistor, inductor, and capacitor depends on the inductance and capacitance. Radios, cell phones, wireless computer connections, etc., function by changing the capacitance of an internal circuit so that the resonant frequency equals the frequency of the desired signal (transmitted by an electromagnetic wave, which will be discussed in Chapter 9).

Capacitors can be used in *power-factor correction* in the AC electrical power grid. We will talk about AC circuits in Chapter 8, but the main point is that by adding capacitors (or inductors) to a load, a larger fraction of the power delivered can actually be used, instead of sloshing back and forth between the power station and the load. This sloshing wastes energy by heating up the transmission lines.

A *condenser microphone* makes use of the fact that the capacitance of a parallel-plate capacitor depends on the plate separation. ("Condenser" is simply another name for a capacitor.) A small capacitor consists of a fixed plate and a movable diaphragm. The pressure from the sound waves in the air moves the diaphragm back and forth, changing the separation and hence the capacitance. This movement is extremely small, but is large enough to affect a circuit and generate an electric signal that can be sent to a speaker. Due to the large resistance of the circuit, the charge on the capacitor remains essentially constant as the diaphragm vibrates back and forth. So the voltage changes only because the capacitance changes, that is, $\phi = Q_0/C$.

A *supercapacitor* has a capacitance vastly larger than what can be produced by a pair of parallel plates. A supercapacitor with the dimensions of a standard D-cell battery can have a capacitance of well over a farad (and even up to the kilofarad range). Two square plates separated by 1 mm would need to be 10 km on a side to have a $\epsilon_0 A/s$ capacitance of 1 farad! A supercapacitor works by effectively making $A$ very large and $s$ very small. Two pieces of carbon foam are separated by an insulating membrane and immersed in an electrolyte solution. (Other types, for example ones using graphene layers, also exist.) The effective area $A$ is large due to the porous nature of the foam, and the effective distance $s$ is small (on the order of an atomic length) due to the fact that the electrolyte touches the foam. The voltage is generally only a few volts, so a supercapacitor is used when there is a need for a steady supply of energy, as opposed to a burst of energy. That is, it is used as a battery (although often short-lived, on the order of a minute) instead of, say, a flashbulb capacitor. The charging time of a supercapacitor is also on the order of a minute – much faster than a conventional battery.

# CHAPTER SUMMARY

- Assuming there are no other forces involved, the electric field is zero within the material of a *conductor* (in the stationary state). Equivalently, the conductor is an *equipotential*. Just outside the conductor, the field is perpendicular to the surface and has magnitude $E = \sigma/\epsilon_0$, from Gauss's law.

- The *uniqueness theorem* states that for a set of conductors at given potentials, the solution for the potential $\phi(x, y, z)$ is unique. This implies that once we have found a solution (by whatever means), we know that it must be *the* solution. A quick corollary is that if the space inside a hollow conductor of any shape is empty of charge, the electric field there is zero.

- If a given charge $q$ is located inside a conducting shell, then a total charge of $-q$ resides on the inner surface of the shell. Any additional charge resides on the outer surface, and it distributes itself in the same manner as if neither the given charge $q$ nor the inner-surface charge $-q$ were present. These results follow from Gauss's law and the fact that the electric field is zero inside the material of the conductor.

- The method of *image charges* is useful for finding the electric field that satisfies a given set of boundary conditions at conductors. In the case of a point charge $q$ and an infinite conducting plane, the image charge $-q$ is located on the other side of the plane, an equal distance from it. As expected from Gauss's law, the total charge on the plane is $-q$.

- The *capacitance C*, defined by $Q = C\phi$, gives a measure of how much charge a conductor can hold, for a given potential $\phi$. The capacitances of a sphere and a parallel-plate *capacitor* are

$$C_{\text{sphere}} = 4\pi\epsilon_0 r \quad \text{and} \quad C_{\text{plates}} = \frac{\epsilon_0 A}{s}. \tag{3.36}$$

If there are many conductors in a system, the charge on each is a linear function of the various potentials, with the *coefficients of capacitance* being the constants of proportionality.

- The energy stored in a capacitor can be written in several ways:

$$U = \frac{1}{2}C\phi^2 = \frac{Q^2}{2C} = \frac{1}{2}Q\phi. \tag{3.37}$$

## Problems

3.1   *Inner-surface charge density* **

A positive point charge $q$ is located off-center inside a conducting spherical shell, as shown in Fig. 3.22. (You can assume that the shell is neutral, although this doesn't matter.) We know from Gauss's law that the total charge on the inner surface of the shell

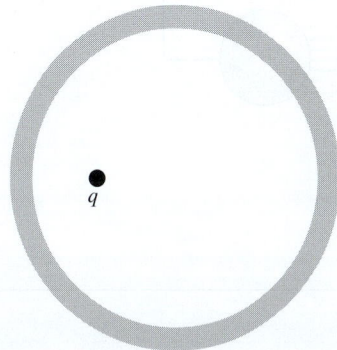

**Figure 3.22.**

(a)

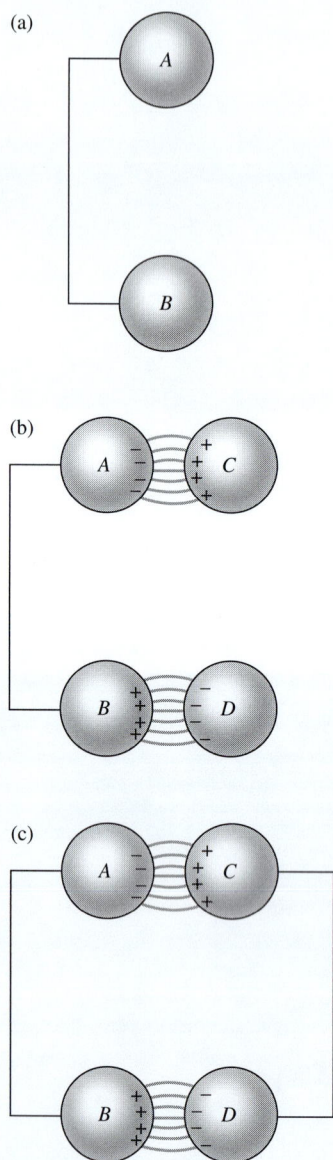

(b)

(c)

**Figure 3.23.**

is $-q$. Is the surface charge density negative over the entire inner surface? Or can it be positive on the far side of the inner surface if the point charge $q$ is close enough to the shell so that it attracts enough negative charge to the near side? Justify your answer. *Hint: Think about field lines.*

3.2   *Holding the charge in place* **

The two metal spheres in Fig. 3.23(a) are connected by a wire; the total charge is zero. In Fig. 3.23(b) two oppositely charged conducting spheres have been brought into the positions shown, inducing charges of opposite sign in $A$ and in $B$. If now $C$ and $D$ are connected by a wire as in Fig. 3.23(c), it could be argued that something like the charge distribution in Fig. 3.23(b) ought to persist, each charge concentration being held in place by the attraction of the opposite charge nearby. What about that? Can you prove it won't happen?

3.3   *Principal radii of curvature* **

Consider a point on the surface of a conductor. The *principal radii of curvature* of the surface at that point are defined to be the largest and smallest radii of curvature there. To find the radii of curvature, consider a plane that contains the normal to the surface at the given point. Rotate this plane around the normal, and look at the curve representing the intersection of the plane and the surface. The radius of curvature is defined to be the radius of the circle that locally matches up with the curve. For example, a sphere has its principal radii everywhere equal to the radius $R$. A cylinder has one principal radius equal to the cross-sectional radius $R$, and the other equal to infinity.

It turns out that the spatial derivative (in the direction of the normal) of the electric field just outside a conductor can be written in terms of the principal radii, $R_1$ and $R_2$, as follows:

$$\frac{dE}{dx} = -\left(\frac{1}{R_1} + \frac{1}{R_2}\right)E. \qquad (3.38)$$

(a) Verify this expression for a sphere, a cylinder, and a plane.
(b) Prove this expression. Use Gauss's law with a wisely chosen pillbox just outside the surface. Remember that near the surface, the electric field is normal to it.

3.4   *Charge distribution on a conducting disk* **

There is a very sneaky way of finding the charge distribution on a conducting circular disk with radius $R$ and charge $Q$. Our goal is to find a charge distribution such that the electric field at any point in the disk has zero component parallel to the disk. From Problem 1.17 we know that the field at any point $P$ inside a spherical shell with uniform surface charge density is zero. Consider the projection of this shell onto the equatorial plane containing $P$. Explain why this setup is relevant, and use it to find the desired charge density on a conducting disk.

3.5  *Charge distribution on a conducting stick* ****

If we put some charge in a 3D conducting ball, it all heads to the surface; the volume charge density is zero inside. If we put some charge on a 2D conducting "ball" (that is, a disk), then we found in Problem 3.4 that the resulting surface charge density is nonzero throughout the disk, but that it increases toward the edge. If we put some charge on a 1D conducting "ball" (that is, a stick), then it turns out that the same strategy used in Problem 3.4 can be used to show that the resulting linear charge density on the stick is essentially uniform; see Good (1997). At first glance, this seems absurd, because if we consider a little piece of charge at an off-center position, there is more charge on one side than on the other. So the electric field at the little piece isn't zero, as we know it must be in a conductor.

Your task is to explain what is meant by the above phrase, *essentially uniform,* by considering a setup with a very large number $N$ of point charges, each with initial value $Q/N$, that are evenly spaced on the stick, a fixed small distance $L/N$ apart. Determine roughly (in an order-of-magnitude sense) how much charge needs to be added to an adjacent point charge so that the field felt by a given off-center point charge is zero. Then take the $N \to \infty$ limit. Consider the cases where the given point charge is, or isn't, very close to an end. (This problem is partly quantitative and partly qualitative. Feel free to drop all factors of order 1 and just look at the dependence of various quantities on the given parameters, in particular $N$.)

3.6  *A charge inside a shell* *

Is the following reasoning correct or incorrect (if incorrect, state the error). A point charge $q$ lies at an off-center position inside a conducting spherical shell. The surface of the conductor is at constant potential, so, by the uniqueness theorem, the potential is constant inside. The field inside is therefore zero, so the charge experiences no force.

3.7  *Inside/outside asymmetry* **

If a point charge is located *outside* a hollow conducting shell, there is an electric field outside, but no electric field inside. On the other hand, if a point charge is located *inside* a hollow conducting shell, there is an electric field both inside and outside (although the external field would be zero in the special case where the shell happened to have charge exactly equal and opposite to the point charge). The situation is therefore not symmetric with respect to inside and

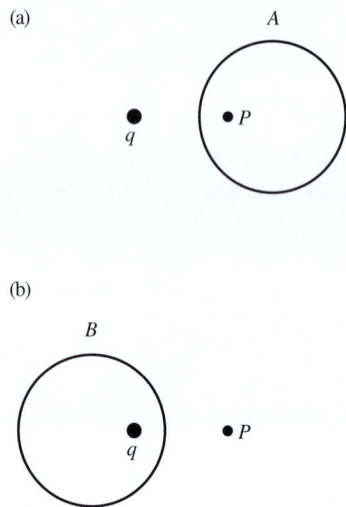

(a)

(b)

**Figure 3.24.**

outside. Explain why this is the case, by considering where electric field lines can begin and end.

3.8 *Inside or outside* **

A setup consists of a spherical metal shell and a point charge $q$. We are interested in the electric field at a given point $P$. In Fig. 3.24(a), if the shell is placed in position $A$ around point $P$, with the charge $q$ outside, then we know that the field at $P$ is zero by the uniqueness theorem. On the other hand, if the shell is placed in position $B$ around the charge $q$, with point $P$ outside, then we know that the field at $P$ is nonzero (see the example in Section 3.2).

However, we can transition continuously from one of these cases to the other by increasing the size of shell $A$ until the left part of it becomes an infinite plane between $q$ and $P$, and then considering this plane to be the right part of an infinite shell $B$, and then shrinking this shell down to the given size. During this process the point $P$ goes from being inside the shell to being outside. What's going on here? How can we transition from zero field to nonzero field at point $P$?

3.9 *Grounding a shell* **

A conducting spherical shell has charge $Q$ and radius $R_1$. A larger concentric conducting spherical shell has charge $-Q$ and radius $R_2$. If the outer shell is grounded, explain why nothing happens to the charge on it. If instead the inner shell is grounded, find its final charge.

3.10 *Why leave?* ***

In the setup in Problem 3.9, let the inner shell be grounded by connecting it to a large conducting neutral object very far away via a very thin wire that passes through a very small hole in the outer shell. If you think in terms of potentials (as you probably did if you solved Problem 3.9), then you can quickly see why some of the charge on the inner shell flows off to infinity. The potential of the inner shell is initially higher than the potential at infinity.

However, if you think in terms of forces on the positive charges on the inner shell, then things aren't as clear. A small bit of positive charge will certainly want to hop on the wire and follow the electric field across the gap to the larger shell. But when it gets to the larger shell, it seems like it has no reason to keep going to infinity, because the field is zero outside. And, even worse, the field will point *inward* once some positive charge has moved away from the shells. So it seems like the field will drag back any positive charge that has left. What's going on? Does charge actually leave the inner shell? If so, what is wrong with the above reasoning?

3.11 *How much work?* ∗

A charge $Q$ is located a distance $h$ above a conducting plane, just as in Fig. 3.10(a). Asked to predict the amount of work that would have to be done to move this charge out to infinite distance from the plane, one student says that it is the same as the work required to separate to infinite distance two charges $Q$ and $-Q$ that are initially $2h$ apart, hence $W = Q^2/4\pi\epsilon_0(2h)$. Another student calculates the force that acts on the charge as it is being moved and integrates $F\,dx$, but gets a different answer. What did the second student get, and who is right?

3.12 *Image charges for two planes* ∗∗

A point charge $q$ is located between two parallel infinite conducting planes, a distance $d$ from one and $\ell - d$ from the other. Where should image charges be located so that the electric field is everywhere perpendicular to the planes?

3.13 *Image charge for a grounded spherical shell* ∗∗∗

(a) A point charge $-q$ is located at $x = a$, and a point charge $Q$ is located at $x = A$. Show that the locus of points with $\phi = 0$ is a circle in the $xy$ plane (and hence a spherical shell in space).

(b) What must be the relation among $q$, $Q$, $a$, and $A$ so that the center of the circle is located at $x = 0$?

(c) Assuming that the relation you found in part (b) holds, what is the radius of the circle in terms of $a$ and $A$?

(d) Explain why the previous results imply the following statement: if a charge $Q$ is *externally* located a distance $A > R$ from the center of a grounded conducting spherical shell with radius $R$, then the *external* field due to the shell is the same as the field of an image point charge $-q = -QR/A$ located a distance $a = R^2/A$ from the center of the shell; see Fig. 3.25. The total external field is the sum of this field plus the field from $Q$. (The *internal* field is zero, by the uniqueness theorem.)

(e) Likewise for the following statement: if a charge $-q$ is *internally* located a distance $a < R$ from the center of a grounded conducting spherical shell with radius $R$, then the *internal* field due to the shell is the same as the field of an image point charge $Q = qR/a$ located a distance $A = R^2/a$ from the center of the shell; see Fig. 3.26. The total internal field is the sum of this field plus the field from $q$. (The *external* field is zero, because otherwise the shell would not have the same potential as infinity. Evidently a charge $+q$ flows onto the grounded shell.)

3.14 *Force from a conducting shell* ∗∗

A charge $Q$ is located a distance $r > R$ from the center of a grounded conducting spherical shell with radius $R$. Using the result

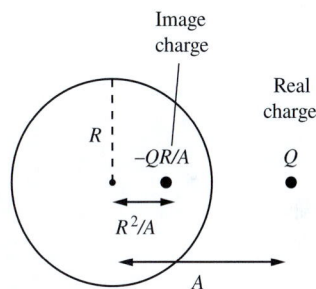

**Figure 3.25.**

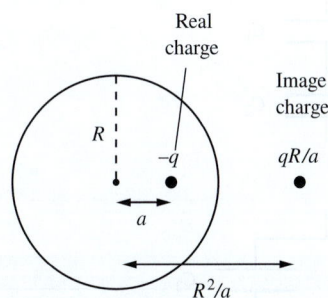

**Figure 3.26.**

from Problem 3.13, find the force from the shell on the charge $Q$. Consider the $r \approx R$ and $r \to \infty$ limits.

**3.15** *Dipole from a shell in a uniform field* ***

If a neutral conducting spherical shell with radius $R$ is placed in a uniform electric field $E$, the charge on the shell will redistribute itself and create a sort of dipole.

(a) Show that the external field due to the redistributed charge on the shell is in fact *exactly* equal to the field due to an idealized dipole at the center of the shell. What is the strength $p$ of the dipole?

(b) Using the form of the dipole field given in Eq. (2.36), verify that the total external field ($E$ plus the field from the shell) is perpendicular to the shell at the surface.

(c) What is the surface charge density as a function of position on the shell?

*Hint:* Use the result from Problem 3.13, and consider the uniform field $E$ to be generated by a charge $Q$ at position $x = -A$, plus a charge $-Q$ at position $x = A$. In the limit where both $Q$ and $A$ go to infinity (in an appropriate manner), the field at the location of the shell is finite, essentially uniform, and points in the positive $x$ direction.

**3.16** *Image charge for a nongrounded spherical shell* **

A charge $Q$ is located a distance $r > R$ from the center of a *nongrounded* conducting spherical shell with radius $R$ and total charge $q_s$. The field external to the shell can be mimicked by the combination of the image charge discussed in Problem 3.13, plus a second image charge. What is this second charge, and where is it located?

**3.17** *Capacitance of raindrops* *

$N$ charged raindrops with radius $a$ all have the same potential. Assume that they are far enough apart so that the charge distribution on each isn't affected by the others (that is, it is spherically symmetric). What is the total capacitance of this system? How does this capacitance compare with the capacitance in the case where the drops are combined into one big drop?

**3.18** *Adding capacitors* **

(a) Two capacitors, $C_1$ and $C_2$, are connected in series, as shown in Fig. 3.27(a). Show that the effective capacitance $C$ of the system is given by

$$\frac{1}{C} = \frac{1}{C_1} + \frac{1}{C_2}. \tag{3.39}$$

Check the $C_1 \to 0$ and $C_1 \to \infty$ limits.

(a)

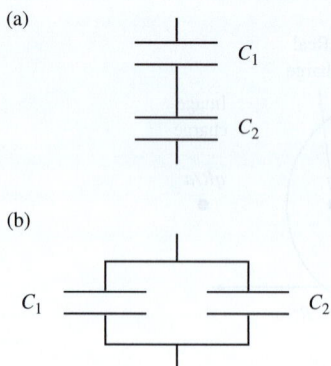

$C_1$

$C_2$

(b)

$C_1$

$C_2$

**Figure 3.27.**

(b) If the capacitors are instead connected in parallel, as shown in Fig. 3.27(b), show that the effective capacitance is given by

$$C = C_1 + C_2. \tag{3.40}$$

Again check the $C_1 \to 0$ and $C_1 \to \infty$ limits.

These two rules are the opposites of the rules for adding resistors (see Problem 4.3) and inductors (see Problem 7.13).

**3.19** *Uniform charge on a capacitor* **

Problem 3.4 shows that the charge distribution on an isolated conducting disk is not uniform. But when two oppositely charged disks (or any other planar shape) are placed very close to each other to form a capacitor, the charge distribution on each is essentially uniform, assuming the separation is small. Can you prove this?

**3.20** *Distribution of charge on a capacitor* **

Consider a parallel-plate capacitor with *different* magnitudes of charge on the two plates. Let the charges be $Q_1$ and $Q_2$ (which we normally set equal to $Q$ and $-Q$). Find the four amounts of charge on the inner and outer surfaces of the two plates.

**3.21** *A four-plate capacitor* **

Consider a capacitor made of four parallel plates with large area $A$, evenly spaced with small separation $s$. The first and third are connected by a wire, as are the second and fourth. What is the capacitance of this system?

**3.22** *A three-cylinder capacitor* **

A capacitor consists of three concentric cylindrical shells with radii $R$, $2R$, and $3R$. The inner and outer shells are connected by a wire, so they are at the same potential. The shells start neutral, and then a battery transfers charge from the middle shell to the inner/outer shells.

(a) If the final charge per unit length on the middle shell is $-\lambda$, what are the charges per unit length on the inner and outer shells?

(b) What is the capacitance per unit length of the system?

(c) If the battery is disconnected, what happens to the three charges-per-length on the shells if $\lambda_{\text{new}}$ is added to the outer shell?

**3.23** *Capacitance coefficients and C* **

Consider the setup in the example in Section 3.6. Explain why the relations in Eq. (3.24), which contain four capacitance coefficients, reduce properly to the simple $Q = C\phi$ statement for a parallel-plate capacitor, with $C$ given in Eq. (3.15).

3.24 *Human capacitance* ∗

Make a rough estimate of the capacitance of an isolated human body. (*Hint:* It must lie somewhere between that of an inscribed sphere and that of a circumscribed sphere.) By shuffling over a nylon rug on a dry winter day, you can easily charge yourself up to a couple of kilovolts – as shown by the length of the spark when your hand comes too close to a grounded conductor. How much energy would be dissipated in such a spark?

3.25 *Energy of a disk* ∗

Given that the capacitance of an isolated conducting disk of radius $a$ is $8\epsilon_0 a$, what is the energy stored in the electric field of such a disk when the net charge on the disk is $Q$? Compare this with the energy in the field of a nonconducting disk of the same radius that has an equal charge $Q$ distributed with uniform density over its surface. (See Exercise 2.56.) Which ought to be larger? Why?

3.26 *Force on a capacitor plate* ∗∗∗

A parallel-plate capacitor consists of a fixed plate and a movable plate that is allowed to slide in the direction parallel to the plates. Let $x$ be the distance of overlap, as shown in Fig. 3.28. The separation between the plates is fixed.

(a) Assume that the plates are electrically isolated, so that their charges $\pm Q$ are constant. In terms of $Q$ and the (variable) capacitance $C$, derive an expression for the leftward force on the movable plate. *Hint:* Consider how the energy of the system changes with $x$.

(b) Now assume that the plates are connected to a battery, so that the potential difference $\phi$ is held constant. In terms of $\phi$ and the capacitance $C$, derive an expression for the force.

(c) If the movable plate is held in place by an opposing force, then either of the above two setups could be the relevant one, because nothing is moving. So the forces in (a) and (b) should be equal. Verify that this is the case.

3.27 *Force on a capacitor plate, again* ∗∗∗

Repeat the three parts of Problem 3.26, but don't use the word "capacitance" in your solution. Instead find the stored energy by considering the energy density of the electric field. Write the force in terms of the overlap $x$, the separation $s$, the width of the plates $\ell$ (in the direction perpendicular to the page), and either the charge $Q$ (for part (a)) or the density $\sigma$ (for part (b)).

3.28 *Maximum energy storage between spheres* ∗∗

We want to design a spherical vacuum capacitor, with a given radius $a$ for the outer spherical shell, that will be able to store the greatest amount of electrical energy subject to the constraint that

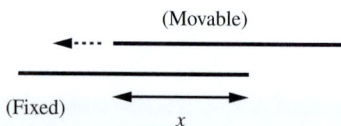

(Movable)

(Fixed)

$x$

**Figure 3.28.**

the electric field strength at the surface of the inner sphere may not exceed $E_0$. What radius $b$ should be chosen for the inner spherical conductor, and how much energy can be stored?

3.29  *Compressing a sphere* **

A spherical conducting shell has radius $R$ and potential $\phi$. If you want, you can consider it to be part of a capacitor with the other shell at infinity. You compress the shell down to essentially zero size (always keeping it spherical) while a battery holds the potential constant at $\phi$. By calculating the initial and final energies stored in the system, and also the work done by (or on) you and the battery, verify that energy is conserved. (Be sure to specify clearly what your conservation-of-energy statement is, paying careful attention to the signs of the various quantities.)

3.30  *Two ways of calculating energy* ***

A capacitor consists of two arbitrarily shaped conducting shells, with one inside the other. The inner conductor has charge $Q$, the outer has charge $-Q$. We know of two ways of calculating the energy $U$ stored in this system. We can find the electric field $E$ and then integrate $\epsilon_0 E^2/2$ over the volume between the conductors. Or if we know the potential difference $\phi$, we can write $U = Q\phi/2$ (or equivalently $U = C\phi^2/2$).

(a)  Show that these two methods give the same energy in the case of two concentric shells.

(b)  By using the identity $\nabla \cdot (\phi \nabla \phi) = (\nabla \phi)^2 + \phi \nabla^2 \phi$, show that the two methods give the same energy for conductors of any shape.

## Exercises

3.31  *In or out* *

A positive charge $q$ is placed at the center of each of the neutral cylindrical-ish hollow conducting shells whose cross sections are shown in Fig. 3.29. (White areas on the page denote vacuum; the shaded curves denote the metal of the conducting shells.) For each case, indicate roughly the induced charge distribution on the conductor. Be sure to indicate which part of the surface the charge lies on. Are your distributions consistent with the fact that there is no electric field in an empty cavity inside a conducting shell?

3.32  *Gravity screen* *

What is wrong with the idea of a gravity screen, something that will "block" gravity the way a metal sheet seems to "block" the electric field? Think about the difference between the gravitational source and electrical sources. Note that the walls of the box in

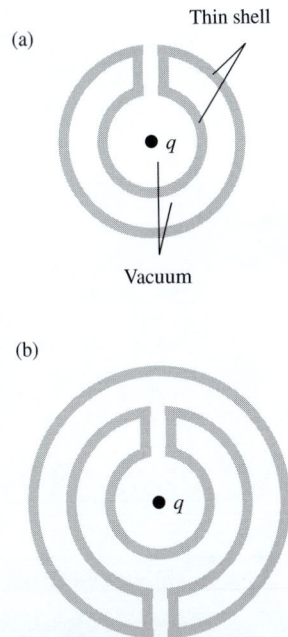

(a)

Thin shell

$q$

Vacuum

(b)

$q$

**Figure 3.29.**

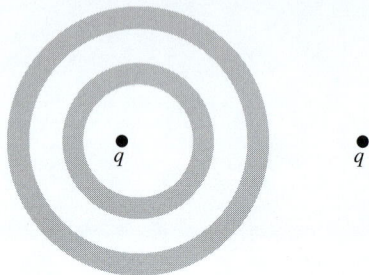

**Figure 3.30.**

Fig. 3.8 do not block the field of the outside sources but merely allow the surface charges to set up a compensating field. Why can't something of this sort be contrived for gravity? What would you need to accomplish it?

**3.33** *Two concentric shells* $**$

(a) The shaded regions in Fig. 3.30 represent two neutral concentric conducting spherical shells. The white regions represent vacuum. Two point charges $q$ are located as shown; the interior one is off-center. Draw a reasonably accurate picture of the field lines everywhere, and indicate the various charge densities. What quantities are spherically symmetric? (There are two possible cases for what your picture can look like, depending on how close the exterior point charge is; see Exercise 3.49. Take your pick.)

(b) Repeat the above tasks in the case where the two shells are connected by a wire, so that they are at the same potential.

**3.34** *Equipotentials* $**$

A point charge is located in the vicinity of a neutral conducting sphere. Make a rough sketch of a few equipotential surfaces; you need only indicate the qualitative features. How do the surfaces make the transition from very small circles (or spheres, in space) around the point charge to very large circles around the whole system? Explain why there must be points on the surface of the sphere where the electric field is zero.

**3.35** *Electric field at a corner* $***$

A very long conducting tube has a square cross section. The charge per unit length in the longitudinal direction is $\lambda$. Explain why the external electric field diverges at the corners of the tube. Does this result depend on the specific shape of the tube? What if the cross section is triangular or hexagonal? Or what if the point in question is at the tip of a cone or at a kink in a wire?

**3.36** *Zero flow* $***$

If you did Exercise 2.50, you found that the charge on each sphere is proportional to the radius $r$. As mentioned in that exercise, if the spheres are connected by a wire, no charge will flow in the wire. Imagine that one sphere is much smaller than the other. Then, since the electric field is proportional to $1/r^2$, the field is much larger at the surface of the smaller sphere than the larger sphere, because the charge is proportional only to $r$. So why doesn't the charge get repelled from the smaller sphere and flow through the wire to the larger sphere? *Hint:* See Problem 3.10.

**3.37** *A charge between two plates* **

Two parallel plates are connected by a wire so that they remain at the same potential. Let one plate coincide with the $xz$ plane and the other with the plane $y = s$. The distance $s$ between the plates is much smaller than the lateral dimensions of the plates. A point charge $Q$ is located between the plates at $y = b$ (see Fig. 3.31). What is the magnitude of the total surface charge on the inner surface of each plate?

Here are some helpful thoughts. The total surface charge on the inner surface of both plates must of course be $-Q$ (why?), and we can guess that a larger fraction of it will be found on the nearer plate. If the charge were very close to the left plate, $b \ll s$, the presence of the plate on the right couldn't make much difference. However, we want to know exactly how the charge divides. If you try to use an image method you will discover that you need an infinite chain of images, rather like the images you see in a barbershop with mirrors on both walls (see Problem 3.12). It is not easy to calculate the resultant field at any point on one of the surfaces (see Exercise 3.45). Nevertheless, the question we asked can be answered by a very simple calculation based on superposition. (*Hint:* Adding another charge $Q$ anywhere on the plane $y = b$ would just double the surface charge on each plate. In fact the total surface charge induced by any number of charges is independent of their position on the plane $y = b$. If only we had a sheet of uniform charge on this plane the electric fields would be simple, and we could use Gauss's law. Take it from there.)

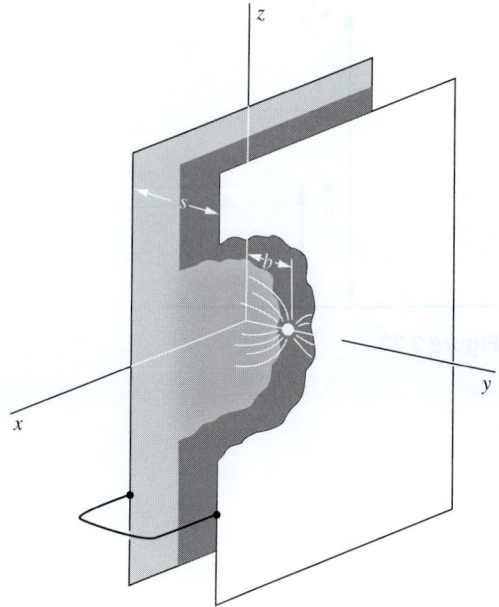

**Figure 3.31.**

**3.38** *Two charges and a plane* *

A positive point charge $Q$ is fixed a distance $\ell$ above a horizontal conducting plane. An equal negative charge $-Q$ is to be located somewhere along the perpendicular dropped from $Q$ to the plane. Where can $-Q$ be placed so that the total force on it will be zero?

**3.39** *A wire above the earth* *

By solving the problem of the point charge and the plane conductor, we have, in effect, solved every problem that can be constructed from it by superposition. For instance, suppose we have a straight wire 200 meters long, uniformly charged with $10^{-5}$ C per meter of length, running parallel to the earth at a height of 5 meters. What is the field strength at the surface of the earth, immediately below the wire? (For steady fields the earth behaves like a good conductor.) You may work in the approximation where the length of the wire is much greater than its height. What is the electrical force acting on the wire?

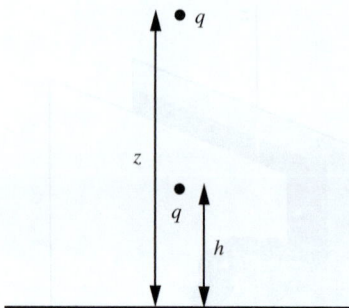

**Figure 3.32.**

3.40 *Direction of the force*  ∗∗

A point charge $q$ is located a fixed height $h$ above an infinite horizontal conducting plane, as shown in Fig. 3.32. Another point charge $q$ is located a height $z$ (with $z > h$) above the plane. The two charges lie on the same vertical line. If $z$ is only slightly larger than $h$, then the force on the top charge is clearly upward. But for larger values of $z$, is the force still always upward? Try to solve this without doing any calculations. *Hint:* Think dipole.

3.41 *Horizontal field line*  ∗∗

In the field of the point charge over the plane (Fig. 3.11), if you follow a field line that starts out from the point charge in a horizontal direction, that is, parallel to the plane, where does it meet the surface of the conductor? (You'll need Gauss's law and a simple integration.)

3.42 *Point charge near a corner*  ∗∗

Locate two charges $q$ each and two charges $-q$ each on the corners of a square, with like charges diagonally opposite one another. Show that there are two equipotential surfaces that are planes. In this way sketch qualitatively the field of the system where a single point charge is located symmetrically in the inside corner formed by bending a metal sheet through a right angle. Which configurations of conducting planes and point charges can be solved this way and which can't? How about a point charge located on the bisector of a 120° dihedral angle between two conducting planes?

3.43 *Images from three planes*  ∗∗

Imagine the $xy$ plane, the $xz$ plane, and the $yz$ plane all made of metal and soldered together at the intersections. A single point charge $Q$ is located a distance $d$ from each of the planes. Sketch the configuration of image charges you need to satisfy the boundary conditions. What is the direction and magnitude of the force that acts on the charge $Q$?

3.44 *Force on a charge between two planes*  ∗∗

A point charge $q$ is located between two parallel infinite conducting planes, a distance $b$ from one and $\ell - b$ from the other. Using the results from Problem 3.12 and Section 2.7, find an approximate expression for the force on the charge in the case where the charge is very close to one of the planes (that is, $b \ll \ell$).

3.45 *Charge on each plane*  ∗∗∗∗

(a) Consider a point charge $q$ located between two parallel infinite conducting planes. The planes are a distance $\ell$ apart, and the point charge is distance $b$ from the right plane. Using the result from Problem 3.12, show that the electric field on the inside surface of the right plane, at a point a distance $r$ from the axis containing all the image charges, is given by

$$4\pi\epsilon_0 E = \frac{2qb}{(b^2 + r^2)^{3/2}} + \sum_{n=1}^{\infty}\left( -\frac{2q(2n\ell - b)}{\left((2n\ell - b)^2 + r^2\right)^{3/2}} \right.$$
$$\left. + \frac{2q(2n\ell + b)}{\left((2n\ell + b)^2 + r^2\right)^{3/2}} \right).$$
$$(3.41)$$

(b) Since $\sigma = -\epsilon_0 E$, the above expression (divided by $-4\pi$) gives the density on the right plane.[9] The task of Exercise 3.37 is to determine (via a slick method) the total charge on each of the two planes. Complete the same task here (in a much more complicated manner) by directly integrating the density $\sigma$ over the entire right plane. This task is tricky due to the following complication.

If you integrate each term in the above sum separately, you will run into difficulty because every term in the (infinite) sum gives $\pm q$. (This is expected, because the result in Eq. (3.5) does not depend on the distance from the charge to the plane.) So the sum is not well defined. What you will need to do is group the sum in the pairs indicated above, and integrate out to a fixed large value of $r$; call it $R$. After *first* combining the terms in each pair (making suitable approximations with $b \ll R$), you will *then* obtain a sum over $n$ that converges. This sum can be calculated by converting it to an integral (which is a valid step in the large-$R$ limit). The process is quite involved, but the end result is very clean. Your answer will be independent of $R$, so letting $R \to \infty$ won't affect it.

3.46 *Sphere and plane image charges* ∗

Using the result from Problem 3.13, show that in the case where the real charge is very close to the grounded spherical shell (either inside or outside), the setup reduces correctly (that is, yields the correct value and location of the image charge) to the image-charge setup for the infinite plane discussed in Section 3.4.

3.47 *Bump on a plane* ∗∗

An infinite conducting plane has a hemispherical bump on it with radius $R$. A point charge $Q$ is located a distance $R$ above the top of the hemisphere, as shown in Fig. 3.33. With the help of Problem 3.13, find the image charges needed to make the electric field perpendicular to the plane and the hemisphere at all points.

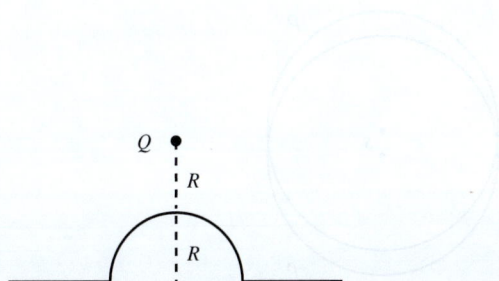

Figure 3.33.

[9] The minus sign comes from the fact that a positive field produces flux *into* a Gaussian surface containing the right plane.

3.48  *Density at top of bump on a plane* ***

If you solved Exercise 3.47, here's an extension. Assume that the distance $A$ from the given charge $Q$ to the plane is much larger than the radius $R$ of the hemisphere. Show that in the $A \gg R$ limit, the surface charge density at the top of the hemisphere is three times as large as the density would be on the plane (at the foot of the perpendicular from $Q$) if we simply had a flat plane without the bump.

3.49  *Positive or negative density* **

A positive point charge $Q$ is located outside a nongrounded conducting spherical shell with radius $R$. The net charge on the shell is also $Q$. If the point charge is located very close to the shell, the surface charge density on the near part of the shell is negative. But if the point charge is located far away, the surface density is positive everywhere (and essentially uniform). Using the results from Problems 3.13 and 3.16, show that the cutoff between these two cases occurs when the point charge is a distance $R(3 + \sqrt{5})/2$ from the center of the shell, or equivalently $R(1 + \sqrt{5})/2$ from the surface; this factor is the golden ratio.

3.50  *Attractive or repulsive?* **

A point charge $Q$ is located a distance $r > R$ from the center of a nongrounded conducting spherical shell with radius $R$ and net charge that is also $Q$. If the point charge is very far from the shell, then the shell looks essentially like a point charge $Q$, so the force between the two objects is repulsive. But if the point charge is very close to the shell, then the excess negative charge on the near side of the shell dominates, so the force is attractive. Using the results from Problems 3.13 and 3.16, show that the value of $r$ where the force makes the transition from repulsive to attractive is $r = R(1 + \sqrt{5})/2 \approx (1.618)R$. The factor here is the golden ratio. (In your solution, don't panic if you end up with a quintic equation. Just show that it has a factor of the form $x^2 - x - 1$.)

3.51  *Conducting sphere in a uniform field* ****

A neutral conducting spherical shell with radius $R$ is placed in a uniform electric field $E$. Problem 3.15 presents one method of finding the resulting surface charge density, which comes out to be $\sigma = 3\epsilon_0 E \cos\theta$, where $\theta$ is measured relative to the direction of the uniform field $E$. This problem presents another method.

(a) Consider two solid *non*conducting spheres with uniform volume charge densities $\pm\rho$. Imagine that they are initially placed right on top of each other, and that one is then manually moved a distance $s$, as shown in Fig. 3.34. (Assume that they can somehow freely pass through each other.) With the centers

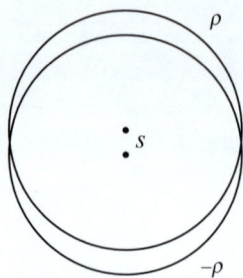

**Figure 3.34.**

now $s$ apart, find the electric field due to the two spheres in the region of overlap (where the net density is zero). Ignore the additional uniform field $E$ for now. The technique from Problem 1.27 will be useful.

(b) With the centers $s$ apart, find the force that one shell exerts on the other. You may assume that $s \ll R$. *Hint:* In Fig. 3.35, the part of the top sphere that lies within the dashed-line sphere experiences no force from the bottom sphere, because this part is symmetric with respect to the center of the bottom sphere. So we care only about the force from the bottom sphere on the shaded part of the top sphere. For small $s$, all of this region lies essentially right on the surface of the bottom sphere, so if you can find the thickness as a function of position, then you should be able to find the total force.

(c) Now let's introduce the uniform field $E$. If the two spheres are placed in this field, equilibrium will be reached when the mutual attraction balances the repulsion due to the field $E$. What is the resulting separation $s$ between the centers?

(d) Given the $s$ you just found, show that the total electric field (due to the two spheres plus the uniform field $E$) in the overlap region is zero. For small $s$, it then follows that the surface of the sphere is at constant potential, which means that we have re-created the conducting-shell boundary condition.

(e) By the uniqueness theorem, the field and surface charge density for our two-sphere system is the same (for small $s$) as the field and surface charge density for a conducting shell. By considering Fig. 3.34, show that the surface charge density on a conducting shell in a uniform field $E$ is $\sigma = 3\epsilon_0 E \cos\theta$.

**3.52** *Aluminum capacitor* ✶

Two aluminized optical flats 15 cm in diameter are separated by a gap of 0.04 mm, forming a capacitor. What is the capacitance in picofarads?

**3.53** *Inserting a plate* ✶✶

If the capacitance in Fig. 3.36(a) is $C$, what is the capacitance in Fig. 3.36(b), where a third plate is inserted and the outer plates are connected by a wire?

**3.54** *Dividing the surface charge* ✶✶

Three conducting plates are placed parallel to one another as shown in Fig. 3.37. The outer plates are connected by a wire. The inner plate is isolated and has a net surface charge density of $\sigma$ (the combined value from the top and bottom faces of the plate). What are the surface densities, $\sigma_1$ and $\sigma_2$, on the top and bottom faces of the inner plate?

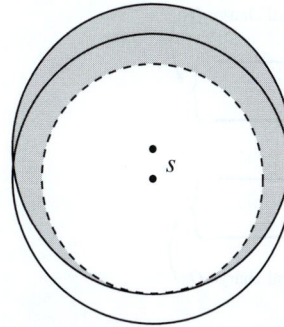

**Figure 3.35.**

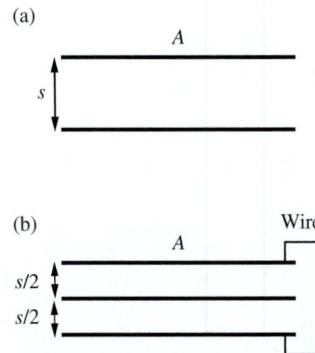

**Figure 3.36.**

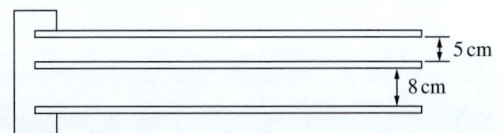

**Figure 3.37.**

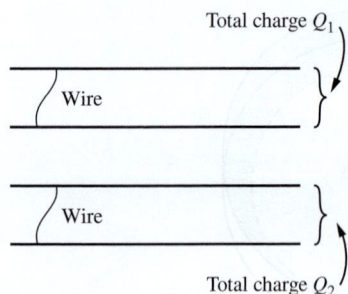

Total charge $Q_1$

Wire

Wire

Total charge $Q_2$

**Figure 3.38.**

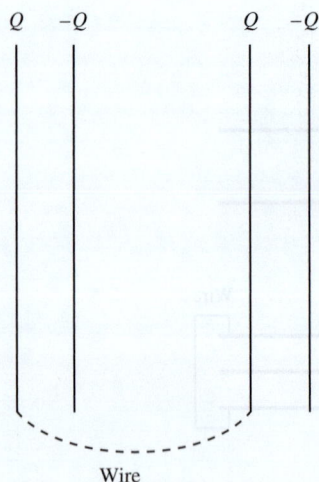

$Q$   $-Q$        $Q$   $-Q$

Wire

**Figure 3.39.**

**3.55  *Two pairs of plates*** **

Four conducting plates lie parallel to each other, as shown in Fig. 3.38. The spacings between them are arbitrary (but small compared with the lateral dimensions). The top two plates are connected by a wire so that they are at the same potential, and likewise for the bottom two. A total charge $Q_1$ resides on the top two plates, and a total charge $Q_2$ on the bottom two. What is the charge on each of the four plates?

**3.56  *Field just outside a capacitor*** **

A capacitor consists of two disks with radius $R$, small separation $s$, and surface charge densities $\pm\sigma$. Find the electric field just outside the capacitor, an infinitesimal distance from the center of the positive disk.

**3.57  *A 2N-plate capacitor*** **

Consider the setup in Problem 3.21, but now with $2N$ parallel plates instead of four. The first, third, fifth, etc. plates are connected by wires, and likewise for the second, fourth, sixth, etc. plates. What is the capacitance of this system? What does it equal in the $N \to \infty$ limit?

**3.58  *Capacitor paradox*** **

Two capacitors with the same capacitance $C$ and charge $Q$ are placed next to each other, as shown in Fig. 3.39. The two positive plates are then connected by a wire. Will charge flow in the wire? Consider two possible reasonings:

(A) Before the plates are connected, the potential differences of the two capacitors are the same (because $Q$ and $C$ are the same). So the potentials of the two positive plates are equal. Therefore, no charge will flow in the wire when the plates are connected.

(B) Number the plates 1 through 4, from left to right. Before the plates are connected, there is zero electric field in the region between the capacitors, so plate 3 must be at the same potential as plate 2. But plate 2 is at a lower potential than plate 1. Therefore, plate 3 is at a lower potential than plate 1, so charge will flow in the wire when the plates are connected.

Which reasoning is correct, and what is wrong with the wrong reasoning?

**3.59  *Coaxial capacitor*** **

A capacitor consists of two coaxial cylinders of length $L$, with outer and inner radii $a$ and $b$. Assume $L \gg a - b$, so that end corrections may be neglected. Show that the capacitance is $C = 2\pi\epsilon_0 L/\ln(a/b)$. Verify that if the gap between the cylinders, $a - b$, is very small compared with the radius, this result reduces to one that could have been obtained by using the formula for the parallel-plate capacitor.

**3.60  *A three-shell capacitor* ∗∗**
A capacitor consists of three concentric spherical shells with radii $R$, $2R$, and $3R$. The inner and outer shells are connected by a wire (passing through a hole in the middle shell, without touching it), so they are at the same potential. The shells start neutral, and then a battery transfers charge from the middle shell to the inner and outer shells.

(a) If the final charge on the middle shell is $-Q$, what are the charges on the inner and outer shells?

(b) What is the capacitance of the system?

(c) If the battery is disconnected, what happens to the three charges on the shells if charge $q$ is added to the outer shell?

**3.61  *Capacitance of a spheroid* ∗∗**
Here is the exact formula for the capacitance $C$ of a conductor in the form of a prolate spheroid of length $2a$ and diameter $2b$:

$$C = \frac{8\pi\epsilon_0 a\epsilon}{\ln\left(\dfrac{1+\epsilon}{1-\epsilon}\right)}, \qquad \text{where} \qquad \epsilon = \sqrt{1 - \frac{b^2}{a^2}}. \qquad (3.42)$$

First verify that the formula reduces to the correct expression for the capacitance of a sphere if $b \to a$. Now imagine that the spheroid is a charged water drop. If this drop is deformed at constant volume and constant charge $Q$ from a sphere to a prolate spheroid, will the energy stored in the electric field increase or decrease? (The volume of the spheroid is $(4/3)\pi ab^2$.)

**3.62  *Deriving C for a spheroid* ∗∗∗**
If you worked Exercise 2.44, use that result to derive the formula given in Exercise 3.61 for the capacitance of an isolated conductor of prolate spheroidal shape.

**3.63  *Capacitance coefficients for shells* ∗∗**
A capacitor consists of two concentric spherical shells. Label the inner shell, of radius $b$, as conductor 1; and label the outer shell, of radius $a$, as conductor 2. For this two-conductor system, find $C_{11}$, $C_{22}$, and $C_{12}$.

**3.64  *Capacitance-coefficient symmetry* ∗∗**
Here are some suggestions that should enable you to construct a proof that $C_{12}$ must always equal $C_{21}$. We know that, when an element of charge $dQ$ is transferred from zero potential to a conductor at potential $\phi$, some external agency has to supply an amount of energy $\phi\, dQ$. Consider a two-conductor system in which the two

conductors have been charged so that their potentials are, respectively, $\phi_{1f}$ and $\phi_{2f}$ ("f" for "final"). This condition might have been brought about, starting from a state with all charges and potentials zero, in many different ways. Two possible ways are of particular interest.

(a) Keep $\phi_2$ at zero while raising $\phi_1$ gradually from zero to $\phi_{1f}$. Then raise $\phi_2$ from zero to $\phi_{2f}$ while holding $\phi_1$ constant at $\phi_{1f}$.

(b) Carry out a similar program with the roles of 1 and 2 exchanged, that is, raise $\phi_2$ from zero to $\phi_{2f}$ first, and so on.

Compute the total work done by external agencies, for each of the two charging programs. Then complete the argument.

**3.65  *Capacitor energy*  ∗**

We found in Section 3.7 that the energy stored in a capacitor is $U = Q\phi/2$. Show that Eq. (2.32) yields this same result.

**3.66  *Adding a capacitor*  ∗∗**

A 100 pF capacitor is charged to 100 volts. After the charging battery is disconnected, the capacitor is connected in parallel with another capacitor. If the final voltage is 30 volts, what is the capacitance of the second capacitor? How much energy was lost, and what happened to it?

**3.67  *Energy in coaxial tubes*  ∗∗**

Two coaxial aluminum tubes are 30 cm long. The outer diameter of the inner tube is 3 cm, the inner diameter of the outer tube is 4 cm. When these are connected to a 45 volt battery, how much energy is stored in the electric field between the tubes?

**3.68  *Maximum energy storage between cylinders*  ∗∗**

We want to design a cylindrical vacuum capacitor, with a given radius $a$ for the outer cylindrical shell, that will be able to store the greatest amount of electrical energy per unit length, subject to the constraint that the electric field strength at the surface of the inner cylinder may not exceed $E_0$. What radius $b$ should be chosen for the inner cylindrical conductor, and how much energy can be stored per unit length?

**3.69  *Force, and potential squared*  ∗**

(a) In Gaussian units, show that the square of a potential difference $(\phi_2 - \phi_1)^2$ has the same dimensions as force. (In SI units, $\epsilon_0(\phi_2 - \phi_1)^2$ has the same units as force.) This tells us that the electrostatic forces between bodies will largely be determined, as to order of magnitude, by the potential differences involved. Dimensions will enter only in ratios, and there may be some constants like $4\pi$. What is the order of magnitude of force you expect with 1 statvolt potential difference between something and something else?

(b) Practically achievable potential differences are rather severely limited, for reasons having to do with the structure of matter. The highest man-made difference of electric potential is about $10^7$ volts, achieved by a Van de Graaff electrostatic generator operating under high pressure. (Billion-volt accelerators do not involve potential differences that large.) How many pounds force are you likely to find associated with a "square megavolt"? These considerations may suggest why electrostatic motors have not found much application.

**3.70** *Force and energy for two plates* ∗∗

Calculate the electrical force that acts on one plate of a parallel-plate capacitor. The potential difference between the plates is 10 volts, and the plates are squares 20 cm on a side with a separation of 3 cm. If the plates are insulated so the charge cannot change, how much external work could be done by letting the plates come together? Does this equal the energy that was initially stored in the electric field?

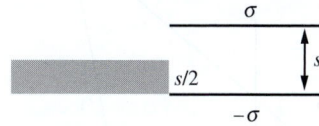

**Figure 3.40.**

**3.71** *Conductor in a capacitor* ∗∗

(a) The plates of a capacitor have area $A$ and separation $s$ (assumed to be small). The plates are isolated, so the charges on them remain constant; the charge densities are $\pm\sigma$. A neutral conducting slab with the same area $A$ but thickness $s/2$ is initially held outside the capacitor; see Fig. 3.40. The slab is released. What is its kinetic energy at the moment it is completely inside the capacitor? (The slab will indeed get drawn into the capacitor, as evidenced by the fact that the kinetic energy you calculate will be positive.)

(b) Same question, but now let the plates be connected to a battery that maintains a constant potential difference. The charge densities are initially $\pm\sigma$. (Don't forget to include the work done by the battery, which you will find to be nonzero.)

**3.72** *Force on a capacitor sheet* ∗∗∗

The aluminum sheet $A$ shown in Fig. 3.41 is suspended by an insulating thread between the surfaces formed by the bent aluminum sheet $B$. The sheets $A$ and $B$ are oppositely charged; the difference in potential is $V$. This causes a force $F$, in addition to the weight of $A$, pulling $A$ downward. If we can measure $F$ and know the various dimensions, we should be able to infer $V$. As an application of Eq. (3.32), work out a formula giving $V$ in terms of $F$ and the relevant dimensions.

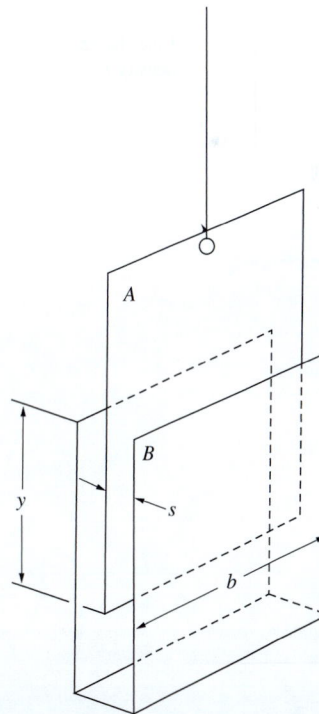

**Figure 3.41.**

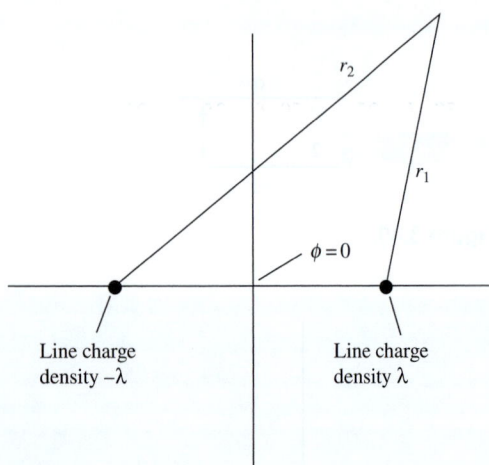

**Figure 3.42.**

Line charge
density $-\lambda$

Line charge
density $\lambda$

$\phi = 0$

$r_2$

$r_1$

**3.73** *Force on a coaxial capacitor* ∗∗∗

A cylinder with 4 cm outer diameter hangs, with its axis vertical, from one arm of a beam balance. The lower portion of the hanging cylinder is surrounded by a stationary cylinder, coaxial, with inner diameter 6 cm. Calculate the magnitude of the force tending to pull the hanging cylinder further down when the potential difference between the two cylinders is held constant at 5 kilovolts.

**3.74** *Equipotentials for two pipes* ∗∗∗

A typical two-dimensional boundary-value problem is that of two parallel circular conducting cylinders, such as two metal pipes, of infinite length and at different potentials. These two-dimensional problems happen to be much more tractable than three-dimensional problems, mathematically. In fact, the key to all problems of the "two-pipe" class is given by the field around two parallel line charges of equal and opposite linear density; see Fig. 3.42. All equipotential surfaces in this field are circular cylinders! See if you can prove this. (And all field lines are circular too, but you don't have to prove that here.) It is easiest to work with the potential, but you must note that one cannot set the potential zero at infinity in a two-dimensional system. Let zero potential be at the line midway between the two line charges, that is, at the origin in the cross-sectional diagram. The potential at any point is the sum of the potentials calculated for each line charge separately. This should lead you quickly to the discovery that the potential is simply proportional to $\ln(r_2/r_1)$ and is therefore constant on a curve traced by a point whose distances from two points are in a constant ratio. Make a sketch showing some of the equipotentials.

**3.75** *Average of six points* ∗

Let $\phi(x, y, z)$ be any function that can be expanded in a power series around a point $(x_0, y_0, z_0)$. Write a Taylor series expansion for the value of $\phi$ at each of the six points $(x_0 + \delta, y_0, z_0)$, $(x_0 - \delta, y_0, z_0)$, $(x_0, y_0 + \delta, z_0)$, $(x_0, y_0 - \delta, z_0)$, $(x_0, y_0, z_0 + \delta)$, $(x_0, y_0, z_0 - \delta)$, which symmetrically surround the point $(x_0, y_0, z_0)$ at a distance $\delta$. Show that, if $\phi$ satisfies Laplace's equation, the average of these six values is equal to $(x_0, y_0, z_0)$ through terms of the third order in $\delta$.

**3.76** *The relaxation method* ∗∗

Here's how to solve Laplace's equation approximately, for given boundary values, using nothing but arithmetic. The method is the relaxation method mentioned in Section 3.8, and it is based on the result of Exercise 3.75. For simplicity we take a two-dimensional

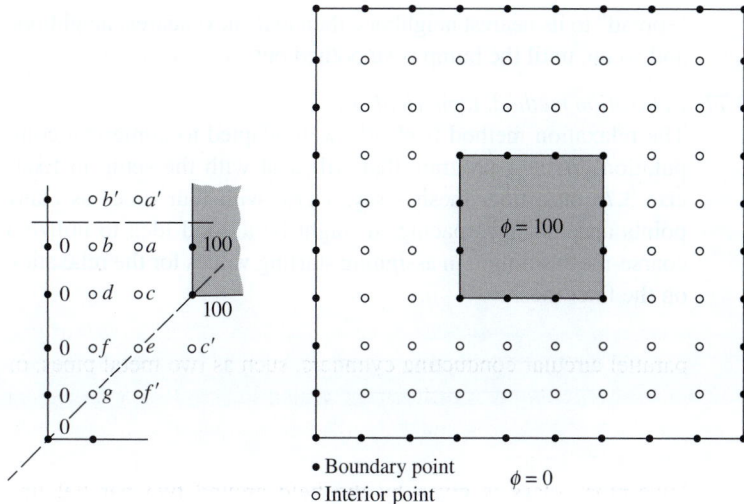

**Figure 3.43.**
Replace value at an interior point by $1/4 \times$ sum of its four neighbors: $c \to (100 + a + d + e)/4$; keep $a' = a$, $b' = b$, $c' = c$, and $f' = f$. Suggested starting values: $a = 50$, $b = 25$, $c = 50$, $d = 25$, $e = 50$, $f = 25$, $g = 25$.

• Boundary point
○ Interior point

$\phi = 0$

$\phi = 100$

example. In Fig. 3.43 there are two square equipotential boundaries, one inside the other. This might be a cross section through a capacitor made of two sizes of square metal tubing. The problem is to find, for an array of discrete points, numbers that will be a good approximation to the values at those points of the exact two-dimensional potential function $\phi(x, y)$. We make the array rather coarse, to keep the labor within bounds.

Let us assign, arbitrarily, potential 100 to the inner boundary and zero to the outer. All points on these boundaries retain those values. You could start with any values at the interior points, but time will be saved by a little judicious guesswork. We know the correct values must lie between 0 and 100, and we expect that points closer to the inner boundary will have higher values than those closer to the outer boundary. Some reasonable starting values are suggested in the figure. Obviously, you should take advantage of the symmetry of the configuration: only seven different interior values need to be computed. Now you simply go over these seven interior lattice points in some systematic manner, replacing the value at each interior point by the average of its four neighbors. Repeat until all changes resulting from a sweep over the array are acceptably small. For this exercise, let us agree that it will be time to quit when no change larger in absolute magnitude than one unit occurs in the course of the sweep. Enter your final values on the array, and sketch the approximate course that two equipotentials, for $\phi = 25$ and $\phi = 50$, would have in the actual continuous $\phi(x, y)$.

The relaxation of the values toward an eventually unchanging distribution is closely related to the physical phenomenon of *diffusion*. If you start with much too high a value at one point, it will "spread" to its nearest neighbors, then to its next nearest neighbors, and so on, until the bump is smoothed out.

**3.77** *Relaxation method, numerical* ***

The relaxation method is clearly well adapted to numerical computation. Write a program that will deal with the setup in Exercise 3.76 on a finer mesh – say, a grid with four times as many points and half the spacing. It might be a good idea to utilize a coarse-mesh solution in assigning starting values for the relaxation on the finer mesh.

**4**

# Electric currents

**Overview** In this chapter we discuss charge in motion, or electric *current*. The *current density* is defined as the current per cross-sectional area. It is related to the charge density by the *continuity equation*. In most cases, the current density is proportional to the electric field; the constant of proportionality is called the *conductivity,* with the inverse of the conductivity being the *resistivity.* *Ohm's law* gives an equivalent way of expressing this proportionality. We show in detail how the conductivity arises on a molecular level, by considering the drift velocity of the charge carriers when an electric field is applied. We then look at how this applies to *metals* and *semiconductors.* In a *circuit*, an *electromotive force (emf)* drives the current. A battery produces an emf by means of chemical reactions. The current in a circuit can be found either by reducing the circuit via the series and parallel rules for resistors, or by using *Kirchhoff's rules.* The *power* dissipated in a resistor depends on the resistance and the current passing through it. Any circuit can be reduced to a *Thévenin equivalent* circuit involving one resistor and one emf source. We end the chapter by investigating how the current changes in an *RC* circuit.

## 4.1 Electric current and current density

An electric current is charge in motion. The carriers of the charge can be physical particles like electrons or protons, which may or may not be attached to larger objects, atoms or molecules. Here we are not concerned with the nature of the charge carriers but only with the net transport of electric charge their motion causes. The electric current in a wire is the amount of charge passing a fixed mark on the wire in unit time. The SI unit of current is the coulomb/second, which is called an *ampere* (amp, or A):

$$1 \text{ ampere} = 1 \, \frac{\text{coulomb}}{\text{second}}. \tag{4.1}$$

In Gaussian units current is expressed in esu/second. A current of 1 A is the same as a current of $2.998 \cdot 10^9$ esu/s, which is equivalent to $6.24 \cdot 10^{18}$ elementary electronic charges per second.

It is the net charge transport that counts, with due regard to sign. Negative charge moving east is equivalent to positive charge moving west. Water flowing through a hose could be said to involve the transport of an immense amount of charge – about $3 \cdot 10^{23}$ electrons per gram of water! But since an equal number of protons move along with the electrons (every water molecule contains ten of each), the electric current is zero. On the other hand, if you were to charge negatively a nylon thread and pull it steadily through a nonconducting tube, that would constitute an electric current, in the direction opposite to that of the motion of the thread.

We have been considering current along a well-defined path, like a wire. If the current is *steady* – that is, unchanging in time – it must be the same at every point along the wire, just as with steady traffic the same number of cars must pass, per hour, different points along an unbranching road.

A more general kind of current, or charge transport, involves charge carriers moving around in three-dimensional space. To describe this we need the concept of *current density*. We have to consider average quantities, for charge carriers are discrete particles. We must suppose, as we did in defining the charge density $\rho$, that our scale of distances is such that any small region we wish to average over contains very many particles of any class we are concerned with.

Consider first a special situation in which there are $n$ particles per cubic meter, on the average, all moving with the same vector velocity $\mathbf{u}$ and carrying the same charge $q$. Imagine a small frame of area $\mathbf{a}$ fixed in some orientation, as in Fig. 4.1(a). How many particles pass through the frame in a time interval $\Delta t$? If $\Delta t$ begins the instant shown in Fig. 4.1(a) and (b), the particles destined to pass through the frame in the next $\Delta t$ interval will be just those now located within the oblique prism in Fig. 4.1(b). This prism has the frame area as its base and an edge length $u \, \Delta t$, which is the distance any particle will travel in a time $\Delta t$. Particles outside this prism will either miss the window or fail to reach it. The volume of the prism is the product (base) × (altitude), or $au \, \Delta t \cos \theta$, which can be written $\mathbf{a} \cdot \mathbf{u} \, \Delta t$. On the average, the number of particles found in such a volume will be $n \mathbf{a} \cdot \mathbf{u} \, \Delta t$. Hence the average *rate* at which charge is

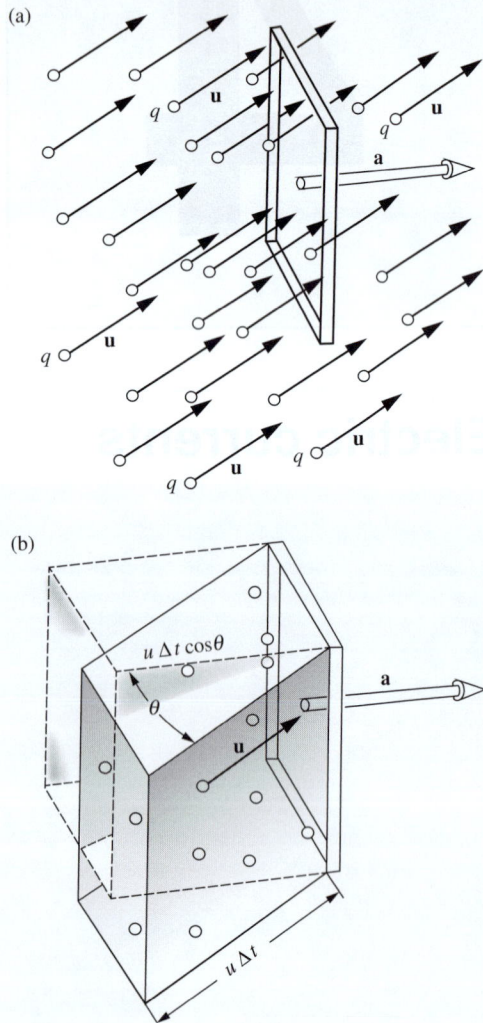

**Figure 4.1.**
(a) A swarm of charged particles all moving with the same velocity $u$. The frame has area $a$. The particles that will pass through the frame in the next $\Delta t$ seconds are those now contained in the oblique prism (b). The prism has base area $a$ and altitude $u \, \Delta t \cos \theta$, hence its volume is $au \, \Delta t \cos \theta$ or $\mathbf{a} \cdot \mathbf{u} \, \Delta t$.

passing through the frame, that is, the current through the frame, which we shall call $I_a$, is

$$I_a = \frac{q(n\mathbf{a} \cdot \mathbf{u}\,\Delta t)}{\Delta t} = nq\mathbf{a} \cdot \mathbf{u}. \tag{4.2}$$

Suppose we had many classes of particles in the swarm, differing in charge $q$, in velocity vector $\mathbf{u}$, or in both. Each would make its own contribution to the current. Let us tag each kind by a subscript $k$. The $k$th class has charge $q_k$ on each particle, moves with velocity vector $\mathbf{u}_k$, and is present with an average population density of $n_k$ such particles per cubic meter. The resulting current through the frame is then

$$I_a = n_1 q_1 \mathbf{a} \cdot \mathbf{u}_1 + n_2 q_2 \mathbf{a} \cdot \mathbf{u}_2 + \cdots = \mathbf{a} \cdot \sum_k n_k q_k \mathbf{u}_k. \tag{4.3}$$

On the right is the scalar product of the vector $\mathbf{a}$ with a vector quantity that we shall call the current density $\mathbf{J}$:

$$\boxed{\mathbf{J} = \sum_k n_k q_k \mathbf{u}_k} \tag{4.4}$$

The SI unit of current density is amperes per square meter $(\mathrm{A/m^2})$,[1] or equivalently coulombs per second per square meter $(\mathrm{C\,s^{-1}m^{-2}})$, although technically the ampere is a fundamental SI unit while the coulomb is not (a coulomb is defined as one ampere-second). The Gaussian unit of current density is esu per second per square centimeter $(\mathrm{esu\,s^{-1}cm^{-2}})$.

Let's look at the contribution to the current density $\mathbf{J}$ from one variety of charge carriers, electrons say, which may be present with many different velocities. In a typical conductor, the electrons will have an almost random distribution of velocities, varying widely in direction and magnitude. Let $N_e$ be the total number of electrons per unit volume, of all velocities. We can divide the electrons into many groups, each of which contains electrons with nearly the same speed and direction. The *average velocity* of all the electrons, like any average, would then be calculated by summing over the groups, weighting each velocity by the number in the group, and dividing by the total number. That is,

$$\bar{\mathbf{u}} = \frac{1}{N_e} \sum_k n_k \mathbf{u}_k. \tag{4.5}$$

We use the bar over the top, as in $\bar{\mathbf{u}}$, to mean the average over a distribution. Comparing Eq. (4.5) with Eq. (4.4), we see that the contribution

---

[1] Sometimes one encounters current density expressed in $\mathrm{A/cm^2}$. Nothing is wrong with that; the meaning is perfectly clear as long as the units are stated. (Long before SI was promulgated, two or three generations of electrical engineers coped quite well with amperes per square inch!)

of the electrons to the current density can be written simply in terms of the average electron velocity. Remembering that the electron charge is $q = -e$, and using the subscript $e$ to show that all quantities refer to this one type of charge carrier, we can write

$$\mathbf{J}_e = -eN_e\bar{\mathbf{u}}_e. \tag{4.6}$$

This may seem rather obvious, but we have gone through it step by step to make clear that the current through the frame depends only on the average velocity of the carriers, which often is only a tiny fraction, in magnitude, of their random speeds. Note that Eq. (4.6) can also be written as $\mathbf{J}_e = \rho_e\bar{\mathbf{u}}_e$, where $\rho_e = -eN_e$ is the volume charge density of the electrons.

## 4.2 Steady currents and charge conservation

The current $I$ flowing through any surface $S$ is just the surface integral

$$I = \int_S \mathbf{J} \cdot d\mathbf{a}. \tag{4.7}$$

We speak of a steady or stationary current system when the current density vector $\mathbf{J}$ remains constant in time everywhere. Steady currents have to obey the law of charge conservation. Consider some region of space completely enclosed by the balloonlike surface $S$. The surface integral of $\mathbf{J}$ over all of $S$ gives the rate at which charge is leaving the volume enclosed. Now if charge forever pours out of, or into, a fixed volume, the charge density inside must grow infinite, unless some compensating charge is continually being created there. But charge creation is just what never happens. Therefore, for a truly time-independent current distribution, the surface integral of $\mathbf{J}$ over *any* closed surface must be zero. This is completely equivalent to the statement that, at every point in space,

$$\text{div}\,\mathbf{J} = 0. \tag{4.8}$$

To appreciate the equivalence, recall Gauss's theorem and our fundamental definition of divergence in terms of the surface integral over a small surface enclosing the location in question.

We can make a more general statement than Eq. (4.8). Suppose the current is not steady, $\mathbf{J}$ being a function of $t$ as well as of $x$, $y$, and $z$. Then, since $\int_S \mathbf{J} \cdot d\mathbf{a}$ is the instantaneous rate at which charge is *leaving* the enclosed volume, while $\int_V \rho\, dv$ is the total charge *inside* the volume at any instant, we have

$$\int_S \mathbf{J} \cdot d\mathbf{a} = -\frac{d}{dt}\int_V \rho\, dv. \tag{4.9}$$

Letting the volume in question shrink down around any point $(x, y, z)$, the relation expressed in Eq. (4.9) becomes:[2]

$$\mathrm{div}\,\mathbf{J} = -\frac{\partial \rho}{\partial t} \qquad \text{(time-dependent charge distribution).} \qquad (4.10)$$

The time derivative of the charge density $\rho$ is written as a partial derivative since $\rho$ will usually be a function of spatial coordinates as well as time. Equations (4.9) and (4.10) express the *(local) conservation of charge:* no charge can flow away from a place without diminishing the amount of charge that is there. Equation (4.10) is known as the *continuity equation.*

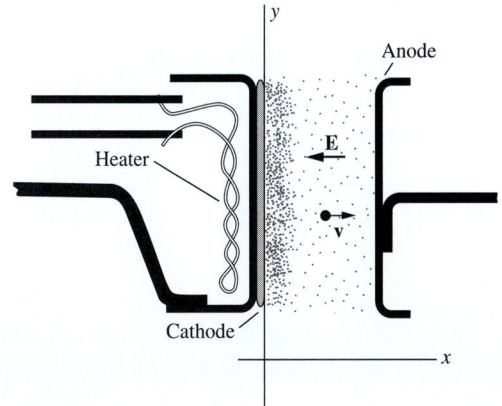

**Figure 4.2.**
A vacuum diode with plane-parallel cathode and anode.

**Example (Vacuum diode)** An instructive example of a stationary current distribution occurs in the plane diode, a two-electrode vacuum tube; see Fig. 4.2. One electrode, the cathode, is coated with a material that emits electrons copiously when heated. The other electrode, the anode, is simply a metal plate. By means of a battery the anode is maintained at a positive potential with respect to the cathode. Electrons emerge from this hot cathode with very low velocities and then, being negatively charged, are accelerated toward the positive anode by the electric field between cathode and anode. In the space between the cathode and anode the electric current consists of these moving electrons. The circuit is completed by the flow of electrons in external wires, possibly by the movement of ions in a battery, and so on, with which we are not here concerned.

In this diode the local density of charge in any region, $\rho$, is simply $-ne$, where $n$ is the local density of electrons, in electrons per cubic meter. The local current density $\mathbf{J}$ is $\rho\mathbf{v}$, where $\mathbf{v}$ is the velocity of electrons in that region. In the plane-parallel diode we may assume $\mathbf{J}$ has no $y$ or $z$ components. If conditions are steady, it follows then that $J_x$ must be independent of $x$, for if $\mathrm{div}\,\mathbf{J} = 0$ as Eq. (4.8) says, $\partial J_x / \partial x$ must be zero if $J_y = J_z = 0$. This is belaboring the obvious; if we have a steady stream of electrons moving in the $x$ direction only, the same number per second have to cross any intermediate plane between cathode and anode. We conclude that $\rho v$ is constant. But observe that $v$ is *not* constant; it varies with $x$ because the electrons are accelerated by the field. Hence $\rho$ is not constant either. Instead, the negative charge density is high near the cathode and low near the anode, just as the density of cars on an expressway is high near a traffic slowdown and low where traffic is moving at high speed.

# 4.3 Electrical conductivity and Ohm's law

There are many ways of causing charge to move, including what we might call "bodily transport" of the charge carriers. In the Van de Graaff

---

[2] If the step between Eqs. (4.9) and (4.10) is not obvious, look back at our fundamental definition of divergence in Chapter 2. As the volume shrinks, we can eventually take $\rho$ outside the volume integral on the right. The volume integral is to be carried out at one instant of time. The time derivative thus depends on the difference between $\rho \int dv$ at $t$ and at $t + dt$. The only difference is due to the change of $\rho$ there, since the boundary of the volume remains in the same place.

electrostatic generator (see Problem 4.1) an insulating belt is given a surface charge, which it conveys to another electrode for removal, much as an escalator conveys people. That constitutes a perfectly good current. In the atmosphere, charged water droplets falling because of their weight form a component of the electric current system of the earth. In this section we shall be interested in a more common agent of charge transport, the force exerted on a charge carrier by an electric field. An electric field $\mathbf{E}$ pushes positive charge carriers in one direction, negative charge carriers in the opposite direction. If either or both can move, the result is an electric current in the direction of $\mathbf{E}$. In most substances, and over a wide range of electric field strengths, we find that the current density is proportional to the strength of the electric field that causes it. The linear relation between current density and field is expressed by

$$\mathbf{J} = \sigma \mathbf{E} \tag{4.11}$$

The factor $\sigma$ is called the *conductivity* of the material. Its value depends on the material in question; it is very large for metallic conductors, extremely small for good insulators. It may depend too on the physical state of the material – on its temperature, for instance. But with such conditions given, it does not depend on the magnitude of $\mathbf{E}$. If you double the field strength, holding everything else constant, you get twice the current density.

After everything we said in Chapter 3 about the electric field being zero inside a conductor, you might be wondering why we are now talking about a nonzero internal field. The reason is that in Chapter 3 we were dealing with static situations, that is, ones in which all the charges have settled down after some initial motion. In such a setup, the charges pile up at certain locations and create a field that internally cancels an applied field. But when dealing with currents in conductors, we are not letting the charges pile up, which means that things can't settle down. For example, a battery feeds in electrons at one end of a wire and takes them out at the other end. If the electrons were *not* taken out at the other end, then they *would* pile up there, and the electric field would eventually (actually very quickly) become zero inside.

The units of $\sigma$ are the units of $\mathbf{J}$ (namely $\mathrm{C\,s^{-1}m^{-2}}$) divided by the units of $\mathbf{E}$ (namely V/m or N/C). You can quickly show that this yields $\mathrm{C^2\,s\,kg^{-1}m^{-3}}$. However, it is customary to write the units of $\sigma$ as the reciprocal of ohm-meter, $(\mathrm{ohm\text{-}m})^{-1}$, where the ohm, which is the unit of resistance, is defined below.

In Eq. (4.11), $\sigma$ may be considered a scalar quantity, implying that the direction of $\mathbf{J}$ is always the same as the direction of $\mathbf{E}$. That is surely what we would expect within a material whose structure has no "built-in" preferred direction. Materials do exist in which the electrical conductivity itself depends on the angle the applied field $\mathbf{E}$ makes with some intrinsic axis in the material. One example is a single crystal of graphite, which has a layered structure on an atomic scale. For another example, see Problem 4.5. In such cases $\mathbf{J}$ may not have the direction of $\mathbf{E}$. But there still are linear relations between the components of $\mathbf{J}$ and the components of $\mathbf{E}$, relations expressed by Eq. (4.11) with $\sigma$ a *tensor* quantity instead of a scalar.[3] From now on we'll consider only *isotropic* materials, those within which the electrical conductivity is the same in all directions.

Equation (4.11) is a statement of *Ohm's law*. It is an *empirical* law, a generalization derived from experiment, not a theorem that must be universally obeyed. In fact, Ohm's law is bound to fail in the case of any particular material if the electric field is too strong. And we shall meet some interesting and useful materials in which "nonohmic" behavior occurs in rather weak fields. Nevertheless, the remarkable fact is the enormous range over which, in the large majority of materials, current density is proportional to electric field. Later in this chapter we'll explain why this should be so. But now, taking Eq. (4.11) for granted, we want to work out its consequences. We are interested in the total current $I$ flowing through a wire or a conductor of any other shape with well-defined ends, or terminals, and in the difference in potential between those terminals, for which we'll use the symbol $V$ (for *voltage*) rather than $\phi_1 - \phi_2$ or $\phi_{12}$.

Now, $I$ is the surface integral of $\mathbf{J}$ over a cross section of the conductor, which implies that $I$ is proportional to $\mathbf{J}$. Also, $V$ is the line integral of $\mathbf{E}$ on a path through the conductor from one terminal to the other, which implies that $V$ is proportional to $\mathbf{E}$. Therefore, if $\mathbf{J}$ is proportional to $\mathbf{E}$ everywhere inside a conductor as Eq. (4.11) states, then $I$ must be proportional to $V$. The relation of $V$ to $I$ is therefore another expression of Ohm's law, which we'll write this way:

$$\boxed{V = IR} \qquad \text{(Ohm's law).} \qquad (4.12)$$

The constant $R$ is the *resistance* of the conductor between the two terminals; $R$ depends on the size and shape of the conductor and the

---

[3] The most general linear relation between the two vectors $\mathbf{J}$ and $\mathbf{E}$ would be expressed as follows. In place of the three equations equivalent to Eq. (4.11), namely, $J_x = \sigma E_x$, $J_y = \sigma E_y$, and $J_z = \sigma E_z$, we would have $J_x = \sigma_{xx} E_x + \sigma_{xy} E_y + \sigma_{xz} E_z$, $J_y = \sigma_{yx} E_x + \sigma_{yy} E_y + \sigma_{yz} E_z$, and $J_z = \sigma_{zx} E_x + \sigma_{zy} E_y + \sigma_{zz} E_z$. These relations can be compactly summarized in the matrix equation,

$$\begin{pmatrix} J_x \\ J_y \\ J_z \end{pmatrix} = \begin{pmatrix} \sigma_{xx} & \sigma_{xy} & \sigma_{xz} \\ \sigma_{yx} & \sigma_{yy} & \sigma_{yz} \\ \sigma_{zx} & \sigma_{zy} & \sigma_{zz} \end{pmatrix} \begin{pmatrix} E_x \\ E_y \\ E_z \end{pmatrix}.$$

The nine coefficients $\sigma_{xx}$, $\sigma_{xy}$, etc., make up a *tensor*, which here is just a matrix. In this case, because of a symmetry requirement, it would turn out that $\sigma_{xy} = \sigma_{yx}$, $\sigma_{yz} = \sigma_{zy}$, $\sigma_{xz} = \sigma_{zx}$. Furthermore, by a suitable orientation of the $x$, $y$, $z$ axes, all the coefficients could be rendered zero except $\sigma_{xx}$, $\sigma_{yy}$, and $\sigma_{zz}$.

**Figure 4.3.**
The resistance of a conductor of length $L$,
uniform cross-sectional area $A$, and
conductivity $\sigma$.

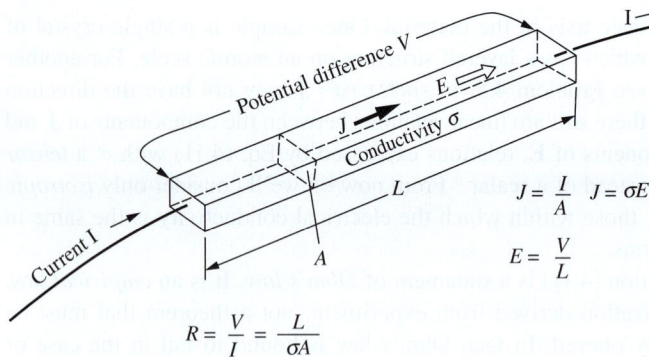

Potential difference $V$

Current $I$

$E \Rightarrow$

$J \rightarrow$

Conductivity $\sigma$

$L$

$A$

$I \rightarrow$

$J = \dfrac{I}{A} \quad J = \sigma E$

$E = \dfrac{V}{L}$

$R = \dfrac{V}{I} = \dfrac{L}{\sigma A}$

conductivity $\sigma$ of the material. The simplest example is a solid rod of cross-sectional area $A$ and length $L$. A steady current $I$ flows through this rod from one end to the other (Fig. 4.3). Of course there must be conductors to carry the current to and from the rod. We consider the terminals of the rod to be the points where these conductors are attached. Inside the rod the current density is given by

$$J = \frac{I}{A},\qquad(4.13)$$

and the electric field strength is given by

$$E = \frac{V}{L}.\qquad(4.14)$$

The resistance $R$ in Eq. (4.12) is $V/I$. Using Eqs. (4.11), (4.13), and (4.14) we easily find that

$$R = \frac{V}{I} = \frac{LE}{AJ} = \frac{L}{A\sigma}.\qquad(4.15)$$

On the way to this simple formula we made some tacit assumptions. First, we assumed the current density is uniform over the cross section of the bar. To see why that must be so, imagine that $J$ is actually greater along one side of the bar than on the other. Then $E$ must also be greater along that side. But then the line integral of **E** from one terminal to the other would be greater for a path along one side than for a path along the other, and that cannot be true for an electrostatic field.

A second assumption was that **J** kept its uniform magnitude and direction right out to the ends of the bar. Whether that is true or not depends on the external conductors that carry current to and from the bar and how they are attached. Compare Fig. 4.4(a) with Fig. 4.4(b). Suppose that the terminal in (b) is made of material with a conductivity much higher than that of the bar. That will make the plane of the end of the bar an equipotential surface, creating the current system to which Eq. (4.15) applies *exactly*. But all we can say in general about such "end effects" is that Eq. (4.15) will give $R$ to a good approximation if the width of the bar is small compared with its length.

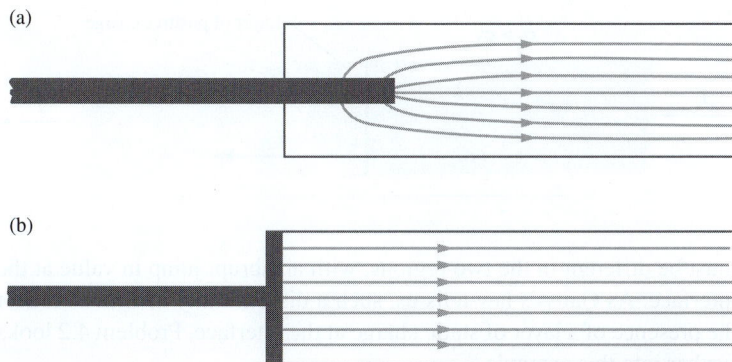

(a)

(b)

**Figure 4.4.**
Different ways in which the current $I$ might be introduced into the conducting bar. In (a) it has to spread out before the current density $\mathbf{J}$ becomes uniform. In (b) if the external conductor has much higher conductivity than the bar, the end of the bar will be an equipotential and the current density will be uniform from the beginning. For long thin conductors, such as ordinary wires, the difference is negligible.

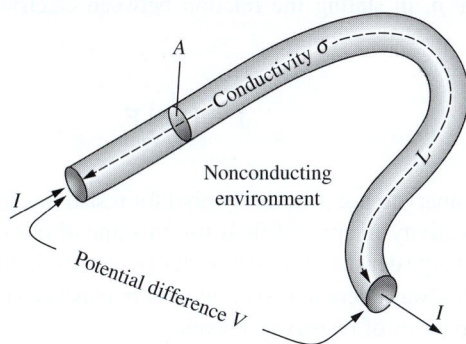

$$R = \frac{L}{\sigma A}$$

**Figure 4.5.**
As long as our conductors are surrounded by a nonconducting medium (air, oil, vacuum, etc.), the resistance $R$ between the terminals doesn't depend on the shape, only on the length of the conductor and its cross-sectional area.

A third assumption is that the bar is surrounded by an electrically nonconducting medium. Without that, we could not even *define* an isolated current path with terminals and talk about *the* current $I$ and *the* resistance $R$. In other words, it is the enormous difference in conductivity between good insulators, including air, and conductors that makes *wires*, as we know them, possible. Imagine the conducting rod of Fig. 4.3 bent into some other shape, as in Fig. 4.5. Because it is embedded in a nonconducting medium into which current cannot leak, the problem presented in Fig. 4.5 is for all practical purposes the same as the one in Fig. 4.3 which we have already solved. Equation (4.15) applies to a bent wire as well as a straight rod, if we measure $L$ along the wire.

In a region where the conductivity $\sigma$ is constant, the steady current condition div $\mathbf{J} = 0$ (Eq. (4.8)) together with Eq. (4.11) implies that div $\mathbf{E} = 0$ also. This tells us that the charge density is zero within that region. On the other hand, if $\sigma$ varies from one place to another in the conducting medium, steady current flow may entail the presence of static charge within the conductor. Figure 4.6 shows a simple example, a bar made of two materials of different conductivity, $\sigma_1$ and $\sigma_2$. The current density $\mathbf{J}$ must be the same on the two sides of the interface; otherwise charge would continue to pile up there. It follows that the electric field $\mathbf{E}$

**Figure 4.6.**
When current flows through this composite conductor, a layer of static charge appears at the interface between the two materials, so as to provide the necessary jump in the electric field **E**. In this example $\sigma_2 < \sigma_1$, hence $E_2$ must be greater than $E_1$.

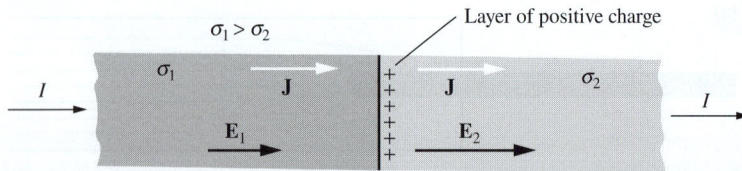

must be different in the two regions, with an abrupt jump in value at the interface. As Gauss's law tells us, such a discontinuity in **E** must reflect the presence of a layer of static charge at the interface. Problem 4.2 looks further into this example.

Instead of the conductivity $\sigma$ we could have used its reciprocal, the *resistivity* $\rho$, in stating the relation between electric field and current density:

$$\mathbf{J} = \left(\frac{1}{\rho}\right)\mathbf{E}. \tag{4.16}$$

It is customary to use $\rho$ as the symbol for resistivity and $\sigma$ as the symbol for conductivity in spite of their use in some of our other equations for volume charge density and surface charge density. In the rest of this chapter $\rho$ will always denote resistivity and $\sigma$ conductivity. Equation (4.15) written in terms of resistivity becomes

$$\boxed{R = \frac{\rho L}{A}} \tag{4.17}$$

The SI unit for resistance is defined to be the *ohm* (denoted by $\Omega$), which is given by Eq. (4.12) as

$$1 \text{ ohm} = 1\,\frac{\text{volt}}{\text{ampere}}. \tag{4.18}$$

In terms of other SI units, you can show that 1 ohm equals $1\,\text{kg}\,\text{m}^2\,\text{C}^{-2}\,\text{s}^{-1}$. If resistance $R$ is in ohms, it is evident from Eq. (4.17) that the resistivity $\rho$ must have units of (ohms) $\times$ (meters). The official SI unit for $\rho$ is therefore the ohm-meter. But another unit of length can be used with perfectly clear meaning. A unit commonly used for resistivity, in both the physics and technology of electrical conduction, is the ohm-centimeter (ohm-cm). If one chooses to measure resistivity in ohm-cm, the corresponding unit for conductivity is written as $\text{ohm}^{-1}\text{cm}^{-1}$, or $(\text{ohm-cm})^{-1}$, and called "reciprocal ohm-cm." It should be emphasized that Eqs. (4.11) through (4.17) are valid for any self-consistent choice of units.

**Example (Lengthening a wire)**   A wire of pure tin is drawn through a die, reducing its diameter by 25 percent and increasing its length. By what factor is its resistance increased? It is then flattened into a ribbon by rolling, which results in a further increase in its length, now twice the original length. What is the overall change in resistance? Assume the density and resistivity remain constant throughout.

**Solution**   Let $A$ be the cross-sectional area, and let $L$ be the length. The volume $AL$ is constant, so $L \propto 1/A$. The resistance $R = \rho L/A$ is therefore proportional to $1/A^2$. If the die reduces the diameter by the factor $3/4$, then it reduces $A$ by the factor $(3/4)^2$. The resistance is therefore multiplied by the factor $1/(3/4)^4 = 3.16$. In terms of the radius $r$, the resistance is proportional to $1/r^4$.

Since $A \propto 1/L$, we can alternatively say that the resistance $R = \rho L/A$ is proportional to $L^2$. An overall increase in $L$ by the factor 2 therefore yields an overall increase in $R$ by the factor $2^2 = 4$.

In Gaussian units, the unit of charge can be expressed in terms of other fundamental units, because Coulomb's law with a dimensionless coefficient yields 1 esu $= 1 \, \mathrm{g^{1/2} \, cm^{3/2} \, s^{-1}}$, as you can verify. You can use this to show that the units of resistance are s/cm. Since Eq. (4.17) still tells us that the resistivity $\rho$ has dimensions of (resistance) $\times$ (length), we see that the Gaussian unit of $\rho$ is simply the second. The analogous statement in the SI system, as you can check, is that the units of $\rho$ are seconds divided by the units of $\epsilon_0$. Hence $\epsilon_0 \rho$ has the dimensions of time. This association of a resistivity with a time has a natural interpretation which will be explained in Section 4.11.

The conductivity and resistivity of a few materials are given in different units for comparison in Table 4.1. The key conversion factor is also given (see Appendix C for the derivation).

**Example (Drift velocity in a copper wire)**   A copper wire $L = 1$ km long is connected across a $V = 6$ V battery. The resistivity of the copper is $\rho = 1.7 \cdot 10^{-8}$ ohm-meter, and the number of conduction electrons per cubic meter is $N = 8 \cdot 10^{28} \, \mathrm{m^{-3}}$. What is the drift velocity of the conduction electrons under these circumstances? How long does it take an electron to drift once around the circuit?

**Solution**   Equation (4.6) gives the magnitude of the current density as $J = Nev$, so the drift velocity is $v = J/Ne$. But $J$ is given by $J = \sigma E = (1/\rho)(V/L)$. Substituting this into $v = J/Ne$ yields

$$v = \frac{V}{\rho L N e} = \frac{6 \, \mathrm{V}}{(1.7 \cdot 10^{-8} \, \mathrm{ohm\text{-}m})(1000 \, \mathrm{m})(8 \cdot 10^{28} \, \mathrm{m^{-3}})(1.6 \cdot 10^{-19} \, \mathrm{C})}$$

$$= 2.8 \cdot 10^{-5} \, \mathrm{m/s}. \tag{4.19}$$

This is *much* slower than the average thermal speed of an electron at room temperature, which happens to be about $10^5$ m/s. The time to drift once around the

**Table 4.1.**
Resistivity and its reciprocal, conductivity, for a few materials

| Material | Resistivity $\rho$ | Conductivity $\sigma$ |
|---|---|---|
| Pure copper, 273 K | $1.56 \cdot 10^{-8}$ ohm-m $1.73 \cdot 10^{-18}$ s | $6.4 \cdot 10^7$ (ohm-m)$^{-1}$ $5.8 \cdot 10^{17}$ s$^{-1}$ |
| Pure copper, 373 K | $2.24 \cdot 10^{-8}$ ohm-m $2.47 \cdot 10^{-18}$ s | $4.5 \cdot 10^7$ (ohm-m)$^{-1}$ $4.0 \cdot 10^{17}$ s$^{-1}$ |
| Pure germanium, 273 K | $2$ ohm-m $2.2 \cdot 10^{-10}$ s | $0.5$ (ohm-m)$^{-1}$ $4.5 \cdot 10^9$ s$^{-1}$ |
| Pure germanium, 500 K | $1.2 \cdot 10^{-3}$ ohm-m $1.3 \cdot 10^{-13}$ s | $830$ (ohm-m)$^{-1}$ $7.7 \cdot 10^{12}$ s$^{-1}$ |
| Pure water, 291 K | $2.5 \cdot 10^5$ ohm-m $2.8 \cdot 10^{-5}$ s | $4.0 \cdot 10^{-6}$ (ohm-m)$^{-1}$ $3.6 \cdot 10^4$ s$^{-1}$ |
| Seawater (varies with salinity) | $0.25$ ohm-m $2.8 \cdot 10^{-11}$ s | $4$ (ohm-m)$^{-1}$ $3.6 \cdot 10^{10}$ s$^{-1}$ |

*Note:* 1 ohm-meter $= 1.11 \cdot 10^{-10}$ s.

circuit is $t = (1000 \, \text{m})/(2.8 \cdot 10^{-5} \, \text{m/s}) = 3.6 \cdot 10^7$ s, which is a little over a year. Note that $v$ is independent of the cross-sectional area. This makes sense, because if we have two separate identical wires connected to the same voltage source, they have the same $v$. If we combine the two wires into one thicker wire, this shouldn't change the $v$.

When dealing with currents in wires, we generally assume that the wire is neutral. That is, we assume that the moving electrons have the same density per unit length as the stationary protons in the lattice. We should mention, however, that an actual current-carrying wire is *not* neutral. There are surface charges on the wire, as explained by Marcus (1941) and demonstrated by Jefimenko (1962). These charges are necessary for three reasons.

First, the surface charges keep the current flowing along the path of the wire. Consider a battery connected to a long wire, and let's say we put a bend in the wire far from the battery. If we then bend the wire in some other arbitrary manner, the battery doesn't "know" that we changed the shape, so it certainly can't be the cause of the electrons taking a new path through space. The cause of the new path must be the electric field due to nearby charges. These charges appear as surface charges on the wire.

Second, the existence of a net charge on the wire is necessary to create the proper flow of energy associated with the current. To get a handle on this energy flow, we will have to wait until we learn about magnetic fields in Chapter 6 and the *Poynting vector* in Chapter 9. But for now we'll just say that to have the proper energy flow, there must be a component of the electric field pointing radially away from the wire. This component wouldn't exist if the net charge on the wire were zero.

Third, the surface charge causes the potential to change along the wire in a manner consistent with Ohm's law. See Jackson (1996) for more discussion on these three roles that the surface charges play.

However, having said all this, it turns out that in most of our discussions of circuits and currents in this book, we won't be interested in the electric field external to the wires. So we can generally ignore the surface charges, with no ill effects.

## 4.4 The physics of electrical conduction
### 4.4.1 Currents and ions

To explain electrical conduction we have to talk first about atoms and molecules. Remember that a neutral atom, one that contains as many electrons as there are protons in its nucleus, is *precisely* neutral (see Section 1.3). On such an object the net force exerted by an electric field is exactly zero. And even if the neutral atom were moved along by some other means, that would not be an electric current. The same holds for neutral molecules. Matter that consists only of neutral molecules ought to have zero electrical conductivity. Here one qualification is in order: we are concerned now with steady electric currents, that is, *direct* currents, not alternating currents. An alternating electric field could cause periodic deformation of a molecule, and that displacement of electric charge would be a true alternating electric current. We shall return to that subject in Chapter 10. For a steady current we need mobile charge carriers, or *ions*. These must be present in the material before the electric field is applied, for the electric fields we shall consider are not nearly strong enough to create ions by tearing electrons off molecules. Thus the physics of electrical conduction centers on two questions: how many ions are there in a unit volume of material, and how do these ions move in the presence of an electric field?

In pure water at room temperature approximately two $H_2O$ molecules in a billion are, at any given moment, dissociated into negative ions, $OH^-$, and positive ions, $H^+$. (Actually the positive ion is better described as $OH_3^+$, that is, a proton attached to a water molecule.) This provides approximately $6 \cdot 10^{13}$ negative ions and an equal number of positive ions in a cubic centimeter of water.[4] The motion of these ions in the

---

4  Students of chemistry may recall that the concentration of hydrogen ions in pure water corresponds to a pH value of 7.0, which means the concentration is $10^{-7.0}$ mole/liter. That is equivalent to $10^{-10.0}$ mole/cm$^3$. A mole of anything is $6.02 \cdot 10^{23}$ things – hence the number $6 \cdot 10^{13}$ given above.

applied electric field accounts for the conductivity of pure water given in Table 4.1. Adding a substance like sodium chloride, whose molecules easily dissociate in water, can increase enormously the number of ions. That is why seawater has electrical conductivity nearly a million times greater than that of pure water. It contains something like $10^{20}$ ions per cubic centimeter, mostly $Na^+$ and $Cl^-$.

In a gas like nitrogen or oxygen at ordinary temperatures there would be no ions at all except for the action of some ionizing radiation such as ultraviolet light, x-rays, or nuclear radiation. For instance, ultraviolet light might eject an electron from a nitrogen molecule, leaving $N_2^+$, a molecular ion with a positive charge $e$. The electron thus freed is a negative ion. It may remain free or it may eventually stick to some molecule as an "extra" electron, thus forming a negative molecular ion. The oxygen molecule happens to have an especially high affinity for an extra electron; when air is ionized, $N_2^+$ and $O_2^-$ are common ion types. In any case, the resulting conductivity of the gas depends on the number of ions present at any moment, which depends in turn on the intensity of the ionizing radiation and perhaps other circumstances as well. So we cannot find in a table *the* conductivity of a gas. Strictly speaking, the conductivity of pure nitrogen shielded from all ionizing radiation would be zero.[5]

Given a certain concentration of positive and negative ions in a material, how is the resulting conductivity, $\sigma$ in Eq. (4.11), determined? Let's consider first a slightly ionized gas. To be specific, suppose its density is like that of air in a room – about $10^{25}$ molecules per cubic meter. Here and there among these neutral molecules are positive and negative ions. Suppose there are $N$ positive ions in unit volume, each of mass $M_+$ and carrying charge $e$, and an equal number of negative ions, each with mass $M_-$ and charge $-e$. The number of ions in unit volume, $2N$, is very much smaller than the number of neutral molecules. When an ion collides with anything, it is almost always a neutral molecule rather than another ion. Occasionally a positive ion does encounter a negative ion and combine with it to form a neutral molecule. Such recombination[6] would steadily deplete the supply of ions if ions were not being continually created by some other process. But in any case the rate of change of $N$ will be so slow that we can neglect it here.

---

[5] But what about thermal energy? Won't that occasionally lead to the ionization of a molecule? In fact, the energy required to ionize, that is, to extract an electron from, a nitrogen molecule is several hundred times the mean thermal energy of a molecule at 300 K. You would not expect to find even one ion so produced in the entire earth's atmosphere!

[6] In calling the process recombination we of course do not wish to imply that the two "recombining" ions were partners originally. Close encounters of a positive ion with a negative ion are made somewhat more likely by their electrostatic attraction. However, that effect is generally not important when the number of ions per unit volume is very much smaller than the number of neutral molecules.

## 4.4.2 Motion in zero electric field

Imagine now the scene, on a molecular scale, before an electric field is applied. The molecules, and the ions too, are flying about with random velocities appropriate to the temperature. The gas is mostly empty space, the mean distance between a molecule and its nearest neighbor being about ten molecular diameters. The mean free path of a molecule, which is the average distance it travels before bumping into another molecule, is much larger, perhaps $10^{-7}$ m, or several hundred molecular diameters. A molecule or an ion in this gas spends 99.9 percent of its time as a free particle. If we could look at a particular ion at a particular instant, say $t = 0$, we would find it moving through space with some velocity **u**.

What will happen next? The ion will move in a straight line at constant speed until, sooner or later, it chances to come close to a molecule, close enough for strong short-range forces to come into play. In this *collision* the total kinetic energy and the total momentum of the two bodies, molecule and ion, will be conserved, but the ion's velocity will be rather suddenly changed in both magnitude and direction to some new velocity **u′**. It will then coast along freely with this new velocity until another collision changes its velocity to **u″**, and so on. After at most a few such collisions the ion is as likely to be moving in any direction as in any other direction. The ion will have "forgotten" the direction it was moving at $t = 0$.

To put it another way, if we picked 10,000 cases of ions moving horizontally south, and followed each of them for a sufficient time of $\tau$ seconds, their final velocity directions would be distributed impartially over a sphere. It may take several collisions to wipe out most of the direction memory or only a few, depending on whether collisions involving small momentum changes or large momentum changes are the more common, and this depends on the nature of the interaction. An extreme case is the collision of hard elastic spheres, which turns out to produce a completely random new direction in just one collision. We need not worry about these differences. The point is that, whatever the nature of the collisions, there will be *some* time interval $\tau$, characteristic of a given system, such that the lapse of $\tau$ seconds leads to substantial loss of *correlation* between the initial velocity direction and the final velocity direction of an ion in that system.[7] This characteristic time $\tau$ will depend on the ion and on the nature of its average environment; it will certainly be shorter the more frequent the collisions, since in our gas nothing happens to an ion between collisions.

---

[7] It would be possible to define $\tau$ precisely for a general system by giving a quantitative measure of the correlation between initial and final directions. It is a statistical problem, like devising a measure of the correlation between the birth weights of rats and their weights at maturity. However, we shall not need a general quantitative definition to complete our analysis.

### 4.4.3 Motion in nonzero electric field

Now we are ready to apply a uniform electric field $\mathbf{E}$ to the system. It will make the description easier if we imagine the loss of direction memory to occur completely at a single collision, as we have said it does in the case of hard spheres. Our main conclusion will actually be independent of this assumption. Immediately after a collision an ion starts off in some random direction. We will denote by $\mathbf{u}^c$ the velocity immediately after a collision. The electric force $\mathbf{E}e$ on the ion imparts momentum to the ion continuously. After time $t$ it will have acquired from the field a momentum increment $\mathbf{E}et$, which simply adds vectorially to its original momentum $M\mathbf{u}^c$. Its momentum is now $M\mathbf{u}^c + \mathbf{E}et$. If the momentum increment is small relative to $M\mathbf{u}^c$, that implies that the velocity has not been affected much, so we can expect the next collision to occur about as soon as it would have in the absence of the electric field. In other words, the average time between collisions, which we shall denote by $\bar{t}$, is independent of the field $\mathbf{E}$ if the field is not too strong.

The momentum acquired from the field is always a vector in the same direction. But it is lost, in effect, at every collision, since the direction of motion after a collision is random, regardless of the direction before.

*What is the average momentum of all the positive ions at a given instant of time?* This question is surprisingly easy to answer if we look at it this way. At the instant in question, suppose we stop the clock and ask each ion how long it has been since its last collision. Suppose we get the particular answer $t_1$ from positive ion 1. Then that ion must have momentum $e\mathbf{E}t_1$ *in addition* to the momentum $M\mathbf{u}_1^c$ with which it emerged from its last collision. The average momentum of all $N$ positive ions is therefore

$$M\bar{\mathbf{u}}_+ = \frac{1}{N} \sum_j \left( M\mathbf{u}_j^c + e\mathbf{E}t_j \right). \tag{4.20}$$

Here $\mathbf{u}_j^c$ is the velocity the $j$th ion had just after its last collision. These velocities $\mathbf{u}_j^c$ are quite random in direction and therefore contribute zero to the average. The second part is simply $\mathbf{E}e$ times the *average of the $t_j$*, that is, times the *average of the time since the last collision*. That must be the same as the average of the time until the *next* collision, and both are the same[8] as the average time between collisions, $\bar{t}$. We conclude

---

[8] You may think the average time between collisions would have to be equal to the *sum* of the *average time since the last collision* and the *average time to the next*. That would be true if collisions occurred at absolutely regular intervals, but they don't. They are independent random events, and for such the above statement, paradoxical as it may seem at first, is true. Think about it. The question does not affect our main conclusion, but if you unravel it you will have grown in statistical wisdom; see Exercise 4.23. (*Hint:* If one collision doesn't affect the probability of having another – that's what *independent* means – it can't matter whether you start the clock at some arbitrary time, or at the time of a collision.)

that the average velocity of a positive ion, in the presence of the steady field **E**, is

$$\bar{\mathbf{u}}_+ = \frac{E e \bar{t}_+}{M_+}.$$ (4.21)

This shows that the average velocity of a charge carrier is proportional to the electric force applied to it. If we observe only the average velocity, it looks as if the medium were resisting the motion with a force proportional to the velocity. This is true because if we write Eq. (4.21) as $\mathbf{E}e - (M_+/\bar{t}_+)\bar{\mathbf{u}}_+ = 0$, we can interpret it as the terminal-velocity statement that the $\mathbf{E}e$ electric force is balanced by a $-b\bar{\mathbf{u}}_+$ drag force, where $b \equiv M_+/\bar{t}_+$. This $-b\mathbf{u}$ force is the kind of frictional drag you feel if you try to stir thick syrup with a spoon, a "viscous" drag. Whenever charge carriers behave like this, we can expect something like Ohm's law, for the following reason.

In Eq. (4.21) we have written $\bar{t}_+$ because the mean time between collisions may well be different for positive and negative ions. The negative ions acquire velocity in the opposite direction, but since they carry negative charge their contribution to the current density **J** adds to that of the positives. The equivalent of Eq. (4.6), with the two sorts of ions included, is now

$$\mathbf{J} = Ne\left(\frac{e\mathbf{E}\bar{t}_+}{M_+}\right) - Ne\left(\frac{-e\mathbf{E}\bar{t}_-}{M_-}\right) = Ne^2\left(\frac{\bar{t}_+}{M_+} + \frac{\bar{t}_-}{M_-}\right)\mathbf{E}.$$ (4.22)

Our theory therefore predicts that the system will obey Ohm's law, for Eq. (4.22) expresses a linear relation between **J** and **E**, the other quantities being constants characteristic of the medium. Compare Eq. (4.22) with Eq. (4.11). The constant $Ne^2(\bar{t}_+/M_+ + \bar{t}_-/M_-)$ appears in the role of $\sigma$, the conductivity.

We made a number of rather special assumptions about this system, but looking back, we can see that they were not essential so far as the linear relation between **E** and **J** is concerned. Any system containing a constant density of free charge carriers, in which the motion of the carriers is frequently "re-randomized" by collisions or other interactions within the system, ought to obey Ohm's law if the field **E** is not too strong. The ratio of **J** to **E**, which is the conductivity $\sigma$ of the medium, will be proportional to the number of charge carriers and to the characteristic time $\tau$, the time for loss of directional correlation. It is *only* through this last quantity that all the complicated details of the collisions enter the problem. The making of a detailed theory of the conductivity of any given system, assuming the number of charge carriers is known, amounts to making a theory for $\tau$. In our particular example this quantity was replaced by $\bar{t}$, and a perfectly definite result was predicted for the conductivity $\sigma$. Introducing the more general quantity $\tau$, and also allowing for the possibility of different numbers of positive and negative carriers, we can summarize our theory as follows:

$$\sigma \approx e^2 \left( \frac{N_+ \tau_+}{M_+} + \frac{N_- \tau_-}{M_-} \right) \qquad (4.23)$$

We use the $\approx$ sign to acknowledge that we did not give $\tau$ a precise definition. That could be done, however.

---

**Example (Atmospheric conductivity)**    Normally in the earth's atmosphere the greatest density of free electrons (liberated by ultraviolet sunlight) amounts to $10^{12}$ per cubic meter and is found at an altitude of about 100 km where the density of air is so low that the mean free path of an electron is about 0.1 m. At the temperature that prevails there, an electron's mean speed is $10^5$ m/s. What is the conductivity in $(\text{ohm-m})^{-1}$?

Solution    We have only one type of charge carrier, so Eq. (4.23) gives the conductivity as $\sigma = Ne^2 \tau / m$. The mean free time is $\tau = (0.1\,\text{m})/(10^5\,\text{m/s}) = 10^{-6}$ s. Therefore,

$$\sigma = \frac{Ne^2 \tau}{m} = \frac{(10^{12}\,\text{m}^{-3})(1.6 \cdot 10^{-19}\,\text{C})^2 (10^{-6}\,\text{s})}{9.1 \cdot 10^{-31}\,\text{kg}} = 0.028\ (\text{ohm-m})^{-1}.$$

$$(4.24)$$

---

To emphasize the fact that electrical conduction ordinarily involves only a slight systematic drift superimposed on the random motion of the charge carriers, we have constructed Fig. 4.7 as an artificial microscopic view of the kind of system we have been talking about. Positive ions are represented by gray dots, negative ions by circles. We assume the latter are electrons and hence, because of their small mass, so much more mobile than the positive ions that we may neglect the motion of the positives altogether. In Fig. 4.7(a) we see a wholly random distribution of particles and of electron speeds. To make the diagram, the location and sign of a particle were determined by a random-number table. The electron velocity vectors were likewise drawn from a random distribution, one corresponding to the "Maxwellian" distribution of molecular velocities in a gas. In Fig. 4.7(b) we have used the same positions, but now the velocities all have a small added increment to the right. That is, Fig. 4.7(b) is a view of an ionized material in which there is a net flow of negative charge to the right, equivalent to a positive current to the left. Figure 4.7(a) illustrates the situation with zero average current. The slightness of the systematic drift is demonstrated by the fact that it is essentially impossible to determine, by looking at the two figures separately, which is the one with zero average current.

Obviously we should not expect the actual average of the velocities of the 46 electrons in Fig. 4.7(a) to be exactly zero, for they are statistically independent quantities. One electron doesn't affect the behavior of another. There will in fact be a randomly fluctuating electric current

(a)

(b)

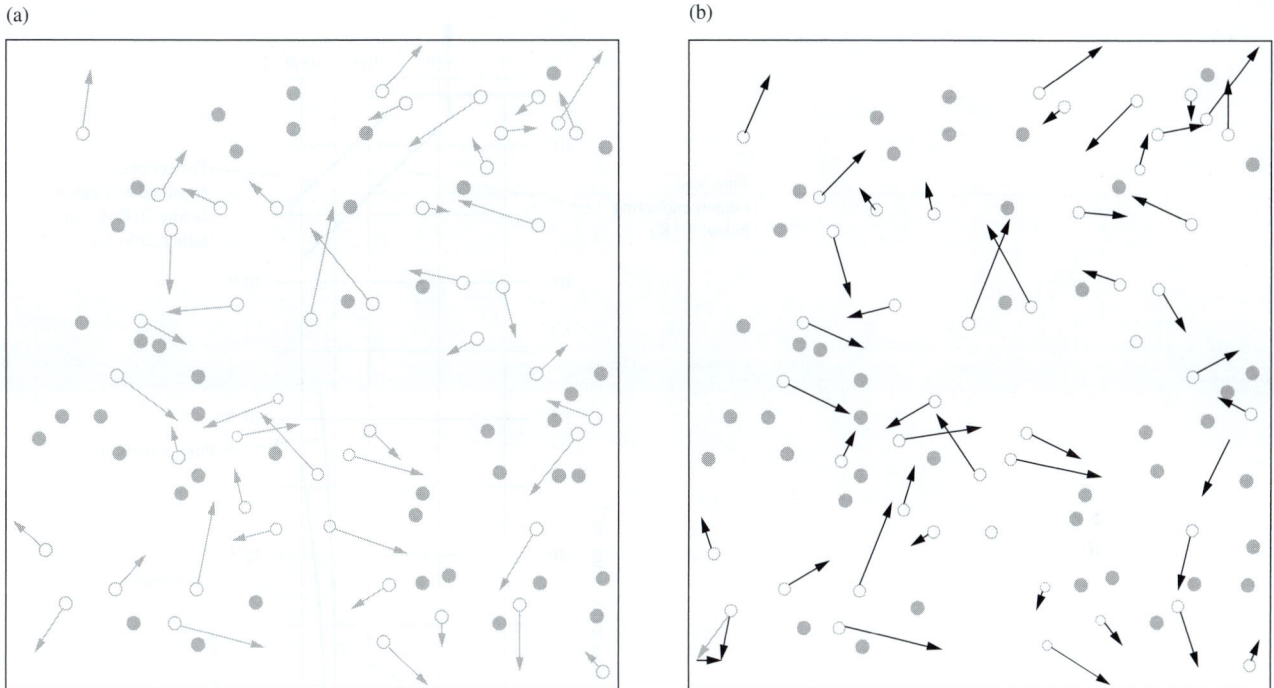

**Figure 4.7.**
(a) A random distribution of electrons and positive ions with about equal numbers of each. Electron velocities are shown as vectors and in (a) are completely random. In (b) a drift toward the right, represented by the velocity vector $\rightarrow$, has been introduced. This velocity was added to each of the original electron velocities, as shown in the case of the electron in the lower left corner.

in the absence of any driving field, simply as a result of statistical fluctuations in the vector sum of the electron velocities. This spontaneously fluctuating current can be measured. It is a source of noise in all electric circuits, and often determines the ultimate limit of sensitivity of devices for detecting weak electric signals.

### 4.4.4 Types of materials

With these ideas in mind, consider the materials whose electrical conductivity is plotted, as a function of temperature, in Fig. 4.8. Glass at room temperature is a good insulator. Ions are not lacking in its internal structure, but they are practically immobile, locked in place. As glass is heated, its structure becomes somewhat less rigid. An ion is able to move now and then, in the direction the electric field is pushing it. That happens in a sodium chloride crystal, too. The ions, in that case $Na^+$ and $Cl^-$, move by infrequent short jumps.[9] Their average rate of progress is proportional to the electric field strength at any given temperature, so Ohm's law is obeyed. In both these materials, the main effect of raising the temperature is to increase the mobility of the charge carriers rather than their number.

Silicon and germanium are called *semiconductors*. Their conductivity, too, depends strongly on the temperature, but for a different reason. At zero absolute temperature, they would be perfect insulators,

---

[9] This involves some disruption of the perfectly orderly array of ions depicted in Fig. 1.7.

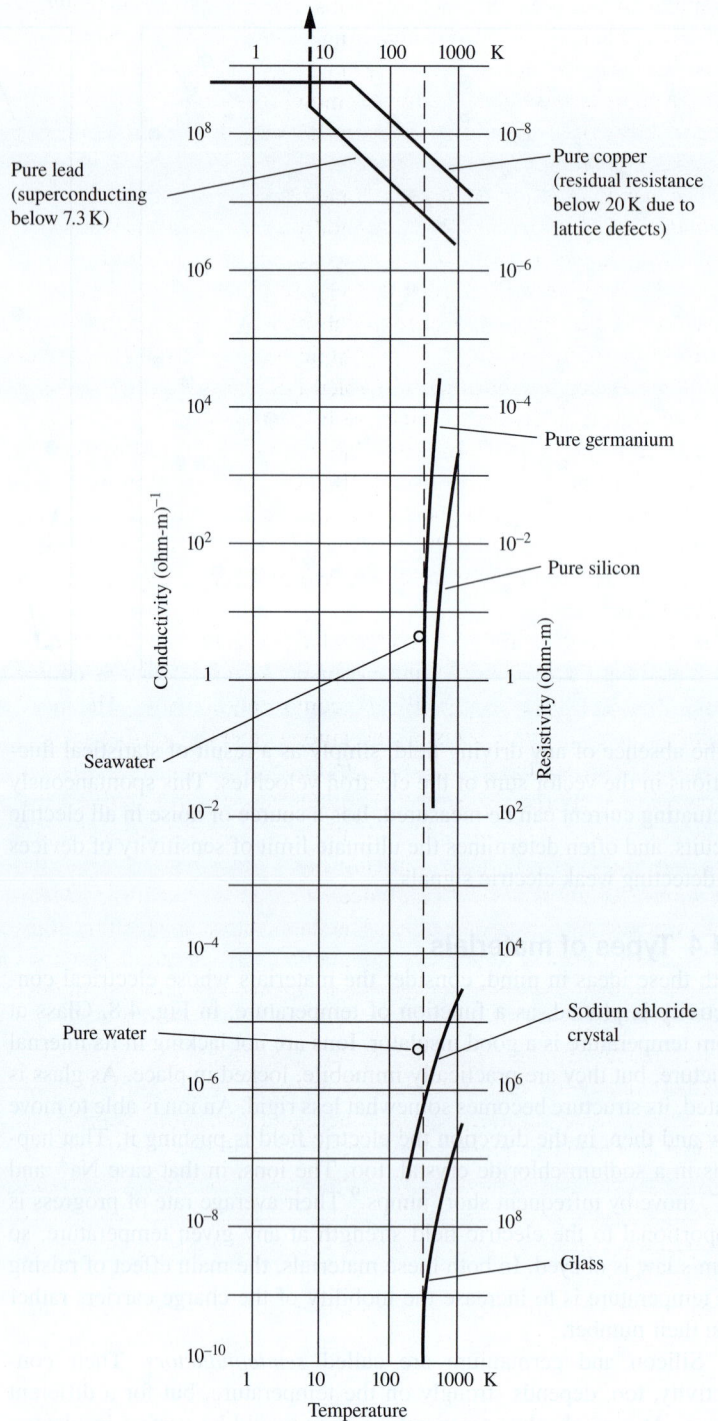

**Figure 4.8.**
The electrical conductivity of some representative substances. Note that logarithmic scales are used for both conductivity and absolute temperature.

containing no ions at all, only neutral atoms. The effect of thermal energy is to create charge carriers by liberating electrons from some of the atoms. The steep rise in conductivity around room temperature and above reflects a great increase in the number of mobile electrons, not an increase in the mobility of an individual electron. We shall look more closely at semiconductors in Section 4.6.

The metals, exemplified by copper and lead in Fig. 4.8, are even better conductors. Their conductivity generally *decreases* with increasing temperature, due to an effect we will discuss in Section 4.5. In fact, over most of the range plotted, the conductivity of a pure metal like copper or lead is inversely proportional to the absolute temperature, as can be seen from the 45° slope of our logarithmic graph. Were that behavior to continue as copper and lead are cooled down toward absolute zero, we could expect an enormous increase in conductivity. At 0.001 K, a temperature readily attainable in the laboratory, we should expect the conductivity of each metal to rise to 300,000 times its room temperature value. In the case of copper, we would be sadly disappointed. As we cool copper below about 20 K, its conductivity ceases to rise and remains constant from there on down. We will try to explain that in Section 4.5.

In the case of lead, normally a somewhat poorer conductor than copper, something far more surprising happens. As a lead wire is cooled below 7.2 K, its resistance abruptly and completely *vanishes*. The metal becomes *superconducting*. This means, among other things, that an electric current, once started flowing in a circuit of lead wire, will continue to flow indefinitely (for years, even!) without any electric field to drive it. The conductivity may be said to be infinite, though the concept really loses its meaning in the superconducting state. Warmed above 7.2 K, the lead wire recovers its normal resistance as abruptly as it lost it. Many metals can become superconductors. The temperature at which the transition from the normal to the superconducting state occurs depends on the material. In *high-temperature superconductors,* transitions as high as 130 K have been observed.

Our model of ions accelerated by the electric field, their progress being continually impeded by collisions, utterly fails us here. Somehow, in the superconducting state all impediment to the electrons' motion has vanished. Not only that, magnetic effects just as profound and mysterious are manifest in the superconductor. At this stage of our study we cannot fully describe, let alone explain, the phenomenon of superconductivity. More will be said in Appendix I, which should be intelligible after our study of magnetism.

Superconductivity aside, all these materials obey Ohm's law. Doubling the electric field doubles the current if other conditions, including the temperature, are held constant. At least that is true if the field is not too strong. It is easy to see how Ohm's law could fail in the case of a partially ionized gas. Suppose the electric field is so strong that the additional velocity an electron acquires between collisions is comparable to its thermal velocity. Then the time between collisions will be shorter than it was before the field was applied, an effect not included in our theory and one that will cause the observed conductivity to depend on the field strength.

A more spectacular breakdown of Ohm's law occurs if the electric field is further increased until an electron gains so much energy between collisions that in striking a neutral atom it can knock another electron loose. The two electrons can now release still more electrons in the same way. Ionization increases explosively, quickly making a conducting path between the electrodes. This is a *spark*. It's what happens when a spark-plug fires, and when you touch a doorknob after walking over a rug on a dry day. There are always a few electrons in the air, liberated by cosmic rays if in no other way. Since one electron is enough to trigger a spark, this sets a practical limit to field strength that can be maintained in a gas. Air at atmospheric pressure will break down at roughly 3 megavolts/meter. In a gas at low pressure, where an electron's free path is quite long, as within the tube of an ordinary fluorescent lamp, a steady current can be maintained with a modest field, with ionization by electron impact occurring at a constant rate. The physics is fairly complex, and the behavior far from ohmic.

## 4.5 Conduction in metals

The high conductivity of metals is due to electrons within the metal that are not attached to atoms but are free to move through the whole solid. Proof of this is the fact that electric current in a copper wire – unlike current in an ionic solution – transports no chemically identifiable substance. A current can flow steadily for years without causing the slightest change in the wire. It could only be electrons that are moving, entering the wire at one end and leaving it at the other.

We know from chemistry that atoms of the metallic elements rather easily lose their outermost electrons.[10] These would be bound to the atom if it were isolated, but become detached when many such atoms are packed close together in a solid. The atoms thus become positive ions, and these positive ions form the rigid lattice of the solid metal, usually in an orderly array. The detached electrons, which we shall call the conduction electrons, move through this three-dimensional lattice of positive ions.

The number of conduction electrons is large. The metal sodium, for instance, contains $2.5 \cdot 10^{22}$ atoms in $1 \, cm^3$, and each atom provides one conduction electron. No wonder sodium is a good conductor! But wait, there is a deep puzzle here. It is brought to light by applying our simple

---

[10] This could even be taken as the property that defines a metallic element, making somewhat tautological the statement that metals are good conductors.

theory of conduction to this case. As we have seen, the mobility of a charge carrier is determined by the time $\tau$ during which it moves freely without bumping into anything. If we have $2.5 \cdot 10^{28}$ electrons of mass $m_e$ per cubic meter, we need only the experimentally measured conductivity of sodium to calculate an electron's mean free time $\tau$. The conductivity of sodium at room temperature is $\sigma = 2.1 \cdot 10^7$ (ohm-m)$^{-1}$. Recalling that 1 ohm $= 1 \, \text{kg m}^2 \, \text{C}^{-2} \, \text{s}^{-1}$, we have $\sigma = 2.1 \cdot 10^7 \, \text{C}^2 \, \text{s} \, \text{kg}^{-1} \, \text{m}^{-3}$. Solving Eq. (4.23) for $\tau_-$, with $N_+ = 0$ as there are no mobile positive carriers, we find

$$\tau_- = \frac{\sigma m_e}{Ne^2} = \frac{\left(2.1 \cdot 10^7 \, \dfrac{\text{C}^2 \, \text{s}}{\text{kg m}^3}\right)\left(9.1 \cdot 10^{-31} \, \text{kg}\right)}{\left(2.5 \cdot 10^{28} \, \dfrac{1}{\text{m}^3}\right)\left(1.6 \cdot 10^{-19} \, \text{C}\right)^2} = 3 \cdot 10^{-14} \, \text{s}. \quad (4.25)$$

This seems a *surprisingly long* time for an electron to move through the lattice of sodium ions without suffering a collision. The thermal speed of an electron at room temperature ought to be about $10^5$ m/s, according to kinetic theory, which in that time should carry it a distance of $3 \cdot 10^{-9}$ m. Now, the ions in a crystal of sodium are practically touching one another. The centers of adjacent ions are only $3.8 \cdot 10^{-10}$ m apart, with strong electric fields and many bound electrons filling most of the intervening space. How could an electron travel nearly ten lattice spaces through these obstacles without being deflected? Why is the lattice of ions so *easily penetrated* by the conduction electrons?

This puzzle baffled physicists until the *wave aspect* of the electrons' motion was recognized and explained by quantum mechanics. Here we can only hint at the nature of the explanation. It goes something like this. We should not now think of the electron as a tiny charged particle deflected by every electric field it encounters. It is *not localized* in that sense. It behaves more like a spread-out wave interacting, at any moment, with a larger region of the crystal. What interrupts the progress of this wave through the crystal is not the regular array of ions, dense though it is, but an *irregularity* in the array. (A light wave traveling through water can be scattered by a bubble or a suspended particle, but not by the water itself; the analogy has some validity.) In a geometrically perfect and flawless crystal the electron wave would never be scattered, which is to say that the electron would never be deflected; our time $\tau$ would be infinite. But real crystals are imperfect in at least two ways. For one thing, there is a random thermal vibration of the ions, which makes the lattice at any moment slightly irregular geometrically, and the more so the higher the temperature. It is this effect that makes the conductivity of a pure metal *decrease* as the temperature is raised. We see it in the sloping portions of the graph of $\sigma$ for pure copper and pure lead in Fig. 4.8. A real crystal can have irregularities, too, in the form of foreign atoms, or impurities, and lattice defects – flaws in the stacking of the atomic array. Scattering by these irregularities limits the free time $\tau$ whatever the temperature. Such defects are responsible for the residual temperature-independent resistivity seen in the plot for copper in Fig. 4.8.

In metals Ohm's law is obeyed exceedingly accurately up to current densities far higher than any that can be long maintained. No deviation has ever been clearly demonstrated experimentally. According to one theoretical prediction, departures on the order of 1 percent might be expected at a current density of $10^{13}$ A/m². That is more than a million times the current density typical of wires in ordinary circuits.

## 4.6 Semiconductors

In a crystal of silicon each atom has four near neighbors. The three-dimensional arrangement of the atoms is shown in Fig. 4.9. Now silicon, like carbon which lies directly above it in the periodic table, has four valence electrons, just the number needed to make each bond between neighbors a shared electron pair – a covalent bond as it is called in chemistry. This neat arrangement makes a quite rigid structure. In fact, this is the way the carbon atoms are arranged in diamond, the hardest known substance. With its bonds all intact, the perfect silicon crystal is a perfect insulator; there are no mobile electrons. But imagine that we could extract an electron from one of these bond pairs and move it a few hundred lattice spaces away in the crystal. This would leave a net positive charge at the site of the extraction and would give us a loose electron. It would also cost a certain amount of energy. We will take up the question of energy in a moment.

First let us note that we have created *two* mobile charges, not just one. The freed electron is mobile. It can move like a conduction electron

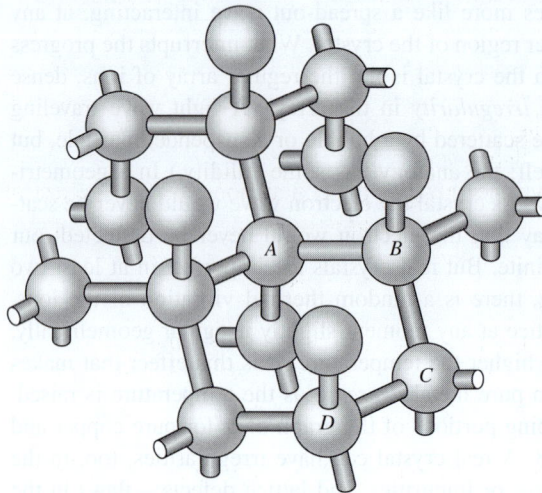

**Figure 4.9.**
The structure of the silicon crystal. The balls are Si atoms. A rod represents a covalent bond between neighboring atoms, made by sharing a pair of electrons. This requires four valence electrons per atom. Diamond has this structure, and so does germanium.

in a metal, like which it is spread out, not sharply localized. The quantum state it occupies we call a state in the *conduction band*. The positive charge left behind is also mobile. If you think of it as an electron missing in the bond between atoms $A$ and $B$ in Fig. 4.9, you can see that this vacancy among the valence electrons could be transferred to the bond between $B$ and $C$, thence to the bond between $C$ and $D$, and so on, just by shifting electrons from one bond to another. Actually, the motion of the hole, as we shall call it henceforth, is even freer than this would suggest. It sails through the lattice like a conduction electron. The difference is that it is a *positive* charge. An electric field $\mathbf{E}$ accelerates the hole in the direction of $\mathbf{E}$, not the reverse. The hole acts as if it had a mass comparable with an electron's mass. This is really rather mysterious, for the hole's motion results from the collective motion of many valence electrons.[11] Nevertheless, and fortunately, it acts so much like a real positive particle that we may picture it as such from now on.

The minimum energy required to extract an electron from a valence state in silicon and leave it in the conduction band is $1.8 \cdot 10^{-19}$ joule, or 1.12 electron-volts (eV). One electron-volt is the work done in moving one electronic charge through a potential difference of one volt. Since 1 volt equals 1 joule/coulomb, we have[12]

$$1 \text{ eV} = \left(1.6 \cdot 10^{-19} \text{ C}\right)\left(1 \text{ J/C}\right) \implies \boxed{1 \text{ eV} = 1.6 \cdot 10^{-19} \text{ J}} \quad (4.26)$$

The above energy of 1.12 eV is the *energy gap* between two bands of possible states, the valence band and the conduction band. States of intermediate energy for the electron simply do not exist. This energy ladder is represented in Fig. 4.10. Two electrons can never have the same quantum state – that is a fundamental law of physics (the *Pauli exclusion principle* which you will learn about in quantum mechanics). States ranging up the energy ladder must therefore be occupied even at absolute zero. As it happens, there are exactly enough states in the valence band to accommodate all the electrons. At $T = 0$, as shown in Fig. 4.10(a), *all* of these valence states are occupied, and *none* of the conduction band states is.

If the temperature is high enough, thermal energy can raise some electrons from the valence band to the conduction band. The effect of temperature on the probability that electron states will be occupied is expressed by the exponential factor $e^{-\Delta E/kT}$, called the Boltzmann factor.

---

[11] This mystery is *not* explained by drawing an analogy, as is sometimes done, with a bubble in a liquid. In a centrifuge, bubbles in a liquid would go in toward the axis; the holes we are talking about would go out. A cryptic but true statement, which only quantum mechanics will make intelligible, is this: the hole behaves dynamically like a positive charge with positive mass because it is a vacancy in states with negative charge and negative mass.

[12] Technically, "eV" should be written as "*e*V," because an electron-volt is the product of two things: the (magnitude of the) electron charge $e$ and one volt V.

(a)                                                        (b)

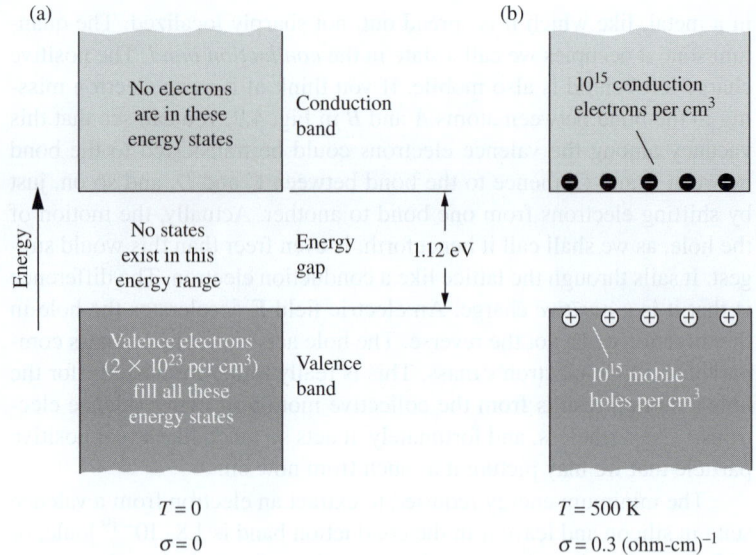

**Figure 4.10.**
A schematic representation of the energy bands in silicon, which are all the possible states for the electrons, arranged in order of energy. Two electrons can't have the same state. (a) At temperature zero the valence band is full; an electron occupies every available state. The conduction band is empty. (b) At $T = 500$ K there are $10^{15}$ electrons in the lowest conduction band states, leaving $10^{15}$ holes in the valence band, in 1 cm$^3$ of the crystal.

Suppose that two states labeled 1 and 2 are available for occupation by an electron and that the electron's energy in state 1 would be $E_1$, while its energy in state 2 would be $E_2$. Let $p_1$ be the probability that the electron will be found occupying state 1, $p_2$ the probability that it will be found in state 2. In a system in thermal equilibrium at temperature $T$, the ratio $p_2/p_1$ depends only on the energy *difference*, $\Delta E = E_2 - E_1$. It is given by

$$\frac{p_2}{p_1} = e^{-\Delta E/kT} \qquad (4.27)$$

The constant $k$, known as Boltzmann's constant, has the value $1.38 \cdot 10^{-23}$ joule/kelvin. This relation holds for any two states. It governs the population of available states on the energy ladder. To predict the resulting number of electrons in the conduction band at a given temperature we would have to know more about the number of states available. But this shows why the number of conduction electrons per unit volume depends so strongly on the temperature. For $T = 300$ K the energy $kT$ is about 0.025 eV. The Boltzmann factor relating states 1 eV apart in energy would be $e^{-40}$, or $4 \cdot 10^{-18}$. In silicon at room temperature the number of electrons in the conduction band, per cubic centimeter, is approximately $10^{10}$. At 500 K one finds about $10^{15}$ electrons per cm$^3$ in the conduction band, and the same number of holes in the valence band (Fig. 4.10(b)). Both holes and electrons contribute to the conductivity, which is 0.3 (ohm-cm)$^{-1}$ at that temperature. Germanium behaves like silicon, but the energy gap is somewhat smaller, 0.7 eV. At any given temperature it has more conduction electrons and holes than

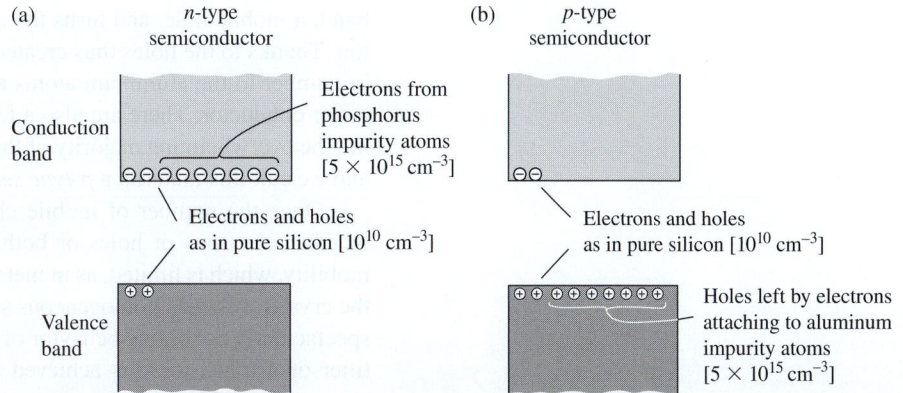

(a)     *n*-type
semiconductor

Conduction
band

Electrons from
phosphorus
impurity atoms
$[5 \times 10^{15}$ cm$^{-3}]$

Electrons and holes
as in pure silicon $[10^{10}$ cm$^{-3}]$

Valence
band

(b)     *p*-type
semiconductor

Electrons and holes
as in pure silicon $[10^{10}$ cm$^{-3}]$

Holes left by electrons
attaching to aluminum
impurity atoms
$[5 \times 10^{15}$ cm$^{-3}]$

**Figure 4.11.**
In an *n*-type semiconductor (a) most of the charge carriers are electrons released from pentavalent impurity atoms such as phosphorus. In a *p*-type semiconductor (b) the majority of the charge carriers are holes. A hole is created when a trivalent impurity atom like aluminum grabs an electron to complete the covalent bonds to its four silicon neighbors. A few carriers of the opposite sign exist in each case. The number densities in brackets refer to our example of $5 \cdot 10^{15}$ impurity atoms per cm$^3$, and room temperature. Under these conditions the number of majority charge carriers is practically equal to the number of impurity atoms, while the number of minority carriers is *very* much smaller.

silicon, consequently higher conductivity, as is evident in Fig. 4.8. Diamond would be a semiconductor, too, if its energy gap weren't so large (5.5 eV) that there are no electrons in the conduction band at any attainable temperature.

With only $10^{10}$ conduction electrons and holes per cubic centimeter, the silicon crystal at room temperature is practically an insulator. But that can be changed dramatically by inserting foreign atoms into the pure silicon lattice. This is the basis for all the marvelous devices of semiconductor electronics. Suppose that some very small fraction of the silicon atoms – for example, 1 in $10^7$ – are replaced by phosphorus atoms. (This "doping" of the silicon can be accomplished in various ways.) The phosphorus atoms, of which there are now about $5 \cdot 10^{15}$ per cm$^3$, occupy regular sites in the silicon lattice. A phosphorus atom has five valence electrons, one too many for the four-bond structure of the perfect silicon crystal. The extra electron easily comes loose. Only 0.044 eV of energy is needed to boost it to the conduction band. What is left behind in this case is not a mobile hole, but an immobile positive phosphorus ion. We now have nearly $5 \cdot 10^{15}$ mobile electrons in the conduction band, and a conductivity of nearly 1 (ohm-cm)$^{-1}$. There are also a few holes in the valence band, but the number is negligible compared with the number of conduction electrons. (It is even smaller than it would be in a pure crystal, because the increase in the number of conduction electrons makes it more likely for a hole to be negated.) Because nearly all the charge carriers are *negative*, we call this "phosphorus-doped" crystal an *n-type semiconductor* (Fig. 4.11(a)).

Now let's dope a pure silicon crystal with aluminum atoms as the impurity. The aluminum atom has three valence electrons, one too few to construct four covalent bonds around its lattice site. That is cheaply remedied if one of the regular valence electrons joins the aluminum atom permanently, completing the bonds around it. The cost in energy is only 0.05 eV, much less than the 1.2 eV required to raise a valence electron up to the conduction band. This promotion creates a vacancy in the valence band, a mobile hole, and turns the aluminum atom into a fixed negative ion. Thanks to the holes thus created – at room temperature nearly equal in number to the aluminum atoms added – the crystal becomes a much better conductor. There are also a few electrons in the conduction band, but the overwhelming majority of the mobile charge carriers are positive, and we call this material a *p-type semiconductor* (Fig. 4.11(b)).

Once the number of mobile charge carriers has been established, whether electrons or holes or both, the conductivity depends on their mobility, which is limited, as in metallic conduction, by scattering within the crystal. A single homogeneous semiconductor obeys Ohm's law. The spectacularly nonohmic behavior of semiconductor devices – as in a rectifier or a transistor – is achieved by combining *n*-type material with *p*-type material in various arrangements.

---

**Example (Mean free time in silicon)** In Fig. 4.10, a conductivity of $30 \ (\text{ohm-m})^{-1}$ results from the presence of $10^{21}$ electrons per m$^3$ in the conduction band, along with the same number of holes. Assume that $\tau_+ = \tau_-$ and $M_+ = M_- = m_e$, the electron mass. What must be the value of the mean free time $\tau$? The rms speed of an electron at 500 K is $1.5 \cdot 10^5$ m/s. Compare the mean free path with the distance between neighboring silicon atoms, which is $2.35 \cdot 10^{-10}$ m.

**Solution** Since we have two types of charge carriers, the electrons and the holes, Eq. (4.23) gives

$$\tau = \frac{m\sigma}{2Ne^2} = \frac{(9.1 \cdot 10^{-31} \ \text{kg})\left(30 \,(\text{ohm-m})^{-1}\right)}{2(10^{21} \ \text{m}^{-3})(1.6 \cdot 10^{-19} \ \text{C})^2} \approx 5.3 \cdot 10^{-13} \ \text{s}. \qquad (4.28)$$

The distance traveled during this time is $v\tau = (1.5 \cdot 10^5 \ \text{m/s})(5.3 \cdot 10^{-13} \ \text{s}) \approx 8 \cdot 10^{-8}$ m, which is more than 300 times the distance between neighboring silicon atoms.

---

## 4.7 Circuits and circuit elements

Electrical devices usually have well-defined terminals to which wires can be connected. Charge can flow into or out of the device over these paths. In particular, if two terminals, and only two, are connected by wires to something outside, and if the current flow is steady with constant potentials everywhere, then obviously the current must be equal and opposite at the two terminals.[13] In that case we can speak of *the* current $I$ that flows through the device, and of *the* voltage $V$ "between the terminals" or "across the terminals," which means their difference in electric

---

[13] It is perfectly possible to have 4 A flowing into one terminal of a two-terminal object with 3 A flowing out at the other terminal. But then the object is accumulating positive charge at the rate of 1 coulomb/second. Its potential must be changing very rapidly – and that can't go on for long. Hence this cannot be a *steady,* or time-independent, current.

potential. The ratio $V/I$ for some given $I$ is a certain number of resistance units (ohms, if $V$ is in volts and $I$ in amps). If Ohm's law is obeyed in all parts of the object through which current flows, that number will be a constant, independent of the current. This one number completely describes the electrical behavior of the object, for steady current flow (DC) between the given terminals. With these rather obvious remarks we introduce a simple idea, the notion of a *circuit element*.

Look at the five boxes in Fig. 4.12. Each has two terminals, and inside each box there is some stuff, different in every box. If any one of these boxes is made part of an electrical circuit by connecting wires to the terminals, the ratio of the potential difference between the terminals to the current flowing in the wire that we have connected to the terminal will be found to be 65 ohms. We say the resistance between the terminals, in each box, is 65 ohms. This statement would surely not be true for all conceivable values of the current or potential difference. As the potential difference or *voltage* between the terminals is raised, various things might happen, earlier in some boxes than in others, to change the *voltage/current* ratio. You might be able to guess which boxes would give trouble first. Still, there is *some* limit below which they all behave linearly; within that range, for *steady* currents, the boxes are alike. They are alike in this sense: if any circuit contains one of these boxes, which box it is makes no difference in the behavior of that circuit. The box is equivalent to a 65 ohm resistor.[14] We represent it by the symbol -\/\/\/- and in the description of the circuit of which the box is one component, we replace the box with this abstraction. An electrical circuit or network is then a collection of such circuit elements joined to one another by paths of negligible resistance.

Taking a network consisting of many elements connected together and selecting two points as terminals, we can regard the whole thing as equivalent, as far as these two terminals are concerned, to a single resistor. We say that the physical network of objects in Fig. 4.13(a) is represented by the diagram of Fig. 4.13(b), and for the terminals $A_1A_2$ the equivalent circuit is Fig. 4.13(c). The equivalent circuit for the terminals at $B_1B_2$ is given in Fig. 4.13(d). If you put this assembly in a box with only that pair of terminals accessible, it will be indistinguishable from a resistor of 57.6 ohm resistance.

There is one very important rule – only *direct-current* measurements are allowed! All that we have said depends on the current and electric fields being constant in time; if they are not, the behavior of a circuit

**Figure 4.12.**
Various devices that are equivalent, for direct current, to a 65 ohm resistor.

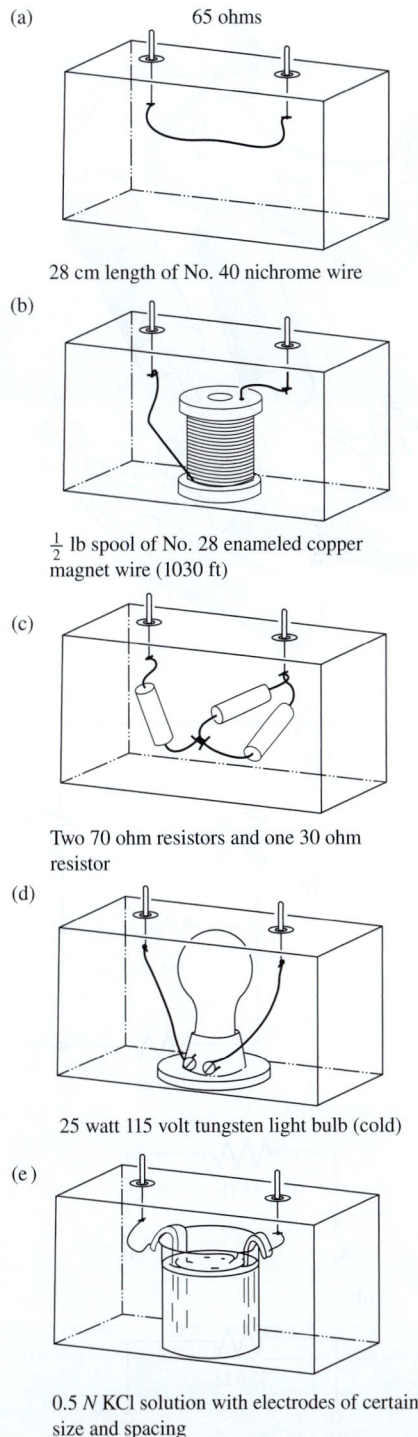

(a) 65 ohms

28 cm length of No. 40 nichrome wire

(b)

$\frac{1}{2}$ lb spool of No. 28 enameled copper magnet wire (1030 ft)

(c)

Two 70 ohm resistors and one 30 ohm resistor

(d)

25 watt 115 volt tungsten light bulb (cold)

(e)

0.5 $N$ KCl solution with electrodes of certain size and spacing

---

[14] We use the term *resistor* for the actual object designed especially for that function. Thus a "200 ohm, 10 watt, wire-wound resistor" is a device consisting of a coil of wire on some insulating base, with terminals, intended to be used in such a way that the average power dissipated in it is not more than 10 watts.

(a)

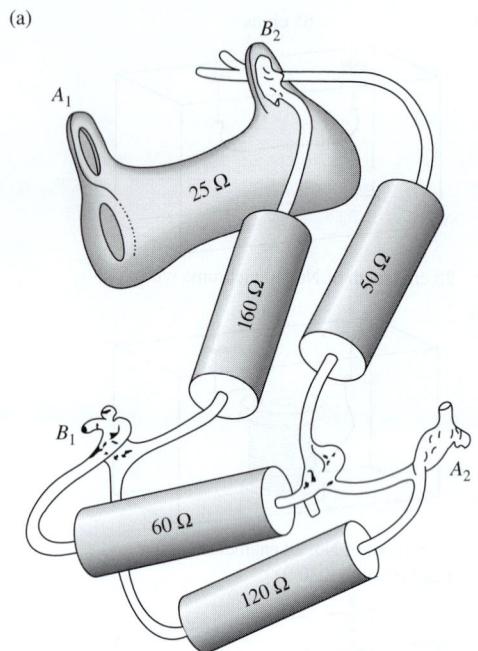

(b)

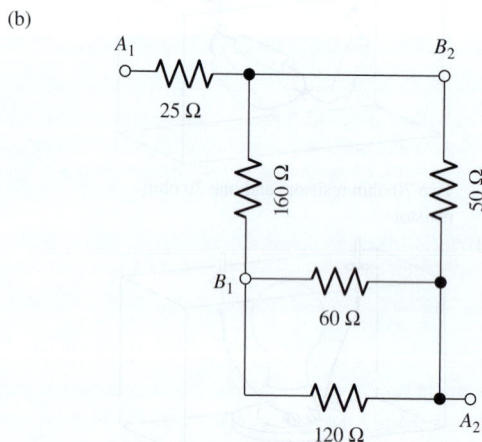

(c)

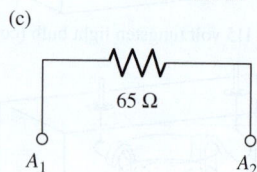

(d)

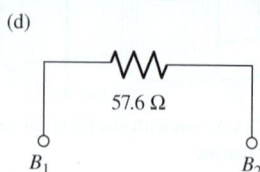

element may not depend on its resistance alone. The concept of equivalent circuits can be extended from these DC networks to systems in which current and voltage vary with time. Indeed, that is where it is most valuable. We are not quite ready to explore that domain.

Little time will be spent here on methods for calculating the equivalent resistance of a network of circuit elements. The cases of series and parallel groups are easy. A combination like that in Fig. 4.14 is two resistors, of value $R_1$ and $R_2$, in series. The equivalent resistance is

$$R = R_1 + R_2 \qquad (4.29)$$

A combination like that in Fig. 4.15 is two resistors in parallel. By an argument that you should be able to give (see Problem 4.3), the equivalent resistance $R$ is found to be

$$\frac{1}{R} = \frac{1}{R_1} + \frac{1}{R_2} \qquad \text{or} \qquad R = \frac{R_1 R_2}{R_1 + R_2}. \qquad (4.30)$$

**Example (Reducing a network)**   Let's use the addition rules in Eqs. (4.29) and (4.30) to reduce the network shown in Fig. 4.16 to an equivalent single resistor. As complicated as this network looks, it can be reduced, step by step, via series or parallel combinations. We assume that every resistor in the circuit has the value 100 ohms.

Using the above rules, we can reduce the network as follows (you should verify all of the following statements). A parallel combination of two 100 ohm resistors is equivalent to 50 ohms. So in the first figure, the top two circled sections are each equivalent to 150 ohms, and the bottom one is equivalent to 50 ohms. In the second figure, the top and bottom circled sections are then equivalent to 160 ohms and 150 ohms. In the third figure, the circled section is then equivalent to 77.4 ohms. The whole circuit is therefore equivalent to $100 + 77.4 + 150 = 327.4$ ohms.

Although Eqs. (4.29) and (4.30) are sufficient to handle the complicated circuit in Fig. 4.16, the simple network of Fig. 4.17 *cannot* be so reduced, so a more general method is required (see Exercise 4.44). Any conceivable network of resistors in which a constant current is flowing has to satisfy these conditions (the first is Ohm's law, the second and third are known as Kirchhoff's rules):

(1) The current through each element must equal the voltage across that element divided by the resistance of the element.

**Figure 4.13.**
Some resistors connected together (a); the circuit diagram (b); and the equivalent resistance between certain pairs of terminals (c) and (d).

(2) At a *node* of the network, a point where three or more connecting wires meet, the algebraic sum of the currents into the node must be zero. (This is our old charge-conservation condition, Eq. (4.8), in circuit language.)

(3) The sum of the potential differences taken in order around a *loop* of the network, a path beginning and ending at the same node, is zero. (This is network language for the general property of the static electric field: $\int \mathbf{E} \cdot d\mathbf{s} = 0$ for any closed path.)

The algebraic statement of these conditions for any network will provide exactly the number of independent linear equations needed to ensure that there is one and only one solution for the equivalent resistance between two selected nodes. We assert this without proving it. It is interesting to note that the structure of a DC network problem depends only on the *topology* of the network, that is, on those features of the diagram of connections that are independent of any distortion of the lines of the diagram. We will give an example of the use of the above three rules in Section 4.10, after we have introduced the concept of electromotive force.

A DC network of resistances is a *linear* system – the voltages and currents are governed by a set of linear equations, the statements of the conditions (1), (2), and (3). Therefore the superposition of different possible states of the network is also a possible state. Figure 4.18 shows a section of a network with certain currents, $I_1, I_2, \ldots$, flowing in the wires and certain potentials, $V_1, V_2, \ldots$, at the nodes. If some other set of currents and potentials, say $I'_1, \ldots, V'_1, \ldots$, is another possible state of affairs in this section of network, then so is the set $(I_1 + I'_1), \ldots, (V_1 + V'_1), \ldots$. These currents and voltages corresponding to the superposition will also satisfy the conditions (1), (2), and (3). Some general theorems about networks, interesting and useful to the electrical engineer, are based on this. One such theorem is *Thévenin's theorem*, discussed in Section 4.10 and proved in Problem 4.13.

## 4.8 Energy dissipation in current flow

The flow of current in a resistor involves the dissipation of energy. If it takes a force $\mathbf{F}$ to push a charge carrier along with average velocity $\mathbf{v}$, any agency that accomplishes this must do work at the rate $\mathbf{F} \cdot \mathbf{v}$. If an electric field $\mathbf{E}$ is driving the ion of charge $q$, then $\mathbf{F} = q\mathbf{E}$, and the rate at which work is done is $q\mathbf{E} \cdot \mathbf{v}$. The energy thus expended shows up eventually as heat. In our model of ionic conduction, the way this comes about is quite clear. The ion acquires some extra kinetic energy, as well as momentum, between collisions. A collision, or at most a few collisions, redirects its momentum at random but does not necessarily restore the kinetic energy to normal. For that to happen the ion has to transfer kinetic energy to the obstacle that deflects it. Suppose the charge carrier has a considerably smaller mass than the neutral atom it collides with. The average transfer

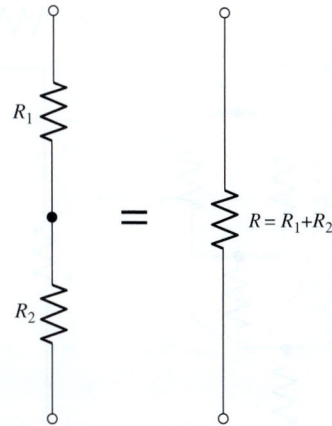

**Figure 4.14.**
Resistances in series.

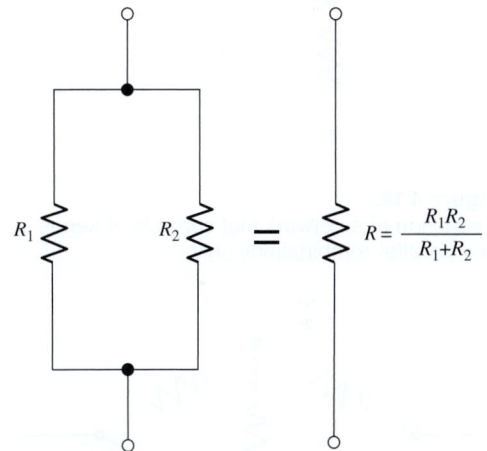

**Figure 4.15.**
Resistances in parallel.

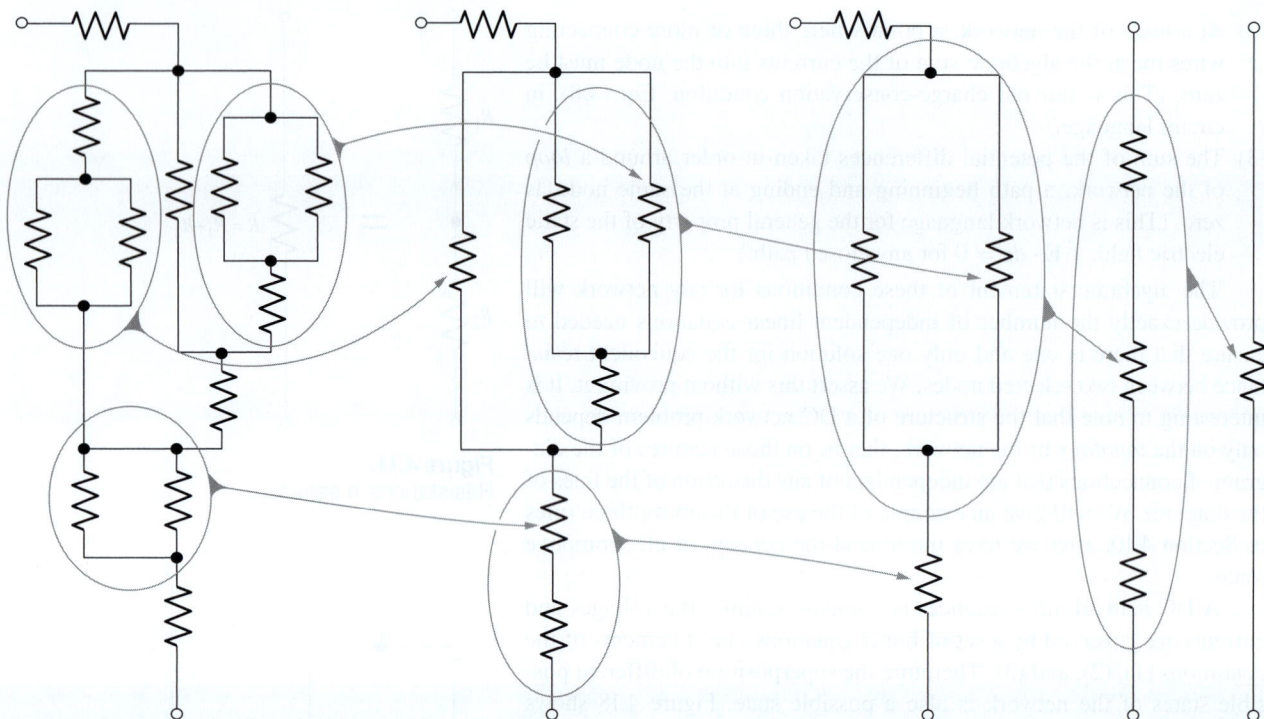

**Figure 4.16.**
Reduction of a network that consists of series and parallel combinations only.

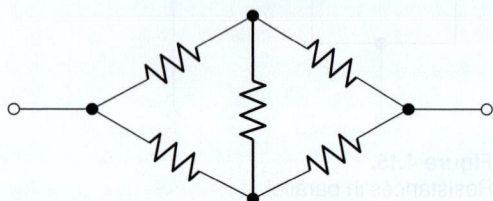

**Figure 4.17.**
A simple bridge network. It can't be reduced in the manner of Fig. 4.16.

of kinetic energy is small when a billiard ball collides with a bowling ball. Therefore the ion (billiard ball) will continue to accumulate extra energy until its average kinetic energy is so high that its average loss of energy in a collision equals the amount gained between collisions. In this way, by first "heating up" the charge carriers themselves, the work done by the electrical force driving the charge carriers is eventually passed on to the rest of the medium as random kinetic energy, or heat.

Suppose a steady current $I$, in amperes, flows through a resistor of $R$ ohms. In a time $\Delta t$, a charge of $I \Delta t$ coulombs is transferred through a potential difference of $V$ volts, where $V = IR$. Hence the work done in time $\Delta t$ is $(I \Delta t)V = I^2 R \Delta t$ in joules (because 1 coulomb $\times$ 1 volt = 1 joule). The rate at which work is done (that is, the power) is therefore

$$P = I^2 R \tag{4.31}$$

The unit of power is the *watt*. In terms of other units, a watt is a joule per second or equivalently a volt-ampere.

Naturally the steady flow of current in a dc circuit requires some source of energy capable of maintaining the electric field that drives the charge carriers. Until now we have avoided the question of the *electromotive force* by studying only parts of entire circuits; we kept the "battery" out of the picture. In Section 4.9 we shall discuss some sources of electromotive force.

# 4.9 Electromotive force and the voltaic cell

The origin of the electromotive force in a direct-current circuit is some mechanism that transports charge carriers in a direction *opposite* to that in which the electric field is trying to move them. A Van de Graaff electrostatic generator (Fig. 4.19) is an example on a large scale. With everything running steadily, we find current in the external resistance flowing in the direction of the electric field **E**, and energy being dissipated there (appearing as heat) at the rate $IV_0$, or $I^2R$. Inside the column of the machine, too, there is a downward-directed electric field. Here charge carriers can be moved against the field if they are stuck to a nonconducting belt. They are stuck so tightly that they can't slide backward along the belt in the generally downward electric field. (They can still be removed from the belt by a much stronger field localized at the brush in the terminal. We need not consider here the means for putting charge on and off the belt near the pulleys.) The energy needed to pull the belt is supplied from elsewhere – usually by an electric motor connected to a power line, but it could be a gasoline engine, or even a person turning a crank. This Van de Graaff generator is in effect a battery with an electromotive force, under these conditions, of $V_0$ volts.

In ordinary batteries it is chemical energy that makes the charge carriers move through a region where the electric field opposes their motion. That is, a *positive* charge carrier may move to a place of *higher* electric potential if by so doing it can engage in a chemical reaction that will yield more energy than it costs to climb the electrical hill.

To see how this works, let us examine one particular voltaic cell. *Voltaic cell* is the generic name for a chemical source of electromotive force. In the experiments of Galvani around 1790 the famous twitching frogs' legs had signaled the chemical production of electric current. It was Volta who proved that the source was not "animal electricity," as Galvani maintained, but the contact of dissimilar metals in the circuit. Volta went on to construct the first battery, a stack of elementary cells, each of which consisted of a zinc disk and a silver disk separated by cardboard moistened with brine. The battery that powers your flashlight comes in a tidier package, but the principle of operation is the same. Several kinds of voltaic cells are in use, differing in their chemistry but having common features: two electrodes of different material immersed in an ionized fluid, or electrolyte.

As an example, we'll describe the lead–sulfuric acid cell which is the basic element of the automobile battery. This cell has the important property that its operation is readily reversible. With a *storage battery* made of such cells, which can be charged and discharged repeatedly, energy can be stored and recovered electrically.

A fully charged lead–sulfuric acid cell has positive plates that hold lead dioxide, $PbO_2$, as a porous powder, and negative plates that hold pure lead of a spongy texture. The mechanical framework, or grid, is made of a lead alloy. All the positive plates are connected together and

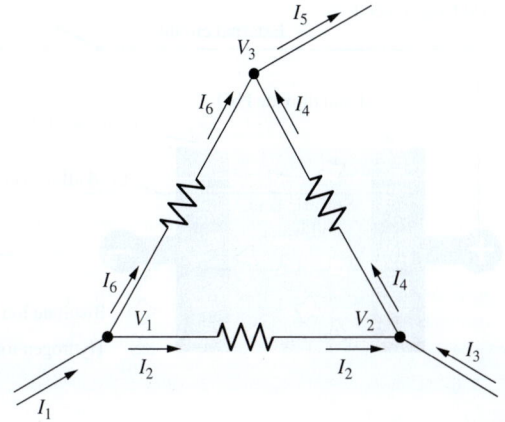

**Figure 4.18.**
Currents and potentials at the nodes of a network.

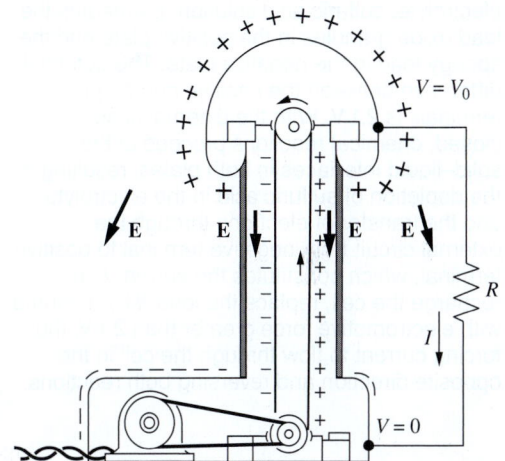

**Figure 4.19.**
In the Van de Graaff generator, charge carriers are mechanically transported in a direction opposite to that in which the electric field would move them.

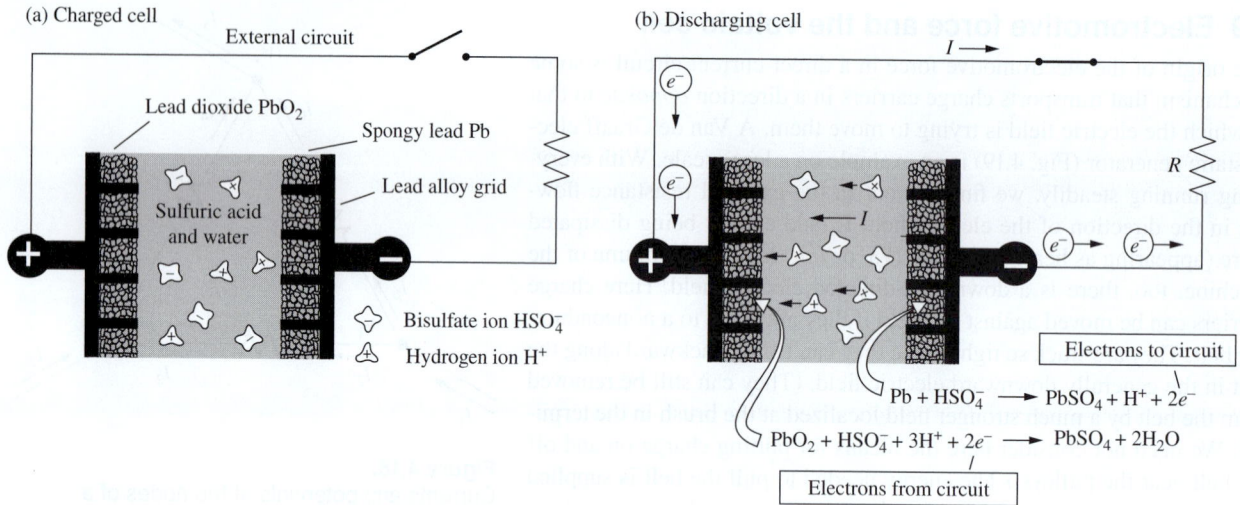

(a) Charged cell

External circuit

Lead dioxide PbO$_2$

Spongy lead Pb

Lead alloy grid

Sulfuric acid
and water

$R$

⬡ Bisulfate ion HSO$_4^-$

△ Hydrogen ion H$^+$

(b) Discharging cell

$I \longrightarrow$

$e^-$

$e^-$

$\longleftarrow I$

$R$

$e^- \rightarrow$ $e^- \rightarrow$

Electrons to circuit

Pb + HSO$_4^-$ $\longrightarrow$ PbSO$_4$ + H$^+$ + 2$e^-$

PbO$_2$ + HSO$_4^-$ + 3H$^+$ + 2$e^-$ $\longrightarrow$ PbSO$_4$ + 2H$_2$O

Electrons from circuit

**Figure 4.20.**
A schematic diagram, not to scale, showing how
the lead–sulfuric acid cell works. The
electrolyte, sulfuric acid solution, permeates the
lead oxide granules in the positive plate and the
spongy lead in the negative plate. The potential
difference between the positive and negative
terminals is 2.1 V. With the external circuit
closed, chemical reactions proceed at the
solid–liquid interfaces in both plates, resulting in
the depletion of sulfuric acid in the electrolyte
and the transfer of electrons through the
external circuit from negative terminal to positive
terminal, which constitutes the current $I$. To
recharge the cell, replace the load $R$ by a source
with electromotive force greater than 2.1 V, thus
forcing current to flow through the cell in the
opposite direction and reversing both reactions.

to the positive terminal of the cell. The negative plates, likewise connected, are interleaved with the positive plates, with a small separation.
The schematic diagram in Fig. 4.20 shows only a small portion of a
positive and a negative plate. The sulfuric acid electrolyte fills the cell,
including the interstices of the active material, the porosity of which provides a large surface area for chemical reaction.

The cell will remain indefinitely in this condition if there is no external circuit connecting its terminals. The potential difference between its
terminals will be close to 2.1 volts. This open-circuit potential difference
is established "automatically" by the chemical interaction of the constituents. This is the *electromotive force* of the cell, for which the symbol
$\mathcal{E}$ will be used. Its value depends on the concentration of sulfuric acid
in the electrolyte, but not at all on the size, number, or separation of
the plates.

Now connect the cell's terminals through an external circuit with
resistance $R$. If $R$ is not too small, the potential difference $V$ between the
cell terminals will drop only a little below its open-circuit value $\mathcal{E}$, and
a current $I = V/R$ will flow around the circuit (Fig. 4.20(b)). Electrons
flow *into* the positive terminal; other electrons flow *out* of the negative
terminal. At each electrode chemical reactions are proceeding, the overall effect of which is to convert lead, lead dioxide, and sulfuric acid into
lead sulfate and water. For every molecule of lead sulfate thus made,
one charge $e$ is passed around the circuit and an amount of energy $e\mathcal{E}$ is
released. Of this energy the amount $eV$ appears as heat in the external
resistance $R$. The difference between $\mathcal{E}$ and $V$ is caused by the resistance
of the electrolyte itself, through which the current $I$ must flow inside the
cell. If we represent this internal resistance by $R_i$, the system can be quite
well described by the equivalent circuit in Fig. 4.21.

As discharge goes on and the electrolyte becomes more diluted with water, the electromotive force $\mathcal{E}$ decreases somewhat. Normally, the cell is considered discharged when $\mathcal{E}$ has fallen below 1.75 volts. To recharge the cell, current must be forced around the circuit in the opposite direction by connecting a voltage source greater than $\mathcal{E}$ across the cell's terminals. The chemical reactions then run backward until all the lead sulfate is turned back into lead dioxide and lead. The investment of energy in charging the cell is somewhat more than the cell will yield on discharge, for the internal resistance $R_i$ causes a power loss $I^2 R_i$ whichever way the current is flowing.

Note in Fig. 4.20(b) that the current $I$ in the electrolyte is produced by a net drift of positive ions toward the positive plate. Evidently the electric field in the electrolyte points toward, not away from, the positive plate. Nevertheless, the line integral of $\mathbf{E}$ around the whole circuit is zero, as it must be for any electrostatic field. The explanation is this: there are two very steep jumps in potential at the interface of the positive plate and the electrolyte and at the interface of the negative plate and the electrolyte. That is where the ions are moved *against* a strong electric field by forces arising in the chemical reactions. It is this region that corresponds to the belt in a Van de Graaff generator.

Every kind of voltaic cell has its characteristic electromotive force, falling generally in the range of 1 to 3 volts. The energy involved, per molecule, in any chemical reaction is essentially the gain or loss in transfer of an outer electron from one atom to a different atom. That is never more than a few electron-volts. We can be pretty sure that no one is going to invent a voltaic cell with a 12 volt electromotive force. The 12 volt automobile battery consists of six separate lead–sulfuric acid cells connected in series. For more discussion of how batteries work, including a helpful analogy, see Roberts (1983).

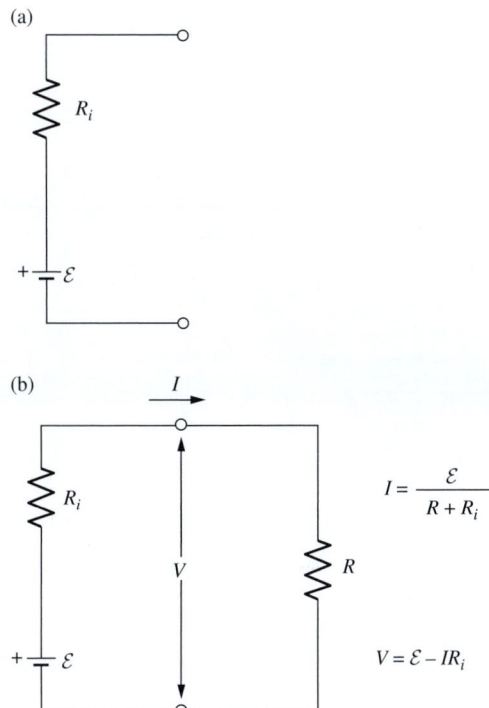

**Figure 4.21.**
(a) The equivalent circuit for a voltaic cell is simply a resistance $R_i$ in series with an electromotive force $\mathcal{E}$ of fixed value.
(b) Calculation of the current in a circuit containing a voltaic cell.

**Example (Lead–acid battery)**   A 12 V lead–acid storage battery with a 20 ampere-hour capacity rating has a mass of 10 kg.

(a)   How many kilograms of lead sulfate are formed when this battery is discharged? (The molecular weight of $PbSO_4$ is 303.)
(b)   How many kilograms of batteries of this type would be required to store the energy derived from 1 kg of gasoline by an engine of 20 percent efficiency? (The heat of combustion of gasoline is $4.5 \cdot 10^7$ J/kg.)

Solution

(a)   The total charge transferred in 20 ampere-hours is $(20\,C/s)(3600\,s) = 72{,}000\,C$. From Fig. 4.20(b), the creation of two electrons is associated with the creation of one molecule of $PbSO_4$. But also the absorption of two electrons is associated with the creation of another molecule of $PbSO_4$. So the travel of two electrons around the circuit is associated with the creation of two molecules of $PbSO_4$. The ratio is thus 1 to 1. The charge transferred per mole of $PbSO_4$ is therefore $(6 \cdot 10^{23})(1.6 \cdot 10^{-19}\,C) = 96{,}000\,C$. The above charge of 72,000 C therefore corresponds to 3/4 of a mole. Since each mole has a mass of 0.303 kg, the desired mass is about 0.23 kg.

(b) At 12 V, the energy output associated with a charge of 72,000 C is $(12 \text{ J/C})(72,000 \text{ C}) = 864,000 \text{ J}$. Also, 1 kg of gasoline burned at 20 percent efficiency yields an energy of $(0.2)(1 \text{ kg})(4.5 \cdot 10^7 \text{ J/kg}) = 9 \cdot 10^6 \text{ J}$. This is equivalent to $(9 \cdot 10^6 \text{ J})/(8.64 \cdot 10^5 \text{ J}) = 10.4$ batteries. Since each battery has a mass of 10 kg, this corresponds to 104 kg of batteries.

## 4.10 Networks with voltage sources

### 4.10.1 Applying Kirchhoff's rules

A network of resistors could contain more than one electromotive force, or voltage source. Consider the following example.

**Example** The circuit in Fig. 4.22 contains two batteries with electromotive force $\mathcal{E}_1$ and $\mathcal{E}_2$, respectively. In each of the conventional battery symbols shown, the longer line indicates the positive terminal. Assume that $R_1$ includes the internal resistance of one battery, $R_2$ that of the other. Supposing the resistances given, what are the currents in this network?

**Solution** Having assigned directions arbitrarily to the currents $I_1$, $I_2$, and $I_3$ in the branches, we can impose the requirements stated in Section 4.7. We have one node and two loops,[15] so we obtain three independent Kirchhoff equations:

$$I_1 - I_2 - I_3 = 0,$$
$$\mathcal{E}_1 - R_1 I_1 - R_3 I_3 = 0,$$
$$\mathcal{E}_2 + R_3 I_3 - R_2 I_2 = 0. \qquad (4.32)$$

To check the signs, note that in writing the two loop equations, we have gone around each loop in the direction current would flow from the battery in that loop. The three equations can be solved for $I_1$, $I_2$, and $I_3$. This is slightly messy by hand, but trivial if we use a computer; the result is

$$I_1 = \frac{\mathcal{E}_1 R_2 + \mathcal{E}_1 R_3 + \mathcal{E}_2 R_3}{R_1 R_2 + R_2 R_3 + R_1 R_3},$$
$$I_2 = \frac{\mathcal{E}_2 R_1 + \mathcal{E}_2 R_3 + \mathcal{E}_1 R_3}{R_1 R_2 + R_2 R_3 + R_1 R_3},$$
$$I_3 = \frac{\mathcal{E}_1 R_2 - \mathcal{E}_2 R_1}{R_1 R_2 + R_2 R_3 + R_1 R_3}. \qquad (4.33)$$

If in a particular case the value of $I_3$ turns out to be negative, it simply means that the current in that branch flows opposite to the direction we had assigned to positive current.

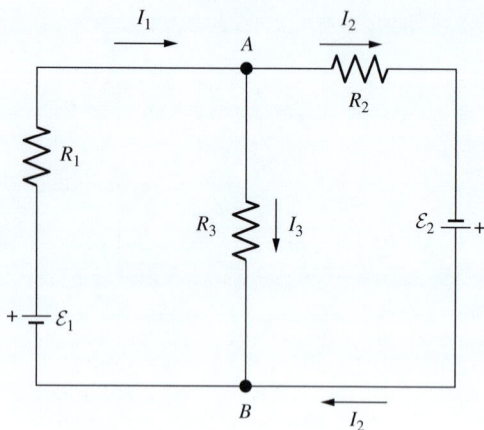

**Figure 4.22.**
A network with two voltage sources.

---

[15] There are actually two nodes, of course, but they give the same information. And there is technically a third loop around the whole network, but the resulting equation is the sum of the two other loop equations.

Alternatively, we can use the "loop" currents shown in Fig. 4.23. The advantages of this method are that (1) the "node" condition in Section 4.7 is automatically satisfied, because whatever current goes into a node also comes out, by construction; and (2) there are only two unknowns to solve for instead of three (although to be fair, the first of the equations in Eq. (4.32) is trivial). The disadvantage is that if we want to find the current in the middle branch ($I_3$ above), we need to take the difference of the loop currents $I_1$ and $I_2$, because with the sign conventions chosen, these currents pass in opposite directions through $R_3$. But this is not much of a burden. The two loop equations are now

$$\mathcal{E}_1 - R_1 I_1 - R_3(I_1 - I_2) = 0,$$
$$\mathcal{E}_2 - R_3(I_2 - I_1) - R_2 I_2 = 0. \qquad (4.34)$$

Of course, these two equations are just the second two equations in Eq. (4.32), with $I_3 = I_1 - I_2$ substituted in from the first equation. So we obtain the same values of $I_1$ and $I_2$ (and hence $I_3$).

**Figure 4.23.**
Loop currents for use in Kirchhoff's rules. Loop currents automatically satisfy the node condition.

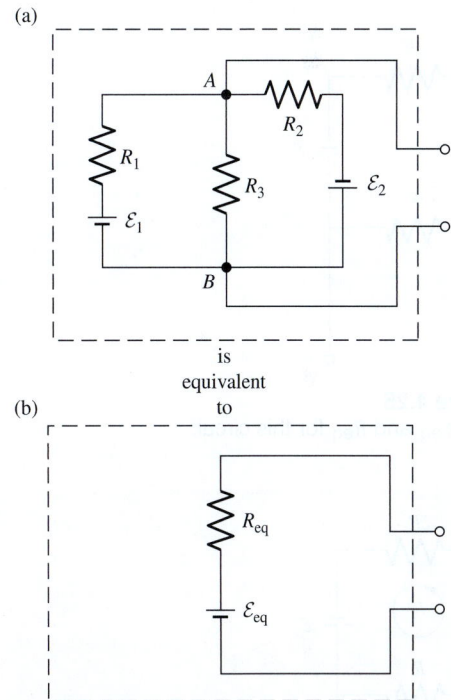

The calculational difference between the two methods in the above example was inconsequential. But in larger networks the second method is often more tractable, because it involves simply writing down a loop equation for every loop you see on the page. This tells you right away how many unknowns (the loop currents) there are. In either case, all of the physics is contained in the equations representing the rules given in Section 4.7. The hardest thing about these equations is making sure all the signs are correct. The actual process of solving them is easy if you use a computer. A larger network is technically no more difficult to solve than a smaller one. The only difference is that the larger network takes more time, because it takes longer to write down the equations (which are all of the same general sort) and then type them into the computer.

### 4.10.2 Thévenin's theorem

Suppose that a network such as the one in Fig. 4.22 forms part of some larger system, to which it is connected at two of its nodes. For example, let us connect wires to the two nodes $A$ and $B$ and enclose the rest in a "black box" with these two wires as the only external terminals, as in Fig. 4.24(a). A general theorem called *Thévenin's theorem* assures us that this two-terminal box is completely equivalent, in its behavior in any other circuit to which it may be connected, to a *single* voltage source $\mathcal{E}_{eq}$ ("eq" for equivalent) with an internal resistance $R_{eq}$. This holds for any network of voltage sources and resistors, no matter how complicated. It is not immediately obvious that such an $\mathcal{E}_{eq}$ and $R_{eq}$ should exist (see Problem 4.13 for a proof), but assuming they do exist, their values can be determined by either experimental measurements or theoretical calculations, in the following ways.

If we *don't* know what is in the box, we can determine $\mathcal{E}_{eq}$ and $R_{eq}$ experimentally by two measurements.

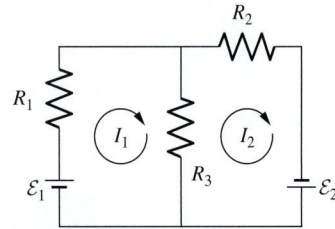

(a)

is
equivalent
to

(b)

**Figure 4.24.**
Make $R_{eq}$ equal to the resistance that would be measured between the terminals in (a) if all electromotive forces were zero. Make $\mathcal{E}_{eq}$ equal to the voltage observed between the terminals in (a) with the external circuit open. Then the circuit in (b) is *equivalent* to the circuit in (a). You can't tell the difference by any direct-current measurement at those terminals.

- Measure the *open-circuit voltage* between the terminals by connecting them via a voltmeter that draws negligible current. (The "infinite" resistance of the voltmeter means that the terminals are effectively unconnected; hence the name "open circuit.") This voltage equals $\mathcal{E}_{eq}$. This is clear from Fig. 4.24(b); if essentially zero current flows through this simple circuit, then there is zero voltage drop across the resistor $R_{eq}$. So the measured voltage equals all of the $\mathcal{E}_{eq}$.
- Measure the *short-circuit current* $I_{sc}$ between the terminals by connecting them via an ammeter with negligible resistance. (The "zero" resistance of the ammeter means that the terminals are effectively connected by a short circuit.) Ohm's law for the short-circuited circuit in Fig. 4.24(b) then yields simply $\mathcal{E}_{eq} = I_{sc}R_{eq}$. The equivalent resistance is therefore given by

$$R_{eq} = \frac{\mathcal{E}_{eq}}{I_{sc}}. \qquad (4.35)$$

If we *do* know what is in the box, we can determine $\mathcal{E}_{eq}$ and $R_{eq}$ by calculating them instead of measuring them.

- For $\mathcal{E}_{eq}$, calculate the open circuit voltage between the two terminals (with nothing connected to them outside the box). In the above example, this is just $I_3R_3$, with $I_3$ given by Eq. (4.33).
- For $R_{eq}$, connect the terminals by a wire with zero resistance, and calculate the short-circuit current $I_{sc}$ through this wire; $R_{eq}$ is then given by $\mathcal{E}_{eq}/I_{sc}$. See Problem 4.14 for how this works in the above example. There is, however, a second method for calculating $R_{eq}$, which is generally much quicker: $R_{eq}$ is the resistance that would be measured between the two terminals with all the internal electromotive forces made zero. In our example that would be the resistance of $R_1$, $R_2$, and $R_3$ all in parallel, which is $R_1R_2R_3/(R_1R_2 + R_2R_3 + R_1R_3)$. The reason why this method works is explained in the solution to Problem 4.13.

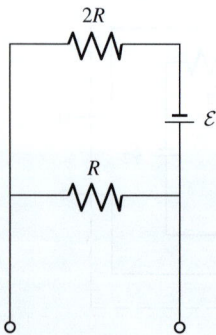

**Figure 4.25.**
Find $\mathcal{E}_{eq}$ and $R_{eq}$ for this circuit.

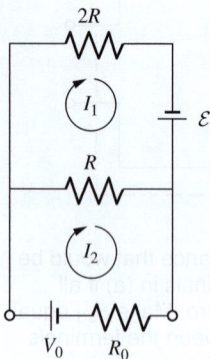

**Figure 4.26.**
Loop currents for use in Kirchhoff's rules.

### Example

(a)  Find the Thévenin equivalent $\mathcal{E}_{eq}$ and $R_{eq}$ for the circuit shown in Fig. 4.25.
(b)  Calculate $\mathcal{E}_{eq}$ and $R_{eq}$ again, but now do it the long way. Use Kirchhoff's rules to find the current passing through the bottom branch of the circuit in Fig. 4.26, and then interpret your result in a way that gives you $\mathcal{E}_{eq}$ and $R_{eq}$.

### Solution

(a)  $\mathcal{E}_{eq}$ is the open-circuit voltage. With nothing connected to the terminals, the current running around the loop is $\mathcal{E}/3R$. The voltage drop across the $R$ resistor is therefore $(\mathcal{E}/3R)(R) = \mathcal{E}/3$. But this is also the open-circuit voltage between the two terminals, so $\mathcal{E}_{eq} = \mathcal{E}/3$.

We can find $R_{eq}$ in two ways. The quick way is to calculate the resistance between the terminals with $\mathcal{E}$ set equal to zero. In that case we have an $R$ and a $2R$ in parallel, so $R_{eq} = 2R/3$.

Alternatively, we can find $R_{eq}$ by calculating the short-circuit current between the terminals. With the short circuit present, no current takes the route through the $R$ resistor, so we just have $\mathcal{E}$ and $2R$ in series. The short-circuit current between the terminals is therefore $I_{sc} = \mathcal{E}/2R$. The equivalent resistance is then given by $R_{eq} = \mathcal{E}_{eq}/I_{sc} = (\mathcal{E}/3)/(\mathcal{E}/2R) = 2R/3$.

(b) The loop equations for the circuit in Fig. 4.26 are

$$0 = \mathcal{E} - R(I_1 - I_2) - (2R)I_1,$$
$$0 = V_0 - R(I_2 - I_1) - R_0 I_2. \qquad (4.36)$$

Solving these equations for $I_2$ gives $I_2 = (\mathcal{E} + 3V_0)/(2R + 3R_0)$ (as you can check), which can be written suggestively as

$$V_0 + \frac{\mathcal{E}}{3} = I_2 \left( R_0 + \frac{2R}{3} \right). \qquad (4.37)$$

But this is exactly the $V = IR$ statement that we would write down for the circuit shown in Fig. 4.27, where the total emf is $V_0 + \mathcal{E}/3$ and the total resistance is $R_0 + 2R/3$. Since the result in Eq. (4.37) holds for any values of $V_0$ and $R_0$, we conclude that the given circuit is equivalent to an emf $\mathcal{E}_{eq} = \mathcal{E}/3$ in series with a resistor $R_{eq} = 2R/3$. Generalizing this method is the basic idea behind the first proof of Thévenin's theorem given in Problem 4.13.

**Figure 4.27.**
The Thévenin equivalent circuit.

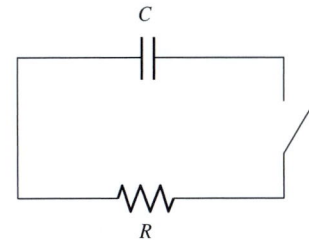

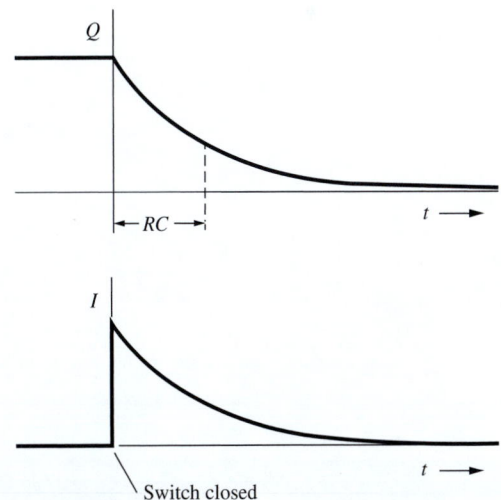

In analyzing a complicated circuit it sometimes helps to replace a two-terminal section by its equivalent $\mathcal{E}_{eq}$ and $R_{eq}$. Thévenin's theorem assumes the linearity of all circuit elements, including the reversibility of currents through batteries. If one of our batteries is a nonrechargeable dry cell with the current through it backward, caution is advisable!

## 4.11 Variable currents in capacitors and resistors

Let a capacitor of capacitance $C$ be charged to some potential $V_0$ and then discharged by suddenly connecting it across a resistance $R$. Figure 4.28 shows the capacitor indicated by the conventional symbol ⊣⊢, the resistor $R$, and a switch which we shall imagine to be closed at time $t = 0$. It is obvious that, as current flows, the capacitor will gradually lose its charge, the voltage across the capacitor will diminish, and this in turn will lessen the flow of current. Let's be quantitative about this.

**Example (RC circuit)** In the circuit in Fig. 4.28, what are the charge $Q$ on the capacitor and the current $I$ in the circuit, as functions of time?

**Solution** To find $Q(t)$ and $I(t)$ we need only write down the conditions that govern the circuit. Let $V(t)$ be the potential difference between the plates, which is also the voltage across the resistor $R$. Let the current $I$ be considered positive

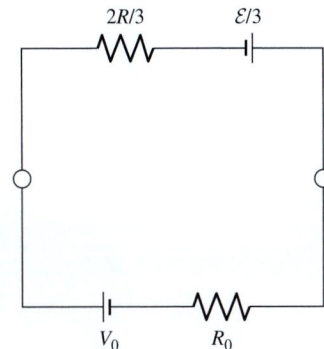

**Figure 4.28.**
Charge and current in an $RC$ circuit. Both quantities decay by the factor $1/e$ in time $RC$.

if it flows away from the positive side of the capacitor. The quantities $Q$, $I$, and $V$, all functions of the time, must be related as follows:

$$Q = CV, \qquad I = \frac{V}{R}, \qquad -\frac{dQ}{dt} = I. \qquad (4.38)$$

Eliminating $I$ and $V$, we obtain the equation that governs the time variation of $Q$:

$$\frac{dQ}{dt} = -\frac{Q}{RC}. \qquad (4.39)$$

Writing this in the form

$$\frac{dQ}{Q} = -\frac{dt}{RC}, \qquad (4.40)$$

we can integrate both sides, obtaining

$$\ln Q = \frac{-t}{RC} + \text{const.} \qquad (4.41)$$

The solution of our differential equation is therefore

$$Q = (\text{another constant}) \cdot e^{-t/RC}. \qquad (4.42)$$

If $V = V_0$ at $t = 0$, then $Q = CV_0$ at $t = 0$. This determines the constant, and we now have the exact behavior of $Q$ after the switch is closed:

$$Q(t) = CV_0 \, e^{-t/RC}. \qquad (4.43)$$

The behavior of the current $I$ is found directly from this:

$$I(t) = -\frac{dQ}{dt} = \frac{V_0}{R} e^{-t/RC}. \qquad (4.44)$$

And the voltage at any time is $V(t) = I(t)R$, or alternatively $V(t) = Q(t)/C$.

At the closing of the switch the current rises at once to the value $V_0/R$ and then decays exponentially to zero. The time that characterizes this decay is the constant $RC$ in the above exponents. People often speak of the "$RC$ time constant" associated with a circuit or part of a circuit. Let's double check that $RC$ does indeed have units of time. In SI units, $R$ is measured in ohms, which from Eq. (4.18) is given by volt/ampere. And $C$ is measured in farads, which from Eq. (3.8) is given by coulomb/volt. So $RC$ has units of coulomb/ampere, which is a second, as desired. If we make the circuit in Fig. 4.28 out of a 0.05 microfarad capacitor and a 5 megohm resistor, both of which are reasonable objects to find around any laboratory, we would have $RC = (5 \cdot 10^6 \text{ ohm})(0.05 \cdot 10^{-6} \text{ farad}) = 0.25 \text{ s}$.

Quite generally, in any electrical system made up of charged conductors and resistive current paths, one time scale – perhaps not the only one – for processes in the system is set by some resistance–capacitance product. This has a bearing on our earlier observation on page 187 that $\epsilon_0\rho$ has the dimensions of time. Imagine a capacitor with plates of area $A$ and separation $s$. Its capacitance $C$ is $\epsilon_0 A/s$. Now imagine the space between the plates suddenly filled with a conductive medium of resistivity $\rho$. To avoid any question of how this might affect the capacitance, let us suppose that the medium is a very slightly ionized gas; a substance of that density will hardly affect the capacitance at all. This new conductive path will discharge the capacitor as effectively as did the external resistor in Fig. 4.28. How quickly will this happen? From Eq. (4.17) the resistance of the path, $R$, is $\rho s/A$. Hence the time constant $RC$ is just $(\rho s/A)(\epsilon_0 A/s) = \epsilon_0 \rho$. For example, if our weakly ionized gas had a resistivity of $10^6$ ohm-meter, the time constant for discharge of the capacitor would be (recalling the units of $\epsilon_0$ and the ohm) $\epsilon_0 \rho = (8.85 \cdot 10^{-12}\, \text{C}^2\,\text{s}^2\,\text{kg}^{-1}\,\text{m}^{-3})(10^6\,\text{kg}\,\text{m}^3\,\text{C}^{-2}\,\text{s}^{-1}) \approx 10$ microseconds. It does not depend on the size or shape of the capacitor.

What we have here is simply the time constant for the relaxation of an electric field in a conducting medium by redistribution of charge. We really don't need the capacitor plates to describe it. Imagine that we could suddenly imbed two sheets of charge, a negative sheet and a positive sheet, opposite one another in a conductor – for instance, in an $n$-type semiconductor (Fig. 4.29(a)). What will make these charges disappear? Do negative charge carriers move from the sheet on the left across the intervening space, neutralizing the positive charges when they arrive at the sheet on the right? Surely not – if that were the process, the time required would be proportional to the distance between the sheets. What happens instead is this. The *entire population* of negative charge carriers that fills the space between the sheets is caused to move by the electric field. Only a *very slight* displacement of this cloud of charge suffices to remove excess negative charge on the left, while providing on the right the extra negative charge needed to neutralize the positive sheet, as indicated in Fig. 4.29(b). Within a conductor, in other words, neutrality is restored by a small readjustment of the entire charge distribution, not by a few charge carriers moving a long distance. That is why the relaxation time can be independent of the size of the system.

For a metal with resistivity typically $10^{-7}$ ohm-meter, the time constant $\epsilon_0 \rho$ is about $10^{-18}$ s, orders of magnitude shorter than the mean free time of a conduction electron in the metal. As a relaxation time this makes no sense. Our theory, at this stage, can tell us nothing about events on a time scale as short as that.

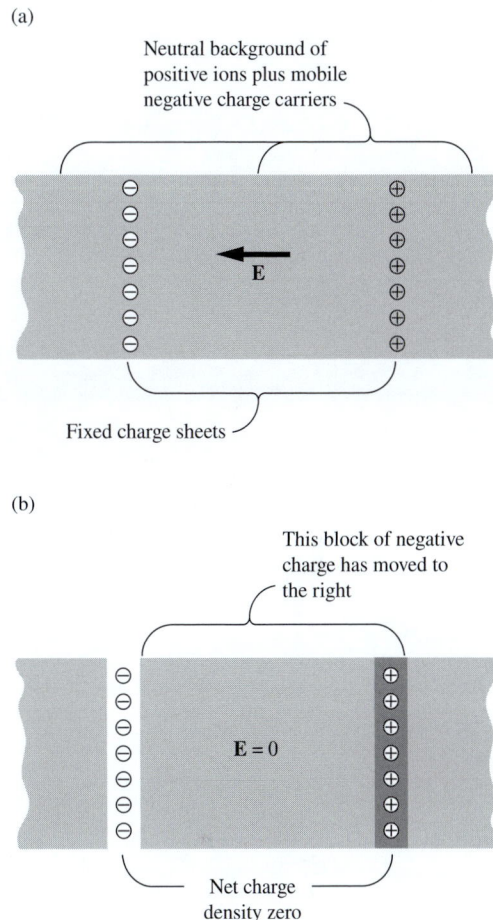

(a)

Neutral background of positive ions plus mobile negative charge carriers

E

Fixed charge sheets

(b)

This block of negative charge has moved to the right

E = 0

Net charge density zero

**Figure 4.29.**
In a conducting medium, here represented by an $n$-type conductor, two fixed sheets of charge, one negative and one positive, can be neutralized by a slight motion of the entire block of mobile charge carriers lying between them. (a) Before the block of negative charge has moved. (b) After the net charge density has been reduced to zero at each sheet.

## 4.12 Applications

The *transatlantic telegraph cable* (see Exercise 4.22) extended about 2000 miles between Newfoundland and Ireland, and was the most expensive and involved electrical engineering project of its time. After many failures, interrupted by a very short-lived success in 1858, it was finally completed in 1866. The initial failures were due partly to the fact that there didn't exist a consistent set of electrical units, in particular a unit for resistance. A byproduct of the project was therefore the hastening of a consistent set of units.

*Electric shocks* can range from barely noticeable, to annoying, to painful, to lethal. The severity of a shock depends on the current, not the applied voltage (although for a given resistance, a higher voltage means a higher current, of course). The duration also matters. The current can be harmful for two reasons: it can cause burns, and it can cause the heart to undergo fibrillation, where the normal coordinated contractions are replaced by uncoordinated ones, which don't pump any blood. Currents as small as 50 mA can cause fibrillation. A *defibrillator* works by passing a large enough current through the heart so that it briefly freezes up. With the uncoordinated contractions halted, the heart is then likely to start beating normally.

A current of 10 mA running through your hand is roughly the "can't let go" threshold, where the induced contractions of the muscles prevent you from letting go of the wire or whatever the voltage source is. And if you can't let go, you will inevitably become sweaty, which will reduce the resistivity of your skin, making things even worse. People who work with electricity often keep one hand in their pocket, to reduce the chance of it touching a grounded object and forming a conductive path, and, even worse, a path that goes across the heart.

If you touch a voltage source, the relation between the voltage and current depends on the resistance involved, and this can vary greatly, depending on the conditions. The main things that the resistance depends on are the resistivity of your skin, and the contact area. Dry skin has a much higher resistivity than wet skin. In a given scenario, the dry-skin resistance might be, say, 100,000 $\Omega$, whereas the wet-skin resistance might be 1000 $\Omega$. The shock from a 120 V wall socket in these two cases will be about 1 mA (hardly noticeable), or about 100 mA (potentially lethal). It's best to assume the latter! The variation of resistivity with the sweatiness of your skin is one of the ingredients in lie-detector machines. The increase in current is quite remarkable, even if you don't think you're sweating much. If you hook yourself up to an apparatus of this sort (which can often be found in science museums and the like), and if you then imagine, say, bungee jumping off a high suspension bridge, the current will rise dramatically. It's like it is magically reading your mind. The wet interior of our bodies has an even lower resistivity, of course, so special precautions must be taken in hospital operating rooms. Even a small voltage across an exposed heart can create a sufficient current to cause fibrillation.

If you shuffle your feet across a carpet on a dry day, you can get charged up to a very high voltage, perhaps 50,000 V. If you then touch a grounded object, you get a shock in the form of a spark. But even if you're quite sweaty, this 50,000 V certainly isn't lethal, whereas the 120 V from the wall socket might very well be. As mentioned in Section 2.18, the reason is that the amount of charge on you after shuffling across the carpet is very small, whereas there is an essentially infinite amount of charge available from the power company. The current in the case of the carpet lasts for a very short time before the charge runs out, and this time interval is too short to do any damage.

A *fuse* is a safety device that protects against large currents in a circuit; 20 A is a typical threshold. Such currents can generate enough heat to start a fire. A fuse consists of a thin strip of metal connected in series with the circuit. If the current becomes sufficiently large, the $I^2R$ resistance heating will melt the strip, producing a gap in the circuit and halting the current flow. A fuse needs to be replaced each time it burns out, but it's better to burn out a fuse than to burn down a house! A similar safety device, a *circuit breaker,* doesn't need to be replaced; it can simply be reset. This will be discussed in Section 6.10.

A *smoke detector* of the "ionization" type works by monitoring a tiny current. This current is created by a small radioactive source that emits alpha particles, which ionize the air. These ions constitute the tiny current. When smoke enters the detector, the ions collect on the smoke particles and are neutralized. This disrupts the current, telling the alarm to go off.

*Electric eels* can generate voltages on the order of 500 V, which is enough to harm or possibly kill a human. The discharge is brief; a current of about 1 ampere lasts for a few milliseconds, but this is long enough. Special *electrocyte* cells, which take up most of the eel's body, are able to generate small voltages of roughly 0.1 V by means of sodium–potassium pumps (a common mechanism in nerve and muscle cells). Thousands of these cells are stacked in series (the same basic idea behind a battery) to produce the 500 V potential, and then many of these stacks are combined in parallel. The tricky thing is for the eel to discharge all the electrocyte cells simultaneously; the mechanism for this isn't entirely understood.

*Sprites* are faint flashes of light that sometimes occur high above thunderstorms. They are predominantly red and generally extend vertically, sometimes up to 100 km. There are still many open questions about how they form, but an important fact is that the air is much less dense at high altitudes, so the breakdown field is lower. (Electrons have more time to accelerate between collisions, so they can achieve larger speeds.) A relatively small field, caused by a dipole distribution of charge in a thundercloud, can be large enough at high altitudes to cause breakdown.

In a conducting metal wire, current can flow in either direction. The two directions are symmetrical; if you flip the wire over, it looks the same. But in a *diode,* current can flow in only one direction. A semiconductor diode consists of an *n*-type semiconducting region adjacent to a *p*-type region (the combination is called a *p-n junction*). The currents in these regions arise from the flow of electrons and holes, respectively. Because of this difference, there is an asymmetry between the two directions of current flow. This setup therefore has at least some chance of allowing current to flow in only one direction. And indeed, it can be shown that (provided that the voltage isn't excessively large) current can flow only in the *p*-to-*n* direction. Or, equivalently, electrons can flow only in the *n*-to-*p* direction. The reasoning involves the diffusion of electrons and holes across the junction, which creates a *depletion* region, where there are no mobile charge carriers. Alternatively, electrons can't flow from the *p* region to the *n* region, because there are no mobile electrons in the *p* region to do the flowing.

When a current flows through a diode, electrons give off energy by dropping from the conduction band down to a hole in the valence band. This energy takes the form of a photon, and for certain materials the energy gap corresponds to a photon whose frequency is in the visible spectrum. Such a diode is called a *light-emitting diode (LED)*. LEDs last much longer than incandescent bulbs (there is no filament that will burn out), and they are much more energy efficient (there is less wasted heat, and fewer wasted photons in nonvisible parts of the spectrum). They are also generally very small, so they can easily be inserted into circuits.

A *transistor* is a device that allows a small signal in one part of a circuit to control a large current in another part. A transistor can be made with, for example, an *n-p-n* junction (although the middle region must be very thin for it to work). This is called a *bipolar junction transistor (BJT)*. If two terminals are connected to the two *n*-type materials on the ends, no current will flow through the transistor, because it effectively consists of two *p-n* junctions pointing in opposite directions, But if a small current is allowed to flow through the middle *p*-type material by way of a third terminal, then it turns out (although this is by no means obvious) that a much larger current will flow through the transistor, that is, through the original two terminals. A transistor can operate in an "all or nothing" mode, that is, as a switch that turns a large current on or off. Or it can operate in a continuous manner, where the resultant current depends on the small current in the middle *p*-type material.

Another type of transistor is the *field effect transistor (FET)*. There are different types of FETs, but in one type a *gate* runs along the side of a *p*-type material. If a positive voltage is applied to the gate, it will attract electrons and effectively turn a thin layer of the *p*-type material into *n*-type. If terminals are connected to two *n*-type regions that are placed at the ends of this layer, then we have a continuous *n*-type region running from one terminal to the other, so current can flow. Increasing the gate voltage increases the thickness of the *n*-type layer, thereby increasing the current flow between the terminals. The result is an amplification of the original signal to the gate. And indeed, the transistor is the main component in the *amplifier* in your sound system. Although BJTs were developed first, FETs are by far the dominant transistor in modern electronics.

*Solar cells* produce power by making use of the *photovoltaic effect*. At a *p-n* junction of two semiconducting materials, an electric field naturally occurs (see Exercise 4.26) due to the diffusion of electrons from the $n$ to $p$ regions, and of holes from the $p$ to $n$ regions. If photons in the sunlight kick electrons up to the conduction band, they (along with the holes created) will move under the influence of the field, and a current will be generated.

# CHAPTER SUMMARY

- The current density associated with a given type of charge carrier is $\mathbf{J} = nq\mathbf{u}$. Conservation of charge is expressed by the *continuity equation*, $\operatorname{div} \mathbf{J} = -\partial \rho / \partial t$, which says that if the charge in a region decreases (or increases), there must be current flowing out of (or into) that region.

- The current density in a conductor is related to the electric field by $\mathbf{J} = \sigma \mathbf{E}$, where $\sigma$ is the *conductivity*. This implies a linear relation between voltage and current, known as *Ohm's law: $V = IR$*. The *resistance* of a wire is $R = L/A\sigma = \rho L/A$, where $\rho$ is the *resistivity*.

- An applied electric field results in a (generally) slow drift velocity of the charge carriers. The conductivity associated with a given type of charge carrier is $\sigma \approx Ne^2\tau/m$, where $\tau$ is the mean time between collisions.

- Conduction in *semiconductors* arises from mobile electrons in the conduction band, or mobile "holes" in the valence band. The former dominate in $n$-type semiconductors, while the latter dominate in $p$-type semiconductors. Doping can dramatically increase the number of electrons or holes.

- It is often possible to reduce a circuit of resistors to a smaller circuit via the rules for adding resistors in series and parallel:

$$R = R_1 + R_2 \quad \text{and} \quad \frac{1}{R} = \frac{1}{R_1} + \frac{1}{R_2}. \tag{4.45}$$

More generally, the currents in a circuit can be found via Ohm's law ($V = IR$) and *Kirchhoff's rules* (zero net current flow into any node, and zero net voltage drop around any loop).

- The *power* dissipated in a resistor is given by $P = I^2 R$, which can also be written as $IV$ or $V^2/R$.

- A voltaic cell, or battery, utilizes chemical reactions to supply an *electromotive force*. Since the line integral of the electric field around a complete circuit is zero, there must be locations where ions move against the electric field. The forces from the chemical reactions are responsible for this.

- *Thévenin's theorem* states that any circuit is equivalent to a single voltage source $\mathcal{E}_{\mathrm{eq}}$ and a single resistor $R_{\mathrm{eq}}$. The *linearity* of circuits is critical in the proof of this theorem.

- If the charge on a capacitor is discharged across a resistor, the charge and current decrease exponentially with a time constant equal to $RC$.

## Problems

4.1   *Van de Graaff current* *

In a Van de Graaff electrostatic generator, a rubberized belt 0.3 m wide travels at a velocity of 20 m/s. The belt is given a surface charge at the lower roller, the surface charge density being high enough to cause a field of $10^6$ V/m on each side of the belt. What is the current in milliamps?

4.2   *Junction charge* **

Show that the total amount of charge at the junction of the two materials in Fig. 4.6 is $\epsilon_0 I(1/\sigma_2 - 1/\sigma_1)$, where $I$ is the current flowing through the junction, and $\sigma_1$ and $\sigma_2$ are the conductivities of the two conductors.

4.3   *Adding resistors* *

(a) Two resistors, $R_1$ and $R_2$, are connected in series, as shown in Fig. 4.30(a). Show that the effective resistance $R$ of the system is given by

$$R = R_1 + R_2. \tag{4.46}$$

Check the $R_1 \to 0$ and $R_1 \to \infty$ limits.

(b) If the resistors are instead connected in parallel, as shown in Fig. 4.30(b), show that the effective resistance is given by

$$\frac{1}{R} = \frac{1}{R_1} + \frac{1}{R_2}. \tag{4.47}$$

Again check the $R_1 \to 0$ and $R_1 \to \infty$ limits.

4.4   *Spherical resistor* **

(a) The region between two concentric spherical shells is filled with a material with resistivity $\rho$. The inner radius is $r_1$, and the outer radius $r_2$ is many times larger (essentially infinite). Show that the resistance between the shells is essentially equal to $\rho/4\pi r_1$.

(b) Without doing any calculations, dimensional analysis suggests that the above resistance should be proportional to $\rho/r_1$, because $\rho$ has units of ohm-meters and $r_1$ has units of meters. But is this reasoning rigorous?

4.5   *Laminated conductor* **

A laminated conductor is made by depositing, alternately, layers of silver 100 angstroms thick and layers of tin 200 angstroms thick (1 angstrom $= 10^{-10}$ m). The composite material, considered on a larger scale, may be considered a homogeneous but anisotropic material with an electrical conductivity $\sigma_\perp$ for currents perpendicular to the planes of the layers, and a different conductivity $\sigma_\parallel$ for

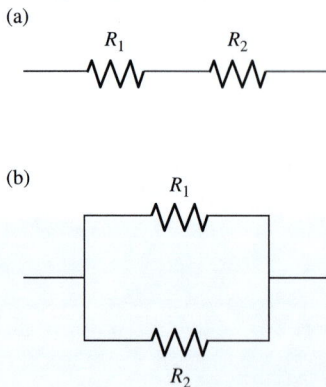

(a)

$R_1$      $R_2$

(b)

$R_1$

$R_2$

**Figure 4.30.**

currents parallel to that plane. Given that the conductivity of silver is 7.2 times that of tin, find the ratio $\sigma_\perp/\sigma_\parallel$.

**4.6** *Validity of tapered-rod approximation*  ✳

(a) The result given in Exercise 4.32 below is only an approximate one, valid in the limit where the taper is slow (that is, where $a - b$ is much smaller than the length of the cone). It can't be universally valid, because in the $b \to \infty$ limit the resistance of the cone would be zero. But the object shown in Fig. 4.31 certainly doesn't have a resistance that approaches zero as $b \to \infty$. Why isn't the result valid?

(b) The technique given in the hint in Exercise 4.32 *is* valid for the object shown in Fig. 4.32, which has spherical endcaps (with a common center) as its end faces. The radial distance between the faces is $\ell$ and their areas are $A_1$ and $A_2$. Find the resistance between the end faces.

**4.7** *Triangles of resistors*  ✳✳

(a) In Fig. 4.33, each segment represents a resistor $R$ (independent of the length on the page). What is the effective resistance $R_{\text{eff}}$ between $A$ and $B$? The numbers you encounter in your calculation should look familiar.

(b) What is $R_{\text{eff}}$ in the limit of a very large number of triangles? You can assume that they spiral out of the page, so that they don't run into each other. Your result should agree with a certain fact you may know concerning the familiar-looking numbers in part (a). *Hint:* In the infinite-triangle limit, if you add on another triangle to the left of $A$, the effective resistance along the new "spoke" must still be $R_{\text{eff}}$.

**4.8** *Infinite square lattice*  ✳✳

Consider a two-dimensional infinite square lattice of $1\,\Omega$ resistors. That is, every lattice point in the plane has four $1\,\Omega$ resistors connected to it. What is the equivalent resistance between two adjacent nodes? This problem is a startling example of the power of symmetry and superposition. *Hint:* If you can determine the voltage drop between two adjacent nodes when a current of, say, 1 A goes in one node and comes out the other, then you are done. Consider this setup as the superposition of two other setups.

**4.9** *Sum of the effective resistances*  ✳✳✳✳

$N$ points in space are connected by a collection of resistors, all with the same value $R$. The network of resistors is arbitrary (and not necessarily planar), except for the one restriction that it is "connected" (that is, it is possible to travel between any two points via an unbroken chain of resistors). Two given points may be connected by multiple resistors (see Fig. 4.34), so the number of resistors emanating from a given point can be any number greater than or equal to 1.

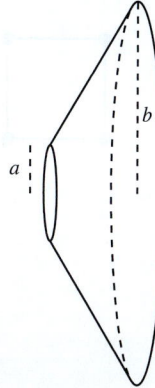

**Figure 4.31.**

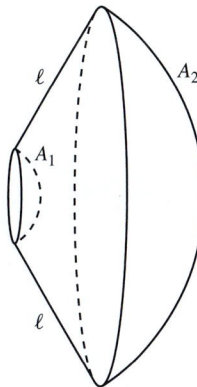

**Figure 4.32.**

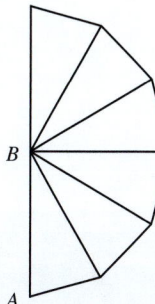

**Figure 4.33.**

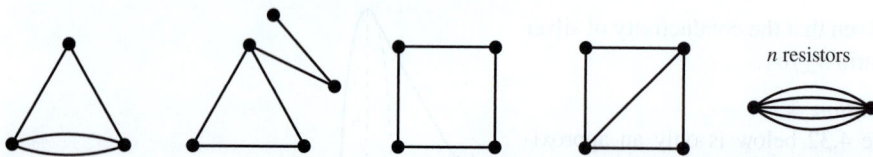

**Figure 4.34.**

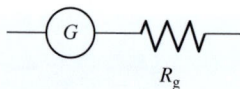

**Figure 4.35.**

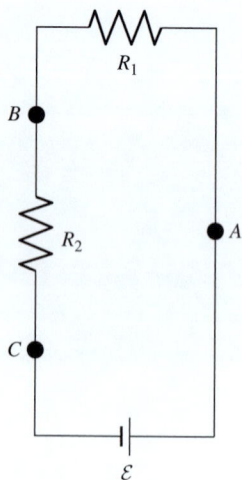

**Figure 4.36.**

Consider a particular resistor. The network produces an effective resistance between the two points at the ends of this resistor. What is the sum of the effective resistances across all the resistors in the network? (The result is known as Foster's theorem.) Solve this in two steps as follows.

(a) Find the desired sum for the various networks shown in Fig. 4.34. For example, in the first network, two of the effective resistances (in units of $R$) are $2/5$ (from a 2 and a 1/2 in parallel) and two are $3/5$ (from a 1 and a 3/2 in parallel), so the sum of all four effective resistances is 2. Remember that the sum is over all *resistors* and not pairs of points, so the two curved resistors each get counted once.

(b) Based on your results in part (a), and perhaps with the help of some other networks you can randomly make up, you should be able to make a conjecture about the general case with $N$ points and an arbitrary configuration of equal resistors (subject to the condition that the network is connected). Prove your conjecture. *Note:* The solution is extremely tricky, so you may want to look at the hint given in the first paragraph of the solution. Even with this hint it is still very tricky.

4.10 *Voltmeter, ammeter* ∗∗
The basic ingredient in voltmeters and ammeters (at least ones from the old days) is the *galvanometer*, which is a device that can measure very small currents. (It works via magnetic effects, but the exact mechanism isn't important here.) Inherent in any galvanometer is some resistance $R_g$, so a physical galvanometer can be represented by the system shown in Fig. 4.35.

Consider a circuit such as the one in Fig. 4.36, with all quantities unknown. Let's say we want to measure experimentally the current flowing across point $A$ (which is the same as everywhere else, in this simple circuit), and also the voltage difference between points $B$ and $C$. Given a galvanometer with known $R_g$, and also a supply of known resistors (ranging from much smaller to much larger than $R_g$), how can you accomplish these two tasks? Explain how to construct your two devices (called an "ammeter" and "voltmeter," respectively), and also how you should connect/insert them in the given circuit. You will need to make sure that you (a) affect the given circuit as little as possible, and (b) don't destroy your galvanometer by passing more current through it than it can handle, which is generally much less than the current in the given circuit.

**4.11** *Tetrahedron resistance* ∗∗

A tetrahedron has equal resistors $R$ along each of its six edges. Find the equivalent resistance between any two vertices. Do this by:

(a) using the symmetry of the tetrahedron to reduce it to an equivalent resistor;

(b) laying the tetrahedron flat on a table, hooking up a battery with an emf $\mathcal{E}$ to two vertices, and writing down the four loop equations. It's easy enough to solve this system of equations by hand, but it's even easier if you use a computer.

**4.12** *Find the voltage difference* ∗∗

What is the potential difference between points $a$ and $b$ in the circuit shown in Fig. 4.37?

**4.13** *Thévenin's theorem* ∗∗∗∗

Consider an arbitrary circuit $A$ and an additional arbitrary circuit $B$ connected to $A$'s external leads, as shown in Fig. 4.38. Prove Thévenin's theorem. That is, show that, as far as $B$ is concerned, $A$ acts the same as a single emf $\mathcal{E}_{eq}$ connected in series with a single resistor $R_{eq}$; explain how to determine these two quantities. Note that this result is independent of the exact nature of $B$. However, feel free to prove the theorem for just the special case where the circuit $B$ is a single emf $\mathcal{E}$. We'll present two proofs. They're a bit tricky, so you may want to look at the first few lines to get started.

**4.14** *Thévenin $R_{eq}$ via $I_{sc}$* ∗∗

Find the Thévenin equivalent resistance $R_{eq}$ for the circuit in Fig. 4.24(a). Do this by calculating the short-circuit current $I_{sc}$ between $A$ and $B$, and then using $R_{eq} = \mathcal{E}_{eq}/I_{sc}$. (You can use the fact that $\mathcal{E}_{eq} = I_3 R_3$, where $I_3$ is given in Eq. (4.33).)

**4.15** *A Thévenin equivalent* ∗∗

Find the Thévenin equivalent resistance and emf for the circuit shown in Fig. 4.39. If a 15 Ω resistor is connected across the terminals, what is the current through it?

**4.16** *Discharging a capacitor* ∗∗

A capacitor initially has charge $Q$. It is then discharged by closing the switch in Fig. 4.40. You might argue that no charge should actually flow in the wire, because the electric field is essentially zero just outside the capacitor; so the charges on the plates feel essentially no force pushing them off the plates onto the wire.

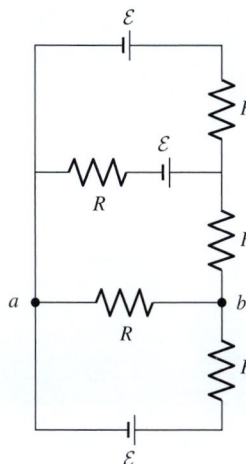

**Figure 4.37.**

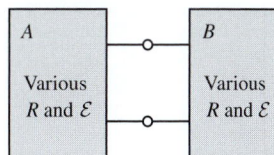

**Figure 4.38.**

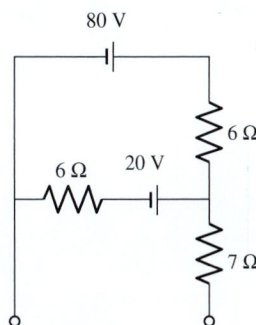

**Figure 4.39.**

**Figure 4.40.**

(a)

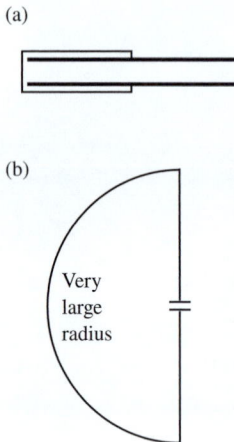

(b)

Very
large
radius

**Figure 4.41.**

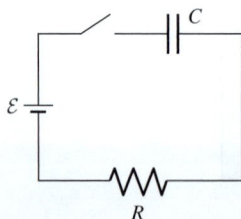

**Figure 4.42.**

$Q = Q_0$    $Q = 0$

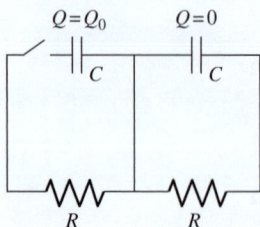

**Figure 4.43.**

$Q = Q_0$    $Q = 0$

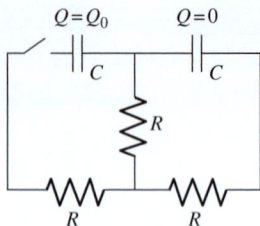

**Figure 4.44.**

Why, then, does the capacitor discharge? State (and justify quantitatively) which parts of the circuits are the (more) relevant ones in the two setups shown in Fig. 4.41.

4.17 *Charging a capacitor* **

A battery is connected to an *RC* circuit, as shown in Fig. 4.42. The switch is initially open, and the charge on the capacitor is initially zero. If the switch is closed at $t = 0$, find the charge on the capacitor, and also the current, as functions of time.

4.18 *A discharge with two capacitors* ***

(a) The circuit in Fig. 4.43 contains two identical capacitors and two identical resistors. Initially, the left capacitor has charge $Q_0$ (with the left plate positive), and the right capacitor is uncharged. If the switch is closed at $t = 0$, find the charges on the capacitors as functions of time. Your loop equations should be simple ones.

(b) Answer the same question for the circuit in Fig. 4.44, in which we have added one more (identical) resistor. What is the maximum (or minimum) charge that the right capacitor achieves? *Note:* Your loop equations should now be more interesting. Perhaps the easiest way to solve them is to take their sum and difference. This allows you to solve for the sum and difference of the charges, from which you can obtain each charge individually.

# Exercises

4.19 *Synchrotron current* *

In a 6 gigaelectron-volt (1 GeV = $10^9$ eV) electron synchrotron, electrons travel around the machine in an approximately circular path 240 meters long. It is normal to have about $10^{11}$ electrons circling on this path during a cycle of acceleration. The speed of the electrons is practically that of light. What is the current? We give this very simple problem to emphasize that nothing in our definition of current as rate of transport requires the velocities of the charge carriers to be nonrelativistic and that there is no rule against a given charged particle getting counted many times during a second as part of the current.

4.20 *Combining the current densities* **

We have $5 \cdot 10^{16}$ doubly charged positive ions per m$^3$, all moving west with a speed of $10^5$ m/s. In the same region there are $10^{17}$ electrons per m$^3$ moving northeast with a speed of $10^6$ m/s. (Don't ask how we managed it!) What are the magnitude and direction of **J**?

**4.21** *Current pulse from an alpha particle* ✶✶✶

The result from Exercise 3.37 can help us to understand the flow of current in a circuit, part of which consists of charged particles moving through space between two electrodes. The question is, what is the nature of the current when only one particle traverses the space? (If we can work that out, we can easily describe any flow involving a larger number arriving on any schedule.)

(a) Consider the simple circuit in Fig. 4.45(a), which consists of two electrodes in vacuum connected by a short wire. Suppose the electrodes are 2 mm apart. A rather slow alpha particle, of charge $2e$, is emitted by a radioactive nucleus in the left plate. It travels directly toward the right plate with a constant speed of $10^6$ m/s and stops at this plate. Make a quantitative graph of the current in the connecting wire, plotting current against time. Do the same for an alpha particle that crosses the gap moving with the same speed but at an angle of 45° to the normal. (Actually for pulses as short as this the inductance of the connecting wire, here neglected, would affect the pulse shape.)

(b) Suppose we had a cylindrical arrangement of electrodes, as shown in Fig. 4.45(b), with the alpha particles being emitted from a thin wire on the axis of a small cylindrical electrode. Would the current pulse have the same shape? (You will need to solve the cylindrical version of Exercise 3.37.)

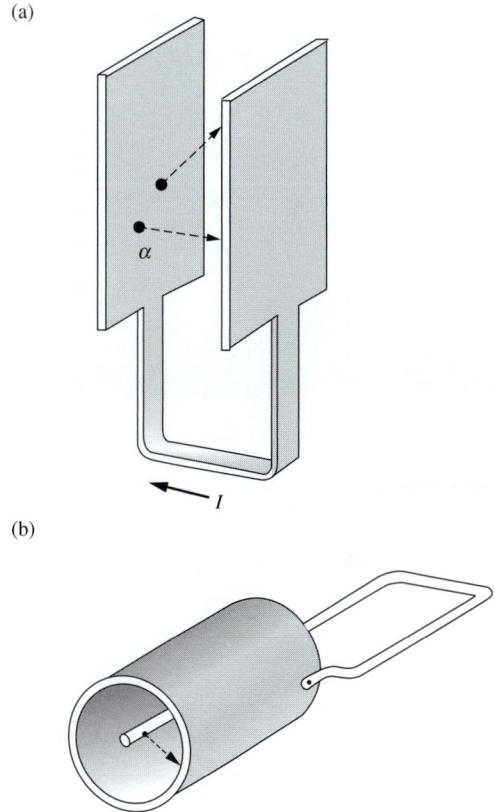

**4.22** *Transatlantic telegraphic cable* ✶✶

The first telegraphic messages crossed the Atlantic in 1858, by a cable 3000 km long laid between Newfoundland and Ireland. The conductor in this cable consisted of seven copper wires, each of diameter 0.73 mm, bundled together and surrounded by an insulating sheath.

(a) Calculate the resistance of the conductor. Use $3 \cdot 10^{-8}$ ohm-meter for the resistivity of the copper, which was of somewhat dubious purity.

(b) A return path for the current was provided by the ocean itself. Given that the resistivity of seawater is about 0.25 ohm-meter, see if you can show that the resistance of the ocean return would have been much smaller than that of the cable. (Assume that the electrodes immersed in the water were spheres with radius, say, 10 cm.)

**4.23** *Intervals between independent events* ✶✶✶

This exercise is more of a math problem than a physics problem, so maybe it doesn't belong in this book. But it's a fun one. Consider a series of events that happen at independent random times, such as the collisions in Section 4.4 that led to the issue discussed in

(a)

(b)

**Figure 4.45.**

Footnote 8. Such a process can be completely characterized by the probability per unit time (call it $p$) of an event happening. The definition of $p$ is that the probability of an event happening in an infinitesimal[16] time $dt$ equals $p\,dt$.

(a) Show that starting at any particular time (not necessarily the time of an event), the probability that the *next* event happens between $t$ and $t + dt$ later equals $e^{-pt}p\,dt$. You can do this by breaking the time interval $t$ into a large number of tiny intervals, and demanding that the event does *not* happen in any of them, but that it *does* happen in the following $dt$. (You will need to use the fact that $(1 - x/N)^N = e^{-x}$ in the $N \to \infty$ limit.) Verify that the integral of $e^{-pt}p\,dt$ correctly equals 1.

(b) Show that starting at any particular time (not necessarily the time of an event), the average waiting time (also called the expectation value of the waiting time) to the next event equals $1/p$. Explain why this is also the average time between events.

(c) Pick a random point in time, and look at the length of the time interval (between successive events) that it belongs to. Explain, using the above results, why the average length of this interval is $2/p$, and not $1/p$.

(d) We have found that the average time between events is $1/p$, and also that the average length of the interval surrounding a randomly chosen point in time is $2/p$. Someone might think that these two results should be the same. Explain intuitively why they are not.

(e) Using the above probability distribution $e^{-pt}p\,dt$ properly, show mathematically why $2/p$ is the correct result for the average length of the interval surrounding a randomly chosen point in time.

4.24 *Mean free time in water* ∗

An ion in a liquid is so closely surrounded by neutral molecules that one can hardly speak of a "free time" between collisions. Still, it is interesting to see what value of $\tau$ is implied by Eq. (4.23) if we take the observed conductivity of pure water from Table 4.1, and if we use $6 \cdot 10^{19}$ m$^{-3}$ for $N_+$ and $N_-$; see Footnote 4. A typical thermal speed for a water molecule is 500 m/s. How far would it travel in that time $\tau$?

4.25 *Drift velocity in seawater* ∗

The resistivity of seawater is about 0.25 ohm-meter. The charge carriers are chiefly Na$^+$ and Cl$^-$ ions, and of each there are about

---

[16] The probability of an event happening in a *non*infinitesimal time $t$ is *not* equal to $pt$. If $t$ is large enough, then $pt$ is larger than 1, so it certainly can't represent a probability. In that case, $pt$ is the *average* number of events that occur in time $t$. But this doesn't equal the probability that an event occurs, because there can be double, triple, etc., events in the time $t$. We don't have to worry about multiple events if $dt$ is infinitesimal.

$3 \cdot 10^{26}$ per $m^3$. If we fill a plastic tube 2 meters long with seawater and connect a 12 volt battery to the electrodes at each end, what is the resulting average drift velocity of the ions?

**4.26** *Silicon junction diode* **

In a silicon junction diode the region of the planar junction between $n$-type and $p$-type semiconductors can be approximately represented as two adjoining slabs of uniform charge density, one negative and one positive. Away from the junction, outside these charge layers, the potential is constant, its value being $\phi_n$ in the $n$-type material and $\phi_p$ in the $p$-type material. Given that the difference between $\phi_p$ and $\phi_n$ is 0.3 volt, and that the thickness of each of the two slabs of charge is $10^{-4}$ m, find the charge density in each of the two slabs, and make a graph of the potential $\phi$ as a function of position $x$ through the junction. What is the strength of the electric field at the midplane?

**4.27** *Unbalanced current* **

As an illustration of the point made in Footnote 13 in Section 4.7, consider a black box that is approximately a 10 cm cube with two binding posts. Each of these terminals is connected by a wire to some external circuits. Otherwise, the box is well insulated from everything. A current of approximately 1 A flows through this circuit element. Suppose now that the current in and the current out differ by one part in a million. About how long would it take, unless something else happens, for the box to rise in potential by 1000 volts?

**4.28** *Parallel resistors* *

By solving the loop equations for the setup shown in Fig. 4.46, derive the rule for adding resistors in parallel.

**4.29** *Keeping the same resistance* *

In the circuit shown in Fig. 4.47, if $R_0$ is given, what value must $R_1$ have in order that the input resistance between the terminals shall be equal to $R_0$?

**4.30** *Automobile battery* *

If the voltage at the terminals of an automobile battery drops from 12.3 to 9.8 volts when a 0.5 ohm resistor is connected across the battery, what is the internal resistance?

**4.31** *Equivalent boxes* **

A black box with three terminals, $a$, $b$, and $c$, contains nothing but three resistors and connecting wire. Measuring the resistance between pairs of terminals, we find $R_{ab} = 30$ ohms, $R_{ac} = 60$ ohms, and $R_{bc} = 70$ ohms. Show that the contents of the box could be either of the configurations shown in Fig. 4.48. Is there any other possibility? Are the two boxes completely equivalent, or

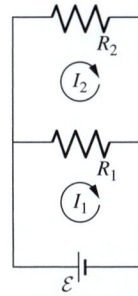

**Figure 4.46.**

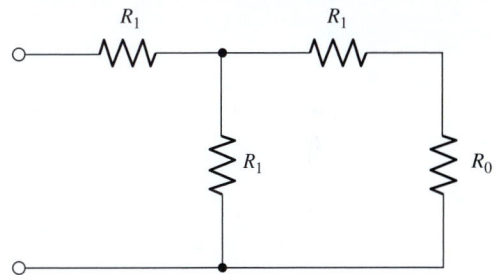

**Figure 4.47.**

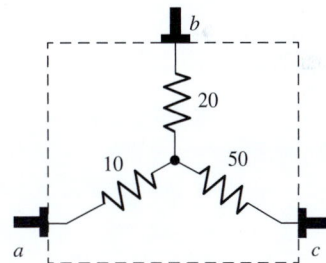

or

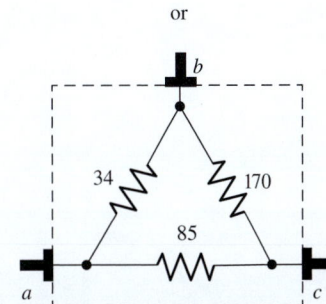

**Figure 4.48.**

is there an external measurement that would distinguish between them?

4.32 *Tapered rod* ∗

Two graphite rods are of equal length. One is a cylinder of radius $a$. The other is conical, tapering (or widening) linearly from radius $a$ at one end to radius $b$ at the other. Show that the end-to-end electrical resistance of the conical rod is $a/b$ times that of the cylindrical rod. *Hint:* Consider the rod to be made up of thin, disk-like slices, all in series. (This result is actually only an approximate one, valid in the limit where the taper is slow. See Problem 4.6 for a discussion of this.)

4.33 *Laminated conductor extremum* ∗∗

(a) Consider the setup in Problem 4.5. For given conductivities of the two materials, show that the ratio $\sigma_\perp/\sigma_\parallel$ is minimum when the layers have the same thickness (independent of what the conductivities are).

Give a physical argument why you would expect $\sigma_\perp/\sigma_\parallel$ to achieve a maximum or minimum somewhere between the two extremes where one thickness is much larger/smaller than the other.

(b) For given layer thicknesses, show that the ratio $\sigma_\perp/\sigma_\parallel$ is maximum when the materials have the same conductivity (independent of what the thicknesses are).

Give a physical argument why you would expect $\sigma_\perp/\sigma_\parallel$ to achieve a maximum or minimum somewhere between the two extremes where one conductivity is much larger/smaller than the other.

4.34 *Effective resistances in lattices* ∗∗

In the spirit of Problem 4.8, find the effective resistance between two adjacent nodes in an infinite (a) 3D cubic lattice, (b) 2D triangular lattice, (c) 2D hexagonal lattice, (d) 1D lattice. (The last of these tasks is a little silly, of course.) Assume that the lattices consist of $1\ \Omega$ resistors.

4.35 *Resistances in a cube* ∗∗

A cube has a resistor $R$ along each edge. Find the equivalent resistance between two nodes that correspond to:

(a) diagonally opposite corners of the cube;
(b) diagonally opposite corners of a face;
(c) adjacent corners.

You do not need to solve a number of simultaneous equations; instead use symmetry arguments. *Hint:* If two vertices are at the same potential, they can be collapsed to one point without changing the equivalent resistance between the two given nodes.

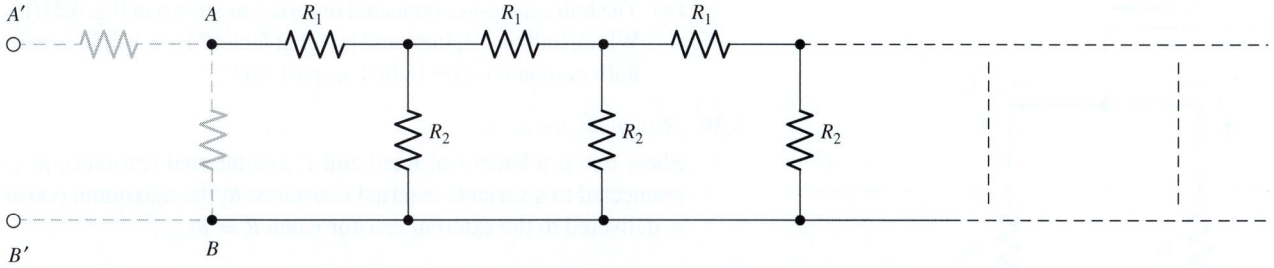

**Figure 4.49.**

**4.36** *Attenuator chain* ∗∗

Some important kinds of networks are infinite in extent. Figure 4.49 shows a chain of series and parallel resistors stretching off endlessly to the right. The line at the bottom is the resistanceless return wire for all of them. This is sometimes called an attenuator chain, or a ladder network.

The problem is to find the "input resistance," that is, the equivalent resistance between terminals $A$ and $B$. Our interest in this problem mainly concerns the method of solution, which takes an odd twist and which can be used in other places in physics where we have an iteration of identical devices (even an infinite chain of lenses, in optics). The point is that the input resistance (which we do not yet know – call it $R$) will not be changed by adding a new set of resistors to the front end of the chain to make it one unit longer. But now, adding this section, we see that this new input resistance is just $R_1$ in series with the parallel combination of $R_2$ and $R$.

Use this strategy to determine $R$. Show that, if voltage $V_0$ is applied at the input to such a chain, the voltage at successive nodes decreases in a geometric series. What should the ratio of the resistors be so that the ladder is an attenuator that halves the voltage at every step? Obviously a truly infinite ladder would not be practical. Can you suggest a way to terminate it after a few sections without introducing any error in its attenuation?

**4.37** *Some golden ratios* ∗

Find the resistance between terminals $A$ and $B$ in each of the infinite chains of resistors shown in Fig. 4.50. All the resistors have the same value $R$. (The strategy from Exercise 4.36 will be useful.)

**4.38** *Two light bulbs* ∗

(a) Two light bulbs are connected in parallel, and then connected to a battery, as shown in Fig. 4.51(a). You observe that bulb 1 is twice as bright as bulb 2. Assuming that the brightness of a bulb is proportional to the power dissipated in the bulb's resistor, which bulb's resistor is larger, and by what factor?

(a)

(b)

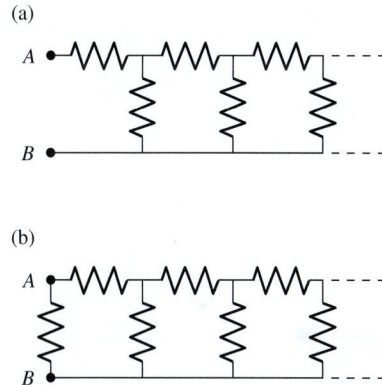

**Figure 4.50.**

(a)

(b)

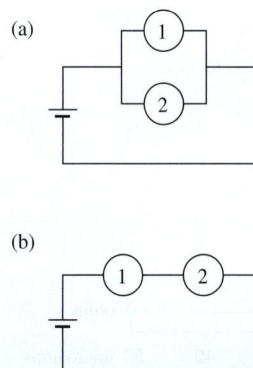

**Figure 4.51.**

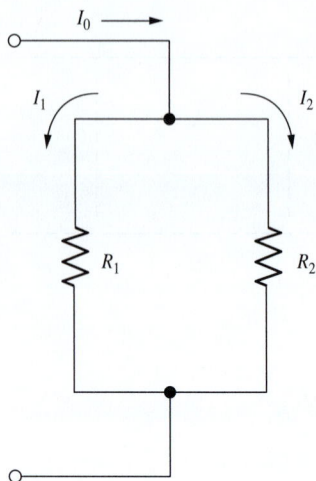

**Figure 4.52.**

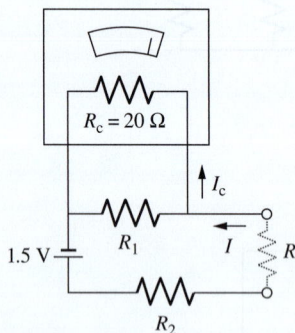

**Figure 4.53.**

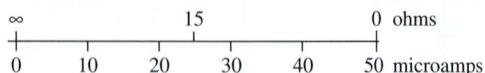

**Figure 4.54.**

(b) The bulbs are now connected in series, as shown in Fig. 4.51(b). Which bulb is brighter, and by what factor? How bright is each bulb compared with bulb 1 in part (a)?

### 4.39 *Maximum power* *

Show that if a battery of fixed emf $\mathcal{E}$ and internal resistance $R_i$ is connected to a variable external resistance $R$, the maximum power is delivered to the external resistor when $R = R_i$.

### 4.40 *Minimum power dissipation* **

Figure 4.52 shows two resistors in parallel, with values $R_1$ and $R_2$. The current $I_0$ divides somehow between them. Show that the condition that $I_1 + I_2 = I_0$, together with the requirement of *minimum power dissipation*, leads to the same current values that we would calculate with ordinary circuit formulas. This illustrates a general variational principle that holds for direct current networks: the distribution of currents within the network, for given input current $I_0$, is always that which gives the *least* total power dissipation.

### 4.41 *D-cell* **

The common 1.5 volt dry cell used in flashlights and innumerable other devices releases its energy by oxidizing the zinc can which is its negative electrode, while reducing manganese dioxide, $MnO_2$, to $Mn_2O_3$ at the positive electrode. (It is called a carbon–zinc cell, but the carbon rod is just an inert conductor.) A cell of size D, weighing 90 g, can supply 100 mA for about 30 hours.

(a) Compare its energy storage, in J/kg, with that of the lead–acid battery described in the example in Section 4.9. Unfortunately the cell is not rechargeable.

(b) How high could you lift yourself with one D-cell powering a 50 percent efficient winch?

### 4.42 *Making an ohmmeter* ***

You have a microammeter that reads 50 µA at full-scale deflection, and the coil in the meter movement has a resistance of 20 ohms. By adding two resistors, $R_1$ and $R_2$, and a 1.5 volt battery as shown in Fig. 4.53, you can convert this into an ohmmeter. When the two outcoming leads of this ohmmeter are connected together, the meter is to register zero ohms by giving exactly full-scale deflection. When the leads are connected across an unknown resistance $R$, the deflection will indicate the resistance value if the scale is appropriately marked. In particular, we want half-scale deflection to indicate 15 ohms, as shown in Fig. 4.54. What values of $R_1$ and $R_2$ are required, and where on the ohm scale will the marks be (with reference to the old microammeter calibration) for 5 ohms and for 50 ohms?

**4.43** *Using symmetry* **

This exercise deals with the equivalent resistance $R_{eq}$ between terminals $T_1$ and $T_2$ for the network of five resistors shown in Fig. 4.55. One way to derive a formula for $R_{eq}$ would be to solve the network for the current $I$ that flows in at $T_1$ for a given voltage difference $V$ between $T_1$ and $T_2$; then $R_{eq} = V/I$. The solution involves rather tedious algebra in which it is easy to make a mistake (although it is quick and painless if you use a computer; see Exercise 4.44), so we'll tell you most of the answer:

$$R_{eq} = \frac{R_1 R_2 R_3 + R_1 R_2 R_4 + [?] + R_2 R_3 R_4 + R_5 (R_1 R_3 + R_2 R_3 + [?] + R_2 R_4)}{R_1 R_2 + R_1 R_4 + [?] + R_3 R_4 + R_5 (R_1 + R_2 + R_3 + R_4)}.$$

(4.48)

By considering the symmetry of the network you should be able to fill in the three missing terms. Now check the formula by directly calculating $R_{eq}$ in four special cases: (a) $R_5 = 0$, (b) $R_5 = \infty$, (c) $R_1 = R_3 = 0$, and (d) $R_1 = R_2 = R_3 = R_4 \equiv R$, and comparing your results with what the formula gives.

**4.44** *Using the loop equations* *

Exercise 4.43 presents the (large) expression for the equivalent resistance between terminals $T_1$ and $T_2$ in Fig. 4.55. Derive this expression by writing down the loop equations involving the currents shown in Fig. 4.56, and then using *Mathematica* so solve for $I_3$. (No need for any messy algebra. Except for some typing, you're basically finished with the problem once you write down the loop equations.)

**4.45** *Battery/resistor loop* **

In the circuit shown in Fig. 4.57, all five resistors have the same value, 100 ohms, and each cell has an electromotive force of 1.5 V. Find the open-circuit voltage and the short-circuit current for the terminals $A$ and $B$. Then find $\mathcal{E}_{eq}$ and $R_{eq}$ for the Thévenin equivalent circuit.

**4.46** *Maximum power via Thévenin* **

A resistor $R$ is to be connected across the terminals $A$ and $B$ of the circuit shown in Fig. 4.58. For what value of $R$ will the power dissipated in the resistor be greatest? To answer this, construct the Thévenin equivalent circuit and then invoke the result from Exercise 4.39. How much power will be dissipated in $R$?

**4.47** *Discharging a capacitor* **

Return to the example of the capacitor $C$ discharging through the resistor $R$, which was worked out in Section 4.11, and show that the total energy dissipated in the resistor agrees with the energy originally stored in the capacitor. Suppose someone objects that

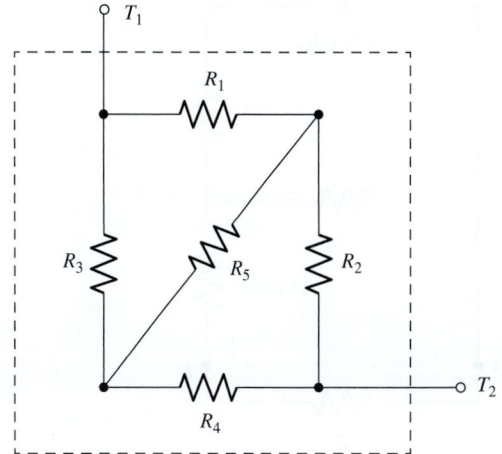

**Figure 4.55.**

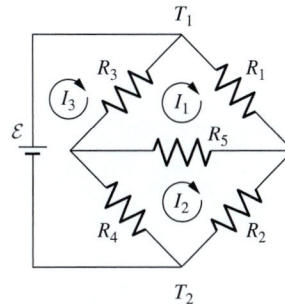

**Figure 4.56.**

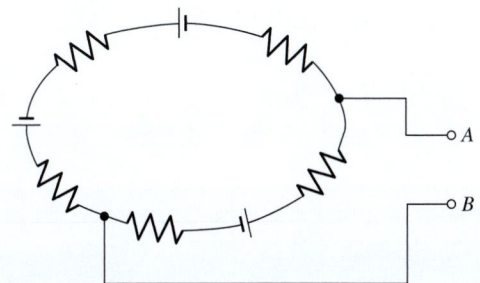

**Figure 4.57.**

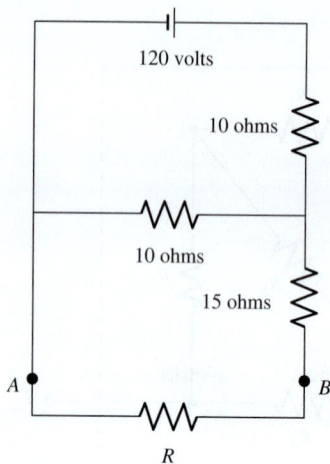

**Figure 4.58.**

the capacitor is never *really* discharged because $Q$ becomes zero only for $t = \infty$. How would you counter this objection? You might find out how long it would take the charge to be reduced to one electron, with some reasonable assumptions.

4.48 *Charging a capacitor* **

Problem 4.17 deals with the charging of a capacitor. Using the results from that problem, show that energy is conserved. That is, show that the total work done by the battery equals the final energy stored in the capacitor plus the energy dissipated in the resistor.

4.49 *Displacing the electron cloud* *

Suppose the conducting medium in Fig. 4.29 is *n*-type silicon with $10^{21}$ electrons per m$^3$ in the conduction band. Assume the initial density of charge on the sheets is such that the electric field strength is $3 \cdot 10^4$ V/m. By what distance must the intervening distribution of electrons be displaced to restore neutrality and reduce the electric field to zero?

# 5

# The fields of moving charges

**Overview** The goal of this chapter is to show that when relativity is combined with our theory of electricity, a necessary conclusion is that a new force, the *magnetic* force, must exist. In nonstatic situations, charge is defined via a surface integral. With this definition, charge is *invariant*, that is, independent of reference frame. Using this invariance, we determine how the electric field transforms between two frames. We then calculate the electric field due to a charge moving with constant velocity; it does *not* equal the spherically symmetric Coulomb field. Interesting field patterns arise in cases where a charge starts or stops.

The main result of this chapter, derived in Section 5.9, is the expression for the force that a moving charge (or a group of moving charges) exerts on another moving charge. On our journey to this result, we will consider setups with increasing complexity. More precisely, in calculating the force on a charge $q$ due to another charge $Q$, there are four basic cases to consider, depending on the charges' motions. (1) If both charges are stationary in a given frame, then we know from Chapter 1 that Coulomb's law gives the force. (2) If the source $Q$ is moving and $q$ is at rest, then we can use the transformation rule for the electric field mentioned above. (3) If the source $Q$ is at rest and $q$ is moving, then we can use the transformation rule for the force, presented in Appendix G, to show that the Coulomb field gives the force, as you would expect. (4) Finally, the case we are most concerned with: if *both* charges are moving, then we will show in Section 5.9 that a detailed consideration of relativistic effects implies that there exists an additional force that must be added to the electrical force; this is the magnetic force.

In short, the magnetic force is a consequence of Coulomb's law, charge invariance, and relativity.

## 5.1 From Oersted to Einstein

In the winter of 1819–1820, Hans Christian Oersted was lecturing on electricity, galvanism, and magnetism to advanced students at the University of Copenhagen. *Electricity* meant electrostatics; *galvanism* referred to the effects produced by continuous currents from batteries, a subject opened up by Galvani's chance discovery and the subsequent experiments of Volta; *magnetism* dealt with the already ancient lore of lodestones, compass needles, and the terrestrial magnetic field. It seemed clear to some that there must be a relation between galvanic currents and electric charge, although there was little more direct evidence than the fact that both could cause shocks. On the other hand, magnetism and electricity appeared to have nothing whatever to do with one another. Still, Oersted had a notion, vague perhaps, but tenaciously pursued, that magnetism, like the galvanic current, might be a sort of "hidden form" of electricity. Groping for some manifestation of this, he tried before his class the experiment of passing a galvanic current through a wire that ran above and at right angles to a compass needle (with the compass held horizontal, so that the needle was free to spin in a horizontal plane). It had no effect. After the lecture, something impelled him to try the experiment with a wire running parallel to the compass needle. The needle swung wide – and when the galvanic current was reversed it swung the other way!

The scientific world was more than ready for this revelation. A ferment of experimentation and discovery followed as soon as the word reached other laboratories. Before long, Ampère, Faraday, and others had worked out an essentially complete and exact description of the magnetic action of electric currents. Faraday's crowning discovery of electromagnetic induction came less than 12 years after Oersted's experiment. In the previous two centuries since the publication in 1600 of William Gilbert's great work *De Magnete,* man's understanding of magnetism had advanced not at all. Out of these experimental discoveries there grew the complete classical theory of electromagnetism. Formulated mathematically by Maxwell in the early 1860s, it was triumphantly corroborated by Hertz's demonstration of electromagnetic waves in 1888.

Special relativity has its historical roots in electromagnetism. Lorentz, exploring the electrodynamics of moving charges, was led very close to the final formulation of Einstein. And Einstein's great paper of 1905 was entitled not "Theory of Relativity," but rather "On the Electrodynamics of Moving Bodies." Today we see in the postulates of relativity and their implications a wide framework, one that embraces all physical laws and not solely those of electromagnetism. We expect any complete physical theory to be relativistically invariant. It ought to tell the same story in all inertial frames of reference. As it happened, physics already *had* one relativistically invariant theory – Maxwell's electromagnetic theory – long before the significance of relativistic invariance was recognized. Whether the ideas of special relativity could have evolved in the absence of a complete theory of the electromagnetic field is a question for the historian of science to speculate about; probably it can't be answered. We can only say that the actual history shows rather plainly a path running from Oersted's compass needle to Einstein's postulates.

Still, relativity is not a branch of electromagnetism, nor a consequence of the existence of light. The central postulate of special relativity, which no observation has yet contradicted, is the equivalence of reference frames moving with constant velocity with respect to one another. Indeed, it is possible, without even mentioning light, to derive the formulas of special relativity from nothing more than that postulate and the assumption that all spatial directions are equivalent.[1] The universal constant $c$ then appears in these formulas as a limiting velocity, approached by an energetic particle but never exceeded. Its value can be ascertained by an experiment that does not involve light or anything else that travels at precisely that speed. In other words, we would have special relativity even if electromagnetic waves could not exist.

Later in this chapter, we are going to follow the historical path from Oersted to Einstein almost in reverse. We will take special relativity as given, and ask how an electrostatic system of charges and fields looks in another reference frame. In this way we shall find the forces that act on electric charges in motion, including the force that acts between electric currents. Magnetism, seen from this viewpoint, is a relativistic aspect of electricity.[2] But first, let's review some of the phenomena we shall be trying to explain.

## 5.2 Magnetic forces

Two wires running parallel to one another and carrying currents in the same direction are drawn together. The force on one of the wires, per unit length of wire, is inversely proportional to the distance between the wires (Fig. 5.1(a)). Reversing the direction of one of the currents changes

---

[1] See Mermin (1984a), in which it is shown that the most general law for the addition of velocities that is consistent with the equivalence of inertial frames must have the form $v = (v_1 + v_2)/(1 + v_1 v_2/c^2)$, identical to our Eq. (G.8) in Appendix G. To discover the value of the constant $c$ in our universe we need only measure with adequate accuracy three lower speeds, $v$, $v_1$, and $v_2$. For references to other articles on the same theme, see also Mermin (1984b).

[2] The earliest exposition of this approach, to our knowledge, is Page (1912). It was natural for Page, writing only seven years after Einstein's revolutionary paper, to consider relativity more in need of confirmation than electrodynamics. His concluding sentence reads: "Viewed from another standpoint, the fact that we have been able, by means of the principle of relativity, to deduce the fundamental relations of electrodynamics from those of electrostatics, may be considered as some confirmation of the principle of relativity."

(a)

(b)

(c)

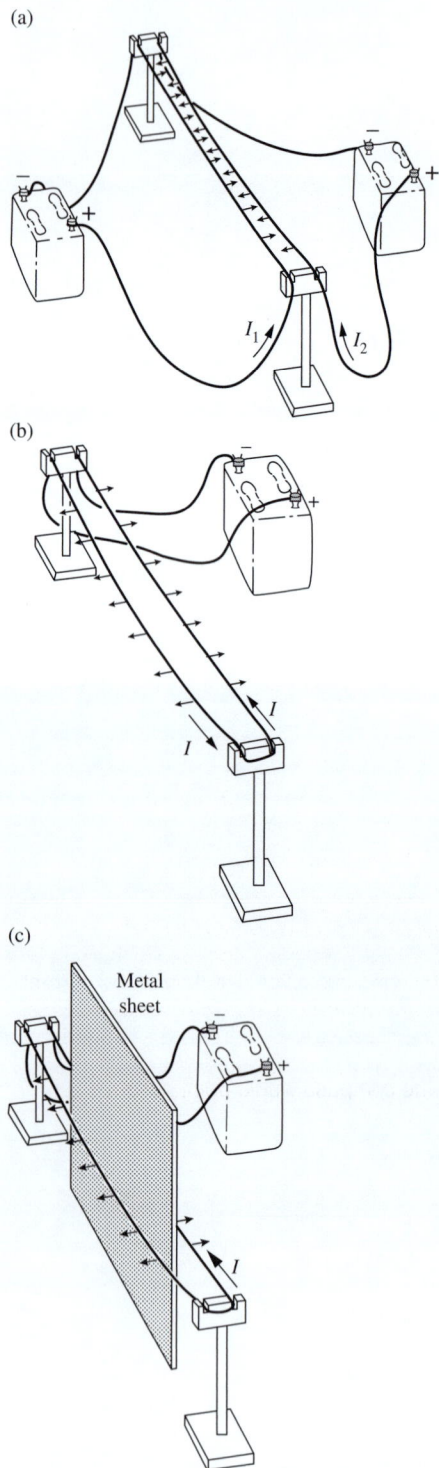

the force to one of repulsion. Thus the two sections of wire in Fig. 5.1(b), which are part of the same circuit, tend to fly apart. There is some sort of "action at a distance" between the two filaments of steady electric current. It seems to have nothing to do with any static electric charge on the surface of the wire. There may be some such charge and the wires may be at different potentials, but the force we are concerned with depends only on the charge *movement* in the wires, that is, on the two currents. You can put a sheet of metal between the two wires without affecting this force at all (Fig. 5.1(c)). These new forces that come into play when charges are moving are called *magnetic*.

Oersted's compass needle (Fig. 5.2(a)) doesn't look much like a direct-current circuit. We now know, however, as Ampère was the first to suspect, that magnetized iron is full of perpetually moving charges – electric currents on an atomic scale; we will talk about this in detail in Chapter 11. A slender coil of wire with a battery to drive current through it (Fig. 5.2(b)) behaves just like the compass needle under the influence of a nearby current.

Observing the motion of a free charged particle, instead of a wire carrying current, we find the same thing happening. In a cathode ray tube, electrons that would otherwise follow a straight path are deflected toward or away from an external current-carrying wire, depending on the relative direction of the current in that wire (Fig. 5.3). This interaction of currents and other moving charges can be described by introducing a *magnetic field*. (The electric field, remember, was simply a way of describing the action at a distance between stationary charges that is expressed in Coulomb's law.) We say that an electric current has associated with it a magnetic field that pervades the surrounding space. Some other current, or any moving charged particle that finds itself in this field, experiences a force proportional to the strength of the magnetic field in that locality. The force is always perpendicular to the velocity, for a charged particle. The entire force on a particle carrying charge $q$ is given by

$$\boxed{\mathbf{F} = q\mathbf{E} + q\mathbf{v} \times \mathbf{B}} \qquad (5.1)$$

where $\mathbf{B}$ is the magnetic field.[3]

**Figure 5.1.**
(a) Parallel wires carrying currents in the same direction are pulled together. (b) Parallel wires carrying currents in opposite directions are pushed apart. (c) These forces are not affected by putting a metal plate between the wires.

---

[3] Here we make use of the vector product, or *cross product,* of two vectors. A reminder: the vector $\mathbf{v} \times \mathbf{B}$ is a vector perpendicular to both $\mathbf{v}$ and $\mathbf{B}$ and of magnitude $vB\sin\theta$, where $\theta$ is the angle between the directions of $\mathbf{v}$ and $\mathbf{B}$. A right-hand rule determines the sense of the direction of the vector $\mathbf{v} \times \mathbf{B}$. In our Cartesian coordinates, $\hat{\mathbf{x}} \times \hat{\mathbf{y}} = \hat{\mathbf{z}}$ and $\mathbf{v} \times \mathbf{B} = \hat{\mathbf{x}}(v_y B_z - v_z B_y) + \hat{\mathbf{y}}(v_z B_x - v_x B_z) + \hat{\mathbf{z}}(v_x B_y - v_y B_x)$.

We shall take Eq. (5.1) as the definition of **B**. All that concerns us now is that the magnetic field strength is a vector that determines the velocity-proportional part of the force on a moving charge. In other words, the command, "Measure the direction and magnitude of the vector **B** at such and such a place," calls for the following operations: Take a particle of known charge $q$. Measure the force on $q$ at rest, to determine **E**. Then measure the force on the particle when its velocity is **v**; repeat with **v** in some other direction. Now find a **B** that will make Eq. (5.1) fit all these results; that is the magnetic field at the place in question.

Clearly this doesn't *explain* anything. Why does Eq. (5.1) work? Why can we always find a **B** that is consistent with this simple relation, for all possible velocities? We want to understand why there is a velocity-proportional force. It is really most remarkable that this force is strictly proportional to $v$, and that the effect of the electric field does not depend on $v$ at all! In the following pages we'll see how this comes about. It will turn out that a field **B** with these properties *must* exist if the forces between electric charges obey the postulates of special relativity. Seen from this point of view, magnetic forces are a relativistic aspect of charge in motion.

A review of the essential ideas and formulas of special relativity is provided in Appendix G. This would be a good time to read through it.

## 5.3 Measurement of charge in motion

How are we going to measure the quantity of electric charge on a moving particle? Until this question is settled, it is pointless to ask what effect motion has on charge itself. A charge can only be measured by the effects it produces. A point charge $Q$ that is at rest can be measured by determining the force that acts on a test charge $q$ a certain distance away (Fig. 5.4(a)). That is based on Coulomb's law. But if the charge we want to measure is moving, we are on uncertain ground. There is now a special direction in space, the instantaneous direction of motion. It could be that the force on the test charge $q$ depends on the *direction* from $Q$ to $q$, as well as on the distance between the two charges. For different positions of the test charge, as in Fig. 5.4(b), we would observe different forces. Putting these into Coulomb's law would lead to different values for the same quantity $Q$. Also we have as yet no assurance that the force will always be in the direction of the radius vector **r**.

To allow for this possibility, let's agree to define $Q$ by averaging over all directions. Imagine a large number of infinitesimal test charges distributed evenly over a sphere (Fig. 5.4(c)). At the instant the moving charge passes the center of the sphere, the radial component of force on each test charge is measured, and the average of these force magnitudes is used to compute $Q$. Now this is just the operation that would be needed

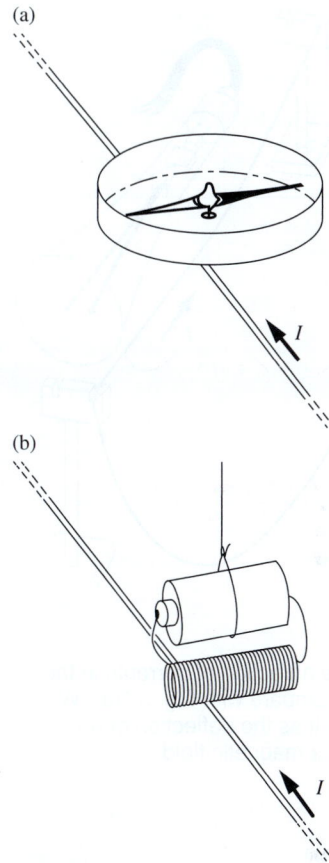

(a)

(b)

**Figure 5.2.**
A compass needle (a) and a coil of wire carrying current (b) are similarly influenced by current in a nearby conductor. The direction of the current $I$ is understood to be that in which positive ions would be moving if they were the carriers of the current. In the earth's magnetic field the black end of the compass would point north.

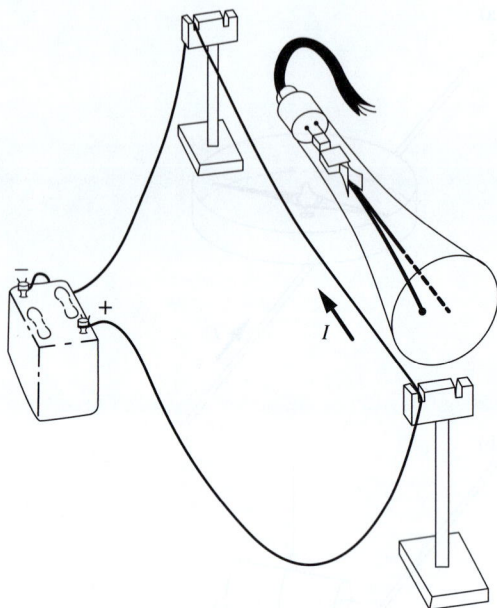

**Figure 5.3.**
An example of the attraction of currents in the same direction. Compare with Fig. 5.1(a). We can also describe it as the deflection of an electron beam by a magnetic field.

to determine the surface integral of the electric field over that sphere, at time $t$. The test charges here are all at rest, remember; the force on $q$ per unit charge gives, by definition, the electric field at that point. This suggests that Gauss's law, rather than Coulomb's law, offers the natural way[4] to define quantity of charge for a moving charged particle, or for a collection of moving charges. We can frame such a definition as follows.

The amount of electric charge in a region is defined by the surface integral of the electric field $\mathbf{E}$ over a surface $S$ enclosing the region. This surface $S$ is fixed in some coordinate frame $F$. The field $\mathbf{E}$ is measured, at any point $(x, y, z)$ and at time $t$ in $F$, by the force on a test charge *at rest in $F$*, at that time and place. The surface integral is to be determined for a particular time $t$. That is, the field values used are those measured simultaneously by observers deployed all over $S$. (This presents no difficulty, for $S$ is stationary in the frame $F$.) Let us denote such a surface integral, over $S$ at time $t$, by $\int_{S(t)} \mathbf{E} \cdot d\mathbf{a}$. We define the amount of charge inside $S$ as $\epsilon_0$ times this integral:

$$Q = \epsilon_0 \int_{S(t)} \mathbf{E} \cdot d\mathbf{a} \qquad (5.2)$$

It would be embarrassing if the value of $Q$ so determined depended on the size and shape of the surface $S$. For a stationary charge it doesn't – that is Gauss's law. But how do we know that Gauss's law holds when charges are moving? Fortunately it does. We can take that as an experimental fact. This fundamental property of the electric field of moving charges permits us to define quantity of charge by Eq. (5.2). From now on we can speak of the amount of charge in a region or on a particle, and that will have a perfectly definite meaning even if the charge is in motion.

Figure 5.5 summarizes these points in an example. Two protons and two electrons are shown in motion, at a particular instant of time. It is a fact that the surface integral of the electric field $\mathbf{E}$ over the surface $S_1$ is precisely equal to the surface integral over $S_2$ evaluated at the same instant, and we may use this integral, as we have always used Gauss's law in electrostatics, to determine the total charge enclosed. Figure 5.6 raises a new question. What if the same particles had some other velocities? For instance, suppose the two protons and two electrons combine to form a hydrogen molecule. Will the total charge appear exactly the same as before?

---

[4] It is not the only *possible* way. You could, for instance, adopt the arbitrary rule that the test charge must always be placed directly ahead (in the direction of motion) of the charge to be measured. Charge so defined would *not* have the simple properties we are about to discuss, and your theory would prove clumsy and complicated.

## 5.4 Invariance of charge

There is conclusive experimental evidence that the total charge in a system is not changed by the motion of the charge carriers. We are so accustomed to taking this for granted that we seldom pause to think how remarkable and fundamental a fact it is. For proof, we can point to the exact electrical neutrality of atoms and molecules. We have already described in Section 1.3 the experimental test of the neutrality of the hydrogen molecule, which proved that the electron and proton carry charges equal in magnitude to better than 1 part in $10^{20}$. A similar experiment was performed with helium atoms. The helium atom contains two protons and two electrons, the same charged particles that make up the hydrogen molecule. In the helium atom their motion is very different. The protons, in particular, instead of revolving slowly 0.7 angstrom apart, are tightly bound into the helium nucleus where they move with kinetic energies in the range of 1 million eV. If *motion* had any effect on the amount of charge, we could not have exact cancelation of nuclear and electronic charge in *both* the hydrogen molecule and the helium atom. In fact, the helium atom was shown to be neutral with nearly the same experimental accuracy.

Another line of evidence comes from the optical spectra of isotopes of the same element, atoms with different nuclear masses but, nominally at least, the same nuclear charge. Here again, we find a marked difference in the motion of the protons within the nucleus, but comparison of the spectral lines of the two species shows no discrepancy that could be attributed to even a slight difference in total nuclear charge.

Mass is *not* invariant in the same way. We know that the energy of a particle is changed by its motion, by the factor $1/(1 - v^2/c^2)^{1/2}$. If the constituents of a composite particle are in motion, then the increase in their energies shows up as an increase in the overall mass of the particle (even though the masses of the constituents remain the same). To emphasize the difference between mass and charge, we show in Fig. 5.7 an imaginary experiment. In the box on the right the two massive charged particles, which are fastened to the end of a pivoted rod, have been set revolving with speed $v$. The mass of the system on the right is *greater* than the mass of the system on the left, as demonstrated by weighing the box on a spring balance or by measuring the force required to accelerate it.[5] The total electric charge, however, is unchanged. A real experiment equivalent to this can be carried out with a mass spectrograph, which can reveal quite plainly a mass difference between an

[5] In general, the total mass $M$ of a system is given by $M^2 c^4 = E^2 - p^2 c^2$, where $E$ is the total energy and $p$ is the total momentum. Assuming that the total momentum is zero, the mass is given in terms of the energy as $M = E/c^2$. This energy can exist in various forms: rest energy, kinetic energy, and potential energy. For the point we are making, we have assumed that the elastic-strain potential energy of the rod in the right-hand box is negligible. If the rod is stiff, this contribution will be small compared with the $v^2/c^2$ term. See if you can show why.

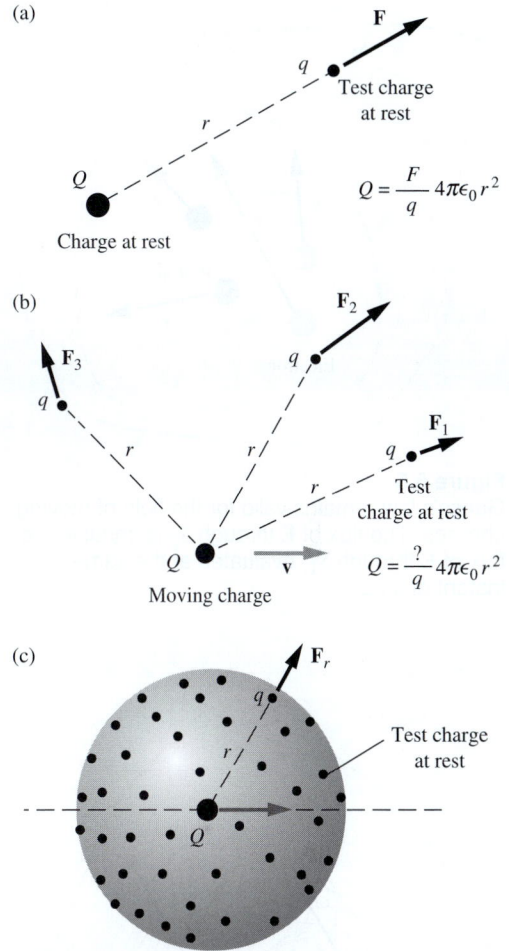

**Figure 5.4.**
(a) The magnitude of a charge at rest is determined by the force on a test charge at rest and Coulomb's law. (b) In the case of a moving charge, the force, for all we know now, may depend on the angular position of the test charge. If so, we can't use procedure (a). (c) At the instant $Q$ passes through the center of the spherical array of test charges, measure the radial force component on each, and use the average value of $F_r$ to determine $Q$. This is equivalent to measuring the surface integral of $\mathbf{E}$.

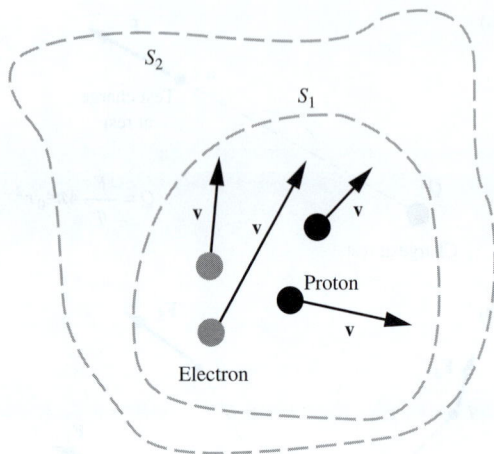

**Figure 5.5.**
Gauss's law remains valid for the field of moving charges. The flux of **E** through $S_2$ is equal to the flux of **E** through $S_1$, evaluated at the same instant of time.

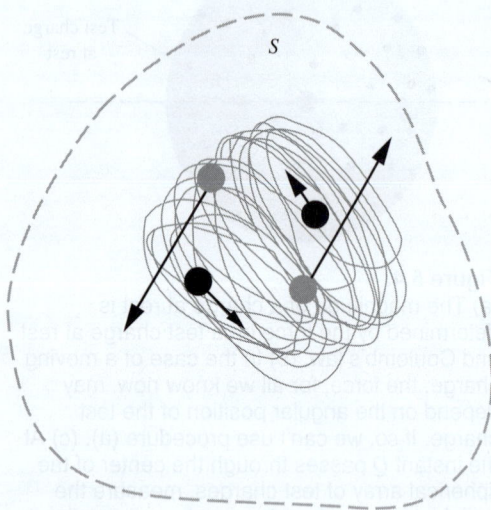

**Figure 5.6.**
Does the flux of **E** through $S$ depend on the state of motion of the charged particles? Is the surface integral of **E** over $S$ the same as in Fig. 5.5? Here the particles are bound together as a hydrogen molecule.

ionized deuterium molecule (two protons, two neutrons, one electron) and an ionized helium atom (also two protons, two neutrons, and one electron). These are two very different structures, within which the component particles are whirling around with very different speeds. The difference in energy shows up as a measurable difference in mass. There is no detectable difference, to very high precision, in the electric charge of the two ions.

This invariance of charge lends a special significance to the fact of charge quantization. We emphasized in Chapter 1 the importance – and the mystery – of the fact that every elementary charged particle has a charge equal in magnitude to that of every other such particle. We now observe that this precise equality holds not only for two particles at rest with respect to one another, but also for *any* state of relative motion.

The experiments we have described, and many others, show that the value of our Gauss's law surface integral $\int_S \mathbf{E} \cdot d\mathbf{a}$ *depends only on the number and variety of charged particles inside S, and not on how they are moving.* According to the postulate of relativity, such a statement must be true for *any* inertial frame of reference if it is true for one. Therefore if $F'$ is some *other* inertial frame, moving with respect to $F$, and if $S'$ is a closed surface in *that* frame which at time $t'$ encloses the same charged bodies that were enclosed by $S$ at time $t$, we must have

$$\int_{S(t)} \mathbf{E} \cdot d\mathbf{a} = \int_{S'(t')} \mathbf{E}' \cdot d\mathbf{a}' \quad \text{(charge invariance).} \quad (5.3)$$

The field $\mathbf{E}'$ is of course measured in $F'$, that is, it is defined by the force on a test charge at rest in $F'$. The distinction between $t$ and $t'$ must not be overlooked. As we know, events that are simultaneous in $F$ need not be simultaneous in $F'$. Each of the surface integrals in Eq. (5.3) is to be evaluated at one instant in *its* frame. If charges lie on the boundary of $S$, or of $S'$, one has to be rather careful about ascertaining that the charges within $S$ at $t$ are the same as those within $S'$ at $t'$. If the charges are well away from the boundary, as in Fig. 5.8 which is intended to illustrate the relation in Eq. (5.3), there is no problem in this respect.

Equation (5.3) is a formal statement of the relativistic invariance of charge. We can choose our Gaussian surface in *any* inertial frame; the surface integral will give a number independent of the frame. Invariance of charge is not the same as charge conservation, which was discussed in Chapter 4 and is expressed mathematically in the equation

$$\operatorname{div} \mathbf{J} = -\frac{\partial \rho}{\partial t}. \quad (5.4)$$

Charge *conservation* implies that, if we take a closed surface fixed in some coordinate system and containing some charged matter, and if no particles cross the boundary, then the total charge inside that surface remains constant. Charge *invariance* implies that, if we look at this

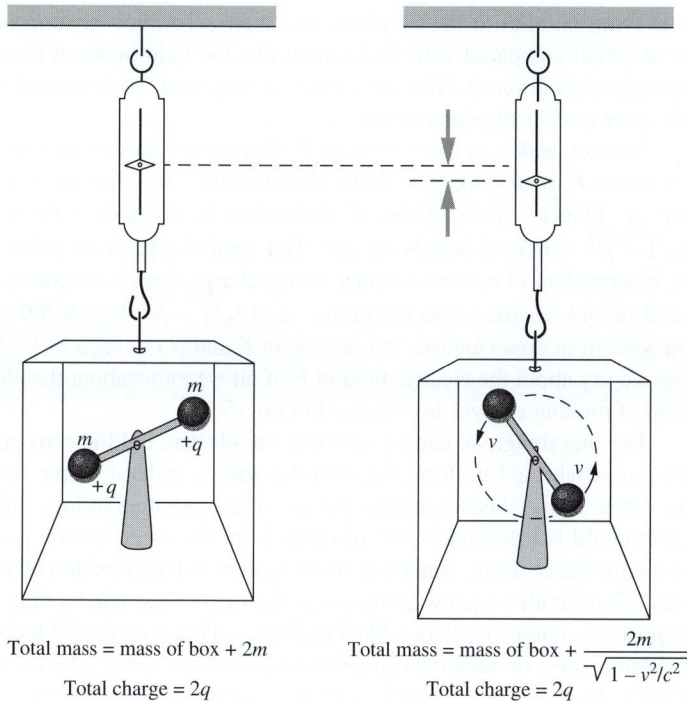

**Figure 5.7.**
An imaginary experiment to show the invariance of charge. The charge in the box is to be measured by measuring the electric field all around the box, or, equivalently, by measuring the force on a distant test charge. Mass is not invariant in the same way; see the comment in Footnote 5.

collection of stuff from any other frame of reference, we will measure exactly the same amount of charge. Energy is conserved, but energy *is not* a relativistic invariant. Charge is conserved, and charge *is* a relativistic invariant. In the language of relativity theory, energy is one component of a four-vector, while charge is a scalar, an invariant number, with respect to the Lorentz transformation. This is an observed fact with far-reaching implications. It completely determines the nature of the field of moving charges.

## 5.5 Electric field measured in different frames of reference

If charge is to be invariant under a Lorentz transformation, the electric field **E** has to transform in a particular way. "Transforming **E**" means answering a question like this: if an observer in a certain inertial frame $F$ measures an electric field **E** as so-and-so-many volts/meter, at a given point in space and time, what field will be measured at the same space-time point by an observer in a different inertial frame $F'$? For a certain class of fields, we can answer this question by applying Gauss's law to some simple systems.

In the frame $F$ (Fig. 5.9(a)) there are two stationary sheets of charge of uniform density $\sigma$ and $-\sigma$, respectively. They are squares $b$ on a

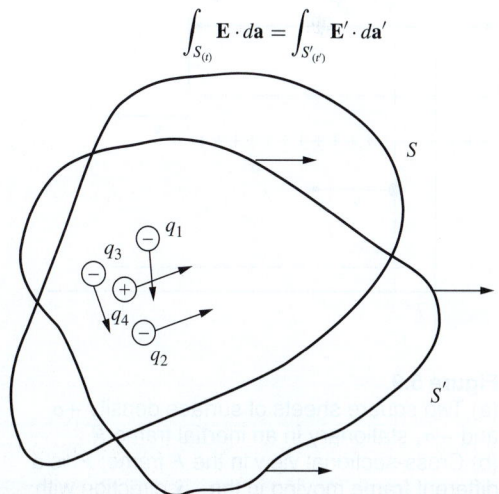

$$\int_{S_{(t)}} \mathbf{E} \cdot d\mathbf{a} = \int_{S'_{(t')}} \mathbf{E}' \cdot d\mathbf{a}'$$

**Figure 5.8.**
The surface integral of **E** over $S$ is equal to the surface integral of **E**' over $S'$. The charge is the same in all frames of reference.

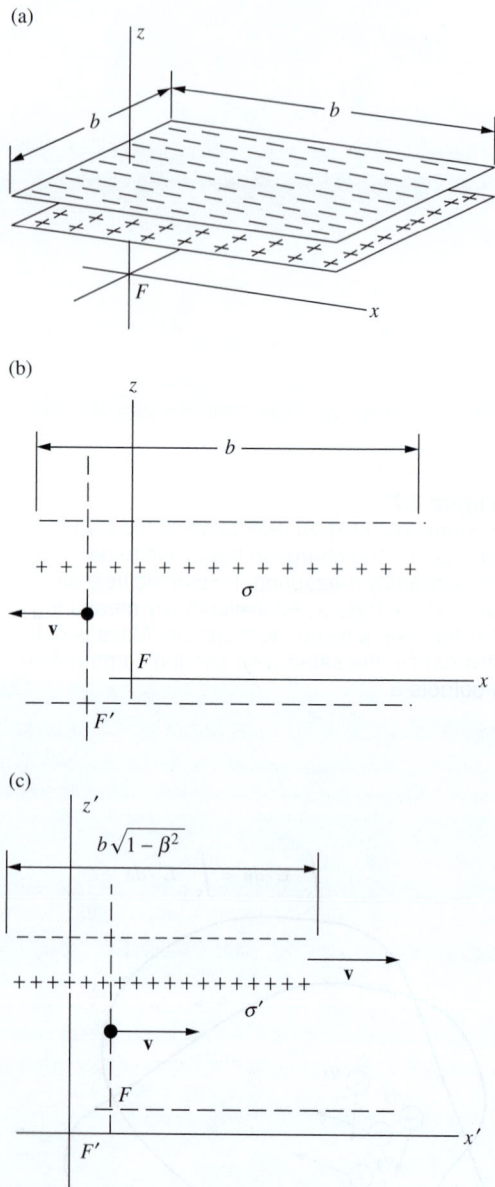

(a)

(b)

(c)

**Figure 5.9.**
(a) Two square sheets of surface density $+\sigma$ and $-\sigma$, stationary in an inertial frame $F$.
(b) Cross-sectional view in the $F$ frame; $F'$ is a different frame moving in the $-\hat{x}$ direction with respect to $F$. (c) Cross section of the charge sheets as seen in frame $F'$. The same charge is on the shorter sheet, so the charge density is greater: $\sigma' = \gamma\sigma$.

side lying parallel to the $xy$ plane, and their separation is supposed to be so small compared with their extent that the field between them can be treated as uniform. The magnitude of this field, as measured by an observer in $F$, is of course just $\sigma/\epsilon_0$.

Now consider an inertial frame $F'$ that moves toward the left, with respect to $F$, with velocity $\mathbf{v}$. To an observer in $F'$, the charged "squares" are no longer square. Their $x'$ dimension is contracted from $b$ to $b\sqrt{1-\beta^2}$, where $\beta$ stands for $v/c$. But total charge is invariant, that is, independent of reference frame, so the charge *density* measured in $F'$ must be *greater* than $\sigma$ in the ratio $\gamma \equiv 1/\sqrt{1-\beta^2}$. Figure 5.9 shows the system in cross section, (b) as seen in $F$ and (c) as seen in $F'$. What can we say about the electric field in $F'$ if all we know about the electric field of moving charges is contained in Eq. (5.3)?

For one thing, we can be sure that the electric field is zero outside the sandwich, and uniform between the sheets, at least in the limit as the extent of the sheets becomes infinite. The field of an infinite uniform sheet could not depend on the distance from the sheet, nor on position along the sheet. There is nothing in the system to fix a position along the sheet. But for all we know at this point, the field of a single moving sheet of positive charge *might* look like Fig. 5.10(a). However, even if it did, the field of a sheet of negative charge moving with the same velocity would have to look like Fig. 5.10(b), and the superposition of the two fields would still give zero field outside our two charged sheets and a uniform perpendicular field between them, as in Fig. 5.10(c). (Actually, as we shall prove before long, the field of a single sheet of charge moving in its own plane is perpendicular to the sheet, unlike the hypothetical fields pictured in Fig. 5.10(a) and (b).)

We can apply Gauss's law to a box stationary in frame $F'$, the box shown in cross section in Fig. 5.10(c). The charge content is determined by $\sigma'$, and the field is zero outside the sandwich. Gauss's law tells us that the magnitude of $E'_z$, which is the only field component inside, must be $\sigma'/\epsilon_0$, or $(\sigma/\sqrt{1-\beta^2})/\epsilon_0$. Hence

$$E'_z = \frac{E_z}{\sqrt{1-\beta^2}} = \gamma E_z. \tag{5.5}$$

Now imagine a different situation with the stationary charged sheets in the frame $F$ oriented perpendicular to the $x$ axis, as in Fig. 5.11. The observer in $F$ now reports a field in the $x$ direction of magnitude $E_x = \sigma/\epsilon_0$. In this case, the surface charge density observed in the frame $F'$ is the *same* as that observed in $F$. The sheets are not contracted; only the distance between them is contracted, but that doesn't enter into the determination of the field. This time we find by applying Gauss's law to the box stationary in $F'$ the following:

$$E'_x = \frac{\sigma'}{\epsilon_0} = \frac{\sigma}{\epsilon_0} = E_x. \tag{5.6}$$

That is all very well for the particularly simple arrangement of charges here pictured; do our conclusions have more general validity? This question takes us to the heart of the meaning of *field*. If the electric field **E** at a point in space-time is to have a unique meaning, then the way **E** appears in other frames of reference, in the same space-time neighborhood, cannot depend on the nature of the sources, wherever they may be, that produced **E**. In other words, observers in $F$, having measured the field in their neighborhood at some time, ought to be able to predict *from these measurements alone* what observers in other frames of reference would measure at the same space-time point. Were this not true, *field* would be a useless concept. The evidence that it *is* true is the eventual agreement of our field theory with experiment.

Seen in this light, the relations expressed in Eqs. (5.5) and (5.6) take on a significance beyond the special case of charges on parallel sheets. Consider *any* charge distribution, all parts of which are at rest with respect to the frame $F$. If observers in $F$ measure a field $E_z$ in the $z$ direction, then observers in frame $F'$ (whose velocity **v** with respect to $F$ is parallel to the $x$ axis) will report, for the same space-time point, a field $E_z' = \gamma E_z$. That is, they will get a number, as the result of their $E_z'$ measurement, that is larger by the factor $\gamma$ than the number the $F$ observers got in their $E_z$ measurement. On the other hand, if observers in $F$ measure a field $E_x$ in the $x$ direction, then observers in $F'$ report a field $E_x'$ *equal* to $E_x$. Obviously the $y$ and the $z$ directions are equivalent, both being transverse to the velocity **v**. Anything we have said about $E_z'$ applies to $E_y'$ too. (For both of the above orientations of the sheets, $E_y$ and $E_y'$ are zero, so $E_y' = \gamma E_y$ does indeed hold, albeit trivially.) Whatever the direction of **E** in the frame $F$, we can treat it as a superposition of fields in the $x$, the $y$, and the $z$ directions, and from the transformation of each of these predict the vector field **E**' at that point in $F'$.

Let's summarize this in words appropriate to relative motion in any direction. Charges at rest in frame $F$ are the source of a field **E**. Let frame $F'$ move with velocity **v** relative to $F$. At any point in $F$, resolve **E** into a longitudinal component $E_\parallel$ parallel to **v** and a transverse component $E_\perp$ perpendicular to **v**. At the same space-time point in $F'$, the field **E**' is to be resolved into $E_\parallel'$ and $E_\perp'$, with $E_\parallel'$ being parallel to **v** and $E_\perp'$ perpendicular thereto. We have now learned that

$$\boxed{\begin{array}{l} E_\parallel' = E_\parallel \\ E_\perp' = \gamma E_\perp \end{array}} \qquad \text{(for charges at rest in frame } F\text{).} \qquad (5.7)$$

If you forget where the $\gamma$ factor goes, just remember the following simple rule, which is a consequence of charge invariance and length contraction:

- *The transverse component of the electric field is* <u>*smaller*</u> *in the frame of the sources than in any other frame.*

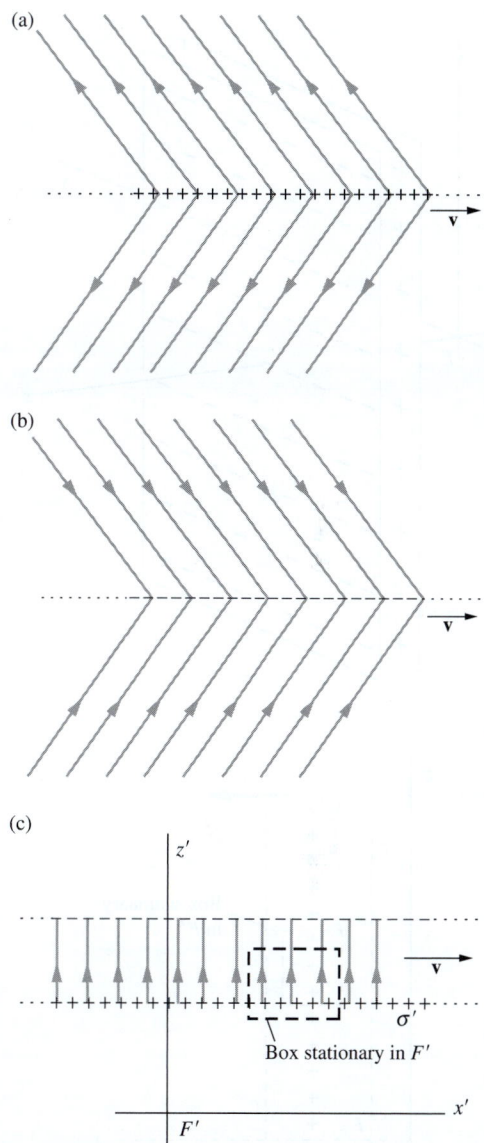

**Figure 5.10.**
(a) Perhaps the field of a single moving sheet of positive charge looks like this. (It really doesn't, but we haven't proved that yet.) (b) If the field of the positive sheet looked like Fig. 5.10(a), the field of a moving negative sheet would look like this. (c) The superposition of the fields of the positive and negative sheets would look like this, even if Figs. 5.10(a) and (b) were correct.

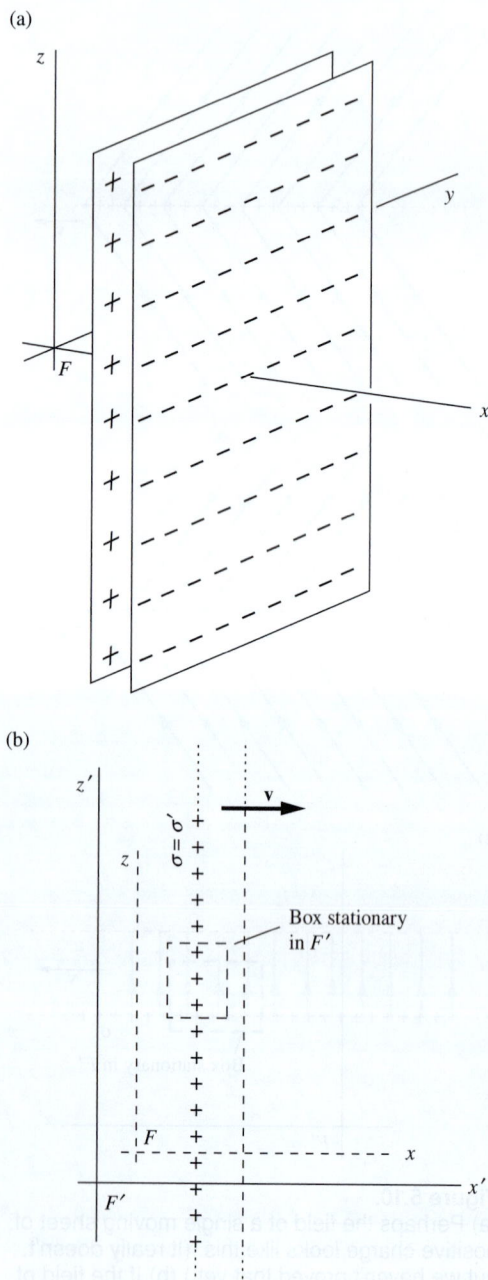

(a)

(b)

**Figure 5.11.**
The electric field in another frame of reference (relative velocity parallel to field direction). (a) In reference frame $F$. (b) Cross-sectional view in reference frame $F'$.

Equation (5.7) holds only for fields that arise from charges stationary in $F$. As we shall see in Section 6.7, if charges in $F$ are moving, the prediction of the electric field in $F'$ involves knowledge of *two* fields in $F$, the electric and the magnetic. But we already have a useful result, one that suffices whenever we can find any inertial frame of reference in which all the charges remain at rest. We shall use it in Section 5.6 to study the electric field of a point charge moving with constant velocity.

**Example (Tilted sheet)**    Fixed in the frame $F$ is a sheet of charge, of uniform surface density $\sigma$, that bisects the dihedral angle formed by the $xy$ and the $yz$ planes. The electric field of this stationary sheet is of course perpendicular to the sheet. How will this setup be described by observers in a frame $F'$ that is moving in the $x$ direction with velocity $0.6c$ with respect to $F$? That is, what is the surface charge density $\sigma'$, and what are the strength and direction of the electric field in $F'$? Find the component of the field perpendicular to the sheet in $F'$, and verify that Gauss's law still holds.

**Solution**    The situations in frames $F$ and $F'$ are shown in Fig. 5.12. Let's first find the surface density $\sigma'$. The $\gamma$ factor associated with $v = 0.6c$ is $\gamma = 5/4$, so in going from $F$ to $F'$, longitudinal distances are decreased by $4/5$. The distance between points $A'$ and $B'$ in $F'$ is therefore shorter than the distance between points $A$ and $B$ in $F$. With the distance $\ell$ shown, the latter distance is $\sqrt{2}\ell$, while the former is $\sqrt{1 + (4/5)^2}\,\ell$. Since charge is invariant, the same amount of charge is contained between $A'$ and $B'$ as between $A$ and $B$. Therefore, $\sigma'\sqrt{1 + (4/5)^2}\ell = \sigma\sqrt{2}\ell \Longrightarrow \sigma' = (1.1043)\sigma$.

Now let's find the electric field $\mathbf{E}'$. The magnitude of the field $\mathbf{E}$ in $F$ is simply $E = \sigma/2\epsilon_0$ (by the standard Gauss's law argument), and it points at a $45°$ angle. So Eq. (5.7) gives the field components in $F'$ as

$$E'_\parallel = E_\parallel = E/\sqrt{2} \quad \text{and} \quad E'_\perp = \gamma E_\perp = \gamma E/\sqrt{2}. \tag{5.8}$$

$\mathbf{E}'$ is shown in Fig. 5.12(b). Its magnitude is $E' = (E/\sqrt{2})\sqrt{1 + (5/4)^2} = (1.1319)E$, and its slope is (negative) $E'_\perp/E'_\parallel = \gamma$. The sheet's slope is also (positive) $\ell/(\ell/\gamma) = \gamma$. So the angles $\theta$ shown are all equal to $\tan^{-1}\gamma = 51.34°$. This means that $\mathbf{E}'$ points at an angle of $2\theta$ with respect to the sheet. Equivalently, it points at an angle of $2\theta - 90° \approx 12.68°$ with respect to the normal to the sheet. The normal component of $E'$ is then

$$E'_n = E'\cos 12.68° = (1.1319E)\cos 12.68° = (1.1043)E. \tag{5.9}$$

Therefore, since the same numerical factor appears in the two equations, $E'_n = (1.1043)E$ and $\sigma' = (1.1043)\sigma$, we can multiply both sides of the relation $E = \sigma/2\epsilon_0$ by 1.1043 to obtain $E'_n = \sigma'/2\epsilon_0$. In other words, Gauss's law holds in frame $F'$. There is also an electric field component parallel to the sheet in $F'$ (unlike in $F$), but this doesn't affect the flux through a Gaussian surface. You can check that it is no coincidence that the numbers worked out here, by solving the problem symbolically in terms of $\gamma$; see Exercise 5.12.

## 5.6 Field of a point charge moving with constant velocity

In the frame $F$ in Fig. 5.13(a) the point charge $Q$ remains at rest at the origin. At every point the electric field $\mathbf{E}$ has the magnitude $Q/4\pi\epsilon_0 r^2$ and is directed radially outward. In the $xz$ plane its components at any point $(x, z)$ are

$$E_x = \frac{Q}{4\pi\epsilon_0 r^2}\cos\theta = \frac{Qx}{4\pi\epsilon_0(x^2 + z^2)^{3/2}},$$

$$E_z = \frac{Q}{4\pi\epsilon_0 r^2}\sin\theta = \frac{Qz}{4\pi\epsilon_0(x^2 + z^2)^{3/2}}. \tag{5.10}$$

Consider another frame $F'$ that is moving in the negative $x$ direction, with speed $v$, with respect to frame $F$. We need the relation between the coordinates of an event in the two frames, for which we turn to the Lorentz transformation given in Eq. (G.2) of Appendix G. It simplifies the description to assume, as we are free to do, that the origins of the two frames coincide at time zero according to observers in both frames. In other words, that event, the coincidence of the origins, can be the event $A$ referred to by Eq. (G.2), with coordinates $(x_A, y_A, z_A, t_A) = (0, 0, 0, 0)$ in frame $F$ and $(x'_A, y'_A, z'_A, t'_A) = (0, 0, 0, 0)$ in frame $F'$. Then event $B$ is the space-time point we are trying to locate. We can omit the tag $B$ and call its coordinates in $F$ just $(x, y, z, t)$, and its coordinates in $F'$ just $(x', y', z', t')$. Then Eq. (G.2) of Appendix G becomes

$$x' = \gamma x - \gamma\beta ct, \quad y' = y, \quad z' = z, \quad t' = \gamma t - \frac{\gamma\beta x}{c}. \tag{5.11}$$

However, *that* transformation was for an $F'$ frame moving in the positive $x$ direction with respect to $F$, as one can quickly verify by noting that, with increasing time $t$, $x'$ gets smaller. To construct the Lorentz transformation for our problem, in which the $F'$ frame moves in the opposite direction, we must either reverse the sign of $\beta$ or switch the primes. We choose to do the latter because we want to express $x$ and $z$ in terms of $x'$ and $z'$. The Lorentz transformation we need is therefore

$$x = \gamma x' - \gamma\beta ct', \quad y = y', \quad z = z', \quad t = \gamma t' - \frac{\gamma\beta x'}{c}. \tag{5.12}$$

According to Eqs. (5.5) and (5.6), $E'_z = \gamma E_z$ and $E'_x = E_x$. Using Eqs. (5.10) and (5.12), we can express the field components $E'_z$ and $E'_x$ in terms of the coordinates in $F'$. For the instant $t' = 0$, when $Q$ passes the origin in $F'$, we have

$$E'_x = E_x = \frac{Q(\gamma x')}{4\pi\epsilon_0[(\gamma x')^2 + z'^2]^{3/2}},$$

$$E'_z = \gamma E_z = \frac{\gamma(Qz')}{4\pi\epsilon_0[(\gamma x')^2 + z'^2]^{3/2}}. \tag{5.13}$$

(a) (Frame $F$)

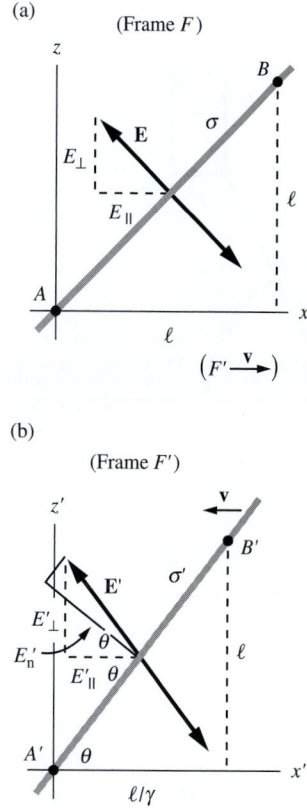

(b) (Frame $F'$)

**Figure 5.12.**
The setup as viewed in frames $F$ and $F'$. The sheet moves to the left in $F'$.

(a)

(b)

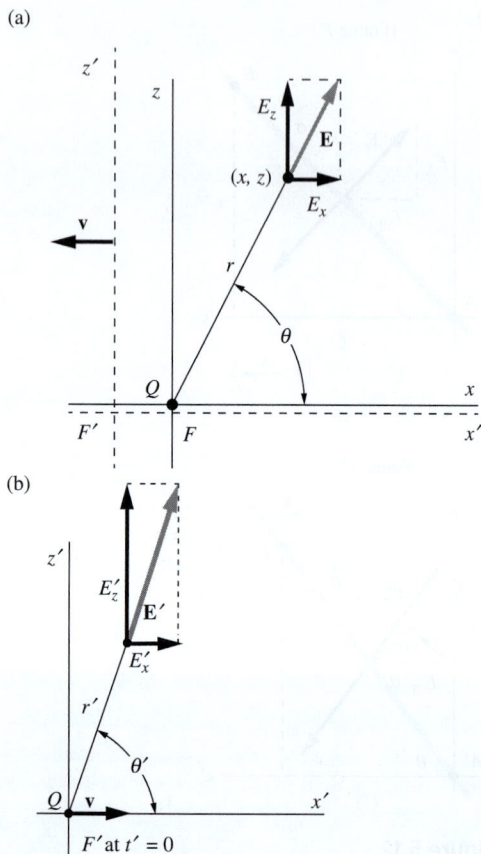

**Figure 5.13.**
The electric field of a point charge: (a) in a frame in which the charge is at rest; (b) in a frame in which the charge moves with constant velocity.

Note first that $E_z'/E_x' = z'/x'$. This tells us that the vector $\mathbf{E}'$ makes the same angle with the $x'$ axis as does the radius vector $\mathbf{r}'$. Hence $\mathbf{E}'$ points radially outward along a line drawn from the instantaneous position of $Q$, as in Fig. 5.13(b). Pause a moment to let this conclusion sink in! It means that, if $Q$ passed the origin of the primed system at precisely 12:00 noon, "prime time," an observer stationed anywhere in the primed system will report that the electric field in his vicinity was pointing, at 12:00 noon, exactly radially from the origin. This sounds at first like instantaneous transmission of information! How can an observer a mile away know where the particle is at the same instant? He can't. That wasn't implied. This particle, remember, has been moving at constant speed forever, on a "flight plan" that calls for it to pass the origin at noon. That information has been available for a long time. It is the *past history* of the particle that determined the field observed, if you want to talk about cause and effect. We will investigate in Section 5.7 what happens when there is an unscheduled change in the flight plan.

To find the strength of the field, we compute $E_x'^2 + E_z'^2$, which is the square of the magnitude of the field, $E'^2$:

$$E'^2 = E_x'^2 + E_z'^2 = \frac{\gamma^2 Q^2 (x'^2 + z'^2)}{(4\pi\epsilon_0)^2 [(\gamma x')^2 + z'^2]^3}$$

$$= \frac{Q^2(x'^2 + z'^2)}{(4\pi\epsilon_0)^2 \gamma^4 [x'^2 + (1-\beta^2)z'^2]^3}$$

$$= \frac{Q^2(1-\beta^2)^2}{(4\pi\epsilon_0)^2 (x'^2 + z'^2)^2 \left(1 - \dfrac{\beta^2 z'^2}{x'^2 + z'^2}\right)^3}. \tag{5.14}$$

(Here, for once, it was neater with $\beta$ worked back into the expression.) Let $r'$ denote the distance from the charge $Q$, which is momentarily at the origin, to the point $(x', z')$ where the field is measured: $r' = (x'^2 + z'^2)^{1/2}$. Let $\theta'$ denote the angle between this radius vector and the velocity of the charge $Q$, which is moving in the positive $x'$ direction in the frame $F'$. Then since $z' = r' \sin\theta'$, the magnitude of the field can be written as

$$\boxed{E' = \frac{Q}{4\pi\epsilon_0 r'^2} \frac{1-\beta^2}{(1 - \beta^2 \sin^2\theta')^{3/2}}} \tag{5.15}$$

There is nothing special about the origin of coordinates, nor about the $x'z'$ plane as compared with any other plane through the $x'$ axis. Therefore we can say quite generally that the electric field of a charge that has been in uniform motion is at a given instant of time directed radially from the instantaneous position of the charge, while its magnitude is given by Eq. (5.15), with $\theta'$ the angle between the direction of motion of the charge and the radius vector from the instantaneous position of the charge to the point of observation.

For low speeds the field reduces simply to $E' \approx Q/4\pi\epsilon_0 r'^2$, and is practically the same, at any instant, as the field of a point charge stationary in $F'$ at the instantaneous location of $Q$. But if $\beta^2$ is not negligible, the field is stronger at right angles to the motion than in the direction of the motion, at the same distance from the charge. If we were to indicate the intensity of the field by the density of field lines, as is often done, the lines tend to concentrate in a pancake perpendicular to the direction of motion. Figure 5.14 shows the density of lines as they pass through a unit sphere, from a charge moving in the $x'$ direction with a speed $v/c = 0.866$. A simpler representation of the field is shown in Fig. 5.15, a cross section through the field with some field lines in the $x'z'$ plane indicated.[6]

This is a remarkable electric field. It is not spherically symmetrical, which is not surprising because in this frame there is a preferred direction, the direction of motion of the charge. However, the field is symmetrical on either side of the plane passing through the charge and perpendicular to the direction of motion of the charge. That, by the way, is sufficient to prove that the field of a uniform sheet of charge moving in its own plane must be perpendicular to the sheet. Think of that field as the sum of the fields of charge elements spread uniformly over the sheet. Since each of these individual fields has the fore-and-aft symmetry of Fig. 5.15 with respect to the direction of motion, their sum could only be perpendicular to the sheet. It could not look like Fig. 5.10(a).

The task of Exercise 5.15 is to verify that the field in Eq. (5.15) satisfies Gauss's law. The inverse-square dependence on $r'$ is a necessary but not sufficient condition for Gauss's law to hold. Additionally, we need the $\beta$ dependence to drop out of the surface integral (because the amount of charge doesn't depend on the velocity), and it is by no means obvious that this happens.

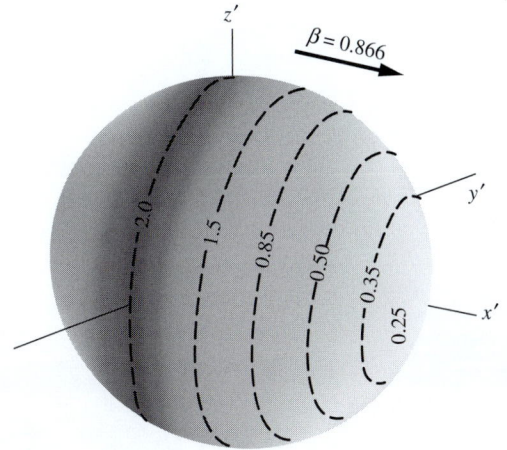

**Figure 5.14.**
The intensity in various directions of the field of a moving charge. At this instant, the charge is passing the origin of the $x'y'z'$ frame. The numbers give the field strength relative to $Q/4\pi\epsilon_0 r'^2$.

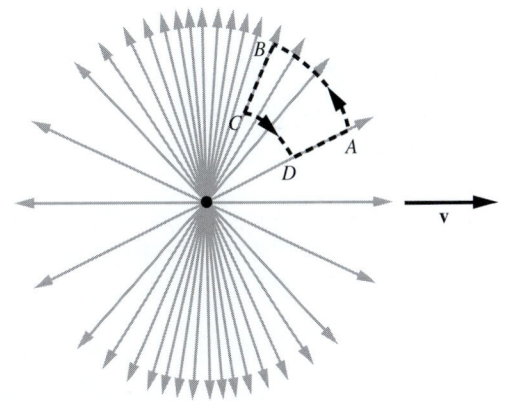

**Figure 5.15.**
Another representation of the field of a uniformly moving charge.

**Example (Transverse and longitudinal fields)**  Let's verify that the electric field in Eq. (5.15) obeys the relations in Eq. (5.7) for the transverse and longitudinal fields. Of course, we know that it must, because we *used* Eq. (5.7) in deriving Eq. (5.15). But it is a good exercise to double check. As above, the unprimed frame $F$ is the frame of the charge $Q$, and the primed frame $F'$ moves to the left with speed $v = \beta c$.

Consider first the transverse field. The electric field in frame $F$ is simply $E = Q/4\pi\epsilon_0 r^2$, in all directions. In frame $F'$, where the charge moves to the right with speed $\beta c$, the transverse field is obtained by setting $\theta = \pi/2$ in Eq. (5.15).

---

[6] A *two-dimensional* diagram like Fig. 5.15 cannot faithfully represent the field intensity by the density of field lines. Unless we arbitrarily break off some of the lines, the density of lines in the picture will fall off as $1/r'$, whereas the intensity of the field we are trying to represent falls off as $1/r'^2$. So Fig. 5.15 gives only a qualitative indication of the variation of $E'$ with $r'$ and $\theta'$.

Using $\gamma \equiv 1/\sqrt{1-\beta^2}$, this gives $E'_{\perp} = \gamma Q/4\pi \epsilon_0 r'^2$. But $r' = r$ because there is no transverse length contraction. Hence $E'_{\perp} = \gamma E_{\perp}$, as desired.

Now consider the longitudinal field. In frame $F'$, the longitudinal field is obtained by setting $\theta = 0$ in Eq. (5.15). This gives $E'_{\parallel} = Q/4\pi \epsilon_0 \gamma^2 r'^2$. According to Eq. (5.7), this should equal the field in frame $F$, namely $E_{\parallel} = Q/4\pi \epsilon_0 r^2$. And indeed it does, because the longitudinal distances are related by $r = \gamma r'$. That is, the distance is longer in frame $F$; see Eq. (5.12) with $t' = 0$.

Let's be more explicit about this $r = \gamma r'$ relation. When we say that $E'_{\parallel} = E_{\parallel}$, we mean that $E'_{\parallel}$ and $E_{\parallel}$ are related this way when measured at the *same point in space-time* by people in the two frames. To visualize this, imagine a longitudinal stick with length $r$ attached to the charge $Q$, with person $P$ sitting on the other end of the stick (and therefore at rest with respect to the charge). Person $P'$ is at rest somewhere on the $x'$ axis in frame $F'$. Each person sees the other fly by with speed $v$. If they both shout out the values of the longitudinal fields they observe when their locations coincide, they will shout the same values, namely $Q/4\pi \epsilon_0 r^2$. Note that $P'$ measures a smaller distance to the charge ($r' = r/\gamma$ instead of $r$, due to the length contraction of the stick), but the field in Eq. (5.15) is suppressed by a factor of $\gamma^2$ in the longitudinal direction, and these two effects exactly cancel.

---

What if we have an essentially continuous stream of particles moving in a line? From Gauss's law we know that the electric field takes the standard form of $\lambda/2\pi \epsilon_0 r$, where $\lambda$ is the charge density of the line, as measured in the given frame. That is, as far as the electric field is concerned, it doesn't matter that the line of charges is moving longitudinally, for a given $\lambda$. (We'll see in Section 5.9 that the moving line also creates a magnetic field, but that doesn't concern us here.) However, it is by no means obvious that the sum of the nonspherically symmetric fields in Eq. (5.15), from all the individual charges, equals $\lambda/2\pi \epsilon_0 r$ for any value of $\beta$. The task of Problem 5.5 is to demonstrate this explicitly.

The field in Fig. 5.15 is a field that *no stationary charge distribution*, whatever its form, could produce. For in this field the line integral of **E'** is in general *not zero* around a closed path. Consider, for example, the closed path $ABCD$ in Fig. 5.15. The circular arcs contribute nothing to the line integral, being perpendicular to the field; on the radial sections, the field is *stronger* along $BC$ than along $DA$, so the *circulation* of **E'** on this path is not zero. But remember, this is not an electro*static* field. In the course of time the electric field **E'** at any point in the frame $F'$ changes as the source charge moves.

Figure 5.16 shows the electric field at certain instants of time observed in a frame of reference through which an electron is moving at constant velocity in the $x$ direction.[7] In Fig. 5.16, the speed of the electron is $0.33c$. Its kinetic energy is therefore about 30,000 eV (30 kiloelectron-volts (keV)). The value of $\beta^2$ is $1/9$, and the electric field does not differ

---

[7] Previously we had the charge at rest in the unprimed frame, moving in the primed frame. Here we adopt $xyz$ for the frame in which the charge is moving, to avoid cluttering the subsequent discussion with primes.

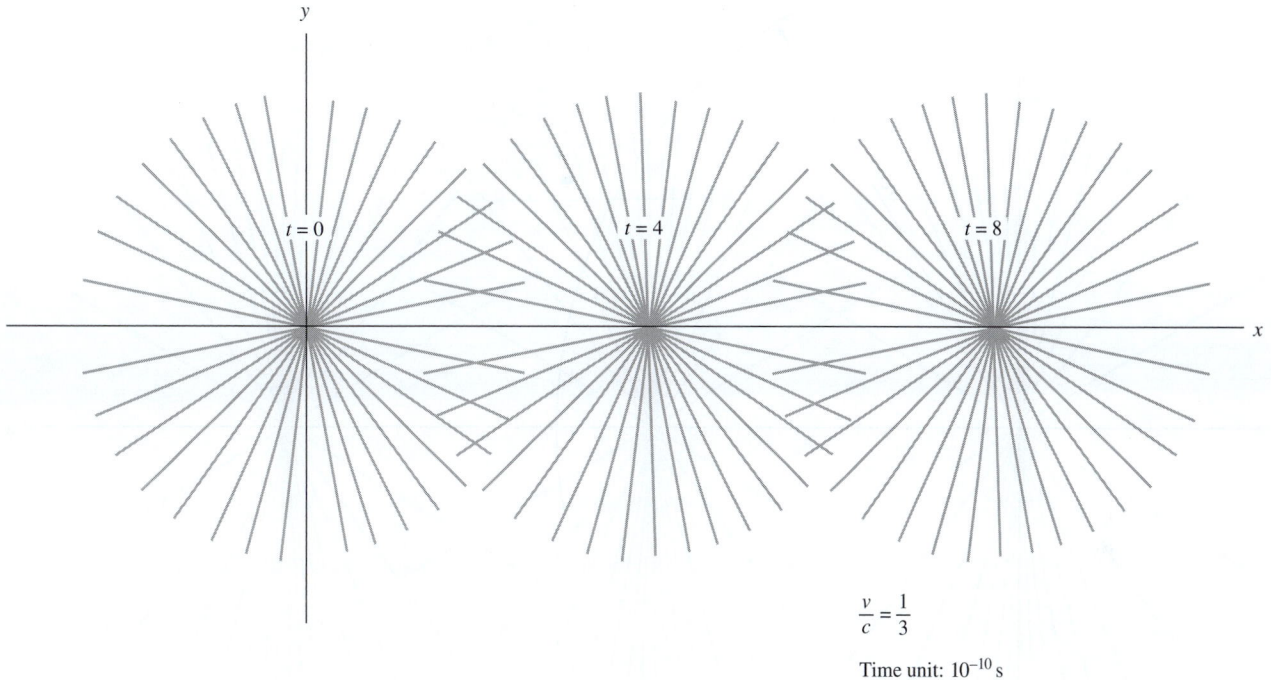

$$\frac{v}{c} = \frac{1}{3}$$

Time unit: $10^{-10}$ s

**Figure 5.16.**
The electric field of a moving charge, shown for three instants of time; $v/c = 1/3$.

greatly from that of a charge at rest. In Fig. 5.17, the speed is $0.8c$, corresponding to a kinetic energy of 335 keV. If the time unit for each diagram is taken as $1.0 \cdot 10^{-10}$ s, the distance scale is life-size, as drawn. Of course, the diagram holds equally well for *any* charged particle moving at the specified fraction of the speed of light. We mention the equivalent energies for an electron merely to remind the reader that relativistic speeds are nothing out of the ordinary in the laboratory.

## 5.7 Field of a charge that starts or stops

It must be clearly understood that *uniform velocity,* as we have been using the term, implies a motion at constant speed in a straight line that has been going on forever. What if our electron had *not* been traveling in the distant past along the negative $x$ axis until it came into view in our diagram at $t = 0$? Suppose it had been sitting quietly at rest at the origin, waiting for the clock to read $t = 0$. Just prior to $t = 0$, something gives the electron a sudden large acceleration, up to the speed $v$, and it moves away along the positive $x$ axis at this speed. Its motion *from then on* precisely duplicates the motion of the electron for which Fig. 5.17 was drawn (assuming $v = 0.8c$). But Fig. 5.17 does *not* correctly represent the field of the electron whose history was just described. To see that it cannot do so, consider the field at the point marked $P$, at time $t = 2$, which means $2 \cdot 10^{-10}$ s. In $2 \cdot 10^{-10}$ s a light signal travels 6 cm. Since this point lies more than 6 cm distant from the origin, it could not

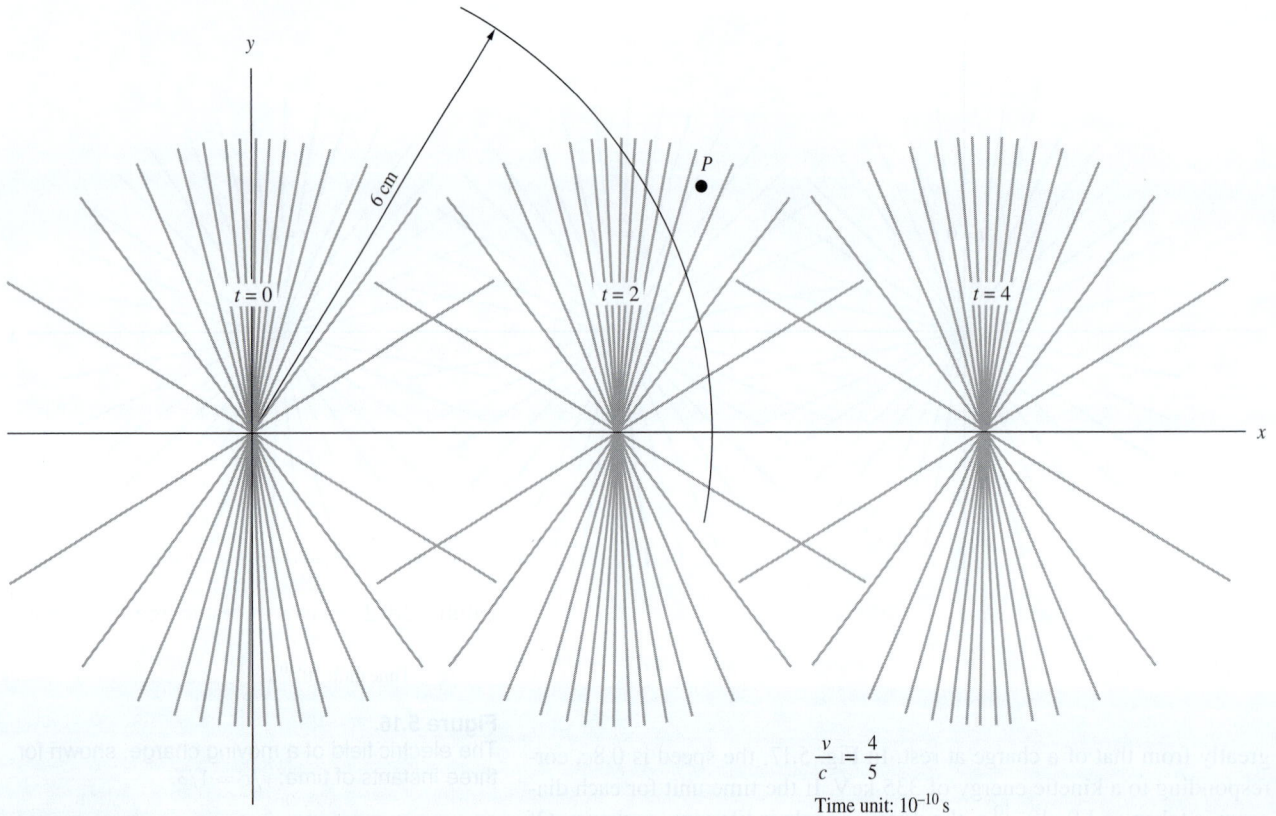

$$\frac{v}{c} = \frac{4}{5}$$

Time unit: $10^{-10}$ s

**Figure 5.17.**
The electric field of a moving charge, shown for
three instants of time; $v/c = 4/5$.

have received the news that the electron had started to move at $t = 0$!
Unless there is a gross violation of relativity – and we are taking the
postulates of relativity as the basis for this whole discussion – the field
at point $P$ at time $t = 2$, and indeed at all points outside the sphere of
radius 6 cm centered on the origin, *must be the field of a charge at rest at
the origin.*

On the other hand, close to the moving charge itself, what happened
in the remote past can't make any difference. The field must somehow
change, as we consider regions farther and farther from the charge, at
the given instant $t = 2$, from the field shown in the second diagram of
Fig. 5.17 to the field of a charge at rest at the origin. We can't deduce more
than this without knowing how fast the news *does* travel. Suppose – just
suppose – it travels as fast as it can without conflicting with the relativity
postulates. Then if the period of acceleration is neglected, we should
expect the field within the entire 6 cm radius sphere, at $t = 2$, to be
the field of a uniformly moving point charge. If that is so, the field of
the electron that starts from rest, suddenly acquiring the speed $v = 0.8c$
at $t = 0$, must look something like Fig. 5.18. There is a thin spherical
shell (whose thickness in an actual case will depend on the duration of

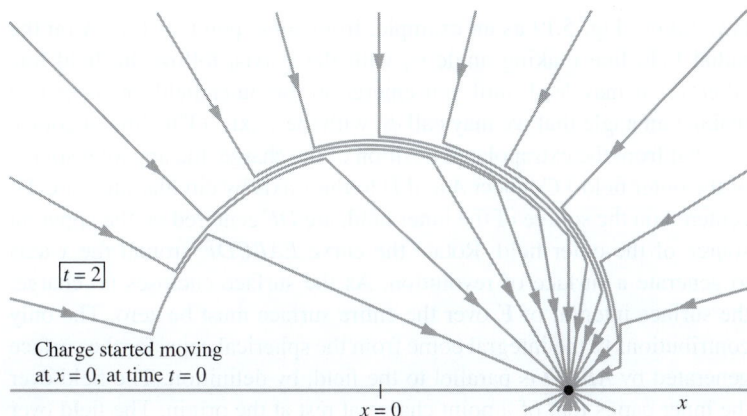

**Figure 5.18.**
An electron initially at rest in the laboratory frame is suddenly accelerated at $t = 0$ and moves with constant velocity thereafter. This is how the electric field looks at the instant $t = 2$ all over the laboratory frame.

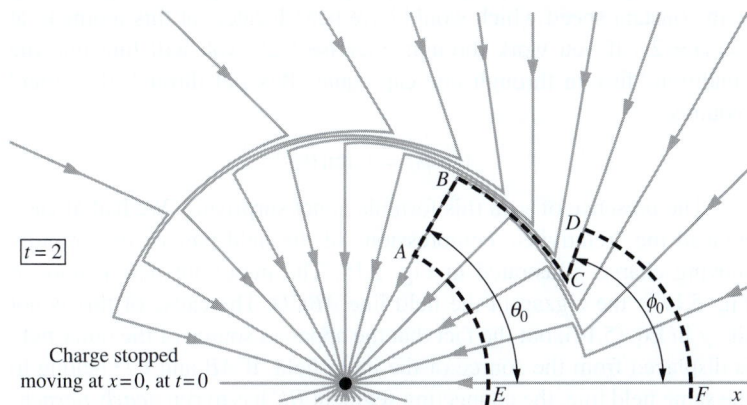

**Figure 5.19.**
An electron that has been moving with constant velocity reaches the origin at $t = 0$, is abruptly stopped, and remains at rest thereafter. This is how the field looks in the laboratory frame at the instant $t = 2$. The dashed outline follows a field line from $A$ to $D$. Rotating the whole outline $EABCDF$ around the $x$ axis generates a closed surface, the total flux through which must be zero. The flux in through the spherical cap $FD$ must equal the flux out through the spherical cap $EA$. This condition suffices to determine the relation between $\theta_0$ and $\phi_0$.

the interval required for acceleration) within which the transition from one type of field to the other takes place. This shell simply expands with speed $c$, its center remaining at $x = 0$. The arrowheads on the field lines indicate the direction of the field when the source is a negative charge, as we have been assuming.

Figure 5.19 shows the field of an electron that had been moving with uniform velocity *until* $t = 0$, at which time it reached $x = 0$ where it was abruptly *stopped*. Now the news that it was stopped cannot reach, by time $t$, any point farther than $ct$ from the origin. The field outside the sphere of radius $R = ct$ must be that which would have prevailed if the electron had kept on moving at its original speed. That is why we see the "brush" of field lines on the right in Fig. 5.19 pointing precisely down to the position where the electron would be if it hadn't stopped. (Note that this last conclusion does not depend on the assumption we introduced in the previous paragraph, that the news travels as fast as it can.) The field almost seems to have a life of its own!

It is a relatively simple matter to connect the inner and outer field lines. There is only one way it can be done that is consistent with Gauss's law. Taking Fig. 5.19 as an example, from some point such as $A$ on the radial field line making angle $\theta_0$ with the $x$ axis, follow the field line wherever it may lead until you emerge in the outer field on some line making an angle that we may call $\phi_0$ with the $x$ axis. (This line of course is radial from the extrapolated position of the charge, the apparent source of the outer field.) Connect $A$ and $D$ to the $x$ axis by circular arcs, arc $AE$ centered on the source of the inner field, arc $DF$ centered on the apparent source of the outer field. Rotate the curve $EABCDF$ around the $x$ axis to generate a surface of revolution. As the surface encloses no charge, the surface integral of $\mathbf{E}$ over the entire surface must be zero. The only contributions to the integral come from the spherical caps, for the surface generated by $ABCD$ is parallel to the field, by definition. The field over the inner cap is that of a point charge at rest at the origin. The field over the outer cap is the field, as given by Eq. (5.15), of a point charge moving with constant speed which would have been located, at this moment, at $x = vt = 2v$. If you work through Exercise 5.20, you will find that the condition "flux in through one cap equals flux out through the other" requires

$$\tan \phi_0 = \gamma \tan \theta_0. \qquad (5.16)$$

The presence of $\gamma$ in this formula is not surprising. We had already noticed the "relativistic compression" of the field pattern of a rapidly moving charge, illustrated in Fig. 5.15. The important new feature in Fig. 5.19 is the zigzag in the field line $ABCD$. The cause of this is not the $\gamma$ in Eq. (5.16), but the fact that the apparent source of the outer field is displaced from the source of the inner field. If $AB$ and $CD$ belong to the same field line, the connecting segment $BC$ has to run *nearly perpendicular* to a radial vector. We have a *transverse* electric field there, and one that, to judge by the crowding of the field lines, is relatively intense compared with the radial field. As time goes on, the zigzag in the field lines will move radially outward with speed $c$. But the thickness of the shell of transverse field will not increase, for that was determined by the duration of the deceleration process.

The ever-expanding shell of transverse electric field would keep on going *even if* at some later time (at $t = 3$, say) we suddenly accelerated the electron back to its original velocity. That would only launch a new outgoing shell, this one looking very much like the field in Fig. 5.18. The field *does* have a life of its own! What has been created here before our eyes is an *electromagnetic wave*. The magnetic field that is also part of it was not revealed in this view. Later, in Chapter 9, we shall learn how the electric and magnetic fields work together in propagating an electrical disturbance through empty space. What we have discovered here is that such waves *must* exist if nature conforms to the postulates of special relativity and if electric charge is a relativistic invariant.

More can be done with our "zigzag-in-the-field-line" analysis. Appendix H shows how to derive, rather simply, an accurate and simple formula for the rate of radiation of energy by an accelerated electric charge. We must return now to the uniformly moving charge, which has more surprises in store.

## 5.8 Force on a moving charge

Equation (5.15) tells us the force experienced by a stationary charge in the field of another charge that is moving at constant velocity. We now ask a different question: what is the force that acts on a moving charge, one that moves in the field of some other charges?

We shall look first into the case of a charged particle moving through the field produced by stationary charges. (Section 5.9 deals with the case where both the charged particle and the sources of the field are moving.) We might have an electron moving between the charged plates of an oscilloscope, or an alpha particle moving through the Coulomb field around an atomic nucleus. The sources of the field, in any case, are all at rest in some frame of reference, $F$, which we shall call the "lab frame." At some place and time in the lab frame we observe a particle carrying charge $q$ that is moving, at that instant, with velocity $\mathbf{v}$ through the electrostatic field. What force appears to act on $q$?

Force means rate of change of momentum, so we are really asking: what is the rate of change of momentum of the particle, $d\mathbf{p}/dt$, at this place and time, as measured in our lab frame of reference, $F$? The answer is contained, by implication, in what we have already learned. Let's look at the system from a coordinate frame $F'$ moving, at the time in question, along with the particle.[8] In this "particle frame" $F'$, the particle will be, at least momentarily, at rest. It is the other charges that are now moving. This is a situation we know how to handle. The charge $q$ has the same value; charge is invariant. The force on the stationary charge $q$ is just $q\mathbf{E}'$, where $\mathbf{E}'$ is the electric field observed in the frame $F'$. We have learned how to find $\mathbf{E}'$ when $\mathbf{E}$ is given; Eq. (5.7) provides our rule. Thus, knowing $\mathbf{E}$, we can find the rate of change of momentum of the particle as observed in $F'$. All that remains is to transform this quantity back to $F$. So our problem hinges on the question: how does force, that is, rate of change of momentum, transform from one inertial frame to another?

The answer to that question is worked out in Eqs. (G.16) and (G.17) in Appendix G. The force component *parallel* to the relative motion of the two frames has the *same* value in the moving frame as it does in the rest frame of the particle. A force component *perpendicular* to the

---

[8] This notation might seem the reverse of the notation we used in Section 5.6, where the unprimed frame $F$ was the particle frame. However, the notation is consistent in the sense that we are taking the source(s) of the electric field we are concerned with to be at rest in the unprimed frame $F$. In Section 5.6 the source was a single charge, whereas in the present case the source is an arbitrary collection of charges. We are concerned with the field due to these charges, not the field due to the "test" particle represented by the frame $F'$.

relative frame velocity is always *smaller*, by $1/\gamma$, than its value in the particle's rest frame. Let us summarize this in Eq. (5.17) using subscripts $\parallel$ and $\perp$ to label momentum components, respectively, parallel to and perpendicular to the relative velocity of $F'$ and $F$, as we did in Eq. (5.7):

$$\frac{dp_\parallel}{dt} = \frac{dp'_\parallel}{dt'}$$

$$\frac{dp_\perp}{dt} = \frac{1}{\gamma}\frac{dp'_\perp}{dt'}$$

(for a particle at rest in frame $F'$). (5.17)

Note that this is not a symmetrical relation between the primed and unprimed quantities. The rest frame of the particle, which we have chosen to call $F'$ in this case (note that we called it $F$ in Appendix G), is special because the particle is the thing upon which the given force is acting. If you forget where the $\gamma$ factor goes, just remember the following simple rule (which can be traced to time dilation).

- *The transverse component of the force on a particle is <u>larger</u> in the frame of the particle than in any other frame.*

   Equipped with the force transformation law, Eq. (5.17), and the transformation law for electric field components, Eq. (5.7), we return now to our charged particle moving through the field **E**, and we discover an astonishingly simple fact. Consider first $E_\parallel$, the component of **E** parallel to the instantaneous direction of motion of our charged particle. Transform to a frame $F'$ moving, at that instant, with the particle. In that frame the longitudinal electric field is $E'_\parallel$, and, according to Eq. (5.7), $E'_\parallel = E_\parallel$. So the force $dp'_\parallel/dt'$ is

$$\frac{dp'_\parallel}{dt'} = qE'_\parallel = qE_\parallel. \qquad (5.18)$$

Back in frame $F$, observers are measuring the longitudinal force, that is, the rate of change of the longitudinal momentum component, $dp_\parallel/dt$. According to Eq. (5.17), $dp_\parallel/dt = dp'_\parallel/dt'$, so in frame $F$ the longitudinal force component they find is

$$\frac{dp_\parallel}{dt} = \frac{dp'_\parallel}{dt'} \implies \frac{dp_\parallel}{dt} = qE_\parallel. \qquad (5.19)$$

Of course, the particle does not *remain* at rest in $F'$ as time goes on. It will be accelerated by the field $\mathbf{E}'$, and so $\mathbf{v}'$, the velocity of the particle in the inertial frame $F'$, will gradually increase from zero. However, as we are concerned with the instantaneous acceleration, only infinitesimal values of $v'$ are involved anyway, and the restriction on Eq. (5.17) is rigorously fulfilled.

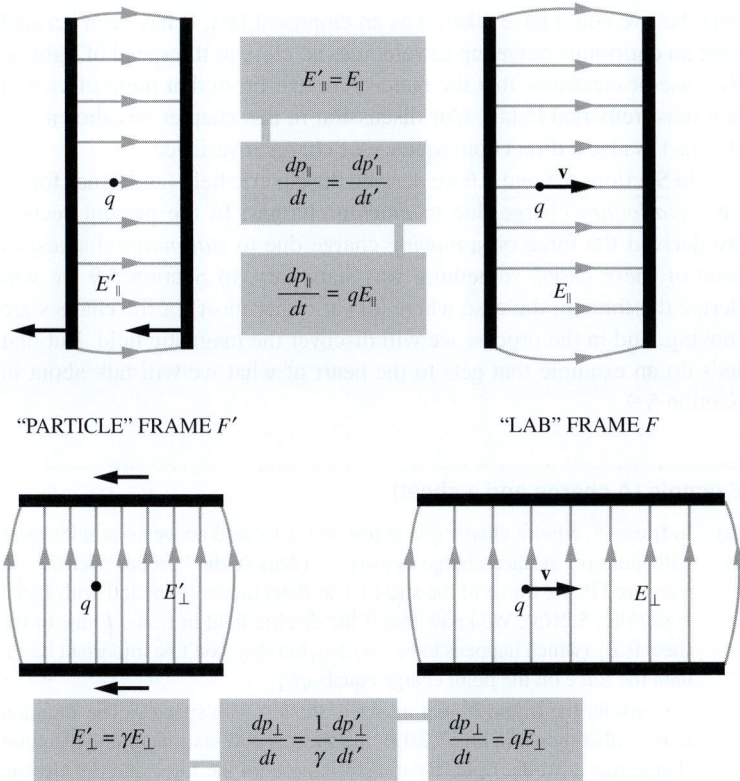

$$E'_\parallel = E_\parallel$$

$$\frac{dp_\parallel}{dt} = \frac{dp'_\parallel}{dt'}$$

$$\frac{dp_\parallel}{dt} = qE_\parallel$$

"PARTICLE" FRAME $F'$          "LAB" FRAME $F$

$$E'_\perp = \gamma E_\perp \qquad \frac{dp_\perp}{dt} = \frac{1}{\gamma}\frac{dp'_\perp}{dt'} \qquad \frac{dp_\perp}{dt} = qE_\perp$$

**Figure 5.20.**
In a frame in which the charges producing the field **E** are at rest, the force on a charge $q$ moving with any velocity is simply $q$**E**.

For $E_\perp$, the transverse field component in $F$, the transformation is $E'_\perp = \gamma E_\perp$, so that

$$\frac{dp'_\perp}{dt'} = qE'_\perp = q\gamma E_\perp. \tag{5.20}$$

But on transforming the force back to frame $F$ we have $dp_\perp/dt = (1/\gamma)(dp'_\perp/dt')$, so the $\gamma$ drops out after all:

$$\frac{dp_\perp}{dt} = \frac{1}{\gamma}\frac{dp'_\perp}{dt'} \implies \frac{dp_\perp}{dt} = \frac{1}{\gamma}(q\gamma E_\perp) \implies \boxed{\frac{dp_\perp}{dt} = qE_\perp}$$

$$\tag{5.21}$$

The message of Eqs. (5.19) and (5.21) is simply this: the force on a charged particle in motion through $F$ is $q$ times the electric field **E** in that frame, *strictly independent* of the velocity of the particle. Figure 5.20 is a reminder of this fact, and of the way we discovered it.

You have already used this result earlier in the book, where you were simply told that the contribution of the electric field to the force on a moving charge is $q$**E**. Because this is familiar and so simple, you may think it is obvious and we have been wasting our time proving it. It is true that we could have taken it as an empirical fact. It has been verified over an enormous range, up to velocities so close to the speed of light, in the case of electrons, that the factor $\gamma$ is $10^4$. From that point of view it is a most remarkable law. Our discussion in this chapter has shown that this fact is also a direct consequence of charge invariance.

In Sections 5.5 and 5.6 we derived the electric field (and hence force) on a *stationary* charge due to *moving* charges. In the present section we derived the force on a *moving* charge due to *stationary* charges. In both of these cases, something was stationary. In Section 5.9 we will derive the force in the case where all (or rather most) of the charges are moving, and in the process we will discover the magnetic field. But first let's do an example that gets to the heart of what we will talk about in Section 5.9.

### Example (A charge and a sheet)

(a) In frame $F$, a point charge $q$ is at rest and is located above an infinite sheet with uniform surface charge density $\sigma$. (This is the "proper" density, as measured in the frame of the sheet.) The sheet moves to the left with speed $v$; see Fig. 5.21(a). We know that if the electric field in frame $F$ due to the sheet is $E_1$ (which happens to be $\gamma\sigma/2\epsilon_0$, but that won't be important here), then the force on the point charge equals $qE_1$.

Consider the frame $F'$ that moves to the left with speed $v$. The situation in $F'$ is also shown in Fig. 5.21(a); the sheet is now stationary and the point charge moves to the right. By transforming both the force and the electric field from $F$ to $F'$, show that the force in $F'$ equals the electric force (as expected).

(b) Now consider a similar scenario where both the sheet and the charge are at rest in $F$, as shown in Fig. 5.21(b). Let the electric field in frame $F$ due to the sheet be $E_2$ (which happens to be $\sigma/2\epsilon_0$, but again, that won't be important). As above, $F'$ moves to the left with speed $v$. By transforming both the force and the electric field from $F$ to $F'$, show that the force in $F'$ does *not* equal the electric force. This implies that there must be some other force in $F'$; it is the magnetic force.

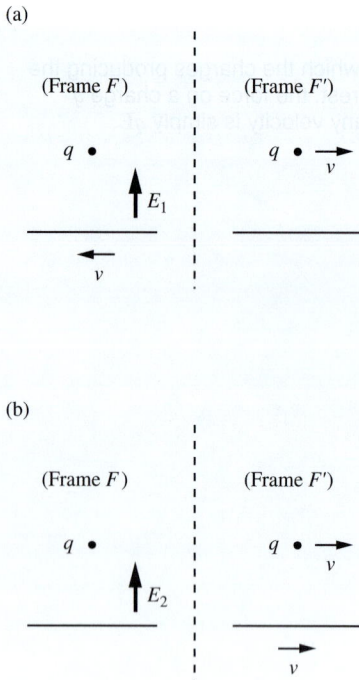

### Solution

(a) From Eq. (5.17) the transverse *force* on a particle is largest in the frame of the *particle* (which is frame $F$ here), so the force in $F'$ is smaller; it equals $qE_1/\gamma$. And from Eq. (5.7) the transverse *field* is smallest in the frame of the *source* ($F'$ here). Since the field equals $E_1$ in $F$, it therefore equals $E_1/\gamma$ in $F'$. The electric force in $F'$ is then $qE_1/\gamma$. We see that the total force in $F'$ is completely accounted for by the electric force.

(b) In this scenario the reasoning with the force is the same. The transverse force on a particle is largest in the frame of the particle ($F$ here), so the force in $F'$ is smaller; it equals $qE_2/\gamma$. However, the reasoning with the field is the reverse of what it was in part (a). The transverse field is smallest in the frame of the source (which is now $F$), so the field in $F'$ is *larger;* it equals $\gamma E_2$. The electric force in $F'$ is therefore $\gamma qE_2$. This is *not* equal

(a)

(Frame $F$)          (Frame $F'$)

$q$ •                $q$ • $\xrightarrow{v}$

$\uparrow E_1$

$\xleftarrow{\phantom{v}}$
$v$

(b)

(Frame $F$)          (Frame $F'$)

$q$ •                $q$ • $\xrightarrow{v}$

$\uparrow E_2$

$\xrightarrow{\phantom{v}}$
$v$

**Figure 5.21.**
(a) A point charge and a sheet moving relative to each other. (b) A point charge and a sheet at rest with respect to each other.

to the total force, $qE_2/\gamma$. Hence there must be some other force in $F'$ that partially cancels the $\gamma qE_2$ electric force and brings it down to the correct value of $qE_2/\gamma$. This force is the magnetic force; it arises when a charge is moving in the vicinity of other moving charges.

The purpose of this example was simply to demonstrate that the transformation rules for the electric field and the force, Eqs. (5.7) and (5.17), lead to the conclusion that *some* other force must be present in $F'$ in this second scenario. Having accomplished this, we'll stop here, but you can work things out quantitatively in Problem 5.8.

## 5.9 Interaction between a moving charge and other moving charges

Equation (5.1) tells us that there can be a velocity-dependent force on a moving charge. That force is associated with a *magnetic field,* the sources of which are electric currents, that is, other charges in motion. Oersted's experiment showed that electric currents could influence magnets, but at that time the nature of a magnet was totally mysterious. Soon Ampère and others unraveled the interaction of electric currents with each other, as in the attraction observed between two parallel wires carrying current in the same direction. This led Ampère to the hypothesis that a magnetic substance contains permanently circulating electric currents. If so, Oersted's experiment could be understood as the interaction of the galvanic current in the wire with the permanent microscopic currents that gave the compass needle its special properties. Ampère gave a complete and elegant mathematical formulation of the interaction of steady currents, and of the equivalence of magnetized matter to systems of permanent currents. His brilliant conjecture about the actual nature of magnetism in iron had to wait a century, more or less, for its ultimate confirmation.

Whether the magnetic manifestations of electric currents arose from anything *more* than the simple transport of charge was not clear to Ampère and his contemporaries. Would the motion of an electrostatically charged object cause effects like those produced by a continuous galvanic current? Later in the century, Maxwell's theoretical work suggested the answer should be *yes*. The first direct evidence was obtained by Henry Rowland, to whose experiment we shall return at the end of Chapter 6.

From our present vantage point, the magnetic interaction of electric currents can be recognized as an inevitable corollary to Coulomb's law. If the postulates of relativity are valid, if electric charge is invariant, and if Coulomb's law holds, then, as we shall now show, the effects we commonly call "magnetic" are bound to occur. They will emerge as soon as we examine the electric interaction between a moving charge and other moving charges. A simple system will illustrate this.

In the lab frame of Fig. 5.22(a), with spatial coordinates $x$, $y$, $z$, there is a line of positive charges, at rest and extending to infinity in both directions. We shall call them ions for short. Indeed, they might represent the

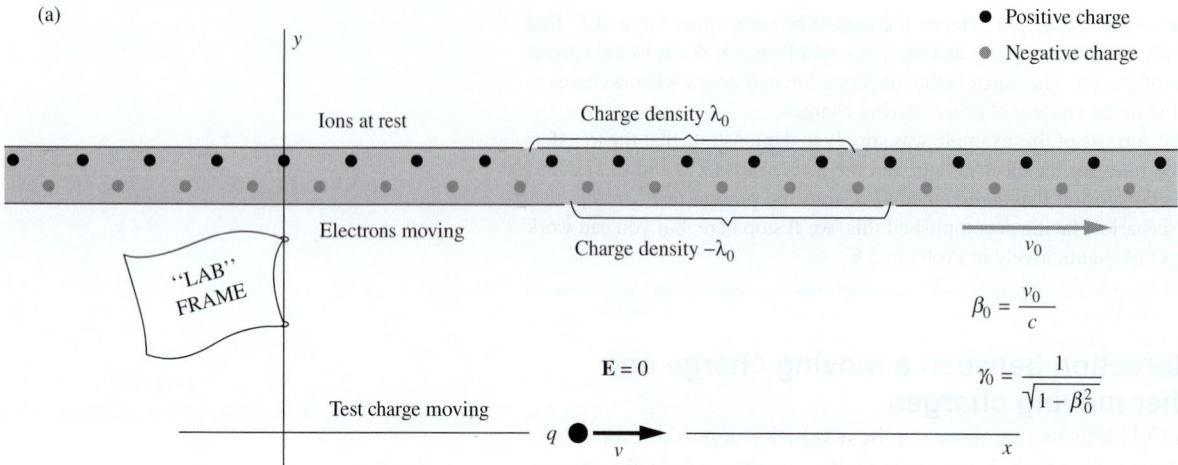

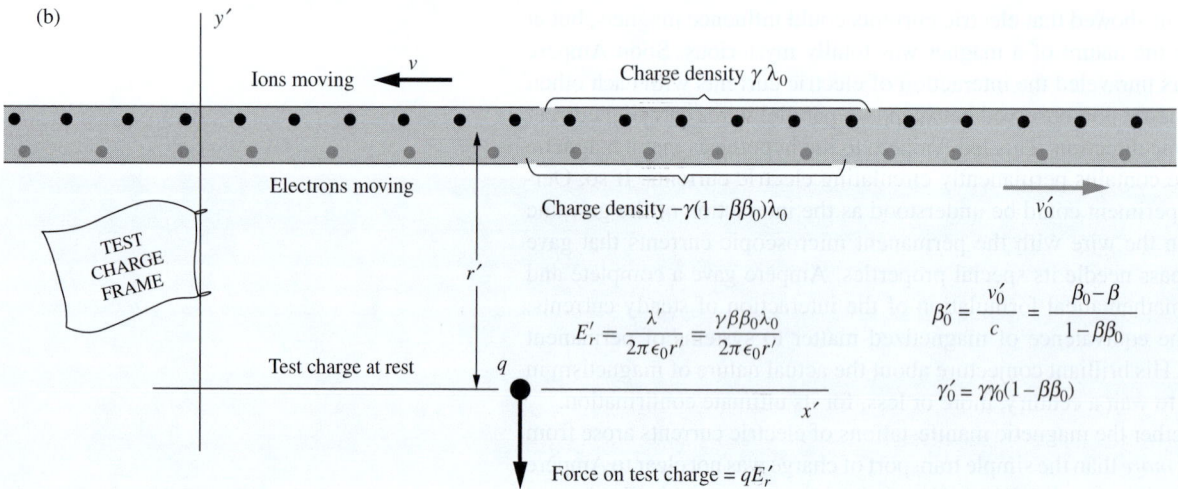

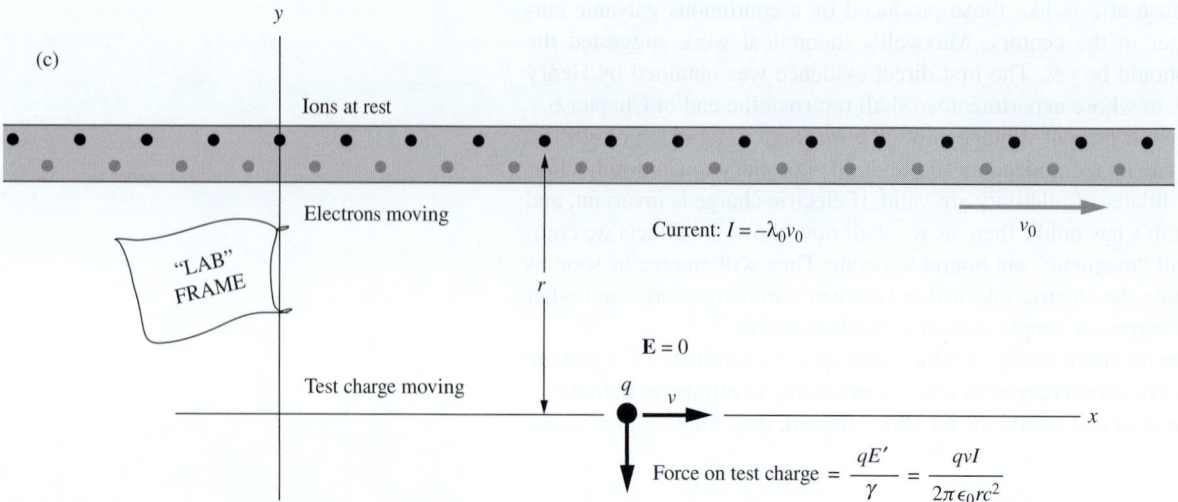

copper ions that constitute the solid substance of a copper wire. There is also a line of negative charges that we shall call electrons. These are all moving to the right with speed $v_0$. In a real wire the electrons would be intermingled with the ions; we've separated them in the diagram for clarity. The linear density of positive charge is $\lambda_0$. It happens that the linear density of negative charge along the line of electrons is exactly equal in magnitude. That is, any given length of "wire" contains at a given instant the same number of electrons and protons.[9] The net charge on the wire is zero. Gauss's law tells us there can be no flux from a cylinder that contains no charge, so the electric field must be zero everywhere outside the wire. A test charge $q$ at rest near this wire experiences no force whatsoever.

Suppose the test charge is not at rest in the lab frame but is moving with speed $v$ in the $x$ direction. Transform to a frame moving with the test charge, the $x'$, $y'$ frame in Fig. 5.22(b). The test charge $q$ is here at rest, but something else has changed: the wire appears to be charged! There are two reasons for that: the positive ions are closer together, and the electrons are farther apart. Because the lab frame in which the positive ions are at rest is moving with speed $v$, the distance between positive ions as seen in the test charge frame is contracted by $\sqrt{1 - v^2/c^2}$, or $1/\gamma$. The linear density of positive charge in this frame is correspondingly greater; it must be $\gamma\lambda_0$. The density of negative charge takes a little longer to calculate, for the electrons were already moving with speed $v_0$ in the lab frame. Hence their linear density in the lab frame, which was $-\lambda_0$, had already been increased by a Lorentz contraction. In the electrons' own rest frame the negative charge density must have been $-\lambda_0/\gamma_0$, where $\gamma_0$ is the *Lorentz factor* that goes with $v_0$.

Now we need the speed of the electrons in the test charge frame in order to calculate their density there. To find that velocity ($v_0'$ in Fig. 5.22(b)) we must add the velocity $-v$ to the velocity $v_0$, remembering

**Figure 5.22.**
A test charge $q$ moving parallel to a current in a wire. (a) In the lab frame, the wire, in which the positive charges are fixed, is at rest. The current consists of electrons moving to the right with speed $v_0$. The net charge on the wire is zero. There is no electric field outside the wire. (b) In a frame in which the test charge is at rest, the positive ions are moving to the left with speed $v$, and the electrons are moving to the right with speed $v_0'$. The linear density of positive charge is greater than the linear density of negative charge. The wire appears positively charged, with an external field $E_r'$, which causes a force $qE_r'$ on the stationary test charge $q$. (c) That force transformed back to the lab frame has the magnitude $qE_r'/\gamma$, which is proportional to the product of the speed $v$ of the test charge and the current in the wire, $-\lambda_0 v_0$.

[9] It doesn't have to, but that equality can always be established, if we choose, by adjusting the number of electrons per unit length. In our idealized setup, we assume this has been done.

to use the relativistic formula for the addition of velocities (Eq. (G.7) in Appendix G). Let $\beta_0' = v_0'/c$, $\beta_0 = v_0/c$, and $\beta = v/c$. Then

$$\beta_0' = \frac{\beta_0 - \beta}{1 - \beta\beta_0}. \tag{5.22}$$

The corresponding Lorentz factor $\gamma_0'$, obtained from Eq. (5.22) with a little algebra (as you can check), is

$$\gamma_0' \equiv (1 - \beta_0'^2)^{-1/2} = \gamma\gamma_0(1 - \beta\beta_0). \tag{5.23}$$

This is the factor by which the linear density of negative charge in the electrons' own rest frame (which was $-\lambda_0/\gamma_0$) is enhanced when it is measured in the test charge frame. The total linear density of charge in the wire in the test charge frame, $\lambda'$, can now be calculated:

$$\lambda' = \gamma\lambda_0 - \frac{\lambda_0}{\gamma_0}\,\underbrace{\gamma\gamma_0(1 - \beta\beta_0)}_{} = \gamma\beta\beta_0\lambda_0. \tag{5.24}$$

| factor for transformation to test charge frame | positive charge density in ions' rest frame | negative charge density in electrons' rest frame | factor for transformation to test charge frame |
|---|---|---|---|

The wire is positively charged. Gauss's law guarantees the existence of a radial electric field whose magnitude $E_r'$ is given by our familiar formula, Eq. (1.39), for the field of any infinite line charge:

$$E_r' = \frac{\lambda'}{2\pi\epsilon_0 r'} = \frac{\gamma\beta\beta_0\lambda_0}{2\pi\epsilon_0 r'}. \tag{5.25}$$

At the location of the test charge $q$ this field points in the $-y'$ direction. The test charge will therefore experience a force

$$F_y' = qE_y' = -\frac{q\gamma\beta\beta_0\lambda_0}{2\pi\epsilon_0 r'}. \tag{5.26}$$

Now let's return to the lab frame, pictured again in Fig. 5.22(c). What is the magnitude of the force on the charge $q$ as measured there? If its value is $qE_y'$ in the rest frame of the test charge, observers in the lab frame will report a force smaller by the factor $1/\gamma$, by Eq. (5.17). Since $r = r'$, the force on our moving test charge, measured in the lab frame, is

$$F_y = \frac{F_y'}{\gamma} = -\frac{q\beta\beta_0\lambda_0}{2\pi\epsilon_0 r}. \tag{5.27}$$

The quantity $-\lambda_0 v_0$, or $-\lambda_0 \beta_0 c$, is just the total current $I$ in the wire, in the lab frame, for it is the amount of charge flowing past a given point per second. We call current positive if it is equivalent to positive charge flowing in the positive $x$ direction. Our current in this example is negative. Our result can be written this way:

$$F_y = \frac{q v_x I}{2\pi \epsilon_0 r c^2} \qquad (5.28)$$

where we have written $v_x$ for $v$ to remind us that the velocity of the test charge $q$ is in the $x$ direction. We have found that, in the lab frame, the moving test charge experiences a force in the (negative) $y$ direction that is proportional to the current in the wire, and to the velocity of the test charge in the $x$ direction. We will see at the beginning of Chapter 6 exactly how this force is related to the magnetic field **B**. But for now we simply note that the force is in the direction of $\mathbf{v} \times \mathbf{B}$ if **B** is a vector in the $\hat{\mathbf{z}}$ direction, pointing at us out of the diagram.

It is a remarkable fact that the force on the moving test charge does not depend separately on the velocity or density of the charge carriers but only on the product, $\beta_0 \lambda_0$ in our example, that determines the charge transport. If we have a certain current $I$, say 1 milliamp, it does not matter whether this current is composed of high-energy electrons moving with 99 percent of the speed of light, or of electrons in a metal executing nearly random thermal motions with a slight drift in one direction, or of charged ions in solution with positive ions moving one way, negatives the other. Or it could be any combination of these, as Exercise 5.30 will demonstrate. Furthermore, the force on the test charge is strictly proportional to the velocity of the test charge $v$. Finally, our derivation was *in no way restricted to small velocities,* either for the charge carriers in the wire or for the moving charge $q$. Equation (5.28) is exact, with no restrictions.

**Example (Repelling wires)**   Let's see how this explains the mutual repulsion of conductors carrying currents in opposite directions, as shown in Fig. 5.1(b) at the beginning of this chapter. Two such wires are represented in the lab frame in Fig. 5.23(a). Assume the wires are uncharged in the lab frame.[10] Then there is no electrical force from the opposite wire on the positive ions which are stationary in the lab frame.

Transferring to a frame in which one set of electrons is at rest (Fig. 5.23(b)), we find that in the other wire the electron distribution is Lorentz-contracted more

---

[10] As mentioned at the end of Section 4.3, there are surface charges on an actual current-carrying wire. Hence there is an electric force between the wires, in addition to the magnetic force we are presently concerned with; see Assis et al. (1999). However, under normal circumstances this electric force is very small compared with the magnetic force, so we can ignore it. It certainly goes to zero in the limit of high conductivity, because essentially zero surface charge is needed to maintain a given current in that case.

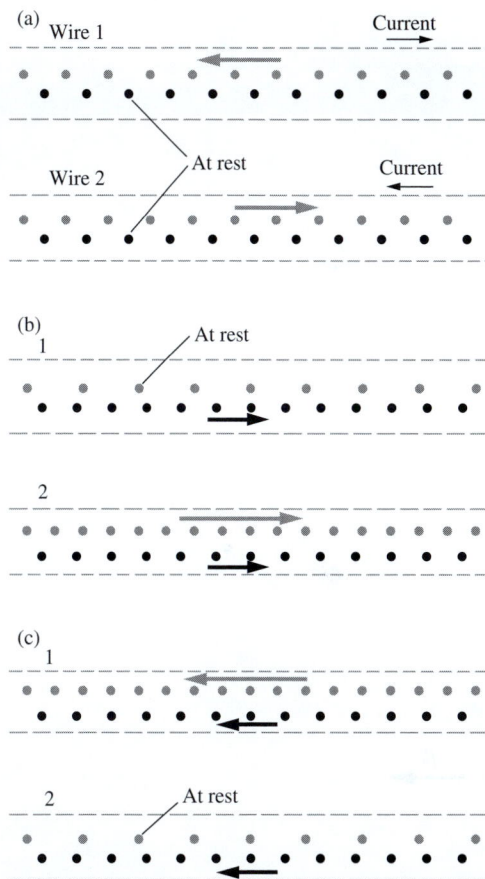

**Figure 5.23.**
(a) Lab frame with two wires carrying current in opposite directions. As in metal wire, the current is due to the motion of negative ions (electrons) only. (b) Rest frame of electrons in wire 1. Note that in wire 2 positive ions are compressed, but electron distribution is contracted even more. (c) Rest frame of electrons in wire 2. Just as in (b), the *other* wire appears to these electrons at rest to be negatively charged.

than the positive ion distribution. (As you can show, Eq. (5.24) now involves the term $1 + \beta\beta_0$, which leads to a negative net density.) Because of that, the electrons at rest in this frame will be repelled by the other wire. And when we transfer to the frame in which those other electrons are at rest (Fig. 5.23(c)), we find the same situation. They too will be repelled. These repulsive forces will be observed in the lab frame as well, modified only by the factor $\gamma$.

We conclude that the two streams of electrons will repel one another in the lab frame. The stationary positive ions, although they feel no direct electrical force from the other wire, will be the indirect bearers of this repulsive force if the electrons remain confined within the wire. So the wires will be pushed apart, as in Fig. 5.1(b), until some external force balances the repulsion.

**Example (Force on protons moving together)**   Two protons are moving parallel to one another a distance $r$ apart, with the same speed $\beta c$ in the lab frame. According to Eq. (5.15), at the instantaneous position of one of the protons, the electric field **E** caused by the other, as measured in the lab frame, has magnitude $\gamma e / 4\pi\epsilon_0 r^2$. But the force on the proton measured in the lab frame is *not* $\gamma e^2/4\pi\epsilon_0 r^2$. Verify this by finding the force in the proton rest frame and transforming that force back to the lab frame. Show that the discrepancy can be accounted for by the second term in Eq. (5.1) if there is a magnetic field **B** that points in the appropriate direction and that has a magnitude $\beta/c = v/c^2$ times the magnitude of the electric field, accompanying the proton as it travels through the lab frame.[11]

**Solution**   In the rest frame of the two protons, the force of repulsion is simply $e^2/4\pi\epsilon_0 r^2$. The force in the lab frame is therefore $(1/\gamma)(e^2/4\pi\epsilon_0 r^2)$. (Remember, the force is always largest in the rest frame of the particle on which it acts.) This *is* the correct total force in the lab frame. But, as mentioned above, the repulsive electrical force $eE$ in the lab frame is $\gamma e^2/4\pi\epsilon_0 r^2$, because Eq. (5.15) tells us that the electric field due to a moving charge is larger by a factor $\gamma$ in the transverse direction. Apparently this must not be the whole force. There must be an extra attractive force that partially cancels the repulsive electric force $\gamma e^2/4\pi\epsilon_0 r^2$, bringing it down to the correct value of $e^2/\gamma 4\pi\epsilon_0 r^2$. This extra attractive force must therefore have magnitude (using $1/\gamma^2 = 1 - \beta^2$)

$$\frac{\gamma e^2}{4\pi\epsilon_0 r^2} - \frac{e^2}{\gamma 4\pi\epsilon_0 r^2} = \gamma\left(1 - \frac{1}{\gamma^2}\right)\frac{e^2}{4\pi\epsilon_0 r^2}$$

$$= \gamma\beta^2\frac{e^2}{4\pi\epsilon_0 r^2} = e(\beta c)\left(\frac{\beta}{c}\frac{\gamma e}{4\pi\epsilon_0 r^2}\right). \tag{5.29}$$

We have chosen to write the force in this way, because we can then interpret it as the $q\mathbf{v} \times \mathbf{B}$ magnetic force in Eq. (5.1), provided that the magnitude of **B** is $(\beta/c)(\gamma e/4\pi\epsilon_0 r^2)$, which is $\beta/c$ times the magnitude of the electric field in the lab frame, and provided that **B** points out of (or into) the page at the location of the top (or bottom) proton in Fig. 5.24. The cross product $\mathbf{v} \times \mathbf{B}$ then points

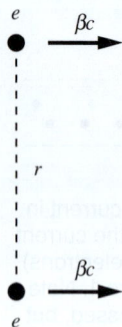

**Figure 5.24.**
Two protons moving parallel to each other, a distance $r$ apart.

---

[11]  The setup in this example is nearly the same as the setup in the example at the end of Section 5.8, but we will follow this one through to completion.

in the proper (attractive) direction. Each proton creates a magnetic field at the location of the other proton. The relative factor of $\beta/c$ between the magnetic and electric fields is consistent with the Lorentz transformations we will derive in Section 6.7.

We see that the magnetic force, through its partial cancelation of the electric force, allows the following two statements to be consistent: (1) the transverse *electric field* due to a charge is *smallest* in the frame of that charge (by a factor of $\gamma$ compared with any other frame), and (2) the transverse *force* on a particle is *largest* in the frame of that particle (by a factor of $\gamma$ compared with any other frame). These two statements imply, respectively, that the *electric* force is larger in the lab frame than in the protons' frame, but the *total* force is smaller in the lab frame than in the protons' frame. These two facts are consistent because the existence of the magnetic force means that the total force isn't equal to just the electric force.

You might be tempted to argue that the proton is not "moving through" the **B** field of the other proton because that field is "moving right along with it." That would be incorrect. In the force law that is the fundamental definition of **B**, namely $\mathbf{F} = q\mathbf{E} + q\mathbf{v} \times \mathbf{B}$, **B** is the field at the position of the charge $q$ at an instant in time, with both position and time measured in the frame in which we are measuring the force on $q$. What the "source" of **B** may be doing at that instant is irrelevant.

Note that the structure of the reasoning in this example is the same as the reasoning in the charge-and-wire example above in the text. In both cases we first found the force in the rest frame of a given point charge. (This was simple in the present example, but involved a detailed length-contraction argument in the charge-and-wire example.) We then transformed the force to the lab frame by a quick division by $\gamma$. And finally we determined what the extra (magnetic) force must be to make the sum of the electric and magnetic forces in the lab frame be correct. (The lab-frame electric force was trivially zero in the charge-and-wire example, but not in the present example.) See Exercise 5.29 for more practice with this type of problem.

---

Moving parallel to a current-carrying conductor, a charged particle experiences a force perpendicular to its direction of motion. What if it moves, instead, at right angles to the conductor? A velocity perpendicular to the wire will give rise to a force parallel to the wire – again, a force perpendicular to the particle's direction of motion. To see how this comes about, let us consider the lab frame of that system and give the test charge a velocity $v$ in the $y$ direction, as in Fig. 5.25(a). Transferring to the rest frame of the test charge (Fig. 5.25(b)), we find the positive ions moving vertically downward. Certainly they cannot cause a horizontal field at the test-charge position. The $x'$ component of the field from an ion on the left will be exactly canceled by the $x'$ component of the field of a symmetrically positioned ion on the right.

The effect we are looking for is caused by the electrons. They are all moving obliquely in this frame, downward and toward the right. Consider the two symmetrically located electrons $e_1$ and $e_2$. Their electric fields, relativistically compressed in the direction of the electrons' motion, have

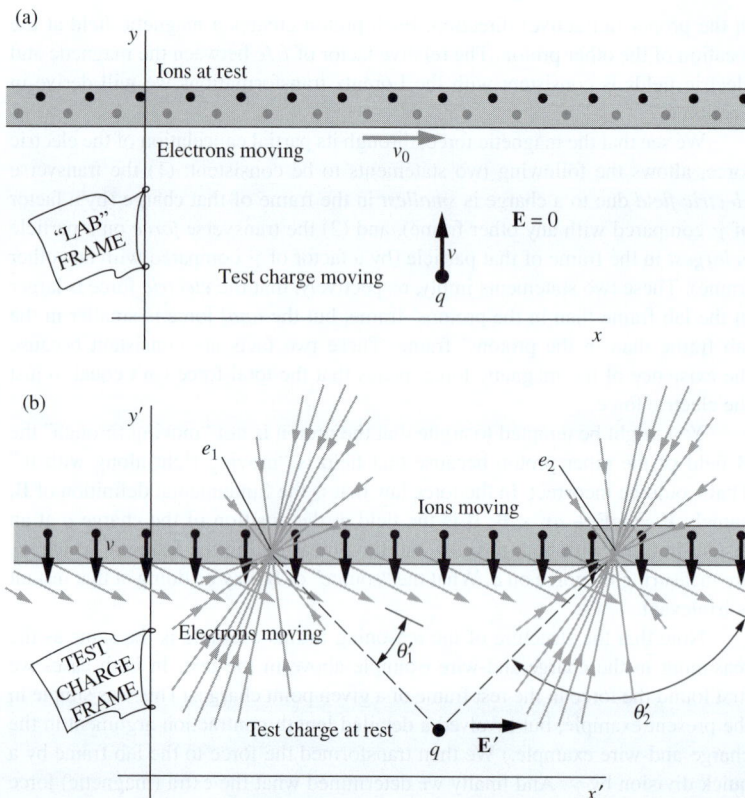

(a)

(b)

**Figure 5.25.**
(a) The "wire" with its current of moving negative charges, or "electrons," is the same as in Fig. 5.22, but now the test charge is moving toward the wire. (b) In the rest frame of the test charge, the positive charges, or "ions," are moving in the $-\hat{y}$ direction. The electrons are moving obliquely. Because the field of a moving charge is stronger in directions more nearly perpendicular to its velocity, an electron on the right, such as $e_2$, causes a stronger field at the position of the test charge than does a symmetrically located electron, $e_1$, on the left. Therefore the vector sum of the fields has in this frame a component in the $\hat{x}'$ direction.

been represented by a brush of field lines in the manner of Fig. 5.15. You can see that, although $e_1$ and $e_2$ are equally far away from the test charge, the field of electron $e_2$ will be *stronger* than the field of electron $e_1$ at that location. That is because the line from $e_2$ to the test charge is more nearly perpendicular to the direction of motion of $e_2$. In other words, the angle $\theta'$ that appears in the denominator of Eq. (5.15) is here different for $e_1$ and $e_2$, so that $\sin^2 \theta'_2 > \sin^2 \theta'_1$. That will be true for any symmetrically located pair of electrons on the line, as you can verify with the aid of Fig. 5.26. The electron on the right always wins. Summing over all the electrons is therefore bound to yield a resultant field $E'$ in the $\hat{x}$ direction. The $y'$ component of the electrons' field will be exactly canceled by the field of the ions. That $E'_y$ is zero is guaranteed by Gauss's law, for the number of charges per unit length of wire is the same as it was in the lab frame. The wire is uncharged in both frames.

The force on our test charge, $qE'_x$, when transformed back into the lab frame, will be a force that is proportional to $v$ and that points in the $\hat{x}$ direction, which (as in the earlier case of motion parallel to the wire) is the direction of $\mathbf{v} \times \mathbf{B}$ if $\mathbf{B}$ is a vector in the $\hat{z}$ direction, pointing at us out of the diagram. We could show that the magnitude of this velocity-dependent force is given here also by Eq. (5.28): $F = qvI/2\pi\epsilon_0 rc^2$.

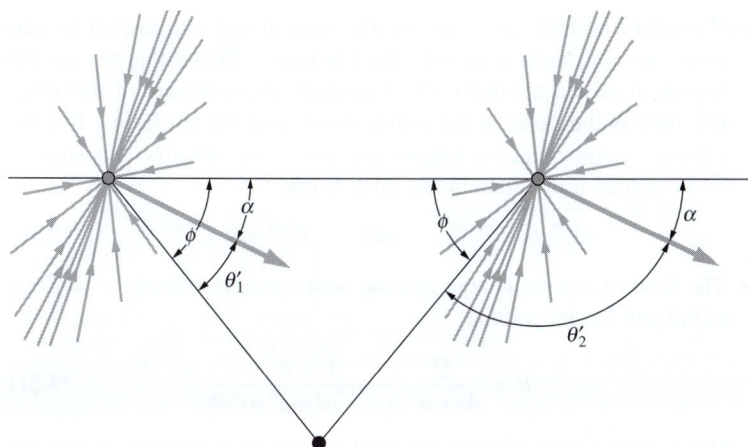

$$\sin \theta_1' = \sin(\phi - \alpha) \qquad\qquad \sin \theta_2' = \sin(\phi + \alpha)$$
$$\sin^2 \theta_2' - \sin^2 \theta_1' = 4 \sin \phi \cos \phi \sin \alpha \cos \alpha \geq 0 \text{ if}$$
$$0 \leq \phi \leq \frac{\pi}{2} \text{ and } 0 \leq \alpha \leq \frac{\pi}{2}$$

**Figure 5.26.**
A closer look at the geometry of Fig. 5.25(b), showing that, for *any* pair of electrons equidistant from the test charge, the one on the right will have a larger value of $\sin^2 \theta'$. Hence, according to Eq. (5.15), it will produce the stronger field at the test charge.

The physics needed is all in Eq. (5.15), but there are many factors to keep straight; see Exercise 5.31.

In this chapter we have seen how the fact of charge invariance implies forces between electric currents. That does not oblige us to look on one fact as the cause of the other. These are simply two aspects of electromagnetism whose relationship beautifully illustrates the more general law: physics is the same in all inertial frames of reference.

If we had to analyze every system of moving charges by transforming back and forth among various coordinate systems, our task would grow both tedious and confusing. There is a better way. The overall effect of one current on another, or of a current on a moving charge, can be described completely and concisely by working with the magnetic field; this is the subject of Chapter 6.

We usually end each chapter with a discussion of applications, but we will save the applications of the magnetic field for Section 6.10, after we have studied magnetism in depth.

## CHAPTER SUMMARY

- Charged particles in a *magnetic field* experience a magnetic force equal to $\mathbf{F} = q\mathbf{v} \times \mathbf{B}$.
- Gauss's law, $Q = \epsilon_0 \int \mathbf{E} \cdot d\mathbf{a}$, holds for moving charges (by the definition of charge) as well as for stationary charges. Charge is *invariant*, that is, the amount of charge in a system is independent of the frame in which it is measured. The total charge enclosed in a volume is independent of the motion of the charge carriers within.

- Consider a collection of charges, which are at rest with respect to each other, moving with respect to the lab frame. These charges are the source of an electric field. The longitudinal component of this electric field is the same in the source frame and the lab frame. But the transverse component is larger in the lab frame. That is, it is smaller in the source frame than in any other frame:

$$E_{\parallel}^{\text{lab}} = E_{\parallel}^{\text{source}} \quad \text{and} \quad E_{\perp}^{\text{lab}} = \gamma E_{\perp}^{\text{source}}. \tag{5.30}$$

- The field of a point charge moving with constant velocity $v = \beta c$ is radial and has magnitude

$$E = \frac{Q}{4\pi\epsilon_0 r^2} \frac{1 - \beta^2}{(1 - \beta^2 \sin^2 \theta)^{3/2}}. \tag{5.31}$$

  If the charge stops moving, the field outside an expanding shell is the same as if the charge had kept moving, as indicated in Fig. 5.19.

- If a particle moves with respect to the lab frame, then the longitudinal component of the force on it is the same in its frame and the lab frame. But the transverse component is smaller in the lab frame. That is, it is larger in the particle frame than in any other frame:

$$F_{\parallel}^{\text{lab}} = F_{\parallel}^{\text{particle}} \quad \text{and} \quad F_{\perp}^{\text{lab}} = \frac{1}{\gamma} F_{\perp}^{\text{particle}}. \tag{5.32}$$

- The force on a charge $q$ moving through the electric field **E** arising from stationary charges is simply $q\mathbf{E}$, as one would expect.

- If a charge is moving with respect to other charges that are also moving in the lab frame (say, in a wire), then the charge experiences a magnetic force. This magnetic force can alternatively be viewed as an electric force in the particle's frame. This nonzero electric field is due to the different length contractions of the positive and negative charges in the wire. In this manner we see that magnetism is a relativistic effect.

## Problems

5.1  *Field from a filament*  ∗

On a nylon filament 0.01 cm in diameter and 4 cm long there are $5.0 \cdot 10^8$ extra electrons distributed uniformly over the surface. What is the electric field strength at the surface of the filament:

(a) in the rest frame of the filament?

(b) in a frame in which the filament is moving at a speed $0.9c$ in a direction parallel to its length?

5.2  *Maximum horizontal force*  ∗∗

A charge $q_1$ is at rest at the origin, and a charge $q_2$ moves with speed $\beta c$ in the $x$ direction, along the line $z = b$. For what angle $\theta$

shown in Fig. 5.27 will the horizontal component of the force on $q_1$ be maximum? What is $\theta$ in the $\beta \approx 1$ and $\beta \approx 0$ limits?

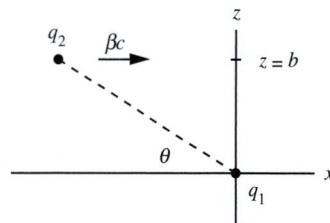

**Figure 5.27.**

5.3  *Newton's third law* **

In the laboratory frame, a proton is at rest at the origin at $t = 0$. At that instant a negative pion (charge $-e$) that has been traveling in along the $x$ axis at a speed of $0.6c$ reaches the point $x = 0.01$ cm. There are no other charges around. What is the magnitude of the force on the pion? What is the magnitude of the force on the proton? What about Newton's third law? (We're getting a little ahead of ourselves with this last question, but see if you can answer it anyway.)

5.4  *Divergence of E* **

(a) Show that the divergence of the **E** field given in Eq. (5.15) is zero (except at the origin). Work with spherical coordinates.

(b) Now show that the divergence of **E** is zero by using Cartesian coordinates and the form of **E** given in Eq. (5.13). (Careful! There's something missing from Eq. (5.13).)

5.5  *E from a line of moving charges* **

An essentially continuous stream of point charges moves with speed $v$ along the $x$ axis. The stream extends from $-\infty$ to $+\infty$. Let the charge density per unit length be $\lambda$, as measured in the lab frame. We know from using a cylindrical Gaussian surface that the electric field a distance $r$ from the $x$ axis is $E = \lambda/2\pi\epsilon_0 r$. Derive this result again by using Eq. (5.15) and integrating over all of the moving charges. You will want to use a computer or the integral table in Appendix K.

5.6  *Maximum field from a passing charge* **

In a colliding beam storage ring an antiproton going east passes a proton going west, the distance of closest approach being $10^{-10}$ m. The kinetic energy of each particle in the lab frame is 93 GeV, corresponding to $\gamma = 100$. In the rest frame of the proton, what is the maximum intensity of the electric field at the proton due to the charge on the antiproton? For about how long, approximately, does the field exceed half its maximum intensity?

5.7  *Electron in an oscilloscope* **

In the lab frame, the electric field in the region between the two plates of an oscilloscope is $E$. An electron enters this region with a relativistic velocity $v_0$ parallel to the plates. If the length of the plates is $\ell$, what are the electron's transverse momentum and transverse deflection distance upon exiting (as measured in the lab frame)? Solve this by working in the lab frame $F$, and then again by working in the inertial frame $F'$ that coincides with the electron's frame when it enters the region. (You will need to use Eqs. (G.11) and (G.12). You may assume that the transverse motion is nonrelativistic.)

5.8  *Finding the magnetic field*  **

Consider the second scenario in the example at the end of Section 5.8. Show that the total force in frame $F'$ equals the sum of the electric and magnetic forces, provided that there is a magnetic field pointing out of the page with magnitude $\gamma v E_2/c^2$.

5.9  *"Twice" the velocity*  **

Suppose that the velocity of the test charge in Fig. 5.22 is chosen so that in its frame the electrons move backward with speed $v_0$.

(a) Show that the $\beta$ associated with the test charge's velocity in the lab frame must be $\beta = 2\beta_0/(1 + \beta_0^2)$.

(b) Using length contraction, find the net charge density in the test-charge frame, and check that it agrees with Eq. (5.24).

## Exercises

5.10  *Capacitor plates in two frames*  *

A capacitor consists of two parallel rectangular plates with a vertical separation of 2 cm. The east–west dimension of the plates is 20 cm, the north–south dimension is 10 cm. The capacitor has been charged by connecting it temporarily to a battery of 300 V. What is the electric field strength between the plates? How many excess electrons are on the negative plate? Now give the following quantities as they would be measured in a frame of reference that is moving eastward, relative to the laboratory in which the plates are at rest, with speed $0.6c$: the three dimensions of the capacitor; the number of excess electrons on the negative plate; the electric field strength between the plates. Answer the same questions for a frame of reference that is moving upward with speed $0.6c$.

5.11  *Electron beam*  *

A beam of 9.5 megaelectron-volt (MeV) electrons ($\gamma \approx 20$), amounting as current to 0.05 μA, is traveling through vacuum. The transverse dimensions of the beam are less than 1 mm, and there are no positive charges in or near it.

(a) In the lab frame, what is the average distance between an electron and the next one ahead of it, measured parallel to the beam? What approximately is the average electric field strength 1 cm away from the beam?

(b) Answer the same questions for the electron rest frame.

**5.12** *Tilted sheet* **

Redo the "Tilted sheet" example in Section 5.5 in terms of a general $\gamma$ factor, to verify that Gauss's law holds for any choice of the relative speed of the two frames.

**5.13** *Adding the fields* *

A stationary proton is located on the $z$ axis at $z = a$. A negative muon is moving with speed $0.8c$ along the $x$ axis. Consider the total electric field of these two particles, in this frame, at the time when the muon passes through the origin. What are the values at that instant of $E_x$ and $E_z$ at the point $(a, 0, 0)$ on the $x$ axis?

**5.14** *Forgetting relativity* *

Given $\beta$, for what angle $\theta$ does the field in Eq. (5.15) take on the value you would obtain if you forgot about relativity, namely $Q/4\pi\epsilon_0 r'^2$? What is $\theta$ in the $\beta \approx 1$ and $\beta \approx 0$ limits?

**5.15** *Gauss's law for a moving charge* **

Verify that Gauss's law holds for the electric field in Eq. (5.15). That is, verify that the flux of the field, through a sphere centered at the charge, is $q/\epsilon_0$. Of course, we used this fact in deriving Eq. (5.15) in the first place, so we know that it must be true. But it can't hurt to double check. You'll want to use a computer or the integral table in Appendix K.

**5.16** *Cosmic rays* *

The most extremely relativistic charged particles we know about are cosmic rays which arrive from outer space. Occasionally one of these particles has so much kinetic energy that it can initiate in the atmosphere a "giant shower" of secondary particles, dissipating, in total, as much as $10^{19}$ eV of energy (more than 1 joule!). The primary particle, probably a proton, must have had $\gamma \approx 10^{10}$. How far away from such a proton would the field rise to 1 V/m as it passes? Roughly how thick is the "pancake" of field lines at that distance? (You can use the result from either Problem 5.6 or Exercise 5.21 that the angular width of the pancake is on the order of $1/\gamma$.)

**5.17** *Reversing the motion* *

A proton moves in along the $x$ axis toward the origin at a velocity $v_x = -c/2$. At the origin it collides with a massive nucleus, rebounds elastically, and moves outward on the $x$ axis with nearly the same speed. Make a sketch showing approximately how the electric field of which the proton is the source looks at an instant $10^{-10}$ s after the proton reaches the origin.

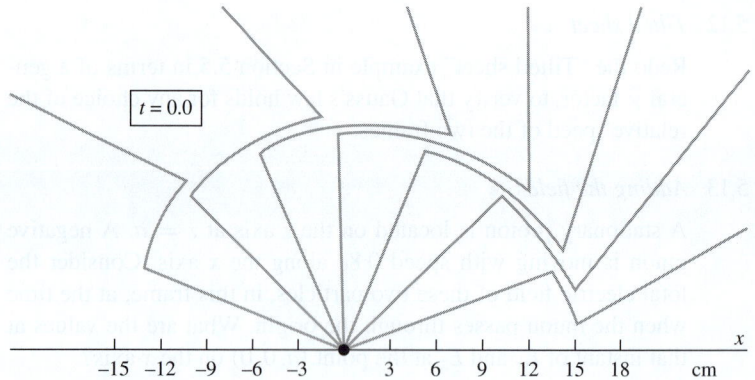

$t = 0.0$

−15 −12 −9 −6 −3    3   6   9   12  15  18   cm    $x$

**Figure 5.28.**

5.18  *A nonuniformly moving electron*  *

In Fig. 5.28 you see an electron at time $t = 0$ and the associated electric field at that instant. Distances in centimeters are given in the diagram.

(a) Describe what *has been* going on. Make your description as complete and quantitative as you can.

(b) Where was the electron at the time $t = -7.5 \cdot 10^{-10}$ s?

(c) What was the strength of the electric field at the origin at that instant?

5.19  *Colliding particles*  **

Figure 5.29 shows a highly relativistic positive particle approaching the origin from the left and a negative particle approaching with equal speed from the right. They collide at the origin at $t = 0$, find some way to dispose of their kinetic energy, and remain there as a neutral entity. What do you think the electric field looks like at some time $t > 0$? Sketch the field lines. How does the field change as time goes on?

5.20  *Relating the angles*  ***

Derive Eq. (5.16) by performing the integration to find the flux of $E$ through each of the spherical caps described in the caption of Fig. 5.19. On the inner cap the field strength is constant, and the element of surface area may be taken as $2\pi r^2 \sin\theta \, d\theta$. On the outer cap the field is described by Eq. (5.15), with the appropriate changes in symbols, and the element of surface area is $2\pi r^2 \sin\phi \, d\phi$.

Feel free to use a computer for the integration. If you want to do it by hand, a hint is to write $\sin^2\phi$ as $1 - \cos^2\phi$ and then let $x \equiv \cos\phi$. The integral you will need is

$$\int \frac{dx}{(a^2 + x^2)^{3/2}} = \frac{x}{a^2(a^2 + x^2)^{1/2}}. \tag{5.33}$$

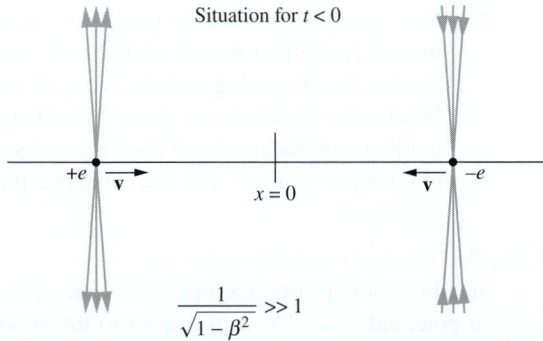

Situation for $t < 0$

$+e$   $v$        $x = 0$        $v$   $-e$

$$\frac{1}{\sqrt{1 - \beta^2}} \gg 1$$

**Figure 5.29.**

**5.21** *Half of the flux* ∗∗∗

In the field of the moving charge $Q$, given by Eq. (5.15), we want to find an angle $\delta$ such that half of the total flux from $Q$ is contained between the two conical surfaces $\theta' = \pi/2 + \delta$ and $\theta' = \pi/2 - \delta$. If you have done Exercise 5.20 you have already done most of the work. You should find that for $\gamma \gg 1$, the angle between the two cones is on the order of $1/\gamma$.

**5.22** *Electron in an oscilloscope* ∗∗

The deflection plates in a high-voltage cathode ray oscilloscope are two rectangular plates, 4 cm long and 1.5 cm wide, and spaced 0.8 cm apart. There is a difference in potential of 6000 V between the plates. An electron that has been accelerated through a potential difference of 250 kV enters this deflector from the left, moving parallel to the plates and halfway between them, initially. We want to find the position of the electron and its direction of motion when it leaves the deflecting field at the other end of the plates. We shall neglect the fringing field and assume the electric field between the plates is uniform right up to the end. The rest energy of the electron may be taken as 500 keV.

(a) First carry out the analysis in the lab frame by answering the following questions:

- What are the values of $\gamma$ and $\beta$?
- What is $p_x$ in units of $mc$?
- How long does the electron spend between the plates? (Neglect the change in horizontal velocity discussed in Exercise 5.25.)
- What is the transverse momentum component acquired, in units of $mc$?
- What is the transverse velocity at exit?
- What is the vertical position at exit?
- What is the direction of flight at exit?

(b) Now describe this whole process as it would appear in an inertial frame that moved with the electron at the moment it entered the deflecting region. What do the plates look like? What is the field between them? What happens to the electron in this coordinate system? Your main object in this exercise is to convince yourself that the two descriptions are completely consistent.

5.23 *Two views of an oscilloscope* **

In a high-voltage oscilloscope the source of electrons is a cathode at potential $-125$ kV with respect to the anode and the enclosed region beyond the anode aperture. Within this region there is a pair of parallel plates 5 cm long in the $x$ direction (the direction of the electron beam) and 8 mm apart in the $y$ direction. An electron leaves the cathode with negligible velocity, is accelerated toward the anode, and subsequently passes between the deflecting plates at a time when the potential of the lower plate is $-120$ V, that of the upper plate $+120$ V.

Fill in the blanks. Use rounded-off constants: electron rest energy $= 5 \cdot 10^5$ eV, etc. When the electron arrives at the anode, its kinetic energy is _____ eV, its total energy has increased by a factor of _____, and its velocity is _____ $c$. Its momentum is _____ kg m/s in the $x$ direction. Beyond the anode the electron passes between parallel metal plates. The field between the plates is _____ V/m; the force on the electron is _____ newtons upward. The electron spends _____ seconds between the plates and emerges, having acquired $y$ momentum of magnitude $p_y =$ _____ kg m/s. Its trajectory now slants upward at an angle $\theta =$ _____ radians.

A fast neutron that just happened to be moving along with the electron when it passed through the anode reported subsequent events as follows. "We were sitting there when this capacitor came flying at us at _____ m/s. It was _____ m long, so it surrounded us for _____ seconds. That didn't bother me, but the electric field of _____ V/m accelerated the electron so that, after the capacitor left us, the electron was moving away from me at _____ m/s, with a momentum of _____ kg m/s."

5.24 *Acquiring transverse momentum* ***

In the rest frame of a particle with charge $q_1$, another particle with charge $q_2$ is approaching, moving with velocity $v$ not small compared with $c$. If it continues to move in a straight line, it will pass a distance $b$ from the position of the first particle. It is so massive that its displacement from the straight path during the encounter is small compared with $b$. Likewise, the first particle is so massive that its displacement from its initial position while the other particle is nearby is also small compared with $b$.

(a) Show that the increment in momentum acquired by each particle as a result of the encounter is perpendicular to **v** and has magnitude $q_1 q_2 / 2\pi \epsilon_0 vb$. (Gauss's law can be useful here.)

(b) Expressed in terms of the other quantities, how large (in order of magnitude) must the masses of the particles be to justify our assumptions?

**5.25** *Decreasing velocity* **

Consider the trajectory of a charged particle that is moving with a speed $0.8c$ in the $x$ direction when it enters a large region in which there is a uniform electric field in the $y$ direction. Show that the $x$ velocity of the particle must actually *decrease*. What about the $x$ component of momentum?

**5.26** *Charges in a wire* *

In Fig. 5.22 the relative spacing of the black and gray dots was designed to be consistent with $\gamma = 1.2$ and $\beta_0 = 0.8$. Calculate $\beta_0'$. Find the value, as a fraction of $\lambda_0$, of the net charge density $\lambda'$ in the test-charge frame.

**5.27** *Equal velocities* *

Suppose that the velocity of the test charge in Fig. 5.22 is made equal to that of the electrons, $v_0$. What would then be the linear densities of positive charge, and of negative charge, in the test-charge frame?

**5.28** *Stationary rod and moving charge* *

A charge $q$ moves with speed $v$ parallel to a long rod with linear charge density $\lambda$, as shown in Fig. 5.30. The rod is at rest. If the charge $q$ is a distance $r$ from the rod, the force on it is simply $F = qE = q\lambda / 2\pi r\epsilon_0$.

Now consider the setup in the frame that moves along with the charge $q$. What is the force on the charge $q$ in this new frame? Solve this by:

(a) transforming the force from the old frame to the new frame, without caring about what causes the force in the new frame;

(b) calculating the electric force in the new frame.

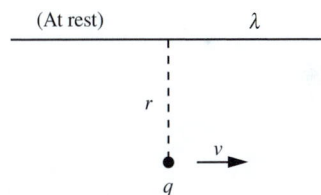

(At rest)     $\lambda$

$r$

$q$   $v$

**Figure 5.30.**

**5.29** *Protons moving in opposite directions* ***

Two protons are moving antiparallel to one another, along lines separated by a distance $r$, with the same speed $\beta c$ in the lab frame. Consider the moment when they have the same horizontal position and are a distance $r$ apart, as shown in Fig. 5.31. According to Eq. (5.15), at the position of each of the protons, the electric field $E$ caused by the other, as measured in the lab frame, is $\gamma e / 4\pi \epsilon_0 r^2$. But the force on each proton measured in the lab frame is *not* $\gamma e^2 / 4\pi \epsilon_0 r^2$. Verify this by finding the force on one of the protons in its own rest frame and transforming that force back to the

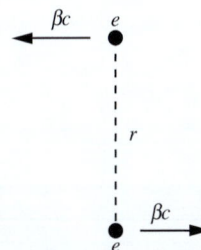

$\beta c$   $e$

$r$

$\beta c$

$e$

**Figure 5.31.**

lab frame. (You will need to use the velocity addition formula to find the speed of one proton as viewed by the other). Show that the discrepancy can be accounted for if each proton is subject to a magnetic field that points in the appropriate direction and that has a magnitude $\beta/c = v/c^2$ times the magnitude of the electric field, at the instant shown.

5.30 *Transformations of $\lambda$ and $I$* ***

Consider a composite line charge consisting of several kinds of carriers, each with its own velocity. For each kind, labeled by $k$, the linear density of charge measured in frame $F$ is $\lambda_k$, and the velocity is $\beta_k c$ parallel to the line. The contribution of these carriers to the current in $F$ is then $I_k = \lambda_k \beta_k c$. How much do these $k$-type carriers contribute to the charge and current in a frame $F'$ that is moving parallel to the line at velocity $-\beta c$ with respect to $F$? By following the steps we took in the transformations in Fig. 5.22, you should be able to show that

$$\lambda'_k = \gamma\left(\lambda_k + \frac{\beta I_k}{c}\right), \qquad I'_k = \gamma(I_k + \beta c \lambda_k). \qquad (5.34)$$

If each component of the linear charge density and current transforms in this way, then so must the total $\lambda$ and $I$:

$$\lambda' = \gamma\left(\lambda + \frac{\beta I}{c}\right), \qquad I' = \gamma(I + \beta c \lambda). \qquad (5.35)$$

You have now derived the Lorentz transformation to a parallel-moving frame for *any* line charge and current, whatever its composition.

5.31 *Moving perpendicular to a wire* ****

At the end of Section 5.9 we discussed the case where a charge $q$ moves perpendicular to a wire. Figures 5.25 and 5.26 show qualitatively why there is a nonzero force on the charge, pointing in the positive $x$ direction. Carry out the calculation to show that the force at a distance $\ell$ from the wire equals $qvI/2\pi\epsilon_0\ell c^2$. That is, use Eq. (5.15) to calculate the force on the charge in its own frame, and then divide by $\gamma$ to transform back to the lab frame.

*Notes:* You can use the fact that in the charge $q$'s frame, the speed of the electrons in the $x$ direction is $v_0/\gamma$ (this comes from the transverse-velocity-addition formula). Remember that the $\beta$ in Eq. (5.15) is the velocity of the electrons in the charge's frame, and this velocity has two components. Be careful with the transverse distance involved. There are many things to keep track of in this problem, but the integration itself is easy if you use a computer (or Appendix K).

# 6

# The magnetic field

**Overview** Having shown in Chapter 5 that the magnetic force must exist, we will now study the various properties of the magnetic field and show how it can be calculated for an arbitrary (steady) current distribution. The *Lorentz force* gives the total force on a charged particle as $\mathbf{F} = q\mathbf{E} + q\mathbf{v} \times \mathbf{B}$. The results from the previous chapter give us the form of the magnetic field due to a long straight wire. This form leads to *Ampère's law*, which relates the line integral of the magnetic field to the current enclosed by the integration loop. It turns out that Ampère's law holds for a wire of any shape. When supplemented with a term involving changing electric fields, this law becomes one of Maxwell's equations (as we will see in Chapter 9). The sources of magnetic fields are currents, in contrast with the sources of electric fields, which are charges; there are no isolated magnetic charges, or *monopoles*. This statement is another of Maxwell's equations.

As in the electric case, the magnetic field can be obtained from a potential, but it is now a *vector potential*; its curl gives the magnetic field. The *Biot–Savart law* allows us to calculate (in principle) the magnetic field due to any steady current distribution. One distribution that comes up often is that of a *solenoid* (a coil of wire), whose field is (essentially) constant inside and zero outside. This field is consistent with an Ampère's-law calculation of the discontinuity of $\mathbf{B}$ across a sheet of current. By considering various special cases, we derive the *Lorentz transformations* of the electric and magnetic fields. The electric (or magnetic) field in one frame depends on *both* the electric and magnetic fields in another frame. The *Hall effect* arises from the $q\mathbf{v} \times \mathbf{B}$ part of the

Lorentz force. This effect allows us, for the first time, to determine the sign of the charge carriers in a current.

## 6.1 Definition of the magnetic field

A charge that is moving parallel to a current of other charges experiences a force perpendicular to its own velocity. We can see it happening in the deflection of the electron beam in Fig. 5.3. We discovered in Section 5.9 that this is consistent with – indeed, is required by – Coulomb's law combined with charge invariance and special relativity. And we found that a force perpendicular to the charged particle's velocity also arises in motion at right angles to the current-carrying wire. For a given current, the magnitude of the force, which we calculated for the particular case in Fig. 5.22(a), is proportional to the product of the particle's charge $q$ and its speed $v$ in our frame. Just as we defined the electric field **E** as the vector force on unit charge at rest, so we can define another field **B** by the *velocity-dependent* part of the force that acts on a charge in motion. The defining relation was introduced at the beginning of Chapter 5. Let us state it again more carefully.

At some instant $t$ a particle of charge $q$ passes the point $(x, y, z)$ in our frame, moving with velocity **v**. At that moment the force on the particle (its rate of change of momentum) is **F**. The electric field at that time and place is known to be **E**. Then the magnetic field at that time and place is defined as the vector **B** that satisfies the following vector equation (for any value of **v**):

$$\mathbf{F} = q\mathbf{E} + q\mathbf{v} \times \mathbf{B} \tag{6.1}$$

This force **F** is called the *Lorentz force*. Of course, **F** here includes only the charge-dependent force and not, for instance, the weight of the particle carrying the charge. A vector **B** satisfying Eq. (6.1) always exists. Given the values of **E** and **B** in some region, we can with Eq. (6.1) predict the force on any particle moving through that region with any velocity. For fields that vary in time and space, Eq. (6.1) is to be understood as a local relation among the instantaneous values of **F**, **E**, **v**, and **B**. Of course, all four of these quantities must be measured in the same inertial frame.

In the case of our "test charge" in the lab frame of Fig. 5.22(a), the electric field **E** was zero. With the charge $q$ moving in the positive $x$ direction, $\mathbf{v} = \hat{\mathbf{x}}v$, we found in Eq. (5.28) that the force on it was in the negative $y$ direction, with magnitude $Iqv/2\pi\epsilon_0 rc^2$:

$$\mathbf{F} = -\hat{\mathbf{y}}\frac{Iqv}{2\pi\epsilon_0 rc^2}. \tag{6.2}$$

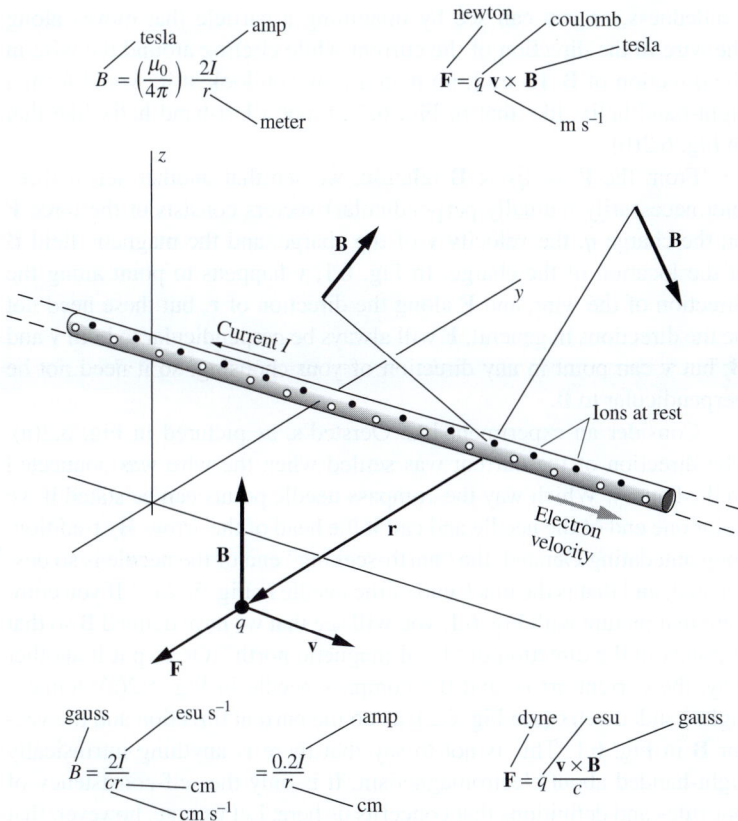

**Figure 6.1.**
The magnetic field of a current in a long straight wire and the force on a charged particle moving through that field.

In this case the magnetic field must be

$$\boxed{\mathbf{B} = \hat{\mathbf{z}}\frac{I}{2\pi\epsilon_0 rc^2}} \qquad (6.3)$$

for then Eq. (6.1) becomes

$$\mathbf{F} = q\mathbf{v} \times \mathbf{B} = (\hat{\mathbf{x}} \times \hat{\mathbf{z}})(qv)\left(\frac{I}{2\pi\epsilon_0 rc^2}\right) = -\hat{\mathbf{y}}\frac{Iqv}{2\pi\epsilon_0 rc^2}, \qquad (6.4)$$

in agreement with Eq. (6.2).

The relation of **B** to **r** and to the current $I$ is shown in Fig. 6.1. Three mutually perpendicular directions are involved: the direction of **B** at the point of interest, the direction of a vector **r** from the wire to that point, and the direction of current flow in the wire. Here questions of *handedness* arise for the first time in our study. Having adopted Eq. (6.1) as the definition of **B** and agreed on the conventional rule for the vector product, that is, $\hat{\mathbf{x}} \times \hat{\mathbf{y}} = \hat{\mathbf{z}}$, etc., in coordinates like those of Fig. 6.1, we have determined the direction of **B**. That relation has a

(a)

RH

(b)

LH

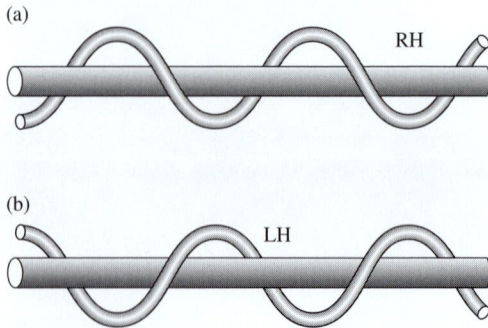

**Figure 6.2.**
A reminder. The helix in (a) is called a
right-handed helix, that in (b) a left-handed helix.

handedness, as you can see by imagining a particle that moves along the wire in the direction of the current while circling around the wire in the direction of **B**. Its trail, no matter how you look at it, would form a right-hand helix, like that in Fig. 6.2(a), not a left-hand helix like that in Fig. 6.2(b).

From the $\mathbf{F} = q\mathbf{v} \times \mathbf{B}$ relation, we see that another set of three (not necessarily mutually perpendicular) vectors consists of the force $\mathbf{F}$ on the charge $q$, the velocity $\mathbf{v}$ of the charge, and the magnetic field $\mathbf{B}$ at the location of the charge. In Fig. 6.1, $\mathbf{v}$ happens to point along the direction of the wire, and $\mathbf{F}$ along the direction of $\mathbf{r}$, but these need not be the directions in general; $\mathbf{F}$ will always be perpendicular to both $\mathbf{v}$ and $\mathbf{B}$, but $\mathbf{v}$ can point in any direction of your choosing, so it need not be perpendicular to $\mathbf{B}$.

Consider an experiment like Oersted's, as pictured in Fig. 5.2(a). The direction of the current was settled when the wire was connected to the battery. Which way the compass needle points can be stated if we color one end of the needle and call it the head of the arrow. By tradition, long antedating Oersted, the "north-seeking" end of the needle is so designated, and that is the black end of the needle in Fig. 5.2(a).[1] If you compare that picture with Fig. 6.1, you will see that we have defined **B** so that it points in the direction of "local magnetic north." Or, to put it another way, the current arrow and the compass needle in Fig. 5.2(a) define a right-handed helix (see Fig. 6.2), as do the current direction and the vector **B** in Fig. 6.1. This is not to say that there is anything intrinsically right-handed about electromagnetism. It is only the self-consistency of our rules and definitions that concerns us here. Let us note, however, that a question of handedness *could never arise* in electrostatics. In this sense the vector **B** differs in character from the vector **E**. In the same way, a vector representing an angular velocity, in mechanics, differs from a vector representing a linear velocity.

The SI units of **B** can be determined from Eq. (6.1). In a magnetic field of unit strength, a charge of one coulomb moving with a velocity of one meter/second perpendicular to the field experiences a force of one newton. The unit of **B** so defined is called the *tesla*:

$$1 \text{ tesla} = 1 \frac{\text{newton}}{\text{coulomb} \cdot \text{meter/second}} = 1 \frac{\text{newton}}{\text{amp} \cdot \text{meter}}. \qquad (6.5)$$

In terms of other units, 1 tesla equals $1 \text{ kg C}^{-1} \text{ s}^{-1}$. In SI units, the relation between field and current in Eq. (6.3) is commonly written as

$$\boxed{\mathbf{B} = \hat{\mathbf{z}} \frac{\mu_0 I}{2\pi r}} \qquad (6.6)$$

[1] We now know that the earth's magnetic field has reversed many times in geologic history. See Problem 7.19 and the reference there given.

where $\mathbf{B}$ is in teslas, $I$ is in amps, and $r$ is in meters. The constant $\mu_0$, like the constant $\epsilon_0$ we met in electrostatics, is a fundamental constant in the SI unit system. Its value is defined to be exactly

$$\boxed{\mu_0 \equiv 4\pi \cdot 10^{-7} \, \frac{\text{kg m}}{\text{C}^2}} \tag{6.7}$$

Of course, if Eq. (6.6) is to agree with Eq. (6.3), we must have

$$\mu_0 = \frac{1}{\epsilon_0 c^2} \implies \boxed{c^2 = \frac{1}{\mu_0 \epsilon_0}} \tag{6.8}$$

With $\epsilon_0$ given in Eq. (1.3), and $c = 2.998 \cdot 10^8$ m/s, you can quickly check that this relation does indeed hold.

REMARK: Given that we already found the $\mathbf{B}$ field due to a current-carrying wire in Eq. (6.3), you might wonder what the point is of rewriting $\mathbf{B}$ in terms of the newly introduced constant $\mu_0$ in Eq. (6.6). The answer is that $\mu_0$ is a product of the historical development of magnetism, which should be contrasted with the special-relativistic development we followed in Chapter 5. The connection between electric and magnetic effects was certainly observed long before the formulation of special relativity in 1905. In particular, as we learned in Section 5.1, Oersted discovered in 1820 that a current-carrying wire produces a magnetic field. And $\mu_0$ was eventually introduced as the constant of proportionality in Eq. (6.6). (Or, more accurately, $\mu_0$ was assigned a given value, and then Eq. (6.6) was used to define the unit of current.) But even with the observed connection between electricity and magnetism, in the mid nineteenth century there was no obvious relation between the $\mu_0$ in the expression for $\mathbf{B}$ and the $\epsilon_0$ in the expression for $\mathbf{E}$. They were two separate constants in two separate theories. But two developments changed this.

First, in 1861 Maxwell wrote down his set of equations that govern all of electromagnetism. He then used these equations to show that electromagnetic waves exist and travel with speed $1/\sqrt{\mu_0 \epsilon_0} \approx 3 \cdot 10^8$ m/s. (We'll study Maxwell's equations and electromagnetic waves in Chapter 9.) This strongly suggested that light is an electromagnetic wave, a fact that was demonstrated experimentally by Hertz in 1888. Therefore, $c = 1/\sqrt{\mu_0 \epsilon_0}$, and hence $\mu_0 = 1/\epsilon_0 c^2$. This line of reasoning shows that the speed of light $c$ is determined by the two constants $\epsilon_0$ and $\mu_0$.

The second development was Einstein's formulation of the special theory of relativity in 1905. Relativity was the basis of our reasoning in Chapter 5 (the main ingredients of which were length contraction and the relativistic velocity-addition formula), which led to the expression for the magnetic field in Eq. (6.3). A comparison of this equation with the historical expression in Eq. (6.6) yields $\mu_0 = 1/\epsilon_0 c^2$. This line of reasoning shows that $\mu_0$ is determined by the two constants $\epsilon_0$ and $c$. Of course, having proceeded the way we did in Chapter 5, there is no need to introduce the constant $\mu_0$ in Eq. (6.6) when we already have Eq. (6.3). Nevertheless, the convention in SI units is to write $\mathbf{B}$ in the form given in Eq. (6.6). If you wish, you can think of $\mu_0$ simply as a convenient shorthand for the more cumbersome expression $1/\epsilon_0 c^2$.

Comparing the previous two paragraphs, it is unclear which derivation of $\mu_0 = 1/\epsilon_0 c^2$ is "better." Is it preferable to take $\epsilon_0$ and $\mu_0$ as the fundamental constants and then derive, with Maxwell's help, the value of $c$, or to take $\epsilon_0$ and

$c$ as the fundamental constants and derive, with Einstein's help, the value of $\mu_0$? The former derivation has the advantage of explaining why $c$ takes on the value $2.998 \cdot 10^8$ m/s, while the latter has the advantage of explaining how magnetic forces arise from electric forces. In the end, it's a matter of opinion, based on what information you want to start with.

In Gaussian units, Eq. (6.1) takes the slightly different form

$$\mathbf{F} = q\mathbf{E} + \frac{q}{c}\mathbf{v} \times \mathbf{B}. \tag{6.9}$$

Note that $\mathbf{B}$ now has the same dimensions as $\mathbf{E}$, the factor $\mathbf{v}/c$ being dimensionless. With force $F$ in dynes and charge $q$ in esu, the unit of magnetic field strength is the dyne/esu. This unit has a name, the *gauss*. There is no special name for the unit dyne/esu when it is used as a unit of electric field strength. It is the same as 1 statvolt/cm, which is the term normally used for unit electric field strength in the Gaussian system. In Gaussian units, the equation analogous to Eq. (6.3) is

$$\mathbf{B} = \hat{\mathbf{z}}\frac{2I}{rc}. \tag{6.10}$$

If you repeat the reasoning of Chapter 5, you will see that this $\mathbf{B}$ is obtained basically by replacing $\epsilon_0$ by $1/4\pi$ and erasing one of the factors of $c$ in Eq. (6.3). $\mathbf{B}$ is in gauss if $I$ is in esu/s, $r$ is in cm, and $c$ is in cm/s.

**Example (Relation between 1 tesla and 1 gauss)**    Show that 1 tesla is equivalent to exactly $10^4$ gauss.

Solution    Consider a setup where a charge of 1 C travels at 1 m/s in a direction perpendicular to a magnetic field with strength 1 tesla. Equations (6.1) and (6.5) tell us that the charge experiences a force of 1 newton. Let us express this fact in terms of the Gaussian force relation in Eq. (6.9). We know that $1\,\text{N} = 10^5$ dyne and $1\,\text{C} = 3 \cdot 10^9$ esu (this "3" isn't actually a 3; see the discussion below). If we let 1 tesla $= n$ gauss, with $n$ to be determined, then the way that Eq. (6.9) describes the given situation is as follows:

$$10^5 \text{ dyne} = \frac{3 \cdot 10^9 \text{ esu}}{3 \cdot 10^{10} \text{ cm/s}} \left(100\,\frac{\text{cm}}{\text{s}}\right)(n \text{ gauss}). \tag{6.11}$$

Since 1 gauss equals 1 dyne/esu, all the units cancel, and we end up with $n = 10^4$, as desired.

Now, the two 3's in Eq. (6.11) are actually 2.998's. This is clear in the denominator because the 3 comes from the factor of $c$. To see why it is the case in the numerator, recall the example in Section 1.4 where we showed that $1\,\text{C} = 3 \cdot 10^9$ esu. If you redo that example and keep things in terms of the constant $k$ given in Eqs. (1.2) and (1.3), you will find that the number $3 \cdot 10^9$ is actually $\sqrt{10^9 k}$ (ignoring the units of $k$). But in view of the definition of $\mu_0$ in Eq. (6.7), the $k = 1/4\pi\epsilon_0$ expression in Eq. (1.3) can be written as $k = 1/(10^7\mu_0\epsilon_0)$. And we know from above that $1/\mu_0\epsilon_0 = c^2$, hence $k = 10^{-7}c^2$ (ignoring the units).

So the number $3 \cdot 10^9$ is really $\sqrt{10^9 k} = \sqrt{10^2 c^2} = 10c$ (ignoring the units), or $2.998 \cdot 10^9$. Since both of the 3's in Eq. (6.11) are modified in the same way, the $n = 10^4$ result is therefore still exact.

Let us use Eqs. (6.1) and (6.6) to calculate the magnetic force between parallel wires carrying current. Let $r$ be the distance between the wires, and let $I_1$ and $I_2$ be the currents which we assume are flowing in the same direction, as shown in Fig. 6.3. The wires are assumed to be infinitely long – a fair assumption in a practical case if they are very long compared with the distance $r$ between them. We want to predict the force that acts on some finite length $l$ of one of the wires, due to the entirety of the other wire. The current in wire 1 causes a magnetic field of strength

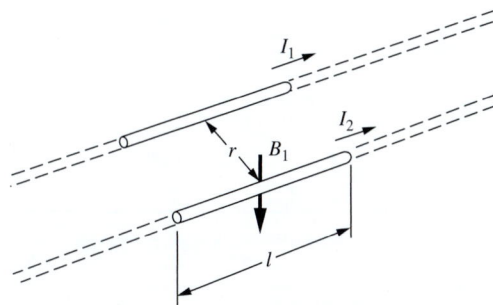

$$B_1 = \frac{\mu_0 I_1}{2\pi r} \qquad (6.12)$$

**Figure 6.3.**
Current $I_1$ produces magnetic field $B_1$ at conductor 2. The force on a length $l$ of conductor 2 is given by Eq. (6.15).

at the location of wire 2. Within wire 2 there are $n_2$ moving charges per meter length of wire, each with charge $q_2$ and speed $v_2$. They constitute the current $I_2$:

$$I_2 = n_2 q_2 v_2. \qquad (6.13)$$

According to Eq. (6.1), the force on each charge is $q_2 v_2 B_1$.[2] The force on each meter length of wire is therefore $n_2 q_2 v_2 B_1$, or simply $I_2 B_1$. The force on a length $l$ of wire 2 is then

$$\boxed{F = I_2 B_1 l} \qquad (6.14)$$

Using the $B_1$ from Eq. (6.12), this becomes

$$\boxed{F = \frac{\mu_0 I_1 I_2 l}{2\pi r}} \qquad (6.15)$$

Here $F$ is in newtons, and $I_1$ and $I_2$ are in amps. As the factor $l/r$ that appears both in Eq. (6.15) and below in Eq. (6.16) is dimensionless, $l$ and $r$ could be in any units.[3]

---

[2] $B_1$ is the field *inside* wire 2, caused by the current in wire 1. When we study magnetic fields inside matter in Chapter 11, we will find that most conductors, including copper and aluminum, but *not* including iron, have very little influence on a magnetic field. For the present, let us agree to avoid things like iron and other ferromagnetic materials. Then we can safely assume that the magnetic field inside the wire is practically what it would be in vacuum with the same currents flowing.

[3] Equation (6.15) has usually been regarded as the primary definition of the ampere in the SI system, $\mu_0$ being *assigned* the value $4\pi \cdot 10^{-7}$. That is to say, one ampere is the current that, flowing in each of two infinitely long parallel wires a distance $r$ apart, will cause a force of exactly $2 \cdot 10^{-7}$ newton on a length $l = r$ of one of the wires. The other SI electrical units are then defined in terms of the ampere. Thus a coulomb is one ampere-second, a volt is one joule/coulomb, and an ohm is one volt/ampere.

The same exercise carried out in Gaussian units, with Eqs. (6.9) and (6.10), will lead to

$$F = \frac{2I_1 I_2 l}{c^2 r}. \qquad (6.16)$$

Equation (6.15) is symmetric in the labels 1 and 2, so the force on an equal length of wire 1 caused by the field of wire 2 must be given by the same formula. We have not bothered to keep track of signs because we know already that currents in the same direction attract one another.

More generally, we can calculate the force on a small piece of current-carrying wire that sits in a magnetic field $\mathbf{B}$. Let the length of the small piece be $dl$, the linear charge density of the moving charges be $\lambda$, and the speed of these charges be $v$. Then the amount of moving charge in the piece is $dq = \lambda \, dl$, and the current is $I = \lambda v$ (in agreement with Eq. (6.13) since $\lambda = nq$). Equation (6.1) tells us that the magnetic force on the piece is

$$d\mathbf{F} = dq \, \mathbf{v} \times \mathbf{B} = (\lambda \, dl)(v\hat{\mathbf{v}}) \times \mathbf{B} = (\lambda v)(dl \, \hat{\mathbf{v}}) \times \mathbf{B}$$

$$\implies \boxed{d\mathbf{F} = I \, d\mathbf{l} \times \mathbf{B}} \qquad (6.17)$$

The vector $d\mathbf{l}$ gives both the magnitude and direction of the small piece. The $F = I_2 B_1 l$ result in Eq. (6.14) is a special case of this result.

---

**Example (Copper wire)**  Let's apply Eqs. (6.13) and (6.15) to the pair of wires in Fig. 6.4(a). They are copper wires 1 mm in diameter and 5 cm apart. In copper the number of conduction electrons per cubic meter, already mentioned in Chapter 4, is $8.45 \cdot 10^{28}$, so the number of electrons per unit length of wire is $n = (\pi/4)(10^{-3} \, \text{m})^2 (8.45 \cdot 10^{28} \, \text{m}^{-3}) = 6.6 \cdot 10^{22} \, \text{m}^{-1}$. Suppose their mean drift velocity $\bar{v}$ is $0.3 \, \text{cm/s} = 0.003 \, \text{m/s}$. (Of course their random speeds are vastly greater.) The current in each wire is then

$$I = nq\bar{v} = (6.6 \cdot 10^{22} \, \text{m}^{-1})(1.6 \cdot 10^{-19} \, \text{C})(0.003 \, \text{m/s}) \approx 32 \, \text{C/s}.$$

The attractive force on a 20 cm length of wire is

$$F = \frac{\mu_0 I^2 l}{2\pi r} = \frac{(4\pi \cdot 10^{-7} \, \text{kg m/C}^2)(32 \, \text{C/s})^2 (0.2 \, \text{m})}{2\pi (0.05 \, \text{m})} \approx 8 \cdot 10^{-4} \, \text{N}. \quad (6.18)$$

This result of $8 \cdot 10^{-4}$ N is not an enormous force, but it is easily measurable. Figure 6.4(b) shows how the force on a given length of conductor could be observed.

---

Recall that the $\mu_0$ in Eq. (6.18) can alternatively be written as $1/\epsilon_0 c^2$. The $c^2$ in the denominator reminds us that, as we discovered in Chapter 5, the magnetic force is a relativistic effect, strictly proportional to $v^2/c^2$ and traceable to a Lorentz contraction. And with the $v$ in the above example less than the speed of a healthy ant, it is causing a quite respectable force! The explanation is the immense amount of negative charge the conduction electrons represent, charge that ordinarily is so precisely neutralized by positive charge that we hardly notice it. To appreciate that,

(a)

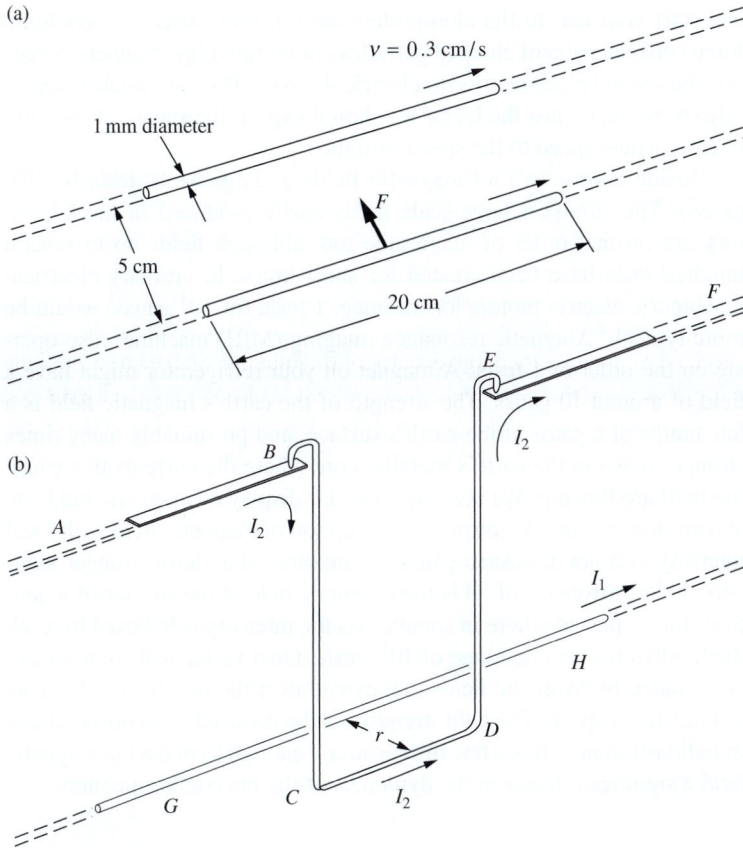

$v = 0.3\,\text{cm/s}$

1 mm diameter

$F$

5 cm

20 cm

$F$

$E$

$I_2$

(b)

$B$

$A$

$I_2$

$I_1$

$H$

$r$

$D$

$I_2$

$G$                $C$                $I_2$

**Figure 6.4.**
(a) The current in each copper wire is 32 amps, and the force $F$ on the 20 cm length of conductor is $8 \cdot 10^{-4}$ newtons. (b) One way to measure the force on a length of conductor. The section $BCDE$ swings like a pendulum below the conducting pivots. The force on the length $CD$ due to the field of the straight conductor $GH$ is the only force deflecting the pendulum from the vertical.

consider the force with which our wires in Fig. 6.4 would repel one another if the charge of the $6.6 \cdot 10^{22}$ electrons per meter were *not* neutralized at all. As an exercise you can show that the force is just $c^2/v^2$ times the force we calculated above, or roughly $4 \times 10^{15}$ tons per meter of wire. So full of electricity is all matter! If all the electrons in just one raindrop were removed from the earth, the whole earth's potential would rise by several million volts.

Matter in bulk, from raindrops to planets, is almost exactly neutral. You will find that any piece of it much larger than a molecule contains nearly the same number of electrons as protons. If it didn't, the resulting electric field would be so strong that the excess charge would be irresistibly blown away. That would happen to electrons in our copper wire even if the excess of negative charge were no more than $10^{-10}$ of the total. A magnetic field, on the other hand, cannot destroy itself in this way. No matter how strong it may be, it exerts no force on a stationary charge. That is why forces that arise from the *motion* of electric charges can dominate the scene. The second term on the right in Eq. (6.1) can be much larger than the first. Thanks to that second term, an electric motor can start your car. In the atomic domain, however, where the coulomb force between pairs of charged particles comes into play, magnetic forces do take second place relative to electrical forces. They are weaker, generally speaking, by just the factor we should expect, the square of the ratio of the particle speed to the speed of light.

Inside atoms we find magnetic fields as large as 10 tesla (or $10^5$ gauss). The strongest large-scale fields easily produced in the laboratory are on that order of magnitude too, although fields up to several hundred tesla have been created for short times. In ordinary electrical machinery, electric motors for instance, 1 tesla (or $10^4$ gauss) would be more typical.[4] Magnetic resonance imaging (MRI) machines also operate on the order of 1 tesla. A magnet on your refrigerator might have a field of around 10 gauss. The strength of the earth's magnetic field is a few tenths of a gauss at the earth's surface, and presumably many times stronger down in the earth's metallic core where the currents that cause the field are flowing. We see a spectacular display of magnetic fields on and around the sun. A *sunspot* is an eruption of magnetic field with local intensity of a few thousand gauss. Some other stars have stronger magnetic fields. Strongest of all is the magnetic field at the surface of a neutron star, or pulsar, where in some cases the intensity is believed to reach the hardly conceivable range of $10^{10}$ tesla. On a vaster scale, our galaxy is pervaded by magnetic fields that extend over thousands of light years of interstellar space. The field strength can be deduced from observations in radioastronomy. It is a few microgauss – enough to make the magnetic field a significant factor in the dynamics of the interstellar medium.

## 6.2 Some properties of the magnetic field

The magnetic field, like the electric field, is a device for describing how charged particles interact with one another. If we say that the magnetic field at the point (4.5, 3.2, 6.0) at 12:00 noon points horizontally in the negative *y* direction and has a magnitude of 5 gauss, we are making a statement about the acceleration a moving charged particle at that point in space-time would exhibit. The remarkable thing is that a statement of this form, giving simply a vector quantity **B**, says all there is to say. With it one can predict uniquely the velocity-dependent part of the force on *any* charged particle moving with *any* velocity. It makes unnecessary any further description of the other charged particles that are the sources of

---

[4] Nikola Tesla (1856–1943), the inventor and electrical engineer for whom the SI unit was named, invented the alternating-current induction motor and other useful electromagnetic devices. Gauss's work in magnetism was concerned mainly with the earth's magnetic field. Perhaps this will help you to remember which is the larger unit. For small magnetic fields, it is generally more convenient to work with gauss than with tesla, even though the gauss technically isn't part of the SI system of units. This shouldn't cause any confusion; you can quickly convert to tesla by dividing by (exactly) $10^4$. If you're wary about leaving the familiar ground of SI units, feel free to think of a gauss as a deci-milli-tesla.

the field. In other words, if two quite different systems of moving charges happen to produce the same **E** and **B** at a particular point, the behavior of any test particle at the point would be exactly the same in the two systems. It is for this reason that the concept of field, as an intermediary in the interaction of particles, is useful. And it is for this reason that we think of the field as an independent entity.

Is the field more, or less, real than the particles whose interaction, as seen from our present point of view, it was invented to describe? That is a deep question which we would do well to set aside for the time being. People to whom the electric and magnetic fields were vividly real – Faraday and Maxwell, to name two – were led thereby to new insights and great discoveries. Let's view the magnetic field as concretely as they did and learn some of its properties.

So far we have studied only the magnetic field of a straight wire or filament of steady current. The field direction, we found, is everywhere perpendicular to the plane containing the filament and the point where the field is observed. The magnitude of the field is proportional to $1/r$. The field lines are circles surrounding the filament, as shown in Fig. 6.5. The sense of direction of **B** is determined by our previously adopted convention about the vector cross-product, by the (arbitrary) decision to write the second term in Eq. (6.1) as $q\mathbf{v} \times \mathbf{B}$, and by the *physical fact* that a positive charge moving in the direction of a positive current is attracted to it rather than repelled. These are all consistent if we relate the direction of **B** to the direction of the current that is its source in the manner shown in Fig. 6.5. Looking in the direction of positive current, we see the **B** lines curling clockwise. Or you may prefer to remember it as a right-hand-thread relation. Point your right thumb in the direction of the current and your fingers will curl in the direction of **B**.

Let's look at the line integral of **B** around a closed path in this field. (Remember that a similar inquiry in the case of the electric field of a point charge led us to a simple and fundamental property of all electrostatic fields, that $\int \mathbf{E} \cdot d\mathbf{s} = 0$ around a closed path, or equivalently that curl **E** = 0.) Consider first the path *ABCD* in Fig. 6.6(a). This lies in a plane perpendicular to the wire; in fact, we need only work in this plane, for **B** has no component parallel to the wire. The line integral of **B** around the path shown is zero, for the following reason. Paths *BC* and *DA* are perpendicular to **B** and contribute nothing. Along *AB*, **B** is stronger in the ratio $r_2/r_1$ than it is along *CD*; but *CD* is longer than *AB* by the same factor, for these two arcs subtend the same angle at the wire. So the two arcs give equal and opposite contributions, and the whole integral is zero.

It follows that the line integral is also zero on any path that can be constructed out of radial segments and arcs, such as the path in Fig. 6.6(b). From this it is a short step to conclude that the line integral is zero around *any* path that does not enclose the wire. To smooth out the corners we would only need to show that the integral around a

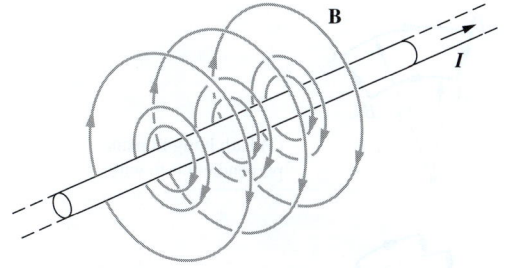

**Figure 6.5.**
Magnetic field lines around a straight wire carrying current.

(a) Path lying in plane perpendicular to wire

(b) Path constructed of radial segments and arcs

(c) Path that does not enclose the wire

(d) Circular path enclosing wire

(e) Crooked path enclosing wire

(f) Circular and crooked path *not* enclosing wire

(g) Loop of $N$ turns enclosing wire

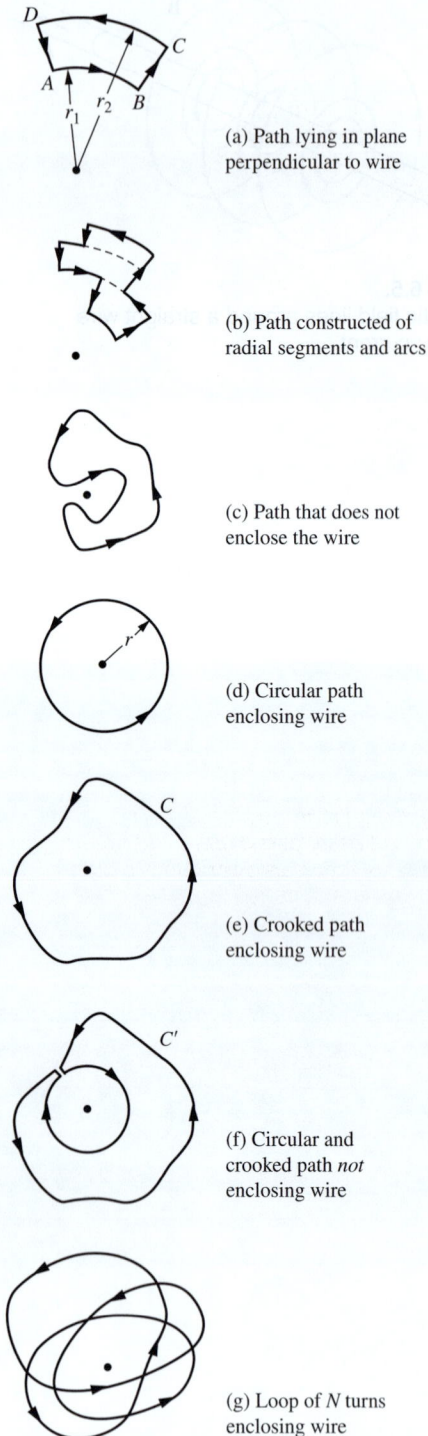

little triangular path vanishes. The same step was involved in the case of the electric field.

A path that does not enclose the wire is one like the path in Fig. 6.6(c), which, if it were made of string, could be pulled free. The line integral around any such path is zero.

Now consider a circular path that encloses the wire, as in Fig. 6.6(d). Here the circumference is $2\pi r$, and the field is $\mu_0 I/2\pi r$ and everywhere parallel to the path, so the value of the line integral around this particular path is $(2\pi r)(\mu_0 I/2\pi r)$, or $\mu_0 I$. We now claim that *any* path looping once around the wire must give the same value. Consider, for instance, the crooked path $C$ in Fig. 6.6(e). Let us construct the path $C'$ in Fig. 6.6(f) made of a path like $C$ and a circular path, but *not* enclosing the wire. The line integral around $C'$ must be zero, and therefore the integral around $C$ must be the negative of the integral around the circle, which we have already evaluated as $\mu_0 I$ in magnitude. The sign will depend in an obvious way on the sense of traversal of the path. Our general conclusion is:

$$\int \mathbf{B} \cdot d\mathbf{s} = \mu_0 \times (\text{current enclosed by path}) \qquad \text{(Ampère's law).}$$

$$(6.19)$$

This is known as *Ampère's law*. It is valid for *steady* currents. In the Gaussian analog of this expression, the $\mu_0$ is replaced with $4\pi/c$, which quickly follows from a comparison of Eqs. (6.6) and (6.10).

Equation (6.19) holds when the path loops the current filament once. Obviously a path that loops it $N$ times, like the one in Fig. 6.6(g), will just give $N$ times as big a result for the line integral.

The magnetic field, as we have emphasized before, depends only on the rate of charge transport, the number of units of charge passing a given point in the circuit, per second. Figure 6.7 shows a circuit with a current of 5 milliamperes. The average velocity of the charge carriers ranges from $10^{-6}$ m/s in one part of the circuit to 0.8 times the speed of light in another. The line integral of $\mathbf{B}$ over a closed path has the same value around every part of this circuit, namely

$$\int \mathbf{B} \cdot d\mathbf{s} = \mu_0 I = \left(4\pi \cdot 10^{-7} \frac{\text{kg m}}{\text{C}^2}\right) \left(0.005 \frac{\text{C}}{\text{s}}\right) = 6.3 \cdot 10^{-9} \frac{\text{kg m}}{\text{C s}}.$$

$$(6.20)$$

You can check that these units are the same as tesla-meter, which they must be, in view of the left-hand side of this equation.

**Figure 6.6.**
The line integral of the magnetic field $\mathbf{B}$ over any closed path depends only on the current enclosed.

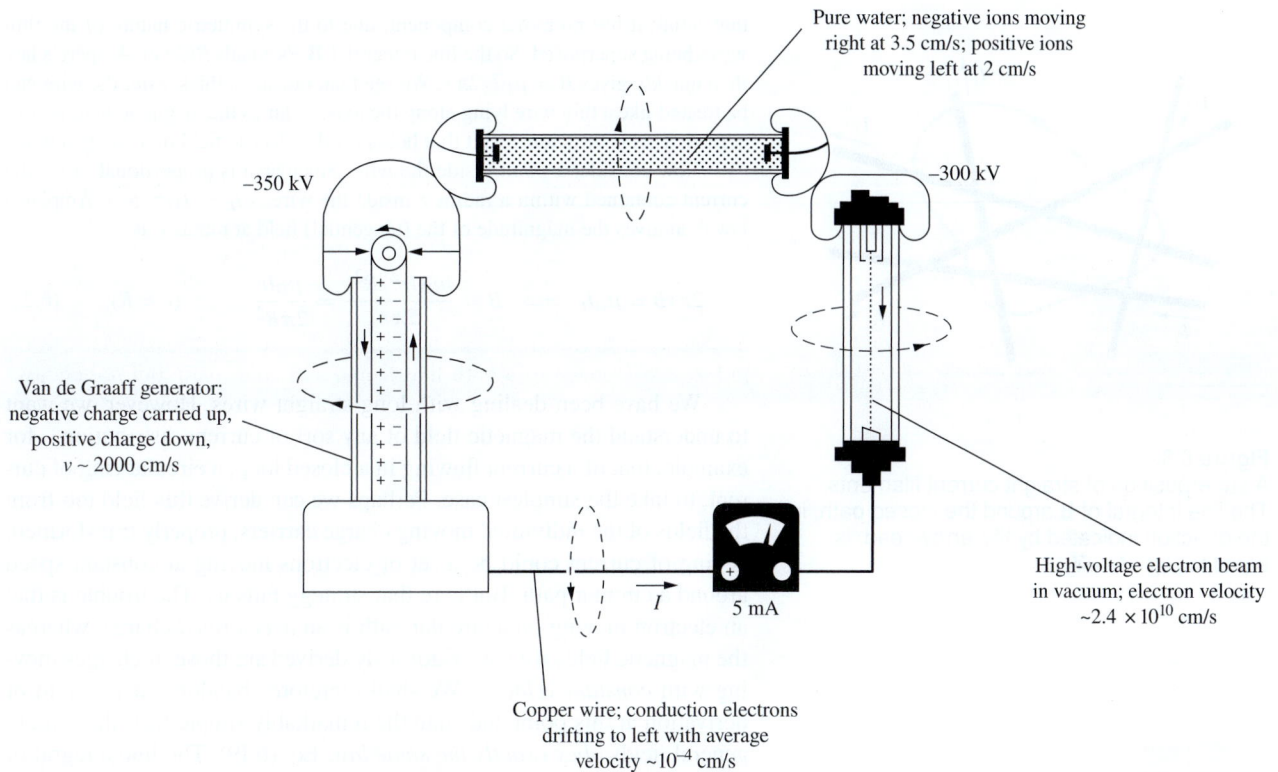

Pure water; negative ions moving
right at 3.5 cm/s; positive ions
moving left at 2 cm/s

−350 kV

−300 kV

Van de Graaff generator;
negative charge carried up,
positive charge down,
$v \sim 2000$ cm/s

$I$

5 mA

High-voltage electron beam
in vacuum; electron velocity
$\sim 2.4 \times 10^{10}$ cm/s

Copper wire; conduction electrons
drifting to left with average
velocity $\sim 10^{-4}$ cm/s

**Figure 6.7.**
The line integral of **B** has precisely the same
value around every part of this circuit, although
the velocity of the charge carriers is quite
different in different parts.

What we have proved for the case of a long straight filament of current clearly holds, by superposition, for the field of any system of straight filaments. In Fig. 6.8 several wires are carrying currents in different directions. If Eq. (6.19) holds for the magnetic field of one of these wires, it must hold for the total field, which is the vector sum, at every point, of the fields of the individual wires. That is a pretty complicated field. Nevertheless, we can predict the value of the line integral of **B** around the closed path in Fig. 6.8 merely by noting which currents the path encircles, and in which sense.

**Example (Magnetic field due to a thick wire)** We know that the magnetic field outside an infinitesimally thin wire points in the tangential direction and has magnitude $B = \mu_0 I / 2\pi r$. But what about a thick wire? Let the wire have radius $R$ and carry current $I$ with uniform current density; the wire may be viewed as the superposition of a large number of thin wires running parallel to each other. Find the field both outside and inside the wire.

**Solution** Consider an Amperian loop (in the spirit of a Gaussian surface) that takes the form of a circle with radius $r$ around the wire. Due to the cylindrical symmetry, **B** has the same magnitude at all points on this loop. Also, **B** is

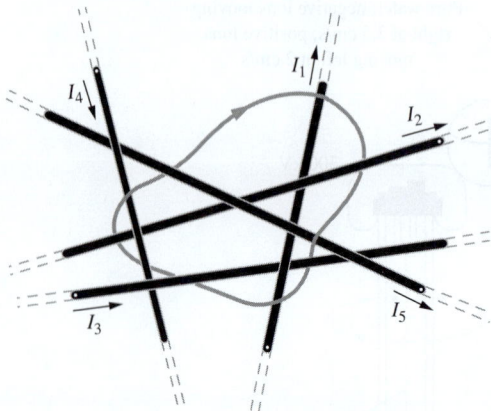

**Figure 6.8.**
A superposition of straight current filaments. The line integral of **B** around the closed path, in the direction indicated by the arrowhead, is equal to $\mu_0(-I_4 + I_5)$.

tangential; it has no radial component, due to the symmetric nature of the thin wires being superposed. So the line integral $\int \mathbf{B} \cdot d\mathbf{s}$ equals $B(2\pi r)$. Ampère's law then quickly gives $B = \mu_0 I / 2\pi r$. We see that, outside a thick wire, the wire can be treated like a thin wire lying along the axis, as far as the magnetic field is concerned. This is the same result that holds for the electric field of a charged wire.

Now consider a point inside the wire. Since area is proportional to $r^2$, the current contained within a radius $r$ inside the wire is $I_r = I(r^2/R^2)$. Ampère's law then gives the magnitude of the (tangential) field at radius $r$ as

$$2\pi r B = \mu_0 I_r \implies B = \frac{\mu_0(Ir^2/R^2)}{2\pi r} = \frac{\mu_0 I r}{2\pi R^2} \qquad (r < R). \qquad (6.21)$$

We have been dealing with long straight wires. However, we want to understand the magnetic field of any sort of current distribution – for example, that of a current flowing in a closed loop, a circular ring of current, to take the simplest case. Perhaps we can derive this field too from the fields of the individual moving charge carriers, properly transformed. A ring of current could be a set of electrons moving at constant speed around a circular path. But here that strategy fails us. The trouble is that an electron moving on a circular path is an *accelerated* charge, whereas the magnetic fields we have rigorously derived are those of charges moving with *constant velocity*. We shall therefore abandon our program of derivation at this point and state the remarkably simple fact: these more general fields *obey exactly the same law*, Eq. (6.19). The line integral of **B** around a bent wire is equal to that around a long straight wire carrying the same current. As this goes beyond anything we have so far deduced, we must look on it here as a postulate confirmed by the experimental tests of its implications.

You may find it unsettling that the validity of Ampère's law applied to an arbitrarily shaped wire simply has to be accepted, given that we have derived everything up to this point. However, this distinction between acceptance and derivation is illusory. As we will see in Chapter 9, Ampère's law is a special case of one of Maxwell's equations. Therefore, accepting Ampère's law is equivalent to accepting one of Maxwell's equations. And considering that Maxwell's equations govern all of electromagnetism (being consistent with countless experimental tests), accepting them is certainly a reasonable thing to do. Likewise, all of our derivations thus far in this book (in particular, the ones in Chapter 5) can be traced back to Coulomb's law, which is equivalent to Gauss's law, which in turn is equivalent to *another* one of Maxwell's equations. Therefore, accepting Coulomb's law is equivalent to accepting this other Maxwell equation. In short, everything boils down to Maxwell's equations sooner or later. Coulomb's law is no more fundamental than Ampère's law. We accepted the former long ago, so we shouldn't be unsettled about accepting the latter now.

To state Ampère's law in the most general way, we must talk about volume distributions of current. A general steady current distribution is described by a current density $\mathbf{J}(x, y, z)$ that varies from place to place but is constant in time. A current in a wire is merely a special case in which $\mathbf{J}$ has a large value within the wire but is zero elsewhere. We discussed volume distributions of current in Chapter 4, where we noted that, for time-independent currents, $\mathbf{J}$ has to satisfy the continuity equation, or conservation-of-charge condition,

$$\text{div }\mathbf{J} = 0. \tag{6.22}$$

Take any closed curve $C$ in a region where currents are flowing. The total current enclosed by $C$ is the flux of $\mathbf{J}$ through the surface spanning $C$, that is, the surface integral $\int_S \mathbf{J} \cdot d\mathbf{a}$ over this surface $S$ (see Fig. 6.9). A general statement of the relation in Eq. (6.19) is therefore

$$\int_C \mathbf{B} \cdot d\mathbf{s} = \mu_0 \int_S \mathbf{J} \cdot d\mathbf{a}. \tag{6.23}$$

Let us compare this with Stokes' theorem, which we developed in Chapter 2:

$$\int_C \mathbf{F} \cdot d\mathbf{s} = \int_S (\text{curl }\mathbf{F}) \cdot d\mathbf{a}. \tag{6.24}$$

We see that a statement equivalent to Eq. (6.23) is this:

$$\boxed{\text{curl }\mathbf{B} = \mu_0 \mathbf{J}} \tag{6.25}$$

This is the differential form of Ampère's law, and it is the simplest and most general statement of the relation between the magnetic field and the moving charges that are its source. As with Eq. (6.19), the Gaussian analog of this expression has the $\mu_0$ replaced by $4\pi/c$. Note that the form of $\mathbf{J}$ in Eq. (6.25) guarantees that Eq. (6.22) is satisfied, because the divergence of the curl is always zero (see Exercise 2.78).

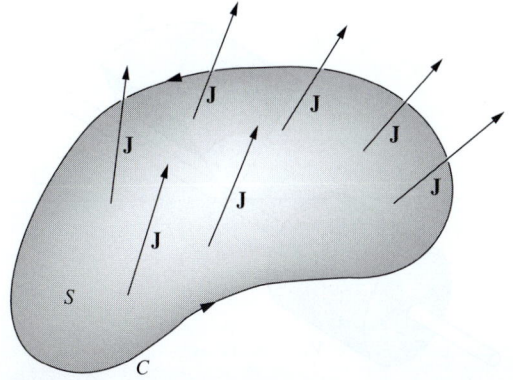

**Figure 6.9.**
$\mathbf{J}$ is the local current density. The surface integral of $\mathbf{J}$ over $S$ is the current enclosed by the curve $C$.

**Example (Curl of B for a thick wire)**  For the above "thick wire" example, verify that curl $\mathbf{B} = \mu_0 \mathbf{J}$ both inside and outside the wire.

**Solution**  We can use the expression for the curl in cylindrical coordinates given in Eq. (F.2) in Appendix F. The only nonzero derivative in the expression is $\partial(rA_\theta)/\partial r$, so outside the wire we have

$$\text{curl }\mathbf{B} = \hat{\mathbf{z}}\frac{1}{r}\frac{\partial(rB_\theta)}{\partial r} = \hat{\mathbf{z}}\frac{1}{r}\frac{\partial}{\partial r}\left(r\frac{\mu_0 I}{2\pi r}\right) = 0, \tag{6.26}$$

which is correct because there is zero current density outside the wire. For the present purposes, the only relevant fact about the external field is that it is proportional to $1/r$.

Inside the wire we have

$$\text{curl }\mathbf{B} = \hat{\mathbf{z}}\frac{1}{r}\frac{\partial(rB_\theta)}{\partial r} = \hat{\mathbf{z}}\frac{1}{r}\frac{\partial}{\partial r}\left(r\frac{\mu_0 I r}{2\pi R^2}\right) = \hat{\mathbf{z}}\mu_0\frac{I}{\pi R^2} = \mu_0(\hat{\mathbf{z}}J) = \mu_0\mathbf{J}, \tag{6.27}$$

as desired.

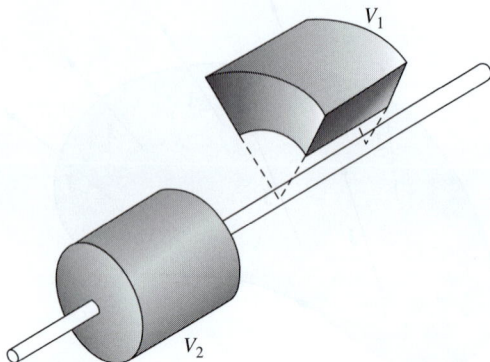

**Figure 6.10.**
There is zero net flux of **B** out of either box.

Equation (6.25) by itself is not enough to determine $\mathbf{B}(x, y, z)$, given $\mathbf{J}(x, y, z)$, because many different vector fields could have the same curl. We need to complete it with another condition. We had better think about the divergence of **B**. Going back to the magnetic field of a single straight wire, we observe that the divergence of that field is zero. You can't draw a little box anywhere, even one enclosing the wire, that will have a net outward or inward flux. It is enough to note that the boxes $V_1$ and $V_2$ in Fig. 6.10 have no net flux and can shrink to zero without developing any. (The $1/r$ dependence of $B$ isn't important here. All that matters is that **B** points in the tangential direction and that its magnitude is independent of $\theta$.) For this field then, div **B** = 0, and hence also for all superpositions of such fields. Again we postulate that the principle can be extended to the field of any distribution of currents, so that a companion to Eq. (6.22) is the condition

$$\boxed{\text{div } \mathbf{B} = 0} \tag{6.28}$$

You can quickly check that this relation holds for the wire in the above example, both inside and outside, by using the cylindrical-coordinate expression for the divergence given in Eq. (F.2) in Appendix F; the only nonzero component of **B** is $B_\theta$, but $\partial B_\theta / \partial \theta = 0$.

We are concerned with fields whose sources lie within some finite region. We won't consider sources that are infinitely remote and infinitely strong. Under these conditions, **B** goes to zero at infinity. With this proviso, we have the following theorem.

**Theorem 6.1** *Assuming that* **B** *vanishes at infinity, Eqs. (6.25) and (6.28) together determine* **B** *uniquely if* **J** *is given.*

*Proof* Suppose both equations are satisfied by two different fields $\mathbf{B}_1$ and $\mathbf{B}_2$. Then their difference, the vector field $\mathbf{D} = \mathbf{B}_1 - \mathbf{B}_2$, is a field with zero divergence and zero curl everywhere. What could it be like? Having zero curl, it must be the gradient[5] of some potential function $f(x, y, z)$, that is, $\mathbf{D} = \nabla f$. But $\nabla \cdot \mathbf{D} = 0$, too, so $\nabla \cdot \nabla f$ or $\nabla^2 f = 0$ everywhere. Over a sufficiently remote enclosing boundary, $f$ must take on some constant value $f_0$, because $\mathbf{B}_1$ and $\mathbf{B}_2$ (and hence **D**) are essentially zero very far away from the sources. Since $f$ satisfies Laplace's equation everywhere inside that boundary, it cannot have a maximum or a minimum anywhere in that region (see Section 2.12), and so it must have the value $f_0$ everywhere. Hence $\mathbf{D} = \nabla f = 0$, and $\mathbf{B}_1 = \mathbf{B}_2$.  $\square$

The fact that a vector field is uniquely determined by its curl and divergence (assuming that it goes to zero at infinity) is known as the

---

[5] This follows from our work in Chapter 2. If curl **D** = 0, then the line integral of **D** around any closed path is zero. This implies that we can uniquely define a potential function $f$ as the line integral of **D** from an arbitrary reference point. It then follows that **D** is the gradient of $f$.

*Helmholtz theorem.* We proved this theorem in the special case where the divergence is zero.

In the case of the electrostatic field, the counterparts of Eqs. (6.25) and (6.28) were

$$\boxed{\operatorname{curl} \mathbf{E} = 0} \qquad \text{and} \qquad \boxed{\operatorname{div} \mathbf{E} = \frac{\rho}{\epsilon_0}} \qquad (6.29)$$

In the case of the electric field, however, we could begin with Coulomb's law, which expressed directly the contribution of each charge to the electric field at any point. Here we shall have to work our way back to some relation of that type.[6] We shall do so by means of a *potential function*.

## 6.3 Vector potential

We found that the scalar potential function $\phi(x, y, z)$ gave us a simple way to calculate the electrostatic field of a charge distribution. If there is some charge distribution $\rho(x, y, z)$, the potential at any point $(x_1, y_1, z_1)$ is given by the volume integral

$$\phi(x_1, y_1, z_1) = \frac{1}{4\pi\epsilon_0} \int \frac{\rho(x_2, y_2, z_2) \, dv_2}{r_{12}}. \qquad (6.30)$$

The integration is extended over the whole charge distribution, and $r_{12}$ is the magnitude of the distance from $(x_2, y_2, z_2)$ to $(x_1, y_1, z_1)$. The electric field $\mathbf{E}$ is obtained as the negative of the gradient of $\phi$:

$$\mathbf{E} = -\operatorname{grad} \phi. \qquad (6.31)$$

The same trick won't work for the magnetic field, because of the essentially different character of $\mathbf{B}$. The curl of $\mathbf{B}$ is *not* necessarily zero, so $\mathbf{B}$ can't, in general, be the gradient of a scalar potential. However, we know another kind of vector derivative, the curl. It turns out that we can usefully represent $\mathbf{B}$, not as the gradient of a scalar function, but as the curl of a *vector* function, like this:

$$\boxed{\mathbf{B} = \operatorname{curl} \mathbf{A}} \qquad (6.32)$$

By obvious analogy, we call $\mathbf{A}$ the *vector potential*. It is *not* obvious, at this point, why this tactic is helpful. That will have to emerge as we proceed. It is encouraging that Eq. (6.28) is automatically satisfied, since $\operatorname{div} \operatorname{curl} \mathbf{A} = 0$, for any $\mathbf{A}$. Or, to put it another way, the fact that $\operatorname{div} \mathbf{B} = 0$ presents us with the opportunity to represent $\mathbf{B}$ as the curl of another vector function.

---

[6] The student may wonder why we couldn't have started from some equivalent of Coulomb's law for the interaction of currents. The answer is that a piece of a current filament, unlike an electric charge, is not an independent object that can be physically isolated. You cannot perform an experiment to determine the field from *part* of a circuit; if the rest of the circuit isn't there, the current can't be steady without violating the continuity condition.

# The magnetic field

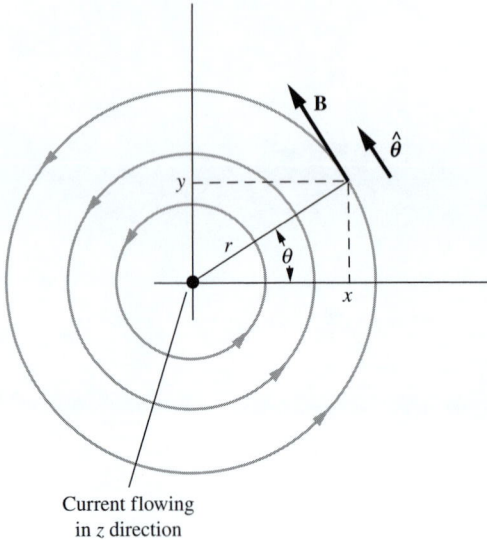

Current flowing
in z direction

**Figure 6.11.**
Some field lines around a current filament.
Current flows toward you (out of the plane of the
paper).

**Example (Vector potential for a wire)**  As an example of a vector potential, consider a long straight wire carrying a current $I$. In Fig. 6.11 we see the current coming toward us out of the page, flowing along the positive $z$ axis. Outside the wire, what is the vector potential $\mathbf{A}$?

**Solution**  We know what the magnetic field of the straight wire looks like. The field lines are circles, as sketched already in Fig. 6.5. A few are shown in Fig. 6.11. The magnitude of $\mathbf{B}$ is $\mu_0 I / 2\pi r$. Using a unit vector $\hat{\theta}$ in the tangential direction, we can write the vector $\mathbf{B}$ as

$$\mathbf{B} = \frac{\mu_0 I}{2\pi r}\hat{\theta}. \tag{6.33}$$

We want to find a vector field $\mathbf{A}$ whose curl equals this $\mathbf{B}$. Equation (F.2) in Appendix F gives the expression for the curl in cylindrical coordinates. In view of Eq. (6.33), we are concerned only with the $\hat{\theta}$ component of the curl expression, which is $(\partial A_r/\partial z - \partial A_z/\partial r)\hat{\theta}$. Due to the symmetry along the $z$ axis, we can't have any $z$ dependence, so we are left with only the $-(\partial A_z/\partial r)\hat{\theta}$ term. Equating this with the $\mathbf{B}$ in Eq. (6.33) gives

$$\nabla \times \mathbf{A} = \mathbf{B} \implies -\frac{\partial A_z}{\partial r} = \frac{\mu_0 I}{2\pi r} \implies \mathbf{A} = -\hat{\mathbf{z}}\frac{\mu_0 I}{2\pi}\ln r. \tag{6.34}$$

This last step can formally be performed by separating variables and integrating. But there is no great need to do this, because we know that the integral of $1/r$ is $\ln r$. The task of Problem 6.4 is to use Cartesian coordinates to verify that the above $\mathbf{A}$ has the correct curl. See also Problem 6.5.

Of course, the $\mathbf{A}$ in Eq. (6.34) is not the only function that could serve as the vector potential for this particular $\mathbf{B}$. To this $\mathbf{A}$ could be added any vector function with zero curl. The above result holds for the space outside the wire. Inside the wire, $\mathbf{B}$ is different, so $\mathbf{A}$ must be different also. It is not hard to find the appropriate vector potential function for the interior of a solid round wire; see Exercise 6.43.

Our job now is to discover a general method of calculating $\mathbf{A}$, when the current distribution $\mathbf{J}$ is given, so that Eq. (6.32) will indeed yield the correct magnetic field. In view of Eq. (6.25), the relation between $\mathbf{J}$ and $\mathbf{A}$ is

$$\text{curl}\,(\text{curl}\,\mathbf{A}) = \mu_0 \mathbf{J}. \tag{6.35}$$

Equation (6.35), being a vector equation, is really three equations. We shall work out one of them, say the $x$-component equation. The $x$ component of curl $\mathbf{B}$ is $\partial B_z/\partial y - \partial B_y/\partial z$. The $z$ and $y$ components of $\mathbf{B}$ are, respectively,

$$B_z = \frac{\partial A_y}{\partial x} - \frac{\partial A_x}{\partial y}, \quad B_y = \frac{\partial A_x}{\partial z} - \frac{\partial A_z}{\partial x}. \tag{6.36}$$

Thus the $x$-component part of Eq. (6.35) reads

$$\frac{\partial}{\partial y}\left(\frac{\partial A_y}{\partial x} - \frac{\partial A_x}{\partial y}\right) - \frac{\partial}{\partial z}\left(\frac{\partial A_x}{\partial z} - \frac{\partial A_z}{\partial x}\right) = \mu_0 J_x. \tag{6.37}$$

We assume our functions are such that the order of partial differentiation can be interchanged. Taking advantage of that and rearranging a little, we can write Eq. (6.37) in the following way:

$$-\frac{\partial^2 A_x}{\partial y^2} - \frac{\partial^2 A_x}{\partial z^2} + \frac{\partial}{\partial x}\left(\frac{\partial A_y}{\partial y}\right) + \frac{\partial}{\partial x}\left(\frac{\partial A_z}{\partial z}\right) = \mu_0 J_x. \qquad (6.38)$$

To make the thing more symmetrical, let's add and subtract the same term, $\partial^2 A_x/\partial x^2$, on the left:[7]

$$-\frac{\partial^2 A_x}{\partial x^2} - \frac{\partial^2 A_x}{\partial y^2} - \frac{\partial^2 A_x}{\partial z^2} + \frac{\partial}{\partial x}\left(\frac{\partial A_x}{\partial x} + \frac{\partial A_y}{\partial y} + \frac{\partial A_z}{\partial z}\right) = \mu_0 J_x. \quad (6.39)$$

We can now recognize the first three terms as the negative of the Laplacian of $A_x$. The quantity in parentheses is the divergence of **A**. Now, we have a certain latitude in the construction of **A**. All we care about is its curl; its divergence can be anything we like. Let us *require* that

$$\text{div } \mathbf{A} = 0. \qquad (6.40)$$

In other words, among the various functions that might satisfy our requirement that curl $\mathbf{A} = \mathbf{B}$, let us consider as candidates only those that also have zero divergence. To see why we are free to do this, suppose we had an **A** such that curl $\mathbf{A} = \mathbf{B}$, but div $\mathbf{A} = f(x, y, z) \neq 0$. We claim that, for any function $f$, we can always find a field **F** such that curl $\mathbf{F} = 0$ and div $\mathbf{F} = -f$. If this claim is true, then we can replace **A** with the new field $\mathbf{A} + \mathbf{F}$. This field has its curl still equal to the desired value of **B**, while its divergence is now equal to the desired value of zero. And the claim *is* indeed true, because if we treat $-f$ like the charge density $\rho$ that generates an electrostatic field, we obviously can find a field **F**, the analog of the electrostatic **E**, such that curl $\mathbf{F} = 0$ and div $\mathbf{F} = -f$; the prescription is given in Fig. 2.29(a), without the $\epsilon_0$.

With div $\mathbf{A} = 0$, the quantity in parentheses in Eq. (6.39) drops away, and we are left simply with

$$\frac{\partial^2 A_x}{\partial x^2} + \frac{\partial^2 A_x}{\partial y^2} + \frac{\partial^2 A_x}{\partial z^2} = -\mu_0 J_x, \qquad (6.41)$$

where $J_x$ is a known scalar function of $x$, $y$, $z$. Let us compare Eq. (6.41) with Poisson's equation, Eq. (2.73), which reads

$$\frac{\partial^2 \phi}{\partial x^2} + \frac{\partial^2 \phi}{\partial y^2} + \frac{\partial^2 \phi}{\partial z^2} = -\frac{\rho}{\epsilon_0}. \qquad (6.42)$$

The two equations are identical in form. We already *know* how to find a solution to Eq. (6.42). The volume integral in Eq. (6.30) is the

---

[7] This equation is the $x$ component of the vector identity, $\nabla \times (\nabla \times \mathbf{A}) = -\nabla^2 \mathbf{A} + \nabla(\nabla \cdot \mathbf{A})$. So in effect, what we've done here is prove this identity. Of course, we could have just invoked this identity and applied it to Eq. (6.35), skipping all of the intermediate steps. But it's helpful to see the proof.

prescription. Therefore a solution to Eq. (6.41) must be given by Eq. (6.30), with $\rho/\epsilon_0$ replaced by $\mu_0 J_x$:

$$A_x(x_1, y_1, z_1) = \frac{\mu_0}{4\pi} \int \frac{J_x(x_2, y_2, z_2)\, dv_2}{r_{12}}. \tag{6.43}$$

The other components must satisfy similar formulas. They can all be combined neatly in one vector formula:

$$\boxed{\mathbf{A}(x_1, y_1, z_1) = \frac{\mu_0}{4\pi} \int \frac{\mathbf{J}(x_2, y_2, z_2)\, dv_2}{r_{12}}} \tag{6.44}$$

In more compact notation we have

$$\mathbf{A} = \frac{\mu_0}{4\pi} \int \frac{\mathbf{J}\, dv}{r} \qquad \text{or} \qquad d\mathbf{A} = \frac{\mu_0}{4\pi} \frac{\mathbf{J}\, dv}{r}. \tag{6.45}$$

There is only one snag. We stipulated that $\operatorname{div} \mathbf{A} = 0$, in order to get Eq. (6.41). If the divergence of the $\mathbf{A}$ in Eq. (6.44) isn't zero, then although this $\mathbf{A}$ will satisfy Eq. (6.41), it won't satisfy Eq. (6.39). That is, it won't satisfy Eq. (6.35). Fortunately, it turns out that the $\mathbf{A}$ in Eq. (6.44) does indeed satisfy $\operatorname{div} \mathbf{A} = 0$, *provided* that the current is steady (that is, $\nabla \cdot \mathbf{J} = 0$), which is the type of situation we are concerned with. You can prove this in Problem 6.6. The proof isn't important for what we will be doing; we include it only for completeness.

Incidentally, the $\mathbf{A}$ for the example above could not have been obtained by Eq. (6.44). The integral would diverge owing to the infinite extent of the wire. This may remind you of the difficulty we encountered in Chapter 2 in setting up a scalar potential for the electric field of a charged wire. Indeed the two problems are very closely related, as we should expect from their identical geometry and the similarity of Eqs. (6.44) and (6.30). We found in Eq. (2.22) that a suitable scalar potential for the line charge problem is $-(\lambda/2\pi\epsilon_0)\ln r + C$, where $C$ is an arbitrary constant. This assigns zero potential to some arbitrary point that is neither on the wire nor an infinite distance away. Both that scalar potential and the vector potential of Eq. (6.34) are singular at the origin and at infinity. However, see Problem 6.5 for a way to get around this issue. For an interesting discussion of the vector potential, including its interpretation as "electromagnetic momentum," see Semon and Taylor (1996).

## 6.4 Field of any current-carrying wire

Figure 6.12 shows a loop of wire carrying current $I$. The vector potential $\mathbf{A}$ at the point $(x_1, y_1, z_1)$ is given according to Eq. (6.44) by the integral over the loop. For current confined to a thin wire we may take as the volume element $dv_2$ a short section of the wire of length $dl$. The current density $J$ is $I/a$, where $a$ is the cross-sectional area and $dv_2 = a\, dl$.

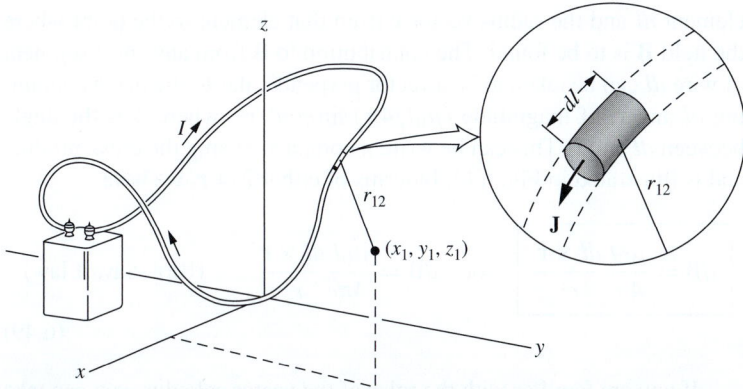

**Figure 6.12.**
Each element of the current loop contributes to the vector potential **A** at the point $(x_1, y_1, z_1)$.

Hence $J\, dv_2 = I\, dl$, and if we make the vector $dl$ point in the direction of positive current, we can simply replace $\mathbf{J}\, dv_2$ by $I\, dl$. Thus for a thin wire or filament, we can write Eq. (6.44) as a line integral over the circuit:

$$\mathbf{A} = \frac{\mu_0 I}{4\pi} \int \frac{dl}{r_{12}}. \tag{6.46}$$

To calculate **A** everywhere and then find **B** by taking the curl of **A** might be a long job. It will be more useful to isolate one contribution to the line integral for **A**, the contribution from the segment of wire at the origin, where the current happens to be flowing in the $x$ direction (Fig. 6.13). We shall denote the length of this segment by $dl$. Let $d\mathbf{A}$ be the contribution of this part of the integral to **A**. Then at the general point $(x, y, z)$, the vector $d\mathbf{A}$, which points in the positive $x$ direction, is

$$d\mathbf{A} = \hat{\mathbf{x}}\, \frac{\mu_0 I}{4\pi}\, \frac{dl}{\sqrt{x^2 + y^2 + z^2}}. \tag{6.47}$$

Let us denote the corresponding part of **B** by $d\mathbf{B}$. If we consider now a point $(x, y, 0)$ in the $xy$ plane, then, when taking the curl of $d\mathbf{A}$ to obtain $d\mathbf{B}$, we find that only one term among the various derivatives survives:

$$d\mathbf{B} = \text{curl}\,(d\mathbf{A}) = \hat{\mathbf{z}} \left( -\frac{\partial A_x}{\partial y} \right)$$
$$= \hat{\mathbf{z}}\, \frac{\mu_0 I}{4\pi}\, \frac{y\, dl}{(x^2 + y^2)^{3/2}} = \hat{\mathbf{z}}\, \frac{\mu_0 I}{4\pi}\, \frac{\sin\phi\; dl}{r^2}, \tag{6.48}$$

where $\phi$ is indicated in Fig. 6.13. You should convince yourself why the symmetry of the $d\mathbf{A}$ in Eq. (6.47) with respect to the $xy$ plane implies that curl $(d\mathbf{A})$ must be perpendicular to the $xy$ plane.

With this result we can free ourselves at once from a particular coordinate system. Obviously all that matters is the relative orientation of the

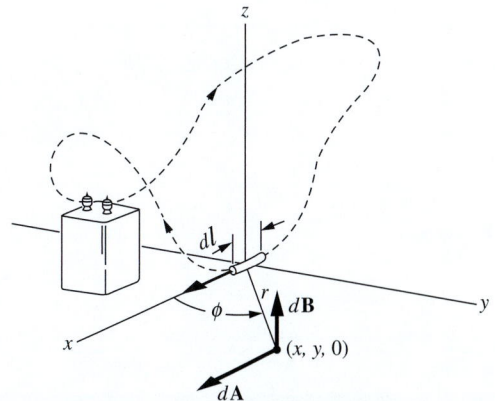

**Figure 6.13.**
If we find $d\mathbf{A}$, the contribution to **A** of the particular element shown, its contribution to **B** can be calculated using $\mathbf{B} = \text{curl}\,\mathbf{A}$.

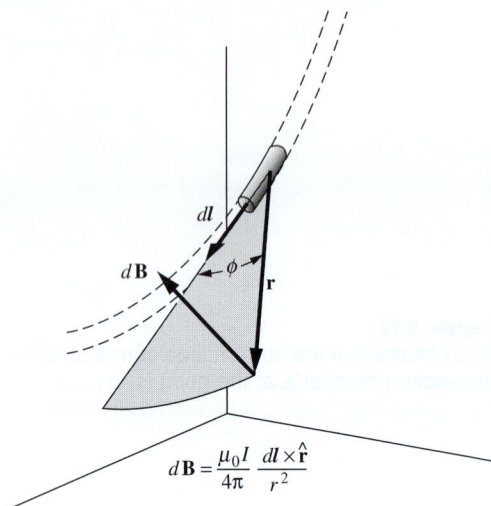

$$dB = \frac{\mu_0 I}{4\pi}\frac{dl \times \hat{\mathbf{r}}}{r^2}$$

**Figure 6.14.**
The Biot–Savart law. The field of any circuit can be calculated by using this relation for the contribution of each circuit element.

element $dl$ and the radius vector $\mathbf{r}$ from that element to the point where the field $\mathbf{B}$ is to be found. The contribution to $\mathbf{B}$ from any short segment of wire $dl$ can be taken to be a vector perpendicular to the plane containing $dl$ and $\mathbf{r}$, of magnitude $(\mu_0 I/4\pi)\sin\phi\, dl/r^2$, where $\phi$ is the angle between $dl$ and $\mathbf{r}$. This can be written compactly using the cross-product and is illustrated in Fig. 6.14. In terms of either $\hat{\mathbf{r}}$ or $\mathbf{r}$, we have

$$\boxed{d\mathbf{B} = \frac{\mu_0 I}{4\pi}\frac{dl \times \hat{\mathbf{r}}}{r^2}} \quad \text{or} \quad d\mathbf{B} = \frac{\mu_0 I}{4\pi}\frac{dl \times \mathbf{r}}{r^3} \quad \text{(Biot–Savart law)}.$$

$$(6.49)$$

If you are familiar with the rules of the vector calculus, you can take a shortcut from Eq. (6.46) to Eq. (6.49) without making reference to a coordinate system. Writing $d\mathbf{B} = \nabla \times d\mathbf{A}$, with $d\mathbf{A} = (\mu_0 I/4\pi)\, dl/r$, we can use the vector identity $\nabla \times (f\mathbf{F}) = f\nabla \times \mathbf{F} + \nabla f \times \mathbf{F}$ to obtain

$$d\mathbf{B} = \nabla \times \frac{\mu_0 I}{4\pi}\frac{dl}{r} = \frac{\mu_0 I}{4\pi}\left(\frac{1}{r}\nabla \times dl + \nabla\left(\frac{1}{r}\right)\times dl\right). \quad (6.50)$$

But $dl$ is a constant, so the first term on the right-hand side is zero. And recall that $\nabla(1/r) = -\hat{\mathbf{r}}/r^2$ (as in going from the Coulomb potential to the Coulomb field). Thus

$$d\mathbf{B} = \frac{\mu_0 I}{4\pi}\left(-\frac{\hat{\mathbf{r}}}{r^2}\right)\times dl = \frac{\mu_0 I}{4\pi}\frac{dl \times \hat{\mathbf{r}}}{r^2}. \quad (6.51)$$

Historically, Eq. (6.49) is known as the Biot–Savart law. The meaning of Eq. (6.49) is that, if $\mathbf{B}$ is computed by integrating over the *complete circuit*, taking the contribution from each element to be given by this formula, the resulting $\mathbf{B}$ will be correct. As we remarked in Footnote 6, the contribution of part of a circuit is not physically identifiable. In fact, Eq. (6.49) is not the only formula that could be used to get a correct result for $\mathbf{B}$ – to it could be added any function that would give zero when integrated around a closed path.

The Biot–Savart law is valid for *steady* currents (or for sufficiently slowly changing currents).[8] There is no restriction on the speed of the charges that make up the steady current, provided that they are essentially continuously distributed. The speeds can be relativistic, and the Biot–Savart law still works fine. If the current isn't steady (and is changing rapidly enough) then, although the Biot–Savart law isn't valid, a somewhat similar law that involves the so-called "retarded time" is valid. We won't get into that here, but see Problem 6.28 if you want to get a sense of what the retarded time is all about.

[8] There are actually a few other conditions under which the law is valid, but we won't worry about those here. See Griffiths and Heald (1991) for everything you might want to know about the conditions under which the various laws of electricity and magnetism are valid.

We seem to have discarded the vector potential as soon as it performed one essential service for us. Indeed, it is often easier, as a practical matter, to calculate the field of a current system directly, now that we have Eq. (6.49), than to find the vector potential first.[9] We shall practice on some examples in Section 6.5. However, the vector potential is important for deeper reasons. For one thing, it has revealed to us a striking parallel between the relation of the electrostatic field $\mathbf{E}$ to its sources (static electric charges) and the relation of the magnetostatic field $\mathbf{B}$ to its sources (steady electric currents; that's what magnetostatic means). Its greatest usefulness becomes evident in more advanced topics, such as electromagnetic radiation and other time-varying fields.

## 6.5 Fields of rings and coils

We will now do two examples where we use Eq. (6.49) to calculate a magnetic field. The second example will build on the result of the first.

**Example (Circular ring)**  A current filament in the form of a circular ring of radius $b$ is shown in Fig. 6.15(a). We could predict without any calculation that the magnetic field of this source must look something like Fig. 6.15(b), where we have sketched some field lines in a plane through the axis of symmetry. The field as a whole must be rotationally symmetrical about this axis, the $z$ axis in Fig. 6.15(a), and the field lines themselves (ignoring their direction) must be symmetrical with respect to the plane of the loop, the $xy$ plane. Very close to the filament the field will resemble that near a long straight wire, since the distant parts of the ring are there relatively unimportant.

It is easy to calculate the field on the axis, using Eq. (6.49). Each element of the ring of length $dl$ contributes a $d\mathbf{B}$ perpendicular to $\mathbf{r}$. We need only include the $z$ component of $d\mathbf{B}$, for we know the total field on the axis must point in the $z$ direction. This brings in a factor of $\cos\theta$, so we obtain

$$dB_z = \frac{\mu_0 I}{4\pi} \frac{dl}{r^2} \cos\theta = \frac{\mu_0 I}{4\pi} \frac{dl}{r^2} \frac{b}{r}. \tag{6.52}$$

Integrating over the whole ring, we have simply $\int dl = 2\pi b$, so the field on the axis at any point $z$ is

$$B_z = \frac{\mu_0 I}{4\pi} \frac{2\pi b^2}{r^3} = \frac{\mu_0 I b^2}{2(b^2 + z^2)^{3/2}} \qquad \text{(field on axis).} \tag{6.53}$$

At the center of the ring, $z = 0$, the magnitude of the field is

$$B_z = \frac{\mu_0 I}{2b} \qquad \text{(field at center).} \tag{6.54}$$

Note that the field points in the same direction (upward) everywhere along the $z$ axis.

[9] The main reason for this is that if we want to use $\mathbf{A}$ to calculate $\mathbf{B}$ at a given point, we need to know what $\mathbf{A}$ is at nearby points too. That is, we need to know $\mathbf{A}$ as a function of the coordinates so that we can calculate the derivatives in the curl. On the other hand, if we calculate $\mathbf{B}$ via Eq. (6.49), we simply need to find $\mathbf{B}$ at the one given point.

(a)

(b)

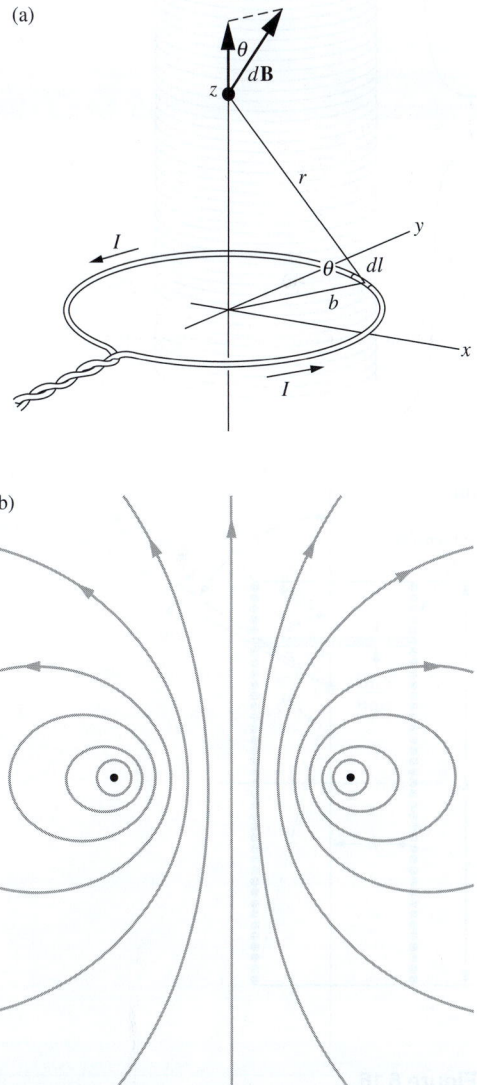

**Figure 6.15.**
The magnetic field of a ring of current.
(a) Calculation of the field on the axis. (b) Some field lines.

(a)

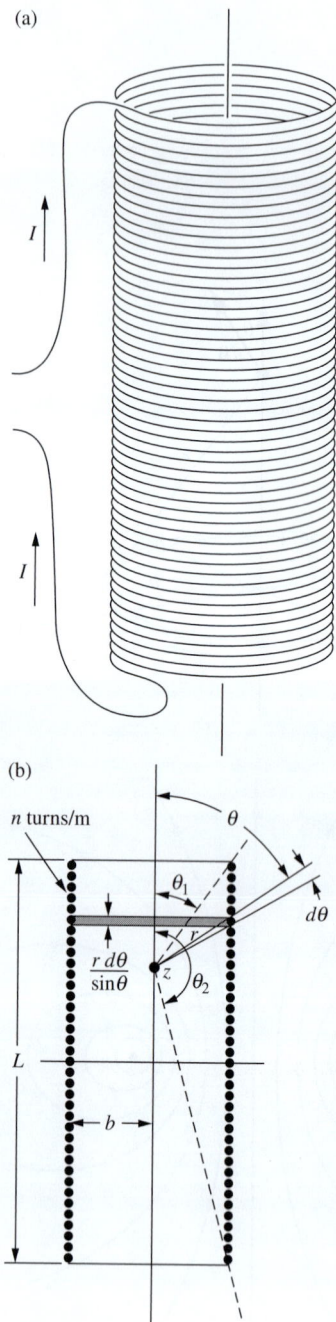

(b)

*n* turns/m

$\theta_1$

$\theta$

$d\theta$

$\dfrac{r\,d\theta}{\sin\theta}$

$z$  $\theta_2$

$L$

$\leftarrow b \rightarrow$

**Figure 6.16.**
(a) Solenoid. (b) Calculation of the field on the axis of a solenoid.

**Example (Solenoid)** The cylindrical coil of wire shown in Fig. 6.16(a) is usually called a *solenoid*. We assume the wire is closely and evenly spaced so that the number of turns in the winding, per meter length along the cylinder, is a constant, *n*. Now, the current path is actually helical, but if the turns are many and closely spaced, we can ignore this and regard the whole solenoid as equivalent to a stack of current rings. Then we can use Eq. (6.53) as a basis for calculating the field at any point, such as the point *z*, on the axis of the coil.

Consider the contribution from the current rings included between radii from the point *z* that make angles $\theta$ and $\theta + d\theta$ with the axis. The length of this segment of the solenoid, indicated in Fig. 6.16(b), is $r\,d\theta/\sin\theta$ (because $r\,d\theta$ is the tilted angular span of the segment, and the factor of $1/\sin\theta$ gives the vertical span). It is therefore equivalent to a ring carrying a current $dI = In(r\,d\theta/\sin\theta)$. Since $r = b/\sin\theta$, Eq. (6.53) gives, for the contribution of this ring to the axial field:

$$dB_z = (dI)\frac{\mu_0 b^2}{2r^3} = \left(\frac{nIr\,d\theta}{\sin\theta}\right)\frac{\mu_0 b^2}{2r^3} = \frac{\mu_0 nI}{2}\sin\theta\,d\theta. \tag{6.55}$$

Carrying out the integration between the limits $\theta_1$ and $\theta_2$, gives

$$B_z = \frac{\mu_0 nI}{2}\int_{\theta_1}^{\theta_2}\sin\theta\,d\theta = \frac{\mu_0 nI}{2}(\cos\theta_1 - \cos\theta_2). \tag{6.56}$$

We have used Eq. (6.56) to make a graph, in Fig. 6.17, of the field strength on the axis of a coil, the length of which is four times its diameter. The ordinate is the field strength $B_z$ relative to the field strength in a coil of infinite length with the same number of turns per meter and the same currents in each turn. For the infinite coil, $\theta_1 = 0$ and $\theta_2 = \pi$, so

$$\boxed{B_z = \mu_0 nI} \quad \text{(infinitely long solenoid).} \tag{6.57}$$

At the center of the "four-to-one" coil the field is very nearly as large as this, and it stays pretty nearly constant until we approach one of the ends. Equation (6.57) actually holds for all points inside an infinite solenoid, not just for points on the axis; see Problem 6.19.

Figure 6.18 shows the magnetic field lines in and around a coil of these proportions. Note that some field lines actually penetrate the winding. The cylindrical sheath of current is a surface of discontinuity for the magnetic field. Of course, if we were to examine the field very closely in the neighborhood of the wires, we would not find any infinitely abrupt kinks, but we would find a very complicated, ripply pattern around and through the individual wires.

It is quite possible to make a long solenoid with a *single* turn of a thin ribbonlike conductor, as in Fig. 6.19. To this, our calculation and the diagram in Fig. 6.18 apply exactly, the quantity *nI* being merely replaced by the current per meter flowing in the sheet. Now the change in direction of a field line that penetrates the wall occurs entirely within the thickness of the sheet, as suggested in the inset in Fig. 6.19.

In calculating the field of the solenoid in Fig. 6.16, we treated it as a stack of rings, ignoring the longitudinal current that must exist in

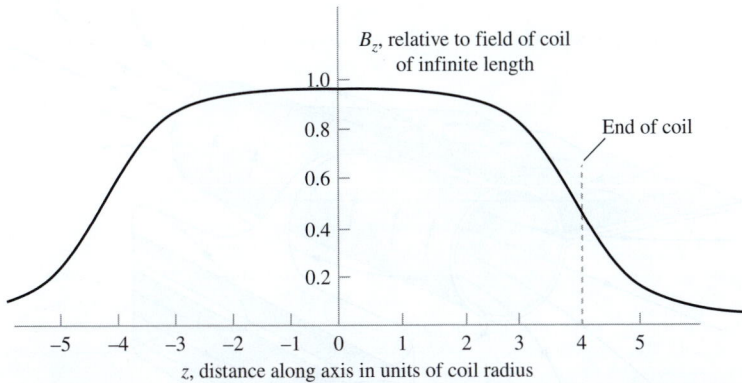

**Figure 6.17.**
Field strength $B_z$ on the axis, for the solenoid shown in Fig. 6.18.

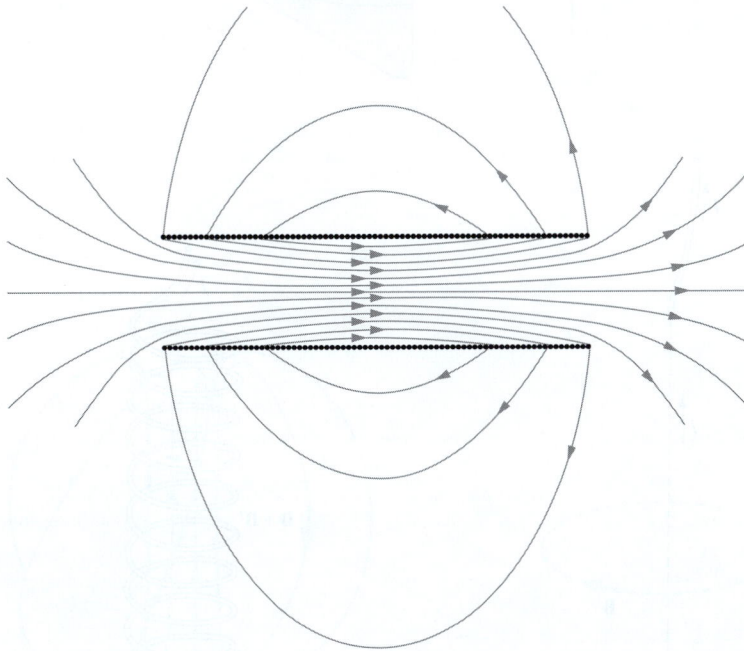

**Figure 6.18.**
Field lines in and around a solenoid.

any coil in which the current enters at one end and leaves at the other. Let us see how the field is modified if that is taken into account. The helical coil in Fig. 6.20(c) is equivalent, so far as the field external to the solenoid is concerned, to the superposition of the stack of current rings in Fig. 6.20(a) and a single axial conductor in Fig. 6.20(b). Adding the field of the latter, $\mathbf{B}'$, to the field $\mathbf{B}$ of the former, we get the external field of the coil. It has a helical twist. Some field lines have been sketched in Fig. 6.20(c). As for the field inside the solenoid, the longitudinal current $I$ flows, in effect, on the cylinder itself. Such a current distribution, a uniform hollow tube of current, produces zero field inside the cylinder

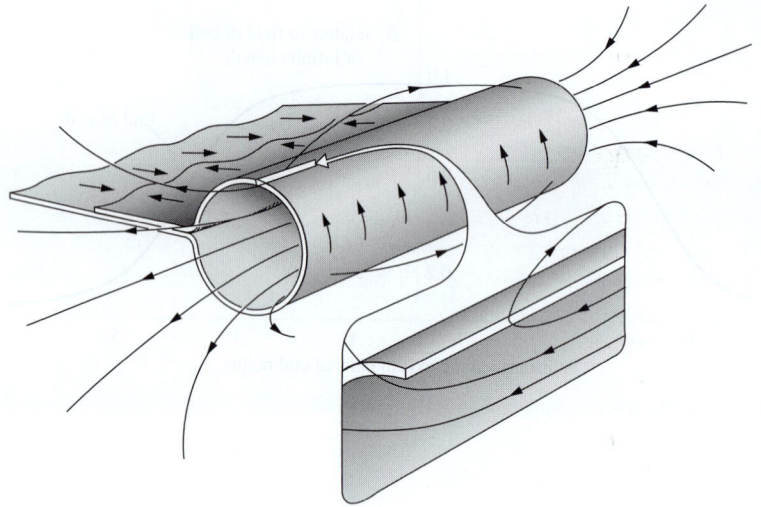

**Figure 6.19.**
A solenoid formed by a single cylindrical conducting sheet. Inset shows how the field lines change direction inside the current-carrying conductor.

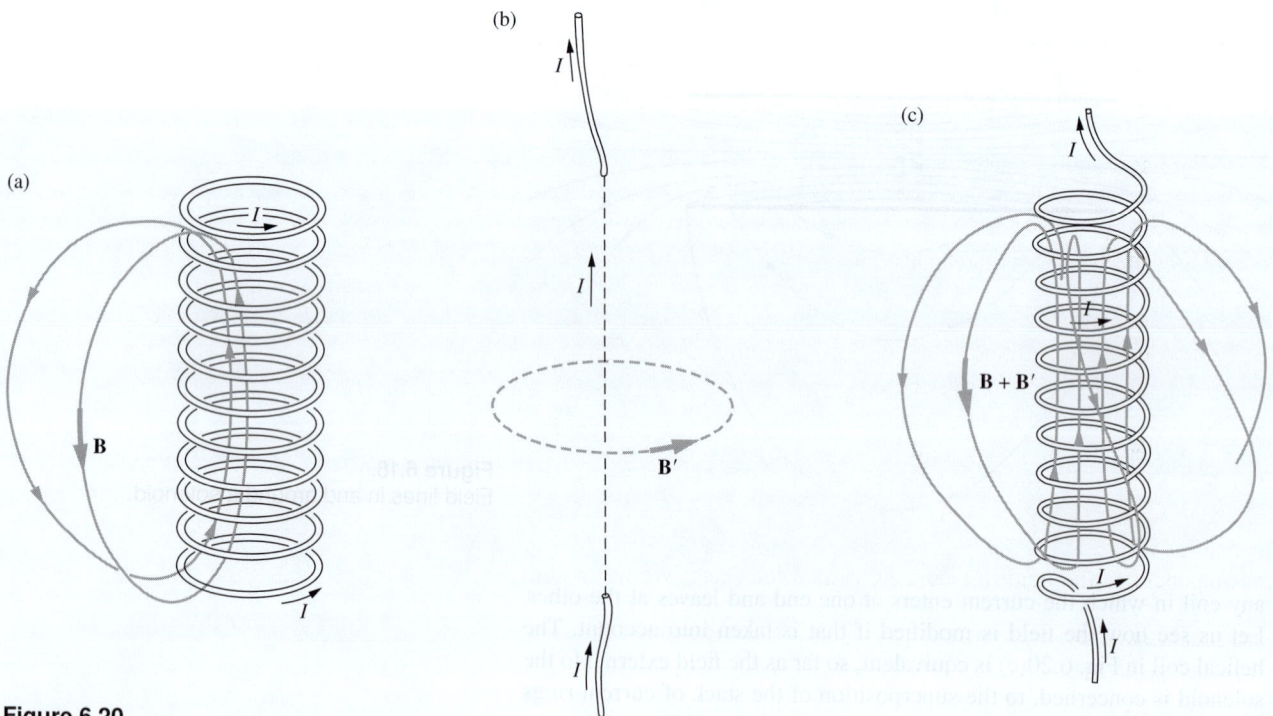

(a)

(b)

(c)

**Figure 6.20.**
The helical coil (c) is equivalent to a stack of circular rings, each carrying current $I$ and shown in (a), plus a current $I$ parallel to the axis of the coil as shown in (b). A path around the coil encloses the current $I$, the field of which, $\mathbf{B'}$, must be added to the field $\mathbf{B}$ of the rings to form the external field of the helical coil.

(due to Eq. (6.19) and the fact that a circular path inside the tube encloses no current), leaving unmodified the interior field we calculated before. If you follow a looping field line from inside to outside to inside again, you will discover that it does *not* close on itself. Field lines generally don't. You might find it interesting to figure out how this picture would be changed if the wire that leads the current $I$ away from the coil were brought down along the axis of the coil to emerge at the bottom.

## 6.6 Change in B at a current sheet

In the setup of Fig. 6.19 we had a solenoid constructed from a single curved sheet of current. Let's look at something even simpler, a flat, unbounded current sheet. You may think of this as a sheet of copper of uniform thickness in which a current flows with constant density and direction everywhere within the metal. In order to refer to directions, let us locate the sheet in the $xz$ plane and let the current flow in the $x$ direction. As the sheet is supposed to be of infinite extent with no edges, it is hard to draw a picture of it! We show a broken-out fragment of the sheet in Fig. 6.21, in order to have something to draw; you must imagine the rest of it extending over the whole plane. The thickness of the sheet will not be very important, finally, but we may suppose that it has some definite thickness $d$.

If the current density inside the metal is $J$ in $C\,s^{-1}\,m^{-2}$, then every length $l$ of height, in the $z$ direction, includes a ribbon of current amounting to $J(ld)$ in $C/s$. We call $Jd$ the *surface current density* or *sheet current density* and use the symbol $\mathcal{J}$ to distinguish it from the volume current density $\mathbf{J}$. The units[10] of $\mathcal{J}$ are amps/meter; multiplying $\mathcal{J}$ by the length $l$ of a line segment (perpendicular to the current flow) on the surface gives the current crossing that segment. If we are not concerned with what goes on inside the sheet itself, $\mathcal{J}$ is a useful quantity. It is $\mathcal{J}$ that determines the *change* in the magnetic field from one side of the sheet to the other, as we shall see.

The field in Fig. 6.21 is not merely that due to the sheet alone. Some other field in the $z$ direction is present, from another source. The total field, including the effect of the current sheet, is represented by the $\mathbf{B}$ vectors drawn in front of and behind the sheet.

Consider the line integral of $\mathbf{B}$ around the rectangle 12341 in Fig. 6.21. One of the long sides is in front of the surface, the other behind it, with the short sides piercing the sheet. Let $B_z^+$ denote the $z$ component of the magnetic field immediately in front of the sheet, $B_z^-$ the $z$ component of the field immediately behind the sheet. We mean here the field of *all* sources that may be around, including the sheet itself. The line integral of $\mathbf{B}$ around the long rectangle is simply $l(B_z^+ - B_z^-)$. (Even if there were some other source that caused a field component parallel to the short legs of the rectangle, these legs themselves can be kept much shorter than the long sides, since we assume the sheet is

---

[10] The terms "surface" current density and "volume" current density indicate the dimension of the space in which the current flows. Since the units of $\mathcal{J}$ and $\mathbf{J}$ are A/m and A/m$^2$, respectively, we are using these terms in a different sense than we use them when talking about surface and volume *charge* densities, which have dimensions C/m$^2$ and C/m$^3$. For example, to obtain a current from a *volume* current density, we multiply it by an *area*.

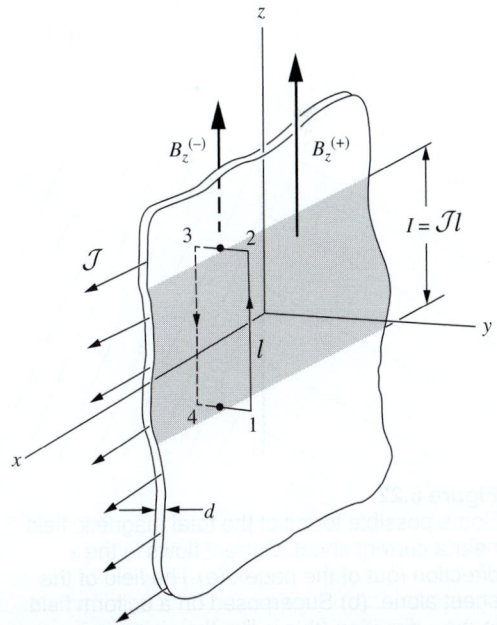

**Figure 6.21.**
At a sheet of surface current there must be a change in the parallel component of $\mathbf{B}$ from one side to the other.

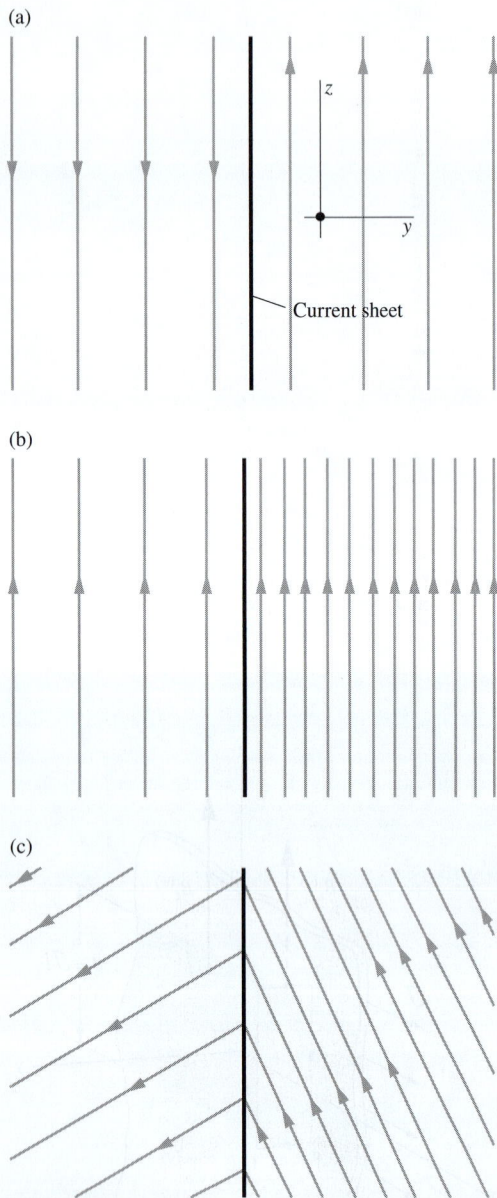

(a)

(b)

(c)

**Figure 6.22.**
Some possible forms of the total magnetic field near a current sheet. Current flows in the $x$ direction (out of the page). (a) The field of the sheet alone. (b) Superposed on a uniform field in the $z$ direction (this is like the situation in Fig. 6.21). (c) Superposed on a uniform field in another direction. In every case the component $B_z$ changes by $\mu_0 \mathcal{J}$, on passing through the sheet, with no change in $B_y$.

thin, in any case, compared with the scale of any field variation.) The current enclosed by the rectangle is just $\mathcal{J}l$. Hence Eq. (6.19) yields $l(B_z^+ - B_z^-) = \mu_0 \mathcal{J}l$, or

$$B_z^+ - B_z^- = \mu_0 \mathcal{J}. \qquad (6.58)$$

A current sheet of density $\mathcal{J}$ gives rise to a jump in the component of **B** that is parallel to the surface and perpendicular to $\mathcal{J}$. This may remind you of the change in electric field at a sheet of charge. There, the *perpendicular* component of **E** is discontinuous, the magnitude of the jump depending on the density of surface charge.

If the sheet is the only current source we have, then of course the field is symmetrical about the sheet; $B_z^+$ is $\mu_0 \mathcal{J}/2$, and $B_z^-$ is $-\mu_0 \mathcal{J}/2$. This is shown in Fig. 6.22(a). Some other situations, in which the effect of the current sheet is superposed on a field already present from another source, are shown in Fig. 6.22(b) and (c). Suppose there are two sheets carrying equal and opposite surface currents, as shown in cross section in Fig. 6.23, with no other sources around. The direction of current flow is perpendicular to the plane of the paper, out on the left and in on the right. The field between the sheets is $\mu_0 \mathcal{J}$, and there is no field at all outside. Something like this is found when current is carried by two parallel ribbons or slabs, close together compared with their width, as sketched in Fig. 6.24. Often *bus bars* for distributing heavy currents in power stations are of this form.

**Example (Field from a cylinder of current)**    A cylindrical shell has radius $R$ and carries uniformly distributed current $I$ parallel to its axis. Find the magnetic field outside the shell, an infinitesimal distance away from it. Do this in the following way.

(a)    Slice the shell into infinitely long "rods" parallel to the axis, and then integrate the field contributions from all the rods. You should obtain $B = \mu_0 I/4\pi R$. However, this is not the correct field, because we know from Ampère's law that the field outside a wire (or a cylinder) takes the form of $\mu_0 I/2\pi r$. And $r = R$ here.

(b)    What is wrong with the above reasoning? Explain how to modify it to obtain the correct result of $\mu_0 I/2\pi R$. *Hint:* You could have very well performed the above integral in an effort to obtain the magnetic field an infinitesimal distance *inside* the cylinder, where we know the field is zero.

Solution

(a)    A cross section of the cylinder is shown in Fig. 6.25. A small piece of the circumference of the circle represents a rod pointing into and out of the page. Let the rods be parameterized by the angle $\theta$ relative to the point $P$ at which we are calculating the field. If a rod subtends an angle $d\theta$, then it contains a fraction $d\theta/2\pi$ of the total current $I$. So the current in the rod is $I(d\theta/2\pi)$. The rod is a distance $2R \sin(\theta/2)$ from $P$, which is infinitesimally close to the top of the cylinder.

If the current heads into the page, then the field due to the rod shown is directed up and to the right at $P$. Only the horizontal (tangential) component of this field survives, because the vertical component cancels with that from the corresponding rod on the left side of the cylinder. This brings in a factor of $\sin(\theta/2)$, as you can verify. Using the fact that the field from a straight rod takes the form of $\mu_0 I/2\pi r$, we find that the field at point $P$ is directed to the right and has magnitude (apparently) equal to

$$B = 2 \int_0^\pi \frac{\mu_0 (I\, d\theta/2\pi)}{2\pi (2R\sin(\theta/2))} \sin(\theta/2) = \frac{\mu_0 I}{4\pi^2 R} \int_0^\pi d\theta = \frac{\mu_0 I}{4\pi R}. \quad (6.59)$$

(b) (You should try to solve this on your own before reading further. You may want to take a look at Exercise 1.67.) As noted in the statement of the problem, it is no surprise that the above result is incorrect, because the same calculation would supposedly yield the field just inside the cylinder too. But we know that the field there is zero. The calculation does, however, yield the next best thing, namely the average of the correct fields inside and outside (zero and $\mu_0 I/2\pi R$). We'll see why shortly.

The reason why the calculation is invalid is that it doesn't correctly describe the field due to rods on the cylinder very close to the given point $P$, that is, for rods characterized by $\theta \approx 0$. It is incorrect for two reasons. The closeup view in Fig. 6.26 (with an exaggerated distance from $P$ to the cylinder, for clarity) shows that the distance from a rod to $P$ is *not* equal to $2R\sin(\theta/2)$. Additionally, it shows that the field at $P$ does *not* point perpendicular to the line from the rod to the top of the cylinder. It points more horizontally, so the extra factor of $\sin(\theta/2)$ in Eq. (6.59) is not correct.

What *is* true is that if we remove a thin strip from the top of the cylinder (so we now have a gap in the circle representing the cross-sectional view), then the above integral is valid for the remaining part of the cylinder. The strip contributes negligibly to the integral in Eq. (6.59) (assuming it subtends a very small angle), so we can say that the field due to the remaining part of the cylinder is equal to the above result of $\mu_0 I/4\pi R$. By superposition, the total field due to the entire cylinder is this field of $\mu_0 I/4\pi R$ plus the field due to the thin strip. But if the point in question is infinitesimally close to the cylinder, then the thin strip looks like an infinite sheet of current, the field of which we know is $\mu_0 \mathcal{J}/2 = \mu_0(I/2\pi R)/2 = \mu_0 I/4\pi R$. The desired total field is then

$$B_{\text{outside}} = B_{\text{cylinder minus strip}} + B_{\text{strip}} = \frac{\mu_0 I}{4\pi R} + \frac{\mu_0 I}{4\pi R} = \frac{\mu_0 I}{2\pi R}. \quad (6.60)$$

The relative sign here is indeed a plus sign, because right above the strip, the strip's field points to the right, which is the same as the direction of the field due to the rest of the cylinder. By superposition we also obtain the correct field just inside the cylinder:

$$B_{\text{inside}} = B_{\text{cylinder minus strip}} - B_{\text{strip}} = \frac{\mu_0 I}{4\pi R} - \frac{\mu_0 I}{4\pi R} = 0. \quad (6.61)$$

The relative minus sign comes from the fact that right below the strip, the strip's field points to the left. For an alternative way of solving this problem, see Bose and Scott (1985).

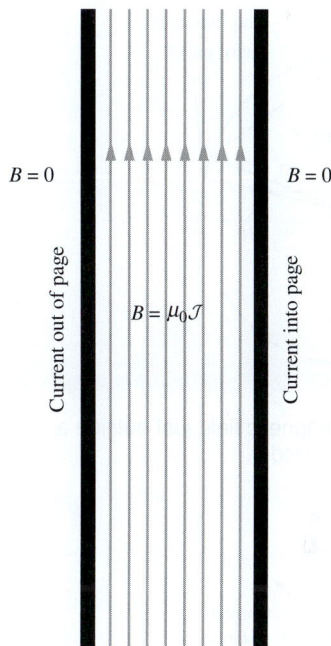

$B = 0$    $B = 0$

Current out of page    $B = \mu_0 \mathcal{J}$    Current into page

**Figure 6.23.**
The magnetic field between plane-parallel current sheets.

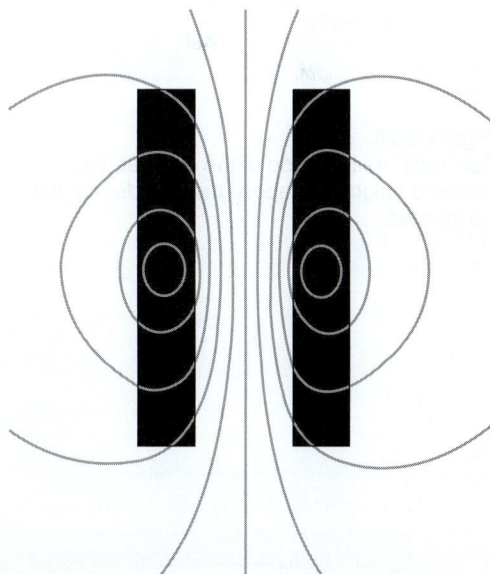

**Figure 6.24.**
The magnetic field of a pair of copper bus bars, shown in cross section, carrying current in opposite directions.

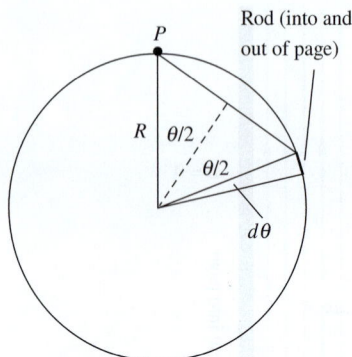

**Figure 6.25.**
Calculating the magnetic field just outside a current-carrying cylinder.

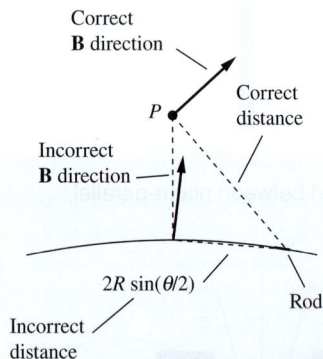

**Figure 6.26.**
For "rods" near the top of the cylinder, the nonzero height of $P$ above the cylinder cannot be ignored.

The change in **B** from one side of a sheet to the other takes place within the sheet, as we already remarked in connection with Fig. 6.19. For the same $\mathcal{J}$, the thinner the sheet, the more abrupt the transition. We looked at a situation very much like this in Chapter 1 when we examined the discontinuity in the perpendicular component of **E** that occurs at a sheet of surface charge. It was instructive then to ask about the force on the surface charge, and we shall ask a similar question here.

Consider a square portion of the sheet, a length $\ell$ on a side (a rectangle would work fine too). The current included is equal to $\mathcal{J}\ell$, the length of current path is $\ell$, and the *average* field that acts on this current is $(B_z^+ + B_z^-)/2$. The force on a length $\ell$ of current-carrying wire equals $IB\ell$ (see Eq. (6.14)), so the force on this portion of the current distribution is

$$\text{force on } \ell^2 \text{ of sheet} = IB_{\text{avg}}\ell = (\mathcal{J}\ell)\left(\frac{B_z^+ + B_z^-}{2}\right)\ell. \quad (6.62)$$

In view of Eq. (6.58), we can substitute $(B_z^+ - B_z^-)/\mu_0$ for $\mathcal{J}$, so that the force per unit area can be expressed in this way:

$$\text{force per unit area} = \left(\frac{B_z^+ - B_z^-}{\mu_0}\right)\left(\frac{B_z^+ + B_z^-}{2}\right)$$

$$= \frac{1}{2\mu_0}\left[(B_z^+)^2 - (B_z^-)^2\right]. \quad (6.63)$$

The force is perpendicular to the surface and proportional to the area, like the stress caused by hydrostatic pressure. To make sure of the sign, we can figure out the direction of the force in a particular case, such as that in Fig. 6.23. The force is *outward* on each conductor. It is as if the high-field region were the region of high pressure. The repulsion of any two conductors carrying current in opposite directions, as in Fig. 6.24, can be seen as an example of that.

We have been considering an infinite flat sheet, but things are much the same in the immediate neighborhood of any surface where there is a change in **B**. Wherever the component of **B** parallel to the surface changes from $B_1$ to $B_2$, from one side of the surface to the other, we may conclude not only that there is a sheet of current flowing in the surface, but also that the surface must be under a perpendicular stress of $(B_1^2 - B_2^2)/2\mu_0$, measured in N/m². This is one of the controlling principles in *magnetohydrodynamics*, the study of electrically conducting fluids, a subject of interest both to electrical engineers and to astrophysicists.

## 6.7 How the fields transform

A sheet of surface charge, if it is moving parallel to itself, constitutes a surface current. If we have a uniform charge density of $\sigma$ on the surface, with the surface itself sliding along at speed $v$, the surface current density is just $\mathcal{J} = \sigma v$. This is true because the area that slides past a transverse line with length $\ell$ during time $dt$ is $(v\,dt)\ell$, which yields a current of $\sigma(v\,dt\,\ell)/dt = (\sigma v)\ell$. And since the current is also $\mathcal{J}\ell$ by definition, we have $\mathcal{J} = \sigma v$. This simple idea of a sliding surface will help us to see how the electric and magnetic field quantities must change when we transform from one inertial frame of reference to another. We will deal first with the transverse fields, and then with the longitudinal fields.

Let's imagine two plane sheets of surface charge, parallel to the $xz$ plane and moving with speed $v_0$ as in Fig. 6.27. Again, we show fragments of surfaces only in the sketch; the surfaces are really infinite in extent. In the inertial frame $F$ with coordinates $x$, $y$, and $z$, where the sheets move with speed $v_0$, the density of surface charge is $\sigma$ on one sheet and $-\sigma$ on the other. Here $\sigma$ means the amount of charge within unit area when area is measured by observers stationary in $F$. (It is not the density of charge in the rest frame of the charges themselves, which would be smaller by $1/\gamma_0$.) In the frame $F$ the uniform electric field $\mathbf{E}$ points in the positive $y$ direction, and Gauss's law assures us, as usual, that its strength is

$$E_y = \frac{\sigma}{\epsilon_0}. \tag{6.64}$$

In this frame $F$ the sheets are both moving in the positive $x$ direction with speed $v_0$, so that we have a pair of current sheets. The density of surface current is $\mathcal{J}_x = \sigma v_0$ in one sheet, the negative of that in the other. As in the arrangement in Fig. 6.23, the field between two such current sheets is

$$B_z = \mu_0 \mathcal{J}_x = \mu_0 \sigma v_0. \tag{6.65}$$

The inertial frame $F'$ is one that moves, as seen from $F$, with a speed $v$ in the positive $x$ direction. *What fields will an observer in $F'$ measure?* To answer this we need only find out what the sources look like in $F'$.

In $F'$ the $x'$ velocity of the charge-bearing sheets is $v_0'$, given by the velocity addition formula

$$v_0' = \frac{v_0 - v}{1 - v_0 v/c^2} = c\frac{\beta_0 - \beta}{1 - \beta_0 \beta}. \tag{6.66}$$

There is a different Lorentz contraction of the charge density in this frame, exactly as in our earlier example of the moving line charge in Section 5.9. We can repeat the argument we used then: the density in the rest frame of the charges themselves is $\sigma(1 - v_0^2/c^2)^{1/2}$, or $\sigma/\gamma_0$, and therefore the density of surface charge in the frame $F'$ is

$$\sigma' = \sigma\frac{\gamma_0'}{\gamma_0}. \tag{6.67}$$

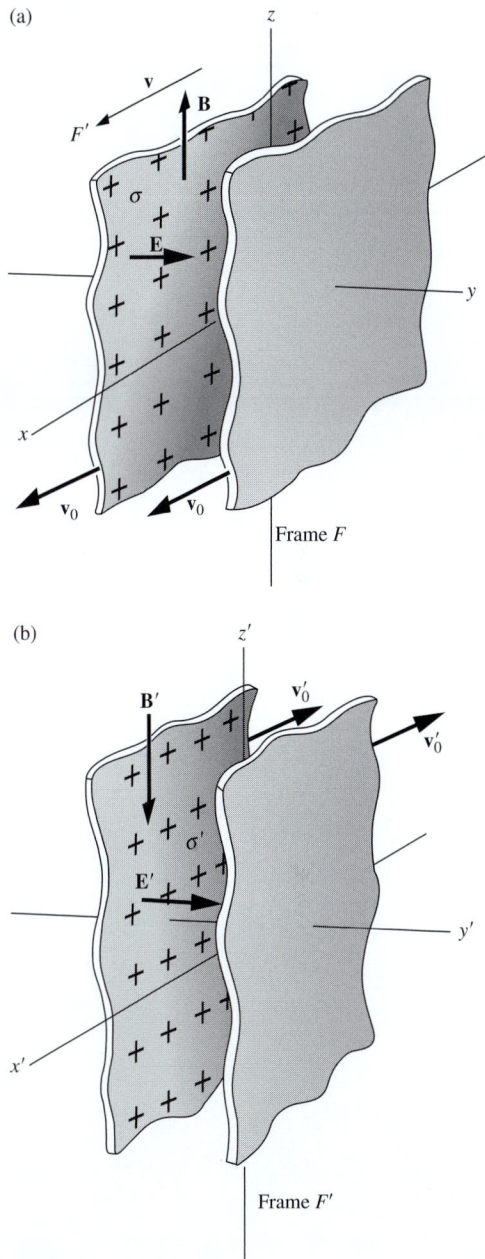

**Figure 6.27.**
(a) As observed in frame $F$, the surface charge density is $\sigma$ and the surface current density is $\sigma v_0$. (b) Frame $F'$ moves in the $x$ direction with speed $v$ as seen from $F$. In $F'$ the surface charge density is $\sigma'$ and the current density is $\sigma' v_0'$.

As usual, $\gamma_0'$ stands for $(1 - v_0'^2/c^2)^{-1/2}$. Using Eq. (6.66), you can show that $\gamma_0' = \gamma_0 \gamma (1 - \beta_0 \beta)$. Hence,

$$\sigma' = \sigma \gamma (1 - \beta_0 \beta). \tag{6.68}$$

The surface current density in the frame $F'$ is (charge density) × (charge velocity):

$$\mathcal{J}' = \sigma' v_0' = \sigma \gamma (1 - \beta_0 \beta) \cdot c \frac{\beta_0 - \beta}{1 - \beta_0 \beta} = \sigma \gamma (v_0 - v). \tag{6.69}$$

We now know how the sources appear in frame $F'$, so we know what the fields in that frame must be. In saying this, we are again invoking the postulate of relativity. The laws of physics must be the same in all inertial frames, and that includes the formulas connecting electric field with surface charge density, and magnetic field with surface current density. It follows then that

$$E_y' = \frac{\sigma'}{\epsilon_0} = \gamma \left[ \frac{\sigma}{\epsilon_0} - \frac{\sigma}{\epsilon_0} \left( \frac{v_0}{c} \right) \left( \frac{v}{c} \right) \right] = \gamma \left[ \frac{\sigma}{\epsilon_0} - \frac{v}{\mu_0 \epsilon_0 c^2} \cdot \mu_0 \sigma v_0 \right],$$

$$B_z' = \mu_0 \mathcal{J}' = \gamma \left[ \mu_0 \sigma v_0 - \mu_0 \sigma v \right] = \gamma \left[ \mu_0 \sigma v_0 - \mu_0 \epsilon_0 v \cdot \frac{\sigma}{\epsilon_0} \right]. \tag{6.70}$$

(These expressions might look a bit scary, but don't worry, they'll simplify!) We have chosen to write $E_y'$ and $B_z'$ in this way because if we look back at the values of $E_y$ and $B_z$ in Eqs. (6.64) and (6.65), we see that our result can be written as follows:

$$E_y' = \gamma \left( E_y - \frac{v}{\mu_0 \epsilon_0 c^2} \cdot B_z \right),$$

$$B_z' = \gamma \left( B_z - \mu_0 \epsilon_0 v \cdot E_y \right). \tag{6.71}$$

We can further simplify these expressions by using the relation $1/\mu_0 \epsilon_0 = c^2$ from Eq. (6.8). We finally obtain

$$E_y' = \gamma \left( E_y - v B_z \right),$$

$$B_z' = \gamma \left( B_z - \frac{v}{c^2} E_y \right), \tag{6.72}$$

or equivalently

$$E_y' = \gamma \left( E_y - \beta (c B_z) \right),$$

$$c B_z' = \gamma \left( (c B_z) - \beta E_y \right). \tag{6.73}$$

You will note that these are exactly the same *Lorentz transformations* that apply to $x$ and $t$ (see Eq. (G.2) in Appendix G). They are symmetric in $E_y$ and $c B_z$.

If the sandwich of current sheets had been oriented parallel to the *xy* plane instead of the *xz* plane, we would have obtained relations connecting $E'_z$ with $E_z$ and $B_y$, and connecting $B'_y$ with $B_y$ and $E_z$. Of course, they would have the same form as the relations above, but if you trace the directions through, you will find that there are differences in sign, following from the rules for the direction of **B**.

Now we must learn how the field components in the direction of motion change. We discovered in Section 5.5 that a longitudinal component of **E** has the same magnitude in the two frames. That this is true also of a longitudinal component of **B** can be seen as follows. Suppose a longitudinal component of **B**, a $B_x$ component in the arrangement in Fig. 6.27, is produced by a solenoid around the *x* axis in frame *F* (at rest in *F*). The field strength inside a solenoid, as we know from Eq. (6.57), depends only on the current in the wire, *I*, which is charge per second, and *n*, the number of turns of wire per meter of axial length. In the frame *F'* the solenoid will be Lorentz-contracted, so the number of turns per meter in that frame will be greater. But the current, as reckoned by observers in *F'*, will be reduced, since, from their point of view, the *F* observers who measured the current by counting the number of electrons passing a point on the wire, per second, were using a slow-running watch. The time dilation just cancels the length contraction in the product *nI*. Indeed any quantity of the dimensions (longitudinal length)$^{-1}$ × (time)$^{-1}$ is unchanged in a Lorentz transformation. So $B'_x = B_x$.

Remember the point made early in Chapter 5, in the discussion following Eq. (5.6): the transformation properties of the field are *local* properties. The values of **E** and **B** at some space-time point in one frame must uniquely determine the field components observed in any other frame at that same space-time point. Therefore the fact that we have used an especially simple kind of source (the parallel uniformly charged sheets, or the solenoid) in our derivation in no way compromises the generality of our result. We have in fact arrived at the general laws for the transformation of all components of the electric and magnetic field, of whatever origin or configuration.

We give below the full list of transformations. All primed quantities are measured in the frame *F'*, which is moving in the positive *x* direction with speed *v* as seen from *F*. Unprimed quantities are the numbers that are the results of measurement in *F*. As usual, $\beta$ stands for $v/c$ and $\gamma$ for $(1 - \beta^2)^{-1/2}$.

$$
\begin{aligned}
E'_x &= E_x & E'_y &= \gamma(E_y - vB_z) & E'_z &= \gamma(E_z + vB_y) \\
B'_x &= B_x & B'_y &= \gamma\big(B_y + (v/c^2)E_z\big) & B'_z &= \gamma\big(B_z - (v/c^2)E_y\big)
\end{aligned}
$$

$$(6.74)$$

When these equations are written in the alternative form given in Eq. (6.73), the symmetry between **E** and *c***B** is evident. If the printer had mistakenly interchanged $E$'s with $cB$'s, and $y$'s with $z$'s, the equations would come out exactly the same. Certainly magnetic phenomena as we find them in Nature are distinctly different from electrical phenomena. The world around us is by no means symmetrical with respect to electricity and magnetism. Nevertheless, with the sources out of the picture, we find that the fields themselves, $\mathbf{E}$ and $c\mathbf{B}$, are connected to one another in a highly symmetrical way.

It appears too that the electric and magnetic fields are in some sense aspects, or components, of a single entity. We can speak of the *electromagnetic* field, and we may think of $E_x$, $E_y$, $E_z$, $cB_x$, $cB_y$, and $cB_z$ as six components of the electromagnetic field. The *same* field viewed in different inertial frames will be represented by different sets of values for these components, somewhat as a vector is represented by different components in different coordinate systems rotated with respect to one another. However, the electromagnetic field so conceived is not a vector, mathematically speaking, but rather something called a *tensor*. The totality of the equations in the box on page 309 forms the prescription for transforming the components of such a tensor when we shift from one inertial frame to another. We are not going to develop that mathematical language here. In fact, we shall return now to our old way of talking about the electric field as a vector field, and the magnetic field as another vector field coupled to the first in a manner to be explored further in Chapter 7. To follow up on this brief hint of the unity of the electromagnetic field as represented in four-dimensional space-time, you will have to wait for a more advanced course.

We can express the transformation of the fields, Eq. (6.74), in a more elegant way which is often useful. Let $\mathbf{v}$ be the velocity of a frame $F'$ as seen from a frame $F$. We can always resolve the fields in both $F$ and $F'$ into vectors parallel to and perpendicular to, respectively, the direction of $\mathbf{v}$. Thus, using an obvious notation:

$$\mathbf{E} = \mathbf{E}_\parallel + \mathbf{E}_\perp, \qquad \mathbf{E}' = \mathbf{E}'_\parallel + \mathbf{E}'_\perp,$$
$$\mathbf{B} = \mathbf{B}_\parallel + \mathbf{B}_\perp, \qquad \mathbf{B}' = \mathbf{B}'_\parallel + \mathbf{B}'_\perp. \tag{6.75}$$

Then the transformation can be written like this (as you can verify):

$$
\boxed{
\begin{aligned}
&\mathbf{E}'_\parallel = \mathbf{E}_\parallel \qquad \mathbf{E}'_\perp = \gamma\left(\mathbf{E}_\perp + \mathbf{v} \times \mathbf{B}_\perp\right) \\
&\mathbf{B}'_\parallel = \mathbf{B}_\parallel \qquad \mathbf{B}'_\perp = \gamma\left(\mathbf{B}_\perp - (\mathbf{v}/c^2) \times \mathbf{E}_\perp\right) \\
&(\mathbf{v} \text{ is the velocity of } F' \text{ with respect to } F)
\end{aligned}
}
\tag{6.76}
$$

In the special case that led us to Eq. (6.72), $\mathbf{v}$ was $v\hat{\mathbf{x}}$, $\mathbf{E}_\perp$ was $(\sigma/\epsilon_0)\hat{\mathbf{y}}$, and $\mathbf{B}_\perp$ was $\mu_0\sigma v_0\hat{\mathbf{z}}$. You can check that these vectors turn the "$\perp$" equations in Eq. (6.76) into the equations in Eq. (6.70); you will need to use $1/\mu_0\epsilon_0 = c^2$.

In Gaussian units, with $\mathbf{E}$ in statvolts/cm and $\mathbf{B}$ in gauss, the Lorentz transformation of the fields reads as follows (with $\boldsymbol{\beta} \equiv \mathbf{v}/c$):

$$\boxed{\begin{aligned} \mathbf{E}_\parallel' &= \mathbf{E}_\parallel & \mathbf{E}_\perp' &= \gamma \left( \mathbf{E}_\perp + \boldsymbol{\beta} \times \mathbf{B}_\perp \right) \\ \mathbf{B}_\parallel' &= \mathbf{B}_\parallel & \mathbf{B}_\perp' &= \gamma \left( \mathbf{B}_\perp - \boldsymbol{\beta} \times \mathbf{E}_\perp \right) \end{aligned}}$$

(6.77)

You can derive these relations by working through the above procedure in Gaussian units. But a quicker method is to note that, when changing formulas from SI to Gaussian units, we must replace $\epsilon_0$ with $1/4\pi$ (by looking at the analogous expressions for Coulomb's law, or equivalently Gauss's law), and $\mu_0$ with $4\pi/c$ (by looking at the analogous expressions for Ampère's law). The product $\mu_0\epsilon_0$ therefore gets replaced with $1/c$, and Eq. (6.71) then leads to Eq. (6.77).

One advantage of the Gaussian system of units is that the transformations in Eq. (6.77) are more symmetrical than those in Eq. (6.76). This can be traced to the fact that $\mathbf{E}$ and $\mathbf{B}$ have the same units in the Gaussian system. In the SI system, unfortunately, the use of different units for $\mathbf{E}$ and $\mathbf{B}$ (due to the definition of $\mathbf{B}$ in Eq. (6.1)) tends to obscure the essential electromagnetic symmetry of the vacuum. The electric and magnetic fields are after all components of one tensor. The Lorentz transformation is something like a rotation, turning $\mathbf{E}$ partly into $\mathbf{B}'$, and $\mathbf{B}$ partly into $\mathbf{E}'$. It seems quite natural and appropriate that the only parameter in Eq. (6.77) is the dimensionless ratio $\beta$. To draw an analogy that is not altogether unfair, imagine that it has been decreed that east–west displacement components must be expressed in meters while north–south components are to be in feet. The transformation effecting a rotation of coordinate axes would be, to say the least, aesthetically unappealing. Nor is symmetry restored to Eq. (6.76) when $\mathbf{B}$ is replaced, as is often done, by a vector $\mathbf{H}$, which we shall meet in Chapter 11, and which in the vacuum is simply $\mathbf{B}/\mu_0$.

Having said this, however, we should note that there isn't anything disastrous about the extra factor of $c$ that appears in the SI Lorentz transformation. The Lorentz symmetry between $\mathbf{E}$ and $c\mathbf{B}$ is similar to the Lorentz symmetry between $x$ and $ct$. The coordinates $x$ and $t$ have different dimensions, but are still related by a Lorentz transformation with an extra factor of $c$ thrown in.

**Example (Stationary charge and rod)**   A charge $q$ is at rest a distance $r$ from a long rod with linear charge density $\lambda$, as shown in Fig. 6.28. The charges in the rod are also at rest. The electric field due to the rod takes the standard form of $E = \lambda/2\pi\epsilon_0 r$, so the force on the charge $q$ in the lab frame is simply $F = qE = q\lambda/2\pi\epsilon_0 r$. This force is repulsive, assuming $q$ and $\lambda$ have the same sign.

(Lab frame)

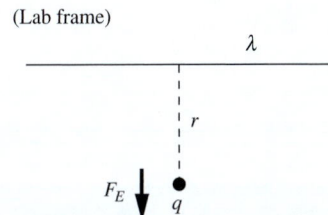

**Figure 6.28.**
A point charge at rest with respect to a charged rod.

Now consider the setup in the frame that moves to the left with speed $v$. In this frame both the charge $q$ and the charges in the rod move to the right with speed $v$. What is the force on the charge $q$ in this new frame? Solve this in three different ways.

(a) Transform the force from the lab frame to the new frame, without caring about where it comes from in the new frame.

(b) Directly calculate the electric and magnetic forces in the new frame, by considering the charges in the rod.

(c) Transform the fields using the Lorentz transformations.

### Solution

(a) The force on a particle is always largest in the rest frame of the particle. It is smaller in any other frame by the $\gamma$ factor associated with the speed $v$ of the particle. The force in the particle frame (the lab frame) is $q\lambda/2\pi\epsilon_0 r$, so the force in the new frame is $q\lambda/2\gamma\pi\epsilon_0 r$.

(b) In the new frame (call it $F'$), the linear charge density in the rod is increased to $\gamma\lambda$, due to length contraction. So the electric field is $E' = \gamma\lambda/2\pi\epsilon_0 r$. This field produces a repulsive electric force of $F_E = \gamma q\lambda/2\pi\epsilon_0 r$.

   In $F'$ the current produced by the rod is the density times the speed, so $I = (\gamma\lambda)v$. The magnetic field is then $B' = \mu_0 I/2\pi r = \mu_0\gamma\lambda v/2\pi r$, directed into the page in Fig. 6.29 (assuming $\lambda$ is positive). The magnetic force is therefore attractive and has magnitude (using $\mu_0 = 1/\epsilon_0 c^2$)

$$F_B = qvB' = qv \cdot \frac{\mu_0\gamma\lambda v}{2\pi r} = \frac{\gamma q\lambda v^2}{2\pi\epsilon_0 rc^2}. \tag{6.78}$$

The net repulsive force acting on the charge $q$ in the new frame is therefore

$$F_E - F_B = \frac{\gamma q\lambda}{2\pi\epsilon_0 r} - \frac{\gamma q\lambda v^2}{2\pi\epsilon_0 rc^2} = \frac{\gamma q\lambda}{2\pi\epsilon_0 r}\left(1 - \frac{v^2}{c^2}\right) = \frac{q\lambda}{2\gamma\pi\epsilon_0 r}, \tag{6.79}$$

where we have used $1 - v^2/c^2 \equiv 1/\gamma^2$. This net force agrees with the result in part (a).

(c) In the lab frame, the charges in the rod aren't moving, so $\mathbf{E}_\perp$ is the only nonzero field in the Lorentz transformations in Eq. (6.76). It is directed away from the rod with magnitude $\lambda/2\pi\epsilon_0 r$. Equation (6.76) immediately gives the electric field in the new frame as $\mathbf{E}'_\perp = \gamma\mathbf{E}_\perp$. So $\mathbf{E}'_\perp$ has magnitude $E'_\perp = \gamma\lambda/2\pi\epsilon_0 r$ and is directed away from the rod, in agreement with the electric field we found in part (b).

Equation (6.76) gives the magnetic field in the new frame as $\mathbf{B}'_\perp = -\gamma(\mathbf{v}/c^2) \times \mathbf{E}_\perp$. The velocity $\mathbf{v}$ of $F'$ with respect to the lab frame $F$ points to the *left* with magnitude $v$. We therefore find that $\mathbf{B}'_\perp$ points into the page with magnitude $B'_\perp = \gamma(v/c^2)(\lambda/2\pi\epsilon_0 r)$. In terms of $\mu_0 = 1/\epsilon_0 c^2$, this can be written as $B'_\perp = \mu_0\gamma\lambda v/2\pi r$, in agreement with the magnetic field we found in part (b). We therefore arrive at the same net force, $F_E - F_B$, as in part (b).

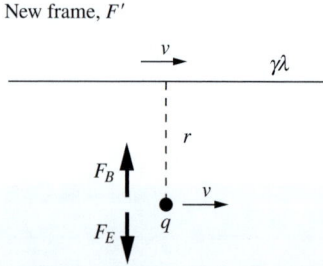

New frame, $F'$

**Figure 6.29.**
The electric and magnetic forces in the new frame.

There is a remarkably simple relation between the electric and magnetic field vectors in a special but important class of cases. Suppose a frame exists – let's call it the unprimed frame – in which $\mathbf{B}$ is zero in some region (as in the above example). Then in *any* other frame $F'$ that moves with velocity $\mathbf{v}$ relative to that special frame, we have, according to Eq. (6.76),

$$\mathbf{E}'_{\parallel} = \mathbf{E}_{\parallel}, \qquad\qquad \mathbf{E}'_{\perp} = \gamma \mathbf{E}_{\perp},$$
$$\mathbf{B}'_{\parallel} = 0, \qquad\qquad \mathbf{B}'_{\perp} = -\gamma(\mathbf{v}/c^2) \times \mathbf{E}_{\perp}. \qquad (6.80)$$

In the last of these equations we can replace $\mathbf{B}'_{\perp}$ with $\mathbf{B}'$, because $\mathbf{B}'_{\parallel} = 0$. We can also replace $\gamma \mathbf{E}_{\perp}$ with $\mathbf{E}'_{\perp}$, which we can in turn replace with $\mathbf{E}'$ because $\mathbf{v} \times \mathbf{E}'_{\parallel} = 0$ (since $\mathbf{E}'_{\parallel}$ is parallel to $\mathbf{v}$ by definition). The last equation therefore becomes a simple relation between the full $\mathbf{E}'$ and $\mathbf{B}'$ fields:

$$\mathbf{B}' = -(\mathbf{v}/c^2) \times \mathbf{E}' \qquad \text{(if } \mathbf{B} = 0 \text{ in one frame).} \qquad (6.81)$$

This holds in every frame if $\mathbf{B} = 0$ in one frame. Remember that $\mathbf{v}$ is the velocity of the frame in question (the primed frame) with respect to the special frame in which $\mathbf{B} = 0$.

In the same way, we can deduce from Eq. (6.76) that, if there exists a frame in which $\mathbf{E} = 0$, then in any other frame

$$\mathbf{E}' = \mathbf{v} \times \mathbf{B}' \qquad \text{(if } \mathbf{E} = 0 \text{ in one frame).} \qquad (6.82)$$

As before, $\mathbf{v}$ is the velocity of the frame $F'$ with respect to the special frame $F$ in which, in this case, $\mathbf{E} = 0$.

Because Eqs. (6.81) and (6.82) involve only quantities measured in the same frame of reference, they are easy to apply, whenever the restriction is met, to fields that vary in space. A good example is the field of a point charge $q$ moving with constant velocity, the problem studied in Chapter 5. Take the unprimed frame to be the frame in which the charge is at rest. In this frame, of course, there is no magnetic field. Equation (6.81) tells us that in the lab frame, where we find the charge moving with speed $v$, there must be a magnetic field perpendicular to the electric field and to the direction of motion. We have already worked out the exact form of the electric field in this frame: we know the field is radial from the instantaneous position of the charge, with a magnitude given by Eq. (5.15). The magnetic field lines must be circles around the direction of motion, as indicated crudely in Fig. 6.30. When the velocity of the charge is high ($v \approx c$), so that $\gamma \gg 1$, the radial "spokes" that are the electric field lines are folded together into a thin disk. The circular magnetic field lines are likewise concentrated in this disk. The magnitude of $\mathbf{B}$ is then nearly equal to the magnitude of $\mathbf{E}/c$. That is, the magnitude of the magnetic field in tesla is almost exactly the same as $1/c$ times the magnitude of the electric field in volts/meter, at the same point and instant of time.

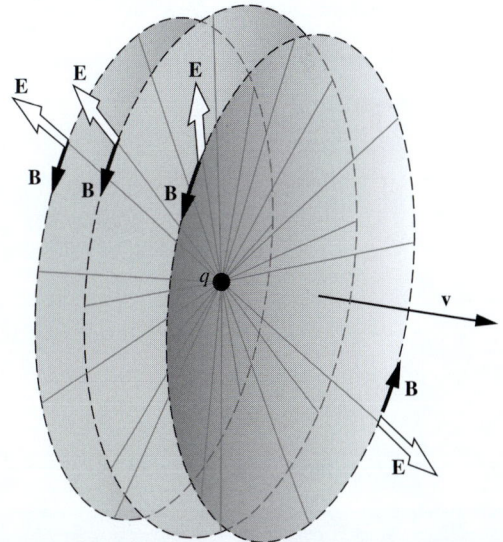

**Figure 6.30.**
The electric and magnetic fields, at one instant of time, of a charge in uniform motion.

We have come a long way from Coulomb's law in the last two chapters. Yet with each step we have only been following out consistently the requirements of relativity and of the invariance of electric charge. We can begin to see that the existence of the magnetic field and its curiously symmetrical relationship to the electric field is a necessary consequence of these general principles. We remind the reader again that this was not at all the historical order of discovery and elucidation of the laws of electromagnetism. One aspect of the coupling between the electric and magnetic fields, which is implicit in Eq. (6.74), came to light in Michael Faraday's experiments with changing electric currents, which will be described in Chapter 7. That was 75 years before Einstein, in his epochal paper of 1905, first wrote out our Eq. (6.74).

## 6.8 Rowland's experiment

As we remarked in Section 5.9, it was not obvious 150 years ago that a current flowing in a wire and a moving electrically charged object are essentially alike as sources of magnetic field. The unified view of electricity and magnetism that was then emerging from Maxwell's work suggested that any moving charge ought to cause a magnetic field, but experimental proof was hard to come by.

That the motion of an electrostatically charged sheet produces a magnetic field was first demonstrated by Henry Rowland, the great American physicist renowned for his perfection of the diffraction grating. Rowland made many ingenious and accurate electrical measurements, but none that taxed his experimental virtuosity as severely as the detection and measurement of the magnetic field of a rotating charged disk. The field to be detected was something like $10^{-5}$ of the earth's field in magnitude – a formidable experiment, even with today's instruments! In Fig. 6.31, you will see a sketch of Rowland's apparatus and a reproduction of the first page of the paper in which he described his experiment. Ten years before Hertz's discovery of electromagnetic waves, Rowland's result gave independent, if less dramatic, support to Maxwell's theory of the electromagnetic field.

## 6.9 Electrical conduction in a magnetic field: the Hall effect

When a current flows in a conductor in the presence of a magnetic field, the force $q\mathbf{v} \times \mathbf{B}$ acts directly on the moving charge carriers. Yet we observe a force on the conductor as a whole. Let's see how this comes about. Figure 6.32(a) shows a section of a metal bar in which a steady current is flowing. Driven by a field $\mathbf{E}$, electrons are drifting to the left with average speed $\bar{v}$, which has the same meaning as the $\bar{u}$ in our discussion of conduction in Chapter 4. The conduction electrons are indicated, very schematically, by the gray dots. The black dots are the positive

ON THE MAGNETIC EFFECT OF ELECTRIC CONVECTION[1]

*[American Journal of Science* [3], *XV*, 30–38, 1878]

The experiments described in this paper were made with a view of determining whether or not an electrified body in motion produces magnetic effects.
There seems to be no theoretical ground upon which we can settle the question,
seeing that the magnetic action of a conducted electric current may be ascribed
to some mutual action between the conductor and the current. Hence an experiment is of value. Professor Maxwell, in his 'Treatise on Electricity,' Art. 770, has
computed the magnetic action of a moving electrified surface, but that the action
exists has not yet been proved experimentally or theoretically.

The apparatus employed consisted of a vulcanite disc 21·1 centimetres in
diameter and ·5 centimetre thick which could be made to revolve around a vertical
axis with a velocity of 61· turns per second. On either side of the disc at a distance
of ·6 cm. were fixed glass plates having a diameter of 38·9 cm. and a hole in
the centre of 7·8 cm. The vulcanite disc was gilded on both sides and the glass
plates had an annular ring of gilt on one side, the outside and inside diameters
being 24·0 cm. and 8·9 cm. respectively. The gilt sides could be turned toward
or from the revolving disc but were usually turned toward it so that the problem
might be calculated more readily and there should be no uncertainty as to the
electrification. The outside plates were usually connected with the earth; and
the inside disc with an electric battery, by means of a point which approached
within one-third of a millimetre of the edge and turned toward it. As the edge
was broad, the point would not discharge unless there was a difference of
potential between it and the edge. Between the electric battery and the disc, . . .

---

[1]  The experiments described were made in the laboratory of the Berlin University
     through the kindness of Professor Helmholtz, to whose advice they are greatly
     indebated for their completeness. The idea of the experiment first occurred to me
     in 1868 and was recorded in a note book of that date.

**Figure 6.31.**
The essential parts of Rowland's apparatus. In
the tube at the left, two short magnetized
needles are suspended horizontally.

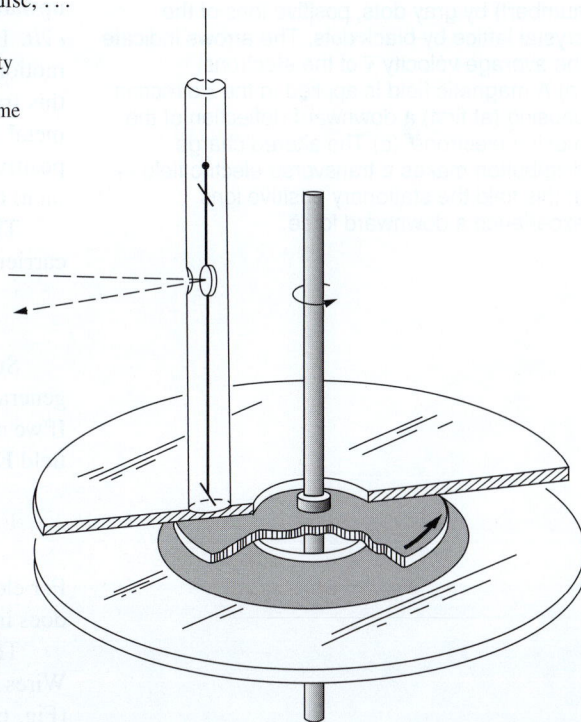

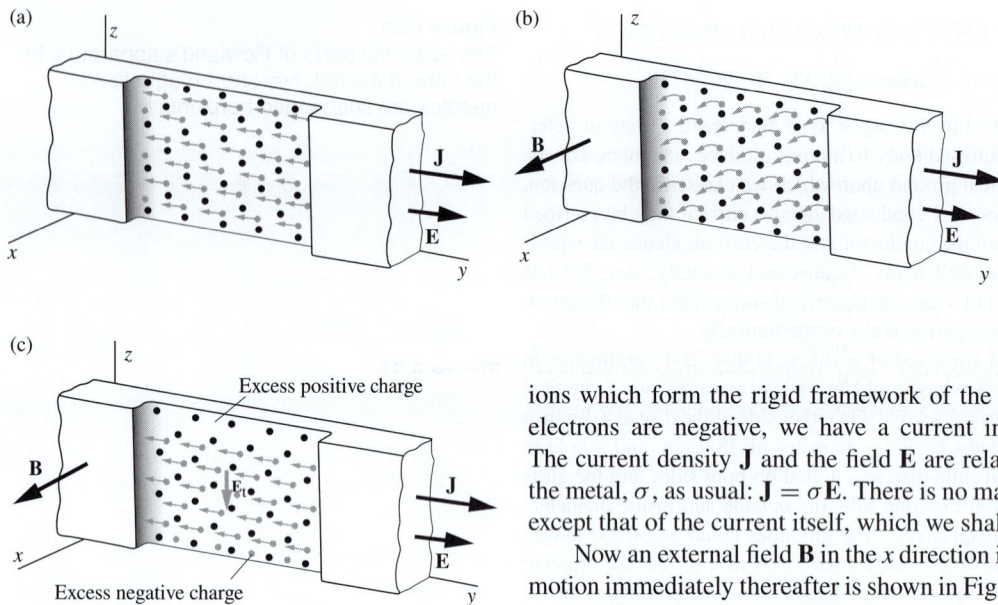

(a)

(b)

(c)

Excess positive charge

Excess negative charge

**Figure 6.32.**
(a) A current flows in a metal bar. Only a short
section of the bar is shown. Conduction
electrons are indicated (not in true size and
number!) by gray dots, positive ions of the
crystal lattice by black dots. The arrows indicate
the average velocity $\bar{v}$ of the electrons.
(b) A magnetic field is applied in the $x$ direction,
causing (at first) a downward deflection of the
moving electrons. (c) The altered charge
distribution makes a transverse electric field $E_t$.
In this field the stationary positive ions
experience a downward force.

ions which form the rigid framework of the solid metal bar. Since the
electrons are negative, we have a current in the positive $y$ direction.
The current density $\mathbf{J}$ and the field $\mathbf{E}$ are related by the conductivity of
the metal, $\sigma$, as usual: $\mathbf{J} = \sigma \mathbf{E}$. There is no magnetic field in Fig. 6.32(a)
except that of the current itself, which we shall ignore.

Now an external field $\mathbf{B}$ in the $x$ direction is switched on. The state of
motion immediately thereafter is shown in Fig. 6.32(b). The electrons are
being deflected downward. But since they cannot escape at the bottom of
the bar, they simply pile up there, until the surplus of negative charge
at the bottom of the bar and the corresponding excess of positive charge
at the top create a downward transverse electric field $\mathbf{E}_t$ in which the
upward force, of magnitude $eE_t$, exactly balances the downward force
$e\bar{v}B$. In the steady state (which is attained very quickly!) the average
motion is horizontal again, and there exists in the interior of the metal
this transverse electric field $\mathbf{E}_t$, as observed in coordinates fixed in the
metal lattice (Fig. 6.32(c)). This field causes a downward force on the
positive ions. That is how the force, $-e\bar{v} \times \mathbf{B}$, on the electrons is passed
on to the solid bar. The bar, of course, pushes on whatever is holding *it*.

The condition for zero average transverse force on the moving charge
carriers is

$$\mathbf{E}_t + \bar{\mathbf{v}} \times \mathbf{B} = 0. \qquad (6.83)$$

Suppose there are $n$ mobile charge carriers per m$^3$ and, to be more
general, denote the charge of each by $q$. Then the current density $\mathbf{J}$ is $nq\bar{\mathbf{v}}$.
If we now substitute $\mathbf{J}/nq$ for $\bar{\mathbf{v}}$ in Eq. (6.83), we can relate the transverse
field $\mathbf{E}_t$ to the directly measurable quantities $\mathbf{J}$ and $\mathbf{B}$:

$$\mathbf{E}_t = \frac{-\mathbf{J} \times \mathbf{B}}{nq}. \qquad (6.84)$$

For electrons $q = -e$, so $\mathbf{E}_t$ has in that case the direction of $\mathbf{J} \times \mathbf{B}$, as it
does in Fig. 6.32(c).

The existence of the transverse field can easily be demonstrated.
Wires are connected to points $P_1$ and $P_2$ on opposite edges of the bar
(Fig. 6.33), the junction points being carefully located so that they are

at the same potential when current is flowing in the bar and **B** is zero. The wires are connected to a voltmeter. After the field **B** is turned on, $P_1$ and $P_2$ are no longer at the same potential. The potential difference is $E_t$ times the width of the bar, and in the case illustrated $P_1$ is positive relative to $P_2$. A steady current will flow around the external circuit from $P_1$ to $P_2$, its magnitude determined by the resistance of the voltmeter. Note that the potential difference would be reversed if the current **J** consisted of positive carriers moving to the right rather than electrons moving to the left. This is true because the positive charge carriers would be deflected downward, just as the electrons were (because *two* things in the $q\mathbf{v} \times \mathbf{B}$ force have switched signs, namely the $q$ and the **v**). The field $E_t$ would therefore have the opposite sign, being now directed upward. Here for the first time we have an experiment that promises to tell us the *sign* of the charge carriers in a conductor.

The effect was discovered in 1879 by E. H. Hall, who was studying under Rowland at Johns Hopkins. In those days no one understood the mechanism of conduction in metals. The electron itself was unknown. It was hard to make much sense of the results. Generally the sign of the "Hall voltage" was consistent with conduction by negative carriers, but there were exceptions even to that. A complete understanding of the Hall effect in metallic conductors came only with the quantum theory of metals, about 50 years after Hall's discovery.

The Hall effect has proved to be especially useful in the study of semiconductors. There it fulfills its promise to reveal directly both the concentration and the sign of the charge carriers. The *n*-type and *p*-type semiconductors described in Chapter 4 give Hall voltages of opposite sign, as we should expect. As the Hall voltage is proportional to $B$, an appropriate semiconductor in the arrangement of Fig. 6.33 can serve, once calibrated, as a simple and compact device for measuring an unknown magnetic field. An example is described in Exercise 6.73.

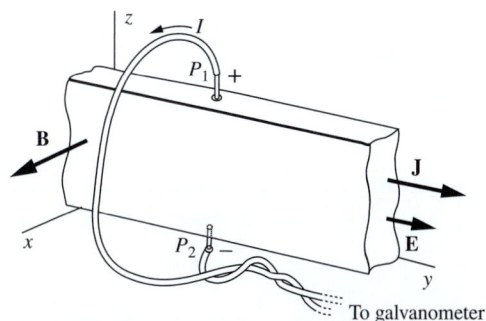

**Figure 6.33.**
The Hall effect. When a magnetic field is applied perpendicular to a conductor carrying current, a potential difference is observed between points on opposite sides of the bar – points that, in the absence of the field, would be at the same potential. This is consistent with the existence of the field $E_t$ inside the bar. By measuring the "Hall voltage" one can determine the number of charge carriers per unit volume, and their sign.

## 6.10 Applications

A *mass spectrometer* is used to determine the chemical makeup of a substance. Its operation is based on the fact that if a particle moves perpendicular to a magnetic field, the radius of curvature of the circular path depends on the particle's mass (see Exercise 6.29). In the spectrometer, molecules in a sample are first positively ionized, perhaps by bombarding them with electrons, which knocks electrons free. The ions are accelerated through a voltage difference and then sent through a magnetic field. Lighter ions have a smaller radius of curvature. (Even though lighter ions are accelerated to higher speeds, you can show that the magnetic field still bends them more compared with heavier ions.) By observing the final positions, the masses (or technically, the mass-to-charge ratios) of the various ions can be determined. Uses of mass spectroscopy include forensics, drug testing, testing for contaminants in food, and determining the composition of the atmosphere of planets.

The image in an old *television set* (made prior to the early 2000s) is created by a *cathode ray tube*. The wide end of the tube is the television screen. Electrons are fired toward the screen and are deflected by the magnetic field produced by current-carrying coils. This current is varied in such a way that the impact point of the electrons on the screen traces out the entire screen (many times per second) via a sequence of horizontal lines. When the electrons hit the screen, they cause phosphor in the screen to emit light. A particular image appears, depending on which locations are illuminated; the intensity of the electron beam is modulated to create the desired shading at each point (black, white, or something in between). Color TVs have three different electron beams for the three primary colors.

Consider two coaxial solenoids, with their currents oriented the same way, separated by some distance in the longitudinal direction. The magnetic field lines will diverge as they leave one solenoid, then reach a maximum width halfway between the solenoids, and then converge as they approach the other solenoid. Consider a charged particle moving approximately in a circle perpendicular to the field lines (under the influence of the Lorentz force), while also drifting in the direction of the field lines. It turns out that the drifting motion will be reversed in regions where the field lines converge, provided that they converge quickly enough. The particle can therefore be trapped in what is called a *magnetic bottle*, bouncing back and forth between the ends. A bottle-type effect can be used, for example, to contain plasma in fusion experiments.

A magnetic bottle can also be created by the magnetic field due to a current ring. The ring effectively produces a curved magnetic bottle; the field lines expand near the plane of the ring and converge near the axis. This is basically what the magnetic field of the earth looks like. The regions of trapped charged particles are called the *Van Allen belts*. If particles approach the ends of the belt, that is, if they approach the earth's atmosphere, the collisions with the air molecules cause the molecules to emit light. We know this light as the *northern (or southern) lights*, or alternatively as the *aurora borealis (or australis)*. The various colors come from the different atomic transitions in oxygen and nitrogen. The sources of the charged particles are *solar wind* and *cosmic rays*. However, the source can also be man-made: a 1962 high-altitude hydrogen bomb test, code-named "*Starfish Prime*," gave a wide area of the Pacific Ocean quite a light show. But at least people were "warned" about the test; the headline of the *Honolulu Advertiser* read, "N-Blast Tonight May Be Dazzling: Good View Likely."

The earth's atmosphere protects us significantly from solar wind (consisting mostly of protons and electrons) and from the steady background of cosmic rays (consisting mostly of protons). But the earth's magnetic field also helps out. The charged particles are deflected away from the earth by the Lorentz force from the magnetic field. In the event of a severe solar flare, however, a larger number of particles make it through the field, disrupting satellites and other electronics. Providing a way to shield astronauts from radiation is one of the main obstacles to extended space travel. Generating a magnetic field via currents in superconductors (see Appendix I) is a potential solution. Although this would certainly be expensive, propelling thick heavy shielding (the analog of the earth's atmosphere) into space would also be very costly.

A *railgun* is a device that uses the $q\mathbf{v} \times \mathbf{B}$ Lorentz force, instead of an explosive, to accelerate a projectile. The gun consists of two parallel conducting rails with a conducting object spanning the gap between them. This object (which is the projectile, or perhaps a larger object holding the projectile) is free to slide along the rails. A power source sends current down one rail, across the projectile, and back along the other rail. The current in the rails produces a magnetic field, and you can use the right-hand rule to show that the resulting Lorentz force on the projectile is directed away from the power source. Large-scale rail guns can achieve projectile speeds of a few kilometers per second and have a range of hundreds of kilometers.

In one form of *electric motor*, a DC current flows through a coil that is free to rotate between the poles of a fixed magnet. The Lorentz force on the charges moving in the coil produces a torque (see Exercise 6.34), so the coil is made to rotate. (Alternatively, the coil behaves just like a magnet with north and south poles, and this magnet interacts with the field of the fixed magnet.) The coil can then apply a torque to whatever object the motor is attached to. There are both AC and DC motors, and different kinds of each. In a *brushed* DC motor, a *commutator* causes the current to change direction every half cycle, so that the torque is always in the same direction. If for some reason the motor gets stuck and stops rotating, the *back emf* (which we will learn about in Chapter 7) drops to zero. The current through the coil then increases, causing it to heat up. The accompanying smell will let you know that you are in the process of burning out your motor.

Solenoids containing *superconducting* wires can be used to create very large fields, on the order of 20 T. The superconducting wires allow a large current to flow with no resistive heating. A *magnetic resonance imaging (MRI)* machine uses the physics described in Appendix J to make images of the inside of your body. Its magnetic field is usually about one or two tesla. The magnets in the Large Hadron Collider at CERN are also superconducting. If a defect causes the circuit to become nonsuperconducting (a result known as a quench), then the circuit will heat up rapidly and a chain reaction of very bad things is likely. As we will see in Chapter 7, solenoids store energy, and this energy needs to go somewhere. In 2008 at CERN, a quench caused major damage and took the accelerator offline for a year.

The size of the magnetic field in a superconducting solenoid is limited by the fact that a superconducting wire can't support magnetic fields or electric currents above certain critical values. The largest sustained magnetic fields in a laboratory are actually created with resistive conductors. The optimal design has the coils of the solenoid replaced by *Bitter plates* arranged in a helical pattern. These plates operate in basically the same manner as coils, but they allow for water cooling via well-placed holes. A Bitter magnet, which can achieve a field of about 35 T, requires a serious amount of water cooling – around 100 gallons per second!

If a coil of wire is wrapped around an iron core, the coil's magnetic field is magnified by the iron, for reasons we will see in Chapter 11. The magnification factor can be 100 or 1000, or even larger. This magnification effect is used in relays, circuit breakers, junkyard magnets (all discussed below), and many other devices. The combination of a coil and an iron core (or even a coil without a core) is called an *electromagnet*. The main advantage of an electromagnet over a permanent magnet is that the field can be turned on and off.

An electric *relay* is a device that uses a small signal to switch on (or off) a larger signal. (So a relay and a transistor act in the same manner.) A small current in one circuit passes through an electromagnet. The magnetic field of this electromagnet pulls on a spring-mounted iron lever that closes (or opens) a second circuit. The power source in this second circuit is generally much larger than in the first. A relay is used, for example, in conjunction with a thermostat. A small current in the thermostat's temperature sensor switches on a much larger current in the actual heating system, which involves, say, a hot-water pump. In years past, relays were used as telegraph repeaters. The relay took a weak incoming signal at the end of a long wire and automatically turned it into a strong outgoing signal, thereby eliminating the need for a human being to receive and retransmit the information.

A *circuit breaker* is similar to a fuse (see Section 4.12), in that it prevents the current in a circuit from becoming too large; 15 A or 20 A are typical thresholds for a household circuit breaker. If you are running many appliances in your home at the same time, the total current may be large enough to cause a wire somewhere inside a wall to overheat and start a fire. Similarly, a sustained short circuit will almost certainly cause a fire. One type of circuit breaker contains an electromagnet whose magnetic field pulls on an iron lever. The lever is held in place by a spring, but if the current in the electromagnet becomes sufficiently large, the force on the lever is large enough to pull it away from its resting position. This movement breaks the circuit in one way or another, depending on the design. The circuit breaker can be reset by simply flipping a switch that manually moves the lever back to its resting position. This should be contrasted with a fuse, which must be replaced each time it burns out.

*Junkyard magnets* can have fields in the 1 tesla range. This is fairly large, but the real key to the magnet's strength is its large area, which is on the order of a square meter. If you've ever played with a small rare-earth magnet (with a field of around 1 tesla) and felt how strongly it can stick to things, imagine playing with one that is a meter in diameter. It's no wonder it can pick up a car!

*Doorbells* (or at least those of the "ding dong" type) consist of a piston located inside a solenoid. Part of the piston is a permanent magnet. A spring holds the piston off to one side (say, to the left), where its left end rests on a sound bar. When the doorbell button is pressed, a circuit is completed and current flows through the solenoid. The resulting magnetic field pulls the piston through the solenoid and causes the right end to hit another sound bar located off to the right (this is the "ding" bar). As long as the button is held down, the piston stays there. But when the button is released, the current stops flowing in the solenoid, and the spring pushes the piston back to its initial position, where it strikes the left sound bar (the "dong" bar).

The *speakers* in your sound system are simple devices in principle, although they require a great deal of engineering to produce a quality sound. A speaker converts an electrical signal into sound waves. A coil of wire is located behind a movable cone (the main part of the speaker that you see, also called the diaphragm) and attached to its middle. The coil surrounds one pole of a permanent magnet, and is in turn surrounded by the other pole (imagine a stamp/cutter for making doughnuts; the coil is free to slide along the inner cylinder of the stamp). Depending on the direction of the current in the coil, the permanent magnet pushes the coil (and hence the cone) one way or the other. The movement of the cone produces the sound waves that travel to your ears. If the correct current (with the proper time-varying amplitude to control the volume, and frequency to control the pitch) is fed through the coil, the cone will oscillate in exactly the manner needed to produce the desired sound. Audible frequencies range from roughly 20 Hz to 20 kHz, so the oscillations will be quick. The necessary current originates in a *microphone*, which operates in the same way as a speaker, but in reverse. That is, a microphone converts sound waves into an electrical signal. There are many different types of microphones; we talked about one type in Section 3.9, and we will talk about another in Section 7.11, after we have covered electromagnetic induction.

*Maglev trains* (short for "magnetic levitation") are vertically supported, laterally stabilized, and longitudinally accelerated by magnetic fields. The train has no contact with the track, which means that it can go faster than a conventional train; speeds can reach 500 km/hr, or 300 mph. There is also less wear and tear. There are two main types of maglev trains: *electromagnetic suspension (EMS)* and *electrodynamic suspension (EDS)*. The EMS system makes use of both permanent magnets and electromagnets. The lateral motion of the train is unstable, so sensitive computerized correction is required. The EDS system makes use of electromagnets (in some cases involving superconductors), along with magnetic induction (discussed in Chapter 7). An advantage of EDS is that the lateral motion is stable, but a disadvantage is that the stray magnetic fields inside the cabin can be fairly large. Additionally, an EDS train requires a minimum speed to levitate, so wheels are needed at low speeds. The propulsion mechanism in both systems involves using alternating current to create magnetic fields that continually accelerate (or decelerate) the train. All maglev trains require a specially built track, which is a large impediment to adoption.

## CHAPTER SUMMARY

- The *Lorentz force* on a charged particle in an electromagnetic field is

$$\mathbf{F} = q\mathbf{E} + q\mathbf{v} \times \mathbf{B}. \tag{6.85}$$

- The magnetic field due to a current in a long straight wire points in the tangential direction and has magnitude

$$B = \frac{I}{2\pi \epsilon_0 r c^2} = \frac{\mu_0 I}{2\pi r}, \tag{6.86}$$

where

$$\mu_0 \equiv 4\pi \cdot 10^{-7} \frac{\text{kg m}}{\text{C}^2} \quad \text{and} \quad c^2 = \frac{1}{\mu_0 \epsilon_0}. \tag{6.87}$$

If a wire carrying current $I_2$ lies perpendicular to a magnetic field $B_1$, then the magnitude of the force on a length $l$ of the wire is $F = I_2 B_1 l$. The SI and Gaussian units of magnetic field are the *tesla* and *gauss*, respectively. One tesla equals exactly $10^4$ gauss.

- *Ampère's law* in integral and differential form is

$$\int \mathbf{B} \cdot d\mathbf{s} = \mu_0 I \quad \Longleftrightarrow \quad \text{curl}\,\mathbf{B} = \mu_0 \mathbf{J}. \tag{6.88}$$

The magnetic field also satisfies

$$\text{div}\,\mathbf{B} = 0 \tag{6.89}$$

This is the statement that there are no magnetic *monopoles*, or equivalently that magnetic field lines have no endings.

- The *vector potential* $\mathbf{A}$ is defined by

$$\mathbf{B} = \text{curl}\,\mathbf{A}, \tag{6.90}$$

which leads to div $\mathbf{B} = 0$ being identically true. Given the current density $\mathbf{J}$, the vector potential can be found via

$$\mathbf{A} = \frac{\mu_0}{4\pi} \int \frac{\mathbf{J}\,dv}{r} \quad \text{or} \quad \mathbf{A} = \frac{\mu_0 I}{4\pi} \int \frac{d\mathbf{l}}{r} \quad \text{(for a thin wire)}. \tag{6.91}$$

- The contribution to the magnetic field from a piece of a wire carrying current $I$ is given by the *Biot–Savart law*:

$$d\mathbf{B} = \frac{\mu_0 I}{4\pi} \frac{d\mathbf{l} \times \hat{\mathbf{r}}}{r^2} \quad \text{or} \quad d\mathbf{B} = \frac{\mu_0 I}{4\pi} \frac{d\mathbf{l} \times \mathbf{r}}{r^3}. \quad (6.92)$$

  This law is valid for steady currents.

- The field due to an infinitely long *solenoid* is zero outside and has magnitude $B = \mu_0 n I$ inside, where $n$ is the number of turns per unit length. If a sheet of current has current density $\mathcal{J}$, then the change in $B$ across the sheet is $\Delta B = \mu_0 \mathcal{J}$.

- The *Lorentz transformations* give the relations between the $\mathbf{E}$ and $\mathbf{B}$ fields in two different frames:

$$\mathbf{E}'_{\parallel} = \mathbf{E}_{\parallel}, \qquad \mathbf{E}'_{\perp} = \gamma\left(\mathbf{E}_{\perp} + \mathbf{v} \times \mathbf{B}_{\perp}\right),$$
$$\mathbf{B}'_{\parallel} = \mathbf{B}_{\parallel}, \qquad \mathbf{B}'_{\perp} = \gamma\left(\mathbf{B}_{\perp} - (\mathbf{v}/c^2) \times \mathbf{E}_{\perp}\right), \quad (6.93)$$

  where $\mathbf{v}$ is the velocity of frame $F'$ with respect to frame $F$. If there exists a frame in which $\mathbf{B} = 0$ (for example, if all the charges are at rest in one frame), then $\mathbf{B}' = -(\mathbf{v}/c^2) \times \mathbf{E}'$ in *all* frames. Similarly, if there exists a frame in which $\mathbf{E} = 0$ (for example, the frame of a neutral current-carrying wire), then $\mathbf{E}' = \mathbf{v} \times \mathbf{B}'$ in *all* frames.

- Henry Rowland demonstrated that a magnetic field is produced not only by a current in a wire, but also by the overall motion of an electrostatically charged object.

- In the *Hall effect*, an external magnetic field causes the charge carriers in a current-carrying wire to pile up on one side of the wire. This causes a transverse electric field inside the wire given by $\mathbf{E}_t = -(\mathbf{J} \times \mathbf{B})/nq$. For most purposes, a current of negative charges moving in one direction acts the same as a current of positive charges moving in the other direction. But the Hall effect can be used to determine the sign of the actual charge carriers.

## Problems

6.1    *Interstellar dust grain*  **

This problem concerns the electrically charged interstellar dust grain that was the subject of Exercise 2.38. Its mass, which was not involved in that problem, may be taken as $10^{-16}$ kg. Suppose it is moving quite freely, with speed $v \ll c$, in a plane perpendicular to the interstellar magnetic field, which in that region has a strength of $3 \cdot 10^{-6}$ gauss. How many years will it take to complete a circular orbit?

6.2    *Field from power lines*  *

A 50 kV direct-current power line consists of two wire conductors 2 m apart. When this line is transmitting 10 MW of power, how strong is the magnetic field midway between the conductors?

6.3    *Repelling wires* **

Suppose that the current $I_2$ in Fig. 6.4(b) is equal to $I_1$, but reversed, so that $CD$ is repelled by $GH$. Suppose also that $AB$ and $EF$ lie vertically above $GH$, that the lengths $BC$ and $CD$ are 30 and 15 cm, respectively, and that the conductor $BCDE$, which is 1 mm diameter copper wire as in Fig. 6.4(a), has a weight of 0.08 N/m. In equilibrium the deflection of the hanging frame from the vertical is such that $r = 0.5$ cm. How large is the current? Is the equilibrium stable?

6.4    *Vector potential for a wire* *

Consider the example in Section 6.3, concerning the vector potential for a long straight wire. Rewrite Eqs. (6.33) and (6.34) in terms of Cartesian coordinates, and verify that $\nabla \times \mathbf{A} = \mathbf{B}$.

6.5    *Vector potential for a finite wire* **

(a) Recall the example in Section 6.3 dealing with a thin infinite wire carrying current $I$. We showed that the vector potential $\mathbf{A}$ given in Eq. (6.34), or equivalently in Eq. (12.272) in the solution to Problem 6.4, correctly produced the desired magnetic field $\mathbf{B}$. However, although they successfully produced $\mathbf{B}$, there is something fundamentally wrong with those expressions for $\mathbf{A}$. What is it? (The infinities at $r = 0$ and $r = \infty$ are technically fine.)

(b) As mentioned at the end of Section 6.3, if you use Eq. (6.44) to calculate $\mathbf{A}$ for an infinite wire, you will obtain an infinite result. Your task here is instead to calculate $\mathbf{A}$ for a finite wire of length $2L$ (ignore the return path for the current), at a distance $r$ from the center. You can then find an approximate expression for $\mathbf{A}$ for large $L$ (is the issue from part (a) fixed?), and then take the curl to obtain $\mathbf{B}$, and then take the $L \to \infty$ limit to obtain the $\mathbf{B}$ field for a truly infinite wire.

6.6    *Zero divergence of A* ***

Show that the vector potential given by Eq. (6.44) satisfies $\nabla \cdot \mathbf{A} = 0$, provided that the current is steady (that is, $\nabla \cdot \mathbf{J} = 0$). *Hints:* Use the divergence theorem, and be careful about the two types of coordinates in Eq. (6.44) (the 1's and 2's). You will need to show that $\nabla_1(1/r_{12}) = -\nabla_2(1/r_{12})$, where the subscript denotes the set of coordinates with respect to which the derivative is taken. The vector identity $\nabla \cdot (f\mathbf{F}) = f\nabla \cdot \mathbf{F} + \mathbf{F} \cdot \nabla f$ will come in handy.

6.7    *Vector potential on a spinning sphere* ****

A spherical shell with radius $R$ and uniform surface charge density $\sigma$ rotates with angular speed $\omega$ around the $z$ axis. Calculate the vector potential at a point on the surface of the sphere. Do this in three steps as follows.

(a) By direct integration, calculate **A** at the point $(R, 0, 0)$. You will want to slice the shell into rings whose points are equidistant from $(R, 0, 0)$. The calculation isn't so bad once you realize that only one component of the velocity survives.

(b) Find **A** at the point $(x, 0, z)$ in Fig. 6.34 by considering the setup to be the superposition of two shells rotating with the angular velocity vectors $\boldsymbol{\omega}_1$ and $\boldsymbol{\omega}_2$ shown. (This works because angular velocity vectors simply add.)

(c) Finally, determine **A** at a general point $(x, y, z)$ on the surface of the sphere.

**6.8** *The field from a loopy wire* ∗

A current $I$ runs along an arbitrarily shaped wire that connects two given points, as shown in Fig. 6.35 (it need not lie in a plane). Show that the magnetic field at distant locations is essentially the same as the field due to a straight wire with current $I$ running between the two points.

**6.9** *Scaled-up ring* ∗

Consider two circular rings of copper wire. One ring is a scaled-up version of the other, twice as large in all regards (radius, cross-sectional radius). If currents around the rings are driven by equal voltage sources, how do the magnetic fields at the centers compare?

**6.10** *Rings with opposite currents* ∗∗

Two parallel rings have the same axis and are separated by a small distance $\epsilon$. They have the same radius $a$, and they carry the same current $I$ but in opposite directions. Consider the magnetic field at points on the axis of the rings. The field is zero midway between the rings, because the contributions from the rings cancel. And the field is zero very far away. So it must reach a maximum value at some point in between. Find this point. Work in the approximation where $\epsilon \ll a$.

**6.11** *Field at the center of a sphere* ∗∗

A spherical shell with radius $R$ and uniform surface charge density $\sigma$ spins with angular frequency $\omega$ around a diameter. Find the magnetic field at the center.

**6.12** *Field in the plane of a ring* ∗∗

A ring with radius $R$ carries a current $I$. Show that the magnetic field due to the ring, at a point in the plane of the ring, a distance $a$ from the center (either inside or outside the ring), is given by

$$B = 2 \cdot \frac{\mu_0 I}{4\pi} \int_0^\pi \frac{(R - a\cos\theta)R\,d\theta}{(a^2 + R^2 - 2aR\cos\theta)^{3/2}}. \tag{6.94}$$

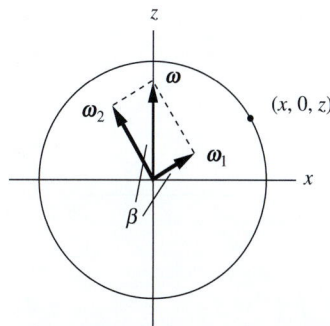

**Figure 6.34.**

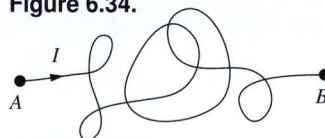

**Figure 6.35.**

*Hint:* The easiest way to handle the cross product in the Biot–Savart law is to write the Cartesian coordinates of $dl$ and $\mathbf{r}$ in terms of an angle $\theta$ in the ring.

This integral can't be evaluated in closed form (except in terms of elliptic functions), but it can always be evaluated numerically if desired. For the special case of $a = 0$ at the center of the ring, the integral is easy to do; verify that it yields the result given in Eq. (6.54).

6.13 *Magnetic dipole* **

Consider the result from Problem 6.12. In the $a \gg R$ limit (that is, very far from the ring), make suitable approximations and show that the magnitude of the magnetic field in the plane of the ring is approximately equal to $(\mu_0/4\pi)(m/a^3)$, where $m \equiv \pi R^2 I =$ (area)$I$ is the *magnetic dipole moment* of the ring. This is a special case of a result we will derive in Chapter 11.

6.14 *Far field from a square loop* ***

Consider a square loop with current $I$ and side length $a$. The goal of this problem is to determine the magnetic field at a point a large distance $r$ (with $r \gg a$) from the loop.

**Figure 6.36.**

(a) At the distant point $P$ in Fig. 6.36, the two vertical sides give essentially zero Biot–Savart contributions to the field, because they are essentially parallel to the radius vector to $P$. What are the Biot–Savart contributions from the two horizontal sides? These are easy to calculate because every little interval in these sides is essentially perpendicular to the radius vector to $P$. Show that the sum (or difference) of these contributions equals $\mu_0 I a^2/2\pi r^3$, to leading order in $a$.

(b) This result of $\mu_0 I a^2/2\pi r^3$ is not the correct field from the loop at point $P$. The correct field is half of this, or $\mu_0 I a^2/4\pi r^3$. We will eventually derive this in Chapter 11, where we will show that the general result is $\mu_0 I A/4\pi r^3$, where $A$ is the area of a loop with arbitrary shape. But we *should* be able to calculate it via the Biot–Savart law. Where is the error in the reasoning in part (a), and how do you go about fixing it? This is a nice one – don't peek at the answer too soon!

6.15 *Magnetic scalar "potential"* **

(a) Consider an infinite straight wire carrying current $I$. We know that the magnetic field outside the wire is $\mathbf{B} = (\mu_0 I/2\pi r)\,\hat{\boldsymbol{\theta}}$. There are no currents outside the wire, so $\nabla \times \mathbf{B} = 0$; verify this by explicitly calculating the curl.

(b) Since $\nabla \times \mathbf{B} = 0$, we should be able to write $\mathbf{B}$ as the gradient of a function, $\mathbf{B} = \nabla \psi$. Find $\psi$, but then explain why the usefulness of $\psi$ as a potential function is limited.

6.16 *Copper solenoid* **

A solenoid is made by winding two layers of No. 14 copper wire on a cylindrical form 8 cm in diameter. There are four turns per centimeter in each layer, and the length of the solenoid is 32 cm. From the wire tables we find that No. 14 copper wire, which has a diameter of 0.163 cm, has a resistance of 0.010 ohm/m at 75°C. (The coil will run hot!) If the solenoid is connected to a 50 V generator, what will be the magnetic field strength at the center of the solenoid in gauss, and what is the power dissipation in watts?

6.17 *A rotating solid cylinder* **

(a) A very long cylinder with radius $R$ and uniform volume charge density $\rho$ spins with frequency $\omega$ around its axis. What is the magnetic field at a point on the axis?

(b) How would your answer change if all the charge were concentrated on the surface?

6.18 *Vector potential for a solenoid* **

A solenoid has radius $R$, current $I$, and $n$ turns per unit length. Given that the magnetic field is $B = \mu_0 nI$ inside and $B = 0$ outside, find the vector potential **A** both inside and outside. Do this in two ways as follows.

(a) Use the result from Exercise 6.41.

(b) Use the expression for the curl in cylindrical coordinates given in Appendix F to find the forms of **A** that yield the correct values of $\mathbf{B} = \nabla \times \mathbf{A}$ in the two regions.

6.19 *Solenoid field, inside and outside* ***

Consider an infinite solenoid with circular cross section. The current is $I$, and there are $n$ turns per unit length. Show that the magnetic field is zero outside and $B = \mu_0 nI$ (in the longitudinal direction) everywhere inside. Do this in three steps as follows.

(a) Show that the field has only a longitudinal component. *Hint:* Consider the contributions to the field from rings that are symmetrically located with respect to a given point.

(b) Use Ampère's law to show that the field has a uniform value outside and a uniform value inside, and that these two values differ by $\mu_0 nI$.

(c) Show that $B \to 0$ as $r \to \infty$. There are various ways to do this. One is to obtain an upper bound on the field contribution due to a given ring by unwrapping the ring into a straight wire segment, and then finding the field due to this straight segment.

6.20 *A slab and a sheet* **

A volume current density $\mathbf{J} = J\hat{z}$ exists in a slab between the infinite planes at $x = -b$ and $x = b$. (So the current is coming out

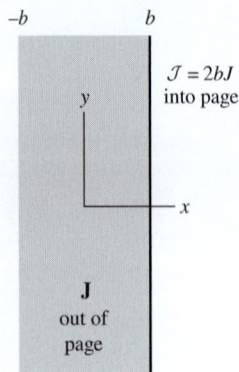

**Figure 6.37.**

of the page in Fig. 6.37.) Additionally, a surface current density $\mathcal{J} = 2bJ$ points in the $-\hat{\mathbf{z}}$ direction on the plane at $x = b$.

(a) Find the magnetic field as a function of $x$, both inside and outside the slab.

(b) Verify that $\nabla \times \mathbf{B} = \mu_0 \mathbf{J}$ inside the slab. (Don't worry about the boundaries.)

**6.21** *Maximum field in a cyclotron* **

For some purposes it is useful to accelerate negative hydrogen ions in a cyclotron. A negative hydrogen ion, H$^-$, is a hydrogen atom to which an extra electron has become attached. The attachment is fairly weak; an electric field of only $4.5 \cdot 10^8$ V/m in the frame of the ion (a rather small field by atomic standards) will pull an electron loose, leaving a hydrogen atom. If we want to accelerate H$^-$ ions up to a kinetic energy of 1 GeV ($10^9$ eV), what is the highest magnetic field we dare use to keep them on a circular orbit up to final energy? (To find $\gamma$ for this problem you only need the rest energy of the H$^-$ ion, which is of course practically the same as that of the proton, approximately 1 GeV.)

**6.22** *Zero force in any frame* **

A neutral wire carries current $I$. A stationary charge is nearby. There is no electric field from the neutral wire, so the electric force on the charge is zero. And although there is a magnetic field, the charge isn't moving, so the magnetic force is also zero. The total force on the charge is therefore zero. Hence it must be zero in every other frame. Verify this, in a particular case, by using the Lorentz transformations to find the **E** and **B** fields in a frame moving parallel to the wire with velocity **v**.

**6.23** *No magnetic shield* **

A student said, "You almost convinced me that the force between currents, which I thought was magnetism, is explained by electric fields of moving charges. But if so, why doesn't the metal plate in Fig. 5.1(c) shield one wire from the influence of the other?" Can you explain it?

**6.24** *E and B for a point charge* **

(a) Use the Lorentz transformations to show that the **E** and **B** fields due to a point charge moving with constant velocity **v** are related by $\mathbf{B} = (\mathbf{v}/c^2) \times \mathbf{E}$.

(b) If $v \ll c$, then **E** is essentially obtained from Coulomb's law, and **B** can be calculated from the Biot–Savart law. Calculate **B** this way, and then verify that it satisfies $\mathbf{B} = (\mathbf{v}/c^2) \times \mathbf{E}$. (It may be helpful to think of the point charge as a tiny rod of charge, in order to get a handle on the $d\mathbf{l}$ in the Biot–Savart law.)

6.25 *Force in three frames* ***

A charge $q$ moves with speed $v$ parallel to a wire with linear charge density $\lambda$ (as measured in the lab frame). The charges in the wire move with speed $u$ in the opposite direction, as shown in Fig. 6.38. If the charge $q$ is a distance $r$ from the wire, find the force on it in (a) the given lab frame, (b) its own rest frame, (c) the rest frame of the charges in the wire. Do this by calculating the electric and magnetic forces in the various frames. Then check that the force in the charge's rest frame relates properly to the forces in the other two frames. You can use the fact that the $\gamma$ factor associated with the relativistic addition of $u$ and $v$ is $\gamma_u \gamma_v (1 + \beta_u \beta_v)$.

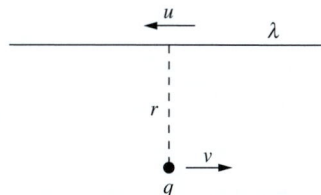

**Figure 6.38.**

6.26 *Motion in E and B fields* ***

The task of Exercise 6.29 is to show that if a charged particle moves in the $xy$ plane in the presence of a uniform magnetic field in the $z$ direction, the path will be a circle. What does the path look like if we add on a uniform electric field in the $y$ direction? Let the particle have mass $m$ and charge $q$. And let the magnitudes of the electric and magnetic fields be $E$ and $B$. Assume that the velocity is nonrelativistic, so that $\gamma \approx 1$ (this assumption isn't necessary in Exercise 6.29, because $v$ is constant there). Be careful, the answer is a bit counterintuitive.

6.27 *Special cases of Lorentz transformations* ***

Figure 6.39 shows four setups involving two infinite charged sheets in a given frame $F$. Another frame $F'$ moves to the right with speed $v$. Explain why these setups demonstrate the six indicated special cases (depending on which field is set equal to zero) of the Lorentz transformations in Eq. (6.76). (Note: one of the sheets has been drawn shorter to indicate length contraction. This is purely symbolic; all of the sheets have infinite length.)

6.28 *The retarded potential* ****

A point charge $q$ moves with speed $v$ along the line $y = r$ in the $xy$ plane. We want to find the magnetic field at the origin at the moment the charge crosses the $y$ axis.

(a) Starting with the electric field in the charge's frame, use the Lorentz transformation to show that, in the lab frame, the magnitude of the magnetic field at the origin (at the moment the charge crosses the $y$ axis) equals $B = (\mu_0/4\pi)(\gamma q v / r^2)$.

(b) Use the Biot–Savart law to calculate the magnetic field at the origin. For the purposes of obtaining the current, you may assume that the "point" charge takes the shape of a very short stick. You should obtain an incorrect answer, lacking the $\gamma$ factor in the above correct answer.

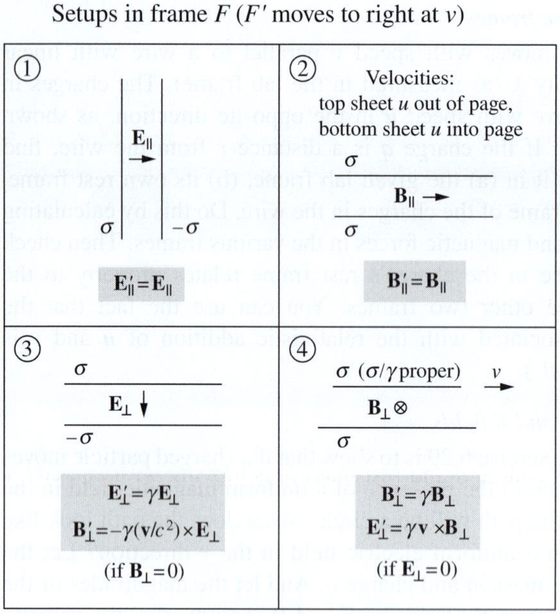

Setups in frame $F$ ($F'$ moves to right at $v$)

**Figure 6.39.**
Setups in frame $F$ ($F'$ moves to right at $v$).

(c) The Biot–Savart method is invalid because the Biot–Savart law holds for *steady* currents (or slowly changing ones, but see Footnote 8). But the current due to the point charge is certainly not steady. At a given location along the line of the charge's motion, the current is zero, then nonzero, then zero again.

For non-steady currents, the validity of the Biot–Savart law can be restored if we use the so-called "retarded time."[11] The basic idea with the retarded time is that, since information can travel no faster than the speed of light, the magnetic field at the origin, at the moment the charge crosses the $y$ axis, must be related to what the charge was doing *at an earlier time*. More precisely, this earlier time (the "retarded time") is the time such that if a light signal were emitted from the charge at this time, then it would reach the origin at the same instant the charge crosses the $y$ axis. Said in another way, if someone standing at the origin takes a photograph of the surroundings at the moment the charge crosses the $y$ axis, then the position of the charge in the photograph (which will *not* be on the $y$ axis) is the charge's location we are concerned with.[12]

---

[11] However, there is an additional term in the modified Biot–Savart law, which makes things more complicated. So we'll work instead with the vector potential $\mathbf{A}$, which still has only one term in its modified form. We can then obtain $\mathbf{B}$ by taking the curl of $\mathbf{A}$.

[12] Due to the finite speed of light, this is quite believable, so we will just accept it as true. But intuitive motivations aside, the modified retarded-time forms of $\mathbf{B}$ and $\mathbf{A}$ can be

Your tasks are to: find the location of the charge in the photograph; explain why the length of the little stick representing the charge has a greater length in the photograph than you might naively think; find this length. For the purposes of calculating the vector potential **A** at the origin, we therefore see that *the current extends over a greater length* than in the incorrect calculation above in part (b). Show that this effect produces the necessary extra $\gamma$ factor in **A**, and hence also in **B**. (Having taken into account the retarded time, the expression for **A** in Eq. (6.46) remains valid.)[13]

## Exercises

**6.29** *Motion in a B field* ✱✱

A particle of charge $q$ and rest mass $m$ is moving with velocity **v** where the magnetic field is **B**. Here **B** is perpendicular to **v**, and there is no electric field. Show that the path of the particle is a curve with radius of curvature $R$ given by $R = p/qB$, where $p$ is the momentum of the particle, $\gamma mv$. (*Hint:* Note that the force $q\mathbf{v} \times \mathbf{B}$ can only change the direction of the momentum, not the magnitude. By what angle $\Delta\theta$ is the direction of **p** changed in a short time $\Delta t$?) If **B** is the same everywhere, the particle will follow a circular path. Find the time required to complete one revolution.

**6.30** *Proton in space* ✱

A proton with kinetic energy $10^{16}$ eV ($\gamma = 10^7$) is moving perpendicular to the interstellar magnetic field, which in that region of the galaxy has a strength $3 \cdot 10^{-6}$ gauss. What is the radius of curvature of its path and how long does it take to complete one revolution? (Use the results from Exercise 6.29.)

**6.31** *Field from three wires* ✱

Three long straight parallel wires are located as shown in Fig. 6.40. One wire carries current $2I$ into the paper; each of the others carries current $I$ in the opposite direction. What is the strength of the magnetic field at the point $P_1$ and at the point $P_2$?

**6.32** *Oersted's experiment* ✱

Describing the experiment in which he discovered the influence of an electric current on a nearby compass needle, H. C. Oersted wrote: "If the distance of the connecting wire does not exceed three-quarters of an inch from the needle, the declination of the

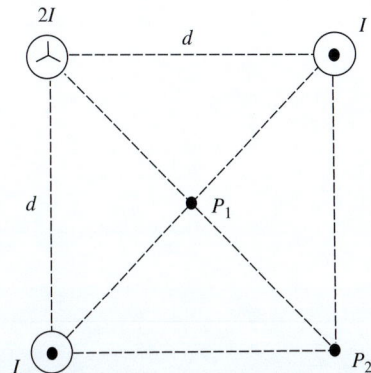

**Figure 6.40.**

---

rigorously derived from Maxwell's equations, as must be the case for any true statement about electromagnetic fields.

[13] If you want to solve this problem by working with the modified Biot–Savart law, it's a bit trickier. You'll need to use Eq. (14) in the article mentioned in Footnote 8. And you'll need to be *very* careful with all of the lengths involved.

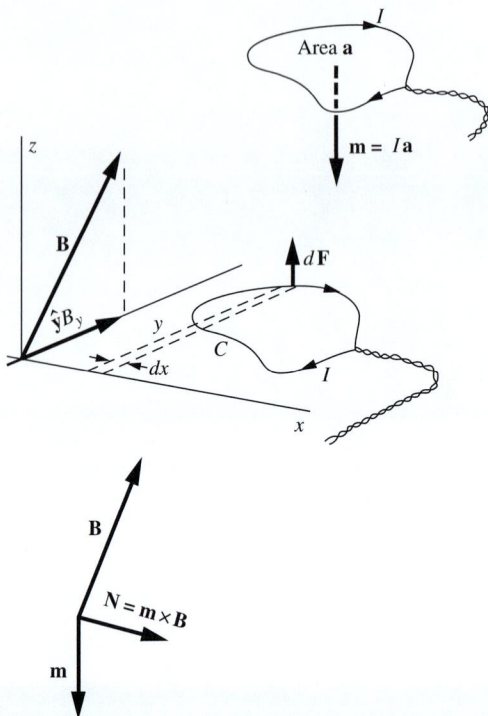

**Figure 6.41.**

needle makes an angle of about 45°. If the distance is increased the angle diminishes proportionally. The declination likewise varies with the power of the battery." About how large a current must have been flowing in Oersted's "connecting wire"? Assume the horizontal component of the earth's field in Copenhagen in 1820 was the same as it is today, 0.2 gauss.

**6.33 *Force between wires* ∗**

Suppose the current $I$ that flows in the circuit in Fig. 5.1(b) is 20 amperes. The distance between the wires is 5 cm. How large is the force, per meter of length, that pushes horizontally on one of the wires?

**6.34 *Torque on a loop* ∗∗∗**

The main goal of this problem is to find the torque that acts on a planar current loop in a uniform magnetic field. The uniform field **B** points in some direction in space. We shall orient our coordinates so that **B** is perpendicular to the $x$ axis, and our current loop lies in the $xy$ plane, as shown in Fig. 6.41. (You should convince yourself that this is always possible.) The shape and size of the (planar) loop are arbitrary; we may think of the current as being supplied by twisted leads on which any net force will be zero. Consider some small element of the loop, and work out its contribution to the torque about the $x$ axis. Only the $z$ component of the force on it will be involved, and hence only the $y$ component of the field **B**, which we have indicated as $\hat{\mathbf{y}}B_y$ in the diagram. Set up the integral that will give the total torque. Show that this integral will give, except for constant factors, the *area* of the loop.

The *magnetic moment* of a current loop is defined as a vector **m** whose magnitude is $Ia$, where $I$ is the current and $a$ is the area of the loop, and whose direction is normal to the loop with a right-hand-thread relation to the current, as shown in the figure. (We will meet the current loop and its magnetic moment again in Chapter 11.) Show now that your result implies that the torque **N** on any current loop is given by the vector equation

$$\mathbf{N} = \mathbf{m} \times \mathbf{B}. \qquad (6.95)$$

What about the net *force* on the loop?

**6.35 *Determining c* ∗∗∗∗**

The value of $1/\sqrt{\mu_0 \epsilon_0}$ (or equivalently the value of $c$) can be determined by electrical experiments involving low-frequency fields only. Consider the arrangement shown in Fig. 6.42. The force between capacitor plates is balanced against the force between parallel wires carrying current in the same direction. A voltage alternating sinusoidally at a frequency $f$ (in cycles per second) is applied to the parallel-plate capacitor $C_1$ and also to the capacitor $C_2$.

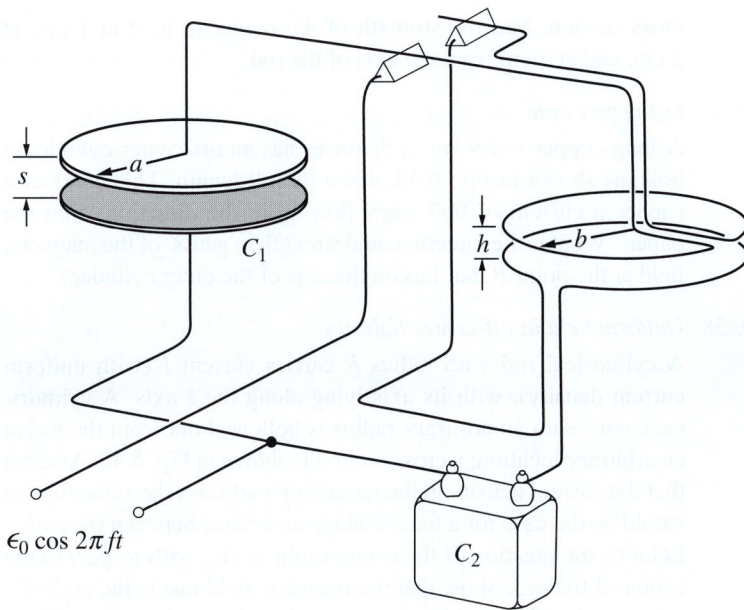

**Figure 6.42.**

The charge flowing into and out of $C_2$ constitutes the current in the rings.

Suppose that $C_2$ and the various distances involved have been adjusted so that the time-average downward force on the upper plate of $C_1$ exactly balances the time-average downward force on the upper ring. (Of course, the *weights* of the two sides should be adjusted to balance with the voltage turned off.) Show that under these conditions the constant $1/\sqrt{\mu_0\epsilon_0}$ $(= c)$ can be computed from measured quantities as follows:

$$\frac{1}{\sqrt{\mu_0\epsilon_0}} = (2\pi)^{3/2} a \left(\frac{b}{h}\right)^{1/2} \left(\frac{C_2}{C_1}\right) f. \qquad (6.96)$$

(If you work with Gaussian instead of SI units, you will end up solving for $c$ instead of $1/\sqrt{\mu_0\epsilon_0}$.) Assume $s \ll a$ and $h \ll b$.

Note that only measurements of *distance* and *time* (or frequency) are required, apart from a measurement of the ratio of the two capacitances $C_1$ and $C_2$. Electrical units, as such, are not involved in the result. (The experiment is actually feasible at a frequency as low as 60 cycles/second if $C_2$ is made, say, $10^6$ times $C_1$ and the current rings are made with several turns to multiply the effect of a small current.)

6.36 *Field at different radii* ∗

A current of 8000 amperes flows through an aluminum rod 4 cm in diameter. Assuming the current density is uniform through the

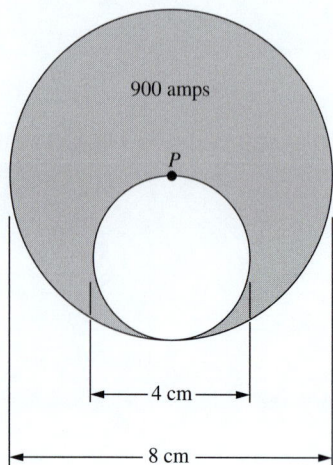

900 amps

*P*

|← 4 cm →|

|← 8 cm →|

**Figure 6.43.**

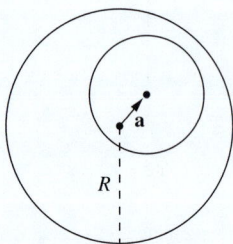

*R*

**a**

**Figure 6.44.**

cross section, find the strength of the magnetic field at 1 cm, at 2 cm, and at 3 cm from the axis of the rod.

6.37  *Off-center hole*  *

A long copper rod 8 cm in diameter has an off-center cylindrical hole, as shown in Fig. 6.43, down its full length. This conductor carries a current of 900 amps flowing in the direction "into the paper." What is the direction, and strength in gauss, of the magnetic field at the point *P* that lies on the axis of the outer cylinder?

6.38  *Uniform field in off-center hole*  **

A cylindrical rod with radius *R* carries current *I* (with uniform current density), with its axis lying along the *z* axis. A cylindrical cavity with an arbitrary radius is hollowed out from the rod at an arbitrary location; a cross section is shown in Fig. 6.44. Assume that the current density in the remaining part stays the same (which would be the case for a fixed voltage difference between the ends). Let **a** be the position of the center of the cavity with respect to the center of the rod. Show that the magnetic field inside the cylindrical cavity is uniform (in both magnitude and direction). *Hint:* Show that the field inside a solid cylinder can be written in the form $\mathbf{B} = (\mu_0 I/2\pi R^2)\hat{\mathbf{z}} \times \mathbf{r}$, and then use superposition with another appropriately chosen cylinder.

6.39  *Constant magnitude of B*  **

How should the current density inside a thick cylindrical wire depend on *r* so that the magnetic field has constant magnitude inside the wire?

6.40  *The pinch effect*  **

Since parallel current filaments attract one another, one might think that a current flowing in a solid rod like the conductor in Problem 6.36 would tend to concentrate near the axis of the rod. That is, the conduction electrons, instead of distributing themselves evenly as usual over the interior of the metal, would crowd in toward the axis and most of the current would be there. What do you think prevents this from happening? Ought it happen to any extent at all? Can you suggest an experiment to detect such an effect, if it should exist?

6.41  *Integral of A, flux of B*  *

Show that the line integral of the vector potential **A** around a closed curve *C* equals the magnetic flux Φ through a surface *S* bounded by the curve. This result is very similar to Ampère's law, which says that the line integral of the magnetic field **B** around a closed curve *C* equals (up to a factor of $\mu_0$) the current flux *I* through a surface *S* bounded by the curve.

**6.42** *Finding the vector potential* ∗

See if you can devise a vector potential that will correspond to a uniform field in the $z$ direction: $B_x = 0$, $B_y = 0$, $B_z = B_0$.

**6.43** *Vector potential inside a wire* ∗∗

A round wire of radius $r_0$ carries a current $I$ distributed uniformly over the cross section of the wire. Let the axis of the wire be the $z$ axis, with $\hat{\mathbf{z}}$ the direction of the current. Show that a vector potential of the form $\mathbf{A} = A_0\hat{\mathbf{z}}(x^2 + y^2)$ will correctly give the magnetic field $\mathbf{B}$ of this current at all points inside the wire. What is the value of the constant, $A_0$?

**6.44** *Line integral along the axis* ∗∗

Consider the magnetic field of a circular current ring, at points on the axis of the ring, given by Eq. (6.53). Calculate explicitly the line integral of the field along the axis from $-\infty$ to $\infty$, to check the general formula

$$\int \mathbf{B} \cdot d\mathbf{s} = \mu_0 I. \qquad (6.97)$$

Why may we ignore the "return" part of the path which would be necessary to complete a closed loop?

**6.45** *Field from an infinite wire* ∗∗

Use the Biot–Savart law to calculate the magnetic field at a distance $b$ from an infinite straight wire carrying current $I$.

**6.46** *Field from a wire frame* ∗

(a) Current $I$ flows around the wire frame in Fig. 6.45(a). What is the direction of the magnetic field at $P$, the center of the cube?

(b) Show by using superposition that the field at $P$ is the same as if the frame were replaced by the single square loop shown in Fig. 6.45(b).

**6.47** *Field at the center of an orbit* ∗

An electron is moving at a speed $0.01c$ on a circular orbit of radius $10^{-10}$ m. What is the strength of the resulting magnetic field at the center of the orbit? (The numbers given are typical, in order of magnitude, for an electron in an atom.)

**6.48** *Fields from two rings* ∗

A ring with radius $r$ and linear charge density $\lambda$ spins with frequency $\omega$. A second ring with radius $2r$ has the same density $\lambda$ and frequency $\omega$. Each ring produces a magnetic field at its center. How do the magnitudes of these fields compare?

**6.49** *Field at the center of a disk* ∗

A disk with radius $R$ and surface charge density $\sigma$ spins with angular frequency $\omega$. What is the magnetic field at the center?

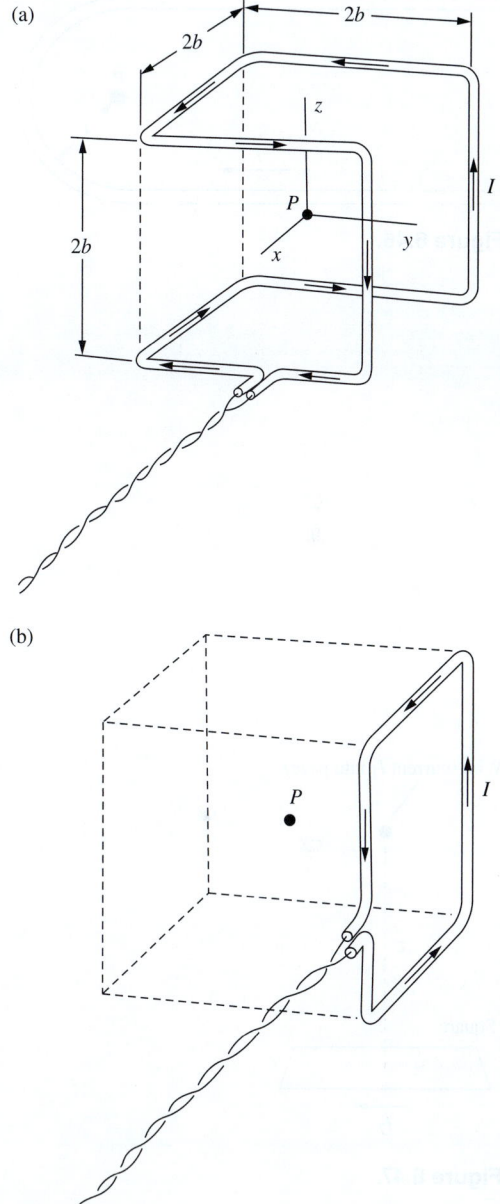

Figure 6.45.

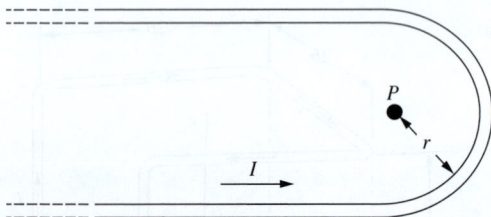

**Figure 6.46.**

**6.50** *Hairpin field* *

A long wire is bent into the hairpin-like shape shown in Fig. 6.46. Find an exact expression for the magnetic field at the point $P$ that lies at the center of the half-circle.

**6.51** *Current in the earth* *

The earth's metallic core extends out to 3000 km, about half the earth's radius. Imagine that the field we observe at the earth's surface, which has a strength of roughly 0.5 gauss at the north magnetic pole, is caused by a current flow in a ring around the "equator" of the core. How big would that current be?

**6.52** *Right-angled wire* **

A wire carrying current $I$ runs down the $y$ axis to the origin, thence out to infinity along the positive $x$ axis. Show that the magnetic field at any point in the $xy$ plane (except right on one of the axes) is given by

$$B_z = \frac{\mu_0 I}{4\pi}\left(\frac{1}{x} + \frac{1}{y} + \frac{x}{y\sqrt{x^2+y^2}} + \frac{y}{x\sqrt{x^2+y^2}}\right). \qquad (6.98)$$

**6.53** *Superposing right angles* **

Use the result from Exercise 6.52, along with superposition, to derive the magnetic field due to an infinite straight wire. (This is certainly an inefficient way of obtaining this field!)

**6.54** *Force between a wire and a loop* **

Figure 6.47 shows a horizontal infinite straight wire with current $I_1$ pointing into the page, passing a height $z$ above a square horizontal loop with side length $\ell$ and current $I_2$. Two of the sides of the square are parallel to the wire. As with a circular ring, this square produces a magnetic field that points upward on its axis. The field fans out away from the axis. From the right-hand rule, you can show that the magnetic force on the straight wire points to the right. By Newton's third law, the magnetic force on the square must therefore point to the left.

Your tasks: explain qualitatively, by drawing the fields and forces, why the force on the square does indeed point to the left; then show that the net force equals $\mu_0 I_1 I_2 \ell^2/2\pi R^2$, where $R = \sqrt{z^2 + (\ell/2)^2}$ is the distance from the wire to the right and left sides of the square. (The calculation of the force on the wire is a bit more involved. We'll save that for Exercise 11.20, after we've discussed magnetic dipoles.)

**6.55** *Helmholtz coils* **

One way to produce a very uniform magnetic field is to use a very long solenoid and work only in the middle section of its interior. This is often inconvenient, wasteful of space and power. Can you

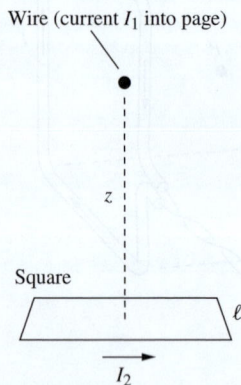

**Figure 6.47.**

suggest ways in which two short coils or current rings might be arranged to achieve good uniformity over a limited region? *Hint:* Consider two coaxial current rings of radius $a$, separated axially by a distance $b$. Investigate the uniformity of the field in the vicinity of the point on the axis midway between the two coils. Determine the magnitude of the coil separation $b$ that for given coil radius $a$ will make the field in this region as nearly uniform as possible.

6.56 *Field at the tip of a cone* **
A hollow cone (like a party hat) has vertex angle $2\theta$, slant height $L$, and surface charge density $\sigma$. It spins around its symmetry axis with angular frequency $\omega$. What is the magnetic field at the tip?

6.57 *A rotating cylinder* *
An infinite cylinder with radius $R$ and surface charge density $\sigma$ spins around its symmetry axis with angular frequency $\omega$. Find the magnetic field inside the cylinder.

6.58 *Rotating cylinders* **
Two long coaxial aluminum cylinders are charged to a potential difference of 15 kV. The inner cylinder (assumed to be the positive one) has an outer diameter of 6 cm, the outer cylinder an inner diameter of 8 cm. With the outer cylinder stationary the inner cylinder is rotated around its axis at a constant speed of 30 revolutions per second. Describe the magnetic field this produces and determine its intensity in gauss. What if both cylinders are rotated in the same direction at 30 revolutions per second?

6.59 *Scaled-down solenoid* **
Consider two solenoids, one of which is a tenth-scale model of the other. The larger solenoid is 2 meters long, 1 meter in diameter, and is wound with 1 cm diameter copper wire. When the coil is connected to a 120 V direct-current generator, the magnetic field at its center is 1000 gauss. The scaled-down model is exactly one-tenth the size in every linear dimension, including the diameter of the wire. The number of turns is the same, and it is designed to provide the same central field.
(a) Show that the voltage required is the same, namely 120 V.
(b) Compare the coils with respect to the power dissipated and the difficulty of removing this heat by some cooling means.

6.60 *Zero field outside a solenoid* ***
We showed in the solution to Problem 6.19 that the magnetic field is zero outside an infinite solenoid with arbitrary (uniform) cross-sectional shape. We can demonstrate this fact in another way, similar in spirit to Problem 1.17.

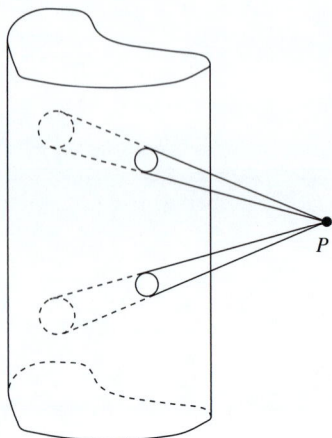

**Figure 6.48.**

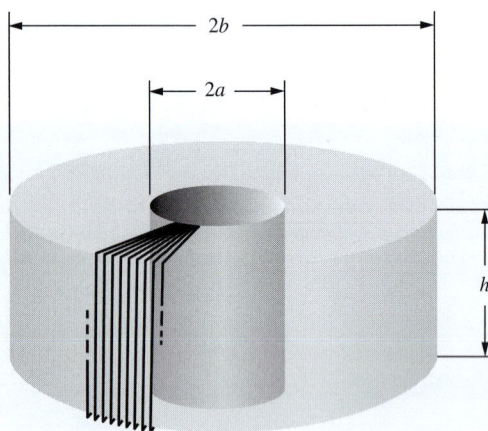

**Figure 6.49.**

Consider a thin cone emanating from an exterior point $P$, and look at the two patches where it intersects the solenoid. Consider also the thin cone symmetrically located on the other side of $P$ (as shown in Fig. 6.48), along with its two associated patches. Show that the sum of the field contributions due to these four patches is zero at $P$.

6.61 *Rectangular torus* ∗∗∗

A coil is wound evenly on a torus of rectangular cross section. There are $N$ turns of wire in all. Only a few are shown in Fig. 6.49. With so many turns, we shall assume that the current on the surface of the torus flows exactly radially on the annular end faces, and exactly longitudinally on the inner and outer cylindrical surfaces. First convince yourself that on this assumption, symmetry requires that the magnetic field everywhere should point in a "circumferential" direction, that is, that all field lines are circles about the axis of the torus. Second, prove that the field is zero at all points outside the torus, including the interior of the central hole. Third, find the magnitude of the field inside the torus, as a function of radius.

6.62 *Creating a uniform field* ∗∗∗

For a delicate magnetic experiment, a physicist wants to cancel the earth's field over a volume roughly $30 \times 30 \times 30$ cm in size, so that the residual field in this region will not be greater than 10 milligauss at any point. The strength of the earth's field in this location is 0.55 gauss, making an angle of 30° with the vertical. It may be assumed constant to a milligauss or so over the volume in question. (The earth's field itself would hardly vary that much over a foot or so, but in a laboratory there are often local perturbations.) Determine roughly what solenoid dimensions would be suitable for the task, and estimate the number of ampere turns (that is, the current $I$ multiplied by the number of turns $N$) required in your compensating system.

6.63 *Solenoids and superposition* ∗∗∗

A number of simple facts about the fields of solenoids can be found by using superposition. The idea is that two solenoids of the same diameter, and length $L$, if joined end to end, make a solenoid of length $2L$. Two semi-infinite solenoids butted together make an infinite solenoid, and so on. (A semi-infinite solenoid is one that has one end here and the other infinitely far away.) Here are some facts you can prove this way.

(a) In the finite-length solenoid shown in Fig. 6.50(a), the magnetic field on the axis at the point $P_2$ at one end is approximately half the field at the point $P_1$ in the center. (Is it slightly more than half, or slightly less than half?)

(b) In the semi-infinite solenoid shown in Fig. 6.50(b), the field line *FGH*, which passes through the very end of the winding, is a straight line from *G* out to infinity.

(c) The flux of **B** through the end face of the semi-infinite solenoid is just half the flux through the coil at a large distance back in the interior.

(d) Any field line that is a distance $r_0$ from the axis far back in the interior of the coil exits from the end of the coil at a radius $r_1 = \sqrt{2}\, r_0$, assuming that $r_0 < $ (solenoid radius)$/\sqrt{2}$.

Show that these statements are true. What else can you find out?

**6.64  Equal magnitudes** **

Suppose we have a situation in which the component of the magnetic field parallel to the plane of a sheet has the same *magnitude* on both sides, but changes *direction* by 90° in going through the sheet. What is going on here? Would there be a force on the sheet? Should our formula for the force on a current sheet apply to cases like this?

**6.65  Proton beam** **

A high-energy accelerator produces a beam of protons with kinetic energy 2 GeV (that is, $2 \cdot 10^9$ eV per proton). You may assume that the rest energy of a proton is 1 GeV. The current is 1 milliamp, and the beam diameter is 2 mm. As measured in the laboratory frame:

(a) what is the strength of the electric field caused by the beam 1 cm from the central axis of the beam?

(b) What is the strength of the magnetic field at the same distance?

(c) Now consider a frame $F'$ that is moving along with the protons. What fields would be measured in $F'$?

**6.66  Fields in a new frame** **

In the neighborhood of the origin in the coordinate system $x$, $y$, $z$, there is an electric field **E** of magnitude 100 V/m, pointing in a direction that makes angles of 30° with the $x$ axis, 60° with the $y$ axis. The frame $F'$ has its axes parallel to those just described, but is moving, relative to the first frame, with a speed $0.6c$ in the positive $y$ direction. Find the direction and magnitude of the electric field that will be reported by an observer in the frame $F'$. What magnetic field does this observer report?

**6.67  Fields from two ions** **

According to observers in the frame $F$, the following events occurred in the $xy$ plane. A singly charged positive ion that had been moving with the constant velocity $v = 0.6c$ in the $\hat{\mathbf{y}}$ direction passed through the origin at $t = 0$. At the same instant a similar

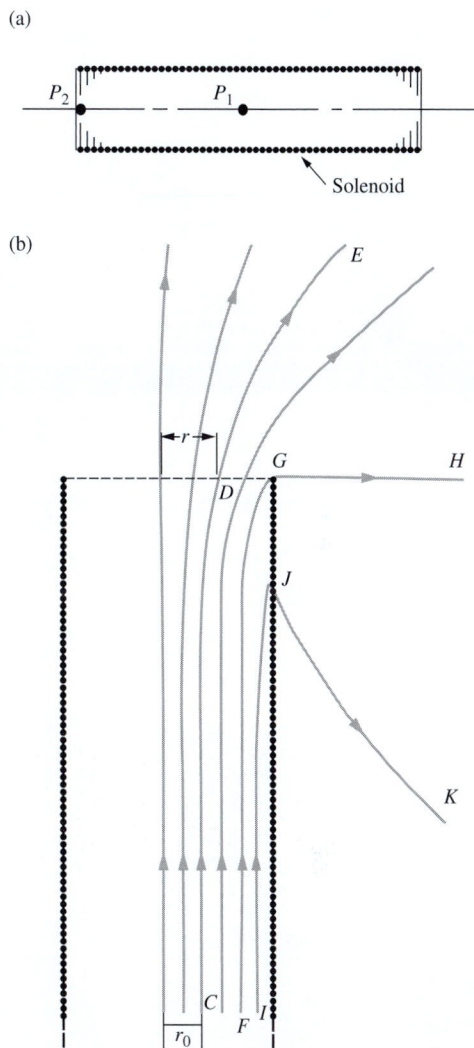

(a)

(b)

**Figure 6.50.**

ion that had been moving with the same speed, but in the $-\hat{\mathbf{y}}$ direction, passed the point $(2, 0, 0)$ on the $x$ axis. The distances are in meters.

(a) What are the strength and direction of the electric field, at $t = 0$, at the point $(3, 0, 0)$?

(b) What are the strength and direction of the magnetic field at the same place and time?

6.68 *Force on electrons moving together* **

Consider two electrons in a cathode ray tube that are moving on parallel paths, side by side, at the same speed $v$. The distance between them, a distance measured at right angles to their velocity, is $r$. What is the force that acts on one of them, owing to the presence of the other, as observed in the laboratory frame? If $v$ were very small compared with $c$, you could answer $e^2/4\pi\epsilon_0 r^2$ and let it go at that. But $v$ isn't small, so you have to be careful.

(a) The easiest way to get the answer is as follows. Go to a frame of reference moving with the electrons. In that frame the two electrons are at rest, the distance between them is still $r$ (why?), and the force *is* just $e^2/4\pi\epsilon_0 r^2$. Now transform the force into the laboratory frame, using the force transformation law, Eq. (5.17). (Be careful about which is the primed system; is the force in the lab frame greater or less than the force in the electron frame?)

(b) It should be possible to get the same answer working entirely in the lab frame. In the lab frame, at the instantaneous position of electron 1, there are both electric and magnetic fields arising from electron 2 (see Fig. 6.30). Calculate the net force on electron 1, which is moving through these fields with speed $v$, and show that you get the same result as in (a). Make a diagram to show the directions of the fields and forces.

(c) In the light of this, what can you say about the force between two side-by-side moving electrons, in the limit $v \to c$?

6.69 *Relating the forces* **

Two very long sticks each have uniform linear proper charge density $\lambda$. ("Proper" means as measured in the rest frame of the given object.) One stick is stationary in the lab frame, while the other stick moves to the left with speed $v$, as shown in Fig. 6.51. They are $2r$ apart, and a stationary point charge $q$ lies midway between them. Find the electric and magnetic forces on the charge $q$ in the lab frame, and also in the frame of the bottom stick. (Be sure to specify the directions.) Then verify that the total forces in the two frames relate properly.

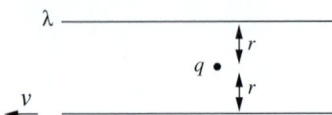

**Figure 6.51.**

**6.70** *Drifting motion* **

Figure 6.52 shows the path of a positive ion moving in the $xy$ plane. There is a uniform magnetic field of 6000 gauss in the $z$ direction. Each period of the ion's cycloidal motion is completed in 1 microsecond. What are the magnitude and the direction of the electric field that must be present? *Hint:* Think about a frame in which the electric field is zero.

**6.71** *Rowland's experiment* ***

Calculate approximately the magnetic field to be expected just above the rotating disk in Rowland's experiment. Take the relevant data from the description on the page of his paper that is reproduced in Fig. 6.31. You will need to know also that the potential of the rotating disk, with respect to the grounded plates above and below it, was around 10 kilovolts in most of his runs. This information is of course given later in his paper, as is a description of a crucial part of the apparatus, the "astatic" magnetometer shown in the vertical tube on the left. This is an arrangement in which two magnetic needles, oppositely oriented, are rigidly connected together on one suspension so that the torques caused by the earth's field cancel one another. The field produced by the rotating disk, acting mainly on the nearer needle, can then be detected in the presence of a very much stronger uniform field. That is by no means the only precaution Rowland had to take. In solving this problem, you can make the simplifying assumption that the charges in the disk all travel with their average speed.

**6.72** *Transverse Hall field* *

Show that the Gaussian version of Eq. (6.84) must read $\mathbf{E}_t = -\mathbf{J} \times \mathbf{B}/nqc$, where $E_t$ is in statvolts/cm, $B$ is in gauss, $n$ is in cm$^{-3}$, and $q$ is in esu.

**6.73** *Hall voltage* **

A Hall probe for measuring magnetic fields is made from arsenic-doped silicon, which has $2 \cdot 10^{21}$ conduction electrons per m$^3$ and a resistivity of 0.016 ohm-m. The Hall voltage is measured across a ribbon of this $n$-type silicon that is 0.2 cm wide, 0.005 cm thick, and 0.5 cm long between thicker ends at which it is connected into a 1 V battery circuit. What voltage will be measured across the 0.2 cm dimension of the ribbon when the probe is inserted into a field of 1 kilogauss?

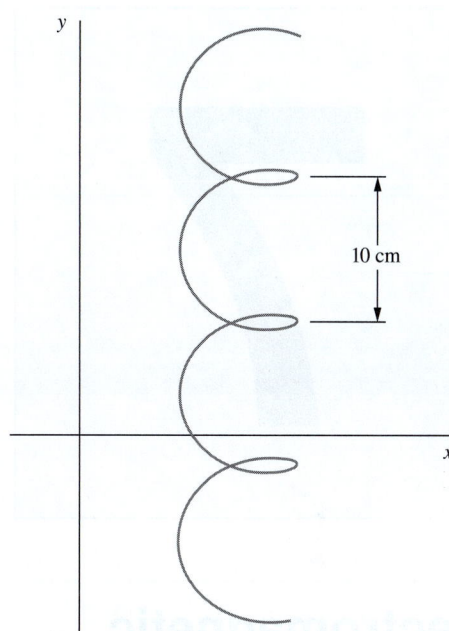

**Figure 6.52.**

# 7

# Electromagnetic induction

**Overview** In this chapter we study the effects of magnetic fields that change with time. Our main result will be that a changing magnetic field causes an electric field. We begin by using the Lorentz force to calculate the emf around a loop moving through a magnetic field. We then make the observation that this emf can be written in terms of the rate of change of the *magnetic flux* through the loop. The sign of the induced emf is determined by *Lenz's law*. If we shift frames so that the loop is now stationary and the source of the magnetic field is moving, we obtain the same result for the emf in terms of the rate of change of flux, as expected. *Faraday's law of induction* states that this result holds independent of the cause of the flux change. For example, it applies to the case in which we turn a dial to decrease the magnetic field while keeping all objects stationary. The differential form of Faraday's law is one of Maxwell's equations. *Mutual inductance* is the effect by which a changing current in one loop causes an emf in another loop. This effect is symmetrical between the two loops, as we will prove. *Self-inductance* is the effect by which a changing current in a loop causes an emf in itself. The most commonly used object with self-inductance is a solenoid, which we call an *inductor*, symbolized by $L$. The current in an $RL$ circuit changes in a specific way, as we will discover. The energy stored in an inductor equals $LI^2/2$, which parallels the $CV^2/2$ energy stored in a capacitor. Similarly, the energy density in a magnetic field equals $B^2/2\mu_0$, which parallels the $\epsilon_0 E^2/2$ energy density in an electric field.

## 7.1 Faraday's discovery

Michael Faraday's account of the discovery of electromagnetic induction begins as follows:

1. The power which electricity of tension possesses of causing an opposite electrical state in its vicinity has been expressed by the general term Induction; which, as it has been received into scientific language, may also, with propriety, be used in the same general sense to express the power which electrical currents may possess of inducing any particular state upon matter in their immediate neighbourhood, otherwise indifferent. It is with this meaning that I purpose using it in the present paper.

2. Certain effects of the induction of electrical currents have already been recognised and described: as those of magnetization; Ampère's experiments of bringing a copper disc near to a flat spiral; his repetition with electromagnets of Arago's extraordinary experiments, and perhaps a few others. Still it appeared unlikely that these could be all the effects which induction by currents could produce; especially as, upon dispensing with iron, almost the whole of them disappear, whilst yet an infinity of bodies, exhibiting definite phenomena of induction with electricity of tension, still remain to be acted upon by the induction of electricity in motion.

3. Further: Whether Ampère's beautiful theory were adopted, or any other, or whatever reservation were mentally made, still it appeared very extraordinary, that as every electric current was accompanied by a corresponding intensity of magnetic action at right angles to the current, good conductors of electricity, when placed within the sphere of this action, should not have any current induced through them, or some sensible effect produced equivalent in force to such a current.

4. These considerations, with their consequence, the hope of obtaining electricity from ordinary magnetism, have stimulated me at various times to investigate experimentally the inductive effect of electric currents. I lately arrived at positive results; and not only had my hopes fulfilled, but obtained a key which appeared to me to open out a full explanation of Arago's magnetic phenomena, and also to discover a new state, which may probably have great influence in some of the most important effects of electric currents.

5. These results I purpose describing, not as they were obtained, but in such a manner as to give the most concise view of the whole.

This passage was part of a paper Faraday presented in 1831. It is quoted from his "Experimental Researches in Electricity," published in London in 1839 (Faraday, 1839). There follows in the paper a description of a dozen or more experiments, through which Faraday brought to light every essential feature of the production of electric effects by magnetic action.

By "electricity of tension" Faraday meant electrostatic charges, and the induction he refers to in the first sentence involves nothing more than we have studied in Chapter 3: that the presence of a charge causes a redistribution of charges on conductors nearby. Faraday's question was, why does an electric current not cause another current in nearby conductors?

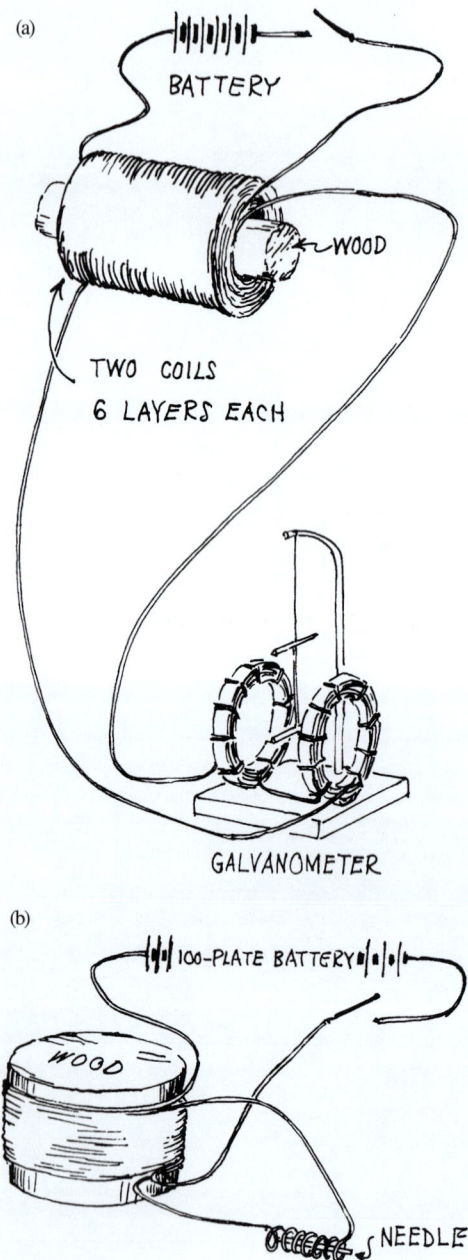

(a)

BATTERY

WOOD

TWO COILS
6 LAYERS EACH

GALVANOMETER

(b)

100-PLATE BATTERY

WOOD

NEEDLE

**Figure 7.1.**
Interpretation by the author of some of Faraday's
experiments described in his "Experimental
Researches in Electricity," London, 1839.

The production of magnetic fields by electric currents had been thoroughly investigated after Oersted's discovery in 1820. The familiar laboratory source of these "galvanic" currents was the voltaic battery. The most sensitive detector of such currents was a galvanometer. It consisted of a magnetized needle pivoted like a compass needle or suspended by a weak fiber between two coils of wire. Sometimes another needle, outside the coil but connected rigidly to the first needle, was used to compensate for the influence of the earth's magnetic field (Fig. 7.1(a)). The sketches in Fig. 7.1(b)–(e) represent a few of Faraday's induction experiments. You must read his own account, one of the classics of experimental science, to appreciate the resourcefulness with which he pressed the search, the alert and open mind with which he viewed the evidence.

In his early experiments Faraday was puzzled to find that a steady current had no detectable effect on a nearby circuit. He constructed various coils of wire, of which Fig. 7.1(a) shows an example, winding two conductors so that they should lie very close together while still separated by cloth or paper insulation. One conductor would form a circuit with the galvanometer. Through the other he would send a strong current from a battery. There was, disappointingly, no deflection of the galvanometer. But in one of these experiments he noticed a very slight disturbance of the galvanometer when the current was switched on and another when it was switched off. Pursuing this lead, he soon established beyond doubt that currents in other conductors are induced, not by a *steady* current, but by a *changing* current. One of Faraday's brilliant experimental tactics at this stage was to replace his galvanometer, which he realized was not a good detector for a brief pulse of current, by a simple small coil in which he put an unmagnetized steel needle; see Fig. 7.1(b). He found that the needle was left magnetized by the pulse of current induced when the primary current was switched on – and it could be magnetized in the opposite sense by the current pulse induced when the primary circuit was broken.

Here is his own description of another experiment:

> In the preceding experiments the wires were placed near to each other, and the contact of the inducing one with the battery made when the inductive effect was required; but as the particular action might be supposed to be exerted only at the moments of making and breaking contact, the induction was produced in another way. Several feet of copper wire were stretched in wide zigzag forms, representing the letter W, on one surface of a broad board; a second wire was stretched in precisely similar forms on a second board, so that when brought near the first, the wires should everywhere touch, except that a sheet of thick paper was interposed. One of these wires was connected with the galvanometer, and the other with a voltaic battery. The first wire was then moved towards the second, and as it approached, the needle was deflected. Being then removed, the needle was deflected in the opposite direction. By first making the wires approach and then recede, simultaneously with the vibrations of the needle, the latter soon became very extensive; but when the wires ceased to move from or towards each other, the galvanometer needle soon came to its usual position.

As the wires approximated, the induced current was in the *contrary* direction to the inducing current. As the wires receded, the induced current was in the *same* direction as the inducing current. When the wires remained stationary, there was no induced current.

In this chapter we study the electromagnetic interaction that Faraday explored in those experiments. From our present viewpoint, induction can be seen as a natural consequence of the force on a charge moving in a magnetic field. In a limited sense, we can derive the induction law from what we already know. In following this course we again depart from the historical order of development, but we do so (borrowing Faraday's own words from the end of the passage first quoted) "to give the most concise view of the whole."

## 7.2 Conducting rod moving through a uniform magnetic field

Figure 7.2(a) shows a straight piece of wire, or a slender metal rod, supposed to be moving at constant velocity **v** in a direction perpendicular to its length. Pervading the space through which the rod moves there is a uniform magnetic field **B**, constant in time. This could be supplied by a large solenoid enclosing the entire region of the diagram. The reference frame $F$ with coordinates $x$, $y$, $z$ is the one in which this solenoid is at rest. In the absence of the rod there is no electric field in that frame, only the uniform magnetic field **B**.

The rod, being a conductor, contains charged particles that will move if a force is applied to them. Any charged particle that is carried along with the rod, such as the particle of charge $q$ in Fig. 7.2(b), necessarily moves through the magnetic field **B** and therefore experiences a force

$$\mathbf{f} = q\mathbf{v} \times \mathbf{B}. \qquad (7.1)$$

With **B** and **v** directed as shown in Fig. 7.2, the force is in the positive $x$ direction if $q$ is a positive charge, and in the opposite direction for the negatively charged electrons that are in fact the mobile charge carriers in most conductors. The consequences will be the same, whether negatives or positives, or both, are mobile.

When the rod is moving at constant speed and things have settled down to a steady state, the force **f** given by Eq. (7.1) must be balanced, at every point inside the rod, by an equal and opposite force. This can only arise from an electric field in the rod. The electric field develops in this way: the force **f** pushes negative charges toward one end of the rod, leaving the other end positively charged. This goes on until these separated charges themselves cause an electric field **E** such that, everywhere in the interior of the rod,

$$q\mathbf{E} = -\mathbf{f}. \qquad (7.2)$$

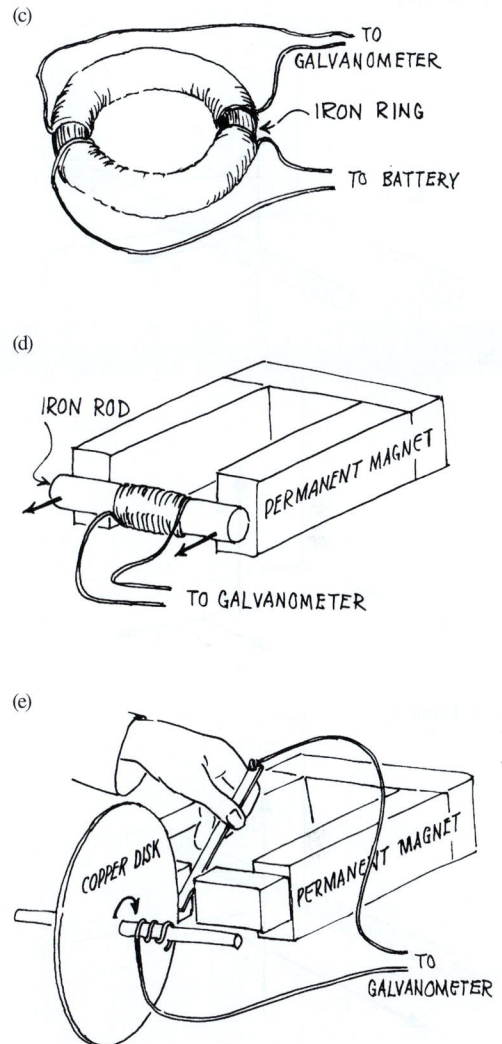

(c)

TO GALVANOMETER

IRON RING

TO BATTERY

(d)

IRON ROD

PERMANENT MAGNET

TO GALVANOMETER

(e)

COPPER DISK

PERMANENT MAGNET

TO GALVANOMETER

**Figure 7.1.**
*(Continued)*

(a) Frame $F$

$z$

$\mathbf{B}$

$\mathbf{E} = 0$

$x$

$\mathbf{v}$

$y$

(b)

$z$

$\mathbf{B}$

$q$

$x$

$\mathbf{v}$

$y$

$f$

(c) Frame $F'$

$z'$

$\mathbf{B}'$

$x'$

$y'$

$\mathbf{E}'$

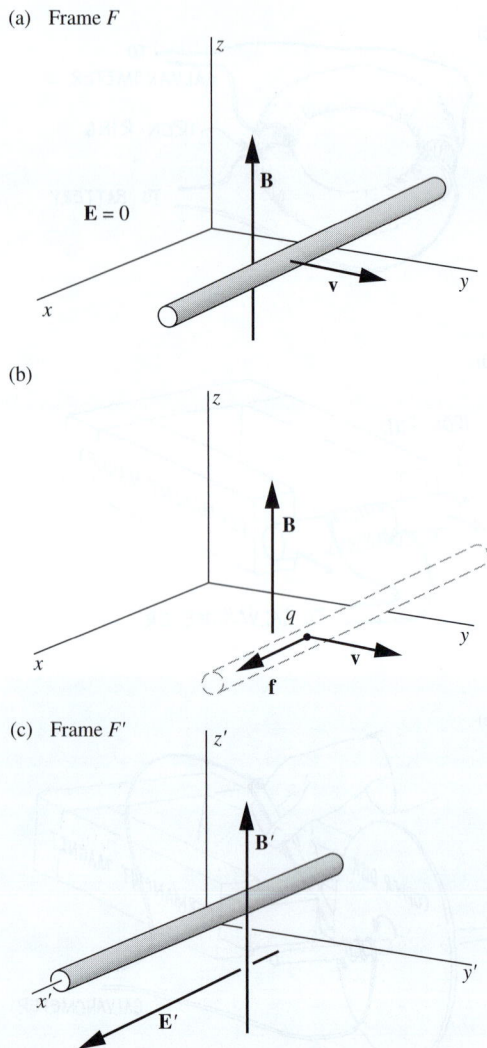

**Figure 7.2.**
(a) A conducting rod moves through a magnetic field. (b) Any charge $q$ that travels with the rod is acted upon by the force $q\mathbf{v} \times \mathbf{B}$. (c) The reference frame $F'$ moves with the rod; in this frame there is an electric field $\mathbf{E}'$.

Then the motion of charge relative to the rod ceases. This charge distribution causes an electric field outside the rod, as well as inside. The field outside looks something like that of separated positive and negative charges, with the difference that the charges are not concentrated entirely at the ends of the rod but are distributed along it. The external field is sketched in Fig. 7.3(a). Figure 7.3(b) is an enlarged view of the positively charged end of the rod, showing the charge distribution on the surface and some field lines both outside and inside the conductor. That is the way things look, at any instant of time, in frame $F$.

Let us observe this system from a frame $F'$ that moves with the rod. Ignoring the rod for the moment, we see in this frame $F'$, indicated in Fig. 7.2(c), a magnetic field $\mathbf{B}'$ (not much different from $\mathbf{B}$ if $v$ is small, by Eq. (6.76)) together with a uniform electric field, as given by Eq. (6.82),

$$\mathbf{E}' = \mathbf{v} \times \mathbf{B}'. \tag{7.3}$$

This is valid for any value of $v$. When we add the rod to this system, all we are doing is putting a stationary conducting rod into a uniform electric field. There will be a redistribution of charge on the surface of the rod so as to make the electric field zero inside, as in the case of the metal box of Fig. 3.8, or of any other conductor in an electric field. The presence of the magnetic field $\mathbf{B}'$ has no influence on this static charge distribution. Figure 7.4(a) shows some electric field lines in the frame $F'$. This field is the sum of the uniform field in Eq. (7.3) and the field due to the separated positive and negative charges. In the magnified view of the end of the rod in Fig. 7.4(b), we observe that the electric field *inside* the rod is zero.

Except for the Lorentz contraction, which is second order in $v/c$, the charge distribution seen at one instant in frame $F$, Fig. 7.3(b), is the same as that seen in $F'$. The electric fields differ because the field in Fig. 7.3 is that of the surface charge distribution alone, while the electric field we see in Fig. 7.4 is the field of the surface charge distribution *plus* the uniform electric field that exists in that frame of reference. An observer in $F$ says, "Inside the rod there has developed an electric field $\mathbf{E} = -\mathbf{v} \times \mathbf{B}$, exerting a force $q\mathbf{E} = -q\mathbf{v} \times \mathbf{B}$ which just balances the force $q\mathbf{v} \times \mathbf{B}$ that would otherwise cause any charge $q$ to move along the rod." An observer in $F'$ says, "Inside the rod there is no electric field, because the redistribution of charge on the rod causes there to be zero net internal field, as usual in a conductor. And although there is a uniform magnetic field here, no force arises from it because no charges are moving." Each account is correct.

## 7.3 Loop moving through a nonuniform magnetic field

What if we made a rectangular loop of wire, as shown in Fig. 7.5, and moved it at constant speed through the uniform field $\mathbf{B}$? To predict what will happen, we need only ask ourselves – adopting the frame $F'$ – what would happen if we put such a loop into a uniform electric field. Obviously two opposite sides of the rectangle would acquire some charge, but that would be all. Suppose, however, that the field $\mathbf{B}$ in the frame $F$, though constant in time, is *not uniform* in space. To make this vivid, we show in Fig. 7.6 the field $\mathbf{B}$ with a short solenoid as its source. This solenoid, together with the battery that supplies its constant current, is fixed near the origin in the frame $F$. (We said earlier there is no electric field in $F$; if we really use a solenoid of finite resistance to provide the field, there will be an electric field associated with the battery and this circuit. It is irrelevant to our problem and can be ignored. Or we can pack the whole solenoid, with its battery, inside a metal box, making sure the total charge is zero.)

Now, with the loop moving with speed $v$ in the $y$ direction, in the frame $F$, let its position at some instant $t$ be such that the magnetic field strength is $B_1$ at the left side of the loop and $B_2$ along the right side (Fig. 7.6). Let $\mathbf{f}$ denote the force that acts on a charge $q$ that rides along with the loop. This force is a function of position on the loop, at this instant of time. Let's evaluate the line integral of $\mathbf{f}$, taken around the whole loop (counterclockwise as viewed from above). On the two sides of the loop that lie parallel to the direction of motion, $\mathbf{f}$ is perpendicular to the path element $d\mathbf{s}$, so these give nothing. Taking account of the contributions from the other two sides, each of length $w$, we have

$$\int \mathbf{f} \cdot d\mathbf{s} = qv(B_1 - B_2)w. \tag{7.4}$$

If we imagine a charge $q$ to move all around the loop, in a time short enough so that the position of the loop has not changed appreciably, then Eq. (7.4) gives the work done by the force $\mathbf{f}$. The work done *per unit charge* is $(1/q) \int \mathbf{f} \cdot d\mathbf{s}$. We call this quantity *electromotive force*. We use the symbol $\mathcal{E}$ for it, and often shorten the name to emf. So we have

$$\boxed{\mathcal{E} \equiv \frac{1}{q} \int \mathbf{f} \cdot d\mathbf{s}} \tag{7.5}$$

$\mathcal{E}$ has the same dimensions as electric potential, so the SI unit is the volt, or joule per coulomb. In the Gaussian system, $\mathcal{E}$ is measured in statvolts, or ergs per esu.

We have noted that the force $\mathbf{f}$ does work. However, $\mathbf{f}$ is a magnetic force, and we know that magnetic forces do no work, because the force is always perpendicular to the velocity. So we seem to have an issue here. Is the magnetic force somehow doing work? If not, then what is? This is the subject of Problem 7.2.

The term *electromotive force* was introduced earlier, in Section 4.9. It was defined as the work per unit charge involved in moving a charge

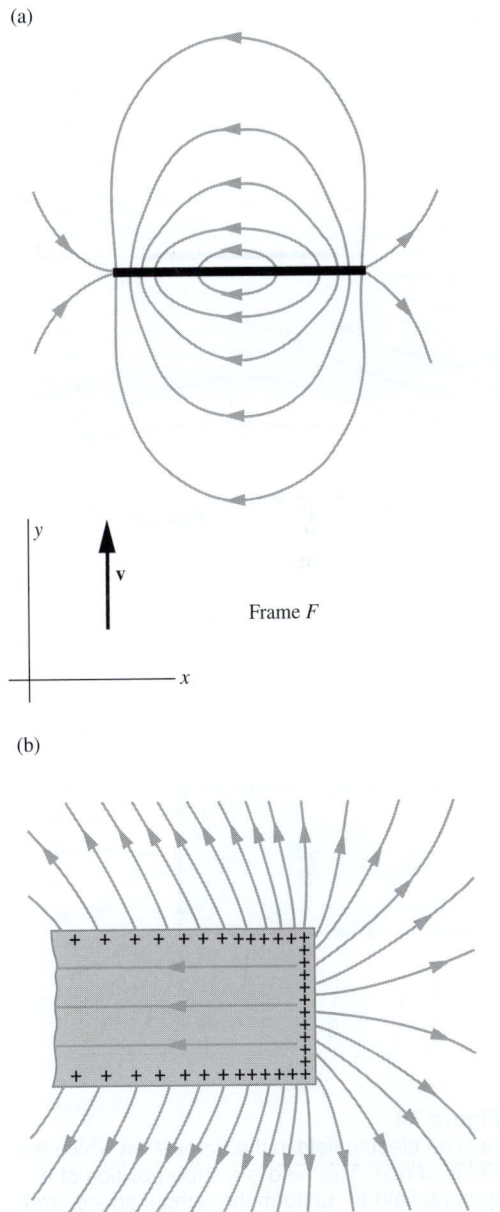

(a)

Frame $F$

(b)

**Figure 7.3.**
(a) The electric field, as seen at one instant of time, in the frame $F$. There is an electric field in the vicinity of the rod, and also inside the rod. The sources of the field are charges on the surface of the rod, as shown in (b), the enlarged view of the right-hand end of the rod.

(a)

(b)

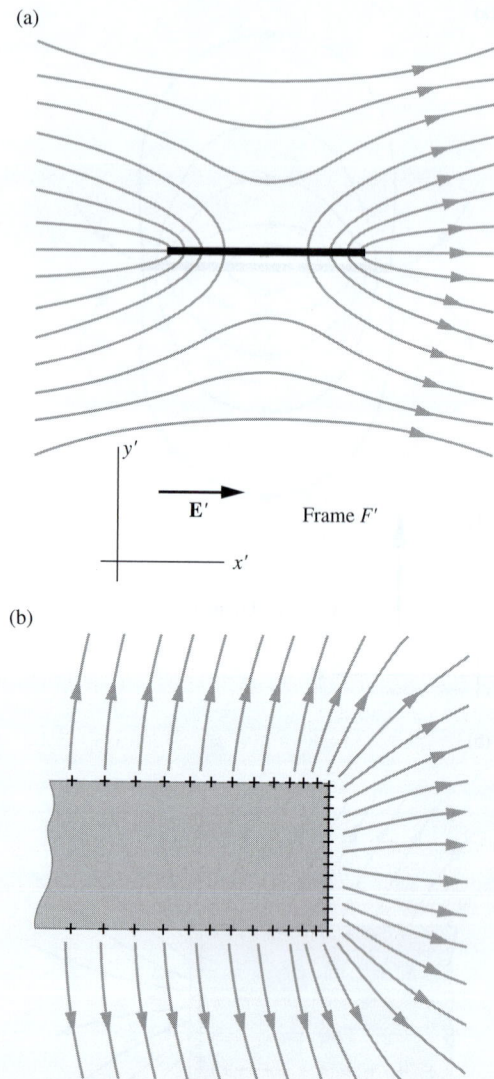

**Figure 7.4.**
(a) The electric field in the frame $F'$ in which the rod is at rest. This field is a superposition of a general field $\mathbf{E}'$, uniform throughout space, and the field of the surface charge distribution. The result is zero electric field inside the rod, shown in magnified detail in (b). Compare with Fig. 7.3.

around a circuit containing a voltaic cell. We now broaden the definition of emf to include any influence that causes charge to circulate around a closed path. If the path happens to be a physical circuit with resistance $R$, then the emf $\mathcal{E}$ will cause a current to flow according to Ohm's law: $I = \mathcal{E}/R$. Note that since curl $\mathbf{E} = 0$ for an electrostatic field, such a field cannot cause a charge to circulate around a closed path. By our above definition of electromotive force, an emf must therefore be *nonelectrostatic* in origin. See Varney and Fisher (1980) for a discussion of electromotive force.

In the particular case we are considering, $\mathbf{f}$ is the force that acts on a charge moving in a magnetic field, and $\mathcal{E}$ has the magnitude

$$\mathcal{E} = vw(B_1 - B_2). \tag{7.6}$$

The electromotive force given by Eq. (7.6) is related in a very simple way to the *rate of change of magnetic flux* through the loop. (We will be quantitative about this in Theorem 7.1.) By the magnetic flux through a loop we mean the surface integral of $\mathbf{B}$ over a surface that has the loop for its boundary. The flux $\Phi$ through the closed curve or loop $C$ in Fig. 7.7(a) is given by the surface integral of $\mathbf{B}$ over $S_1$:

$$\Phi_{S_1} = \int_{S_1} \mathbf{B} \cdot d\mathbf{a}_1. \tag{7.7}$$

We could draw infinitely many surfaces bounded by $C$. Figure 7.7(b) shows another one, $S_2$. Why don't we have to specify which surface to use in computing the flux? It *doesn't make any difference* because $\int \mathbf{B} \cdot d\mathbf{a}$ will have the same value for all surfaces. Let's take a minute to settle this point once and for all.

The flux through $S_2$ will be $\int_{S_2} \mathbf{B} \cdot d\mathbf{a}_2$. Note that we let the vector $d\mathbf{a}_2$ stick out from the upper side of $S_2$, to be consistent with our choice of side of $S_1$. This will give a positive number if the net flux through $C$ is upward:

$$\Phi_{S_2} = \int_{S_2} \mathbf{B} \cdot d\mathbf{a}_2. \tag{7.8}$$

We learned in Section 6.2 that the magnetic field has zero divergence: div $\mathbf{B} = 0$. It follows then from Gauss's theorem that, if $S$ is any *closed* surface ("balloon") and $V$ is the volume inside it, we have

$$\int_S \mathbf{B} \cdot d\mathbf{a} = \int_V \text{div } \mathbf{B} \, dv = 0. \tag{7.9}$$

Apply this to the closed surface, rather like a kettledrum, formed by joining our $S_1$ to $S_2$, as in Fig. 7.7(c). On $S_2$ the outward normal is *opposite* the vector $d\mathbf{a}_2$ we used in calculating the flux through $C$. Thus

$$0 = \int_S \mathbf{B} \cdot d\mathbf{a} = \int_{S_1} \mathbf{B} \cdot d\mathbf{a}_1 + \int_{S_2} \mathbf{B} \cdot (-d\mathbf{a}_2), \tag{7.10}$$

or

$$\int_{S_1} \mathbf{B} \cdot d\mathbf{a}_1 = \int_{S_2} \mathbf{B} \cdot d\mathbf{a}_2. \qquad (7.11)$$

This shows that it doesn't matter which surface we use to compute the flux through $C$.

This is all pretty obvious if you realize that div $\mathbf{B} = 0$ implies a kind of spatial conservation of flux. As much flux enters any volume as leaves it. (We are considering the situation in the whole space at one instant of time.) It is often helpful to visualize "tubes" of flux. A flux tube (Fig. 7.8) is a surface at every point on which the magnetic field line lies in the plane of the surface. It is a surface through which no flux passes, and we can think of it as containing a certain amount of flux, as a fiber optic cable contains fibers. Through any closed curve drawn tightly around a flux tube, the same flux passes. This could be said about the electric field $\mathbf{E}$ only for regions where there is no electric charge, since div $\mathbf{E} = \rho/\epsilon_0$. The magnetic field always has zero divergence everywhere.

Returning now to the moving rectangular loop, let us find the *rate of change* of flux through the loop. In time $dt$ the loop moves a distance $v\,dt$. This changes in two ways the total flux through the loop, which is $\int \mathbf{B} \cdot d\mathbf{a}$ over a surface spanning the loop. As you can see in Fig. 7.9, flux is gained at the right, in amount $B_2 wv\,dt$, while an amount of flux $B_1 wv\,dt$ is lost at the left. Hence $d\Phi$, the change in flux through the loop in time $dt$, is

$$d\Phi = -(B_1 - B_2)wv\,dt. \qquad (7.12)$$

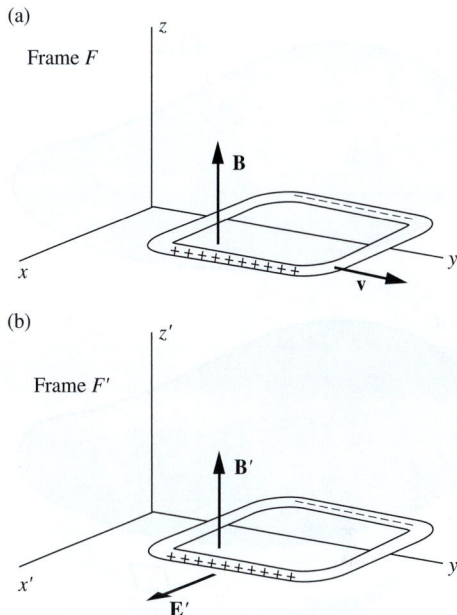

**Figure 7.5.**
(a) Here the wire loop is moving in a uniform magnetic field $\mathbf{B}$. (b) Observed in the frame $F'$ in which the loop is at rest, the fields are $\mathbf{B}'$ and $\mathbf{E}'$.

**Figure 7.6.**
Here the field $\mathbf{B}$, observed in $F$, is not uniform. It varies in both direction and magnitude from place to place.

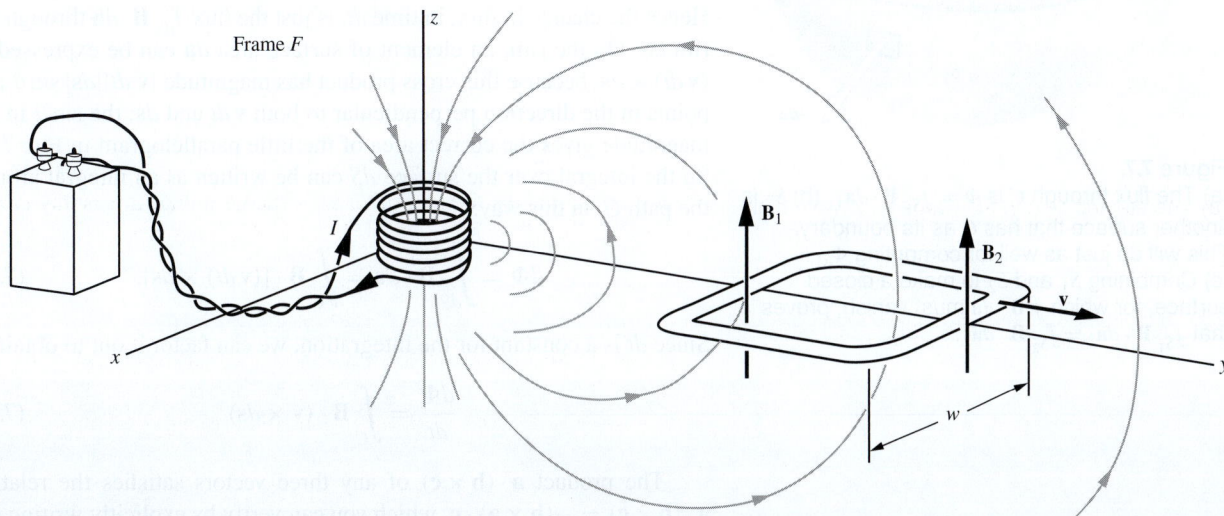

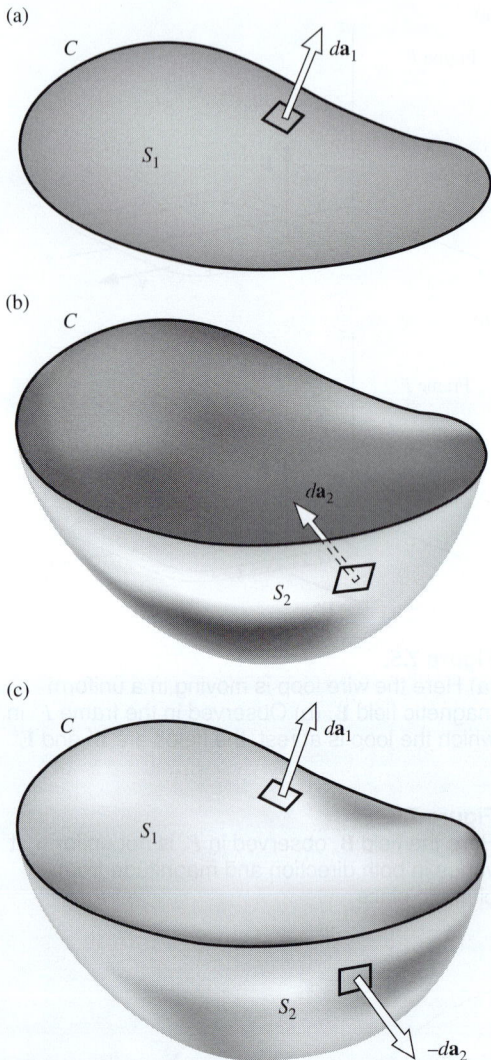

(a)

(b)

(c)

**Figure 7.7.**
(a) The flux through $C$ is $\Phi = \int_{S_1} \mathbf{B} \cdot d\mathbf{a}_1$. (b) $S_2$ is another surface that has $C$ as its boundary. This will do just as well for computing $\Phi$. (c) Combining $S_1$ and $S_2$ to make a closed surface, for which $\int \mathbf{B} \cdot d\mathbf{a}$ must vanish, proves that $\int_{S_1} \mathbf{B} \cdot d\mathbf{a}_1 = \int_{S_2} \mathbf{B} \cdot d\mathbf{a}_2$.

Comparing Eq. (7.12) with Eq. (7.6), we see that, in this case at least, the electromotive force can be expressed as $\mathcal{E} = -d\Phi/dt$. It turns out that this is a general result, as the following theorem states.

**Theorem 7.1**  *If the magnetic field in a given frame is constant in time, then for a loop of any shape moving in any manner, the emf $\mathcal{E}$ around the loop is related to the magnetic flux $\Phi$ through the loop by*

$$\mathcal{E} = -\frac{d\Phi}{dt} \tag{7.13}$$

*Proof*  The loop $C$ in Fig. 7.10 occupies the position $C_1$ at time $t$, and it is moving so that it occupies the position $C_2$ at time $t + dt$. A particular element of the loop $ds$ has been transported with velocity $\mathbf{v}$ to its new position. $S$ indicates a surface that spans the loop at time $t$. The flux through the loop at this instant of time is

$$\Phi(t) = \int_S \mathbf{B} \cdot d\mathbf{a}. \tag{7.14}$$

The magnetic field $\mathbf{B}$ comes from sources that are stationary in our frame of reference and remains constant in time, at any point fixed in this frame. At time $t + dt$ a surface that spans the loop is the original surface $S$, left fixed in space, augmented by the "rim" $dS$. (Remember, we are allowed to use *any* surface spanning the loop to compute the flux through it.) Thus

$$\Phi(t + dt) = \int_{S+dS} \mathbf{B} \cdot d\mathbf{a} = \Phi(t) + \int_{dS} \mathbf{B} \cdot d\mathbf{a}. \tag{7.15}$$

Hence the change in flux, in time $dt$, is just the flux $\int_{dS} \mathbf{B} \cdot d\mathbf{a}$ through the rim $dS$. On the rim, an element of surface area $d\mathbf{a}$ can be expressed as $(\mathbf{v}\, dt) \times d\mathbf{s}$, because this cross product has magnitude $|\mathbf{v}\, dt||d\mathbf{s}| \sin\theta$ and points in the direction perpendicular to both $\mathbf{v}\, dt$ and $d\mathbf{s}$; the $\sin\theta$ in the magnitude gives the correct area of the little parallelogram in Fig. 7.10. So the integral over the surface $dS$ can be written as an integral around the path $C$, in this way:

$$d\Phi = \int_{dS} \mathbf{B} \cdot d\mathbf{a} = \int_C \mathbf{B} \cdot [(\mathbf{v}\, dt) \times d\mathbf{s}]. \tag{7.16}$$

Since $dt$ is a constant for the integration, we can factor it out to obtain

$$\frac{d\Phi}{dt} = \int_C \mathbf{B} \cdot (\mathbf{v} \times d\mathbf{s}). \tag{7.17}$$

The product $\mathbf{a} \cdot (\mathbf{b} \times \mathbf{c})$ of any three vectors satisfies the relation $\mathbf{a} \cdot (\mathbf{b} \times \mathbf{c}) = -(\mathbf{b} \times \mathbf{a}) \cdot \mathbf{c}$, which you can verify by explicitly writing out each side in Cartesian components. Using this identity to rearrange the integrand in Eq. (7.17), we have

$$\frac{d\Phi}{dt} = -\int_C (\mathbf{v} \times \mathbf{B}) \cdot d\mathbf{s}. \qquad (7.18)$$

Now, the force on a charge $q$ that is carried along by the loop is just $q\mathbf{v} \times \mathbf{B}$, so the electromotive force, which is the line integral around the loop of the force per unit charge, is just

$$\mathcal{E} = \int_C (\mathbf{v} \times \mathbf{B}) \cdot d\mathbf{s}. \qquad (7.19)$$

Comparing Eq. (7.18) with Eq. (7.19), we get the simple relation given in Eq. (7.13), valid for arbitrary shape and motion of the loop. (We did not even have to assume that $\mathbf{v}$ is the same for all parts of the loop!) In summary, the line integral around a moving loop of $\mathbf{f}/q$, the force per unit charge, is just the negative of the rate of change of flux through the loop. $\qquad \square$

The sense of the line integral and the direction in which flux is called positive are to be related by a right-hand-thread rule. For instance, in Fig. 7.6, the flux is *upward* through the loop and is *decreasing*. Taking the minus sign in Eq. (7.13) into account, our rule would predict an electromotive force that would tend to drive a positive charge around the loop in a counterclockwise direction, as seen looking down on the loop (Fig. 7.11).

There is a better way to look at this question of sign and direction. Note that if a current should flow in the direction of the induced electromotive force, in the situation shown in Fig. 7.11, this current itself would create some flux through the loop in a direction to *counteract* the assumed flux change (because the Biot–Savart law, Eq. (6.49), tells us that the contributions from this current to the $\mathbf{B}$ field inside the loop all point upward in Fig. 7.11). That is an essential physical fact, and not the consequence of an arbitrary convention about signs and directions. It is a manifestation of the tendency of systems to resist change. In this context it is traditionally called *Lenz's law*.

**Lenz's law**   *The direction of the induced electromotive force is such that the induced current creates a magnetic field that opposes the change in flux.*

Another example of Lenz's law is illustrated in Fig. 7.12. The conducting ring is falling in the magnetic field of the coil. The flux through the ring is *downward* and is *increasing* in magnitude. To counteract this change, some new flux upward is needed. It would take a current flowing around the ring in the direction of the arrows to produce such flux. Lenz's law assures us that the induced emf will be in the correct direction to cause such a current.

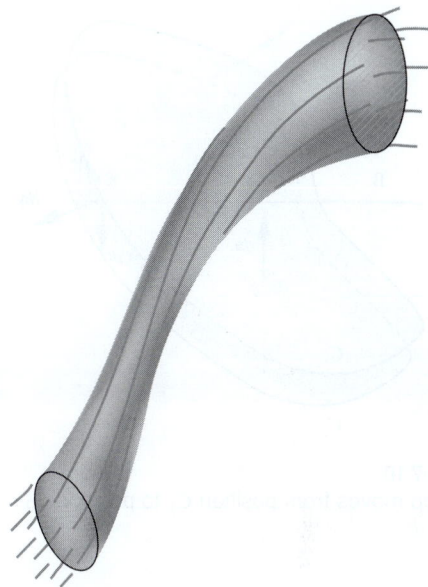

**Figure 7.8.**
A flux tube. Magnetic field lines lie in the surface of the tube. The tube encloses a certain amount of flux $\Phi$. No matter where you chop it, you will find that $\int \mathbf{B} \cdot d\mathbf{a}$ over the section has the same value $\Phi$. A flux tube doesn't have to be round. You can start somewhere with any cross section, and the course of the field lines will determine how the section changes size and shape as you go along the tube.

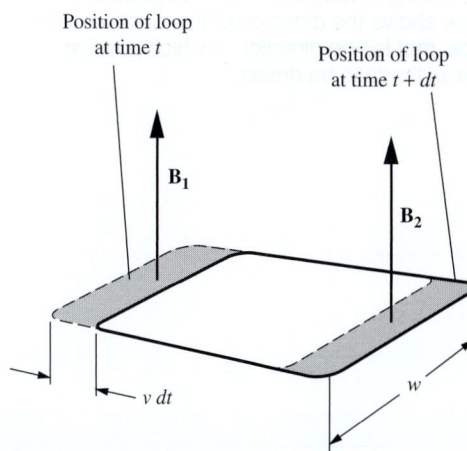

**Figure 7.9.**
In the interval $dt$, the loop gains an increment of flux $B_2 wv\,dt$ and loses an increment $B_1 wv\,dt$.

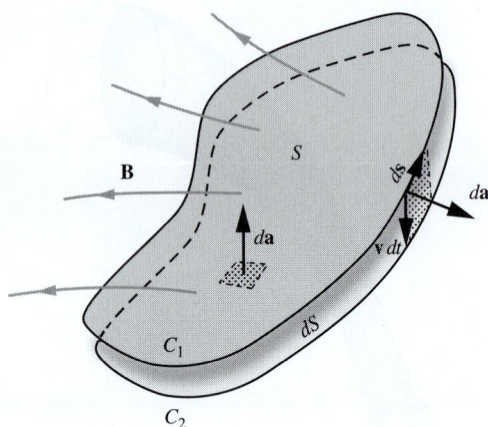

**Figure 7.10.**
The loop moves from position $C_1$ to position $C_2$ in time $dt$.

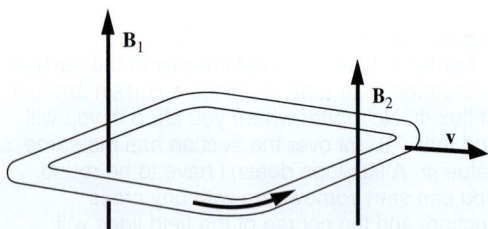

**Figure 7.11.**
The flux through the loop is upward and is decreasing in magnitude as time goes on. The arrow shows the direction of the electromotive force, that is, the direction in which positive charge tends to be driven.

If the electromotive force causes current to flow in the loop that is shown in Figs. 7.6 and 7.11, as it will if the loop has a finite resistance, some energy will be dissipated in the wire. What supplies this energy? To answer that, consider the force that acts on the current in the loop if it flows in the sense indicated by the arrow in Fig 7.11. The side on the right, in the field $B_2$, will experience a force toward the right, while the opposite side of the loop, in the field $B_1$, will be pushed toward the left. But $B_1$ is greater than $B_2$, so the net force on the loop is toward the left, *opposing the motion*. To keep the loop moving at constant speed, some external agency has to do work, and the energy thus invested eventually shows up as heat in the wire (see Exercise 7.30). Imagine what would happen if Lenz's law were violated, or if the force on the loop were to act in a direction to assist the motion of the loop!

**Example (Sinusoidal $\mathcal{E}$)** A very common element in electrical machinery and electrical instruments is a loop or coil that rotates in a magnetic field. Let's apply what we have just learned to the system shown in Fig. 7.13, a single loop rotating at constant speed in a magnetic field that is approximately uniform. The mechanical essentials, shaft, bearings, drive, etc., are not drawn. The field **B** is provided by the two fixed coils. Suppose the loop rotates with angular velocity $\omega$, in radians/second. If its position at any instant is specified by the angle $\theta$, then $\theta = \omega t + \alpha$, where the constant $\alpha$ is simply the position of the loop at $t = 0$. The component of **B** perpendicular to the plane of the loop is $B \sin \theta$. Therefore the flux through the loop at time $t$ is

$$\Phi(t) = SB \sin(\omega t + \alpha), \tag{7.20}$$

where $S$ is the area of the loop. For the induced electromotive force we then have

$$\mathcal{E} = -\frac{d\Phi}{dt} = -SB\omega \cos(\omega t + \alpha). \tag{7.21}$$

If the loop instead of being closed is connected through slip rings to external wires, as shown in Fig. 7.13, we can detect at these terminals a sinusoidally alternating potential difference.

A numerical example will show how the units work out. Suppose the area of the loop in Fig. 7.13 is 80 cm², the field strength $B$ is 50 gauss, and the loop is rotating at 30 revolutions per second. Then $\omega = 2\pi \cdot 30$, or 188 radians/second. The amplitude, that is, the maximum magnitude of the oscillating electromotive force induced in the loop, is

$$\mathcal{E}_0 = SB\omega = \left(0.008 \text{ m}^2\right)\left(0.005 \text{ tesla}\right)\left(188 \text{ s}^{-1}\right) = 7.52 \cdot 10^{-3} \text{ V}. \tag{7.22}$$

You should verify that 1 m²· tesla/s is indeed equivalent to 1 volt.

## 7.4 Stationary loop with the field source moving

We can, if we like, look at the events depicted in Fig. 7.6 from a frame of reference that is moving with the loop. That can't change the physics, only the words we use to describe it. Let $F'$, with coordinates $x'$, $y'$, $z'$, be the frame attached to the loop, which we now regard as stationary (Fig. 7.14). The coil and battery, stationary in frame $F$, are moving in the $-y'$ direction with velocity $\mathbf{v}' = -\mathbf{v}$. Let $B_1'$ and $B_2'$ be the magnetic field measured at the two ends of the loop by observers in $F'$ at some instant $t'$. At these positions there will be an electric field in $F'$. Equation (6.82) tells us that

$$\mathbf{E}_1' = \mathbf{v} \times \mathbf{B}_1' \quad \text{and} \quad \mathbf{E}_2' = \mathbf{v} \times \mathbf{B}_2'. \tag{7.23}$$

For observers in $F'$ this is a genuine electric field. It is not an electrostatic field; the line integral of $\mathbf{E}'$ around any closed path in $F'$ is not generally zero. In fact, from Eq. (7.23) the line integral of $\mathbf{E}'$ around the rectangular loop is

$$\int \mathbf{E}' \cdot d\mathbf{s}' = wv(B_1' - B_2'). \tag{7.24}$$

We can call the line integral in Eq. (7.24) the electromotive force $\mathcal{E}'$ on this path. If a charged particle moves once around the path, $\mathcal{E}'$ is the work done on it, per unit charge. $\mathcal{E}'$ is related to the rate of change of flux through the loop. To see this, note that, while the loop itself is stationary, the *magnetic field pattern* is now moving with the velocity $-\mathbf{v}$ of the source. Hence for the flux lost or gained at either end of the loop, in a time interval $dt'$, we get a result similar to Eq. (7.12), and we conclude that

$$\mathcal{E}' = -\frac{d\Phi'}{dt'}. \tag{7.25}$$

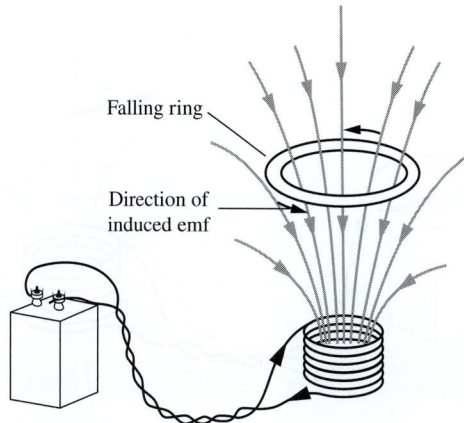

**Figure 7.12.**
As the ring falls, the downward flux through the ring is increasing. Lenz's law tells us that the induced emf will be in the direction indicated by the arrows, for that is the direction in which current must flow to produce upward flux through the ring. The system reacts so as to oppose the change that is occurring.

**Figure 7.13.**
The two coils produce a magnetic field **B** that is approximately uniform in the vicinity of the loop. In the loop, rotating with angular velocity $\omega$, a sinusoidally varying electromotive force is induced.

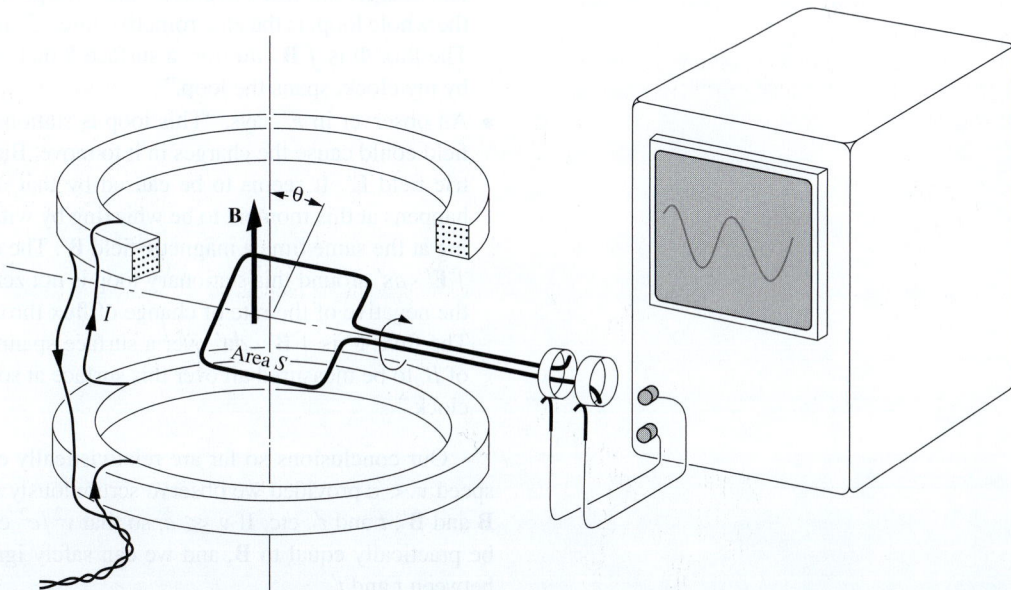

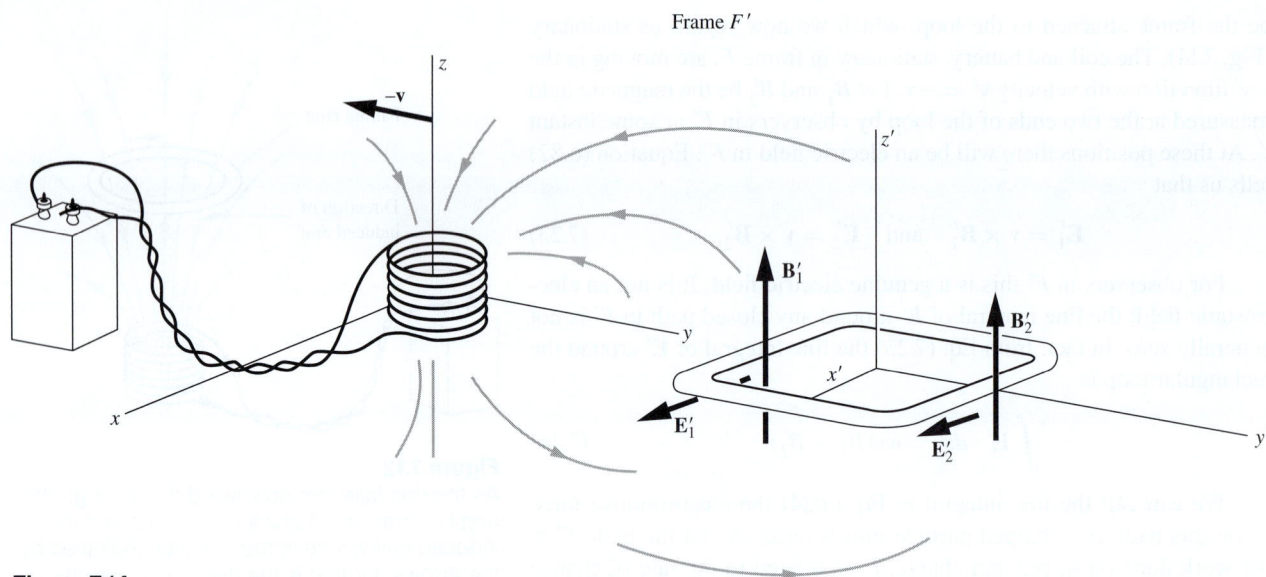

**Figure 7.14.**
As observed in the frame $F'$, the loop is at rest and the field source is moving. The fields $\mathbf{B}'$ and $\mathbf{E}'$ are both present and are functions of both position and time.

We can summarize as follows the descriptions in the two frames of reference, $F$, in which the source of $\mathbf{B}$ is at rest, and $F'$, in which the loop is at rest.

- An observer in $F$ says, "We have here a magnetic field that, though it is not uniform spatially, is constant in time. There is no electric field. That wire loop over there is moving with velocity $\mathbf{v}$ through the magnetic field, so the charges in it are acted on by a force $\mathbf{v} \times \mathbf{B}$ per unit charge. The line integral of this force per unit charge, taken around the whole loop, is the electromotive force $\mathcal{E}$, and it is equal to $-d\Phi/dt$. The flux $\Phi$ is $\int \mathbf{B} \cdot d\mathbf{a}$ over a surface $S$ that, at some instant of time $t$ by my clock, spans the loop."

- An observer in $F'$ says, "This loop is stationary, and only an electric field could cause the charges in it to move. But there is in fact an electric field $\mathbf{E}'$. It seems to be caused by that magnetlike object which happens at this moment to be whizzing by with a velocity $-\mathbf{v}$, producing at the same time a magnetic field $\mathbf{B}'$. The electric field is such that $\int \mathbf{E}' \cdot d\mathbf{s}'$ around this stationary loop is not zero but instead is equal to the negative of the rate of change of flux through the loop, $-d\Phi'/dt'$. The flux $\Phi'$ is $\int \mathbf{B}' \cdot d\mathbf{a}'$ over a surface spanning the loop, the values of $B'$ to be measured all over this surface at some one instant $t'$, by my clock."

Our conclusions so far are relativistically exact. They hold for any speed $v < c$ provided we observe scrupulously the distinctions between $\mathbf{B}$ and $\mathbf{B}'$, $t$ and $t'$, etc. If $v \ll c$, so that $v^2/c^2$ can be neglected, $\mathbf{B}'$ will be practically equal to $\mathbf{B}$, and we can safely ignore also the distinction between $t$ and $t'$.

## 7.5 Universal law of induction

Let's carry out three experiments with the apparatus shown in Fig. 7.15. The tables are on wheels so that they can be easily moved. A sensitive galvanometer has been connected to our old rectangular loop, and to increase any induced electromotive force we put several turns of wire in the loop rather than one. Frankly though, our sensitivity might still be marginal, with the feeble source of magnetic field pictured. Perhaps you can devise a more practical version of the experiment.

**Experiment I.** With constant current in the coil and table 1 stationary, table 2 moves toward the right (away from table 1) with speed $v$. The *galvanometer deflects*. We are not surprised; we have already analyzed this situation in Section 7.3.

**Experiment II.** With constant current in the coil and table 2 stationary, table 1 moves to the left (away from table 2) with speed $v$. The *galvanometer deflects*. This doesn't surprise us either. We have just discussed in Section 7.4 the equivalence of Experiments I and II, an equivalence that is an example of Lorentz invariance or, for the low speeds of our tables, Galilean invariance. We know that in both experiments the deflection of the galvanometer can be related to the rate of change of flux of **B** through the loop.

**Experiment III.** Both tables remain at rest, but we vary the current $I$ in the coil by sliding the contact $K$ along the resistance strip. We do this in such a way that the *rate of decrease* of the field **B** at the loop is the same as it was in Experiments I and II. *Does the galvanometer deflect?*

**Figure 7.15.**
We imagine that either table can move or, with both tables fixed, the current $I$ in the coil can be gradually changed.

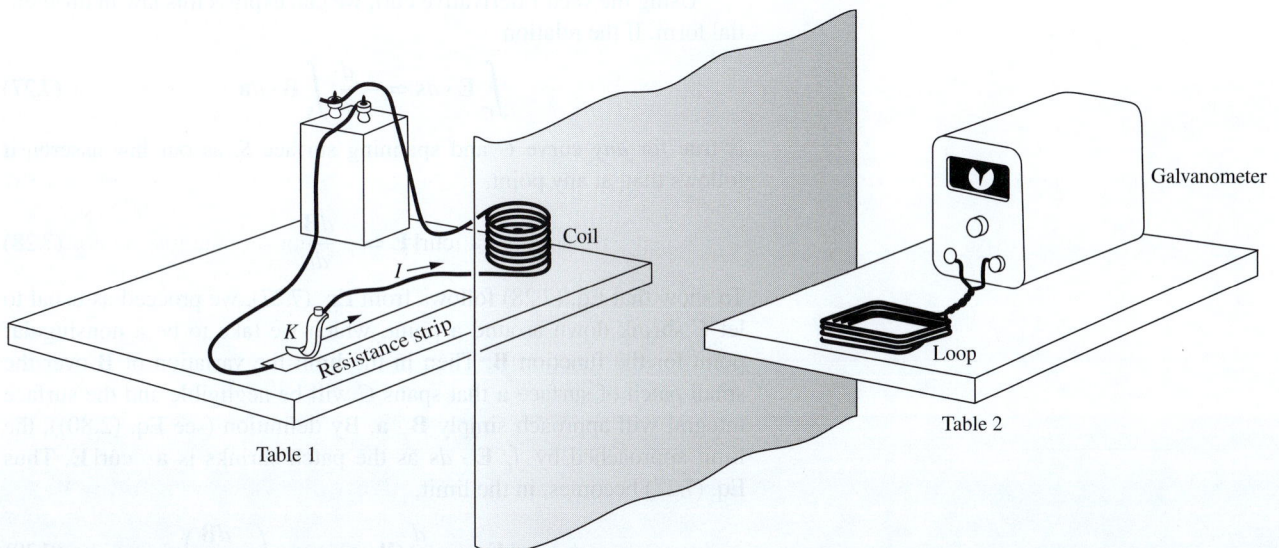

For an observer stationed at the loop on table 2 and measuring the magnetic field in that neighborhood as a function of time and position, there is no way to distinguish among Experiments I, II, and III. Imagine a black cloth curtain between the two tables. Although there might be minor differences between the field configurations for II and III, an observer who did not know what was behind the curtain could not decide, on the basis of local **B** measurements alone, which case it was. Therefore if the galvanometer did *not* respond with the same deflection in Experiment III, it would mean that the relation between the magnetic and electric fields in a region depends on the nature of a remote source. Two magnetic fields essentially similar in their local properties would have associated electric fields with different values of $\int \mathbf{E} \cdot d\mathbf{s}$.

We find by experiment that III *is* equivalent to I and II. The galvanometer deflects, by the same amount as before. Faraday's experiments were the first to demonstrate this fundamental fact. The electromotive force we observe depends only on the rate of change of the flux of **B**, and not on anything else. We can state as a universal relation *Faraday's law of induction:*

> If $C$ is some closed curve, stationary in coordinates $x$, $y$, $z$; if $S$ is a surface spanning $C$; and if $\mathbf{B}(x, y, z, t)$ is the magnetic field measured in $x$, $y$, $z$, at any time $t$, then
>
> $$\mathcal{E} = \int_C \mathbf{E} \cdot d\mathbf{s} = -\frac{d}{dt} \int_S \mathbf{B} \cdot d\mathbf{a} = -\frac{d\Phi}{dt} \qquad \text{(Faraday's law)}$$
>
> $$(7.26)$$

Using the vector derivative curl, we can express this law in differential form. If the relation

$$\int_C \mathbf{E} \cdot d\mathbf{s} = -\frac{d}{dt} \int_S \mathbf{B} \cdot d\mathbf{a} \qquad (7.27)$$

is true for *any* curve $C$ and spanning surface $S$, as our law asserts, it follows that, at any point,

$$\text{curl } \mathbf{E} = -\frac{d\mathbf{B}}{dt}. \qquad (7.28)$$

To show that Eq. (7.28) follows from Eq. (7.27), we proceed as usual to let $C$ shrink down around a point, which we take to be a nonsingular point for the function **B**. Then in the limit the variation of **B** over the small patch of surface **a** that spans $C$ will be negligible and the surface integral will approach simply $\mathbf{B} \cdot \mathbf{a}$. By definition (see Eq. (2.80)), the limit approached by $\int_C \mathbf{E} \cdot d\mathbf{s}$ as the patch shrinks is $\mathbf{a} \cdot \text{curl } \mathbf{E}$. Thus Eq. (7.27) becomes, in the limit,

$$\mathbf{a} \cdot \text{curl } \mathbf{E} = -\frac{d}{dt}(\mathbf{B} \cdot \mathbf{a}) = \mathbf{a} \cdot \left(-\frac{d\mathbf{B}}{dt}\right). \qquad (7.29)$$

Since this holds for *any* infinitesimal **a**, it must be that[1]

$$\text{curl } \mathbf{E} = -\frac{d\mathbf{B}}{dt}. \tag{7.30}$$

Recognizing that **B** may depend on position as well as time, we write $\partial \mathbf{B}/\partial t$ in place of $d\mathbf{B}/dt$. We have then these two entirely equivalent statements of the law of induction:

$$\int_C \mathbf{E} \cdot d\mathbf{s} = -\frac{d}{dt} \int_S \mathbf{B} \cdot d\mathbf{a}$$

$$\text{curl } \mathbf{E} = -\frac{\partial \mathbf{B}}{\partial t} \tag{7.31}$$

With Faraday's law of induction, we are one step closer to the complete set of Maxwell's equations. We will obtain the last piece to the puzzle in Chapter 9.

In Eq. (7.31) the electric field **E** is to be expressed in our SI units of volts/meter, with **B** in teslas, $d\mathbf{s}$ in meters, and $d\mathbf{a}$ in m$^2$. The electromotive force $\mathcal{E} = \int_C \mathbf{E} \cdot d\mathbf{s}$ will then be given in volts. In Gaussian units the relation expressed by Eq. (7.31) looks like this:

$$\int_C \mathbf{E} \cdot d\mathbf{s} = -\frac{1}{c}\frac{d}{dt} \int_S \mathbf{B} \cdot d\mathbf{a},$$

$$\text{curl } \mathbf{E} = -\frac{1}{c}\frac{\partial \mathbf{B}}{\partial t}. \tag{7.32}$$

Here **E** is in statvolts/cm, **B** is in gauss, $d\mathbf{s}$ and $d\mathbf{a}$ are in cm and cm$^2$, respectively, and $c$ is in cm/s. The electromotive force $\mathcal{E} = \int_C \mathbf{E} \cdot d\mathbf{s}$ will be given in statvolts.

The magnetic flux $\Phi$, which is $\int_C \mathbf{B} \cdot d\mathbf{a}$, is expressed in tesla-m$^2$ in our SI units, and in gauss-cm$^2$, a unit exactly $10^8$ times smaller, in Gaussian units (because 1 m$^2 = 10^4$ cm$^2$ and 1 tesla $= 10^4$ gauss, exactly). The SI flux unit is assigned a name of its own, the *weber*.

When in doubt about the units, you may find one of the following equivalent statements helpful:

- Electromotive force in statvolts equals:
  $1/c$ times rate of change of flux in gauss-cm$^2$/s.
- Electromotive force in volts equals:
  rate of change of flux in tesla-m$^2$/s.
- Electromotive force in volts equals:
  $10^{-8}$ times rate of change of flux in gauss-cm$^2$/s.

---

[1] If that isn't obvious, note that choosing **a** in the $x$ direction will establish that $(\text{curl } \mathbf{E})_x = -dB_x/dt$, and so on.

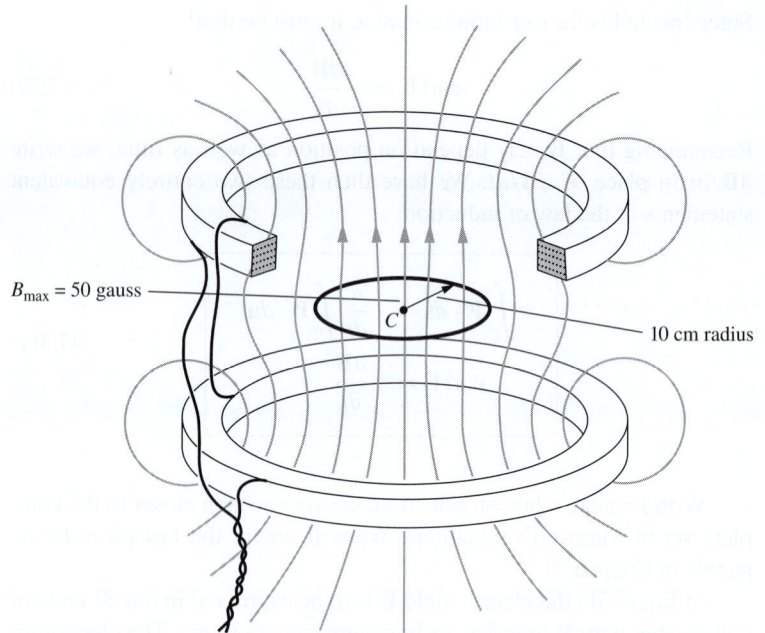

**Figure 7.16.**
Alternating current in the coils produces a magnetic field which, at the center, oscillates between 50 gauss upward and 50 gauss downward. At any instant the field is approximately uniform within the circle $C$.

If these seem confusing, don't try to remember them. Just remember that you can look them up on this page.

The differential expression, $\operatorname{curl} \mathbf{E} = -\partial\mathbf{B}/\partial t$, brings out rather plainly the point we tried to make earlier about the local nature of the field relations. The variation in time of $\mathbf{B}$ in a neighborhood completely determines $\operatorname{curl} \mathbf{E}$ there – nothing else matters. That does not completely determine $\mathbf{E}$ itself, of course. Without affecting this relation, any electrostatic field with $\operatorname{curl} \mathbf{E} = 0$ could be superposed.

**Example (Sinusoidal B field)**   As a concrete example of Faraday's law, suppose coils like those in Fig. 7.13 are supplied with 60 cycles per second alternating current, instead of direct current. The current and the magnetic field vary as $\sin(2\pi \cdot 60\,\mathrm{s}^{-1} \cdot t)$, or $\sin(377\,\mathrm{s}^{-1} \cdot t)$. Suppose the amplitude of the current is such that the magnetic field $\mathbf{B}$ in the central region reaches a maximum value of 50 gauss, or 0.005 tesla. We want to investigate the induced electric field, and the electromotive force, on the circular path 10 cm in radius shown in Fig. 7.16. We may assume that the field $B$ is practically uniform in the interior of this circle, at any instant of time. So we have

$$B = (0.005\ \mathrm{T})\sin(377\,\mathrm{s}^{-1} \cdot t). \tag{7.33}$$

The flux through the loop $C$ is

$$\Phi = \pi r^2 B = \pi \cdot (0.1\ \mathrm{m})^2 \cdot (0.005\ \mathrm{T})\sin(377\,\mathrm{s}^{-1} \cdot t)$$
$$= 1.57 \cdot 10^{-4} \sin(377\,\mathrm{s}^{-1} \cdot t)\ \mathrm{T\,m}^2. \tag{7.34}$$

Using Eq. (7.26) to calculate the electromotive force, we obtain

$$\mathcal{E} = -\frac{d\Phi}{dt} = -(377 \, \mathrm{s}^{-1}) \cdot 1.57 \cdot 10^{-4} \cos(377 \, \mathrm{s}^{-1} \cdot t) \, \mathrm{T \, m^2}$$

$$= -0.059 \cos(377 \, \mathrm{s}^{-1} \cdot t) \, \mathrm{V}. \tag{7.35}$$

The maximum attained by $\mathcal{E}$ is 59 millivolts. The minus sign will ensure that Lenz's law is respected, if we have defined our directions consistently. The variation of both $\Phi$ and $\mathcal{E}$ with time is shown in Fig. 7.17.

What about the electric field itself? Usually we cannot deduce $\mathbf{E}$ from a knowledge of curl $\mathbf{E}$ alone. However, our path $C$ is here a circle around the center of a symmetrical system. *If there are no other* electric fields around, we may assume that, on the circle $C$, $\mathbf{E}$ lies in that plane and has a constant magnitude. Then it is a trivial matter to predict its magnitude, since $\int_C \mathbf{E} \cdot d\mathbf{s} = 2\pi r E = \mathcal{E}$, which we have already calculated. In this case, the electric field on the circle might look like Fig. 7.18(a) at a particular instant. But if there are other field sources, it could look quite different. If there happened to be a positive and a negative charge located on the axis as shown in Fig. 7.18(b), the electric field in the vicinity of the circle would be the superposition of the electrostatic field of the two charges and the induced electric field.

A consequence of Faraday's law of induction is that Kirchhoff's loop rule (which states that $\int \mathbf{E} \cdot d\mathbf{s} = 0$ around a closed path) is no longer valid in situations where there is a changing magnetic field. Faraday has taken us beyond the comfortable realm of conservative electric fields. The voltage difference between two points now depends on the path between them. Problem 7.4 provides an instructive example of this fact.

A note on the terminology: the term "*potential* difference" is generally reserved for electrostatic fields, because it is only for such fields that we can uniquely define a potential function $\phi$ at all points in space, with the property that $\mathbf{E} = -\nabla\phi$. For these fields, the potential difference between points $a$ and $b$ is given by $\phi_b - \phi_a = -\int_a^b \mathbf{E} \cdot d\mathbf{s}$. The term "*voltage* difference" applies to *any* electric field, not necessarily electrostatic, and it is defined similarly as $V_b - V_a = -\int_a^b \mathbf{E} \cdot d\mathbf{s}$. If there are changing magnetic fields involved, this line integral will depend on the path between $a$ and $b$. The voltage difference is what a voltmeter measures, and we can hook up a voltmeter to any type of circuit, of course, no matter what kinds of electric fields it involves. But if there are changing magnetic fields, Problem 7.4 shows that it matters *how* we hook it up. See Romer (1982) for more discussion of this issue.

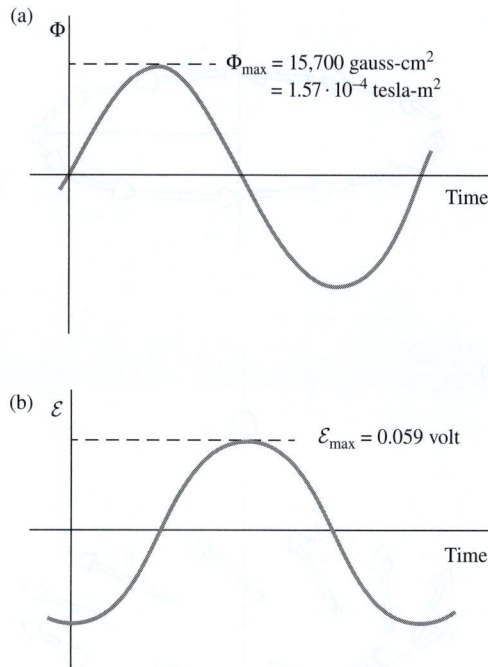

**Figure 7.17.**
(a) The flux through the circle $C$. (b) The electromotive force associated with the path $C$.

## 7.6 Mutual inductance

Two circuits, or loops, $C_1$ and $C_2$ are fixed in position relative to one another (Fig. 7.19). By some means, such as a battery and a variable resistance, a controllable current $I_1$ is caused to flow in circuit $C_1$. Let

(a)

(b)

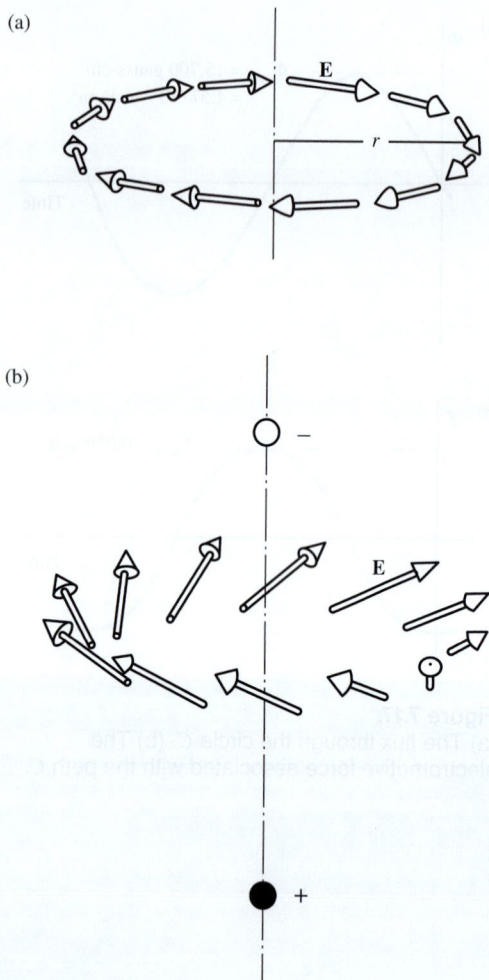

**Figure 7.18.**
The electric field on the circular path $C$. (a) In the absence of sources other than the symmetrical, oscillating current. (b) Including the electrostatic field of two charges on the axis.

$\mathbf{B}_1(x, y, z)$ be the magnetic field that would exist if the current in $C_1$ remained constant at the value $I_1$, and let $\Phi_{21}$ denote the flux of $\mathbf{B}_1$ through the circuit $C_2$. Thus

$$\Phi_{21} = \int_{S_2} \mathbf{B}_1 \cdot d\mathbf{a}_2, \tag{7.36}$$

where $S_2$ is a surface spanning the loop $C_2$. With the shape and relative position of the two circuits fixed, $\Phi_{21}$ will be proportional to $I_1$:

$$\frac{\Phi_{21}}{I_1} = \text{constant} \equiv M_{21}. \tag{7.37}$$

Suppose now that $I_1$ changes with time, but *slowly enough* so that the field $\mathbf{B}_1$ at any point in the vicinity of $C_2$ is related to the current $I_1$ in $C_1$ (at the same instant of time) in the same way as it would be related for a steady current. (To see why such a restriction is necessary, imagine that $C_1$ and $C_2$ are 10 meters apart and we cause the current in $C_1$ to double in value in 10 nanoseconds!) The flux $\Phi_{21}$ will change in proportion as $I_1$ changes. There will be an electromotive force induced in circuit $C_2$, of magnitude

$$\mathcal{E}_{21} = -\frac{d\Phi_{21}}{dt} \implies \mathcal{E}_{21} = -M_{21}\frac{dI_1}{dt}. \tag{7.38}$$

In Gaussian units there is a factor of $c$ in the denominator here. But we can define a new constant $M'_{21} \equiv M_{21}/c$ so that the relation between $\mathcal{E}_{21}$ and $dI_1/dt$ remains of the same form.

We call the constant $M_{21}$ the coefficient of *mutual inductance*. Its value is determined by the geometry of our arrangement of loops. The units will of course depend on our choice of units for $\mathcal{E}$, $I$, and $t$. In SI

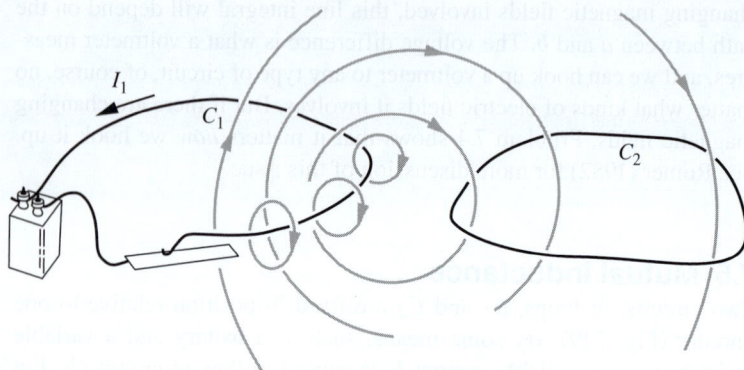

**Figure 7.19.**
Current $I_1$ in loop $C_1$ causes a certain flux $\Phi_{21}$ through loop $C_2$.

units, with $\mathcal{E}$ in volts and $I$ in amperes, the unit for $M_{21}$ is volt $\cdot$ amp$^{-1} \cdot$ s, or ohm $\cdot$ s. This unit is called the *henry*;[2]

$$1 \text{ henry} = 1 \frac{\text{volt} \cdot \text{second}}{\text{amp}} = 1 \text{ ohm} \cdot \text{second}. \qquad (7.39)$$

That is, the mutual inductance $M_{21}$ is one henry if a current $I_1$ changing at the rate of 1 ampere/second induces an electromotive force of 1 volt in circuit $C_2$. In Gaussian units, with $\mathcal{E}$ in statvolts and $I$ in esu/second, the unit for $M_{21}$ is statvolt $\cdot$ (esu/second)$^{-1} \cdot$ second. Since 1 statvolt equals 1 esu/cm, this unit can also be written as second$^2$/cm.

**Example (Concentric rings)** Figure 7.20 shows two coplanar, concentric rings: a small ring $C_2$ and a much larger ring $C_1$. Assuming $R_2 \ll R_1$, what is the mutual inductance $M_{21}$?

**Solution** At the center of $C_1$, with $I_1$ flowing, the field $B_1$ is given by Eq. (6.54) as

$$B_1 = \frac{\mu_0 I_1}{2R_1}. \qquad (7.40)$$

Since we are assuming $R_2 \ll R_1$, we can neglect the variation of $B_1$ over the interior of the small ring. The flux through the small ring is then

$$\Phi_{21} = (\pi R_2^2) \frac{\mu_0 I_1}{2R_1} = \frac{\mu_0 \pi I_1 R_2^2}{2R_1}. \qquad (7.41)$$

The mutual inductance $M_{21}$ in Eq. (7.37) is therefore

$$M_{21} = \frac{\Phi_{21}}{I_1} = \frac{\mu_0 \pi R_2^2}{2R_1}, \qquad (7.42)$$

and the electromotive force induced in $C_2$ is

$$\mathcal{E}_{21} = -M_{21} \frac{dI_1}{dt} = -\frac{\mu_0 \pi R_2^2}{2R_1} \frac{dI_1}{dt}. \qquad (7.43)$$

Since $\mu_0 = 4\pi \cdot 10^{-7}$ kg m/C$^2$, we can write $M_{21}$ alternatively as

$$M_{21} = \frac{(2\pi^2 \cdot 10^{-7} \text{ kg m/C}^2) R_2^2}{R_1}. \qquad (7.44)$$

The numerical value of this expression gives $M_{21}$ in henrys. In Gaussian units, you can show that the relation corresponding to Eq. (7.43) is

$$\mathcal{E}_{21} = -\frac{1}{c} \frac{2\pi^2 R_2^2}{cR_1} \frac{dI_1}{dt}, \qquad (7.45)$$

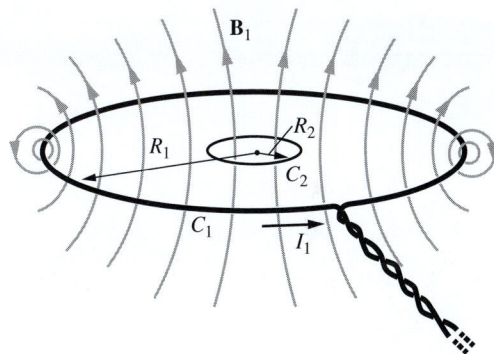

**Figure 7.20.**
Current $I_1$ in ring $C_1$ causes field $\mathbf{B}_1$, which is approximately uniform over the region of the small ring $C_2$.

[2] The unit is named after Joseph Henry (1797–1878), the foremost American physicist of his time. Electromagnetic induction was discovered independently by Henry, practically at the same time as Faraday conducted his experiments. Henry was the first to recognize the phenomenon of self-induction. He developed the electromagnet and the prototype of the electric motor, invented the electric relay, and all but invented telegraphy.

with $\mathcal{E}_{21}$ in statvolts, the $R$'s in cm, and $I_1$ in esu/second. $M_{21}$ is the coefficient of the $dI_1/dt$ term, namely $2\pi^2 R_2^2/c^2 R_1$ (in second$^2$/cm). Appendix C states, and derives, the conversion factor from henry to second$^2$/cm.

Incidentally, the minus sign we have been carrying along doesn't tell us much at this stage. If you want to be sure which way the electromotive force will tend to drive current in $C_2$, Lenz's law is your most reliable guide.

If the circuit $C_1$ consisted of $N_1$ turns of wire instead of a single ring, the field $B_1$ at the center would be $N_1$ times as strong, for a given current $I_1$. Also, if the small loop $C_2$ consisted of $N_2$ turns, all of the same radius $R_2$, the electromotive force in each turn would add to that in the next, making the total electromotive force in that circuit $N_2$ times that of a single turn. Thus for *multiple turns* in each coil the mutual inductance will be given by

$$M_{21} = \frac{\mu_0 \pi N_1 N_2 R_2^2}{2R_1}. \tag{7.46}$$

This assumes that the turns in each coil are neatly bundled together, the cross section of the bundle being small compared with the coil radius. However, the mutual inductance $M_{21}$ has a well-defined meaning for two circuits of any shape or distribution. As we wrote in Eq. (7.38), $M_{21}$ is the (negative) ratio of the electromotive force in circuit 2, caused by changing current in circuit 1, to the rate of change of current $I_1$. That is,

$$M_{21} = -\frac{\mathcal{E}_{21}}{dI_1/dt}. \tag{7.47}$$

## 7.7 A reciprocity theorem

In considering the circuits $C_1$ and $C_2$ in the preceding example, we might have inquired about the electromotive force induced in circuit $C_1$ by a changing current in circuit $C_2$. That would involve another coefficient of mutual inductance, $M_{12}$, given by (ignoring the sign)

$$M_{12} = \frac{\mathcal{E}_{12}}{dI_2/dt}. \tag{7.48}$$

$M_{12}$ is related to $M_{21}$ by the following remarkable theorem.

**Theorem 7.2**  *For any two circuits,*

$$\boxed{M_{12} = M_{21}} \tag{7.49}$$

This theorem is not a matter of geometrical symmetry. Even the simple example in Fig. 7.20 is not symmetrical with respect to the two circuits. Note that $R_1$ and $R_2$ enter in different ways into the expression for $M_{21}$; Eq. (7.49) asserts that, for these two dissimilar circuits, if

$$M_{21} = \frac{\pi\mu_0 N_1 N_2 R_2^2}{2R_1}, \qquad \text{then} \qquad M_{12} = \frac{\pi\mu_0 N_1 N_2 R_2^2}{2R_1} \qquad (7.50)$$

also – and *not* what we would get by switching 1's and 2's everywhere!

*Proof* In view of the definition of mutual inductance in Eq. (7.37), our goal is to show that $\Phi_{12}/I_2 = \Phi_{21}/I_1$, where $\Phi_{12}$ is the flux through some circuit $C_1$ due to a current $I_2$ in another circuit $C_2$, and $\Phi_{21}$ is the flux through $C_2$ due to a current $I_1$ in $C_1$. We will use the vector potential. Stokes' theorem tells us that

$$\int_C \mathbf{A} \cdot d\mathbf{s} = \int_S (\text{curl } \mathbf{A}) \cdot d\mathbf{a}. \qquad (7.51)$$

In particular, if $\mathbf{A}$ is the vector potential of a magnetic field $\mathbf{B}$, in other words, if $\mathbf{B} = \text{curl } \mathbf{A}$, then we have

$$\boxed{\int_C \mathbf{A} \cdot d\mathbf{s} = \int_S \mathbf{B} \cdot d\mathbf{a} = \Phi_S} \qquad (7.52)$$

*That is, the line integral of the vector potential around a loop is equal to the flux of $\mathbf{B}$ through the loop.*

Now, the vector potential is related to its current source as follows, according to Eq. (6.46):

$$\mathbf{A}_{21} = \frac{\mu_0 I_1}{4\pi} \int_{C_1} \frac{d\mathbf{s}_1}{r_{21}}, \qquad (7.53)$$

where $\mathbf{A}_{21}$ is the vector potential, at some point $(x_2, y_2, z_2)$, of the magnetic field caused by current $I_1$ flowing in circuit $C_1$; $d\mathbf{s}_1$ is an element of the loop $C_1$; and $r_{21}$ is the magnitude of the distance from that element to the point $(x_2, y_2, z_2)$.

Figure 7.21 shows the two loops $C_1$ and $C_2$, with current $I_1$ flowing in $C_1$. Let $(x_2, y_2, z_2)$ be a point on the loop $C_2$. Then Eqs. (7.52) and (7.53) give the flux through $C_2$ due to current $I_1$ in $C_1$ as

$$\Phi_{21} = \int_{C_2} d\mathbf{s}_2 \cdot \mathbf{A}_{21} = \int_{C_2} d\mathbf{s}_2 \cdot \frac{\mu_0 I_1}{4\pi} \int_{C_1} \frac{d\mathbf{s}_1}{r_{21}}$$

$$= \frac{\mu_0 I_1}{4\pi} \int_{C_2} \int_{C_1} \frac{d\mathbf{s}_2 \cdot d\mathbf{s}_1}{r_{21}}. \qquad (7.54)$$

Similarly, the flux through $C_1$ due to current $I_2$ flowing in $C_2$ is given by the same expression with the labels 1 and 2 reversed:

$$\Phi_{12} = \frac{\mu_0 I_2}{4\pi} \int_{C_1} \int_{C_2} \frac{d\mathbf{s}_1 \cdot d\mathbf{s}_2}{r_{12}}. \qquad (7.55)$$

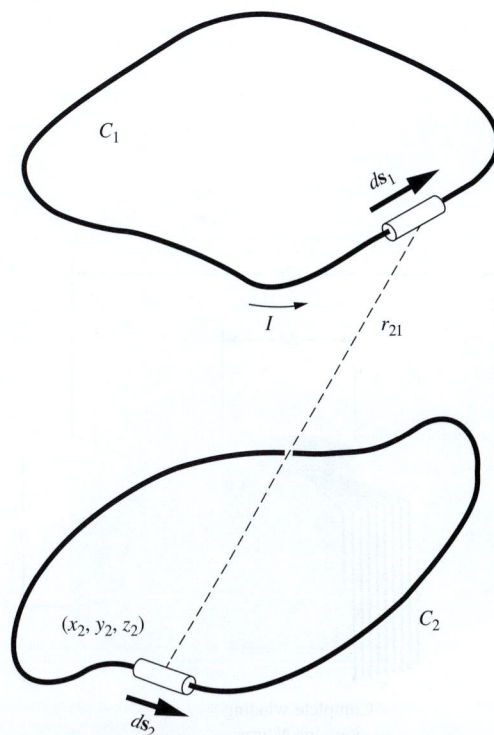

**Figure 7.21.**
Calculation of the flux $\Phi_{21}$ that passes through $C_2$ as a result of current $I_1$ flowing in $C_1$.

Now $r_{12} = r_{21}$, for these are just distance magnitudes, not vectors. The meaning of each of the integrals above is as follows: take the scalar product of a pair of line elements, one on each loop, divide by the distance between them, and sum over all pairs. The only difference between Eqs. (7.54) and (7.55) is the *order* in which this operation is carried out, and that cannot affect the final sum. Hence $\Phi_{21}/I_1 = \Phi_{12}/I_2$, as desired. Thanks to this theorem, we need make no distinction between $M_{12}$ and $M_{21}$. We may speak, henceforth, of *the* mutual inductance $M$ of any two circuits.                                                                 □

Theorems of this sort are often called "reciprocity" theorems. There are some other reciprocity theorems on electric circuits not unrelated to this one. This may remind you of the relation $C_{jk} = C_{kj}$ mentioned in Section 3.6 and treated in Exercise 3.64. (In the spirit of that exercise, see Problem 7.10 for a second proof of the above $M_{12} = M_{21}$ theorem.) A reciprocity relation usually expresses some general symmetry law that is *not* apparent in the superficial structure of the system.

## 7.8 Self-inductance

When the current $I_1$ is changing, there is a change in the flux through circuit $C_1$ itself, and consequently an electromotive force is induced. Call this $\mathcal{E}_{11}$. The induction law holds, whatever the source of the flux:

$$\mathcal{E}_{11} = -\frac{d\Phi_{11}}{dt}, \tag{7.56}$$

where $\Phi_{11}$ is the flux through circuit 1 of the field $B_1$ due to the current $I_1$ in circuit 1. The minus sign expresses the fact that the electromotive force is always directed so as to *oppose* the *change* in current – Lenz's law, again. Since $\Phi_{11}$ will be proportional to $I_1$ we can write

$$\frac{\Phi_{11}}{I_1} = \text{constant} \equiv L_1. \tag{7.57}$$

Equation (7.56) then becomes

$$\mathcal{E}_{11} = -L_1 \frac{dI_1}{dt}. \tag{7.58}$$

The constant $L_1$ is called the *self-inductance* of the circuit. We usually drop the subscript "1."

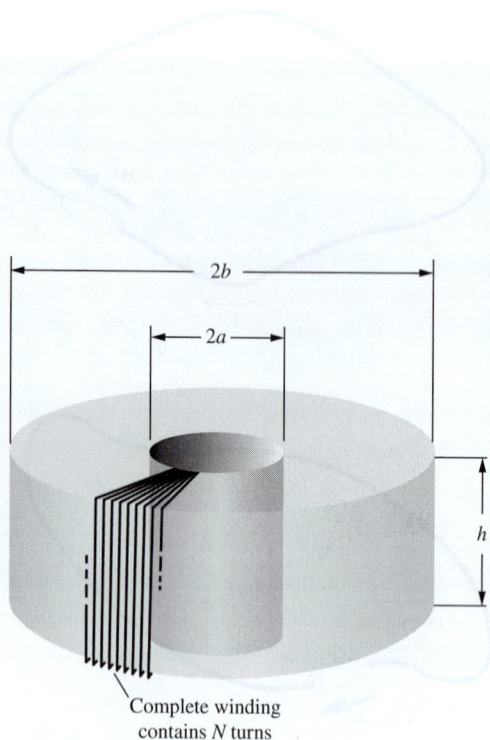

**Figure 7.22.**
Toroidal coil of rectangular cross section. Only a few turns are shown.

**Example (Rectangular toroidal coil)**   As an example of a circuit for which $L$ can be calculated, consider the rectangular toroidal coil of Exercise 6.61, shown here again in Fig. 7.22. You found (if you worked that exercise) that a current $I$ flowing in the coil of $N$ turns produces a field, the strength of which, at a radial distance $r$ from the axis of the coil, is given by $B = \mu_0 NI/2\pi r$. The total flux through one turn of the coil is the integral of this field over the cross section of the coil:

$$\Phi(\text{one turn}) = h \int_a^b \frac{\mu_0 NI}{2\pi r} \, dr = \frac{\mu_0 NIh}{2\pi} \ln\left(\frac{b}{a}\right). \tag{7.59}$$

The flux threading the circuit of $N$ turns is $N$ times as great:

$$\Phi = \frac{\mu_0 N^2 Ih}{2\pi} \ln\left(\frac{b}{a}\right). \tag{7.60}$$

Hence the induced electromotive force $\mathcal{E}$ is

$$\mathcal{E} = -\frac{d\Phi}{dt} = -\frac{\mu_0 N^2 h}{2\pi} \ln\left(\frac{b}{a}\right) \frac{dI}{dt}. \tag{7.61}$$

Thus the self-inductance of this coil is given by

$$L = \frac{\mu_0 N^2 h}{2\pi} \ln\left(\frac{b}{a}\right). \tag{7.62}$$

Since $\mu_0 = 4\pi \cdot 10^{-7}$ kg m/C$^2$, we can rewrite this in a form similar to Eq. (7.44):

$$L = (2 \cdot 10^{-7} \text{ kg m/C}^2) N^2 h \ln\left(\frac{b}{a}\right). \tag{7.63}$$

The numerical value of this expression gives $L$ in henrys. In Gaussian units, you can show that the self-inductance is

$$L = \frac{2N^2 h}{c^2} \ln\left(\frac{b}{a}\right). \tag{7.64}$$

You may think that one of the rings we considered earlier would have made a simpler example to illustrate the calculation of self-inductance. However, if we try to calculate the inductance of a simple circular loop of wire, we encounter a puzzling difficulty. It seems a good idea to simplify the problem by assuming that the wire has zero diameter. But we soon discover that, if finite current flows in a filament of zero diameter, the flux threading a loop made of such a filament is infinite! The reason is that the field $B$, in the neighborhood of a filamentary current, varies as $1/r$, where $r$ is the distance from the filament, and the integral of $B \times$ (area) diverges as $\int (dr/r)$ when we extend it down to $r = 0$. To avoid this we may let the radius of the wire be finite, not zero, which is more realistic anyway. This may make the calculation a bit more complicated, in a given case, but that won't worry us. The real difficulty is that different parts of the wire (at different distances from the center of the loop) now appear as *different circuits*, linked by different amounts of flux. We are no longer sure what we mean by *the* flux through *the* circuit. In fact, because the electromotive force is different in the different filamentary loops into which the circuit can be divided, some *redistribution* of current density must occur when rapidly changing currents flow in the ring. Hence the inductance of the circuit may depend somewhat on the rapidity of change of $I$, and thus not be strictly a constant as Eq. (7.58) would imply.

(a)

(b)

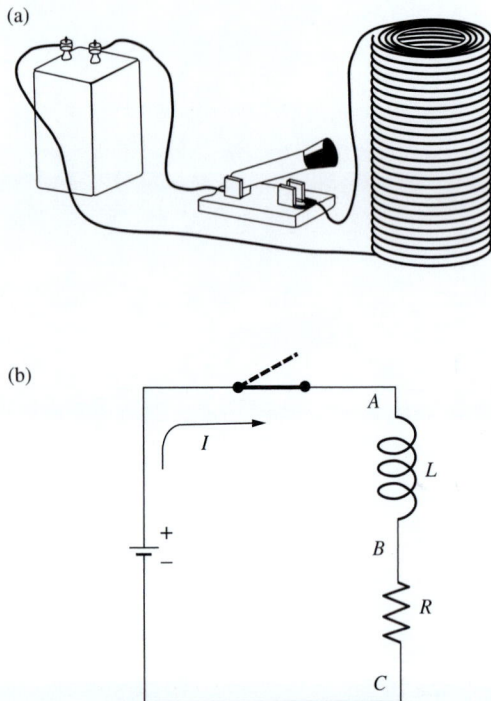

**Figure 7.23.**
A simple circuit with inductance (a) and resistance (b).

We avoided this embarrassment in the toroidal coil example by ignoring the field in the immediate vicinity of the individual turns of the winding. Most of the flux does *not* pass through the wires themselves, and whenever that is the case the effect we have just been worrying about will be unimportant.

## 7.9 Circuit containing self-inductance

Suppose we connect a battery, providing electromotive force $\mathcal{E}_0$, to a coil, or *inductor*, with self-inductance $L$, as in Fig. 7.23(a). The coil itself, the connecting wires, and even the battery will have some resistance. We don't care how this is distributed around the circuit. It can all be lumped together in one resistance $R$, indicated on the circuit diagram of Fig. 7.23(b) by a resistor symbol with this value. Also, the rest of the circuit, especially the connecting wires, contribute a bit to the self-inductance of the whole circuit; we assume that this is included in $L$. In other words, Fig. 7.23(b) represents an idealization of the physical circuit. The inductor $L$, symbolized by $\sim\!\ell\ell\ell\sim$, has no resistance; the resistor $R$ has no inductance. It is this idealized circuit that we shall now analyze.

If the current $I$ in the circuit is changing at the rate $dI/dt$, an electromotive force $L\,dI/dt$ will be induced, in a direction to oppose the change. Also, there is the constant electromotive force $\mathcal{E}_0$ of the battery. If we define the positive current direction as the one in which the battery tends to drive current around the circuit, then the net electromotive force at any instant is $\mathcal{E}_0 - L\,dI/dt$. This drives the current $I$ through the resistor $R$. That is,

$$\mathcal{E}_0 - L\frac{dI}{dt} = RI. \tag{7.65}$$

We can also describe the situation in this way: the voltage difference between points $A$ and $B$ in Fig. 7.23(b), which we call the *voltage across the inductor*, is $L\,dI/dt$, with the upper end of the inductor positive if $I$ in the direction shown is *increasing*. The voltage difference between $B$ and $C$, the voltage across the resistor, is $RI$, with the upper end of the resistor positive. Hence the sum of the voltage across the inductor and the voltage across the resistor is $L\,dI/dt + RI$. This is the same as the potential difference between the battery terminals, which is $\mathcal{E}_0$ (our idealized battery has no internal resistance). Thus we have

$$\mathcal{E}_0 = L\frac{dI}{dt} + RI, \tag{7.66}$$

which is merely a restatement of Eq. (7.65).

Before we look at the mathematical solution of Eq. (7.65), let's predict what ought to happen in this circuit if the switch is closed at $t = 0$. Before the switch is closed, $I = 0$, necessarily. A long time after the switch has been closed, some steady state will have been attained, with current practically constant at some value $I_0$. Then and thereafter, $dI/dt \approx 0$, and Eq. (7.65) reduces to

$$\mathcal{E}_0 = RI_0. \tag{7.67}$$

The transition from zero current to the steady-state current $I_0$ cannot occur abruptly at $t = 0$, for then $dI/dt$ would be infinite. In fact, just after $t = 0$, the current $I$ will be so small that the $RI$ term in Eq. (7.65) can be ignored, giving

$$\frac{dI}{dt} = \frac{\mathcal{E}_0}{L}. \tag{7.68}$$

The inductance $L$ limits the rate of rise of the current.

What we now know is summarized in Fig. 7.24(a). It only remains to find how the whole change takes place. Equation (7.65) is a differential equation very much like Eq. (4.39) in Chapter 4. The constant $\mathcal{E}_0$ term complicates things slightly, but the equation is still straightforward to solve. In Problem 7.14 you can show that the solution to Eq. (7.65) that satisfies our initial condition, $I = 0$ at $t = 0$, is

$$I(t) = \frac{\mathcal{E}_0}{R}\left(1 - e^{-(R/L)t}\right). \tag{7.69}$$

The graph in Fig. 7.24(b) shows the current approaching its asymptotic value $I_0$ exponentially. The "time constant" of this circuit is the quantity $L/R$. If $L$ is measured in henrys and $R$ in ohms, this comes out in seconds, since henrys = volt · amp$^{-1}$ · second, and ohms = volt · amp$^{-1}$.

What happens if we open the switch after the current $I_0$ has been established, thus forcing the current to drop abruptly to zero? That would make the term $L\,dI/dt$ negatively infinite! The catastrophe can be more than mathematical. People have been killed opening switches in highly inductive circuits. What happens generally is that a very high induced voltage causes a spark or arc across the open switch contacts, so that the current continues after all. Let us instead remove the battery from the circuit by closing a conducting path *across* the $LR$ combination, as in Fig. 7.25(a), at the same time disconnecting the battery. We now have a circuit described by the equation

$$0 = L\frac{dI}{dt} + RI, \tag{7.70}$$

with the initial condition $I = I_0$ at $t = t_1$, where $t_1$ is the instant at which the short circuit was closed. The solution is the simple exponential decay function

$$I(t) = I_0 e^{-(R/L)(t-t_1)} \tag{7.71}$$

with the same characteristic time $L/R$ as before.

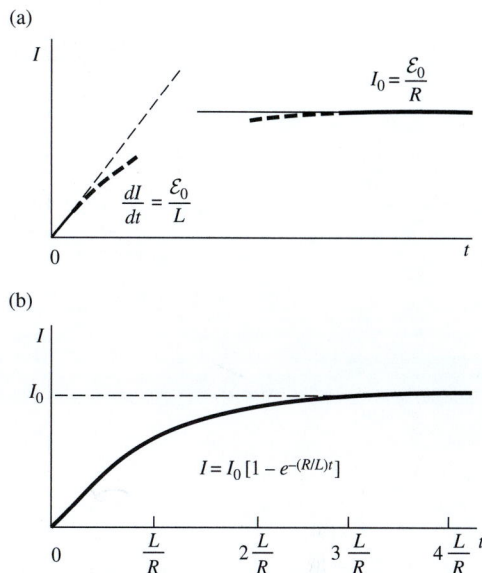

**Figure 7.24.**
(a) How the current must behave initially, and after a very long time has elapsed. (b) The complete variation of current with time in the circuit of Fig. 7.23.

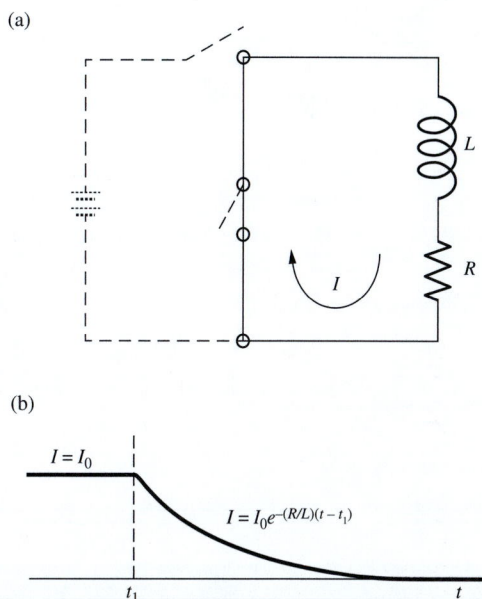

**Figure 7.25.**
(a) *LR* circuit. (b) Exponential decay of current in the *LR* circuit.

## 7.10 Energy stored in the magnetic field

During the decay of the current described by Eq. (7.71) and Fig. 7.25(b), energy is dissipated in the resistor $R$. Since the energy $dU$ dissipated in any short interval $dt$ is $RI^2\,dt$, the total energy dissipated after the closing of the switch at time $t_1$ is given by

$$U = \int_{t_1}^{\infty} RI^2\,dt = \int_{t_1}^{\infty} RI_0^2 e^{-(2R/L)(t-t_1)}\,dt$$

$$= -RI_0^2 \left(\frac{L}{2R}\right) e^{-(2R/L)(t-t_1)} \bigg|_{t_1}^{\infty} = \frac{1}{2} LI_0^2. \qquad (7.72)$$

The source of this energy was the inductor with its magnetic field. Indeed, exactly that amount of work had been done by the battery to build up the current in the first place – over and above the energy dissipated in the resistor between $t = 0$ and $t = t_1$, which was also provided by the battery. To see that this is a general relation, note that, if we have an increasing current in an inductor, work must be done to drive the current $I$ against the induced electromotive force $L\,dI/dt$. Since the electromotive force is defined to be the work done per unit charge, and since a charge $I\,dt$ moves through the inductor in time $dt$, the work done in time $dt$ is

$$dW = L\frac{dI}{dt}(I\,dt) = LI\,dI = \frac{1}{2}L\,d(I^2). \qquad (7.73)$$

Therefore, we may assign a total energy

$$\boxed{U = \frac{1}{2}LI^2} \qquad (7.74)$$

to an inductor carrying current $I$. With the eventual decay of this current, that amount of energy will appear somewhere else.

It is natural to regard this as energy stored in the magnetic field of the inductor, just as we have described the energy of a charged capacitor as stored in its electric field. The energy of a capacitor charged to potential difference $V$ is $(1/2)CV^2$ and is accounted for by assigning to an element of volume $dv$, where the electric field strength is $E$, an amount of energy $(\epsilon_0/2)E^2\,dv$. It is pleasant, but hardly surprising, to find that a similar relation holds for the energy stored in an inductor. That is, we can ascribe to the magnetic field an energy density $(1/2\mu_0)B^2$, and summing the energy of the whole field will give the energy $(1/2)LI^2$.

---

**Example (Rectangular toroidal coil)**   To show how the energy density $B^2/2\mu_0$ works out in one case, we can go back to the toroidal coil whose inductance $L$ we calculated in Section 7.8. We found in Eq. (7.62) that

$$L = \frac{\mu_0 N^2 h}{2\pi} \ln\left(\frac{b}{a}\right). \qquad (7.75)$$

The magnetic field strength $B$, with current $I$ flowing, was given by

$$B = \frac{\mu_0 NI}{2\pi r}. \qquad (7.76)$$

To calculate the volume integral of $B^2/2\mu_0$ we can use a volume element consisting of the cylindrical shell sketched in Fig. 7.26, with volume $2\pi rh\,dr$. As this shell expands from $r = a$ to $r = b$, it sweeps through all the space that contains magnetic field. (The field $B$ is zero everywhere outside the torus, remember.) So,

$$\frac{1}{2\mu_0}\int B^2\,dv = \frac{1}{2\mu_0}\int_a^b \left(\frac{\mu_0 NI}{2\pi r}\right)^2 2\pi rh\,dr = \frac{\mu_0 N^2 hI^2}{4\pi}\ln\left(\frac{b}{a}\right). \qquad (7.77)$$

Comparing this result with Eq. (7.75), we see that, indeed,

$$\frac{1}{2\mu_0}\int B^2\,dv = \frac{1}{2}LI^2. \qquad (7.78)$$

The task of Problem 7.18 is to show that this result holds for an arbitrary circuit with inductance $L$.

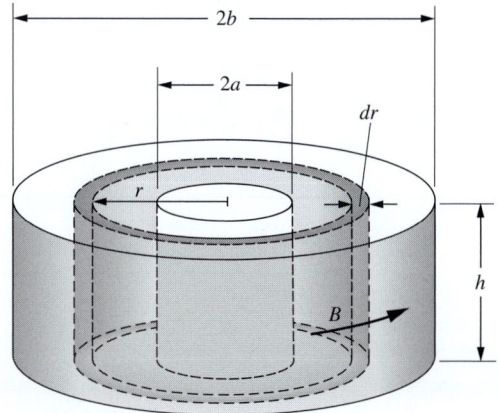

**Figure 7.26.**
Calculation of energy stored in the magnetic field of the toroidal coil of Fig. 7.22.

The more general statement, the counterpart of our statement for the electric field in Eq. (1.53), is that the energy $U$ to be associated with any magnetic field $B(x, y, z)$ is given by

$$U = \frac{1}{2\mu_0}\int_{\substack{\text{entire} \\ \text{field}}} B^2\,dv \qquad (7.79)$$

With $B$ in tesla and $v$ in m$^3$, the energy $U$ will be given in joules, as you can check. In Eq. (7.74), with $L$ in henrys and $I$ in amperes, $U$ will also be given in joules. The Gaussian equivalent of Eq. (7.79) for $U$ in ergs, $B$ in gauss, and $v$ in cm$^3$ is

$$U = \frac{1}{8\pi}\int_{\substack{\text{entire} \\ \text{field}}} B^2\,dv. \qquad (7.80)$$

The Gaussian equivalent of Eq. (7.74) remains $U = LI^2/2$, because the reasoning leading up to that equation is unchanged.

## 7.11 Applications

An *electrodynamic tether* is a long (perhaps 20 km) straight conducting wire that has one end connected to a satellite. The other end hangs down toward (or up away from) the earth. As the satellite and tether orbit the earth, they pass through the earth's magnetic field. Just as with the moving rod in Section 7.2, an emf is generated along the wire. If this were the whole story, charge would simply pile up on the ends. But the satellite is moving through the ionosphere, which contains enough ions to yield a return path for the charge. A complete circuit is therefore formed, so the emf can be used to provide power to the satellite. However, the current in the wire will experience a Lorentz force, and you can show that the direction is opposite to the satellite's motion. On the other hand, if a power source on the satellite drives current in the opposite direction along the tether, the Lorentz force will be in the same direction as the satellite's motion. So the tether can serve as a (gentle) propulsion device.

If you drop a wire hoop into a region containing a horizontal magnetic field, the changing flux will induce a current in the hoop. From Lenz's law and the right-hand rule, the resulting Lorentz force on the current is upward, independent of whether the hoop is entering or leaving the region of the magnetic field. So the direction of the force is always opposite to that of the velocity. If you drop a solid metal sheet into the region, loops (or *eddies*) of current will develop in the sheet, and the same braking effect, known as *eddy-current braking,* will occur. This braking effect has many applications, from coin vending machines to trains to amusement park rides. The loss in kinetic energy shows up as resistive heating. Eddy currents are also used in *metal detectors,* both of the airport security type and the hunting-for-buried-treasure type. The metal detector sends out a changing magnetic field, which induces eddy currents in any metal present. These currents produce their own changing magnetic field, which is then detected by the metal detector.

An *electric guitar* generates its sound via magnetic induction. The strings are made of a material that is easily magnetized, and they vibrate back and forth above *pickups.* A pickup is a coil of wire wrapped around a permanent magnet. This magnet causes the string to become magnetized, and the string's magnetic field then produces a flux through the coil. Because the string is vibrating, this flux is changing, so an emf is induced in the coil. This sends an oscillating current (with the same frequency as the string's vibration) to an amplifier, which then amplifies the sound. Without the amplifier, the sound is barely audible. Electric guitars with nylon (or otherwise nonmagnetic) strings won't work!

A *shake flashlight* is a nice application of Faraday's law. As you shake the flashlight, a permanent magnet passes back and forth through a coil. The induced emf in the coil produces a current that deposits charge on a capacitor, where it can be stored. A *bridge rectifier* (consisting of a certain configuration of four diodes) changes the alternating-current emf from your shaking motion to a direct-current emf, so that positive charge always flows toward the positive side of the capacitor, independent of which way the magnet is moving through the coil. When you flip the light switch, you allow current to flow from the capacitor through the light bulb.

An *electric generator* is based on the circuit in Fig. 7.13. An external torque causes the loop of wire to rotate, and the changing flux through the loop induces an oscillating (that is, alternating) emf. In practice, however, in most generators the rotating part (which is effectively the turbine) contains a permanent magnet that produces a changing flux through coils of wire arranged around the perimeter. In any case, it is Faraday's law at work. The force causing the turbine to rotate can come from various sources: water pressure from a dam, air pressure on a windmill, steam pressure from a coal plant or nuclear reactor, etc.

The *alternator* in a car engine is simply a small electric generator. In addition to turning the wheels, the engine also turns a small magnet inside the alternator, producing an alternating current. However, the car's battery requires direct current, so a rectifier converts the ac to dc. On an even smaller scale, the magnet in one type of bicycle-light generator is made to rotate due to the friction force from the tire. The power generated is usually only a few watts – a small fraction of the total power output from the rider. But don't think that you could make much money by harnessing all of your power on a stationary bike. If a professional cyclist sold the power he could generate during one hour of hard pedaling, he would earn about 5 cents.

*Hybrid cars,* which are powered by both gasoline and a battery, use *regenerative braking* to capture the kinetic energy of the car when the brakes are applied. The motor acts as a generator. More precisely, the friction force between the tires and the ground provides a torque on the gears in the motor, which in turn provide a torque on the magnet inside the generator.

A *microphone* is basically the opposite of a speaker (see Section 6.10). It converts sound waves into an electrical signal. A common type, called a *dynamic microphone,* makes use of electromagnetic induction. A coil of wire is attached to a diaphragm and surrounds one pole of a permanent magnet, in the same manner as in a speaker. When a sound wave causes the diaphragm to vibrate, the coil likewise vibrates. Its motion through the field of the magnet causes changing flux through the coil. An emf, and hence current, are therefore induced in the coil. This current signal is sent to a speaker, which reverses the process, turning the electrical signal back into sound waves. Alternatively, the signal is sent to a device that stores the information; see the discussion of cassette tapes and hard disks in Section 11.12.

A *ground-fault circuit interrupter (GFCI)* helps prevent (or at least mitigate) electric shocks. Under normal conditions, the current coming out of the "hot" slot in a wall socket (the short slot) equals the current flowing into the neutral slot (the tall slot). Now let's say you are receiving an electric shock. In a common type of shock, some of the current is taking an alternate route to ground – through you instead of through the neutral wire. The GFCI monitors the difference between the currents in the hot and neutral wires, and if it detects a difference of more than 5 or 10 mA, it trips the circuit (quickly, in about 30 ms). This monitoring is accomplished by positioning a toroidal coil around the two wires to and from the slots. The currents in these wires travel in opposite directions, so when the currents agree, there is zero net current in the pair. But if there is a mismatch in the (oscillating) currents, then the nonzero net current will produce a changing magnetic field circling around the wires. So there will be a changing magnetic field in the toroidal coil, which will induce a detectable current in the coil. A signal is then sent to a mechanism that trips the circuit. Having survived the shock, you can reset the GFCI by pressing the reset button on the outlet. As with a circuit breaker (see Section 6.10), nothing needs to be replaced after the circuit is tripped. In contrast, a fuse (see Section 4.12) needs to be replaced after it burns out. However, the purpose of a GFCI is different. A GFCI protects *people* by preventing tiny currents from traveling through them (even 50 mA can disrupt the functioning of a heart), while a fuse or a circuit breaker protects *buildings* by preventing large currents (on the order of 20 A), which can generate heat and cause fires.

A *transformer* changes the voltage in a circuit. Imagine two solenoids, $A$ and $B$, both wound around the same cylinder. Let $B$ (the secondary winding) have ten times the number of turns as $A$ (the primary winding). If a sinusoidal voltage source is connected to $A$, it will cause a changing flux through $A$, and hence also through $B$, with the latter flux being ten times the former. The induced emf in $B$ will therefore be ten times the emf in $A$. By adjusting the ratio of the number of turns, the voltage can be stepped up or stepped down by any factor. In practice, the solenoids in a transformer aren't actually right on top of each other, but instead wrapped around different parts of an iron core, which funnels the magnetic field lines along the core from one solenoid to the other. The ease with which ac voltages can be stepped up or down is the main reason why the electric grid uses ac. It is necessary to step up the voltage for long-distance transmission (and hence step down the current, for a given value of the power $P = IV$), because otherwise there would be prohibitively large energy losses due to the $I^2R$ resistance heating in the wires.

An *ignition system coil* in your car converts the 12 volts from the battery into the 30,000 or so volts needed to cause the arcing (the spark) across the spark plugs. Like a transformer, the ignition coil has primary and secondary windings, but the mechanism is slightly different. Instead of producing an oscillating voltage, it produces a one-time surge in voltage, whose original source was the 12 volt dc battery. It does this in two steps. First, the battery produces a steady current through a circuit containing the primary winding. A switch is then opened, and the current drops rapidly to zero (the rate is controlled by inserting a capacitor in the circuit, in parallel with the switch). This changing current creates a large back emf in the primary coil, say 300 volts. Second, if the secondary winding has, say, 100 times as many turns as the primary, then the transformer reasoning in the preceding paragraph leads to 30,000 volts in the secondary coil. This is enough to cause arcing in the spark plug. This two-step process means that we don't need to have $30{,}000/12 \approx 3000$ as many turns in the secondary coil!

A *boost converter* (or step-up converter) increases the voltage in a dc circuit. It is used, for example, to power a 3 volt LED lamp with a

1.5 volt battery. The main idea behind the converter is the fact that inductors resist sudden changes in current. Consider a circuit where current flows from a battery through an inductor. If a switch is opened downstream from the inductor, and if an alternative path is available through a capacitor, then current will still flow for a brief time through the inductor onto the capacitor. Charge will therefore build up on the capacitor. This process is repeated at a high frequency, perhaps 50 kHz. Even if the back voltage from the capacitor is higher than the forward voltage from the battery, a positive current will still flow briefly onto the capacitor each time the switch is opened. (Backward current can be prevented with a diode.) The capacitor then serves as a higher-voltage effective battery for powering the LED.

The magnetic field of the earth cannot be caused by a permanent magnet, because the interior temperature is far too hot to allow the iron core to exist in a state of permanent magnetization. Instead, the field is caused by the *dynamo effect* (see Problem 7.19 and Exercise 7.47). A source of energy is needed to drive the dynamo, otherwise the field would decay on a time scale of 20,000 years or so. This source isn't completely understood; possibilities include tidal forces, gravitational setting, radioactivity, and the buoyancy of lighter elements. The dynamo mechanism requires a fluid region inside the earth (this region is the outer core) and also a means of charge separation (perhaps friction between layers) so that currents can exist. It also requires that the earth be rotating, so that the Coriolis force can act on the fluid. Computer models indicate that the motion of the fluid is extremely complicated, and also that the reversal of the field (which happens every 200,000 years, on average) is likewise complicated. The poles don't simply rotate into each other. Rather, all sorts of secondary poles appear on the surface of the earth during the process, which probably takes a few thousand years. Who knows where all the famous explorers and their compasses would have ended up if a reversal had been taking place during the last thousand years! In recent years, the magnetic north pole has been moving at the brisk rate of about 50 km per year. This speed isn't terribly unusual, though, so it doesn't necessarily imply that a reversal is imminent.

# CHAPTER SUMMARY

- Faraday discovered that a current in one circuit can be induced by a *changing* current in another circuit.

- If a loop moves through a magnetic field, the *induced emf* equals $\mathcal{E} = vw(B_1 - B_2)$, where $w$ is the length of the transverse sides, and the $B$'s are the fields at these sides. This emf can be viewed as a consequence of the Lorentz force acting on the charges in the transverse sides.

• More generally, the emf can be written in terms of the magnetic flux as

$$\mathcal{E} = -\frac{d\Phi}{dt}. \tag{7.81}$$

This is known as *Faraday's law of induction,* and it holds in all cases: the loop can be moving, or the source of the magnetic field can be moving, or the flux can be changed by some other arbitrary means. The sign of the induced emf is determined by *Lenz's law:* the induced current flows in the direction that produces a magnetic field that opposes the change in flux. The differential form of Faraday's law is

$$\nabla \times \mathbf{E} = -\frac{\partial \mathbf{B}}{\partial t}. \tag{7.82}$$

This is one of Maxwell's equations.

• If we have two circuits $C_1$ and $C_2$, a current $I_1$ in one circuit will produce a flux $\Phi_{21}$ through the other. The *mutual inductance* $M_{21}$ is defined by $M_{21} = \Phi_{21}/I_1$. It then follows that a changing $I_1$ produces an emf in $C_2$ equal to $\mathcal{E}_{21} = -M_{21}\, dI_1/dt$. The two coefficients of mutual inductance are symmetric: $M_{12} = M_{21}$.

• The *self-inductance* $L$ is defined analogously. A current $I$ in a circuit will produce a flux $\Phi$ through the circuit, and the self-inductance is defined by $L = \Phi/I$. The emf is then $\mathcal{E} = -L\, dI/dt$.

• If a circuit contains an *inductor,* and if a switch is opened (or closed), the current can't change discontinuously, because that would create an infinite value of $\mathcal{E} = -L\, dI/dt$. The current must therefore gradually change. If a switch is closed in an $RL$ circuit, the current takes the form

$$I = \frac{\mathcal{E}_0}{R}\left(1 - e^{-(R/L)t}\right). \tag{7.83}$$

The quantity $L/R$ is the *time constant* of the circuit.

• The energy stored in an inductor equals $U = LI^2/2$. It can be shown that this is equivalent to the statement that a magnetic field contains an energy density of $B^2/2\mu_0$ (just as an electric field contains an energy density of $\epsilon_0 E^2/2$).

## Problems

7.1 *Current in a bottle* ✱✱

An ocean current flows at a speed of 2 knots (approximately 1 m/s) in a region where the vertical component of the earth's magnetic field is 0.35 gauss. The conductivity of seawater in that region is 4 (ohm-m)$^{-1}$. On the assumption that there is no other horizontal component of **E** than the motional term $\mathbf{v} \times \mathbf{B}$, find the density $J$ of the horizontal electric current. If you were to carry a bottle of seawater through the earth's field at this speed, would such a current be flowing in it?

7.2   *What's doing work?* ✳✳✳

In Fig. 7.27 a conducting rod is pulled to the right at speed $v$ while maintaining contact with two rails. A magnetic field points into the page. From the reasoning in Section 7.3, we know that an induced emf will cause a current to flow in the counterclockwise direction around the loop. Now, the magnetic force $q\mathbf{u} \times \mathbf{B}$ is perpendicular to the velocity $\mathbf{u}$ of the moving charges, so it can't do work on them. However, the magnetic force $\mathbf{f}$ in Eq. (7.5) certainly looks like it is doing work. What's going on here? Is the magnetic force doing work or not? If not, then what is? There is definitely *something* doing work because the wire will heat up.

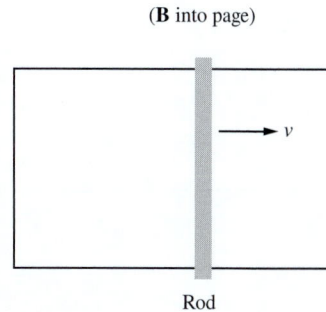

(**B** into page)

Rod

**Figure 7.27.**

7.3   *Pulling a square frame* ✳✳

A square wire frame with side length $\ell$ has total resistance $R$. It is being pulled with speed $v$ out of a region where there is a uniform $\mathbf{B}$ field pointing out of the page (the shaded area in Fig. 7.28). Consider the moment when the left corner is a distance $x$ inside the shaded area.

(a) What force do you need to apply to the square so that it moves with constant speed $v$?

(b) Verify that the work you do from $x = x_0$ (which you can assume is less than $\ell/\sqrt{2}$) down to $x = 0$ equals the energy dissipated in the resistor.

7.4   *Loops around a solenoid* ✳✳

We can think of a voltmeter as a device that registers the line integral $\int \mathbf{E} \cdot d\mathbf{s}$ along a path $C$ *from* the clip at the end of its (+) lead, *through* the voltmeter, *to* the clip at the end of its (−) lead. Note that part of $C$ lies inside the voltmeter itself. Path $C$ may also be part of a loop that is completed by some external path from the (−) clip to the (+) clip. With that in mind, consider the arrangement in Fig. 7.29. The solenoid is so long that its external magnetic field is negligible. Its cross-sectional area is 20 cm$^2$, and the field inside is toward the right and increasing at the rate of 100 gauss/s. Two identical voltmeters are connected to points on a loop that encloses the solenoid and contains two 50 ohm resistors, as shown. The voltmeters are capable of reading microvolts and have high internal resistance. What will each voltmeter read? Make sure your answer is consistent, from every point of view, with Eq. (7.26).

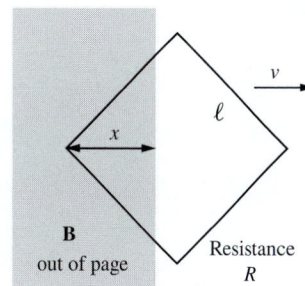

**B**
out of page

Resistance
$R$

**Figure 7.28.**

7.5   *Total charge* ✳✳

A circular coil of wire, with $N$ turns of radius $a$, is located in the field of an electromagnet. The magnetic field is perpendicular to the coil (that is, parallel to the axis of the coil), and its strength has the constant value $B_0$ over that area. The coil is connected by a pair of twisted leads to an external resistance. The total resistance of

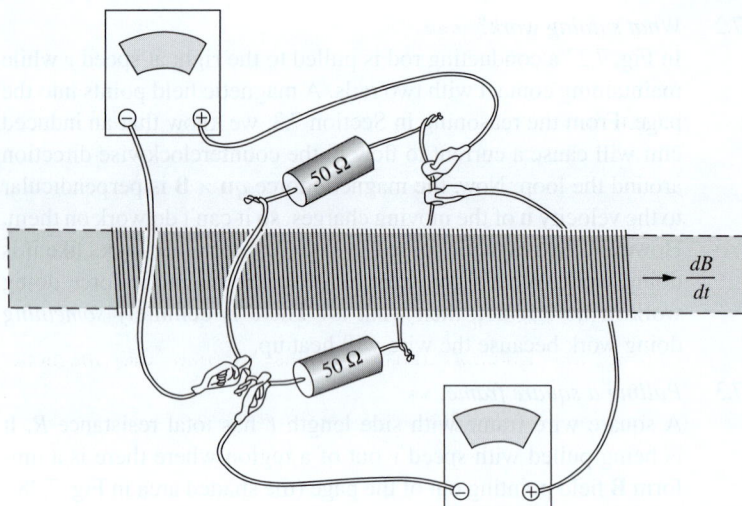

**Figure 7.29.**

this closed circuit, including that of the coil itself, is $R$. Suppose the electromagnet is turned off, its field dropping more or less rapidly to zero. The induced electromotive force causes current to flow around the circuit. Derive a formula for the total charge $Q = \int I \, dt$ that passes through the resistor, and explain why it does not depend on the rapidity with which the field drops to zero.

**7.6 Growing current in a solenoid** **

An infinite solenoid has radius $R$ and $n$ turns per unit length. The current grows linearly with time, according to $I(t) = Ct$. Use the integral form of Faraday's law to find the electric field at radius $r$, both inside and outside the solenoid. Then verify that your answers satisfy the differential form of the law.

**7.7 Maximum emf for a thin loop** ***

A long straight stationary wire is parallel to the $y$ axis and passes through the point $z = h$ on the $z$ axis. A current $I$ flows in this wire, returning by a remote conductor whose field we may neglect. Lying in the $xy$ plane is a thin rectangular loop with two of its sides, of length $\ell$, parallel to the long wire. The length $b$ of the other two sides is very small. The loop slides with constant speed $v$ in the $\hat{x}$ direction. Find the magnitude of the electromotive force induced in the loop at the moment the center of the loop has position $x$. For what values of $x$ does this emf have a local maximum or minimum? (Work in the approximation where $b \ll x$, so that you can approximate the relevant difference in $B$ fields by a derivative.)

7.8  *Faraday's law for a moving tilted sheet* ∗∗∗∗
Recall the "tilted sheet" example in Section 5.5, in which a charged
sheet was tilted at 45° in the lab frame. We calculated the electric
field in the frame $F'$ moving to the right with speed $v$ (which was
$0.6c$ in the example). The goal of this problem is to demonstrate
that Faraday's law holds in this setup.

(a) For a general speed $v$, find the component of the electric field
that is parallel to the sheet in frame $F'$ (in which the sheet
moves to the left with speed $v$). If you solved Exercise 5.12,
you've already done most of the work.

(b) Use the Lorentz transformations to find the magnetic
field in $F'$.

(c) In $F'$, verify that $\int \mathbf{E} \cdot d\mathbf{s} = -d\Phi/dt$ holds for the rectangle
shown in Fig. 7.30 (this rectangle is fixed in $F'$).

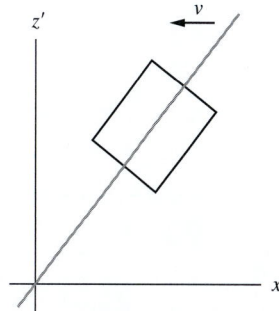

**Figure 7.30.**

7.9  *Mutual inductance for two solenoids* ∗∗
Figure 7.31 shows a solenoid of radius $a_1$ and length $b_1$ located
inside a longer solenoid of radius $a_2$ and length $b_2$. The total
number of turns is $N_1$ on the inner coil, $N_2$ on the outer. Work
out an approximate formula for the mutual inductance $M$.

7.10  *Mutual-inductance symmetry* ∗∗
In Section 7.7 we made use of the vector potential to prove that
$M_{12} = M_{21}$. We can give a second proof, this time in the spirit of
Exercise 3.64. Imagine increasing the currents in two circuits gradually from zero to the final values of $I_{1f}$ and $I_{2f}$ ("f" for "final").
Due to the induced emfs, some external agency has to supply power
to increase (or maintain) the currents. The final currents can be
brought about in many different ways. Two possible ways are of
particular interest.

(a) Keep $I_2$ at zero while raising $I_1$ gradually from zero to $I_{1f}$.
Then raise $I_2$ from zero to $I_{2f}$ while holding $I_1$ constant at $I_{1f}$.

(b) Carry out a similar program with the roles of 1 and 2 exchanged,
that is, raise $I_2$ from zero to $I_{2f}$ first, and so on.

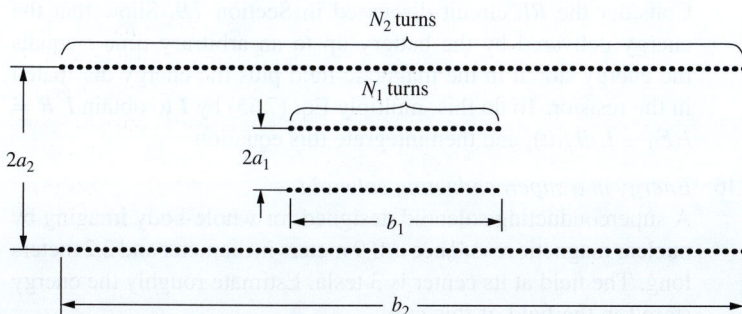

**Figure 7.31.**

Compute the total work done by external agencies, for each of the two programs. Then complete the argument. See Crawford (1992) for further discussion.

7.11 *L for a solenoid* ∗
Find the self-inductance of a long solenoid with radius $r$, length $\ell$, and $N$ turns.

7.12 *Doubling a solenoid* ∗
(a) Two identical solenoids are connected end-to-end to make a solenoid of twice the length. By what factor is the self-inductance increased? The answer quickly follows from the formula for a solenoid's $L$, but you should also explain in words why the factor is what it is.
(b) Same question, but now with the two solenoids placed right on top of one another. (Imagine that one solenoid is slightly wider and surrounds the other.) They are connected so that the current flows in the same direction in each.

7.13 *Adding inductors* ∗
(a) Two inductors, $L_1$ and $L_2$, are connected in series, as shown in Fig. 7.32(a). Show that the effective inductance $L$ of the system is given by

$$L = L_1 + L_2. \tag{7.84}$$

Check the $L_1 \to 0$ and $L_1 \to \infty$ limits.
(b) If the inductors are instead connected in parallel, as shown in Fig. 7.32(b), show that the effective inductance is given by

$$\frac{1}{L} = \frac{1}{L_1} + \frac{1}{L_2}. \tag{7.85}$$

Again check the $L_1 \to 0$ and $L_1 \to \infty$ limits.

7.14 *Current in an RL circuit* ∗∗
Show that the expression for the current in an $RL$ circuit given in Eq. (7.69) follows from Eq. (7.65).

7.15 *Energy in an RL circuit* ∗
Consider the $RL$ circuit discussed in Section 7.9. Show that the energy delivered by the battery up to an arbitrary time $t$ equals the energy stored in the magnetic field plus the energy dissipated in the resistor. To do this, multiply Eq. (7.65) by $I$ to obtain $I^2 R = I(\mathcal{E}_0 - L\, dI/dt)$, and then integrate this equation.

7.16 *Energy in a superconducting solenoid* ∗
A superconducting solenoid designed for whole-body imaging by nuclear magnetic resonance is 0.9 meters in diameter and 2.2 meters long. The field at its center is 3 tesla. Estimate roughly the energy stored in the field of this coil.

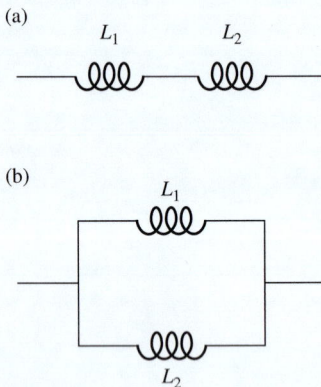

(a)

$L_1 \qquad L_2$

(b)

$L_1$

$L_2$

**Figure 7.32.**

7.17 *Two expressions for the energy* ∗

Two different expressions for the energy stored in a long solenoid are $LI^2/2$ and $(B^2/2\mu_0)$(volume). Show that these expressions are consistent.

7.18 *Two expressions for the energy (general)* ∗∗∗

The task of Problem 2.24 was to demonstrate that two different expressions for the electrostatic energy, $\int(\epsilon_0 E^2/2)\,dv$ and $\int(\rho\phi/2)\,dv$, are equivalent (as they must be, if they are both valid). The latter expression can quickly be converted to $C\phi^2/2$ in the case of oppositely charged conductors in a capacitor (see Exercise 3.65).

The task of this problem is to demonstrate the analogous relation for the magnetic energy, that is, to show that if a circuit (of finite extent) with self-inductance $L$ contains current $I$, then $\int(B^2/2\mu_0)\,dv$ equals $LI^2/2$. This is a bit trickier than the electrostatic case, so here are some hints: (1) a useful vector identity is $\nabla\cdot(\mathbf{A}\times\mathbf{B})=\mathbf{B}\cdot(\nabla\times\mathbf{A})-\mathbf{A}\cdot(\nabla\times\mathbf{B})$, (2) the vector potential and magnetic field satisfy $\nabla\times\mathbf{A}=\mathbf{B}$, (3) $\nabla\times\mathbf{B}=\mu_0\mathbf{J}$, (4) $\Phi=\int\mathbf{A}\cdot d\mathbf{l}$ from Eq. (7.52), and (5) $L$ is defined by $\Phi=LI$.

7.19 *Critical frequency of a dynamo* ∗∗∗

A dynamo like the one in Exercise 7.47 has a certain critical speed $\omega_0$. If the disk revolves with an angular velocity less than $\omega_0$, nothing happens. Only when that speed is attained is the induced $\mathcal{E}$ large enough to make the current large enough to make the magnetic field large enough to induce an $\mathcal{E}$ of that magnitude. The critical speed can depend only on the size and shape of the conductors, the conductivity $\sigma$, and the constant $\mu_0$. Let $d$ be some characteristic dimension expressing the size of the dynamo, such as the radius of the disk in our example.

(a) Show by a dimensional argument that $\omega_0$ must be given by a relation of this form: $\omega_0 = K/\mu_0\sigma d^2$, where $K$ is some dimensionless numerical factor that depends only on the arrangement and *relative* size of the various parts of the dynamo.

(b) Demonstrate this result again by using physical reasoning that relates the various quantities in the problem ($R$, $\mathcal{E}$, $E$, $I$, $B$, etc.). You can ignore all numerical factors in your calculations and absorb them into the constant $K$.

Additional comments: for a dynamo of modest size made wholly of copper, the critical speed $\omega_0$ would be practically unattainable. It is ferromagnetism that makes possible the ordinary dc generator by providing a magnetic field much stronger than the current in the coils, unaided, could produce. For an earth-sized dynamo, however, with $d$ measured in hundreds of kilometers rather than meters, the critical speed is very much smaller. The earth's magnetic field is almost certainly produced by a nonferromagnetic dynamo involving motions in the fluid metallic core. That fluid happens to be molten iron, but it is not even slightly ferromagnetic because it is too hot. (That will be explained in Chapter 11.) We don't know how the conducting fluid moves, or what configuration of electric currents and magnetic fields its motion generates in the core. The magnetic field we observe at the earth's surface is the external field of the dynamo in the core. The direction of the earth's field a million years ago is preserved in the magnetization of rocks that solidified at that time. That magnetic record shows that the field has reversed its direction nearly 200 times in the last 100 million years. Although a reversal cannot have been instantaneous (see Exercise 7.46), it was a relatively sudden event on the geological time scale. The immense value of *paleomagnetism* as an indelible record of our planet's history is well explained in Chapter 18 of Press and Siever (1978).

## Exercises

7.20 *Induced voltage from the tides* *

Faraday describes in the following words an unsuccessful attempt to detect a current induced when part of a circuit consists of water moving through the earth's magnetic field (Faraday, 1839, p. 55):

> I made experiments therefore (by favour) at Waterloo Bridge, extending a copper wire nine hundred and sixty feet in length upon the parapet of the bridge, and dropping from its extremities other wires with extensive plates of metal attached to them to complete contact with the water. Thus the wire and the water made one conducting circuit; and as the water ebbed or flowed with the tide, I hoped to obtain currents analogous to those of the brass ball. I constantly obtained deflections at the galvanometer, but they were irregular, and were, in succession, referred to other causes than that sought for. The different condition of the water as to purity on the two sides of the river; the difference in temperature; slight differences in the plates, in the solder used, in the more or less perfect contact made by twisting or otherwise; all produced effects in turn: and though I experimented on the water passing through the middle arches only; used platina plates instead of copper; and took every other precaution, I could not after three days obtain any satisfactory results.

Assume the vertical component of the field was 0.5 gauss, make a reasonable guess about the velocity of tidal currents in the Thames, and estimate the magnitude of the induced voltage Faraday was trying to detect.

7.21 *Maximum emf* *

What is the maximum electromotive force induced in a coil of 4000 turns, average radius 12 cm, rotating at 30 revolutions per second in the earth's magnetic field where the field intensity is 0.5 gauss?

7.22 *Oscillating E and B* *

In the central region of a solenoid that is connected to a radio-frequency power source, the magnetic field oscillates at $2.5 \cdot 10^6$ cycles per second with an amplitude of 4 gauss. What is the amplitude of the oscillating electric field at a point 3 cm from the axis? (This point lies within the region where the magnetic field is nearly uniform.)

7.23 *Vibrating wire* *

A taut wire passes through the gap of a small magnet (Fig. 7.33), where the field strength is 5000 gauss. The length of wire within the gap is 1.8 cm. Calculate the amplitude of the induced alternating voltage when the wire is vibrating at its fundamental frequency of 2000 Hz with an amplitude of 0.03 cm, transverse to the magnetic field.

7.24 *Pulling a frame* **

The shaded region in Fig. 7.34 represents the pole of an electromagnet where there is a strong magnetic field perpendicular to the plane of the paper. The rectangular frame is made of a 5 mm diameter aluminum rod, bent and with its ends welded together. Suppose that by applying a steady force of 1 newton, starting at the position shown, the frame can be pulled out of the magnet in 1 second. Then, if the force is doubled, to 2 newtons, the frame will be pulled out in _____ seconds. Brass has about twice the resistivity of aluminum. If the frame had been made of a 5 mm brass rod, the force needed to pull it out in 1 second would be _____ newtons. If the frame had been made of a 1 cm diameter aluminum rod, the force required to pull it out in 1 second would be _____ newtons. You may neglect in all cases the inertia of the frame.

7.25 *Sliding loop* **

A long straight stationary wire is parallel to the $y$ axis and passes through the point $z = h$ on the $z$ axis. A current $I$ flows in this wire, returning by a remote conductor whose field we may neglect. Lying in the $xy$ plane is a square loop with two of its sides, of length $b$, parallel to the long wire. This loop slides with constant speed $v$ in the $\hat{\mathbf{x}}$ direction. Find the magnitude of the electromotive force induced in the loop at the moment when the center of the loop crosses the $y$ axis.

7.26 *Sliding bar* **

A metal crossbar of mass $m$ slides without friction on two long parallel conducting rails a distance $b$ apart; see Fig. 7.35. A resistor $R$ is connected across the rails at one end; compared with $R$, the resistance of bar and rails is negligible. There is a uniform field $\mathbf{B}$ perpendicular to the plane of the figure. At time $t = 0$ the crossbar is given a velocity $v_0$ toward the right. What happens afterward?

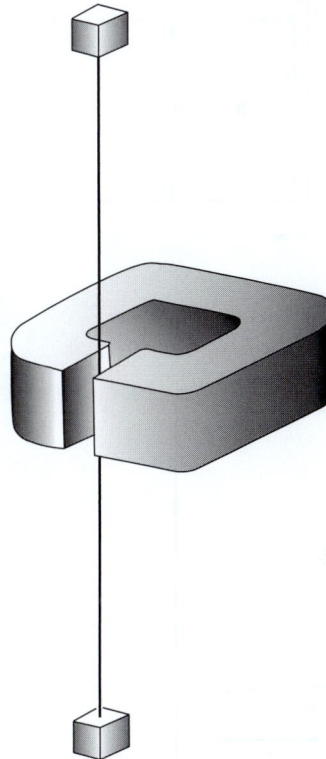

**Figure 7.33.**

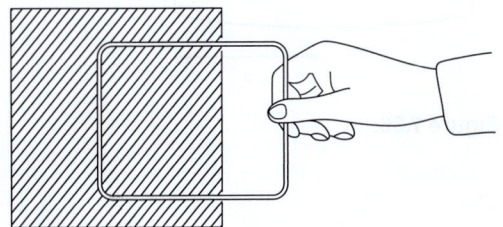

**Figure 7.34.**

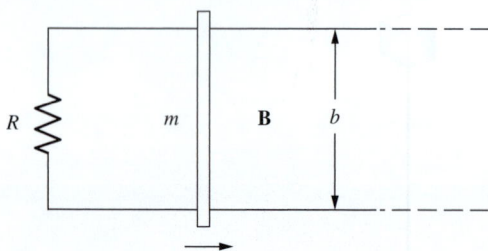

**Figure 7.35.**

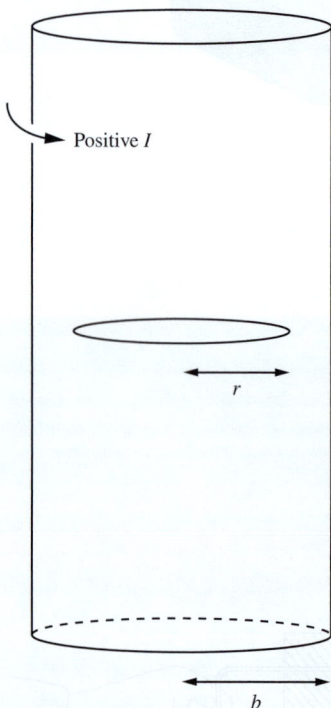

**Figure 7.36.**

(a)  Does the rod ever stop moving? If so, when?

(b)  How far does it go?

(c)  How about conservation of energy?

7.27  *Ring in a solenoid*  **

An infinite solenoid with radius $b$ has $n$ turns per unit length. The current varies in time according to $I(t) = I_0 \cos \omega t$ (with positive defined as shown in Fig. 7.36). A ring with radius $r < b$ and resistance $R$ is centered on the solenoid's axis, with its plane perpendicular to the axis.

(a)  What is the induced current in the ring?

(b)  A given little piece of the ring will feel a magnetic force. For what values of $t$ is this force maximum?

(c)  What is the effect of the force on the ring? That is, does the force cause the ring to translate, spin, flip over, stretch/shrink, etc.?

7.28  *A loop with two surfaces*  **

Consider the loop of wire shown in Fig. 7.37. Suppose we want to calculate the flux of **B** through this loop. *Two* surfaces bounded by the loop are shown in parts (a) and (b) of the figure. What is the essential difference between them? Which, if either, is the correct surface to use in performing the surface integral $\int \mathbf{B} \cdot d\mathbf{a}$ to find the flux? Describe the corresponding surface for a three-turn coil. Show that this is all consistent with our previous assertion that, for a compact coil of $N$ turns, the electromotive force is just $N$ times what it would be for a single loop of the same size and shape.

7.29  *Induced emf in a loop*  ***

Calculate the electromotive force in the moving loop in Fig. 7.38 at the instant when it is in the position shown. Assume the resistance of the loop is so great that the effect of the current in the loop itself is negligible. Estimate very roughly how large a resistance would be safe, in this respect. Indicate the direction in which current would flow in the loop, at the instant shown.

7.30  *Work and dissipated energy*  **

Suppose the loop in Fig. 7.6 has a resistance $R$. Show that whoever is pulling the loop along at constant speed does an amount of work during the interval $dt$ that agrees precisely with the energy dissipated in the resistance during this interval, assuming that the self-inductance of the loop can be neglected. What is the source of the energy in Fig. 7.14 where the loop is stationary?

**7.31** *Sinusoidal emf* **

Does the prediction of a simple sinusoidal variation of electromotive force for the rotating planar loop in Fig. 7.13 depend on the loop being rectangular, on the magnetic field being uniform, or on both? Explain. Can you suggest an arrangement of rotating loop and stationary coils that will give a definitely nonsinusoidal emf? Sketch the voltage–time curve you would expect to see on the oscilloscope, with that arrangement.

**7.32** *Emfs and voltmeters* **

The circular wire in Fig. 7.39(a) encircles a solenoid in which the magnetic flux $d\Phi/dt$ is increasing at a constant rate $\mathcal{E}_0$ out of the page. So the clockwise emf around the loop is $\mathcal{E}_0$.

In Fig. 7.39(b) the solenoid has been removed, and a capacitor has been inserted in the loop. The upper plate is positive. The voltage difference between the plates is $\mathcal{E}_0$, and this voltage is maintained by someone physically dragging positive charges from the negative plate to the positive plate (or rather, dragging electrons the other way). So this person is the source of the emf.

In Fig. 7.39(c) the above capacitor has been replaced by $N$ little capacitors, each with a voltage difference of $\mathcal{E}_0/N$. The figure shows $N = 12$, but assume that $N$ is large, essentially infinite. As above, the emf is maintained by people dragging charges from one plate to the other in every capacitor. This setup is similar to the setup in Fig. 7.39(a), in that the emf is evenly distributed around the circuit.

By definition, the voltage difference between two points is given by $V_b - V_a \equiv -\int_a^b \mathbf{E} \cdot d\mathbf{s}$. This is what a voltmeter measures. For each of the above three setups, find the voltage difference $V_b - V_a$ along path 1 (shown in part (a) of the figure), and also the voltage difference $V_a - V_b$ along path 2. Comment on the similarities and differences in your results, and also on each of the total voltage drops in a complete round trip.

**7.33** *Getting a ring to spin* **

A nonconducting thin ring of radius $a$ carries a static charge $q$. This ring is in a magnetic field of strength $B_0$, parallel to the ring's axis, and is supported so that it is free to rotate about that axis. If the field is switched off, how much angular momentum will be added to the ring? Supposing the mass of the ring to be $m$, show that the ring, if initially at rest, will acquire an angular velocity $\omega = qB_0/2m$. Note that, as in Problem 7.5, the result depends only on the initial and final values of the field strength, and not on the rapidity of change.

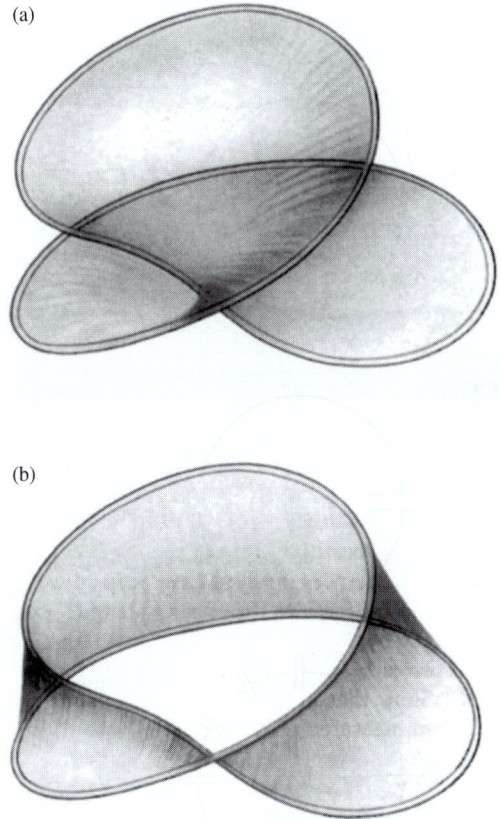

(a)

(b)

**Figure 7.37.**

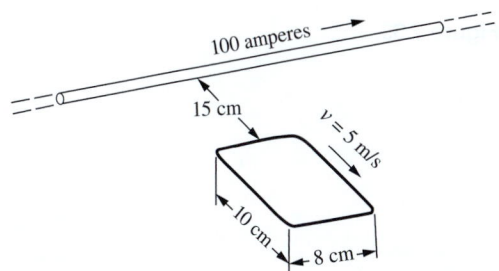

100 amperes

15 cm

$v = 5$ m/s

10 cm

8 cm

**Figure 7.38.**

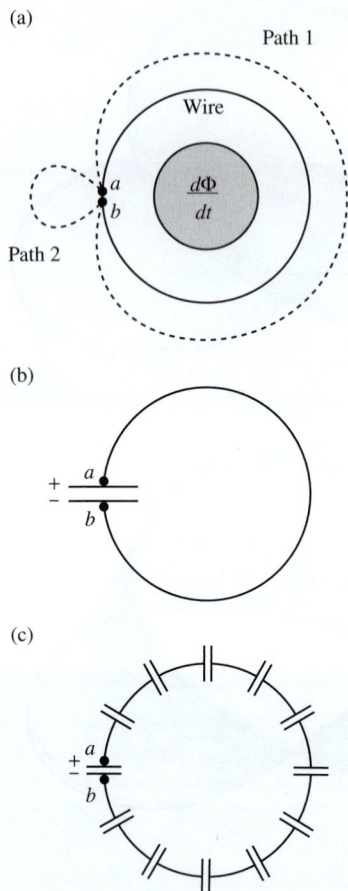

(a)

(b)

(c)

**Figure 7.39.**

**7.34** *Faraday's experiment* ∗∗∗

The coils that first produced a slight but detectable kick in Faraday's galvanometer he describes as made of 203 feet of copper wire each, wound around a large block of wood; see Fig. 7.1(a). The turns of the second spiral (that is, single-layer coil) were interposed between those of the first, but separated from them by twine. The diameter of the copper wire itself was 1/20 inch. He does not give the dimensions of the wooden block or the number of turns in the coils. In the experiment, one of these coils was connected to a "battery of 100 plates." (Assume that one plate is roughly 1 volt.) See if you can make a rough estimate of the duration in seconds (it will be small) and magnitude in amperes of the pulse of current that passed through his galvanometer.

**7.35** *M for two rings* ∗∗

Derive an approximate formula for the mutual inductance of two circular rings of the same radius $a$, arranged like wheels on the same axle with their centers a distance $b$ apart. Use an approximation good for $b \gg a$.

**7.36** *Connecting two circuits* ∗∗

Part (a) of Fig. 7.40 shows two coils with self-inductances $L_1$ and $L_2$. In the relative position shown, their mutual inductance is $M$. The positive current direction and the positive electromotive force direction in each coil are defined by the arrows in the figure. The equations relating currents and electromotive forces are

$$\mathcal{E}_1 = -L_1 \frac{dI_1}{dt} \pm M \frac{dI_2}{dt} \quad \text{and} \quad \mathcal{E}_2 = -L_2 \frac{dI_2}{dt} \pm M \frac{dI_1}{dt}. \quad (7.86)$$

(a) Given that $M$ is always to be taken as a positive constant, how must the signs be chosen in these equations? What if we had chosen, as we might have, the other direction for positive current, and for positive electromotive force, in the lower coil?

(b) Now connect the two coils together, as in part (b) of the figure, to form a single circuit. What is the self-inductance $L'$ of this circuit, expressed in terms of $L_1$, $L_2$, and $M$? What is the self-inductance $L''$ of the circuit formed by connecting the coils as shown in (c)? Which circuit, (b) or (c), has the greater self-inductance?

(c) Considering that the self-inductance of any circuit must be a positive quantity (why couldn't it be negative?), see if you can draw a general conclusion, valid for any conceivable pair of coils, concerning the relative magnitude of $L_1$, $L_2$, and $M$.

**7.37** *Flux through two rings* ∗∗

Discuss the implications of the theorem $\Phi_{21}/I_1 = \Phi_{12}/I_2$ in the case of the large and small concentric rings in Fig. 7.20. With fixed current $I_1$ in the outer ring, obviously $\Phi_{21}$, the flux through the inner ring, decreases if $R_1$ is increased, simply because the field at the center gets weaker. But with fixed current in the inner ring, why should $\Phi_{12}$, the flux through the outer ring, *decrease* as $R_1$ increases, holding $R_2$ constant? It must do so to satisfy our theorem.

**7.38** *Using the mutual inductance for two rings* ***
Can you devise a way to use the theorem $\Phi_{21}/I_1 = \Phi_{12}/I_2$ to find the magnetic field strength due to a ring current at points in the plane of the ring at a distance from the ring much greater than the ring radius? (*Hint:* Consider the effect of a small change $\Delta R_1$ in the radius of the outer ring in Fig. 7.20; it must have the same effect on $\Phi_{12}/I_2$ as on $\Phi_{21}/I_1$.)

**7.39** *Small L* *
How could we wind a resistance coil so that its self-inductance would be *small*?

**7.40** *L for a cylindrical solenoid* **
Calculate the self-inductance of a cylindrical solenoid 10 cm in diameter and 2 m long. It has a single-layer winding containing a total of 1200 turns. Assume that the magnetic field inside the solenoid is approximately uniform right out to the ends. Estimate roughly the magnitude of the error you will thereby incur. Is the true $L$ larger or smaller than your approximate result?

**7.41** *Opening a switch* **
In the circuit shown in Fig. 7.41 the 10 volt battery has negligible internal resistance. The switch $S$ is closed for several seconds, then opened. Make a graph with the abscissa time in milliseconds, showing the potential of point $A$ with respect to ground, just before and then for 5 milliseconds after the opening of switch $S$. Show also the variation of the potential at point $B$ in the same period of time.

**7.42** *RL circuit* **
A coil with resistance of 0.01 ohm and self-inductance 0.50 millihenry is connected across a large 12 volt battery of negligible internal resistance. How long after the switch is closed will the current reach 90 percent of its final value? At that time, how much energy, in joules, is stored in the magnetic field? How much energy has been withdrawn from the battery up to that time?

**7.43** *Energy in an RL circuit* **
Consider the *RL* circuit discussed in Section 7.9. Show that the energy delivered by the battery up to an arbitrary time $t$ equals the energy stored in the magnetic field plus the energy dissipated in the resistor. Do this by using the expression for $I(t)$ in Eq. (7.69)

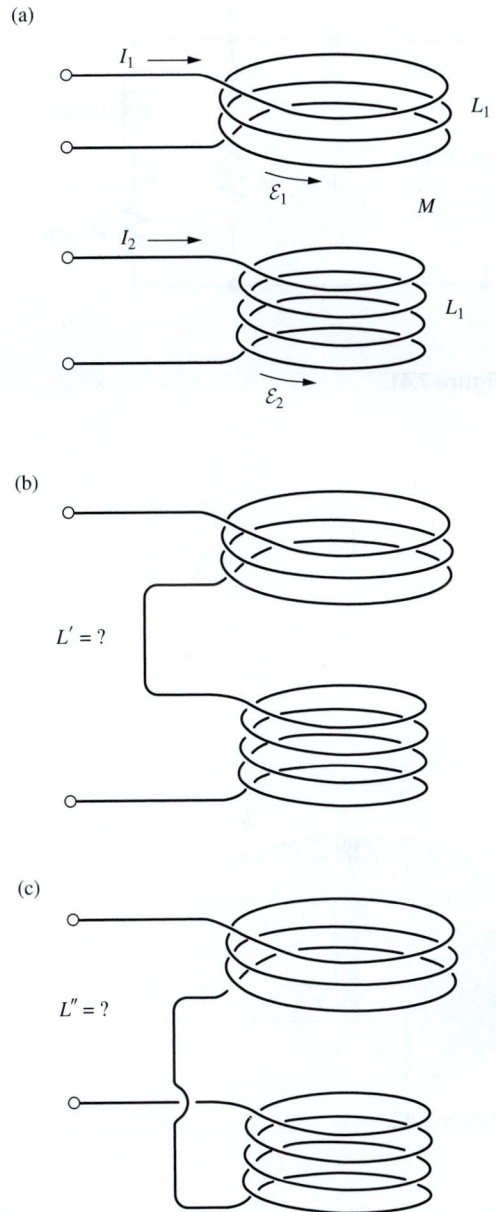

**Figure 7.40.**

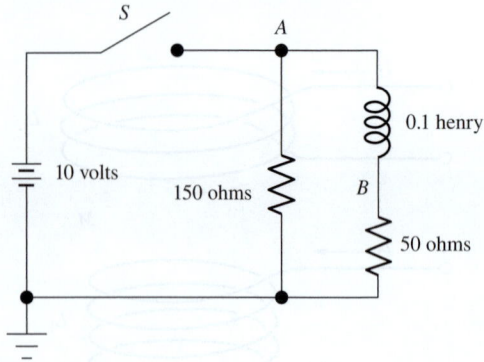

**Figure 7.41.**

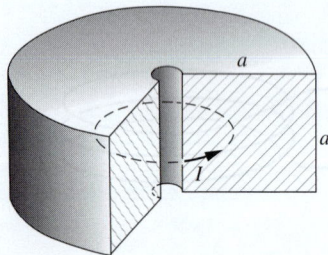

**Figure 7.42.**

and explicitly calculating the relevant integrals. This method is rather tedious, so feel free to use a computer to evaluate the integrals. See Problem 7.15 for a much quicker method.

7.44 *Magnetic energy in the galaxy* *

A magnetic field exists in most of the interstellar space in our galaxy. There is evidence that its strength in most regions is between $10^{-6}$ and $10^{-5}$ gauss. Adopting $3 \cdot 10^{-6}$ gauss as a typical value, find, in order of magnitude, the total energy stored in the magnetic field of the galaxy. For this purpose you may assume the galaxy is a disk roughly $10^{21}$ m in diameter and $10^{19}$ m thick. To see whether the magnetic energy amounts to much, on that scale, you might consider the fact that all the stars in the galaxy are radiating about $10^{37}$ joules/second. How many years of starlight is the magnetic energy worth?

7.45 *Magnetic energy near a neutron star* *

It has been estimated that the magnetic field strength at the surface of a neutron star, or *pulsar,* may be as high as $10^{10}$ tesla. What is the energy density in such a field? Express it, using the mass–energy equivalence, in kilograms per m$^3$.

7.46 *Decay time for current in the earth* **

Magnetic fields inside good conductors cannot change quickly. We found that current in a simple inductive circuit decays exponentially with characteristic time $L/R$; see Eq. (7.71). In a large conducting body such as the metallic core of the earth, the "circuit" is not easy to identify. Nevertheless, we can find the order of magnitude of the decay time, and what it depends on, by making some reasonable approximations.

Consider a solid doughnut of square cross section, as shown in Fig. 7.42, made of material with conductivity $\sigma$. A current $I$ flows around it. Of course, $I$ is spread out in some manner over the cross section, but we shall assume the resistance is that of a wire of area $a^2$ and length $\pi a$, that is, $R \approx \pi/a\sigma$. For the field $B$ we adopt the field at the center of a ring with current $I$ and radius $a/2$. For the stored energy $U$, a reasonable estimate would be $B^2/2\mu_0$ times the volume of the doughnut. Since $dU/dt = -I^2R$, the decay time of the energy $U$ will be $\tau \approx U/I^2R$. Show that, except for some numerical factor depending on our various approximations, $\tau \approx \mu_0 a^2 \sigma$. The radius of the earth's core is 3000 km, and its conductivity is believed to be $10^6$ (ohm-m)$^{-1}$, roughly one-tenth that of iron at room temperature. Evaluate $\tau$ in centuries.

7.47 *A dynamo* **

In this question the term *dynamo* will be used for a generator that works in the following way. By some external agency – the shaft of a steam turbine, for instance – a conductor is driven through a magnetic field, inducing an electromotive force in a circuit of which that conductor is part. The source of the magnetic field is the current that is caused to flow in that circuit by that electromotive force. An electrical engineer would call it a self-excited dc generator. One of the simplest dynamos conceivable is sketched in Fig. 7.43. It has only two essential parts. One part is a solid metal disk and axle which can be driven in rotation. The other is a two-turn "coil" which is stationary but is connected by sliding contacts, or "brushes," to the axle and to the rim of the revolving disk. One of the two devices pictured is, at least potentially, a dynamo. The other is not. Which is the dynamo?

Note that the answer to this question cannot depend on any convention about handedness or current directions. An intelligent extraterrestrial being inspecting the sketches could give the answer, provided only that it knows about arrows! What do you think determines the direction of the current in such a dynamo? What will determine the magnitude of the current?

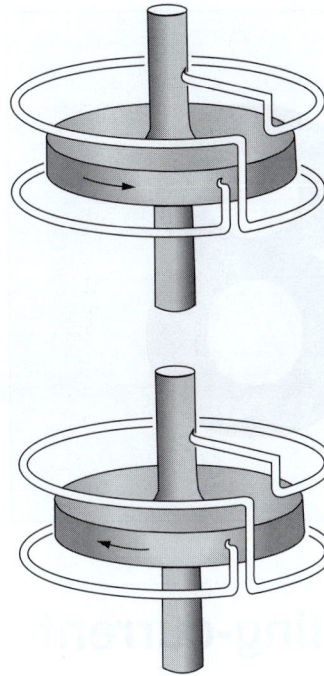

**Figure 7.43.**

# 8

# Alternating-current circuits

**Overview** In earlier chapters we encountered resistors, capacitors, and inductors. We will now study circuits containing all three of these elements. If such a circuit contains no emf source, the current takes the form of a *decaying oscillation* (in the case of small damping). The rate of decay is described by the *Q factor*. If we add on a sinusoidally oscillating emf source, then the current will reach a *steady state* with the same frequency of oscillation as the emf source. However, in general there will be a *phase difference* between the current and the emf. This phase, along with the amplitude of the current, can be determined by three methods. The first method is to guess a *sinusoidal* solution to the differential equation representing the Kirchhoff loop equation. The second is to guess a *complex exponential* solution and then take the real part to obtain the actual current. The third is to use complex voltages, currents, and impedances. These *complex impedances* can be combined via the same series and parallel rules that work for resistors. As we will see, the third method is essentially the same as the second method, but with better bookkeeping; this makes it far more tractable in the case of complicated circuits. Finally, we derive an expression for the power dissipated in a circuit, which reduces to the familiar $V^2/R$ result if the circuit is purely resistive.

## 8.1 A resonant circuit

A mass attached to a spring is a familiar example of an oscillator. If the amplitude of oscillation is not too large, the motion will be a sinusoidal function of the time. In that case, we call it a *harmonic oscillator*.

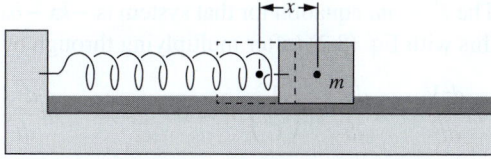

**Figure 8.1.**
A mechanical damped harmonic oscillator.

The characteristic feature of any mechanical harmonic oscillator is a restoring force proportional to the displacement of a mass $m$ from its position of equilibrium, $F = -kx$ (Fig. 8.1). In the absence of other external forces, the mass, if initially displaced, will oscillate with unchanging amplitude at the angular frequency $\omega = \sqrt{k/m}$. But usually some kind of friction will bring it eventually to rest. The simplest case is that of a retarding force proportional to the velocity of the mass, $dx/dt$. Motion in a viscous fluid provides an example. A system in which the restoring force is proportional to some displacement $x$ and the retarding force is proportional to the time derivative $dx/dt$ is called a *damped harmonic oscillator.*

An electric circuit containing capacitance and inductance has the essentials of a harmonic oscillator. Ohmic resistance makes it a damped harmonic oscillator. Indeed, thanks to the extraordinary linearity of actual electric circuit elements, the electrical damped harmonic oscillator is more nearly ideal than most mechanical oscillators. The system we shall study first is the "series $RLC$" circuit shown in Fig. 8.2. Note that there is no emf in this circuit. We will introduce an $\mathcal{E}$ (an oscillating one) in Section 8.2.

Let $Q$ be the charge, at time $t$, on the capacitor in this circuit. The potential difference, or voltage across the capacitor, is $V$, which obviously is the same as the voltage across the series combination of inductor $L$ and resistor $R$. We take $V$ to be positive when the upper capacitor plate is positively charged, and we define the positive current direction by the arrow in Fig. 8.2. With the signs chosen that way, the relations connecting charge $Q$, current $I$, and voltage across the capacitor $V$ are

$$I = -\frac{dQ}{dt}, \quad Q = CV, \quad V = L\frac{dI}{dt} + RI. \quad (8.1)$$

We want to eliminate two of the three variables $Q$, $I$, and $V$. Let us write $Q$ and $I$ in terms of $V$. From the first two equations we obtain $I = -C\,dV/dt$, and the third equation becomes $V = -LC(d^2V/dt^2) - RC(dV/dt)$, or

$$\frac{d^2V}{dt^2} + \left(\frac{R}{L}\right)\frac{dV}{dt} + \left(\frac{1}{LC}\right)V = 0. \quad (8.2)$$

This equation takes exactly the same form as the $F = ma$ equation for a mass on the end of a spring immersed in a fluid in which the damping force is $-bv$, where $b$ is the damping coefficient and $v$ is the velocity.

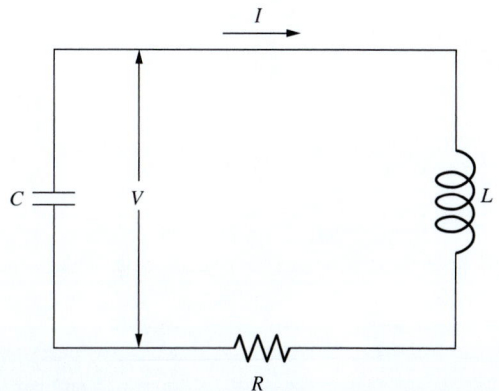

**Figure 8.2.**
A "series $RLC$" circuit.

The $F = ma$ equation for that system is $-kx - b\dot{x} = m\ddot{x}$. We can compare this with Eq. (8.2) (after multiplying through by $L$):

$$L\frac{d^2V}{dt^2} + R\frac{dV}{dt} + \left(\frac{1}{C}\right)V = 0 \quad\Longleftrightarrow\quad m\frac{d^2x}{dt^2} + b\frac{dx}{dt} + kx = 0. \quad (8.3)$$

We see that the inductance $L$ is the analog of the mass $m$; this element provides the inertia that resists change. The resistance $R$ is the analog of the damping coefficient $b$; this element causes energy dissipation. And the inverse of the capacitance, $1/C$, is the analog of the spring constant $k$; this element provides the restoring force. (There isn't anything too deep about the reciprocal form of $1/C$ here; we could have just as easily defined a quantity $C' \equiv 1/C$, with $V = C'Q$.)

Equation (8.2) is a second-order differential equation with constant coefficients. We shall try a solution of the form

$$V(t) = Ae^{-\alpha t}\cos\omega t, \quad (8.4)$$

where $A$, $\alpha$, and $\omega$ are constants. (See Problem 8.3 for an explanation of where this form comes from.) The first and second derivatives of this function are

$$\frac{dV}{dt} = Ae^{-\alpha t}\big[-\alpha\cos\omega t - \omega\sin\omega t\big],$$

$$\frac{d^2V}{dt^2} = Ae^{-\alpha t}\big[(\alpha^2 - \omega^2)\cos\omega t + 2\alpha\omega\sin\omega t\big]. \quad (8.5)$$

Substituting back into Eq. (8.2), we cancel out the common factor $Ae^{-\alpha t}$ and are left with

$$(\alpha^2 - \omega^2)\cos\omega t + 2\alpha\omega\sin\omega t - \frac{R}{L}(\alpha\cos\omega t + \omega\sin\omega t)$$

$$+ \frac{1}{LC}\cos\omega t = 0. \quad (8.6)$$

This will be satisfied for all $t$ if, and only if, the coefficients of $\sin\omega t$ and $\cos\omega t$ are both zero. That is, we must require

$$2\alpha\omega - \frac{R\omega}{L} = 0 \quad\text{and}\quad \alpha^2 - \omega^2 - \alpha\frac{R}{L} + \frac{1}{LC} = 0. \quad (8.7)$$

The first of these equations gives a condition on $\alpha$:

$$\boxed{\alpha = \frac{R}{2L}} \quad (8.8)$$

while the second equation requires that

$$\omega^2 = \frac{1}{LC} - \alpha\frac{R}{L} + \alpha^2 \quad\Longrightarrow\quad \boxed{\omega^2 = \frac{1}{LC} - \frac{R^2}{4L^2}} \quad (8.9)$$

We are assuming that the $\omega$ in Eq. (8.4) is a real number, so $\omega^2$ cannot be negative. Therefore we succeed in obtaining a solution of the form assumed in Eq. (8.4) only if $R^2/4L^2 \leq 1/LC$. In fact, it is the case of "light damping," that is, low resistance, that we want to examine, so we shall assume that the values of $R$, $L$, and $C$ in the circuit are such that the inequality $R < 2\sqrt{L/C}$ holds. However, see the end of this section for a brief discussion of the $R = 2\sqrt{L/C}$ and $R > 2\sqrt{L/C}$ cases.

The function $Ae^{-\alpha t} \cos \omega t$ is not the only possible solution; $Be^{-\alpha t} \sin \omega t$ works just as well, with the same requirements, Eqs. (8.8) and (8.9), on $\alpha$ and $\omega$, respectively. The general solution is the sum of these:

$$V(t) = e^{-\alpha t}(A \cos \omega t + B \sin \omega t) \qquad (8.10)$$

The arbitrary constants $A$ and $B$ could be adjusted to fit initial conditions. That is not very interesting. Whether the solution in any given case involves the sine or the cosine function, or some superposition, is a trivial matter of how the clock is set. The essential phenomenon is a damped sinusoidal oscillation.

The variation of voltage with time is shown in Fig. 8.3(a). Of course, this cannot really hold for all *past* time. At some time in the past the circuit must have been provided with energy somehow, and then left running. For instance, the capacitor might have been charged, with the circuit open, and then connected to the coil.

In Fig. 8.3(b) the time scale has been expanded, and the dashed curve showing the variation of the current $I$ has been added. For $V$ let us take the damped cosine, Eq. (8.4). Then the current as a function of time is given by

$$I(t) = -C\frac{dV}{dt} = AC\omega \left( \sin \omega t + \frac{\alpha}{\omega} \cos \omega t \right) e^{-\alpha t}. \qquad (8.11)$$

The ratio $\alpha/\omega$ is a measure of the damping. This is true because if $\alpha/\omega$ is very small, many oscillations occur while the amplitude is decaying only a little. For Fig. 8.3 we chose a case in which $\alpha/\omega \approx 0.04$. Then the cosine term in Eq. (8.11) doesn't amount to much. All it does, in effect, is shift the phase by a small angle, $\tan^{-1}(\alpha/\omega)$. So the current oscillation is almost exactly one-quarter cycle out of phase with the voltage oscillation.

The oscillation involves a transfer of energy back and forth from the capacitor to the inductor, or from electric field to magnetic field. At the times marked 1 in Fig. 8.3(b) all the energy is in the electric field. A quarter-cycle later, at 2, the capacitor is discharged and nearly all this energy is found in the magnetic field of the coil. Meanwhile, the circuit resistance $R$ is taking its toll, and as the oscillation goes on, the energy remaining in the fields gradually diminishes.

The relative damping in an oscillator is often expressed by giving a number called $Q$. This number $Q$ (not to be confused with the charge on the capacitor!) is said to stand for *quality* or *quality factor*. In fact, no one calls it that; we just call it $Q$. The less the damping, the larger the number $Q$. For an oscillator with frequency $\omega$, $Q$ is the dimensionless ratio formed as follows:

$$Q = \omega \cdot \frac{\text{energy stored}}{\text{average power dissipated}} \qquad (8.12)$$

Or you may prefer to remember $Q$ as follows:

- $Q$ is the number of radians of the argument $\omega t$ (that is, $2\pi$ times the number of cycles) required for the energy in the oscillator to diminish by the factor $1/e$.

In our circuit the stored energy is proportional to $V^2$ or $I^2$ and, therefore, to $e^{-2\alpha t}$. So the energy decays by $1/e$ in a time $t = 1/2\alpha$, which covers $\omega t = \omega/2\alpha$ radians. Hence, for our *RLC* circuit, using Eq. (8.8),

$$Q = \frac{\omega}{2\alpha} = \frac{\omega L}{R}. \qquad (8.13)$$

You should verify that Eq. (8.12) gives the same result.

What is $Q$ for the oscillation represented in Fig. 8.3? The energy decreases by a factor $1/e$ when $V$ decreases by a factor $1/\sqrt{e} \approx 0.6$. As a rough estimate, this decrease occurs after about two oscillations, which is roughly 13 radians. So $Q \approx 13$.

A special case of the above circuit is where $R = 0$. In this case we have the completely undamped oscillator, whose frequency $\omega_0$ is given by Eq. (8.9) as

$$\omega_0 = \frac{1}{\sqrt{LC}} \qquad (8.14)$$

Mostly we deal with systems in which the damping is small enough to be ignored in calculating the frequency. As we can see from Eq. (8.9), and as Problem 8.5 and Exercise 8.18 will demonstrate, light damping has only a second-order effect on $\omega$. Note that in view of Eq. (8.3), the $1/\sqrt{LC}$ frequency for our undamped resonant circuit is the analog of the familiar $\sqrt{k/m}$ frequency for an undamped mechanical oscillator.

For completeness we review briefly what goes on in the overdamped circuit, in which $R > 2\sqrt{L/C}$. Equation (8.2) then has a solution of the form $V = Ae^{-\beta t}$ for two values of $\beta$, the general solution being

$$V(t) = Ae^{-\beta_1 t} + Be^{-\beta_2 t}. \qquad (8.15)$$

**Figure 8.3.**
(a) The damped sinusoidal oscillation of voltage in the *RLC* circuit. (b) A portion of (a) with the time scale expanded and the graph of the current $I$ included. (c) The periodic transfer of energy from electric field to magnetic field and back again. Each picture represents the condition at times marked by the corresponding number in (b).

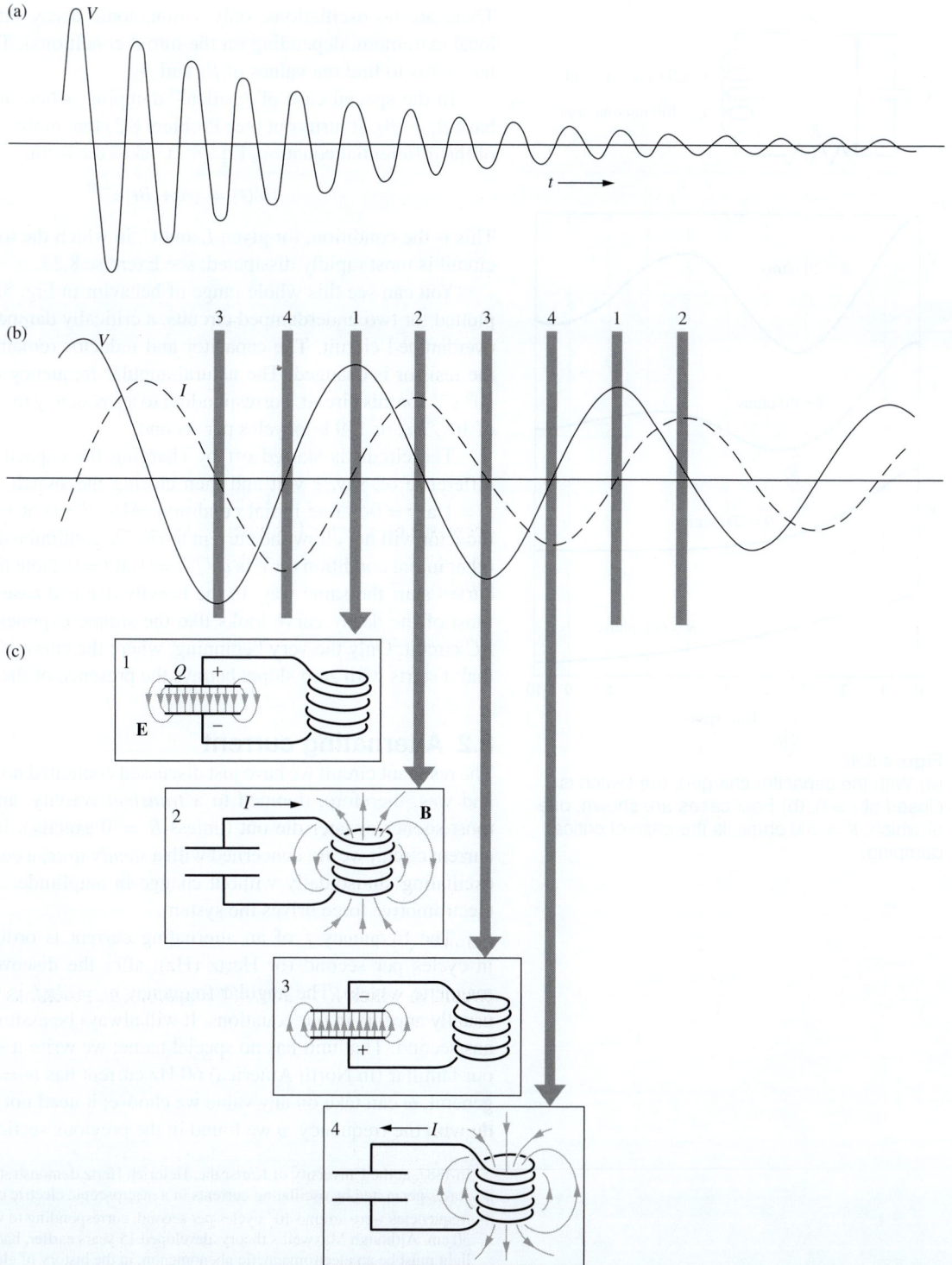

(a)

(b)

(c)

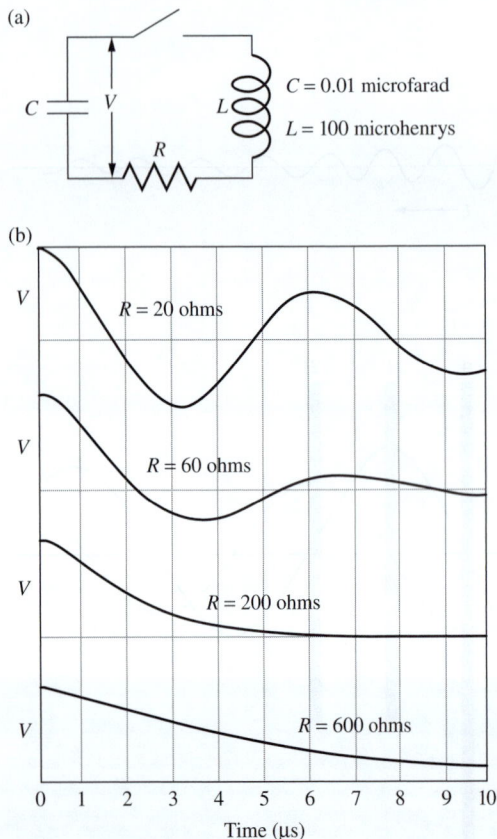

(a)

$C = 0.01$ microfarad

$L = 100$ microhenrys

(b)

$V$    $R = 20$ ohms

$V$    $R = 60$ ohms

$V$    $R = 200$ ohms

$V$    $R = 600$ ohms

0  1  2  3  4  5  6  7  8  9  10

Time (µs)

**Figure 8.4.**
(a) With the capacitor charged, the switch is closed at $t = 0$. (b) Four cases are shown, one of which, $R = 200$ ohms, is the case of critical damping.

There are no oscillations, only a monotonic decay (after perhaps one local extremum, depending on the initial conditions). The task of Problem 8.4 is to find the values of $\beta_1$ and $\beta_2$.

In the special case of "critical" damping, where $R = 2\sqrt{L/C}$, we have $\beta_1 = \beta_2$. It turns out (see Problem 8.2) that in this case the solution of the differential equation, Eq. (8.2), takes the form,

$$V(t) = (A + Bt)e^{-\beta t}. \tag{8.16}$$

This is the condition, for given $L$ and $C$, in which the total energy in the circuit is most rapidly dissipated; see Exercise 8.23.

You can see this whole range of behavior in Fig. 8.4, where $V(t)$ is plotted for two underdamped circuits, a critically damped circuit, and an overdamped circuit. The capacitor and inductor remain the same; only the resistor is changed. The natural angular frequency $\omega_0 = 1/\sqrt{LC}$ is $10^6$ s$^{-1}$ for this circuit, corresponding to a frequency in cycles per second of $10^6/2\pi$, or 159 kilocycles per second.

The circuit is started off by charging the capacitor to a potential difference of, say, 1 volt and then closing the switch at $t = 0$. That is, $V = 1$ at $t = 0$ is one initial condition. Also, $I = 0$ at $t = 0$, because the inductor will not allow the current to rise discontinuously. Therefore, the other initial condition on $V$ is $dV/dt = 0$, at $t = 0$. Note that all four decay curves start the same way. In the heavily damped case ($R = 600$ ohms) most of the decay curve looks like the simple exponential decay of an $RC$ circuit. Only the very beginning, where the curve is rounded over so that it starts with zero slope, betrays the presence of the inductance $L$.

## 8.2 Alternating current

The resonant circuit we have just discussed contained no source of energy and was, therefore, doomed to a *transient* activity, an oscillation that must sooner or later die out (unless $R = 0$ exactly). In an alternating-current circuit we are concerned with a *steady state,* a current and voltage oscillating sinusoidally without change in amplitude. Some oscillating electromotive force drives the system.

The frequency $f$ of an alternating current is ordinarily expressed in cycles per second (or Hertz (Hz), after the discoverer[1] of electromagnetic waves). The angular frequency $\omega = 2\pi f$ is the quantity that usually appears in our equations. It will always be assumed to be in radians/second. That unit has no special name; we write it simply s$^{-1}$. Thus our familiar (in North America) 60 Hz current has $\omega = 377$ s$^{-1}$. But, in general, $\omega$ can take on any value we choose; it need not have anything to do with the frequency $\omega$ we found in the previous section in Eq. (8.9).

[1] In 1887, at the University of Karlsruhe, Heinrich Hertz demonstrated electromagnetic waves generated by oscillating currents in a macroscopic electric circuit. The frequencies were around $10^9$ cycles per second, corresponding to wavelengths around 30 cm. Although Maxwell's theory, developed 15 years earlier, had left little doubt that light must be an electromagnetic phenomenon, in the history of electromagnetism Hertz's experiments were an immensely significant turning point.

Our goal in this section is to determine how the current behaves in a series *RLC* circuit with an oscillating voltage source. To warm up, we consider a few simpler circuits first. In Section 8.3 we provide an alternative method for solving the *RLC* circuit. This method uses complex exponentials in a rather slick way. In Sections 8.4 and 8.5 we generalize this complex-exponential method in a manner that allows us to treat an alternating-current circuit (involving resistors, inductors, and capacitors) in essentially the same simple way that we treat a direct-current circuit involving only resistors.

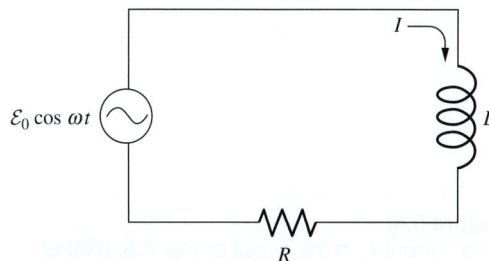

**Figure 8.5.**
A circuit with inductance, driven by an alternating electromotive force.

### 8.2.1 *RL* circuit

Let us apply an electromotive force $\mathcal{E} = \mathcal{E}_0 \cos \omega t$ to a circuit containing inductance and resistance. We might generate $\mathcal{E}$ by a machine schematically like the one in Fig. 7.13, having provided some engine or motor to turn the shaft at the constant angular speed $\omega$. The symbol at the left in Fig. 8.5 is a conventional way to show the presence of an alternating electromotive force in a circuit. It suggests a generator connected in series with the rest of the circuit. But you need not think of an electromotive force as located at a particular place in the circuit. It is only the line integral around the whole circuit that matters. Figure 8.5 could just as well represent a circuit in which the electromotive force arises from a changing magnetic field over the whole area enclosed by the circuit.

We set the sum of voltage drops over the elements of this circuit equal to the electromotive force $\mathcal{E}$, exactly as we did in developing Eq. (7.66). The equation governing the current is then

$$L\frac{dI}{dt} + RI = \mathcal{E}_0 \cos \omega t. \qquad (8.17)$$

There may be some transient behavior, depending on the initial conditions, that is, on how and when the generator is switched on. But we are interested only in the steady state, when the current is oscillating obediently at the frequency of the driving force, with the amplitude and phase necessary to keep Eq. (8.17) satisfied. To show that this is possible, consider a current described by

$$\boxed{I(t) = I_0 \cos(\omega t + \phi)} \qquad (8.18)$$

To determine the constants $I_0$ and $\phi$, we put this into Eq. (8.17):

$$-LI_0\omega \sin(\omega t + \phi) + RI_0 \cos(\omega t + \phi) = \mathcal{E}_0 \cos \omega t. \qquad (8.19)$$

The functions $\sin \omega t$ and $\cos \omega t$ can be separated out:

$$-LI_0\omega(\sin \omega t \cos \phi + \cos \omega t \sin \phi)$$
$$+ RI_0(\cos \omega t \cos \phi - \sin \omega t \sin \phi) = \mathcal{E}_0 \cos \omega t. \qquad (8.20)$$

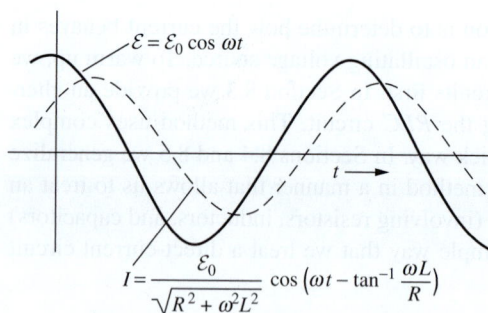

$\mathcal{E} = \mathcal{E}_0 \cos \omega t$

$I = \dfrac{\mathcal{E}_0}{\sqrt{R^2 + \omega^2 L^2}} \cos\left(\omega t - \tan^{-1}\dfrac{\omega L}{R}\right)$

**Figure 8.6.**
The current $I_1$ in the circuit of Fig. 8.5, plotted along with the electromotive force $\mathcal{E}$ on the same time scale. Note the phase difference.

Setting the coefficients of $\sin \omega t$ and $\cos \omega t$ separately equal to zero gives, respectively,

$$-LI_0\omega \cos \phi - RI_0 \sin \phi = 0 \implies \boxed{\tan \phi = -\frac{\omega L}{R}} \tag{8.21}$$

and

$$-LI_0\omega \sin \phi + RI_0 \cos \phi - \mathcal{E}_0 = 0, \tag{8.22}$$

which gives

$$I_0 = \frac{\mathcal{E}_0}{R\cos \phi - \omega L \sin \phi}$$
$$= \frac{\mathcal{E}_0}{R(\cos \phi + \tan \phi \sin \phi)} = \frac{\mathcal{E}_0 \cos \phi}{R}. \tag{8.23}$$

Since Eq. (8.21) implies[2]

$$\cos \phi = \frac{R}{\sqrt{R^2 + \omega^2 L^2}}, \tag{8.24}$$

we can write $I_0$ as

$$\boxed{I_0 = \frac{\mathcal{E}_0}{\sqrt{R^2 + \omega^2 L^2}}} \tag{8.25}$$

In Fig. 8.6 the oscillations of $\mathcal{E}$ and $I$ are plotted on the same graph. Since $\phi$ is a negative angle, the current reaches its maximum a bit *later* than the electromotive force. One says, "The current lags the voltage in an inductive circuit." The quantity $\omega L$, which has the dimensions of resistance and can be expressed in ohms, is called the *inductive reactance*.

---

[2] The $\tan \phi$ expression in Eq. (8.21) actually gives only the magnitude of $\cos \phi$ and not the sign, since $\phi$ could lie in the second or fourth quadrants. But since the convention is to take $I_0$ and $\mathcal{E}_0$ positive, Eq. (8.23) tells us that $\cos \phi$ is positive. The angle $\phi$ therefore lies in the fourth quadrant, at least for an $RL$ circuit.

## 8.2.2 *RC* circuit

If we replace the inductor $L$ by a capacitor $C$, as in Fig. 8.7, we have a circuit governed by the equation

$$-\frac{Q}{C} + RI = \mathcal{E}_0 \cos \omega t, \qquad (8.26)$$

where we have defined $Q$ to be the charge on the bottom plate of the capacitor, as shown. We again consider the steady-state solution

$$I(t) = I_0 \cos(\omega t + \phi). \qquad (8.27)$$

Since $I = -dQ/dt$, we have

$$Q = -\int I \, dt = -\frac{I_0}{\omega} \sin(\omega t + \phi). \qquad (8.28)$$

Note that, in going from $I$ to $Q$ by integration, there is no question of adding a constant of integration, for we know that $Q$ must oscillate symmetrically about zero in the steady state. Substituting $Q$ back into Eq. (8.26) leads to

$$\frac{I_0}{\omega C} \sin(\omega t + \phi) + RI_0 \cos(\omega t + \phi) = \mathcal{E}_0 \cos \omega t. \qquad (8.29)$$

Just as before, we obtain conditions on $\phi$ and $I_0$ by requiring that the coefficients of $\sin \omega t$ and $\cos \omega t$ separately vanish. Alternatively, we can avoid this process by noting that, in going from Eq. (8.19) to Eq. (8.29), we have simply traded $-\omega L$ for $1/\omega C$. The results analogous to Eqs. (8.21) and (8.25) are therefore

$$\boxed{\tan \phi = \frac{1}{R \omega C}} \quad \text{and} \quad \boxed{I_0 = \frac{\mathcal{E}_0}{\sqrt{R^2 + (1/\omega C)^2}}} \qquad (8.30)$$

Note that the phase angle is now positive, that is, it lies in the first quadrant. (The result in Eq. (8.23) is unchanged, so $\cos \phi$ is again positive. But $\tan \phi$ is now also positive.) As the saying goes, the current "leads the voltage" in a capacitive circuit. What this means is apparent in the graph of Fig. 8.8.

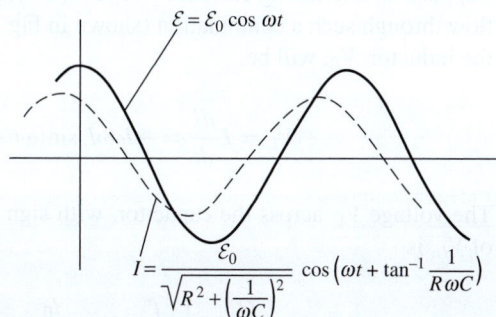

**Figure 8.7.**
An alternating electromotive force in a circuit containing resistance and capacitance.

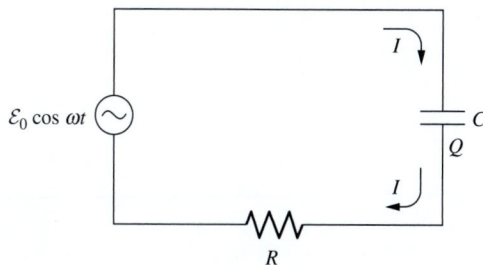

**Figure 8.8.**
The current in the *RC* circuit. Compare the phase shift here with the phase shift in the inductive circuit in Fig. 8.6. The maximum in $I$ occurs here a little earlier than the maximum in $\mathcal{E}$.

$$\mathcal{E} = \mathcal{E}_0 \cos \omega t$$

$$I = \frac{\mathcal{E}_0}{\sqrt{R^2 + \left(\frac{1}{\omega C}\right)^2}} \cos\left(\omega t + \tan^{-1} \frac{1}{R \omega C}\right)$$

### 8.2.3 Transients

Mathematically speaking, the solution for the $RL$ circuit,

$$I(t) = \frac{\mathcal{E}_0}{\sqrt{R^2 + \omega^2 L^2}} \cos\left(\omega t - \tan^{-1} \frac{\omega L}{R}\right), \qquad (8.31)$$

is a *particular integral* of the differential equation, Eq. (8.17). To this could be added a *complementary function,* that is, any solution of the homogeneous differential equation,

$$L\frac{dI}{dt} + RI = 0. \qquad (8.32)$$

This is true because Eq. (8.17) is linear in $I$, so the superposition of the particular and complementary functions is still a solution; the complementary function simply increases the right-hand side of Eq. (8.17) by zero, and therefore doesn't affect the equality. Now, Eq. (8.32) is just Eq. (7.70) of Chapter 7, whose solution we found, in Section 7.9, to be an exponentially decaying function,

$$I(t) \sim e^{-(R/L)t}. \qquad (8.33)$$

The physical significance is this: a transient, determined by some initial conditions, is represented by a decaying component of $I(t)$, of the form of Eq. (8.33). After a time $t \gg L/R$, this will have vanished, leaving only the steady sinusoidal oscillation at the driving frequency, represented by the particular integral, Eq. (8.31). This oscillation is entirely independent of the initial conditions; all memory of the initial conditions is lost.

### 8.2.4 *RLC* circuit

To solve for the current in a series $RLC$ circuit, a certain observation will be helpful. The similarity of our results for the $RL$ circuit and the $RC$ circuit suggests a way to look at the inductor and capacitor in series. Suppose an alternating current $I = I_0 \cos(\omega t + \phi)$ is somehow caused to flow through such a combination (shown in Fig. 8.9). The voltage across the inductor, $V_L$, will be

$$V_L = L\frac{dI}{dt} = -I_0 \omega L \sin(\omega t + \phi). \qquad (8.34)$$

The voltage $V_C$ across the capacitor, with sign consistent with the sign of $V_L$, is

$$V_C = -\frac{Q}{C} = \frac{1}{C}\int I\, dt = \frac{I_0}{\omega C}\sin(\omega t + \phi). \qquad (8.35)$$

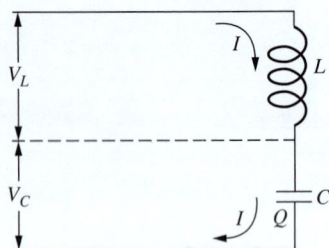

**Figure 8.9.**
The inductor and capacitor in series are equivalent to a single reactive element that is either an inductor or a capacitor, depending on whether $\omega^2 LC$ is greater or less than 1.

The voltage across the combination is then

$$V_L + V_C = -\left(\omega L - \frac{1}{\omega C}\right) I_0 \sin(\omega t + \phi). \qquad (8.36)$$

*For a given* $\omega$, the combination is evidently equivalent to a single element, either an inductor or a capacitor, depending on whether the quantity $\omega L - 1/\omega C$ is positive or negative. Suppose, for example, that $\omega L > 1/\omega C$. Then the combination is equivalent to an inductor $L'$ such that

$$\omega L' = \omega L - \frac{1}{\omega C}. \qquad (8.37)$$

*Equivalence* means *only* that the relation between current and voltage, for steady oscillation at the particular frequency $\omega$, is the same. This allows us to replace $L$ and $C$ by $L'$ in any circuit driven at this frequency. The main point here is that the voltages across the inductor and capacitor are both proportional to $\sin(\omega t + \phi)$, so they are always in phase with each other (or rather, exactly out of phase).

This can be applied to the simple *RLC* circuit in Fig. 8.10. We need only recall Eqs. (8.21) and (8.25), the solution for the *RL* circuit driven by the electromotive force $\mathcal{E}_0 \cos \omega t$, and replace $\omega L$ by $\omega L - 1/\omega C$:

$$\boxed{I(t) = \frac{\mathcal{E}_0}{\sqrt{R^2 + (\omega L - 1/\omega C)^2}} \cos(\omega t + \phi)} \qquad (8.38)$$

where

$$\boxed{\tan\phi = \frac{1}{R\omega C} - \frac{\omega L}{R}} \qquad (8.39)$$

These expressions are also correct if $1/\omega C > \omega L$, in which case we equivalently have a capacitor $C'$ such that $1/\omega C' = 1/\omega C - \omega L$.

Of course, we could have just solved the *RLC* circuit from scratch. The loop equation is

$$L\frac{dI}{dt} - \frac{Q}{C} + RI = \mathcal{E}_0 \cos \omega t. \qquad (8.40)$$

Instead of either Eq. (8.19) or Eq. (8.29), we now have all three types of terms (involving $L$, $C$, and $R$) on the left-hand side. The coefficient of the $\sin(\omega t + \phi)$ term is $-I_0(\omega L - 1/\omega C)$, so we see that we can simply use our results for the *RL* circuit, with $\omega L$ replaced by $\omega L - 1/\omega C$, as we observed above.

### 8.2.5 Resonance

For fixed amplitude $\mathcal{E}_0$ of the electromotive force, and for given circuit elements $L$, $C$, and $R$, Eq. (8.38) tells us that we get the greatest current when the driving frequency $\omega$ is such that

$$\omega L - \frac{1}{\omega C} = 0, \qquad (8.41)$$

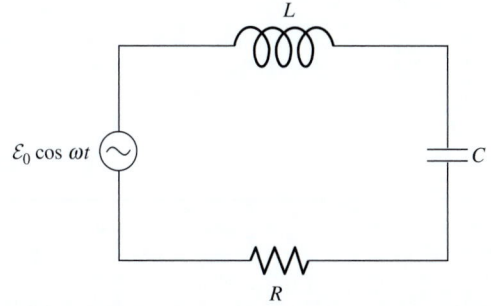

**Figure 8.10.**
The *RLC* circuit driven by a sinusoidal electromotive force.

which is the same as saying that $\omega = 1/\sqrt{LC} = \omega_0$, the resonant frequency of the undamped $LC$ circuit. In that case Eq. (8.38) reduces to

$$I(t) = \frac{\mathcal{E}_0 \cos \omega t}{R}. \tag{8.42}$$

That is exactly the current that would flow if the circuit contained the resistor alone. The reason for this is that when $\omega = 1/\sqrt{LC}$, the voltages across the inductor and capacitor are always equal and opposite. Since they cancel, they are effectively not present, and we simply have a circuit consisting of a resistor and the applied emf $\mathcal{E}_0 \cos \omega t$.

**Example** Consider the circuit of Fig. 8.4(a), connected now to a source or generator of alternating emf, $\mathcal{E} = \mathcal{E}_0 \cos \omega t$. The driving frequency $\omega$ may be different from the resonant frequency $\omega_0 = 1/\sqrt{LC}$, which, for the given capacitance (0.01 microfarads) and inductance (100 microhenrys), is $10^6$ radians/s (or $10^6/2\pi$ cycles per second). Figure 8.11 shows the amplitude of the oscillating current as a function of the driving frequency $\omega$, for three different values of the circuit resistance $R$. It is assumed that the amplitude $\mathcal{E}_0$ of the emf is 100 volts in each case. Note the resonance peak at $\omega = \omega_0$, which is most prominent and sharp for the lowest resistance value, $R = 20$ ohms. This is the same value of $R$ for which, running as a damped oscillator without any driving emf, the circuit behaved as shown in the top graph of Fig. 8.4(b).

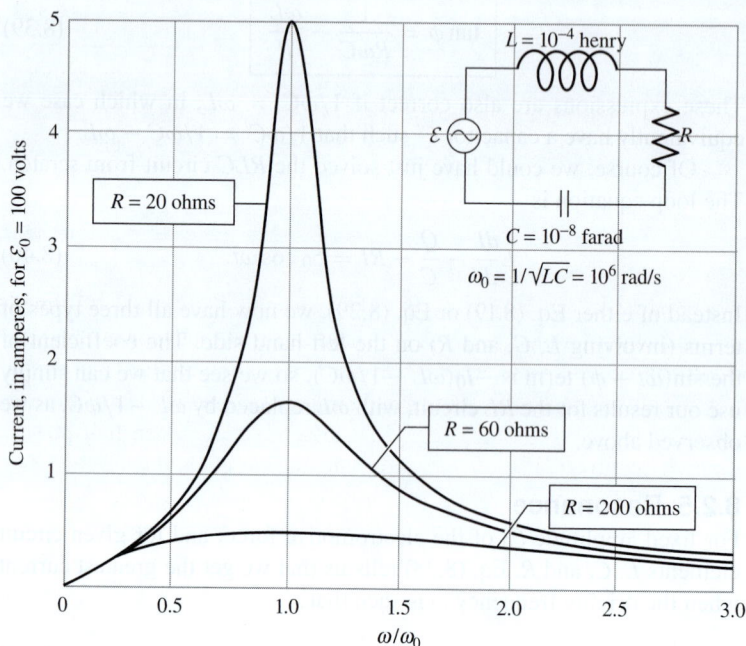

**Figure 8.11.**
An emf of 100 volts amplitude is applied to a series $RLC$ circuit. The circuit elements are the same as in the example of the damped circuit in Fig. 8.4. Circuit amplitude is calculated by Eq. (8.38) and plotted, as a function of $\omega/\omega_0$, for three different resistance values.

Note that we have encountered three (generally different) frequencies up to this point:

- the frequency of the applied oscillating emf, which can take on any value we choose;
- the resonant frequency, $\omega_0 = 1/\sqrt{LC}$, for which the amplitude of the oscillating current is largest;
- the frequency (in the underdamped case) of the transient behavior, given by Eq. (8.9). For light damping, this frequency is approximately equal to the resonant frequency, $\omega_0 = 1/\sqrt{LC}$.

### 8.2.6 Width of the $I_0(\omega)$ curve

The $Q$ factor of the circuit in the above example with $R = 20$ ohms, given in Eq. (8.13) as[3] $\omega_0 L/R$, is $(10^6 \cdot 10^{-4})/20$, or 5, in this case. Generally speaking, the higher the $Q$ of a circuit, the narrower and higher the peak of its response as a function of driving frequency $\omega$. To be more precise, consider frequencies in the neighborhood of $\omega_0$, writing $\omega = \omega_0 + \Delta\omega$. Then, to first order in $\Delta\omega/\omega_0$, the expression $\omega L - 1/\omega C$ that occurs in the denominator in Eq. (8.38) can be approximated this way:

$$\omega L - \frac{1}{\omega C} = \omega_0 L \left(1 + \frac{\Delta\omega}{\omega_0}\right) - \frac{1}{\omega_0 C(1 + \Delta\omega/\omega_0)}, \qquad (8.43)$$

and since $\omega_0$ is $1/\sqrt{LC}$, this becomes

$$\omega_0 L \left(1 + \frac{\Delta\omega}{\omega_0} - \frac{1}{1 + \Delta\omega/\omega_0}\right) \approx \omega_0 L \left(2\frac{\Delta\omega}{\omega_0}\right), \qquad (8.44)$$

where we have used the approximation, $1/(1 + \epsilon) \approx 1 - \epsilon$. Exactly at resonance, the quantity inside the square root sign in Eq. (8.38) is just $R^2$. As $\omega$ is shifted away from resonance, the quantity under the square root will have doubled when $|\omega L - 1/\omega C| = R$, or when, approximately,

$$\frac{2|\Delta\omega|}{\omega_0} = \frac{R}{\omega_0 L} = \frac{1}{Q}. \qquad (8.45)$$

This means that the current amplitude will have fallen to $1/\sqrt{2}$ times the peak when $|\Delta\omega|/\omega_0 = 1/2Q$. These are the "half-power" points, because the energy or power is proportional to the amplitude squared, as we shall explain in Section 8.6. One often expresses the width of a resonance peak by giving the full width, $2\Delta\omega$, between half-power points. Evidently that is just $1/Q$ times the resonant frequency itself. Circuits with very much higher $Q$ than this one are quite common. A radio receiver may select a particular station and discriminate against others

---

[3] The $\omega$ in Eq. (8.13) is the frequency of the freely decaying damped oscillator, practically the same as $\omega_0$ for moderate or light damping. We use $\omega_0$ here in the expression for $Q$. In the present discussion, $\omega$ is *any* frequency we may choose to apply to this circuit.

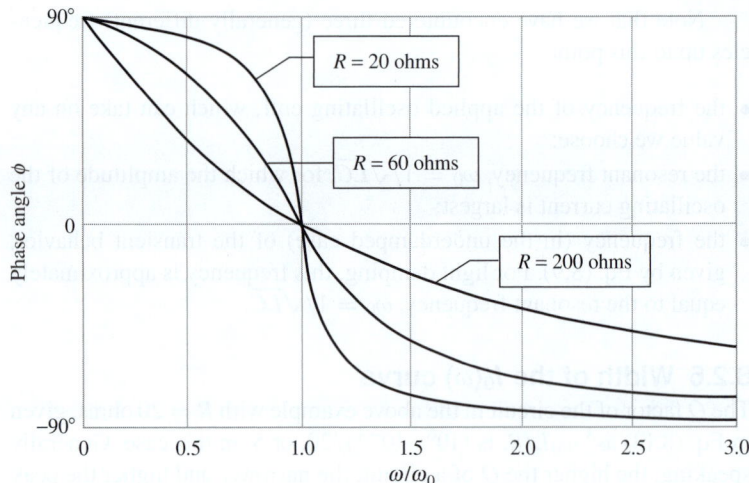

**Figure 8.12.**
The variation of phase angle with frequency, in
the circuit of Fig. 8.11.

by means of a resonant circuit with a $Q$ of several hundred. It is quite
easy to make a microwave resonant circuit with a $Q$ of $10^4$, or even $10^5$.

The angle $\phi$, which expresses the relative phase of the current and
emf oscillations, varies with frequency in the manner shown in Fig. 8.12.
At a very low frequency the capacitor is the dominant hindrance to current flow, and $\phi$ is positive. At resonance, $\phi = 0$. The higher the $Q$, the
more abruptly $\phi$ shifts from positive to negative angles as the frequency
is raised through $\omega_0$.

To summarize what we know about $Q$, we have encountered two
different meanings:

- In an $RLC$ circuit with an applied oscillating emf, $1/Q$ gives a measure of the width of the current and power curves, as functions of $\omega$.
  The higher the $Q$, the narrower the curves. More precisely, the width
  (at half maximum) of the power curve is $\omega_0/Q$.
- If we remove the emf source, the current and energy will decay; $Q$
  gives a measure of how slow this decay is. The higher the $Q$, the more
  oscillations it takes for the amplitude to decrease by a given factor.
  More precisely, the energy decreases by a factor $1/e$ after $Q$ radians
  (or $Q/2\pi$ cycles). Equivalently, as Exercise 8.17 shows, the current
  decreases by a factor of $e^{-\pi}$ after $Q$ cycles. (It's hard to pass up a
  chance to mention a result of $e^{-\pi}$!)

## 8.3 Complex exponential solutions

In Section 8.2 we solved for the current in the series $RLC$ circuit (including a voltage source $\mathcal{E}_0 \cos \omega t$) in Fig. 8.10 by guessing a sinusoidal form
for the current $I(t)$. In the present section we will solve for the current
in a different way, using complex numbers. This method is extremely

powerful, and it forms the basis of what we will do in the remainder of this chapter.

Our strategy will be the following. We will write down the Kirchhoff loop equation as we did above, but instead of solving it directly, we will solve a slightly modified equation in which the $\mathcal{E}_0 \cos \omega t$ voltage source is replaced by $\mathcal{E}_0 e^{i\omega t}$. We will guess an exponential solution of the form $\tilde{I}(t) = \tilde{I} e^{i\omega t}$ and solve for $\tilde{I}$, which will turn out to be a complex number.[4] Of course, our solution for $\tilde{I}(t)$ cannot possibly be the current we are looking for, because $\tilde{I}(t)$ is complex, whereas an actual current must be real. However, if we take the real part of $\tilde{I}(t)$, we will obtain (for reasons we will explain) the desired current $I(t)$ that actually flows in the circuit. Let's see how all this works. Our goal is to reproduce the $I(t)$ in Eqs. (8.38) and (8.39).

The Kirchhoff loop equation for the series $RLC$ circuit in Fig. 8.10 is[5]

$$L\frac{dI(t)}{dt} + RI(t) + \frac{Q(t)}{C} = \mathcal{E}_0 \cos \omega t. \tag{8.46}$$

If we take clockwise current to be positive, then $Q(t)$ is the integral of $I(t)$, that is, $Q(t) = \int I(t)\,dt$. Consider now the modified equation where $\cos \omega t$ is replaced by $e^{i\omega t}$,

$$L\frac{d\tilde{I}(t)}{dt} + R\tilde{I}(t) + \frac{\tilde{Q}(t)}{C} = \mathcal{E}_0 e^{i\omega t}. \tag{8.47}$$

If $\tilde{I}(t)$ is a (complex) solution to this equation, then if we take the real part of the entire equation, we obtain (using the facts that differentiation and integration with respect to $t$ commute with taking the real part)

$$L\frac{d}{dt}\mathrm{Re}[\tilde{I}(t)] + R\,\mathrm{Re}[\tilde{I}(t)] + \frac{1}{C}\int \mathrm{Re}[\tilde{I}(t)]\,dt = \mathcal{E}_0 \cos \omega t. \tag{8.48}$$

We have used the remarkable mathematical identity, $e^{i\theta} = \cos\theta + i\sin\theta$, which tells us that $\cos \omega t$ is the real part of $e^{i\omega t}$. (See Appendix K for a review of complex numbers.)

Equation (8.48) is simply the statement that $I(t) \equiv \mathrm{Re}[\tilde{I}(t)]$ is a solution to our original differential equation in Eq. (8.46). Our goal is therefore to find a complex function $\tilde{I}(t)$ that satisfies Eq. (8.47), and then take the real part. Note the critical role that linearity played here.

---

[4] The tilde on the $I$ terms denotes a complex number. Note that $\tilde{I}(t)$ has time dependence, whereas $\tilde{I}$ does not. More precisely, $\tilde{I} = \tilde{I}(0)$. When writing $\tilde{I}(t)$, be careful not to drop the $t$ argument, because that will change the meaning to $\tilde{I}$ (although the meaning is generally clear from the context). There will actually be a total of four different versions of the letter $I$ that we will encounter in this method. They are summarized in Fig. 8.13.

[5] We are now taking $Q$ to be the charge on the top plate of the capacitor (for no deep reason). You should verify that if we instead took $Q$ to be the charge on the bottom plate, then two minus signs would end up canceling, and we would still arrive at Eq. (8.48). After all, that equation for $\tilde{I}(t)$ can't depend on our arbitrary convention for $Q$.

If our differential equation were modified to contain a term that wasn't linear in $I(t)$, for example $RI(t)^2$, then this method wouldn't work, because $\text{Re}[\tilde{I}(t)^2]$ is *not* equal to $\big(\text{Re}[\tilde{I}(t)]\big)^2$. The modified form of Eq. (8.48) would *not* be the statement that $I(t) \equiv \text{Re}[\tilde{I}(t)]$ satisfies the modified form of Eq. (8.46).

A function of the form $\tilde{I}(t) = \tilde{I}e^{i\omega t}$ will certainly yield a solution to Eq. (8.47), because the $e^{i\omega t}$ factor will cancel through the whole equation, yielding an equation with no time dependence. Now, if $\tilde{I}(t) = \tilde{I}e^{i\omega t}$, then $\tilde{Q}(t)$, which is the integral of $\tilde{I}(t)$, equals $\tilde{I}e^{i\omega t}/i\omega$. (There is no need for a constant of integration because we know that $Q$ oscillates around zero.) So Eq. (8.47) becomes

$$ Li\omega \tilde{I}e^{i\omega t} + R\tilde{I}e^{i\omega t} + \frac{\tilde{I}e^{i\omega t}}{i\omega C} = \mathcal{E}_0 e^{i\omega t}. \tag{8.49} $$

Canceling the $e^{i\omega t}$, solving for $\tilde{I}$, and getting the $i$ out of the denominator by multiplying by 1 in the form of the complex conjugate divided by itself, yields

$$ \tilde{I} = \frac{\mathcal{E}_0}{i\omega L + R + 1/i\omega C} = \frac{\mathcal{E}_0\big[R - i(\omega L - 1/\omega C)\big]}{R^2 + (\omega L - 1/\omega C)^2}. \tag{8.50} $$

The term in the square brackets is a complex number written in $a + bi$ form, but it will be advantageous to write it in "polar" form, that is, as a magnitude times a phase, $Ae^{i\phi}$. The magnitude is $A = \sqrt{a^2 + b^2}$, and the phase is $\phi = \tan^{-1}(b/a)$; see Problem 8.7. So we have

$$ \tilde{I} = \frac{\mathcal{E}_0}{R^2 + (\omega L - 1/\omega C)^2} \cdot \sqrt{R^2 + (\omega L - 1/\omega C)^2}\, e^{i\phi} $$

$$ = \frac{\mathcal{E}_0}{\sqrt{R^2 + (\omega L - 1/\omega C)^2}}\, e^{i\phi} \equiv I_0 e^{i\phi}, \tag{8.51} $$

where

$$ I_0 = \frac{\mathcal{E}_0}{\sqrt{R^2 + (\omega L - 1/\omega C)^2}} \quad \text{and} \quad \tan\phi = \frac{1}{R\omega C} - \frac{\omega L}{R}. \tag{8.52} $$

The actual current $I(t)$ is obtained by taking the real part of the full $\tilde{I}(t) = \tilde{I}e^{i\omega t}$ solution:

$$ I(t) = \text{Re}\big[\tilde{I}e^{i\omega t}\big] = \text{Re}\big[I_0 e^{i\phi} e^{i\omega t}\big] = I_0 \cos(\omega t + \phi) $$

$$ = \frac{\mathcal{E}_0}{\sqrt{R^2 + (\omega L - 1/\omega C)^2}} \cos(\omega t + \phi), \tag{8.53} $$

in agreement with Eqs. (8.38) and (8.39). $I_0$ is the amplitude of the current, and $\phi$ is the phase relative to the applied voltage.

As mentioned above, there are four different types of $I$'s that appear in this procedure: $\tilde{I}(t)$, $\tilde{I}$, $I(t)$, and $I_0$. These are related to each other in the following ways (summarized in Fig. 8.13).

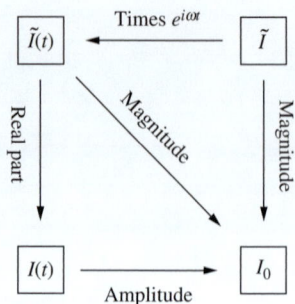

**Figure 8.13.**
Relations among the various usages of the letter "$I$."

- The two complex quantities, $\tilde{I}(t)$ and $\tilde{I}$, are related by a simple factor of $e^{i\omega t}$: $\tilde{I}(t) = \tilde{I}e^{i\omega t}$; $\tilde{I}$ equals $\tilde{I}(0)$.
- $I(t)$, which is the actual current, equals the real part of $\tilde{I}(t)$: $I(t) = \text{Re}[\tilde{I}(t)]$.
- $I_0$ is the magnitude of both $\tilde{I}(t)$ and $\tilde{I}$: $I_0 = |\tilde{I}(t)|$ and $I_0 = |\tilde{I}|$.
- $I_0$ is the amplitude of $I(t)$: $I(t) = I_0 \cos(\omega t + \phi)$.

Although the above method involving complex exponentials might take some getting used to, it is much cleaner and quicker than the method involving trig functions that we used in Section 8.2. Recall the system of equations that we needed to solve in Eqs. (8.21)–(8.25). We had to demand that the coefficients of $\sin \omega t$ and $\cos \omega t$ in Eq. (8.20) were independently zero. That involved a fair bit of algebra. In the present complex-exponential method, the $e^{i\omega t}$ terms cancel in Eq. (8.49), so we are left with only one equation, which we can quickly solve. The point here is that the derivative of an exponential gives back an exponential, whereas sines and cosines flip flop under differentiation. Of course, from the relation $e^{i\theta} = \cos\theta + i\sin\theta$, we know that exponentials can be written in terms of trig functions, and vice versa via $\cos\theta = (e^{i\theta} + e^{-i\theta})/2$ and $\sin\theta = (e^{i\theta} - e^{-i\theta})/2i$. So any task that can be accomplished with exponential functions can also be accomplished with trig functions. But exponentials invariably make the calculations much easier.

In the event that the applied voltage isn't a nice sinusoidal function, our method of guessing exponentials (or trig functions) is still applicable, due to two critical things: (1) Fourier analysis and (2) the linearity of the differential equation in Eq. (8.46). You will study the all-important subject of Fourier analysis in your future math and physics courses, but for now we simply note that Fourier analysis tells us that any reasonably well-behaved function for the voltage source can be written as the (perhaps infinite) sum of exponentials, or equivalently trig functions. And then linearity tells us that we can just add up the solutions for all these exponential voltage sources to obtain the solution for the original voltage source. In effect, this is what we did when we took the real part of $\tilde{I}(t)$ to obtain the actual current $I(t)$. We would have arrived at the same answer if we wrote the applied voltage $\mathcal{E}_0 \cos \omega t$ as $\mathcal{E}_0(e^{i\omega t} + e^{-i\omega t})/2$, then found the solutions for these two exponential voltages, and then added them together. So the strategy of taking the real part is just a special case of the strategy of superposing solutions via Fourier analysis.

## 8.4 Alternating-current networks

In this section we will generalize the results from Section 8.3, where our circuit involved only one loop. Complex numbers provide us with a remarkably efficient way of dealing with arbitrary alternating-current networks. An alternating-current network is any collection of resistors,

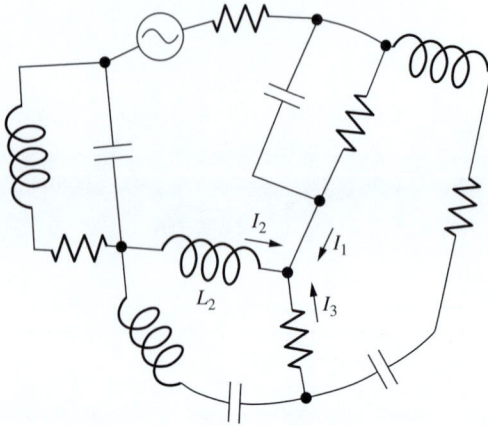

**Figure 8.14.**
An alternating-current network.

capacitors, and inductors in which currents flow that are oscillating steadily at the constant frequency $\omega$. One or more electromotive forces, at this frequency, drive the oscillation. Figure 8.14 is a diagram of one such network. The source of alternating electromotive force is represented by the symbol $-\!\bigcirc\!-$. In a branch of the network, for instance the branch that contains the inductor $L_2$, the current as a function of time is

$$I_2(t) = I_{02} \cos(\omega t + \phi_2). \tag{8.54}$$

Since the frequency is a constant for the whole network, two numbers, such as the amplitude $I_{02}$ and the phase constant $\phi_2$ above, are enough to determine for all time the current in a particular branch. Similarly, the voltage across a branch oscillates with a certain amplitude and phase:

$$V_2(t) = V_{02} \cos(\omega t + \theta_2). \tag{8.55}$$

If we have determined the currents and voltages in all branches of a network, we have analyzed it completely. To find them by constructing and solving all the appropriate differential equations is possible, of course; and if we were concerned with the transient behavior of the network, we might have to do something like that. For the steady state at some given frequency $\omega$, we can use a far simpler and more elegant method. It is based on two ideas:

(1) An alternating current or voltage can be represented by a complex number;
(2) Any one branch or element of the circuit can be characterized, at a given frequency, by the relation between the voltage and current in that branch.

As we saw above, the first idea exploits the identity, $e^{i\theta} = \cos\theta + i\sin\theta$. To carry it out we adopt the following *rule* for the representation:

---

An alternating current $I(t) = I_0 \cos(\omega t + \phi)$ is to be *represented by* the complex number $I_0 e^{i\phi}$, that is, the number whose real part is $I_0 \cos\phi$ and whose imaginary part is $I_0 \sin\phi$.

Going the other way, if the complex number $x + iy$ *represents* a current $I(t)$, then the current as a function of time is given by the real part of the product $(x + iy)e^{i\omega t}$. Equivalently, if $I_0 e^{i\phi}$ represents a current $I(t)$, then $I(t)$ is given by the real part of the product $I_0 e^{i\phi} e^{i\omega t}$, which is $I_0 \cos(\omega t + \phi)$.

---

Figure 8.15 is a reminder of this two-way correspondence. Since a complex number $z = x + iy$ can be graphically represented on the two-dimensional plane, it is easy to visualize the phase constant as the angle $\tan^{-1}(y/x)$ and the amplitude $I_0$ as the modulus $\sqrt{x^2 + y^2}$.

CURRENT AS A
FUNCTION OF TIME

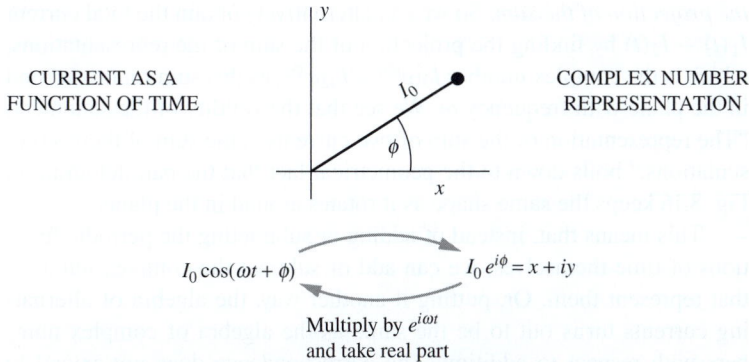

COMPLEX NUMBER
REPRESENTATION

$I_0 \cos(\omega t + \phi)$               $I_0 e^{i\phi} = x + iy$

Multiply by $e^{i\omega t}$
and take real part

**Figure 8.15.**
Rules for representing an alternating current by
a complex number.

What makes all this useful is the following fact. *The representation of the sum of two currents is the sum of their representations.* Consider the sum of two currents $I_1(t)$ and $I_2(t)$ that meet at a junction of wires in Fig. 8.14. At any instant of time $t$, the sum of the currents is given by

$$I_1(t) + I_2(t) = I_{01} \cos(\omega t + \phi_1) + I_{02} \cos(\omega t + \phi_2)$$
$$= (I_{01} \cos \phi_1 + I_{02} \cos \phi_2) \cos \omega t$$
$$- (I_{01} \sin \phi_1 + I_{02} \sin \phi_2) \sin \omega t. \qquad (8.56)$$

On the other hand, the sum of the complex numbers that, according to our rule, represent $I_1(t)$ and $I_2(t)$ is

$$I_{01}e^{i\phi_1} + I_{02}e^{i\phi_2} = (I_{01} \cos \phi_1 + I_{02} \cos \phi_2) + i(I_{01} \sin \phi_1 + I_{02} \sin \phi_2). \qquad (8.57)$$

If you multiply the right-hand side of Eq. (8.57) by $\cos \omega t + i \sin \omega t$ and take the real part of the result, you will get just what appears on the right in Eq. (8.56). This is no surprise, of course, because what we've just done is show (the long way) that

$$\text{Re}\left[I_{01}e^{i(\omega t + \phi_1)} + I_{02}e^{i(\omega t + \phi_2)}\right] = \text{Re}\left[\left(I_{01}e^{i\phi_1} + I_{02}e^{i\phi_2}\right)\left(e^{i\omega t}\right)\right]. \qquad (8.58)$$

The left-hand side of this equation is what appears in Eq. (8.56), and the right-hand side is the result of multiplying Eq. (8.57) by $e^{i\omega t} = \cos \omega t + i \sin \omega t$ and taking the real part.

Figure 8.16 shows geometrically what is going on. The real part of a number in the complex plane is its projection onto the $x$ axis. So the current $I_1(t) = I_{01} \cos(\omega t + \phi_1)$ is the horizontal projection of the complex number $I_{01}e^{i(\omega t + \phi_1)}$, and this complex number can be visualized as the vector $I_{01}e^{i\phi_1}$ rotating around in the plane with angular frequency $\omega$ (because the angle increases according to $\omega t$). Likewise for the current $I_2(t) = I_{02} \cos(\omega t + \phi_2)$. Now, *the sum of the projections of two vectors is*

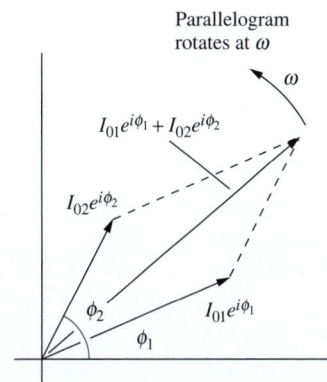

**Figure 8.16.**
As these three vectors rotate around in the plane with the same frequency $\omega$, the horizontal projection of the long vector (the sum) always equals the sum of the horizontal projections of the other two vectors.

*the projection of the sum.* So we can alternatively obtain the total current $I_1(t) + I_2(t)$ by finding the projection of the sum of the representations, which is the complex number $I_{01}e^{i\phi_1} + I_{02}e^{i\phi_2}$, as this sum rotates around in the plane with frequency $\omega$. We see that the validity of the statement, "The representation of the sum of two currents is the sum of their representations," boils down to the geometrical fact that the parallelogram in Fig. 8.16 keeps the same shape as it rotates around in the plane.

This means that, instead of adding or subtracting the periodic functions of time themselves, we can add or subtract the complex numbers that represent them. Or, putting it another way, the algebra of alternating currents turns out to be the same as the algebra of complex numbers with respect to addition. The correspondence does *not* extend to multiplication. The complex number $I_{01}I_{02}e^{i(\phi_1+\phi_2)}$ does *not* represent the product of the two current functions in Eq. (8.56), because the real part of the product of two complex numbers is not equal to the product of the real parts (the latter omits the contribution from the product of the imaginary parts).

However, it is only addition of currents and voltages that we need to carry out in analyzing the network. For example, at the junction where $I_1(t)$ meets $I_2(t)$ in Fig. 8.14, there is the physical requirement that *at every instant* the net flow of current into the junction shall be zero. Hence the condition

$$I_1(t) + I_2(t) + I_3(t) = 0 \qquad (8.59)$$

must hold, where $I_1(t)$, $I_2(t)$, and $I_3(t)$ are the *actual periodic functions of time.* Thanks to our correspondence, this can be expressed in the simple algebraic statement that the sum of three complex numbers is zero. Voltages can be handled in the same way. Instantaneously, the sum of voltage drops around any loop in the network must equal the electromotive force in the loop at that instant. This condition relating periodic voltage functions can likewise be replaced by a statement about the sum of some complex numbers, the representations of the various oscillating functions, $V_1(t)$, $V_2(t)$, etc.

## 8.5 Admittance and impedance

The relation between current flow in a circuit element and the voltage across the element can be expressed as a relation between the complex numbers that represent the voltage and the current. Look at the inductor–resistor combination in Fig. 8.5. The voltage oscillation is represented by[6] $\tilde{V} = \mathcal{E}_0$ and the current by $\tilde{I} = I_0e^{i\phi}$, where $I_0 = \mathcal{E}_0/\sqrt{R^2 + \omega^2L^2}$ and $\tan\phi = -\omega L/R$. The phase difference $\phi$ and the ratio of current

[6] As in Section 8.3, we will indicate complex voltages (and currents) by putting a tilde over them, to avoid confusion with the actual voltages (or currents) $V(t)$ which, as we have noted, are given by the real part of $\tilde{V}e^{i\omega t}$.

amplitude to voltage amplitude are properties of the circuit at this frequency. We define a complex number $Y$ as follows:

$$Y = \frac{e^{i\phi}}{\sqrt{R^2 + \omega^2 L^2}}, \quad \text{with} \quad \phi = \tan^{-1}\left(-\frac{\omega L}{R}\right). \quad (8.60)$$

Then the relation

$$\tilde{I} = Y\tilde{V} \quad (8.61)$$

holds, where $\tilde{V}$ is the complex number (which happens to be just the real number $\mathcal{E}_0$ in the present case) that represents the voltage across the series combination of $R$ and $L$, and $\tilde{I}$ is the complex number that represents the current. $Y$ is called the *admittance*. The same relation can be expressed with the reciprocal of $Y$, denoted by[7] $Z$ and called the *impedance*:

$$\tilde{V} = \left(\frac{1}{Y}\right)\tilde{I} \implies \tilde{V} = Z\tilde{I} \quad (8.62)$$

In Eqs. (8.61) and (8.62) we do make use of the product of two complex numbers, but only one of the numbers is the representation of an alternating current or voltage. The other is the impedance or admittance. Our algebra thus contains two categories of complex numbers, those that represent admittances and impedances, and those that represent currents and voltages. The product of two "impedance numbers," like the product of two "current numbers," doesn't represent anything.

The impedance is measured in ohms. Indeed, if the circuit element had consisted of the resistance $R$ alone, the impedance would be real and equal simply to $R$, so that Eq. (8.62) would resemble Ohm's law for a direct-current circuit: $V = RI$.

The admittance of a resistanceless inductor is the imaginary quantity $Y = -i/\omega L$. This can be seen by letting $R$ go to zero in Eq. (8.60), which yields $\phi = -\pi/2 \Rightarrow e^{i\phi} = -i$. The factor $-i$ means that the current oscillation lags the voltage oscillation by $\pi/2$ in phase. On the complex number diagram, if the voltage is represented by $\tilde{V}$ (Fig. 8.17(b)), the current might be represented by $\tilde{I}$, located as shown there. For the capacitor, we have $Y = i\omega C$, as can be seen from the expression for the current in Eq. (8.30). In this case $\tilde{V}$ and $\tilde{I}$ are related as indicated in Fig. 8.17(c); the current leads the voltage by $\pi/2$. The inset in each of the figures shows how the relative sign of $\tilde{V}$ and $\tilde{I}$ is to be specified. Unless that is done consistently, *leading* and *lagging* are meaningless. Note that we always define the positive current direction so that a positive voltage applied to a

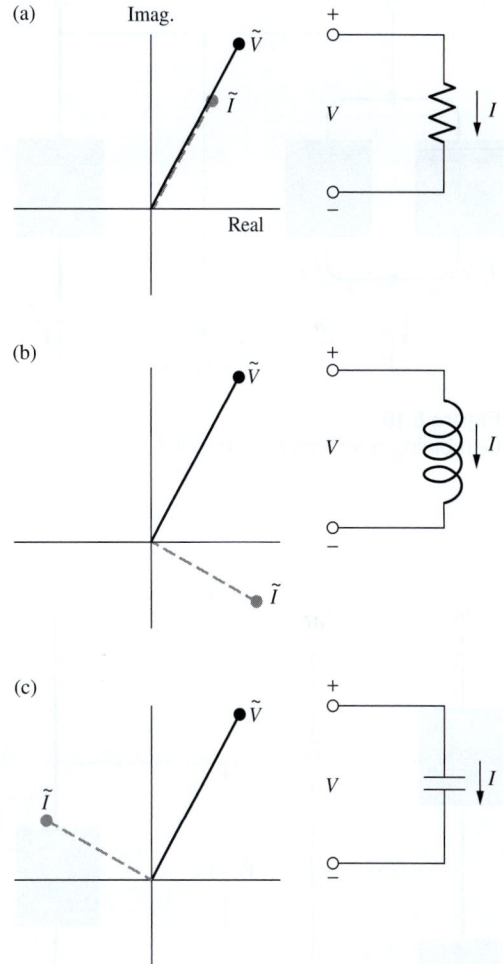

**Figure 8.17.**
$\tilde{V}$ and $\tilde{I}$ are complex numbers that represent the voltage across a circuit element and the current through it. The relative phase of current and voltage oscillation is manifest here in the angle between the "vectors." (a) In the resistor, current and voltage are in phase. (b) In the inductor, current lags the voltage. (c) In the capacitor, current leads the voltage.

[7] We won't put a tilde over $Y$ or $Z$, even though they are complex numbers, because we will rarely have the need to take their real parts (except when finding the phase $\phi$). So we won't need to worry about confusion between two different types of impedances.

**Table 8.1.**
Complex impedances

| Symbol | Admittance, $Y$ | Impedance, $Z = 1/Y$ |
|---|---|---|
| $R$ ⎍ | $\dfrac{1}{R}$ | $R$ |
| $L$ ⎍ | $\dfrac{1}{i\omega L}$ | $i\omega L$ |
| $C$ ⊣⊢ | $i\omega C$ | $\dfrac{1}{i\omega C}$ |
| | $I = YV$ | $V = ZI$ |

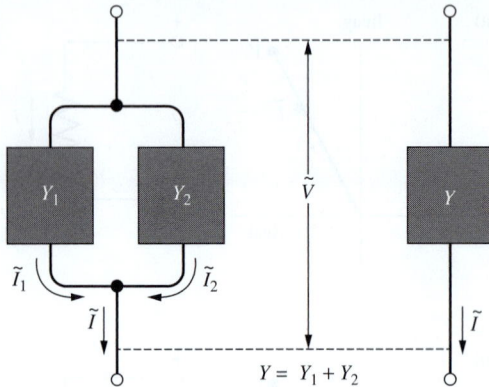

**Figure 8.18.**
Combining admittances in parallel.

$$Y = Y_1 + Y_2$$

resistor causes positive current (Fig. 8.17(a)). The properties of the three basic circuit elements are summarized in Table 8.1.

We can build up any circuit from these elements. When elements or combinations of elements are connected in *parallel*, it is convenient to use the *admittance*, for in that case admittances add. In Fig. 8.18 two black boxes with admittances $Y_1$ and $Y_2$ are connected in parallel. Since the voltages across each box are the same and since the currents add, we have

$$\tilde{I} = \tilde{I}_1 + \tilde{I}_2 = Y_1\tilde{V} + Y_2\tilde{V} = (Y_1 + Y_2)\tilde{V}, \tag{8.63}$$

which implies that the equivalent single black box has an admittance $Y = Y_1 + Y_2$. From Fig. 8.19 we see that the *impedances* add for elements connected in *series*, because the currents are the same and the voltages add:

$$\tilde{V} = \tilde{V}_1 + \tilde{V}_2 = Z_1\tilde{I} + Z_2\tilde{I} = (Z_1 + Z_2)\tilde{I}, \tag{8.64}$$

which implies that the equivalent single black box has an impedance $Z = Z_1 + Z_2$. It sounds as if we are talking about a direct-current network! In fact, we have now reduced the ac network problem to the dc network problem, with only this difference: the numbers we deal with are complex numbers.

**Example (Parallel *RLC* circuit)** Consider the "parallel *RLC*" circuit in Fig. 8.20. The combined admittance of the three parallel branches is

$$Y = \frac{1}{R} + i\omega C - \frac{i}{\omega L}. \tag{8.65}$$

The voltage is simply $\mathcal{E}_0$, so the complex current is

$$\tilde{I} = Y\tilde{V} = \left[\frac{1}{R} + i\left(\omega C - \frac{1}{\omega L}\right)\right]\mathcal{E}_0. \tag{8.66}$$

The amplitude $I_0$ of the current oscillation $I(t)$ is the modulus of the complex number $\tilde{I}$, and the phase angle relative to the voltage is $\tan^{-1}[\mathrm{Im}(Y)/\mathrm{Re}(Y)]$.

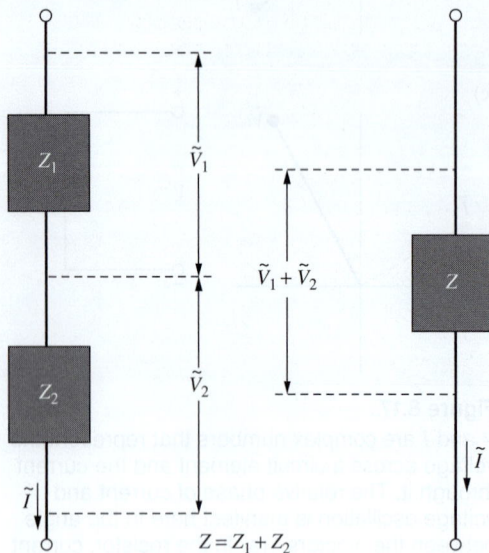

**Figure 8.19.**
Combining impedances in series.

$$Z = Z_1 + Z_2$$

Assuming that the voltage is given as usual by $\mathcal{E}_0 \cos \omega t$ (that is, with no phase), we have

$$I(t) = \mathcal{E}_0 \sqrt{(1/R)^2 + (\omega C - 1/\omega L)^2} \, \cos(\omega t + \phi),$$

$$\tan \phi = R \omega C - \frac{R}{\omega L}. \tag{8.67}$$

You can compare these results with the results in Eqs. (8.38) and (8.39) for the series $RLC$ circuit. For both of these circuits, you are encouraged to check limiting cases for the $R$, $L$, and $C$ values.

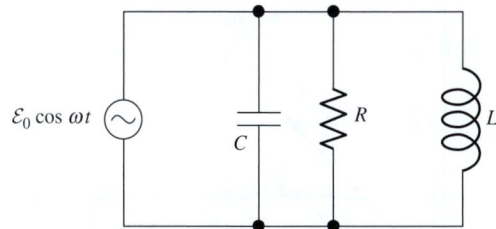

**Figure 8.20.**
A parallel resonant circuit. Add the complex admittances of the three elements, as in Eq. (8.65).

Let's now analyze a more complicated circuit. We will examine in detail what the various complex voltages and currents look like in the complex plane and how they relate to each other.

**Example**  Consider the circuit in Fig. 8.21. Our goal will be to find the complex voltage across, and current through, each of the three elements. We will then draw the associated vectors in the complex plane and verify that the relations among them are correct. To keep the calculations from getting out of hand, we will arrange for all three of the complex impedances to have magnitude $R$. If we take $R$ and $\omega$ as given, this can be arranged by letting $L = R/\omega$ and $C = 1/\omega R$. The three impedances are then

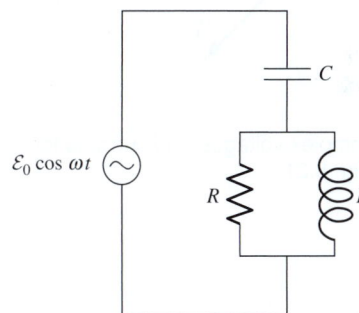

**Figure 8.21.**
What are the complex voltages and currents across each of the three elements in this circuit?

$$Z_R = R, \quad Z_L = i\omega L = iR, \quad Z_C = 1/i\omega C = -iR. \tag{8.68}$$

With these values, the impedance of the entire circuit is

$$Z = Z_C + \frac{Z_R Z_L}{Z_R + Z_L} = R\left(-i + \frac{1 \cdot i}{1 + i}\right) = R\frac{1 - i}{2}. \tag{8.69}$$

Assuming that the applied voltage is given as usual by $\mathcal{E}_0 \cos \omega t$ (with no extra phase), the applied complex voltage $\tilde{V}_\mathcal{E}$ is simply the real number $\mathcal{E}_0$. The total complex current $\tilde{I}$ (which is also the complex current $\tilde{I}_C$ through the capacitor) is therefore given by

$$\tilde{V}_\mathcal{E} = \tilde{I}Z \implies \tilde{I} = \frac{\mathcal{E}_0}{Z} = \frac{\mathcal{E}_0}{R}\frac{2}{1 - i} = \frac{\mathcal{E}_0}{R}(1 + i). \tag{8.70}$$

The complex voltage across the capacitor is then

$$\tilde{V}_C = \tilde{I}_C Z_C = \frac{\mathcal{E}_0}{R}(1 + i) \cdot (-iR) = \mathcal{E}_0(1 - i). \tag{8.71}$$

The complex voltages across the resistor and inductor are the same, and their common value equals $\mathcal{E}_0$ minus the complex voltage across the capacitor:

$$\tilde{V}_R = \tilde{V}_L = \mathcal{E}_0 - \tilde{V}_C = \mathcal{E}_0 - \mathcal{E}_0(1 - i) = i\mathcal{E}_0. \tag{8.72}$$

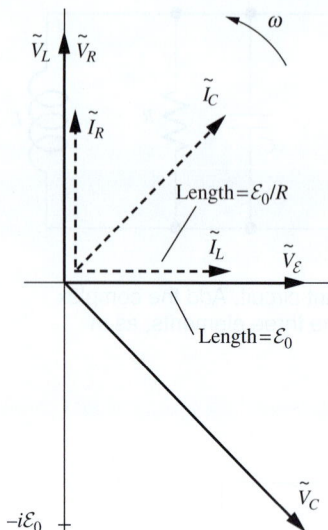

**Figure 8.22.**
The various complex voltages and currents for
the circuit in Fig. 8.21.

The complex current through the resistor is therefore

$$\tilde{I}_R = \frac{\tilde{V}_R}{Z_R} = \frac{i\mathcal{E}_0}{R}, \tag{8.73}$$

and the complex current through the inductor is

$$\tilde{I}_L = \frac{\tilde{V}_R}{Z_L} = \frac{i\mathcal{E}_0}{iR} = \frac{\mathcal{E}_0}{R}. \tag{8.74}$$

Our results for the three complex voltages (along with the $\mathcal{E}_0$ source) and the three complex currents are drawn in the complex plane in Fig. 8.22. (The $\tilde{V}$'s and $\tilde{I}$'s have different units, so the relative size of the two groups of vectors is meaningless.) There are various true statements we can make about the vectors: (1) $\mathcal{E}_0$ equals the sum of $\tilde{V}_C$ and either $\tilde{V}_L$ or $\tilde{V}_R$, (2) $\tilde{I}_C$ equals the sum of $\tilde{I}_R$ and $\tilde{I}_L$, (3) $\tilde{I}_L$ is 90° behind $\tilde{V}_L$ as the vectors rotate counterclockwise around in the plane, (4) $\tilde{I}_R$ is in phase with $\tilde{V}_R$, and (5) $\tilde{I}_C$ is 90° ahead of $\tilde{V}_C$.

As time goes on, the vectors in Fig. 8.22 all rotate around in the complex plane with the same angular speed $\omega$. The vectors keep the same rigid shape with respect to each other. The horizontal projections (the real parts) are the actual quantities that exist in the real world. Equivalently, the actual quantities are given by $I_R(t) = \text{Re}[\tilde{I}_R e^{i\omega t}]$, etc. The $e^{i\omega t}$ factor increases the phase by $\omega t$, so this is what causes the vectors to rotate around in the plane. Figure 8.22 gives the vectors at $t = 0$ (assuming the applied voltage equals $\mathcal{E}_0 \cos \omega t$ with no extra phase), or at any time for which $\omega t$ is a multiple of $2\pi$.

As mentioned in Section 8.4, the critical thing to realize about this rotation around in the plane is that since, for example, the vector $\tilde{I}_C$ always equals the sum of vectors $\tilde{I}_R$ and $\tilde{I}_L$ (because the system rotates as a rigid "object"), the horizontal projections also always satisfy this relation. That is, $I_C(t) = I_R(t) + I_L(t)$. In other words, the Kirchhoff node condition is satisfied at the node below the capacitor. Likewise, since the applied voltage $\tilde{V}_\mathcal{E}$ always equals $\tilde{V}_C$ plus $\tilde{V}_R$ (or $\tilde{V}_L$), we have $V_\mathcal{E}(t) = V_C(t) + V_R(t)$. So the Kirchhoff loop condition is satisfied. In short, if the complex voltages and currents satisfy Kirchhoff's rules at a *particular* time, then the actual voltages and currents satisfy Kirchhoff's rules at *all* times.

As noted earlier in this section, the $i$'s in $Z_L$ and $Z_C$ in Table 8.1 are consistent with the $\pm\pi/2$ phases between the voltages and currents. Let's verify this for Fig. 8.22. In the case of the inductor, we have

$$\tilde{V}_L = \tilde{I}_L Z_L \implies \tilde{V}_L = \tilde{I}_L(i\omega L) \implies \tilde{V}_L = \tilde{I}_L(e^{i\pi/2}\omega L), \tag{8.75}$$

which means that $\tilde{V}_L$ is $\pi/2$ ahead of $\tilde{I}_L$. The opposite is true for the capacitor. More generally, we can write $\tilde{V} = \tilde{I}Z$ for the entire circuit or any subpart, just as we can for a network containing only resistors.

If the complex voltage $\tilde{V}$, complex current $\tilde{I}$, and impedance $Z$ are written in polar form as[8]

$$\tilde{V} = V_0 e^{i\phi_V}, \quad \tilde{I} = I_0 e^{i\phi_I}, \quad Z = |Z| e^{i\phi_Z}, \tag{8.76}$$

then, by looking at the modulus and phase of the two sides of the $\tilde{V} = \tilde{I}Z$ equation, we obtain

$$\boxed{V_0 = I_0 |Z|} \quad \text{and} \quad \boxed{\phi_V = \phi_I + \phi_Z} \tag{8.77}$$

The former of these statements looks just like Ohm's law, $V = IR$. The latter says that the voltage is $\phi_Z$ ahead of the current. You are encouraged at this point to solve Problem 8.9, the task of which is to draw all the complex voltages and currents for the series and parallel $RLC$ circuits in Figs. 8.10 and 8.20.

We should emphasize that the above methods are valid only for *linear* circuit elements, elements in which the current is proportional to the voltage. In other words, our circuit must be described by a linear differential equation. You can't even define an impedance for a nonlinear element. Nonlinear circuit elements are very important and interesting devices. If you have studied some in the laboratory, you can see why they will not yield to this kind of analysis.

This is all predicated, too, on continuous oscillation at constant frequency. The transient behavior of the circuit is a different problem. However, for linear circuits the tools we have just developed have some utility, even for transients. The reason, as we noted at the end of Section 8.3, is that by superposing steady oscillations of many frequencies we can represent a nonsteady behavior, and the response to each of the individual frequencies can be calculated as if that frequency were present alone.

We have encountered three different methods for dealing with steady states in circuits containing a sinusoidal voltage source. Let's summarize them.

## Method 1 (Trig functions)
This is the method we used in Section 8.2. The steps are as follows.

- Write down the differential equation expressing the fact that the voltage drop around each loop in a circuit is zero. The various voltage drops take the form of $IR$, $L\,dI/dt$, and $Q/C$. Write the differential equation in terms of only one quantity, say the current $I(t)$.
- Guess a trig solution of the form $I(t) = I_0 \cos(\omega t + \phi)$. There will be many such currents if there are many loops.

---

[8] We have written the modulus of $Z$ as $|Z|$ rather than $Z_0$ to signify that $Z$ isn't the same type of quantity as $\tilde{V}$ and $\tilde{I}$. The quantities $V_0$ and $I_0$ are the amplitudes of the actual voltage and current oscillations, and we don't want to give the impression that $Z$ represents an oscillatory function.

- Use the trig sum formulas to expand $\cos(\omega t + \phi)$ and $\sin(\omega t + \phi)$, and then demand that the coefficients of $\cos \omega t$ and $\sin \omega t$ are separately identically equal to zero. This yields solutions for $I_0$ and $\phi$.

## Method 2 (Exponential functions)

This is the method we used in Section 8.3. The steps are as follows.

- As in Method 1, write down the differential equation for the voltage drop around each loop, and then write it in terms of only, say, the current $I(t)$.
- Replace the $\mathcal{E}_0 \cos \omega t$ voltage source with $\mathcal{E}_0 e^{i\omega t}$, and then guess a complex solution for the current of the form $\tilde{I}(t) \equiv \tilde{I} e^{i\omega t}$. The actual current in the circuit will be given by the real part of this. That is, $I(t) = \mathrm{Re}[\tilde{I}(t)]$. There will be many such currents if there are many loops.
- The solution for $\tilde{I}$ can be written in the general polar form, $\tilde{I} = I_0 e^{i\phi}$, The actual current is then

$$I(t) = \mathrm{Re}[\tilde{I}(t)] = \mathrm{Re}[\tilde{I} e^{i\omega t}] = \mathrm{Re}[I_0 e^{i\phi} e^{i\omega t}] = I_0 \cos(\omega t + \phi).$$
(8.78)

$I_0$ is the amplitude of the current, and $\phi$ is the phase relative to the voltage source.

## Method 3 (Complex impedances)

This is the method we used in Sections 8.4 and 8.5. The steps are as follows.

- Assign impedances of $R$, $i\omega L$, and $1/i\omega C$ to the resistors, inductors, and capacitors in the circuit, and then use the standard rules for adding impedances in series and in parallel (the same rules as for simple resistors).
- Write down $\tilde{V} = \tilde{I} Z$ for the entire circuit or any subpart, just as you would for a network containing only resistors. With the complex quantities written in polar form, $\tilde{V} = \tilde{I} Z$ quickly yields $V_0 = I_0 |Z|$ and $\phi_V = \phi_I + \phi_Z$. The former of these statements looks just like Ohm's law, $V = IR$. The latter says that the voltage is $\phi_Z$ ahead of the current.
- The $\tilde{V}$ and $\tilde{I}$ vectors rotate around in the complex plane with the same angular speed $\omega$. The horizontal projections (the real parts) are the actual quantities that exist in the real world. Since the vectors keep the same rigid shape with respect to each other, it follows that if the complex voltages and currents satisfy Kirchhoff's rules at a given time, the actual voltages and currents satisfy Kirchhoff's rules at all times.
- This third method is actually just a more systematic version of the second method. But for circuits involving more than one loop, the third method is vastly more tractable than the second, which in turn is much more tractable than the first.

## 8.6 Power and energy in alternating-current circuits

If the voltage across a resistor $R$ is $V_0 \cos \omega t$, the current is $I = (V_0/R) \cos \omega t$. The instantaneous power, that is, the instantaneous rate at which energy is being dissipated in the resistor, is given by

$$P_R = RI^2 = \frac{V_0^2}{R} \cos^2 \omega t. \tag{8.79}$$

Since the average of $\cos^2 \omega t$ over many cycles is $1/2$ (because it has the same average as $\sin^2 \omega t$, and $\sin^2 \omega t + \cos^2 \omega t = 1$), the average power dissipated in the resistor is

$$\overline{P}_R = \frac{1}{2} \frac{V_0^2}{R}. \tag{8.80}$$

It is customary to express voltage and current in ac circuits by giving not the amplitude but $1/\sqrt{2}$ times the amplitude. This is often called the *root-mean-square* (rms) value: $V_{\text{rms}} = V_0/\sqrt{2}$. That takes care of the factor $1/2$ in Eq. (8.80), so that

$$\boxed{\overline{P}_R = \frac{V_{\text{rms}}^2}{R}} \tag{8.81}$$

For example, the common domestic line voltage in North America is 120 volts, which corresponds to an *amplitude* $120\sqrt{2} = 170$ volts. The potential difference between the terminals of the electric outlet in your room (if the voltage is up to normal) is

$$V(t) = 170 \cos(377 \, \text{s}^{-1} \cdot t), \tag{8.82}$$

where we have used the fact that the frequency is 60 Hz. An ac ammeter is calibrated to read 1 amp when the current amplitude is 1.414 amps.

Equation (8.81) holds in the case of a single resistor. More generally, the instantaneous rate at which energy is delivered to a circuit element (or a combination of circuit elements) is $VI$, the product of the total instantaneous voltage across the element(s) and the current, with due regard to sign. Consider this aspect of the current flow in the simple $LR$ circuit in Fig. 8.5. In Fig. 8.23 we have redrawn the current and voltage graphs and added a curve proportional to the product $VI$. Positive $VI$ means energy is being transferred into the $LR$ combination from the source of electromotive force, or generator. Note that $VI$ is negative in certain parts of the cycle. In those periods some energy is being returned to the generator. This is explained by the oscillation in the energy stored in the magnetic field of the inductor. This stored energy, $LI^2/2$, goes through a maximum twice in each full cycle.

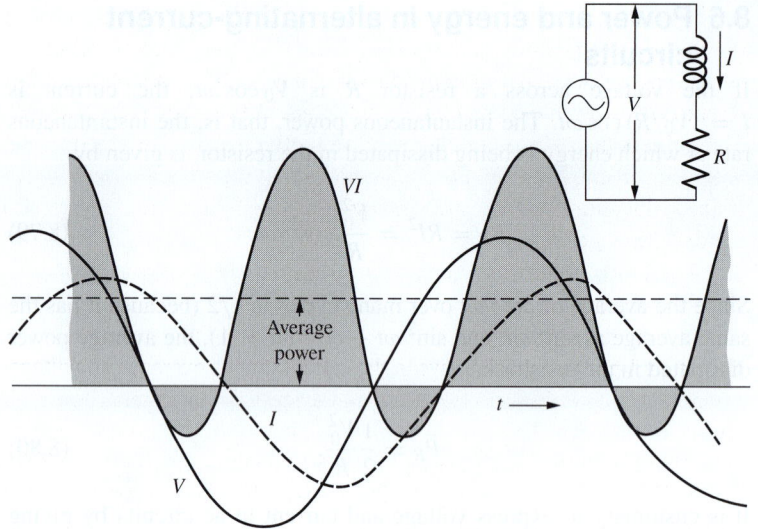

**Figure 8.23.**
The instantaneous power $VI$ is the rate at which energy is being transferred from the source of electromotive force on the left to the circuit elements on the right. The time average of this is indicated by the horizontal dashed line.

The *average* power $\overline{P}$ delivered to the $LR$ circuit corresponds to the horizontal dashed line. To calculate its value, let's take a look at the product $VI$, with $V = \mathcal{E}_0 \cos \omega t$ and $I = I_0 \cos(\omega t + \phi)$:

$$VI = \mathcal{E}_0 I_0 \cos \omega t \cos(\omega t + \phi)$$

$$= \mathcal{E}_0 I_0 (\cos^2 \omega t \cos \phi - \cos \omega t \sin \omega t \sin \phi). \qquad (8.83)$$

The term proportional to $\cos \omega t \sin \omega t$ has a time average zero, as is obvious if you write it as $(1/2) \sin 2\omega t$, while the average of $\cos^2 \omega t$ is $1/2$. Thus for the time average we have

$$\overline{P} = \overline{VI} = \frac{1}{2} \mathcal{E}_0 I_0 \cos \phi. \qquad (8.84)$$

If both current and voltage are expressed as rms values, in volts and amps, respectively, then

$$\boxed{\overline{P} = V_{\mathrm{rms}} I_{\mathrm{rms}} \cos \phi} \qquad (8.85)$$

In this circuit all the energy dissipated goes into the resistance $R$. Naturally, any real inductor has some resistance. For the purpose of analyzing the circuit, we included that with the resistance $R$. Of course, the heat evolves at the actual site of the resistance.

The power $P$ equals the product of the actual voltage $V(t)$ and actual current $I(t)$. These quantities in turn are the real parts of the complex voltage $\tilde{V}(t)$ and complex current $\tilde{I}(t)$. Does this mean that the power equals the real part of the product $\tilde{V}(t)\tilde{I}(t)$? Definitely not, because the product of the real parts doesn't equal the real part of the product; the real part of the product also has a contribution from the product of the *imaginary* parts of $\tilde{V}(t)$ and $\tilde{I}(t)$. As we mentioned in Section 8.4, it doesn't make any sense to form the product of two complex quantities (excluding products with impedances and admittances, which are a different type of number; they aren't functions of time that we solve for). The point is that, since our original differential equations were linear in voltages and currents, we must keep things that way. The product of two of these quantities doesn't have anything to do with the actual solution to the differential equation.

There was nothing special about our *LR* circuit, so Eq. (8.85) holds for a general circuit (or subpart of a circuit), provided that $V_{rms}$ is the total rms voltage across the circuit, $I_{rms}$ is the rms current through the circuit, and $\phi$ is the phase between the instantaneous current and voltage. Equation (8.85) reduces to Eq. (8.81) in the special case where the circuit consists of a single resistor. In that case, the current across the resistor is in phase with the voltage, so $\phi = 0$. Additionally, $I_{rms} = V_{rms}/R$, so Eq. (8.85) simplifies to Eq. (8.81). In the case where a resistor is part of a larger circuit, remember that the $V_{rms}$ in Eq. (8.85) is the voltage across the entire circuit (or whatever part we're concerned with), while the $V_{rms}$ in Eq. (8.81) is the voltage across only the resistor; see Problem 8.14.

**Example**  To get some more practice with the methods we developed in Section 8.5, we'll analyze the circuit in Fig. 8.24(a). A 10,000 ohm, 1 watt resistor (this rating gives the maximum power the resistor can safely absorb) has been connected up with two capacitors of capacitance 0.2 and 0.5 microfarads. We propose to plug this into the 120 volt, 60 Hz outlet. *Question:* Will the 1 watt resistor get too hot? In the course of finding out whether the average power dissipated in $R$ exceeds the 1 watt rating, we'll calculate some of the currents and voltages we might expect to measure in this circuit. One way to work through the circuit is outlined below.

(a)

Admittance of $C_2 = i\omega C_2 = i(377)(2 \cdot 10^{-7}) = 0.754 \cdot 10^{-4} i$ ohm$^{-1}$

Admittance of the resistor $= \dfrac{1}{R} = 10^{-4}$ ohm$^{-1}$

Admittance of ⧌ $= 10^{-4}(1 + 0.754i)$ ohm$^{-1}$

Impedance of ⧌ $= \dfrac{1}{10^{-4}(1 + 0.754i)} = \dfrac{10^4(1 - 0.754i)}{1^2 + 0.754^2}$

$= (6380 - 4810i)$ ohms

Impedance of $C_1 = -\dfrac{i}{\omega C} = -\dfrac{i}{(377)(5 \cdot 10^{-7})} = -5300i$ ohms

Impedance of entire circuit $= (6380 - 10,110i)$ ohms

$I_1 = \dfrac{120}{6380 - 10,110i} = \dfrac{120(6380 + 10,110i)}{(6380)^2 + (10,110)^2} = (5.36 + 8.49i) \cdot 10^{-3}$ amp

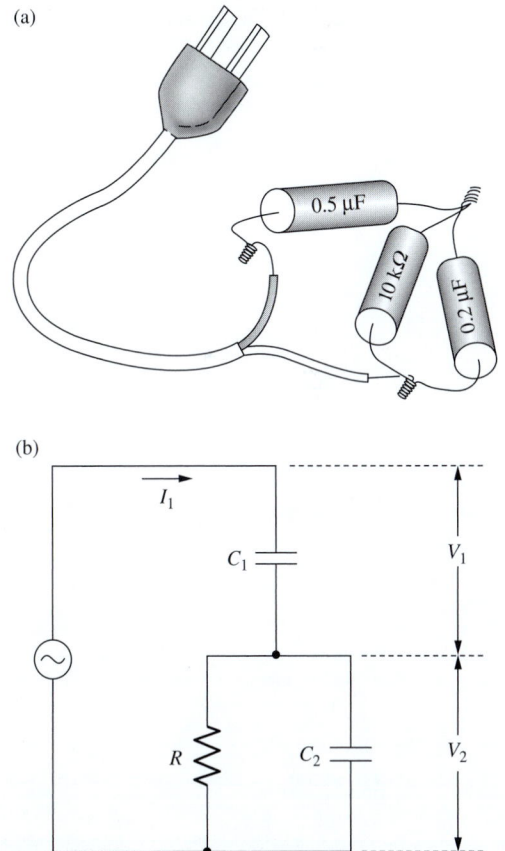

(b)

$I_1$

$C_1$　$V_1$

$R$　$C_2$　$V_2$

**Figure 8.24.**
An actual network (a) ready to be connected to a source of electromotive force, and (b) the circuit diagram.

Since 120 volts is the rms voltage, we obtain the rms current. That is, the modulus of the complex number $I_1$, which is $[(5.36)^2 + (8.49)^2]^{1/2} \cdot 10^{-3}$ amp or 10.0 milliamps, is the rms current. An ac milliammeter inserted in series with the line would read 10 milliamps. This current has a phase angle $\phi = \tan^{-1}(0.849/0.536)$ or 1.01 radians with respect to the line voltage. From Eq. (8.85), the average power delivered to the entire circuit is then

$$\overline{P} = (120 \text{ volts})(0.010 \text{ amp}) \cos 1.01 = 0.64 \text{ watt}. \qquad (8.86)$$

In this circuit the resistor is the only dissipative element, so this must be the average power dissipated in it. Just as a check, we can find the voltage $V_2$ across the resistor. If $V_1$ is the voltage across $C_1$, we have

$$V_1 = I_1 \left( \frac{-i}{\omega C} \right) = (5.36 + 8.49i)(-5300i) 10^{-3} = (45.0 - 28.4i) \text{ volts};$$

$$V_2 = 120 - V_1 = (75.0 + 28.4i) \text{ volts}. \qquad (8.87)$$

The current $I_2$ in $R$ will be in phase with $V_2$, of course, so the average power in $R$ will be

$$\overline{P} = \frac{V_2^2}{R} = \frac{(75.0)^2 + (28.4)^2}{10^4} = 0.64 \text{ watt}, \qquad (8.88)$$

which checks. Thus the rating of the resistor isn't exceeded, for what that assurance is worth. Actually, whether the resistor will get too hot depends not only on the average power dissipated in it, but also on how easily it can get rid of the heat. The power rating of a resistor is only a rough guide.

## 8.7 Applications

The *resonance* of electrical circuits has numerous applications in the modern world. Our lives wouldn't be the same without it. Any wireless communication, from radios to cell phones to computers to GPS systems, is made possible by resonance. If you have a radio sitting on your desk, it is being bombarded by electromagnetic waves (discussed in Chapter 9) with all sorts of frequencies. If you want to pick out a particular frequency emitted by a radio station, you can "tune" your radio to that frequency by adjusting the radio's resonant frequency. This is normally done by adjusting the capacitance of the internal circuit by using *varactors* – diodes whose capacitance can be controlled by an applied voltage. Assuming that the resistance of the circuit is small, two things will happen when the resonant frequency matches the frequency of the radio station: there will be a large oscillation in the circuit at the radio station's frequency, and there will also be a negligible oscillation at all the other frequencies that are bombarding the radio. A high $Q$ value of the circuit leads to both of these effects, due to the facts that the height of the peak in Fig. 8.11 is proportional to $Q$ (as you can show) and that the width is proportional to $1/Q$. The oscillation in the circuit can then be demodulated (see the AM/FM discussion in Section 9.8) and amplified and sent to the speakers, creating the sound that you hear. Resonance provides us with an astonishingly simple and automatic mechanism for finding needles in haystacks.

The microwaves in a *microwave oven* are created by a *magnetron*. This device consists of a ring-like chamber with a number (often eight) of cavities around the perimeter (Fig. 8.25). These cavities have both a capacitance and an inductance (and also a small resistance), so they act like little resonant *LC* circuits. Their size is chosen so that the resonant frequency is about 2.5 GHz. The charge on the tips of the little *LC* cavities alternates in sign around the perimeter of the ring. Charge (and hence energy) is added to the system by emitting electrons from the center of the ring. These electrons are attracted toward the positive tips. If this were the whole story, the effect would be to *reduce* the charge in the system. But there is a clever way of reversing the effect: by applying an appropriate magnetic field, the paths of the electrons can be bent by just the right amount to make them hit the *negative* tips. Charge is therefore added to the system instead of subtracted. The microwave radiation can be extracted by, say, inducing a current in small coils contained in the *LC* cavities.

**Figure 8.25.**
A magnetron. The cavities have both a capacitance and an inductance.

The electricity that comes out of your wall socket is alternating current (ac) as opposed to direct current (dc). The rms voltage in North America is 120 V, and the frequency is 60 Hz. (In Europe the values are 230 V and 50 Hz, respectively.) The fundamental reason we use ac instead of dc is that, in the case of ac, it is easy to increase or decrease the voltage via a transformer. This is critical for the purpose of transmitting power over long distances, because for a given power $P = IV$ supplied by a power plant, a large $V$ implies a small $I$, which in turn implies a small $I^2R$ power loss in the long transmission lines. It is much more difficult to change the voltage in the case of dc. This was the deciding factor during the "War of Currents" in the 1880s, when ac and dc power were battling for dominance. Because dc power had to be shipped at the same low voltage at which it was used, dc power plants needed to be located within a few miles of the load. This had obvious disadvantages: cities would need to contain many power plants, and conversely a dam located far from a populated area would be useless. However, modern developments have made the conversion of dc voltages easier, so *high-voltage, direct current (HVDC)* power transmission is used in some instances. For both ac and dc, the long-haul voltages are on the order of a few hundred kilovolts. The War of Currents pitted (among many other people) Thomas Edison on the dc side against Nikola Tesla on the ac side.

Most of the electricity produced in power plants is *three-phase*. That is, there are three separate wires carrying voltages that are 120° out of phase. This can be achieved, for example, by having three loops of wire in Fig. 7.13 instead of just the one shown. There are various advantages to three-phase power, one of which is that it delivers a more steady power compared with single-phase, which has two moments during each cycle when the voltage is zero. However, this is mainly relevant for large machinery. Most households are connected to only one of the phases (or between two of them) in the power grid.

The ac power delivered to your home works fine for many electrical devices. For example, a toaster and an incandescent light bulb require only the generation of $I^2R$ power, which is created by either ac or dc. But many other devices require dc, because the direction of the current in the electronic circuits matters. A *power adapter* converts ac to dc, while generally also lowering the voltage. The voltage is lowered by a transformer, and then the conversion to dc is accomplished by a *bridge rectifier*, which consists of a combination of four diodes that lets the current flow in only one direction. Additionally, a capacitor helps smooth out the dc voltage by storing charge and then releasing it when the voltage would otherwise dip.

As mentioned in Section 3.9, it is advantageous to perform *power-factor correction* in the ac electrical power grid. The larger the imaginary part of an impedance of, say, an electrical motor, the larger the phase angle $\phi$, and hence the smaller the $\cos\phi$ factor in Eq. (8.85), which is known as the *power factor*. At first glance, this doesn't seem to present a problem, because the unused power simply sloshes back and forth between the power station and the motor. However, for a given amount of net power consumed, a smaller power factor means that the current $I$ will need to be larger. This in turn means that there will be larger $I^2R$ power losses in the (generally long) transmission lines. For this reason, industries are usually charged a higher rate if their power factor is below 0.95. In an inductive circuit (for example, a motor with its many windings), the power factor can be increased by adding capacitance to the circuit, because this will reduce the magnitude of the imaginary part of the impedance.

## CHAPTER SUMMARY

- The loop equation for a series $RLC$ circuit (with no emf source) yields a linear differential equation involving three terms, one for each element. In the underdamped case, the solution for the voltage across the capacitor is

$$V(t) = e^{-\alpha t}(A\cos\omega t + B\sin\omega t), \tag{8.89}$$

where

$$\alpha = \frac{R}{2L} \quad \text{and} \quad \omega^2 = \frac{1}{LC} - \frac{R^2}{4L^2}. \tag{8.90}$$

The solutions for the overdamped and critically damped cases take other forms. The *quality factor* of a circuit is given by

$$Q = \omega \cdot \frac{\text{energy stored}}{\text{average power dissipated}}. \qquad (8.91)$$

- If we add to the series *RLC* circuit a sinusoidal emf source, $\mathcal{E}(t) = \mathcal{E}_0 \cos \omega t$, then the solution for the current is $I(t) = I_0 \cos(\omega t + \phi)$, where

$$I_0 = \frac{\mathcal{E}_0}{\sqrt{R^2 + (\omega L - 1/\omega C)^2}} \quad \text{and} \quad \tan \phi = \frac{1}{R\omega C} - \frac{\omega L}{R}. \qquad (8.92)$$

This is the *steady-state* solution that survives after the *transient* solution from Section 8.1 has decayed away. $I_0$ is maximum when $\omega$ equals the resonant frequency, $\omega_0 = 1/\sqrt{LC}$. The width of the $I_0(\omega)$ curve around the resonance peak is on the order of $\omega_0/Q$.

- The series *RLC* circuit can also be solved by replacing the $\mathcal{E}_0 \cos \omega t$ term in the Kirchhoff differential equation with $\mathcal{E}_0 e^{i\omega t}$, and then guessing an *exponential solution* of the form $\tilde{I}(t) = \tilde{I} e^{i\omega t}$. The actual current $I(t)$ is obtained by taking the real part of $\tilde{I}(t)$.

- In alternating-current *networks,* currents and voltages can be represented by *complex numbers*. The real part of the complex number is the actual current or voltage. The complex current and voltage are related to each other via the complex *admittance* or *impedance*: $\tilde{I} = Y\tilde{V}$ or $\tilde{V} = Z\tilde{I}$. The admittances and impedances for the three circuit elements $R, L, C$ are given in Table 8.1. Admittances add in parallel, and impedances add in series.

- We have presented three different methods for solving alternating-current networks. See the summary at the end of Section 8.5.

- The average *power* delivered to a circuit is

$$\overline{P} = \frac{1}{2}\mathcal{E}_0 I_0 \cos \phi = V_{\text{rms}} I_{\text{rms}} \cos \phi, \qquad (8.93)$$

where the rms values are $1/\sqrt{2}$ times the peak values. This reduces to $\overline{P}_R = V_{\text{rms}}^2/R$ in the case of a single resistor.

## Problems

8.1 *Linear combinations of solutions* ∗

Homogeneous linear differential equations have the property that the sum, or any linear combination, of two solutions is again a solution. ("Homogeneous" means there's a zero on one side of the equation.) Consider, for example, the second-order equation (although the property holds for any order),

$$A\ddot{x} + B\dot{x} + Cx = 0. \qquad (8.94)$$

Show that if $x_1(t)$ and $x_2(t)$ are solutions, then the sum $x_1(t) + x_2(t)$ is also a solution. Show that this property does *not* hold for the nonlinear differential equation $A\ddot{x} + B\dot{x}^2 + Cx = 0$.

8.2 *Solving linear differential equations* **

Consider the $n$th-order homogeneous linear differential equation

$$a_n \frac{d^n x}{dt^n} + a_{n-1} \frac{d^{n-1} x}{dt^{n-1}} + \cdots + a_1 \frac{dx}{dt} + a_0 x = 0. \tag{8.95}$$

Show that the solutions take the form of $x(t) = A_i e^{r_i t}$, where the $r_i$ depend on the $a_j$ coefficients. *Hint:* If the $(d/dt)$ derivatives were replaced by the letter $z$, then we would have an $n$th-order polynomial in $z$, which we know can be factored, by the fundamental theorem of algebra. (You can assume that the roots of this polynomial are distinct. Things are a little more complicated if there are double roots; this is discussed in the solution.)

8.3 *Underdamped motion* ***

A second-order homogeneous linear differential equation can be written in the general form of

$$\ddot{x} + 2\alpha \dot{x} + \omega_0^2 x = 0, \tag{8.96}$$

where $\alpha$ and $\omega_0$ are constants. (For the series $RLC$ circuit in Section 8.1, Eq. (8.2) gives these constants as $\alpha = R/2L$ and $\omega_0^2 = 1/LC$.) From Problem 8.2 we know that there are two independent exponential solutions to this equation. Find these two solutions, and then show that, in the underdamped case where $\alpha < \omega_0$, the general solution can be written in the form of Eq. (8.10).

8.4 *Overdamped RLC circuit* **

Find the constants $\beta_1$ and $\beta_2$ in Eq. (8.15) by plugging an exponential trial solution into Eq. (8.2). If $R$ is very large, what does the solution look like for large $t$?

8.5 *Change in frequency* **

For the decaying signal shown in Exercise 8.19, estimate the percentage by which the frequency differs from the natural frequency $1/\sqrt{LC}$ of the circuit.

8.6 *Limits of an RLC circuit* ***

(a) In the $R \to 0$ limit, verify that the solution in Eq. (8.4) correctly reduces to the solution for an $LC$ circuit. That is, show that the voltage behaves like $\cos \omega_0 t$.

(b) In the $L \to 0$ limit, verify that the solution in Eq. (8.15) correctly reduces to the solution for an $RC$ circuit. That is, show that the voltage behaves like $e^{-t/RC}$. You will need to use the results from Problem 8.4.

(c) In the $C \to \infty$ limit, verify that the solution in Eq. (8.15) correctly reduces to the solution for an $RL$ circuit. That is, show that the voltage behaves like $e^{-(R/L)t}$, up to an additive constant. What is the physical meaning of this constant?

**8.7**  *Magnitude and phase*  *

Show that $a + bi$ can be written as $I_0 e^{i\phi}$, where $I_0 = \sqrt{a^2 + b^2}$ and $\phi = \tan^{-1}(b/a)$.

**8.8**  *RLC circuit via vectors*  ***

(a)  The loop equation for the series $RLC$ in Fig. 8.26 is

$$L\frac{dI}{dt} + RI + \frac{Q}{C} = \mathcal{E}_0 \cos \omega t, \qquad (8.97)$$

where we have taken positive $I$ to be clockwise and $Q$ to be the charge on the right plate of the capacitor. If $I$ takes the form of $I(t) = I_0 \cos(\omega t + \phi)$, show that Eq. (8.97) can be written as

$$\omega L I_0 \cos(\omega t + \phi + \pi/2) + R I_0 \cos(\omega t + \phi)$$
$$+ \frac{I_0}{\omega C} \cos(\omega t + \phi - \pi/2) = \mathcal{E}_0 \cos \omega t. \qquad (8.98)$$

(b)  At any given time, the four terms in Eq. (8.98) can be considered to be the real parts of four vectors in the complex plane. Draw the appropriate quadrilateral that represents the fact that the sum of the three terms on the left side of the equation equals the term on the right side.

(c)  Use your quadrilateral to determine the amplitude $I_0$ and phase $\phi$ of the current, and check that they agree with the values in Eqs. (8.38) and (8.39).

**8.9**  *Drawing the complex vectors*  **

For the series and parallel $RLC$ circuits in Figs. 8.10 and 8.20, draw the vectors representing all of the complex voltages and currents. For the sake of making a concrete picture, assume that $R = |Z_L| = 2|Z_C|$. The vectors all rotate around in the complex plane, so you can draw them at whatever instant in time you find most convenient.

**8.10**  *Real impedance*  *

Is it possible to find a frequency at which the impedance at the terminals of the circuit in Fig. 8.27 will be purely real?

**8.11**  *Light bulb*  *

A 120 volt (rms), 60 Hz line provides power to a 40 watt light bulb. By what factor will the brightness decrease if a 10 $\mu$F capacitor is connected in series with the light bulb? (Assume that the brightness is proportional to the power dissipated in the bulb's resistor.)

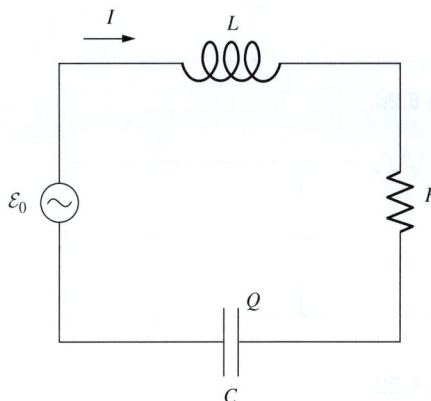

**Figure 8.26.**

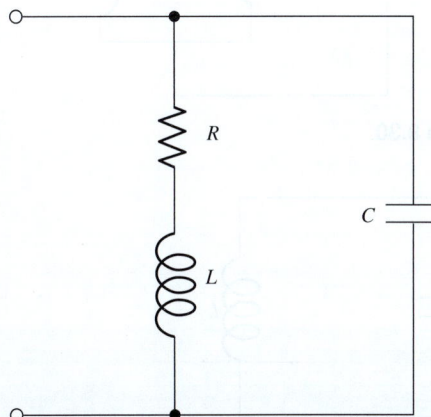

**Figure 8.27.**

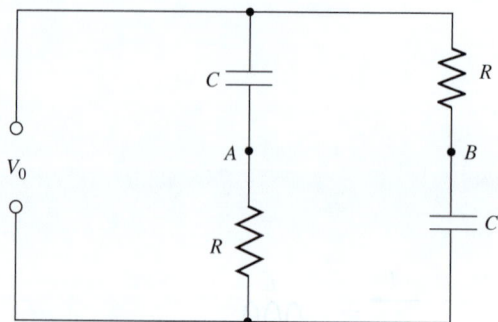

**Figure 8.28.**

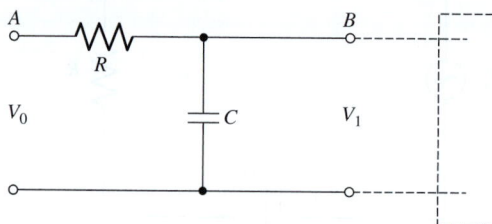

**Figure 8.29.**

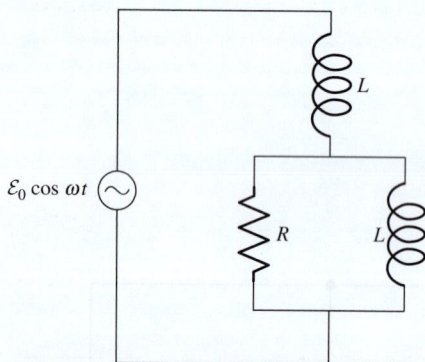

**Figure 8.30.**

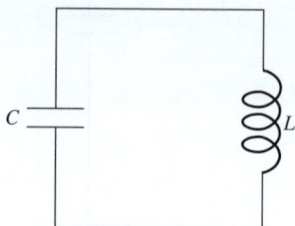

**Figure 8.31.**

**8.12** *Fixed voltage magnitude* **

Let $V_{AB} \equiv V_B - V_A$ in the circuit in Fig. 8.28. Show that $|V_{AB}|^2 = V_0^2$ for any frequency $\omega$. Find the frequency for which $V_{AB}$ is 90° out of phase with $V_0$.

**8.13** *Low-pass filter* **

In Fig. 8.29 an alternating voltage $V_0 \cos \omega t$ is applied to the terminals at $A$. The terminals at $B$ are connected to an audio amplifier of very high input impedance. (That is, current flow into the amplifier is negligible.) Calculate the ratio $|\tilde{V}_1|^2/V_0^2$. Here $|\tilde{V}_1|$ is the absolute value of the complex voltage amplitude at terminals $B$. Choose values for $R$ and $C$ to make $|\tilde{V}_1|^2/V_0^2 = 0.1$ for a 5000 Hz signal. This circuit is the most primitive of "low-pass" filters, providing attenuation that increases with increasing frequency. Show that, for sufficiently high frequencies, the signal power is reduced by a factor $1/4$ for every doubling of the frequency. Can you devise a filter with a more drastic cutoff – such as a factor $1/16$ per octave?

**8.14** *Series RLC power* **

Consider the series *RLC* circuit in Fig. 8.10. Show that the average power delivered to the circuit, which is given in Eq. (8.84), equals the average power dissipated in the resistor, which is given in Eq. (8.80). (These equations are a little easier to work with than the equivalent rms equations, Eqs. (8.85) and (8.81).)

**8.15** *Two inductors and a resistor* **

The circuit in Fig. 8.30 has two equal inductors $L$ and a resistor $R$. The frequency of the emf source, $\mathcal{E}_0 \cos \omega t$, is chosen to be $\omega = R/L$.

(a) What is the total complex impedance of the circuit? Give it in terms of $R$ only.

(b) If the total current through the circuit is written as $I_0 \cos(\omega t + \phi)$, what are $I_0$ and $\phi$?

(c) What is the average power dissipated in the circuit?

## Exercises

**8.16** *Voltages and energies* *

Consider the *LC* circuit in Fig. 8.31. Initial conditions have been set up so that the voltage change across the capacitor (proceeding around the loop in a clockwise manner) equals $V_0 \cos \omega t$, where $\omega = 1/\sqrt{LC}$. At $t = 0$, what are the voltage changes (proceeding clockwise) across the capacitor and inductor? Where is the energy stored? Answer the same questions for $t = \pi/2\omega$.

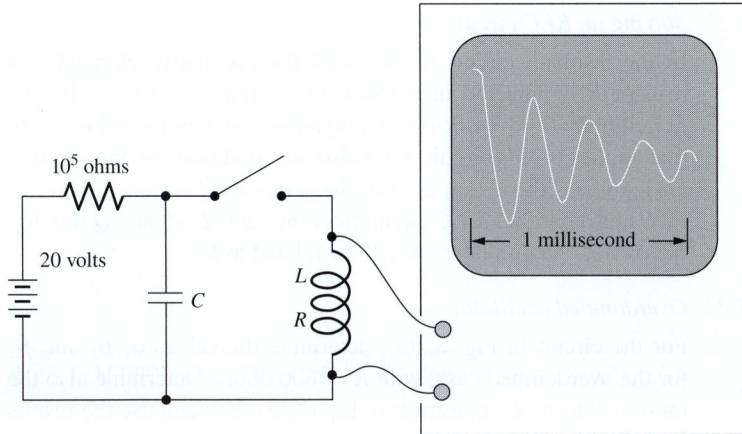

**Figure 8.32.**

8.17 *Amplitude after Q cycles* ∗
In the *RLC* circuit in Section 8.1, show that the current (or voltage) amplitude decreases by a factor of $e^{-\pi} \approx 0.043$ after $Q$ cycles.

8.18 *Effect of damping on frequency* ∗∗
Using Eqs. (8.9) and (8.13), express the effect of damping on the frequency of a series *RLC* circuit, by writing $\omega$ in terms of $Q$ and $\omega_0 = 1/\sqrt{LC}$. Suppose enough resistance is added to bring $Q$ from $\infty$ down to 1000. By what percentage is the frequency $\omega$ thereby shifted from $\omega_0$? How about if $Q$ is brought from $\infty$ down to 5?

8.19 *Decaying signal* ∗∗
The coil in the circuit shown in Fig. 8.32 is known to have an inductance of 0.01 henry. When the switch is closed, the oscilloscope sweep is triggered. The $10^5$ ohm resistor is large enough (as you will discover) so that it can be treated as essentially infinite for parts (a) and (b) of this problem.

(a) Determine as well as you can the value of the capacitance $C$.
(b) Estimate the value of the resistance $R$ of the coil.
(c) What is the magnitude of the voltage across the oscilloscope input a long time, say 1 second, after the switch has been closed?

8.20 *Resonant cavity* ∗∗
A resonant cavity of the form illustrated in Fig. 8.33 is an essential part of many microwave oscillators. It can be regarded as a simple *LC* circuit. The inductance is that of a rectangular toroid with one turn; see Eq. (7.62). This inductor is connected directly to a parallel-plate capacitor. Find an expression for the resonant frequency of this circuit, and show by a rough sketch the configuration of the magnetic and electric fields.

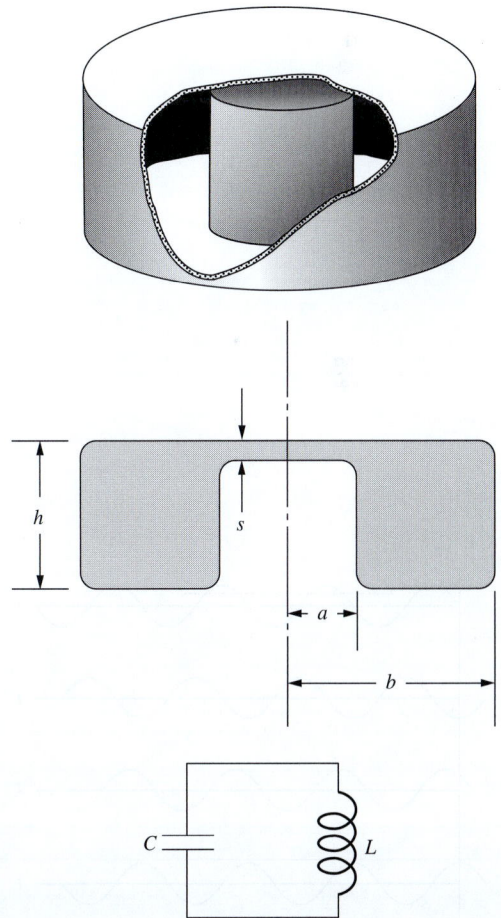

**Figure 8.33.**

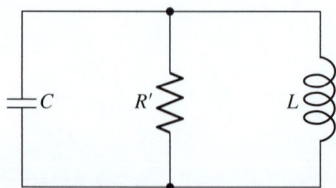

**Figure 8.34.**

**8.21  Solving an RLC circuit  ✷✷✷**

In the resonant circuit in Fig. 8.34 the dissipative element is a resistor $R'$ connected in parallel, rather than in series, with the $LC$ combination. Work out the equation, analogous to Eq. (8.2), that applies to this circuit. Find also the conditions on the solution analogous to those that hold in the series $RLC$ circuit. If a series $RLC$ and a parallel $R'LC$ circuit have the same $L$, $C$, and $Q$ (quality factor, not charge), how must $R'$ be related to $R$?

**8.22  Overdamped oscillator  ✷✷**

For the circuit in Fig. 8.4(a), determine the values of $\beta_1$ and $\beta_2$ for the overdamped case, with $R = 600$ ohms. Determine also the ratio of $B$ to $A$, the constants in Eq. (8.15). You can use the results from Problem 8.4.

**8.23  Energy in an RLC circuit  ✷✷✷**

For the damped $RLC$ circuit of Fig. 8.2, work out an expression for the total energy stored in the circuit (the energy in the capacitor plus the energy in the inductor) at any time $t$, for all three of the underdamped, overdamped, and critically damped cases; you need not simplify your answers. If $R$ is varied while $L$ and $C$ are kept fixed, show that the critical damping condition, $R = 2\sqrt{L/C}$, is the one in which the total energy is most quickly dissipated. (The exponential behavior is all that matters here.) The results from Problem 8.4 will be useful.

**8.24  RC circuit with a voltage source  ✷✷**

A voltage source $\mathcal{E}_0 \cos \omega t$ is connected in series with a resistor $R$ and a capacitor $C$. Write down the differential equation expressing Kirchhoff's law. Then guess an exponential form for the current, and take the real part of your solution to find the actual current. Determine how the amplitude and phase of the current behave for very large and very small $\omega$, and explain the results physically.

**8.25  Light bulb  ✷✷**

How large an inductance should be connected in series with a 120 volt (rms), 60 watt light bulb if it is to operate normally when the combination is connected across a 240 volt, 60 Hz line? (First determine the inductive reactance required. You may neglect the resistance of the inductor and the inductance of the light bulb.)

**8.26  Label the curves  ✷✷**

The four curves in Fig. 8.35 are plots, in some order, of the applied voltage and the voltages across the resistor, inductor, and capacitor of a series $RLC$ circuit. Which is which? Whose impedance is larger, the inductor's or the capacitor's?

$R, L, C, \mathcal{E}$

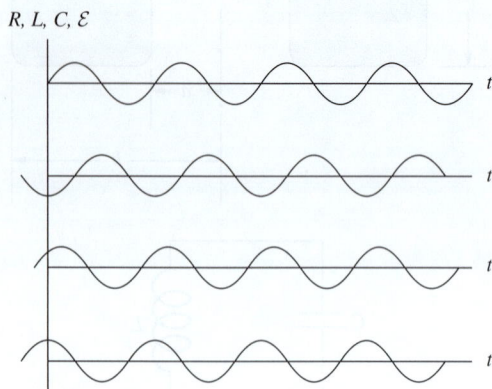

**Figure 8.35.**

**8.27** *RLC parallel circuit* **

A 1000 ohm resistor, a 500 picofarad capacitor, and a 2 millihenry inductor are connected in parallel. What is the impedance of this combination at a frequency of 10 kilocycles per second? At a frequency of 10 megacycles per second? What is the frequency at which the absolute value of the impedance is greatest?

**8.28** *Small impedance* *

Consider the circuit in Fig. 8.36. The frequency is chosen to be $\omega = 1/\sqrt{LC}$. Given $L$ and $C$, how should you pick $R$ so that the impedance of the circuit is small?

**8.29** *Real impedance* *

Is it possible to find a frequency at which the impedance at the terminals of the circuit in Fig. 8.37 will be purely real?

**8.30** *Equal impedance?* *

Do there exist values of $R$, $L$, and $C$ for which the two circuits in Fig. 8.38 have the same impedance? (The resistor $R$ has the same value in both.) Can you give a physical explanation why or why not?

**8.31** *Zero voltage difference* **

Show that, if the condition $R_1 R_2 = L/C$ is satisfied by the components of the circuit in Fig. 8.39, the difference in voltage between points $A$ and $B$ will be zero at any frequency. Discuss the suitability of this circuit as an ac bridge for measurement of an unknown inductance.

**8.32** *Finding L* **

In the laboratory you find an inductor of unknown inductance $L$ and unknown internal resistance $R$. Using a dc ohmmeter, an ac voltmeter of high impedance, a 1 microfarad capacitor, and a 1000 Hz signal generator, determine $L$ and $R$ as follows. According to the ohmmeter, $R$ is 35 ohms. You connect the capacitor in series with the inductor and the signal generator. The voltage across both is 10.1 volts. The voltage across the capacitor alone is 15.5 volts. You note also, as a check, that the voltage across the inductor alone is 25.4 volts. How large is $L$? Is the check consistent?

**8.33** *Equivalent boxes* ***

Show that the impedance $Z$ at the terminals of each of the two circuits in Fig. 8.40 is (ignoring the units)

$$Z = \frac{5000 + 16 \cdot 10^{-3}\omega^2 - 16i\omega}{1 + 16 \cdot 10^{-6}\omega^2}. \qquad (8.99)$$

Since they present, at any frequency, the identical impedance, the two black boxes are completely equivalent and indistinguishable

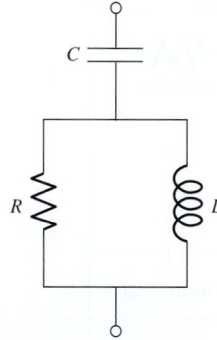

Figure 8.36.

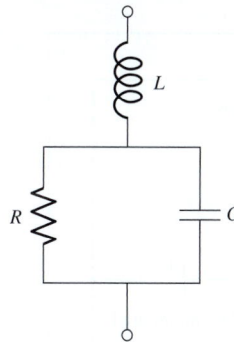

Figure 8.37.

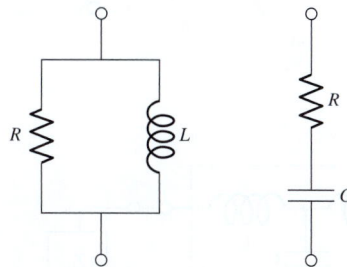

Figure 8.38.

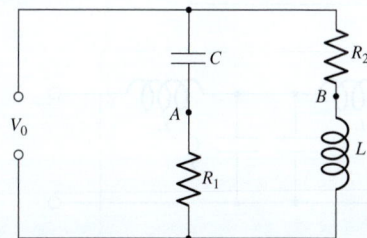

Figure 8.39.

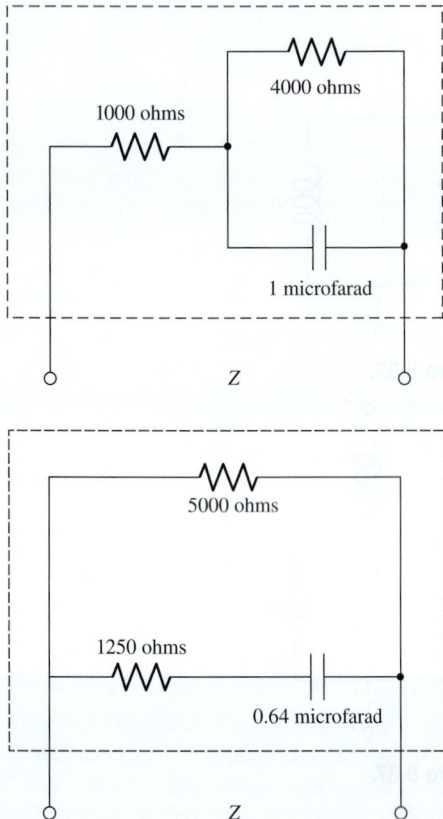

**Figure 8.40.**

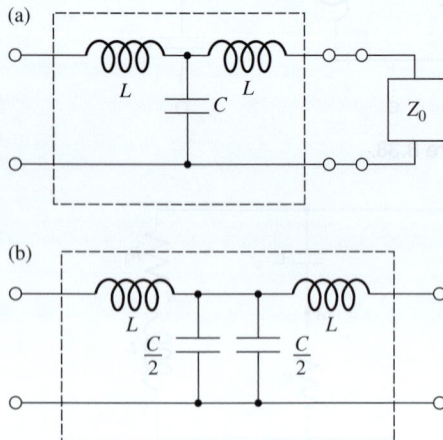

**Figure 8.41.**

from the outside. See if you can discover the general rules for finding the resistances and capacitance in the bottom box, given the resistances and capacitance in the top box.

**8.34** *LC chain* **∗∗**

The box in Fig. 8.41(a) with four terminals contains a capacitor $C$ and two inductors of equal inductance $L$ connected as shown. An impedance $Z_0$ is to be connected to the terminals on the right. For given frequency $\omega$, find the value that $Z_0$ must have if the resulting impedance between the terminals on the left (the "input" impedance) is to be equal to $Z_0$.

(You will find that the required value of $Z_0$ is a pure resistance $R_0$ provided that $\omega^2 < 2/LC$. A chain of such boxes could be connected together to form a ladder network resembling the ladder of resistors in Exercise 4.36. If the chain is terminated with a resistor of the correct value $R_0$, its input impedance at frequency $\omega$ will be $R_0$, no matter how many boxes make up the chain.)

What is $Z_0$ in the special case $\omega = \sqrt{2/LC}$? It helps in understanding that case to note that the contents of the box (a) can be equally well represented by box (b).

**8.35** *RC circuit* **∗∗**

A 2000 ohm resistor and a 1 microfarad capacitor are connected in series across a 120 volt (rms), 60 Hz line.

(a) What is the total impedance?
(b) What is the rms value of the current?
(c) What is the average power dissipated in the circuit?
(d) What will be the reading of an ac voltmeter connected across the resistor? Across the capacitor?
(e) The left and right plates of a cathode ray tube are connected across the resistor, and the top and bottom plates are connected across the capacitor. The horizontal and vertical axes of the tube's screen therefore indicate the voltages across the resistor and capacitor, respectively. Sketch the pattern that you expect to see on the screen. From the given information, is it possible to determine the direction in which the pattern is traced out?

**8.36** *High-pass filter* **∗∗**

Consider the setup in Problem 8.13, but with the capacitor replaced by an inductor. Calculate the ratio $|\tilde{V}_1|^2/V_0^2$. Choose values for $R$ and $L$ to make $|\tilde{V}_1|^2/V_0^2 = 0.1$ for a 100 Hz signal. This circuit is the most primitive of "high-pass" filters, providing attenuation that increases with decreasing frequency. Show that, for sufficiently low frequencies, the signal power is reduced by a factor $1/4$ for every halving of the frequency.

8.37 *Parallel RLC power* **

Repeat the task of Problem 8.14, but now for the parallel *RLC* circuit in Fig. 8.20.

8.38 *Two resistors and a capacitor* **

The circuit in Fig. 8.42 has two equal resistors $R$ and a capacitor $C$. The frequency of the emf source, $\mathcal{E}_0 \cos \omega t$, is chosen to be $\omega = 1/RC$.

(a) What is the total complex impedance of the circuit? Give it in terms of $R$ only.

(b) If the total current through the circuit is written as $I_0 \cos(\omega t + \phi)$, what are $I_0$ and $\phi$?

(c) What is the average power dissipated in the circuit?

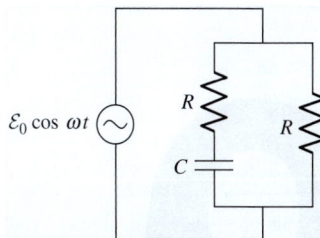

**Figure 8.42.**

# 9

# Maxwell's equations and electromagnetic waves

**Overview** In the course of our study of electromagnetism, we have been gradually putting together the pieces of the puzzle of *Maxwell's equations.* In this chapter we find the final piece, known as the *displacement current.* We do this by exposing a contradiction in our present theory and then resolving it. Once we write down the full set of Maxwell's equations, we quickly discover that in vacuum they lead to *wave* solutions with a set of specific properties. These waves are light waves, and unlike other waves you are familiar with, they require no medium to support their propagation. *Traveling* electromagnetic waves carry energy, and more generally the *Poynting vector* describes the energy flow in an arbitrary electromagnetic field. *Standing* electromagnetic waves, which are the superposition of traveling waves, carry no net energy. By examining how the electric and magnetic fields transform between frames, we find that a light wave in one frame looks like a light wave in any other frame.

## 9.1 "Something is missing"

Let us review the relations between charges and fields. As we learned in Chapter 2, a statement equivalent to Coulomb's law is the differential form of Gauss's law,

$$\operatorname{div} \mathbf{E} = \frac{\rho}{\epsilon_0} \tag{9.1}$$

connecting the electric charge density $\rho$ and the electric field $\mathbf{E}$. This holds for moving charges as well as stationary charges. That is, $\rho$ can be a function of time as well as position. As we emphasized in Chapter 5, the fact that Eq. (9.1) holds for moving charges is consistent with *charge invariance:* no matter how an isolated charged particle may be moving, its charge, as measured by the integral of **E** over a surface surrounding it, appears the same in every frame of reference.

Electric charge in motion is electric current. Because charge is never created or destroyed, the charge density $\rho$ and the current density **J** always satisfy the condition

$$\text{div } \mathbf{J} = -\frac{\partial \rho}{\partial t} \tag{9.2}$$

We first wrote down this "equation of continuity" as Eq. (4.10).

If the current density **J** is constant in time, we call it a *stationary current distribution.* The magnetic field of such a current satisfies the equation

$$\text{curl } \mathbf{B} = \mu_0 \mathbf{J} \qquad \text{(stationary current distribution).} \tag{9.3}$$

We worked with this relation in Chapter 6.

Now we are interested in charge distributions and fields that are changing in time. Suppose we have a charge distribution $\rho(x, y, z, t)$ with $\partial \rho / \partial t \neq 0$. For instance, we might have a capacitor that is discharging through a resistor. According to Eq. (9.2), $\partial \rho / \partial t \neq 0$ implies

$$\text{div } \mathbf{J} \neq 0. \tag{9.4}$$

But according to Eq. (9.3), since the divergence of the curl of *any* vector function is identically zero (see Exercise 2.78),

$$\text{div } \mathbf{J} = \frac{1}{\mu_0} \text{div (curl } \mathbf{B}) = 0. \tag{9.5}$$

The contradiction shows that Eq. (9.3) *cannot be correct* for a system in which the charge density is varying in time. Of course, no one claimed it was; a stationary current distribution, for which Eq. (9.3) *does* hold, is one in which not even the current density **J**, let alone the charge density $\rho$, is time-dependent.

The problem can be posed in somewhat different terms by considering the line integral of magnetic field around the wire that carries charge away from the capacitor plate in Fig. 9.1. According to Stokes' theorem,

$$\int_C \mathbf{B} \cdot d\mathbf{l} = \int_S \text{curl } \mathbf{B} \cdot d\mathbf{a}. \tag{9.6}$$

The surface $S$ passes right through the conductor in which a current $I$ is flowing. Inside this conductor, curl **B** has a finite value, namely $\mu_0 \mathbf{J}$, and the integral on the right comes out equal to $\mu_0 I$. That is to say, if the curve $C$ is close to the wire and well away from the capacitor gap, the

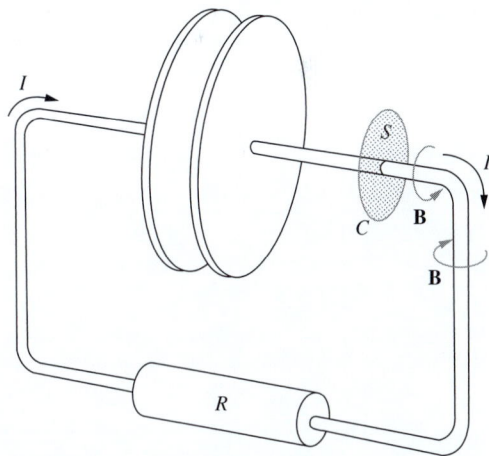

**Figure 9.1.**
Having been charged with the right-hand plate positive, the capacitor is being discharged through the resistor. There is a magnetic field **B** around the wire. The integral of curl **B**, over the surface $S$ that passes through the wire, has the value $\mu_0 I$.

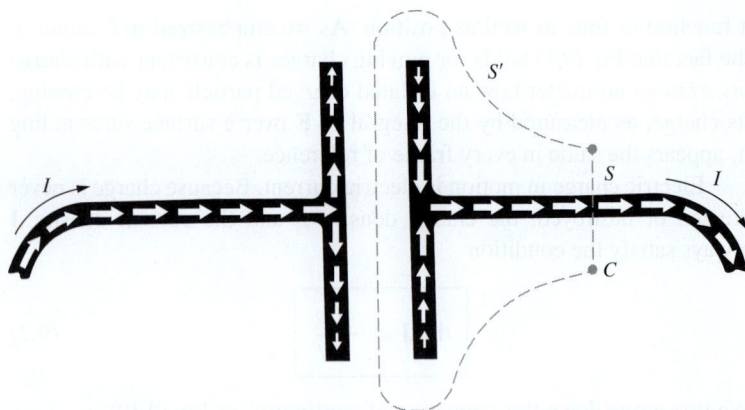

**Figure 9.2.**
The white arrows show the current flow in the conductors. The surface $S'$, which like $S$ has the curve $C$ for its edge, has no current passing through it.

magnetic field there is not different from the field around any wire carrying the same current. Now, the surface $S'$ in Fig. 9.2 is also a surface spanning $C$, and has an equally good claim to be used in the statement of Stokes' theorem, Eq. (9.6). Through this surface, however, there flows *no current at all!* Nevertheless, curl **B** cannot be zero over all of $S'$ without violating Stokes' theorem. Therefore, on $S'$, curl **B** must depend on something other than the current density **J**.

We can only conclude that Eq. (9.3) has to be replaced by some other relation, in the more general situation of changing charge distributions. Let's write instead

$$\text{curl } \mathbf{B} = \mu_0 \mathbf{J} + (?) \tag{9.7}$$

and see if we can discover what (?) must be.

Another line of thought suggests the answer. Remember that the Lorentz-transformation laws of the electromagnetic field, Eq. (6.73), are symmetrical in **E** and $c\mathbf{B}$. Now, in Faraday's induction phenomenon, a *changing magnetic field* is accompanied by an *electric field,* in a manner described by Eq. (7.31):

$$\boxed{\text{curl } \mathbf{E} = -\frac{\partial \mathbf{B}}{\partial t}} \tag{9.8}$$

This is a local relation connecting the electric and magnetic fields in empty space – charges are not directly involved. If symmetry with respect to **E** and $c\mathbf{B}$ is to prevail, we must expect that a *changing electric field* can give rise to a *magnetic field*. There ought to be an induction phenomenon described by an equation like Eq. (9.8), but with the roles of **E** and $c\mathbf{B}$ switched. Writing Eq. (9.8) as curl $\mathbf{E} = -(1/c)\partial(c\mathbf{B})/\partial t$ and then reversing the roles of **E** and $c\mathbf{B}$, we obtain curl $(c\mathbf{B}) = -(1/c)\partial \mathbf{E}/\partial t \implies$ curl $\mathbf{B} = -(1/c^2)\partial \mathbf{E}/\partial t$. It will turn out that we need to change the sign in order for Eq. (9.13) below to work out correctly, but that is all:

$$\text{curl }\mathbf{B} = \frac{1}{c^2}\frac{\partial \mathbf{E}}{\partial t} \implies \text{curl }\mathbf{B} = \mu_0 \epsilon_0 \frac{\partial \mathbf{E}}{\partial t}, \qquad (9.9)$$

where we have used the relation $c^2 = 1/\mu_0\epsilon_0$ from Eq. (6.8). The second of the expressions in Eq. (9.9) is the standard way of writing the relation in SI units.

This provides the missing term that is called for in Eq. (9.7). To try it out, write

$$\boxed{\text{curl }\mathbf{B} = \mu_0\mathbf{J} + \mu_0\epsilon_0 \frac{\partial \mathbf{E}}{\partial t}} \qquad (9.10)$$

and take the divergence of both sides:

$$\text{div (curl }\mathbf{B}) = \text{div }(\mu_0\mathbf{J}) + \text{div}\left(\mu_0\epsilon_0 \frac{\partial \mathbf{E}}{\partial t}\right). \qquad (9.11)$$

The left side is necessarily zero, as already remarked. In the second term on the right we can interchange the order of differentiation with respect to space coordinates and time. Thus

$$\text{div}\left(\mu_0\epsilon_0 \frac{\partial \mathbf{E}}{\partial t}\right) = \mu_0\epsilon_0\frac{\partial}{\partial t}(\text{div }\mathbf{E}) = \mu_0\epsilon_0\frac{\partial}{\partial t}\left(\frac{\rho}{\epsilon_0}\right) = \mu_0\frac{\partial \rho}{\partial t}, \quad (9.12)$$

by Eq. (9.1). The right-hand side of Eq. (9.11) now becomes

$$\mu_0\text{div }\mathbf{J} + \mu_0\frac{\partial \rho}{\partial t}, \qquad (9.13)$$

which is zero by virtue of the continuity condition, Eq. (9.2).

The new term resolves the difficulty raised in Fig. 9.2. As charge flows out of the capacitor, the electric field, which at any instant has the configuration in Fig. 9.3, *diminishes* in intensity. In this case, $\partial \mathbf{E}/\partial t$ points opposite to $\mathbf{E}$. The vector function $\mu_0\epsilon_0(\partial \mathbf{E}/\partial t)$ is represented by the black arrows in Fig. 9.4. With curl $\mathbf{B} = \mu_0\mathbf{J} + \mu_0\epsilon_0(\partial \mathbf{E}/\partial t)$, the integral of curl $\mathbf{B}$ over $S'$ now has the same value as it does over $S$. On $S'$ the second term contributes everything; on $S$ the first term, the term with $\mathbf{J}$, is practically all that counts.

## 9.2 The displacement current

Observe that the vector field $\mu_0\epsilon_0(\partial \mathbf{E}/\partial t)$ appears to form a *continuation* of the conduction current distribution. Maxwell called it the *displacement current,* and the name has stuck although it no longer seems very appropriate. To be precise, we can define a *displacement current density* $\mathbf{J}_\text{d}$, to be distinguished from the conduction current density $\mathbf{J}$, by writing Eq. (9.10) this way:

$$\text{curl }\mathbf{B} = \mu_0(\mathbf{J} + \mathbf{J}_\text{d}), \qquad (9.14)$$

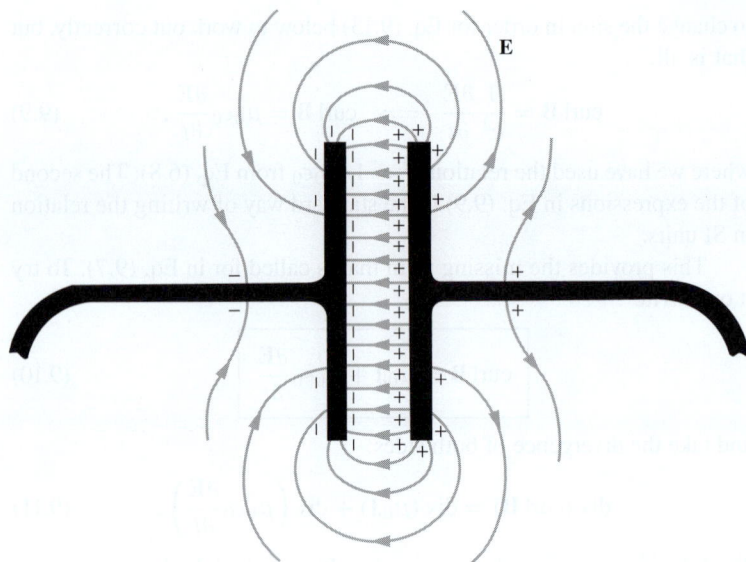

**Figure 9.3.**
The electric field at a particular instant. The magnitude of **E** is decreasing everywhere as time goes on.

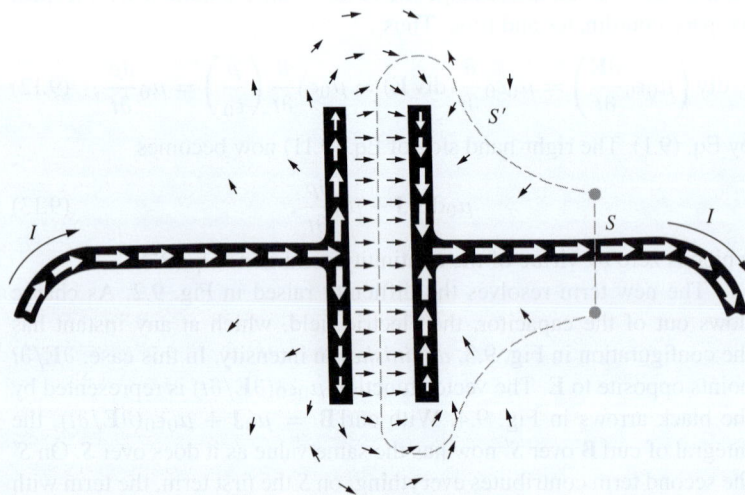

**Figure 9.4.**
The conduction current (white arrows) and the displacement current (black arrows).

where we have used the relations given in Eq. (9.5) to record
of the expressions in Eq. (9.14) as a current density, by writing the relation
in SI units.

This gives the displacement term called for in Eq. (9.7). To try

$$\oint \mathbf{B} \cdot d\mathbf{s} = $$

And then for an ordinary surface $S$ the

$$\oint \mathbf{B} \cdot d\mathbf{s} = \left( \right)$$

by Eq. (9.1). The right-hand side of Eq. (9.11) now becomes

$$\oint \qquad = \qquad \epsilon_0 \qquad \qquad (9.13)$$

and defining

$$\mathbf{J}_\mathrm{d} \equiv \epsilon_0 \frac{\partial \mathbf{E}}{\partial t}. \qquad (9.15)$$

We needed the new term to make the relation between current and magnetic field consistent with the continuity equation, in the case of conduction currents changing in time. If it belongs there, it implies the existence of a new induction effect in which a changing electric field is accompanied by a magnetic field. If the effect is real, why didn't

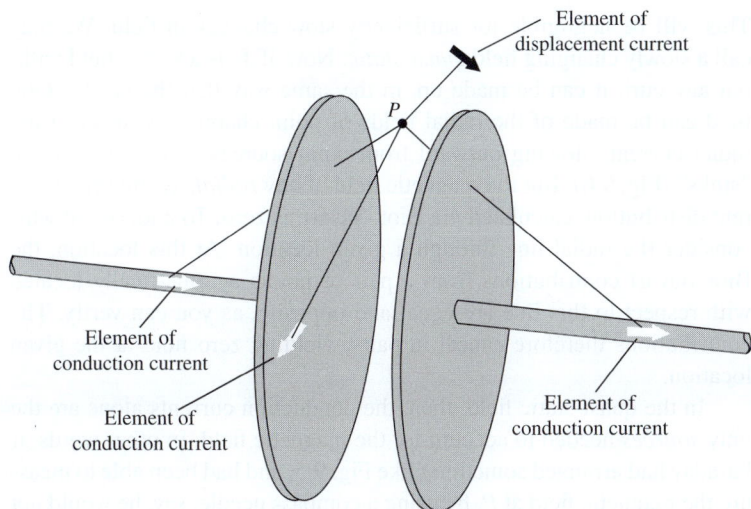

Element of
displacement current

P

Element of
conduction current

Element of
conduction current

Element of
conduction current

**Figure 9.5.**
In the case of slowly varying fields, the total contribution to the magnetic field at any point, from all displacement currents, is zero. The magnetic field at $P$ can be calculated by the Biot–Savart formula applied to conduction current elements only.

Faraday discover it? For one thing, he wasn't looking for it, but there is a more fundamental reason why experiments like Faraday's could not have revealed any new effects attributable to the last term in Eq. (9.10). In any apparatus in which there are changing electric fields, there are present, at the same time, conduction currents, charges in motion. The magnetic field **B**, everywhere around the apparatus, is just about what you would expect those conduction currents to produce. In fact, it is almost exactly the field you would calculate if, ignoring the fact that the circuits may not be continuous, you use the Biot–Savart formula, Eq. (6.49), to find the contribution of each conduction current element to the field at some point in space.

Consider, for example, the point $P$ in the space between our discharging capacitor plates, Fig. 9.5. Each element of conduction current, in the wires and on the surface of the plates, contributes to the field at $P$, according to the Biot–Savart formula. Must we include also the elements of displacement current density $\mathbf{J_d}$? The answer is rather surprising. We *may* include $\mathbf{J_d}$, but if we are careful to include the *entire* displacement current distribution, its net effect will be *zero* for relatively slowly varying fields.

To see why this is so, note that the vector function $\mathbf{J_d}$, indicated by the black arrows in Fig. 9.4, has the same form as the electric field **E** in Fig. 9.3. This electric field is practically an electrostatic field, except that it is slowly dying away. We expect therefore that its curl is practically zero, which would imply that curl $\mathbf{J_d}$ must be practically zero. More precisely, we have curl $\mathbf{E} = -\partial\mathbf{B}/\partial t$, and with the displacement current $\mathbf{J_d} = \epsilon_0(\partial\mathbf{E}/\partial t)$, we get, by interchanging the order of differentiation,

$$\text{curl } \mathbf{J_d} = \epsilon_0 \text{ curl} \left(\frac{\partial\mathbf{E}}{\partial t}\right) = \epsilon_0 \frac{\partial}{\partial t}(\text{curl }\mathbf{E}) = -\epsilon_0 \frac{\partial^2\mathbf{B}}{\partial t^2}. \tag{9.16}$$

(a)

(b)

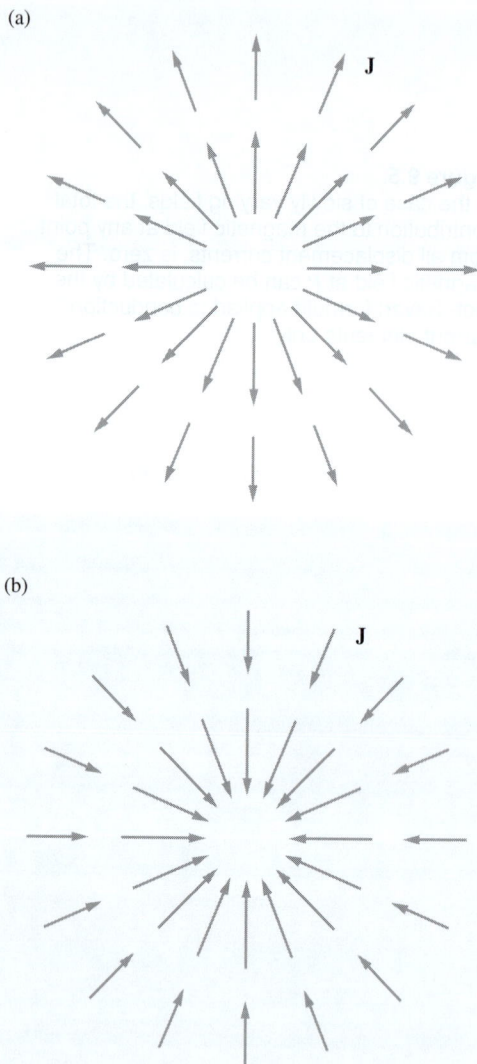

**Figure 9.6.**
Showing what is meant by a radial current distribution. The current density **J** for the point source in (a), or for the point "sink" in (b), is like the electric field of a point charge. Any current distribution with curl **J** = 0 could be made by superposing such sources and sinks, and must therefore have zero magnetic field.

This will be negligible for sufficiently slow changes in field. We may call a slowly changing field *quasi-static*. Now, if $\mathbf{J}_d$ is a vector field without any curl, it can be made up, in the same way that the electrostatic field can be made of the radial fields of point charges, by superposing radial currents flowing outward from point sources or in toward point "sinks" (Fig. 9.6). But the magnetic field of any *radial,* symmetrical current distribution, calculated via Biot–Savart, is zero. To understand why, consider the radial line through a given location. At this location, the Biot–Savart contributions from a pair of points symmetrically located with respect to this line are equal and opposite, as you can verify. The contributions therefore cancel in pairs, yielding zero field at the given location.

In the quasi-static field, then, the conduction currents alone are the only *sources* needed to account for the magnetic field. In other words, if Faraday had arranged something like Fig. 9.5, and had been able to measure the magnetic field at *P*, by using a compass needle, say, he would not have been surprised. He would not have needed to invent a displacement current to explain it.

To see this new induction effect, we need rapidly changing fields. In fact, we need changes to occur in the time it takes light to cross the apparatus. That is why the direct demonstration had to wait for Hertz, whose experiment came roughly 25 years after the law itself had been worked out by Maxwell.

## 9.3 Maxwell's equations

James Clerk Maxwell (1831–1879), after immersing himself in the accounts of Faraday's electrical researches, set out to formulate mathematically a theory of electricity and magnetism. Maxwell could not exploit relativity – that came 50 years later. The electrical constitution of matter was a mystery, the relation between light and electromagnetism unsuspected. Many of the arguments that we have used to make our next step seem obvious were unthinkable then. Nevertheless, as Maxwell's theory developed, the term we have been discussing, $\partial \mathbf{E}/\partial t$, appeared quite naturally in his formulation. He called it the displacement current. Maxwell was concerned with electric fields in solid matter as well as in vacuum, and when he talks about a displacement current he is often including some charge in motion, too. We'll clarify that point in Chapter 10 when we study electric fields in matter. Indeed, Maxwell thought of space itself as a medium, the "aether," so that even in the absence of solid matter, the displacement current was occurring *in* something. But never mind – his mathematical equations were perfectly clear and unambiguous, and his introduction of the displacement current was a *theoretical* discovery of the first rank.

Maxwell's description of the electromagnetic field was essentially complete. We have arrived by different routes at various pieces of it, which we shall now assemble in the form traditionally called *Maxwell's equations:*

$$\operatorname{curl}\mathbf{E} = -\frac{\partial\mathbf{B}}{\partial t}$$

$$\operatorname{curl}\mathbf{B} = \mu_0\epsilon_0\frac{\partial\mathbf{E}}{\partial t} + \mu_0\mathbf{J}$$

$$\operatorname{div}\mathbf{E} = \frac{\rho}{\epsilon_0}$$

$$\operatorname{div}\mathbf{B} = 0$$

(9.17)

These are written for fields in the presence of electric charge of density $\rho$ and electric current, that is, charge in motion, of density $\mathbf{J}$.

The first equation is Faraday's *law of induction.* The second expresses the dependence of the magnetic field on the *displacement current* density, or rate of change of electric field, and on the *conduction current* density, or rate of motion of charge. (If $\partial\mathbf{E}/\partial t = 0$, this equation reduces to Ampère's law.) The third equation is equivalent to Coulomb's law; it is the differential form of Gauss's law. The fourth equation states that there are no sources of magnetic field *except* currents; that is, there are no magnetic monopoles. We shall have more to say about this aspect of Nature in Chapter 11.

Note that the lack of symmetry in these equations, with respect to $\mathbf{B}$ and $\mathbf{E}$ (or rather $c\mathbf{B}$ and $\mathbf{E}$; see Eq. (9.19)), is entirely due to the presence of electric charge and electric conduction current. In empty space, the terms with $\rho$ and $\mathbf{J}$ are zero, and Maxwell's equations become

$$\operatorname{curl}\mathbf{E} = -\frac{\partial\mathbf{B}}{\partial t} \qquad \operatorname{div}\mathbf{E} = 0$$

$$\operatorname{curl}\mathbf{B} = \mu_0\epsilon_0\frac{\partial\mathbf{E}}{\partial t} \qquad \operatorname{div}\mathbf{B} = 0$$

(9.18)

Remembering that $\mu_0\epsilon_0 = 1/c^2$, we can write the two "induction" equations as

$$\operatorname{curl}\mathbf{E} = -\frac{1}{c}\frac{\partial(c\mathbf{B})}{\partial t} \qquad \text{and} \qquad \operatorname{curl}(c\mathbf{B}) = \frac{1}{c}\frac{\partial\mathbf{E}}{\partial t},$$

(9.19)

where the symmetry between $c\mathbf{B}$ and $\mathbf{E}$ is clear. This symmetry, after all, is what led us to the displacement current in the first place; see the paragraph preceding Eq. (9.9).

In Eq. (9.18) the displacement current term is all important. Its presence, along with its counterpart in the first equation, implies the possibility of *electromagnetic waves,* as we will see in Section 9.4. Recognizing this, Maxwell went on to develop with brilliant success an electromagnetic theory of light.

In Gaussian units Maxwell's equations look like this:

$$\text{curl } \mathbf{E} = -\frac{1}{c}\frac{\partial \mathbf{B}}{\partial t}$$

$$\text{curl } \mathbf{B} = \frac{1}{c}\frac{\partial \mathbf{E}}{\partial t} + \frac{4\pi}{c}\mathbf{J} \qquad (9.20)$$

$$\text{div } \mathbf{E} = 4\pi\rho$$

$$\text{div } \mathbf{B} = 0$$

And in empty space, with $\rho$ and $\mathbf{J}$ equal to zero, these become

$$\text{curl } \mathbf{E} = -\frac{1}{c}\frac{\partial \mathbf{B}}{\partial t} \qquad \text{div } \mathbf{E} = 0$$

$$\text{curl } \mathbf{B} = \frac{1}{c}\frac{\partial \mathbf{E}}{\partial t} \qquad \text{div } \mathbf{B} = 0 \qquad (9.21)$$

## 9.4 An electromagnetic wave

We are going to construct a rather simple electromagnetic field that will satisfy Maxwell's equations for empty space, Eq. (9.18). Suppose there is an electric field $\mathbf{E}$, everywhere parallel to the $z$ axis, whose intensity depends only on the space coordinate $y$ and the time $t$. Let the dependence have this particular form:[1]

$$\mathbf{E} = \hat{\mathbf{z}}E_0 \sin(y - vt), \qquad (9.22)$$

in which $E_0$ and $v$ are simply constants. This field fills all space – at least all the space we are presently concerned with. We'll need a magnetic field, too. We shall assume that it has an $x$ component only, with a dependence on $y$ and $t$ similar to that of $E_z$:

$$\mathbf{B} = \hat{\mathbf{x}}B_0 \sin(y - vt), \qquad (9.23)$$

where $B_0$ is another constant.

Figure 9.7 may help you to visualize these fields. It is difficult to represent graphically two such fields filling all space. Remember that nothing varies with $x$ or $z$; whatever is happening at a point on the $y$ axis is happening everywhere on the perpendicular plane through that point. As time goes on, the entire field pattern slides steadily to the right,

---

[1] There is technically an issue with the units here, because the argument of the sine function should be dimensionless. We should really be writing it as $\sin(ky - \omega t)$ or something similar; see the example in Section 9.5. However, the present form makes things a little less cluttered, without affecting the final results.

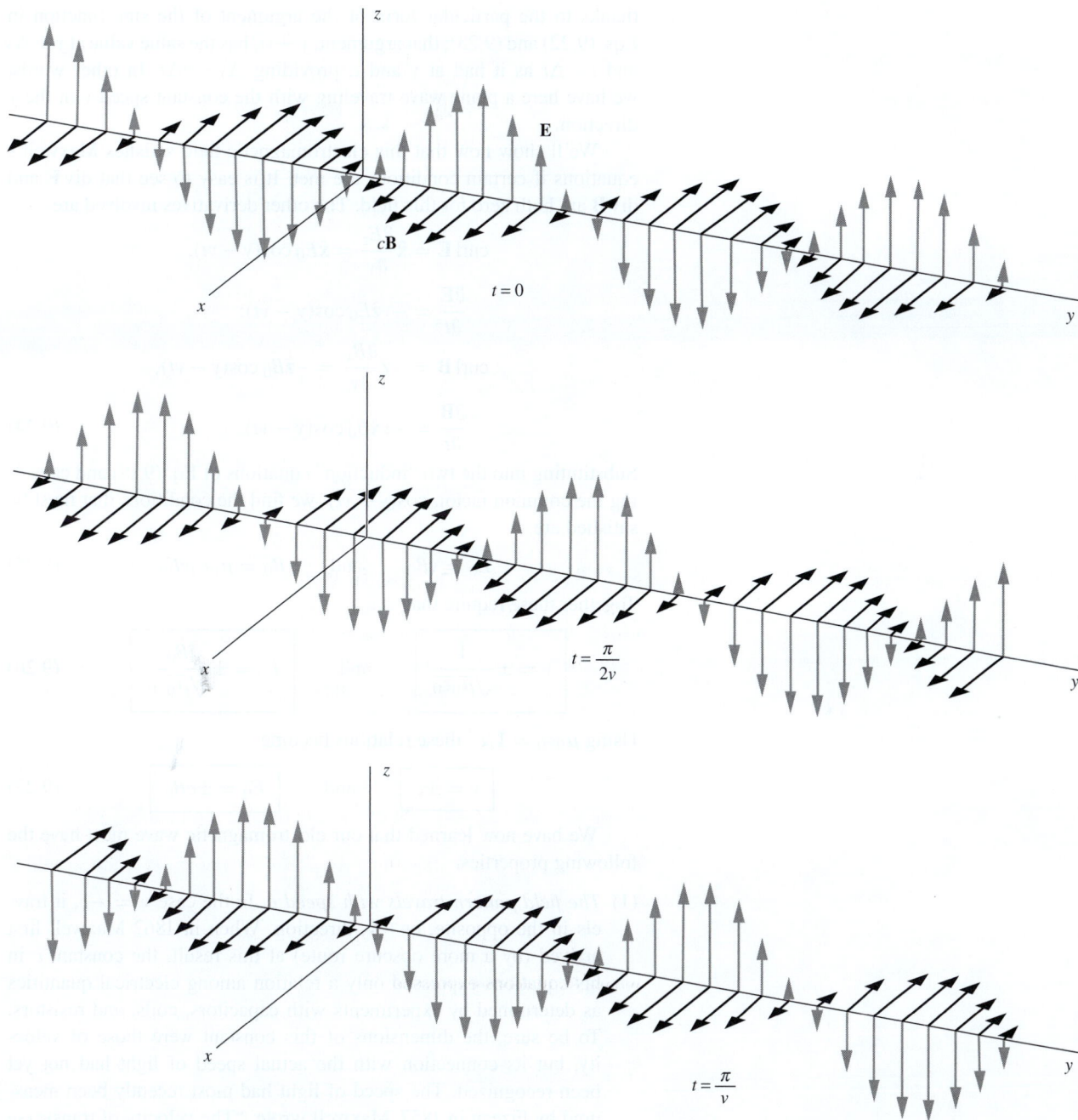

**Figure 9.7.**
The wave described by Eqs. (9.22) and (9.23) is shown at three different times. It is traveling to the right, in the positive $y$ direction.

thanks to the particular form of the argument of the sine function in Eqs. (9.22) and (9.23); that argument, $y - vt$, has the same value at $y + \Delta y$ and $t + \Delta t$ as it had at $y$ and $t$, providing $\Delta y = v\Delta t$. In other words, we have here a plane wave traveling with the constant speed $v$ in the $\hat{\mathbf{y}}$ direction.

We'll show now that this electromagnetic field satisfies Maxwell's equations if certain conditions are met. It is easy to see that div $\mathbf{E}$ and div $\mathbf{B}$ are both zero for this field. The other derivatives involved are

$$\text{curl } \mathbf{E} = \hat{\mathbf{x}}\frac{\partial E_z}{\partial y} = \hat{\mathbf{x}}E_0 \cos(y - vt),$$

$$\frac{\partial \mathbf{E}}{\partial t} = -v\hat{\mathbf{z}}E_0 \cos(y - vt);$$

$$\text{curl } \mathbf{B} = -\hat{\mathbf{z}}\frac{\partial B_x}{\partial y} = -\hat{\mathbf{z}}B_0 \cos(y - vt),$$

$$\frac{\partial \mathbf{B}}{\partial t} = -v\hat{\mathbf{x}}B_0 \cos(y - vt). \tag{9.24}$$

Substituting into the two "induction" equations of Eq. (9.18) and canceling the common factor, $\cos(y - vt)$, we find the conditions that must be satisfied are

$$E_0 = vB_0 \quad \text{and} \quad B_0 = \mu_0\epsilon_0 vE_0. \tag{9.25}$$

Together these require that

$$\boxed{v = \pm\frac{1}{\sqrt{\mu_0\epsilon_0}}} \quad \text{and} \quad \boxed{E_0 = \pm\frac{B_0}{\sqrt{\mu_0\epsilon_0}}} \tag{9.26}$$

Using $\mu_0\epsilon_0 = 1/c^2$ these relations become

$$\boxed{v = \pm c} \quad \text{and} \quad \boxed{E_0 = \pm cB_0} \tag{9.27}$$

We have now learned that our electromagnetic wave must have the following properties.

(1) *The field pattern travels with speed c.* In the case $v = -c$, it travels in the opposite, or $-\hat{\mathbf{y}}$, direction. When in 1862 Maxwell first arrived (by a more obscure route) at this result, the constant $c$ in his equations expressed only a relation among electrical quantities as determined by experiments with capacitors, coils, and resistors. To be sure, the dimensions of this constant were those of velocity, but its connection with the actual speed of light had not yet been recognized. The speed of light had most recently been measured by Fizeau in 1857. Maxwell wrote, "The velocity of transverse undulations in our hypothetical medium, calculated from the electromagnetic experiments of MM. Kohlrausch and Weber, agrees so exactly with the velocity of light calculated from the optical experiments of M. Fizeau, that we can scarcely avoid the inference that *light consists in the transverse undulations of the same medium which*

*is the cause of electric and magnetic phenomena."* The italics are Maxwell's.

(2) *At every point in the wave at any instant of time, the electric field strength equals c times the magnetic field strength.* In our SI units, $B$ is expressed in tesla and $E$ in volts/meter. If the electric field strength is 1 volt/meter, the associated magnetic field strength is $1/(3 \cdot 10^8) = 3.33 \cdot 10^{-9}$ tesla. (In Gaussian units, the electric and magnetic field strengths are equal, with no need for the factor of $c$.)

(3) *The electric field and the magnetic field are perpendicular to one another and to the direction of travel, or propagation.* To be sure, we had already assumed this when we constructed our example, but it is not hard to show that it is a necessary condition, given that the fields do not depend on the coordinates perpendicular to the direction of propagation. Note that, if $v = -c$, which would make the direction of propagation $-\hat{\mathbf{y}}$, we must have $E_0 = -cB_0$. This preserves the handedness of the essential triad of directions, the direction of $\mathbf{E}$, the direction of $\mathbf{B}$, and the direction of propagation. We can describe this without reference to a particular coordinate frame as follows: the wave always travels in the direction of the vector $\mathbf{E} \times \mathbf{B}$.

Any plane electromagnetic wave in empty space has these three properties.

## 9.5 Other waveforms; superposition of waves

In the example we have just studied, the function $\sin(y - vt)$ was chosen merely for its simplicity. The "waviness" of the sinusoidal function has *nothing to do* with the essential property of wave motion, which is the propagation unchanged of a form or pattern – *any* pattern. It was not the nature of the function but the way $y$ and $t$ were combined in its argument that caused the pattern to propagate. If we replace the sine function by *any* other function, $f(y - vt)$, we obtain a pattern that travels with speed $v$ in the $\hat{\mathbf{y}}$ direction. Moreover, Eq. (9.25) will apply as before (as you should check by working out the steps analogous to those in Eq. (9.24)), and our wave will have the three general properties just listed.

Here is another example, the plane electromagnetic wave pictured in Fig. 9.8, which is described mathematically as follows:

$$\mathbf{E} = \frac{E_0\hat{\mathbf{y}}}{1 + \frac{(x + ct)^2}{\ell^2}} \, , \qquad \mathbf{B} = \frac{-(E_0/c)\hat{\mathbf{z}}}{1 + \frac{(x + ct)^2}{\ell^2}} \, , \qquad (9.28)$$

where $\ell$ is a fixed length that we have chosen as $\ell = 1$ foot for the purposes of drawing Fig. 9.8. (The speed of light is very nearly 1 foot/nanosecond.) This electromagnetic field satisfies Maxwell's equations, Eq. (9.18). It is a *plane* wave because nothing depends on $y$ or $z$. It is traveling in the direction $-\hat{\mathbf{x}}$, as we recognize at once from the $+$ sign in the argument $x + ct$. That is indeed the direction of $\mathbf{E} \times \mathbf{B}$. In this

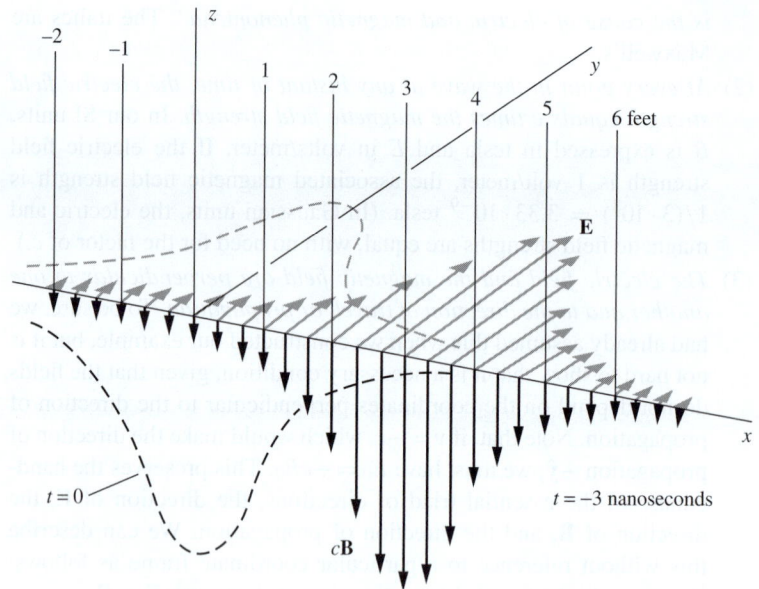

**Figure 9.8.**
The wave described by Eq. (9.28) is traveling in the negative $x$ direction. It is shown 3 nanoseconds before its peak passes the origin.

wave nothing is oscillating or alternating; it is simply an electromagnetic pulse with long tails. At time $t = 0$, the maximum field strengths, $E = E_0$ (in volts/meter) and $B = E_0/c$ (which correctly has units of tesla) will be experienced by an observer at the origin, or at any other point on the $yz$ plane. In Fig. 9.8 we have shown the field as it was at $t = -3$ nanoseconds, with the distances marked off in feet.

Maxwell's equations for **E** and **B** in empty space are linear. The superposition of two solutions is also a solution. Any number of electromagnetic waves can propagate through the same region without affecting one another. The field **E** at a space-time point is the vector sum of the electric fields of the individual waves, and the same goes for **B**.

**Example (Standing wave)**    An important example is the superposition of two similar plane waves traveling in opposite directions. Consider a wave traveling in the $\hat{\mathbf{y}}$ direction, described by

$$\mathbf{E}_1 = \hat{\mathbf{z}}E_0 \sin \frac{2\pi}{\lambda}(y - ct), \qquad \mathbf{B}_1 = \hat{\mathbf{x}}\frac{E_0}{c} \sin \frac{2\pi}{\lambda}(y - ct). \qquad (9.29)$$

This wave differs in only minor ways from the wave in Eqs. (9.22) and (9.23). We have introduced the wavelength $\lambda$ of the periodic function, and we have used $B_0 = E_0/c$.

Now consider another wave:

$$\mathbf{E}_2 = \hat{\mathbf{z}}E_0 \sin \frac{2\pi}{\lambda}(y + ct), \qquad \mathbf{B}_2 = -\hat{\mathbf{x}}\frac{E_0}{c} \sin \frac{2\pi}{\lambda}(y + ct). \qquad (9.30)$$

This is a wave with the same amplitude and wavelength, but propagating in the $-\hat{\mathbf{y}}$ direction. With the two waves both present, Maxwell's equations are still satisfied, the electric and magnetic fields now being

$$\mathbf{E} = \mathbf{E}_1 + \mathbf{E}_2 = \hat{\mathbf{z}}E_0 \left[ \sin\left(\frac{2\pi y}{\lambda} - \frac{2\pi ct}{\lambda}\right) + \sin\left(\frac{2\pi y}{\lambda} + \frac{2\pi ct}{\lambda}\right) \right],$$

$$\mathbf{B} = \mathbf{B}_1 + \mathbf{B}_2 = \hat{\mathbf{x}}\frac{E_0}{c} \left[ \sin\left(\frac{2\pi y}{\lambda} - \frac{2\pi ct}{\lambda}\right) - \sin\left(\frac{2\pi y}{\lambda} + \frac{2\pi ct}{\lambda}\right) \right]. \quad (9.31)$$

Remembering the formula for the sine of the sum of two angles, you can easily reduce Eq. (9.31) to

$$\mathbf{E} = 2\hat{\mathbf{z}}E_0 \sin\frac{2\pi y}{\lambda} \cos\frac{2\pi ct}{\lambda}, \qquad \mathbf{B} = -2\hat{\mathbf{x}}\frac{E_0}{c} \cos\frac{2\pi y}{\lambda} \sin\frac{2\pi ct}{\lambda}. \quad (9.32)$$

The field described by Eq. (9.32) is called a *standing wave*. Figure 9.9 suggests what it looks like at different times. The factor $c/\lambda$ is the *frequency* (in time) with which the field oscillates at any position $x$, and $2\pi c/\lambda$ is the corresponding angular frequency. According to Eq. (9.32), whenever $2ct/\lambda$ equals an integer, which happens every half-period, we have $\sin 2\pi ct/\lambda = 0$, and the magnetic field $\mathbf{B}$ vanishes *everywhere*. On the other hand, whenever $2ct/\lambda$ equals an integer plus one-half, we have $\cos 2\pi ct/\lambda = 0$, and the electric field vanishes everywhere. The maxima of $\mathbf{B}$ and the maxima of $\mathbf{E}$ occur at different places as well as at different times. In contrast with the traveling wave, the standing wave has its electric and magnetic fields "out of step" in both space and time.

In the above standing wave, note that $\mathbf{E} = 0$ *at all times* on the plane $y = 0$ and on every other plane for which $y$ equals an integral number of half-wavelengths. Imagine that we could cover the $xz$ plane at $y = 0$ with a sheet of perfectly conducting metal. At the surface of a perfect conductor, the electric field component parallel to the surface must be zero – otherwise an infinite current would flow. That imposes a drastic *boundary condition* on any electromagnetic field in the surrounding space. But our standing wave, which is described by Eq. (9.32), *already satisfies* that condition, as well as satisfying Maxwell's equations in the entire space $y > 0$. Therefore it provides a ready-made solution to the problem of a plane electromagnetic wave reflected, at normal incidence, from a flat conducting mirror (see Fig. 9.10). The incident wave is described by Eq. (9.30), for $y > 0$, the reflected wave by Eq. (9.29). There is no field at all behind the mirror, or if there is, it has nothing to do with the field in front. Immediately in front of the mirror there is a magnetic field parallel to the surface, given by Eq. (9.32): $\mathbf{B} = -2\hat{\mathbf{x}}(E_0/c)\sin(2\pi ct/\lambda)$. The jump in $\mathbf{B}$ from this value in front of the conducting sheet to zero behind shows that an alternating current must be flowing in the sheet (see Section 6.6).

You could install a conducting sheet at any other plane where $\mathbf{E}$, as given by Eq. (9.32), is permanently zero, and thus trap an electromagnetic standing wave between two mirrors. That arrangement has many applications, including lasers. In fact, with an understanding of the

**Figure 9.9 (see p. 444).**
A standing wave, resulting from the superposition of a wave traveling in the positive $y$ direction, Eq. (9.29), and a similar wave traveling in the negative $y$ direction, Eq. (9.30). Beginning with the top figure, the fields are shown at four different times, separated successively by one-eighth of a full period.

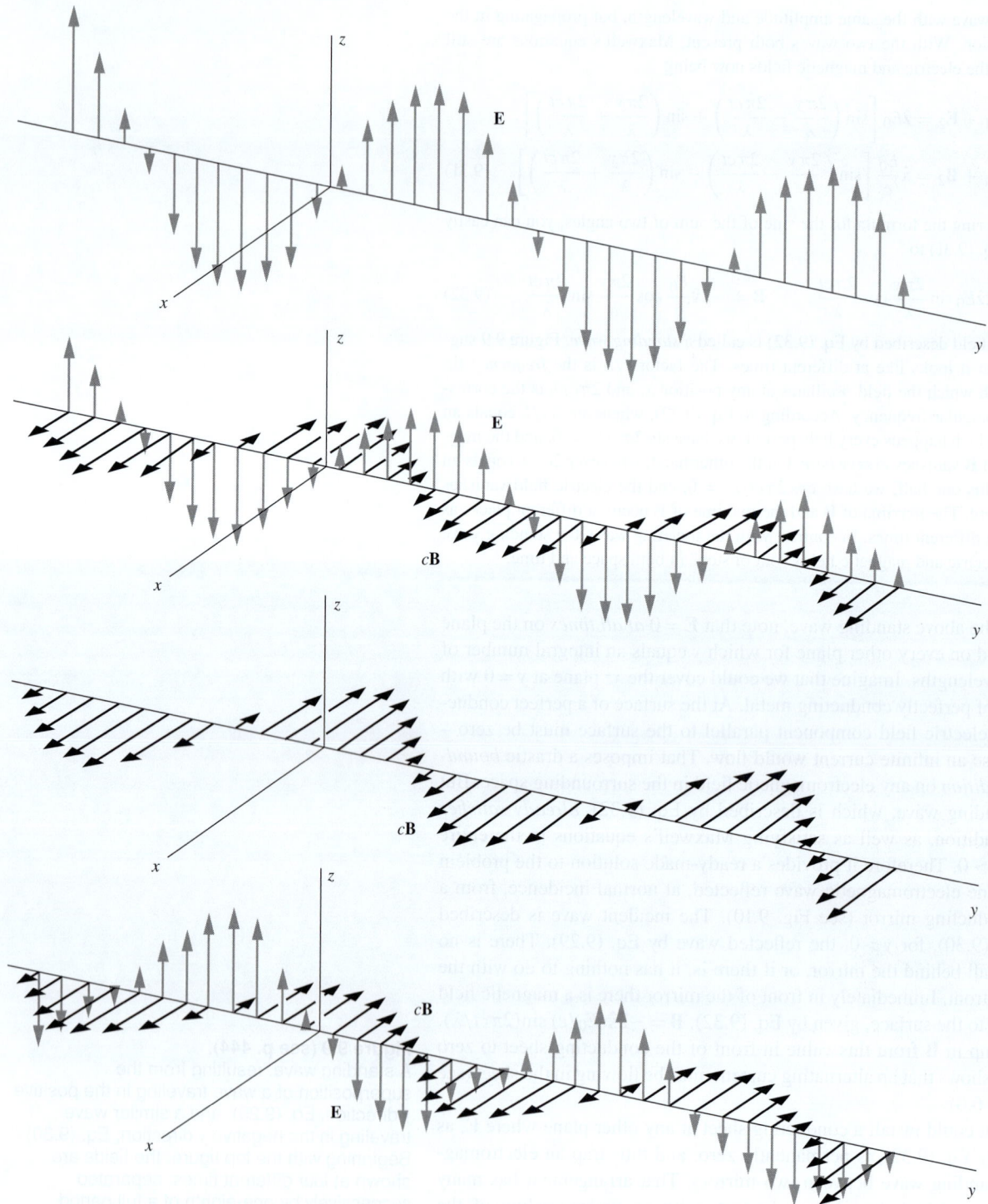

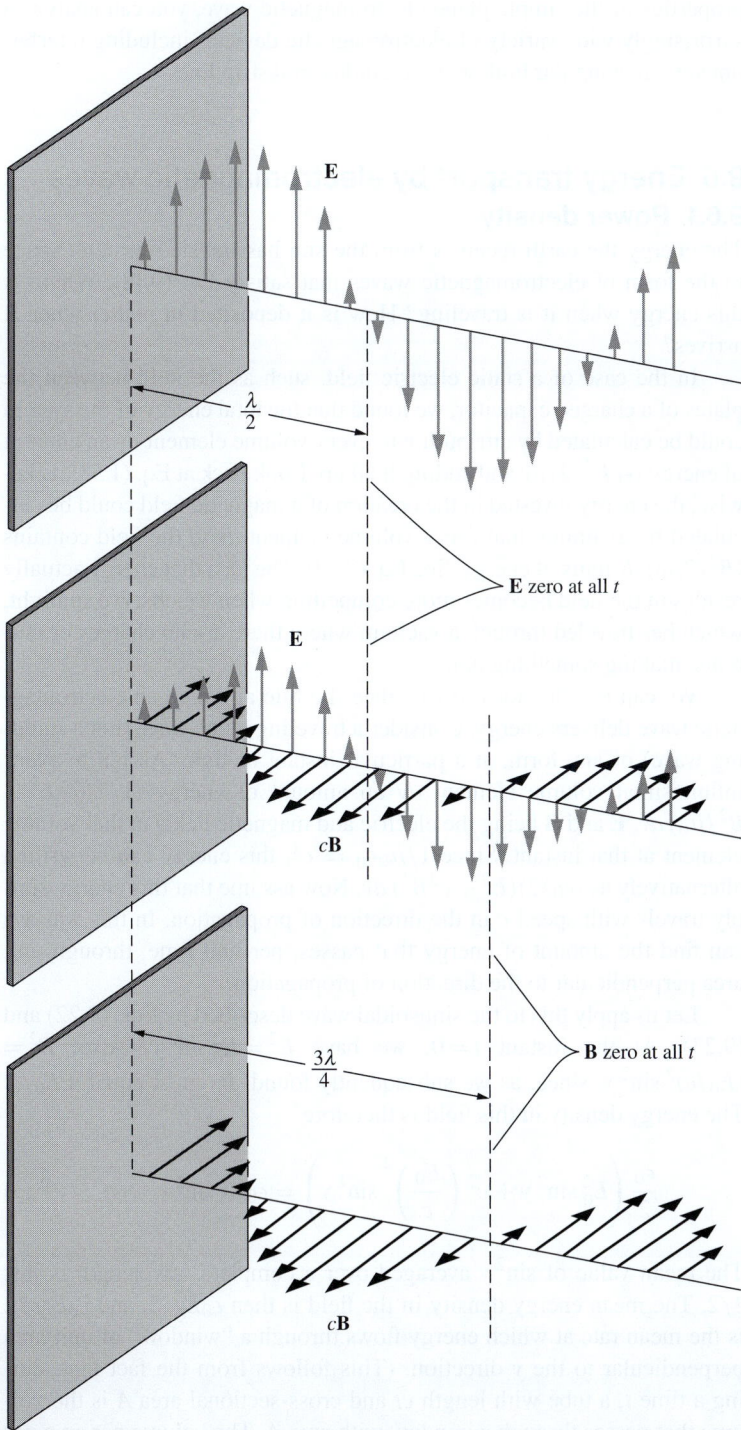

**E** zero at all *t*

**B** zero at all *t*

**Figure 9.10.**
A standing wave produced by reflection at a
perfectly conducting sheet.

properties of the simple plane electromagnetic wave, you can analyze a surprisingly wide variety of electromagnetic devices, including interferometers, rectangular hollow wave guides, and strip lines.

## 9.6 Energy transport by electromagnetic waves

### 9.6.1 Power density

The energy the earth receives from the sun has traveled through space in the form of electromagnetic waves that satisfy Eq. (9.18). Where *is* this energy when it is traveling? How is it deposited in matter when it arrives?

In the case of a static electric field, such as the field between the plates of a charged capacitor, we found that the total energy of the system could be calculated by attributing to every volume element $dv$ an amount of energy $(\epsilon_0 E^2/2)\, dv$ and adding it all up. Look back at Eq. (1.53). Likewise, the energy invested in the creation of a magnetic field could be calculated by assuming that every volume element $dv$ in the field contains $(B^2/2\mu_0)\, dv$ units of energy. See Eq. (7.79). The idea that energy actually resides in the field becomes more compelling when we observe sunlight, which has traveled through a vacuum where there are no charges or currents, making something hot.

We can use this idea to calculate the rate at which an electromagnetic wave delivers energy. Consider a traveling plane wave (not a standing wave) of any form, at a particular instant of time. Assign to every infinitesimal volume element $dv$ an amount of energy $(1/2)(\epsilon_0 E^2 + B^2/\mu_0)\, dv$, **E** and **B** being the electric and magnetic fields in that volume element at that instant. Since $1/\mu_0\epsilon_0 = c^2$, this energy can be written alternatively as $(\epsilon_0/2)(E^2 + c^2 B^2)\, dv$. Now assume that this energy simply travels with speed $c$ in the direction of propagation. In this way we can find the amount of energy that passes, per unit time, through unit area perpendicular to the direction of propagation.

Let us apply this to the sinusoidal wave described by Eqs. (9.22) and (9.23). At the instant $t = 0$, we have $E^2 = E_0^2 \sin^2 y$. Also, $B^2 = (E_0/c)^2 \sin^2 y$, since, as we subsequently found, $B_0$ must equal $\pm E_0/c$. The energy density in this field is therefore

$$\frac{\epsilon_0}{2}\left(E_0^2 \sin^2 y + c^2 \left(\frac{E_0}{c}\right)^2 \sin^2 y\right) = \epsilon_0 E_0^2 \sin^2 y. \tag{9.33}$$

The mean value of $\sin^2 y$ averaged over a complete wavelength is just $1/2$. The mean energy density in the field is then $\epsilon_0 E_0^2/2$, and $\epsilon_0 E_0^2 c/2$ is the mean rate at which energy flows through a "window" of unit area perpendicular to the $y$ direction. (This follows from the fact that, during a time $t$, a tube with length $ct$ and cross-sectional area $A$ is the volume that passes through a window with area $A$. The volume per area per time is therefore $(ct)A/At = c$.) We can say more generally that, for any continuous, repetitive wave, whether sinusoidal or not, the rate of energy flow per unit area, which we call the *power density S*, is given by

$$S = \epsilon_0 \overline{E^2} c \qquad (9.34)$$

Here $\overline{E^2}$ is the mean square electric field strength, which was $E_0^2/2$ for the sinusoidal wave of amplitude $E_0$. $S$ will be in joules per second per square meter, or equivalently watts per square meter, if $E$ is in volts per meter and $c$ is in meters per second.

In Gaussian units the formula for power density is

$$S = \frac{\overline{E^2} c}{4\pi}, \qquad (9.35)$$

where $S$ is in ergs per second per square centimeter if $E$ is in statvolts per centimeter and $c$ is in centimeters per second.

If you want to write Eq. (9.34) without reference to $c$, then substituting $c = 1/\sqrt{\mu_0 \epsilon_0}$ yields

$$S = \frac{\overline{E^2}}{\sqrt{\mu_0/\epsilon_0}} \qquad (9.36)$$

This expression for $S$ is based only on the physics that was known in 1861 when Maxwell wrote down his set of equations. That is, it invokes nothing about the nature of light; you can repeat the above derivation by using the expression for $v$ in Eq. (9.26) without introducing the speed of light, $c$. The fact that $1/\sqrt{\mu_0 \epsilon_0}$ can indeed be replaced by $c$ was conjectured by Maxwell in 1862, demonstrated experimentally by Hertz in 1888, and explained theoretically by Einstein in 1905 through his special theory of relativity. The last of these routes was the one we took in Chapters 5 and 6, where we showed that $\mu_0 = 1/\epsilon_0 c^2$.

The constant $\sqrt{\mu_0/\epsilon_0}$ in Eq. (9.36) has the dimensions of resistance, and its value is 376.73 ohms. Rounding it off to 377 ohms, we have a convenient and easily remembered formula:

$$S(\text{watts/meter}^2) = \frac{\overline{E^2}(\text{volts/meter})^2}{377 \text{ ohms}} \qquad (9.37)$$

The units here reduce to: watts $=$ volt$^2$/ohm, which are the same as in the standard $P = V^2/R$ expression for the power in an ordinary resistor. If you need help in remembering the number 377, it happens to be the number of radians per second in 60 hertz, and also the 14th Fibonacci number.

When the electromagnetic wave encounters an electrical conductor, the electric field causes currents to flow. This generally results in energy being dissipated within the conductor at the expense of the energy in the wave. The total reflection of the incident wave in Fig. 9.10 was a special case in which the conductivity of the reflecting surface was infinite.

If the resistivity of the reflector is not zero, the amplitude of the reflected wave will be less than that of the incident wave. Aluminum, for example, reflects visible light, at normal incidence, with about 92 percent efficiency. That is, 92 percent of the incident energy is reflected, the amplitude of the reflected wave being $\sqrt{0.92}$ or 0.96 times that of the incident wave. The lost 8 percent of the incident energy ends up as heat in the aluminum, where the current driven by the electric field of the wave encounters ohmic resistance. What counts, of course, is the resistivity of aluminum at the frequency of the light wave, in this case about $5 \cdot 10^{14}$ Hz. That may be somewhat different from the dc or low-frequency resistivity of the metal. Still, the reflectivity of most metals for visible light is essentially due to the same highly mobile conduction electrons that make metals good conductors of steady current. It is no accident that good conductors are generally shiny. But why clean copper looks reddish while aluminum looks "silvery" can't be explained without a detailed theory of each metal's electronic structure.

Energy can also be absorbed when an electromagnetic wave meets nonconducting matter. Little of the light that strikes a black rubber tire is reflected, although the rubber is an excellent insulator for low-frequency electric fields. Here the dissipation of the electromagnetic energy involves the action of the high-frequency electric field on the electrons in the molecules of the material. In the broadest sense, that applies to the absorption of light in everything around us, including the retina of the eye.

Some insulators transmit electromagnetic waves with very little absorption. The transparency of glass for visible light, with which we are so familiar, is really a remarkable property. In the purest glass fibers used for optical transmission of audio and video signals, a wave travels as much as a hundred kilometers, or more than $10^{11}$ wavelengths, before most of the energy is lost. However transparent a material medium may be, the propagation of an electromagnetic wave within the medium differs in essential ways from propagation through the vacuum. The matter interacts with the electromagnetic field. To take that interaction into account, Eq. (9.18) must be modified in a way that will be explained in Chapter 10.

### 9.6.2 The Poynting vector

With the help of Maxwell's equations, we can produce a more general version of the power density given in Eq. (9.34). That result was valid only for traveling waves. The present result will be valid for arbitrary electromagnetic fields. Furthermore, it will be valid as a function of time (and space), and not just as a time average. As above, our starting point will be the fact that the energy density of an electromagnetic field, which we label $\mathcal{U}$, is given by $\epsilon_0 E^2/2 + B^2/2\mu_0$. Consider the rate of change of $\mathcal{U}$. If we write $E^2$ and $B^2$ as $\mathbf{E} \cdot \mathbf{E}$ and $\mathbf{B} \cdot \mathbf{B}$, then

$$\frac{\partial \mathcal{U}}{\partial t} = \epsilon_0 \frac{\partial \mathbf{E}}{\partial t} \cdot \mathbf{E} + \frac{1}{\mu_0} \frac{\partial \mathbf{B}}{\partial t} \cdot \mathbf{B}. \tag{9.38}$$

The product rule works for vectors just as it does for regular functions, as you can check by explicitly writing out the Cartesian components. We can rewrite the time derivatives here with the help of the two "induction" Maxwell equations in free space, $\nabla \times \mathbf{B} = \mu_0 \epsilon_0 \, \partial \mathbf{E}/\partial t$ and $\nabla \times \mathbf{E} = -\partial \mathbf{B}/\partial t$. This yields

$$\frac{\partial \mathcal{U}}{\partial t} = \frac{1}{\mu_0}(\nabla \times \mathbf{B}) \cdot \mathbf{E} - \frac{1}{\mu_0}(\nabla \times \mathbf{E}) \cdot \mathbf{B}. \qquad (9.39)$$

The right-hand side of this expression conveniently has the same form as the right-hand side of the vector identity

$$\nabla \cdot (\mathbf{C} \times \mathbf{D}) = (\nabla \times \mathbf{C}) \cdot \mathbf{D} - (\nabla \times \mathbf{D}) \cdot \mathbf{C}. \qquad (9.40)$$

Hence $\partial \mathcal{U}/\partial t = (1/\mu_0)\nabla \cdot (\mathbf{B} \times \mathbf{E})$. For reasons that will become clear, let's switch the order of $\mathbf{B}$ and $\mathbf{E}$, which brings in a minus sign. We then have

$$\frac{\partial \mathcal{U}}{\partial t} = -\frac{1}{\mu_0}\nabla \cdot (\mathbf{E} \times \mathbf{B}). \qquad (9.41)$$

If we now define the *Poynting vector* $\mathbf{S}$ by

$$\boxed{\mathbf{S} \equiv \frac{\mathbf{E} \times \mathbf{B}}{\mu_0}} \qquad \text{(Poynting vector)}, \qquad (9.42)$$

then we can write our result as

$$-\frac{\partial \mathcal{U}}{\partial t} = \nabla \cdot \mathbf{S}. \qquad (9.43)$$

This equation should remind you of another one we have encountered. It has exactly the same form as the continuity equation,

$$-\frac{\partial \rho}{\partial t} = \nabla \cdot \mathbf{J}. \qquad (9.44)$$

Therefore, just as $\mathbf{J}$ gives the current density (the flow of charge per time per area), we can likewise say that $\mathbf{S}$ gives the power density (the flow of energy per time per area). Equivalently, Eqs. (9.43) and (9.44) are the statements of conservation of energy and charge, respectively. Energy (or charge) can't just disappear; if the energy in a given region decreases, it must be the case that energy flowed out of that region, and into another region.

   If you don't trust the analogy with $\mathbf{J}$, you can work with the integral form of Eq. (9.43). The integral of the energy density $\mathcal{U}$ over a given volume $V$ is simply the total energy $U$ contained in that volume. So we have

$$\frac{dU}{dt} = \frac{d}{dt}\int_V \mathcal{U}\,dv = \int_V \frac{\partial \mathcal{U}}{\partial t}\,dv = -\int_V \nabla \cdot \mathbf{S}\,dv = -\int_S \mathbf{S} \cdot d\mathbf{a}, \quad (9.45)$$

where we have used the divergence theorem. This shows that the rate of change of the energy in a given volume $V$ equals the negative of the flux of the vector $\mathbf{S}$ outward through the closed surface $S$ that bounds $V$. (Remember that $d\mathbf{a}$ is defined to be the outward-pointing normal.) The minus sign in Eq. (9.45) makes sense; a positive outward flux of $\mathbf{S}$ means that $U$ is decreasing. Since Eq. (9.45) holds for an arbitrary closed volume, the natural interpretation of $\mathbf{S}$ is that it gives the rate of energy flow per area through any surface, closed or not.

The Poynting vector $\mathbf{S}$ gives the power density for an arbitrary electromagnetic field, not just for the special case of a traveling wave. For any electromagnetic field, at any given point at any instant in time, the direction of $\mathbf{S}$ gives the direction of the energy flow, and the magnitude of $\mathbf{S}$ gives the energy per time per area flowing through a small frame. The units of $\mathbf{S}$ are joules per second per square meter, or watts per square meter.

In the special case of a traveling wave (sinusoidal or not), we know from the third property listed in Section 9.4 that the velocity points in the direction of $\mathbf{E} \times \mathbf{B}$. This equals the direction of $\mathbf{S}$, as must be the case. We also know that a traveling wave has $\mathbf{B}$ perpendicular to $\mathbf{E}$, with $B = E/c$. The magnitude of $\mathbf{S}$ is therefore $S = E(E/c)/\mu_0$. Using $\mu_0 = 1/\epsilon_0 c^2$, we obtain $S = \epsilon_0 E^2 c$. This is the instantaneous power density. Its average value is simply $\overline{S} = \epsilon_0 \overline{E^2} c$, in agreement with Eq. (9.34). (In that equation we were using $S$, without the line over it, to denote the average power density.)

Interestingly, there can also be energy flow in a static electromagnetic field. Consider a very long stick with uniform linear charge density $\lambda$, moving with speed $v$ in the longitudinal direction, say, rightward. Close to the stick and not too close to the ends, the stick creates $\mathbf{E}$ and $\mathbf{B}$ fields that are essentially static, with $\mathbf{E}$ pointing radially and $\mathbf{B}$ pointing tangentially. Their cross product is therefore nonzero, so the Poynting vector is nonzero. Hence there is energy flow, and it moves in the same direction as the stick moves (for either sign of $\lambda$), as you can show with the right-hand rule. The energy density at a given point (not too close to the ends) doesn't change, because energy flows into a given volume from the left at the same rate it flows out to the right. However, near the ends the fields are changing, so there *is* a net energy flow into or out of a given volume. (Think of a uniform caravan of cars moving along the highway. The density of cars changes only at points near the ends of the caravan.) The rightward flow of energy is consistent with the fact that the whole system is moving to the right.

The Poynting vector (named after John Henry Poynting) falls into a wonderful class of phonetically accurate theorems/results. Others are the Low energy theorem (after F. E. Low) dealing with low-energy photons, and the Schwarzschild radius of a black hole (after Karl Schwarzschild, whose last name means "black shield" in German).

**Example (Energy flow into a capacitor)** A capacitor has circular plates with radius $R$ and is being charged by a constant current $I$. The electric field $E$ between the plates is increasing, so the energy density is also increasing. This implies that there must be a flow of energy into the capacitor. Calculate the Poynting vector at radius $r$ inside the capacitor (in terms of $r$ and $E$), and verify that its flux equals the rate of change of the energy stored in the region bounded by radius $r$.

**Solution** If the Poynting vector is to be nonzero, there must be a nonzero magnetic field inside the capacitor. And indeed, because the electric field is changing, there is an induced magnetic field due to the $\nabla \times \mathbf{B} = \epsilon_0 \mu_0 \, \partial \mathbf{E}/\partial t$ Maxwell equation. If we integrate this equation over the area of a disk with radius $r$ inside the capacitor (see Fig. 9.11) and use Stokes' theorem on the left-hand side, we obtain

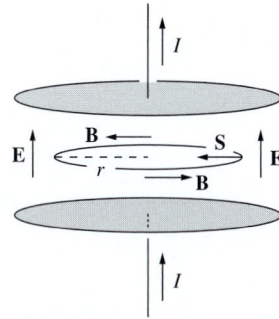

$$\int \mathbf{B} \cdot d\mathbf{s} = \epsilon_0 \mu_0 \frac{\partial E}{\partial t} \, (\text{area}) \implies B(2\pi r) = \epsilon_0 \mu_0 \frac{\partial E}{\partial t} (\pi r^2)$$

$$\implies B = \frac{\epsilon_0 \mu_0 r}{2} \frac{\partial E}{\partial t}. \tag{9.46}$$

**Figure 9.11.**
The changing vertical electric field inside the capacitor induces a tangential magnetic field. The cross product of $\mathbf{E}$ and $\mathbf{B}$ yields an inward-pointing Poynting vector, consistent with the increasing energy density.

This magnetic field points tangentially around the circle of radius $r$. Since $\mathbf{E}$ is increasing upward, $\mathbf{B}$ is directed counterclockwise when viewed from above, as you can check via the right-hand rule. The Poynting vector $\mathbf{S} = (\mathbf{E} \times \mathbf{B})/\mu_0$ then points radially inward everywhere on the circle of radius $r$. So the direction is correct; energy is flowing into the region bounded by radius $r$.

Let's now find the magnitude of $\mathbf{S}$. Since $\mathbf{E}$ is perpendicular to $\mathbf{B}$, the magnitude of $\mathbf{S}$ is

$$S = \frac{EB}{\mu_0} = \frac{E}{\mu_0} \left( \frac{\epsilon_0 \mu_0 r}{2} \frac{\partial E}{\partial t} \right) = \frac{\epsilon_0 r}{2} E \frac{\partial E}{\partial t}. \tag{9.47}$$

To find the total energy per time (that is, the power) flowing past radius $r$, we must multiply $S$ by the lateral area of the cylinder of radius $r$; that is, we must find the flux of $S$. If the separation between the plates is $h$, the lateral area is $2\pi rh$. The total power flowing into the cylinder of radius $r$ is then

$$P = \left( \frac{\epsilon_0 r}{2} E \frac{\partial E}{\partial t} \right) 2\pi rh = (\pi r^2 h) \epsilon_0 E \frac{\partial E}{\partial t} = \frac{d}{dt} \left( (\text{volume}) \frac{\epsilon_0 E^2}{2} \right) = \frac{dU}{dt}. \tag{9.48}$$

So the Poynting-vector flux does indeed equal the rate of change of the stored energy. In the special case where $r$ equals the radius of the capacitor, $R$, we obtain the total power flowing into the capacitor. Note that $S$ and $P$ are largest at $r = R$, and zero at $r = 0$, as expected.

REMARK: You might be worried that although we found there to be a nonzero magnetic field inside the capacitor, we didn't take into account the resulting magnetic energy density, $B^2/2\mu_0$. We used only the electric $\epsilon_0 E^2/2$ part of the density. However, a constant current $I$ implies a constant $d\sigma/dt$ (where $\pm\sigma$ are the charge densities on the plates), which in turn implies a constant $\partial E/\partial t$, which in turn implies a constant $B$, from Eq. (9.46). The magnetic energy density is therefore constant and thus doesn't affect the $dU/dt$ in Eq. (9.48). We can therefore rightfully ignore it. On the other hand, if $I$ *isn't* constant, then things are more complicated. However, for "everyday" rates of change of $I$, it is a very good approximation to say that the magnetic energy density in a capacitor is much smaller than the electric energy density; see Exercise 9.30.

At the end of Section 4.3 we mentioned that the energy flow in a circuit is due to the Poynting vector. We can now say more about this. There are two important parts to the energy flow. The first is the flow that yields the resistance heating. The current in a conducting wire is caused by a longitudinal $\mathbf{E}$ field inside the wire; recall $\mathbf{J} = \sigma\mathbf{E}$. Since the curl of $\mathbf{E}$ is zero, this same longitudinal $\mathbf{E}$ component must also exist right outside the surface of the wire. As you can show in Exercise 9.28, the Poynting-vector flux through a cylinder right outside the wire exactly accounts for the $IV$ resistance heating.

The second part is the energy flow along the wire. As discussed at the end of Section 4.3, there are surface charges on the wire. These create an electric field perpendicular to the wire, which in turn creates a Poynting vector parallel to the wire, as you can verify. This gives an energy flow along the wire; see Galili and Goihbarg (2005). More generally, the energy flow need not be constrained to lie near the wire if the wire loops around in space. Energy can flow across open space too, from one part of a circuit to another; see Jackson (1996).

If there are other electric fields present in the system, there can be a third part to the energy flow, now *away* from the wire. See Problem 9.10.

## 9.7 How a wave looks in a different frame

A plane electromagnetic wave is traveling through the vacuum. Let $\mathbf{E}$ and $\mathbf{B}$ be the electric and magnetic fields measured at some place and time in $F$, by an observer in $F$. What field will be measured by an observer in a different frame who happens to be passing that point at that time? Suppose that frame $F'$ is moving with speed $v$ in the $\hat{\mathbf{x}}$ direction relative to $F$, with its axes parallel to those of $F$. We can turn to Eq. (6.74) for the transformations of the field components. Let us write them out again:

$$E'_x = E_x, \qquad E'_y = \gamma(E_y - vB_z), \qquad E'_z = \gamma(E_z + vB_y);$$
$$B'_x = B_x, \qquad B'_y = \gamma\left(B_y + (v/c^2)E_z\right), \qquad B'_z = \gamma\left(B_z - (v/c^2)E_y\right).$$

$$(9.49)$$

The key to our problem is the way two particular scalar quantities transform, namely, $\mathbf{E} \cdot \mathbf{B}$ and $E^2 - B^2$. Let us use Eq. (9.49) to calculate $\mathbf{E}' \cdot \mathbf{B}'$ and see how it is related to $\mathbf{E} \cdot \mathbf{B}$:

$$
\begin{aligned}
\mathbf{E}' \cdot \mathbf{B}' &= E'_x B'_x + E'_y B'_y + E'_z B'_z \\
&= E_x B_x + \gamma^2 \left[ E_y B_y + \cancel{(v/c^2)E_y E_z} - \cancel{vB_y B_z} - (v/c)^2 E_z B_z \right] \\
&\quad + \gamma^2 \left[ E_z B_z - \cancel{(v/c^2)E_y E_z} + \cancel{vB_y B_z} - (v/c)^2 E_y B_y \right] \\
&= E_x B_x + \gamma^2 (1 - \beta^2)(E_y B_y + E_z B_z) = \mathbf{E} \cdot \mathbf{B}. \qquad (9.50)
\end{aligned}
$$

The scalar product $\mathbf{E} \cdot \mathbf{B}$ is *not changed* in the Lorentz transformation of the fields; it is an invariant. A similar calculation, which will be left to the reader as Exercise 9.32, shows that $E_x^2 + E_y^2 + E_z^2 - c^2(B_x^2 + B_y^2 + B_z^2)$ is also unchanged by the Lorentz transformation. We therefore have

$$
\boxed{\mathbf{E}' \cdot \mathbf{B}' = \mathbf{E} \cdot \mathbf{B}} \quad \text{and} \quad \boxed{E'^2 - c^2 B'^2 = E^2 - c^2 B^2} \qquad (9.51)
$$

The invariance of these two quantities is an important general property of any electromagnetic field, not just the field of an electromagnetic wave with which we are concerned at the moment. For the wave field, its implications are especially simple and direct. We know that the plane wave has $\mathbf{B}$ perpendicular to $\mathbf{E}$, and $cB = E$. Each of our two invariants, $\mathbf{E} \cdot \mathbf{B}$ and $E^2 - c^2 B^2$, is therefore zero. And if an invariant is zero in one frame, it must be zero in all frames. We see that *any* Lorentz transformation of the wave will leave $\mathbf{E}$ and $c\mathbf{B}$ perpendicular and equal in magnitude. *A light wave looks like a light wave in any inertial frame of reference.* That should not surprise us. It could be said that we have merely come full circle, back to the postulates of relativity, Einstein's starting point. Indeed, according to Einstein's own autobiographical account, he had begun 10 years earlier (at age 16!) to wonder what one would observe if one could "catch up" with a light wave. With the transformations in Eq. (9.49), which were given in Einstein's 1905 paper, the question can be answered. Consider a traveling wave with amplitudes given by $E_y = E_0, E_x = E_z = 0, B_z = E_0/c, B_x = B_y = 0$. This is a wave traveling in the $\hat{\mathbf{x}}$ direction, as we can tell from the fact that $\mathbf{E} \times \mathbf{B}$ points in that direction. Using Eq. (9.49) and the relation $\gamma = 1/\sqrt{1 - \beta^2}$, we find that

$$
E'_y = E_0 \sqrt{\frac{1 - \beta}{1 + \beta}}, \qquad B'_z = \frac{E_0}{c} \sqrt{\frac{1 - \beta}{1 + \beta}}. \qquad (9.52)
$$

As observed in $F'$ the amplitude of the wave is reduced. The wave velocity, of course, is $c$ in $F'$, as it is in $F$. The electromagnetic wave has no rest frame. In the limit $\beta = 1$, the amplitudes $E'_y$ and $B'_z$ observed in $F'$ are reduced to zero. The wave has vanished!

## 9.8 Applications

The power density of sunlight when it reaches the earth (or rather, the top of the atmosphere) is about $1360 \, \text{W/m}^2$, on average. You can show that this implies that the total power output of the sun is about $4 \cdot 10^{26} \, \text{W}$. If the power in one square kilometer of sunlight were converted to electricity at 15 percent efficiency, the result would be 200 megawatts. However, assuming that the sun is shining for only 6 hours per day on average, this would yield an average of 50 megawatts. Effects of atmosphere absorption and latitude would further reduce the result somewhat, but there would still be enough electrical power for a city of 25,000 people.

The *cosmic microwave background (CMB) radiation* (see Exercise 9.25) was discovered by Penzias and Wilson in 1965. This radiation is left over from the big bang and fills all of space. About 300,000 years after the big bang, the universe became transparent to photons, shortly after the hot plasma of electrons and ions cooled to the point where stable atoms could form. The CMB photons have been traveling freely ever since. The wavelength was shorter back then, but it has been continually expanding along with the expansion of the universe. The radiation consists of a distribution of wavelengths, but the peak is around 2 mm. It looks nearly the same in all directions, but its slight anisotropies yield information about what the early universe looked like.

*Comets* generally have two kinds of tails. The *dust tail* consists of dust that is pushed away from the comet by the radiation pressure from the sunlight. (The sunlight carries energy, so it also carries momentum; see Problem 9.11.) The dust drifts relatively slowly away from the comet, so this results in the tail curving and drifting behind the comet. The *ion tail* consists of ions that are blown away from the comet by the sun's solar wind (consisting of charged particles). These ions move very quickly away from the comet, so the ion tail always points essentially radially away from the sun, independent of the comet's location around the sun.

*Radio frequency identification (RFID) tags* have many uses: anti-theft tags, inventory tracking, tollbooth transponders, chip timing in road races, library books, and so on. Although some RFID tags contain their own power source, most (called "passive RFID") do not. They are powered by *resonant inductive coupling*: a small coil and capacitor in the tag constitute an *LC* circuit with a particular resonant frequency. A "reader" transmits a radio wave with this frequency, and the (changing) magnetic field in this wave induces a current in the tag's circuit. This powers a small microchip, which then transmits a specific identification message back to the reader.

Cell phones, radios, and many other communication devices make use of electromagnetic waves in the *radio frequency* part of the spectrum, usually from about 1 MHz to a few GHz. A pure sinusoidal wave at a given frequency contains minimal information, so if we want to transmit useful information, we must modify the wave in some manner. The two simplest ways of modifying the wave are *amplitude modulation (AM)* and *frequency modulation (FM)*. In the case of AM, the carrier wave (with a frequency in the 1 MHz range) has its *amplitude* modulated by the sound wave (with a much smaller frequency in the 1 kHz range) that is being sent. The larger the value of the sound wave at a given instant, the larger the amplitude of the transmitted wave. The sound wave is in some sense the envelope of the transmitted wave. The receiver is able to extract the amplitude information and can then reconstruct the original sound wave; a plot of the amplitude of the transmitted wave as a function of time is effectively a plot of the original sound wave as a function of time.

In the case of FM, the carrier wave (with a frequency of around 100 MHz, as you know from your FM radio dial) has its *frequency* modulated by the sound wave that is being sent. The larger the value of the sound wave at a given instant, the more the carrier-wave frequency shifts relative to a particular value. The receiver is able to extract the frequency information and can then reconstruct the original sound wave; a plot of the frequency as a function of time is effectively a plot of the original sound wave as a function of time. Note that this frequency is well defined, because even if a time interval is fairly short on the time scale of the sound wave, a very large number of the carrier-wave oscillations still fit into it. One method of extracting the frequency information is called *slope detection*. In this method, the resonant frequency of the receiver is chosen to be slightly shifted from that of the carrier wave, so that the span of the carrier wave's frequencies lies on the steep side part of the resonance peak. If the span lies, say, on the left side of the peak, then the response of the receiver's circuit increases (approximately linearly) as the frequency of the carrier wave increases. So we simply need to measure the amplitude of the current in the circuit, and we will obtain the frequency of the carrier wave (up to some factor). The thing that makes all of this possible is *resonance*, which enables the receiver to respond to a narrow range of frequencies and ignore all others.

## CHAPTER SUMMARY

- Because the divergence of the curl of a vector is identically zero, the differential form of Ampère's law, curl $\mathbf{B} = \mu_0 \mathbf{J}$, implies that div $\mathbf{J}$ is identically zero. This is inconsistent with the continuity equation, div $\mathbf{J} = -\partial \rho / \partial t$ (which follows from conservation of charge), in situations where $\rho$ changes with time. Therefore, curl $\mathbf{B} = \mu_0 \mathbf{J}$ cannot be correct. The correct expression has an extra term, $\mu_0 \epsilon_0 \, \partial \mathbf{E} / \partial t$, on the right-hand side. With this term, div $\mathbf{J}$ correctly equals $-\partial \rho / \partial t$.

- The quantity $\epsilon_0 \, \partial \mathbf{E}/\partial t$ is called the *displacement current*. This is the last piece of the puzzle, and we can now write down the complete set of *Maxwell's equations:*

$$\text{curl } \mathbf{E} = -\frac{\partial \mathbf{B}}{\partial t},$$

$$\text{curl } \mathbf{B} = \mu_0 \epsilon_0 \frac{\partial \mathbf{E}}{\partial t} + \mu_0 \mathbf{J},$$

$$\text{div } \mathbf{E} = \frac{\rho}{\epsilon_0},$$

$$\text{div } \mathbf{B} = 0. \tag{9.53}$$

These are, respectively, (1) Faraday's law, (2) Ampère's law with the addition of the displacement current, (3) Gauss's law, and (4) the statement that there are no magnetic monopoles.

- A possible form of a *traveling electromagnetic wave* is

$$\mathbf{E} = \hat{\mathbf{z}} E_0 \sin(y - vt) \quad \text{and} \quad \mathbf{B} = \hat{\mathbf{x}} B_0 \sin(y - vt), \tag{9.54}$$

where

$$v = \pm \frac{1}{\sqrt{\mu_0 \epsilon_0}} = \pm c \quad \text{and} \quad E_0 = \pm \frac{B_0}{\sqrt{\mu_0 \epsilon_0}} = \pm c B_0. \tag{9.55}$$

In general, we can produce a traveling wave by replacing the $\sin(y - vt)$ function with *any* function $f(y - vt)$, provided that (1) $v = \pm c$, (2) $E_0 = \pm c B_0$, and (3) $\mathbf{E}$ and $\mathbf{B}$ are perpendicular to each other and also to the direction of propagation.

- A *standing* wave is formed by adding two waves traveling in opposite directions. In a standing wave there are (unlike in a traveling wave) positions where $\mathbf{E}$ is zero at all times, and times when $\mathbf{E}$ is zero at all positions. Likewise for $\mathbf{B}$.

- The *power density* (energy per unit area per unit time) of a sinusoidal electromagnetic wave can be written in various forms:

$$S = \epsilon_0 \overline{E^2} c = \frac{\overline{E^2}}{\sqrt{\mu_0/\epsilon_0}} = \frac{\overline{E^2}(\text{volts/meter})^2}{377 \text{ ohms}}. \tag{9.56}$$

More generally, the *Poynting vector,*

$$\mathbf{S} = \frac{\mathbf{E} \times \mathbf{B}}{\mu_0}, \tag{9.57}$$

gives the power density of an arbitrary electromagnetic field at every point.

- Using the fact that the $\mathbf{E}$ and $\mathbf{B}$ fields transform according to the Lorentz transformations, we can derive two invariants:

$$\mathbf{E}' \cdot \mathbf{B}' = \mathbf{E} \cdot \mathbf{B} \quad \text{and} \quad E'^2 - c^2 B'^2 = E^2 - c^2 B^2. \tag{9.58}$$

These imply that if in one frame $\mathbf{E}$ and $\mathbf{B}$ are perpendicular and $E = cB$, then these two relations are also true in any other frame. That is, a light wave in one inertial frame looks like a light wave in any other inertial frame.

## Problems

9.1 *The missing term* ∗∗

Due to the contradiction between Eqs. (9.2) and (9.5), we know that there must be an extra term in the $\nabla \times \mathbf{B}$ relation, as we found in Eq. (9.10). Call this term $\mathbf{W}$. In the text, we used the Lorentz transformations to motivate a guess for $\mathbf{W}$. Find $\mathbf{W}$ here by taking the divergence of both sides of $\nabla \times \mathbf{B} = \mu_0 \mathbf{J} + \mathbf{W}$. Assume that the only facts you are allowed to work with are (1) $\nabla \cdot \mathbf{E} = \rho/\epsilon_0$, (2) $\nabla \cdot \mathbf{B} = 0$, (3) $\nabla \cdot \mathbf{J} = -\partial\rho/\partial t$, and (4) $\nabla \times \mathbf{B} = \mu_0 \mathbf{J}$ in the case of steady currents.

9.2 *Spherically symmetric current* ∗

A spherically symmetric (and constant) current density flows radially inward to a spherical shell, causing the charge on the shell to increase at the constant rate $dQ/dt$. Verify that Maxwell's equation, $\nabla \times \mathbf{B} = \mu_0 \mathbf{J} + \mu_0 \epsilon_0 \, \partial\mathbf{E}/\partial t$, is satisfied at points outside the shell.

9.3 *A charge and a half-infinite wire* ∗∗

A half-infinite wire carries current $I$ from negative infinity to the origin, where it builds up at a point charge with increasing $q$ (so $dq/dt = I$). Consider the circle shown in Fig. 9.12, which has radius $b$ and subtends an angle $2\theta$ with respect to the charge. Calculate the integral $\int \mathbf{B} \cdot d\mathbf{s}$ around this circle. Do this in three ways.

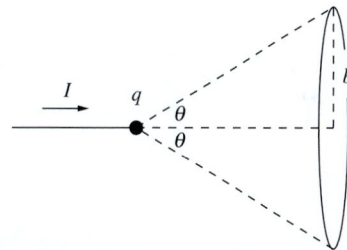

**Figure 9.12.**

(a) Find the $\mathbf{B}$ field at a given point on the circle by using the Biot–Savart law to add up the contributions from the different parts of the wire.

(b) Use the integrated form of Maxwell's equation (that is, the generalized form of Ampère's law including the displacement current),

$$\int_C \mathbf{B} \cdot d\mathbf{s} = \mu_0 I + \mu_0 \epsilon_0 \int_S \frac{\partial \mathbf{E}}{\partial t} \cdot d\mathbf{a}, \qquad (9.59)$$

with $S$ chosen to be a surface that is bounded by the circle and doesn't intersect the wire, but is otherwise arbitrary. (You can invoke the result from Problem 1.15.)

(c) Use the same strategy as in (b), but now let $S$ intersect the wire.

9.4 *B in a discharging capacitor, via conduction current* ∗∗

As mentioned in Exercise 9.15, the magnetic field inside a discharging capacitor can be calculated by summing the contributions from all elements of conduction current. This calculation is extremely tedious. We can, however, get a handle on the contribution from the plates' conduction current in a much easier way that doesn't involve a nasty integral. If we make the usual assumption that the distance $s$ between the plates is small compared with their radius $b$, then any point $P$ inside the capacitor is close enough to the plates so that they look essentially like infinite planes, with a surface current density equal to the density at the nearest point.

(a) Determine the current that crosses a circle of radius $r$ in the capacitor plates, and then use this to find the surface current density. *Hint:* The charge on each plate is essentially uniformly distributed at all times.

(b) Combine the field contributions from the wires and the plates to show that the field at a point $P$ inside the capacitor, a distance $r$ from the axis of symmetry, equals $B = \mu_0 I r / 2\pi b^2$. (Assume $s \ll r$, so that you can approximate the two wires as a complete infinite wire.)

9.5 *Maxwell's equations for a moving charge* ***

In part (b) of Problem 6.24 we dealt with approximate expressions for the electric and magnetic fields due to a slowly moving charge, valid in the limit $v \ll c$. In this problem we will use the exact forms. The exact **E**, for any value of $v$, is given in Eq. (5.13) or Eq. (5.15). The Lorentz transformation then gives[2] the exact **B** as $\mathbf{B} = (1/c^2)\mathbf{v} \times \mathbf{E}$; see Problem 6.24(a). Verify that these exact expressions for **E** and **B** satisfy Maxwell's equations in vacuum. That is:

(a) Show that $\nabla \cdot \mathbf{B} = 0$. (We already showed in Problem 5.4 that $\nabla \cdot \mathbf{E} = 0$.) The vector identity for $\nabla \cdot (\mathbf{A} \times \mathbf{B})$ in Appendix K will come in handy.

(b) Show that $\nabla \times \mathbf{E} = -\partial \mathbf{B}/\partial t$. (The $\nabla \times \mathbf{B} = \partial \mathbf{E}/\partial t$ calculation is nearly the same, so you can skip that.) *Note:* Although the calculation is doable if you use the spherical-coordinate expression for $E$ in Eq. (5.15) (don't forget that both $r$ and $\theta$ vary with time), it's a bit easier if you use the Cartesian-coordinate expression in Eq. (5.13).

9.6 *Oscillating field in a solenoid* ***

A solenoid with radius $R$ has $n$ turns per unit length. The current varies with time according to $I(t) = I_0 \cos \omega t$. The magnetic field inside the solenoid, $B(t) = \mu_0 n I(t)$, therefore changes with time. In this problem you will need to make use of wisely chosen Faraday/Ampère loops.

---

[2] If you want to derive the magnetic field for a fast-moving charge via the Biot–Savart law, you need to incorporate the so-called "retarded time" arising from the finite speed of light. We won't get into that here, but see Problem 6.28 for a special case.

(a) Changing $B$ fields cause $E$ fields. Assuming that the $B$ field is given by $B_0(t) \equiv \mu_0 n I_0 \cos \omega t$, find the electric field at radius $r$ inside the solenoid.

(b) Changing $E$ fields cause $B$ fields. Find the $B$ field (at radius $r$ inside the solenoid) caused by the changing $E$ field that you just found. More precisely, find the difference between the $B$ at radius $r$ and the $B$ on the axis. Label this difference as $\Delta B(r, t)$.

(c) The total $B$ field does not equal $\mu_0 n I_0 \cos \omega t$ throughout the solenoid, due to the $\Delta B(r, t)$ difference you just found.[3] What is the ratio $\Delta B(r, t)/B_0(t)$? Explain why we are justified in making the statement, "The magnetic field inside the solenoid is essentially equal to the naive $\mu_0 n I_0 \cos \omega t$ value, provided that the changes in the current occur on a time scale that is long compared with the time it takes light to travel across the width of the solenoid." (This time is very short, so for an "everyday" value of $\omega$, the field is essentially equal to $\mu_0 n I_0 \cos \omega t$.)

**9.7** *Traveling and standing waves* **

Consider the two oppositely traveling electric-field waves,

$$\mathbf{E}_1 = \hat{\mathbf{x}} E_0 \cos(kz - \omega t) \quad \text{and} \quad \mathbf{E}_2 = \hat{\mathbf{x}} E_0 \cos(kz + \omega t). \quad (9.60)$$

The sum of these two waves is the standing wave, $2\hat{\mathbf{x}} E_0 \cos kz \cos \omega t$.

(a) Find the magnetic field associated with this standing electric wave by finding the **B** fields associated with each of the above traveling **E** fields, and then adding them.

(b) Find the magnetic field by instead using Maxwell's equations to find the **B** field associated with the standing electric wave, $2\hat{\mathbf{x}} E_0 \cos kz \cos \omega t$.

**9.8** *Sunlight* *

The power density in sunlight, at the earth, is roughly 1 kilowatt/m². How large is the rms magnetic field strength?

**9.9** *Energy flow for a standing wave* **

(a) Consider the standing wave in Eq. (9.32). Draw plots of the energy density $\mathcal{U}(y, t)$ at $\omega t$ values of $0, \pi/4, \pi/2, 3\pi/4$, and $\pi$.

(b) Make a plot of the $y$ component of the Poynting vector, $S_y(y, t)$, at $\omega t$ values of $\pi/4, \pi/2$, and $3\pi/4$. Explain why these plots are consistent with how the energy sloshes back and forth between the different energy plots.

**9.10** *Energy flow from a wire* **

A very thin straight wire carries a constant current $I$ from infinity radially inward to a spherical conducting shell with radius $R$. The

---

[3] Of course, this $\Delta B(r, t)$ difference causes another $E$ field, and so on. So we would get an infinite series of corrections if we kept going. But as long as the current doesn't change too quickly, the higher-order terms are negligible.

increase in the charge on the shell causes the electric field in the surrounding space to increase, which means that the energy density increases. This implies that there must be a flow of energy from somewhere. This "somewhere" is the wire. Verify that the total flux of the Poynting vector away from a thin tube surrounding the wire equals the rate of change of the energy stored in the electric field. (You can assume that the radius of the wire is much smaller than the radius of the tube, which in turn is much smaller than the radius of the shell.)

9.11 *Momentum in an electromagnetic field* **

We know from Section 9.6 that traveling electromagnetic waves carry energy. But the theory of relativity tells us that anything that transports energy must also transport momentum. Since light may be considered to be made of massless particles (photons), the relation $p = E/c$ must hold; see Eq. (G.19). In terms of **E** and **B**, find the momentum density of a traveling electromagnetic wave. That is, find the quantity that, when integrated over a given volume, yields the momentum contained in the wave in that volume.

Although we won't prove it here, the result that you just found for traveling waves is a special case of the more general result that the momentum density equals $1/c^2$ times the energy flow per area per time. This holds for *any* type of energy flow (matter or field). In particular, it holds for any type of electromagnetic field; even a static field with a nonzero $\mathbf{E} \times \mathbf{B}/\mu_0$ Poynting vector carries momentum. For a nice example of this, see Problem 9.12.

9.12 *Angular momentum paradox* ***

A setup consists of three very long coaxial cylindrical objects: a nonconducting cylindrical shell with radius $a$ and total (uniform) charge $Q$, another nonconducting cylindrical shell with radius $b > a$ and total (uniform) charge $-Q$, and a solenoid with radius $R > b$; see Fig. 9.13. (This setup is a variation of the setup in Boos (1984).) The current in the solenoid produces a uniform magnetic field $B_0$ in its interior. The solenoid is fixed, but the two cylinders are free to rotate (independently) around the axis. They are initially at rest. Imagine that the current in the solenoid is then decreased to zero. (If you want to be picky about keeping the system isolated from external torques, you can imagine the current initially flowing in a superconductor which becomes a normal conductor when heated up.) The changing $B$ field inside the solenoid will induce an $E$ field at the locations of the two cylinders.

(a) Find the angular momentum gained by each cylinder by the time the magnetic field has decreased to zero.

(b) You should find that the total change in angular momentum of the cylinders is not zero. Does this mean that angular

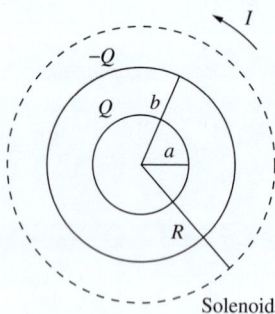

**Figure 9.13.**

momentum isn't conserved? If it *is* conserved, verify this quantitatively. You may assume that the two cylinders are massive
enough so that they don't end up spinning very quickly, which
means that we can ignore the *B* fields they generate. *Hint:* See
Problem 9.11.

## Exercises

9.13 *Displacement-current flux* *

The flux of the real current through the surface *S* in Fig. 9.4 is
simply *I*. Verify explicitly that the flux of the displacement current,
$\mathbf{J}_d \equiv \epsilon_0(\partial \mathbf{E}/\partial t)$, through the surface *S'* also equals *I*. What about
the sign of the flux? As usual, work in the approximation where
the spacing between the capacitor plates is small.

9.14 *Sphere with a hole* **

A current *I* flows along a wire toward a point charge, causing the
charge to increase with time. Consider a spherical surface *S* centered at the charge, with a tiny hole where the wire is, as shown in
Fig. 9.14. The circumference *C* of this hole is the boundary of the
surface *S*. Verify that the integral form of Maxwell's equation,

$$\int_C \mathbf{B} \cdot d\mathbf{s} = \int_S \left( \mu_0 \epsilon_0 \frac{\partial \mathbf{E}}{\partial t} + \mu_0 \mathbf{J} \right) \cdot d\mathbf{a}, \qquad (9.61)$$

is satisfied.

9.15 *Field inside a discharging capacitor* **

The magnetic field inside the discharging capacitor shown in
Fig. 9.1 can in principle be calculated by summing the contributions from all elements of conduction current, as indicated in
Fig. 9.5. That might be a long job. If we can assume symmetry
about this axis, it is very much easier to find the field **B** at a point
by using the integral law,

$$\int_C \mathbf{B} \cdot d\mathbf{s} = \int_S \left( \mu_0 \epsilon_0 \frac{\partial \mathbf{E}}{\partial t} + \mu_0 \mathbf{J} \right) \cdot d\mathbf{a}, \qquad (9.62)$$

applied to a circular path through the point. Use this to show that
the field at *P*, which is midway between the capacitor plates in
Fig. 9.15, and a distance *r* from the axis of symmetry, equals $B = \mu_0 I r / 2\pi b^2$. You may assume that the distance *s* between the plates
is small compared with their radius *b*. (Compare this with the cal-
culation of the induced electric field **E** in the example of Fig. 7.16.)

9.16 *Changing flux from a moving charge* **

In terms of the electric field **E** of a point charge moving with
constant velocity **v**, the Lorentz transformation gives the magnetic

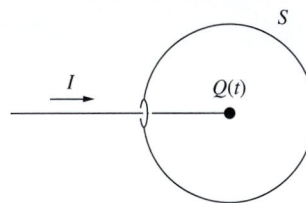

**Figure 9.14.**

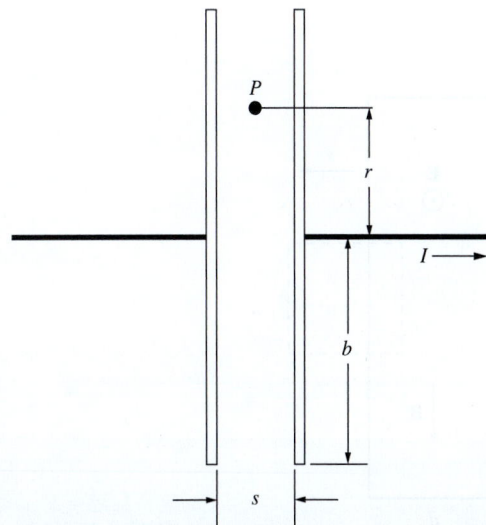

**Figure 9.15.**

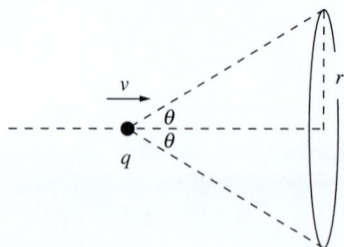

**Figure 9.16.**

field as $\mathbf{B} = (\mathbf{v}/c^2) \times \mathbf{E}$. Verify that Maxwell's equation in integral form, $\int \mathbf{B} \cdot d\mathbf{s} = (1/c^2)(d\Phi_E/dt)$, holds for the circle shown in Fig. 9.16. (We can therefore think of the magnetic field as being induced by the changing electric field of the moving charge.) *Hint:* Indicate geometrically the new electric flux that passes through the circle after the charge has moved a small distance to the right.

**9.17** *Gaussian conditions* *
Start with the source-free, or "empty-space," Maxwell's equations in Gaussian units in Eq. (9.21). Consider a wave described by Eqs. (9.22) and (9.23), but now with $E_0$ in statvolts/cm and $B_0$ in gauss. What conditions must $E_0, B_0$, and $v$ meet to satisfy Maxwell's equations?

**9.18** *Associated B field* *
If the electric field in free space is $\mathbf{E} = E_0(\hat{\mathbf{x}} + \hat{\mathbf{y}}) \sin[(2\pi/\lambda)(z + ct)]$ with $E_0 = 20$ volts/m, then the magnetic field, not including any static magnetic field, must be what?

**9.19** *Find the wave* *
Write out formulas for $\mathbf{E}$ and $\mathbf{B}$ that specify a plane electromagnetic sinusoidal wave with the following characteristics. The wave is traveling in the direction $-\hat{\mathbf{x}}$; its frequency is 100 megahertz (MHz), or $10^8$ cycles per second; the electric field is perpendicular to the $\hat{\mathbf{z}}$ direction.

**9.20** *Kicked by a wave* **
A free proton was at rest at the origin before the wave described by Eq. (9.28) came past. Let the amplitude $E_0$ equal 100 kilovolts/m. Where would you expect to find the proton at time $t = 1$ microsecond? The proton mass is $1.67 \cdot 10^{-27}$ kg. *Hint:* Since the duration of the pulse is only a few nanoseconds, you can neglect the displacement of the proton during the passage of the pulse. Also, if the velocity of the proton is not too large, you may ignore the effect of the magnetic field on its motion. The first thing to calculate is the momentum acquired by the proton during the pulse.

**9.21** *Effect of the magnetic field* **
Suppose that in Exercise 9.20 the effect of the magnetic field was not entirely negligible. How would it change the direction of the proton's final velocity? (It suffices to give the dependence on the various parameters; you can ignore any numerical factors.)

**9.22** *Plane-wave pulse* **
Consider the plane-wave pulse of $\mathbf{E}$ and $\mathbf{B}$ fields shown in Fig. 9.17; $\mathbf{E}$ points out of the page, and $\mathbf{B}$ points downward. The fields are uniform inside a "slab" region and are zero outside. The slab has length $d$ in the $x$ direction and large (essentially infinite) lengths in the $y$ and $z$ directions. It moves with speed $v$ (to be determined)

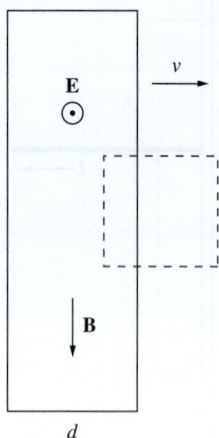

**Figure 9.17.**

in the $x$ direction. This slab can be considered to be a small section of the transition shell in Appendix H. However, you need not worry about how these fields were generated. All that matters is that the electromagnetic field is self-sustaining via the two "induction" Maxwell equations.

(a) With the dashed rectangular loop shown (which is fixed in space, while the slab moves), use the integral form of one of Maxwell's equations to obtain a relation between $E$ and $B$.

(b) Make a similar argument, with a loop perpendicular to the plane of the page, to obtain another relation between $E$ and $B$. (Be careful with the signs.) Then solve for $v$.

9.23 *Field in a box* ***

Show that the electromagnetic field described by

$$\mathbf{E} = E_0\hat{\mathbf{z}}\cos kx\cos ky\cos \omega t,$$
$$\mathbf{B} = B_0(\hat{\mathbf{x}}\cos kx\sin ky - \hat{\mathbf{y}}\sin kx\cos ky)\sin \omega t \qquad (9.63)$$

will satisfy the empty-space Maxwell equations in Eq. (9.18) if $E_0 = \sqrt{2}cB_0$ and $\omega = \sqrt{2}ck$. This field can exist inside a square metal box enclosing the region $-\pi/2k < x < \pi/2k$ and $-\pi/2k < y < \pi/2k$, with arbitrary height in the $z$ direction. Roughly what do the electric and magnetic fields look like?

9.24 *Satellite signal* *

From a satellite in stationary orbit, a signal is beamed earthward with a power of 10 kilowatts and a beam width covering a region roughly circular and 1000 km in diameter. What is the electric field strength at the receivers, in millivolts/meter?

9.25 *Microwave background radiation* **

Of all the electromagnetic energy in the universe, by far the largest amount is in the form of waves with wavelengths in the millimeter range. This is the cosmic microwave background radiation discovered by Penzias and Wilson in 1965. It apparently fills all space, including the vast space between galaxies, with an energy density of $4\cdot 10^{-14}$ joule/m$^3$. Calculate the rms electric field strength in this radiation, in volts/m. Roughly how far away from a 1 kilowatt radio transmitter would you find a comparable electromagnetic wave intensity?

9.26 *An electromagnetic wave* **

Here is a particular electromagnetic field in free space:

$$E_x = 0, \quad E_y = E_0\sin(kx + \omega t), \quad E_z = 0; \qquad (9.64)$$
$$B_x = 0, \quad B_y = 0, \qquad\qquad\qquad B_z = -(E_0/c)\sin(kx + \omega t).$$

(a) Show that this field can satisfy Maxwell's equations if $\omega$ and $k$ are related in a certain way.

(b) Suppose $\omega = 10^{10}\,\text{s}^{-1}$ and $E_0 = 1\,\text{kV/m}$. What is the wavelength? What is the energy density in joules per cubic meter, averaged over a large region? From this calculate the power density, the energy flow in joules per square meter per second.

**9.27** *Reflected wave* $**$

A sinusoidal wave is reflected at the surface of a medium whose properties are such that half the incident energy is absorbed. Consider the field that results from the superposition of the incident and the reflected wave. An observer stationed somewhere in this field finds the local electric field oscillating with a certain amplitude $E$. What is the ratio of the largest such amplitude noted by any observer to the smallest amplitude noted by any observer? (This is called the *voltage standing wave ratio*, or, in laboratory jargon, VSWR.)

**9.28** *Poynting vector and resistance heating* $**$

A longitudinal $\mathbf{E}$ field inside a wire causes a current via $\mathbf{J} = \sigma\mathbf{E}$. And since the curl of $\mathbf{E}$ is zero, this same longitudinal $\mathbf{E}$ component must also exist right outside the surface of the wire. Show that the Poynting vector flux through a cylinder right outside the wire accounts for the $IV$ resistance heating.

**9.29** *Energy flow in a capacitor* $**$

A capacitor is charged by having current flow in a thin straight wire from the middle of one circular plate to the middle of the other (as opposed to wires coming in from infinity, as in the example in Section 9.6.2). The electric field inside the capacitor increases, so the energy density also increases. This implies that there must be a flow of energy from somewhere. As in Problem 9.10, this "somewhere" is the wire. Verify that the flux of the Poynting vector away from the wire equals the rate of change of the energy stored in the field. (Of course, we would need to place a battery somewhere along the wire to produce the current flow, and this battery is where the energy flow originates. See Galili and Goihbarg (2005).)

**9.30** *Comparing the energy densities* $**$

Consider the capacitor example in Section 9.6.2, but now let the current change in a way that makes the electric field inside the capacitor take the form of $E(t) = E_0 \cos \omega t$. The induced magnetic field is given in Eq. (9.46). Show that the energy density of the magnetic field is much smaller than the energy density of the electric field, provided that the time scale of $\omega$ (namely $2\pi/\omega$) is much longer than the time it takes light to travel across the diameter of the capacitor disks. (As in Problem 9.6, we are ignoring higher-order effects.)

9.31 *Field momentum of a moving charge* ∗∗∗

Consider a charged particle in the shape of a small spherical shell with radius $a$ and charge $q$. It moves with a nonrelativistic speed $v$. The electric field due to the shell is essentially given by the simple Coulomb field, and the magnetic field is then given by Eq. (6.81). Using the result from Problem 9.11, integrate the momentum density over all space. Show that the resulting total momentum of the electromagnetic field can be written as $mv$, where $m \equiv (4/3)(q^2/8\pi\epsilon_0 a)/c^2$.

An interesting aside: for nonrelativistic speeds, the total energy in the electromagnetic field is dominated by the electric energy. So from Problem 1.32 the total energy in the field equals $U = (q^2/8\pi\epsilon_0 a)$. Using the above value of $m$, we therefore find that $U = (3/4)mc^2$. This doesn't agree with Einstein's $U = \gamma mc^2$ result, with $\gamma \approx 1$ for a nonrelativistic particle. The qualitative resolution to this puzzle is that, although we correctly calculated the electromagnetic energy, this isn't the total energy. There must be other forces at play, of course, because otherwise the Coulomb repulsion would cause the particle to fly apart.

9.32 *A Lorentz invariant* ∗∗∗

Starting from the field transformation given by Eq. (6.76), show that the scalar quantity $E^2 - c^2 B^2$ is invariant under the transformation. In other words, show that $E'^2 - c^2 B'^2 = E^2 - c^2 B^2$. You can do this using only vector algebra, without writing out $x$, $y$, $z$ components of anything. (The resolution into parallel and perpendicular vectors is convenient for this, since $\mathbf{E}_\perp \cdot \mathbf{E}_\parallel = 0$, $\mathbf{B}_\parallel \times \mathbf{E}_\parallel = 0$, etc.)

# 10

# Electric fields in matter

**Overview** In this chapter we study how electric fields affect, and are affected by, matter. We concern ourselves with insulators, or *dielectrics,* characterized by a *dielectric constant.* The study of electric fields in matter is largely the study of *dipoles.* We discussed these earlier in Chapter 2, but we will derive their properties in more generality here, showing in detail how the *multipole expansion* comes about. The net dipole moment induced in matter by an electric field can come about in two ways. In some cases the electric field polarizes the molecules; the *atomic polarizability* quantifies this effect. In other cases a molecule has an inherent dipole moment, and the external field serves to align these moments. In any case, a material can be described by a *polarization density P.* The *electric susceptibility* gives (up to a factor of $\epsilon_0$) the ratio of $P$ to the electric field. The effect of the polarization density is to create a surface charge density on a dielectric material. This explains why the capacitance of a capacitor is increased when it is filled with a dielectric; the surface charge on the dielectric partially cancels the free charge on the capacitor plates.

We study the special case of a uniformly polarized sphere, which interestingly has a uniform electric field in its interior. We then extend this result to the case of a dielectric sphere placed in a uniform electric field. By considering separately the *free charge* and *bound charge,* we are led to the *electric displacement vector D,* whose divergence involves only the free charge (unlike the electric field, whose divergence involves *all* the charge, by Gauss's law). We look at the effects of temperature on the polarization density, how the polarization responds to rapidly changing fields, and how the bound-charge current affects the "curl **B**" Maxwell equation. Finally, we consider an electromagnetic wave in a dielectric. We find that only a slight modification to the vacuum case is needed.

## 10.1 Dielectrics

The capacitor we studied in Chapter 3 consisted of two conductors, insulated from one another, with nothing in between. The system of two conductors was characterized by a certain capacitance $C$, a constant relating the magnitude of the charge $Q$ on the capacitor (positive charge $Q$ on one plate, equal negative charge on the other) to the difference in electric potential between the two conductors, $\phi_1 - \phi_2$. Let's denote the potential difference by $\phi_{12}$:

$$C = \frac{Q}{\phi_{12}}. \tag{10.1}$$

For the parallel-plate capacitor, two flat plates each of area $A$ and separated by a distance $s$, we found that the capacitance is given by

$$C = \frac{\epsilon_0 A}{s}. \tag{10.2}$$

Capacitors like this can be found in some electrical apparatus. They are called *vacuum capacitors* and consist of plates enclosed in a highly evacuated bottle. They are used chiefly where extremely high and rapidly varying potentials are involved. Far more common, however, are capacitors in which the space between the plates is filled with some non-conducting solid or liquid substance. Most of the capacitors you have worked with in the laboratory are of that sort; there are dozens of them in any television screen. For conductors embedded in a material medium, Eq. (10.2) does not agree with experiment. Suppose we fill the space between the two plates shown in Fig. 10.1(a) with a slab of plastic, as in Fig. 10.1(b). Experimenting with this new capacitor, we still find a simple proportionality between charge and potential difference, so that we can still *define* a capacitance by Eq. (10.1). But we find $C$ to be substantially *larger* than Eq. (10.2) would have predicted. That is, we find more charge on each of the plates, for the same potential difference, plate area, and distance of separation. The plastic slab must be the cause of this.

It is not hard to understand in a general way how this comes about. The plastic slab consists of molecules, the molecules are composed of atoms, which in turn are made of electrically charged particles – electrons and atomic nuclei. The electric field between the capacitor plates acts on those charges, pulling the negative charges up, if the upper plate is positive as in Fig. 10.2, and pushing the positive charges down. Nothing moves very far. (There are no free electrons around, already detached

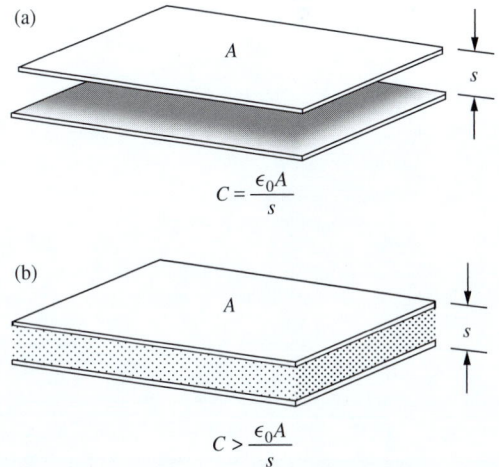

(a)

$$C = \frac{\epsilon_0 A}{s}$$

(b)

$$C > \frac{\epsilon_0 A}{s}$$

**Figure 10.1.**
(a) A capacitor formed by parallel conducting plates. (b) The same plates with a slab of insulator in between.

(a)

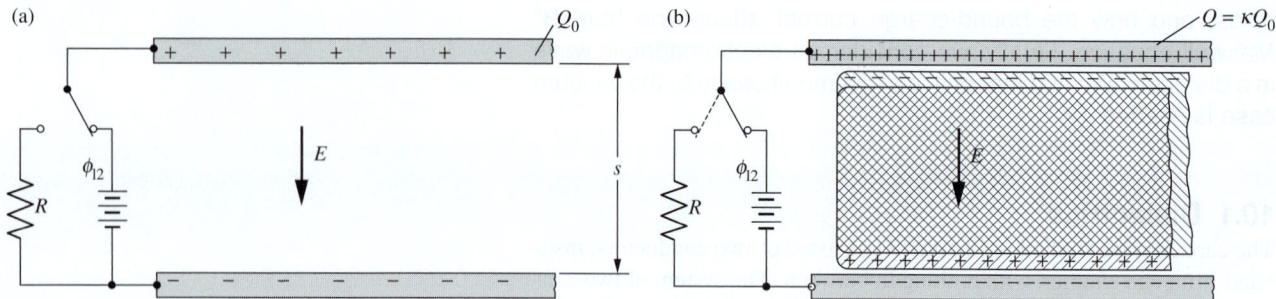

(b)

**Figure 10.2.**
How a dielectric increases the charge on the plates of a capacitor. (a) Space between the plates empty; $Q_0 = C_0\phi_{12}$. (b) Space between the plates filled with a nonconducting material, that is, a dielectric. Electric field pulls negative charges up and pushes positive charges down, exposing a layer of uncompensated negative charge on the upper surface of the dielectric and a layer of uncompensated positive charge on the lower surface. The total charge at the top, *including* charge $Q$ on the upper plate, is the same as in (a). $Q$ itself is now greater than $Q_0$; $Q = \kappa Q_0$. This $Q$ is the amount of charge that will flow through the resistor $R$ if the capacitor is discharged by throwing the switch.

from atoms and ready to travel, as there would be in a metallic conductor.) There will be some slight displacement of the charges nevertheless, for an atom is not an infinitely rigid structure. The effect of this within the plastic slab is that the negative charge distribution, viewed as a whole, and the total positive charge distribution (the atomic nuclei) are very slightly displaced relative to one another, as indicated in Fig. 10.2(b). The interior of the block remains electrically neutral, but a thin layer of uncompensated negative charge has emerged at the top, with a corresponding layer of uncompensated positive charge at the bottom.

In the presence of the induced layer of negative charge below the upper plate, the charge $Q$ on the plate itself will increase. In fact, $Q$ must increase until the total charge at the top, the algebraic sum of $Q$ and the induced charge layer, equals $Q_0$ (the charge on the upper plate before the plastic was inserted). We shall be able to prove this when we return to this problem in Section 10.8 after settling some questions about the electric field inside matter. The important point now is that the charge $Q$ in Fig. 10.2(b) is *larger* than $Q_0$ and that this $Q$ *is* the charge of the capacitor in the relation $Q = C\phi_{12}$. It is the charge that came out of the battery, and it is the amount of charge that would flow through the resistor $R$ were we to discharge the capacitor by throwing the switch in the diagram. If we did that, the induced charge layer, which is not part of $Q$, would simply disappear into the slab.

According to this explanation, the ability of a particular material to increase the capacitance ought to depend on the amount of electric charge in its structure and the ease with which the electrons can be displaced with respect to the atomic nuclei. The factor by which the capacitance is increased when an empty capacitor is filled with a particular material, $Q/Q_0$ in our example, is called the *dielectric constant* of that material. The symbol $\kappa$ is usually used for it:

$$Q = \kappa Q_0 \qquad \Longleftrightarrow \qquad C = \kappa C_0 \qquad (10.3)$$

The material itself is often called a *dielectric* when we are talking about its behavior in an electric field. But any homogeneous nonconducting

**Table 10.1.**
Dielectric constants of various substances

| Substance | Conditions | Dielectric constant ($\kappa$) |
|---|---|---|
| Air | gas, 0 °C, 1 atm | 1.00059 |
| Methane, $CH_4$ | gas, 0 °C, 1 atm | 1.00088 |
| Hydrogen chloride, HCl | gas, 0 °C, 1 atm | 1.0046 |
| Water, $H_2O$ | gas, 110 °C, 1 atm | 1.0126 |
| | liquid, 20 °C | 80.4 |
| Benzene, $C_6H_6$ | liquid, 20 °C | 2.28 |
| Methanol, $CH_3OH$ | liquid, 20 °C | 33.6 |
| Ammonia, $NH_3$ | liquid, −34 °C | 22.6 |
| Mineral oil | liquid, 20 °C | 2.24 |
| Sodium chloride, NaCl | solid, 20 °C | 6.12 |
| Sulfur, S | solid, 20 °C | 4.0 |
| Silicon, Si | solid, 20 °C | 11.7 |
| Polyethylene | solid, 20 °C | 2.25–2.3 |
| Porcelain | solid, 20 °C | 6.0-8.0 |
| Paraffin wax | solid, 20 °C | 2.1–2.5 |
| Pyrex glass 7070 | solid, 20 °C | 4.00 |

substance can be so characterized. Table 10.1 lists the measured values of the dielectric constants for a miscellaneous assortment of substances.

Every dielectric constant in the table is larger than 1. We should expect that if our explanation is correct. The presence of a dielectric could *reduce* the capacitance below that of the empty capacitor only if its electrons moved, when the electric field was applied, in a direction opposite to the resulting force. For oscillating electric fields, by the way, some such behavior would not be absurd. But for the steady fields we are considering here it can't work that way.

The dielectric constant of a perfect vacuum is, of course, exactly 1.0 by our definition. For gases under ordinary conditions, $\kappa$ is only a little larger than 1.0, simply because a gas is mostly empty space. Ordinary solids and liquids usually have dielectric constants ranging from 2 to 6 or so. Note, however, that liquid ammonia is an exception to this rule, and water is a spectacular exception. Actually liquid water is slightly conductive, but that, as we shall have to explain later, does not prevent our defining and measuring its dielectric constant. The ionic conductivity of the liquid is not the reason for the gigantic dielectric constant of water. You can discern this extraordinary property of water in the dielectric constant of the vapor if you remember that it is really the *difference* between $\kappa$ and 1 that reveals the electrical influence of the material. Compare the values of $\kappa$ given in the table for water vapor and for air.

Once the dielectric constant of a particular material has been determined, perhaps by measuring the capacitance of one capacitor filled with it, we are able to predict the behavior, not merely of two-plate capacitors, but of *any* electrostatic system made up of conductors and pieces of that dielectric of any shape. That is, we can predict all electric fields that will exist in the vacuum outside the dielectrics for given charges or potentials on the conductors in the system.

The theory that enables us to do this was fully worked out by the physicists of the nineteenth century. Lacking a complete picture of the atomic structure of matter, they were more or less obliged to adopt a macroscopic description. From that point of view, the interior of a dielectric is a featureless expanse of perfectly smooth "mathematical jelly" whose single electrical property distinguishing it from a vacuum is a dielectric constant different from unity.

If we develop only a macroscopic description of matter in an electric field, we shall find it hard to answer some rather obvious-sounding questions – or, rather, hard to ask these questions in such a way that they can be meaningfully answered. For instance, what is the strength of the electric field *inside* the plastic slab of Fig. 10.1(b) when there are certain charges on the plates? Electric field strength is defined by the force on a test charge. How can we put a test charge inside a perfectly dense solid, without disturbing anything, and measure the force on it? What would that force mean if we did measure it? You might think of boring a hole and putting the test charge in the hole with some room to move around, so that you can measure the force on it as on a free particle. But then you will be measuring not the electric field in the dielectric, but the electric field in a cavity in the dielectric, which is quite a different thing.

Fortunately another line of attack is available to us, one that leads up from the microscopic or *atomic* level. We know that matter is made of atoms and molecules; these in turn are composed of elementary charged particles. We know something about the size and structure of these atoms, and we know something about their arrangement in crystals and fluids and gases. Instead of describing our dielectric slab as a volume of structureless but nonvacuous jelly, we shall describe it as a collection of molecules inhabiting a vacuum. If we can find out what the electric charges in *one* molecule do when that molecule is all by itself in an electric field, we should be able to understand the behavior of two such molecules a certain distance apart in a vacuum. It will only be necessary to include the influence, on each molecule, of any electric field arising from the other. This is a vacuum problem. Now all we have to do is extend this to a population of, say, $10^{20}$ molecules occupying a cubic centimeter or so of vacuum, and we have our real dielectric. We hope to do this without generating $10^{20}$ separate problems.

This program, if carried through, will reward us in two ways. We shall be able at last to say something meaningful about the electric and magnetic fields inside matter, answering questions such as the one raised above. What is more valuable, we shall understand how the macroscopic electric and magnetic phenomena in matter arise from, and therefore reveal, the nature of the underlying atomic structure. We are going to study electric and magnetic effects separately. We begin with dielectrics.

Since our first goal is to describe the electric field produced by an atom or molecule, it will help to make some general observations about the electrostatic field external to any small system of charges.

## 10.2 The moments of a charge distribution

An atom or molecule consists of some electric charges occupying a small volume, perhaps a few cubic angstroms ($10^{-30}$ m$^3$) of space. We are interested in the electric field outside that volume, which arises from this rather complicated charge distribution. We shall be particularly concerned with the field far away from the source, by which we mean far away compared with the size of the source itself. What features of the charge structure mainly determine the field at remote points? To answer this, let's look at some arbitrary distribution of charges and see how we might go about computing the field at a point outside it. The discussion in this and the following section generalizes our earlier discussion of dipoles in Section 2.7.

Figure 10.3 shows a charge distribution of some sort located in the neighborhood of the origin of coordinates. It might be a molecule consisting of several positive nuclei and quite a large number of electrons. In any case we shall suppose it is described by a given charge density function $\rho(x, y, z)$; $\rho$ is negative where the electrons are and positive where the nuclei are. To find the electric field at distant points we can begin by computing the potential of the charge distribution. To illustrate, let's take some point $A$ out on the $z$ axis. (Since we are not assuming any special symmetry in the charge distribution, there is nothing special about the $z$ axis.) Let $r$ be the distance of $A$ from the origin. The electric potential at $A$, denoted by $\phi_A$, is obtained as usual by adding the contributions from all elements of the charge distribution:

$$\phi_A = \frac{1}{4\pi\epsilon_0} \int \frac{\rho(x', y', z')\, dv'}{R}. \qquad (10.4)$$

In the integrand, $dv'$ is an element of volume within the charge distribution, $\rho(x', y', z')$ is the charge density there, and $R$ in the denominator is the distance from $A$ to this particular charge element. The integration is carried out in the coordinates $x'$, $y'$, $z'$, of course, and is extended over all the region containing charge. We can express $R$ in terms of $r$ and the distance $r'$ from the origin to the charge element. Using the law of cosines with $\theta$ the angle between $\mathbf{r}'$ and the axis on which $A$ lies, we have

$$R = (r^2 + r'^2 - 2rr'\cos\theta)^{1/2}. \qquad (10.5)$$

With this substitution for $R$, the integral becomes

$$\phi_A = \frac{1}{4\pi\epsilon_0} \int \rho\, dv'(r^2 + r'^2 - 2rr'\cos\theta)^{-1/2}. \qquad (10.6)$$

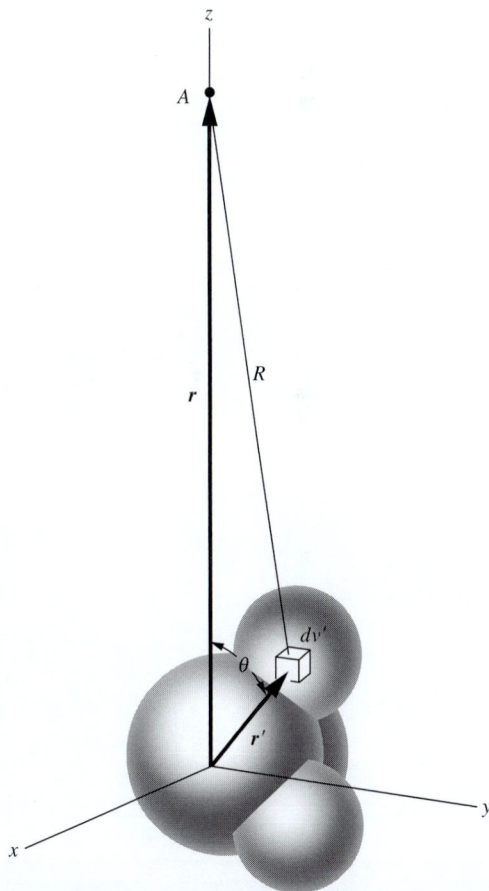

**Figure 10.3.**
Calculation of the potential, at a point $A$, of a molecular charge distribution.

Now we want to take advantage of the fact that, for a distant point like $A$, $r'$ is much smaller than $r$ for all parts of the charge distribution. This suggests that we should expand the square root in Eq. (10.5) in powers of $r'/r$. Writing

$$(r^2 + r'^2 - 2rr' \cos\theta)^{-1/2} = \frac{1}{r} \left[ 1 + \left( \frac{r'^2}{r^2} - \frac{2r'}{r} \cos\theta \right) \right]^{-1/2} \quad (10.7)$$

and using the expansion $(1 + \delta)^{-1/2} = 1 - \delta/2 + 3\delta^2/8 - \cdots$, we get, after collecting together terms of the same power in $r'/r$, the following:

$$(r^2 + r'^2 - 2rr' \cos\theta)^{-1/2}$$
$$= \frac{1}{r} \left[ 1 + \frac{r'}{r} \cos\theta + \left( \frac{r'}{r} \right)^2 \frac{(3\cos^2\theta - 1)}{2} + \mathcal{O}\left[ \left( \frac{r'}{r} \right)^3 \right] \right], \quad (10.8)$$

where the last term here indicates terms of order at least $(r'/r)^3$. These are very small if $r' \ll r$. Now, $r$ is a constant in the integration, so we can take it outside and write the prescription for the potential at $A$ as follows:

$$\phi_A = \frac{1}{4\pi\epsilon_0} \left[ \frac{1}{r} \underbrace{\int \rho \, dv'}_{K_0} + \frac{1}{r^2} \underbrace{\int r' \cos\theta \, \rho \, dv'}_{K_1} \right. \quad (10.9)$$
$$\left. + \frac{1}{r^3} \underbrace{\int r'^2 \frac{(3\cos^2\theta - 1)}{2} \rho \, dv'}_{K_2} + \cdots \right].$$

Each of the integrals above, $K_0$, $K_1$, $K_2$, and so on, has a value that depends only on the structure of the charge distribution, not on the distance to point $A$. Hence the potential for all points along the $z$ axis can be written as a power series in $1/r$ with constant coefficients:

$$\phi_A = \frac{1}{4\pi\epsilon_0} \left[ \frac{K_0}{r} + \frac{K_1}{r^2} + \frac{K_2}{r^3} + \cdots \right]. \quad (10.10)$$

This power series is called the *multipole expansion* of the potential, although we have calculated it only for a point on the $z$ axis here. To finish the problem we would have to get the potential $\phi$ at all other points, in order to calculate the electric field as $-\text{grad}\,\phi$. We have gone far enough, though, to bring out the essential point: *The behavior of the potential at large distances from the source will be dominated by the first term in the above series whose coefficient is not zero.*

Let us look at these coefficients more closely. The coefficient $K_0$ is $\int \rho \, dv'$, which is simply the total charge in the distribution. If we have equal amounts of positive and negative charge, as in a neutral molecule,

$K_0$ will be zero. For a singly ionized molecule, $K_0$ will have the value $e$. If $K_0$ is not zero, then no matter how large $K_1$, $K_2$, etc., may be, if we go out to a sufficiently large distance, the term $K_0/r$ will win out. Beyond that, the potential will approach that of a point charge at the origin and so will the field. This is hardly surprising.

Suppose we have a neutral molecule, so that $K_0$ is equal to zero. Our interest now shifts to the second term, with coefficient $K_1 = \int r' \cos \theta \, \rho \, dv'$. Since $r' \cos \theta$ is simply $z'$, this term measures the relative displacement, in the direction toward $A$, of the positive and negative charge. It has a nonzero value for the distributions sketched in Fig. 10.4, where the densities of positive and negative charge have been indicated separately. In fact, all the distributions shown have approximately the same value of $K_1$. Furthermore – and this is a crucial point – *if any charge distribution is neutral, the value of $K_1$ is independent of the position chosen as origin.* That is, if we replace $z'$ by $z' + z_0'$, in effect shifting the origin, the value of the integral is not changed: $\int (z' + z_0') \rho \, dv' = \int z' \rho \, dv' + z_0' \int \rho \, dv'$, and the latter integral is always zero for a neutral distribution.

Evidently, if $K_0 = 0$ and $K_1 \neq 0$, the potential along the $z$ axis will vary asymptotically (that is, with ever-closer approximation as we go out to larger distances) as $1/r^2$. We recognize this dependence on $r$ from the dipole discussion in Section 2.7. We expect the electric field strength to behave asymptotically like $1/r^3$, in contrast with the $1/r^2$ dependence of the field from a point charge. Of course, we have discussed only the potential on the $z$ axis. We will return to the question of the exact form of the field after getting a general view of the situation.

If $K_0$ and $K_1$ are both zero, and $K_2$ is not, the potential will behave like $1/r^3$ at large distances, and the field strength will fall off with the inverse fourth power of the distance. Figure 10.5 shows a charge distribution for which $K_0$ and $K_1$ are both zero (and would be zero no matter what direction we had chosen for the $z$ axis), while $K_2$ is not zero.

The quantities $K_0$, $K_1$, $K_2$,... are related to what are called the *moments* of the charge distribution. Using this language, we call $K_0$, which is simply the net charge, the *monopole moment,* or *monopole strength.* $K_1$ is one component of the *dipole moment* of the distribution. The dipole moment has the dimensions (charge) × (displacement); it is a *vector,* and our $K_1$ is its $z$ component. The third constant $K_2$ is related to the *quadrupole moment* of the distribution, the next to the *octupole moment,* and so on. The quadrupole moment is not a vector, but a tensor. The charge distribution shown in Fig. 10.5 has a nonzero quadrupole moment. You can quickly show that $K_2 = 3ea^2$, where $a$ is the distance from each charge to the origin.

**Figure 10.4.**
Some charge distributions with $K_0 = 0$, $K_1 \neq 0$. That is, each has net charge zero, but nonzero dipole moment.

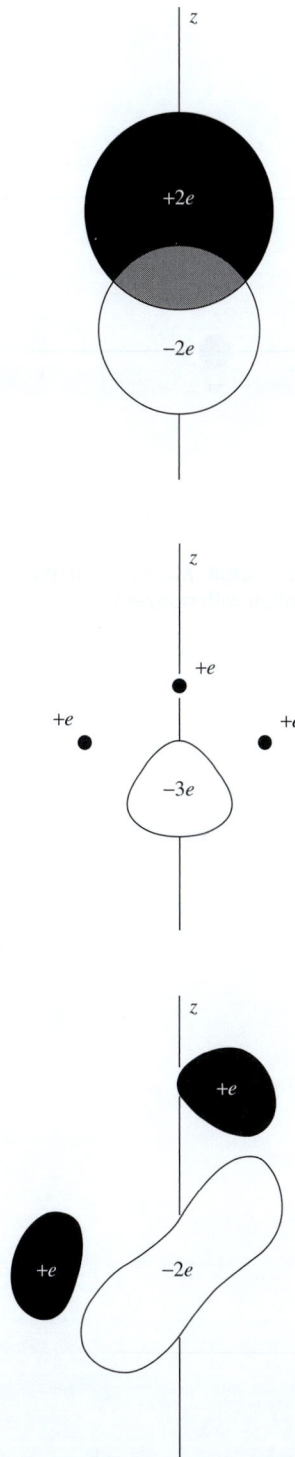

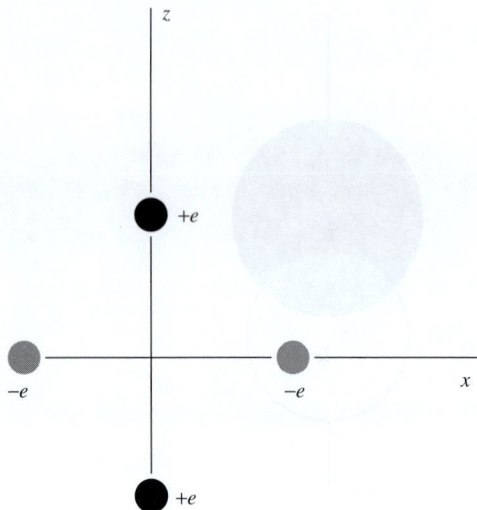

**Figure 10.5.**
For this distribution of charge, $K_0 = K_1 = 0$, but $K_2 \neq 0$. It is a distribution with nonzero quadrupole moment.

**Example (Sphere monopole)**    The external potential due to a spherical shell with uniform surface charge density is $Q/4\pi\epsilon_0 r$. Therefore, its only nonzero moment is the monopole moment. That is, all of the $K_i$ terms except $K_0$ in Eq. (10.10) are zero. Using the integral forms given in Eq. (10.9), verify that $K_1$ and $K_2$ are zero.

**Solution**    For a surface charge density, the $\rho\,dv'$ in the $K_i$ integrals turns into $\sigma\,da' = \sigma(2\pi R \sin\theta)(R\,d\theta)$. Since we're trying to show that the integrals are zero, the various constants in $\sigma\,da'$ don't matter. Only the angular dependence, $\sin\theta\,d\theta$, is relevant. So we have

$$K_1 \propto \int_0^\pi \cos\theta \sin\theta\,d\theta = -\frac{1}{2}\cos^2\theta \Big|_0^\pi = 0,$$

$$K_2 \propto \int_0^\pi (3\cos^2\theta - 1)\sin\theta\,d\theta = \left(-\cos^3\theta + \cos\theta\right)\Big|_0^\pi = 0, \quad (10.11)$$

as desired. Intuitively, it is clear from symmetry that $K_1$ is zero; for every bit of charge with height $z'$, there is a corresponding bit of charge with height $-z'$. But it isn't as intuitively obvious that $K_2$ vanishes.

As mentioned above, $K_1$ and $K_2$ are only components of the complete dipole vector and quadrupole tensor. But the other components can likewise be shown to equal zero, as we know they must. If you want to calculate the general form of the complete quadrupole tensor, one way is to write the $R$ in Eq. (10.5) as $R = \sqrt{(x-x')^2 + (y-y')^2 + (z-z')^2}$, and then perform a Taylor expansion as we did above. See Problem 10.6.

The advantage to us of describing a charge distribution by this hierarchy of moments is that it singles out just those features of the charge distribution that determine the field at a great distance. If we were concerned only with the field in the immediate neighborhood of the distribution, it would be a fruitless exercise. For our main task, understanding what goes on in a dielectric, it turns out that *only* the monopole strength (the net charge) and the dipole strength of the molecular building blocks are important. We can ignore all other moments. And if the building blocks are neutral, we have only their dipole moments to consider.

## 10.3 The potential and field of a dipole

The dipole contribution to the potential at the point $A$, at distance $r$ from the origin, is given by $(1/4\pi\epsilon_0 r^2) \int r' \cos\theta\, \rho\,dv'$. We can write $r' \cos\theta$, which is just the projection of $\mathbf{r}'$ on the direction toward $A$, as $\hat{\mathbf{r}} \cdot \mathbf{r}'$. Thus we can write the potential without reference to any arbitrary axis as

$$\phi_A = \frac{1}{4\pi\epsilon_0 r^2} \int \hat{\mathbf{r}} \cdot \mathbf{r}' \rho\,dv' = \frac{\hat{\mathbf{r}}}{4\pi\epsilon_0 r^2} \cdot \int \mathbf{r}' \rho\,dv', \quad (10.12)$$

which will serve to give the potential at any point with location $r\hat{\mathbf{r}}$. The integral on the right in Eq. (10.12) is the *dipole moment* of the charge

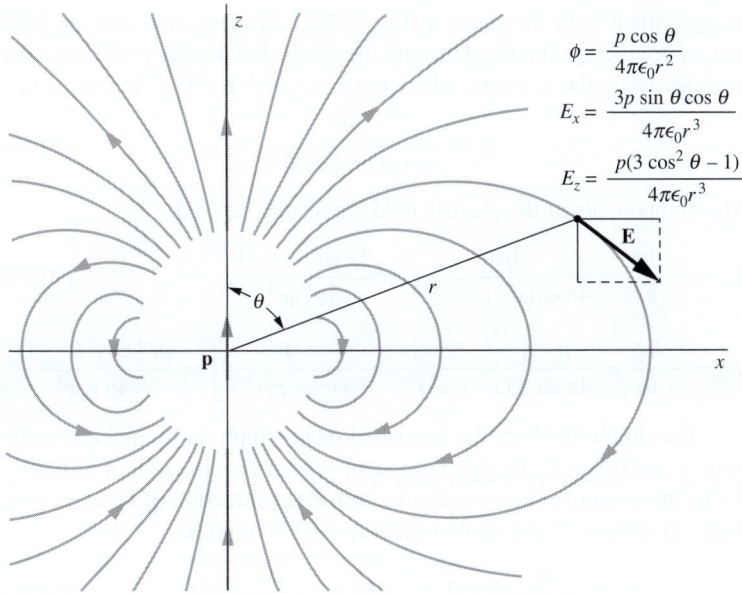

$$\phi = \frac{p\cos\theta}{4\pi\epsilon_0 r^2}$$

$$E_x = \frac{3p\sin\theta\cos\theta}{4\pi\epsilon_0 r^3}$$

$$E_z = \frac{p(3\cos^2\theta - 1)}{4\pi\epsilon_0 r^3}$$

**Figure 10.6.**
The electric field of a dipole, indicated by some field lines.

distribution. It is a vector, obviously, with the dimensions (charge) × (distance). We shall denote the dipole moment vector by **p**:

$$\mathbf{p} = \int \mathbf{r}'\rho\,dv' \tag{10.13}$$

The dipole moment $p = q\ell$ in Section 2.7 is a special case of this result. If we have two point charges $\pm q$ located at positions $z = \pm\ell/2$, then $\rho$ is nonzero only at these two points. So the integral in Eq. (10.13) becomes a discrete sum: $\mathbf{p} = q(\hat{\mathbf{z}}\ell/2) + (-q)(-\hat{\mathbf{z}}\ell/2) = (q\ell)\hat{\mathbf{z}}$, which agrees with the $p = q\ell$ result in Eq. (2.35). The dipole vector points in the direction from the negative charge to the positive charge.

Using the dipole moment **p**, we can rewrite Eq. (10.12) as

$$\phi(\mathbf{r}) = \frac{\hat{\mathbf{r}}\cdot\mathbf{p}}{4\pi\epsilon_0 r^2}. \tag{10.14}$$

The electric field is the negative gradient of this potential. To see what the dipole field is like, locate a dipole **p** at the origin, pointing in the $z$ direction (Fig. 10.6). With this arrangement,

$$\phi = \frac{p\cos\theta}{4\pi\epsilon_0 r^2} \tag{10.15}$$

in agreement with the result in Eq. (2.35).[1] The potential and the field are, of course, symmetrical around the $z$ axis. Let's work with Cartesian coordinates in the $xz$ plane, where $\cos\theta = z/(x^2 + z^2)^{1/2}$. In that plane,

$$\phi = \frac{pz}{4\pi\epsilon_0(x^2 + z^2)^{3/2}}. \tag{10.16}$$

The components of the electric field are readily derived:

$$E_x = -\frac{\partial\phi}{\partial x} = \frac{3pxz}{4\pi\epsilon_0(x^2 + z^2)^{5/2}} = \frac{3p\sin\theta\cos\theta}{4\pi\epsilon_0 r^3}, \tag{10.17}$$

$$E_z = -\frac{\partial\phi}{\partial z} = \frac{p}{4\pi\epsilon_0}\left[\frac{3z^2}{(x^2 + z^2)^{5/2}} - \frac{1}{(x^2 + z^2)^{3/2}}\right] = \frac{p(3\cos^2\theta - 1)}{4\pi\epsilon_0 r^3}.$$

The dipole field can be described more simply in the polar coordinates $r$ and $\theta$. Let $E_r$ be the component of $\mathbf{E}$ in the direction of $\hat{\mathbf{r}}$, and let $E_\theta$ be the component perpendicular to $\hat{\mathbf{r}}$ in the direction of increasing $\theta$. You can show in Problem 10.4 that Eq. (10.17) implies

$$E_r = \frac{p}{2\pi\epsilon_0 r^3}\cos\theta, \qquad E_\theta = \frac{p}{4\pi\epsilon_0 r^3}\sin\theta, \tag{10.18}$$

in agreement with the result in Eq. (2.36). Alternatively, you can quickly derive Eq. (10.18) directly by working in polar coordinates and taking the negative gradient of the potential given by Eq. (10.15). This is the route we took in Section 2.7.

Proceeding out in any direction from the dipole, we find the electric field strength falling off as $1/r^3$, as we had anticipated. Along the $z$ axis the field is parallel to the dipole moment $\mathbf{p}$, with magnitude $p/2\pi\epsilon_0 r^3$; that is, it has the value $\mathbf{p}/2\pi\epsilon_0 r^3$. In the equatorial plane the field points antiparallel to $\mathbf{p}$ and has the value $-\mathbf{p}/4\pi\epsilon_0 r^3$. This field may remind you of the field in the setup with a point charge over a conducting plane, with its image charge, from Section 3.4. That of course is just the two-charge dipole we discussed in Section 2.7. In Fig. 10.7 we show the field of this pair of charges, mainly to emphasize that the field near the charges is *not* a dipole field. This charge distribution has many multipole moments, indeed infinitely many, so it is only the far field at distances $r \gg s$ that can be represented as a dipole field.

To generate a complete dipole field right into the origin we would have to let $s$ shrink to zero while increasing $q$ without limit so as to keep $p = qs$ finite. This highly singular abstraction is not very interesting. We know that our molecular charge distribution will have complicated near fields, so we could not easily represent the near region in any case. Fortunately we shall not need to.

---

[1] Note that the angle $\theta$ here has a different meaning from the angle $\theta$ in Fig. 10.3 and Eqs. (10.5)–(10.9), where it indicated the position of a point in the charge distribution. The present $\theta$ indicates the position of a given point (at which we want to calculate $\phi$ and $\mathbf{E}$) with respect to the dipole direction.

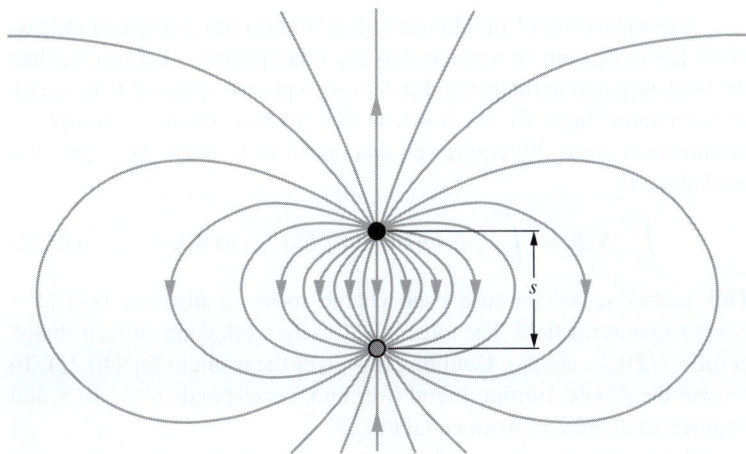

**Figure 10.7.**
The electric field of a pair of equal and opposite point charges approximates the field of a dipole for distances large compared with the separation $s$.

## 10.4 The torque and the force on a dipole in an external field

Suppose two charges, $q$ and $-q$, are mechanically connected so that $s$, the distance between them, is fixed. You may think of the charges as stuck on the end of a short nonconducting rod of length $s$. We shall call this object a dipole. Its dipole moment $p$ is simply $qs$. Let us put the dipole in an external electric field, that is, the field from some other source. The field of the dipole itself does not concern us now. Consider first a uniform electric field, as in Fig. 10.8(a). The positive end of the dipole is pulled toward the right, the negative end toward the left, by a force of strength $qE$. The net force on the object is zero, and so is the torque, in this position.

A dipole that makes some angle $\theta$ with the field direction, as in Fig. 10.8(b), obviously experiences a torque. In general, the torque $\mathbf{N}$ around an axis through some chosen origin is $\mathbf{r} \times \mathbf{F}$, where $\mathbf{F}$ is the force applied at a position $\mathbf{r}$ relative to the origin. Taking the origin in the center of the dipole, so that $r = s/2$, we have

$$\mathbf{N} = \mathbf{r} \times \mathbf{F}_+ + (-\mathbf{r}) \times \mathbf{F}_-. \qquad (10.19)$$

$\mathbf{N}$ is a vector perpendicular to the figure, and its magnitude is given by

$$N = \frac{s}{2} qE \sin\theta + \frac{s}{2} qE \sin\theta = sqE \sin\theta = pE \sin\theta. \qquad (10.20)$$

This can be written simply as

$$\boxed{\mathbf{N} = \mathbf{p} \times \mathbf{E}} \qquad (10.21)$$

When the total force on the dipole is zero, as it is in this case, the torque is independent of the choice of origin (as you should verify), which therefore need not be specified.

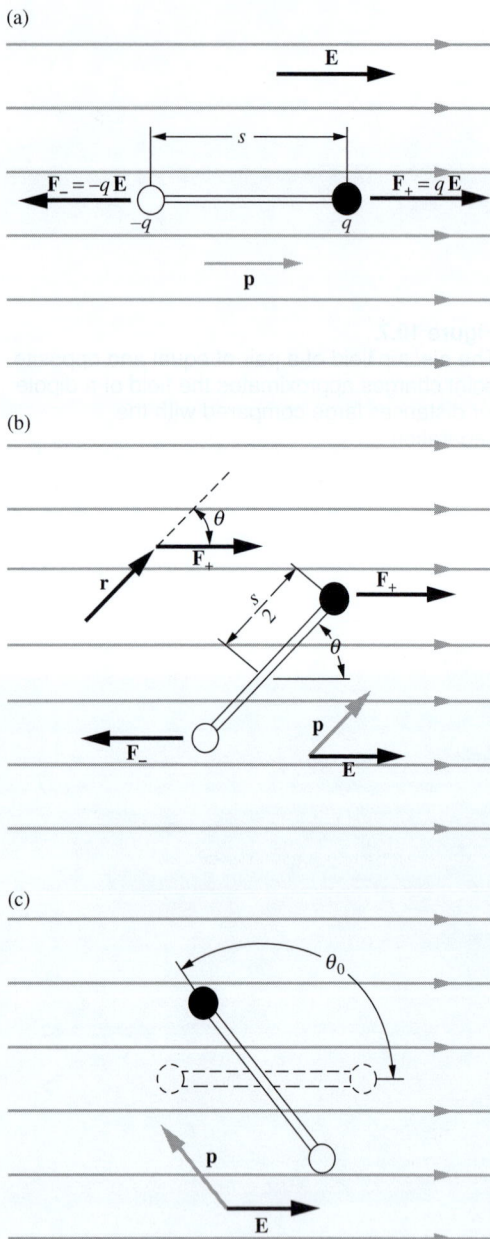

(a)

(b)

(c)

**Figure 10.8.**
(a) A dipole in a uniform field. (b) The torque on the dipole is $\mathbf{N} = \mathbf{p} \times \mathbf{E}$; the vector $\mathbf{N}$ points into the page. (c) The work done in turning the dipole from an orientation parallel to the field to the orientation shown is $pE(1 - \cos\theta_0)$.

The orientation of the dipole in Fig. 10.8(a) has the lowest energy. Work has to be done to rotate it into any other position. Let us calculate the work required to rotate the dipole from a position parallel to the field, through some angle $\theta_0$, as shown in Fig. 10.8(c). Rotation through an infinitesimal angle $d\theta$ requires an amount of work $N\,d\theta$. Thus the total work done is

$$\int_0^{\theta_0} N\,d\theta = \int_0^{\theta_0} pE \sin\theta\,d\theta = pE(1 - \cos\theta_0). \tag{10.22}$$

This makes sense, because each charge moves a distance $(s/2)(1 - \cos\theta_0)$ against the field. The force is $qE$, so the work done on each charge is $(qE)(s/2)(1 - \cos\theta_0)$. Doubling this gives the result in Eq. (10.22). To reverse the dipole, turning it end over end, corresponds to $\theta_0 = \pi$ and requires an amount of work equal to $2pE$.

The net force on the dipole in any *uniform* field is zero, obviously, regardless of its orientation. In a nonuniform field the forces on the two ends of the dipole will generally not be exactly equal and opposite, and there will be a net force on the object. A simple example is a dipole in the field of a point charge $Q$. If the dipole is oriented radially, as in Fig. 10.9(a), with the positive end nearer the positive charge $Q$, the net force will be outward, and its magnitude will be

$$F = (q)\frac{Q}{4\pi\epsilon_0 r^2} + (-q)\frac{Q}{4\pi\epsilon_0 (r+s)^2}. \tag{10.23}$$

For $s \ll r$, we need only evaluate this to first order in $s/r$:

$$F = \frac{qQ}{4\pi\epsilon_0 r^2}\left[1 - \frac{1}{(1 + \frac{s}{r})^2}\right] \approx \frac{qQ}{4\pi\epsilon_0 r^2}\left[1 - \frac{1}{1 + \frac{2s}{r}}\right]$$

$$\approx \frac{qQ}{4\pi\epsilon_0 r^2}\left[1 - \left(1 - \frac{2s}{r}\right)\right] = \frac{sqQ}{2\pi\epsilon_0 r^3}. \tag{10.24}$$

In terms of the dipole moment $p$, this is simply

$$F = \frac{pQ}{2\pi\epsilon_0 r^3}. \tag{10.25}$$

With the dipole at right angles to the field, as in Fig. 10.9(b), there is also a force. Now the forces on the two ends, though equal in magnitude, are not exactly opposite in direction. In this case there is a net upward force.

It is not hard to work out a general formula for the force on a dipole in a nonuniform electric field. The force depends essentially on the *gradients* of the various components of the field. In general, the $x$ component of the force on a dipole of moment $\mathbf{p}$ is

$$\boxed{F_x = \mathbf{p} \cdot \text{grad}\,E_x} \tag{10.26}$$

with corresponding formulas for $F_y$ and $F_z$; see Problem 10.7. All three components can be collected into the concise statement, $\mathbf{F} = (\mathbf{p} \cdot \nabla)\mathbf{E}$.

## 10.5 Atomic and molecular dipoles; induced dipole moments

Consider the simplest atom, the hydrogen atom, which consists of a nucleus and one electron. If you imagine the negatively charged electron revolving around the positive nucleus like a planet around the sun – as in the original atomic model of Niels Bohr – you will conclude that the atom has, at any one instant of time, an electric dipole moment. The dipole moment vector **p** points parallel to the electron–proton radius vector, and its magnitude is $e$ times the electron–proton distance. The direction of this vector will be continually changing as the electron, in this picture of the atom, circles around its orbit. To be sure, the *time average* of **p** will be zero for a circular orbit, but we should expect the periodically changing dipole moment components to generate rapidly oscillating electric fields and electromagnetic radiation.

The absence of such radiation in the normal hydrogen atom was one of the baffling paradoxes of early quantum physics. Modern quantum mechanics tells us that it is better to think of the hydrogen atom in its lowest energy state (the usual condition of most of the hydrogen atoms in the universe) as a spherically symmetrical structure with the electronic charge distributed, in the time average, over a cloud surrounding the nucleus. Nothing is revolving in a circle or oscillating. If we could take a snapshot with an exposure time shorter than $10^{-16}$ s, we might discern an electron localized some distance away from the nucleus. But for processes involving times much longer than that, we have, in effect, a smooth distribution of negative charge surrounding the nucleus and extending out in all directions with steadily decreasing density. The total charge in this distribution is just $-e$, the charge of one electron. Roughly half of it lies within a sphere of radius 0.5 angstrom ($0.5 \cdot 10^{-10}$ m). The density decreases exponentially outward; a sphere only 2.2 angstroms in radius contains 99 percent of the charge. The electric field in the atom is just what a stationary charge distribution of this form, together with the positive nucleus, would produce.

A similar picture is the best one to adopt for other atoms and molecules. We can treat the nuclei in molecules as point charges; for our present purposes their size is too small to matter. The entire electronic structure of the molecule is to be pictured as a single cloud of negative charge of smoothly varying density. The shape of this cloud, and the variation of charge density within it, will of course be different for different molecules. But at the fringes of the cloud the density will always fall off exponentially, so that it makes some sense to talk of the size and shape of the molecular charge distribution.

Quantum mechanics makes a crucial distinction between stationary states and time-dependent states of an atom. The state of lowest energy is a time-independent structure, a stationary state. It has to be, according to the laws of quantum mechanics. It is that state of the atom or

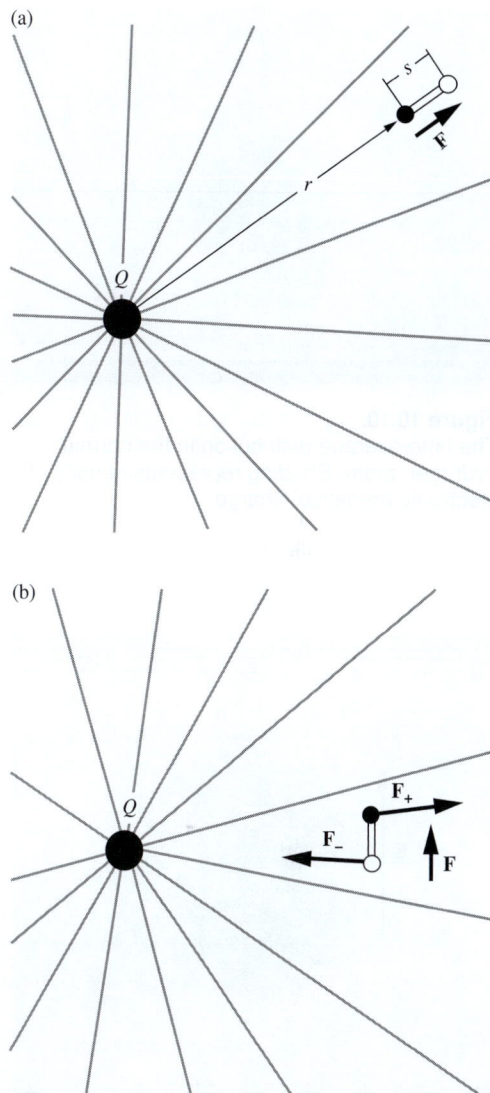

**Figure 10.9.**
The force on a dipole in a nonuniform field.
(a) The net force on the dipole in this position is radially outward. (b) The net force on the dipole in this position is upward.

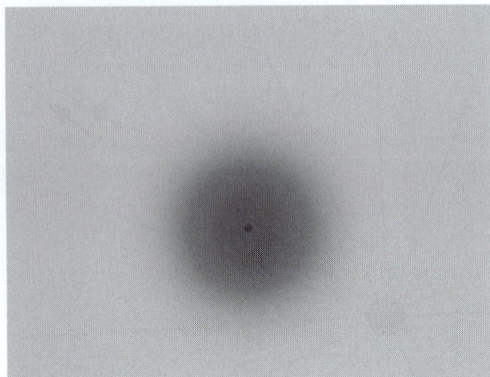

**Figure 10.10.**
The time-average distribution in the normal hydrogen atom. Shading represents density of electronic (negative) charge.

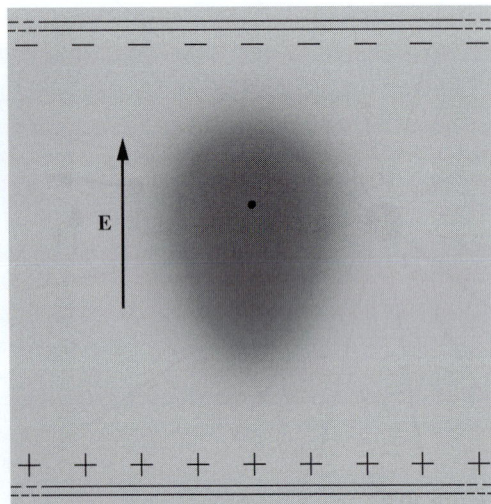

**Figure 10.11.**
In an electric field, the negative charge is pulled one way and the positive nucleus is pulled the other way. The distortion is grossly exaggerated in this picture. To distort the atom that much would require a field of $10^{10}$ volts/m.

molecule that concerns us here. Of course, atoms *can* radiate electromagnetic energy. That happens with the atom in a nonstationary state in which there is an oscillating electric charge.

Figure 10.10 represents the charge distribution in the normal hydrogen atom. It is a cross section through the spherically symmetrical cloud, with the density suggested by shading. Obviously the dipole moment of such a distribution is zero. The same is true of any atom in its state of lowest energy, no matter how many electrons it contains, for in all such states the electron distribution has spherical symmetry. It is also true of any ionized atom, though an ion of course has a monopole moment, that is, a net charge.

So far we have found nothing very interesting. But now let us put the hydrogen atom in an electric field supplied by some external source, as in Fig. 10.11. The electric field distorts the atom, pulling the negative charge down and pushing the positive nucleus up. The distorted atom will have an electric dipole moment because the "center of gravity" of the negative charge will no longer coincide with the positive nucleus, but will be displaced from the nucleus by some small distance $\Delta z$. The electric dipole moment of the atom is now $e\,\Delta z$.

How much distortion will be caused by a field of given strength $E$? Remember that electric fields already exist in the unperturbed atom, of strength $e/4\pi\epsilon_0 a^2$ in order of magnitude, where $a$ is a typical atomic dimension. We should expect the relative distortion of the atom's structure, measured by the ratio $\Delta z/a$, to have the same order of magnitude as the ratio of the perturbing field $E$ to the internal fields that hold the atom together. We predict, in other words, that

$$\frac{\Delta z}{a} \approx \frac{E}{e/4\pi\epsilon_0 a^2}. \qquad (10.27)$$

If you don't trust this reasoning, Exercise 10.30 gives an alternative method for finding the relation between $\Delta z$ and $E$.

Now $a$ is a length of order $10^{-10}$ m, and $e/4\pi\epsilon_0 a^2$ is approximately $10^{11}$ volts/m, a field thousands of times more intense than any large-scale steady field we could make in the laboratory. Evidently the distortion of the atom is going to be very slight indeed, in any practical case. If Eq. (10.27) is correct, it follows that the dipole moment $p$ of the distorted atom, which is just $e\,\Delta z$, will be

$$p = e\,\Delta z \approx 4\pi\epsilon_0 a^3 E. \qquad (10.28)$$

Since the atom was spherically symmetrical before the field **E** was applied, the dipole moment vector **p** will be in the direction of **E**. The factor that relates **p** to **E** is called the *atomic polarizability,* and is usually denoted by $\alpha$:

$$\boxed{\mathbf{p} = \alpha\mathbf{E}} \qquad (10.29)$$

**Table 10.2.**
Atomic polarizabilities ($\alpha/4\pi\epsilon_0$), in units of $10^{-30}\,\mathrm{m}^3$

| Element | H | He | Li | Be | C | Ne | Na | Ar | K |
|---|---|---|---|---|---|---|---|---|---|
| $\alpha/4\pi\epsilon_0$ | 0.66 | 0.21 | 12 | 9.3 | 1.5 | 0.4 | 27 | 1.6 | 34 |

It is common to work instead with the quantity $\alpha/4\pi\epsilon_0$, which has the dimensions of volume. The reason for this is that a direct comparison between $\mathbf{p}$ and $\mathbf{E}$ isn't quite a fair one, because electric fields contain a somewhat arbitrary factor of $1/4\pi\epsilon_0$ multiplying the factors of charge and distance in Coulomb's law. A more reasonable comparison would therefore involve $\mathbf{p}$ and $4\pi\epsilon_0\mathbf{E}$. These quantities have dimensions of (charge) × (distance) and (charge)/(distance)$^2$, respectively. Equation (10.29) then yields $\mathbf{p}/(4\pi\epsilon_0\mathbf{E}) = \alpha/4\pi\epsilon_0$. This quantity is often also called the atomic polarizability, so the term is a little ambiguous. It is best to say explicitly whether you are working with $\alpha$ or $\alpha/4\pi\epsilon_0$.

According to our estimate in Eq. (10.28), we have $\alpha \approx 4\pi\epsilon_0 a^3$, so $\alpha/4\pi\epsilon_0$ is in order of magnitude an atomic volume, something like $a^3 \approx 10^{-30}\,\mathrm{m}^3$. Its value for a particular atom will depend on the details of the atom's electronic structure. An exact quantum-mechanical calculation of the polarizability of the hydrogen atom predicts $\alpha/4\pi\epsilon_0 = (9/2)a_0^3$, where $a_0$ is the *Bohr radius*, $0.52 \cdot 10^{-10}\,\mathrm{m}$, the characteristic distance in the H-atom structure in its normal state. The values of $\alpha/4\pi\epsilon_0$ for several species of atoms, experimentally determined, are given in Table 10.2. The examples given are arranged in order of increasing number of electrons. Note the wide variations in $\alpha/4\pi\epsilon_0$. If you are acquainted with the periodic table of the elements, you may discern something systematic here. Hydrogen and the alkali metals lithium, sodium, and potassium, which occupy the first column of the periodic table, have large values of $\alpha/4\pi\epsilon_0$, and these increase steadily with increasing atomic number, from hydrogen to potassium. The noble gases have much smaller atomic polarizabilities, but these also increase as we proceed, within the family, from helium to neon to krypton. Apparently the alkali atoms, as a class, are easily deformed by an electric field, whereas the electronic structure of a noble gas atom is much stiffer. It is the loosely bound outer, or "valence," electron in the alkali atom structure that is responsible for the easy polarizability.

A molecule, too, develops an induced dipole moment when an electric field is applied to it. The methane molecule depicted in Fig. 10.12 is made from four hydrogen atoms arranged at the corners of a tetrahedron around the central carbon atom. This object has an electrical polarizability, determined experimentally, of

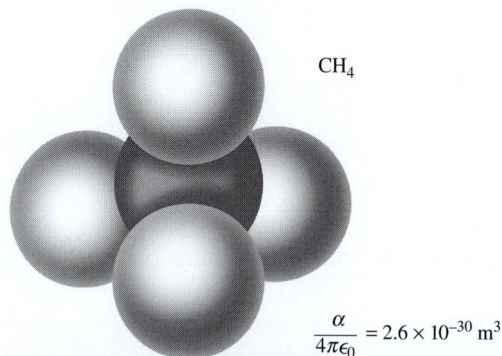

$CH_4$

$$\frac{\alpha}{4\pi\epsilon_0} = 2.6 \times 10^{-30}\,\mathrm{m}^3$$

$$\frac{\alpha}{4\pi\epsilon_0} = 2.6 \cdot 10^{-30}\,\mathrm{m}^3. \tag{10.30}$$

**Figure 10.12.**
The methane molecule, made of four hydrogen atoms and a carbon atom.

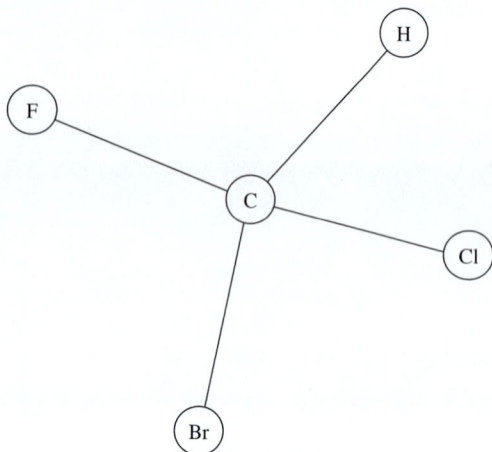

**Figure 10.13.**
A molecule with no symmetry whatsoever, bromochloroflouromethane. This is methane with three different halogens substituted for three of the hydrogens. The bond lengths and the tetrahedron edges are all a bit different.

It is interesting to compare this with the sum of the polarizabilities of a carbon atom and four isolated hydrogen atoms. Taking the data from Table 10.2, we find $\alpha_C/4\pi\epsilon_0 + 4\alpha_H/4\pi\epsilon_0 = 4.1 \cdot 10^{-30}$ m$^3$. Evidently the binding of the atoms into a molecule has somewhat altered the electronic structure. Measurements of atomic and molecular polarizabilities have long been used by chemists as clues to molecular structure.

## 10.6 Permanent dipole moments

Some molecules are so constructed that they have electric dipole moments even in the absence of an electric field. They are unsymmetrical in their normal state. The molecule shown in Fig. 10.13 is an example. A simpler example is provided by any diatomic molecule made out of dissimilar atoms, such as hydrogen chloride, HCl. There is no point on the axis of this molecule about which the molecule is symmetrical fore and aft; the two ends of the molecule are physically different. It would be a pure accident if the center of gravity of the positive charge and that of the negative charge happened to fall at the same point along the axis. When the HCl molecule is formed from the originally spherical H and Cl atoms, the electron of the H atom shifts partially over to the Cl structure, leaving the hydrogen nucleus partially denuded. So there is some excess of positive charge at the hydrogen end of the molecule and a corresponding excess of negative charge at the chlorine end. The magnitude of the resulting electric dipole moment, $p = 3.4 \cdot 10^{-30}$ coulomb-meter, is equivalent to shifting one electron about 0.2 angstrom (using $s = p/e$).

By contrast, the hydrogen atom in a field of 1 megavolt per meter, with the polarizability listed in Table 10.2, acquires an induced moment less than $10^{-34}$ coulomb-meter. Permanent dipole moments, when they exist, are as a rule enormously larger than any moment that can be induced by ordinary laboratory electric fields.[2] Because of this, the distinction between *polar* molecules, as molecules with "built-in" dipole moments are called, and *nonpolar* molecules is very sharp.

We said at the beginning of Section 10.5 that the hydrogen atom had, at any instant of time, a dipole moment. But then we dismissed it as being zero in the time average, on account of the rapid motion of the electron. Now we seem to be talking about molecular dipole moments as if a molecule were an ordinary stationary object like a baseball bat whose ends could be examined at leisure to see which was larger! Molecules move more slowly than electrons, but their motion is rapid by ordinary standards. Why can we credit them with "permanent" electric dipole moments? If this inconsistency was bothering you, you are to be commended. The full answer can't be given without some quantum

[2] There is a good reason for this. The internal electric fields in atoms and molecules, as we remarked in Section 10.5, are naturally on the order of $e/4\pi\epsilon_0(10^{-10}$ m$)^2$, which is roughly $10^{11}$ volts/m! We cannot apply such a field to matter in the laboratory for the closely related reason that it would tear the matter to bits.

mechanics, but the difference essentially involves the time scale of the motion. The time it takes a molecule to interact with its surroundings is generally *shorter* than the time it takes the intrinsic motion of the molecule to average out the dipole moment smoothly. Hence the molecule *really acts* as if it had the moment we have been talking about. A very short time qualifies as permanent in the world of one molecule and its neighbors.

Some common polar molecules are shown in Fig. 10.14, with the direction and magnitude of the permanent dipole moment indicated for each. The water molecule has an electric dipole moment because it is bent in the middle, the O–H axes making an angle of about 105° with one another. This is a structural oddity with the most far-reaching consequences. The dipole moment of the molecule is largely responsible for the properties of water as a solvent, and it plays a decisive role in chemistry that goes on in an aqueous environment. It is hard to imagine what the world would be like if the $H_2O$ molecule, like the $CO_2$ molecule, had its parts arranged in a straight line; probably we wouldn't be here to observe it. We hasten to add that the shape of the $H_2O$ molecule is not a capricious whim of Nature. Quantum mechanics has revealed clearly why a molecule made of an eight-electron atom joined to two one-electron atoms must prefer to be bent.

The behavior of a polar substance as a dielectric is strikingly different from that of material composed of nonpolar molecules. The dielectric constant of water is about 80, that of methyl alcohol 33, while a typical nonpolar liquid might have a dielectric constant around 2. In a nonpolar substance the application of an electric field induces a slight dipole moment in each molecule. In the polar substance dipoles are already present in great strength but, in the absence of a field, are pointing in random directions so that they have no large-scale effect. An applied electric field merely *aligns* them to a certain degree. In either process, however, the macroscopic effects will be determined by the net amount of polarization per unit volume.

## 10.7 The electric field caused by polarized matter
### 10.7.1 The field outside matter

Suppose we build up a block of matter by assembling a very large number of molecules in a previously empty region of space. Suppose too that each of these molecules is polarized in the same direction. For the present we need not concern ourselves with the nature of the molecules or with the means by which their polarization is maintained. We are

**Figure 10.14.**
Some well-known polar molecules. The observed value of the permanent dipole moment $p$ is given in units of $10^{-30}$ coulomb-meters.

Hydrogen chloride

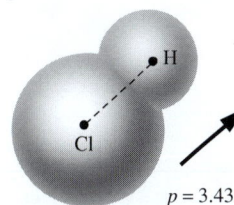

$p = 3.43$

Ammonia

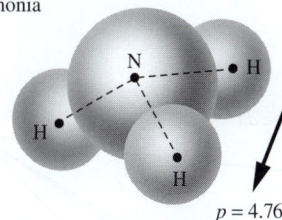

$p = 4.76$

Carbon monoxide

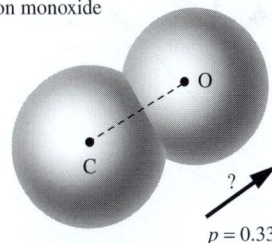

$p = 0.33$

Water

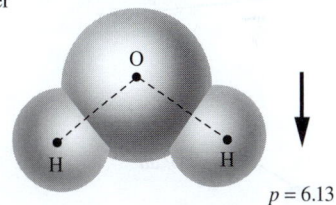

$p = 6.13$

Methanol

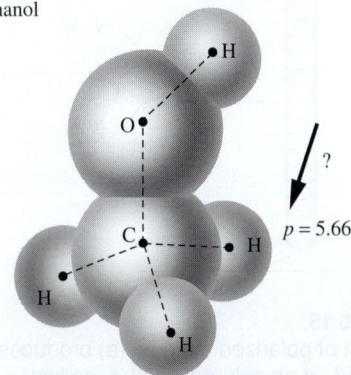

$p = 5.66$

interested only in the electric field *they* produce when they are in this condition; later we can introduce any fields from other sources that might be around. If you like, you can imagine that these are molecules with permanent dipole moments that have been lined up neatly, all pointing the same way, and frozen in position. All we need to specify is $N$, the number of dipoles per cubic meter, and the moment of each dipole **p**. We shall assume that $N$ is so large that any macroscopically small volume $dv$ contains quite a large number of dipoles. The total dipole strength in such a volume is $\mathbf{p}N\,dv$. At any point far away from this volume element compared with its size, the electric field from these particular dipoles would be practically the same if they were replaced by a single dipole moment of strength $\mathbf{p}N\,dv$. We shall call $\mathbf{p}N$ the density of polarization, and denote it by **P**, a vector quantity with the units C-m/m$^3$ (or C/m$^2$):

$$\mathbf{P} \equiv \mathbf{p}N = \frac{\text{dipole moment}}{\text{volume}}. \tag{10.31}$$

$\mathbf{P}\,dv$ is the dipole moment to be associated with any small-volume element $dv$ for the purpose of computing the electric field at a distance. By the way, our matter has been assembled from neutral molecules only; there is no net charge in the system or on any molecule, so we have *only* the dipole moments to consider as sources of a distant field.

Figure 10.15 shows a slender column, or cylinder, of this polarized material. Its cross section is $da$, and it extends vertically from $z_1$ to $z_2$. The polarization density **P** within the column is uniform over the length and points in the positive $z$ direction. We are about to calculate the electric potential, at some external point, due to this column of polarization. An element of the cylinder, of height $dz$, has a dipole moment $\mathbf{P}\,dv = \mathbf{P}\,da\,dz$. Its contribution to the potential at the point $A$ can be written down by referring back to our formula Eq. (10.15) for the potential of a dipole, that is,

$$d\phi_A = \frac{P\,da\,dz\cos\theta}{4\pi\epsilon_0 r^2}. \tag{10.32}$$

The potential due to the entire column is

$$\phi_A = \frac{P\,da}{4\pi\epsilon_0} \int_{z_1}^{z_2} \frac{dz\cos\theta}{r^2}. \tag{10.33}$$

This is simpler than it looks: $dz\cos\theta$ is just $-dr$, so that the integrand is a perfect differential, $d(1/r)$. The result of the integration is then

$$\boxed{\phi_A = \frac{P\,da}{4\pi\epsilon_0}\left(\frac{1}{r_2} - \frac{1}{r_1}\right)} \tag{10.34}$$

Equation (10.34) is precisely the same as the expression for the potential at $A$ that would be produced by two point charges, a positive charge of magnitude $P\,da$ sitting on top of the column at a distance $r_2$ from $A$, and a negative charge of the same magnitude at the bottom of

(a)

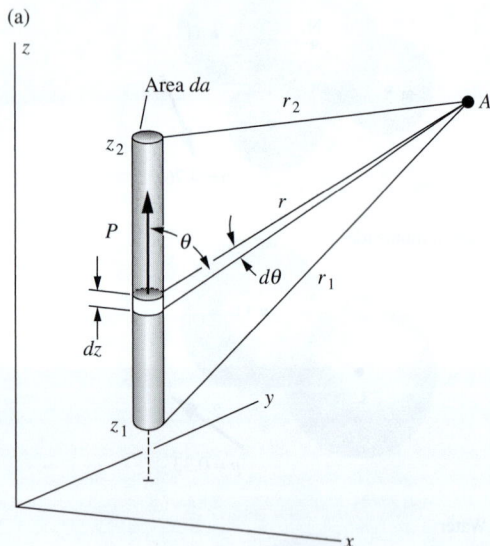

(b)

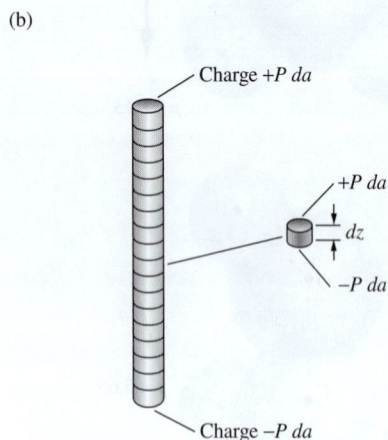

**Figure 10.15.**
A column of polarized material (a) produces the same field, at an external point $A$, as two charges, one at each end of the column (b).

(a)

(b)

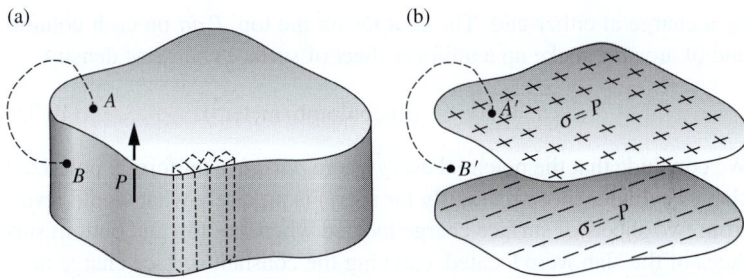

**Figure 10.16.**
A block of polarized material (a) is equivalent to two sheets of charge (b), as far as the field outside is concerned.

the column. The source consisting of a column of uniformly polarized matter is equivalent, at least so far as its field at all *external* points is concerned, to two concentrated charges. Note that nowhere have we assumed that $A$ is far away from the column, that is, that $r_1$ and $r_2$ are much larger than the height of the column, $z_2 - z_1$. All that is required is that the distance from $A$ to any point in the column is much larger than the size of the dipoles (assumed to be very small) and also much larger than the width of the column (also assumed to be small), for then Eq. (10.32) will be valid.

We can prove Eq. (10.34) in another way without any mathematics. Consider a small section of the column of height $dz$, containing a dipole moment $P\,da\,dz$. Let us make an imitation or substitute for this by taking an unpolarized insulator of the same size and shape and sticking a charge $P\,da$ on top of it and a charge $-P\,da$ on the bottom. This little block now has the same dipole moment as that bit of our original column, and therefore it will make an identical contribution to the field at any remote point $A$. (The field inside our substitute, or very close to it, may be different from the field of the original – we don't care about that.) Now make a whole set of such blocks and stack them up to imitate the polarized column; see Fig. 10.15(b). They must give the same field at $A$ as the whole column does, for each block gave the same contribution as its counterpart in the original. Now see what we have! At every joint the positive charge on the top of one block coincides with the negative charge on the bottom of the block above it, making the charge equal zero. The only charges left uncompensated are the negative charge $-P\,da$ on the bottom of the bottom block and the positive charge $P\,da$ on the top of the top block. Seen from a distant point such as $A$ ("distant" compared with the size of a block, not necessarily the whole column), these look like point charges. We conclude, as before, that two such charges produce at $A$ exactly the same field as does our whole column of polarized material.

With no further calculation we can extend this to a slab, or right cylinder, of any proportions uniformly polarized in a direction perpendicular to its parallel faces; see Fig. 10.16(a). The slab can simply be subdivided into a bundle of columns, and the potential outside will be the sum of the contributions of the columns, each of which can be replaced by a charge at either end. The charges on the top, *P da* on each column end of area *da*, make up a uniform sheet of surface charge of density

$$\boxed{\sigma = P} \qquad \text{(in coulomb/meter}^2\text{)}. \qquad (10.35)$$

We conclude that the potential everywhere *outside* a uniformly polarized slab or cylinder (not necessarily far away) is precisely what would result from two sheets of surface charge located where the top and bottom surfaces of the slab were located, carrying the constant surface charge density $\sigma = P$ and $\sigma = -P$, respectively; see Fig. 10.16(b).

We are not quite ready to say anything about the field *inside* the slab. However, we do know the potential at all points on the surface of the slab – top, bottom, or sides. Any two such points, *A* and *B*, can be connected by a path running entirely through the external field, so that the line integral $\int \mathbf{E} \cdot d\mathbf{s}$ is entirely determined by the external field. It must be the same as the integral along the path $A'B'$ in Fig. 10.16(b). A point literally on the surface of the dielectric might be within range of the intense molecular fields, the near field of the molecule that we have left out of our account. Let's agree to define the boundary of the dielectric as a surface far enough out from the outermost atomic nucleus – 10 or 20 angstroms would be margin enough – so that at any point outside this boundary, the near fields of the individual atoms make a negligible contribution to the whole line integral from *A* to *B*.

With this in mind, let's look at a rather thin, wide plate of polarized material, of thickness *t*, shown in cross section in Fig. 10.17(a). Figure 10.17(b) shows, likewise in cross section, the equivalent sheets of charge. For the system of two charge sheets, we know the field, of course, in the space both outside and between the sheets. The field strength inside, well away from the edges, must be just $\sigma/\epsilon_0$, pointing down, and the potential difference between points $A'$ and $B'$ is therefore $\sigma t/\epsilon_0$. The *same potential difference* must exist between corresponding points *A* and *B* on our polarized slab, because the entire *external* field is the same in the two systems.

## 10.7.2 The field inside matter

We can now address the field inside polarized matter. Is the internal field the same in the two systems in Fig. 10.17? Certainly *not*, because the slab is full of positive nuclei and electrons, with fields on the order of $10^{11}$ volts per meter pointing in one direction here, another direction there. But one thing *is* the same: the line integral of the field, reckoned over *any* internal path from *A* to *B*, must be just $\phi_B - \phi_A$, which, as we have seen, is the same as $\phi_{B'} - \phi_{A'}$, which is equal to $\sigma t/\epsilon_0$, or $Pt/\epsilon_0$. This must be so because the introduction of atomic charges, no matter what their distribution, cannot destroy the conservative property of the electric field, expressed in the statement that $\int \mathbf{E} \cdot d\mathbf{s}$ is independent of path, or curl $\mathbf{E} = 0$.

(a)

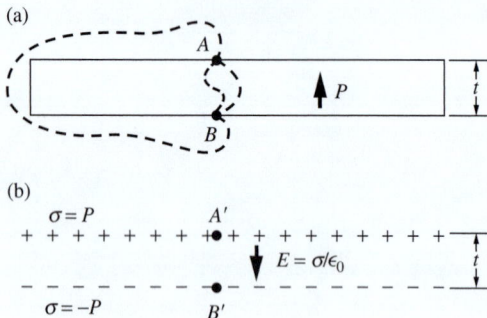

(b)

**Figure 10.17.**
(a) The line integral of **E** from *A* to *B* must be the same over all paths, internal or external, because the internal microscopic or atomic electric fields also are conservative (curl **E** = 0). The equivalent charge sheets (b) have the same external field.

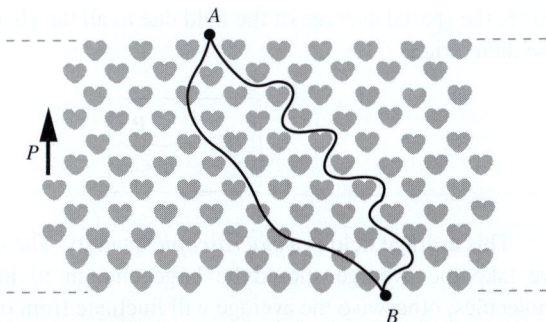

**Figure 10.18.**
Over any path from $A$ to $B$, the line integral of the actual microscopic field is the same.

We know that in Fig. 10.17(b) the potential difference between the top and bottom sheets is nearly constant, except near the edges, because the interior electric field is practically uniform. Therefore in the central area of our polarized plate the potential difference between top and bottom must likewise be constant. In this region the line integral $\int_A^B \mathbf{E} \cdot d\mathbf{s}$ taken from *any* point $A$ on top of the slab to *any* point $B$ on the bottom, by *any* path, must always yield the same value $Pt/\epsilon_0$. Figure 10.18 is a "magnified view" of the central region of the slab, in which the polarized molecules have been made to look something like $H_2O$ molecules all pointing the same way. We have not attempted to depict the very intense fields that exist between and inside the molecules. (The field ten angstroms away from a water molecule is on the order of a hundred megavolts per meter, as you can discover from Fig. 10.14 and Eq. (10.18).) You must imagine some rather complicated field configurations in the neighborhood of each molecule. Now, the $\mathbf{E}$ in $\int \mathbf{E} \cdot d\mathbf{s}$ represents the *total electric field* at a given point in space, inside or outside a molecule; it includes these complicated and intense fields just mentioned. We have reached the remarkable conclusion that *any* path through this welter of charges and fields, whether it dodges molecules or penetrates them, must yield the same value for the path integral, namely the value we find in the system of Fig. 10.17(b), where the field is quite uniform and has the strength $P/\epsilon_0$.

This tells us that the *spatial average* of the electric field within our polarized slab must be $-\mathbf{P}/\epsilon_0$. By the spatial average of a field $\mathbf{E}$ over some volume $V$, which we might denote by $\langle \mathbf{E} \rangle_V$, we mean precisely this:

$$\langle \mathbf{E} \rangle_V = \frac{1}{V} \int_V \mathbf{E} \, dv. \tag{10.36}$$

One way to sample impartially the field in many equal volumes $dv$ into which $V$ might be divided would be to measure the field along each line in a bundle of closely spaced parallel lines. We have just seen that the line integral of $\mathbf{E}$ along any or all such paths is the same as if we were in a constant electric field of strength $-\mathbf{P}/\epsilon_0$. That is the justification for the conclusion that, within the polarized dielectric slab of Figs. 10.17 and

10.18, the spatial average of the field due to all the charges that belong to the dielectric is

$$\langle \mathbf{E} \rangle = -\frac{\mathbf{P}}{\epsilon_0} \qquad (10.37)$$

This average field is a *macroscopic* quantity. The volume over which we take the average should be large enough to include very many molecules, otherwise the average will fluctuate from one such volume to the adjoining one. The average field $\langle \mathbf{E} \rangle$ defined by Eq. (10.36) is really the only kind of *macroscopic* electric field in the interior of a dielectric that we can talk about. It provides the only satisfactory answer, in the context of a macroscopic description of matter, to the question, What is the electric field inside a dielectric material?

We may call the $\mathbf{E}$ in the integrand on the right, in Eq. (10.36), the *microscopic* field. If we imagine that we could measure the field values we need for the path integral, we will be measuring electric fields in vacuum, in the presence, of course, of electric charge. We will need very tiny instruments, for we may be called on to measure the field at a particular point just inside one end of a certain molecule. Have we any right to talk in this way about taking the line integral of $\mathbf{E}$ along some path that skirts the southwest corner of a particular molecule and then tunnels through its neighbor? Yes. The justification is the massive evidence that the laws of electromagnetism work down to a scale of distances much smaller than atomic size. We can even describe an experiment that would serve to measure the average of the microscopic electric field along a path defined well within the limits of atomic dimensions. All we have to do is shoot an energetic charged particle, an alpha particle for example, through the material. From the net change in its momentum, the average electric field that acted on it, over its whole path, could be inferred.

Let us review the properties of the average, or macroscopic, field $\langle \mathbf{E} \rangle$ defined by Eq. (10.36). Its line integral $\int_A^B \langle \mathbf{E} \rangle \cdot d\mathbf{s}$ between any two points $A$ and $B$ that are reasonably far apart is independent of the path. It follows that curl $\langle \mathbf{E} \rangle = 0$ and that $\langle \mathbf{E} \rangle$ is the negative gradient of a potential $\langle \phi \rangle$. This potential function $\langle \phi \rangle$ is itself a smoothed-out average, in the sense of Eq. (10.36), of the microscopic potential $\phi$. (The latter rises to several million volts in the interior of every atomic nucleus!) The surface integral of $\langle \mathbf{E} \rangle$, $\int \langle \mathbf{E} \rangle \cdot d\mathbf{a}$, over any surface that encloses a reasonably large volume, is equal to $1/\epsilon_0$ times the charge within that volume.[3] That is to say, $\langle \mathbf{E} \rangle$ obeys Gauss's law, a statement we can also make in differential

---

[3]  We state this without proof, postponing consideration of the relation of the surface integral of an average field to the average of surface integrals of the microscopic field to Chapter 11, where the question arises in Section 11.8 in connection with the magnetic field inside matter. (See Fig. 11.18.)

form: div $\langle \mathbf{E} \rangle = \langle \rho \rangle / \epsilon_0$, with the understanding that $\langle \rho \rangle$ too is a local average over a suitably macroscopic volume. In short, the spatial average quantities $\langle \mathbf{E} \rangle$, $\langle \phi \rangle$, and $\langle \rho \rangle$ are related to one another in the same way as are the microscopic electric field, potential, and charge density in vacuum.

From now on, when we speak of the electric field $\mathbf{E}$ inside any piece of matter much larger than a molecule, we will mean an average, or macroscopic, field as defined by Eq. (10.36), even when the brackets $\langle \ \rangle$ are omitted.

## 10.8 Another look at the capacitor

At the beginning of this chapter we explained in a qualitative way how the presence of a dielectric between the plates of a capacitor increases its capacitance. Now we are ready to analyze quantitatively the dielectric-filled capacitor. What we have just learned about the electric field inside matter is the key to the problem. We identified as the macroscopic field $\mathbf{E}$, the spatial average of the microscopic field. The line integral of that macroscopic $\mathbf{E}$ between any two points $A$ and $B$ is path-independent and equal to the potential difference. Looking back at Fig. 10.2(a) we observe that the field $\mathbf{E}$ in the empty capacitor must have had the value $\phi_{12}/s$. But the potential difference between the plates, $\phi_{12}$, which was established by the battery, was exactly the same in the dielectric-filled capacitor in Fig. 10.2(b). Hence the field $\mathbf{E}$ in the dielectric, understood now as the macroscopic field, must have had the same value too, for it extends and is uniform over the same distance $s$. (The layers in the diagram are actually negligible in thickness compared with $s$.)

The fact that the $\mathbf{E}$ fields are the same implies that the total charge on and near the top plate in the dielectric-filled capacitor must be the same as it was in the empty capacitor, namely $Q_0$. To prove that, we need only invoke Gauss's law for a suitable imaginary box enclosing the charge layers, as indicated in Fig. 10.19. Now, the charge is made up of two parts, the charge on the plate $Q$ (which will flow off when the capacitor is discharged) and $Q'$, the charge that belongs to the dielectric. The charge on the plate is given by $Q = \kappa Q_0$. That was our definition of $\kappa$. Therefore, if $Q + Q' = Q_0$ as we have just concluded, we must have

$$Q' = Q_0 - Q = Q_0(1 - \kappa). \tag{10.38}$$

We can think of this system as the superposition of a vacuum capacitor and a polarized dielectric slab, Fig. 10.19(b) and (c). In the vacuum capacitor with charge $\kappa Q_0$, the electric field $E''$ would be $\kappa$ times the field $E$. In the isolated polarized dielectric slab the field $E'$ is $-P/\epsilon_0$, as stated in Eq. (10.37). The superposition of these two objects creates the actual field $E$. That is,

$$E = E'' + E' = \kappa E - \frac{P}{\epsilon_0}, \tag{10.39}$$

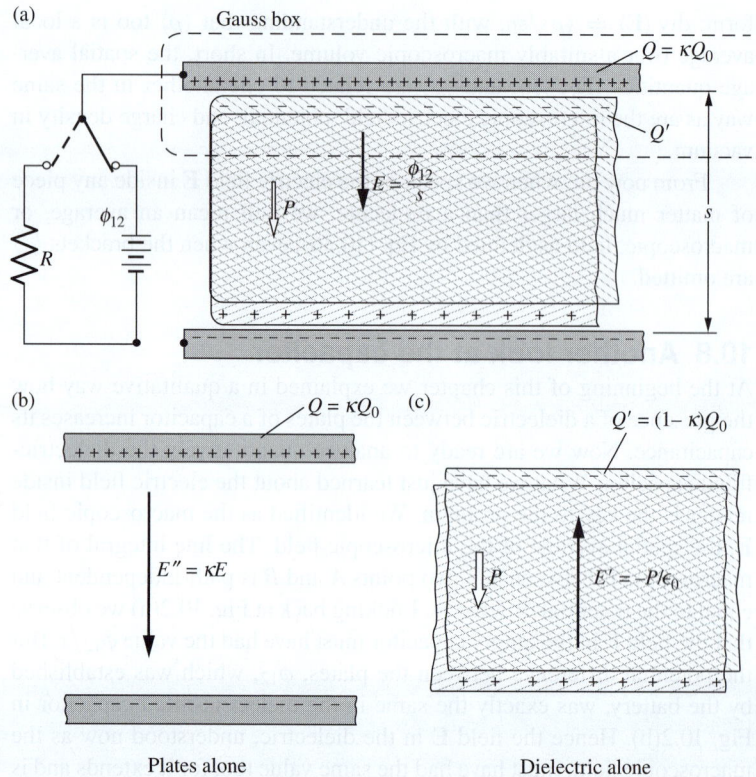

**Figure 10.19.**
The dielectric-filled capacitor of Fig. 10.2(b). The field $E$, which is the average, or macroscopic, field in the dielectric, is $\phi_{12}/s$, equal to the field in the empty capacitor of Fig. 10.2(a). The charge inside the Gauss box must equal $Q_0$, the charge on the plate of the empty capacitor. The system can be regarded as the superposition of a vacuum capacitor (b) and a polarized dielectric (c).

which can be rearranged like this:

$$\frac{P}{\epsilon_0 E} = \kappa - 1 \tag{10.40}$$

The ratio $P/\epsilon_0 E$ (which is dimensionless) is called the electric susceptibility of the dielectric material and is denoted by $\chi_e$ (Greek chi):

$$\chi_e \equiv \frac{P}{\epsilon_0 E} \implies \boxed{P = \chi_e \epsilon_0 E} \tag{10.41}$$

From Eq. (10.40) we have

$$\chi_e = \kappa - 1 \implies \boxed{\kappa = 1 + \chi_e} \tag{10.42}$$

In most materials under ordinary circumstances, it is the field **E** in the dielectric that *causes* **P**. The relation is quite linear. That is to say, the electric susceptibility $\chi_e$ is a constant characteristic of the particular material and is not dependent on the strength of the electric field or the size or shape of the electrodes. We call such materials, in which **P** is proportional to **E**, *linear dielectrics*. Cases are known, however, usually involving materials composed of polar molecules, in which polarization can be literally frozen in. A block of ice polarized by an externally applied electric field and then cooled in liquid helium will retain its polarization indefinitely after the external field is removed, thus providing a real example of the hypothetical polarized slab in Fig. 10.18.

In addition to frozen-in polarization, there are two other cases where the $\mathbf{P} \propto \mathbf{E}$ relation doesn't hold. First, we can have a nonisotropic crystal, that is, one in which the polarization responds differently to electric fields in different directions. Each component of $\mathbf{P}$ is then a (usually linear) function of, in general, all three components of $\mathbf{E}$. In other words, the $\mathbf{P}$ and $\mathbf{E}$ vectors are related by a full matrix instead of a simple constant of proportionality. So they need not point in the same direction; see the discussion in Footnote 3 in Chapter 4 dealing with the analogous case involving $\mathbf{J}$ and $\mathbf{E}$. However, we will assume that the dielectrics in this chapter are isotropic, unless otherwise stated.

Second, the $\mathbf{P} \propto \mathbf{E}$ relation doesn't hold if, as mentioned above, the proportionality factor $\chi_e \epsilon_0$ depends on the *strength* $E$ of the electric field. In this case $\chi_e \epsilon_0$ is a function of $E$, which means that $P$ is a *nonlinear* function of $E$. If you want, you can consider the linear-dielectric $P = \chi_e \epsilon_0 E$ relation to be the first term in the Taylor series of $P$ as a function of $E$. But the nice thing is that, in most materials, this first term is all we need. As with isotropy, we will assume that our dielectrics are linear, unless otherwise stated. Note that in using definite values of $\kappa$ (that is, ones that are independent of $E$) to describe the various materials in Table 10.1, we are already assuming (correctly) that they are linear dielectrics. Furthermore, we are assuming (again correctly) that the materials are homogeneous and isotropic (on a macroscopic scale); that is, they have the same properties at all points and in all directions.

Strictly speaking, filling a vacuum capacitor with dielectric material increases its capacitance by the precise factor $\kappa$ characteristic of that material only if we fill all the surrounding space too, or at least all the space where there is any electric field. In the example we discussed, it was tacitly assumed that the plates were so large compared with their distance of separation that "edge effects," including the small amount of charge that would be on the outside of the plates near the edge (see Fig. 3.14(b)), could be neglected. A quite general statement can be made about a system of conductors of any shape or arrangement that is entirely immersed in a homogeneous dielectric – for instance, in a large tank of oil. With any charges, $Q_1$, $Q_2$, etc., on the various conductors, the macroscopic electric field $\mathbf{E}_{med}$ at any location in the medium is just $1/\kappa$ times the field $\mathbf{E}_{vac}$ that would exist at that location with the same charges on the same conductors in vacuum (Fig. 10.20). This has important consequences in semiconductors. When silicon, for example, is doped with phosphorus to make an $n$-type semiconductor, the high dielectric constant of the silicon crystal (see Table 10.1) greatly reduces the electrical attraction between the outermost electron of the phosphorus atom and the

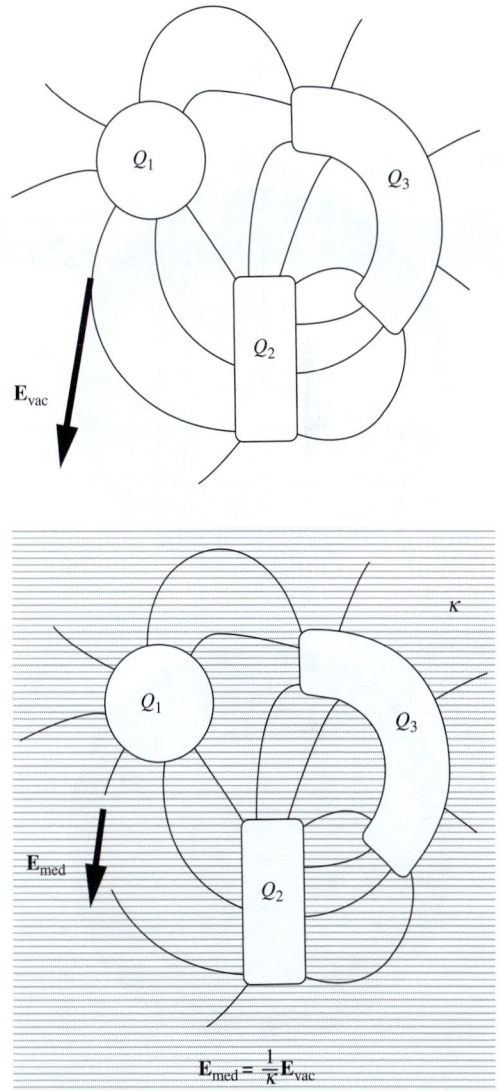

**Figure 10.20.**
For the same charges on the conductors, the presence of the dielectric medium reduces all electric field intensities (and hence all potential differences) by the factor $1/\kappa$. The charges $Q_1$, $Q_2$, and $Q_3$ are the charges that would actually flow off the conductors if we were to discharge the system.

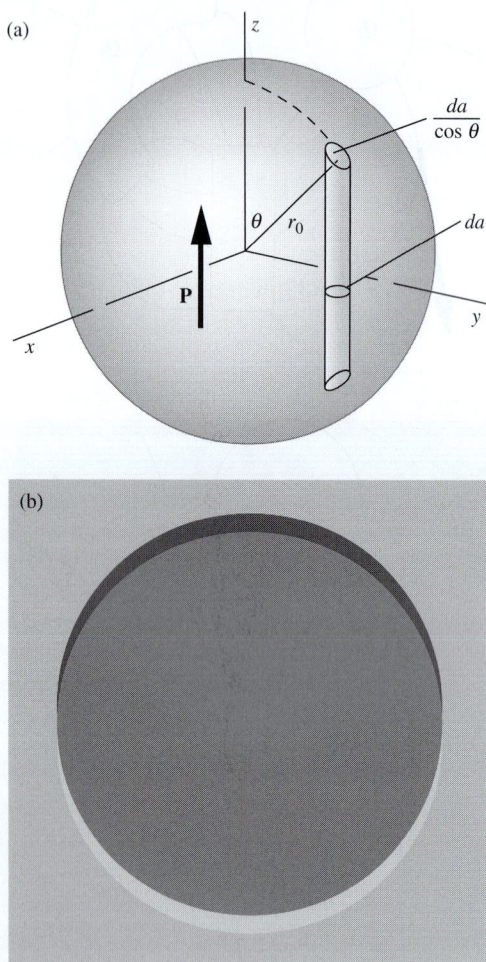

(a)

(b)

**Figure 10.21.**
(a) Divide the polarized sphere into polarized rods, and replace each rod by patches of charge on the surface of the sphere. (b) A ball of positive volume charge density and a ball of negative volume charge density, slightly displaced, are equivalent to a distribution of charge on the spherical surface.

rest of the atom. This makes it easy for the electron to leave the residual $P^-$ ion and join the conduction band, as in Fig. 4.11(a).

This brings us to a more general problem. What if the space in our system is partly filled with dielectric and partly empty, with electric fields in both parts? We'll begin with a somewhat artificial but instructive example, a polarized solid sphere in otherwise empty space.

## 10.9 The field of a polarized sphere

The solid sphere in Fig. 10.21(a) is supposed to be uniformly polarized, as if it had been carved out of the substance of the slab in Fig. 10.16(a). What must the electric field be like, both inside and outside the sphere? We take **P** as usual to denote the density of polarization, constant in magnitude and direction throughout the volume of the sphere. The polarized material could be divided, like the slab in Fig. 10.16(a), into columns parallel to **P**, and each of these replaced by a charge of magnitude $P \times$ (column cross section) at top and bottom. Thus the field we seek is that of a surface charge distribution spread over a sphere with density $\sigma = P \cos \theta$. The factor $\cos \theta$ enters, as should be evident from the figure, because a column of cross section $da$ intercepts on the sphere a patch of surface of area $da / \cos \theta$. Figure 10.21(b) is a cross section through this shell of equivalent surface charge in which the density of charge has been indicated by the varying thickness of the black semicircle above (positive charge density) and the light semicircle below (negative charge density).

If it has not already occurred to you, this figure may suggest that we think of the polarization **P** as having arisen from the slight upward displacement of a ball filled uniformly with positive charge of volume density $\rho$, relative to a ball of negative charge of density $-\rho$. That would leave uncompensated positive charge poking out at the top and negative charge showing at the bottom, varying in amount precisely as $\cos \theta$ over the whole boundary.[4] In the interior, where the positive and negative charge densities still overlap, they would exactly cancel one another. Taking this view, we see a very easy way to calculate the field *outside* the shell of surface charge. Any spherical charge distribution, as we know, has an external field the same as if its entire charge were concentrated at the center. So the superposition of two spheres of total charge $Q$ and $-Q$, with their centers separated by a small displacement $s$, will produce an external field the same as that of two point charges $Q$ and $-Q$, a distance $s$ apart. This is just a dipole with dipole moment $p_0 = Qs$.

A microscopic description of the polarized substance leads us to the same conclusion. In Fig. 10.22(a) the molecular dipoles actually responsible for the polarization **P** have been crudely represented as consisting individually of a pair of charges $q$ and $-q$, a distance $s$ apart, to make

---

[4] This follows from the fact that the thickness of the "semicircle" at a given point is the radial component of the vertical vector representing the displacement $s$ of the top sphere relative to the bottom sphere. You can quickly show that this radial component is $s \cos \theta$.

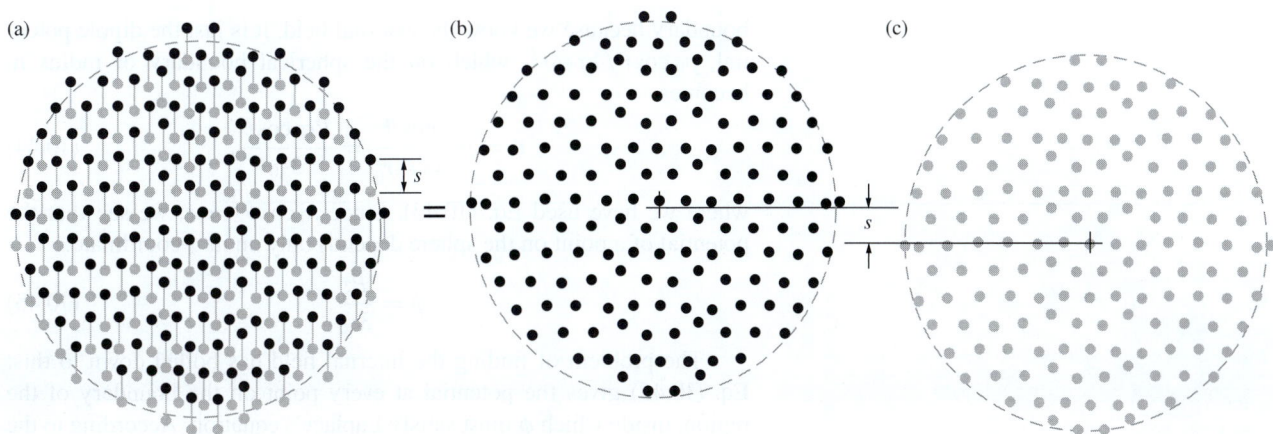

**Figure 10.22.**
A sphere of lined-up molecular dipoles (a) is equivalent to superposed, slightly displaced, spheres of positive (b) and negative (c) charges.

a dipole moment $p = qs$. With $N$ of these per cubic meter, we have $P = Np = Nqs$, and the total number of such dipoles in the sphere is $(4\pi/3)r_0^3N$. The positive charges, considered separately (Fig. 10.22(b)), are distributed throughout a sphere with total charge content $Q = (4\pi/3)r_0^3Nq$, and the negative charges occupy a similar sphere with its center displaced (Fig. 10.22(c)). Clearly each of these charge distributions can be replaced by a point charge at its center, if we are concerned with the field well outside the distribution. "Well outside" means far enough away from the surface so that the actual graininess of the charge distribution doesn't matter, and of course that is something we always have to ignore when we speak of the macroscopic fields.

So, for present purposes, the picture of overlapping spheres of uniform charge density and the description in terms of actual dipoles in a vacuum are equivalent,[5] and show that the field outside the distribution is the same as that of a single dipole located at the center. The moment of this dipole $p_0$ is simply the total polarization in the sphere:

$$p_0 = Qs = \frac{4\pi}{3}r_0^3Nqs = \frac{4\pi}{3}r_0^3P. \tag{10.43}$$

The quantities $Q$ and $s$ have, separately, no significance and may now be dropped from the discussion.

The external field of the polarized sphere is that of a central dipole $p_0$, not only at a great distance from the sphere but also right down to the surface, macroscopically speaking. All we had to do to construct Fig. 10.23, a representation of the external field lines, was to block out a circular area from Fig. 10.6.

The internal field is a different matter. Let's look at the electric potential, $\phi(x, y, z)$. We know the potential at all points on the spherical

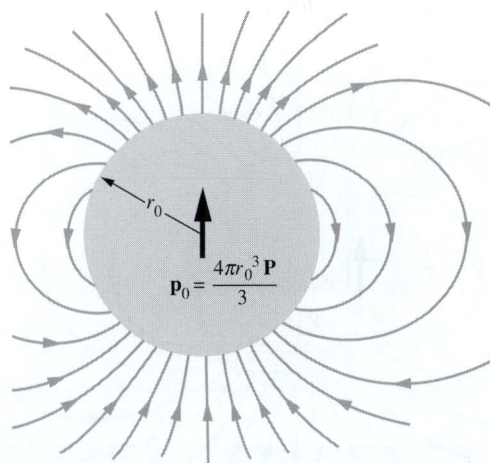

**Figure 10.23.**
The field outside a uniformly polarized sphere is exactly the same as that of a dipole located at the center of the sphere.

---

[5]  This may have been obvious enough, but we have labored the details in this one case to allay any suspicion that the "smooth-charge-ball" picture, which is so different from what we know the interior of a real substance to be like, might be leading us astray.

boundary because we know the external field. It is just the dipole potential, $p_0 \cos \theta / 4 \pi \epsilon_0 r^2$, which on the spherical boundary of radius $r_0$ becomes

$$\phi = p_0 \frac{\cos \theta}{4 \pi \epsilon_0 r_0^2} = \frac{P r_0 \cos \theta}{3 \epsilon_0}, \qquad (10.44)$$

where we have used Eq. (10.43). Since $r_0 \cos \theta = z$, we see that the potential of a point on the sphere depends only on its $z$ coordinate:

$$\phi = \frac{P z}{3 \epsilon_0}. \qquad (10.45)$$

The problem of finding the internal field has boiled down to this: Eq. (10.45) gives the potential at every point on the boundary of the region, inside which $\phi$ must satisfy Laplace's equation. According to the uniqueness theorem we proved in Chapter 3, that suffices to determine $\phi$ throughout the interior. If we can find *a* solution, it must be *the* solution. Now the function $Cz$, where $C$ is any constant, satisfies Laplace's equation, so Eq. (10.45) has actually handed us the solution to the potential *in the interior* of the sphere. That is, $\phi_{\text{in}} = P z / 3 \epsilon_0$. The electric field associated with this potential is uniform and points in the $-z$ direction:

$$E_z = -\frac{\partial \phi_{\text{in}}}{\partial z} = -\frac{\partial}{\partial z} \left( \frac{P z}{3 \epsilon_0} \right) = -\frac{P}{3 \epsilon_0}. \qquad (10.46)$$

As the direction of $\mathbf{P}$ was the only thing that distinguished the $z$ axis, we can write our result in more general form:

$$\boxed{\mathbf{E}_{\text{in}} = -\frac{\mathbf{P}}{3 \epsilon_0}} \qquad (10.47)$$

This is the macroscopic field $\mathbf{E}$ in the polarized material.

Figure 10.24 shows both the internal and external fields. At the upper pole of the sphere, the strength of the upward-pointing external field is, from Eq. (10.17) or Eq. (10.18) for the field of a dipole,

$$E_z = \frac{2 p_0}{4 \pi \epsilon_0 r^3} = \frac{2 (4 \pi r_0^3 P / 3)}{4 \pi \epsilon_0 r_0^3} = \frac{2 P}{3 \epsilon_0} \qquad \text{(outside, at top)}, \quad (10.48)$$

which is just twice the magnitude of the downward-pointing internal field.

This example illustrates the general rules for the behavior of the field components at the surface of a polarized medium. $\mathbf{E}$ is discontinuous at the boundary of a polarized medium, exactly as it would be at a surface in vacuum that carried a surface charge density $\sigma = P_\perp$. The symbol $P_\perp$ stands for the component of $\mathbf{P}$ normal to the surface outward (which in the present case is $P_\perp = P \cos \theta$). It follows that $E_\perp$, the normal component of $\mathbf{E}$, must change abruptly by an amount $P_\perp / \epsilon_0$; whereas $E_\parallel$, the component of $\mathbf{E}$ parallel to the boundary, remains continuous, that is, has the same value on both sides of the boundary (Fig. 10.25). Indeed,

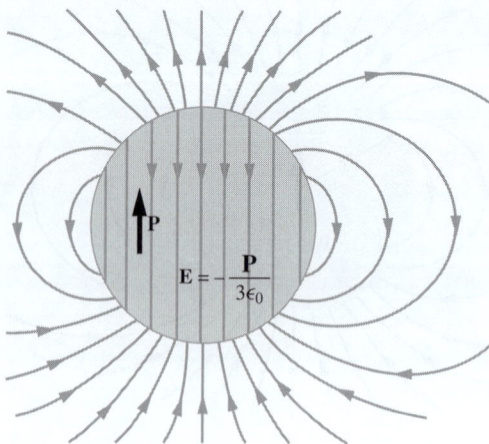

**Figure 10.24.**
The field of the uniformly polarized sphere, both inside and outside.

at the north pole of our sphere, the net change in $E_z$ is $2P/3\epsilon_0 -$ $(-P/3\epsilon_0)$, or $P/\epsilon_0$.

---

**Example (Continuity of $E_\parallel$)**   For our polarized sphere, let's check that the component of **E** parallel to the surface is continuous from inside to outside everywhere on the sphere. From Eq. (10.47) the internal field has magnitude $P/3\epsilon_0$ and points downward, so $E_\parallel^{\text{in}}$ is obtained by simply tacking on a factor of $\sin\theta$. That is, $E_\parallel^{\text{in}} = P\sin\theta/3\epsilon_0$. The tangential component of the external dipole field is given by the $E_\theta$ in Eq. (10.18):

$$E_\parallel^{\text{out}} = \frac{p_0 \sin\theta}{4\pi\epsilon_0 r^3} = \frac{(4\pi r_0^3 P/3)\sin\theta}{4\pi\epsilon_0 r_0^3} = \frac{P\sin\theta}{3\epsilon_0}, \qquad (10.49)$$

which equals $E_\parallel^{\text{in}}$, as desired.

Note that, for $0 < \theta < \pi$, the $\sin\theta$ factor is positive, so $E_\parallel^{\text{in}}$ and $E_\parallel^{\text{out}}$ point in the positive $\hat{\boldsymbol{\theta}}$ direction, that is, away from the north pole. Similarly, for $\pi < \theta < 2\pi$, $E_\parallel^{\text{in}}$ and $E_\parallel^{\text{out}}$ point in the *negative* $\hat{\boldsymbol{\theta}}$ direction, which again is away from the north pole (because positive $\hat{\boldsymbol{\theta}}$ is directed clockwise around the full circle). A quick glance at Fig. 10.24 shows that the field lines are consistent with these facts.

The task of Exercise 10.36 is to use the explicit forms of the internal and external fields to show that $E_\perp$ has a discontinuity of $P_\perp/\epsilon_0$ everywhere on the surface of the sphere.

---

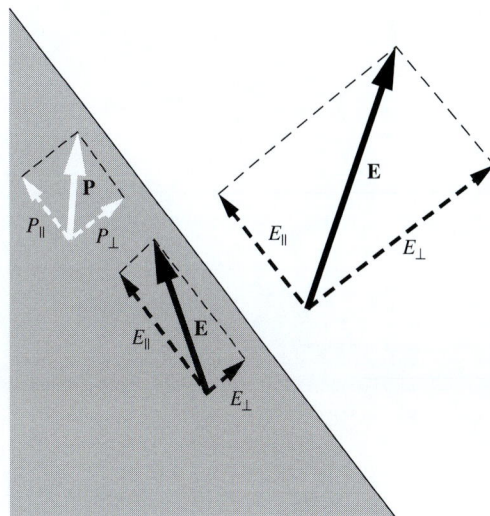

**Figure 10.25.**
The change in $E$ at the boundary of a polarized dielectric: $E_\parallel$ is the same on both sides of the boundary; $E_\perp$ increases by $P_\perp/\epsilon_0$ in going from dielectric to vacuum. (Note that $E$ and $P/\epsilon_0$ are not drawn to the same scale.)

None of these conclusions depends on how the polarization of the sphere was caused. Assuming any sphere *is* uniformly polarized, Fig. 10.24 shows *its* field. Onto this can be superposed any field from other sources, thus representing many possible systems. This will not affect the discontinuity in **E** at the boundary of the polarized medium. The above rules therefore apply in any system, the discontinuity in **E** being determined solely by the existing polarization.

## 10.10 A dielectric sphere in a uniform field

As an example, let us put a sphere of dielectric material characterized by a dielectric constant $\kappa$ into a homogeneous electric field $\mathbf{E}_0$ like the field between the parallel plates of a vacuum capacitor, Fig. 10.26. Let the sources of this field, the charges on the plates, be far from the sphere so that they do not shift as the sphere is introduced. Then whatever the field may be in the vicinity of the sphere, it will remain practically $\mathbf{E}_0$ at a great distance. This is what is meant by putting a sphere into a uniform field. The total field **E** is no longer uniform in the neighborhood of the sphere. It is the *sum* of the uniform field $\mathbf{E}_0$ of the distant sources and a field $\mathbf{E}'$ generated by the polarized matter itself:

$$\mathbf{E} = \mathbf{E}_0 + \mathbf{E}'. \qquad (10.50)$$

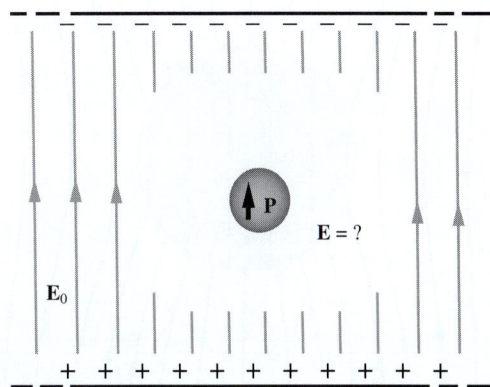

**Figure 10.26.**
The sources of the field $\mathbf{E}_0$ remain fixed. The dielectric sphere develops some polarization **P**. The total field **E** is the superposition of $\mathbf{E}_0$ and the field of this polarized sphere.

This relation is valid both inside and outside the sphere. The field $\mathbf{E}'$ depends on the polarization $\mathbf{P}$ of the dielectric, which in turn depends on the value of $\mathbf{E}$ inside the sphere:

$$\mathbf{P} = \chi_e \epsilon_0 \mathbf{E}_{\text{in}} = (\kappa - 1)\epsilon_0 \mathbf{E}_{\text{in}}. \tag{10.51}$$

Remember that the $\mathbf{E}$ that appears in this expression involving $\chi_e$ is the *total* electric field.

We don't know yet what the total field $\mathbf{E}$ is; we know only that Eq. (10.51) has to hold at any point inside the sphere. *If* the sphere becomes uniformly polarized, an assumption that will need to be justified by our results, the relation between the polarization $\mathbf{P}$ of the sphere and its own field at points inside, $\mathbf{E}'_{\text{in}}$, is given by Eq. (10.47):[6]

$$\mathbf{E}'_{\text{in}} = -\frac{\mathbf{P}}{3\epsilon_0}. \tag{10.52}$$

Substituting the $\mathbf{P}$ from Eq. (10.51) into Eq. (10.52) quickly gives $\mathbf{E}'_{\text{in}}$ in terms of $\mathbf{E}_{\text{in}}$; we obtain $\mathbf{E}'_{\text{in}} = -(\kappa - 1)\mathbf{E}_{\text{in}}/3$. Substituting this into Eq. (10.50) gives the total field inside the sphere as

$$\mathbf{E}_{\text{in}} = \mathbf{E}_0 - \frac{\kappa - 1}{3}\mathbf{E}_{\text{in}} \implies \boxed{\mathbf{E}_{\text{in}} = \left(\frac{3}{2 + \kappa}\right)\mathbf{E}_0} \tag{10.53}$$

Because $\kappa$ is greater than 1, the factor $3/(2 + \kappa)$ will be less than 1; the field inside the dielectric is weaker than $\mathbf{E}_0$. The polarization is

$$\mathbf{P} = (\kappa - 1)\epsilon_0 \mathbf{E}_{\text{in}} \implies \boxed{\mathbf{P} = 3\left(\frac{\kappa - 1}{\kappa + 2}\right)\epsilon_0 \mathbf{E}_0} \tag{10.54}$$

The assumption of uniform polarization is now seen to be self-consistent.[7] To compute the total field $\mathbf{E}_{\text{out}}$ outside the sphere we must add vectorially to $\mathbf{E}_0$ the field of a central dipole with dipole moment equal to $\mathbf{P}$ times the volume of the sphere. Some field lines of $\mathbf{E}$, both inside and outside the dielectric sphere, are shown in Fig. 10.27.

To summarize, we found $\mathbf{E}_{\text{in}}$ by effectively equating two different expressions for the field $\mathbf{E}'_{\text{in}}$ caused by the polarized matter. One expression is simply the statement of superposition, $\mathbf{E}'_{\text{in}} = \mathbf{E}_{\text{in}} - \mathbf{E}_0$. The other expression is $\mathbf{E}'_{\text{in}} = -(\kappa - 1)\mathbf{E}_{\text{in}}/3$, which comes from the facts that $\mathbf{E}'_{\text{in}}$ is proportional to $\mathbf{P}$ (in the case of a sphere) and that $\mathbf{P}$ is proportional to $\mathbf{E}_{\text{in}}$ (in a linear dielectric).

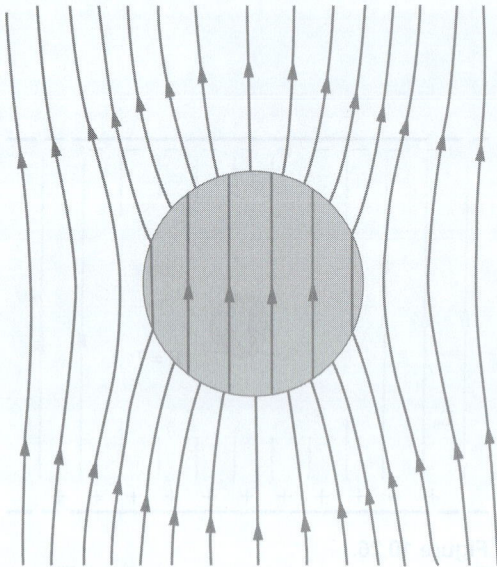

**Figure 10.27.**
The total field **E**, both inside and outside the dielectric sphere.

---

[6] In Eq. (10.47) we were using the symbol $\mathbf{E}_{\text{in}}$, without the prime, for this field. In that case it was the only field present.

[7] That is what makes this system easy to deal with. For a dielectric cylinder of finite length in a uniform electric field, the assumption would not work. The field $\mathbf{E}'$ of a uniformly polarized cylinder – for instance one with its length about equal to its diameter – is *not* uniform inside the cylinder. (What must it look like?) Therefore $\mathbf{E}_{\text{in}} = \mathbf{E}_0 + \mathbf{E}'_{\text{in}}$ cannot be uniform – but in that case $\mathbf{P} = \chi_e \mathbf{E}_{\text{in}}$ could not be uniform after all. In fact, it is only dielectrics of ellipsoidal shape, of which the sphere is a special case, that acquire uniform polarization in a uniform field.

## 10.11 The field of a charge in a dielectric medium, and Gauss's law

Suppose that a very large volume of homogeneous linear dielectric has somewhere within it a concentrated charge $Q$, not part of the regular molecular structure of the dielectric. Imagine, for instance, that a small metal sphere has been charged and then dropped into a tank of oil. As was stated at the end of Section 10.8, the electric field in the oil is simply $1/\kappa$ times the field that $Q$ would produce in a vacuum:

$$E = \frac{Q}{4\pi\epsilon_0\kappa r^2}. \tag{10.55}$$

The product $\epsilon_0\kappa$ is commonly denoted by $\epsilon$, so we can write

$$\boxed{E = \frac{Q}{4\pi\epsilon r^2}} \quad \text{where} \quad \boxed{\epsilon \equiv \kappa\epsilon_0} \implies \kappa = \frac{\epsilon}{\epsilon_0}. \tag{10.56}$$

The quantity $\epsilon$ is known as the *permittivity* of the dielectric. The vacuum permittivity, also called the *permittivity of free space,* is simply $\epsilon_0$.

It is interesting to see how Gauss's law works out. The surface integral of $\mathbf{E}$ (which is the macroscopic, or space average, field, remember) taken over a sphere surrounding $Q$, gives $Q/\kappa\epsilon_0$, or $Q/\epsilon$, if we believe Eq. (10.55), and *not* $Q/\epsilon_0$. Why not? The answer is that $Q$ is not the only charge inside the sphere. There are also all the charges that make up the atoms and molecules of the dielectric. Ordinarily any volume of the oil would be electrically neutral. But now the oil is radially polarized, which means that the charge $Q$, assuming it is positive, has pulled in toward itself the negative charge in the oil molecules and pushed away the positive charges. Although the displacement may be only very slight in each molecule, still on the average any sphere we draw around $Q$ will contain more oil-molecule negative charge than oil-molecule positive charge. Hence the net charge in the sphere, including the "foreign" charge $Q$ at the center, is *less* than $Q$. In fact, it is $Q/\kappa$.

It is often useful to distinguish between the foreign charge $Q$ and the charges that make up the dielectric itself. Over the former we have some degree of control – charge can be added to or removed from an object, such as the plate of a capacitor. This is often called *free* charge. The other charges, which are integral parts of the atoms or molecules of the dielectric, are usually called *bound* charge. *Structural* charge might be a better name. These charges are not mobile; they are more or less elastically bound, contributing, by their slight displacement, to the polarization.

One can devise a vector quantity that is related by something like Gauss's law to the free charge only. In the system we have just examined (a point charge $Q$ immersed in a dielectric), the vector $\kappa\mathbf{E}$ has this property. That is, $\int \kappa\mathbf{E} \cdot d\mathbf{a}$, taken over some closed surface $S$, equals $Q/\epsilon_0$ if $S$ encloses $Q$, and zero if it does not. By superposition, this must hold for any collection of free charges described by a free-charge density $\rho_{\text{free}}(x, y, z)$ in an infinite homogeneous linear dielectric medium:

$$\int_S \kappa \mathbf{E} \cdot d\mathbf{a} = \frac{1}{\epsilon_0} \int_V \rho_{\text{free}} \, dv, \qquad (10.57)$$

where $V$ is the volume enclosed by the surface $S$. An integral relation like this implies a "local" relation between the divergence of the vector field $\kappa \mathbf{E}$ and the free charge density:

$$\text{div} \, (\kappa \mathbf{E}) = \frac{\rho_{\text{free}}}{\epsilon_0}. \qquad (10.58)$$

Since $\kappa$ has been assumed to be constant throughout the medium, Eq. (10.58) tells us nothing new. However, it can help us to isolate the role of the bound charge. In any system whatsoever, the fundamental relation (namely Gauss's law) between electric field $\mathbf{E}$ and total charge density $\rho_{\text{free}} + \rho_{\text{bound}}$ remains valid:

$$\text{div} \, \mathbf{E} = \frac{1}{\epsilon_0} (\rho_{\text{free}} + \rho_{\text{bound}}). \qquad (10.59)$$

Subtracting Eq. (10.59) from Eq. (10.58) yields

$$\text{div} \, (\kappa - 1) \mathbf{E} = -\frac{\rho_{\text{bound}}}{\epsilon_0}. \qquad (10.60)$$

According to Eq. (10.40), $(\kappa - 1)\mathbf{E} = \mathbf{P}/\epsilon_0$ for a linear dielectric, so Eq. (10.60) implies that

$$\boxed{\text{div} \, \mathbf{P} = -\rho_{\text{bound}}} \qquad (10.61)$$

Equation (10.61) states a local relation. It cannot depend on conditions elsewhere in the system, nor on how the particular arrangement of bound charges is maintained. Any arrangement of bound charge that has a certain local excess, per unit volume, of nuclear protons over atomic electrons must represent a polarization with a certain divergence. So, although we derived Eq. (10.61) by using relations pertaining to linear dielectrics, *it must in fact hold universally, not just in an unbounded linear dielectric.* It doesn't matter how the polarization comes about. (See Problem 10.11 for a general proof.) You can get a feeling for the identity expressed in Eq. (10.61) by imagining a few polar molecules arranged to give a polarization with a positive divergence (Fig. 10.28). The dipoles point outward, which necessarily leaves a little concentration of negative charge in the middle. Of course, Eq. (10.61) refers to averages over volume elements so large that $\mathbf{P}$ and $\rho_{\text{bound}}$ can be treated as smoothly varying quantities.

From Eqs. (10.59) and (10.61), both of which are true in any system whatsoever, we get the relation

$$\text{div} \, (\epsilon_0 \mathbf{E} + \mathbf{P}) = \rho_{\text{free}}. \qquad (10.62)$$

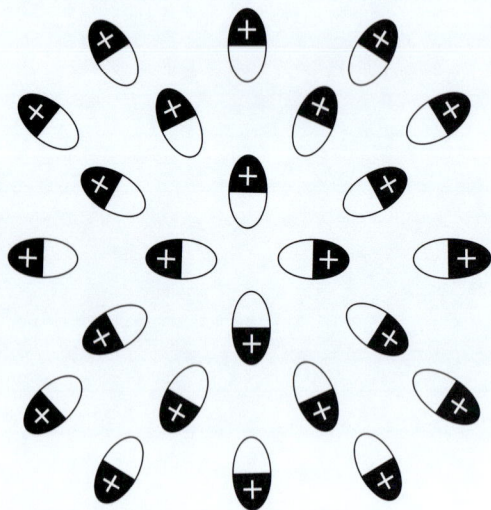

**Figure 10.28.**
Molecular dipoles arranged so that div $\mathbf{P} > 0$. Note the concentration of negative charge in the middle, consistent with Eq. (10.61).

This is quite independent of any relation between **E** and **P**; it is not limited to linear dielectrics (where **P** is proportional to **E**).

It is customary to give the combination $\epsilon_0 \mathbf{E} + \mathbf{P}$ a special name, the *electric displacement* vector, and its own symbol, **D**. That is, we define **D** by

$$\boxed{\mathbf{D} \equiv \epsilon_0 \mathbf{E} + \mathbf{P}} \tag{10.63}$$

and Eq. (10.62) becomes

$$\boxed{\text{div } \mathbf{D} = \rho_{\text{free}}} \tag{10.64}$$

This relation, or equivalently Eq. (10.62), holds in any situation in which the macroscopic quantities **P**, **E**, and $\rho$ can be defined.

If additionally we are dealing with a linear dielectric, then by comparing Eqs. (10.58) and (10.64) we see that **D** is simply $\kappa \epsilon_0 \mathbf{E}$, or

$$\boxed{\mathbf{D} = \epsilon \mathbf{E}} \quad \text{(for a linear dielectric)}. \tag{10.65}$$

This alternatively follows from Eq. (10.63) by using Eq. (10.41) to write **P** as $\chi_e \epsilon_0 \mathbf{E}$, and then using Eq. (10.42) to write $1 + \chi_e$ as $\kappa$.

The appearance of Eq. (10.64) may suggest that we should look on **D** as a vector field whose source is the free charge distribution $\rho_{\text{free}}$ (up to a factor of $\epsilon_0$), in the same sense that the total charge distribution $\rho$ is the source of **E**. That would be wrong. The electrostatic field **E** is uniquely determined – except for the addition of a constant field – by the charge distribution $\rho$ because, supplementing the law div $\mathbf{E} = \rho/\epsilon_0$, there is another universal condition, curl $\mathbf{E} = 0$. It is *not* true, in general, that curl $\mathbf{D} = 0$. Thus the distribution of free charge is not sufficient to determine **D** through Eq. (10.64). Something else is needed, such as the boundary conditions at various dielectric surfaces. The boundary conditions on **D** are of course merely an alternative way of expressing the boundary conditions involving **E** and **P**, already stated near the end of Section 10.9 and in Fig. 10.25.

---

**Example (Continuity of $D_\perp$)** For our polarized sphere in Section 10.9, we saw that $E_\parallel$ was continuous across the boundary whereas $E_\perp$ was not. These boundary conditions hold for any shape of polarized material. It turns out that the opposite conditions are true for **D**. That is, $D_\perp$ is continuous across the boundary whereas $D_\parallel$ is not. You can derive these boundary conditions in Problem 10.12. For now, let's just verify that $D_\perp$ is continuous across the boundary of our polarized sphere.

Inside the sphere, we have $\mathbf{E} = -\mathbf{P}/3\epsilon_0$, so the displacement vector is $\mathbf{D} = \epsilon_0(-\mathbf{P}/3\epsilon_0) + \mathbf{P} = 2\mathbf{P}/3$. The radial component of this is

$$D_\perp^{\text{in}} \equiv D_r^{\text{in}} = \frac{2P \cos \theta}{3}. \tag{10.66}$$

Outside the sphere, **E** is the field due to a dipole with $\mathbf{p}_0 = (4\pi R^3/3)\mathbf{P}$. The radial component of the dipole field is $E_r = p_0 \cos\theta/2\pi\epsilon_0 R^3$. In terms of $P$ this becomes $E_r = 2P\cos\theta/3\epsilon_0$. Since $\mathbf{P} = 0$ outside the sphere, the external $\mathbf{D}$ is obtained by simply multiplying the external $\mathbf{E}$ by $\epsilon_0$. Therefore

$$D_\perp^{\text{out}} \equiv D_r^{\text{out}} = \frac{2P\cos\theta}{3}. \tag{10.67}$$

This equals the above $D_\perp^{\text{in}}$, as desired.

The task of Exercise 10.41 is to use the explicit forms of the internal and external fields to find the discontinuity in $D_\parallel$ everywhere on the surface of the sphere.

In the approach we have taken to electric fields in matter, the introduction of $\mathbf{D}$ is an artifice that is not, on the whole, very helpful. We have mentioned $\mathbf{D}$ because it is hallowed by tradition, beginning with Maxwell,[8] and the student is sure to encounter it in other books, many of which treat it with more respect than it deserves.

Our essential conclusions about electric fields in matter can be summarized as follows:

(1) Matter can be polarized, its condition being described completely, so far as the macroscopic field is concerned, by a polarization density $\mathbf{P}$, which is the dipole moment per unit volume. The contribution of such matter to the electric field $\mathbf{E}$ is the same as that of a charge distribution $\rho_{\text{bound}}$, existing in vacuum and having the density $\rho_{\text{bound}} = -\text{div}\,\mathbf{P}$. In particular, at the surface of a polarized substance, where there is a discontinuity in $\mathbf{P}$, this reduces to a surface charge of density $\sigma = -\Delta P_\perp$. Add any free charge distribution that may be present, and the electric field is the field that this *total* charge distribution would produce in vacuum. This is the macroscopic field $\mathbf{E}$ both inside and outside matter, with the understanding that inside matter it is the spatial average of the true microscopic field.

(2) If $\mathbf{P}$ is proportional to $\mathbf{E}$ in a material, we call the material a linear dielectric. We define the electric susceptibility $\chi_e$ and the dielectric constant $\kappa$ characteristic of that material as $\chi_e = \mathbf{P}/\epsilon_0\mathbf{E}$ and $\kappa = 1 + \chi_e$. Free charges immersed in a linear dielectric give rise to electric fields that are $1/\kappa$ times as strong as the same charges would produce in vacuum.

## 10.12 A microscopic view of the dielectric

The polarization $\mathbf{P}$ in the dielectric is simply the large-scale manifestation of the electric dipole moments of the atoms or molecules of which

---

[8] The prominence of $\mathbf{D}$ in Maxwell's formulation of electromagnetic theory, and his choice of the name *displacement*, can perhaps be traced to his inclination toward a kind of mechanical model of the "aether." Whittaker has pointed out in his classic text (Whittaker, 1960) that this inclination may have led Maxwell himself astray at one point in the application of his theory to the problem of reflection of light from a dielectric.

the material is composed. **P** is the mean dipole moment density, the total vector dipole moment per unit volume – averaged, of course, over a region large enough to contain an enormous number of atoms. If there is no electric field to establish a preferred direction, **P** will be zero. That will surely be true for an ordinary liquid or a gas, and for solids too if we ignore the possibility of "frozen-in" polarization mentioned in Section 10.8. In the presence of an electric field in the medium, polarization can arise in two ways. (1) Every atom or molecule will acquire an induced dipole moment proportional to, and in the direction of, the field **E** that acts on that atom or molecule. (2) If molecules with permanent dipole moments are present in the medium, their orientations will no longer be perfectly random; alignment of their dipole moments in the field direction will be favored slightly over alignment in the opposite direction. Both effects (1) and (2) lead to polarization in the direction **E**, that is, to a positive value of $\chi_e \equiv \mathbf{P}/\epsilon_0 \mathbf{E}$, the electric susceptibility.

Let us consider first the induced atomic moments in a medium in which the atoms or molecules are rather far apart. An example is a gas at atmospheric density, in which there are something like $3 \cdot 10^{25}$ molecules per cubic meter. We shall assume that the field **E** that acts on an individual molecule is the same as the average, or macroscopic, field **E** in the medium. In making this assumption, we are neglecting the field at a molecule that is produced by the induced dipole moment of a nearby molecule. Let $\alpha$ be the polarizability of every molecule and $N$ the mean number of molecules per cubic meter. The dipole moment induced in each molecule is $\mathbf{p} = \alpha \mathbf{E}$, and the resulting polarization of the medium, **P**, is simply

$$\mathbf{P} = N\mathbf{p} = N\alpha\mathbf{E}. \tag{10.68}$$

This gives us at once the electric susceptibility $\chi_e$,

$$\chi_e = \frac{\mathbf{P}}{\epsilon_0 \mathbf{E}} = \frac{N\alpha}{\epsilon_0}, \tag{10.69}$$

and the dielectric constant $\kappa$,

$$\kappa = 1 + \chi_e = 1 + \frac{N\alpha}{\epsilon_0}. \tag{10.70}$$

The methane molecule in Fig. 10.12 has a polarizability value (or rather an $\alpha/4\pi\epsilon_0$ value) of $2.6 \cdot 10^{-30}$ m$^3$. At standard conditions of $0\,^\circ$C and atmospheric pressure there are approximately $2.8 \cdot 10^{25}$ molecules in 1 m$^3$. According to Eq. (10.70), the dielectric constant of methane at that density ought to have the value

$$\kappa = 1 + \frac{N\alpha}{\epsilon_0} = 1 + \frac{1}{\epsilon_0}(2.8 \cdot 10^{25}\text{ m}^{-3})(4\pi\epsilon_0 \cdot 2.6 \cdot 10^{-30}\text{ m}^3)$$

$$= 1.00091. \tag{10.71}$$

Up to rounding errors in the numbers we used, this agrees with the value of $\kappa$ listed for methane in Table 10.1. The agreement is hardly surprising, for the value of $\alpha/4\pi\epsilon_0$ given in Fig. 10.12 was probably deduced originally by applying the simple theory we have just developed to an experimentally measured dielectric constant.

We have already noted in Section 10.5 that the atomic polarizability $\alpha/4\pi\epsilon_0$, which has the dimensions of volume, is in order of magnitude about equal to the volume of an atom. That being so, the product $N\alpha/4\pi\epsilon_0$, which is just $\chi_e/4\pi$ according to Eq. (10.69), is about equal to the fraction of the volume of the medium that is taken up by atoms. Now the density of a gas under standard conditions is roughly one-thousandth of the density of the same substance condensed to liquid or solid. In the case of methane the ratio is close to $1/1000$; in the case of air, $1/700$. The gas is about 99.9 percent empty space. In the solid or liquid, on the other hand, the molecules are practically touching one another. The fraction of the volume they occupy is not much less than unity. This tells us that, in condensed matter generally, the induced polarization will result in a value of $\chi_e/4\pi$ of order of magnitude unity. In fact, as our brief list in Table 10.1 suggests, and as a more extensive tabulation would confirm, the value of $\chi_e/4\pi = (\kappa - 1)/4\pi$ for most nonpolar liquids and solids ranges from about 0.1 to 1. We can now see why.

We can see, too, why an exact theory of the susceptibility $\chi_e$ of a solid or liquid is not so easy to develop. When the atoms are crowded together until they almost "touch," the effect of one atom on its neighbors cannot be neglected. The distance $b$ between nearest neighbors is approximately $N^{-1/3}$. Let an electric field $E$ induce a dipole moment $p = E\alpha$ in each atom. This dipole $p$ on one atom will cause a field of strength $E' \sim p/4\pi\epsilon_0 b^3$ at the location of the next atom. But $1/b^3 \approx N$, hence $E' \sim E\alpha N/4\pi\epsilon_0$ . As we have just explained, in condensed matter $\alpha N/4\pi\epsilon_0$ is necessarily of order unity. Hence $\mathbf{E}'$ is *not* small, and certainly not negligible, compared with $\mathbf{E}$. Just what the effective field is that polarizes an atom in this situation is a question with no very obvious answer.[9]

Molecules with permanent electric dipole moments, *polar* molecules, respond to an electric field by trying to line up parallel to it. So long as the dipole moment $\mathbf{p}$ is not pointing in the direction of $\mathbf{E}$, there is a torque $\mathbf{p} \times \mathbf{E}$ tending to turn $\mathbf{p}$ into the direction of $\mathbf{E}$. (Look back at Eq. (10.21) and Fig. 10.8(b).) Of course, the torque is zero if $\mathbf{p}$ happens to be pointing exactly opposite to $\mathbf{E}$, but that condition is unstable. Torque on the electric dipole is torque on the molecule itself. A state of lowest energy will have been attained if and when all the polar molecules have rotated to bring their dipole moments into the $\mathbf{E}$ direction. While settling down to that state of perfect alignment they will have given off some energy,

---

[9] An elementary, approximate, treatment of this problem, leading to what is called the Clausius–Mossotti relation, can be found in Section 9.13 in the first edition of this book.

through rotational friction, to their surroundings. The resulting polarization would be gigantic. In water there are about $3.3 \cdot 10^{28}$ molecules/m³; the dipole moment of each (Fig. 10.14) is $6.1 \cdot 10^{-30}$ C-m. With complete alignment of the dipoles, **P** would be 0.2 C/m². If Fig. 10.24 were a picture of a water droplet thus polarized, the field strength just outside the drop, of order $P/\epsilon_0$ from Eq. (10.48), would exceed $10^{10}$ V/m!

This does *not* happen. Nothing approaching complete alignment is attained in any reasonable applied field **E**. Why not? The reason is essentially the same as the reason why the molecules of air in a room are not found all lying on the floor – which is, after all, the arrangement of lowest potential energy. We must think about *temperature* and about the energy of thermal agitation that every molecule exhibits at a given absolute temperature $T$. In magnitude, that energy is $kT$, where $k$ is the universal constant called *Boltzmann's constant*. At room temperature, $kT$ amounts to $4 \cdot 10^{-21}$ joule. In a system all at temperature $T$, the mean translational energy of a molecule – or, for that matter, of any object small or large – is $(3/2)kT$. More to the point here, the mean rotational energy of a molecule is just $kT$. Now, the air molecules do not all gather near the floor because the change in gravitational potential energy in elevating by a couple of meters a molecule of mass $5 \cdot 10^{-26}$ kg is only, as you can readily compute, about $10^{-24}$ joule, less than $1/1000$ of $kT$. On the other hand, the air near the floor *is* slightly more dense than the air near the ceiling, even when there is no temperature gradient. That is just the well-known change of barometric pressure with height. Air near the floor is fractionally more dense (when the difference is slight) by just $mgh/kT$, $mgh$ being the difference in gravitational potential energy between the two levels.

Similarly, in our dielectric we shall find a slight excess of molecular dipoles in the orientation of lower potential energy, that is, pointing in the direction of **E**, or with a component in that direction. The fractional excess in the favored directions will be, in order of magnitude, $pE/kT$. The numerator represents the difference in potential energy. Actually the work required to turn a dipole from the direction of **E** to the opposite direction is $2pE$ (see Eq. (10.22)), but averaging over angles would bring in other numerical factors that we are leaving out. With $N$ dipoles per unit volume, the polarization $P$, which would be $Np$ if they were totally aligned, will be smaller by something like the factor $pE/kT$. The polarization to be expected is therefore, in order of magnitude,

$$P \approx Np \left( \frac{pE}{kT} \right) = \frac{Np^2}{kT} E, \qquad (10.72)$$

and the susceptibility is

$$\chi_e = \frac{P}{\epsilon_0 E} \approx \frac{Np^2}{\epsilon_0 kT}. \qquad (10.73)$$

For water at room temperature, the quantity on the right in Eq. (10.73) is about 35, whereas with $\kappa = 80$, the actual value of $\chi_e$ is 79. Evidently a factor of roughly 2.3 is needed on the right in Eq. (10.73), in this case, to convert our order-of-magnitude estimate into a correct prediction. Deriving that factor theoretically is quite difficult, for the interactions of neighboring molecules complicate matters even more than in the case of the nonpolar dielectric.

If you apply an electric field of $10^4$ V/m to water, the resulting polarization, $P = \chi_e \epsilon_0 E = 7 \cdot 10^{-6}$ C/m$^2$, is equivalent to the alignment of $1.1 \cdot 10^{24}$ H$_2$O dipoles per cubic meter, or about one molecule in 30,000. Even so, this polarization is an order of magnitude greater than that caused by the same field in any nonpolar dielectric.

## 10.13 Polarization in changing fields

So far we have considered only electrostatic fields in matter. We need to look at the effects of electric fields that are varying in time, like the field in a capacitor used in an alternating-current circuit. The important question is, will the changes in polarization keep up with the changes in the field? Will the ratio of **P** to **E**, at any instant, be the same as in a static electric field? For very slow changes we should expect no difference but, as always, the criterion for slowness depends on the particular physical process. It turns out that induced polarization and the orientation of permanent dipoles are two processes with quite different response times.

The induced polarization of atoms and molecules occurs by the distortion of the electronic structure. Little mass is involved, and the structure is very stiff; its natural frequencies of vibration are extremely high. To put it another way, the motions of the electrons in atoms and molecules are characterized by periods on the order of $10^{-16}$ s – something like the period of a visible light wave. To an atom, $10^{-14}$ s is a *long* time. It has no trouble readjusting its electronic structure in a time like that. Because of this, strictly nonpolar substances behave practically the same from direct current (zero frequency) up to frequencies close to those of visible light. The polarization keeps in step with the field, and the susceptibility $\chi_e = P/\epsilon_0 E$ is independent of frequency.

The orientation of a polar molecule is a process quite different from the mere distortion of the electron cloud. The whole molecular framework has to rotate. On a microscopic scale, it is rather like turning a peanut end for end in a bag of peanuts. The frictional drag tends to make the rotation lag behind the torque and to reduce the amplitude of the resulting polarization. Where on the time scale this effect sets in varies enormously from one polar substance to another. In water, the "response time" for dipole reorientation is something like $10^{-11}$ s. The dielectric constant remains around 80 up to frequencies on the order of $10^{10}$ Hz. Above $10^{11}$ Hz, $\kappa$ falls to a modest value typical of a nonpolar liquid. The dipoles simply cannot follow so rapid an alternation of the field.

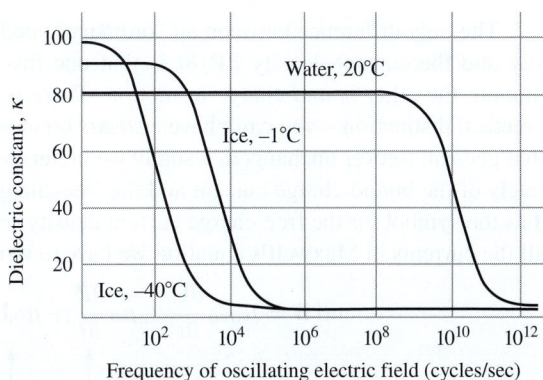

**Figure 10.29.**
The variation with frequency of the dielectric constant of water and ice. Based on information from Smyth (1955) for water data, and Auty and Cole (1952) for ice data.

In other substances, especially solids, the characteristic time can be much longer. In ice just below the freezing point, the response time for electrical polarization is around $10^{-5}$ s. Figure 10.29 shows some experimental curves of dielectric constant versus frequency for water and ice.

## 10.14 The bound-charge current

Wherever the polarization in matter changes with time, there is an electric current, a genuine motion of charge. Suppose there are $N$ dipoles per cubic meter of dielectric, and that in the time interval $dt$ each changes from $\mathbf{p}$ to $\mathbf{p} + d\mathbf{p}$. Then the macroscopic polarization density $\mathbf{P}$ changes from $\mathbf{P} = N\mathbf{p}$ to $\mathbf{P} + d\mathbf{P} = N(\mathbf{p} + d\mathbf{p})$. Suppose the change $d\mathbf{p}$ was effected by moving a charge $q$ through a distance $d\mathbf{s}$, in each atom: $q\,d\mathbf{s} = d\mathbf{p}$. Then during the time $dt$ there was actually a charge cloud of density $\rho = Nq$, moving with velocity $\mathbf{v} = d\mathbf{s}/dt$. This is a conduction current of a certain density $\mathbf{J}$ in coulombs per second per square meter:

$$\mathbf{J} = \rho\mathbf{v} = Nq\frac{d\mathbf{s}}{dt} = N\frac{d\mathbf{p}}{dt} \implies \boxed{\mathbf{J} = \frac{d\mathbf{P}}{dt}} \tag{10.74}$$

The connection between rate of change of polarization and current density, $\mathbf{J} = d\mathbf{P}/dt$, is independent of the details of the model. A changing polarization *is* a conduction current, not essentially different from any other. Note that if we take the divergence of both sides of Eq. (10.74) and use Eq. (10.61), we obtain $\operatorname{div}\mathbf{J} = d(\operatorname{div}\mathbf{P})/dt = -d\rho_{\text{bound}}/dt$, which is consistent with the continuity equation in Eq. (4.10).

Naturally, such a current is a source of magnetic field. If there are no other currents around, we should write Maxwell's equation, $\operatorname{curl}\mathbf{B} = \mu_0\epsilon_0(\partial\mathbf{E}/\partial t) + \mu_0\mathbf{J}$, as

$$\operatorname{curl}\mathbf{B} = \mu_0\epsilon_0\frac{\partial\mathbf{E}}{\partial t} + \mu_0\frac{\partial\mathbf{P}}{\partial t}. \tag{10.75}$$

The only difference between an "ordinary" conduction current density and the current density $\partial \mathbf{P}/\partial t$ is that one involves *free* charge in motion, the other *bound* charge in motion. There is one rather obvious practical distinction – you can't have a *steady* bound-charge current, one that goes on forever unchanged. Usually we prefer to keep account separately of the bound-charge current and the free-charge current, retaining $\mathbf{J}$ as the symbol for the free-charge current density only. Then to include all the currents in Maxwell's equation we have to write it this way:

$$\text{curl } \mathbf{B} = \mu_0 \epsilon_0 \frac{\partial \mathbf{E}}{\partial t} + \mu_0 \frac{\partial \mathbf{P}}{\partial t} + \mu_0 \mathbf{J}. \qquad (10.76)$$

$$\underset{\substack{\text{bound-charge} \\ \text{current density}}}{\uparrow} \quad \underset{\substack{\text{free-charge} \\ \text{current density}}}{\uparrow}$$

In a linear dielectric medium, we can write $\epsilon_0 \mathbf{E} + \mathbf{P} = \epsilon \mathbf{E}$, allowing a shorter version of Eq. (10.76):

$$\text{curl } \mathbf{B} = \mu_0 \epsilon \frac{\partial \mathbf{E}}{\partial t} + \mu_0 \mathbf{J}. \qquad (10.77)$$

More generally, Eq. (10.76) can also be abbreviated by introducing the vector $\mathbf{D}$, previously defined in any medium as $\epsilon_0 \mathbf{E} + \mathbf{P}$ (which reduces to $\epsilon \mathbf{E}$ in a linear dielectric):

$$\boxed{\text{curl } \mathbf{B} = \mu_0 \frac{\partial \mathbf{D}}{\partial t} + \mu_0 \mathbf{J}} \qquad (10.78)$$

The term $\partial \mathbf{D}/\partial t$ is usually referred to as the displacement current. Actually, the part of it that involves $\partial \mathbf{P}/\partial t$ represents, as we have seen, an honest conduction current, real charges in motion. The only part of the total current density that is not simply charge in motion is the $\partial \mathbf{E}/\partial t$ part, the true vacuum displacement current which we discussed in Chapter 9. Incidentally, if we want to express all components of the full current density in units corresponding to those of $\mathbf{J}$, we can factor out the $\mu_0$ and write Eq. (10.76) as follows:

$$\text{curl } \mathbf{B} = \mu_0 \left( \epsilon_0 \frac{\partial \mathbf{E}}{\partial t} + \frac{\partial \mathbf{P}}{\partial t} + \mathbf{J} \right). \qquad (10.79)$$

$$\underset{\substack{\text{vacuum} \\ \text{displacement} \\ \text{current} \\ \text{density}}}{\uparrow} \quad \underset{\substack{\text{bound-} \\ \text{charge} \\ \text{current} \\ \text{density}}}{\uparrow} \quad \underset{\substack{\text{free-} \\ \text{charge} \\ \text{current} \\ \text{density}}}{\uparrow}$$

Involved in the distinction between bound charge and free charge is a question we haven't squarely faced: can one always identify unambiguously the "molecular dipole moments" in matter, especially solid matter? The answer is no. Let us take a microscopic view of a thin wafer of sodium chloride crystal. The arrangement of the positive sodium ions

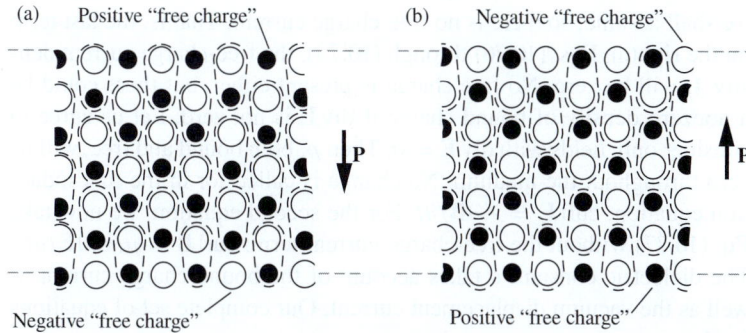

**Figure 10.30.**
An ionic lattice, with charges grouped in pairs as "molecules," in two ways: polarization vector directed downward (a), or upward (b). The systems are physically identical; the difference is only in the description.

and the negative chlorine ions was shown in Fig. 1.7. Figure 10.30 is a cross section through the crystal, which extends on out to the right and the left. If we choose to, we may consider an adjacent pair of ions as a neutral molecule with a dipole moment. Grouping them as in Fig. 10.30(a), we describe the medium as having a uniform macroscopic polarization density **P**, a vector directed downward. At the same time, we observe that there is a layer of positive charge over the top of the crystal and a layer of negative charge over the bottom, which, not having been included in our molecules, must be accounted *free charge.*

Now we might just as well have chosen to group the ions as in Fig. 10.30(b). According to that description, **P** is a vector *upward,* but we have a negative free-charge layer on top of the crystal and a positive free-charge layer beneath. *Either description is correct.* You will have no trouble finding another one, also correct, in which **P** is zero and there is no free charge. Each description predicts **E** = 0. The macroscopic field **E** is an observable physical quantity. It can depend only on the charge distribution, not on how we choose to *describe* the charge distribution.

This example teaches us that in the real atomic world the distinction between bound charge and free charge is more or less arbitrary, and so, therefore, is the concept of polarization density **P**. The molecular dipole is a well-defined notion only where molecules as such are identifiable – where there is some physical reason for saying, "This atom belongs to this molecule and not to that." In many crystals such an assignment is meaningless. An atom or ion may interact about equally strongly with all its neighbors; one can only speak of the whole crystal as a single molecule.

## 10.15 An electromagnetic wave in a dielectric

In Eq. (9.17) we wrote out Maxwell's equations for the electric and magnetic fields in vacuum, including source terms – charge density $\rho$ and current density **J**. Now we want to consider an electromagnetic field in an unbounded dielectric medium. The dielectric is a perfect insulator, we shall assume, so there is no free-charge current. That is, the last term on the right in Eqs. (10.76) through (10.79), the free-charge current density **J**, will be zero. No free charge is present either, but there could be a nonzero density of bound charge if div **E** is not zero. Let us agree to consider only fields with div **E** = 0. Then $\rho$, both bound and free, will be zero throughout the medium. No change is called for in the first induction equation, curl **E** = $-\partial$**B**/$\partial t$. For the second equation, we now take Eq. (10.77) without the free-charge-current term: curl **B** = $\mu_0\epsilon(\partial$**E**/$\partial t)$. The dielectric constant $\epsilon$ takes account of the bound-charge current as well as the vacuum displacement current. Our complete set of equations has become

$$\text{curl }\mathbf{E} = -\frac{\partial \mathbf{B}}{\partial t}, \qquad\qquad \text{div }\mathbf{E} = 0;$$

$$\text{curl }\mathbf{B} = \mu_0\epsilon\frac{\partial \mathbf{E}}{\partial t}, \qquad\qquad \text{div }\mathbf{B} = 0. \qquad (10.80)$$

These differ from Eq. (9.18) only in the replacement of $\epsilon_0$ with $\epsilon$ in the second induction equation.

As we did in Section 9.4, let us construct a wavelike electromagnetic field that can be made to satisfy Maxwell's equations. This time we give our trial wave function a slightly more general form:

$$\mathbf{E} = \hat{\mathbf{z}}E_0\sin(ky - \omega t),$$
$$\mathbf{B} = \hat{\mathbf{x}}B_0\sin(ky - \omega t). \qquad (10.81)$$

The angle $(ky - \omega t)$ is called the *phase* of the wave. For a point that moves in the positive $y$ direction with speed $\omega/k$, the phase $ky - \omega t$ remains constant. In other words, $\omega/k$ is the *phase velocity* of this wave. This term is used when it is necessary to distinguish between two velocities, phase velocity and group velocity. There is no difference in the case we are considering, so we shall call $\omega/k$ simply the wave velocity, the same as $v$ in our discussion in Section 9.4. At any fixed location, such as $y = y_0$, the fields oscillate in time with angular frequency $\omega$. At any instant of time, such as $t = t_0$, the phase differs by $2\pi$ at planes one wavelength $\lambda$ apart, where $\lambda = 2\pi/k$.

The divergence equations in Eq. (10.80) are quickly seen to be satisfied by the wave in Eq. (10.81). For the curl equations, the space and time derivatives we need are those listed in Eq. (9.24) with small alterations:

$$\text{curl }\mathbf{E} = \hat{\mathbf{x}}E_0k\cos(ky - \omega t), \qquad \frac{\partial \mathbf{E}}{\partial t} = -\hat{\mathbf{z}}E_0\omega\cos(ky - \omega t),$$

$$\text{curl }\mathbf{B} = -\hat{\mathbf{z}}B_0k\cos(ky - \omega t), \qquad \frac{\partial \mathbf{B}}{\partial t} = -\hat{\mathbf{x}}B_0\omega\cos(ky - \omega t).$$
$$(10.82)$$

Substituting these into Eq. (10.80), we find that the curl equations are satisfied if

$$\frac{\omega}{k} = \pm\frac{1}{\sqrt{\mu_0\epsilon}} \quad \text{and} \quad E_0 = \pm\frac{B_0}{\sqrt{\mu_0\epsilon}} \tag{10.83}$$

The wave velocity $v = \omega/k$ differs from the velocity of light in vacuum (which is $c = 1/\sqrt{\mu_0\epsilon_0}$) by the factor $\sqrt{\epsilon_0/\epsilon} = 1/\sqrt{\kappa}$. Since $\kappa > 1$, we have $v < c$. The electric and magnetic field amplitudes, $E_0$ and $B_0$, which were related by $E_0 = cB_0$ for the wave in vacuum, here are related by $E_0 = vB_0$, where $v = 1/\sqrt{\mu_0\epsilon}$. For a given magnetic amplitude $B_0$, the electric amplitude $E_0$ is smaller in the dielectric than in vacuum. In other respects the wave resembles our plane wave in vacuum: $\mathbf{B}$ is perpendicular to $\mathbf{E}$, and the wave travels in the direction of $\mathbf{E} \times \mathbf{B}$. Of course, if we compare a wave in a dielectric with a wave of the same frequency in vacuum, the wavelength $\lambda$ in the dielectric will be less than the vacuum wavelength by $1/\sqrt{\kappa}$ since *frequency* $\times$ *wavelength* $=$ *velocity*.

Light traveling through glass provides an example of the wave just described. In optics it is customary to define $n$, the *index of refraction* of a medium, as the ratio of the speed of light in vacuum to the speed of light in that medium. We have now discovered that $n$ is nothing more than $\sqrt{\kappa}$. In fact, we have now laid most of the foundation for a classical theory of optics. Of course, we must be careful to use the appropriate value of $\kappa$. Take water, for example. If we use the $\kappa = 80$ value from Table 10.1, we obtain $n \approx 9$. But the actual index of refraction of water is $n = 1.33$. What's going on here? *Hint*: The answer is contained in a figure in this chapter.

## 10.16 Applications

The *pollination* of flowers by bees is helped by polarization effects. When bees travel through the air, they become positively charged due to triboelectric effects with the air; air molecules strip off electrons from the bee when its wings collide with the molecules. When the bee gets close to the pollen on the anther of the flower, the bee's charge polarizes the pollen, which then experiences a dipole attraction toward the bee. The pollen jumps to the bee and lands on the bee's hairs (while maintaining zero net charge, because the hairs aren't conductive). When the bee then gets close to the stigma of a flower, it induces a negative charge on the stigma. The electric field from the somewhat pointed stigma wins out over the field from the somewhat rounded bee, so the pollen jumps to the stigma. Note the lack of symmetry between the anther and the stigma; the anther gives up the pollen, while the stigma attracts it. This lack of symmetry arises mainly from the fact that the stigma has a more conductive path to ground than the anther has, so it can acquire a net negative charge when the bee is near. The stigma's attraction to the pollen is therefore of the monopole–dipole type, whereas the anther's attraction to the pollen is of the smaller dipole–dipole type.

The ability of the *gecko* lizard to stick to a window or walk upside down on a ceiling is due to the *van der Waals* force. This force is an interaction between dipoles in the gecko's feet and dipoles on the surface. The force falls off rapidly with distance, being proportional to $1/r^7$. Equivalently, the potential energy is proportional to $1/r^6$. (In short, quantum fluctuations create random dipole moments in a given molecule. The resulting $E \propto 1/r^3$ field induces a dipole moment $p \propto E \propto 1/r^3$ in a neighboring molecule. The resulting $(\mathbf{p} \cdot \nabla)E$ force on this molecule is then proportional to $1/r^7$.) The key for geckos, therefore, is to make the distance $r$ be as small as possible. They do this by having hundreds of thousands of tiny hairs (called *setae*) on their feet, each of which contains hundreds of even tinier hairs called *spatulae*. These spatulae are able to penetrate the nooks and crannies on the surface, making the various distances $r$ extremely small. If all of a gecko's spatulae are engaged, it could walk on a ceiling with a few hundred pounds strapped to it!

The main ingredient in soaps and detergents is a *surfactant*. This is a molecule in the form of a long chain with a polar end and a nonpolar end. The polar end is attracted to the polar water molecules (it is called *hydrophilic*), whereas the nonpolar end isn't (it is called *hydrophobic*). The nonpolar end is instead attracted to other hydrophobic molecules, such as oils and other grime on you or your clothes. More precisely, the hydrophobic ends/molecules aren't actually attracted to each other. Rather, the attraction of all the polar molecules to themselves has the effect of forcing all the hydrophobic ends/molecules into little clumps, called *micelles*. This gives the appearance of an attraction. (The same reasoning leads to the everyday fact that oil and water don't mix.) The clumps (with the oils and nonpolar surfactant ends in the interior, and the polar surfactant ends on the surface) float around in the water and can be eliminated by discarding the water, that is, by rinsing. So your laundry detergent won't work without water!

The large dipole moment of a water molecule is what allows a *microwave oven* to heat your food. The alternating electric field of the microwave radiation (created by a magnetron; see Section 8.7) causes the water dipoles to rotate back and forth. This jiggling of the molecules causes them to bump into each other, which results in the thermal energy (the heat) that you observe. The specific microwave frequency that is used (usually about 2.5 GHz, which corresponds to a wavelength of about 12 cm) has nothing to do with the resonant vibrational frequency of a free water molecule in vapor, which is about 20 GHz. There isn't anything special about the 12 cm wavelength, although if it were much shorter the waves wouldn't penetrate the food as well. The water molecules in ice can't rotate as easily, so that's why it takes a while to defrost frozen food. You can't just crank up the power because, if one part of the food thaws first, then it will absorb energy much faster than the remaining frozen part. You will then end up with, for example, thoroughly cooked meat right next to frozen meat. The defrost cycle in a microwave oven functions by simply shutting off for periods of time, allowing the heat to diffuse.

When some materials, such as *quartz,* are put under stress and bent, a voltage difference arises between different parts. This is known as the *piezoelectric effect.* (Conversely, if a voltage difference is applied to different parts, the material will bend.) What happens is that the molecules are stretched or squashed, and for certain configurations this results in the molecules acquiring a dipole moment. The material therefore behaves like a polarized dielectric; there are net surface charges, so the result is effectively a capacitor with a voltage difference between the plates. The piezoelectric property of quartz allows your wristwatch to keep time. A tiny quartz crystal, which is the analog of the pendulum in a pendulum clock, is cut so that it vibrates with a specific resonant frequency, usually $2^{15} = 32{,}768\,\text{Hz}$. (This power of 2 makes it easy for a chain of frequency dividers to generate the desired 1 Hz frequency.) Via the piezoelectric effect, an electric signal with this 32,768 Hz frequency is sent to a circuit. The circuit then amplifies the signal and sends it back to the crystal, providing the necessary driving force to keep the resonant oscillation going. Quartz has a very high $Q$ value, so only a tiny amount of input energy is needed. That is why the battery can last so long. The oscillations are initially produced by the random ac noise in the circuit. This ac noise contains at least a little bit of the resonant frequency, which the crystal responds to.

## CHAPTER SUMMARY

- If a capacitor is filled with a *dielectric* (an insulator), the capacitance increases by a factor $\kappa$, known as the *dielectric constant*. This is a consequence of the fact that the polarization of the molecules in the dielectric causes layers of charge to form near the capacitor plates, partially canceling the free charge.

- The potential due to a charge distribution can be written as the sum of terms with increasing powers of $1/r$. The coefficients of these terms are called *moments*. A net charge has a monopole moment. Two opposite monopoles create a dipole. Two opposite dipoles create a quadrupole, and so on. The potential and field due to a dipole are given by

$$\phi(r,\theta) = \frac{p\cos\theta}{4\pi\epsilon_0 r^2}, \qquad \mathbf{E}(r,\theta) = \frac{p}{4\pi\epsilon_0 r^3}\left(2\cos\theta\,\hat{\mathbf{r}} + \sin\theta\,\hat{\boldsymbol{\theta}}\right).$$

$$(10.84)$$

- The torque on an electric dipole is $\mathbf{N} = \mathbf{p}\times\mathbf{E}$. The force is $F_x = \mathbf{p}\cdot\nabla E_x$, and likewise for the $y$ and $z$ components.

- An external electric field will cause an atom to become polarized. The *atomic polarizability* $\alpha$ is defined by $\mathbf{p} = \alpha\mathbf{E}$. However, the quantity $\alpha/4\pi\epsilon_0$, with has the dimensions of volume, is also commonly called the atomic polarizability. In order of magnitude, $\alpha/4\pi\epsilon_0$ equals an atomic volume.

- Some molecules have a permanent dipole moment; the moment exists even in the absence of an external electric field. These are called *polar molecules*. An external electric field causes the dipoles to align (at least partially), which leads to an overall polarization of the material.

- The *polarization* per unit volume is given by $\mathbf{P} = \mathbf{p}N$. Uniformly polarized matter is equivalent to a surface charge density $\sigma = P$, or $\sigma = P\cos\theta$ if the surface is tilted with respect to the direction of $\mathbf{P}$.

- When we talk about the electric field inside matter, we mean the spatial average, $\langle\mathbf{E}\rangle = (1/V)\int\mathbf{E}\,dv$. Inside a uniformly polarized slab, this average is $-\mathbf{P}/\epsilon_0$.

- The *electric susceptibility* $\chi_e$ is defined by

$$\chi_e \equiv \frac{P}{\epsilon_0 E} \quad\Longrightarrow\quad \chi_e = \kappa - 1. \tag{10.85}$$

- The field inside a uniformly polarized sphere is $-\mathbf{P}/3\epsilon_0$. If a dielectric sphere is placed in a uniform electric field $\mathbf{E}_0$, the resulting polarization is uniform and is given by $\mathbf{P} = 3\epsilon_0(\kappa - 1)\mathbf{E}_0/(\kappa + 2)$.

- For any material, $\text{div}\,\mathbf{P} = -\rho_{\text{bound}}$. Combining this with Gauss's law, $\text{div}\,\mathbf{E} = \rho_{\text{total}}/\epsilon_0$, gives

$$\text{div}\,\mathbf{D} = \rho_{\text{free}}, \qquad \text{where} \quad \mathbf{D} \equiv \epsilon_0\mathbf{E} + \mathbf{P}. \tag{10.86}$$

$\mathbf{D}$ is known as the *electric displacement vector*. If additionally we are dealing with a linear dielectric, then

$$\mathbf{D} \equiv \epsilon\mathbf{E}, \qquad \text{where} \quad \epsilon \equiv \kappa\epsilon_0. \tag{10.87}$$

- If an external electric field is applied to a dielectric containing polar molecules, the polarizations tend to align with the field, but thermal energy generally prevents the alignment from being large. The susceptibility is given roughly by $\chi_e \approx Np^2/\epsilon_0 kT$.

- In a rapidly changing electric field, the induced polarization of atoms and molecules can keep up with the field at high frequencies. However, the polarization arising from polar molecules cannot, because it is much more difficult to rotate a molecule as a whole than simply to stretch it.

- The *bound-charge* current density satisfies $\mathbf{J}_{\text{bound}} = d\mathbf{P}/dt$. This represents a true current, but when writing the "curl $\mathbf{B}$" Maxwell equation it is often convenient to separate this current from the free-charge current:

$$\text{curl}\,\mathbf{B} = \mu_0\left(\epsilon_0\frac{\partial\mathbf{E}}{\partial t} + \frac{\partial\mathbf{P}}{\partial t} + \mathbf{J}_{\text{free}}\right) \equiv \mu_0\left(\frac{\partial\mathbf{D}}{\partial t} + \mathbf{J}_{\text{free}}\right). \tag{10.88}$$

• An electromagnetic wave in a dielectric travels with speed $v = 1/\sqrt{\mu_0 \epsilon}$. This is slower than the speed $c = 1/\sqrt{\mu_0 \epsilon_0}$ in vacuum by the factor $1/\sqrt{\kappa}$. The *index of refraction* equals the inverse of this: $n = \sqrt{\kappa}$. The **E** and **B** fields are still perpendicular to each other and to the direction of travel. Their amplitudes are related by $E_0 = vB_0$.

## Problems

10.1  *Leaky cell membrane* **

In Section 4.11 we discussed the relaxation time of a capacitor filled with a material having a resistivity $\rho$. If you look back at that discussion you will notice that we dodged the question of the dielectric constant of the material. Now you can repair that omission, by introducing $\kappa$ properly into the expression for the time constant. A leaky capacitor important to us all is formed by the wall of a living cell, an insulator (among its many other functions!) that separates two conducting fluids. Its electrical properties are of particular interest in the case of the nerve cell, for the propagation of a nerve impulse is accompanied by rapid changes in the electric potential difference between interior and exterior.

(a) The cell membrane typically has a capacitance around 1 microfarad per square centimeter of membrane area. It is believed the membrane consists of material having a dielectric constant about 3. What thickness does this imply?

(b) Other electrical measurements have indicated that the resistance of 1 cm$^2$ of cell membrane, measured from the conducting fluid on one side to that on the other, is around 1000 ohms. Show that the time constant of such a leaky capacitor is independent of the area of the capacitor. How large is it in this case? Where would the resistivity $\rho$ of such membrane material fall on the chart of Fig. 4.8?

10.2  *Force on a dielectric* **

A rectangular capacitor with side lengths $a$ and $b$ has separation $s$, with $s$ much smaller than $a$ and $b$. It is partially filled with a dielectric with dielectric constant $\kappa$. The overlap distance is $x$; see Fig. 10.31. The capacitor is isolated and has constant charge $Q$.

(a) What is the energy stored in the system? (Treat the capacitor like two capacitors in parallel.)

(b) What is the force on the dielectric? Does this force pull the dielectric into the capacitor or push it out?

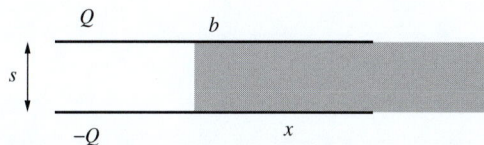

**Figure 10.31.**

10.3  *Energy of dipoles* **

Find the potential energy of the first and third dipole configurations in Exercise 10.29 (the second and fourth require only a slight modification), by explicitly looking at the potential energy of the

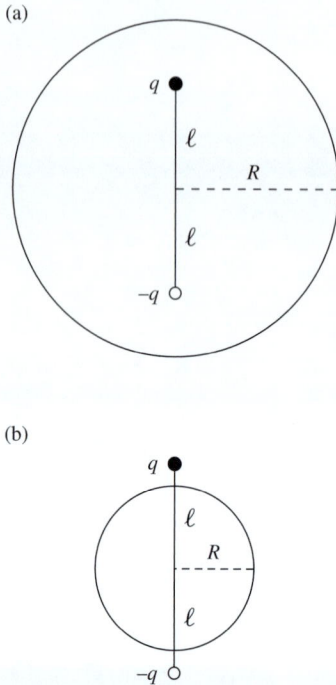

(a)

(b)

**Figure 10.32.**

relevant pairs of point charges in the dipoles, and then making suitable approximations. Let the charges $q$ and $-q$ in each dipole be separated by a distance $\ell$, and let the centers of the dipoles be separated by a distance $d$ (with $\ell \ll d$).

**10.4** *Dipole polar components* **

Show that Eq. (10.18) follows from Eq. (10.17). *Hint:* You can write the Cartesian unit vectors in terms of the polar unit vectors, or you can project the vector $(E_x, E_z)$ onto the radial and tangential directions.

**10.5** *Average field* **

(a) (This problem builds on the results from Problem 1.28.) Given an arbitrary collection of charges inside a sphere of radius $R$, show that the average electric field over the volume of the sphere is given by $\mathbf{E}_{\mathrm{avg}} = -\mathbf{p}/4\pi\epsilon_0 R^3$, where $\mathbf{p}$ is the total dipole moment, measured relative to the center.

(b) For the specific case of the dipole shown in Fig. 10.32(a), find the average electric field over both the *surface* and the *volume* of the sphere of radius $R$.

(c) Repeat for the case shown in Fig. 10.32(b).

**10.6** *Quadrupole tensor* ***

You should see the quadrupole tensor at least once in your life, so here it is. Calculate the general form of the quadrupole tensor by writing the $R$ in Eq. (10.5) in Cartesian form as

$$R = \sqrt{(x_1 - x_1')^2 + (x_2 - x_2')^2 + (x_3 - x_3')^2},  \qquad (10.89)$$

and then performing a Taylor expansion as we did in Section 10.2. (It's a little cleaner to work with $x_1, x_2, x_3$ instead of $x, y, z$.) Your goal is to write the $1/r^3$ part of the potential in a form that has all of its dependence on the primed coordinates collected into a matrix. (By analogy, the $1/r^2$ part of the potential in Eq. (10.12) has all of its dependence on the primed coordinates collected into the vector $\mathbf{p}$ given by Eq. (10.13).)

**10.7** *Force on a dipole* **

Derive Eq. (10.26). As usual, work in the approximation where the dipole length $s$ is small. *Hint:* Let the two charges be at positions $\mathbf{r}$ and $\mathbf{r} + \mathbf{s}$.

**10.8** *Force from an induced dipole* **

Between any ion and any neutral atom there is a force that arises as follows. The electric field of the ion polarizes the atom; the field of that induced dipole reacts on the ion. Show that this force is always attractive, and that it varies with the inverse fifth power of the distance of separation $r$. Derive an expression for the associated potential energy, with zero energy corresponding to infinite separation. For what distance $r$ does this potential energy have the same magnitude as $kT$ at room temperature (which is $4 \cdot 10^{-21}$ joule) if the ion is singly charged and the atom is a sodium atom? (See Table 10.2.)

**10.9** *Polarized water* **

The electric dipole moment of the water molecule is given in Fig. 10.14. Imagine that all the molecular dipoles in a cup of water could be made to point down. Calculate the magnitude of the resulting surface charge density at the upper water surface, and express it in electrons per square centimeter.

**10.10** *Tangent field lines* ***

Assume that the uniform field $\mathbf{E}_0$ that causes the electric field in Fig. 10.27 is produced by large capacitor plates very far away. Consider the special set of field lines that are tangent to the sphere. These lines hit each of the distant capacitor plates in a circle of radius $r$. What is $r$ in terms of the radius $R$ and dielectric constant $\kappa$ of the sphere? *Hint:* Consider a well-chosen Gaussian surface that has the horizontal great circle of the sphere as part of its boundary.

**10.11** *Bound charge and divergence of P* ***

Derive Eq. (10.61) by considering the volume integral of both sides of the equation. Assume that the dipoles consist of charges $\pm q$ separated by a distance $s$. *Hint:* Consider a small patch of the surface bounding a given volume. What causes there to be a net bound charge inside the volume due to the dipoles near this patch?

**10.12** *Boundary conditions on D* **

Using $\mathbf{D} \equiv \epsilon_0 \mathbf{E} + \mathbf{P}$ and div $\mathbf{D} = \rho_{\text{free}}$, derive the general rules for the discontinuities (if any) in $D_\parallel$ and $D_\perp$ across the surface of an arbitrarily shaped polarized material (with no free charge).

**10.13** *Q for a leaky capacitor* ***

Consider an oscillating electric field, $E_0 \cos \omega t$, inside a dielectric medium that is not a perfect insulator. The medium has dielectric constant $\kappa$ and conductivity $\sigma$. This could be the electric field in some leaky capacitor that is part of a resonant circuit, or it could be the electric field at a particular location in an electromagnetic wave. Show that the $Q$ factor, as defined by Eq. (8.12), is $\omega \epsilon / \sigma$ for this system, and evaluate it for seawater at a frequency of 1000 MHz. You will need to use the result from Exercise 10.42. (The conductivity is given in Table 4.1, and the dielectric constant may be assumed to be the same as that of pure water at the same frequency. See Fig. 10.29.) What does your result suggest about the propagation of decimeter waves through seawater?

**10.14** *Boundary conditions on E and B* **

Find the boundary conditions on $E_\parallel$, $E_\perp$, $B_\parallel$, and $B_\perp$ across the interface between two linear dielectrics. Assume that there are no free charges or free currents present.

## Exercises

**10.15** *Densities on a capacitor* **

Consider the setup of Problem 10.2. In terms of the various parameters given there, find the charge densities on the left and right parts of the capacitor. You should find that as $x$ increases, the charge densities on *both* parts of plates decrease. At first glance this seems a bit absurd, so try to explain intuitively how it is possible.

**10.16** *Leyden jar* **

In 1746 a Professor Musschenbroek in Leiden charged water in a bottle by touching a wire, projecting from the neck of the bottle, to his electrostatic machine. When his assistant, who was holding the bottle in one hand, tried to remove the wire with the other, he got a violent shock. Thus did the simple capacitor force itself on the attention of electrical scientists. The discovery of the "Leyden jar" revolutionized electrical experimentation. In 1747 Benjamin Franklin was already writing about his experiments with "Mr. Musschenbroek's wonderful bottle." The jar was really nothing but glass with a conductor on each side of it. To see why it caused such a sensation, estimate roughly the capacitance of a jar made of a 1 liter bottle with walls 2 mm thick, the glass having a dielectric constant 4. What diameter sphere, in air, would have the same capacitance?

**10.17** *Maximum energy storage* **

Materials to be used as insulators or dielectrics in capacitors are rated with respect to *dielectric strength,* defined as the maximum internal electric field the material can support without danger of electrical breakdown. It is customary to express the dielectric strength in kilovolts per mil. (One mil is 0.001 inch, or 0.00254 cm.) For example, Mylar (a Dupont polyester film) is rated as having a dielectric strength of 14 kilovolts/mil when it is used in a thin sheet – as it would be in a typical capacitor. The dielectric constant $\kappa$ of Mylar is 3.25. Its density is 1.40 g/cm$^3$. Calculate the maximum amount of energy that can be stored in a Mylar-filled capacitor, and express it in joules/kg of Mylar. Assuming the electrodes and case account for 25 percent of the capacitor's weight, how high could the capacitor be lifted by the energy stored in it? Compare the capacitor as an energy storage device with the battery in Exercise 4.41.

**10.18** *Partially filled capacitors* **
Figure 10.33 shows three capacitors of the same area and plate separation. Call the capacitance of the vacuum capacitor $C_0$. Each of the others is half-filled with a dielectric, with the same dielectric constant $\kappa$, but differently disposed, as shown. Find the capacitance of each of these two capacitors. (Neglect edge effects.)

**10.19** *Capacitor roll* **
You have a supply of polyethylene tape, with dielectric constant 2.3, that is 2.25 inches wide and 0.001 inch thick; you also have a supply of aluminum tape that is 2 inches wide and 0.0005 inch thick. You want to make a capacitor of about 0.05 microfarad capacitance, in the form of a compact cylindrical roll. Describe how you might do this, estimating the amount of tape of each kind that would be needed, and the overall diameter of the finished capacitor. (It may help to look at Problem 3.21 and Exercise 3.57.)

**10.20** *Work in a dipole field* *
How much work is done in moving unit positive charge from $A$ to $B$ in the field of the dipole **p** shown in Fig. 10.34?

**10.21** *A few dipole moments* *
What is the magnitude and direction of the dipole moment vector **p** of each of the charge distributions in parts (a), (b), and (c) of Fig. 10.35?

**10.22** *Fringing field from a capacitor* *
A parallel-plate capacitor, with a measured capacitance $C = 250$ picofarads ($250 \cdot 10^{-12}$ F), is charged to a potential difference of 2000 volts. The plates are 1.5 cm apart. We are interested in the field outside the capacitor, the "fringing" field which we usually ignore. In particular, we would like to know the field at a distance from the capacitor large compared with the size of the capacitor itself. This can be found by treating the charge distribution on the capacitor as a dipole. Estimate the electric field strength

(a) at a point 3 meters from the capacitor in the plane of the plates;
(b) at a point the same distance away, in a direction perpendicular to the plates.

**10.23** *Dipole field plus uniform field* **
A dipole of strength $p = 6 \cdot 10^{-10}$ C-m is located at the origin, pointing in the $\hat{\mathbf{z}}$ direction. To its field is added a uniform electric field of strength $150\,\text{kV/m}$ in the $\hat{\mathbf{y}}$ direction. At how many places, located where, is the total field zero?

**10.24** *Field lines* **
A field line in the dipole field is described in polar coordinates by the very simple equation $r = r_0 \sin^2 \theta$, in which $r_0$ is the radius

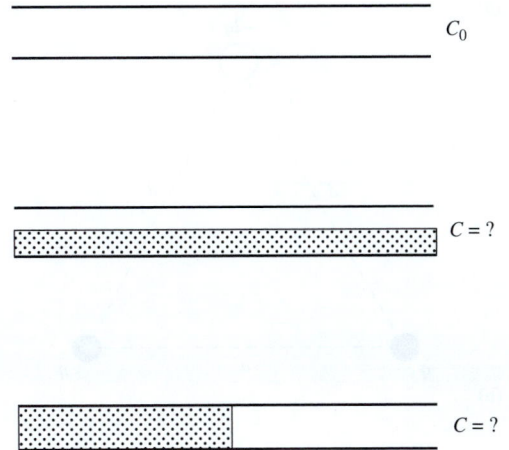

**Figure 10.33.**

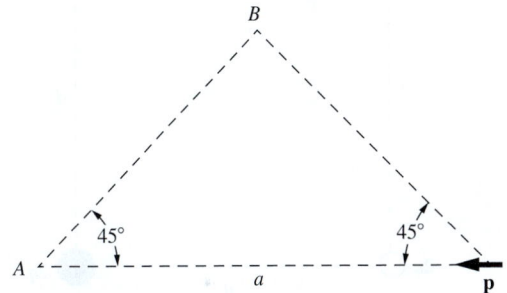

**Figure 10.34.**

(a)

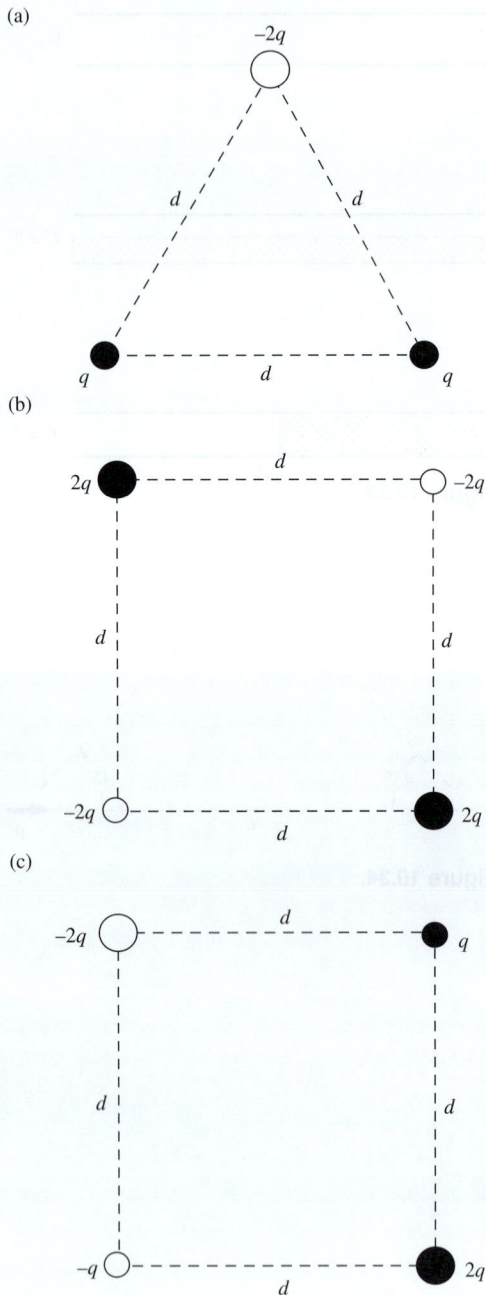

(b)

(c)

**Figure 10.35.**

at which the field line passes through the equatorial plane of the dipole. Show that this is true by demonstrating that at any point on that curve the tangent has the same direction as the dipole field.

**10.25** *Average dipole field on a sphere* **

By direct integration, show that the average of the dipole field, over the surface of a sphere centered at the dipole, is zero. You will want to work with the Cartesian components of **E**, but feel free to write these components in terms of spherical coordinates, as in Eq. (10.17).

**10.26** *Quadrupole for a square* **

Calculate the quadrupole matrix **Q** (see the result of Problem 10.6) for the configuration of charges in Fig. 10.5. Then show that the potential at a point at (large) radius $r$ on the $z \equiv x_3$ axis equals $3ea^2/4\pi\epsilon_0 r^3$, where $a$ is the distance from all the charges to the origin. What is the potential at the point $(r/\sqrt{2})(1, 0, 1)$?

**10.27** *Pascal's triangle and the multipole expansion* **

(a) If two monopoles with opposite sign are placed near each other, they make a dipole. Likewise, if two dipoles with opposite sign are placed near each other, they make a quadrupole, and so on. Explain how this fact was used in Fig. 10.36 to obtain each of the configurations from the one above it. Note that the magnitudes of the charges form Pascal's triangle. This triangle provides a very simple way of generating successively higher terms in the multipole expansion.

(b) If the bottom configuration in Fig. 10.36 is indeed an octupole, then the leading-order term in the potential at the point $P$ in Fig. 10.37 must be of order $1/r^4$. (This is two orders of $1/r$ higher than the $1/r^2$ dipole potential.) Verify this. That is, use a Taylor series to show that the order 1, $1/r$, $1/r^2$, and $1/r^3$ terms in the potential vanish. Define $r$ to be the distance to the rightmost charge, and assume $r \gg a$. This problem is easy if you use the Series operation in *Mathematica*, but you should work it out by hand. It is interesting to see how each of the terms vanishes.

As you will discover, this result is related to a nice little theorem about the sum $\sum_{k=0}^{N} \binom{N}{k} k^m (-1)^k$, where $m$ can take on any value from 0 to $N - 1$. You are encouraged to think about this, although you don't need to prove it here. *Hint:* Expand $(1 - x)^N$ with the binomial theorem, and take the derivative. Then multiply by $x$ and take the derivative again. Repeat this process as needed, and then set $x = 1$.

**10.28** *Force on a dipole* **

What are the magnitude and direction of the force on the central dipole caused by the field of the other two dipoles in Fig. 10.38?

**10.29** *Energy of dipole pairs* ✶✶

Shown in Fig. 10.39 are four different arrangements of the electric dipole moments of two neighboring polar molecules. Find the potential energy of each arrangement, the potential energy being defined as the work done in bringing the two molecules together from infinite separation while keeping their moments in the specified orientation. That is not necessarily the easiest way to calculate it. You can always bring them together one way and then rotate them.

**10.30** *Polarized hydrogen* ✶✶

A hydrogen atom is placed in an electric field $E$. The proton and the electron cloud are pulled in opposite directions. Assume simplistically (since we are concerned only with a rough result here) that the electron cloud takes the form of a uniform sphere with radius $a$, with the proton a distance $\Delta z$ from the center, as shown in Fig. 10.40. Find $\Delta z$, and show that your result agrees with Eq. (10.27).

**10.31** *Mutually induced dipoles* ✶✶

Two polarizable atoms $A$ and $B$ are a fixed distance apart. The polarizability of each atom is $\alpha$. Consider the following intriguing possibility. Atom $A$ is polarized by an electric field, the source of which is the electric dipole moment $\mathbf{p}_B$ of atom $B$. This dipole moment is induced in atom $B$ by an electric field, the source of which is the dipole moment $\mathbf{p}_A$ of atom $A$. Can this happen? If so, under what conditions? If not, why not?

**10.32** *Hydration* ✶

The phenomenon of *hydration* is important in the chemistry of aqueous solutions. This refers to the fact that an ion in solution gathers around itself a cluster of water molecules, which cling to it rather tightly. The force of attraction between a dipole and a point charge is responsible for this. Estimate the energy required to separate an ion carrying a single charge $e$ from a water molecule, assuming that initially the ion is located 1.5 angstroms from the effective location of the $H_2O$ dipole. (This distance is actually a rather ill-defined quantity, since the water molecule, viewed from close up, is a charge distribution, not an infinitesimal dipole.) Which part of the water molecule will be found nearest to a negative ion? See Fig. 10.14 for the dipole moment of the water molecule.

Monopole

1

Dipole

−1        1

Quadrupole

1        −2        1

Octupole

−1        3        −3        1

**Figure 10.36.**

Octupole

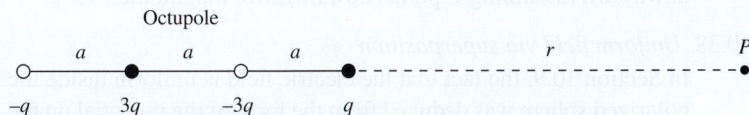

$a$     $a$     $a$      $r$    $P$

−$q$    $3q$    −$3q$    $q$

**Figure 10.37.**

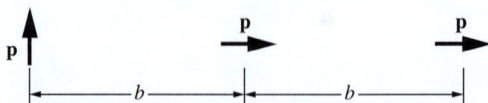

**Figure 10.38.**

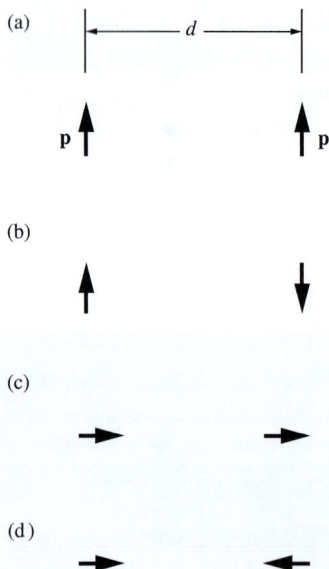

(a)

(b)

(c)

(d)

**Figure 10.39.**

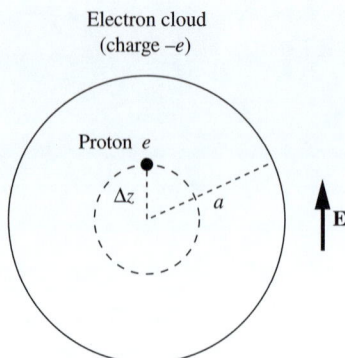

Electron cloud
(charge $-e$)

Proton $e$

$\Delta z$

$a$

$E$

**Figure 10.40.**

**10.33** *Field from hydrogen chloride*  *

A hydrogen chloride molecule is located at the origin, with the H–Cl line along the $z$ axis and Cl uppermost. What is the direction of the electric field, and its strength in volts/m, at a point 10 angstroms up from the origin, on the $z$ axis? At a point 10 angstroms out from the origin, on the $y$ axis? ($p$ is given in Fig. 10.14.)

**10.34** *Hydrogen chloride dipole moment*  **

In the hydrogen chloride molecule the distance between the chlorine nucleus and the proton is 1.28 angstroms. Suppose the electron from the hydrogen atom is transferred entirely to the chlorine atom, joining with the other electrons to form a spherically symmetrical negative charge that is centered on the chlorine nucleus. How does the electric dipole moment of this model compare with the actual HCl dipole moment given in Fig. 10.14? Where must the actual "center of gravity" of the negative charge distribution be located in the real molecule? (The chlorine nucleus has a charge $17e$, and the hydrogen nucleus has a charge $e$.)

**10.35** *Some electric susceptibilities*  **

From the values of $\kappa$ given for water, ammonia, and methanol in Table 10.1, we know that the electric susceptibility $\chi_e$ for each liquid is given by $\chi_e = \kappa - 1$. Our theoretical prediction in Eq. (10.73) can be written $\chi_e = CNp^2/\epsilon_0 kT$, with the factor $C$ as yet unknown, but expected to have order of magnitude unity. The densities of the liquids are 1.00, 0.82, and 1.33 g/cm$^3$, respectively; their molecular weights are 18, 17, and 32. Taking the value of the dipole moment from Fig. 10.14, find for each case the value of $C$ required to fit the observed value of $\chi_e$.

**10.36** *Discontinuity in $E_\perp$*  **

Consider the polarized sphere from Section 10.9. Using the forms of the internal and external electric fields, show that $E_\perp$ has a discontinuity of $P_\perp/\epsilon_0 = P\cos\theta/\epsilon_0$ everywhere on the surface of the sphere.

**10.37** *E at the center of a polarized sphere*  **

If you don't trust the $\mathbf{E} = -\mathbf{P}/3\epsilon_0$ result we obtained in Section 10.9 for the field inside a uniformly polarized sphere, you will find it more believable if you check it in a special case. By direct integration of the contributions from the $\sigma = P\cos\theta$ surface charge density, show that the field at the center is directed downward (assuming $\mathbf{P}$ points upward) with magnitude $P/3\epsilon_0$.

**10.38** *Uniform field via superposition*  **

In Section 10.9, the fact that the electric field is uniform inside the polarized sphere was deduced from the form of the potential on the boundary. You can also prove it by superposing the internal fields of two balls of charge whose centers are separated.

(a) Show that, inside a spherical uniform charge distribution, $\mathbf{E}$ is proportional to $\mathbf{r}$.

(b) Now take two spherical distributions with density $\rho$ and $-\rho$, centers at $C_1$ and $C_2$, and show that the resultant field is constant and parallel to the line from $C_1$ to $C_2$. Verify that the field can be written as $-\mathbf{P}/3\epsilon_0$.

(c) Analyze in the same way the field of a long cylindrical rod that is polarized perpendicular to its axis.

10.39 *Conducting-sphere limit* **
Our formula for the dielectric sphere in Section 10.10 can actually serve to describe a metal sphere in a uniform field. To demonstrate this, investigate the limiting case $\kappa \to \infty$, and show that the external field then takes on a form that satisfies the perfect-conductor boundary conditions. What about the internal field? Make a sketch of some field lines for this limiting case. What is the radius of a conducting sphere with polarizability equal to that of the hydrogen atom, given in Table 10.2?

10.40 *Continuity of D* *
Use the definition of $\mathbf{D}$, namely $\mathbf{D} \equiv \epsilon_0\mathbf{E}+\mathbf{P}$, to show that $\mathbf{D}$ is continuous across the faces of a uniformly polarized slab. Assume that the polarization is perpendicular to the faces, and that the thickness of the slab is small compared with the other two dimensions.

10.41 *Discontinuity in $D_\parallel$* **
Consider the polarized sphere from Section 10.9. Using the forms of the internal and external electric fields, find the discontinuity in $D_\parallel$ across the surface of the sphere, as a function of $\theta$.

10.42 *Energy density in a dielectric* **
By considering how the introduction of a dielectric changes the energy stored in a capacitor, show that the correct expression for the energy density in a dielectric must be $\epsilon E^2/2$. Then compare the energy stored in the electric field with that stored in the magnetic field in the wave studied in Section 10.15.

10.43 *Reflected wave* ***
A block of glass, refractive index $n = \sqrt{\kappa}$, fills the space $y > 0$, its surface being the $xz$ plane. A plane wave traveling in the positive $y$ direction through the empty space $y < 0$ is incident upon this surface. The electric field in this wave is $\hat{\mathbf{z}}E_i \sin(ky-\omega t)$. There is a transmitted wave inside the glass block, with an electric field given by $\hat{\mathbf{z}}E_t \sin(k'y - \omega t)$. There is also a reflected wave in the space $y < 0$, traveling away from the glass in the negative $y$ direction.

Its electric field is $\hat{\mathbf{z}}E_r \sin(ky + \omega t)$. Of course, each wave has its magnetic field, of amplitude, respectively, $B_i$, $B_t$, and $B_r$.

The total magnetic field must be continuous at $y = 0$; and the total electric field, being parallel to the surface, must be continuous also (see Problem 10.14). Show that these requirements, and the relation of $B_t$ to $E_t$ given in Eq. (10.83) (with the "0" subscript changed to "t"), suffice to determine the ratio of $E_r$ to $E_i$. When a light wave is incident normally at an air–glass interface, what fraction of the energy is reflected if the index $n$ is 1.6?

# 11

# Magnetic fields in matter

**Overview** Magnetic fields in matter are a bit more involved than electric fields in matter. Our main goal in this chapter is to understand the three types of magnetic materials: *diamagnetic* materials, which are weakly repelled by a solenoid; *paramagnetic* materials, which are somewhat strongly attracted; and *ferromagnetic* materials, which are very strongly attracted. As was the case in Chapter 10, we will need to understand dipoles. The far field of a *magnetic dipole* has the same form as that of an electric dipole, with the *magnetic dipole moment* replacing the electric dipole moment. However, the near fields are fundamentally different due to the absence of magnetic charge. We will find that diamagnetism is due to the fact that an applied magnetic field causes the magnetic dipole moment arising from the *orbital* motion of electrons in atoms to pick up a contribution pointing *opposite* to the applied field. In contrast, in the case of paramagnetism, the *spin* dipole moment is the relevant one, and it picks up a contribution pointing in the *same* direction as the applied field. Ferromagnetism is similar to paramagnetism, although a certain quantum phenomenon makes the overall effect much larger; a ferromagnetic dipole moment can exist in the absence of an external magnetic field. Magnetized materials can be described by the *magnetization* $M$, the curl of which gives the *bound* currents (which arise from both orbital motion and spin). By considering separately the free and bound currents, we are led to the field $H$ (also called the "magnetic field") whose curl involves only the free current (unlike the magnetic field $B$, whose curl involves *all* the current, by Ampère's law).

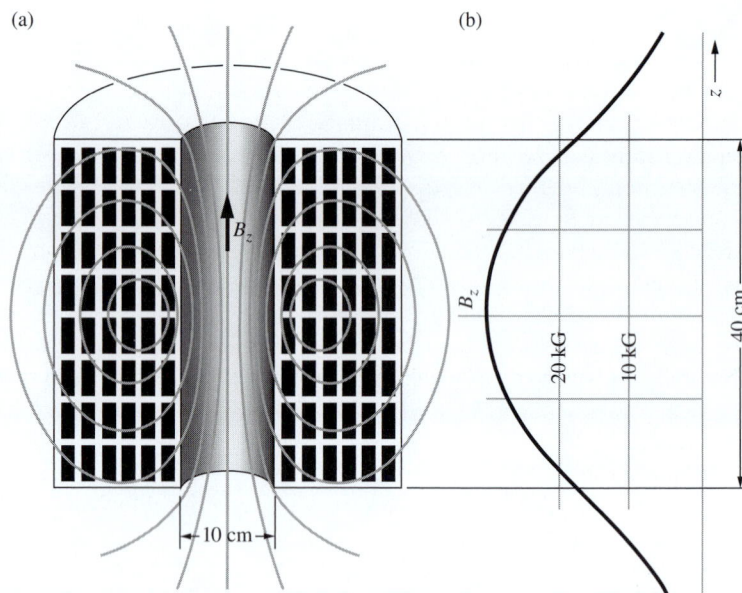

**Figure 11.1.**
(a) A coil designed to produce a strong magnetic field. The water-cooled winding is shown in cross section. (b) A graph of the field strength $B_z$ on the axis of the coil.

## 11.1 How various substances respond to a magnetic field

Imagine doing some experiments with a very intense magnetic field. To be definite, suppose we have built a solenoid of 10 cm inside diameter, 40 cm long, like the one shown in Fig. 11.1. Its outer diameter is 40 cm, most of the space being filled with copper windings. This coil will provide a steady field of 3.0 tesla, or 30,000 gauss, at its center if supplied with 400 kilowatts of electric power – and something like 30 gallons of water per minute, to carry off the heat. We mention these practical details to show that our device, though nothing extraordinary, is a pretty respectable laboratory magnet. The field strength at the center is nearly $10^5$ times the earth's field, and probably 5 or 10 times stronger than the field near any iron bar magnet or horseshoe magnet you may have experimented with, although some rare-earth magnets can have fields of around 1 tesla.

The field will be fairly uniform near the center of the solenoid, falling, on the axis at either end, to roughly half its central value. It will be rather less uniform than the field of the solenoid in Fig. 6.18, since our coil is equivalent to a "nested" superposition of solenoids with length–diameter ratio varying from 4:1 to 1:1. In fact, if we analyze our coil in that way and use the formula that we derived for the field on the axis of a solenoid with a single-layer winding (see Eq. (6.56)), it is not hard to calculate the axial field exactly. A graph of the field strength on the axis, with the central field taken as 3.0 tesla = 30 kilogauss, is included in Fig. 11.1. The intensity just at the end of the coil is 1.8 tesla, and in that neighborhood the field is changing with a gradient of approximately 17 tesla/m, or 1700 gauss/cm.

Let's put various substances into this field and see if a force acts on them. Generally, we do detect a force. It vanishes when the current in the coil is switched off. We soon discover that the force is strongest not when our sample of substance is at the center of the coil where the magnetic field $B_z$ is strongest, but when it is located near the end of the coil where the gradient $dB_z/dz$ is large. From now on let us support each sample just inside the upper end of the coil. Figure 11.2 shows one such sample, contained in a test tube suspended by a spring which can be calibrated to indicate the extra force caused by the magnetic field. Naturally we have to do a "blank" experiment with the test tube and suspension alone, to allow for the magnetic force on everything other than the sample.

We find in such an experiment that the force on a particular substance – metallic aluminum, for instance – is proportional to the mass of the sample and independent of its shape, as long as the sample is not too large. (Experiments with a small sample in this coil show that the force remains practically constant over a region a few centimeters in extent, inside the end of the coil; if we use samples no more than 1 to 2 cm$^3$ in volume, they can be kept well within this region.) We can express our quantitative results, for a given substance, as so many newtons force per kilogram of sample, under the conditions $B_z = 1.8$ tesla, $dB_z/dz = 17$ tesla/m.

But first the qualitative results, which are a bit bewildering. For a large number of quite ordinary pure substances, the force observed, although easily measurable, seems ridiculously small, despite all our effort to provide an intense magnetic field. Typically, the force is 0.1 or 0.2 newtons per kilogram, that is, no more than a few percent of the weight of the sample (which is 9.8 newtons per kilogram). It is directed upward for some samples, downward for others. This has nothing to do with the *direction* of the magnetic field, as we can verify by reversing the current in the coil. Instead, it appears that some substances are always pulled in the direction of *increasing* field intensity, others in the direction of *decreasing* field intensity, irrespective of the field direction.

We do find some substances that are attracted to the coil with considerably greater force. For instance, copper chloride crystals are pulled downward with a force of 2.8 newtons per kilogram of sample. Liquid oxygen behaves spectacularly in this experiment; it is pulled into the coil with a force nearly eight times its weight. In fact, if we were to bring an uncovered flask of liquid oxygen up to the bottom end of our coil, the liquid would be lifted right out of the flask. (Where do you think it would end up?) Liquid nitrogen, on the other hand, proves to be quite unexciting; it is pushed away from the coil with the feeble force of 0.1 newtons per kilogram. In Table 11.1 we have listed some results that one might obtain in such an experiment. The substances, including those

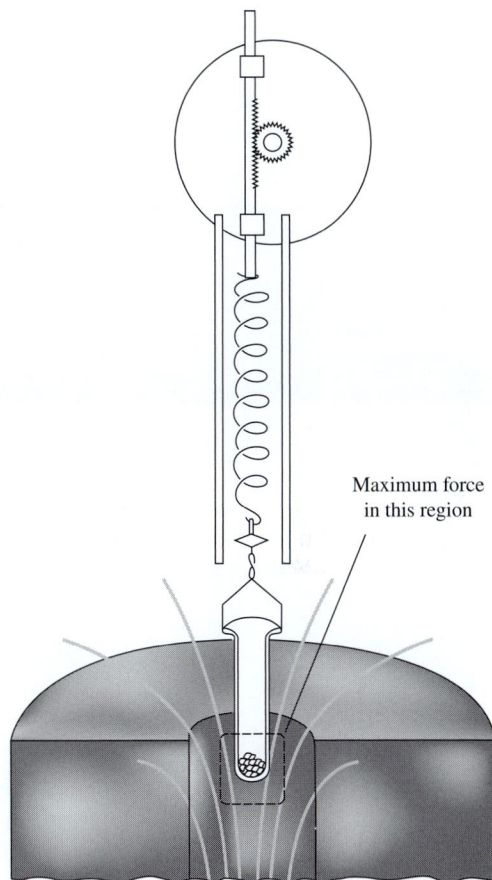

Maximum force in this region

**Figure 11.2.**
An arrangement for measuring the force on a substance in a magnetic field.

**Table 11.1.**
Force per kilogram near the upper end of the coil in our experiment, where $B_z = 1.8$ tesla and $dB_z/dz = 17$ tesla/m

| Substance | Formula | Force (newtons) |
|---|---|---|
| **Diamagnetic** | | |
| Water | $H_2O$ | −0.22 |
| Copper | Cu | −0.026 |
| Sodium chloride | NaCl | −0.15 |
| Sulfur | S | −0.16 |
| Diamond | C | −0.16 |
| Graphite | C | −1.10 |
| Liquid nitrogen | $N_2$ | −0.10 (78 K) |
| **Paramagnetic** | | |
| Sodium | Na | 0.20 |
| Aluminum | Al | 0.17 |
| Copper chloride | $CuCl_2$ | 2.8 |
| Nickel sulfate | $NiSO_4$ | 8.3 |
| Liquid oxygen | $O_2$ | 75  (90 K) |
| **Ferromagnetic** | | |
| Iron | Fe | 4000 |
| Magnetite | $Fe_3O_4$ | 1200 |

Direction of force: downward (into coil), +; upward, −.
All measurements were made at a temperature of 20°C unless otherwise stated.
The three types of magnetism are defined in the text.

already mentioned, have been chosen to suggest, as best one can with a sparse sampling, the wide range of magnetic behavior we find in ordinary materials. Note that our convention for the sign of the force is that a positive force is directed into the coil.

As you know, a few substances, of which the most familiar is metallic iron, seem far more "magnetic" than any others. In Table 11.1 we give the force per kilogram that would act on a piece of iron put in the same position in the field as the other samples. Since 1 newton is about 0.22 pounds, the force per kilogram is roughly 900 pounds, or nearly 1 pound for a 1 gram sample! (We would not have been so naive as to approach our magnet with a gram of iron suspended in a test tube from a delicate spring – a different suspension would have to be used.) Observe that there is a factor of more than $10^5$ between the force per kilogram on iron and the force per kilogram on copper, elements not otherwise radically different. Incidentally, this suggests that reliable magnetic measurements on a substance like copper may not be easy. A few parts per million contamination by metallic iron particles would utterly falsify the result.

There is another essential difference between the behavior of the iron and the magnetite and that of the other substances in the table. Suppose we make the obvious test, by varying the field strength of the magnet, to ascertain whether the force on a sample is proportional to the field. For instance, we might reduce the solenoid current by half, thereby halving both the field intensity $B_z$ and its gradient $dB_z/dz$. We would find, in the case of every substance above iron in the table, that the force is reduced to *one-fourth* its former value, whereas the force on the iron sample, and that on the magnetite, would be reduced only to one-half or perhaps a little less. Evidently the force, under these conditions at least, is proportional to the square of the field strength for all the other substances listed, but nearly proportional to the field strength itself for Fe and $Fe_3O_4$.

It appears that we may be dealing with several different phenomena here, and complicated ones at that. As a small step toward understanding, we can introduce some classification.

(1) Diamagnetism  First, those substances that are feebly *repelled* by our magnet – water, sodium chloride, diamond, etc. – are called *diamagnetic*. The majority of inorganic compounds and practically all organic compounds are diamagnetic. It turns out, in fact, that diamagnetism is a property of *every* atom and molecule. When the opposite behavior is observed, it is because the diamagnetism is outweighed by a different and stronger effect, one that leads to attraction.

(2) Paramagnetism  Substances that are *attracted* toward the region of stronger field are called *paramagnetic*. In some cases, notably metals such as aluminum, sodium, and many others, the paramagnetism is not much stronger than the common diamagnetism. In other materials, such as the $NiSO_4$ and the $CuCl_2$ on our list, the paramagnetic effect is much stronger. In these substances also, it *increases* as the temperature is lowered, leading to quite large effects at temperatures near absolute zero. The increase of paramagnetism with lowering temperature is responsible in part for the large force recorded for liquid oxygen. If you think all this is going to be easy to explain, observe that copper is diamagnetic while copper chloride is paramagnetic, but sodium is paramagnetic while sodium chloride is diamagnetic.

(3) Ferromagnetism  Finally, substances that behave like iron and magnetite are called *ferromagnetic*. In addition to the common metals of this class – iron, cobalt, and nickel – quite a number of ferromagnetic alloys and crystalline compounds are known. Indeed current research in ferromagnetism is steadily lengthening the list.

In this chapter we have two tasks. One is to develop a treatment of the large-scale phenomena involving magnetized matter, in which the material itself is characterized by a few parameters and the experimentally determined relations among them. It is like a treatment of dielectrics based on some observed relation between electric field and bulk polarization. We sometimes call such a theory *phenomenological;* it is more of a description than an explanation. Our second task is to try to understand, at least in a general way, the atomic origin of the various magnetic effects. Even more than dielectric phenomena, the magnetic effects, once understood, reveal some basic features of atomic structure.

One general fact stands out in Table 11.1. Very little energy, on the scale of molecular energies, is involved in diamagnetism and paramagnetism. Take the extreme example of liquid oxygen. To pull 1 kilogram (although the sample size would certainly be much smaller) of liquid oxygen away from our magnet, energy would have to be expended amounting, in joules, to 75 newtons times a distance of roughly 0.1 meters (since the field strength falls off substantially over a distance of a few centimeters). In order of magnitude, let us say the energy is 10 joules. There are $2 \cdot 10^{25}$ molecules in 1 kilogram of the liquid, so this is less than $10^{-24}$ joules per molecule. Just to vaporize 1 kilogram of liquid oxygen requires 50,000 calories, or about $10^{-20}$ joules per molecule, using 1 calorie $= 4.18$ joules. (Most of that energy is used in separating the molecules from one another.) Whatever may be happening in liquid oxygen at the molecular level as a result of the magnetic field, it is apparently a very minor affair in terms of energy.

Even a strong magnetic field has hardly any effect on chemical processes, including biochemical. You could put your hand and forearm into our 3 tesla solenoid without experiencing any significant sensation or consequence. It is hard to predict whether your arm would prove to be paramagnetic or diamagnetic, but the force on it would be no more than a fraction of an ounce in any case. Conversely, the presence of someone's hand close to the sample in Fig. 11.2 would perturb the field and change the force on the sample by no more than a few parts in a million. In whole-body imaging with nuclear magnetic resonance, the body is pervaded by a magnetic field of a few tesla in strength with no physiological effects whatsoever. It appears that the only hazards associated with large-scale, strong, steady magnetic fields arise from metal objects in the vicinity. For example, implants containing metal may heat up, move within the body, or malfunction. And there is also the danger that a loose iron object will be snatched by the fringing field and hurled into the magnet. Be careful what you bring into a magnetic resonance imaging (MRI) room!

In its interaction with matter, the magnetic field plays a role utterly different from that of the electric field. The reason is simple and fundamental. Atoms and molecules are made of electrically charged particles that move with velocities generally small compared with the speed of light. A magnetic field exerts no force at all on a stationary electric charge; on a moving charged particle the force is proportional to $v^2/c^2$.[1]

---

[1] This factor of $v^2/c^2$ follows from Eq. (5.28). The current $I$ in that equation involves the velocity of charges, and we are assuming that all velocities here are of the same order of magnitude. Of course, if we have a charged particle moving in a region where the electric field is zero and the magnetic field is nonzero, then the magnetic field dominates. But for general random motions of the charges (both the charges creating the fields and the charges affected by the fields), the magnetic force is smaller by a factor of $v^2/c^2$.

Said in a sloppier way, in SI units the $1/4\pi\epsilon_0$ factor in Coulomb's law for the electric field is large, while the $\mu_0/4\pi$ factor in the Biot–Savart law for the magnetic field is small. Electric forces overwhelmingly dominate the atomic scene. As we have remarked before, magnetism appears, in our world at least, to be a relativistic effect. The story would be different if matter were made of magnetically charged particles. We must explain now what *magnetic charge* means and what its apparent absence signifies.

## 11.2 The absence of magnetic "charge"

The magnetic field outside a magnetized rod such as a compass needle looks very much like the electric field outside an electrically polarized rod, a rod that has an excess of positive charge at one end, negative charge at the other (Fig. 11.3). It is conceivable that the magnetic field has sources that are related to it in the same way as electric charge is related to the electric field. Then the north pole of the compass needle would be the location of an excess of one kind of magnetic charge, and the south pole would be the location of an excess of the opposite kind. We might call "north charge" positive and "south charge" negative, with magnetic field directed from positive to negative, a rule like that adopted for electric field and electric charge. Historically, that is how our convention about the positive direction of magnetic field was established.[2] What we have called *magnetic charge* has usually been called *magnetic pole strength.*

This idea is perfectly sound as far as it goes. It becomes even more plausible when we recall that the fundamental equations of the electromagnetic field are symmetrical in $\mathbf{E}$ and $c\mathbf{B}$. Why, then, should we not expect to find symmetry in the sources of the field? With magnetic charge as a possible source of the static magnetic field $\mathbf{B}$, we would have div $\mathbf{B} \propto \eta$, where $\eta$ stands for the density of magnetic charge, in complete analogy to the electric charge density $\rho$. Two positive magnetic charges (or north poles) would repel one another, and so on.

The trouble is, that is not the way things are. Nature for some reason has not made use of this opportunity. The world around us appears totally asymmetrical in the sense that we find *no magnetic charges at all.* No one has yet observed an isolated excess of one kind of magnetic charge – an isolated north pole for example. If such a *magnetic monopole* existed it could be recognized in several ways. Unlike a magnetic dipole,

---

[2] In Chapter 6, remember, we established the positive direction of $\mathbf{B}$ by reference to a current direction (direction of motion of positive charge) and a right-hand rule. Now *north pole* means "north-seeking pole" of the compass needle. We know of no reason why the earth's magnetic polarity should be one way rather than the other. Franklin's designation of "positive" electricity had nothing to do with any of this. So the fact that it takes a right-hand rule rather than a left-hand rule to make this all consistent is the purest accident.

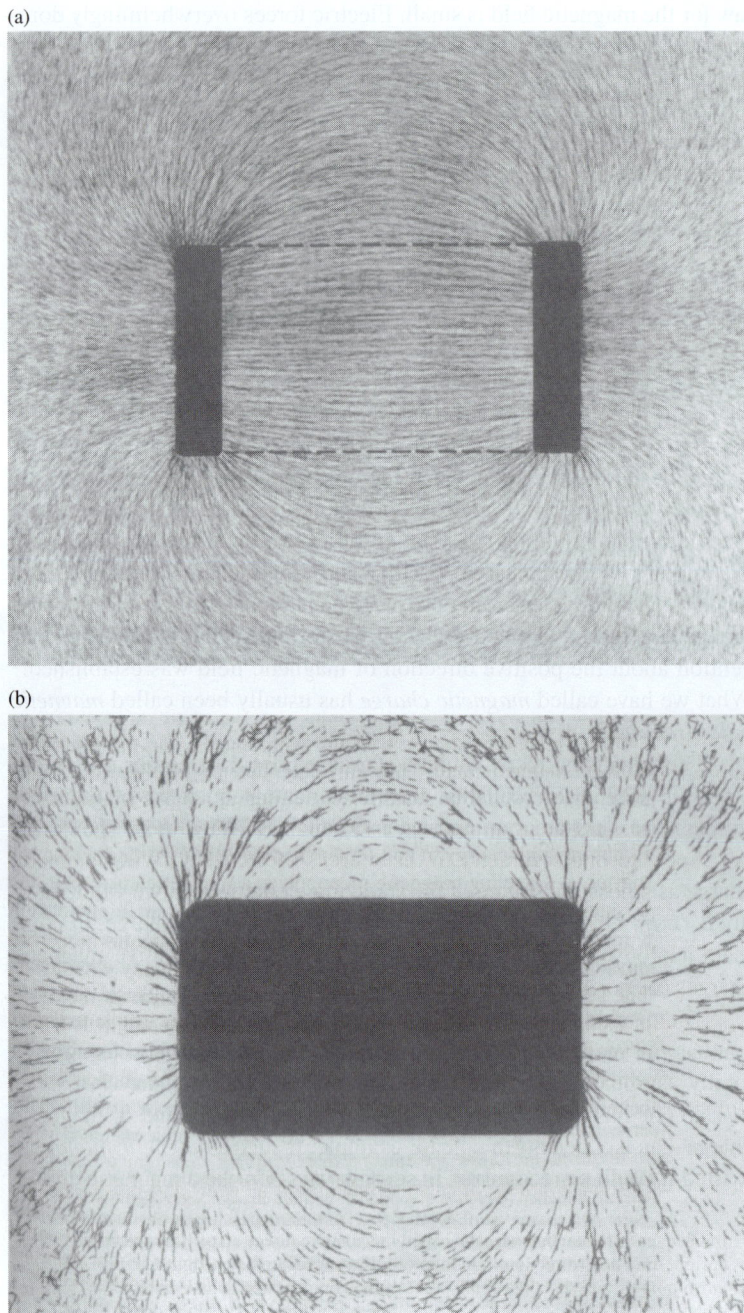

**Figure 11.3.**
(a) Two oppositely charged disks (the electrodes showing in cross section as solid black bars) have an electric field that is the same as that of a polarized rod. That is, if you imagine such a rod to occupy the region within the dashed boundary, its external field would be like that shown. The electric field here was made visible by a multitude of tiny black fibers, suspended in oil, which oriented themselves along the field direction. This elegant method of demonstrating electric field configurations is due to Harold M. Waage, Palmer Physical Laboratory, Princeton University, who kindly prepared the original photograph for this illustration (Waage, 1964).
(b) The magnetic field around a magnetized cylinder, shown by the orientation of small pieces of nickel wire, immersed in glycerine. (This attempt to improve on the traditional iron filings demonstration by an adaptation of Waage's technique was not very successful – the nickel wires tend to join in long strings that are then pulled in toward the magnet.) Theoretically constructed diagrams of the fields in the two systems are shown later in Fig. 11.22.

it would experience a force if placed in a uniform magnetic field. Thus an elementary particle carrying a magnetic charge would be steadily accelerated in a static magnetic field, as a proton or an electron is steadily accelerated in an electric field. Reaching high energy, it could then be detected by its interaction with matter. A traveling magnetic monopole is a magnetic current; it must be encircled by an electric field, as an electric current is encircled by a magnetic field. With strategies based on these unique properties, physicists have looked for magnetic monopoles in many experiments. The search was renewed when a development in the theory of elementary particles suggested that the universe ought to contain at least a few magnetic monopoles, left over from the "big bang" in which it presumably began. But not one magnetic monopole has yet been detected, and it is now evident that if they exist at all they are exceedingly rare. Of course, the proven existence of even one magnetically charged particle would have profound implications, but it would not alter the fact that in matter as we know it, the only sources of the magnetic field are electric currents. As far as we know,

$$\text{div } \mathbf{B} = 0 \quad (\textit{everywhere}) \tag{11.1}$$

This takes us back to the hypothesis of Ampère, his idea that magnetism in matter is to be accounted for by a multitude of tiny rings of electric current distributed through the substance. We begin by studying the magnetic field created by a single current loop at points relatively far from the loop.

## 11.3 The field of a current loop

A closed conducting loop, not necessarily circular, lies in the $xy$ plane encircling the origin, as in Fig. 11.4(a). A steady current $I$ flows around the loop. We are interested in the magnetic field this current creates – not near the loop, but at distant points like $P_1$ in the figure. We assume that $r_1$, the distance to $P_1$, is much larger than any dimension of the loop. To simplify the diagram we have located $P_1$ in the $yz$ plane; it will turn out that this is no restriction. This is a good place to use the vector potential. We shall compute first the vector potential $\mathbf{A}$ at the location $P_1$, that is, $\mathbf{A}(0, y_1, z_1)$. From this it will be obvious what the vector potential is at any other point $(x, y, z)$ far from the loop. Then by taking the curl of $\mathbf{A}$ we can get the magnetic field $\mathbf{B}$.

For a current confined to a wire, Eq. (6.46) gives $\mathbf{A}$ as

$$\mathbf{A}(0, y_1, z_1) = \frac{\mu_0 I}{4\pi} \int_{\text{loop}} \frac{d\mathbf{l}_2}{r_{12}}. \tag{11.2}$$

When we used this equation in Section 6.4, we were concerned only with the contribution of a small segment of the circuit; now we have to integrate around the entire loop. Consider the variation in the denominator

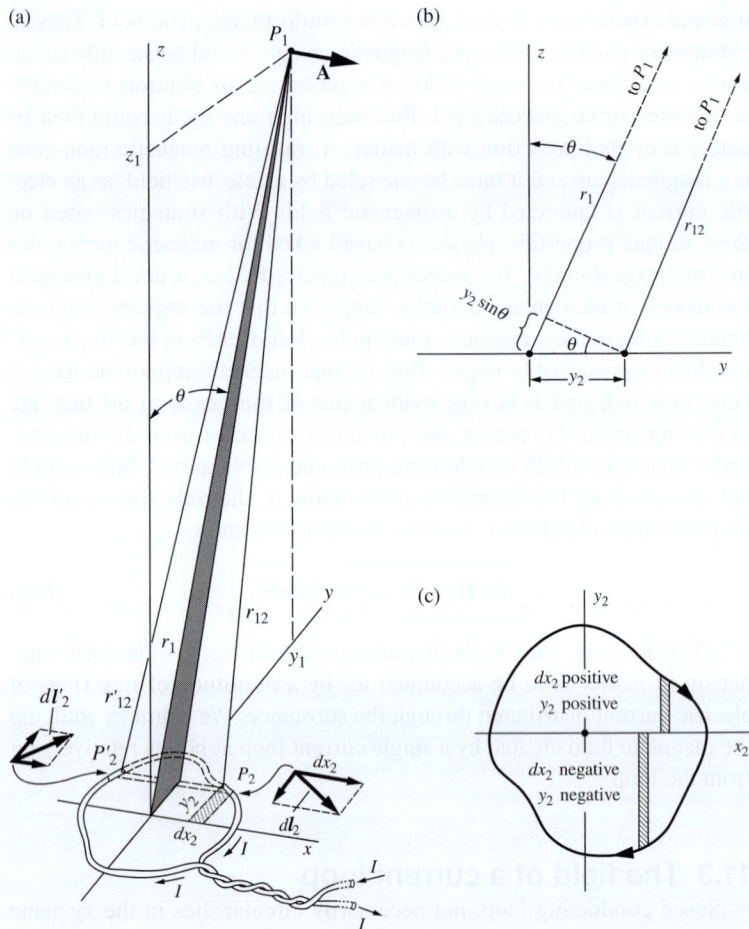

**Figure 11.4.**
(a) Calculation of the vector potential **A** at a point far from the current loop. (b) Side view, looking in along the $x$ axis, showing that $r_{12} \approx r_1 - y_2 \sin\theta$ if $r_1 \gg y_2$. (c) Top view, to show that $\int_{\mathrm{loop}} y_2 \, dx_2$ is the area of the loop.

$r_{12}$ as we go around the loop. If $P_1$ is far away, the first-order variation in $r_{12}$ depends only on the coordinate $y_2$ of the segment $d\mathbf{l}_2$, and not on $x_2$. This is true because, from the Pythagorean theorem, the contribution to $r_{12}$ from $x_2$ is of second order, whereas the side view in Fig. 11.4(b) shows the first-order contribution from $y_1$. Thus, neglecting quantities proportional to $(x_2/r_{12})^2$, we may treat $r_{12}$ and $r'_{12}$, which lie on top of one another in the side view, as equal. And in general, to first order in the ratio (loop dimension/distance to $P_1$), we have

$$r_{12} \approx r_1 - y_2 \sin\theta. \tag{11.3}$$

Look now at the two elements of the path $d\mathbf{l}_2$ and $d\mathbf{l}'_2$ shown in Fig. 11.4(a). For these the $dy_2$ displacements are equal and opposite, and as we have already pointed out, the $r_{12}$ distances are equal to first order. To this order then, the $dy_2$ contributions to the line integral will cancel, and this will be true for the whole loop. Hence $\mathbf{A}$ at $P_1$ will not have a $y$ component. Obviously it will not have a $z$ component either, for $dz_2$ is always zero since the current path itself has nowhere a $z$ component.

However, $\mathbf{A}$ at $P_1$ *will* have an $x$ component. The $x$ component of the vector potential comes from the $dx_2$ part of the path integral:

$$\mathbf{A}(0, y_1, z_1) = \hat{\mathbf{x}} \frac{\mu_0 I}{4\pi} \int \frac{dx_2}{r_{12}}. \tag{11.4}$$

Without spoiling our first-order approximation, we can turn Eq. (11.3) into

$$\frac{1}{r_{12}} = \frac{1}{r_1 \left(1 - (y_2/r_1) \sin \theta\right)} \approx \frac{1}{r_1} \left(1 + \frac{y_2 \sin \theta}{r_1}\right), \tag{11.5}$$

and using this for the integrand, we have

$$\mathbf{A}(0, y_1, z_1) = \hat{\mathbf{x}} \frac{\mu_0 I}{4\pi r_1} \int \left(1 + \frac{y_2 \sin \theta}{r_1}\right) dx_2. \tag{11.6}$$

In the integration, $r_1$ and $\theta$ are constants. Obviously $\int dx_2$ around the loop vanishes. Now $\int y_2 \, dx_2$ around the loop is just the area of the loop, regardless of its shape; see Fig. 11.4(c). So we get finally

$$\mathbf{A}(0, y_1, z_1) = \hat{\mathbf{x}} \frac{\mu_0 I \sin \theta}{4\pi r_1^2} \times (\text{area of loop}). \tag{11.7}$$

The intuitive reason why this result is nonzero is that the parts of the loop that are closer to $P_1$ give larger contributions to the integral, because they have a smaller $r_{12}$. There is partial, but not complete, cancelation from corresponding pieces of the loop with the same $x_2$ value but opposite $dx_2$ values.

Here is a simple but crucial point: since the *shape* of the loop hasn't mattered, our restriction on $P_1$ to the $yz$ plane cannot make any essential difference. Therefore we must have in Eq. (11.7) the general result we seek, if only we *state* it generally: the vector potential of a current loop of any shape, at a distance $r$ from the loop that is much greater than the size of the loop, is a vector perpendicular to the plane containing $\mathbf{r}$ and the normal to the plane of the loop, of magnitude

$$A = \frac{\mu_0 I a \sin \theta}{4\pi r^2}, \tag{11.8}$$

where $a$ stands for the area of the loop.

This vector potential is symmetrical around the axis of the loop, which implies that the field $\mathbf{B}$ will be symmetrical also. The explanation is that we are considering regions so far from the loop that the details of the shape of the loop have negligible influence. All loops with the same *current $\times$ area* product produce the same far field. We call the product $Ia$

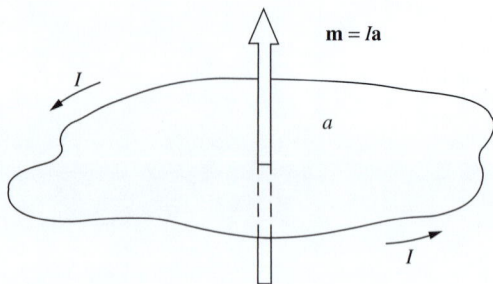

**Figure 11.5.**
By definition, the magnetic moment vector is
related to the current by a right-hand-screw rule
as shown here.

the *magnetic dipole moment* of the current loop, and denote it by **m**. Its
units are amp-m$^2$. The magnetic dipole moment is a vector, its direction
defined to be that of the normal to the loop, or that of the vector **a**, the
directed area of the region surrounded by the loop:

$$\boxed{\mathbf{m} = I\mathbf{a}} \tag{11.9}$$

As for sign, let us agree that the direction of **m** and the sense of positive
current flow in the loop are to be related by a right-hand-screw rule,
illustrated in Fig. 11.5. (The dipole moment of the loop in Fig. 11.4(a)
points downward, according to this rule.) The vector potential for the
field of a magnetic dipole **m** can now be written neatly with vectors:

$$\boxed{\mathbf{A} = \frac{\mu_0}{4\pi} \frac{\mathbf{m} \times \hat{\mathbf{r}}}{r^2}} \tag{11.10}$$

where $\hat{\mathbf{r}}$ is a unit vector in the direction *from* the loop *to* the point for
which **A** is being computed. You can check that this agrees with our
convention about sign. Note that the direction of **A** will always be that of
the current in the *nearest* part of the loop.

Figure 11.6 shows a magnetic dipole located at the origin, with the
dipole moment vector **m** pointed in the positive $z$ direction. To express
the vector potential at any point $(x, y, z)$, we observe that $r^2 = x^2+y^2+z^2$,
and $\sin\theta = \sqrt{x^2 + y^2}/r$. The magnitude $A$ of the vector potential at that
point is given by

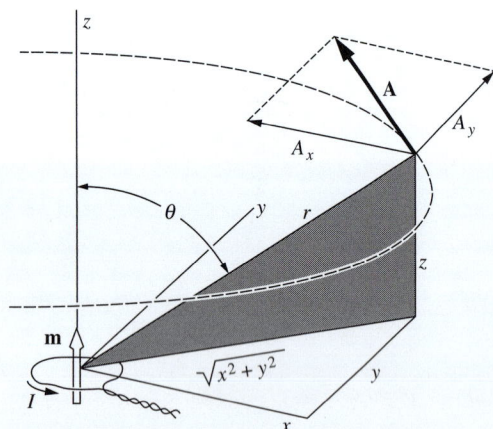

**Figure 11.6.**
A magnetic dipole located at the origin. At every
point far from the loop, **A** is a vector parallel to
the $xy$ plane, tangent to a circle around the $z$
axis.

$$A = \frac{\mu_0}{4\pi} \frac{m\sin\theta}{r^2} = \frac{\mu_0}{4\pi} \frac{m\sqrt{x^2 + y^2}}{r^3}. \tag{11.11}$$

Since **A** is tangent to a horizontal circle around the $z$ axis, its components are

$$A_x = A\left(\frac{-y}{\sqrt{x^2 + y^2}}\right) = -\frac{\mu_0}{4\pi}\frac{my}{r^3},$$

$$A_y = A\left(\frac{x}{\sqrt{x^2 + y^2}}\right) = \frac{\mu_0}{4\pi}\frac{mx}{r^3},$$

$$A_z = 0. \tag{11.12}$$

Let's evaluate **B** for a point in the $xz$ plane, by finding the components of curl **A** and then (not before!) setting $y = 0$:

$$B_x = (\nabla \times \mathbf{A})_x = \frac{\partial A_z}{\partial y} - \frac{\partial A_y}{\partial z} = -\frac{\mu_0}{4\pi}\frac{\partial}{\partial z}\frac{mx}{(x^2 + y^2 + z^2)^{3/2}} = \frac{\mu_0}{4\pi}\frac{3mxz}{r^5},$$

$$B_y = (\nabla \times \mathbf{A})_y = \frac{\partial A_x}{\partial z} - \frac{\partial A_z}{\partial x} = \frac{\mu_0}{4\pi}\frac{\partial}{\partial z}\frac{-my}{(x^2 + y^2 + z^2)^{3/2}} = \frac{\mu_0}{4\pi}\frac{3myz}{r^5},$$

$$B_z = (\nabla \times \mathbf{A})_z = \frac{\partial A_y}{\partial x} - \frac{\partial A_x}{\partial y}$$

$$= \frac{\mu_0}{4\pi} m \left[ \frac{-2x^2 + y^2 + z^2}{(x^2 + y^2 + z^2)^{5/2}} + \frac{x^2 - 2y^2 + z^2}{(x^2 + y^2 + z^2)^{5/2}} \right] = \frac{\mu_0}{4\pi} \frac{m(3z^2 - r^2)}{r^5}.$$

$$(11.13)$$

In the $xz$ plane, we have $y = 0$, $\sin\theta = x/r$, and $\cos\theta = z/r$. The field components at any point in that plane are thus given by

$$B_x = \frac{\mu_0}{4\pi} \frac{3m\sin\theta\cos\theta}{r^3},$$

$$B_y = 0,$$

$$B_z = \frac{\mu_0}{4\pi} \frac{m(3\cos^2\theta - 1)}{r^3}. \qquad (11.14)$$

Now turn back to Section 10.3, where in Eq. (10.17) we expressed the components in the $xz$ plane of the field $\mathbf{E}$ of an electric dipole $\mathbf{p}$, which was situated exactly like our magnetic dipole $\mathbf{m}$. The expressions are essentially identical, the only changes being $p \to m$ and $1/\epsilon_0 \to \mu_0$. We have thus found that the magnetic field of a small current loop has, at remote points, the same form as the electric field of two separated charges. We already know what that field, the electric dipole field, looks like. Figure 11.7 is an attempt to suggest the three-dimensional form of the magnetic field $\mathbf{B}$ arising from our current loop with dipole moment $\mathbf{m}$. As in the case of the electric dipole, the field is described somewhat more simply in spherical polar coordinates:

$$B_r = \frac{\mu_0 m}{2\pi r^3} \cos\theta, \quad B_\theta = \frac{\mu_0 m}{4\pi r^3} \sin\theta, \quad B_\phi = 0. \qquad (11.15)$$

The magnetic field *close* to a current loop is entirely different from the electric field close to a pair of separated positive and negative charges, as the comparison in Fig. 11.8 shows. Note that between the charges the electric field points down, while inside the current ring the magnetic field points up, although the far fields are alike. This reflects the fact that our magnetic field satisfies $\nabla \cdot \mathbf{B} = 0$ everywhere, *even inside the source*. The magnetic field lines don't end. By *near* and *far* we mean, of course, relative to the size of the current loop or the separation of the charges. If we imagine the current ring shrinking in size, the current meanwhile increasing so that the dipole moment $m = Ia$ remains constant, we approach the infinitesimal magnetic dipole, the counterpart of the infinitesimal electric dipole described in Chapter 10.

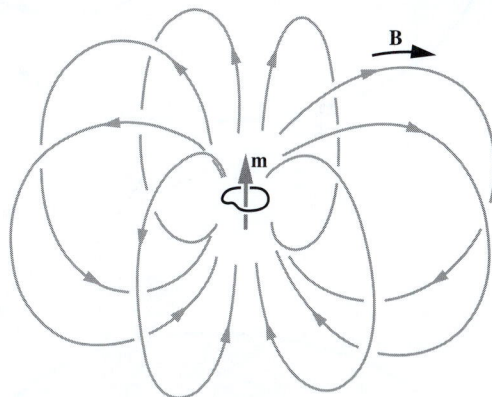

**Figure 11.7.**
Some magnetic field lines in the field of a magnetic dipole, that is, a small loop of current.

## 11.4 The force on a dipole in an external field

Consider a small circular current loop of radius $r$, placed in the magnetic field of some other current system, such as a solenoid. In Fig. 11.9, a field

**Magnetic fields in matter**

(a)

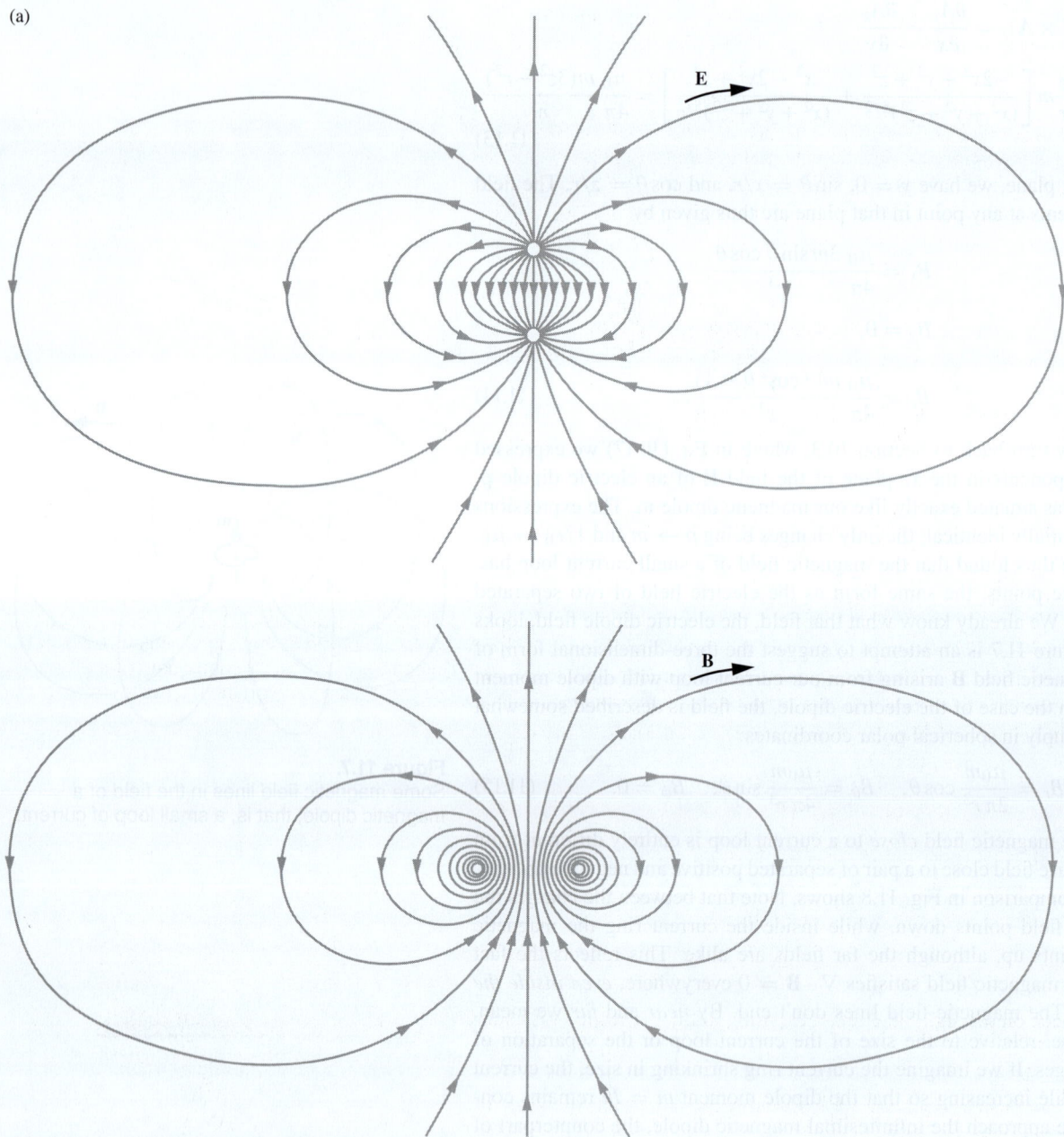

**Figure 11.8.**
(a) The electric field of a pair of equal and
opposite charges. Far away it becomes the field
of an electric dipole. (b) The magnetic field of a
current ring. Far away it becomes the field of a
magnetic dipole.

**B** is drawn that is generally in the $z$ direction. It is not a uniform field. Instead, it gets weaker as we proceed in the $z$ direction; this is evident from the fanning out of the field lines. Let us assume, for simplicity, that the field is symmetric about the $z$ axis. Then it resembles the field near the upper end of the solenoid in Fig. 11.1. The field represented in Fig. 11.9 does *not* include the magnetic field of the current ring itself. We want to find the force on the current ring caused by the other field, which we shall call, for want of a better name, the *external field*. The net force on the current ring due to its *own* field is certainly zero, so we are free to ignore its own field in this discussion.

If you study the situation in Fig. 11.9, you will soon conclude that there is a net force on the current ring. It arises because the external field **B** has an outward component $B_r$ everywhere around the ring. (The vertical component $B_z$ produces a force in the horizontal plane that simply stretches or compresses the ring – negligibly, assuming the ring is fairly rigid.) Therefore if the current flows in the direction indicated, each element of the loop, $dl$, must be experiencing a downward force of magnitude $IB_r \, dl$ (see Eq. (6.14)). If $B_r$ has the same magnitude at all points on the ring, as it must in the symmetrically spreading field assumed, the total downward force will have the magnitude

$$F = 2\pi rIB_r. \tag{11.16}$$

Now, $B_r$ can be directly related to the gradient of $B_z$. Since div $\mathbf{B} = 0$ at all points, the net flux of magnetic field out of any volume is zero. Consider a pancake-like cylinder of radius $r$ and height $\Delta z$ (Fig. 11.10). The outward flux from the side is $2\pi r(\Delta z)B_r$ and the net outward flux from the end surfaces is

$$\pi r^2[-B_z(z) + B_z(z + \Delta z)], \tag{11.17}$$

which to the first order in the small distance $\Delta z$ is $\pi r^2(\partial B_z/\partial z)\Delta z$. Setting the total outward flux equal to zero gives $0 = \pi r^2(\partial B_z/\partial z)\Delta z + 2\pi rB_r\Delta z$, or

$$B_r = -\frac{r}{2}\frac{\partial B_z}{\partial z}. \tag{11.18}$$

As a check on the sign, note that, according to Eq. (11.18), $B_r$ is positive when $B_z$ is decreasing upward; a glance at the figure shows that to be correct.

The force on the dipole (with upward taken to be positive) can now be expressed in terms of the gradient of the component $B_z$ of the external field:

$$F = -2\pi rI\left(-\frac{r}{2}\frac{\partial B_z}{\partial z}\right) = \pi r^2I\frac{\partial B_z}{\partial z}. \tag{11.19}$$

In the present case, $\partial B_z/\partial z$ is negative, so the force is correctly downward. In the factor $\pi r^2I$ we recognize the magnitude $m$ of the magnetic

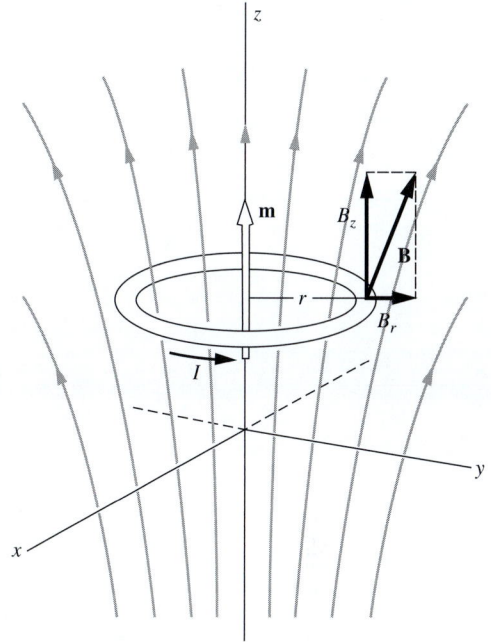

**Figure 11.9.**
A current ring in an inhomogeneous magnetic field. (The field of the ring itself is not shown.) Because of the radial component of the field, $B_r$, there is a force on the ring as a whole.

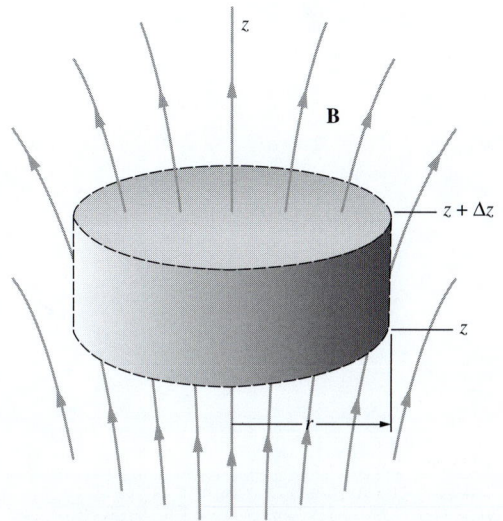

**Figure 11.10.**
Gauss's theorem can be used to relate $B_r$ and $\partial B_z/\partial z$, leading to Eq. (11.18).

dipole moment of our current ring. So the force on the ring can be expressed very simply in terms of the dipole moment:

$$F = m \frac{\partial B_z}{\partial z} \qquad (11.20)$$

We haven't proved it, but you will not be surprised to hear that for small loops of any other shape the force depends only on the *current × area* product, that is, on the dipole moment. The shape doesn't matter. Of course, we are discussing only loops small enough so that only the first-order variation of the external field, over the span of the loop, is significant.

Our ring in Fig. 11.9 has a magnetic dipole moment **m** pointing upward, and the force on it is downward. Obviously, if we could reverse the current in the ring, thereby reversing **m**, the force would reverse its direction. The situation can be summarized as follows.

- Dipole moment *parallel* to external field: force acts in direction of *increasing* field strength.
- Dipole moment *antiparallel* to external field: force acts in direction of *decreasing* field strength.
- *Uniform* external field: *zero* force.

Quite obviously, this is not the most general situation. The moment **m** could be pointing at some odd angle with respect to the field **B**, and the different components of **B** could be varying, spatially, in different ways. Given all the similarities between electric and magnetic dipoles, it is tempting to say that the force on a magnetic dipole should take the same form as the force on an electric dipole, given in Eq. (10.26). That is, the *x* component of the force on a magnetic dipole **m** should be given by

$$F_x = \mathbf{m} \cdot \nabla B_x \quad \text{(incorrect)}, \qquad (11.21)$$

with corresponding formulas for $F_y$ and $F_z$. All three components can be combined into the compact relation,

$$\mathbf{F} = (\mathbf{m} \cdot \nabla)\mathbf{B} \quad \text{(incorrect)}. \qquad (11.22)$$

You can check in Problem 11.4 that in the above setup with the ring, this force reduces to the force in Eq. (11.20).

However, this argument by analogy is risky, because, although the fields due to electric and magnetic dipoles look the same at large distances, the dipoles themselves look very different up close. One consists of two point charges, the other of a loop of current. The far field is irrelevant when dealing with the force *on* a dipole. It turns out that, although Eq. (11.22) gives the correct force on a magnetic dipole in many cases, it is *not* correct in general. The correct expression for the force turns out to be

$$\mathbf{F} = \nabla(\mathbf{m} \cdot \mathbf{B}) \qquad (11.23)$$

You can check in Problem 11.4 that in the above setup, this also reduces to the force in Eq. (11.20). At first glance it might seem like Eq. (11.23) comes out of the blue, but there is actually very good motivation for it. We will see in Section 11.6 that the energy of a magnetic dipole in a magnetic field is $-\mathbf{m} \cdot \mathbf{B}$. (But see Feynman *et al.* (1977), chap. 15, for a discussion of a subtlety about this energy.) So Eq. (11.23) is the familiar statement that the force equals the negative gradient of the energy.

Under what conditions are the force expressions in Eqs. (11.22) and (11.23) equal? Using the "$\nabla(\mathbf{A} \cdot \mathbf{B})$" vector identity in Appendix K, along with the fact that $\mathbf{m}$ has no spatial dependence, we find

$$\nabla(\mathbf{m} \cdot \mathbf{B}) = (\mathbf{m} \cdot \nabla)\mathbf{B} + \mathbf{m} \times (\nabla \times \mathbf{B}). \qquad (11.24)$$

Our two expressions for the force are therefore equal if $\nabla \times \mathbf{B} = 0$.[3] If we deal only with static setups where $\partial \mathbf{E}/\partial t = 0$, then the relevant Maxwell equation reduces to Ampère's law, $\nabla \times \mathbf{B} = \mu_0 \mathbf{J}$. So we see that the two expressions for the force agree if the setup involves no currents at the location of the dipole (other than the current in the dipole loop itself). This was the case in the above example. However, Problem 11.4 presents a setup where Eqs. (11.22) and (11.23) yield different forces; the task of that problem is to calculate the force explicitly and show that it agrees with Eq. (11.23).

In Eqs. (11.20) and (11.23) the force is in newtons, with the magnetic field gradient in tesla/meter and the magnetic dipole moment $m$ given by Eq. (11.9): $m = Ia$, where $I$ is in amps and $a$ in $m^2$. There are several equivalent ways to express the units of $m$. From Eq. (11.9) the units are

$$[m] = \text{amp-m}^2. \qquad (11.25)$$

But, as you can see from Eq. (11.20), we also have

$$[m] = \frac{\text{newtons}}{\text{tesla/m}} = \frac{\text{newton-m}}{\text{tesla}} = \frac{\text{joules}}{\text{tesla}}. \qquad (11.26)$$

Looking back at the three cases summarized on p. 538, we can begin to see what must be happening in the experiments described at the beginning of this chapter. A substance located at the position of the sample in Fig. 11.2 would be attracted *into* the solenoid if it contained magnetic dipoles *parallel* to the field $\mathbf{B}$ of the coil. It would be pushed *out* of the solenoid if it contained dipoles pointing in the opposite direction, *antiparallel* to the field. The force would depend on the gradient of the axial field strength, and would be zero at the midpoint of the solenoid. Also, if the total strength of dipole moments in the sample were proportional to the field strength $\mathbf{B}$, then in a given position the force would be proportional to $B$ times $\partial B/\partial z$, and hence to the square of the solenoid current. This is the observed behavior in the case of the diamagnetic

---

[3] This is a sufficient condition, but not necessary. Technically all we need is for $\nabla \times \mathbf{B}$ to be parallel to $\mathbf{m}$. But if we want the two expressions to be equal for any orientation of $\mathbf{m}$, then we need $\nabla \times \mathbf{B} = 0$.

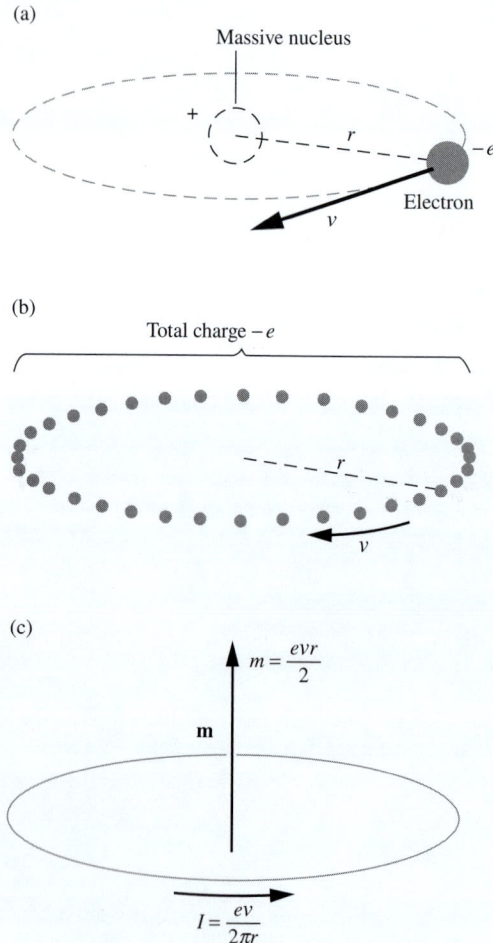

and the paramagnetic substances. It looks as if the ferromagnetic samples must have possessed a magnetic moment nearly independent of field strength, but we must set them aside for a special discussion anyway.

How does the application of a magnetic field to a substance evoke in the substance magnetic dipole moments with total strength proportional to the applied field? And why should they be parallel to the field in some substances, and oppositely directed in others? If we can answer these questions, we shall be on the way to understanding the physics of diamagnetism and paramagnetism.

## 11.5 Electric currents in atoms

We know that an atom consists of a positive nucleus surrounded by negative electrons. To describe it fully we would need the concepts of quantum physics. Fortunately, a simple and easily visualized model of an atom is very helpful for understanding diamagnetism. It is a planetary model with the electrons in orbits around the nucleus, like the model in Bohr's first quantum theory of the hydrogen atom.

We begin with one electron moving at constant speed on a circular path. Since we are not attempting here to explain atomic structure, we shall not inquire into the reasons why the electron has this particular orbit. We ask only, if it does move in such an orbit, what magnetic effects are to be expected? In Fig. 11.11 we see the electron, visualized as a particle carrying a concentrated electric charge $-e$, moving with speed $v$ on a circular path of radius $r$. In the middle is a positive nuclear charge, making the system electrically neutral. But the nucleus, because of its relatively great mass, moves so slowly that its magnetic effects can be neglected.

At any instant, the electron and the positive charge would appear as an electric dipole, but on the time average the electric dipole moment is zero, producing no steady electric field at a distance. We discussed this point in Section 10.5. The *magnetic* field of the system, far away, is *not* zero on the time average. Instead, it is just the field of a current ring. This is because, when considering the time average, it can't make any difference whether we have all the negative charge gathered into one lump, going around the track, or distributed in bits, as in Fig. 11.11(b), to make a uniform endless procession. The current is the amount of charge that passes a given point on the ring, per second. Since the electron makes $v/2\pi r$ revolutions per second, the current is

$$I = \frac{ev}{2\pi r}. \tag{11.27}$$

The orbiting electron is equivalent to a ring current of this magnitude with the direction of positive flow opposite to $v$, as shown in Fig. 11.11(c). Its far field is therefore that of a magnetic dipole, of strength

(a)

Massive nucleus

$v$

Electron $-e$

(b)

Total charge $-e$

$r$

$v$

(c)

$m = \dfrac{evr}{2}$

**m**

$I = \dfrac{ev}{2\pi r}$

**Figure 11.11.**
(a) A model of an atom in which one electron moves at speed $v$ in a circular orbit.
(b) Equivalent procession of charge. The average electric current is the same as if the charge $-e$ were divided into small bits, forming a rotating ring of charge. (c) The magnetic moment is the product of current and area.

$$m = \pi r^2 I = \frac{evr}{2}. \qquad (11.28)$$

Let us note in passing a simple relation between the magnetic moment **m** associated with the electron orbit, and the orbital angular momentum **L**. The angular momentum is a vector of magnitude $L = m_e vr$, where $m_e$ denotes the mass of the electron,[4] and it points downward if the electron is revolving in the sense shown in Fig. 11.11(a). Note that the product $vr$ occurs in both $m$ and $L$. With due regard to direction, we can write

$$\boxed{\mathbf{m} = \frac{-e}{2m_e} \mathbf{L}} \qquad (11.29)$$

This relation involves nothing but fundamental constants, which should make you suspect that it holds quite generally. Indeed that is the case, although we shall not prove it here. It holds for elliptical orbits, and it holds even for the rosette-like orbits that occur in a central field that is not inverse-square. Remember the important property of any orbit in a central field: angular momentum is a constant of the motion. It follows then, from the general relation expressed by Eq. (11.29) (derived by us only for a special case), that wherever angular momentum is conserved, the magnetic moment also remains constant in magnitude and direction. The factor

$$\frac{-e}{2m_e} \quad \text{or} \quad \frac{\text{magnetic moment}}{\text{angular momentum}} \qquad (11.30)$$

is called the *orbital magnetomechanical ratio* for the electron.[5] The intimate connection between magnetic moment and angular momentum is central to any account of atomic magnetism.

Why don't we notice the magnetic fields of all the electrons orbiting in all the atoms of every substance? The answer must be that there is a mutual cancelation. In an ordinary lump of matter there must be as many electrons going one way as the other. This is to be expected, for there is nothing to make one sense of rotation intrinsically easier than another, or otherwise to distinguish any unique axial direction. There would have to be something in the structure of the material to single out not merely an axis, but a *sense of rotation around that axis!*

We may picture a piece of matter, in the absence of any external magnetic field, as containing revolving electrons with their various orbital angular momentum vectors and associated orbital magnetic moments

---

[4] Our choice of the symbol **m** for magnetic moment makes it necessary, in this chapter, to use a different symbol for the electron mass. For angular momentum we choose the symbol **L**, because **L** is traditionally used in atomic physics for orbital angular momentum, which is what we consider here. We shall be dealing with speeds $v$ much less than $c$, so the nonrelativistic expression for **L** will suffice.

[5] Many people use the term *gyromagnetic ratio* for this quantity. Some call it the *magnetogyric ratio*. Whatever the name, it is understood that the magnetic moment is the numerator.

(a) Initial state

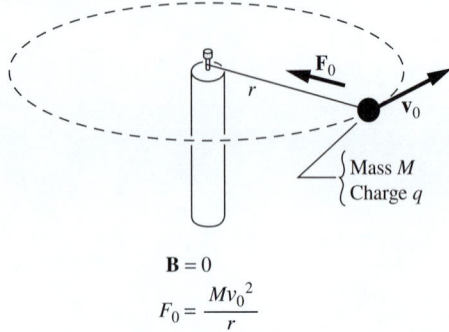

$$\mathbf{B} = 0$$

$$F_0 = \frac{Mv_0^2}{r}$$

(b) Intermediate state, **B** increasing in downward direction

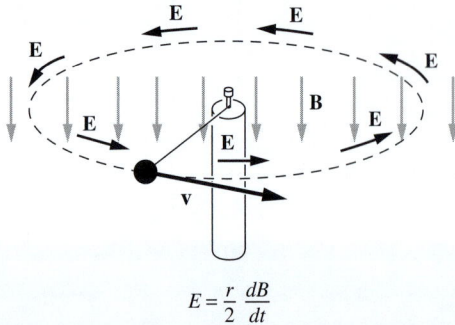

$$E = \frac{r}{2}\frac{dB}{dt}$$

(c) Final state, after time $\Delta t$

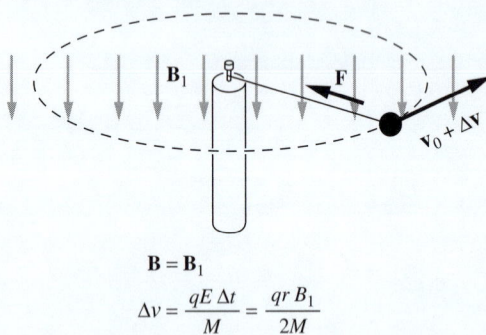

$$\mathbf{B} = \mathbf{B}_1$$

$$\Delta v = \frac{qE\,\Delta t}{M} = \frac{qr\,B_1}{2M}$$

**Figure 11.12.**
The growth of the magnetic field **B** induces an electric field **E** that accelerates the revolving charged body.

distributed evenly over all directions in space. Consider those orbits that happen to have their planes approximately parallel to the $xy$ plane, of which there will be about equal numbers with **m** up and **m** down. Let's find out what happens to one of these orbits when we switch on an external magnetic field in the $z$ direction.

We will start by analyzing an electromechanical system that doesn't look much like an atom. In Fig. 11.12 there is an object of mass $M$ and electric charge $q$, tethered to a fixed point by a cord of fixed length $r$. This cord provides the centripetal force that holds the object in its circular orbit. The magnitude of that force $F_0$ is given, as we know, by

$$F_0 = \frac{Mv_0^2}{r}. \tag{11.31}$$

In the initial state, Fig. 11.12(a), there is no external magnetic field. Now, by means of some suitable large solenoid, we begin creating a field **B** in the negative $z$ direction, uniform over the whole region at any given time. While this field is growing at the rate $dB/dt$, there will be an induced electric field **E** all around the path, as indicated in Fig. 11.12(b). To find the magnitude of this field **E** we note that the rate of change of flux through the circular path is

$$\frac{d\Phi}{dt} = \pi r^2 \frac{dB}{dt}. \tag{11.32}$$

This determines the line integral of the electric field, which is really all that matters (we only assume for symmetry and simplicity that it is the same all around the path). Faraday's law, $\mathcal{E} = -d\Phi/dt$, gives (ignoring the signs)

$$\int \mathbf{E} \cdot d\mathbf{l} = \pi r^2 \frac{dB}{dt}. \tag{11.33}$$

The left-hand side equals $2\pi rE$, so we find that

$$E = \frac{r}{2}\frac{dB}{dt}. \tag{11.34}$$

We have ignored signs so far, but if you apply to Fig. 11.12 your favorite rule for finding the direction of an induced electromotive force, you will see that **E** must be in a direction to accelerate the body, if $q$ is a positive charge. The acceleration along the path, $dv/dt$, is determined by the force $qE$:

$$M\frac{dv}{dt} = qE = \frac{qr}{2}\frac{dB}{dt}, \tag{11.35}$$

so that we have a relation between the change in $v$ and the change in $B$:

$$dv = \frac{qr}{2M}dB. \tag{11.36}$$

(a)

(c)

**Initial states**
(a), (c)

$$m_0 = \frac{qr}{2} v_0$$

(b)

(d)

**Final states**
(b), (d)

**B** = **B**$_1$ (downward)

$\Delta$**m** upward in both cases

$$\Delta m = \mathbf{B}_1 \frac{q^2 r^2}{4M}$$

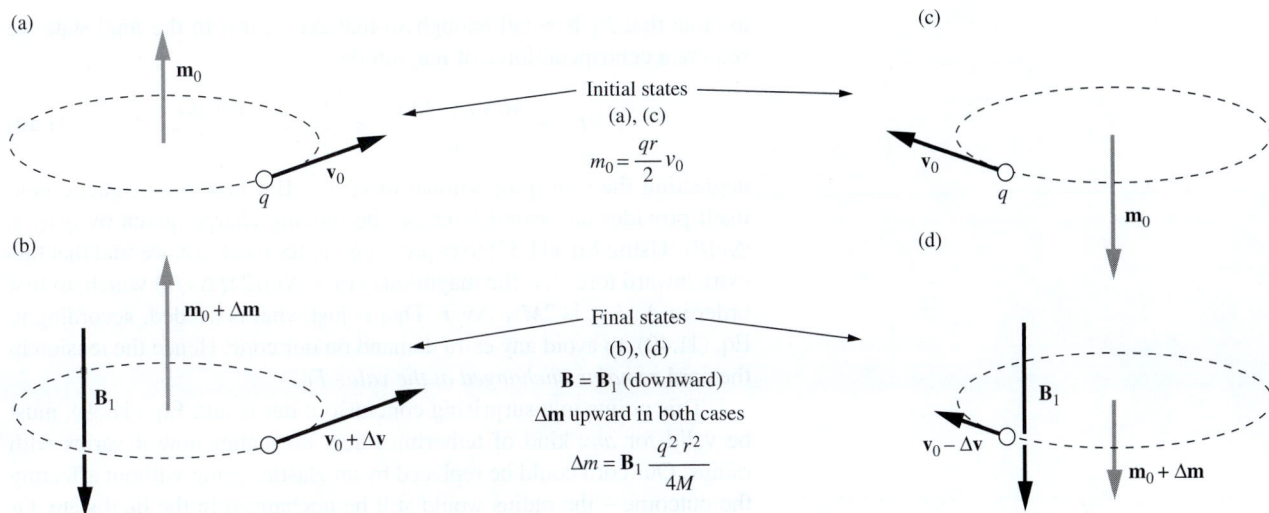

**Figure 11.13.**
The change in the magnetic moment vector is opposite to the direction of **B**, for both directions of motion.

The radius $r$ being fixed by the length of the cord, the factor $qr/2M$ is a constant. Let $\Delta v$ denote the net change in $v$ in the whole process of bringing the field up to the final value $B_1$. Then

$$\Delta v = \int_{v_0}^{v_0 + \Delta v} dv = \frac{qr}{2M} \int_0^{B_1} dB = \frac{qrB_1}{2M}. \qquad (11.37)$$

Note that the time has dropped out – the final velocity is the same whether the change is made slowly or quickly.

The increased speed of the charge in the final state means an increase in the upward-directed magnetic moment **m**. A *negatively* charged body would have been *decelerated* under similar circumstances, which would have *decreased* its *downward* moment. In either case, then, the application of the field **B**$_1$ has brought about a change in magnetic moment opposite to the field. From Eq. (11.28), the magnitude of the change in magnetic moment $\Delta m$ is

$$\Delta m = \frac{qr}{2} \Delta v = \frac{q^2 r^2}{4M} B_1. \qquad (11.38)$$

Likewise for charges, either positive or negative, revolving in the other direction, the induced change in magnetic moment is opposite to the change in applied magnetic field. Figure 11.13 shows this for a positive charge. It appears that the following relation holds for either sign of charge and either direction of revolution:

$$\boxed{\Delta \mathbf{m} = -\frac{q^2 r^2}{4M} \mathbf{B}_1} \qquad (11.39)$$

In this example we forced $r$ to be constant by using a cord of fixed length. Let us see how the tension in the cord has changed. We shall assume that $B_1$ is small enough so that $\Delta v \ll v_0$. In the final state we require a centripetal force of magnitude

$$F_1 = \frac{M(v_0 + \Delta v)^2}{r} \approx \frac{Mv_0^2}{r} + \frac{2Mv_0\Delta v}{r}, \qquad (11.40)$$

neglecting the term proportional to $(\Delta v)^2$. But now the magnetic field itself provides an inward force on the moving charge, given by $q(v_0 + \Delta v)B_1$. Using Eq. (11.37) to express $qB_1$ in terms of $\Delta v$, we find that this extra inward force has the magnitude $(v_0 + \Delta v)(2M\Delta v/r)$ which, to first order in $\Delta v/v_0$, is $2Mv_0 \, \Delta v/r$. That is just what is needed, according to Eq. (11.40), to avoid any extra demand on our cord! Hence the tension in the cord *remains unchanged at the value $F_0$.*

This points to a surprising conclusion: our result, Eq. (11.39), must be valid for *any* kind of tethering force, no matter how it varies with radius. Our cord could be replaced by an elastic spring without affecting the outcome – the radius would still be unchanged in the final state. Or to go at once to a system we are interested in, it could be replaced by the Coulomb attraction of a nucleus for an electron. Or it could be the effective force that acts on one electron in an atom containing many electrons, which has a still different dependence on radius.

Let us apply this to an electron in an atom, substituting the electron mass $m_e$ for $M$, and $e^2$ for $q^2$. Now $\Delta m$ is the magnetic moment induced by the application of a field $B_1$ to the atom. In other words, $\Delta m/B_1$ is a magnetic polarizability, defined in the same way as the electric polarizability $\alpha$ we introduced in Section 10.5. Remember that $\alpha/4\pi\epsilon_0$ had the dimensions of volume and turned out to be, in order of magnitude, $10^{-30} \text{ m}^3$, roughly the volume of an atom. By Eq. (11.39), the magnetic polarizability due to one electron in an orbit of radius $r$ is

$$\frac{\Delta m}{B_1} = -\frac{e^2 r^2}{4m_e}. \qquad (11.41)$$

Taking the orbit radius $r$ to be the Bohr radius, $0.53 \cdot 10^{-10}$ m, we find

$$\frac{\Delta m}{B_1} = -\frac{(1.6 \cdot 10^{-19} \text{ C})^2 (0.53 \cdot 10^{-10} \text{ m})^2}{4(9.1 \cdot 10^{-31} \text{ kg})} = -2 \cdot 10^{-29} \, \frac{\text{C}^2 \, \text{m}^2}{\text{kg}}. \qquad (11.42)$$

However, this comparison between $\Delta m$ and $B_1$ isn't quite a fair one, because magnetic fields contain a somewhat arbitrary factor of $\mu_0/4\pi$ multiplying the factors of current and distance; see the Biot–Savart law in Eq. (6.49). (An analogous issue arose with the electric polarizability in Section 10.5.) A more reasonable comparison would therefore involve $(\mu_0/4\pi)\Delta m$ and $B_1$. Since $\mu_0/4\pi = 1 \cdot 10^{-7} \text{ kg m/C}^2$, the numerical value of the ratio is simply modified by a factor of $10^{-7}$, and we have

$$\frac{\mu_0}{4\pi} \frac{\Delta m}{B_1} = -2 \cdot 10^{-36} \text{ m}^3. \qquad (11.43)$$

This has the dimensions of volume, just like the electric polarizability $\alpha/4\pi\epsilon_0$. However, it is five or six orders of magnitude smaller than typical electric polarizabilities, as sampled in Table 10.2. We can be a little more precise about this disparity. Using Eq. (11.41), we have (recalling $\mu_0 = 1/\epsilon_0 c^2$)

$$\frac{\mu_0}{4\pi}\frac{\Delta m}{B_1} = -\frac{r^2}{4}\frac{\mu_0 e^2}{4\pi m_e} = -\frac{r^2}{4}\left(\frac{1}{4\pi\epsilon_0}\frac{e^2}{m_e c^2}\right) \equiv -\frac{r^2}{4}r_0. \quad (11.44)$$

The quantity in parentheses has the dimensions of length and is known as the *classical electron radius*,[6] $r_0 = 2.8 \cdot 10^{-15}$ m. Since the electric polarizability is given by $\alpha/4\pi\epsilon_0 \approx r^3$, where $r$ is an atomic radius, we see that the magnetic polarizability is smaller than the electric polarizability by the ratio (roughly, up to a factor of 4 and other factors of order 1 that our models ignore) of the classical electron radius $r_0$ to an atomic radius $r$.

---

**Example** Let us see if Eq. (11.41) will account for the force on our diamagnetic samples listed in Table 11.1. The total number of electrons is about the same in one gram of almost anything. It is about one electron for every two nucleons (recall that the atomic weight is about twice the atomic number for most of the elements), or $N = 3 \cdot 10^{26}$ electrons per kilogram of matter. This follows from the fact that the mass of a nucleon is $1.67 \cdot 10^{-27}$ kg, so there are $6 \cdot 10^{26}$ nucleons in a kilogram. Equivalently, hydrogen has an atomic weight of 1, so the number of nucleons in one gram is Avogadro's number, $6.02 \cdot 10^{23}$.

Of course, the $r^2$ in Eq. (11.41) must now be replaced by a mean square orbit radius $\langle r^2 \rangle$, where the average is taken over all the electrons in the atom, some of which have larger orbits than others. Actually $\langle r^2 \rangle$ varies remarkably little from atom to atom through the whole periodic table, and $a_0^2$, the square of the Bohr radius which we have just used, remains a surprisingly good estimate. Adopting that, we would predict, using Eq. (11.42), that a field of 1.8 tesla would induce in 1 kg of substance a magnetic moment of magnitude

$$\Delta m = N\frac{e^2 r^2}{4m_e}B_1 = (3 \cdot 10^{26})(2 \cdot 10^{-29}\,\text{C}^2\,\text{m}^2/\text{kg})(1.8\,\text{tesla})$$

$$= 1.1 \cdot 10^{-2}\,\frac{\text{joule}}{\text{tesla}} \quad (\text{or amp-m}^2), \quad (11.45)$$

which in a gradient of 17 tesla/m would give rise to a force of magnitude

$$F = \Delta m\frac{\partial B_z}{\partial z} = \left(1.1 \cdot 10^{-2}\,\frac{\text{joule}}{\text{tesla}}\right)\left(17\,\frac{\text{tesla}}{\text{m}}\right) = 0.18\,\text{newtons}. \quad (11.46)$$

This agrees quite well, indeed better than we had any right to expect, with the values for the several purely diamagnetic substances listed in Table 11.1. As far as the sign of the force goes, we know from Eq. (11.39) that the magnetic moment is

---

[6] This radius is obtained by setting the rest energy of the electron, $m_e c^2$, equal to (in order of magnitude) the electrostatic potential energy of a ball with radius $r_0$ and charge $e$. See Exercise 1.62.

antiparallel to the external magnetic field. The discussion in Section 11.4 therefore tells us that the force acts in the direction of decreasing field strength, that is, outward from the solenoid. This agrees with the behavior of the diamagnetic samples, because the convention in Table 11.1 was that "−" meant outward.

We can see now why diamagnetism is a universal phenomenon, and a rather inconspicuous one. It is about the same in molecules as in atoms. The fact that a molecule can be a much larger structure than an atom – it may be built of hundreds or thousands of atoms – does not generally increase the effective mean-square orbit radius. The reason is that in a molecule any given electron is pretty well localized on an atom. There are some interesting exceptions, and we included one in Table 11.1 – graphite. The anomalous diamagnetism of graphite is due to an unusual structure that permits some electrons to circulate rather freely within a planar group of atoms in the crystal lattice. For these electrons $\langle r^2 \rangle$ is extraordinarily large.

As mentioned at the beginning of this section, diamagnetism (and likewise paramagnetism and ferromagnetism) can be explained only with quantum mechanics. A purely classical theory of diamagnetism does not exist; see O'Dell and Zia (1986). Nevertheless, the above discussion is helpful for understanding the critical property of diamagnetism, namely that the change in the magnetic moment is directed opposite to the applied magnetic field.

## 11.6 Electron spin and magnetic moment

In addition to its orbital angular momentum, the electron possesses another kind of angular momentum that has nothing to do with its orbital motion. It behaves in many ways as if it were continually rotating around an axis of its own. This property is called *spin*. While diamagnetism is a result of the orbital angular momentum of electrons, paramagnetism is a result of their spin angular momentum (as is ferromagnetism, which we will discuss in Section 11.11). A consequence of these origins is that a diamagnetic moment points antiparallel to the external magnetic field, whereas a paramagnetic moment points parallel to the external field (in an average sense, as we will see).

When the magnitude of the spin angular momentum is measured, the same result is always obtained: $h/4\pi$, where $h$ is Planck's constant, $6.626 \cdot 10^{-34}$ kg m$^2$/s. Electron spin is a quantum phenomenon. Its significance for us now lies in the fact that there is associated with this intrinsic, or "built-in," angular momentum a *magnetic moment,* likewise of invariable magnitude. This magnetic moment points in the direction you would expect if you visualize the electron as a ball of negative charge spinning around its axis. That is, the magnetic moment vector points antiparallel to the spin angular momentum vector, as indicated in

Fig. 11.14. The magnetic moment, however, is twice as large, relative to the angular momentum, as it is in the case of orbital motion.

There is no point in trying to devise a classical model of this object; its properties are essentially quantum mechanical. We need not even go so far as to say it *is* a current loop. What matters is only that it behaves like one in the following respects: (1) it produces a magnetic field that, at a distance, is that of a magnetic dipole; (2) in an external field **B** it experiences a torque equal to that which would act on a current loop of equivalent dipole moment; (3) within the space occupied by the electron, div **B** = 0 everywhere, as in the ordinary sources of magnetic field with which we are already familiar.

Since the magnitude of the spin magnetic moment is always the same, the only thing an external field can influence is its direction. (Contrast this with the changing magnitude of the orbital magnetic moment in Section 11.5.) A magnetic dipole in an external field experiences a torque. If you worked through Exercise 6.34, you proved that the torque **N** on a current loop of any shape, with dipole moment **m**, in a field **B**, is given by

$$\boxed{\mathbf{N} = \mathbf{m} \times \mathbf{B}} \tag{11.47}$$

For those who have not been through that demonstration, let's take time out to calculate the torque in a simple special case. In Fig. 11.15 we see a rectangular loop of wire carrying current $I$. The loop has a magnetic moment **m**, of magnitude $m = Iab$. The torque on the loop arises from the forces $\mathbf{F}_1$ and $\mathbf{F}_2$ that act on the horizontal wires. Each of these forces has the magnitude $F = IbB$, and its moment arm is the distance $(a/2) \sin \theta$. We see that the magnitude of the torque on the loop is

$$N = 2(IbB)\frac{a}{2}\sin\theta = (Iab)B\sin\theta = mB\sin\theta. \tag{11.48}$$

The torque acts in a direction to bring **m** parallel to **B**; it is represented by a vector **N** in the positive $x$ direction, in the situation shown. All this is consistent with the general formula, Eq. (11.47). Note that Eq. (11.47) corresponds exactly to the formula we derived in Section 10.4 for the torque on an electric dipole **p** in an external field **E**, namely, $\mathbf{N} = \mathbf{p} \times \mathbf{E}$. The orientation with **m** in the direction of **B**, like that of the electric dipole **p** parallel to **E**, is the orientation of lowest energy. Similarly, the work required to rotate a dipole **m** from an orientation parallel to **B**, through an angle $\theta_0$, is $mB(1 - \cos\theta_0)$; see Eq. (10.22). This equals $2mB$ for a rotation from parallel to antiparallel.

If the electron spin moments in a substance are free to orient themselves, we expect them to prefer the orientation in the direction of any applied field **B**, the orientation of lowest energy. Suppose every electron in a kilogram of material takes up this orientation. We have already calculated that there are roughly $3 \cdot 10^{26}$ electrons in a kilogram of anything. The spin magnetic moment of an electron, $m_s$, is given in Fig. 11.14 as

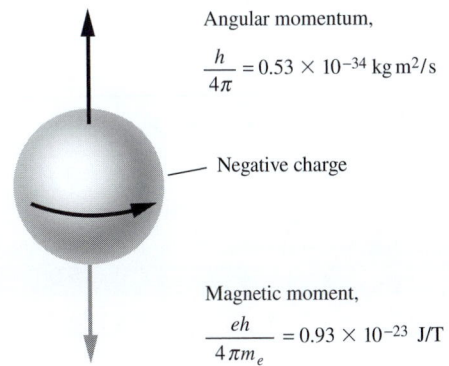

Angular momentum,

$$\frac{h}{4\pi} = 0.53 \times 10^{-34} \text{ kg m}^2/\text{s}$$

Negative charge

Magnetic moment,

$$\frac{eh}{4\pi m_e} = 0.93 \times 10^{-23} \text{ J/T}$$

**Figure 11.14.**
The intrinsic angular momentum, or spin, and the associated magnetic moment, of the electron. Note that the ratio of magnetic moment to angular momentum is $e/m_e$, not $e/2m_e$ as it is for orbital motion; see Eq. (11.29). This has no classical explanation.

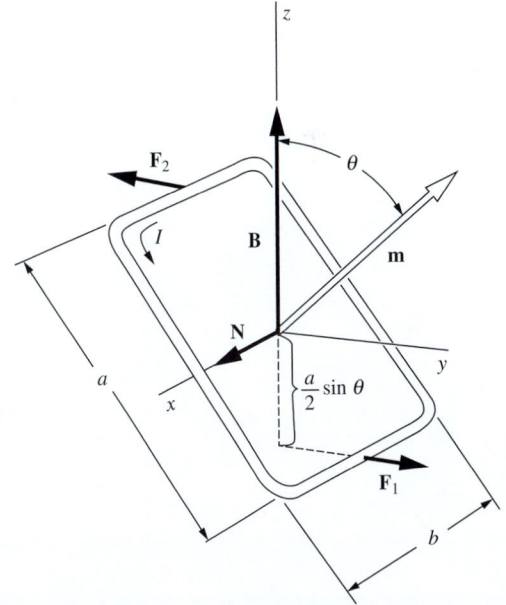

**Figure 11.15.**
Calculation of the torque on a current loop in a magnetic field **B**. The magnetic moment of the current loop is **m**.

$$m_{\mathrm{s}} = 9.3 \cdot 10^{-24} \frac{\text{joules}}{\text{tesla}} \quad \text{(or amp-m}^2\text{).} \qquad (11.49)$$

The total magnetic moment of our lined-up spins in one kilogram will be $(3 \cdot 10^{26}) \times (9.3 \cdot 10^{-24})$ or 2800 joules/tesla. From Eq. (11.20), the force per kilogram, in our coil where the field gradient is 17 tesla/m, would be $4.7 \cdot 10^4$ newtons. This is a little over 10,000 pounds, or equivalently 10 pounds for a tiny 1 gram sample!

Obviously this is much greater than the force recorded for any of the paramagnetic samples. Our assumptions were wrong in two ways. First, the electron spin moments are not all free to orient themselves. Second, thermal agitation prevents perfect alignment of any spin moments that are free. Let us look at these two issues in turn.

In most atoms and molecules, the electrons are associated in pairs, with the spins in each pair constrained to point in opposite directions regardless of the applied magnetic field. As a result, the magnetic moments of such a pair of electrons exactly cancel one another. All that is left is the diamagnetism of the orbital motion which we have already explored. The vast majority of molecules are purely diamagnetic. A few molecules (really *very* few) contain an odd number of electrons. In such a molecule, total cancelation of spin moments in pairs is clearly impossible. Nitric oxide, NO, with 15 electrons in the molecule is an example; it is paramagnetic. The oxygen molecule $O_2$ contains 16 electrons, but its electronic structure happens to favor noncancelation of two of the electron spins. In single atoms the inner electrons are generally paired, and if there is an outer unpaired electron, its spin is often paired off with that of a neighbor when the atom is part of a compound or crystal. Certain atoms, however, do contain unpaired electron spins which remain relatively free to orient in a field even when the atom is packed in with others. Important examples are the elements ranging from chromium to copper in the periodic table, a sequence that includes iron, cobalt, and nickel. Another group of elements with this property is the rare earth sequence around gadolinium. Compounds or alloys of these elements are generally paramagnetic, and in some cases ferromagnetic. The number of free electron spins involved in paramagnetism is typically one or two per atom. We can think of each paramagnetic atom as equipped with one freely swiveling magnetic moment **m**, which in a field **B** would be found pointing, like a tiny compass needle, in the direction of the field – if it were not for thermal disturbances.

Thermal agitation tends always to create a random distribution of spin axis directions. The degree of alignment that eventually prevails represents a compromise between the preference for the direction of lowest energy and the disorienting influence of thermal motion. We have met this problem before. In Section 10.12 we considered the alignment, by an electric field **E**, of the electric dipole moments of polar molecules. It turned out to depend on the ratio of two energies: $pE$, the energetic advantage of orientation of a dipole moment **p** parallel to **E** as compared with an average over completely random orientations, and $kT$, the mean thermal energy associated with any form of molecular motion at absolute temperature $T$. Only if $pE$ were much larger than $kT$ would nearly complete alignment of the dipole moments be attained. If $pE$ is much smaller than $kT$, the equilibrium polarization is equivalent to perfect alignment of a small fraction, approximately $pE/kT$, of the dipoles. We can take this result over directly for paramagnetism. We need only replace $pE$ by $mB$, the energy involved in the orientation of a magnetic dipole moment **m** in a magnetic field **B**. Providing $mB/kT$ is small, it follows that the total magnetic moment, per unit volume, resulting from application of the field **B** to $N$ dipoles per unit volume will be approximately

$$M \approx Nm \left( \frac{mB}{kT} \right) = \frac{Nm^2}{kT} B. \qquad (11.50)$$

The induced moment is proportional to $B$ and inversely proportional to the temperature.

For one electron spin moment ($m = 9.3 \cdot 10^{-24}$ joule/tesla) in our field of 1.8 tesla, $mB$ is $1.7 \cdot 10^{-23}$ joule. For room temperature, $kT$ is $4 \cdot 10^{-21}$ joule; in that case $mB/kT$ is indeed small. But if we could lower the temperature to 1 K in the same field, $mB/kT$ would be about unity. With further lowering of the temperature we could expect to approach complete alignment, with total moment approaching $Nm$. These conditions are quite frequently achieved in low-temperature experiments. Indeed, paramagnetism is both more impressive and more interesting at very low temperatures, in contrast to dielectric polarization. Molecular electric dipoles would be totally frozen in position, incapable of any reorientation. The electron spin moments are still remarkably free.

## 11.7 Magnetic susceptibility

We have seen that both diamagnetic and paramagnetic substances develop a magnetic moment proportional to the applied field. At least, that is true under most conditions. At very low temperatures, in fairly strong fields, the induced paramagnetic moment can be observed to approach a limiting value as the field strength is increased, as we have noted. Setting this "saturation" effect aside, the relation between moment and applied field is linear, so that we can characterize the magnetic properties of a substance by the ratio of induced moment to applied field. The ratio is called the *magnetic susceptibility*. Depending on whether we choose the moment of 1 kg of material, of 1 m$^3$ of material, or of 1 mole, we define the *specific* susceptibility, the *volume* susceptibility, or the *molar* susceptibility. Our discussion in Section 11.5 suggests that for diamagnetic substances the specific susceptibility, based on the induced moment per kilogram, should be nearly the same from one substance to another. However, the volume susceptibility, based on the induced magnetic moment per cubic meter, is more relevant to our present concerns.

The magnetic moment per unit volume we shall call the *magnetic polarization,* or the *magnetization,* using for it the symbol $\mathbf{M}$:

$$\mathbf{M} = \frac{\text{magnetic moment}}{\text{volume}}. \qquad (11.51)$$

Since the magnetic moment $\mathbf{m}$ has units of amps $\times$ meter$^2$, the magnetization $\mathbf{M}$ has units of amps/meter. Now, magnetization $\mathbf{M}$ and the quotient $\mathbf{B}/\mu_0$ have the same dimensions. One way to see this is to recall that the dimensions of the field $\mathbf{B}$ of a magnetic dipole are given in Eq. (11.14) by $\mu_0$(magnetic dipole moment)/(distance)$^3$, while $\mathbf{M}$, as we have just defined it, has the dimensions (magnetic dipole moment)/(volume). If we now define the volume magnetic susceptibility, denoted by $\chi_m$, through the relation

$$\mathbf{M} = \chi_m \frac{\mathbf{B}}{\mu_0} \quad (\textit{warning: see remarks below}), \qquad (11.52)$$

the susceptibility will be a dimensionless number, negative for diamagnetic substances, positive for paramagnetic. This is exactly analogous to the procedure, expressed in Eq. (10.41), by which we defined the electric susceptibility $\chi_e$ as the ratio of electric polarization $P$ to the product $\epsilon_0 E$. For the paramagnetic contribution, if any, to the susceptibility (let us denote it $\chi_{pm}$), we can use Eq. (11.50) to write down a formula analogous to Eq. (10.73):

$$\chi_{pm} = \frac{M}{B/\mu_0} \approx \frac{\mu_0 N m^2}{kT}, \qquad (11.53)$$

where $N$ refers to the number of spin dipoles per unit volume. Of course, the full susceptibility $\chi_m$ includes the ever-present diamagnetic contribution, which is negative, and derivable from Eq. (11.41).

Unfortunately, Eq. (11.52) is *not* the customary definition of volume magnetic susceptibility. In the usual definition, another field $\mathbf{H}$, which we shall meet in Section 11.10, appears instead of $\mathbf{B}$. Although illogical, the definition in terms of $\mathbf{H}$ has a certain practical justification, and the tradition is so well established that we shall eventually have to bow to it. But in this chapter we want to follow as long as we can a path that naturally and consistently parallels the description of the electric fields in matter. A significant parallel is this: the macroscopic field $\mathbf{B}$ inside matter will turn out to be the average of the microscopic $\mathbf{B}$, just as the macroscopic $\mathbf{E}$ turned out to be the average of the microscopic $\mathbf{E}$.

The difference in definition is of no practical consequence as long as $\chi_m$ is a number very small compared with unity. The values of $\chi_m$ for purely diamagnetic substances, solid or liquid, lie typically between $-0.5 \cdot 10^{-5}$ and $-1.0 \cdot 10^{-5}$. Even for oxygen under the conditions given in Table 11.1, the paramagnetic susceptibility is less than $10^{-2}$. This means that the magnetic field caused by the dipole moments in the substance, at least as a large-scale average, is very much weaker than the applied field **B**. That gives us some confidence that in such systems we may assume the field that acts on the atomic dipole to orient them is the same as the field that would exist there in the absence of the sample. However, we shall be interested in other systems in which the field of the magnetic moments is *not* small. Therefore we must study, just as we did in the case of electric polarization, the magnetic fields that magnetized matter itself produces, both inside and outside the material.

## 11.8 The magnetic field caused by magnetized matter

A block of material that contains, evenly distributed through its volume, a large number of atomic magnetic dipoles all pointing in the same direction, is said to be *uniformly magnetized*. The magnetization vector **M** is simply the product of the number of oriented dipoles per unit volume and the magnetic moment **m** of each dipole. We don't care how the alignment of these dipoles is maintained. There may be some field applied from another source, but we are not interested in that. We want to study only the field produced by the dipoles themselves.

Consider first a slab of material of thickness $dz$, sliced out perpendicular to the direction of magnetization, as shown in Fig. 11.16(a). The slab can be divided into little tiles, as indicated in Fig. 11.16(b). One such tile, which has a top surface of area $da$, contains a total dipole moment amounting to $M\,da\,dz$, since $M$ is the dipole moment per unit volume. The magnetic field this tile produces at all *distant* points – distant compared with the size of the tile – is just that of any dipole with the same magnetic moment. We could construct a dipole of that strength by bending a conducting ribbon of width $dz$ into the shape of the tile, and sending around this loop a current $I = M\,dz$; see Fig. 11.16(c). That will give the loop a dipole moment:

$$m = I \times \text{area} = (M\,dz)\,da, \qquad (11.54)$$

which is the same as that of the tile.

Let us substitute such a current loop for every tile in the slab, as indicated in Fig. 11.16(d). The current is the same in all of these, and therefore, at every interior boundary we find equal and opposite currents, equivalent to zero current. Our "egg-crate" of loops is therefore equivalent to a single ribbon running around the outside, carrying the current $I = M\,dz$; see Fig. 11.16(e). Now, these tiles can be made quite small, so long as we don't subdivide all the way down to molecular size. They must be large enough so that their magnetization does not vary appreciably from one tile to the next. Within that limitation, we can state that the field at any *external* point, even close to the slab, is the same as that of the current ribbon.

It remains only to reconstruct a whole block from such laminations, or slabs, as in Fig. 11.17(a). The entire block is then equivalent to the

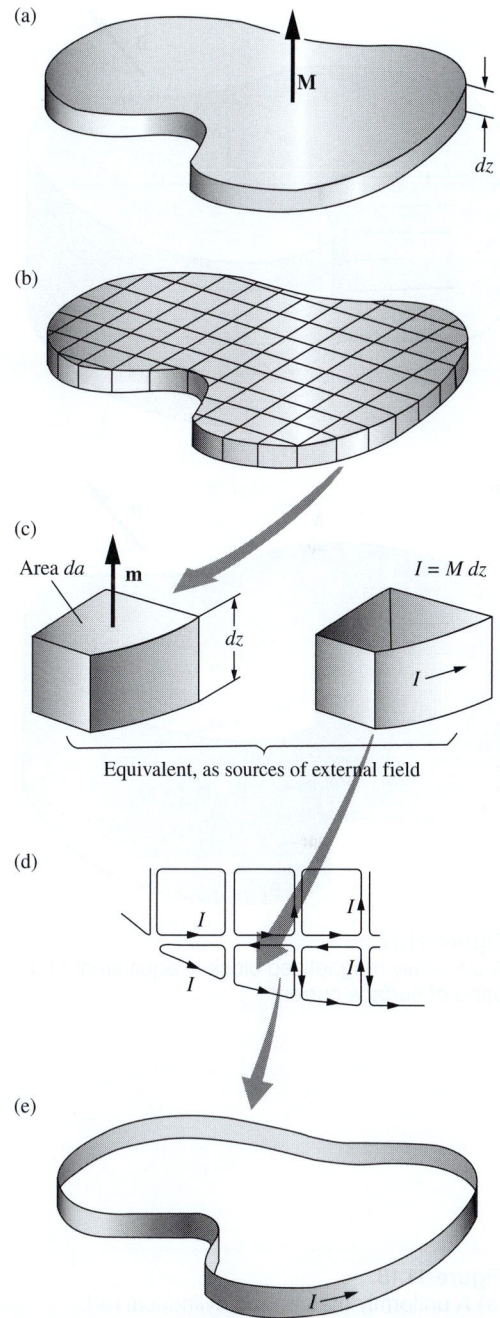

(a)

(b)

(c)
Area $da$   **m**   $I = M\,dz$

Equivalent, as sources of external field

(d)

(e)

**Figure 11.16.**
The thin slab, magnetized perpendicular to its broad surface, is equivalent to a ribbon of current so far as its external field is concerned.

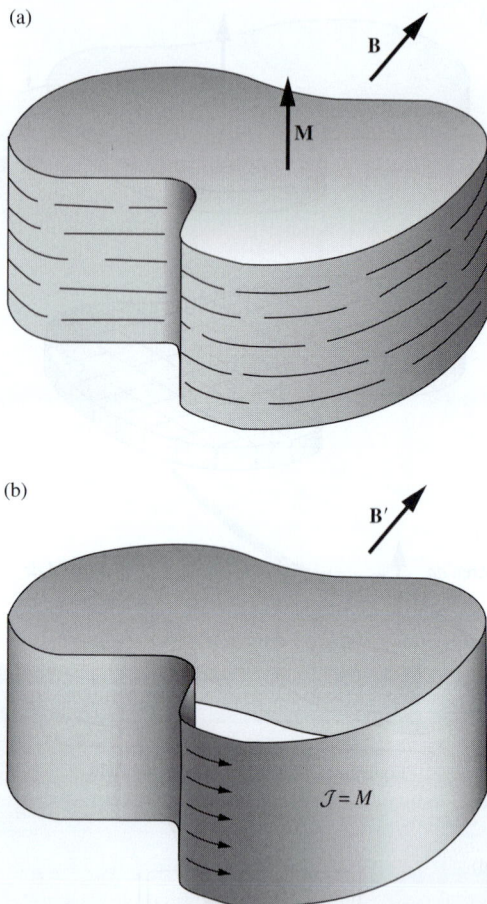

(a)

**B**

**M**

(b)

**B'**

$\mathcal{J} = M$

**Figure 11.17.**
A uniformly magnetized block is equivalent to a
band of surface current.

**Figure 11.18.**
(a) A uniformly magnetized cylindrical rod.
(b) The equivalent hollow cylinder, or sheath, of
current. Its field is **B'**. (c) We can sample the
interior of the rod, and thus obtain a spatial
average of the microscopic field, by closely
spaced parallel surfaces, $S_1, S_2, \ldots$

wide ribbon in Fig. 11.17(b), around which flows a current $M\,dz$, in C/s,
in every strip $dz$, or, stated more simply, a surface current of density $\mathcal{J}$,
in C/(s-m), given by

$$\mathcal{J} = M \qquad (11.55)$$

The magnetic field **B** at any point outside the magnetized block in
Fig. 11.17(a), and even close to the block provided we don't approach
within molecular distances, is the same as the field **B'** at the corresponding point in the neighborhood of the wide current ribbon in Fig. 11.17(b).

But what about the field inside the magnetized block? Here we face
a question like the one we met in Chapter 10. Inside matter the magnetic
field is not at all uniform if we observe it on the atomic scale, which we
have been calling microscopic. It varies sharply in both magnitude and
direction between points only a few angstroms apart. This *microscopic*
field **B** is simply a magnetic field in vacuum, for from the microscopic
viewpoint, as we emphasized in Chapter 10, matter is a collection of
particles and electric charge in otherwise empty space. The only large-
scale field that can be uniquely defined inside matter is the spatial average
of the microscopic field.

Because of the absence of effects attributable to magnetic charge,
we believe that the microscopic field itself satisfies div **B** $= 0$. If that
is true, it follows quite directly that the spatial average of the internal
microscopic field in our block is equal to the field **B'** inside the equivalent
hollow cylinder of current.

To demonstrate this, consider the long rod uniformly magnetized
parallel to its length, shown in Fig. 11.18(a). We have just shown that

(a)                              (b)

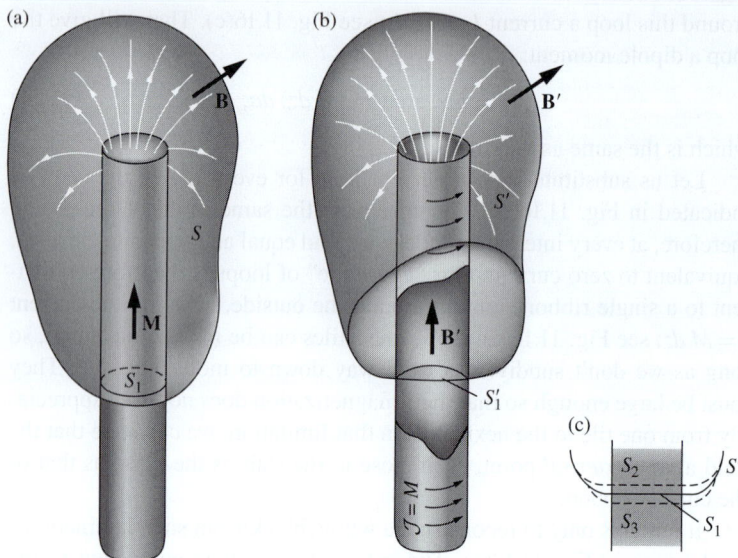

**B**

**B'**

$S$

$S'$

$S_1$

$S_1'$

**M**

**B'**

$\mathcal{J} = M$

(c)

$S_2$        $S'$

$S_3$        $S_1$

the external field will be the same as that of the long cylinder of current (practically equivalent to a single-layer solenoid) shown in Fig. 11.18(b). The surface $S$ in Fig. 11.18(a) indicates a closed surface that includes a portion $S_1$ passing through the interior of the rod. Because div $\mathbf{B} = 0$ for the internal microscopic field, as well as for the external field, div $\mathbf{B}$ is zero throughout the entire volume enclosed by $S$. It then follows from Gauss's theorem that the surface integral of $\mathbf{B}$ over $S$ must be zero. The surface integral of $\mathbf{B}'$ over the closed surface $S'$ in Fig. 11.18(b) is zero also. Over the portions of $S$ and $S'$ *external* to the cylinders, $\mathbf{B}$ and $\mathbf{B}'$ are identical. Therefore the surface integral of $\mathbf{B}$ over the internal disk $S_1$ must be equal to the surface integral of $\mathbf{B}'$ over the internal disk $S_1'$. This must hold also for any one of a closely spaced set of parallel disks, such as $S_2$, $S_3$, etc., indicated in Fig. 11.18(c), because the field outside the cylinder in this neighborhood is negligibly small, so that the outside parts don't change anything. Now, taking the surface integral over a series of equally spaced planes like this is a perfectly good way to compute the volume average of the field $\mathbf{B}$ in that neighborhood, for it samples all volume elements impartially. It follows that the spatial average of the microscopic field $\mathbf{B}$ inside the magnetized rod is equal to the field $\mathbf{B}'$ inside the current sheath of Fig. 11.18(b).

It is instructive to compare the arguments we have just developed with our analysis of the corresponding questions in Chapter 10. Figure 11.19 displays these developments side by side. You will see that they run logically parallel, but that at each stage there is a difference that reflects the essential asymmetry epitomized in the observation that *electric charges* are the source of *electric fields,* while *moving electric charges* are the source of *magnetic fields.* For example, in the arguments about the average of the microscopic field, the key to the problem in the electric case is the assumption that curl $\mathbf{E} = 0$ for the microscopic electric field. In the magnetic case, the key is the assumption that div $\mathbf{B} = 0$ for the microscopic magnetic field.

If the magnetization $\mathbf{M}$ within a volume of material is not uniform but instead varies with position as $\mathbf{M}(x, y, z)$, the equivalent current distribution is given simply by

$$\boxed{\mathbf{J} = \text{curl}\,\mathbf{M}} \tag{11.56}$$

Let's see how this comes about in one situation. Suppose there is a magnetization in the $z$ direction that gets stronger as we proceed in the $y$ direction. This is represented in Fig. 11.20(a), which shows a small region in the material subdivided into little blocks. The blocks are supposed to be so small that we may consider the magnetization uniform within a single block. Then we can replace each block by a current ribbon, with surface current density $\mathcal{J} = M_z$. The current $I$ carried by such a ribbon, if the block is $\Delta z$ in height, is $\mathcal{J}\Delta z$ or $M_z\Delta z$. Now, each ribbon has a bit more

(a) As a source of *external* electric field **E**

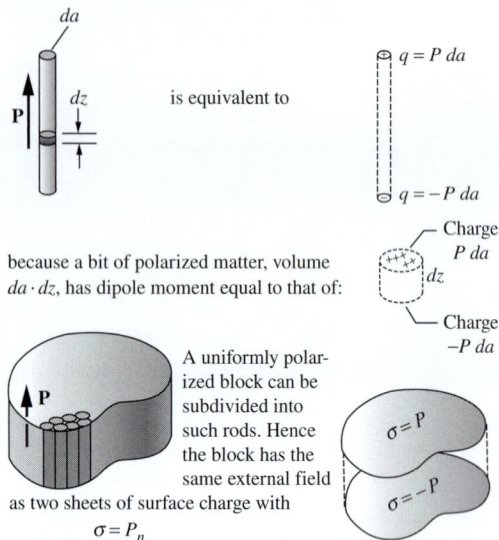

is equivalent to

$q = P\,da$

$q = -P\,da$

because a bit of polarized matter, volume $da \cdot dz$, has dipole moment equal to that of:

Charge $P\,da$

Charge $-P\,da$

A uniformly polarized block can be subdivided into such rods. Hence the block has the same external field as two sheets of surface charge with

$\sigma = P_n$

(More generally, for nonuniform polarization, polarized matter is equivalent to a charge distribution $\rho = -\operatorname{div} \mathbf{P}$.)

(b) As a source of *external* magnetic field **B**

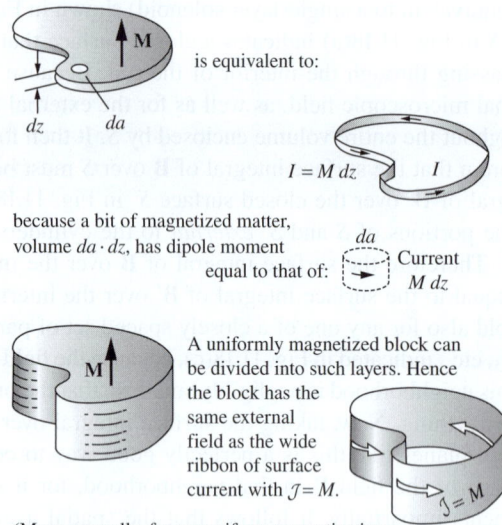

is equivalent to:

$I = M\,dz$

because a bit of magnetized matter, volume $da \cdot dz$, has dipole moment equal to that of:

Current $M\,dz$

A uniformly magnetized block can be divided into such layers. Hence the block has the same external field as the wide ribbon of surface current with $\mathcal{J} = M$.

(More generally, for nonuniform magnetization, magnetized matter is equivalent to a current distribution $\mathbf{J} = \operatorname{curl} \mathbf{M}$.)

PROOF THAT THE EQUIVALENCE EXTENDS TO
THE SPATIAL AVERAGE OF THE INTERNAL FIELDS

Consider a wide, thin, uniformly polarized slab and its equivalent sheets of surface charge.

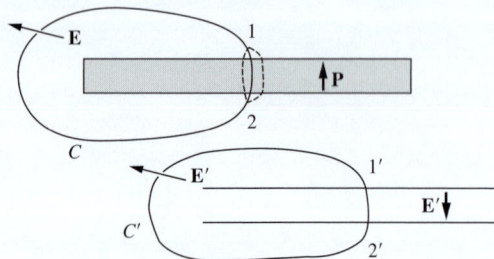

Near the middle the external field is slight and **E**′ is uniform. If $\nabla \times \mathbf{E} = 0$ for the internal field, then $\oint_C \mathbf{E} \cdot d\mathbf{l} = 0$. But $\mathbf{E} = \mathbf{E}'$ on the external path. Hence $\int_1^2 \mathbf{E} \cdot d\mathbf{l} = \int_{1'}^{2'} \mathbf{E}' \cdot d\mathbf{l}'$ for all internal paths.

Conclusion: $\langle \mathbf{E} \rangle = \mathbf{E}'$; the spatial average of the internal electric field is equal to the field $\mathbf{E}'$ that would be produced at that point in empty space by the equivalent charge distribution described above (together with any external sources).

Consider a long uniformly magnetized column and its equivalent cylinder of surface current.

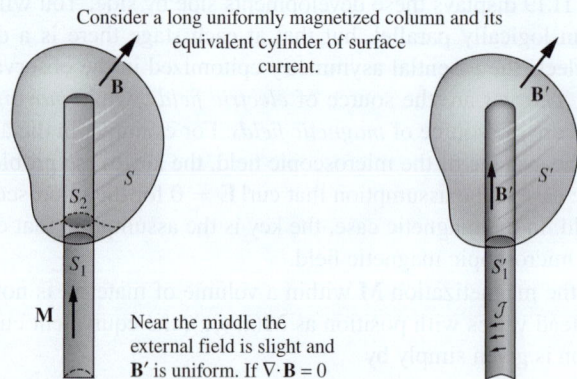

Near the middle the external field is slight and **B**′ is uniform. If $\nabla \cdot \mathbf{B} = 0$ for the internal field, then $\int_S \mathbf{B} \cdot d\mathbf{a} = 0$. But $\mathbf{B} = \mathbf{B}'$ on the surface external to the column. Hence $\int_{S_1} \mathbf{B} \cdot d\mathbf{a} = \int_{S_1'} \mathbf{B}' \cdot d\mathbf{a}'$ over any interior portion of surface, like $S_1$, $S_2$, etc.

Conclusion: $\langle \mathbf{B} \rangle = \mathbf{B}'$; the spatial average of the internal magnetic field is equal to the field $\mathbf{B}'$ that would be produced at that point in empty space by the equivalent charge distribution described above (together with any external sources).

**Figure 11.19.**
The electric (a) and magnetic (b) cases compared.

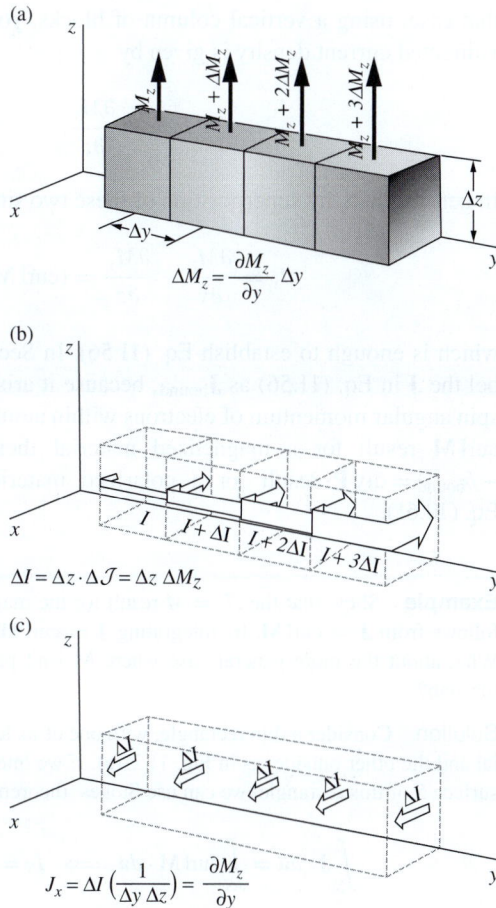

**Figure 11.20.**
Nonuniform magnetization is equivalent to a volume current density.

current density than the one to the left of it. The current in each loop is greater than the current in the loop to the left by

$$\Delta I = \Delta z \, \Delta M_z = \Delta z \frac{\partial M_z}{\partial y} \Delta y. \tag{11.57}$$

At every interface in this row of blocks there is a net current in the $x$ direction of magnitude $\Delta I$; see Fig. 11.20(c). To get the current per unit area flowing in the $x$ direction we have to multiply by the number of blocks per unit area, which is $1/(\Delta y \, \Delta z)$. Thus

$$J_x = \Delta I \left( \frac{1}{\Delta y \, \Delta z} \right) = \frac{\partial M_z}{\partial y}. \tag{11.58}$$

Another way of getting an $x$-directed current is to have a $y$ component of magnetization that varies in the $z$ direction. If you trace through that case, using a vertical column of blocks, you will find that the net $x$-directed current density is given by

$$J_x = -\frac{\partial M_y}{\partial z}. \tag{11.59}$$

In general then, by superposition of these two situations,

$$J_x = \frac{\partial M_z}{\partial y} - \frac{\partial M_y}{\partial z} = (\text{curl } \mathbf{M})_x, \tag{11.60}$$

which is enough to establish Eq. (11.56). In Section 11.10 we will relabel the $\mathbf{J}$ in Eq. (11.56) as $\mathbf{J}_{\text{bound}}$, because it arises from the orbital and spin angular momentum of electrons within atoms. The present $\mathbf{J}_{\text{bound}} = \text{curl } \mathbf{M}$ result for a magnetized material then clearly parallels the $-\rho_{\text{bound}} = \text{div } \mathbf{P}$ result for a polarized material that we derived in Eq. (10.61).

---

**Example** Show that the $\mathcal{J} = M$ result for the magnetized slab in Fig. 11.17 follows from $\mathbf{J} = \text{curl } \mathbf{M}$, by integrating $\mathbf{J} = \text{curl } \mathbf{M}$ over an appropriate area. What about the more general case where $\mathbf{M}$ isn't parallel to the boundary of the slab?

**Solution** Consider a thin rectangle, with one of its long sides inside the material and the other outside, as in Fig. 11.21(a). If we integrate $\mathbf{J} = \text{curl } \mathbf{M}$ over the surface $S$ of this rectangle, we can use Stokes' theorem to write

$$\int_S \mathbf{J} \cdot d\mathbf{a} = \int_S \text{curl } \mathbf{M} \cdot d\mathbf{a} \implies I_S = \int_C \mathbf{M} \cdot d\mathbf{s}, \tag{11.61}$$

where $I_S$ is the current passing through $S$. This current can be written as $\mathcal{J}\ell$, where $\ell$ is the height of the rectangle. This is true because we can make the rectangle arbitrarily thin, so any current passing through it must arise from a surface current density $\mathcal{J}$. The integral $\int_C \mathbf{M} \cdot d\mathbf{s}$ simply equals $M\ell$ (if the integral runs around the loop in the clockwise direction), because $\mathbf{M}$ is nonzero only along the left side of the rectangle. Equation (11.61) therefore gives $\mathcal{J}\ell = M\ell \implies \mathcal{J} = M$, as desired. If $\mathbf{M}$ points upward, the surface current density flows into the page.

If we have the more general case shown in Fig. 11.21(b), where the surface is tilted with respect to the direction of $\mathbf{M}$, then integrating over the area of the thin rectangle still gives $I_S = \int_C \mathbf{M} \cdot d\mathbf{s}$. But now only the component of $\mathbf{M}$ parallel to the long side of the rectangle survives in the dot product. Call this component $M_\parallel$. The above reasoning then quickly yields $\mathcal{J} = M_\parallel$. (Compare this with the surface charge density $\sigma = P_\perp$ for an electrically polarized material.) You can also arrive at this result by taking into account all the tiny current loops, as we did in Fig. 11.16. In short, the same number of current loops fit into a given height, but the relevant surface area is larger if the surface is tilted. So the surface current density is smaller.

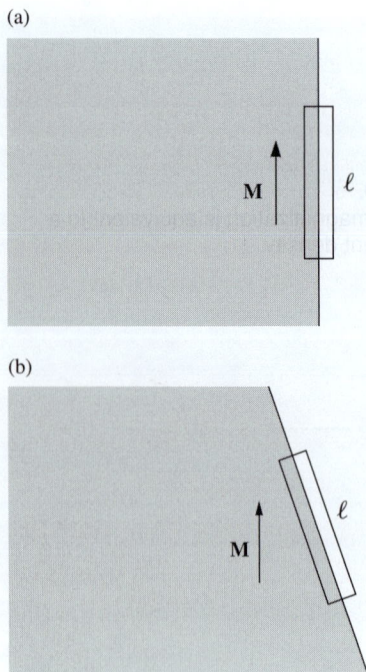

(a)

(b)

**Figure 11.21.**
The surface integral over a thin rectangle at the boundary, combined with Stokes' theorem, shows that $\mathcal{J} = M$ follows from $\mathbf{J} = \text{curl } \mathbf{M}$.

## 11.9 The field of a permanent magnet

The uniformly polarized spheres and rods we talked about in Chapter 10 are seldom seen, even in the laboratory. Frozen-in electric polarization can occur in some substances, although it is usually disguised by some accumulation of free charge. To make Fig. 11.3(a), which shows how the field of a polarized rod *would* look, it was necessary to use two charged disks. On the other hand, materials with permanent magnetic polarization, that is, permanent *magnetization,* are familiar and useful. Permanent magnets can be made from many alloys and compounds of ferromagnetic substances. What makes this possible is a question we leave for Section 11.11, where we dip briefly into the physics of ferromagnetism. In this section, taking the existence of permanent magnets for granted, we want to study the magnetic field **B** of a uniformly magnetized cylindrical rod and compare it carefully with the electric field **E** of a uniformly polarized rod of the same shape.

Figure 11.22 shows each of these solid cylinders in cross section. The polarization, in each case, is parallel to the axis, and it is uniform. That is, the polarization **P** and the magnetization **M** have uniform magnitude and direction everywhere within their respective cylinders. In the magnetic case this implies that every cubic millimeter of the permanent magnet has the same number of lined-up electron spins, pointing in the same direction. (A very good approximation to this can be achieved with modern permanent magnet materials.)

By the field inside the cylinder we mean, of course, the macroscopic field defined as the space average of the microscopic field. With this understanding, we show in Fig. 11.22 the field lines both inside and

**Figure 11.22.**
(a) The electric field **E** outside and inside a uniformly polarized cylinder. (b) The magnetic field **B** outside and inside a uniformly magnetized cylinder. In each case, the interior field shown is the macroscopic field, that is, the local average of the atomic or microscopic field.

(a)

(b)

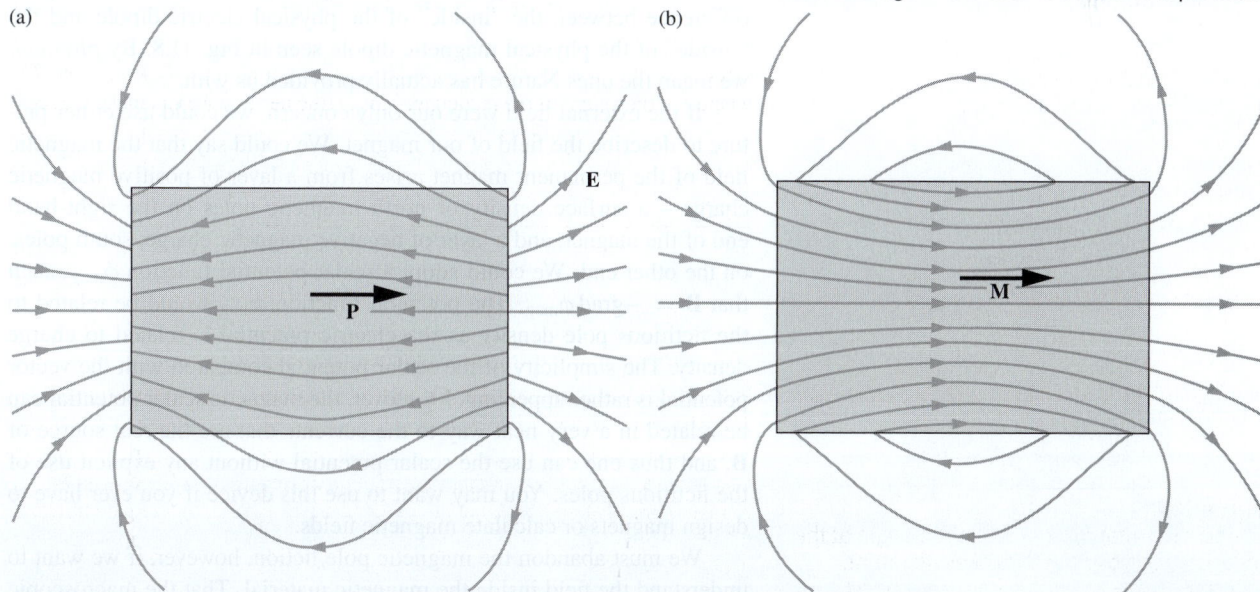

outside the rods. By the way, these rods are not supposed to be near one another; we only put the diagrams together for convenient comparison. Each rod is isolated in otherwise field-free space. (Which do you think would more seriously disturb the field of the other, if they *were* close together?)

Outside the rods the fields **E** and **B** *look alike*. In fact the field lines follow precisely the same course. That should not surprise you if you recall that the electric dipole and the magnetic dipole have similar far fields. Each little chunk of the magnet is a magnetic dipole, each little chunk of the polarized rod (sometimes called an *electret*) is an electric dipole, and the field outside is the superposition of all their far fields.

The field **B**, inside and out, is the same as that of a cylindrical sheath of current. In fact, if we were to wind very evenly, on a cardboard cylinder, a single-layer solenoid of fine wire, we could hook a battery up to it and duplicate the exterior and interior field **B** of the permanent magnet. (The coil would get hot and the battery would run down, whereas electron spins provide the current free and frictionless!) The electric field **E**, both inside and outside the polarized rod, is that of two disks of charge, one at each end of the cylinder.

Observe that the *interior* fields **E** and **B** are essentially different in form: **B** points to the right, is continuous at the ends of the cylinder, and suffers a sharp change in direction at the cylindrical surface. (These three facts are consistent with the **B** field that arises from a cylindrical sheath of current.) On the other hand, **E** points to the left, passes through the cylindrical surface as if it weren't there, but is discontinuous at the end surfaces. (These three facts are consistent with the **E** field that arises from two disks of charge.) These differences arise from the essential difference between the "inside" of the physical electric dipole and the "inside" of the physical magnetic dipole seen in Fig. 11.8. By *physical*, we mean the ones Nature has actually provided us with.

If the external field were our only concern, we could use either picture to describe the field of our magnet. We could say that the magnetic field of the permanent magnet arises from a layer of positive magnetic charge – a surface density of north magnetic poles on the right-hand end of the magnet, and a layer of negative magnetic charge, south poles, on the other end. We could adopt a scalar potential function $\phi_{\text{mag}}$, such that $\mathbf{B} = -\operatorname{grad}\phi_{\text{mag}}$. The potential function $\phi_{\text{mag}}$ would be related to the fictitious pole density as the electric potential is related to charge density. The simplicity of the scalar potential compared with the vector potential is rather appealing. Moreover, the magnetic scalar potential can be related in a very neat way to the currents that are the real source of **B**, and thus one can use the scalar potential without any explicit use of the fictitious poles. You may want to use this device if you ever have to design magnets or calculate magnetic fields.

We must abandon the magnetic pole fiction, however, if we want to understand the field inside the magnetic material. That the macroscopic magnetic field inside a permanent magnet is, in a very real sense, like the field in Fig. 11.22(b) rather than the field in Fig. 11.22(a) has been demonstrated experimentally by deflecting energetic charged particles in magnetized iron, as well as by the magnetic effects on slow neutrons, which pass even more easily through the interior of matter.

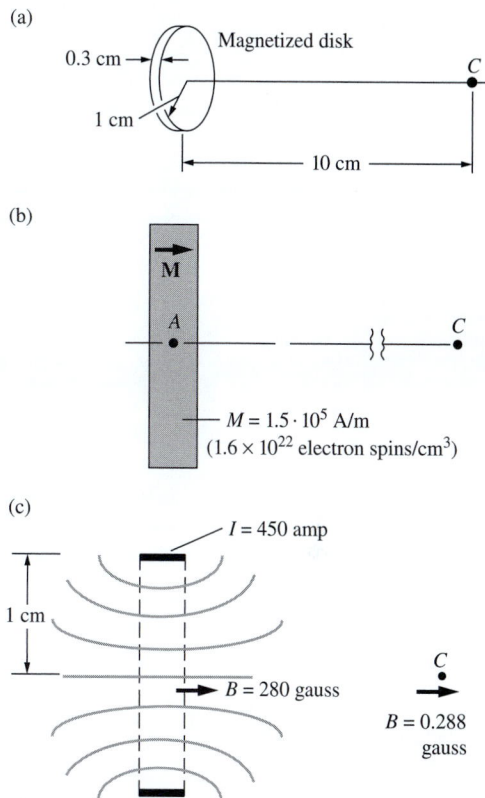

**Example (Disk magnet)** Figure 11.23(a) shows a small disk-shaped permanent magnet, in which the magnetization is parallel to the axis of symmetry. Although many permanent magnets take the form of bars or horseshoes made of iron, flat disk magnets of considerable strength can be made with certain rare-earth elements. The magnetization $M$ is given as $1.5 \cdot 10^5$ joules/(tesla-m$^3$), or equivalently amps/meter. The magnetic moment of the electron is $9.3 \cdot 10^{-24}$ joule/tesla, so this value of $M$ corresponds to $1.6 \cdot 10^{28}$ lined-up electron spins per cubic meter. The disk is equivalent to a band of current around its rim, of surface density $\mathcal{J} = M$. The rim being $\ell = 0.3$ cm wide, the current $I$ amounts to

$$I = \mathcal{J}\ell = M\ell = (1.5 \cdot 10^5 \text{ amp/m})(3 \cdot 10^{-3} \text{ m}) = 450 \text{ amps.} \quad (11.62)$$

This is rather more current than you draw by short-circuiting an automobile battery! The field **B** at any point in space, including points inside the disk, is simply the field of this band of current. For instance, near the center of the disk, **B** is approximately (using Eq. (6.54))

$$B = \frac{\mu_0 I}{2r} = \frac{(4\pi \cdot 10^{-7} \text{ kg m/C}^2)(450 \text{ C/s})}{2(0.01 \text{ m})} = 2.8 \cdot 10^{-2} \text{ tesla,} \quad (11.63)$$

or 280 gauss. The approximation consists in treating the 0.3 cm wide band of current as if it were concentrated in a single thin ring. (The corresponding approximation in an electrical setup would be to treat the equivalent charge sheets as large compared with their separation.) As for the field at a distant point, it would be easy to compute it for the ring current, but we could also, for an approximate calculation, proceed as we did in the electrical example. That is, we could find the total magnetic moment of the object, and find the distant field of a single dipole of that strength.

**Figure 11.23.**
(a) A disk uniformly magnetized parallel to its axis. (b) Cross-sectional view of disk. (c) The equivalent current is a band of current amounting to 450 amps flowing around the rim of the disk. The magnetic field $B$ is the same as the magnetic field of a very short solenoid, or approximately that of a simple ring of current of 1 cm radius.

## 11.10 Free currents, and the field H

It is often useful to distinguish between bound currents and free currents. *Bound* currents are currents associated with molecular or atomic magnetic moments, including the intrinsic magnetic moment of particles with spin. These are the molecular current loops envisioned by Ampère, the source of the magnetization we have just been considering. *Free* currents are ordinary conduction currents flowing on macroscopic paths – currents that can be started and stopped with a switch and measured with an ammeter.

The current density $\mathbf{J}$ in Eq. (11.56) is the macroscopic average of the bound currents, so let us henceforth label it $\mathbf{J}_{bound}$:

$$\mathbf{J}_{bound} = \mathrm{curl}\,\mathbf{M}. \tag{11.64}$$

At a surface where $\mathbf{M}$ is discontinuous, such as the side of the magnetized block in Fig. 11.17, we have a surface current density $\mathcal{J}$ which also represents bound current.

We found that $\mathbf{B}$, both outside matter and, as a space average, inside matter, is related to $\mathbf{J}_{bound}$ just as it is to any current density. That is, $\mathrm{curl}\,\mathbf{B} = \mu_0 \mathbf{J}_{bound}$. But that was in the absence of free currents. If we bring these into the picture, the field they produce simply adds on to the field caused by the magnetized matter and we have

$$\mathrm{curl}\,\mathbf{B} = \mu_0(\mathbf{J}_{bound} + \mathbf{J}_{free}) = \mu_0 \mathbf{J}_{total}. \tag{11.65}$$

Let us express $\mathbf{J}_{bound}$ in terms of $\mathbf{M}$, through Eq. (11.64). Then Eq. (11.65) becomes

$$\mathrm{curl}\,\mathbf{B} = \mu_0(\mathrm{curl}\,\mathbf{M}) + \mu_0 \mathbf{J}_{free}, \tag{11.66}$$

which can be rearranged as

$$\mathrm{curl}\left(\frac{\mathbf{B}}{\mu_0} - \mathbf{M}\right) = \mathbf{J}_{free}. \tag{11.67}$$

If we now *define* a vector function $\mathbf{H}(x, y, z)$ at every point in space by the relation

$$\boxed{\mathbf{H} \equiv \frac{\mathbf{B}}{\mu_0} - \mathbf{M}} \tag{11.68}$$

then Eq. (11.67) can be written as

$$\boxed{\mathrm{curl}\,\mathbf{H} = \mathbf{J}_{free}} \tag{11.69}$$

In other words, the vector $\mathbf{H}$, defined by Eq. (11.68), is related (up to a factor of $\mu_0$) to the *free* current in the way $\mathbf{B}$ is related to the total current, *bound* plus *free*. The parallel is not complete, however, for we always have $\mathrm{div}\,\mathbf{B} = 0$, whereas our vector function $\mathbf{H}$ does not necessarily have zero divergence.

This surely has reminded you of the vector $\mathbf{D}$ which we introduced, a bit grudgingly, in Chapter 10. Recall that $\mathbf{D}$ is related (up to a factor of $\epsilon_0$) to the free charge as $\mathbf{E}$ is related to the total charge. Although we rather disparaged $\mathbf{D}$, the vector $\mathbf{H}$ is really useful, for a practical reason that is worth understanding. In short, the reason is that in our two equations, $\mathrm{div}\,\mathbf{D} = \rho_{free}$ and $\mathrm{curl}\,\mathbf{H} = \mathbf{J}_{free}$, the charge density $\rho_{free}$ is difficult to measure, while the current density $\mathbf{J}_{free}$ is easy. Let's look at this in more detail.

In electrical systems, what we can easily control and measure are the potential differences of bodies, and not the amounts of free charge on them. Thus we control the electric field **E** directly. **D** is out of our direct control, and since it is not a fundamental quantity in any sense, what happens to it is not of much concern. In magnetic systems, however, it is precisely the free currents that we can most readily control. We lead them through wires, measure them with ammeters, channel them in well-defined paths with insulation, and so on. We have much less direct control, as a rule, over magnetization, and hence over **B**. So the auxiliary vector **H** *is* useful, even if **D** is not.

The integral relation equivalent to Eq. (11.69) is

$$\int_C \mathbf{H} \cdot d\mathbf{l} = \int_S \mathbf{J}_{\text{free}} \cdot d\mathbf{a} = I_{\text{free}} \tag{11.70}$$

where $I_{\text{free}}$ is the total free current enclosed by the path $C$. Suppose we wind a coil around a piece of iron and send through this coil a certain current $I$, which we can measure by connecting an ammeter in series with the coil. This is the free current, and it is the only free current in the system. Therefore one thing we know for sure is the line integral of **H** around any closed path, whether that path goes through the iron or not. The integral depends only on the number of turns of our coil that are linked by the path, and not on the magnetization in the iron. The determination of **M** and **B** in this system may be rather complicated. It helps to have singled out one quantity that we can determine quite directly. Figure 11.24 illustrates this property of **H** by an example, and is a reminder of the units we may use in a practical case. **H** has the same units as $\mathbf{B}/\mu_0$, or equivalently the same units as **M**, which are amps/meter. This is consistent with the fact that curl **H** equals $\mathbf{J}_{\text{free}}$, which has units of amps/meter$^2$. It is also consistent with the fact that $\int \mathbf{H} \cdot d\mathbf{l}$ equals $I_{\text{free}}$, which has units of amps.

We consider **B** the fundamental magnetic field vector because the absence of magnetic charge, which we discussed in Section 11.2, implies div **B** = 0 everywhere, even inside atoms and molecules. From div **B** = 0 it follows, as we showed in Section 11.8, that the average macroscopic field inside matter is **B**, not **H**. The implications of this have not always been understood or heeded in the past. However, **H** has the practical advantage we have already explained. In some older books you will find **H** introduced as the primary magnetic field. **B** is then defined as $\mu_0(\mathbf{H} + \mathbf{M})$, and given the name *magnetic induction*. Even some writers who treat **B** as the primary field feel obliged to call it the magnetic induction because the name *magnetic field* was historically preempted by **H**. This seems clumsy and pedantic. If you go into the laboratory and ask a physicist what causes the pion trajectories in his bubble chamber to curve, you will probably receive the answer "magnetic field," not "magnetic induction." You will seldom hear a geophysicist refer to the earth's magnetic induction, or an astrophysicist talk about the magnetic induction in the galaxy. We propose to keep on calling **B** the magnetic field.

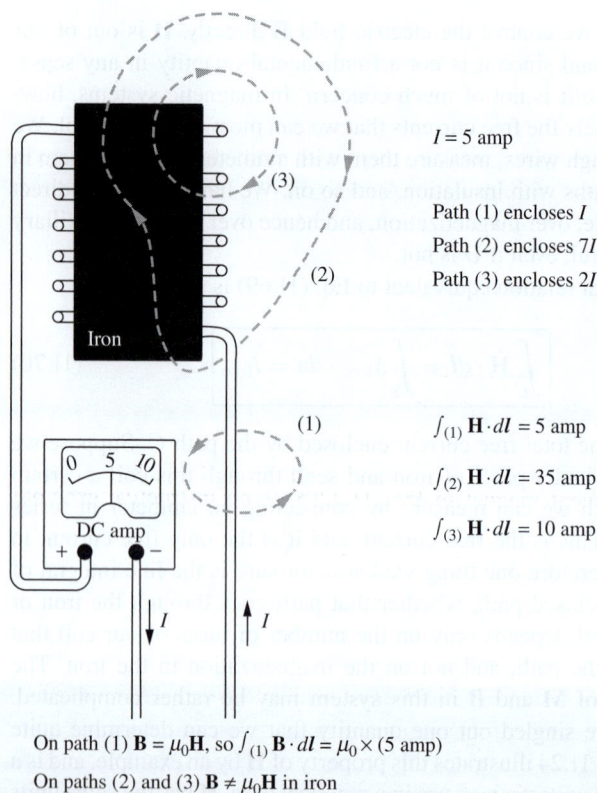

$I = 5$ amp

Path (1) encloses $I$
Path (2) encloses $7I$
Path (3) encloses $2I$

$\int_{(1)} \mathbf{H} \cdot d\mathbf{l} = 5$ amp
$\int_{(2)} \mathbf{H} \cdot d\mathbf{l} = 35$ amp
$\int_{(3)} \mathbf{H} \cdot d\mathbf{l} = 10$ amp

**Figure 11.24.**
Illustrating the relation between free current and the line integral of **H**.

On path (1) $\mathbf{B} = \mu_0\mathbf{H}$, so $\int_{(1)} \mathbf{B} \cdot d\mathbf{l} = \mu_0 \times (5$ amp$)$
On paths (2) and (3) $\mathbf{B} \neq \mu_0\mathbf{H}$ in iron

As for **H**, although other names have been invented for it, we shall call it *the field* **H**, or even *the magnetic field* **H**.

It is only the names that give trouble, not the symbols. Everyone agrees that in the SI system the relation connecting **B**, **M**, and **H** is that stated in Eq. (11.68). In empty space we have $\mathbf{H} = \mathbf{B}/\mu_0$, for **M** must be zero where there is no matter.

In the description of an electromagnetic wave, it is common to use **H** and **E**, rather than **B** and **E**, for the magnetic and electric fields. For the plane wave in free space that we studied in Section 9.4, the relation between the magnetic amplitude $H_0$ in amps/meter and the electric amplitude $E_0$ in volts/meter involves the constant $\sqrt{\mu_0/\epsilon_0}$ which has the dimensions of resistance and the approximate value 377 ohms. For its exact value, see Appendix E. We met this constant before in Section 9.6, where it appeared in the expression for the power density in the plane wave, Eq. (9.36). The condition that corresponds to $E_0$ and $B_0$, as stated by Eq. (9.26), becomes

$$E_0(\text{volt/meter}) = H_0(\text{amp/meter}) \times 377 \text{ ohms}. \qquad (11.71)$$

This makes a convenient system of units for dealing with electromagnetic fields in vacuum whose sources are macroscopic alternating currents and voltages. But remember that the basic magnetic field *inside* matter is **B**, not **H**, as we found in Section 11.9. That is not a matter of mere definition, but a consequence of the absence of magnetic charge.

The way in which **H** is related to **B** and **M** is reviewed in Fig. 11.25, for both the SI and the Gaussian systems of units. These relations hold whether **M** is proportional to **B** or not. However, if **M** is proportional to **B**, then it will also be proportional to **H**. In fact, the traditional definition of the volume magnetic susceptibility $\chi_m$ is not the logically preferable one given in Eq. (11.52), but rather

$$\boxed{\mathbf{M} = \chi_m \mathbf{H}} \qquad (\text{if } \mathbf{M} \propto \mathbf{B}), \qquad (11.72)$$

which we shall reluctantly adopt from here on. If $\chi_m \ll 1$, which is commonly the case, there is negligible difference between the two definitions; see Exercise 11.38.

The permanent magnet in Fig. 11.22(b) is an instructive example of the relation of **H** to **B** and **M**. To obtain **H** at some point inside the magnetized material, we have to add vectorially to $\mathbf{B}/\mu_0$ at that point the vector $-\mathbf{M}$. Figure 11.26 depicts this for a particular point $P$. It turns out that the lines of **H** inside the magnet look just like the lines of **E** inside the polarized cylinder of Fig. 11.22(a). The reason for this is the following. In the permanent magnet there are no free currents at all. Consequently, the line integral of **H**, according to Eq. (11.70), must be zero around any closed path. You can see that this will be the case if the **H** lines look like the **E** lines in Fig. 11.22(a), for we know the line integral of that electrostatic field is zero around any closed path.

Said in a different way, if magnetic poles, rather than electric currents, really were the source of the magnetization, then the macroscopic magnetic field inside the magnetized material would look just like the macroscopic electric field inside the polarized material (because that field is produced by electric poles). The similarity of magnetic polarization and electric polarization would be complete. The **B** field in a (hypothetical) setup with magnetic poles looks like the **H** field in a (real) setup with current loops.

In the example of the permanent magnet, Eq. (11.72) does not apply. The magnetization vector **M** is not proportional to **H** but is determined, instead, by the previous treatment of the material. How this can come about will be explained in the following section.

For any material in which **M** is proportional to **H**, so that Eq. (11.72) applies as well as the basic relation, Eq. (11.68), we have

$$\mathbf{B} = \mu_0(\mathbf{H} + \mathbf{M}) = \mu_0(1 + \chi_m)\mathbf{H}. \qquad (11.73)$$

Hence **B** is proportional to **H**. The factor of proportionality, $\mu_0(1 + \chi_m)$, is called the *magnetic permeability* and denoted usually by $\mu$:

$$\boxed{\mathbf{B} = \mu\mathbf{H}} \qquad \text{where} \quad \mu \equiv \mu_0(1 + \chi_m). \qquad (11.74)$$

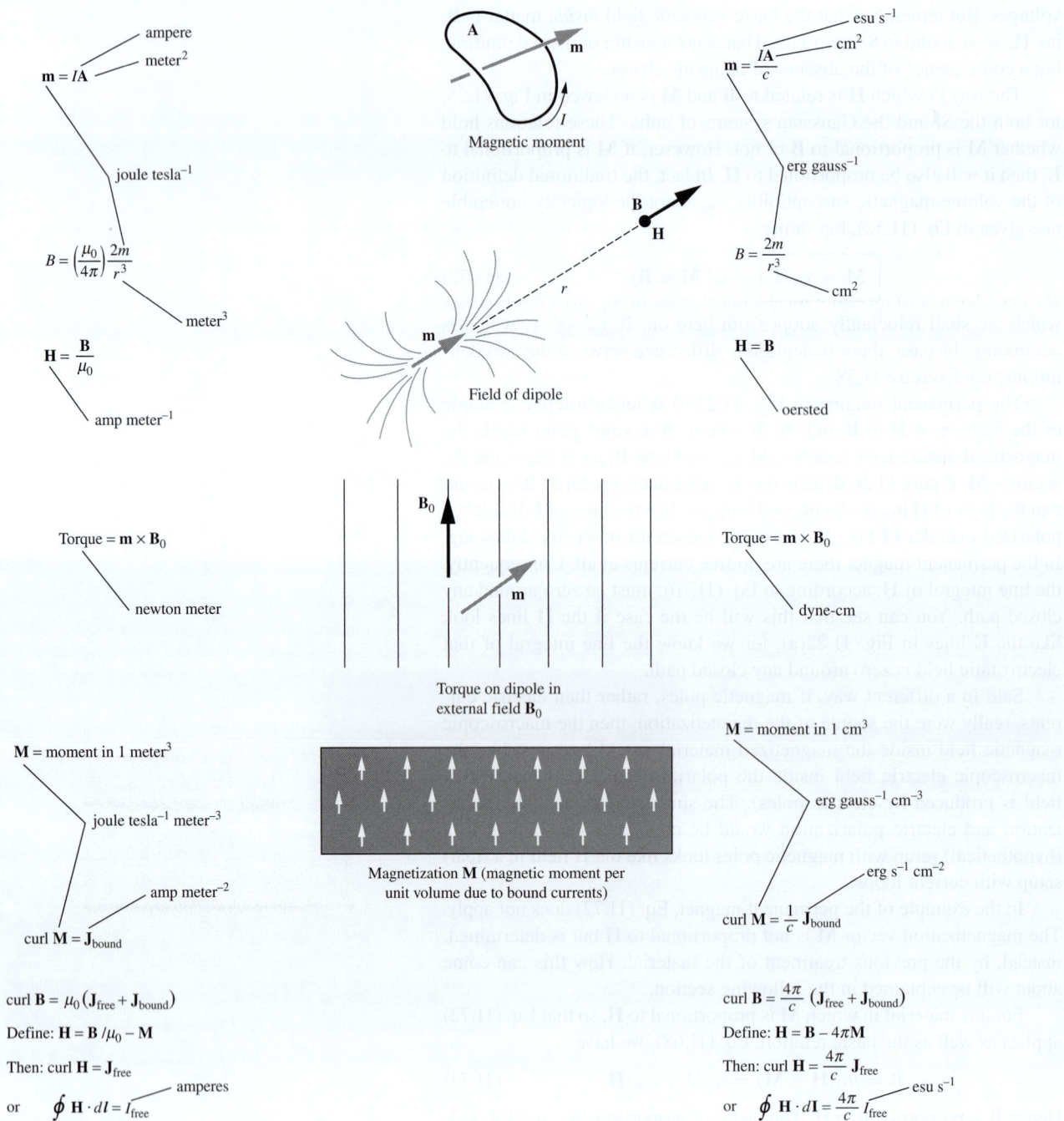

$$\mathbf{m} = I\mathbf{A}$$ — ampere — meter$^2$

$$B = \left(\frac{\mu_0}{4\pi}\right)\frac{2m}{r^3}$$ — joule tesla$^{-1}$ — meter$^3$

$$\mathbf{H} = \frac{\mathbf{B}}{\mu_0}$$ — amp meter$^{-1}$

$$\text{Torque} = \mathbf{m} \times \mathbf{B}_0$$ — newton meter

$$\mathbf{M} = \text{moment in } 1 \text{ meter}^3$$ — joule tesla$^{-1}$ meter$^{-3}$ — amp meter$^{-2}$

$$\text{curl } \mathbf{M} = \mathbf{J}_{\text{bound}}$$

$$\text{curl } \mathbf{B} = \mu_0\left(\mathbf{J}_{\text{free}} + \mathbf{J}_{\text{bound}}\right)$$

Define: $\mathbf{H} = \mathbf{B}/\mu_0 - \mathbf{M}$

Then: $\text{curl } \mathbf{H} = \mathbf{J}_{\text{free}}$

or    $\oint \mathbf{H} \cdot d\mathbf{l} = I_{\text{free}}$ — amperes

Magnetic moment

Field of dipole

Torque on dipole in external field $\mathbf{B}_0$

Magnetization $\mathbf{M}$ (magnetic moment per unit volume due to bound currents)

$$\mathbf{m} = \frac{I\mathbf{A}}{c}$$ — esu s$^{-1}$ — cm$^2$

$$B = \frac{2m}{r^3}$$ — erg gauss$^{-1}$ — cm$^2$

$$\mathbf{H} = \mathbf{B}$$ — oersted

$$\text{Torque} = \mathbf{m} \times \mathbf{B}_0$$ — dyne-cm

$$\mathbf{M} = \text{moment in } 1 \text{ cm}^3$$ — erg gauss$^{-1}$ cm$^{-3}$ — erg s$^{-1}$ cm$^{-2}$

$$\text{curl } \mathbf{M} = \frac{1}{c}\mathbf{J}_{\text{bound}}$$

$$\text{curl } \mathbf{B} = \frac{4\pi}{c}\left(\mathbf{J}_{\text{free}} + \mathbf{J}_{\text{bound}}\right)$$

Define: $\mathbf{H} = \mathbf{B} - 4\pi\mathbf{M}$

Then: $\text{curl } \mathbf{H} = \frac{4\pi}{c}\mathbf{J}_{\text{free}}$

or    $\oint \mathbf{H} \cdot d\mathbf{l} = \frac{4\pi}{c}I_{\text{free}}$ — esu s$^{-1}$

**Figure 11.25.**
Summary of relations involving **B**, **H**, **M**, **m**, $\mathbf{J}_{\text{free}}$, and $\mathbf{J}_{\text{bound}}$.

The permeability $\mu$, rather than the susceptibility $\chi$, is customarily used in describing ferromagnetism.

## 11.11 Ferromagnetism

Ferromagnetism has served and puzzled man for a long time. The *lodestone* (magnetite) was known in antiquity, and the influence on history of iron in the shape of compass needles was perhaps second only to that of iron in the shape of swords. For a century our electrical technology depended heavily on the circumstance that one abundant metal happens to possess this peculiar property. Nevertheless, it was only with the development of quantum mechanics that anything like a fundamental understanding of ferromagnetism was achieved.

We have already described some properties of ferromagnets. In a very strong magnetic field the force on a ferromagnetic substance is in such a direction as to pull it into a stronger field, as for paramagnetic materials, but instead of being proportional to the product of the field **B** and its gradient, the force is proportional to the gradient itself. As we remarked at the end of Section 11.4, this suggests that, if the field is strong enough, the magnetic moment acquired by the ferromagnet reaches some limiting magnitude. The direction of the magnetic moment vector must still be controlled by the field, for otherwise the force would not always act in the direction of increasing field intensity.

In permanent magnets we observe a magnetic moment even in the absence of any externally applied field, and it maintains its magnitude and direction even when external fields are applied, if they are not too strong. The field of the permanent magnet itself is always present, of course, and you may wonder whether it could not keep its own sources lined up. However, if you look again at Fig. 11.22(b) and Fig. 11.26, you will notice that **M** is generally not parallel to either **B** or **H**. This suggests that the magnetic dipoles must be clamped in direction by something other than purely magnetic forces.

The magnetization observed in ferromagnetic materials is much larger than we are used to in paramagnetic substances. Permanent magnets quite commonly have fields in the range of a few thousand gauss. A more characteristic quantity is the limiting value of the magnetization, the magnetic moment per unit volume, that the material acquires in a very strong field. This is called the *saturation* magnetization.

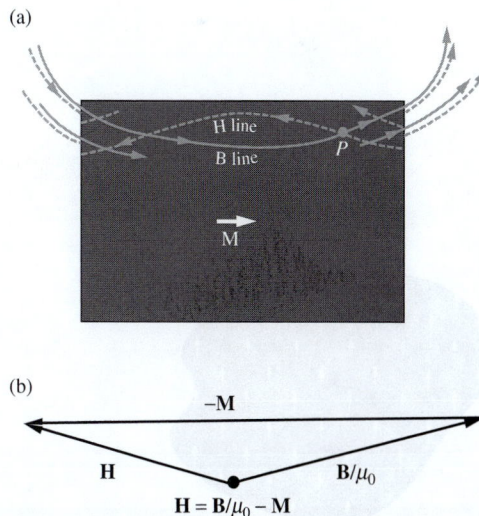

(a)

(b)

**Figure 11.26.**
(a) The relation of **B**, **H**, and **M** at a point inside the magnetized cylinder of Fig. 11.22(b).
(b) Relation of vectors at point $P$.

**Example**   We can deduce the saturation magnetization of iron from the data in Table 11.1. In a field with a gradient of 17 tesla/m, the force on 1 kg of iron was 4000 newtons. From Eq. (11.20), which relates the force on a dipole to the field gradient, we find

$$m = \frac{F}{dB/dz} = \frac{4000 \text{ newtons}}{17 \text{ tesla/m}} = 235 \text{ joules/tesla} \quad \text{(for 1 kg).} \quad (11.75)$$

To get the moment per cubic meter we multiply $m$ by the density of iron, 7800 kg/m$^3$. The magnetization $M$ is thus

$$M = (235 \text{ joules/tesla-kg})(7800 \text{ kg/m}^3)$$
$$= 1.83 \cdot 10^6 \text{ joules/(tesla-m}^3). \tag{11.76}$$

It is $\mu_0 M$, not $M$, that we should compare with field strengths in tesla. In the present case, $\mu_0 M$ has the value of 2.3 tesla.

It is more interesting to see how many electron spin moments this magnetization corresponds to. Dividing $M$ by the electron moment given in Fig. 11.14, namely $9.3 \cdot 10^{-24}$ joule/tesla, we get about $2 \cdot 10^{29}$ spin moments per cubic meter. Now, 1 m$^3$ of iron contains about $10^{29}$ atoms. The limiting magnetization seems to correspond to about two lined-up spins per atom. As most of the electrons in the atom are paired off and have no magnetic effect at all, this indicates that we are dealing with substantially complete alignment of those few electron spins in the atom's structure that are at liberty to point in the same direction.

A very suggestive fact about ferromagnets is this: a given ferromagnetic substance, pure iron for example, loses its ferromagnetic properties quite abruptly if heated to a certain temperature. Above 770 °C, pure iron acts like a paramagnetic substance. Cooled below 770 °C, it immediately recovers its ferromagnetic properties. This transition temperature, called the *Curie point* after Pierre Curie who was one of its early investigators, is different for different substances. For pure nickel it is 358 °C.

What is this ferromagnetic behavior that so sharply distinguishes iron below 770 °C from iron above 770 °C, and from copper at any temperature? It is the *spontaneous* lining up in one direction of the atomic magnetic moments, which implies alignment of the spin axes of certain electrons in each iron atom. By *spontaneous,* we mean that no external magnetic field need be involved. Over a region in the iron large enough to contain millions of atoms, the spins and magnetic moments of nearly all the atoms are pointing in the same direction. Well below the Curie point – at room temperature, for instance, in the case of iron – the alignment is nearly perfect. If you could magically look into the interior of a crystal of metallic iron and see the elementary magnetic moments as vectors with arrowheads on them, you might see something like Fig. 11.27.

It is hardly surprising that a high temperature should destroy this neat arrangement. Thermal energy is the enemy of order, so to speak. A crystal, an orderly arrangement of atoms, changes to a liquid, a much less orderly arrangement, at a sharply defined temperature, the melting point. The melting point, like the Curie point, is different for different substances. Let us concentrate here on the ordered state itself. Two or three questions are obvious.

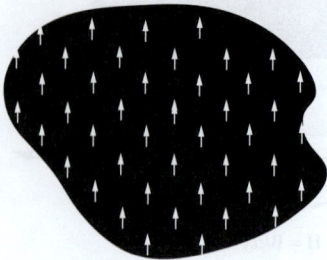

**Figure 11.27.**
The orderliness of the spin directions in a small region in a crystal of iron. Each arrow represents the magnetic moment of one iron atom.

**Question 1** What makes the spins line up and keeps them lined up?

**Question 2** How, if there is no external field present, can the spins choose one direction rather than another? Why didn't all the moments in Fig. 11.27 point down, or to the right, or to the left?

**Question 3** If the atomic moments *are* all lined up, why isn't every piece of iron at room temperature a strong magnet?

The answers to these three questions will help us to understand, in a general way at least, the behavior of ferromagnetic materials when an external field, neither very strong nor very weak, is applied. That includes a very rich variety of phenomena that we haven't even described yet.

**Answer 1** For some reason connected with the quantum mechanics of the structure of the iron atom, it is energetically favorable for the spins of adjacent iron atoms to be parallel. This is *not* due to their magnetic interaction. It is a stronger effect than that, and moreover it favors parallel spins whether like this ↑↑ or like this →→ (dipole interactions don't work that way – see Exercise 10.29). Now if atom A (Fig. 11.28) wants to have its spin in the same direction as that of its neighbors, atoms B, C, D, and E, and each of *them* prefers to have its spin in the same direction as the spin of *its* neighbors, including atom A, you can readily imagine that if a local majority ever develops there will be a strong tendency to "make it unanimous," and then the fad will spread.

**Answer 2** Accident somehow determines which of the various equivalent directions in the crystal is chosen, if we commence from a disordered state – as, for example, if the iron is cooled through its Curie point without any external field applied. Pure iron consists of body-centered cubic crystals. Each atom has eight nearest neighbors. The symmetry of the environment imposes itself on every physical aspect of the atom, including the coupling between spins. In iron the cubic axes happen to be the axes of easiest magnetization. That is, the spins like to point in the same direction, but they like it even better if that direction is one of the six directions $\pm\hat{\mathbf{x}}$, $\pm\hat{\mathbf{y}}$, $\pm\hat{\mathbf{z}}$ (Fig. 11.29). This is important because it means that the spins cannot easily swivel around *en masse* from one of the easy directions to an equivalent one at right angles. To do so, they would have to swing through *less* favorable orientations on the way. It is just this hindrance that makes permanent magnets possible.

**Answer 3** An apparently unmagnetized piece of iron is actually composed of many *domains,* in each of which the spins are all lined up one way, but in a direction different from that of the spins in neighboring domains. On the average over the whole piece of "unmagnetized" iron, all directions are equally represented, so no large-scale magnetic field results. Even in a single crystal the magnetic domains establish themselves. The domains are usually microscopic in the everyday sense of the

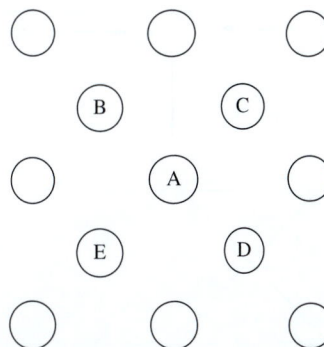

**Figure 11.28.**
An atom A and its nearest neighbors in the crystal lattice. (Of course, the lattice is really three-dimensional.)

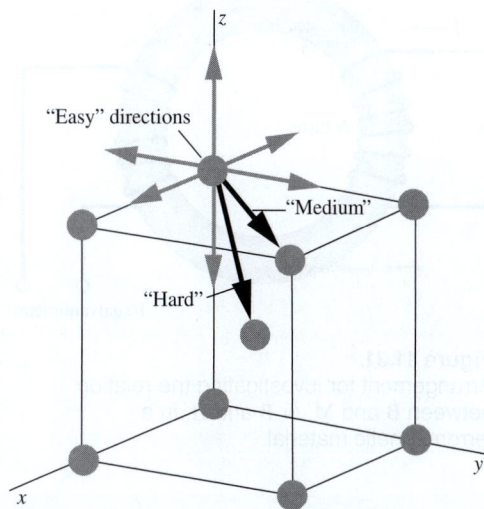

**Figure 11.29.**
In iron the energetically preferred direction of magnetization is along a cubic axis of the crystal.

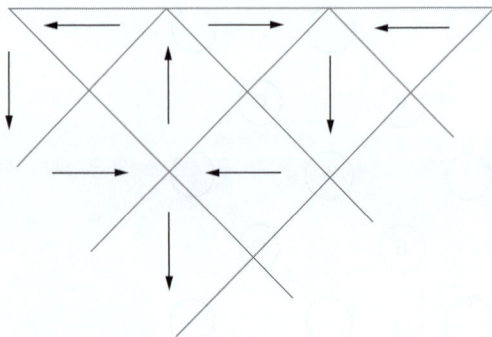

**Figure 11.30.**
Possible arrangement of magnetic domains in a single uniform crystal of iron.

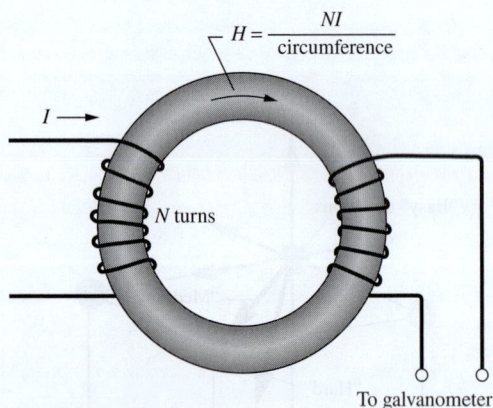

**Figure 11.31.**
Arrangement for investigating the relation between **B** and **M**, or **B** and **H**, in a ferromagnetic material.

word. In fact, they can be made visible under a low-power microscope. That is still enormous, of course, on an atomic scale, so a magnetic domain typically includes billions of elementary magnetic moments. Figure 11.30 depicts a division into domains. The division comes about because it is cheaper in energy than an arrangement with all the spins pointing in one direction. The latter arrangement would be a permanent magnet with a strong field extending out into the space around it. The energy stored in this exterior field is larger than the energy needed to turn some small fraction of the spins in the crystal, namely those at a domain boundary, out of line with their immediate neighbors. The domain structure is thus the outcome of an energy-minimization contest.

If we wind a coil of wire around an iron rod, we can apply a magnetic field to the material by passing a current through the wire. In this field, moments pointing parallel to the field will have a lower energy than those pointing antiparallel, or in some other direction. This favors some domains over others; those that happen to have a favorably oriented moment direction[7] will tend to grow at the expense of the others, if that is possible. A domain grows like a club, that is, by expanding its membership. This happens at the boundaries. Spins belonging to an unfavored domain, but located next to the boundary with a favored domain, simply switch allegiance by adopting the favored direction. That merely shifts the domain boundary, which is nothing more than the dividing surface between the two classes of spins. This happens rather easily in single crystals. That is, a very weak applied field can bring about, through boundary movement, a very large domain growth, and hence a large overall change in magnetization. Depending on the grain structure of the material, however, the movement of domain boundaries can be difficult.

If the applied field does not happen to lie along one of the "easy" directions (in the case of a cubic crystal, for example), the exhaustion of the unfavored domains still leaves the moments not pointing exactly parallel to the field. It may now take a considerably stronger field to pull them into line with the field direction so as to create, finally, the maximum magnetization possible.

Let us look at the large-scale consequences of this, as they appear in the magnetic behavior of a piece of iron under various applied fields. A convenient experimental arrangement is an iron torus, around which we wind two coils (Fig. 11.31). This affords a practically uniform field within the iron, with no end effects to complicate matters. By measuring

---

[7] We tend to use *spins* and *moments* almost interchangeably in this discussion. The moment is an intrinsic aspect of the spin, and if one is lined up so is the other. To be meticulous, we should remind the reader that in the case of the electron the magnetic moment and angular momentum vectors point in opposite directions (Fig. 11.14).

the voltage induced in one of the coils, we can determine changes in flux $\Phi$, and hence in **B** inside the iron. If we keep track of the *changes* in **B**, starting from $B = 0$, we always know what **B** is. A current through the other coil establishes **H**, which we take as the independent variable. If we know **B** and **H**, we can always compute **M**. It is more usual to plot **B** rather than **M**, as a function of **H**. A typical *B-H* curve for iron is shown in Fig. 11.32. Note the different units on the axes; $B$ is measured in tesla while $H$ is measured in amps/meter. If there were no iron in the coil, **B** would equal $\mu_0$**H**, so $H = 1$ amp/meter would be worth exactly $B = 4\pi \cdot 10^{-7}$ tesla. Or equivalently, $H = 300$ amps/meter would yield $B \approx 4 \cdot 10^{-4}$ tesla. But with the iron present, the resulting $B$ field is much larger. We see from the figure that when $H = 300$ amps/meter, $B$ has risen to more than 1 tesla. Of course, $B$ and $H$ here refer to an average throughout the whole iron ring; the fine domain structure as such never exhibits itself.

Starting with unmagnetized iron, $B = 0$ and $H = 0$, increasing $H$ causes $B$ to rise in a conspicuously nonlinear way, slowly at first, then more rapidly, then very slowly, finally flattening off. What actually becomes constant in the limit is not $B$ but $M$. In this graph, however, since $\mathbf{M} = \mathbf{B}/\mu_0 - \mathbf{H}$, and $H \ll B/\mu_0$, the difference between $B$ and $\mu_0 M$ is not appreciable.

The lower part of the *B-H* curve is governed by the motion of domain boundaries, that is, by the growth of "right-pointing" domains at the expense of "wrong-pointing" domains. In the upper flattening part of the curve, the atomic moments are being pulled by "brute force" into line with the field. The iron here is an ordinary polycrystalline metal, so only a small fraction of the microcrystals will be fortunate enough to have an easy direction lined up with the field direction.

If we now slowly decrease the current in the coil, thus lowering $H$, the curve *does not retrace itself.* Instead, we find the behavior given by the dashed curve in Fig. 11.32. This irreversibility is called *hysteresis.* It is largely due to the domain boundary movements being partially irreversible. The reasons are not obvious from anything we have said, but are well understood by physicists who work on ferromagnetism. The irreversibility is a nuisance, and a cause of energy loss in many technical applications of ferromagnetic materials – for instance, in alternating-current transformers. But it is indispensable for permanent magnetization, and for such applications, one wants to enhance the irreversibility. Figure 11.33 shows the corresponding portion of the *B-H* curve for a good permanent magnet alloy. Note that $H$ has to become about 50,000 amps/meter in the *reverse* direction before $B$ is reduced to zero. If the coil is simply switched off and removed, we are left with $B$ at 1.3 tesla, called the *remanence.* Since $H$ is zero, this is essentially the same as $\mu_0 M$. The alloy has acquired a permanent magnetization, that is, one that will persist indefinitely if it is exposed only to weak magnetic fields.

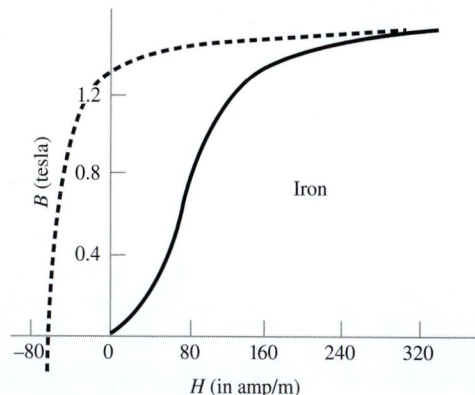

**Figure 11.32.**
Magnetization curve for fairly pure iron. The dashed curve is obtained as $H$ is reduced from a high positive value.

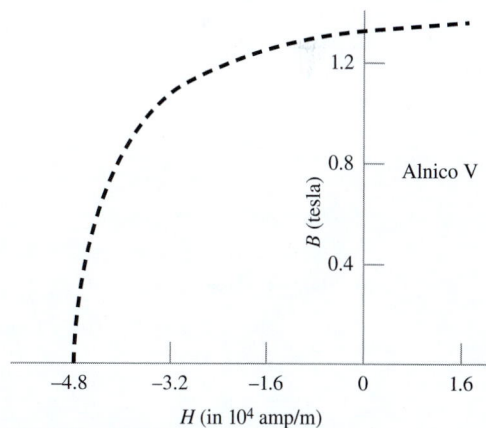

**Figure 11.33.**
Alnico V is an alloy of aluminum, nickel, and cobalt that is used for permanent magnets. Compare this portion of its magnetization curve with the corresponding portion of the characteristic for a "soft" magnetic material, shown in Fig. 11.32.

All the information that is stored on magnetic tapes and disks owes its permanence to this physical phenomenon.

## 11.12 Applications

*Lodestone* is magnetite (which is a type of iron oxide) that has been magnetized. Most magnetite isn't lodestone; the earth's magnetic field is too weak to create the observed magnetization. (And additionally, a particular crystal structure in the magnetite is needed.) A much stronger field is required, and this is the field produced by the brief but large electric current in a lightning strike. So without lightning, there would be no magnetic compasses, and the explorers throughout history would have had a much more difficult time!

Red pigments in paint often contain iron oxide grains. When the paint is wet, the grains' magnetic dipole moments are free to rotate and align themselves with the earth's magnetic field. But when the paint dries, they are locked in place. The direction of the dipole moments in a painting therefore gives the direction of the earth's magnetic field at the time the painting was made. By studying paintings whose positions are fixed, such as murals, we can gather information on how the earth's magnetic field has changed over time, assuming that the creation dates are known. (This doesn't work with framed paintings that have been moved around, or with murals that have undergone restoration!) Conversely, knowing the direction of the dipoles in a given painting can help determine the creation date; this can be very useful for archaeological purposes.

A common application of ferromagnetic materials is the magnification of the magnetic field from a coil of wire. If a coil is wrapped around a ferromagnetic core (usually iron-based), the coil's magnetic field causes the spins in the core material to align with the field, thereby producing a much larger field than the coil would produce by itself. The magnification factor (which is just the relative permeability, $\mu/\mu_0$) depends on various things, but for ferromagnetic materials it can be 100 or 1000, or even larger. This magnification effect is used in the electromagnets in relays, circuit breakers, junkyard magnets, and many other devices (see Section 6.10).

Many *data storage* devices, from cassette tapes to computer hard disks to credit cards, rely on the existence of ferromagnetic materials. (There are many other types of storage devices too. For example, CDs and DVDs operate by reflecting laser light off tiny pits in a plastic disk coated with aluminum. And flash memory makes use of a special kind of transistor.) A *cassette tape* (RIP) was the dominant format for music recording in the 1980s. The cassette consists of a long plastic tape coated with iron oxide, wound around two small reels. The *read/write head* is a tiny electromagnet. In the writing (recording) mode, a current passes through the coil of the electromagnet, producing a magnetic field that aligns the magnetic dipoles on the tape as it moves past. The tape therefore encodes the information in the current in the original signal.[8] Conversely, in the reading (playback) mode, the magnetic field of the dipoles in the tape induces a current in the electromagnet as the magnetic domains in the tape move past. This induced current contains the same information as what is on the tape. It can then be amplified and sent to a speaker.

A computer *hard disk* operates under the same general principle as a cassette tape, although there are some differences. Instead of a long tape, the hard disk, as the name suggests, takes the form of a disk on which the magnetic data is encoded in very narrow circular tracks. The magnetic domains are much smaller, and the average linear speed relative to the read/write head is much larger. The information at a given location on a disk can be accessed much more quickly than on a linear tape, because there is no need to run through the tape to gain access to a given point. Another important difference is that the cassette tape is an *analog* device, whereas the hard disk is a *digital* one. That is, as the current in the cassette's write-head electromagnet changes smoothly, the induced magnetization of the tape also changes smoothly, and this smoothly varying function contains the desired information. But in a hard disk, the magnetization is "all or nothing"; the write-head electromagnet either does nothing or completely saturates a magnetic domain. The abrupt changes (or lack thereof) in the magnetization are what is detected by the read head. So the information that comes out is just a string of yes's or no's, that is, 1's or 0's. In modern computers the reading is done with a separate *magnetoresistive* head, whose resistance depends on the magnetic field from the disk. This type of head is better able to read the tiny magnetic features on the disk. As incredibly useful as magnetic data storage is, beware that it isn't permanent. Hard-disk data can degrade on the order of 10 years. If you really want your data to last, you can use CDs that are made with a layer of gold. These are designed to last 300 years, although the supporting hardware will certainly be replaced long before then.

As discussed in Exercise 11.25, *magnetic bacteria* contain crystals of iron (often in the form of magnetite) that keep the bacteria aligned with the earth's magnetic field. The alignment is passive, that is, it would work even if a bacterium were dead. The point of the alignment is not to indicate which way is north, but rather which way is *down*. Except near the magnetic equator, the earth's magnetic field is inclined vertically. In the simplest scenario, bacteria follow the downward field lines to the oxygen-deprived regions (which they like) in the mud. (However,

---

[8] The process is complicated due to hysteresis; the magnetization tends to keep its original value and doesn't respond linearly to the applied magnetic field. This is remedied by adding a *bias signal* to the original signal. The purpose of this bias signal (whose frequency is high – usually around 100 kHz) is basically to average out the positive and negative effects of the hysteresis.

other mechanisms allow them also to seek out oxygen-deprived layers of water higher up.) The orientation of the magnetization of the crystals is opposite in bacteria in the northern and southern hemispheres, so they all do indeed swim downward. If a bacterium is transported to the other hemisphere, it will swim upward. Near the magnetic equator, there are roughly equal numbers of each type, and presumably another navigation mechanism takes over.

Many other creatures, such as homing pigeons, sea turtles, and rainbow trout, also use the earth's magnetic field to navigate. The exact mechanism isn't known, but with pigeons, for example, the beak most likely contains tiny crystals of magnetite that transmit a signal to the brain. The mechanism differs from the one in bacteria in that it is active (that is, a signal needs to be sent) rather than passive; in bacteria the torque on the magnetic crystals simply rotates the bacteria. Also, for pigeons the purpose of the magnetite is to indicate which way is north (or south, etc.), and not which way is down, as in the case of bacteria. A bacterium wants to make a beeline along a magnetic field line to the mud, whereas a pigeon has no great desire to make a beeline for the ground!

A *ferrofluid* consists of tiny (on the order of a nanometer) ferromagnetic particles suspended in a fluid. A surfactant (discussed in Section 10.16) keeps the particles evenly distributed in the fluid, so that they don't all clump together in the presence of a magnetic field. When a magnetic field is applied, a ferrofluid can take on bizarre shapes, with spikes and valleys brought about by the complicated balancing of magnetic, gravitational, and surface-tension forces. Aside from making cool shapes, ferrofluids have many useful applications – in magnetic resonance imaging, as seals in devices with rotating parts, and as a means of heat transfer. The ability to transfer heat efficiently is due to the temperature dependence of the magnetic susceptibility. In a speaker, for example, heat is generated near the voice coil, and the hotter ferrofluid (with a smaller magnetization than the colder ferrofluid) will experience a smaller magnetic force, and can therefore be displaced toward a heat sink.

*Magnetorheological fluids* are similar to ferrofluids, except that the magnetic particles are larger. When a magnetic field is applied, the particles tend to line up along the field lines, which increases the viscosity in the orthogonal direction. One application is in shock absorbers, where an electromagnet constantly adjusts the stiffness of the fluid, depending on the instantaneous road condition.

If you hold one permanent magnet above another, you can pick up the bottom one. So the magnetic force from the top magnet does work on the bottom magnet. However, isn't it the case that magnetic forces do no work, because the $q\mathbf{v} \times \mathbf{B}$ Lorentz force is always perpendicular to the velocity $\mathbf{v}$? Actually, no. Magnetic forces *can* do work in some cases. More precisely, the no-magnetic-work conclusion holds only if the force is governed by the $q\mathbf{v} \times \mathbf{B}$ Lorentz force law, which applies to point charges moving through space. This force law has nothing to do with the force on a magnetic dipole arising from a quantum mechanical spin, which *cannot* be viewed as a tiny current loop. The force on a dipole is instead given by $\nabla(\mathbf{m} \cdot \mathbf{B})$. Although we motivated this force law in Section 11.4 by considering a current loop, it can in fact be derived from fundamental physical principles (relativity, quantum mechanics, and a few other things), with no reference to moving charges or the Lorentz force. The $\nabla(\mathbf{m} \cdot \mathbf{B})$ force has a potential energy associated with it, and the change in this potential energy equals the work done on the dipole. If you picked up an actual current ring by holding another current ring above it, then, as the bottom ring moved upward, you can show that the current in it would decrease (which the spin of an electron can't do). A battery (or something) would be needed to maintain the current, and this battery would be the agency doing the work, because the no-magnetic-work conclusion *would* hold in this case. See Problem 7.2 for a related setup and a detailed discussion of why the magnetic Lorentz force ends up doing zero net work on moving charges.

Most permanent magnets are made of a combination of iron and other elements, such as nickel, cobalt, and oxygen. But in recent years permanent magnets made of *rare-earth* elements such as neodymium and samarium (combined with other more common elements) have come into wide usage. Rare-earth magnets are generally a bit stronger than iron-based ones. The fields are often in the 1–1.5 tesla range, whereas the field of an iron magnet is usually less than 0.5 tesla (although it can reach 1 tesla for some types). Additionally, the crystal structure of a rare-earth magnet is highly anisotropic. As discussed in the answer to Question 2 in Section 11.11, this means that it is difficult to change the direction of the individual magnetic dipole moments. It is therefore difficult to demagnetize a rare-earth magnet. Rare-earth magnets have replaced standard iron magnets in many electronic devices.

## CHAPTER SUMMARY

- There are three types of magnetism. (1) *Diamagnetism* arises from the *orbital* angular momentum of electrons; a diamagnetic moment points antiparallel to the external magnetic field. (2) *Paramagnetism* arises from the *spin* angular momentum of electrons; a paramagnetic moment points parallel to the external magnetic field (in an average sense). (3) *Ferromagnetism* also arises from the spin angular momentum of electrons, but the interactions involved can be explained only with quantum mechanics; a ferromagnetic moment can exist in the absence of an external magnetic field.

- The force on a diamagnetic material points in the direction of decreasing magnetic field strength, while the force on a paramagnetic material points in the direction of increasing magnetic field strength. The force on a ferromagnetic material can point either way, but in a field strong enough to wash out any initial magnetization that may have existed, the force points in the direction of increasing magnetic field strength.

- Unlike electric fields, magnetic fields are caused by currents, not poles. The absence of magnetic poles can be stated as div $\mathbf{B} = 0$.
- The *magnetic moment* of a current loop is $\mathbf{m} = I\mathbf{a}$. The vector potential due to the loop can be written as $\mathbf{A} = (\mu_0/4\pi)\mathbf{m} \times \hat{\mathbf{r}}/r^2$, and the *magnetic dipole* field in spherical coordinates is

$$B_r = \frac{\mu_0 m}{2\pi r^3}\cos\theta, \quad B_\theta = \frac{\mu_0 m}{4\pi r^3}\sin\theta. \quad (11.77)$$

- The force on a magnetic dipole is $\mathbf{F} = \nabla(\mathbf{m}\cdot\mathbf{B})$. This takes a different form compared with the force on an electric dipole.
- The magnetic moment due to the *orbital* motion of an electron is $\mathbf{m} = -(e/2m_e)\mathbf{L}$, where $\mathbf{L}$ is the orbital angular momentum. The diamagnetic moment induced in an external field $\mathbf{B}$ is $\mathbf{m} = -(e^2 r^2/4m_e)\mathbf{B}$.
- Electrons also contain *spin* angular momentum and a spin dipole moment. Paramagnetism arises from the (partial) alignment of the spins. The torque on a dipole moment is $\mathbf{N} = \mathbf{m} \times \mathbf{B}$.
- In many cases the diamagnetic and paramagnetic dipole *magnetizations* $\mathbf{M}$ are proportional to the applied external field $\mathbf{B}$. The *magnetic susceptibility* can be defined by $\mathbf{M} = \chi_m \mathbf{B}/\mu_0$, although a different definition in terms of $\mathbf{H}$ (see below) is the conventional one. The paramagnetic susceptibility is given approximately by $\chi_{pm} = \mu_0 N m^2/kT$.
- The magnetic field due to a uniformly magnetized slab is equivalent, both inside and outside, to the magnetic field due to a current on the "side" surface with density $\mathcal{J} = M$. If the magnetization is not constant, then we more generally have $\mathbf{J} = \text{curl}\,\mathbf{M}$.
- The magnetic field $\mathbf{H}$ is defined by

$$\mathbf{H} \equiv \frac{\mathbf{B}}{\mu_0} - \mathbf{M}. \quad (11.78)$$

This field satisfies

$$\text{curl}\,\mathbf{H} = \mathbf{J}_{\text{free}} \quad \Longleftrightarrow \quad \int_C \mathbf{H}\cdot d\mathbf{l} = I_{\text{free}}. \quad (11.79)$$

If $\mathbf{M}$ is proportional to $\mathbf{B}$, then the accepted definition of the magnetic susceptibility is $\mathbf{M} = \chi_m \mathbf{H}$. The *magnetic permeability* is defined by

$$\mathbf{B} = \mu\mathbf{H}, \quad \text{where} \quad \mu \equiv \mu_0(1 + \chi_m). \quad (11.80)$$

- Like paramagnetism, ferromagnetism arises from the spin angular momentum of electrons, but in a more inherently *quantum-mechanical* way that allows magnetization to exist even in the absence of an external magnetic field. A ferromagnetic material is divided into different *domains* where the spins are aligned. Increasing the external magnetic field can shift the boundaries of these domains. However, this process is not exactly reversible – an effect called *hysteresis*.

# Problems

**11.1** *Maxwell's equations with magnetic charge* ∗∗∗

Write out Maxwell's equations as they would appear if we had magnetic charge and magnetic currents as well as electric charge and electric currents. Invent any new symbols you need and define carefully what they stand for. Be particularly careful about + and − signs.

**11.2** *Magnetic dipole* ∗∗

In Chapter 6 we calculated the field at a point on the axis of a current ring of radius $b$; see Eq. (6.53). Show that for $z \gg b$ this approaches the field of a magnetic dipole, and find how far out on the axis one has to go before the field has come within 1 percent of the field that an infinitesimal dipole of the same dipole moment would produce at that point.

**11.3** *Dipole in spherical coordinates* ∗∗

Derive Eq. (11.15) by using spherical coordinates to take the curl of the expression for **A** in Eq. (11.10). You will want to use a vector identity from Appendix K.

**11.4** *Force on a dipole* ∗∗

(a) In Section 11.4 we found that the force on the magnetic dipole in the setup shown in Fig. 11.9 was $F_z = m_z(\partial B_z/\partial z)$. Verify that this force is what the more general expression in Eq. (11.22), $\mathbf{F} = (\mathbf{m} \cdot \nabla)\mathbf{B}$, reduces to for our ring setup. The motivation for this form of **F** is that it parallels the force on an electric dipole.

(b) Verify that $F_z = m_z(\partial B_z/\partial z)$ is also what an alternative force expression, $\mathbf{F} = \nabla(\mathbf{m} \cdot \mathbf{B})$, reduces to for our ring setup. The motivation for this form of **F** is that $-\mathbf{m} \cdot \mathbf{B}$ is the energy of a dipole in a magnetic field.

(c) Only one of the preceding forms of **F** can be the correct general expression for all possible setups. Determine the correct one by finding the force on the current loop shown in Fig. 11.34. The magnetic field points in the $z$ direction (perpendicular to the page) and is proportional to $x$, that is, $\mathbf{B} = \hat{\mathbf{z}}B_0 x$.

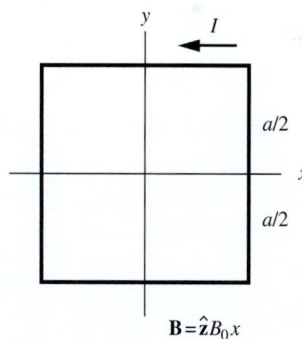

**Figure 11.34.**

**11.5** *Converting $\chi_m$* ∗∗

In SI units, the magnetic moment is defined by $m = Ia$, and the magnetic susceptibility is defined by $\chi_m = \mu_0 M/B$. (The non-accepted definition of $\chi_m$ in Eq. (11.52) will suffice for the present purposes.) In Gaussian units, the corresponding definitions are $m = Ia/c$ and $\chi_m = M/B$. In both systems $\chi_m$ is dimensionless. Show that, for a given setup, the $\chi_m$ in SI units equals $4\pi$ times the $\chi_m$ in Gaussian units.

**11.6 *Paramagnetic susceptibility of liquid oxygen*** **

How large is the magnetic moment of 1 g of liquid oxygen in a field of 1.8 tesla, according to the data in Table 11.1? Given that the density of liquid oxygen is 850 kg/m$^3$ at 90 K, what is its paramagnetic susceptibility $\chi_m$?

**11.7 *Rotating shell*** **

A sphere with radius $R$ has uniform magnetization $\mathbf{M}$. Show that the surface current density is the same as that generated by a spherical shell with radius $R$ and uniform surface charge density $\sigma$, rotating with a specific angular speed $\omega$. How must the various parameters be related? (Note that the result of Problem 11.8 below then tells us what the field of a rotating charged spherical shell is, both inside and outside.)

**11.8 *B inside a magnetized sphere*** ***

In Section 10.9 we determined the electric field inside a uniformly polarized sphere by finding the potential on the surface and then using the uniqueness theorem to find the potential inside. We can use a similar strategy to find the magnetic field inside a uniformly magnetized sphere. Here are the steps.

(a) The field due to magnetic dipole, given in Eq. (11.15), takes the same form as the field due to an electric dipole, given in Eq. (10.18). Explain why this implies that the *external* magnetic field due to a uniformly magnetized sphere with radius $R$ is the same as the field due to a magnetic dipole $\mathbf{m}_0$ located at the center, with magnitude $m_0 = (4\pi R^3/3)M$.

(b) If $\mathbf{m}_0$ points in the $z$ direction, then Eq. (11.12) gives the Cartesian components of the vector potential $\mathbf{A}$ on the surface of the sphere. After looking back at Section 6.3, explain why the uniqueness theorem applies, and find $\mathbf{A}$ inside the sphere. Then take the curl to find $\mathbf{B}$. How do the features of this $\mathbf{B}$ compare with those of the $\mathbf{E}$ inside a polarized sphere?

**11.9 *B at the north pole of a solid rotating sphere*** **

A solid sphere with radius $R$ has uniform volume charge density $\rho$ and rotates with angular speed $\omega$. Use the results from Problems 11.7 and 11.8 to show that the magnetic field at the "north pole" of the sphere equals $2\mu_0\rho\omega R^2/15$.

**11.10 *Surface current on a cube*** **

A cube of magnetite 5 cm on an edge is magnetized to saturation in a direction perpendicular to two of its faces. Find the magnitude of the ribbon of bound-charge current that flows around the circuit consisting of the other four faces of the cube. The saturation magnetization in magnetite is $4.8 \cdot 10^5$ joules/tesla-m$^3$. Would the field of this cubical magnet seriously disturb a compass 2 meters away?

**11.11** *An iron torus* ∗

An iron torus of inner diameter 10 cm, outer diameter 12 cm, has 20 turns of wire wound on it. Use the *B-H* curve in Fig. 11.32 to estimate the current required to produce a field of 1.2 tesla in the iron.

## Exercises

**11.12** *Earth dipole* ∗∗

At the north magnetic pole the earth's magnetic field is vertical and has a strength of 0.62 gauss. The earth's field at the surface and further out is approximately that of a central dipole.

(a) What is the magnitude of the dipole moment in joules/tesla?

(b) Imagine that the source of the field is a current ring on the "equator" of the earth's metallic core, which has a radius of 3000 km, about half the earth's radius. How large would the current have to be?

**11.13** *Disk dipole* ∗∗

A disk with radius $R$ has uniform surface charge density $\sigma$ and spins with angular speed $\omega$. Far away, it looks like a magnetic dipole. What is the magnetic dipole moment?

**11.14** *Sphere dipole* ∗∗

A sphere of radius $R$ carries charge $Q$ distributed uniformly over the surface, with density $\sigma = Q/4\pi R^2$. This shell of charge is rotating about an axis of the sphere with angular speed $\omega$. Find its magnetic moment. (Divide the sphere into narrow bands of rotating charge; find the current to which each band is equivalent, and its dipole moment, and then integrate over all bands.)

**11.15** *A solenoid as a dipole* ∗∗

A solenoid like the one described in Section 11.1 is located in the basement of a physics laboratory. Physicists on the top floor of the building, 60 feet higher and displaced horizontally by 80 feet, complain that its field is disturbing their measurements. Assuming that the solenoid is operating under the conditions described, and treating it as a simple magnetic dipole, give a rough estimate (order of magnitude is fine) of the field strength at the location of the complaining physicists. Comment, if you see any grounds for doing so, on the merit of their complaint.

**11.16** *Dipole in a uniform field* ∗∗

A magnetic dipole of strength $m$ is placed in a homogeneous magnetic field of strength $B_0$, with the dipole moment directed

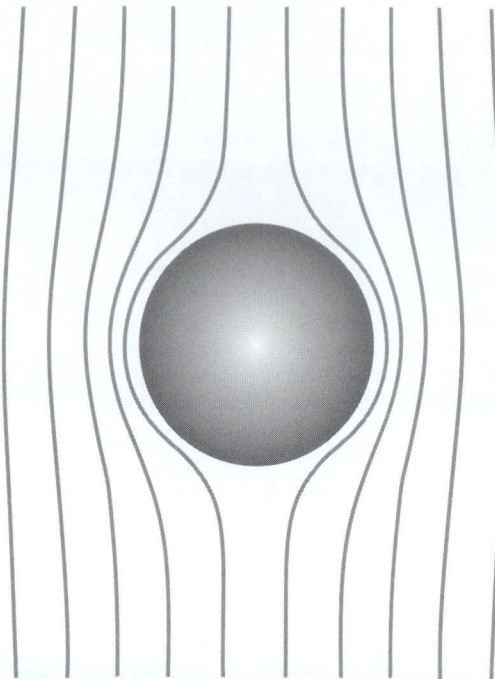

**Figure 11.35.**

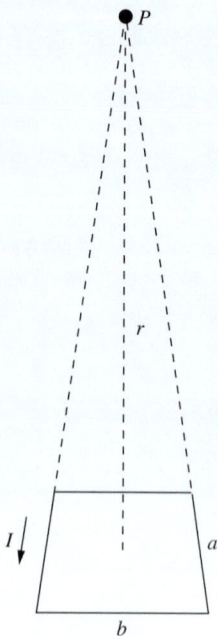

**Figure 11.36.**

opposite to the field. Show that, in the combined field, there is a certain spherical surface, centered on the dipole, through which no field lines pass. The external field, one may say, has been "pushed out" of this sphere. The field lines outside the sphere have been drawn in Fig. 11.35. Roughly what do the field lines inside the sphere look like? What is the strength of the field immediately outside the sphere, at the equator?

So far as its effect on the external field is concerned, the dipole could be replaced by currents flowing in the spherical surface, if we could provide just the right current distribution. (See Problems 11.7 and 11.8, although the exact distribution isn't necessary for this problem.) What is the field inside the sphere in this case? Why can you be sure? (This is an important configuration in the study of superconductivity. A superconducting sphere, in fact, does push out all field from its interior.)

11.17 *Trapezoid dipole* ∗∗

A current $I$ runs around the trapezoidal loop shown in Fig. 11.36. The left and right sides are nearly parallel, and they point toward a distant point $P$. Using the Biot–Savart law, find the magnetic field at $P$, and check that it is consistent with the field in Eq. (11.15). Work in the approximation where $a$ and $b$ are much smaller than $r$. (If you think that it's silly to consider a trapezoid when we could alternatively consider the seemingly simpler shape of a square, you should look at Problem 6.14.)

11.18 *Field somewhat close to a solenoid* ∗∗

A solenoid has length $\ell$ and radius $R$ (with $\ell \gg R$). Consider a point $P$ a distance $\ell$ off to the side, as shown in Fig. 11.37. Show that up to numerical factors, the magnetic field at $P$ behaves like $B_0 R^2/\ell^2$, where $B_0$ is the field inside the solenoid.

11.19 *Using reciprocity* ∗∗

The magnetic dipole $\mathbf{m}$ in Fig. 11.38 oscillates at frequency $\omega$ and has amplitude $\mathbf{m}_0$. Some of its flux links the nearby circuit $C_1$, inducing in $C_1$ an electromotive force, $\mathcal{E}_1 \sin \omega t$. It would be easy to compute $\mathcal{E}_1$ if we knew how much flux from the dipole is enclosed by $C_1$, but that might be hard to calculate. Suppose that all we know about $C_1$ is this: if a current $I_1$ were flowing in $C_1$, it would produce a magnetic field $\mathbf{B}_1$ at the location of $\mathbf{m}$. We are told the value of $\mathbf{B}_1/I_1$, but nothing more about $C_1$, not even its shape or location. Show that this information suffices to relate $\mathcal{E}_1$ to $\mathbf{m}_0$ by the simple formula $\mathcal{E}_1 = (\omega/I_1)\mathbf{B}_1 \cdot \mathbf{m}_0$. *Hint:* Represent $\mathbf{m}$ as a small loop of area $A$ carrying current $I_2$. Call this circuit $C_2$. Consider the voltage induced in $C_2$ by a varying current in $C_1$; then invoke the reciprocity of mutual inductance, which we proved in Section 7.7.

**11.20** *Force between a wire and a loop* ∗∗∗

In Exercise 6.54 we calculated the force on the square loop in Fig. 6.47 due to the magnetic field from the infinite wire. Verify that Newton's third law holds by calculating the force on the wire due to the magnetic field from the square loop. You can assume that the objects are far apart, so that you can use the dipole approximation for the field from the loop.

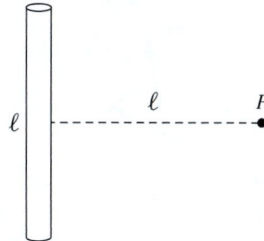

**Figure 11.37.**

**11.21** *Dipoles on a chessboard* ∗∗∗

Imagine that a magnetic dipole of strength $m$ is located at the center of every square on a chessboard, with dipoles on white squares pointing up, dipoles on black squares pointing down. The side of a square is $s$. To answer the following questions you will have to write a little program.

(a) Compute the work required to remove any particular one of the dipoles to infinity, leaving the other 63 fixed in position and orientation. Thus determine which of the dipoles are in this respect most tightly bound.

(b) How much work must be done to disperse all 64 dipoles to infinite separation from one another?

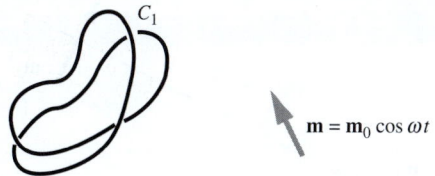

**Figure 11.38.**

**11.22** *Potential momentum* ∗∗

Someone who knows a little about the quantum theory of the atom might be troubled by one point in our analysis in Section 11.5 of the effect of a magnetic field on the orbital velocity of an atomic electron. When the velocity changes, while $r$ remains constant, the angular momentum $mvr$ changes. But the angular momentum of an electron orbit is supposed to be precisely an integral multiple of the constant $h/2\pi$, $h$ being the universal quantum constant, Planck's constant. How can $mvr$ change without violating this fundamental quantum law?

The resolution of this paradox is important for the quantum mechanics of charged particles, but it is not peculiar to quantum theory. When we consider conservation of energy for a particle carrying charge $q$, moving in an external electrostatic field $\mathbf{E}$, we always include, along with the kinetic energy $mv^2/2$, the potential energy $q\phi$, where $\phi$ is the scalar electric potential at the location of the particle. We should not be surprised to find that, when we consider conservation of momentum, we must consider not only the ordinary momentum $M\mathbf{v}$, but also a quantity involving the vector potential of the magnetic field, $\mathbf{A}$.

It turns out that the momentum must be taken as $M\mathbf{v} + q\mathbf{A}$, where $\mathbf{A}$ is the vector potential of the external field evaluated at the location of the particle. We might call $M\mathbf{v}$ the *kinetic momentum* and $q\mathbf{A}$ the *potential momentum*. (In relativity the inclusion of the $q\mathbf{A}$ term is an obvious step because, just as energy and momentum

(a)

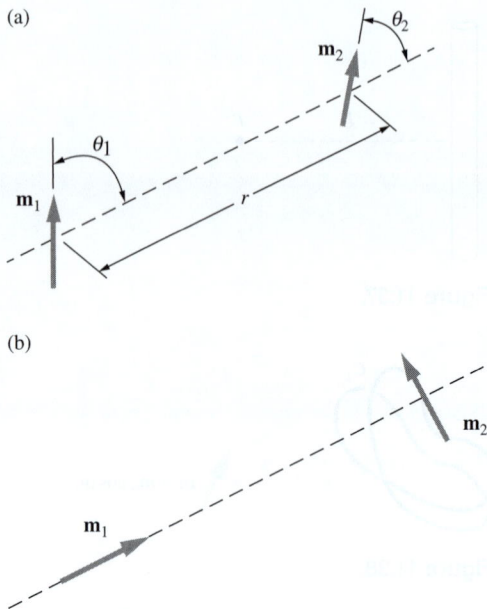

(b)

**Figure 11.39.**

(times $c$) make up a "four-vector," so do $\phi$ and $c\mathbf{A}$, the scalar and vector potentials of the field.) The angular momentum that concerns us here must then be, not just

$$\mathbf{r} \times (M\mathbf{v}), \quad \text{but} \quad \mathbf{r} \times (M\mathbf{v} + q\mathbf{A}). \tag{11.81}$$

Go back now to the case of the charge revolving at the end of the cord in Fig. 11.12. Check first that a vector potential appropriate to a field $\mathbf{B}$ in the negative $z$ direction is $\mathbf{A} = (B/2)(\hat{\mathbf{x}}y - \hat{\mathbf{y}}x)$. Then find what happens to the angular momentum $\mathbf{r} \times (M\mathbf{v} + q\mathbf{A})$ as the field is turned on.

**11.23** *Energy of a dipole configuration* ✶✶✶

We want to find the energy required to bring two dipoles from infinite separation into the configuration shown in Fig. 11.39(a), defined by the distance apart $r$ and the angles $\theta_1$ and $\theta_2$. Both dipoles lie in the plane of the paper. Perhaps the simplest way to compute the energy is this: bring the dipoles in from infinity while keeping them in the orientation shown in Fig. 11.39(b). This takes no work, for the force on each dipole is zero. Now calculate the work done in rotating $\mathbf{m}_1$ into its final orientation while holding $\mathbf{m}_2$ fixed. Then calculate the work required to rotate $\mathbf{m}_2$ into its final orientation. Thus show that the total work done, which we may call the potential energy of the system, is equal to $(\mu_0 m_1 m_2 / 4\pi r^3)(\sin\theta_1 \sin\theta_2 - 2\cos\theta_1 \cos\theta_2)$.

**11.24** *Octahedron energy* ✶✶

Two opposite vertices of a regular octahedron of edge length $b$ are located on the $z$ axis. At each of these vertices, and also at each of the other four vertices, is a dipole of strength $m$ pointing in the $\hat{\mathbf{z}}$ direction. Using the result for Exercise 11.23, calculate the potential energy of this system.

**11.25** *Rotating a bacterium* ✶

In magnetite, $Fe_3O_4$, the saturation magnetization $M_0$ is $4.8 \cdot 10^5$ joule/(tesla-m$^3$). The magnetic bacteria discovered in 1975 by R. P. Blakemore contain crystals of magnetite, approximately cubical, of dimension $5 \cdot 10^{-8}$ m. A bacterium, itself about $10^{-6}$ m in size, may contain from 10 to 20 such crystals strung out as a chain. This magnet keeps the whole cell aligned with the earth's magnetic field, and thus controls the direction in which the bacterium swims; see Blakemore and Frankel (1981). Calculate the energy involved in rotating a cell containing such a magnet through 90° in the earth's field (assuming initial alignment), and compare it with the energy of thermal agitation, $kT$.

**11.26** *Electric vs. magnetic dipole moments* ✶✶

The electric dipole moment of a polar molecule is typically $10^{-30}$ or $10^{-29}$ C-m in order of magnitude (see Fig. 10.14). The magnetic

moment of an atom or molecule with an unpaired electron spin
is $10^{-23}$ amp-m$^2$ (see Fig. 11.14). What (roughly) is the ratio of
the forces from these two moments on a charge $q$ at a given distance moving with speed $v = c/100$? Your result should provide a
reminder that on the atomic scale, magnetism is a relatively feeble
effect.

**11.27** *Diamagnetic susceptibility of water* **

From the data in Table 11.1, determine the diamagnetic susceptibility of water.

**11.28** *Paramagnetic susceptibility of water* **

The water molecule $H_2O$ contains ten electrons with spins paired
off and, consequently, zero magnetic moment. Its electronic structure is purely diamagnetic. However, the hydrogen nucleus, the
proton, is a particle with intrinsic spin and magnetic moment. The
magnetic moment of the proton is about 700 times smaller than
that of the electron. In water the two proton spins in a molecule are
not locked antiparallel but are practically free to orient individually, subject only to thermal agitation.

(a) Using Eq. (11.53), calculate the resulting paramagnetic susceptibility of water at 20 °C.

(b) How large is the magnetic moment induced in 1 liter of water
in a field of 1.5 tesla?

(c) If you wrapped a single turn of wire around a 1 liter flask,
about how large a current, in microamps, would produce an
equivalent magnetic moment?

**11.29** *Work on a paramagnetic material* **

Show that the work done per kilogram in pulling a paramagnetic
material from a region where the magnetic field strength is $B$ to
a region where the field strength is negligibly small is $\chi B^2/2\mu_0$,
$\chi$ being the specific susceptibility. Show that the work on a given
sample can be written as $FB/(2\,\partial B/\partial z)$, where $F$ is the force at the
initial location. Then calculate exactly how much work would be
required to remove one gram of liquid oxygen from the position
referred to in Section 11.1. (Of course, this applies only if $\chi$ is a
constant over the range of field strengths involved.)

**11.30** *Greatest force in a solenoid* ***

A cylindrical solenoid has a single-layer winding of radius $r_0$. It
is so long that near one end the field may be taken to be that of
a semi-infinite solenoid. Show that the point on the axis of the
solenoid where a small paramagnetic sample will experience the
greatest force is located a distance $r_0/\sqrt{15}$ in from the end.

**11.31** *Boundary conditions for B* ∗∗

Use the result from Problem 11.8 to show that the radial component of **B** is continuous across the boundary of a uniformly magnetized sphere, while the tangential component has a discontinuity of $\mu_0 \mathcal{J}_\theta$, where $\mathcal{J}_\theta$ is the surface current density at position $\theta$ on the sphere.

**11.32** *B at the center of a solid rotating sphere* ∗∗

A solid sphere with radius $R$ has uniform volume charge density $\rho$ and rotates with angular speed $\omega$. Use the results from Problems 11.7 and 11.8 to show that the magnetic field at the center of the sphere equals $\mu_0 \rho \omega R^2 / 3$.

**11.33** *Spheres of frozen magnetization* ∗∗

A remarkable permanent magnet alloy of samarium and cobalt has a saturation magnetization of $7.5 \cdot 10^5$ joule/tesla-m$^3$, which it retains undiminished in external fields up to 1.5 tesla. It provides a good approximation to rigidly frozen magnetization. Consider a sphere of uniformly magnetized samarium-cobalt 1 cm in radius.

(a) What is the strength of its magnetic field **B** just outside the sphere at one of its poles? You can invoke the result from Problem 11.8.

(b) What is the strength of its magnetic field **B** at its magnetic equator?

(c) Imagine two such spheres magnetically stuck together with unlike poles touching. How much force must be applied to separate them?

**11.34** *Muon deflection* ∗∗

An iron plate 0.2 m thick is magnetized to saturation in a direction parallel to the surface of the plate. A 10 GeV muon moving perpendicular to that surface enters the plate and passes through it with relatively little loss of energy. Calculate approximately the angular deflection of the muon's trajectory, given that the rest-mass energy of the muon is 200 MeV and that the saturation magnetization in iron is equivalent to $1.5 \cdot 10^{29}$ electron moments per cubic meter.

**11.35** *Volume integral of near field* ∗

In the case of an electric dipole made of two charges $Q$ and $-Q$ separated by a distance $s$, the volume of the near region, where the field is essentially different from the ideal dipole field, is proportional to $s^3$. The field strength in this region is proportional to $Q/s^2$, at similar points as $s$ is varied. The dipole moment is $p = Qs$, so that if we shrink $s$ while holding $p$ constant, the product of volume and field strength does what? Carry through the corresponding argument for the magnetic field of a current loop. The moral is: if we are concerned with the space average field in any volume containing dipoles, the essential difference between the insides of electric and magnetic dipoles *cannot* be ignored, even when we are treating the dipoles otherwise as infinitesimal.

**11.36** *Equilibrium orientations* **

Three magnetic compasses are placed at the corners of a horizontal equilateral triangle. As in any ordinary compass, each compass needle is a magnetic dipole constrained to rotate in a horizontal plane. In this case the earth's magnetic field has been precisely annulled. The only field that acts on each dipole is that of the other two dipoles. What orientation will they eventually assume? (Use symmetry arguments!) Can your answer be generalized for $N$ compasses at the vertices of an $N$-gon?

**11.37** *B inside a magnetized sphere* **

In Problem 11.8 we found the magnetic field **B** inside a sphere with uniform magnetization **M**. The task of this exercise is to rederive that result by making use of the result from Section 10.9 for a uniformly polarized sphere, namely $\mathbf{E} = -\mathbf{P}/3\epsilon_0$. To do this, consider the following equations that are valid for static fields:

$$\nabla \cdot (\epsilon_0 \mathbf{E} + \mathbf{P}) = \rho_{\text{free}}, \qquad \nabla \cdot \mathbf{B} = 0,$$
$$\nabla \times \mathbf{E} = 0, \qquad \nabla \times (\mathbf{B}/\mu_0 - \mathbf{M}) = \mathbf{J}_{\text{free}}. \quad (11.82)$$

(The first and last of these are Eqs. (10.62) and (11.67).) If additionally $\rho_{\text{free}} = 0$ and $\mathbf{J}_{\text{free}} = 0$, which is the case for our polarized and magnetized spheres, the right-hand sides of all the equations are zero. Rewrite the two magnetic equations in terms of **H**, and then take advantage of the resulting similarity with the electric equations.

**11.38** *Two susceptibilities* *

Let us denote by $\chi_m'$ the magnetic susceptibility defined by Eq. (11.52), to distinguish it from the susceptibility $\chi_m$ in the conventional definition, Eq. (11.72). Show that

$$\chi_m = \chi_m'/(1 - \chi_m'). \quad (11.83)$$

**11.39** *Magnetic moment of a rock* **

The direction of the earth's magnetic field in geological ages past can be deduced by studying the remanent magnetization in rocks. The magnetic moment of a rock specimen can be determined by rotating it inside a coil and measuring the alternating voltage thereby induced. The two coils in Fig. 11.40 are connected in series. Each has 1500 turns and a mean radius of 6 cm. The rock is rotated at 1740 revolutions per minute by a shaft perpendicular to the plane of the diagram. Assume that the magnetic moment lies in the plane of the page.

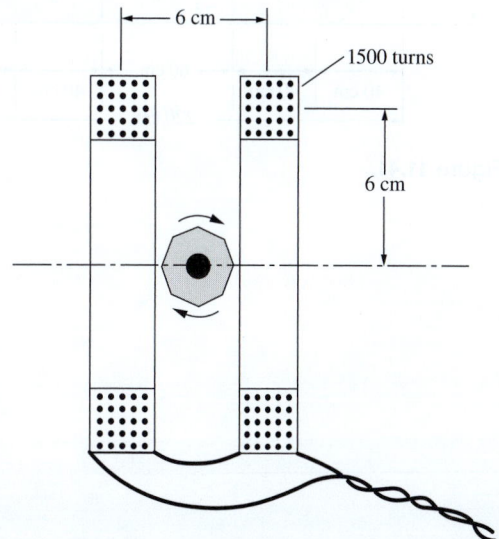

**Figure 11.40.**

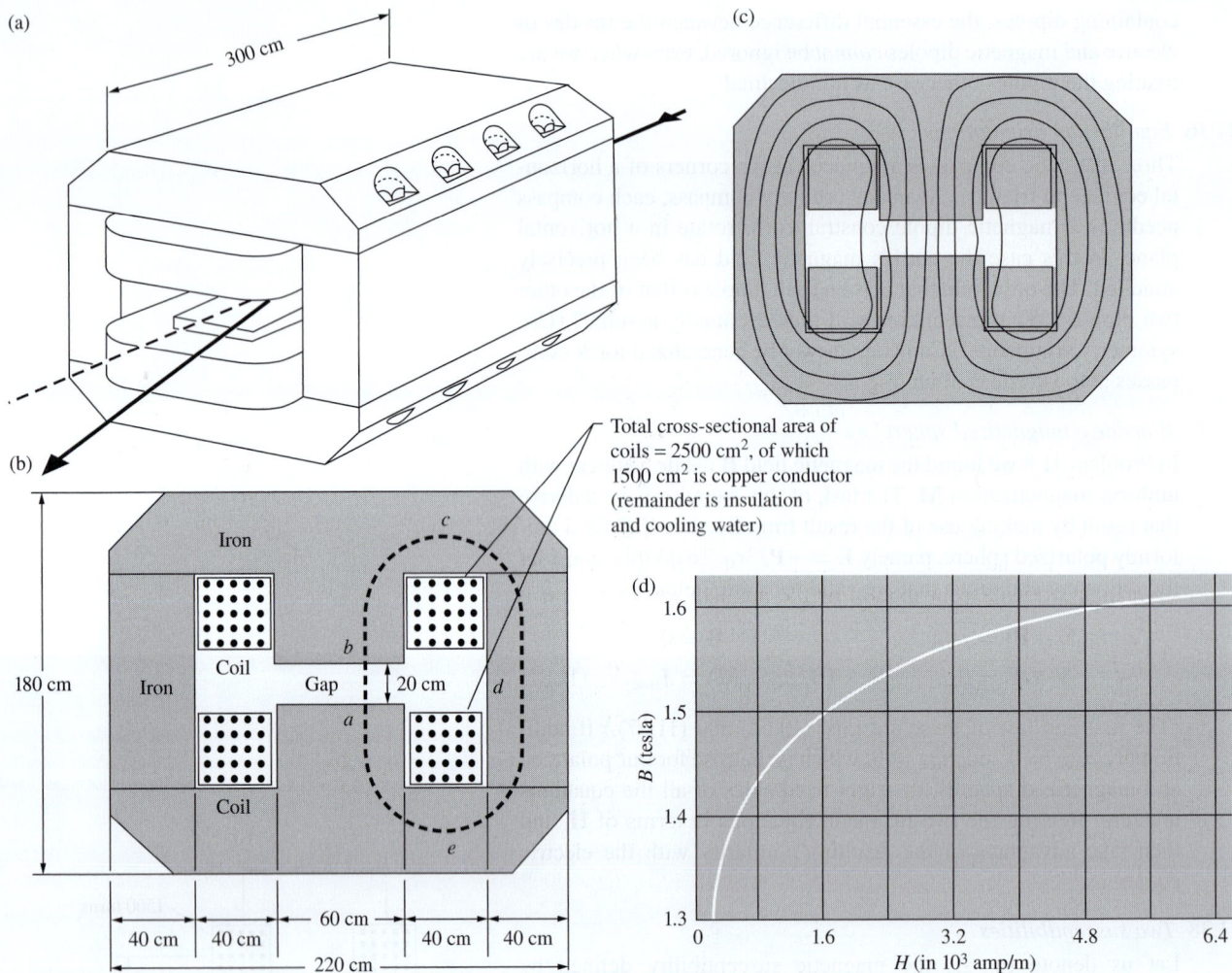

(a)

300 cm

(b)

Iron

Iron

180 cm

Coil

Gap

20 cm

Coil

40 cm   40 cm   60 cm   40 cm   40 cm

220 cm

**Figure 11.41.**

(c)

Total cross-sectional area of
coils = 2500 cm², of which
1500 cm² is copper conductor
(remainder is insulation
and cooling water)

(d)

$B$ (tesla)

1.6

1.5

1.4

1.3

0        1.6       3.2       4.8       6.4

$H$ (in $10^3$ amp/m)

(a) How large is the magnetic moment of the rock if the amplitude
of the induced electromotive force is 1 millivolt? The formula
derived in Exercise 11.19 is useful here.

(b) In order of magnitude, what is the minimum amount of ferromagnetic material required to produce an effect that large?

11.40 *Deflecting high-energy particles* ∗∗∗

For deflecting a beam of high-energy particles in a certain experiment, one requires a magnetic field of 1.6 tesla intensity, maintained over a rectangular region 3 m long in the beam direction, 60
cm wide, and 20 cm high. A suitable magnet might be designed
along the lines indicated in parts (a) and (b) of Fig. 11.41; part
(b) shows the cross section of two horizontal coils. Taking the

dimensions as given (you can make rough estimates for any other lengths you need), and referring to the additional comments below, determine:

(a) the total number of ampere turns required in the two coils to produce a 1.6 tesla field in the gap;

(b) the power in kilowatts that must be supplied;

(c) the number of turns that each coil should contain, and the corresponding cross-sectional area of the wire, so that the desired field will be attained when the coils are connected in series to a 400 volt dc power supply.

For use in (a), a portion of the B-H curve for Armco magnet iron is shown in Fig. 11.41(d). All that you need to determine is the line integral of H around a path like *abcdea*. In the gap, $H = B/\mu_0$. In the iron, you may assume that B has the same intensity as in the gap. The field lines will look something like those in Fig. 11.41(c). You can estimate roughly the length of path in the iron. This is not very critical, for you will find that the long path *bcdea* contributes a relatively small amount to the line integral, compared with the contribution of the air path *ab*. In fact, it is not a bad approximation, at lower field strengths, to neglect H in the iron.

For (b), let each coil contain N turns, and assume the resistivity of copper is $\rho = 2.0 \cdot 10^{-8}$ ohm-m. You will find that the power required for a given number of ampere turns is independent of N; that is, it is the same for many turns of fine wire or a few turns of thick wire, provided that the total cross section of copper is fixed as specified (1500 cm$^2$ in our setup). The designer therefore selects N and conductor cross section to match the magnet to the voltage of the intended power source.

# 12

## Solutions to the problems

Solutions to the
problems

### 12.1  Chapter 1

1.1 *Gravity vs. electricity*

  (a) The general expressions for the magnitudes of the gravitational and electrical forces are

$$F_{\text{g}} = \frac{Gm_1 m_2}{r^2} \quad \text{and} \quad F_{\text{e}} = \frac{q_1 q_2}{4\pi \epsilon_0 r^2}. \tag{12.1}$$

For two protons, the ratio of these forces is

$$\frac{F_{\text{g}}}{F_{\text{e}}} = \frac{4\pi \epsilon_0 G m^2}{q^2}$$

$$= \frac{4\pi \left(8.85 \cdot 10^{-12} \, \frac{\text{s}^2 \text{C}^2}{\text{kg m}^3}\right) \left(6.67 \cdot 10^{-11} \, \frac{\text{m}^3}{\text{kg s}^2}\right) \left(1.67 \cdot 10^{-27} \, \text{kg}\right)^2}{(1.6 \cdot 10^{-19} \, \text{C})^2}$$

$$= 8.1 \cdot 10^{-37} \approx 10^{-36}, \tag{12.2}$$

which is extremely small. Note that this ratio is independent of $r$. To get a sense of how large the number $10^{36}$ is, imagine forming a row of neutrons stretching from the earth to the sun. If you made ten billion copies of this, you would have about $10^{36}$ neutrons.

  (b) If $r = 10^{-15}$ m, the electrical force is

$$F_{\text{e}} = \frac{1}{4\pi \epsilon_0} \frac{q^2}{r^2} = \left(9 \cdot 10^9 \, \frac{\text{kg m}^3}{\text{s}^2 \text{C}^2}\right) \frac{(1.6 \cdot 10^{-19} \, \text{C})^2}{(10^{-15} \, \text{m})^2} = 230 \, \text{N}. \tag{12.3}$$

Since 1 N is equivalent to about 0.22 pounds, this force is roughly 50 pounds! This is balanced by the similarly huge "strong" force that holds the nucleus together.

1.2   *Zero force from a triangle*

First note that the desired point cannot be located in the interior of the triangle, because the components of the fields along the symmetry axis would all point in the same direction (toward the negative ion). Let the sides of the triangle be 2 units long. Consider a point $P$ that lies a distance $y$ (so $y$ is defined to be a positive number) beyond the side containing the two positive ions, as shown in Fig. 12.1. $P$ is a distance $y + \sqrt{3}$ from the negative ion, and $\sqrt{1 + y^2}$ from each of the positive ions. If the electric field equals zero at $P$, then the upward field due to the negative ion must cancel the downward field due to the two positive ions. This gives (ignoring the factor of $e/4\pi\epsilon_0$)

$$\frac{1}{(y + \sqrt{3})^2} = 2 \cdot \frac{1}{1^2 + y^2} \left( \frac{y}{\sqrt{1^2 + y^2}} \right) \implies y = \frac{(1 + y^2)^{3/2}}{2(y + \sqrt{3})^2},$$

(12.4)

where the $y/\sqrt{1 + y^2}$ factor in the first equality arises from taking the vertical component of the titled field lines due to the positive ions. Equation (12.4) can be solved numerically, and the result is $y \approx 0.1463$. It can also be solved by iteration: evaluate the right side for some guessed initial $y$, then replace $y$ with that calculated value. For this equation the process converges rapidly to $y \approx 0.1463$.

A second point with $E = 0$ lies somewhere beyond the negative ion. To locate it, let $y$ now be the distance (so $y$ is still a positive quantity) from the same origin as before (the midpoint of the side connecting the two positive ions). We obtain the same equation as above, except that $+\sqrt{3}$ is replaced with $-\sqrt{3}$. The numerical solution is now $y \approx 6.2045$. This corresponds to a distance $6.2045 - \sqrt{3} \approx 4.4724$ beyond the negative ion.

The existence of each of these points with zero field follows from a continuity argument. For the upper point: the electric field just above the negative ion points downward. But the field at a large distance above the setup points upward, because the triangle looks effectively like a point charge with net charge $+e$ from afar. Therefore, by continuity there must be an intermediate point where the field makes the transition from pointing downward to pointing upward. So $E = 0$ at this point. A similar continuity argument holds for the lower point.

1.3   *Force from a cone*

(a) Consider a thin ring around the cone, located a slant distance $x$ away from the tip, with width $dx$, as shown in Fig. 12.2. If we look at all the bits of charge in this ring, the horizontal components of their forces cancel in pairs from diametrically opposite points. So we are left with only the vertical components, which brings in a factor of $\cos\theta$. (Equivalently, we can just say that the field is directed vertically, from symmetry.) The vertical component of the force due to a small piece of charge $dQ$ in the ring is $q(dQ)\cos\theta/4\pi\epsilon_0 x^2$. Integrating over the entire ring simply turns the $dQ$ into the total charge in the ring. The total (vertical) force due to the ring is therefore $q(Q_{\text{ring}})\cos\theta/4\pi\epsilon_0 x^2$.

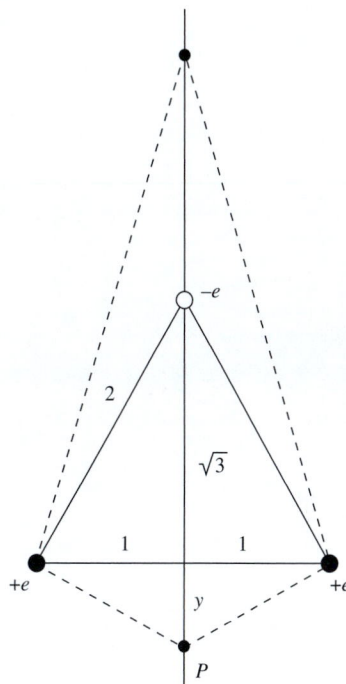

**Figure 12.1.**

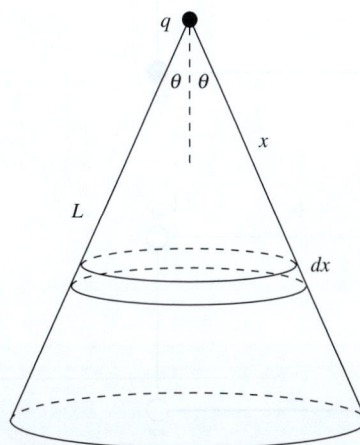

**Figure 12.2.**

The radius of the ring is $x \sin \theta$, so its area is $2\pi (x \sin \theta) dx$. The charge in the ring is then $Q_{\text{ring}} = \sigma 2\pi x \sin \theta \, dx$. Integrating over all the rings from $x = 0$ to $x = L$ gives a total force on $q$ equal to

$$F = \int_0^L \frac{q(\sigma 2\pi x \sin \theta \, dx) \cos \theta}{4\pi \epsilon_0 x^2} = \frac{q\sigma \sin \theta \cos \theta}{2\epsilon_0} \int_0^L \frac{dx}{x}. \quad (12.5)$$

But this integral diverges, so the force is infinite. In short, for small $x$ the largeness of the $1/x^2$ factor in Coulomb's law wins out over the smallness of the $x$ factor in the area of a ring (but just barely; the above integral diverges very slowly like a log).

(b) The only difference now is that the integral starts at $L/2$ instead of zero. So we have

$$F = \frac{q\sigma \sin \theta \cos \theta}{2\epsilon_0} \int_{L/2}^L \frac{dx}{x} = \frac{q\sigma \sin \theta \cos \theta}{2\epsilon_0} (\ln 2). \quad (12.6)$$

Since $\sin \theta \cos \theta = (1/2) \sin 2\theta$, this force is maximized when $2\theta = 90° \implies \theta = 45°$, in which case it equals $q\sigma (\ln 2)/4\epsilon_0$. The force correctly equals zero when $\theta = 90°$ ($q$ is at the center of a hole in a flat disk) and also when $\theta = 0$ (the cone is infinitesimally thin and hence contains zero charge).

Note that the force $F$ is independent of $L$. If we imagine scaling the size by a factor of, say, 5, then a patch of the expanded cone has $5^2$ times the charge as the corresponding patch of the original cone (because areas are proportional to distances squared), and this exactly cancels the factor of $1/5^2$ from the $1/r^2$ in Coulomb's law.

1.4 *Work for a rectangle*

The two basic arrangements are shown in Fig. 12.3. In each case there are six pairs of charges. Two of the separations are $a$, two are $b$, and two are $\sqrt{a^2 + b^2}$. In the first arrangement, the energy of the system (which equals the work required to bring the charges together) is given by

$$U = \frac{e^2}{4\pi \epsilon_0} \left( -2 \cdot \frac{1}{a} - 2 \cdot \frac{1}{b} + 2 \cdot \frac{1}{\sqrt{a^2 + b^2}} \right). \quad (12.7)$$

Each of the first two terms has a larger magnitude than the third, so $U$ is negative for any values of $a$ and $b$. In the second arrangement, the energy is given by

$$U = \frac{e^2}{4\pi \epsilon_0} \left( 2 \cdot \frac{1}{a} - 2 \cdot \frac{1}{b} - 2 \cdot \frac{1}{\sqrt{a^2 + b^2}} \right). \quad (12.8)$$

We quickly see that this is negative if $b = a$, and positive if $b \gg a$. So the energy can be positive if $b$ is large enough compared with $a$. For convenience, let $a = 1$. Then $U = 0$ when

$$1 - \frac{1}{b} = \frac{1}{\sqrt{1 + b^2}} \implies \frac{(b-1)^2}{b^2} = \frac{1}{1 + b^2}$$

$$\implies b^4 - 2b^3 + b^2 - 2b + 1 = 0. \quad (12.9)$$

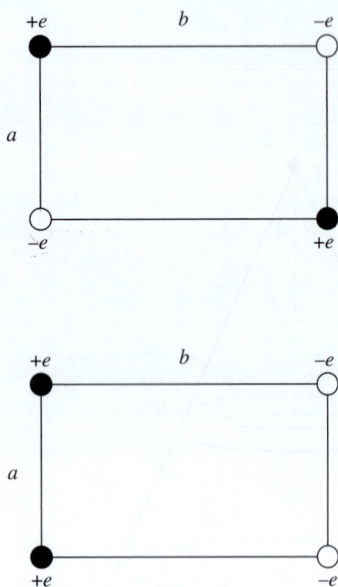

**Figure 12.3.**

We can solve this numerically, and the root we are concerned with is $b \approx 1.883$, or more generally $b = (1.883)a$. If $b$ is larger than this, then $U$ is positive.

1.5 *Stable or unstable?*

Let the four corners be located at the points $(\pm a, \pm a)$. Then the four distances from these points to the point $(x, y)$ are $\sqrt{(\pm a - x)^2 + (\pm a - y)^2}$. Ignoring the common factor of $-Qq/4\pi\epsilon_0$, the code for the *Mathematica* Series expansion (to second order in $x$ and $y$) for the energy $U(x, y)$ of the $-Q$ charge is as follows:

```
Series[
1/Sqrt[(a-x)^2+(a-y)^2]+
1/Sqrt[(a-x)^2+(-a-y)^2]+
1/Sqrt[(-a-x)^2+(a-y)^2]+
1/Sqrt[(-a-x)^2+(-a-y)^2],
{x,0,2},{y,0,2}]
```

This yields an energy of (to second order)

$$U(x, y) = \frac{-Qq}{4\pi\epsilon_0 a}\left(2\sqrt{2} + \frac{x^2 + y^2}{2\sqrt{2}\,a^2}\right). \tag{12.10}$$

We see that the energy decreases with $x$ and $y$, so the equilibrium is unstable for any direction of motion of the $-Q$ charge in the $xy$ plane. This instability is consistent with a theorem that we will prove in Section 2.12.

Note that $U(x, y)$ reduces properly when $x = y = 0$, in which case the $-Q$ charge is $\sqrt{2}\,a$ away from the four $q$ charges. Note also that the lack of terms linear in $x$ or $y$ is consistent with the fact that the force on the $-Q$ charge is zero at the center of the square (the force involves the derivative of the energy). As far as motion in the $xy$ plane near the origin is concerned, you can think of the charge roughly as a ball sitting on top of an inverted bowl.

1.6 *Zero potential energy for equilibrium*

(a) By symmetry, the force on $Q$ is zero, so we need only worry about the force on the $q$ charges. And again by symmetry, we need only worry about one of these. The force on the right $q$ (ignoring the $1/4\pi\epsilon_0$ since it will cancel) equals $qQ/d^2 + q^2/(2d)^2$. Setting this equal to zero gives $Q = -q/4$.

(b) As in part (a), we need only worry about the force on one of the $q$ charges. The force on the top $q$ (ignoring the $1/4\pi\epsilon_0$) equals $qQ/d^2 + 2(q^2/(\sqrt{3}d)^2)(\sqrt{3}/2)$, where the last factor is the cos 30° involved in taking the vertical component of the force. We have used the fact that a side of the equilateral triangle has length $\sqrt{3}d$. Setting the force equal to zero gives $Q = -q/\sqrt{3}$.

(c) We have three pairs of charges in part (a), so the potential energy is

$$\frac{1}{4\pi\epsilon_0}\left(\frac{q^2}{2d} + 2\cdot\frac{qQ}{d}\right) = \frac{1}{4\pi\epsilon_0}\left(\frac{q^2}{2d} + 2\cdot\frac{q(-q/4)}{d}\right) = 0.$$

$$\tag{12.11}$$

We have six pairs of charges in part (b), so the potential energy is

$$\frac{1}{4\pi\epsilon_0}\left(3\cdot\frac{q^2}{\sqrt{3}d}+3\cdot\frac{qQ}{d}\right)=\frac{1}{4\pi\epsilon_0}\left(3\cdot\frac{q^2}{\sqrt{3}d}+3\cdot\frac{q(-q/\sqrt{3})}{d}\right)$$

$$=0. \tag{12.12}$$

These are both equal to zero, as desired.

(d) Consider an arbitrary set of charges in equilibrium, and imagine moving them out to infinity by uniformly expanding the size of the configuration, so that all relative distances stay the same. For example, in part (b) we will simply expand the equilateral triangle until it becomes infinitely large. At a later time, let $f$ be the factor by which all distances have increased. Then because the electrostatic force is proportional to $1/r^2$, the forces between all pairs of charges have decreased by a factor $1/f^2$. So the net force on any charge is $1/f^2$ of what it was at the start. But it was zero at the start, so it is zero at any later time. Therefore, since the force on any charge is always zero, zero work is needed to bring it out to infinity. The initial potential energy of the system is thus zero, as desired. (You can quickly show with a counterexample that the converse of our result is *not* true.)

From this reasoning, we see that the particular inverse-square nature of the electrostatic force is irrelevant. Any power-law force leads to the same result. But a force such as $e^{-\alpha r}$ does not. For further discussion of this topic, see Crosignani and Di Porto (1977).

1.7 *Potential energy in a two-dimensional crystal*
Consider the potential energy of a given ion due to the full infinite plane. Call it $U_0$. If we sum over all ions (or a very large number $N$) to find the total $U$ of these ions, we obtain $NU_0$. However, we have counted each pair twice, so we must divide by 2 to obtain the actual total energy. Dividing by $N$ then gives the energy per ion as $(NU_0/2)/N=U_0/2$.

Equivalently, we can calculate the potential energy of a given ion due to only the half-plane above it and the half-line to the right of it; see Fig. 12.4. Looking at only half of the ions like this is equivalent to the above division by 2. Physically, this strategy corresponds to how you might actually build up the lattice. Imagine that the half-plane above and the half-line to the right have already been put in position. The question is, how much energy is involved in bringing in a new ion? This is simply the potential energy due to the ions already in place. We can then continue adding on new ions in the same line, onward to the left (as in Exercise 1.42), and then eventually we can move down to the next line.

If we index the ions by the coordinates $(m,n)$, then the potential energy of the ion at $(0,0)$ due to the half-line to the right of it and the half-plane above it is given by

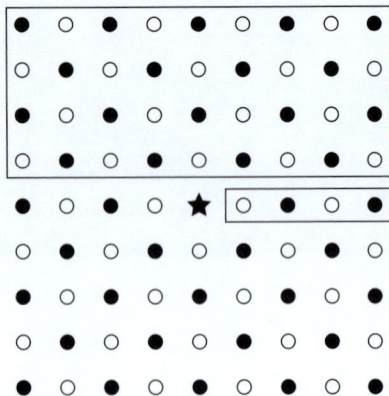

Figure 12.4.

$$U=\frac{e^2}{4\pi\epsilon_0 a}\left(\sum_{m=1}^{\infty}\frac{(-1)^m}{m}+\sum_{n=1}^{\infty}\sum_{m=-\infty}^{\infty}\frac{(-1)^{m+n}}{\sqrt{m^2+n^2}}\right). \tag{12.13}$$

Taking the limits to be 1000 instead of $\infty$ yields decent enough results via *Mathematica*. We obtain

$$U = \frac{e^2}{4\pi\epsilon_0 a}(-0.693 - 0.115) = -\frac{(0.808)e^2}{4\pi\epsilon_0 a}. \tag{12.14}$$

This result is negative, which means that it requires energy to move the ions away from each other. This makes sense, because the four nearest neighbors are of the opposite sign.

1.8     *Oscillating in a ring*

Consider a small piece of the ring with length $R\,d\theta$. Using the law of cosines in Fig. 12.5, the distance from the point $(r, 0)$ to this piece is $\sqrt{R^2 + r^2 - 2Rr\cos\theta}$. The potential energy of the charge $q$, as a function of $r$, is therefore

$$U(r) = 2\int_0^\pi \frac{1}{4\pi\epsilon_0} \frac{q(\lambda R\,d\theta)}{\sqrt{R^2 + r^2 - 2Rr\cos\theta}}$$

$$= \frac{q\lambda}{2\pi\epsilon_0} \int_0^\pi \frac{d\theta}{\sqrt{1 + r^2/R^2 - 2(r/R)\cos\theta}}. \tag{12.15}$$

Using the given Taylor series with $\epsilon \equiv r^2/R^2 - 2(r/R)\cos\theta$, and keeping terms only to order $r^2$, yields

$$U(r) = \frac{q\lambda}{2\pi\epsilon_0} \int_0^\pi \left[ 1 - \frac{1}{2}\left(\frac{r^2}{R^2} - \frac{2r}{R}\cos\theta\right) \right.$$

$$\left. + \frac{3}{8}\left(\left(-\frac{2r}{R}\cos\theta\right)^2 + \cdots\right) \right] d\theta$$

$$= \frac{q\lambda}{2\pi\epsilon_0} \int_0^\pi \left(1 + \frac{r^2}{2R^2}\left(3\cos^2\theta - 1\right)\right) d\theta, \tag{12.16}$$

where we have used the fact that the term linear in $\cos\theta$ integrates to zero. As far as $\cos^2\theta$ goes, we can simply replace it with $1/2$, because that is its average value. We therefore obtain

$$U(r) = \frac{q\lambda}{2\epsilon_0} + \frac{q\lambda r^2}{8\epsilon_0 R^2}. \tag{12.17}$$

The force on the charge $q$ is then

$$F(r) = -\frac{dU}{dr} = -\frac{q\lambda r}{4\epsilon_0 R^2}, \tag{12.18}$$

which is a Hooke's law type force, being proportional to the displacement. The $F = ma$ equation for the charge is

$$F = ma \implies -\frac{q\lambda r}{4\epsilon_0 R^2} = m\ddot{r} \implies \ddot{r} = -\left(\frac{q\lambda}{4\epsilon_0 m R^2}\right)r. \tag{12.19}$$

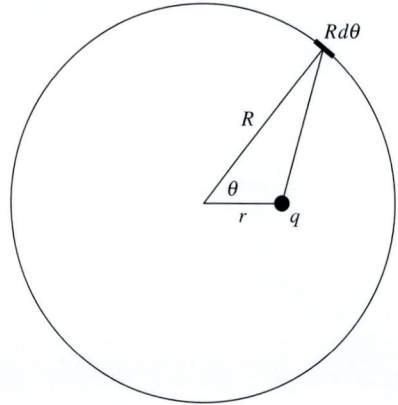

**Figure 12.5.**

The angular frequency of small oscillations is the square root of the (negative of the) coefficient of $r$, so $\omega = \sqrt{q\lambda/4\epsilon_0 mR^2}$. In terms of the charge $Q$ on the ring, we have $\lambda = Q/2\pi R$, so $\omega = \sqrt{qQ/8\pi\epsilon_0 mR^3}$. If $r = 0.1$ m, $m = 0.01$ kg, and $q$ and $Q$ are both one microcoulomb, then you can show that $\omega = 21$ s$^{-1}$, which is a little over 3 Hz.

1.9 *Field from two charges*

(a) The field cannot be zero anywhere *between* two charges of opposite sign, or anywhere closer to the greater charge than to the lesser charge. Hence the point we seek must lie to the right of the $-q$ charge; that is, its $x$ value must satisfy $x > a$. It is important to be clear about this before plunging into the algebra. The field will vanish there if (ignoring the $4\pi\epsilon_0$)

$$\frac{2q}{x^2} - \frac{q}{(x-a)^2} = 0 \implies x^2 - 4xa + 2a^2 = 0$$

$$\implies x = (2 \pm \sqrt{2})a. \qquad (12.20)$$

The positive root locates the point of vanishing field at $x = (2.414)a$. The other root lies between the charges, and gives a second location where the *magnitudes* of the two fields are equal. But in this case the fields point in the same direction instead of canceling.

Note that in order to have any chance of the field being zero at a given point, the point must lie on the $x$ axis. This follows from the fact that if the point does not lie on the $x$ axis, then the fields from the $2q$ and $-q$ charges point in different directions, so it is impossible for them to cancel each other. The $x = (2.414)a$ point is therefore the only point at which the field is zero.

(b) At the point $(a, y)$ in Fig. 12.6, with $y$ positive, the field component $E_y$ has the value (ignoring the $4\pi\epsilon_0$)

$$E_y = \frac{2q}{a^2 + y^2}\left(\frac{y}{\sqrt{a^2 + y^2}}\right) - \frac{q}{y^2}, \qquad (12.21)$$

where the factor in parentheses yields the vertical component of the titled field due to the $2q$ charge. $E_y$ vanishes when $2y^3 = (a^2+y^2)^{3/2}$, which can be written as $2^{2/3}y^2 = a^2+y^2$. Hence $y = a/\sqrt{2^{2/3} - 1} = (1.305)a$. The value $y = -(1.305)a$ also works, by symmetry. (Alternatively, if $y$ is negative, there is a plus sign in front of the second term in Eq. (12.21).)

The existence of such a point where $E_y = 0$ follows from a continuity argument: at a point on the line $x = a$ just above the $-q$ charge, this charge dominates, so the field points downward. But for large positive values of $y$, the $2q$ charge dominates, so the field points upward. By continuity, the field must make the transition between these two directions and point horizontally at some intermediate $y$ value. This reasoning does *not* apply if we consider instead the field at points on a vertical line through the $2q$ charge. Indeed, the field points upward (for positive $y$) everywhere on that line.

1.10 *45-degree field line*

Let's parameterize the line by the angle $\theta$ shown in Fig. 12.7. A little piece of the line that subtends an angle $d\theta$ is a distance $\ell/\cos\theta$ from the given

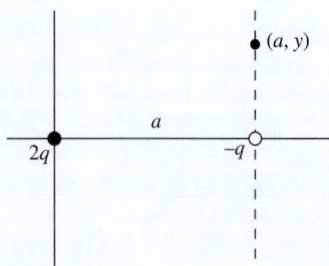

**Figure 12.6.**

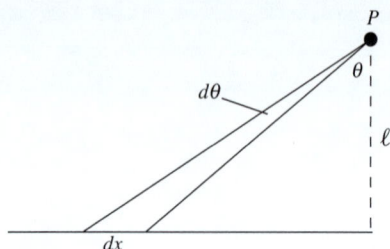

**Figure 12.7.**

point $P$, and its length is $dx = d(\ell \tan \theta) = \ell\, d\theta / \cos^2 \theta$. The magnitude of the field contribution at point $P$ is therefore

$$dE = \frac{1}{4\pi \epsilon_0} \frac{(\ell\, d\theta / \cos^2 \theta)\lambda}{(\ell / \cos \theta)^2} = \frac{1}{4\pi \epsilon_0} \frac{\lambda\, d\theta}{\ell}. \qquad (12.22)$$

The horizontal component of this is obtained by multiplying by $\sin \theta$. The total horizontal component of the field at $P$ is then

$$E_x = \int_0^{\pi/2} \frac{1}{4\pi \epsilon_0} \frac{\lambda\, d\theta}{\ell} \sin \theta = \frac{\lambda}{4\pi \epsilon_0 \ell} \int_0^{\pi/2} \sin \theta\, d\theta = \frac{\lambda}{4\pi \epsilon_0 \ell}.$$

$$(12.23)$$

Similarly, the vertical component of the field contribution in Eq. (12.22) is obtained by multiplying by $\cos \theta$. And since $\int_0^{\pi/2} \cos \theta\, d\theta$ equals 1 just like the $\sin \theta$ integral, we see that $E_x$ and $E_y$ at point $P$ both equal $\lambda/4\pi \epsilon_0 \ell$. The field therefore points up at a 45° angle, as desired.

The $E_y = \lambda/4\pi \epsilon_0 \ell$ result is consistent with the fact that the field from a full infinite line is $\lambda/2\pi \epsilon_0 \ell$, which follows from a direct integration or a quick application of Gauss's law. By superposition, two half-infinite lines placed end-to-end yield a full infinite line, so the $E_y$ from the latter must be twice the $E_y$ from the former.

Note that since $\ell$ is the only length scale in the problem, both components must be proportional to $\lambda/\epsilon_0 \ell$ (assuming that they are finite). Their ratio, and hence the angle of the field, is therefore independent of $\ell$. But it takes a calculation to show that the angle is 45°.

1.11   *Field at the end of a cylinder*

(a) We will solve this problem by slicing up the cylindrical shell into a series of rings stacked on top of each other. (As an exercise, you can also calculate the field by slicing up the cylindrical shell into half-infinite parallel strips.) In Fig. 12.8 each little piece of a ring gives a field contribution of $dq/4\pi \epsilon_0 r^2$. By symmetry, only the vertical component survives, and this brings in a factor of $\sin \theta$. Integrating over $dq$ simply yields the total charge $q$ in the ring, so we find that the vertical component of the field due to a ring is

$$E_{\text{ring}} = \frac{q}{4\pi \epsilon_0 r^2} \sin \theta. \qquad (12.24)$$

But $q$ is given by $\sigma(\text{area}) = \sigma(2\pi R\, dy)$, where (see Fig. 12.9 for a zoomed-in view) $dy = r\, d\theta / \cos \theta = r\, d\theta / (R/r) = r^2\, d\theta / R$. The field from a ring subtending an angle $d\theta$ is therefore

$$E_{\text{ring}} = \frac{\sigma(2\pi R)(r^2\, d\theta / R)}{4\pi \epsilon_0 r^2} \sin \theta = \frac{\sigma \sin \theta\, d\theta}{2\epsilon_0}. \qquad (12.25)$$

Integrating this from $\theta = 0$ to $\theta = \pi/2$ quickly gives a total field of $E = \sigma/2\epsilon_0$. Note that this result is independent of $R$; see the final paragraph of the solution to Problem 1.3.

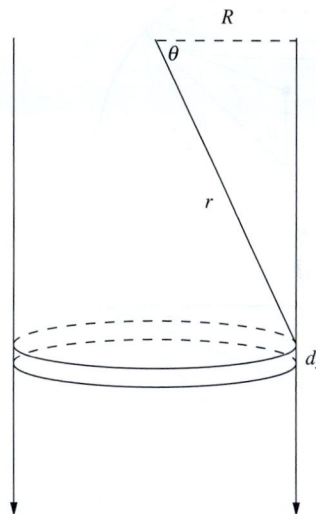

**Figure 12.8.**

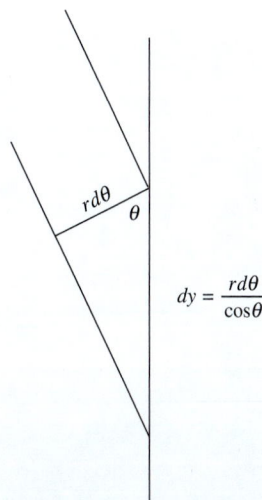

**Figure 12.9.**

An interesting corollary of this $E = \sigma/2\epsilon_0$ result is that if we cap the end of the cylinder with a flat disk with radius $R$ and the same charge density $\sigma$, then the field inside the cylinder, just below the center of the disk, is zero. This follows from the fact that the disk looks essentially like an infinite plane from up close, and the field due to an infinite plane is $\sigma/2\epsilon_0$ on either side.

(b) If we slice up the solid cylinder into concentric cylindrical shells with thickness $dR$, then the effective surface charge density of each shell is $\rho\, dR$. The above result then tells us that the field from each shell is $(\rho\, dR)/2\epsilon_0$. Integrating over $R$ simply turns the $dR$ into an $R$, so the total field is $\rho R/2\epsilon_0$.

1.12 *Field from a hemispherical shell*

Consider a ring defined by the angle $\theta$ down from the top of the hemisphere, subtending an angle $d\theta$. Its area is $2\pi(R\sin\theta)(R\,d\theta)$, so its charge is $\sigma(2\pi R^2 \sin\theta\, d\theta)$. From the law of cosines, the length $r$ in Fig. 12.10 is $r = \sqrt{R^2 + z^2 - 2Rz\cos\theta}$. The $x$ and $y$ components of the fields from the various parts of the ring will cancel in pairs, so we need only worry about the $z$ component from each little piece. From Fig. 12.11 the field from each piece points downward at an angle $\phi$ below the horizontal. So the $z$ component involves a factor of $-\sin\phi$, where $\sin\phi = (R\cos\theta - z)/r$. Adding up the $z$ components from all the parts of the ring gives the field from the ring as

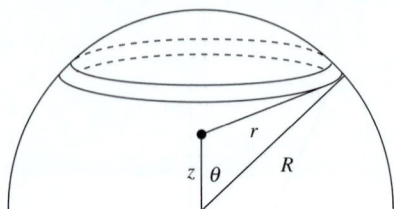

**Figure 12.10.**

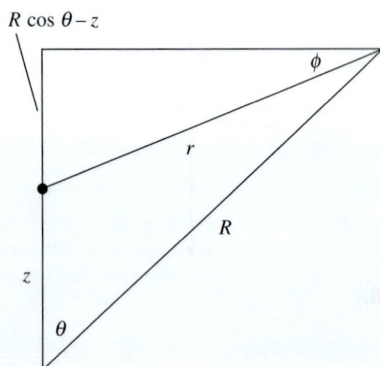

**Figure 12.11.**

$$dE_z = -\frac{\sigma(2\pi R^2 \sin\theta\, d\theta)}{4\pi\epsilon_0 r^2} \cdot \frac{(R\cos\theta - z)}{r} \tag{12.26}$$

The angle $\theta$ runs from 0 to $\pi/2$, so integrating over all the rings and using $r = \sqrt{R^2 + z^2 - 2Rz\cos\theta}$ gives a total field of

$$E_z(z) = -\frac{\sigma R^2}{2\epsilon_0}\int_0^{\pi/2} \frac{\sin\theta(R\cos\theta - z)d\theta}{(R^2 + z^2 - 2Rz\cos\theta)^{3/2}}. \tag{12.27}$$

This integral happens to be doable in closed form. From *Mathematica* or Appendix K we obtain

$$E_z(z) = \frac{\sigma R^2}{2\epsilon_0} \cdot \frac{R - z\cos\theta}{z^2\sqrt{R^2 + z^2 - 2Rz\cos\theta}}\Bigg|_0^{\pi/2}$$

$$= \frac{\sigma R^2}{2\epsilon_0 z^2}\left(\frac{R}{\sqrt{R^2 + z^2}} - \frac{R - z}{\sqrt{(R-z)^2}}\right). \tag{12.28}$$

There are two possibilities for the value of the second term, depending on the sign of $R - z$. We find

$$E_z(z) = \frac{\sigma R^2}{2\epsilon_0 z^2}\left(\frac{1}{\sqrt{1 + z^2/R^2}} - 1\right) \qquad (z < R),$$

$$E_z(z) = \frac{\sigma R^2}{2\epsilon_0 z^2}\left(\frac{1}{\sqrt{1 + z^2/R^2}} + 1\right) \qquad (z > R). \tag{12.29}$$

The first of these results is always negative, so the field always points downward if $z < R$ (assuming $\sigma$ is positive). Intuitively, it isn't so obvious which way the field points in the case where $z$ is slightly smaller than $R$. On the other hand, if $z > R$ then the field obviously points upward. $E_z$ is discontinuous at $z = R$, with a jump of $\sigma/\epsilon_0$. This is consistent with an application of Gauss's law with a pillbox spanning the surface.

In terms of the total charge $Q = 2\pi R^2 \sigma$ on the hemisphere, the factor out front in Eq. (12.29) equals $Q/4\pi\epsilon_0 z^2$. So the above fields correctly approach $\pm Q/4\pi\epsilon_0 z^2$ in the $z \to \pm\infty$ limits (the hemisphere looks like a point charge from far away).

For $z \to 0$, you can use $1/\sqrt{1+\epsilon} \approx 1 - \epsilon/2$ to Taylor-expand the $1/\sqrt{1 + z^2/R^2}$ term in the first result in Eq. (12.29). This yields a field of $-\sigma/4\epsilon_0 = -Q/8\pi\epsilon_0 R^2$ at the center of the hemisphere, which agrees with the result from Exercise 1.50, if you have solved that. This field of $\sigma/4\epsilon_0$ is half the size of the $\sigma/2\epsilon_0$ field from an infinite sheet. You should convince yourself why it must be smaller (consider the amount of charge subtended by a given solid angle), although the factor of $1/2$ isn't obvious.

Note that for $-R < z < R$ the field is an even function of $z$. There is a quick way of seeing this. *Hint:* Superpose a complete spherical shell with charge density $-\sigma$ on top of the given hemispherical shell, and use the fact that a complete spherical shell has no field inside.

1.13   *A very uniform field*

(a) At height $z$ above the lower ring, the field due to a charge $dQ$ on the lower ring has magnitude $dQ/4\pi\epsilon_0(r^2 + z^2)$. But only the $z$ component survives, by symmetry, and this brings in a factor of $z/\sqrt{r^2 + z^2}$. Integrating over the entire ring simply turns the $dQ$ into $Q$, so the upward vertical field due to the bottom ring is $E_z = Qz/4\pi\epsilon_0(r^2 + z^2)^{3/2}$.

The same type of reasoning holds with the top (negative) ring, with $z$ replaced by $h - z$. And this field also points upward (at locations below it). So the total field at height $z$ is

$$E_z = \frac{Q}{4\pi\epsilon_0} \left( \frac{z}{(r^2 + z^2)^{3/2}} + \frac{h - z}{(r^2 + (h-z)^2)^{3/2}} \right). \qquad (12.30)$$

(b) $E_z$ doesn't change if $z$ is replaced by $h - z$. Equivalently, if we define $z' \equiv z - h/2$ to be the coordinate relative to the midpoint, then the field doesn't change when we replace $z'$ with $(h-z) - h/2 = h/2 - z = -z'$. This means that the field is an even function of $z'$. So it is symmetric with respect to $z = h/2$, as desired. The point $z = h/2$ is therefore a local extremum. As an exercise, you can also demonstrate this fact by imagining flipping the whole setup upside down and then negating the charge on each ring (which brings you back to the original setup).

Via *Mathematica*, the second derivative of $E_z$ is

$$\frac{d^2 E_z}{dz^2} = \frac{15z^3}{(r^2 + z^2)^{7/2}} - \frac{9z}{(r^2 + z^2)^{5/2}}$$

$$+ \frac{15(h-z)^3}{(r^2 + (h-z)^2)^{7/2}} - \frac{9(h-z)}{(r^2 + (h-z)^2)^{5/2}}. \qquad (12.31)$$

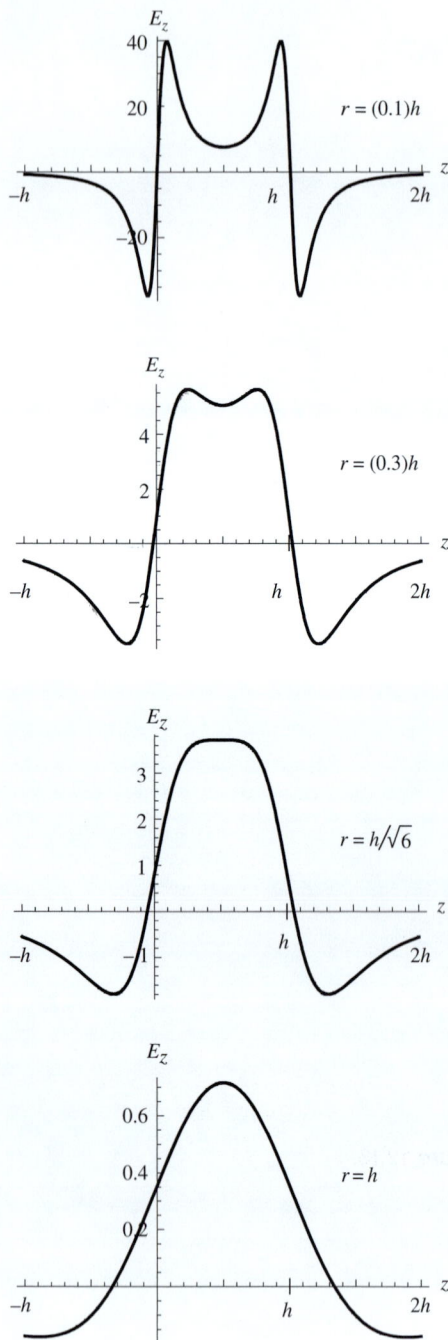

**Figure 12.12.**

As stated in the problem, we want this to be zero at $z = h/2$. Things simplify fairly quickly when setting it equal to zero with $z = h/2$. You can show that the result is $r = h/\sqrt{6} \approx (0.41)h$. So the diameter should be chosen to be about 0.82 times the separation between the rings. Plots of $E_z$ are shown in Fig. 12.12 in units of $Q/4\pi\epsilon_0 h^2$ for values of $r$ that are much smaller, smaller, equal to, and larger than $h/\sqrt{6}$. (Note the different scales on the vertical axes.) The $r = h/\sqrt{6}$ value marks the transition between $z = h/2$ being a local minimum or a local maximum of the field. For very small values of $r$, the field is very large near the rings, because the rings look essentially like point charges at distances larger than a few multiples of $r$.

1.14 *Hole in a plane*

(a) By symmetry, only the component of the electric field perpendicular to the plane survives. A small piece of charge $dq$ at radius $r$ in the plane produces a field with magnitude $dq/4\pi\epsilon_0(r^2 + z^2)$ at the given point. To obtain the component perpendicular to the plane, we must multiply this by $z/\sqrt{r^2 + z^2}$. Slicing up the plane into rings with charge $dq = (2\pi r\,dr)\sigma$, we find that the total field from the plane (minus the hole) is

$$E(z) = \int_R^\infty \frac{2\pi\sigma zr\,dr}{4\pi\epsilon_0(r^2 + z^2)^{3/2}} = -\frac{2\pi\sigma z}{4\pi\epsilon_0\sqrt{r^2 + z^2}}\Big|_{r=R}^{r=\infty}$$

$$= \frac{\sigma z}{2\epsilon_0\sqrt{R^2 + z^2}}. \tag{12.32}$$

Note that if $R = 0$ (so that we have a uniform plane without a hole), then $E = \sigma/2\epsilon_0$, which is the familiar field due to an infinite plane.

(b) If $z \ll R$, then Eq. (12.32) gives $E(z) \approx \sigma z/2\epsilon_0 R$. So $F = ma$ for the charge $-q$ yields

$$(-q)E = m\ddot{z} \implies \ddot{z} + \left(\frac{q\sigma}{2\epsilon_0 mR}\right)z = 0. \tag{12.33}$$

This equation represents simple harmonic motion. The frequency of small oscillations is the square root of the coefficient of the $z$ term:

$$\omega = \sqrt{\frac{q\sigma}{2\epsilon_0 mR}}. \tag{12.34}$$

For the parameters given in the problem, this frequency equals

$$\omega = \sqrt{\frac{(10^{-8}\,\text{C})(10^{-6}\,\text{C/m}^2)}{2\left(8.85 \cdot 10^{-12}\,\frac{\text{s}^2\,\text{C}^2}{\text{kg m}^3}\right)(10^{-3}\,\text{kg})(0.1\,\text{m})}} = 2.4\,\text{s}^{-1}, \tag{12.35}$$

which is about 0.4 Hz. The charged particle would have to be constrained to lie on the line $L$ because the equilibrium is unstable in the transverse directions.

(c) Integrating the magnitude of the force, $qE$, to obtain the difference in potential energy between the center of the hole and position $z$ (technically, $U = -\int F\,dz$ and $F = (-q)E$ if you want to worry about the signs) gives

$$U(z) = \int_0^z qE(z')dz' = \int_0^z \frac{q\sigma z'\,dz'}{2\epsilon_0\sqrt{R^2 + z'^2}}$$

$$= \frac{q\sigma}{2\epsilon_0}\sqrt{R^2 + z'^2}\,\Big|_0^z = \frac{q\sigma}{2\epsilon_0}\left(\sqrt{R^2 + z^2} - R\right). \qquad (12.36)$$

By conservation of energy, the speed at the center of the hole is given by $mv^2/2 = U(z)$. Therefore,

$$v = \sqrt{\frac{q\sigma}{m\epsilon_0}\left(\sqrt{R^2 + z^2} - R\right)}. \qquad (12.37)$$

For large $z$ this reduces to $v = \sqrt{q\sigma z/m\epsilon_0}$. We can also obtain this last result by noting that, for large $z$, the magnitude of the force arising from the field in Eq. (12.32) reduces to $F = q\sigma/2\epsilon_0$. This is constant, so the acceleration has the constant magnitude of $a = q\sigma/2m\epsilon_0$. The standard 1D kinematic result of $v = \sqrt{2az}$ then gives $v = \sqrt{q\sigma z/m\epsilon_0}$, as above.

1.15 *Flux through a circle*

(a) The claim stated in the problem (that the flux is the same through any surface that is bounded by the circle and that stays to the right of the origin) is true because otherwise there would be nonzero net flux into or out of the closed surface formed by the union of two such surfaces. This would violate Gauss's law, because the closed surface contains no charge.

For the case of the flat disk, the magnitude of the field at the angle $\beta$ shown in Fig. 12.13 is $q/4\pi\epsilon_0 r^2 = q/4\pi\epsilon_0(\ell/\cos\beta)^2$. Only the horizontal component is relevant for the flux, and this brings in a factor of $\cos\beta$. The radius of a constant-$\beta$ ring on the disk is $\ell\tan\beta$, so if the ring subtends an angle $d\beta$, its area is $da = 2\pi(\ell\tan\beta)\,d(\ell\tan\beta) = 2\pi\ell\tan\beta(\ell\,d\beta/\cos^2\beta)$. The total flux through the disk is therefore

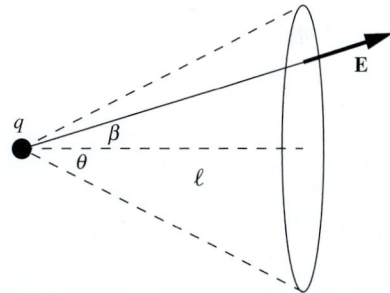

**Figure 12.13.**

$$\int E_x\,da = \int E\cos\beta\,da = \int_0^\theta \frac{q\cos^2\beta}{4\pi\epsilon_0\ell^2}\cos\beta \cdot \frac{2\pi\ell^2\tan\beta\,d\beta}{\cos^2\beta}$$

$$= \frac{q}{2\epsilon_0}\int_0^\theta \sin\beta\,d\beta = \frac{q}{2\epsilon_0}(1 - \cos\theta). \qquad (12.38)$$

For $\theta \to 0$ this correctly equals zero, and for $\theta \to \pi/2$ it equals $q/2\epsilon_0$, which is correctly half of the total flux of $q/\epsilon_0$ due to $q$. Note that the flux is independent of $\ell$; this is the familiar consequence of the $1/r^2$ nature of Coulomb's law. Only the angle $\theta$ matters.

(b) The field everywhere on the spherical cap is $q/4\pi\epsilon_0 R^2$, where $R = \ell/\cos\theta$ is the radius of the sphere. The field is normal to the sphere, so we don't have to worry about taking a component. The radius of a constant-$\beta$ ring on the cap is $R \sin \beta$, so the area of the ring is $2\pi(R \sin \beta)(R \, d\beta)$. The total flux through the spherical cap is therefore

$$\int E \, da = \int_0^\theta \frac{q}{4\pi\epsilon_0 R^2} \cdot 2\pi R^2 \sin \beta \, d\beta$$

$$= \frac{q}{2\epsilon_0} \int_0^\theta \sin \beta \, d\beta = \frac{q}{2\epsilon_0}(1 - \cos \theta), \qquad (12.39)$$

in agreement with the result in part (a). The same limiting cases work out, and we can now also consider the $\theta \to \pi$ limit (without the surface crossing the charge, as would happen with the flat disk). In this case we have a very small circle on the left side of the origin. Our sphere is nearly a complete sphere, except for a small hole where the circle is. Equation (12.39) gives a flux of $q/\epsilon_0$ for $\theta \to \pi$, which is correctly the total flux due to the charge $q$.

1.16 *Gauss's law and two point charges*

(a) The field at position $x$ on the $x$ axis is (dropping terms of order $x^2$)

$$E_x(x) = \frac{q}{4\pi\epsilon_0(\ell + x)^2} - \frac{q}{4\pi\epsilon_0(\ell - x)^2}$$

$$\approx \frac{q}{4\pi\epsilon_0\ell^2}\left(\frac{1}{1 + 2x/\ell} - \frac{1}{1 - 2x/\ell}\right)$$

$$\approx \frac{q}{4\pi\epsilon_0\ell^2}\left((1 - 2x/\ell) - (1 + 2x/\ell)\right)$$

$$= -\frac{qx}{\pi\epsilon_0\ell^3}. \qquad (12.40)$$

To find the field at position $y$ on the $y$ axis, we must take the vertical component of the fields from the two charges, which brings in a factor of $y/\sqrt{\ell^2 + y^2}$. The field is therefore (dropping terms of order $y^2$)

$$E_y(y) = 2 \cdot \frac{q}{4\pi\epsilon_0(\ell^2 + y^2)} \cdot \frac{y}{\sqrt{\ell^2 + y^2}} \approx \frac{qy}{2\pi\epsilon_0\ell^3}. \qquad (12.41)$$

The $y$ axis can be chosen to be any axis perpendicular to the line of the charges, so this result holds for any point on the perpendicular-bisector plane of the charges.

(b) Although we found the field components in part (a) only for points on the $x$ axis or on the perpendicular-bisector plane, the results are actually valid for all points in space near the origin. That is, $E_x(x, y) \approx -qx/\pi\epsilon_0\ell^3$, independent of $y$. And $E_y(x, y) \approx qy/2\pi\epsilon_0\ell^3$, independent of $x$. You can check these facts by writing out the exact expressions for the fields. For example, in Eq. (12.41) the $\ell$ values for the two charges become $\ell \pm x$, and this doesn't change the result, to leading order. Alternatively, note that, due to symmetry, $E_x(x, y)$ is an even function of $y$. This means that $E_x(x, y)$ has no linear dependence on $y$. The variation with $y$ therefore starts only at order $y^2$, which is negligible for small $y$. So $E_x$ is essentially independent of $y$ near the $x$ axis. Similar reasoning works with $E_y$ as a function of $x$.

For convenience, define $C \equiv q/2\pi\epsilon_0\ell^3$. Then the longitudinal and transverse field components have magnitudes $2Cx$ and $Cy$, respectively. The two circular faces of the small cylinder have a combined area of $a_{\text{circ}} = 2\pi r_0^2$. And the cylindrical boundary has an area of $a_{\text{cyl}} = (2\pi r_0)(2x_0) = 4\pi r_0 x_0$. There is inward flux through the circles; this flux comes from $E_x$ only, which has magnitude $2Cx_0$. And there is outward flux through the cylindrical part; this flux comes from $E_y$ only, which has magnitude $Cr_0$. The net outward flux is therefore

$$-(2\pi r_0^2)(2Cx_0) + (4\pi r_0 x_0)(Cr_0) = 0, \qquad (12.42)$$

as desired.

1.17 *Zero field inside a spherical shell*

Let $a$ be the distance from point $P$ to patch $A$, and let $b$ be the distance from $P$ to patch $B$; see Fig. 12.14. (Since the cones are assumed to be thin, it doesn't matter exactly which points in the patches we use to define these distances.) Draw the "perpendicular" bases of the cones, and call them $A'$ and $B'$, as shown. The ratio of the areas of $A'$ and $B'$ is $a^2/b^2$, because areas are proportional to lengths squared. The key point is that the angle between the planes of $A$ and $A'$ is the same as the angle between the planes of $B$ and $B'$. This is true because the chord between $A$ and $B$ (that is, the line perpendicular to $A'$ and $B'$) meets the circle at equal angles at its ends. The ratio of the areas of $A$ and $B$ is therefore also equal to $a^2/b^2$. So the charge on patch $A$ is $a^2/b^2$ times the charge on patch $B$.

The magnitudes of the fields due to the two patches take the general form of $q/4\pi\epsilon_0 r^2$. We just found that the $q$ for $A$ is $a^2/b^2$ times the $q$ for $B$. But we also know that the $r^2$ for $A$ is $a^2/b^2$ times the $r^2$ for $B$. So the values of $q/4\pi\epsilon_0 r^2$ for the two patches are equal. The fields at $P$ due to $A$ and $B$ (which can be treated essentially like point charges, because the cones are assumed to be thin) are therefore equal in magnitude (and opposite in direction, of course). If we draw enough cones to cover the whole shell, the contributions to the field from little patches over the whole shell cancel in pairs, so we are left with zero field at $P$. This holds for any point $P$ inside the shell.

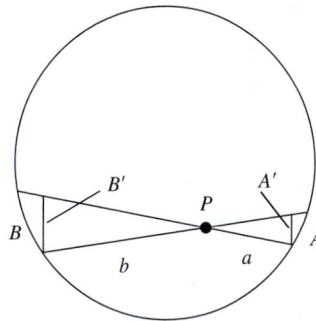

**Figure 12.14.**

1.18 *Fields at the surfaces*

(a) From Gauss's law, the field due to the sphere is the same as if all of the charge were concentrated at the center, in a point charge $q = (4\pi R^3/3)\rho$. Therefore,

$$E = \frac{q}{4\pi\epsilon_0 R^2} = \frac{(4\pi R^3/3)\rho}{4\pi\epsilon_0 R^2} = \frac{R\rho}{3\epsilon_0}. \qquad (12.43)$$

(b) Again from Gauss's law, the field due to the cylinder is the same as if all of the charge were concentrated on the axis with linear charge

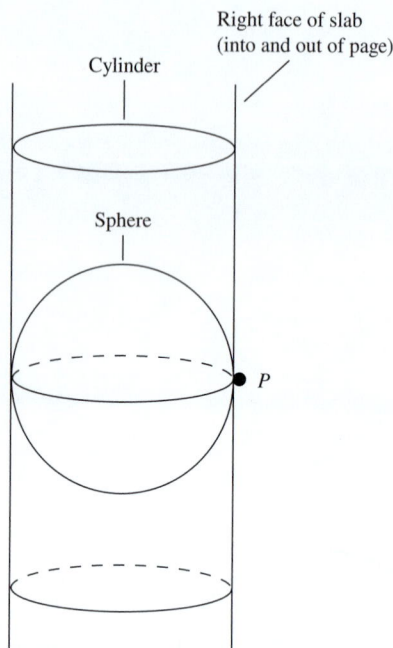

Cylinder

Right face of slab
(into and out of page)

Sphere

• P

**Figure 12.15.**

density $\lambda = \pi R^2 \rho$. This $\lambda$ follows from the fact that the amount of charge in a length $L$ of the cylinder can be written as both $\lambda L$ (by definition) and $\pi R^2 L \rho$ (because $\pi R^2 L$ is the relevant volume). Therefore,

$$E = \frac{\lambda}{2\pi \epsilon_0 R} = \frac{\pi R^2 \rho}{2\pi \epsilon_0 R} = \frac{R\rho}{2\epsilon_0}. \qquad (12.44)$$

(c) Again from Gauss's law, the field due to the slab is the same as if all of the charge were concentrated on a sheet with surface charge density $\sigma = 2R\rho$. As above, this $\sigma$ follows from the fact that the amount of charge in an area $A$ of the slab can be written as both $\sigma A$ (by definition) and $2RA\rho$ (because $2R \cdot A$ is the relevant volume). Therefore,

$$E = \frac{\sigma}{2\epsilon_0} = \frac{2R\rho}{2\epsilon_0} = \frac{R\rho}{\epsilon_0}. \qquad (12.45)$$

The fields of the sphere, cylinder, and slab are therefore in the ratio of $1/3$ to $1/2$ to $1$. (These numbers can be traced to the dimensionality of the sphere's volume, the cylinder's cross-sectional area, and the slab's thickness.) The size order makes sense, because in Fig. 12.15 the slab completely contains the cylinder, which in turn completely contains the sphere. So at point $P$, the field from the slab must be greater than the field from the cylinder, which in turn must be greater than the field from the sphere (because in each case the extra charge creates a nonzero field pointing to the right).

1.19 *Sheet on a sphere*

The electric field due to the sheet is $\sigma/2\epsilon_0$, where $\sigma = \rho x$ is the effective charge per area. This follows from exactly the same Gauss's law argument as in the case of a thin sheet. The field at point $B$ on top of the sheet is the sum of this $\sigma/2\epsilon_0$ field plus the Coulomb field due to the sphere, but at radius $R + x$. The field at point $A$ underneath the sheet is the difference (because now the sheet's field points downward) between the $\sigma/2\epsilon_0$ field and the Coulomb field due to the sphere, at radius $R$. The charge in the sphere is $(4/3)\pi R^3 \rho_0$, so the field is larger above the sheet if

$$\frac{(4/3)\pi R^3 \rho_0}{4\pi \epsilon_0 (R+x)^2} + \frac{\rho x}{2\epsilon_0} > \frac{(4/3)\pi R^3 \rho_0}{4\pi \epsilon_0 R^2} - \frac{\rho x}{2\epsilon_0}$$

$$\Longleftrightarrow \quad \rho x > \frac{R\rho_0}{3}\left(1 - \frac{1}{(1 + x/R)^2}\right)$$

$$\Longleftrightarrow \quad \rho x > \frac{R\rho_0}{3}\left(\frac{2x}{R}\right)$$

$$\Longleftrightarrow \quad \rho > \frac{2}{3}\rho_0, \qquad (12.46)$$

where we have used $1/(1 + \epsilon)^2 \approx 1/(1 + 2\epsilon) \approx 1 - 2\epsilon$ to go from the second to the third line.

This problem is basically the same problem as the "mine shaft" problem: if you descend in a mine shaft, does the gravitational field increase or decrease? The answer is that it decreases if $\rho_{\text{crust}} > (2/3)\rho_{\text{avg}}$, where $\rho_{\text{crust}}$ is the mass density of the earth's crust (assumed to be roughly constant) and $\rho_{\text{avg}}$ is the average mass density of the entire earth (of which the crust is a negligible part). The two problems are equivalent because the gravitational and electrical fields both fall off as $1/r^2$, and because the nearby crust behaves essentially like a large flat plane.

We know that there must exist a cutoff value of $\rho_{\text{crust}}$ for which the field does not depend on the depth, for the following reason. If $\rho_{\text{crust}}$ is very small (imagine the limit where the crust is essentially massless; equivalently, pretend that the boundary of the earth is a few miles up in the air), then descending in a mine shaft will decrease the $r$ in the gravitational force $F = GmM/r^2$, while barely decreasing the $M$ (this $M$ is the mass contained inside radius $r$); so $F$ will increase. On the other hand, if $\rho_{\text{crust}}$ is very large (imagine the limit of a thin spherical shell), then descending will barely decrease the $r$ in $GmM/r^2$, while significantly decreasing the $M$; so $F$ will decrease. By continuity, there must be a value of $\rho_{\text{crust}}$ for which $F$ does not change as you descend. However, the above factor of $2/3$ is by no means obvious.

1.20   *Thundercloud*

(a) Assuming that the cloud is large enough to be treated roughly like an infinite plane, an opposite charge will be induced on the ground, so the field is $E = \sigma/\epsilon_0$, where $\sigma$ is the charge per area in the cloud. Therefore,

$$\sigma = \epsilon_0 E = \left(8.85 \cdot 10^{-12} \, \frac{\text{s}^2 \, \text{C}^2}{\text{kg} \, \text{m}^3}\right)\left(3000 \, \frac{\text{V}}{\text{m}}\right) \approx 2.7 \cdot 10^{-8} \, \frac{\text{C}}{\text{m}^2}.$$

(12.47)

You can check that the units work out by using $1 \, \text{V/m} = 1 \, \text{N/C}$.

(b) Let $h$ be the rainfall depth, which is $h = 2.5 \cdot 10^{-3}$ m here. If $r$ is the drop radius, then the number of drops that land on a patch of area $A$ on the ground is $N = Ah/(4\pi r^3/3)$. The total charge initially in the cloud above this patch is $\sigma A$, so the charge $q$ on each drop is $q = (\sigma A)/N = \sigma A/(3Ah/4\pi r^3) = 4\pi r^3 \sigma/3h$. This charge causes a radial field of strength $q/4\pi \epsilon_0 r^2$ at the surface of each drop, which equals

$$\frac{q}{4\pi \epsilon_0 r^2} = \frac{4\pi r^3 \sigma/3h}{4\pi \epsilon_0 r^2} = \frac{\sigma}{\epsilon_0} \frac{r}{3h} = E \frac{r}{3h}$$

$$= \left(3000 \, \frac{\text{V}}{\text{m}}\right) \frac{5 \cdot 10^{-4} \, \text{m}}{3(2.5 \cdot 10^{-3} \, \text{m})} = 200 \, \frac{\text{V}}{\text{m}}.$$  (12.48)

This is *only* the field caused by the net charge on the drop. Even an uncharged drop has, in the field $E$, charges on its surface, of opposite sign on top and bottom. The associated field is described in Chapter 10; see Fig. 10.27.

Note that the field at the surface of a drop is proportional to $r^3/r^2 = r$, from Eq. (12.48); this is the standard result for a sphere with uniform volume density. So the larger the drop, the larger the field. Theoretically, if the field just outside a drop is large enough, electrons can be ripped away from air molecules. This causes the air to be conducting, and a spark could jump from a drop when it gets near the ground. How large a raindrop would be required for this "arcing" to happen? The electrical breakdown of air is about $3 \cdot 10^6$ V/m. This is $1.5 \cdot 10^4$ times the above field of 200 V/m, so we need the radius to be $1.5 \cdot 10^4$ times the above $5 \cdot 10^{-4}$ m radius, which gives 7.5 m. Needless to say, if raindrops were this large we would have more to worry about than sparks jumping from them!

1.21   *Field in the end face*

Gauss's law, combined with cylindrical symmetry, tells us that the field inside a full infinite (in both directions) cylindrical shell is zero. A full infinite cylinder is the superposition of two half-infinite cylinders placed end-to-end. Assume (in search of a contradiction) that the field in the end face of a half-infinite cylinder has a nonzero radial component. Then when we put two half-infinite cylinders together to make a full infinite cylinder, the radial components of the fields from the two half-infinite cylinders will add, yielding a nonzero field inside the resulting full infinite cylinder. This contradicts the fact that the field inside a full infinite cylindrical shell is zero. We therefore conclude that the field in the end face of a half-infinite cylinder must have zero radial component, as desired.

Figure 12.16 shows a rough sketch of a few field lines due to the half-infinite cylinder. In the inside of the (half imaginary) full infinite cylinder, the field lines must be symmetric with respect to the end face, because otherwise they wouldn't cancel with the field from the other half-infinite cylinder when it is placed end-to-end.

1.22   *Field from a spherical shell, right and wrong*

(a) Let the rings be parameterized by the angle $\theta$ down from the top of the sphere, as shown in Fig. 12.17. The width of a ring is $R\,d\theta$, and its circumference is $2\pi(R\sin\theta)$. So its area is $2\pi R^2 \sin\theta\,d\theta$. All points on the ring are a distance $2R\sin(\theta/2)$ from the given point $P$, which is infinitesimally close to the top of the shell. Only the vertical component of the field survives, and this brings in a factor of $\sin(\theta/2)$, as you should check. The total field at the top of the shell is therefore apparently equal to (writing $\sin\theta$ as $2\sin(\theta/2)\cos(\theta/2)$)

$$\frac{1}{4\pi\epsilon_0} \int_0^\pi \frac{\sigma 2\pi R^2 \sin\theta\,d\theta}{\left(2R\sin(\theta/2)\right)^2} \sin(\theta/2) = \frac{\sigma}{4\epsilon_0} \int_0^\pi \cos(\theta/2)\,d\theta$$

$$= \frac{\sigma}{2\epsilon_0} \sin(\theta/2)\bigg|_0^\pi = \frac{\sigma}{2\epsilon_0}.$$

$$(12.49)$$

(b) As noted in the statement of the problem, it is no surprise that the above result is incorrect, because the same calculation would

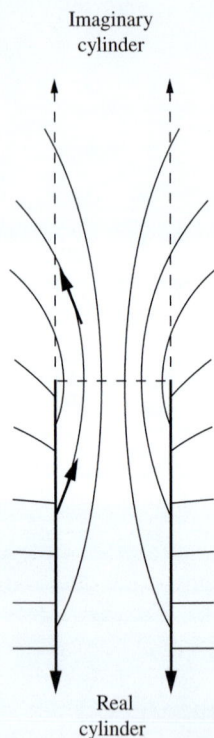

Imaginary cylinder

Real cylinder

**Figure 12.16.**

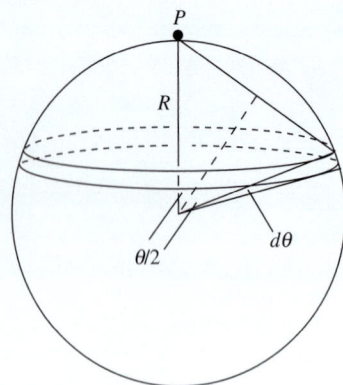

$P$

$R$

$\theta/2$

$d\theta$

**Figure 12.17.**

supposedly yield the field just inside the shell too, where we know it equals zero instead of $\sigma/\epsilon_0$. The calculation does, however, give the next best thing, namely the average of these two values. We'll see why shortly.

The reason why the calculation is invalid is that it doesn't correctly describe the field arising from points on the shell very close to the point $P$, that is, for rings characterized by $\theta \approx 0$. It is incorrect for two reasons. The closeup view in Fig. 12.18 shows that the distance from a ring to the given point $P$ is *not* equal to $2R\sin(\theta/2)$. Additionally, it shows that the field does *not* point along the line from the particular point on the ring to the top of the shell. It points more vertically, toward $P$, so the extra factor of $\sin(\theta/2)$ in Eq. (12.49) is not correct. No matter how close $P$ is to the shell, we can always zoom in close enough so that the picture looks like the one in Fig. 12.18. The only difference is that the more we need to zoom in, the straighter the arc of the circle is. In the limit where $P$ is very close to the shell, the arc is essentially a straight line (we drew it curved for the sake of the illustration).

What *is* true is that if we remove a tiny circular patch from the top of the shell (whose radius is much larger than the distance from $P$ to the shell, but much smaller than the radius of the shell), then the integral in part (a) is valid for the remaining part of the shell. From the form of the integrand in Eq. (12.49), we see that the tiny patch contributes negligibly to the integral. So we can say that the field due to the remaining part of the shell is essentially equal to the above result of $\sigma/2\epsilon_0$. By superposition, the total field due to the entire shell equals this field of $\sigma/2\epsilon_0$ plus the field due to the tiny circular patch. But if the point in question is infinitesimally close to the shell, then this tiny patch looks like an infinite plane, the field of which we know is $\sigma/2\epsilon_0$. The desired total field is therefore

$$E_{\text{outside}} = E_{\text{shell minus patch}} + E_{\text{patch}} = \frac{\sigma}{2\epsilon_0} + \frac{\sigma}{2\epsilon_0} = \frac{\sigma}{\epsilon_0}. \quad (12.50)$$

By superposition we also obtain the correct field just inside the shell:

$$E_{\text{inside}} = E_{\text{shell minus patch}} - E_{\text{patch}} = \frac{\sigma}{2\epsilon_0} - \frac{\sigma}{2\epsilon_0} = 0. \quad (12.51)$$

The relative minus sign arises because the field from the shell-minus-patch is continuous across the hole, but the field from the patch is not; it points in different directions on either side of the patch.

1.23 *Field near a stick*

A piece of the stick with length $dr$ at a distance $r$ from the given point $P$ produces a contribution to $E_{\parallel}$ equal to $(\lambda\,dr)/4\pi\epsilon_0 r^2$. When this is integrated over the parts of the stick on either side of $P$, both integrals diverge. However, the divergences cancel because the contribution from the short piece with length $(1-\eta)\ell$ in Fig. 12.19 cancels the contribution from the closest $(1-\eta)\ell$ part of the long piece with length $(1+\eta)\ell$. So in

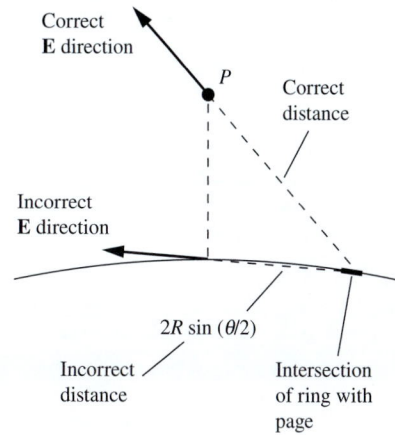

**Figure 12.18.**

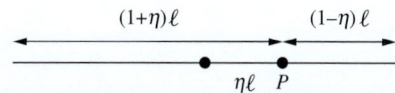

**Figure 12.19.**

the end we just need to integrate over the long piece from $r = (1 - \eta)\ell$ to $r = (1 + \eta)\ell$.[1] The result is

$$E_\parallel = \int_{(1-\eta)\ell}^{(1+\eta)\ell} \frac{\lambda \, dr}{4\pi\epsilon_0 r^2} = \frac{\lambda}{4\pi\epsilon_0 \ell} \left( \frac{1}{1-\eta} - \frac{1}{1+\eta} \right)$$

$$= \frac{\lambda}{4\pi\epsilon_0 \ell} \cdot \frac{2\eta}{1 - \eta^2}. \tag{12.52}$$

This equals zero when $\eta = 0$, as it should. It diverges as $\eta \to 1$ near the end of the stick. If we let $\eta \equiv 1 - \epsilon$ (so $\epsilon\ell$ is the distance from the end), then to leading order in $\epsilon$ we have $1 - \eta^2 = 1 - (1 - \epsilon)^2 \approx 2\epsilon$. So $E_\parallel \approx \lambda/4\pi\epsilon_0(\epsilon\ell)$. It makes sense that this diverges like $1/\epsilon\ell$, because, as we get close to the end, the short piece in Fig. 12.19 cancels only a small part of the long piece. So we need to integrate the $1/r^2$ field from the long piece almost down to $r = 0$ (more precisely, down to $\epsilon\ell$). And the integral of $1/r^2$ diverges like $1/r$ near $r = 0$.

1.24 *Potential energy of a cylinder*
Consider the setup at an intermediate stage when the cylinder has radius $r$. The charge per unit length is $\lambda_r = \rho\pi r^2$, and the electric field at radius $r'$ external to the cylinder is $E = \lambda_r/2\pi\epsilon_0 r'$. So the work done in bringing a charge $dq$ in from radius $R$ down to radius $r$ is (the minus sign here comes from the fact that the external agency opposes the field)

$$dW = -\int_R^r (dq) E \, dr' = -\int_R^r dq \frac{\lambda_r}{2\pi\epsilon_0 r'} \, dr' = \frac{\lambda_r \, dq}{2\pi\epsilon_0} \ln\left( \frac{R}{r} \right). \tag{12.53}$$

As we build up the cylinder, the charge increments $dq$ (the cylindrical shells) are equal to $(2\pi r \, dr)\ell\rho$, where $\ell$ is the length of the part of the cylinder we are considering. The total work done in building up the cylinder from $r = 0$ to $r = a$ is therefore (using the integral table in Appendix K)

$$W = \int dW = \int_0^a \frac{\lambda_r \, dq}{2\pi\epsilon_0} \ln\left( \frac{R}{r} \right) = \int_0^a \frac{(\rho\pi r^2)(2\pi r \ell\rho \, dr)}{2\pi\epsilon_0} \ln\left( \frac{R}{r} \right)$$

$$= \frac{\pi\rho^2\ell}{\epsilon_0} \int_0^a r^3 \ln\left( \frac{R}{r} \right) dr = \frac{\pi\rho^2\ell}{\epsilon_0} \left( \frac{r^4}{16} + \frac{r^4}{4} \ln\left( \frac{R}{r} \right) \right) \Big|_0^a$$

$$= \frac{\pi\rho^2\ell}{\epsilon_0} \left( \frac{a^4}{16} + \frac{a^4}{4} \ln\left( \frac{R}{a} \right) \right). \tag{12.54}$$

Finding the potential energy per unit length means simply erasing the $\ell$. If we define $\lambda \equiv \lambda_a$ as the charge per unit length in the finished cylinder,

---

[1] If the given point doesn't lie exactly on the stick, then there is actually no divergence. But it is still worth mentioning the canceling divergences, because $E_\parallel$ is well defined even if the point does lie exactly on the stick (whereas $E_\perp$ isn't, for an infinitely thin stick).

then $\rho = \lambda/\pi a^2$. Substituting this into Eq. (12.54) gives the energy per unit length (relative to the cylinder at radius $R$) as

$$\frac{\lambda^2}{4\pi\epsilon_0}\left(\frac{1}{4} + \ln\left(\frac{R}{a}\right)\right). \tag{12.55}$$

As mentioned in the statement of the problem, this diverges as $R \to \infty$. If $R = a$, so that all of the charge is initially on the surface of the given cylinder, we see that the energy per unit length needed to turn the initial surface charge density into a uniform volume density is $\lambda^2/16\pi\epsilon_0$. In the more general case where $R > a$, the log term in the result therefore represents the energy needed to bring the charge in from a shell of radius $R$ to a shell of radius $a$. As an exercise, you can verify this directly by gradually bringing in infinitesimally thin shells of charge.

1.25 *Two equal fields*

Let $L$ be the line from the top of the shell to the rings. The short thick segments in Fig. 12.20 represent cross-sectional slices of the rings on the shell and the sheet. The important point to note is that the surfaces of both rings lie at the *same* angle, $\alpha = 90° - \theta$, with respect to the line $L$ (because, in short, vertical is perpendicular to horizontal, and tangential is perpendicular to radial). If this angle were $90°$, then the width of each ring would simply be $\ell\,d\theta$ (we haven't indicated $d\theta$ in the figure, lest it get too cluttered), where $\ell$ is the distance from the top of the shell to the given ring (on either the shell or the sheet). But a general $\alpha$ causes this width to increase to $\ell\,d\theta/\sin\alpha = \ell\,d\theta/\cos\theta$. The radius of the ring is $\ell\sin\theta$, so the area of the ring is $(\ell\,d\theta/\cos\theta)(2\pi\ell\sin\theta)$. We care only about the vertical component of the fields from the various pieces of the ring, and this brings in a factor of $\cos\theta$. So the vertical component of the field at the top of the shell due to a given ring is

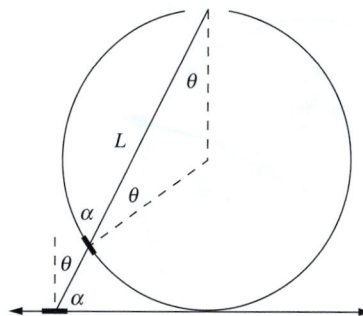

**Figure 12.20.**

$$dE = \frac{(\ell\,d\theta/\cos\theta)(2\pi\ell\sin\theta)\sigma}{4\pi\epsilon_0\ell^2}\cos\theta = \frac{\sigma\sin\theta\,d\theta}{2\epsilon_0}. \tag{12.56}$$

This is independent of $\ell$, so it is the same for the ring on the shell and the ring on the sheet. Integrating over $\theta$, from 0 to $\pi/2$, gives the same result of $\sigma/2\epsilon_0$ for the total field due to either the shell or the sheet.

To sum up, the fields from the rings on the shell and sheet that are associated with the same $\theta$ and $d\theta$ are equal for two reasons: (1) the fields are independent of $\ell$, because area is proportional to length squared, so the $\ell^2$ in the area cancels the $\ell^2$ in Coulomb's law; and (2) the fields have the same $\theta$ dependence, because both rings cut the line $L$ at the same angle.

1.26 *Stable equilibrium in electron jelly*

The only things we need to know about the equilibrium positions of the protons are (1) they are located along a diameter, because otherwise the force from the other proton wouldn't cancel the force from the jelly, which is directed toward the center, (2) they are on opposite sides of the center (and at equal radii, as you can quickly show), because otherwise the force on the proton closer to the center would point radially inward, and (3) they are inside the jelly sphere, because otherwise the negative force from the effective $-2e$ electron charge at the center of the sphere would be larger

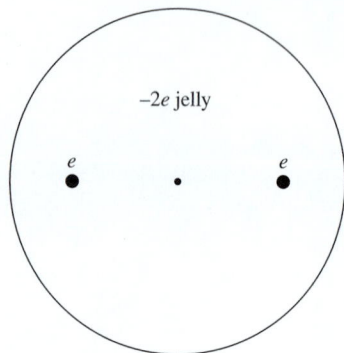

**Figure 12.21.**

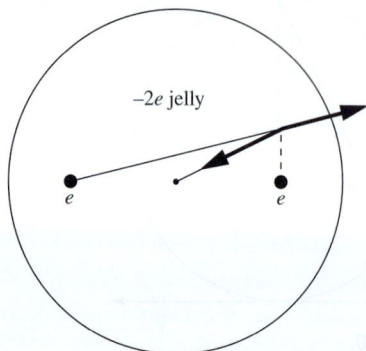

**Figure 12.22.**

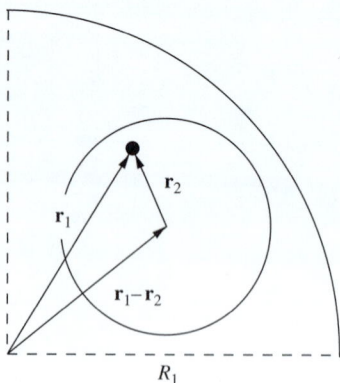

**Figure 12.23.**

than the force from the more distant $+e$ proton charge. So the setup looks something like the one shown in Fig. 12.21.

To show that the equilibrium is stable, we must consider both radial and transverse displacements. If we move one of the protons radially outward, then the repulsive force from the other proton decreases, while the attractive force from the jelly increases (because it grows like $r$, since the charge inside radius $r$ is proportional to $r^3$, and there is a $1/r^2$ in Coulomb's law). The proton is therefore pulled back toward the equilibrium position. Similar reasoning holds if the proton is moved radially inward. Hence the equilibrium is stable in the radial direction.

If we move one of the protons transversely, as shown in Fig. 12.22, then both forces (from the jelly and the other proton) don't change in magnitude, to first order in the small distance moved (because the Pythagorean theorem involves the square of the small distance). The magnitudes of the forces are therefore still equal, just as they were at the equilibrium position (by definition). But the slope of the jelly force is twice as large, so its negative $y$ component is twice as large as the proton force's positive $y$ component. The net force is therefore negative, so the proton is pulled back toward the equilibrium position. Hence the equilibrium is stable in the transverse direction also.

Note that if the sphere of electron jelly were replaced by a negative point charge at the center (which would have to be $-e/4$ if an equilibrium configuration is to exist), then the equilibrium would still be stable under transverse displacements, but not under radial ones, as you can check. This is consistent with the "no stable electrostatic equilibrium" theorem we will prove in Section 2.12. This theorem doesn't apply to the original jelly setup because that setup has a nonzero volume charge density, and the theorem holds only in empty space.

**1.27   Uniform field in a cavity**
At radius $r$ inside a sphere with density $\rho$, the electric field is due to the charge inside radius $r$. So the field is

$$E = \frac{(4\pi r^3/3)\rho}{4\pi\epsilon_0 r^2} = \frac{\rho r}{3\epsilon_0}. \tag{12.57}$$

The field points radially outward (for positive $\rho$), so we can write the **E** vector compactly as $\mathbf{E} = \rho\mathbf{r}_1/3\epsilon_0$, where $\mathbf{r}_1$ is measured relative to the center of the sphere.

The hollow cavity can be considered to be the result of superposing a sphere with radius $R_2$ and charge density $-\rho$ onto the larger sphere with radius $R_1$ and density $\rho$. If we look at the negative sphere by itself, then the same reasoning as above shows that the field in the interior (due to only this sphere) is $\mathbf{E} = -\rho\mathbf{r}_2/3\epsilon_0$, where $\mathbf{r}_2$ is measured relative to the center of this sphere.

Consider now an arbitrary point in the cavity, as shown in Fig. 12.23. By superposition, the field at this point is

$$E = \frac{\rho\mathbf{r}_1}{3\epsilon_0} - \frac{\rho\mathbf{r}_2}{3\epsilon_0} = \frac{\rho(\mathbf{r}_1 - \mathbf{r}_2)}{3\epsilon_0} = \frac{\rho\mathbf{a}}{3\epsilon_0}, \tag{12.58}$$

where $\mathbf{a} \equiv \mathbf{r}_1 - \mathbf{r}_2$ is the vector from the center of the larger sphere to the center of the cavity. This result is independent of the position inside the cavity, as desired. See Exercise 1.75 for a related problem.

The same reasoning also holds in lower-dimensional analogs. For example, we can carve out an infinite cylindrical cavity from an infinite uniform cylinder of charge (with their axes parallel). Using Gauss's law with a cylinder of length $\ell$, the field at radius $r$ inside a uniform cylinder has magnitude

$$E = \frac{\pi r^2 \ell \rho}{2\pi r \ell \epsilon_0} = \frac{\rho r}{2\epsilon_0}. \tag{12.59}$$

The field points radially outward, so we can again write $\mathbf{E}$ compactly in vector form as $\mathbf{E} = \rho \mathbf{r}/2\epsilon_0$. The only difference between this case and the above case with the sphere is a factor of 2 instead of 3. This doesn't affect the logic leading to the uniform field, so we again find that the field is uniform inside the cavity. The important property of the field is that it is proportional to the vector $\mathbf{r}$.

Dropping down one more dimension, we can carve out a slab cavity from a larger slab, both of which are infinite in the two transverse directions. (So we are simply left with two slabs separated by some space.) Consider a given point located a distance $r$ from the center plane of a uniform slab. Using Gauss's law with a volume of cross-sectional area $A$ that extends a distance $r$ on either side of the center plane of the uniform slab, the field at the given point has magnitude

$$E = \frac{(2r)A\rho}{2A\epsilon_0} = \frac{\rho r}{\epsilon_0}. \tag{12.60}$$

The field points outward from the central plane, so we can again write $\mathbf{E}$ compactly in vector form as $\mathbf{E} = \rho \mathbf{r}/\epsilon_0$. As above, we find that the field is proportional to the vector $\mathbf{r}$, so the same result of uniformity holds inside the slab cavity. Of course, in retrospect this result for the slab is easy to see, because we know that the field from an infinite sheet (or slab) is independent of the distance from the sheet. The field is therefore uniform inside the slab cavity, because the distance from the two remaining slabs on either side doesn't matter.

1.28 *Average field on/in a sphere*

(a) Let us put a total amount of charge $Q$ evenly distributed over the spherical shell of radius $R$. We know that the force that this shell exerts on the internal point charge $q$ is zero, because a spherical shell creates zero field in its interior. By Newton's third law, the point charge also exerts zero force on the spherical shell. But, by definition, this force also equals $Q$ times the average field over the shell. The average field is therefore zero.

Mathematically, the average field over the shell is given by the integral $(1/4\pi R^2) \int \mathbf{E}\, da$. The $da$ area element here is a scalar, and the result of the integral is a vector, which we showed equals zero. This integral should be contrasted with the flux integral $\int \mathbf{E} \cdot d\mathbf{a}$. Here $d\mathbf{a}$ is a vector, and the result of the integral is a scalar, which we know equals $q/\epsilon_0$ from Gauss's law. You should think carefully about what each of these integrals means physically.

(b) Again let us put a total amount of charge $Q$ evenly distributed over the spherical shell of radius $R$. We know that the force that this shell exerts on the external point charge $q$ equals $qQ/4\pi\epsilon_0 r^2$, because a spherical shell looks like a point charge from the outside. By Newton's third law, the point charge also exerts a force of $qQ/4\pi\epsilon_0 r^2$ on the spherical shell. But, by definition, this force also equals $Q$ times the average field over the shell. The average field therefore equals $q/4\pi\epsilon_0 r^2$. It points away from the charge $q$ (if $q$ is positive). As in part (a), you should contrast the integrals $\int \mathbf{E}\,da$ and $\int \mathbf{E}\cdot d\mathbf{a}$.

(c) From part (a), the average electric field over any shell that lies outside radius $r$ is zero, so we can ignore that region. From part (b), the average field over the volume of the sphere inside radius $r$ has magnitude $q/4\pi\epsilon_0 r^2$, because all shells have this common average. We are interested in the average field over the *entire* volume of the larger sphere of radius $R$. Since this volume is $R^3/r^3$ times as large as the volume of the smaller sphere of radius $r$, the average field over the larger sphere is smaller by a factor $r^3/R^3$. The average field over the entire sphere of radius $R$ therefore has magnitude $qr/4\pi\epsilon_0 R^3$, as desired. By symmetry, the field points along the radial line, and it is easy to see that it points inward, because more locations within the sphere have the field pointing generally in that direction. In vector form, the average field can be written as $-q\mathbf{r}/4\pi\epsilon_0 R^3$.

Note that this $qr/4\pi\epsilon_0 R^3$ field has the same magnitude as the field at radius $r$ inside a solid sphere of radius $R$ with charge $q$ uniformly distributed throughout its volume; see the example in Section 1.11.

### 1.29  *Pulling two sheets apart*

(a) We are assuming that $A$ is large and $\ell$ is small, so that we can treat the sheets as effectively infinitely large, which means that we can ignore the complicated nature of the field near the edges of the sheets. The electric field is therefore essentially zero outside the sheets and $\sigma/\epsilon_0$ between them (pointing from the positive sheet to the negative sheet). This follows from superposing the $\sigma/2\epsilon_0$ fields (with appropriate signs) from each sheet individually.

From Eq. (1.49) the attractive force per area on each sheet has magnitude $(1/2)(\sigma/\epsilon_0+0)\sigma = \sigma^2/2\epsilon_0$. (You can also obtain this by simply multiplying the density $\sigma$ of one sheet by the field $\sigma/2\epsilon_0$ from the other.) The force that you must apply to one of the sheets to drag it away from the other is therefore $F = \sigma^2 A/2\epsilon_0$, so the work you do over a distance $x$ is $W = \sigma^2 Ax/2\epsilon_0$.

(b) The field between the sheets takes on the value $\sigma/\epsilon_0$, independent of their separation (assuming that it is small compared with the linear size of the sheets). And the field is zero outside. In moving one of the sheets by a distance $x$, you create a region with volume $Ax$ where the field is now $\sigma/\epsilon_0$ instead of zero. Since the energy density is $\epsilon_0 E^2/2$, the increase in energy stored in the electric field is

$$\Delta U = \frac{\epsilon_0 E^2}{2}(Ax). \tag{12.61}$$

This increase in energy comes from the work you do, so this result must agree with the work calculated in part (a). And indeed, since $E = \sigma/\epsilon_0$ we have

$$\Delta U = \frac{\epsilon_0 (\sigma/\epsilon_0)^2}{2}(Ax) = \frac{\sigma^2 Ax}{2\epsilon_0}, \tag{12.62}$$

as desired.

1.30   *Force on a patch*

The net force on the patch is due to the field $\mathbf{E}^{\text{other}}$ from all the other charges in the system, because an object can't exert a net force on itself. This field $\mathbf{E}^{\text{other}}$ need not be perpendicular to the patch; it can point in any direction. But we know that the perpendicular component will be discontinuous across the surface. We assume that the patch is small enough so that the field is uniform over the location of the patch (on either side).

The force on the patch equals $\mathbf{E}^{\text{other}}$ times the charge on the patch, which is $\sigma A$, where $A$ is the area. So the total force is $\mathbf{F} = \mathbf{E}^{\text{other}}(\sigma A)$. The force per area is then $F/A = \sigma \mathbf{E}^{\text{other}}$. In view of the result stated in the problem, our goal is to show that $\mathbf{E}^{\text{other}} = (\mathbf{E}_1 + \mathbf{E}_2)/2$. That is, we want to show that the field due to all the *other* charges equals the average of the actual fields on either side of the patch with *all* the charges, including the patch, taken into account. We can demonstrate this as follows.

Very close to the patch, the patch looks essentially like an infinite plane, so it produces a perpendicular field with magnitude $\sigma/2\epsilon_0$ on either side, pointing away from the patch (if $\sigma$ is positive). Let the plane of the patch be perpendicular to the $x$ axis. By superposition, the total field $\mathbf{E}_1$ on one side of the patch (due to all the other charges plus the patch) is $\mathbf{E}_1 = \mathbf{E}^{\text{other}} + (\sigma/2\epsilon_0)\hat{\mathbf{x}}$. The total field $\mathbf{E}_2$ on the other side of the patch is $\mathbf{E}_2 = \mathbf{E}^{\text{other}} - (\sigma/2\epsilon_0)\hat{\mathbf{x}}$. Adding these two relations gives

$$\mathbf{E}_1 + \mathbf{E}_2 = 2\mathbf{E}^{\text{other}} \implies \mathbf{E}^{\text{other}} = \frac{\mathbf{E}_1 + \mathbf{E}_2}{2}, \tag{12.63}$$

as desired. In effect, taking the average of the fields is simply a way of getting rid of the discontinuity caused by the patch.

REMARK: If the above derivation seemed a little too quick, that's because it was. There is one issue that we glossed over, although fortunately it doesn't affect the result. The issue involves the force component parallel to the surface (the reasoning involving the perpendicular component was fine). The component of $\mathbf{E}^{\text{other}}$ parallel to the surface (let's call it $\mathbf{E}_{\parallel}^{\text{other}}$) doesn't take on a unique value over the area of the patch, even if the patch is very small. $\mathbf{E}_{\parallel}^{\text{other}}$ actually diverges near the edge of the patch (see Exercise 1.66; the basic idea here is the same). So when we say that the parallel component of the force per area is $\sigma \mathbf{E}_{\parallel}^{\text{other}}$, what $\mathbf{E}_{\parallel}^{\text{other}}$ are we referring to? The value we are concerned with is the *average* of $\mathbf{E}_{\parallel}^{\text{other}}$ over the patch, so we must figure out what this average is. As an exercise, you can show that it simply equals the parallel component of the entire field $\mathbf{E}_{\parallel}$, which *does* take on a unique value over the patch, assuming that the patch is very small. (*Hint:* The patch can't exert a net parallel force on itself.)

But $\mathbf{E}_\parallel$ is continuous across the patch, so it equals both $\mathbf{E}_{1,\parallel}$ and $\mathbf{E}_{2,\parallel}$, and hence trivially $(\mathbf{E}_{1,\parallel} + \mathbf{E}_{2,\parallel})/2$. So the result for the force, involving the average of the fields, is still valid.

1.31  *Decreasing energy?*
No, it doesn't make sense. The point-charge configuration should have more energy, because the charge on the two point charges should want to fly apart and spread itself out over the spherical shell (assuming it is constrained to remain on the shell). And indeed, the point-charge configuration not only has more energy, but *infinitely* more energy, if the charges are true points. The error in the reasoning is that we included only the energy of the point charges due to each other; we forgot to include their self-energies, which are large if the charges occupy small volumes. (From the example in Section 1.15, the energy of a solid sphere with radius $a$ is $(3/5)Q^2/4\pi\epsilon_0 a$.) It takes a large amount of work to compress a charge distribution down to a small size.

1.32  *Energy of a shell*
(a) There is no field inside the sphere, so all of the energy is stored in the external field. This field is $Q/4\pi\epsilon_0 r^2$, so the energy is

$$U = \frac{\epsilon_0}{2} \int_R^\infty \left( \frac{Q}{4\pi\epsilon_0 r^2} \right)^2 4\pi r^2 \, dr = \frac{Q^2}{8\pi\epsilon_0} \int_R^\infty \frac{1}{r^2} \, dr = \frac{Q^2}{8\pi\epsilon_0 R}.$$

(12.64)

Note that this is smaller than the analogous result of $(3/5)Q^2/4\pi\epsilon_0 R$ for a solid sphere, given in the example in Section 1.15. This must be the case, because the external fields are the same in the two setups, but the solid sphere has an additional internal field. This is consistent with the fact that if the solid sphere were suddenly made conducting, all of the charge would flee to the surface because that state has a smaller energy.

(b) At an intermediate stage when the shell has charge $q$ on it, the external field is the same as the field from a point charge $q$ at the center. So the work needed to bring in from infinity an additional charge $dq$ (in the form of an infinitesimally thin shell) is $q \, dq/4\pi\epsilon_0 R$. Integrating this from $q = 0$ to $q = Q$ gives a total energy of

$$U = \int_0^Q \frac{q \, dq}{4\pi\epsilon_0 R} = \frac{Q^2}{8\pi\epsilon_0 R}.$$

(12.65)

We have used the fact that all of the infinitesimally thin shells are located at radius $R$.

1.33  *Deriving the energy density*
In Fig. 12.24 the left proton is located at the origin, and the angle $\theta$ is measured relative to the horizontal. The fields at the position shown are

$$E_1 = \frac{e}{4\pi\epsilon_0 r^2} \qquad \text{and} \qquad E_2 = \frac{e}{4\pi\epsilon_0 R^2},$$

(12.66)

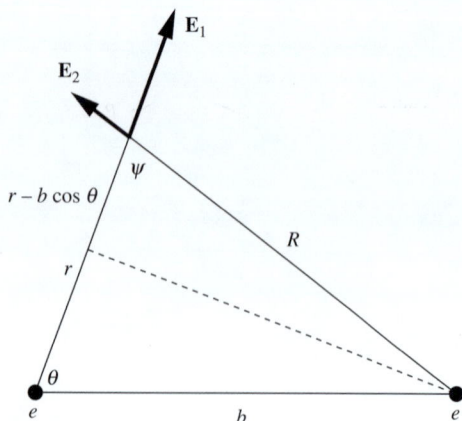

**Figure 12.24.**

where $R$ is given by the law of cosines as $R = (r^2 + b^2 - 2rb \cos \theta)^{1/2}$. The dot product $\mathbf{E}_1 \cdot \mathbf{E}_2$ equals $E_1 E_2 \cos \psi$, where from Fig. 12.24 we have $\cos \psi = (r - b \cos \theta)/R$. The desired integral is therefore

$$\epsilon_0 \int \mathbf{E}_1 \cdot \mathbf{E}_2 \, dv$$

$$= \epsilon_0 \int E_1 E_2 \cos \psi \, dv$$

$$= \epsilon_0 \int_0^{2\pi} \int_0^\pi \int_0^\infty \frac{e}{4\pi\epsilon_0 r^2} \frac{e}{4\pi\epsilon_0 R^2} \frac{r - b\cos\theta}{R} r^2 \sin\theta \, dr \, d\theta \, d\phi$$

$$= \frac{(2\pi)e^2}{16\pi^2\epsilon_0} \int_0^\pi \int_0^\infty \frac{r - b\cos\theta}{(r^2 + b^2 - 2rb\cos\theta)^{3/2}} dr \sin\theta \, d\theta$$

$$= \frac{e^2}{8\pi\epsilon_0} \int_0^\pi \left[ -\frac{1}{(r^2 + b^2 - 2rb\cos\theta)^{1/2}} \Big|_{r=0}^\infty \right] \sin\theta \, d\theta$$

$$= \frac{e^2}{8\pi\epsilon_0} \int_0^\pi \left[ \frac{1}{b} \right] \sin\theta \, d\theta$$

$$= \frac{e^2}{8\pi\epsilon_0 b} \int_0^\pi \sin\theta \, d\theta$$

$$= \frac{e^2}{4\pi\epsilon_0 b}, \tag{12.67}$$

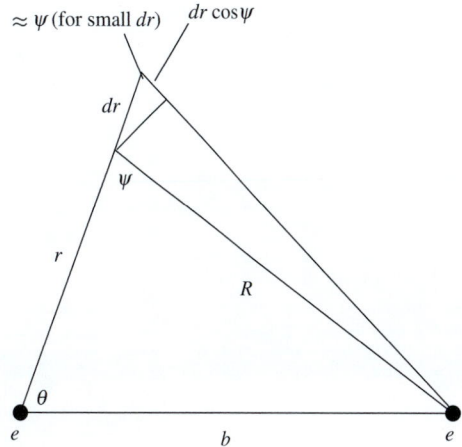

**Figure 12.25.**

as desired. We appear to have gotten lucky with the above $r$ integrand being a perfect differential. However, there is a quick way to see how this comes about. In the fourth line above, we can keep things in terms of $R$ and $\psi$ and write the $r$ integrand as $(\cos\psi/R^2)dr$. But from Fig. 12.25, we see that if we increase $r$ by $dr$ while holding $\theta$ constant, $R$ increases by $dR = dr \cos\psi$. The $r$ integrand can therefore be rewritten as $dR/R^2$. The integral of this is simply $-1/R$, as we found in the fifth line above.

If we have $n$ charges instead of two, then the energy $U$ involves the integral of $\mathbf{E}^2 = (\mathbf{E}_1 + \mathbf{E}_2 + \cdots + \mathbf{E}_n)^2$. As above, we ignore the terms of the form $\mathbf{E}_i^2$, because these give the self-energies of the particles. Each of the cross terms of the form $\mathbf{E}_i \cdot \mathbf{E}_j$ can be handled in exactly the same way as above, and hence yields $e^2/4\pi\epsilon_0 r$, where $r$ is the separation between particles $i$ and $j$.

## 12.2 Chapter 2

2.1 *Equivalent statements*

We are given that $\int \mathbf{E} \cdot d\mathbf{s} = 0$ for any path starting and ending at point $A$ in Fig. 12.26. Let the closed path be divided into paths 1 and 2 by point $B$. Then

$$\int_{A,\,\text{path 1}}^B \mathbf{E} \cdot d\mathbf{s} + \int_{B,\,\text{path 2}}^A \mathbf{E} \cdot d\mathbf{s} = 0. \tag{12.68}$$

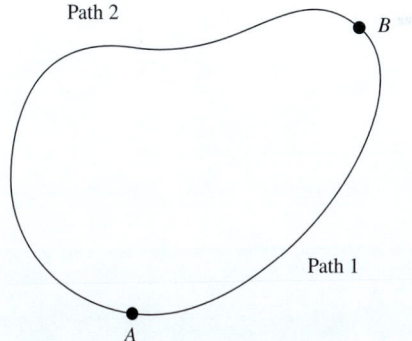

**Figure 12.26.**

It follows that

$$\int_{A,\,\text{path 1}}^{B} \mathbf{E} \cdot d\mathbf{s} = - \int_{B,\,\text{path 2}}^{A} \mathbf{E} \cdot d\mathbf{s}$$

$$= \int_{A,\,\text{path 2}}^{B} \mathbf{E} \cdot d\mathbf{s}, \qquad (12.69)$$

because running the path backwards negates $d\mathbf{s}$. The integral from $A$ to $B$ is therefore independent of the path.

In words, what the above equations say is simply this: if two things (the integral from $A$ to $B$ along path 1, and the integral from $B$ to $A$ along path 2) add up to zero, then they must be negatives of each other. Negating one of them then makes it equal to the other.

2.2  *Combining two shells*
The $Q^2/2\pi\epsilon_0 R$ answer is correct. The thing missing from the $Q^2/4\pi\epsilon_0 R$ answer is the energy of each shell due to the potential of the other. Ignoring the self-energies of the shells, the energy required to build up the second shell, with the first one already in place, is $Q\phi$, where $\phi = Q/4\pi\epsilon_0 R$ is the potential due to the first shell at its surface. This gives $Q^2/4\pi\epsilon_0 R$, which, when added to the incorrect $Q^2/4\pi\epsilon_0 R$ answer involving just the self-energies, gives the correct $Q^2/2\pi\epsilon_0 R$ answer. Alternatively, you can obtain the missing $Q/4\pi\epsilon_0 R$ energy by considering the energies of both shells due to the other, and then dividing by 2, due to the double counting; this is the "2" that appears in Eq. (2.32).

We could have alternatively phrased the question with shells of charge $Q$ and $-Q$. Each of the two self-energies is still $Q^2/8\pi\epsilon_0 R$, so the incorrect reasoning again gives $Q^2/4\pi\epsilon_0 R$. But the total energy must be zero, of course, because in the end we have a shell with zero charge on it. And indeed, if the $Q$ shell is already in place, then the energy required to build up the $-Q$ shell is $(-Q)\phi = -Q^2/4\pi\epsilon_0 R$. When this is added to the $Q^2/4\pi\epsilon_0 R$ answer involving just the self-energies, we correctly obtain zero.

2.3  *Equipotentials from four charges*
The general expression for the potential $\phi(x, y)$ is

$$4\pi\epsilon_0\phi(x, y) = \frac{2q}{\sqrt{x^2 + (y - 2\ell)^2}} + \frac{2q}{\sqrt{x^2 + (y + 2\ell)^2}}$$
$$- \frac{q}{\sqrt{(x - \ell)^2 + y^2}} - \frac{q}{\sqrt{(x + \ell)^2 + y^2}}. \qquad (12.70)$$

The equipotential curve $A$ passes through the point $(0, \ell)$. The potential at this point is (ignoring the factor of $q/4\pi\epsilon_0\ell$)

$$\phi_A = \frac{2}{1} + \frac{2}{3} - \frac{1}{\sqrt{2}} - \frac{1}{\sqrt{2}} = 1.252. \qquad (12.71)$$

Curve $B$ passes through the point $(3.44\ell, 0)$. The potential there is

$$\phi_B = 2 \cdot \frac{2}{\sqrt{3.44^2 + 2^2}} - \frac{1}{2.44} - \frac{1}{4.44} = 0.370. \qquad (12.72)$$

Curve $C$ passes through the origin. The potential there is

$$\phi_C = 2 \cdot \frac{2}{2} - 2 \cdot \frac{1}{1} = 0. \qquad (12.73)$$

At locations where the equipotentials intersect, the slope of $\phi$ is zero in two independent directions, so $\phi$ must locally be a flat plane; the intersection is a saddle point. And since $\mathbf{E}$ is the negative gradient of $\phi$, this implies that $\mathbf{E} = 0$ where the curves intersect. (At the saddle point at the origin, it is clear that $\mathbf{E} = 0$ by symmetry.) Figure 12.27 shows a few more of the equipotential curves. Far away from the charges, the curves are approximately circular.

You can show that $\mathbf{E} = 0$ at the point $(3.44\ell, 0)$ by finding the total $E_x$ component due to the four charges at points on the $x$ axis, and then demanding that $E_x = 0$, which can be solved numerically. Equivalently, you can set $y = 0$ in Eq. (12.70) and then demand that $\partial\phi/\partial x = 0$. The resulting equation will, of course, be the statement that $E_x = 0$, as you should explicitly verify.

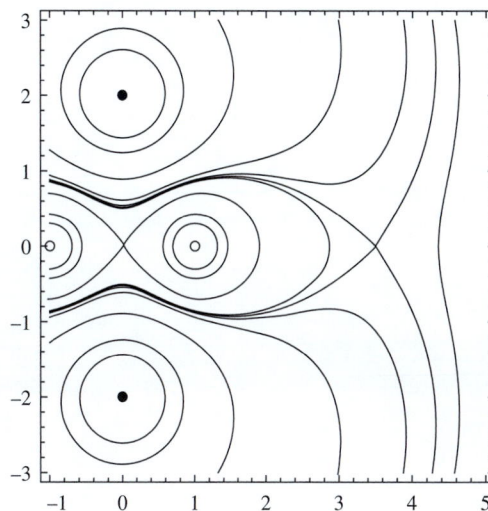

**Figure 12.27.**

2.4 *Center vs. corner of a cube*

Dimensional analysis tells us that for a given charge density $\rho$, the potential $\phi_0$ at the center of a cube of edge $s$ must be proportional to $Q/s$, where $Q$ is the total charge, $\rho s^3$. (This is true because the potential has the dimensions of $q/4\pi\epsilon_0 r$, and $Q$ is the only charge in the setup, and $s$ is the only length scale.) Hence $\phi_0$ is proportional to $\rho s^3/s = \rho s^2$. So for fixed $\rho$, we have $\phi_0 \propto s^2$.

Equivalently, if we imagine increasing the size of the cube by a factor $f$ in each direction, then the integral $\phi \propto \int(\rho\,dv)/r$ picks up a factor of $f^3$ in the $dv$ and a factor of $f$ in the $r$, yielding a net factor of $f^2$ in the numerator.

A cube of edge $2b$ can be considered to be built up from eight cubes of edge $b$. The center of the large cube coincides with a corner of all eight of the smaller cubes. So the potential at the center of the large cube is $8\phi_1$. But, from the above result that $\phi \propto s^2$, this center potential must also be $2^2 = 4$ times the center potential $\phi_0$ of the edge-$b$ cube. Hence $8\phi_1 = 4\phi_0$, or $\phi_0 = 2\phi_1$. It therefore takes twice as much work to bring in a charge from infinity to the center as it does to a corner. From the second example in Section 2.2, the analogous statement for a solid sphere is that it takes $3/2$ as much work to bring in a charge from infinity to the center as it does to the surface. But that problem can't be solved via the above scaling/superposition argument.

The above result for the cube actually holds more generally for any rectangular parallelepiped with uniform charge density. All of the steps in the above logic are still valid, so the potential at the center is twice the potential at a corner.

2.5 *Escaping a cube*

Intuitively, the easiest escape route should be via the midpoint of a face, because that path has the largest closest approach to any of the corners. Let the cube have side length $2\ell$. Then the potential at the center (ignoring the factor of $e/4\pi\epsilon_0\ell$) is $8/\sqrt{3} = 4.6188$, and the potential at the midpoint of a face is $4/\sqrt{2} + 4/\sqrt{6} = 4.4614$. This is smaller than at the center, and it is certainly downhill from there on out, because the fields of all eight charges will be pushing the proton outward. But is it all downhill on the path from the center to the midpoint of the face? At a distance $x$ from the center to the midpoint of a face, the potential is (ignoring the $e/4\pi\epsilon_0\ell$)

$$\phi(x) = \frac{4}{\sqrt{1^2 + 1^2 + (1+x)^2}} + \frac{4}{\sqrt{1^2 + 1^2 + (1-x)^2}}. \qquad (12.74)$$

If you plot this as a function of $x$, you will see that it is indeed a decreasing function, although it is very flat near $x = 0$. So the proton can in fact escape. (See Problem 2.25 for a general theorem.) If you want, you can calculate the derivative of $\phi(x)$ and plot that; you will see that it is always negative (for $x > 0$).

The proton will of course not escape if it moves from the center directly toward a corner of the cube, because the potential is infinite there. But what if it heads directly toward the midpoint of an edge? That is the task of Exercise 2.36.

2.6 *Electrons on a basketball*

Assume that the diameter is about 1 foot (0.3 m), so $r \approx 0.15$ m (the actual value is 0.12 m). With $V_0 = 1000$ V, we are told that

$$\frac{Q}{4\pi\epsilon_0 r} = -V_0 \implies Q = -4\pi\epsilon_0 r V_0. \qquad (12.75)$$

The charge per square meter is then $Q/4\pi r^2 = -\epsilon_0 V_0/r$. The number of extra electrons per square meter is therefore

$$\frac{\epsilon_0 V_0}{er} = \frac{\left(8.85 \cdot 10^{-12}\ \frac{\mathrm{s}^2\,\mathrm{C}^2}{\mathrm{kg}\,\mathrm{m}^3}\right)(1000\ \mathrm{V})}{(1.6 \cdot 10^{-19}\ \mathrm{C})(0.15\ \mathrm{m})} \approx 3.7 \cdot 10^{11}\ \mathrm{m}^{-2}. \qquad (12.76)$$

So the number per square centimeter is $3.7 \cdot 10^7\ \mathrm{cm}^{-2}$.

2.7 *Shell field via direct integration*

Let a given point $P$ be a distance $r$ from the center of the shell, and consider the ring that makes the angle $\theta$ shown in Fig. 12.28. The distance from $P$ to any point on the ring is given by the law of cosines as $\ell = \sqrt{R^2 + r^2 - 2rR\cos\theta}$. The area of the ring is its width (which is $R\,d\theta$) times its circumference (which is $2\pi R\sin\theta$). The charge on the ring is therefore $(R\,d\theta)(2\pi R\sin\theta)\sigma$, so the potential at $P$ due to the ring is (using $\sigma = Q/4\pi R^2$)

$$\phi_{\mathrm{ring}} = \frac{(R\,d\theta)(2\pi R\sin\theta)\sigma}{4\pi\epsilon_0\ell} = \frac{Q\sin\theta\,d\theta}{8\pi\epsilon_0\sqrt{R^2 + r^2 - 2rR\cos\theta}}. \qquad (12.77)$$

The total potential at $P$ is therefore

$$\phi(r) = \frac{Q}{8\pi\epsilon_0} \int_0^\pi \frac{\sin\theta\,d\theta}{\sqrt{R^2 + r^2 - 2rR\cos\theta}}$$

$$= \frac{Q}{8\pi\epsilon_0 rR}\sqrt{R^2 + r^2 - 2rR\cos\theta}\,\Big|_0^\pi. \qquad (12.78)$$

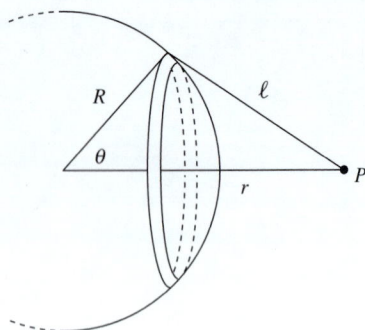

**Figure 12.28.**

The $\sin\theta$ in the numerator is what made this integral easily doable. We must now consider two cases. If $r < R$ (so $P$ is inside the shell), we have

$$\phi(r) = \frac{Q}{8\pi\epsilon_0 rR}\big[(r+R) - (R-r)\big] = \frac{Q}{4\pi\epsilon_0 R}. \qquad (12.79)$$

And if $r > R$ (so $P$ is outside the shell), we have

$$\phi(r) = \frac{Q}{8\pi\epsilon_0 rR}\big[(r+R) - (r-R)\big] = \frac{Q}{4\pi\epsilon_0 r}. \qquad (12.80)$$

We see that if $P$ is inside the shell, then the potential is constant, so the electric field is zero. If $P$ is outside the shell, then

$$E(r) = -\frac{d\phi}{dr} = \frac{Q}{4\pi\epsilon_0 r^2}. \qquad (12.81)$$

Since a solid sphere can be built up from spherical shells, the above results imply that the field outside a solid sphere with charge $Q$ equals $Q/4\pi\epsilon_0 r^2$, and the field inside equals $Q_r/4\pi\epsilon_0 r^2$, where $Q_r$ is the charge that lies inside radius $r$. These results are valid even if the charge distribution varies with $r$, as long as it is spherically symmetric. See Problem 2.8 for an extension of the above method.

2.8    *Verifying the inverse-square law*

   (a) As in Problem 2.7, our strategy for calculating the potential at a point $P$, due to the spherical shell, will be to slice the shell into rings, as shown in Fig. 12.29. The distance from $P$ to any point on the ring is given by the law of cosines as $\ell = \sqrt{R^2 + r^2 - 2rR\cos\theta}$. The area of the ring is $(R\,d\theta)(2\pi R\sin\theta)$, so the charge on it is $dq = \sigma(R\,d\theta)(2\pi R\sin\theta)$.

      If Coulomb's law takes the form of $F(r) = kq_1q_2/r^{2+\delta}$, then the potential due to a point charge $dq$ at distance $\ell$ is $k\,dq/\big((1+\delta)\ell^{1+\delta}\big)$, as you can verify by taking the negative derivative. Using the above value of $dq$, along with $\sigma = Q/4\pi R^2$, we see that the potential at $P$ due to the ring at angle $\theta$ equals

$$\phi_{\text{ring}} = \frac{k\sigma(R\,d\theta)(2\pi R\sin\theta)}{(1+\delta)\ell^{1+\delta}} = \frac{kQ\sin\theta\,d\theta}{2(1+\delta)\ell^{1+\delta}}. \qquad (12.82)$$

Using the above value of $\ell$, the total potential at $P$ is therefore

$$\phi(r) = \int_0^\pi \frac{kQ\sin\theta\,d\theta}{2(1+\delta)\big(R^2+r^2-2rR\cos\theta\big)^{(1+\delta)/2}}$$

$$= \frac{kQ}{2(1-\delta^2)rR}\big(R^2+r^2-2rR\cos\theta\big)^{(1-\delta)/2}\Big|_0^\pi. \qquad (12.83)$$

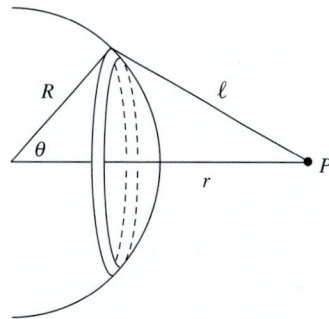

**Figure 12.29.**

The $\sin\theta$ in the numerator is what made this integral easily doable, even with the messy exponent in the denominator. We must now consider two cases. If $r < R$, we have (with $f(x) = x^{1-\delta}$)

$$\phi(r) = \frac{kQ}{2(1-\delta^2)rR}\big[f(R+r)-f(R-r)\big]. \tag{12.84}$$

And if $r > R$ we have

$$\phi(r) = \frac{kQ}{2(1-\delta^2)rR}\big[f(r+R)-f(r-R)\big]. \tag{12.85}$$

If $\delta = 0$ so that $f(x) = x$, these two results reduce to $kQ/R$ and $kQ/r$, respectively, as they should (see Problem 2.7).

(b) The potential at radius $a$ due to the shell of radius $a$ is found from either Eq. (12.84) or Eq. (12.85) to be $(kQ_a/2a^2)f(2a)$, where we have ignored the factor of $(1-\delta^2)$. The potential at radius $a$ due to the smaller shell of radius $b$ is found from Eq. (12.85) to be $(kQ_b/2ab)\big[f(a+b)-f(a-b)\big]$. So we have

$$\phi_a = \frac{kQ_a}{2a^2}f(2a) + \frac{kQ_b}{2ab}\big[f(a+b)-f(a-b)\big]. \tag{12.86}$$

Similarly, the potential at radius $b$ due to the shell of radius $b$ is found from either Eq. (12.84) or Eq. (12.85) to be $(kQ_b/2b^2)f(2b)$. The potential at radius $b$ due to the larger shell of radius $a$ is found from Eq. (12.84) to be $(kQ_a/2ab)\big[f(a+b)-f(a-b)\big]$. So we have

$$\phi_b = \frac{kQ_b}{2b^2}f(2b) + \frac{kQ_a}{2ab}\big[f(a+b)-f(a-b)\big]. \tag{12.87}$$

If $\delta = 0$ so that $f(x) = x$, these two results reduce to

$$\phi_a = \frac{kQ_a}{a} + \frac{kQ_b}{a} \quad \text{and} \quad \phi_b = \frac{kQ_b}{b} + \frac{kQ_a}{a}, \tag{12.88}$$

which are correct, as you can check (note the asymmetry in the denominators).

(c) If we replace the left-hand sides of Eqs. (12.86) and (12.87) with a given common value $\phi$, then we have two equations in the two unknowns, $Q_a$ and $Q_b$. We can eliminate $Q_a$ by multiplying Eq. (12.86) by $a\big[f(a+b)-f(a-b)\big]$ and Eq. (12.87) by $bf(2a)$, and then subtracting the former from the latter. The result is

$$\phi\Big(bf(2a) - a\big[f(a+b)-f(a-b)\big]\Big)$$
$$= \frac{kQ_b}{2b}\Big(f(2a)f(2b) - \big[f(a+b)-f(a-b)\big]^2\Big). \tag{12.89}$$

Therefore (again, with $f(x) = x^{1-\delta}$),

$$Q_b = \frac{2b\phi}{k} \cdot \frac{bf(2a) - a\big[f(a+b)-f(a-b)\big]}{f(2a)f(2b) - \big[f(a+b)-f(a-b)\big]^2}. \tag{12.90}$$

If we had kept the factor of $(1 - \delta^2)$, it would appear in the numerator. As stated in the problem, if $\delta = 0$ so that $f(x) = x$, then $Q_b$ equals zero, as it should. If we pick specific numerical values for $a$ and $b$, for example $a = 1.0$ and $b = 0.5$, then via *Mathematica* we find that to first order in $\delta$, the lengthy fraction in Eq. (12.90) behaves like $\approx (0.26)\delta$. For small $\delta$, the charge on the outer shell is essentially given by the standard Coulomb value of $Q_a = a\phi/k$. So for $a = 1.0$ and $b = 0.5$, the fraction of the charge that ends up on the inner shell is $Q_b/Q_a = [(2b\phi/k) \cdot (0.26)\delta]/(a\phi/k) = (0.26)\delta$.

2.9 *$\phi$ from integration*

(a) All points in a spherical shell at radius $r$ are a distance $r$ from the center, so integrating over all the shells gives

$$\phi_{\text{center}} = \int \frac{dq}{4\pi\epsilon_0 r} = \int_0^R \frac{(4\pi r^2\, dr)\rho}{4\pi\epsilon_0 r} = \frac{R^2\rho}{2\epsilon_0}. \qquad (12.91)$$

(b) Let's find $\phi$ at the "north pole" of the shell. Consider a ring whose points lie at an angle $\theta$ down from the pole. All points in the ring are a distance $2R\sin(\theta/2)$ from the pole; see Fig. 12.30. The area of the ring is $(2\pi R \sin\theta)(R\, d\theta)$, so integrating over all the rings gives (using $\sin\theta = 2\sin(\theta/2)\cos(\theta/2)$)

$$\phi_{\text{surface}} = \int_0^\pi \frac{\sigma(2\pi R\sin\theta)(R\, d\theta)}{4\pi\epsilon_0 2R\sin(\theta/2)} = \frac{\sigma R}{2\epsilon_0} \int_0^\pi \cos(\theta/2)$$

$$= \frac{\sigma R}{\epsilon_0} \sin(\theta/2)\Big|_0^\pi = \frac{\sigma R}{\epsilon_0}. \qquad (12.92)$$

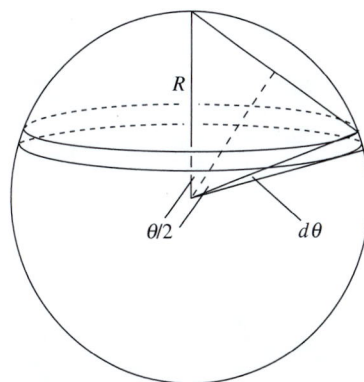

$R$

$\theta/2$   $d\theta$

**Figure 12.30.**

(c) Since $Q = (4\pi R^3/3)\rho$ for the solid sphere in part (a), and $Q = (4\pi R^2)\sigma$ for the shell in part (b), the two results can be written as $\phi_{\text{center}} = (3/2)(Q/4\pi\epsilon_0 R)$ and $\phi_{\text{surface}} = Q/4\pi\epsilon_0 R$. The former is therefore $3/2$ times the latter. This is consistent with the result from the second example in Section 2.2, because we know that $\phi_{\text{surface}}$ for our spherical shell is the same as the potential on the surface of a *solid* sphere with the same charge $Q$ and radius $R$.

2.10 *A thick shell*

(a) • For $0 \le r \le R_1$, the field is $E(r) = 0$, because the electric field inside a shell is zero.

• For $R_1 \le r \le R_2$, the field is $E(r) = Q_r/4\pi\epsilon_0 r^2$, where $Q_r$ is the charge inside radius $r$. Charge is proportional to volume, and volume is proportional to radius cubed, so $Q_r = Q(r^3 - R_1^3)/(R_2^3 - R_1^3)$. Therefore,

$$E(r) = \frac{Q_r}{4\pi\epsilon_0 r^2} = \frac{Q}{4\pi\epsilon_0(R_2^3 - R_1^3)}\left(r - \frac{R_1^3}{r^2}\right). \qquad (12.93)$$

If $R_1 = 0$, in which case we have an entire sphere, the field is proportional to $r$. This agrees with the result from the example in Section 1.11.

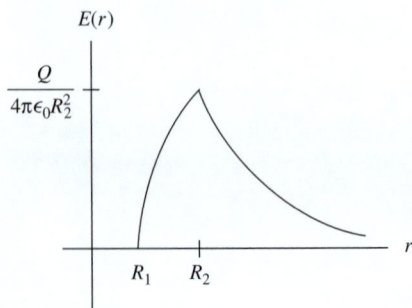

**Figure 12.31.**

- For $R_2 \leq r \leq \infty$, the field is simply $E(r) = Q/4\pi\epsilon_0 r^2$, because the shell looks like a point charge from the outside.

These forms of $E(r)$ agree at the transition points at $R_1$ and $R_2$, as they should. A complete (rough) plot of $E(r)$ is shown in Fig. 12.31. You can show that $E(r)$ is indeed concave downward for $R_1 < r < R_2$ by calculating the second derivative.

(b) With $R_2 = 2R_1 \equiv 2R$, the potential at $r = 0$ is

$$\phi(0) = -\int_\infty^0 E \, dr$$

$$= -\int_\infty^{R_2} \frac{Q}{4\pi\epsilon_0 r^2} \, dr - \int_{R_2}^{R_1} \frac{Q}{4\pi\epsilon_0 (R_2^3 - R_1^3)} \left( r - \frac{R_1^3}{r^2} \right) dr$$

$$\quad - \int_{R_1}^0 (0) dr$$

$$= \frac{Q}{4\pi\epsilon_0 (2R)} - \frac{Q}{4\pi\epsilon_0 ((2R)^3 - R^3)} \left( \frac{r^2}{2} + \frac{R^3}{r} \right) \Bigg|_{2R}^R$$

$$= \frac{Q}{4\pi\epsilon_0 R} \left[ \frac{1}{2} - \frac{1}{7} \left( \frac{3}{2} - \frac{5}{2} \right) \right] = \frac{9}{14} \cdot \frac{Q}{4\pi\epsilon_0 R}. \qquad (12.94)$$

If we write this as $(9/7)\big(Q/4\pi\epsilon_0(2R)\big)$, we see that the factor of $9/7$ for our thick shell correctly lies between the factor of $1$ that is relevant for a thin shell of radius $2R$ and the factor of $3/2$ that is relevant for a solid sphere of radius $2R$ (see the second example in Section 2.2). The external field is the same in all cases, but the solid sphere has a nonzero field extending all the way down to the origin. More work is therefore required in that case to bring in a charge all the way to $r = 0$.

2.11  *E for a line, from a cutoff potential*
A small length $dx$ of the wire has charge $\lambda \, dx$, so the potential (relative to infinity) at a point a distance $r$ from the center of the finite wire equals

$$\phi(r) = \frac{1}{4\pi\epsilon_0} \int_{-L}^L \frac{\lambda \, dx}{\sqrt{x^2 + r^2}}. \qquad (12.95)$$

From Appendix K or *Mathematica*, or by using an $x = r \sinh z$ substitution (although this gets a little messy), this integral becomes

$$\phi(r) = \frac{\lambda}{4\pi\epsilon_0} \ln\left( \sqrt{x^2 + r^2} + x \right) \Bigg|_{-L}^L = \frac{\lambda}{4\pi\epsilon_0} \ln\left( \frac{\sqrt{L^2 + r^2} + L}{\sqrt{L^2 + r^2} - L} \right). \qquad (12.96)$$

In the $L \gg r$ limit we have

$$\sqrt{L^2 + r^2} = L\sqrt{1 + \frac{r^2}{L^2}} \approx L\left( 1 + \frac{r^2}{2L^2} \right) = L + \frac{r^2}{2L}. \qquad (12.97)$$

To leading order in $r/L$, we can then write $\phi(r)$ as

$$\phi(r) = \frac{\lambda}{4\pi\epsilon_0} \ln\left(\frac{2L}{r^2/2L}\right) = \frac{2\lambda}{4\pi\epsilon_0} \ln\left(\frac{2L}{r}\right). \qquad (12.98)$$

The total length of the wire, $2L$, appears in the numerator of the log. However, this quantity is irrelevant because it simply introduces a constant additive term to $\phi(r)$, which yields zero when we take the derivative to find $E(r)$. So $\phi(r)$ is effectively equal to $-(\lambda/2\pi\epsilon_0)\ln(r)$, although technically it makes no sense to take the log of a dimensionful quantity. The (radial) field is therefore

$$E(r) = -\frac{d\phi}{dr} = \frac{\lambda}{2\pi\epsilon_0 r}, \qquad (12.99)$$

as desired. See Exercise 2.49 for the analogous procedure involving a sheet of charge.

This procedure of truncating the wire is valid for the following reason. Since the field from a point charge falls off like $1/r^2$, we know that the field from a very long wire is essentially equal to the field from an infinite wire; the extra infinite pieces at the two ends of the infinite wire give a negligible contribution. (The process of taking the radial component further helps the convergence, but it isn't necessary.) Therefore, the field we found for the above finite wire is essentially the same as the field for an infinite wire (provided that $L$ is large enough to make the approximation in Eq. (12.97) valid). Mathematically, the field in Eq. (12.99) is independent of $L$, so when we finally take the $L \to \infty$ limit, nothing changes.

Equivalently, we showed that, although the potential in Eq. (12.98) depends logarithmically on $L$, the field in Eq. (12.99) is independent of $L$. As far as the field is concerned, it doesn't matter that lengthening the wire changes the potential at every point, because the potentials all change by the same amount. The variation with $r$ remains the same, so $E = -d\phi/dr$ doesn't change. As an analogy, we can measure the gravitational potential energy $mgy$ with respect to the floor. If we shift the origin to be the ceiling instead, then the potential energy at every point changes. But it changes by the same amount everywhere, so the gravitational force is still $mg$.

**2.12** *E and $\phi$ from a ring*

(a) The setup is shown in Fig. 12.32. The magnitude of the field due to a little bit of charge $dQ$ on the ring is $dQ/4\pi\epsilon_0 r^2$. The component of this field that is perpendicular to the $x$ axis will cancel with the analogous component from the diametrically opposite $dQ$. Therefore, we care only about the component along the $x$ axis. This component is

$$E_{dQ} = \frac{dQ}{4\pi\epsilon_0 r^2}\cos\theta = \frac{dQ}{4\pi\epsilon_0 r^2}\cdot\frac{x}{r}. \qquad (12.100)$$

Adding up all the $dQ$ charges simply gives the total charge $Q$, so the total field is

$$E(x) = \frac{Qx}{4\pi\epsilon_0 r^3} = \frac{Qx}{4\pi\epsilon_0(x^2 + R^2)^{3/2}}. \qquad (12.101)$$

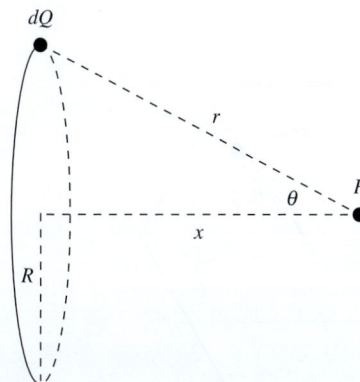

**Figure 12.32.**

If $x \to \infty$ then $E(x) \to Q/4\pi\epsilon_0 x^2$, which is correct because the ring looks like a point charge from far away. And if $x = 0$ then $E(x) = 0$, which is correct because the field components cancel by symmetry.

(b) We don't have to worry about components here, because the potential is a scalar. The potential due to a little piece $dQ$ is $dQ/4\pi\epsilon_0 r$. Adding up all the $dQ$ charges again just gives $Q$, so the total potential is

$$\phi(x) = \frac{Q}{4\pi\epsilon_0 r} = \frac{Q}{4\pi\epsilon_0\sqrt{x^2 + R^2}}. \qquad (12.102)$$

If $x \to \infty$ then $\phi(x) \to dQ/4\pi\epsilon_0 x$, which is correct because again the ring looks like a point charge from far away. And if $x = 0$ then $\phi(x) = Q/4\pi\epsilon_0 R$, which is correct because all points on the ring are a distance $R$ from the origin.

(c) The negative of the derivative of $\phi(x)$ is

$$-\frac{d\phi}{dx} = -\frac{d}{dx}\left(\frac{Q}{4\pi\epsilon_0\sqrt{x^2 + R^2}}\right)$$

$$= -\frac{Q}{4\pi\epsilon_0}\left(-\frac{1}{2}\right)(x^2 + R^2)^{-3/2}(2x)$$

$$= \frac{Qx}{4\pi\epsilon_0(x^2 + R^2)^{3/2}}, \qquad (12.103)$$

which is correctly the field $E(x)$ in Eq. (12.101).

(d) The initial kinetic and potential energies are both zero, so conservation of energy gives the speed at the center via

$$0 + 0 = \frac{1}{2}mv^2 + (-q)\phi(0) \implies 0 = \frac{1}{2}mv^2 - \frac{qQ}{4\pi\epsilon_0 R}$$

$$\implies v = \sqrt{\frac{qQ}{2\pi\epsilon_0 mR}}. \qquad (12.104)$$

If $R \to 0$ then $v \to \infty$ (the ring is essentially a point charge, and we know that the speed would be infinite in that case, because the potential goes to infinity close to the charge). And if $R \to \infty$ then $v \to 0$. This is believable, although not entirely obvious. If we instead had a ring with constant linear charge density (so that $Q \propto R$), then $v$ would be independent of $R$.

2.13 *$\phi$ at the center of an N-gon*

In Fig. 12.33, the area of the little piece of the wedge shown is $r\,dr\,d\theta$. The potential at the center due to this piece is $\sigma(r\,dr\,d\theta)/4\pi\epsilon_0 r = \sigma\,dr\,d\theta/4\pi\epsilon_0$. Integrating this from $r = 0$ up to the length $R$ of the wedge gives a contribution to the potential equal to $\sigma R\,d\theta/4\pi\epsilon_0$. But $R$ equals $a/\cos\theta$, so the potential at the center due to the wedge is $\sigma a\,d\theta/(4\pi\epsilon_0\cos\theta)$. The $N$-gon consists of $2N$ triangles, in each of which $\theta$ runs from 0 to

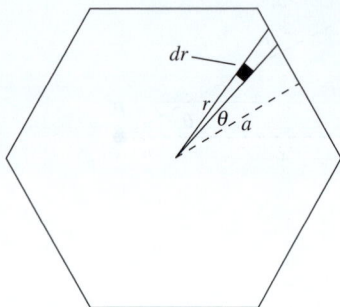

**Figure 12.33.**

$2\pi/2N = \pi/N$. So the potential at the center due to the entire $N$-gon is (using Appendix K for the integral)

$$\phi = 2N \cdot \frac{\sigma a}{4\pi\epsilon_0} \int_0^{\pi/N} \frac{d\theta}{\cos\theta} = \frac{N\sigma a}{2\pi\epsilon_0} \ln\left(\frac{1+\sin\theta}{\cos\theta}\right)\Big|_0^{\pi/N}$$

$$= \frac{N\sigma a}{2\pi\epsilon_0} \ln\left(\frac{1+\sin(\pi/N)}{\cos(\pi/N)}\right). \quad (12.105)$$

In the $N \to \infty$ limit, we can set the cosine term equal to 1 and the sine term equal to $\pi/N$. The Taylor series $\ln(1+\epsilon) \approx \epsilon$ then gives

$$\phi \approx \frac{N\sigma a}{2\pi\epsilon_0} \cdot \frac{\pi}{N} = \frac{\sigma a}{2\epsilon_0}, \quad (12.106)$$

in agreement with Eq. (2.27).

2.14  *Energy of a sphere*

Nonzero $\rho$ exists only inside the sphere, so we need only deal with the potential inside the sphere. The second example in Section 2.2 gives this potential as $\phi = \rho R^2/2\epsilon_0 - \rho r^2/6\epsilon_0$. Equation (2.32) therefore gives

$$U = \frac{1}{2}\int \rho\phi \, dv = \frac{1}{2}\int_0^R \rho\left(\frac{\rho R^2}{2\epsilon_0} - \frac{\rho r^2}{6\epsilon_0}\right) 4\pi r^2 \, dr$$

$$= \frac{\pi\rho^2}{\epsilon_0}\int_0^R \left(R^2 r^2 - \frac{r^4}{3}\right) dr = \frac{\pi\rho^2}{\epsilon_0}\left(\frac{R^5}{3} - \frac{R^5}{15}\right)$$

$$= \frac{4\pi\rho^2 R^5}{15\epsilon_0}. \quad (12.107)$$

In terms of the charge $Q = (4\pi R^3/3)\rho$, you can show that this can be written as $(3/5)Q^2/4\pi\epsilon_0 R$, in agreement with the two other calculations mentioned in the problem.

2.15  *Crossed dipoles*

**First solution**  We can consider the given setup to consist of two dipoles represented by the dashed lines in Fig. 12.34. If the length of the original dipoles is $\ell$, then each of the new dipoles has length $\ell/\sqrt{2}$, and hence a dipole moment $q(\ell/\sqrt{2})$. The total dipole moment is twice this, or $\sqrt{2}q\ell = \sqrt{2}p$. This result can also be obtained by treating each of the original dipoles as a vector with magnitude $p$, pointing from the negative charge to the positive charge, and then simply adding the vectors. We will talk more about the vector nature of the dipole in Chapter 10.

Note that the slight sideways displacement between the two new dipoles doesn't affect the field at large distances. Any changes in $r$ and $\theta$ are higher order in $1/r$, and therefore bring in modifications to the field only at order at least $(1/r)(1/r^3)$.

REMARK:  The more that one of the given dipoles is rotated with respect to the other, the smaller the resulting dipole moment is. If the angle between the given dipoles is $\beta$, you can show that the resulting dipole

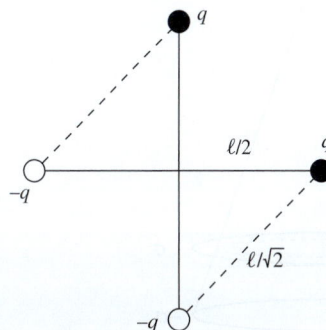

**Figure 12.34.**

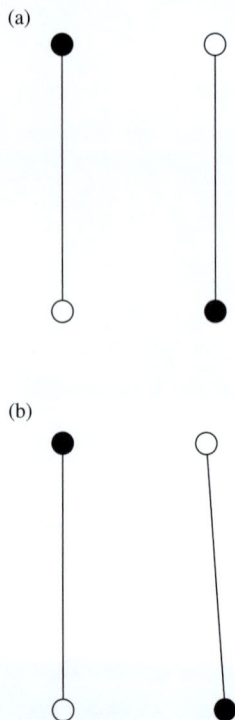

(a)

(b)

**Figure 12.35.**

**Figure 12.36.**

moment is $2p\cos(\beta/2)$. This reduces correctly in the above case where $\beta = 90°$. If $\beta = 180°$ we obtain a dipole moment of zero, as we should because the charges exactly cancel each other. (All higher moments are obviously equal to zero too.) However, if we have $\beta = 180°$ but then shift one of the dipoles sideways, as shown in Fig. 12.35(a), then the dipole moment is still zero, but we end up with a nonzero quadrupole moment. Note that if $\beta$ is very close, but not exactly equal, to $180°$, as in Fig. 12.35(b), then, even though the configuration may look roughly like a quadrupole, it is still a dipole because the dipole moment is nonzero. At sufficiently large distances, the dipole field dominates the quadrupole field.

**Second solution** Alternatively, we can find the net dipole moment by finding the sum of the potentials due to the two given dipoles. If a point is located at angle $\theta$ with respect to the vertical dipole (in the plane of the page), then it is located at angle $\theta - 90°$ with respect to the horizontal dipole. So Eq. (2.35) gives the total potential at the point as (making use of the trig sum formula for cosine)

$$\phi = \frac{p\cos\theta}{4\pi\epsilon_0 r^2} + \frac{p\cos(\theta - 90°)}{4\pi\epsilon_0 r^2} = \frac{p(\cos\theta + \sin\theta)}{4\pi\epsilon_0 r^2}$$

$$= \frac{\sqrt{2}p\cos(\theta - 45°)}{4\pi\epsilon_0 r^2} \equiv \frac{\sqrt{2}p\cos\theta'}{4\pi\epsilon_0 r^2}, \qquad (12.108)$$

where $\theta' \equiv \theta - 45°$. This simply says that we have a dipole with strength $\sqrt{2}p$ oriented along the $\theta' = 0$ direction, which corresponds to the $\theta = 45°$ direction. Of course, this $\phi$ is valid only for points in the plane of the page. So technically all we've done here is show that *if* the setup looks like a dipole from afar, then the dipole moment must be $\sqrt{2}p$.

2.16  *Disks and dipoles*

(a) A little bit of charge $dq$ on the top disk and the corresponding bit $-dq$ below it on the bottom disk form a dipole. The field due to this dipole at a distant point (with $r \gg \ell$) on the axis is $(dq)\ell/2\pi\epsilon_0 r^3$ from Eq. (2.36). All of the dipoles that make up the disks give essentially this same field at a distance point; the slight sideways displacement among them is inconsequential, to leading order. The total field is therefore obtained by replacing $dq$ with the total charge on the disk:

$$E = \frac{(\sigma\pi R^2)\ell}{2\pi\epsilon_0 r^3} = \frac{\sigma R^2 \ell}{2\epsilon_0 r^3}. \qquad (12.109)$$

(b) Consider the corresponding parts of the two disks contained in a common thin cone, as shown in Fig. 12.36. If the distances to the top and bottom parts are $r_t$ and $r_b$, respectively, then the bottom piece has an area, and hence charge, that is larger by a factor $r_b^2/r_t^2$. This factor exactly cancels the effect of the $r^2$ in Coulomb's law, so the two parts give canceling fields. The given cone can be divided into many of these thin cones, all yielding zero net field. Thus the parts of the two disks that are contained within the given cone produce zero net field.

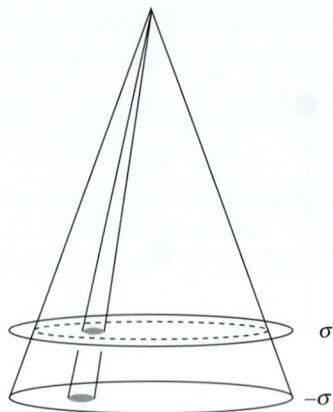

We are therefore concerned only with the leftover ring in the top disk. Let $r$ be measured from the point on the axis that is midway between the disks (although the exact origin won't matter here). From similar triangles in Fig. 12.37, the thickness $b$ of the ring is given by $b/\ell = (R - b)/(r - \ell/2)$. Since $b \ll R$ and $\ell \ll r$, we can ignore the $b$ and $\ell$ on the right-hand side, which leaves us with $b/\ell \approx R/r \implies b \approx R\ell/r$. The charge in the ring is then given by essentially $\sigma(2\pi R)(R\ell/r)$. If $r$ is large, the ring looks like a point charge, so the desired field is

$$E = \frac{\sigma(2\pi R)(R\ell/r)}{4\pi\epsilon_0 r^2} = \frac{\sigma R^2 \ell}{2\epsilon_0 r^3}, \tag{12.110}$$

in agreement with the result in part (a).

2.17 *Linear quadrupole*

(a) Let $\ell$ be the separation between the charges, so the whole quadrupole has length $2\ell$. Along the axis the field is radial, and at a distance $r$ from the center it equals (with $\epsilon \equiv \ell/r$ and $k \equiv 1/4\pi\epsilon_0$)

$$\begin{aligned} E_r &= \frac{kq}{(r-\ell)^2} - \frac{2kq}{r^2} + \frac{kq}{(r+\ell)^2} \\ &= \frac{kq}{r^2}\left[\frac{1}{1-2\epsilon+\epsilon^2} - 2 + \frac{1}{1+2\epsilon+\epsilon^2}\right] \\ &\approx \frac{kq}{r^2}\left[(1+2\epsilon+3\epsilon^2) - 2 + (1-2\epsilon+3\epsilon^2)\right] \\ &= \frac{kq}{r^2}[6\epsilon^2] = \frac{6kq\ell^2}{r^4}. \end{aligned} \tag{12.111}$$

We have indeed correctly inverted the above fractions (at least to order $\epsilon^2$), because $(1-2\epsilon+\epsilon^2)(1+2\epsilon+3\epsilon^2) = 1 + \mathcal{O}(\epsilon^3)$, and likewise for the other term. This field is positive, which makes sense because if we think in terms of the original dipoles, the closer dipole repels more than the farther dipole attracts. The $\ell^2/r^4$ dependence is a factor of $\ell/r$ smaller than the $\ell/r^3$ dependence for a dipole. This is consistent with the multipole-expansion discussion in Section 2.7.

(b) Along the perpendicular bisector, symmetry tells us that the field is again radial. Since the two end charges contribute the same radial field, the total field in Fig. 12.38 at a distance $r$ from the center is (again with $\epsilon \equiv \ell/r$ and $k \equiv 1/4\pi\epsilon_0$)

$$\begin{aligned} E_r &= \frac{2kq}{r_1^2}\cos\theta - \frac{2kq}{r^2} = \frac{2kq}{r^2+\ell^2}\frac{r}{\sqrt{r^2+\ell^2}} - \frac{2kq}{r^2} \\ &= \frac{2kq}{r^2}\left[\frac{1}{(1+\epsilon^2)^{3/2}} - 1\right] \approx \frac{2kq}{r^2}\left[\left(1-\frac{3}{2}\epsilon^2\right) - 1\right] \\ &= \frac{2kq}{r^2}\left[-\frac{3}{2}\epsilon^2\right] = -\frac{3kq\ell^2}{r^4}. \end{aligned} \tag{12.112}$$

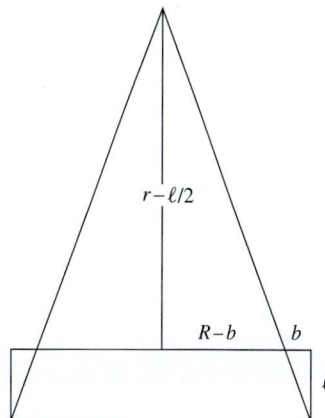

**Figure 12.37.**

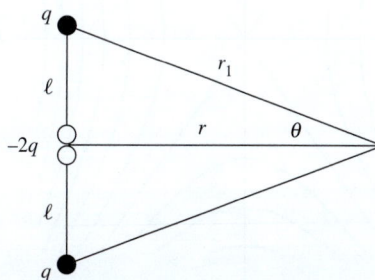

**Figure 12.38.**

This is negative, which makes sense because the point in question is closer to the $-2q$ charge than to the two $q$ charges, and also because the field from the $-2q$ charge points in the radial direction.

Recall that the dipole field points *tangentially* at locations on the perpendicular-bisector plane. The above quadrupole field is fundamentally different in that (among other things) it points *radially* at locations on the perpendicular-bisector plane. The angular dependence of the quadrupole field is more complicated than that of the dipole field, as we will see in Section 10.2.

2.18  *Field lines near the origin*

(a) Setting $a = 1$ and ignoring the factor of $q/4\pi\epsilon_0$, the potential due to the two positive charges, at locations in the $xy$ plane, is

$$\phi(x, y) = \frac{1}{\sqrt{(x+1)^2 + y^2}} + \frac{1}{\sqrt{(x-1)^2 + y^2}}. \tag{12.113}$$

Using the Taylor expansion $1/\sqrt{1+\epsilon} \approx 1 - \epsilon/2 + 3\epsilon^2/8$, and keeping terms up to second order in $x$ and $y$, we have

$$\phi(x, y) = \frac{1}{\sqrt{1 + (2x + x^2 + y^2)}} + \frac{1}{\sqrt{1 + (-2x + x^2 + y^2)}}$$

$$\approx \left(1 - \frac{1}{2}(2x + x^2 + y^2) + \frac{3}{8}(2x + \cdots)^2\right)$$

$$+ \left(1 - \frac{1}{2}(-2x + x^2 + y^2) + \frac{3}{8}(-2x + \cdots)^2\right)$$

$$= 2 + 2x^2 - y^2. \tag{12.114}$$

(Alternatively, you can obtain this from the Series operation in *Mathematica*.) If we had included the $z$ dependence, there would be an extra $-z^2$ tacked on this result. In terms of all the given parameters, you can show that

$$\phi(x, y) \approx \frac{q}{4\pi\epsilon_0 a}\left(2 + \frac{2x^2 - y^2}{a^2}\right). \tag{12.115}$$

Some level surfaces of the function $2x^2 - y^2$ are shown in Fig. 12.39. The origin is a saddle point; it is a minimum with respect to variations in the $x$ direction, and a maximum with respect to variations in the $y$ direction. The constant-$\phi$ lines passing through the equilibrium point are given by $y = \pm\sqrt{2}x$ (near the origin). If we zoom in closer to the origin, the curves keep the same general shape; the picture looks the same, with the only change being the $\phi$ value associated with each curve.

(b) The electric field is the negative gradient of the potential, so we have

$$\mathbf{E} = -\nabla\phi = \frac{q}{4\pi\epsilon_0 a^3}(-4x, 2y). \tag{12.116}$$

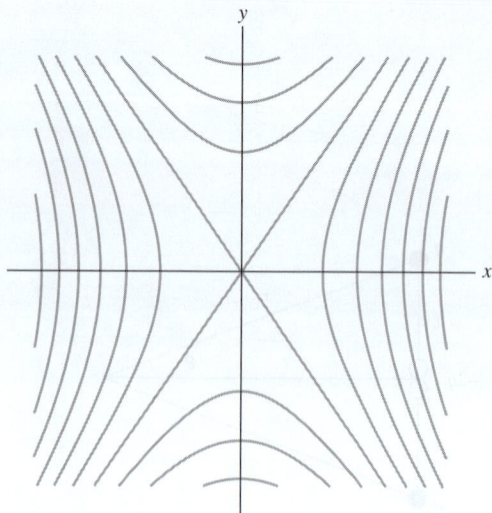

**Figure 12.39.**

The field lines are the curves whose tangents are the **E** field vectors, by definition. Equating the slope of a curve with the slope of the tangent **E** vector, and separating variables and integrating, gives

$$\frac{dy}{dx} = \frac{E_y}{E_x} \implies \frac{dy}{dx} = -\frac{y}{2x} \implies \int \frac{dy}{y} = -\frac{1}{2} \int \frac{dx}{x}$$

$$\implies \ln y = -\frac{1}{2} \ln x + A \implies y = \frac{B}{\sqrt{x}}, \quad (12.117)$$

where $A$ is a constant of integration and $B \equiv e^A$. Different values of $B$ give different field lines. Technically, this $y = B/\sqrt{x}$ result is valid only in the first quadrant. But since the setup is symmetric with respect to the $yz$ plane, and also with respect to rotations around the $x$ axis, the general form of the field lines in the $xy$ plane is shown in Fig. 12.40. If we zoom in closer to the origin, the lines keep the same general shape.

If we include the $z$ dependence, then the correct expression for the field near the origin has the $(-4x, 2y)$ vector in Eq. (12.116) replaced with $(-4x, 2y, 2z)$. As a check on this, the divergence of this vector is zero, which is correct because $\nabla \cdot \mathbf{E} = \rho/\epsilon_0$ and because there are no charges at the origin. Although the abbreviated vector in Eq. (12.116) is sufficient for making a picture of what the field lines look like, it has nonzero divergence, so its utility goes only so far. See Exercise 2.65 for an extension of this problem.

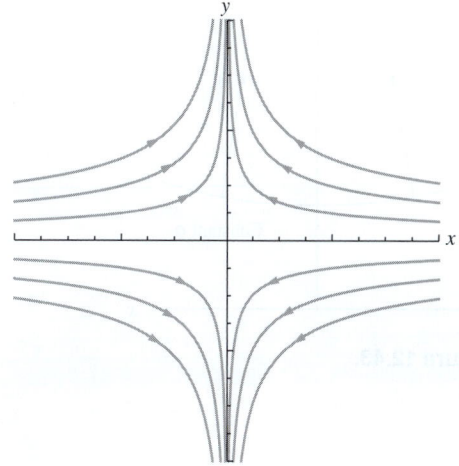

**Figure 12.40.**

2.19 *Equipotentials for a ring*

(a) At a given point on the $z$ axis, the magnitude of the field due to a little piece of charge $dQ$ on the ring in Fig. 12.41 is $dQ/4\pi\epsilon_0 r^2$. The horizontal component of this field will cancel with the horizontal component from the diametrically opposite $dQ$. Therefore, we care only about the vertical component, which brings in a factor of $\cos\theta$. So we have

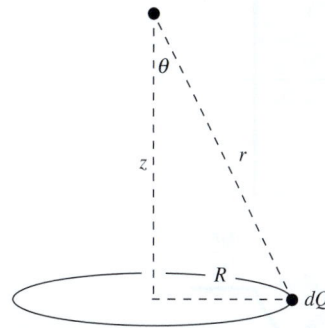

**Figure 12.41.**

$$E_{dQ} = \frac{dQ}{4\pi\epsilon_0 r^2} \cos\theta = \frac{dQ}{4\pi\epsilon_0 r^2} \cdot \frac{z}{r}. \quad (12.118)$$

Adding up all the $dQ$ charges simply gives the total charge $Q$. The total field is then

$$E(z) = \frac{Qz}{4\pi\epsilon_0 r^3} = \frac{Qz}{4\pi\epsilon_0 (z^2 + R^2)^{3/2}}. \quad (12.119)$$

If $z \to \infty$ then $E(z) \to Q/4\pi\epsilon_0 z^2$, which is correct because the ring looks like a point mass from far away. And if $z = 0$ then $E(z) = 0$, which is correct because the field components cancel by symmetry. Taking the derivative of $E(z)$, you can quickly show that the maximum occurs at $z = R/\sqrt{2}$.

(b) A sketch of some equipotential curves is shown in Fig. 12.42. The ring is represented by the two dots; we have chosen $R = 1$. The full surfaces are obtained by rotating the curves around the $z$ axis. Close to

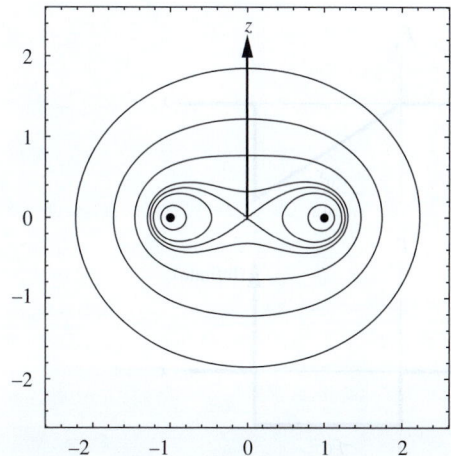

**Figure 12.42.**

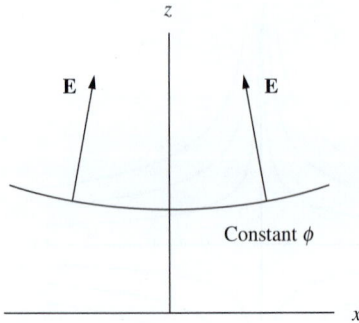

**Figure 12.43.**

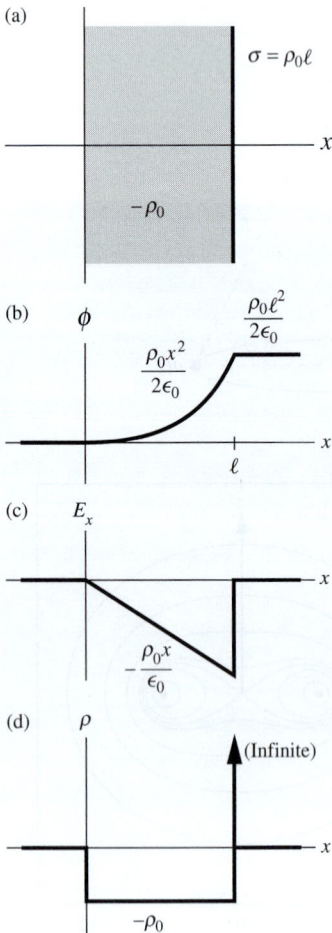

**Figure 12.44.**

the ring, the curves are circles around the dots, which means that the equipotentials are tori in 3D space. Far away, the curves become circles (or spheres in 3D) around the whole setup. The transition between the tori and the spheres occurs where the equipotentials cross at the origin, as shown.

(c) Consider a point on the $z$ axis where the equipotential curve is concave up. Just to the left of this point, the **E** field, which is the negative gradient of $\phi$, points upward and slightly to the right; see Fig. 12.43. So it has a positive $x$ component. Similarly, just to the right, **E** has a negative $x$ component. Therefore $E_x$ decreases as $x$ increases across $x = 0$. In other words, $\partial E_x/\partial x$ is negative at $x = 0$. Likewise, at points where the equipotential curve is concave down, $\partial E_x/\partial x$ is positive. At the transition point from concave up to concave down, $\partial E_x/\partial x$ must make the transition from negative to positive. That is, it must be zero. Since the equipotential surface is symmetric around the $z$ axis, $\partial E_y/\partial y$ must also be zero at the transition point.

There are no charges in the free space at the transition point, so $\nabla \cdot \mathbf{E} = \rho/\epsilon_0$ tells us that $\partial E_x/\partial x + \partial E_y/\partial y + \partial E_z/\partial z = 0$. Since we just found that the $x$ and $y$ derivatives are zero, we see that $\partial E_z/\partial z$ must also be zero. In other words, the transition point is an extremum of $E_z$, so it coincides with the point we found in part (a). See Exercise 2.66 for a variation on this problem.

**2.20** *A one-dimensional charge distribution*
The electric field is given by $\mathbf{E} = -\nabla\phi$, so we quickly find $E_x(x) = -\rho_0 x/\epsilon_0$ (and $E_y = E_z = 0$) in the region $0 < x < \ell$, and $\mathbf{E} = 0$ in the other two regions. Note that $E_x$ is continuous at $x = 0$ but not at $x = \ell$. This implies that there must be a surface charge density on the plane at $x = \ell$.

To obtain the charge distribution, we can use $\rho = -\epsilon_0 \nabla^2 \phi$, or equivalently $\rho = \epsilon_0 \nabla \cdot \mathbf{E}$. Either of these quickly gives $\rho(x) = -\rho_0$ in the region $0 < x < \ell$, and $\rho = 0$ in the other two regions. But as mentioned above, there is also a surface charge density $\sigma$ on the plane at $x = \ell$. This doesn't contradict the volume densities we just found, because those $\rho$ values have nothing to say about the $\rho$ values at the boundaries between the regions. If you tried to calculate $-\epsilon_0 \nabla^2 \phi$ or $\epsilon_0 \nabla \cdot \mathbf{E}$ at $x = 0$ and $x = \ell$, you would obtain results that are, respectively, undefined and infinite. The latter is consistent with the fact that a surface charge occupies zero volume.

To determine the surface density $\sigma$ on the plane at $x = \ell$, we can look at the discontinuity in the field across the plane. The field just to the left is $-\rho_0 \ell/\epsilon_0$, and just to the right is zero. Gauss's law tells us that the change in the field at the surface, which is $\rho_0 \ell/\epsilon_0$, must equal $\sigma/\epsilon_0$. Hence $\sigma = \rho_0 \ell$. Note that this surface density is equal and opposite to the effective surface density of the charged volume with density $-\rho_0$ and thickness $\ell$. The external field (for $x < 0$ and $x > \ell$) is therefore the same as the field from two oppositely charged sheets, which is zero, in agreement with the field we found above. Indeed, working backward from this external field would be another way of concluding that $\sigma = -\rho_0 \ell$.

Figure 12.44(a) shows a view of the slab and sheet; plots of $\phi(x)$, $E_x(x)$, and $\rho(x)$ are shown in Figs. 12.44(b)–(d). As mentioned above, the first derivative of $\phi$ (which is related to the field) is not well defined at $r = \ell$. And the second derivative (which is related to the density) is not well defined at $x = 0$ and is infinite at $x = \ell$, consistent with the fact that we have a finite amount of charge in a zero-volume object (the sheet).

2.21 *A cylindrical charge distribution*

(a) The given $\phi(r)$ is shown in Fig. 12.45(a). The electric field is the negative gradient of $\phi$. For a function that depends only on $r$, the gradient in cylindrical coordinates is simply $\nabla\phi = \hat{\mathbf{r}}(\partial\phi/\partial r)$. Outside the $R \leq r \leq 2R$ region, $\phi$ is constant, so $\mathbf{E}$ is zero. Inside the $R < r < 2R$ region, we have

$$\mathbf{E} = -\nabla\phi = -\hat{\mathbf{r}}\frac{\partial}{\partial r}\left(\frac{\rho_0}{4\epsilon_0}(4R^2 - r^2)\right) = \frac{\rho_0 r}{2\epsilon_0}\hat{\mathbf{r}}. \quad (12.120)$$

The plot of $E_r$ is shown in Fig. 12.45(b). Note the discontinuities at $r = R$ and $r = 2R$.

The charge density is given by $\nabla \cdot \mathbf{E} = \rho/\epsilon_0$, or equivalently $\nabla^2\phi = -\rho/\epsilon_0$. For a function that depends only on $r$, the divergence in cylindrical coordinates is $(1/r)\partial(rE_r/\partial r)$. Outside the $R \leq r \leq 2R$ region, $\mathbf{E}$ is constant (in fact zero), so $\rho$ is zero. Inside the $R < r < 2R$ region, we have

$$\rho = \epsilon_0\nabla \cdot \mathbf{E} = \epsilon_0\frac{1}{r}\frac{\partial}{\partial r}\left(r \cdot \frac{\rho_0 r}{2\epsilon_0}\right) = \rho_0. \quad (12.121)$$

But we're not done. The discontinuities in $E_r$ at $r = R$ and $r = 2R$ imply that there are surface charge densities there. From Gauss's law, the change in the field across a surface equals $\sigma/\epsilon_0$. At $r = R$ the field jumps up by $\rho_0 R/2\epsilon_0$, so the surface density must be $\sigma_R = \rho_0 R/2$. And at $r = 2R$ the field drops down by $\rho_0(2R)/2\epsilon_0$, so the surface density must be $\sigma_{2R} = -\rho_0 R$. Since a surface takes up zero volume, the volume densities of these surface charges are infinite, as indicated by the spikes in Fig. 12.45(c).

(b) The cross-sectional area of the $R < r < 2R$ region is $\pi(2R)^2 - \pi R^2 = 3\pi R^2$. So the volume density $\rho_0$ in this region yields a charge in a length $\ell$ of the cylinder equal to

$$\rho_0\ell(3\pi R^2) = 3\pi R^2\rho_0\ell. \quad (12.122)$$

The surface at $r = R$ yields a charge in length $\ell$ equal to

$$\sigma_R(2\pi R)\ell = (\rho_0 R/2)(2\pi R)\ell = \pi R^2\rho_0\ell. \quad (12.123)$$

And the surface at $r = 2R$ yields a charge in length $\ell$ equal to

$$\sigma_{2R}(2\pi \cdot 2R)\ell = (-\rho_0 R)(4\pi R)\ell = -4\pi R^2\rho_0\ell. \quad (12.124)$$

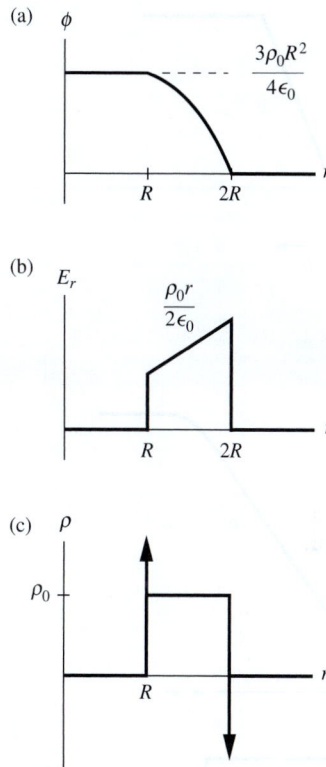

(a) $\phi$

$\frac{3\rho_0 R^2}{4\epsilon_0}$

$R$    $2R$    $r$

(b) $E_r$

$\frac{\rho_0 r}{2\epsilon_0}$

$R$    $2R$    $r$

(c) $\rho$

$\rho_0$

$R$    $r$

**Figure 12.45.**

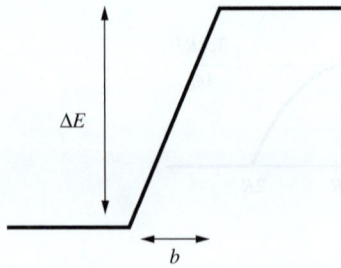

**Figure 12.46.**

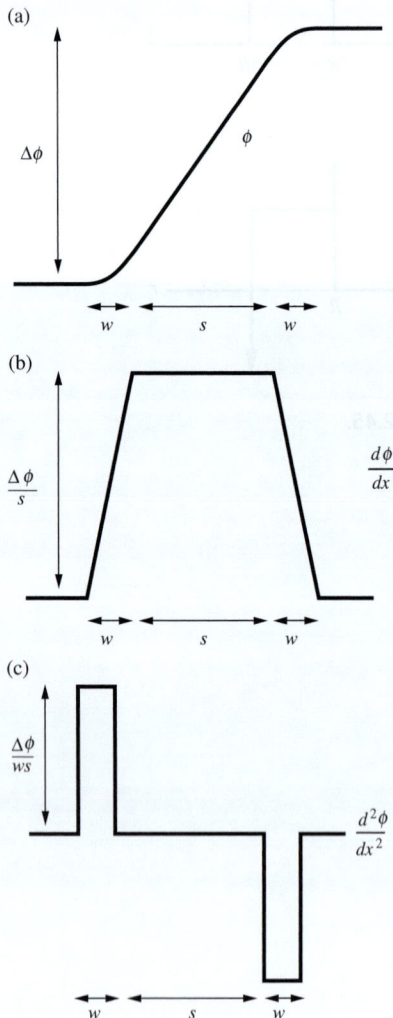

**Figure 12.47.**

Adding up the above three charges and dividing by $\ell$, we see that the total charge per unit length is zero. This makes sense, because otherwise there would be a nonzero field outside the cylinder, whereas we know that the field is zero for $r > 2R$. Note that the $\sigma_R$ surface density on the inner surface yields the same charge that we would obtain if the volume density $\rho_0$ also existed in the $r < R$ region. This is why the $E_r$ in the $R < r < 2R$ region in Fig. 12.45(b) is proportional to $r$ (that is, the slope passes through the origin).

2.22  *Discontinuous E and $\phi$*

(a) We know that if there are no other fields present, a sheet with surface charge density $\sigma$ yields an electric field $\sigma/2\epsilon_0$ pointing away from the sheet on either side. More generally, if there are other fields superposed on the sheet's field, the difference in the normal component of the field on either side of the sheet is $\sigma/\epsilon_0$.

This discontinuity is consistent with the relation $\rho = \epsilon_0 \nabla \cdot \mathbf{E}$. If the field varies only in the $x$ direction, then this relation becomes $\rho = \epsilon_0(dE/dx)$. If $E(x)$ looks like the curve shown in Fig. 12.46, then at the jump we have $dE/dx = \Delta E/b$. So $\rho = \epsilon_0(dE/dx)$ gives $\rho = \epsilon_0(\Delta E/b) \implies \Delta E = \rho b/\epsilon_0$. If we let $b \to 0$ and $\rho \to \infty$ while keeping the product $\rho b$ finite, then we have a sheet with surface density $\sigma = \rho b$, and $\Delta E = \sigma/\epsilon_0$, as desired.

(b) Consider two sheets with surface charge densities $-\sigma$ and $\sigma$, separated by a distance $s$. The electric field between them is $\sigma/\epsilon_0$, so the potential difference is $\phi = Es = \sigma s/\epsilon_0$, with the positive sheet at the higher potential. If we let $s \to 0$ and $\sigma \to \infty$ while keeping the product $\sigma s$ finite, then we have a finite change in $\phi$ over zero distance, that is, a discontinuity.

However, this limiting scenario is much less physical than the one in part (a). In part (a), surface densities with finite $\sigma = \rho b$ and "infinite" $\rho$ exist in approximate form on the surfaces of conductors. To a good approximation the thickness of the charge layer is zero. We can have an infinite $\rho$ with a finite amount of charge. Here in part (b), however, situations with finite $\sigma s$ and "infinite" $\sigma$ actually require an infinite amount of charge in any finite patch of area.[2]

At least mathematically, the discontinuity in $\phi$ is consistent with the relation $\rho = -\epsilon_0 \nabla^2 \phi$. In one dimension this relation is $\rho = -\epsilon_0 \, d^2\phi/dx^2$. If $\phi$ looks like the curve shown in Fig. 12.47(a), then the first derivative $d\phi/dx$ (which is just $-E$) looks something like the curve shown in Fig. 12.47(b). It is zero, then rises to the large value of $\Delta\phi/s$ (if $w \ll s$, essentially all of the change in $\phi$ occurs over the $s$ interval), then decreases back to zero. The second derivative $d^2\phi/dx^2$ therefore looks something like the curve shown in Fig. 12.47(c). It is zero, then jumps to the large value of $\Delta\phi/ws$, then drops back to zero, then drops down to $-\Delta\phi/ws$, then jumps back up to zero.

---

[2]  A line of charge technically has an infinite $\sigma$ (just like a sheet has an infinite $\rho$). Indeed, two oppositely charged lines positioned very close to each other do yield an abrupt change in potential over a short distance. (Two point charges would do the same.) But this isn't so interesting, because all of the important behavior is contained in a very small region.

So the relation $\rho = -\epsilon_0 \, d^2\phi/dx^2$ tells us that the large values of the density have magnitude $\rho = \epsilon_0(\Delta\phi/sw)$, which can be rewritten as $\Delta\phi = ((\rho w)/\epsilon_0)s$. The interpretation of this equation is that we have two very thin sheets with thickness $w$ and effective surface charge density $\pm\sigma = \pm\rho w$ (the negative sheet is on the left), which create a field $E = (\rho w)/\epsilon_0$ between them. When this is multiplied by the sheet separation $s$ we obtain the potential difference $\Delta\phi$.

2.23 *Field due to a distribution*

(a) By symmetry, the electric field points in the $x$ direction and depends only on $x$. So $\nabla \cdot \mathbf{E} = \rho/\epsilon_0$ becomes $dE_x/dx = \rho/\epsilon_0$. Integrating this gives the field inside the slab as $E_x = \rho x/\epsilon_0 + A$, where $A$ is a constant of integration. Physically, the $A$ term is the result of superposing the field of an infinite sheet (or slab) with surface charge density $\sigma$, parallel to the given slab and lying outside it. This sheet creates a constant field $\sigma/2\epsilon_0 \equiv A$ that simply gets added to the field from the given slab. The only thing we assumed in the above reasoning was planar symmetry. An additional sheet satisfies this symmetry, so it is no surprise that its effect is included in the final answer. However, since we are told that there are no other charges present, we must have $A = 0$.

(b) By symmetry, the electric field points in the radial direction and depends only on $r$. In cylindrical coordinates, $\nabla \cdot \mathbf{E}$ equals $(1/r)d(rE_r)/dr$ for a function that depends only on $r$. So $\nabla \cdot \mathbf{E} = \rho/\epsilon_0$ gives the field inside the cylinder as

$$\frac{1}{r}\frac{d(rE_r)}{dr} = \frac{\rho}{\epsilon_0} \implies \frac{d(rE_r)}{dr} = \frac{\rho r}{\epsilon_0}$$

$$\implies rE_r = \frac{\rho r^2}{2\epsilon_0} + B \implies E_r = \frac{\rho r}{2\epsilon_0} + \frac{B}{r}. \tag{12.125}$$

Physically, the $B$ term is the result of superposing the field of a line of charge, with linear charge density $\lambda$, along the axis of the cylinder. This line creates a field $\lambda/2\pi\epsilon_0 r \equiv B/r$ that gets added to the field from the cylinder. The only thing we assumed in the above reasoning was cylindrical symmetry. An additional line of charge along the axis satisfies this symmetry, so again it is no surprise that its effect is included in the final answer. But as in part (a), since we are told that there are no other charges present, we must have $B = 0$.

(c) By symmetry, the electric field points in the radial direction and depends only on $r$. In spherical coordinates, $\nabla \cdot \mathbf{E}$ equals $(1/r^2)d(r^2E_r)/dr$ for a function that depends only on $r$. So $\nabla \cdot \mathbf{E} = \rho/\epsilon_0$ gives the field inside the sphere as

$$\frac{1}{r^2}\frac{d(r^2E_r)}{dr} = \frac{\rho}{\epsilon_0} \implies \frac{d(r^2E_r)}{dr} = \frac{\rho r^2}{\epsilon_0}$$

$$\implies r^2E_r = \frac{\rho r^3}{3\epsilon_0} + C \implies E_r = \frac{\rho r}{3\epsilon_0} + \frac{C}{r^2}. \tag{12.126}$$

Physically, the $C$ term is the result of superposing the field of a point charge $q$ at the center of the sphere. This charge creates a field $q/4\pi\epsilon_0 r^2 \equiv C/r^2$ that gets added to the field from the sphere. The only thing we assumed in the above reasoning was spherical symmetry. An additional point charge at the origin satisfies this symmetry, so again it is no surprise that its effect is included in the final answer. But again, since we are told that there are no other charges present, we must have $C = 0$.

In parts (b) and (c), we actually didn't need to be told that there were no other charges present, because the line charge and the point charge yield a nonzero value of $\rho$ (in fact infinite, since they have a finite amount of charge in zero volume), so they violate our assumption of uniform volume charge density inside the object. In part (a) the additional sheet was located *outside* the slab, so it didn't affect the $\rho$ inside. Of course, we could superpose a spherical or cylindrical shell outside our given sphere or cylinder, but these create zero field in their interiors. You should think about what is different about a "ball" in the 1D case.

In parts (b) and (c), how did we solve an equation involving a uniform density $\rho$ and then end up with a density that wasn't uniform (containing an extra line or point)? This occurred because our calculations in Eqs. (12.125) and (12.126) aren't valid at $r = 0$. Problem 2.26 deals with the complications at $r = 0$ in the spherical case.

(d) There are an infinite number of solutions to the equation $\nabla \cdot \mathbf{E} = \rho/\epsilon_0$, but only a certain subset satisfy the symmetry of the given setup. For example, consider the spherical case in part (c), and imagine superposing additional charges at arbitrary locations outside the sphere. Then the field inside the sphere equals $E_r = \rho r/3\epsilon_0$ plus the standard Coulomb fields from all the charges located outside. This is a perfectly valid solution to $\nabla \cdot \mathbf{E} = \rho/\epsilon_0$ in the interior. But it doesn't respect the spherical symmetry of the original setup. If we keep superposing external charges in a particular manner until we just so happen to create an entire infinite uniform *cylinder* containing the sphere, then the field inside the sphere is now given by the cylindrical solution in part (b).

The point is that when we solved $\nabla \cdot \mathbf{E} = \rho/\epsilon_0$ in the above three cases, we were actually demanding not only that the density equaled $\rho$ inside the object, but also that it equaled *zero outside*. A shortcut to incorporating this information (or at least the important aspect of it) was simply to demand that the solution possessed a certain symmetry. If we had an unsymmetrical object, then we wouldn't be able to take this shortcut. In any case, when determining $\mathbf{E}$ at a given point, the density throughout *all* space matters, even though when determining $\nabla \cdot \mathbf{E}$ at a given point only the *local* density matters. In short, there are many different vector fields that have the same divergence in a given region in space. In 1D, these fields differ by an arbitrary additive constant. But in 2D and 3D, the degeneracy in the solutions is much larger.

2.24 *Two expressions for the energy*

(a) In Cartesian coordinates we have

$$\nabla \cdot (\phi \mathbf{E}) = \frac{\partial(\phi E_x)}{\partial x} + \frac{\partial(\phi E_y)}{\partial y} + \frac{\partial(\phi E_z)}{\partial z}$$

$$= \left( \frac{\partial \phi}{\partial x} E_x + \frac{\partial \phi}{\partial y} E_y + \frac{\partial \phi}{\partial z} E_z \right) + \left( \phi \frac{\partial E_x}{\partial x} + \phi \frac{\partial E_y}{\partial y} + \phi \frac{\partial E_z}{\partial z} \right)$$

$$= \left( \frac{\partial \phi}{\partial x}, \frac{\partial \phi}{\partial y}, \frac{\partial \phi}{\partial z} \right) \cdot (E_x, E_y, E_z) + \phi \left( \frac{\partial E_x}{\partial x} + \frac{\partial E_y}{\partial y} + \frac{\partial E_z}{\partial z} \right)$$

$$= (\nabla \phi) \cdot \mathbf{E} + \phi \nabla \cdot \mathbf{E}, \tag{12.127}$$

as desired.

(b) If $\phi$ and $\mathbf{E}$ are the electric potential and field, then $\nabla \phi = -\mathbf{E}$ and $\nabla \cdot \mathbf{E} = \rho / \epsilon_0$. So the above identity becomes

$$\nabla \cdot (\phi \mathbf{E}) = -\mathbf{E} \cdot \mathbf{E} + \phi \frac{\rho}{\epsilon_0}. \tag{12.128}$$

Let's now integrate both sides over the volume of a very large sphere with radius $R$. On the left-hand side we can use the divergence theorem to write the volume integral $\int_V \nabla \cdot (\phi \mathbf{E})$ as a surface integral $\int_S \phi \mathbf{E} \cdot d\mathbf{a}$. We obtain (using $\mathbf{E} \cdot \mathbf{E} = E^2$)

$$\int_S \phi \mathbf{E} \cdot d\mathbf{a} = - \int_V E^2 \, dv + \int_V \frac{\rho \phi}{\epsilon_0} \, dv. \tag{12.129}$$

If the surface integral is zero, then the resulting equation can be written as

$$\frac{\epsilon_0}{2} \int_V E^2 \, dv = \frac{1}{2} \int_V \rho \phi \, dv, \tag{12.130}$$

which is the desired result. And indeed, the surface integral is zero for the following reason. Since we are assuming that all sources lie within a finite region, we can enclose them in a sphere of some radius $r$. If we let the radius $R$ of our integration surface $S$ go to infinity, the charges inside the sphere of radius $r$ look effectively like a point charge from far away. On $S$, the field $E$ will therefore vanish at least as fast as $1/R^2$ (faster if the net charge of the distribution is zero), and $\phi$ will vanish at least as fast as $1/R$. Since the surface area of $S$ grows like $R^2$, the surface integral over $S$ will vanish at least as fast as $R^2/R^3 = 1/R$. It therefore vanishes as $R \to \infty$. (If the sources were not confined to a finite region, then we could not be sure that any of these integrals would converge when extended over all space.)

Note that the result we have proved holds only as an integral statement over all space. It does *not* hold in differential form, that is, as the local statement that $\epsilon_0 E^2 = \rho \phi$. This is no surprise, because we can easily have a point in space where $\rho = 0$ but $E \neq 0$, in which case the two integrands in our result are certainly not equal at that point.

2.25	*Never trapped*

From Earnshaw's theorem in Section 2.12, we know that there exists a direction in which the potential energy initially decreases. That is, there exists a direction in which the electric field points outward. But if the charge $q$ heads in that direction, how do we know that it won't later encounter a point where heading farther outward would entail increasing the potential energy? We know that there must always be a route for which the potential energy decreases, for the following reason.

Consider the field line (due to all the fixed charges) that the charge $q$ is initially on, for which the potential energy decreases. If we follow this field line, where do we end up? We can't end up back where we started, because that would imply a nonzero curl for the electric field. And we can't end up at one of the fixed positive charges, because all of the field lines point outward near them. The only other option for where we can end up is at infinity. We have therefore constructed an escape path, as desired. See Problem 2.29 for more discussion on where field lines can end.

Note that this reasoning breaks down if there are negative charges among the given fixed charges, because field lines can end at negative charges. And indeed, if a negative charge is sufficiently large, our positive charge $q$ will certainly be trapped.

In Problem 2.5 we explicitly showed that the path through the center of a face of the cube was an escape route. You might think that if you surround the charge $q$ with enough fixed positive charges on the surface of a sphere, then it won't be able to squeeze through a "face." However, if you have a very large number of equal fixed charges, then you essentially have a uniform sphere. And we know that the field inside a sphere is zero. So our charge $q$ has no trouble getting from the center to the vicinity of the midpoint of a face.

2.26	*The delta function*

The Laplacian $\nabla^2$ operator is shorthand for the divergence of the gradient. From Appendix F, the gradient of a function that depends only on $r$ equals $(\partial f/\partial r)\hat{\mathbf{r}}$. So we have $\nabla(1/r) = -\hat{\mathbf{r}}/r^2$. The volume integral of $\nabla^2(1/r)$ is therefore (using the divergence theorem)

$$\int_V \nabla^2 \left(\frac{1}{r}\right) dv = \int_V \nabla \cdot \nabla \left(\frac{1}{r}\right) dv = \int_V \nabla \cdot \frac{-\hat{\mathbf{r}}}{r^2} \, dv = -\int_S \frac{\hat{\mathbf{r}}}{r^2} \cdot d\mathbf{a}.$$

(12.131)

Since $\nabla^2(1/r) = 0$ everywhere away from the origin, any volume containing the origin will yield the same integral. Choosing a sphere of radius $R$ gives

$$\int \nabla^2 \left(\frac{1}{r}\right) dv = -\int \frac{\hat{\mathbf{r}}}{r^2} \cdot d\mathbf{a} = -\int \frac{1}{R^2} \, da = -\frac{4\pi R^2}{R^2} = -4\pi,$$

(12.132)

as desired. A function that is (1) zero everywhere except at one point and (2) infinite enough at that one point so that it has a nonzero integral, is called a *delta function* (up to numerical factors).

Note that for any function $F(\mathbf{r})$, the integral $\int \nabla^2(1/r)F(\mathbf{r})dv$ equals $-4\pi F(0)$. This is true because $\nabla^2(1/r)$ is zero everywhere except the origin, so the only value of $F(\mathbf{r})$ that matters is $F(0)$. We can then pull this constant value outside the integral. The integral of the product of a function $F(\mathbf{r})$ and $\nabla^2(1/r)$ basically picks out the value of the function at the origin.

**2.27** *Relations between $\phi$ and $\rho$*

The derivatives in the $\nabla^2$ operator we are using are with respect to the unprimed coordinates, because the $\phi$ on the left-hand side of the given relation depends on $\mathbf{r}$. Let us write $\nabla^2$ as $\nabla_r^2$, to make this explicit. We have

$$\nabla_r^2 \phi(\mathbf{r}) = \frac{1}{4\pi\epsilon_0} \nabla_r^2 \int \frac{\rho(\mathbf{r}')\,dv'}{|\mathbf{r}'-\mathbf{r}|} = \frac{1}{4\pi\epsilon_0} \int \nabla_r^2 \left(\frac{1}{|\mathbf{r}'-\mathbf{r}|}\right) \rho(\mathbf{r}')\,dv'.$$

$$(12.133)$$

We claim that $\nabla_r^2\big(1/|\mathbf{r}'-\mathbf{r}|\big) = \nabla_{r'}^2\big(1/|\mathbf{r}'-\mathbf{r}|\big)$. That is, the Laplacian $\nabla_r^2$ of $1/|\mathbf{r}'-\mathbf{r}|$ with respect to the unprimed coordinates equals the Laplacian $\nabla_{r'}^2$ of $1/|\mathbf{r}'-\mathbf{r}|$ with respect to the primed coordinates. You can verify this by writing $|\mathbf{r}'-\mathbf{r}|$ as $\sqrt{(x'-x)^2+(y'-y)^2+(z'-z)^2}$ and explicitly calculating the derivatives in Cartesian coordinates. We now have

$$\nabla_r^2 \phi(\mathbf{r}) = \frac{1}{4\pi\epsilon_0} \int \nabla_{r'}^2 \left(\frac{1}{|\mathbf{r}'-\mathbf{r}|}\right) \rho(\mathbf{r}')\,dv'. \qquad (12.134)$$

As mentioned in the solution to Problem 2.26, the integral $\int \nabla^2(1/r)F(\mathbf{r})\,dv$ equals $-4\pi F(0)$. If the $\mathbf{r}$ were not present on the right-hand side of Eq. (12.134), we would have exactly the same type of integral, in which case the right-hand side would equal $(1/4\pi\epsilon_0)\big(-4\pi\rho(0)\big)$. The presence of the $\mathbf{r}$ term simply shifts the origin (equivalently, you can define a new coordinate system with the particular $\mathbf{r}$ value as the origin), so we instead end up with

$$\nabla_r^2 \phi(\mathbf{r}) = \frac{1}{4\pi\epsilon_0}\big(-4\pi\rho(\mathbf{r})\big) = -\frac{\rho(\mathbf{r})}{\epsilon_0}, \qquad (12.135)$$

as desired. Physically, we already knew why the two relations, $\phi = (1/4\pi\epsilon_0)\int (\rho/r)\,dv'$ and $\nabla^2\phi = -\rho/\epsilon_0$, are equivalent: they are both obtained from Coulomb's inverse-square law. The first is obtained by integration and superposition (see Section 2.5), while the second is obtained via Gauss's law (see Section 2.11), which in turn is equivalent to the inverse-square law (see Section 1.10). But it's nice to see how the equivalence can be demonstrated in a strict mathematical sense.

**2.28** *Zero curl*

The curl of $\mathbf{E}$ is given by

$$\text{curl }\mathbf{E} = \begin{vmatrix} \hat{\mathbf{x}} & \hat{\mathbf{y}} & \hat{\mathbf{z}} \\ \partial/\partial x & \partial/\partial y & \partial/\partial z \\ 2xy^2+z^3 & 2x^2y & 3xz^2 \end{vmatrix}$$

$$= \hat{\mathbf{x}}(0-0) + \hat{\mathbf{y}}(3z^2-3z^2) + \hat{\mathbf{z}}(4xy-4xy) = 0. \qquad (12.136)$$

To find the associated potential $\phi$, we could evaluate the (negative) line integral of $\mathbf{E}$ from a given reference point to a general point $(x, y, z)$. But an easier method is simply to find a function $\phi$ that satisfies $\mathbf{E} = -\nabla\phi$. Looking at the $x$ component of this relation, we see that we need $2xy^2 + z^3 = -\partial\phi/\partial x$. Hence $\phi$ must take the form

$$\phi = -x^2 y^2 - xz^3 + f(y, z). \tag{12.137}$$

The arbitrary function $f(y, z)$ won't ruin the $2xy^2 + z^3 = -\partial\phi/\partial x$ equality because $\partial f(y, z)/\partial x = 0$. Similarly, the $y$ component of $\mathbf{E} = -\nabla\phi$ yields $2x^2 y = -\partial\phi/\partial y$, so $\phi$ must take the form

$$\phi = -x^2 y^2 + f(x, z). \tag{12.138}$$

And the $z$ component yields $3xz^2 = -\partial\phi/\partial z$, so $\phi$ must take the form

$$\phi = -xz^3 + f(x, y). \tag{12.139}$$

You can quickly check that the only function consistent with all three of these forms is $\phi = -x^2 y^2 - xz^3 + C$, where $C$ is an arbitrary constant. If the curl of $\mathbf{E}$ weren't zero, then there would exist no function consistent with the three required forms.

2.29 *Ends of the lines*

If an electrostatic field line forms a closed loop, then the field will do nonzero work on a charge during a round trip. But we know that this can't be the case because the electric force is conservative. Equivalently, if a field line forms a closed loop, then the line integral $\int \mathbf{E} \cdot d\mathbf{s}$ around the loop will be nonzero. So the integral $\int (\nabla \times \mathbf{E}) \cdot d\mathbf{a}$ will also be nonzero, by Stokes' theorem. But this contradicts the fact that curl $\mathbf{E}$ is zero for an electrostatic field.

A field line can end only on a charge or at infinity (that is, it never ends), for the following reason. If a field line ends at a point in space where there is no charge, then the divergence of $\mathbf{E}$ will be nonzero there, because a field line goes into a given small volume but doesn't come out. But we are assuming that the charge density is zero at this point, so this violates Gauss's law, $\nabla \cdot \mathbf{E} = \rho/\epsilon_0$.

However, we should promptly make note of the fact that the preceding reasoning is a bit sloppy. A single field line carries no flux, so technically there would be no violation of Gauss's law if a single field line (or a finite number of them) ended in free space. Instead of field lines, we should be talking about narrow bundles of flux (or "flux tubes," but that term is usually reserved for magnetic flux). If a bundle of flux ends in free space, then that would be a true violation of Gauss's law. So a better way of saying things is that a very thin bundle must end up either on a charge or at infinity. Well, that is if it remains a thin bundle...

Admittedly, we're getting a little picky here, but consider the setup in Problem 2.18 dealing with the field from two equal point charges (although the specifics of the setup aren't critical). The field is zero midway between the charges, so if we look at the field line that leaves one of the charges and heads directly toward the other, where does it end up? In some sense, it ends at the $E = 0$ point. But this is mainly a matter of semantics. We could just as well have the line continue outward from the $E = 0$ location along any of the lines fanning out in the perpendicular-bisector plane.

Even if the field line ends at this point, we don't care. As noted above, a single field line is irrelevant; it has no strength; it yields no flux (so it escapes the above Gauss's-law reasoning); it has "measure zero," so there is zero probability that an idealized point particle could actually be on it. If we think (more correctly) in terms of thin bundles of flux, then we can consider a bundle that starts out at one of our two charges, and that has as its axis the line joining the two charges. But then look at what happens to this bundle! From Fig. 12.40 in the solution to Problem 2.18, we see that near the $E = 0$ point the bundle fans out and becomes a pancake spanning the entire perpendicular-bisector plane between the charges. We have a funnel-like surface of revolution generated by rotating one of the field lines around the $x$ axis. No matter how thin the bundle starts off, it ends up as a pancake (assuming that it contains the $x$ axis). In this more physical sense of thinking of field lines in terms of bundles of flux, our problematic field line isn't well defined. As an exercise, you can use the results from Problem 2.18 to show that the flux heading inward through the thin part of the bundle does indeed equal the flux heading outward along the pancake part of the "bundle."

In short, the only statements about an electrostatic field **E** that we have at our disposal are $\nabla \times \mathbf{E} = 0$ and $\nabla \cdot \mathbf{E} = \rho/\epsilon_0$ (along with their integral forms). So we should be wary about making statements that don't involve circulation or flux.

2.30 *Curl of a gradient*

(a) Using the determinant expression for the cross product, we have

$$\nabla \times \mathbf{E} = -\nabla \times \nabla\phi = - \begin{vmatrix} \hat{\mathbf{x}} & \hat{\mathbf{y}} & \hat{\mathbf{z}} \\ \partial/\partial x & \partial/\partial y & \partial/\partial z \\ \partial\phi/\partial x & \partial\phi/\partial y & \partial\phi/\partial z \end{vmatrix} . \quad (12.140)$$

The $x$ component of this is $-\partial^2\phi/\partial y\partial z + \partial^2\phi/\partial z\partial y$. But partial differentiation is commutative (that is, the order doesn't matter), so this component equals zero. Likewise for the $y$ and $z$ components.

(b) Using the given relation $\mathbf{E} = -\nabla\phi$ along with Stokes' theorem, we have

$$\int_S (\nabla \times \mathbf{E}) \cdot d\mathbf{a} = - \int_S (\nabla \times \nabla\phi) \cdot d\mathbf{a} = - \int_C \nabla\phi \cdot d\mathbf{s}, \quad (12.141)$$

where $C$ is the closed curve that bounds the surface $S$. But $\nabla\phi \cdot d\mathbf{s}$ is the change in $\phi$ over the interval $d\mathbf{s}$. When we integrate this we obtain the total change in $\phi$ between the limits of integration. But these limits are the same point because the curve is closed. So the integral is zero.

Hence $\int_S (\nabla \times \mathbf{E}) \cdot d\mathbf{a} = 0$. And since this holds for any surface $S$, it must be the case that $\nabla \times \mathbf{E} = 0$ at all points.

The logic here basically boils down to the mathematical fact that the boundary of a boundary is zero. Curve $C$ is the boundary of surface $S$. And the fact that $C$ itself has no boundary (it is a closed curve with no endpoints) is why $\int_C \nabla\phi \cdot d\mathbf{s}$ equals zero. (Similarly, a surface $S$ that encloses a volume $V$ has no boundary. This fact provides a solution to Exercise 2.78.)

## 12.3 Chapter 3

3.1 *Inner-surface charge density*

The charge density is negative over the entire inner surface. This is true because if there were a location with positive density, then electric field lines would start there, pointing away from it into the spherical cavity. But where could these field lines end? They can't end at infinity, because that's outside the shell. And they can't end at a point in empty space, because that would violate Gauss's law; there would be nonzero flux into a region that contains no charge (see Problem 2.29 for a more detailed discussion of this). They also can't end on the positive point charge $q$, because the field lines point outward from $q$. And finally they can't end on the shell, because that would imply a nonzero line integral of $\mathbf{E}$ (and hence a nonzero potential difference) between two points on the shell. But we know that all points on the conducting shell are at the same potential. Therefore, such a field line (pointing inward from the inner surface) can't exist. So all of the inner surface charge must be negative. Every field line inside the cavity starts at the point charge $q$ and ends on the shell.

3.2 *Holding the charge in place*

Consider a path that runs from conductor $B$ across the gap to conductor $D$, then through the interior of the wire that connects $D$ to $C$, then across the gap to $A$, then finally via the other wire down to $B$. The line integral of $\mathbf{E}$ around any closed path must be zero, if $\mathbf{E}$ is a static electric field. But if the fields are as shown in Fig. 3.23(c), the line integral over the closed path just described is *not* zero. Each gap makes a positive contribution; but in the conductors, including the connecting wires, $\mathbf{E}$ is zero. So the proposed situation cannot represent a static charge distribution.

Although the above reasoning is perfectly valid, you might be looking for a more "cause and effect" reason why the charge redistributes itself. What happens is this: the charge that $C$ induces on $A$ (and likewise that $D$ induces on $B$) isn't enough to keep all of $C$'s charge in place when the wire is connected. The self-repulsion of the charges within $C$ wins out over the attraction from $A$. The quantitative details are contained (for the most part) in Problem 3.13. The main point is that the induced charge on $A$ is *smaller* than the charge on $C$, and the charge that $A$ is able to keep on $C$ is smaller still. So there is a nonzero amount of charge on $C$ that gets repelled away down the wire. Of course, once this happens, then the charge on $A$ decreases, and the whole system cascades down to zero charge everywhere.

3.3    *Principal radii of curvature*

(a) For a sphere with radius $R$, we have $1/R_1 + 1/R_2 = 2/R$. The field outside the sphere is $E = q/4\pi\epsilon_0 r^2$, so $dE/dr = -2q/4\pi\epsilon_0 r^3 = -(2/r)E$. At the surface of the sphere this equals $-(2/R)E$, as desired.

For a cylinder with radius $R$, we have $1/R_1 + 1/R_2 = 1/R + 1/\infty = 1/R$. The field outside the cylinder is $E = \lambda/2\pi\epsilon_0 r$, so $dE/dr = -\lambda/2\pi\epsilon_0 r^2 = -(1/r)E$. At the surface of the cylinder this equals $-(1/R)E$, as desired.

For a plane we have $1/R_1 + 1/R_2 = 2/\infty = 0$. The field from a plane takes on the constant value of $E = \sigma/2\epsilon_0$, so $dE/dx = 0$, as desired.

(b) Consider a pillbox with an approximately rectangular base that lies just outside the surface of the conductor. Let the edges of the base (with lengths $\ell_1$ and $\ell_2$) be aligned parallel to the directions of the principal curvatures. Let the sides of the pillbox (with height $dx$) be normal to the surface, so that they follow the field lines. One of the cross sections is shown in Fig. 12.48.

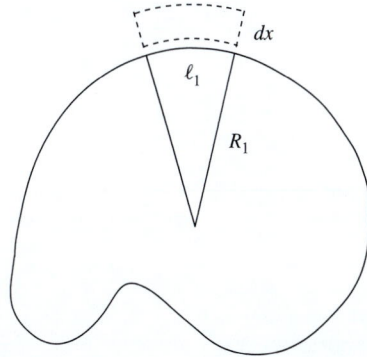

**Figure 12.48.**

Note that the top face of the box is larger than the bottom face (assuming the principal radii are positive). From similar triangles in Fig. 12.48, the top edges are longer than the bottom edges by factors of $(R_1 + dx)/R_1$ and $(R_2 + dx)/R_2$. So the area of the top face is

$$A_{\text{top}} = \left(1 + \frac{dx}{R_1}\right)\ell_1 \cdot \left(1 + \frac{dx}{R_2}\right)\ell_2 \approx \ell_1 \ell_2 \left(1 + dx\left(\frac{1}{R_1} + \frac{1}{R_2}\right)\right),$$

$$(12.142)$$

where we have dropped the $(dx)^2$ term. Now, the net flux through the box is zero because it contains no charges. There is no flux through the sides since they are parallel to the field lines, so equating the flux inward through the bottom with the flux outward through the top gives

$$E_{\text{bot}}A_{\text{bot}} = E_{\text{top}}A_{\text{top}}$$

$$\implies E_{\text{bot}}\ell_1\ell_1 = E_{\text{top}}\ell_1\ell_2\left(1 + dx\left(\frac{1}{R_1} + \frac{1}{R_2}\right)\right)$$

$$\implies E_{\text{top}} = E_{\text{bot}}\left(1 + dx\left(\frac{1}{R_1} + \frac{1}{R_2}\right)\right)^{-1}$$

$$\implies E_{\text{top}} \approx E_{\text{bot}}\left(1 - dx\left(\frac{1}{R_1} + \frac{1}{R_2}\right)\right). \qquad (12.143)$$

The change in the field from bottom to top is $dE = E_{\text{top}} - E_{\text{bot}}$, so we have

$$dE = -E_{\text{bot}}\left(\frac{1}{R_1} + \frac{1}{R_2}\right)dx \implies \frac{dE}{dx} = -\left(\frac{1}{R_1} + \frac{1}{R_2}\right)E,$$

$$(12.144)$$

where we have written $E_{\text{bot}}$ as $E$. This result is valid for negative radii of curvature as well (that is, where the surface is concave). The top surface is now smaller than the bottom surface, but the derivation is exactly the same. We could also have one positive and one negative radius. The above result is also perfectly valid inside a hollow conducting shell, although if there are no charges enclosed, the relation is trivial because both $dE$ and $E$ are zero. Inside the material of a conductor, the relation is trivial in any case, because $dE$ and $E$ are always zero.

3.4  *Charge distribution on a conducting disk*
Recall the argument in Problem 1.17 that showed why the field inside a spherical shell is zero. In Fig. 12.49 the two cones that define the charges $q_1$ and $q_2$ on the surface of the shell are similar, so the areas of the end patches are in the ratio of $r_1^2/r_2^2$. This factor exactly cancels the $1/r^2$ effect in Coulomb's law, so the fields from the two patches are equal and opposite at point $P$. The field contributions from the entire shell therefore cancel in pairs.

Let us now project the charges residing on the upper and lower hemispheres onto the equatorial plane containing $P$. The charges $q_1$ and $q_2$ in the patches mentioned above end up in the shaded patches shown in Fig. 12.49. The (horizontal) fields at point $P$ from these shaded patches have magnitudes $q_1/4\pi\epsilon_0 x_1^2$ and $q_2/4\pi\epsilon_0 x_2^2$. But due to the similar triangles in the figure, $x_1$ and $x_2$ are in the same ratio as $r_1$ and $r_2$. Hence the two forces have equal magnitudes, just as they did in the case of the spherical shell. The forces from all of the various parts of the disk therefore again cancel in pairs, so the horizontal field is zero at $P$. Since $P$ was arbitrary, we see that the horizontal field is zero everywhere in the disk formed by the projection of the charge from the original shell. We have therefore accomplished our goal of finding a charge distribution that produces zero electric field component parallel to the disk.

Since the spherical shell has a larger slope near the sides, more charge is above a given point in the equatorial plane near the edge, compared with at the center. The density of the conducting disk therefore grows with $r$. We can be quantitative about this. In Fig. 12.50, let $\theta$ be measured down from the top of the shell, and let $r$ be the radius of a given point in the disk. Consider a patch with area $A$ at radius $r$ in the plane of the disk. The patch above it on the shell is tilted at an angle $\theta$, so its area is $A/\cos\theta$. The density in the disk is therefore proportional to $1/\cos\theta$. But $\cos\theta = \sqrt{R^2 - r^2}/R$, so the density takes the form of $\sigma R/\sqrt{R^2 - r^2}$, where $\sigma$ is determined by requiring that the total charge be $Q$:

$$Q = \int_0^R \frac{\sigma R}{\sqrt{R^2 - r^2}} 2\pi r\, dr = -2\pi\sigma R\sqrt{R^2 - r^2}\,\Big|_0^R = 2\pi\sigma R^2.$$

$$(12.145)$$

Hence $\sigma = Q/2\pi R^2$, and the desired surface charge density of the conducting disk is

$$\sigma_{\text{disk}} = \frac{Q}{2\pi R\sqrt{R^2 - r^2}}.$$

$$(12.146)$$

Note that the density at the center of the disk is $Q/2\pi R^2$, which is exactly half of the density $Q/\pi R^2$ of a *non*conducting disk with radius $R$ and

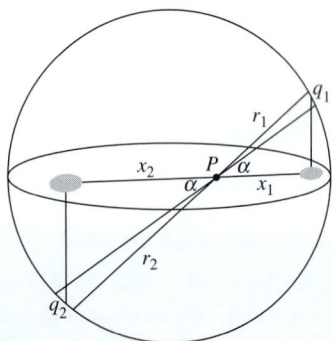

**Figure 12.49.**

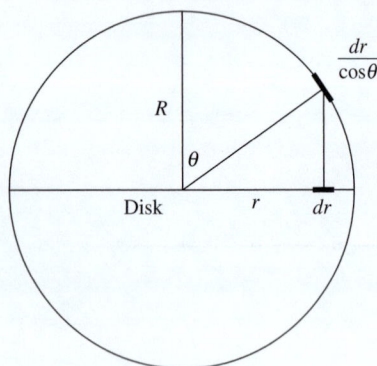

**Figure 12.50.**

charge $Q$. See Friedberg (1993) and Good (1997) for further discussion of this problem.

Another way of obtaining the above $\sigma = Q/2\pi R^2$ result, without doing the integral in Eq. (12.145), is to note that the $\sigma R/\sqrt{R^2 - r^2}$ form of the density implies that the density at the center of the disk is $\sigma$. But this equals the density at the top of the hemispherical shell, because the shell isn't tilted there. And since the shell's density is uniform, we have $(2\pi R^2)\sigma = Q$, because the area of the hemisphere is $2\pi R^2$. Hence $\sigma = Q/2\pi R^2$. (Projecting the bottom hemisphere too wouldn't change the final result for the distribution in the disk, given that the total charge is $Q$.)

The density diverges at the edge of the conducting disk, but the total charge has the finite value $Q$. It is fairly intuitive that the density should grow as $r$ increases, because charges repel each other toward the edge of the disk. However, one should be careful with this type of reasoning. In the lower-dimensional analog involving a one-dimensional rod of charge, the density is actually essentially uniform, all the way out to the end; see Problem 3.5.

3.5   *Charge distribution on a conducting stick*
There are three basic cases, although we can actually group them all together in our reasoning. As shown in Fig. 12.51, a point charge at a given point $P$ can be close to the center, or not close to the center or an end, or close to an end. If the $N$ charges on the line are all equal, then in all three cases the segments of equal length on either side of $P$ produce canceling fields, so the unbalanced field comes from the regions indicated by the shading in the figure.

Let's get a handle on this unbalanced field. Let the point charge at $P$ be the $n$th charge from the left end. Then the unbalanced field comes from the charges (all equal to $Q/N$) that are a distance of at least $n(L/N)$ to the right of point $P$. The unbalanced field is therefore (ignoring the $4\pi\epsilon_0$ since it will cancel throughout this problem)

$$E = \frac{Q/N}{(L/N)^2}\left(\frac{1}{n^2} + \frac{1}{(n+1)^2} + \frac{1}{(n+2)^2} + \cdots\right). \qquad (12.147)$$

This sum can be approximated by an integral (even for small $n$ since we are concerned only with a rough value). In the case where $n \ll N$ (that is, where $P$ is very close to the left end), the sum effectively extends out to infinity, so the integral equals $1/n$. On the other hand, if $n$ is of order $N$, then the integral doesn't extend out to infinity. However, we are concerned with an upper bound on the unbalanced field, and an upper bound on the sum is certainly $1/n$. So, in all cases, the unbalanced field is less than or (roughly) equal to $QN/nL^2$. (For most of the stick, we can say that $n$ is of order $N$, which means that the unbalanced field is bounded by something of order $Q/L^2$. This makes sense because the unbalanced piece has a charge on the order of $Q$, and a distance from $P$ on the order of $L$.)

So the question is: how much charge $dq$ do we need to add to the point charge immediately to the left of $P$, so that its rightward-pointing field increases by an amount on the order of $QN/nL^2$, to balance out the field due to the shaded region in Fig. 12.51? The answer is: not much, due

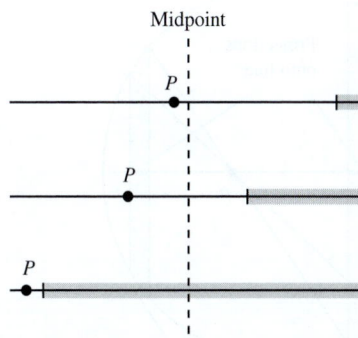

**Figure 12.51.**
The shaded region indicates the part of the stick whose field is left over after the canceling of the fields from the regions of equal length on either side of $P$.

to the close spacing of the charges. The field due to the adjacent point charge will increase by $dq/(L/N)^2$, so if we want this to equal $QN/nL^2$, we need $dq = Q/nN$. This increase is $1/n$ times the original charge $Q/N$.

This modification will of course affect the fields at other locations, so we will need to make iterative corrections. But the general size of the correction to the $n$th charge will be (less than or equal to) of order $Q/nN$. Since the sum of the reciprocals of the integers grows logarithmically, we see that the sum of the changes to all $N$ charges will be of order $Q(\ln N)/N$. In the $N \to \infty$ limit, this goes to zero, which means that we need to add on essentially zero charge, compared with the original charge $Q$ on the stick. In other words, the final charge density is essentially uniform.

Very close to the end of the stick, where $n$ is of order 1, the correction $dq = Q/nN$ is of the same rough size as the original point-charge value $Q/N$. But this region near the end takes up a negligible fraction of the whole length. To see why, let $N$ equal one billion, and consider the end region where the resulting charges differ from $Q/N$ by at most, say, 0.1 percent. Since $dq = Q/nN$, this region extends (roughly) out to the point charge with $n = 1000$. The region therefore has a length that is (roughly) one millionth of the total length. In the $N \to \infty$ limit (however physical that limit may be), the $dq$ corrections vanish at any finite distance away from the end, which means that the stick's density is exactly uniform. The point is that the relative size of the $dq$ correction depends on a given charge's *index n*, as opposed to its *distance* from the end. And compared with an $N$ that heads to infinity, any given number $n$ is infinitesimally small.

It is also possible to demonstrate the general result of this problem by keeping the sizes of the point charges fixed and imagining moving them slightly to balance the fields. The result is that, for large $N$, the charges barely need to move in order to balance out the fields.

The moral of this problem is that due to the $1/r^2$ nature of the field from a point charge (although you can show that any power larger than 1 in the denominator would be sufficient), and due to the very small separation between the charges, it takes only very minor changes in the nearby charge values to create large changes in the local fields, thereby canceling the unbalanced field due to a macroscopic amount of charge a macroscopic distance away. You are encouraged to see how the above reasoning is modified when applied to the 2D case (with a plane built up from lines of charge, whose fields fall off like $1/r$), and also the 3D case (with a volume built up from planes of charge, whose fields don't fall off at all). See Andrews (1997) and Griffiths and Li (1996) for further discussion of this problem.

REMARK: As mentioned in the statement of the problem, the uniform density of the stick can also be demonstrated (in a much quicker manner) by using the technique from Problem 3.4 and Good (1997). If we have a spherical shell with uniform surface charge density, and if we project the two rings in Fig. 12.52 onto the horizontal axis (the stick), then you can check that the two resulting line segments produce canceling fields at point $P$. You can also check that this projection produces a uniform charge distribution on the stick. (In short, for a given horizontal width $dx$ of a

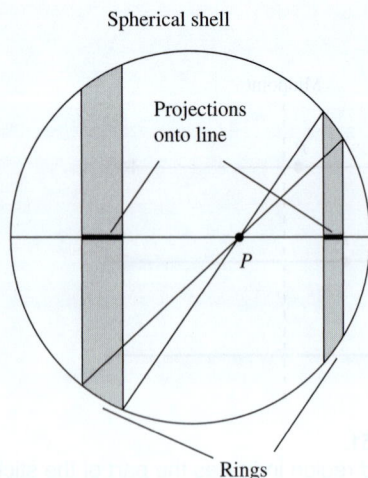

Spherical shell

Projections onto line

$P$

Rings

**Figure 12.52.**

ring, a ring with a smaller radius has a smaller circumference but a larger tilt angle of the surface, and these effects exactly cancel when finding the area of the ring.) Since we can consider the stick to be completely built up from these pairs of line segments (with a smaller piece to the right of $P$, and a larger piece to the left), the field at $P$ is therefore zero. It is amusing how this method of dividing the stick into canceling pairs of little segments is much more useful than the seemingly more natural method involving segments equidistant from $P$, which is the one we used above in Fig. 12.51.

There is actually no need to make use of the above projection of a sphere (which in turn makes use of the $1/r^2$ nature of the electric force) to demonstrate that a uniform stick yields zero net field at any given point $P$. Given a uniform stick, and given a $1/r^d$ force law with $d > 1$, we can divide the stick into pairs of corresponding little segments whose field contributions cancel at $P$. We can start with corresponding little segments at the ends of the stick and then work our way inward to $P$. For a given pair of canceling segments, the segment that is farther from $P$ will be longer (assuming $d > 0$, so that the field falls off with distance). More importantly, if $d > 1$ it will be *enough* longer so that (as you can show) the pairs will eventually be essentially the *same distance* from $P$ when they get very close to $P$; $P$ is therefore located at the *center* of the effective stick that is left over after ignoring all the canceled corresponding pairs. The net field at $P$ is therefore zero. In contrast, if $d = 1$, then, as we work our way inward toward $P$, you can show that the corresponding little segments maintain the *same ratio* of their distances from $P$. Since these distances remain unequal (assuming $P$ isn't located exactly at the center of the original stick), $P$ remains off-center in the effective stick that is left over after ignoring the canceled pairs. The field at $P$ is therefore nonzero. Hence the charge distribution on a conducting stick, in a world with a $1/r$ electric force, is not uniform. Can you determine what it is?

3.6    *A charge inside a shell*
The reasoning is incorrect. The charge will feel a force. The error is that the reasoning takes a solution for one boundary condition and applies it to a situation with another boundary condition. A consequence of the uniqueness theorem is that if we have a surface with constant potential and *no charges* inside, then the constant-$\phi$ solution must be *the* solution. This (true) statement has nothing to do with the given setup containing a point charge. So for the given setup, we can draw no conclusions from the above statement, no matter how true it is. (Physically, there is a nonzero force because the dominant effect is that the charges induced on the nearer part of the shell will attract the given point charge.)

3.7    *Inside/outside asymmetry*
Electrostatic field lines can begin and end only at charges or at infinity. Also, there can be no closed loops since curl $\mathbf{E} = 0$. See Problem 2.29.

If the point charge is located *outside* the shell, the field lines can have their ends at the point charge, the shell, or infinity. There can't be any field lines inside the shell because they would have to start at one point on the shell and end at another. This would imply a nonzero potential difference between these two points, contradicting the fact that all points on the shell have the same potential.

If the point charge is located *inside* the shell, the field lines starting at the point charge must end up on the shell. And there *can* be field lines outside the shell because they can have one end on the shell and the other end at infinity.

We see that the basic difference between inside and outside is that the inside region has only one boundary (the shell), while the outside region has two (the shell and infinity). The former case therefore requires the existence of an additional termination point (a charge) if field lines are to exist.

3.8     *Inside or outside*

As we increase the size of shell *A*, the field at *P* remains zero. When we hit the infinite-plane transition between the two cases, the field is still zero. As we then transition to a very large shell *B* surrounding the charge *q*, the field at *P* is *very small*. One way of seeing this is that there is a large amount of induced negative charge on the shell close to the charge *q* (almost as much as on the infinite plane), and this charge nearly completely shields the field of the charge *q* at point *P*. Alternatively, from the example in Section 3.2, a point charge inside a conducting spherical shell looks like an equal point charge *at the center,* as far as the external field goes. And the center of a very large shell *B* is very far to the left. So the field at *P* due to the effective point charge at the distant center is very small.

As we decrease the size of *B*, the field at *P* increases because not as much induced charge can pile up (due to the mutual repulsion and increased curvature), or equivalently because the center of *B* moves closer. The field at *P* therefore increases to a finite value. In short, the field goes from being zero to nonzero by transitioning in a perfectly reasonable continuous manner.

Although the above reasoning is correct, you might still be wondering about the lack of symmetry between inside and outside. If a charge is outside a conductor, then there is no field on the other side (the inside). But if a charge is inside a conductor, then there *is* a field on the other side (the outside). Why should inside and outside be different? Sure, outside is bigger, but should that really matter? Actually, yes.

Consider a neutral conductor with no other charges around, and let's say we want to put a charge either inside or outside. To create a positive charge, we can take a neutral object and split it into positive and negative pieces, and then get rid of the negative piece. If we do this outside the conductor, then we (1) start with zero field inside the conductor (and outside, too), then (2) create two opposite charges outside and separate them; the field is still zero inside (but not outside now), then (3) get rid of the negative charge by bringing it out to infinity; the field is still zero inside.

However, if we try to do this inside the conductor, then we (1) start with zero field outside the conductor (and inside, too), then (2) create two opposite charges inside and separate them; the field is still zero outside (but not inside now), then (3) get rid of the negative charge by... uh oh, it's stuck inside the conductor, so we can't bring it out to infinity. If we do so, then the charge has to cross the conductor and enter a different region. This breaks the symmetry with the other case, so it's no surprise that we end up with a different result. More precisely, the amount of charge inside the conductor abruptly changes. In the first case, the amount of charge outside the conductor doesn't change when the negative charge is brought out to infinity. See Nan-Xian (1981) for further discussion of this. Problem 3.7 presents another way in which inside and outside fundamentally differ.

3.9 *Grounding a shell*

The field outside the outer shell is zero, so the potential at the outer shell is the same as the potential at infinity. The charge will therefore not move when the outer shell is grounded. If some negative charge did flow off, then there would be a net positive charge on the two shells, and hence an outward-pointing field for $r > R_2$. This would drag the negative charge back onto the outer shell. Likewise, if some positive charge flowed off, then there would be an inward-pointing field for $r > R_2$ which would drag the positive charge back onto the shell.

If the inner shell is grounded, it must end up with the amount of charge that makes its potential equal to the potential at infinity. Let the final charge be $Q_f$. Then the electric field between the shells equals $Q_f/r^2$ (we'll ignore the $1/4\pi\epsilon_0$ in this problem since it will cancel). So the potential of the outer shell relative to the inner shell is $-Q_f(1/R_1 - 1/R_2)$. Similarly, the electric field outside the outer shell equals $(-Q + Q_f)/r^2$, so the potential of the outer shell relative to infinity is $(-Q + Q_f)(1/R_2)$. If the inner shell and infinity are at the same potential, then the previous two potential differences must be equal. This gives

$$Q_f\left(\frac{1}{R_1} - \frac{1}{R_2}\right) = (Q - Q_f)\frac{1}{R_2} \implies Q_f = \frac{R_1}{R_2}Q. \qquad (12.148)$$

Intuitively, if none of the charge leaves (so $Q_f = Q$), then the inner shell is at a *higher* potential than the outer shell, which in turn is at the same potential as infinity in this case. On the other hand, if all of the charge leaves (so $Q_f = 0$), then the inner shell is at the same potential as the outer shell, which in turn is at a *lower* potential than infinity in this case. So, by continuity, there must be a value of $Q_f$ that makes the potential of the inner shell equal to the potential at infinity.

3.10 *Why leave?*

Charge will indeed flow off the inner shell out to infinity, up to the (very short) time when the charge on the shell equals the value calculated in Problem 3.9. The error in the opposing logic is the following. If we consider one small point charge, then this charge would be happy to hang out on the wire right at the hole in the outer shell. However, if we try to do the same thing with another small point charge, then we can't put both of them in the same place, because they will repel each other. So we certainly can't pile up charge right at the hole in the outer shell. What if we stretch out the charge and create a linear density along the wire? Could the charges stabilize in a linear distribution, partially between the shells and partially outside the outer shell?

To answer this, we must use the fact that a very thin wire has essentially *zero capacitance* (see Exercise 3.59; the outer radius in that setup can be assigned some arbitrary fixed value). So charge can't pile up on the wire. If we give the wire a tiny radius, then in a static situation only a tiny bit of charge can pile up. The very strong forces (which go like $1/d^2$, from Coulomb's law) between nearby charges on the wire will make the linear charge density $\lambda$ nearly uniform (see Problem 3.5). So in a sense we have an essentially rigid stick of (a tiny bit of) charge extending from the inner shell out to a very large radius. The question is: what is the direction of the net force on this stick? Will it be drawn inward or be pushed outward? There are competing effects, because the field $E_1(r)$ between the shells points outward, and the field $E_2(r)$ outside the outer shell points inward (at least once some charge has left the inner shell, leaving behind a net negative charge on the two shells). We are ignoring the forces between the charges in the stick here, because they are internal.

Assuming constant density $\lambda$, the outward force on the stick equals $\int_{R_1}^{R_2} E_1 \lambda \, dr$, and the inward force equals $\int_{R_2}^{\infty} E_2 \lambda \, dr$ (this is negative since $E_2$ is negative). The stick won't move if these two forces add up to zero:

$$\int_{R_1}^{R_2} E_1 \lambda \, dr + \int_{R_2}^{\infty} E_2 \lambda \, dr = 0$$

$$\implies \int_{R_1}^{R_2} E_1 \, dr = \int_{\infty}^{R_2} E_2 \, dr$$

$$\implies \phi(R_1) - \phi(R_2) = \phi(\infty) - \phi(R_2)$$

$$\implies \phi(R_1) = \phi(\infty). \tag{12.149}$$

We see that the stick won't move if the potential at the inner shell equals the potential at infinity. So charge will flow off the inner shell until the time when this condition is met, at which point the charge equals the value calculated in Problem 3.9.

If the wire *does* have a nonzero capacitance, then things are different. If we have, say, a little spherical bulge in the wire somewhere outside, then charge will build up there, but only until the potential equals the potential everywhere else along the wire.

As an exercise, you can think about an analogous setup involving two large capacitor sheets separated by a small distance, with charges $\pm Q$, and with one sheet grounded by connecting it to infinity via a very thin wire that passes through a very small hole in the other sheet.

Note that the above reasoning took advantage of the fact that $\int E \, dr$ has two interpretations. It equals the potential difference $\phi$, of course. But if we multiply it by a constant $\lambda$, it also equals the total force on a uniform stick. Basically, multiply $\int E \, dr$ by $q$ and you get the total work done on a charged particle moving between two given points. Multiply it by $\lambda$ and you get the total force on a charged stick lying between the two points.

3.11 *How much work?*

When the charge $Q$ is a distance $x$ above the plane, the force required to balance the electrostatic force and move the charge upward (at constant speed) is $Q^2/4\pi\epsilon_0(2x)^2$, because the force is the same as if the plane were replaced by an image charge $-Q$ at a distance $2x$. The second student therefore calculates the work as

$$W = \int F \, dx = \int_h^\infty \frac{Q^2 \, dx}{4\pi\epsilon_0 (2x)^2} = \frac{Q^2}{4\pi\epsilon_0 (4h)}. \qquad (12.150)$$

This is the correct answer. Concerning the first student's reasoning, if two real charges $Q$ and $-Q$ are pulled apart symmetrically, the *total* work done is $Q^2/4\pi\epsilon_0(2h)$, but the agency moving $Q$ supplies only half of it. The agency moving $-Q$ supplies the other half.

Note that these two real charges must indeed be pulled apart *symmetrically* if we want to mimic the behavior of the conducting plane, because we need the field always to be perpendicular to the given plane. If we instead hold $-Q$ fixed and move only $Q$, then the agency moving $Q$ does in fact do all of the work (which is $Q^2/4\pi\epsilon_0(2h)$). But this setup is not the one we are interested in, because the field isn't perpendicular to the given plane.

Another way to see why the actual work is half of the first student's $Q^2/4\pi\epsilon_0(2h)$ answer is to look at the energy stored in the electric field. If we actually have two real point charges and no plane, then the field exists throughout all space. But in the case of the conducting plane, the field exists only in the half-space on one side of the plane. So the stored energy is half of what it is in the case of the two real point charges.

3.12 *Image charges for two planes*
In the setup in Section 3.4 we had only one conducting plane, so one image charge was sufficient to cause the total electric field to be perpendicular to the plane. Let's see what happens with two planes. In Fig. 12.53 the two given planes are indicated by the bold lines, and the given real charge is labeled $R$. It turns out that we will need an infinite number of image charges, as shown. Solid dots are positive, hollow dots are negative (assuming the given real charge is positive). The reason for all these image charges is the following.

In order to have the $E$ field be perpendicular to the right plane, we need the image charge labeled 1. And likewise, in order to have $E$ be perpendicular to the left plane, we need image charge 2. So far, we just have two copies of the one-plane setup.

However, image charge 1 ruins the orthogonality of the field with the left plane, so we need image charge 3 to remedy this. Likewise, image

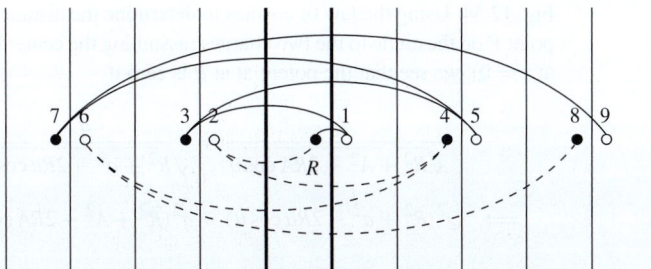

**Figure 12.53.**

charge 2 ruins the orthogonality of the field with the right plane, so we need image charge 4 to remedy this.

But then we need images charges 5 and 6 to remedy the effects of 3 and 4, respectively. And so on. The effects of the charges far away are small, so the process converges. That is, if we have 1000 such charges, the field will be very nearly perpendicular everywhere to the two given planes.

If you want, you can group the charges into two sets – the odds and evens, as indicated by the connecting lines in Fig. 12.53. Each odd charge corrects the effect of the previous odd charge, with respect to alternating planes. Likewise for the evens.

In the special case where the given real charge is located midway between the two planes, all the image charges are similarly located midway between the (imaginary) planes in Fig. 12.53. So the net force on the given charge is zero, as it should be.

3.13 *Image charge for a grounded spherical shell*

(a) The potential at an arbitrary point in the $xy$ plane is (ignoring the $1/4\pi\epsilon_0$)

$$\phi = \frac{Q}{\sqrt{(x-A)^2 + y^2}} - \frac{q}{\sqrt{(x-a)^2 + y^2}}. \tag{12.151}$$

Setting this equal to zero, putting one term on either side of the equation, and squaring gives

$$Q^2(x^2 - 2ax + a^2 + y^2) = q^2(x^2 - 2Ax + A^2 + y^2). \tag{12.152}$$

Since the coefficients of $x^2$ and $y^2$ are equal, this equation describes a circle. More precisely, the equation can be written in the form of $x^2 + y^2 - 2Bx = C$, which in turn can be written as $(x-B)^2 + y^2 = C + B^2$, by completing the square. This equation describes a circle with its center at $(B, 0)$ and with radius $\sqrt{C + B^2}$.

(b) Expanding Eq. (12.152) gives

$$(Q^2 - q^2)x^2 + (Q^2 - q^2)y^2 - 2(Q^2a - q^2A)x = q^2A^2 - Q^2a^2. \tag{12.153}$$

The center of the circle is located at $x = 0$ if the coefficient of $x$ is zero, that is, if $Q^2a = q^2A$.

Alternatively, we can work in terms of the angle $\theta$ shown in Fig. 12.54. Using the law of cosines to determine the distances from a point $P$ on the circle to the two charges (assuming the center is located at $x = 0$), we see that the potential at $P$ is zero if

$$\frac{Q}{\sqrt{R^2 + A^2 - 2RA\cos\theta}} = \frac{q}{\sqrt{R^2 + a^2 - 2Ra\cos\theta}}$$

$$\implies Q^2(R^2 + a^2 - 2Ra\cos\theta) = q^2(R^2 + A^2 - 2RA\cos\theta). \tag{12.154}$$

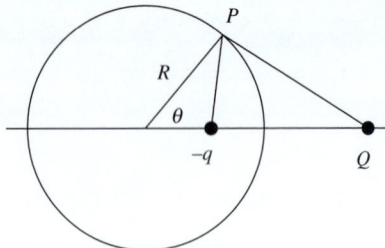

**Figure 12.54.**

If this equation is to be true for *all* values of $\theta$, then the coefficient of $\cos\theta$ must be the same on both sides. This yields $Q^2 a = q^2 A$, as above.

(c) If $Q^2 a = q^2 A$, then dividing Eq. (12.153) by $Q^2 - q^2$ tells us that the radius of the circle is given by

$$R^2 = \frac{q^2 A^2 - Q^2 a^2}{Q^2 - q^2} = \frac{(Q^2 a/A)A^2 - Q^2 a^2}{Q^2 - (Q^2 a/A)} = aA. \qquad (12.155)$$

The radius is therefore the geometric mean of the distances from the two charges to the center of the circle.

Alternatively, we can work with the angle $\theta$ shown in Fig. 12.54. If $Q^2 a = q^2 A$, then Eq. (12.154) gives

$$Q^2(R^2 + a^2) = q^2(R^2 + A^2)$$
$$\Longrightarrow \quad Q^2(R^2 + a^2) = (Q^2 a/A)(R^2 + A^2)$$
$$\Longrightarrow \quad A(R^2 + a^2) = a(R^2 + A^2)$$
$$\Longrightarrow \quad R^2(A - a) = aA(A - a)$$
$$\Longrightarrow \quad R^2 = aA. \qquad (12.156)$$

(d) Having derived $R^2 = aA$, we can eliminate $a$ from the relation $Q^2 a = q^2 A$ to obtain $Q^2(R^2/A) = q^2 A \Longrightarrow q = QR/A$. Putting all the results together, we see that if we have a charge $Q$ at $x = A$ and a charge $-q = -QR/A$ at $x = a = R^2/A$, then the entire surface of the sphere of radius $R$ centered at the origin will be at zero potential. But this is exactly the boundary condition for a grounded conducting sphere. The uniqueness theorem therefore tells us that the two setups (point charge outside grounded conducting sphere, and point charge near image charge) have exactly the same field in the exterior of the sphere. (This reasoning doesn't apply to the interior, because the setups are different there; one contains an image charge, the other doesn't. The uniqueness theorem requires the same charge distribution in the relevant region in both setups.) The results for this problem look a little cleaner if we let $A = nR$, where $n$ is a numerical factor. The image charge then has the value $-q = -Q/n$ and is located at radius $R/n$.

(e) Again using $R^2 = aA$, we can eliminate $A$ from the relation $Q^2 a = q^2 A$ to obtain $Q^2 a = q^2(R^2/a) \Longrightarrow Q = qR/a$. Putting all the results together, we see that if we have a charge $-q$ at $x = a$ and a charge $Q = qR/a$ at $x = A = R^2/a$, then the entire surface of the sphere of radius $R$ centered at the origin will be at zero potential. As above, we conclude that the two setups (point charge inside grounded conducting sphere, and point charge near image charge) have the same field in the interior of the sphere. If we let $a = R/n$, then the image charge has the value $Q = nq$ and is located at radius $nR$.

**3.14** *Force from a conducting shell*

From Problem 3.13, the field at the location of the charge $Q$ is the same as the field of an image charge $-QR/r$ located a distance $a = R^2/r$ from the center of the shell. This image charge is a distance $r - R^2/r$ from the given charge $Q$, so Coulomb's law gives the force on $Q$ as

$$F = \frac{1}{4\pi\epsilon_0} \frac{Q(-QR/r)}{(r - R^2/r)^2} = -\frac{1}{4\pi\epsilon_0} \frac{Q^2 Rr}{(r^2 - R^2)^2}. \qquad (12.157)$$

The minus sign indicates an attractive force.

If $r \approx R$, then the shell looks essentially like a plane from up close, so we should obtain the same force as in the image-charge setup with the infinite plane in Section 3.4. Let $r \equiv R + h$, where $h \ll R$. Then writing $(r^2 - R^2)^2$ as $(r + R)^2(r - R)^2$, the force in Eq. (12.157) becomes

$$F = -\frac{1}{4\pi\epsilon_0} \frac{Q^2 R(R + h)}{(2R + h)^2(h)^2} \approx -\frac{1}{4\pi\epsilon_0} \frac{Q^2}{4h^2}. \qquad (12.158)$$

As expected, this is the force between the real charge $Q$ and an image charge $-Q$, a distance $2h$ apart ($h$ on either side of the plane).

In the $r \to \infty$ limit the force in Eq. (12.157) becomes

$$F \approx -\frac{1}{4\pi\epsilon_0} \frac{Q^2 Rr}{(r^2)^2} \approx -\frac{1}{4\pi\epsilon_0} \frac{Q^2 R}{r^3}. \qquad (12.159)$$

This expression can be understood as follows. Far away from the shell, the shell looks essentially like a point charge, with the charge being that of the image charge, $-QR/r$. And Eq. (12.159) does indeed take the form of $F \approx -Q(QR/r)/4\pi\epsilon_0 r^2$.

3.15 *Dipole from a shell in a uniform field*

(a) From Problem 3.13, the external field due to the shell equals the field due to two image charges: a charge $-QR/A$ at $x = -R^2/A$ and a charge $QR/A$ at $x = R^2/A$. In the $A \to \infty$ limit, the separation $2R^2/A$ between these image charges goes to zero, so the configuration becomes an idealized dipole with dipole moment $p = (QR/A)(2R^2/A) = (2Q/A^2)R^3$. This dipole points in the positive $x$ direction.

At the location of the shell, the total field of the distant $\pm Q$ point charges equals $2Q/4\pi\epsilon_0 A^2$. But we are told that this field equals $E$, so we must have $2Q/A^2 = 4\pi\epsilon_0 E$. We can therefore write the dipole moment as $p = 4\pi\epsilon_0 ER^3$. Hence the external field due to the shell is exactly the same as the field due to an idealized dipole with dipole moment $p = 4\pi\epsilon_0 ER^3$. Note that since $Q/A^2 \propto E$, the charge $Q$ must go to infinity like $A^2$, as the length $A$ goes to infinity.

(b) Using $p = 4\pi\epsilon_0 ER^3$, Eq. (2.36) tells us that the field due to the shell, just outside the shell (that is, at radius $R$), equals $E(2\cos\theta\,\hat{\mathbf{r}} + \sin\theta\,\hat{\boldsymbol{\theta}})$, where $\theta$ is measured with respect to the positive $x$ direction. (Although the dipole field is valid everywhere, our image-charge application of it is valid only outside the shell.) Note that this result is independent of $R$.

The *total* electric field just outside the shell is the sum of the shell's field plus the original uniform field, $\mathbf{E}_u = E\hat{\mathbf{x}}$. You can quickly show that $\hat{\mathbf{x}}$ can be written in terms of the spherical-coordinate unit vectors as $\hat{\mathbf{x}} = \cos\theta\,\hat{\mathbf{r}} - \sin\theta\,\hat{\boldsymbol{\theta}}$. So the uniform field is $\mathbf{E}_u = E(\cos\theta\,\hat{\mathbf{r}} - \sin\theta\,\hat{\boldsymbol{\theta}})$. The total field is therefore $\mathbf{E}_{tot} = 3E\cos\theta\,\hat{\mathbf{r}}$. This correctly has no $\hat{\boldsymbol{\theta}}$ component; the field must be perpendicular to the surface of the conductor.

(c) The surface charge density $\sigma$ is proportional to the (normal) field at the surface. More precisely, Gauss's law tells us that the field just outside the shell is $E_r = \sigma/\epsilon_0$. So $\sigma = \epsilon_0 E_r = 3\epsilon_0 E\cos\theta$.

3.16 *Image charge for a nongrounded spherical shell*

We know from Problem 3.13 that an image charge of $-QR/r$ located at radius $R^2/r$ causes the entire shell to have potential zero. Since the image charge produces the same external field as the shell, Gauss's law implies that the charge on the actual shell is $-QR/r$, whereas we are told that the charge is $q_s$. We can remedy this by placing another image charge of $q_s + QR/r$ at the center. The total charge on the actual shell is now $q_s$, as desired. Furthermore the boundary condition of constant potential on the shell is still satisfied, by symmetry, because the second image charge is located at the center. So by the uniqueness theorem, the field from our two image charges mimics the (external) field from the shell.

Physically, the image charge from Problem 3.13 creates a field that, when combined with the original charge $Q$, is perpendicular to the shell. So if we dump some additional charge ($q_s + QR/r$ in this case) on the shell, it will distribute itself symmetrically, because there is no tangential field that would cause the distribution to be lopsided. And this symmetrical distribution has the same (external) field as a point charge at the center.

3.17 *Capacitance of raindrops*

The capacitance of each of the $N$ drops is $4\pi\epsilon_0 a$. Since we can put $N$ times as much charge on $N$ spheres at a given potential as we can put on one sphere at the same potential, the capacitance of the $N$ spheres is simply $N(4\pi\epsilon_0 a)$. Equivalently, if there is a total charge $Q$ on the $N$ raindrops, then each one has charge $Q/N$. If the potential of each is $\phi$, then

$$\phi = \frac{Q/N}{4\pi\epsilon_0 a} \implies Q = (4\pi\epsilon_0 Na)\phi \implies C = 4\pi\epsilon_0 Na. \quad (12.160)$$

If the drops combine into one big drop, then, since the total volume remains the same, the new radius is given by $(4/3)\pi r^3 = N(4/3)\pi a^3 \implies r = N^{1/3}a$. The capacitance of this single sphere is $4\pi\epsilon_0(N^{1/3}a)$. This is smaller than the capacitance of the system with $N$ drops by the factor $N^{2/3}$.

You should think physically about why the capacitance of the big drop is smaller. *Hint:* Consider a point $P$ on the surface of one of the small drops, and then bring in all the other small drops from far away and put them in a spherical clump, with $P$ arranged to be on the surface of the clump. What happens to the potential at $P$?

3.18 *Adding capacitors*

(a) When capacitors are connected in series, the charges on them are equal, because the charge on the top plate of the bottom capacitor must be the negative of the charge on the bottom plate of the top capacitor (because these two plates are isolated from the rest of the circuit, so the net charge on them must always be zero). Let $\pm Q$ be this common charge, which is also the charge on the overall effective capacitor.

The total voltage (that is, the potential difference) $\phi$ across the effective capacitor is the sum of the voltages across the two capacitors, so $\phi = \phi_1 + \phi_2$. But we know that $\phi = Q/C$, $\phi_1 = Q/C_1$, and $\phi_2 = Q/C_2$, where the same $Q$ appears everywhere here. Plugging these expressions into $\phi = \phi_1 + \phi_2$ gives

$$\frac{Q}{C} = \frac{Q}{C_1} + \frac{Q}{C_2} \implies \frac{1}{C} = \frac{1}{C_1} + \frac{1}{C_2}. \tag{12.161}$$

If $C_1 \to 0$ then $C \to 0$ also. This makes sense because, for a given $Q$ (common to all of the capacitors), the voltage $Q/C_1$ across $C_1$ is huge, which means that the overall voltage $Q/C$ is likewise huge, since it is at least as large. So $C$ must be very small.

If $C_1 \to \infty$ then $C \to C_2$. This makes sense because, for a given $Q$ (common to all of the capacitors), the voltage $Q/C_1$ across $C_1$ is tiny, which means that the overall voltage $Q/C$ is essentially equal to the voltage $Q/C_2$ across $C_2$. So $C \approx C_2$.

(b) When capacitors are connected in parallel, the voltages across them are equal, because the voltage drop from the top of the overall circuit to the bottom can't depend on the path. Let $\phi$ be this common voltage, which is of course also the voltage drop in the overall effective capacitor.

The total charge $Q$ on the effective capacitor is the sum of the charges on the top plates of the two capacitors, so $Q = Q_1 + Q_2$. But we know that $Q = C\phi$, $Q_1 = C_1\phi$, and $Q_2 = C_2\phi$, where the same $\phi$ appears everywhere here. Plugging these expressions into $Q = Q_1 + Q_2$ gives

$$C\phi = C_1\phi + C_2\phi \implies C = C_1 + C_2. \tag{12.162}$$

In short, in the series case the charges on the two capacitors are the same, and the voltages add; whereas in the parallel case the voltages are the same, and the charges add.

If $C_1 \to 0$ then $C \to C_2$. This makes sense because, for a given $\phi$ (common to all of the capacitors), the charge $C_1\phi$ on $C_1$ is tiny, which means that the overall charge $C\phi$ is essentially equal to the charge $C_2\phi$ on $C_2$. So $C \approx C_2$.

If $C_1 \to \infty$ then $C \to \infty$ also. This makes sense because, for a given $\phi$ (common to all of the capacitors), the charge $C_1\phi$ on $C_1$ is huge, which means that the overall charge $C\phi$ is likewise huge, since it is at least as large. So $C$ must be very large.

As remarked in the statement of the problem, the above series/parallel rules are the opposites of the rules for adding resistors and inductors. However, there isn't anything too deep here. If we instead labeled capacitors with the quantity $C'$ defined by $\phi = C'Q$, then the series/parallel rules for adding $C'$'s would be the same as for resistors and inductors.

3.19   *Uniform charge on a capacitor*

Since each plate is an equipotential, the potential difference is the same between any point on one plate and the corresponding point on the other plate. If this difference is $\phi$, then $\phi = Es$, where $E$ is the field (normal to the plates) and $s$ is the plate separation. So $E$ must be the same everywhere between the plates. But $E = \sigma/\epsilon_0$, where $\pm\sigma$ are the local charge densities of the two plates. Hence $\sigma$ must be the same everywhere on each plate (ignoring edge effects).

The above reasoning shows why the density is uniform. However, consider two oppositely charged conducting disks, initially located far apart, with their distributions given by the nonuniform result in Eq. (12.146). If they are brought together to make a capacitor, what actually causes the distribution to shift and become uniform? If you look at Fig. 3.13, you will see that the field lines start out perpendicular to the disk, but then they fan out. So if you bring the disks together with their charges glued in place, the field at each disk, due to the other disk, will have a slight sideways component, for any nonzero separation $s$. If the charges are then unglued, this sideways component will drag the charges in the direction that makes the distribution uniform.

3.20   *Distribution of charge on a capacitor*

First solution   If $E$ is the electric field between the plates, then the charge densities on the inner surfaces must be $\pm\sigma$, where $E = \sigma/\epsilon_0 \implies \sigma = \epsilon_0 E$. This follows from using a Gaussian pillbox at either plate, with one side of the pillbox lying inside the conducting plate where the field is zero. The charges on the inner surfaces are therefore equal and opposite. Alternatively, consider the Gaussian surface indicated by the dashed box in Fig. 12.55. Since there is no flux out of the top or bottom, the net charge enclosed must be zero. Hence there are equal and opposite charges on the inner surfaces.

We now claim that the charges on the outer two surfaces must be equal (with the same sign). Consider a point $P$ inside one of the plates; the field is zero at $P$. The two oppositely charged inner surfaces of the plates produce zero net field at $P$, because they lie on the *same* side of $P$. The two outer surfaces of the plates must therefore also produce zero net field at $P$. Since these two surfaces lie on *opposite* sides of $P$, they must have equal charge densities. The four surface charges therefore take the forms shown in Fig. 12.56 (the fields in the various regions are also shown). So we have

$$Q_1 = Q_{\text{out}} + Q_{\text{in}},$$
$$Q_2 = Q_{\text{out}} - Q_{\text{in}}. \tag{12.163}$$

These equations quickly yield

$$Q_{\text{out}} = \frac{Q_1 + Q_2}{2} \quad \text{and} \quad Q_{\text{in}} = \frac{Q_1 - Q_2}{2}. \tag{12.164}$$

In the special case where $Q_1 = Q_2 \equiv Q$, we have $Q_{\text{out}} = Q$ and $Q_{\text{in}} = 0$; all the charge lies on the outer surfaces. In the special case where $Q_1 = -Q_2 \equiv Q$ (which is the normal case for a capacitor), we

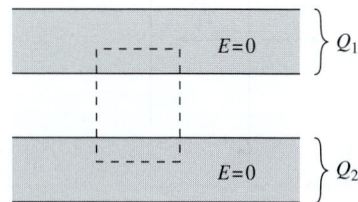

**Figure 12.55.**

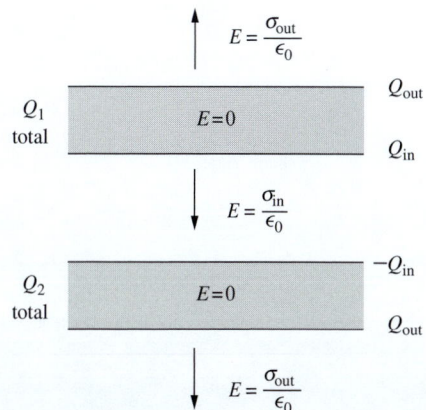

**Figure 12.56.**

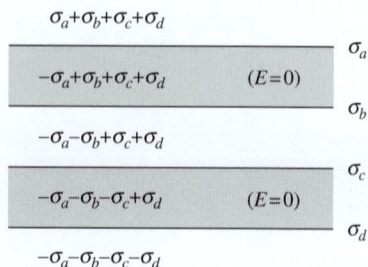

**Figure 12.57.**
The fields in the five regions defined by two capacitor plates, up to a factor of $1/2\epsilon_0$.

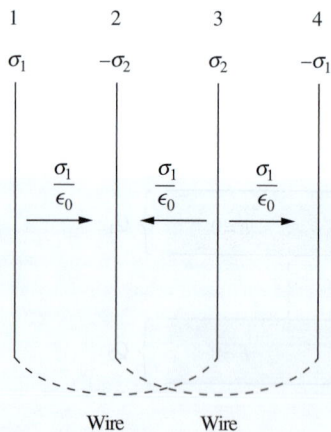

**Figure 12.58.**

have $Q_{\text{out}} = 0$ and $Q_{\text{in}} = Q$; all the charge lies on the inner surfaces. In all cases, the field between the plates is determined solely by $Q_{\text{in}}$ via $E = \sigma/\epsilon_0 = Q_{\text{in}}/A\epsilon_0$.

**Second solution** We can solve for the charges on the four plates by letting the four surface densities be $\sigma_a$, $\sigma_b$, $\sigma_c$, $\sigma_d$, and then explicitly writing down the electric fields in the various regions. The field due to an infinite sheet has magnitude $\sigma/2\epsilon_0$ and points away from the sheet (if $\sigma$ is positive). So up to a factor of $1/2\epsilon_0$, the fields in the five different regions take the forms shown in Fig. 12.57, with upward taken to be positive.

The $E = 0$ regions inside the conductors tell us that $-\sigma_a + \sigma_b + \sigma_c + \sigma_d = 0$ and $-\sigma_a - \sigma_b - \sigma_c + \sigma_d = 0$. Adding these equations gives $\sigma_a = \sigma_d$, and subtracting them gives $\sigma_c = -\sigma_b$. So the charges on the four plates take the form of $Q_{\text{out}}, Q_{\text{in}}, -Q_{\text{in}}, Q_{\text{out}}$, as in the first solution above.

3.21 *A four-plate capacitor*
Assume that the total charge on the first and third plates is positive. An equal and opposite negative charge resides on the second and fourth plates. Let the charge densities on the first two plates be labeled $\sigma_1$ and $-\sigma_2$. Then by left–right symmetry, the charge densities on the third and fourth plates are $\sigma_2$ and $-\sigma_1$. (If we reverse left and right, and then reverse the signs of all charges, we should end up with the same setup.)

The total charge is zero, so there is no field outside the plates. Gauss's law with a pillbox spanning the first plate then tells us that the field between the first and second plates is $\sigma_1/\epsilon_0$. The potential difference between these plates is therefore $\phi = \sigma_1 s/\epsilon_0$. But since the first and third plates have the same potential (due to the connecting wire), the potential difference between the second and third plates must also be $\sigma_1 s/\epsilon_0$. So the field between the second and third plates is also $\sigma_1/\epsilon_0$, but directed to the left. Similarly, the field between the third and fourth plates is also $\sigma_1/\epsilon_0$, directed to the right. The fields are shown in Fig. 12.58.

A Gaussian surface spanning the second plate tells us that its charge density $-\sigma_2$ is given $-\sigma_2 = -2\sigma_1$. The charge density on the third plate is then $\sigma_2 = 2\sigma_1$. The total charge on the two positive plates is therefore $Q = (\sigma_1 + 2\sigma_1)A \implies \sigma_1 = Q/3A$. The $\phi = \sigma_1 s/\epsilon_0$ potential-difference statement between the positive and negative (pairs of) plates in the capacitor can then be written as

$$\phi = \frac{(Q/3A)s}{\epsilon_0} \implies Q = \left(\frac{3A\epsilon_0}{s}\right)\phi \implies C = \frac{3A\epsilon_0}{s}. \tag{12.165}$$

Note that this is larger than the capacitance we would obtain if we juxtaposed the two pairs of plates to create two plates with area $2A$. If we keep the separation $s$, then the capacitance of the resulting standard two-plate capacitor would be $C = \epsilon_0(2A)/s$. The physical reason for the factor of 3 in Eq. (12.165) versus this factor of 2 is the following.

In our original setup with the four parallel plates, the density $\pm 2\sigma_1$ on the interior plates gets split evenly between the two sides of these plates. (This follows from using a Gaussian surface that has one boundary lying inside the conducting plate where the field is zero, and using the fact that the fields on either side of the plate have the same magnitude.) So if we give the plates a small thickness, the setup is shown in Fig. 12.59. We effectively have three identical area-$A$ capacitors, but with the orientation of the middle one reversed. If we had simply juxtaposed the pairs of plates, then we would have the equivalent of only two area-$A$ capacitors (with the same orientation).

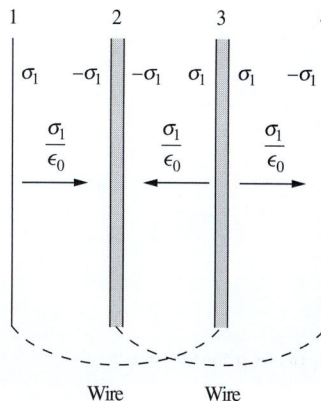

**Figure 12.59.**

3.22   *A three-cylinder capacitor*

(a) Let $\lambda_1$ and $\lambda_3$ be the final charges per length on the inner and outer cylindrical shells, respectively. The outward-pointing field between the inner and middle shells is due only to the inner shell, and it equals $\lambda_1/r$ (ignoring the $1/2\pi\epsilon_0$ since it will cancel). Integrating this gives the potential difference between the inner and middle shells as $\lambda_1 \ln(2R/R) = \lambda_1 \ln 2$, with the inner shell at the higher potential.

   If the inner and outer shells are at the same potential, then $\lambda_1 \ln 2$ must also be the potential difference between the outer and middle shells, with the outer shell at the higher potential. The field between the middle and outer shells must therefore point inward. This field is due to the inner two shells. The charges per length on these shells are $\lambda_1$ and $-\lambda$, so the field points inward with magnitude $(\lambda - \lambda_1)/r$. The potential difference between the outer two shells is then $(\lambda - \lambda_1) \ln(3R/2R) = (\lambda - \lambda_1) \ln(3/2)$, with the outer shell at the higher potential.

   Equating the inner-middle and outer-middle potential differences gives

   $$\lambda_1 \ln 2 = (\lambda - \lambda_1) \ln(3/2)$$
   $$\implies \lambda_1\big(\ln 2 + \ln(3/2)\big) = \lambda \ln(3/2)$$
   $$\implies \lambda_1 = \lambda\frac{\ln(3/2)}{\ln 3} \approx (0.37)\lambda. \qquad (12.166)$$

   And then $\lambda_3 = \lambda - \lambda_1 \approx (0.63)\lambda$ to make the total charge per length on the inner and outer shells equal to $\lambda$.

(b) The potential difference between the inner/outer shells and the middle shell is $\phi = \lambda_1 (\ln 2)/2\pi\epsilon_0$ (bringing the $1/2\pi\epsilon_0$ back in). But $\lambda_1 = \lambda \ln(3/2)/\ln 3$, so we can solve for $\lambda$ in terms of $\phi$. We obtain $\lambda = \phi \cdot 2\pi\epsilon_0 (\ln 3/\ln 2)/\ln(3/2)$. Since $\lambda$ is the charge per unit length, we see that the capacitance per unit length is $2\pi\epsilon_0 (\ln 3/\ln 2)/\ln(3/2) \approx 2\pi\epsilon_0(3.91)$.

(c) Note that $\lambda_3$ didn't appear anywhere in the calculation in part (a). It can therefore take on any value, and the potential differences will still be equal, provided that $\lambda_1 \approx (0.37)\lambda$. So if we add charge per length $\lambda_{\text{new}}$ to the outer shell, it will simply stay there, uniformly distributed on the outside surface of the shell. It will raise the potential everywhere inside by the same amount (which depends on where the $\phi = 0$ point is chosen). But since this change is uniform inside, all differences remain the same. So it actually doesn't matter that the battery was disconnected.

3.23 *Capacitance coefficients and C*

For a standard parallel-plate capacitor, we don't have a box surrounding the plates. Equivalently, the separation $s$ between the plates is much smaller than the distances $r$ and $t$ to the box. So we should be able to recover the $C$ value in Eq. (3.15) by taking the $r \to \infty$ and $t \to \infty$ limits in the four capacitance coefficients from the example in Section 3.6. In these limits, the $1/r$ and $1/t$ terms in $C_{11}$ and $C_{22}$ are negligible, so Eq. (3.24) becomes

$$Q_1 = \frac{\epsilon_0 A}{s} \phi_1 - \frac{\epsilon_0 A}{s} \phi_2,$$

$$Q_2 = -\frac{\epsilon_0 A}{s} \phi_1 + \frac{\epsilon_0 A}{s} \phi_2. \tag{12.167}$$

Now, when we write $Q = C\phi$ for a capacitor, what we really mean is $Q = C \Delta\phi$, where $\Delta\phi$ is the *difference* in the potentials of the two plates (or whatever objects), and where $\pm Q$ are the charges on the plates. We can obtain a potential difference of $\Delta\phi$ if we set $\phi_1 = \Delta\phi/2$ and $\phi_2 = -\Delta\phi/2$ in Eq. (12.167). This gives

$$Q = \frac{\epsilon_0 A}{s}\left(\frac{\Delta\phi}{2}\right) - \frac{\epsilon_0 A}{s}\left(-\frac{\Delta\phi}{2}\right) = \frac{\epsilon_0 A}{s}\Delta\phi,$$

$$-Q = -\frac{\epsilon_0 A}{s}\left(\frac{\Delta\phi}{2}\right) + \frac{\epsilon_0 A}{s}\left(-\frac{\Delta\phi}{2}\right) = -\frac{\epsilon_0 A}{s}\Delta\phi. \tag{12.168}$$

These equations are both identical to the $Q = (\epsilon_0 A/s)\Delta\phi$ statement for a parallel-plate capacitor, as desired.

3.24 *Human capacitance*

Let's assume the capacitance is roughly that of a conducting sphere with 0.5 m radius. Then

$$C = 4\pi \epsilon_0 r = 4\pi \left(8.85 \cdot 10^{-12}\,\frac{s^2\,C^2}{kg\,m^3}\right)(0.5\,m)$$

$$\approx 5.6 \cdot 10^{-11}\,F = 56\,pF. \tag{12.169}$$

If you are charged to, say, 2 kV, the stored energy is

$$U = \frac{1}{2}CV^2 = \frac{1}{2}\left(5.6 \cdot 10^{-11}\,F\right)(2000\,V)^2 = 1.1 \cdot 10^{-4}\,J. \tag{12.170}$$

This is a very small amount of energy, enough to raise the temperature of one milliliter (one gram) of water by only about $2.6 \cdot 10^{-5}\,^\circ C$. (It takes one calorie, or 4.2 joules, to raise one gram of water by $1\,^\circ C$.)

3.25 *Energy of a disk*

The energy stored in the field of the conducting disk is

$$U = \frac{Q^2}{2C} = \frac{Q^2}{2(8\epsilon_0 a)} = \frac{Q^2}{16\epsilon_0 a}. \tag{12.171}$$

For a uniformly charged nonconducting disk, we found in Exercise 2.56 that $U = (2/3\pi^2\epsilon_0)(Q^2/a)$. The ratio of this to the conducting-disk result is $(2/3\pi^2\epsilon_0)/(1/16\epsilon_0) = 32/3\pi^2 = 1.081$. The field of the uniform charge distribution therefore has 8 percent more energy. It makes sense that this energy is larger, because on the conducting disk the charge distributes itself to minimize the energy. If a uniformly charged disk is modified to be conducting, the charge will redistribute itself. But not the other way around.

3.26    *Force on a capacitor plate*

(a) The energy of the capacitor can be written in many ways: $C\phi^2/2$, $Q\phi/2$, or $Q^2/2C$. We want to write the force in terms of $Q$ and $C$, so let's pick the last of these. Since we are assuming that $Q$ is constant, the change in the energy is $dU = (Q^2/2)d(1/C)$. Now, the plates act like a capacitor consisting of just the overlap region with width $x$, because all of the charge will reside in that region (neglecting edge effects). This is true because if there were leftover charge on one plate outside the overlap region, it would be attracted to the corresponding leftover opposite charge on the other plate. All of the charge therefore ends up in the effectively neutral overlap region.

As $x$ increases, the capacitance increases (because the area increases), so $dU$ is negative. This decrease in energy must go somewhere. If nothing is holding the movable plate back, then the energy will show up as kinetic energy of this plate. If the movable plate is attached to another object, then it will do work on this object (increasing the potential and/or kinetic energy, etc.). Equating the magnitude of the change in energy (which is $-dU$) with the work done on another object, we find the magnitude of the force to be

$$-dU = F\,dx \implies \frac{-Q^2}{2}d\left(\frac{1}{C}\right) = F\,dx \implies F = -\frac{Q^2}{2}\frac{d}{dx}\left(\frac{1}{C}\right).$$

$$(12.172)$$

If the movable plate doesn't move at all, then there is no change in energy, but the force is still there, because we could imagine moving the plate by an infinitesimal amount.

(b) Things are a little trickier when the voltage is held constant, because the battery is now part of the system. As $x$ increases, the energy of the capacitor now *increases*. This is true because the energy is $C\phi^2/2$; since we are assuming that $\phi$ is constant, the increase in energy is $(dC)\phi^2/2$. And again, $C$ increases because the area increases.

If this were the whole story, then we would have a violation of conservation of energy, because energy would appear out of nowhere. But the point is that the battery does work. And it does *more* work than $C\phi^2/2$, with the excess being the work that the movable plate can do on an external object. The battery effectively transfers charge from one plate to the other, so the work it does is $(dQ)\phi$. But since $Q = C\phi$, and since we are assuming that $\phi$ is constant, we have $dQ = (dC)\phi$. Hence the work done by the battery is $(dC)\phi^2$.

We found above that the increase in the capacitor's energy is $(dC)\phi^2/2$. So half of the battery's $(dC)\phi^2$ work shows up in the

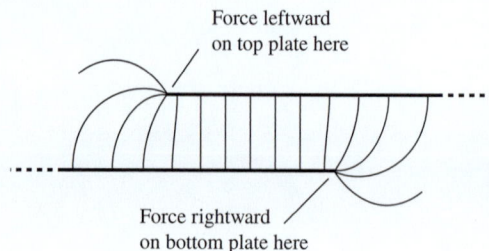

Force leftward
on top plate here

Force rightward
on bottom plate here

**Figure 12.60.**

capacitor. The remaining $(dC)\phi^2/2$ must be the work done on an external object. Therefore,

$$\frac{dC}{2}\phi^2 = F\,dx \implies F = \frac{\phi^2}{2}\frac{dC}{dx}. \tag{12.173}$$

(c) The results in parts (a) and (b) are equal because

$$-\frac{Q^2}{2}\frac{d}{dx}\left(\frac{1}{C}\right) = \frac{Q^2}{2}\left(\frac{1}{C^2}\frac{dC}{dx}\right) = \frac{1}{2}\left(\frac{Q^2}{C^2}\right)\frac{dC}{dx} = \frac{\phi^2}{2}\frac{dC}{dx}. \tag{12.174}$$

Note: End effects do not spoil the accuracy of our results, because the sideways displacement of the top plate leaves the end fields themselves unaltered. It simply lengthens the region where the field is nicely uniform. That is, we can safely ignore the end fields in calculating $dC/dx$, even when they would affect $C$ itself. Nevertheless, the *origin* of the force just calculated lies in the very end fields that our method permits us to ignore, for it is only at the ends that we find sideways components of the electric field; see Fig. 12.60.

3.27 *Force on a capacitor plate, again*

(a) The effective area of the capacitor is $\ell x$, so if $Q$ is the fixed charge in the overlap region (which is where the charge wants to reside; see the solution to Problem 3.26), the charge density is $\sigma = Q/\ell x$. Note that $\sigma$ is a function of $x$. The electric field inside the capacitor is $E = \sigma/\epsilon_0$, and the volume is $V = \ell xs$, so the stored energy as a function of $x$ is

$$U = \frac{\epsilon_0}{2}E^2 V = \frac{\epsilon_0}{2}\left(\frac{Q/\ell x}{\epsilon_0}\right)^2(\ell xs) = \frac{Q^2 s}{2\epsilon_0 \ell x}. \tag{12.175}$$

As mentioned in the solution to Problem 3.26, this decreases with $x$, and we have

$$-dU = F\,dx \implies F = -\frac{dU}{dx} \implies F = \frac{Q^2 s}{2\epsilon_0 \ell x^2}. \tag{12.176}$$

Using $C = \epsilon_0(\ell x)/s$, you can quickly verify that this agrees with the answer to part (a) of Problem 3.26.

(b) The potential difference $\phi$ equals $Es = (\sigma/\epsilon_0)s$. So if $\phi$ is held constant, then in the overlap region the density $\sigma$ and the field $\sigma/\epsilon_0$ remain constant. The stored energy is now

$$U = \frac{\epsilon_0}{2}\left(\frac{\sigma}{\epsilon_0}\right)^2(\ell xs) = \frac{\sigma^2 \ell xs}{2\epsilon_0}, \tag{12.177}$$

which increases with $x$. This increase is made possible by the fact that the battery does work. The battery needs to increase the charge $Q$ on the plates because the density $\sigma$ remains constant while the area $\ell x$ increases. If the battery drags charge $dq$ from the negative plate to the positive plate, the work it does is

$$dW = (dq)\phi = (\sigma \cdot \ell\,dx)\frac{\sigma s}{\epsilon_0} = \frac{\sigma^2 \ell s}{\epsilon_0}\,dx. \tag{12.178}$$

From the above expression for $U$, we see that $dU$ is half of $dW$. Half of the battery's work shows up as stored energy in the capacitor. The other half must be the work done on an external object that the plate pulls on with a force $F$. So we have

$$dW - dU = F\,dx \implies \frac{\sigma^2 \ell s}{2\epsilon_0}\,dx = F\,dx \implies F = \frac{\sigma^2 \ell s}{2\epsilon_0},$$

(12.179)

independent of $x$. Using $C = \epsilon_0(\ell x)/s$ and $\phi = \sigma s/\epsilon_0$, you can show that this agrees with the answer to part (b) of Problem 3.26.

(c) The results in parts (a) and (b) are equal because, for a given value of $x$, we have $Q = \sigma \ell x$.

3.28 *Maximum energy storage between spheres*

First, note that the stored energy should indeed achieve a maximum for some value of $b$ between 0 and $a$, due to the following reasoning. The energy is zero when $b = a$, because there is zero volume containing a nonzero field. And the energy is essentially zero when $b \approx 0$, because the charge on the inner sphere must be very small (otherwise the field at the surface, which is proportional to $1/b^2$, would exceed $E_0$); this implies that the field is very small in the region between the spheres (except very close to the inner sphere, where it is $E_0$). Therefore, since the energy is zero at $b = a$ and $b \approx 0$, it must achieve a maximum at some intermediate value.

For convenience, let $b \equiv ka$. If $E_0$ is the field at radius $ka$, then, since $E \propto 1/r^2$, the field equals $E_0(ka)^2/r^2$ at larger values of $r$ (but less than $a$). So the energy stored in the field is

$$U = \frac{\epsilon_0}{2}\int E^2\,dv = \frac{\epsilon_0}{2}\int_{ka}^{a}\left(E_0\frac{k^2 a^2}{r^2}\right)^2 4\pi r^2\,dr$$

$$= 2\pi\epsilon_0 k^4 a^4 E_0^2 \int_{ka}^{a}\frac{dr}{r^2} = 2\pi\epsilon_0 a^3 E_0^2 (k^3 - k^4).$$

(12.180)

As noted above, this equals zero when $k = 0$ or $k = 1$. Taking the derivative to find the maximum gives $3k^2 - 4k^3 = 0 \implies k = 3/4$. Hence $b = 3a/4$. The stored energy is then

$$U = 2\pi\epsilon_0 a^3 E_0^2\left[\left(\frac{3}{4}\right)^3 - \left(\frac{3}{4}\right)^4\right] = \frac{27}{128}\pi\epsilon_0 a^3 E_0^2.$$

(12.181)

Alternatively, we can solve the problem using capacitance. From the example in Section 3.5, the capacitance of the system is

$$C = 4\pi\epsilon_0 ab/(a - b) = 4\pi\epsilon_0 ak/(1 - k).$$

If $Q$ is the charge on the inner sphere, then $E_0 = Q/4\pi\epsilon_0(ka)^2 \implies Q = 4\pi\epsilon_0(ka)^2 E_0$. The stored energy is then

$$U = \frac{Q^2}{2C} = \frac{(4\pi\epsilon_0(ka)^2 E_0)^2}{2 \cdot 4\pi\epsilon_0 ak/(1 - k)} = 2\pi\epsilon_0 a^3 E_0^2 (k^3 - k^4),$$

(12.182)

in agreement with Eq. (12.180).

3.29 *Compressing a sphere*

The energy of a capacitor is $C\phi^2/2$. Since the capacitance of a sphere is $4\pi\epsilon_0 r$, we see that the initial and final energies stored in the system are (for $r = R$ and $r = 0$, respectively)

$$U_i = \frac{1}{2}(4\pi\epsilon_0 R)\phi^2 = 2\pi\epsilon_0 R\phi^2 \qquad \text{and} \qquad U_f = \frac{1}{2}(4\pi\epsilon_0 \cdot 0)\phi^2 = 0.$$
(12.183)

To find the work done by (or on) the battery, note that the final charge on the shell is zero, because $Q = C\phi$ and because $C = 0$ when $r = 0$. So all of the initial charge $Q_i$ is transferred through a potential difference of $-\phi$ back to the battery. The work done by the battery is therefore

$$W_{batt} = Q_i(-\phi) = -Q_i\phi = -(C_i\phi)\phi = -4\pi\epsilon_0 R\phi^2.$$
(12.184)

Since the work done by the battery is negative, this means that work is actually done *on* the battery; the energy of the battery increases. Basically, every bit of charge $dq$ that leaves the shell gives away (to the battery or in general whatever is maintaining the potential difference) the energy of $(dq)\phi$ that it had.

Now let's find the work that you do. You have to apply a force to the shell to compress it down in size. The force you apply to a given patch must balance the electric force that the rest of the shell exerts on the patch. As we showed in Section 1.14, the electric force per unit area is $\sigma$ times the average of the fields on either side. That is, $F/A = \sigma(E_1 + E_2)/2$. The field is zero inside and $\sigma/\epsilon_0$ just outside, so your force per unit area is $(\sigma)(\sigma/\epsilon_0)/2$. Your total radially inward force over the entire area $A$ of the shell, when the radius is $r$, is then

$$F_{you} = \frac{\sigma^2 A}{2\epsilon_0} = \frac{(Q/A)^2 A}{2\epsilon_0} = \frac{Q^2}{2\epsilon_0 A} = \frac{(C\phi)^2}{2\epsilon_0 A} = \frac{(4\pi\epsilon_0 r\phi)^2}{2\epsilon_0(4\pi r^2)} = 2\pi\epsilon_0\phi^2.$$
(12.185)

Since $\phi$ is constant, this force is constant. Your work over the distance from $r = R$ down to $r = 0$ is therefore

$$W_{you} = F_{you}R = 2\pi\epsilon_0 R\phi^2.$$
(12.186)

Now, the conservation-of-energy statement is

$$U_i + W_{you} + W_{batt} = U_f.$$
(12.187)

That is, the final energy equals the initial energy plus (or minus) the energy that was added to (or subtracted from) the system. And indeed, we have

$$2\pi\epsilon_0 R\phi^2 + 2\pi\epsilon_0 R\phi^2 - 4\pi\epsilon_0 R\phi^2 = 0.$$
(12.188)

Said in another way, the initial energy stored in the shell plus the energy you put into the system all goes into the battery in the end.

3.30 *Two ways of calculating energy*

(a) Let the shells have radii $r_1$ and $r_2$ (with $r_1 < r_2$). The electric field between them is $E = Q/4\pi\epsilon_0 r^2$, and it is zero elsewhere. The first method of calculating the energy of the system therefore gives

$$U = \frac{\epsilon_0}{2} \int E^2 \, dv = \frac{\epsilon_0}{2} \int_{r_1}^{r_2} \left( \frac{Q}{4\pi\epsilon_0 r^2} \right)^2 4\pi r^2 \, dr$$

$$= \frac{Q^2}{8\pi\epsilon_0} \int_{r_1}^{r_2} \frac{dr}{r^2} = \frac{Q^2}{8\pi\epsilon_0} \left( \frac{1}{r_1} - \frac{1}{r_2} \right). \qquad (12.189)$$

The magnitude of the potential difference between the shells is

$$\phi = \int E \, dr = \int_{r_1}^{r_2} \frac{Q}{4\pi\epsilon_0 r^2} \, dr = \frac{Q}{4\pi\epsilon_0} \left( \frac{1}{r_1} - \frac{1}{r_2} \right), \qquad (12.190)$$

which is the familiar result. We then quickly see that the second method of calculating the energy via $Q\phi/2$ also yields the result in Eq. (12.189), as desired.

(b) The electric field is zero both inside the inner conductor (by the uniqueness theorem) and outside the outer conductor (see the discussion in the example in Section 3.2), so the energy is confined to the volume between the conductors. If we integrate the given identity over this volume, we obtain

$$\int_V \nabla \cdot (\phi \nabla \phi) \, dv = \int_V (\nabla \phi)^2 \, dv + \int_V \phi \, \nabla^2 \phi \, dv. \qquad (12.191)$$

In the first term we can use the divergence theorem to write the integral as a surface integral. In the second term (and in the first) we can use $\mathbf{E} = -\nabla \phi$. And the third term equals zero due to Poisson's equation, $\nabla^2 \phi = -\rho/\epsilon_0$, since there is no charge between the conductors. So we have

$$-\int_S \phi \mathbf{E} \cdot d\mathbf{a} = \int_V E^2 \, dv, \qquad (12.192)$$

where the surface $S$ is the boundary of the volume $V$ between the conductors. $S$ consists of two parts: a boundary $S_1$ just outside the inner conductor, and a boundary $S_2$ just inside the outer conductor. The potential $\phi$ takes on a constant value over each of these boundaries, because they coincide with conductors. Let these two constant values be $\phi_1$ and $\phi_2$. Then the left-hand side of Eq. (12.192) becomes

$$-\int_S \phi \mathbf{E} \cdot d\mathbf{a} = -\phi_1 \int_{S_1} \mathbf{E} \cdot d\mathbf{a} - \phi_2 \int_{S_2} \mathbf{E} \cdot d\mathbf{a}. \qquad (12.193)$$

Now, the surface $S_1$ encloses the charge $Q$ on the inner conductor, so Gauss's law tells us that $\int_{S_1} \mathbf{E} \cdot d\mathbf{a} = -Q/\epsilon_0$. (The minus sign comes from the fact that $d\mathbf{a}$ was defined to point outward from our volume $V$, which means inward toward the inner conductor. This is opposite to the direction of $d\mathbf{a}$ if we were considering $S_1$ to be a surface surrounding the inner conductor.) Likewise, the surface $S_2$ also encloses the charge $Q$ on the inner conductor (this time with the standard orientation), so Gauss's law tells us that $\int_{S_2} \mathbf{E} \cdot d\mathbf{a} = Q/\epsilon_0$. Equation (12.193) therefore gives

$$-\int_S \phi \, \mathbf{E} \cdot d\mathbf{a} = \frac{Q}{\epsilon_0}(\phi_1 - \phi_2) \equiv \frac{Q\phi}{\epsilon_0}, \tag{12.194}$$

where $\phi$ is defined to be the potential difference between the conductors. Plugging this result into Eq. (12.192) yields

$$\frac{Q\phi}{\epsilon_0} = \int_V E^2 \, dv \implies \frac{1}{2}Q\phi = \frac{\epsilon_0}{2}\int_V E^2 \, dv, \tag{12.195}$$

as desired. If we instead have two conductors (with charges $Q_1$, $Q_2$ and potentials $\phi_1$, $\phi_2$) inside a third conductor (with charge $-Q_1-Q_2$ and potential $\phi_3$), then you can show that the above procedure leads to

$$\frac{1}{2}Q_1(\phi_1 - \phi_3) + \frac{1}{2}Q_2(\phi_2 - \phi_3) = \frac{\epsilon_0}{2}\int_V E^2 \, dv. \tag{12.196}$$

You should convince yourself that the left-hand side is the energy required to transfer charge from the outer conductor to the two inner conductors, in the correct proportion. If $Q_1 = -Q_2$ we have a standard two-conductor capacitor, with $(1/2)Q\Delta\phi$ on the left.

## 12.4 Chapter 4

### 4.1 Van de Graaff current

The field on either side of the belt is given by $E = \sigma/2\epsilon_0$. Hence

$$\sigma = 2\epsilon_0 E = 2\left(8.85 \cdot 10^{-12}\,\frac{\mathrm{C}}{\mathrm{V\,m}}\right)\left(10^6\,\frac{\mathrm{V}}{\mathrm{m}}\right) = 1.77 \cdot 10^{-5}\,\frac{\mathrm{C}}{\mathrm{m}^2}. \tag{12.197}$$

The current equals $\sigma$ times the area per time that is swept out by the belt. The area swept out in time $dt$ is $\ell(v\,dt)$, where $\ell$ is the belt's width. So the area per time is $\ell v$. The current is therefore

$$I = \sigma \ell v = \left(1.77 \cdot 10^{-5}\,\frac{\mathrm{C}}{\mathrm{m}^2}\right)(0.3\,\mathrm{m})\left(20\,\frac{\mathrm{m}}{\mathrm{s}}\right)$$

$$= 1.06 \cdot 10^{-4}\,\frac{\mathrm{C}}{\mathrm{s}} = 0.106\,\mathrm{mA}. \tag{12.198}$$

### 4.2 Junction charge

Let the junction have area $A$, and let $Q$ be the desired amount of charge on it. Gauss's law tells us that $A(E_2 - E_1) = Q/\epsilon_0$. In the steady state, the current density $J$ is the same in both regions (otherwise charge would continue to pile up at the junction), so we have $E_1 = J/\sigma_1$ and $E_2 = J/\sigma_2$. Therefore,

$$A\left(\frac{J}{\sigma_2} - \frac{J}{\sigma_1}\right) = \frac{Q}{\epsilon_0} \implies Q = \epsilon_0(AJ)\left(\frac{1}{\sigma_2} - \frac{1}{\sigma_1}\right) = \epsilon_0 I\left(\frac{1}{\sigma_2} - \frac{1}{\sigma_1}\right).$$

(12.199)

If $\sigma_1 = \sigma_2$, then $Q$ correctly equals zero. If $\sigma_1 \to \infty$, then $Q = \epsilon_0 I/\sigma_2$. This makes sense because there is essentially no field (external field plus field from junction charge) in the left region (otherwise there would be infinite current), so the charge $Q$ at the junction is the amount necessary to produce the required field in the right region, as you can verify. If $\sigma_1 \to 0$, then $Q \to -\infty$. This makes sense because a huge field is necessary in the left region to create a nonzero $J$, so a large negative charge at the junction is necessary to mostly cancel this huge field and bring it down to the finite value needed in the right region. Physically what happens is that if there is initially no charge at the junction, the huge external field causes charge to flow away from the junction in the right region, leaving behind a negative charge density at the interface, which in turn decreases the field in the right region. This continues until the field in the right region has the value $J/\sigma_2$.

4.3 *Adding resistors*

(a) When resistors are connected in series, the currents across them are equal, because charge can't pile up between them. Let $I$ be this common current. The total voltage $V$ across the effective resistor is the sum of the voltages across the two resistors, so $V = V_1 + V_2$. We know that $V = IR$, $V_1 = IR_1$, and $V_2 = IR_2$, where the same $I$ appears everywhere here. Plugging these expressions into $V = V_1 + V_2$ gives

$$IR = IR_1 + IR_2 \implies R = R_1 + R_2.$$

(12.200)

If $R_1 \to 0$ then $R \to R_2$. This makes sense because, for a given $I$, the voltage $IR_1$ across $R_1$ is tiny, which means that the overall voltage $IR$ is essentially equal to the voltage $IR_2$ across $R_2$. So $R \approx R_2$.

If $R_1 \to \infty$ then $R \to \infty$ also. This makes sense because, for a given $I$, the voltage $IR_1$ across $R_1$ is huge, which means that the overall voltage $IR$ is likewise huge, since it is at least as large. So $R$ must be very large.

(b) When resistors are connected in parallel, the voltages across them are equal, because the voltage drop from the left side of the overall circuit to the right side can't depend on the path. Let $V$ be this common voltage, which is of course also the voltage drop in the overall effective resistor.

The total current $I$ across the effective resistor is the sum of the currents across the two resistors, so $I = I_1 + I_2$. But we know that $I = V/R$, $I_1 = V/R_1$, and $I_2 = V/R_2$, where the same $V$ appears everywhere here. Plugging these expressions into $I = I_1 + I_2$ gives

$$\frac{V}{R} = \frac{V}{R_1} + \frac{V}{R_2} \implies \frac{1}{R} = \frac{1}{R_1} + \frac{1}{R_2}.$$

(12.201)

In short, in the series case the currents across the two resistors are the same, and the voltages add; whereas in the parallel case the voltages are the same, and the currents add.

If $R_1 \to 0$ then $R \to 0$ also. This makes sense because, for a given $V$, the current $V/R_1$ across $R_1$ is huge, which means that the overall current $V/R$ is likewise huge, since it is at least as large. So $R$ must be very large.

If $R_1 \to \infty$ then $R \to R_2$. This makes sense because, for a given $V$, the current $V/R_1$ across $R_1$ is tiny, which means that the overall current $V/R$ is essentially equal to the current $V/R_2$ across $R_2$. So $R \approx R_2$.

4.4    *Spherical resistor*

(a) Using Eq. (4.17), the resistance across a thin shell with radius $r$ and thickness $dr$ is $dR = \rho \, dr / 4\pi r^2$. The shells are in series, so integrating this from $r_1$ to $r_2$ gives a total resistance of

$$R = \int_{r_1}^{r_2} \frac{\rho \, dr}{4\pi r^2} = \frac{\rho}{4\pi} \left( \frac{1}{r_1} - \frac{1}{r_2} \right) \longrightarrow \frac{\rho}{4\pi r_1}, \qquad (12.202)$$

in the limit where $r_2 \gg r_1$.

(b) Without any calculations, it is reasonable to think that the resistance might be proportional to $\rho/r_1$, based on dimensions. However, this isn't completely rigorous, because there is another dimensionful quantity in the problem, namely $r_2$. And the exact answer in Eq. (12.202) *does* depend on $r_2$. But in the $r_2 \to \infty$ limit, any term involving $r_2$ must be either 0 or $\infty$. So all we can say from dimensional analysis is that the desired resistance is 0, $\infty$, or finite and proportional to $\rho/r_1$. A little thought tells us that it can't be zero, because $R$ certainly isn't zero for a finite $r_2$, and $R$ increases with $r_2$. So only the second two options are possible.

In the other extreme where we keep $r_2$ fixed and let $r_1 \to 0$, all of the above reasoning applies, so we know that the resistance must either be $\infty$, or finite and proportional to $\rho/r_2$. It turns out that in this case the $\infty$ option is the correct one, which is evident from the exact answer in Eq. (12.202).

4.5    *Laminated conductor*

The ratio of conductivities, $7.2/1$, and the ratio of layer thicknesses, $1/2$, are the only two things that matter. Let us arbitrarily pick $\sigma_s = 1$ and $\sigma_t = 1/7.2$ (ignoring the units). The resistance of an object with length $L$ and cross-sectional area $A$ can be written in either of the equivalent forms, $\rho L/A$ or $L/\sigma A$. Since we are interested in a relation between $\sigma$ values, we will use the latter.

For perpendicular currents, the layers are in series, so the resistances add:

$$R_\perp = R_s + R_t \implies \frac{L}{\sigma_\perp A} = \frac{L_s}{\sigma_s A} + \frac{L_t}{\sigma_t A}$$

$$\implies \frac{1}{\sigma_\perp} = \frac{1}{\sigma_s} \frac{L_s}{L} + \frac{1}{\sigma_t} \frac{L_t}{L}. \qquad (12.203)$$

$L_s$ and $L_t$ are in the same ratio as the given layer thicknesses, so plugging in the numbers yields

$$\frac{1}{\sigma_\perp} = 1 \cdot \frac{1}{3} + 7.2 \cdot \frac{2}{3} = 5.13 \implies \sigma_\perp = 0.195. \qquad (12.204)$$

For parallel currents, the layers are in parallel, so the resistance add via reciprocals:

$$\frac{1}{R_\parallel} = \frac{1}{R_s} + \frac{1}{R_t} \implies \frac{\sigma_\parallel A}{L} = \frac{\sigma_s A_s}{L} + \frac{\sigma_t A_t}{L}$$

$$\implies \sigma_\parallel = \sigma_s \frac{A_s}{A} + \sigma_t \frac{A_t}{A}. \qquad (12.205)$$

It is now $A_s$ and $A_t$ that are in the same ratio as the given layer thicknesses. Plugging in the numbers yields

$$\sigma_\parallel = 1 \cdot \frac{1}{3} + \frac{1}{7.2} \cdot \frac{2}{3} = 0.426. \qquad (12.206)$$

Therefore,

$$\frac{\sigma_\perp}{\sigma_\parallel} = \frac{0.195}{0.426} = 0.457. \qquad (12.207)$$

Physically, it makes sense that the conductivity is larger in the parallel direction, due to the following reasoning. Consider the case where the layer thicknesses are equal, and where one conductivity is much larger than the other. Let's say they take on the values $\sigma_1 = 1$ and $\sigma_2 = 100$. Then $\sigma_\perp$ is approximately equal to 2, because the "2" region offers essentially no resistance, so the total resistance should be half (equivalently, the conductivity should be twice) what it would be if the conductor were made entirely of the "1" material. And $\sigma_\parallel$ is approximately equal to 50, because the "1" region carries essentially no current, so the total resistance should be twice (equivalently, the conductivity should be half) what it would be if the conductor were made entirely of the "2" material. We see that $\sigma_\parallel$ is indeed (much) larger than $\sigma_\perp$. As an exercise, you can give a general proof that $\sigma_\parallel$ is always greater than or equal to $\sigma_\perp$. Further considerations along these lines are left for Exercise 4.33.

4.6    *Validity of tapered-rod approximation*

(a) The error in the reasoning is that the current doesn't fan out uniformly in the cone. If we are to build up the cone from cross-sectional slices and then add up the resistances from these slices, all parts of a given slice must be at the same potential. Equivalently, if you solve Exercise 4.32 by writing down the resistance of each slice, you will be assuming that the current flows perpendicular to each slice. This is clearly not the case for the slice shown in Fig. 12.61.

(b) The current *does* diverge symmetrically in the object with the spherical endcaps. If we slice the object into pieces with a given radius, then all points in a given slice are the same distance from the left endcap. Let the endcaps be located at radii $r_1$ and $r_2$ from their com-

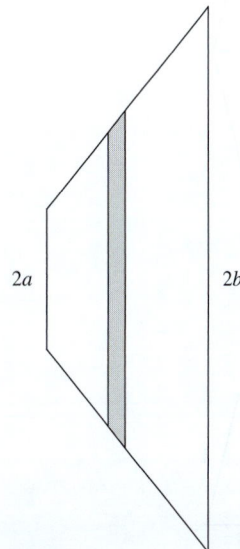

$2a$          $2b$

**Figure 12.61.**

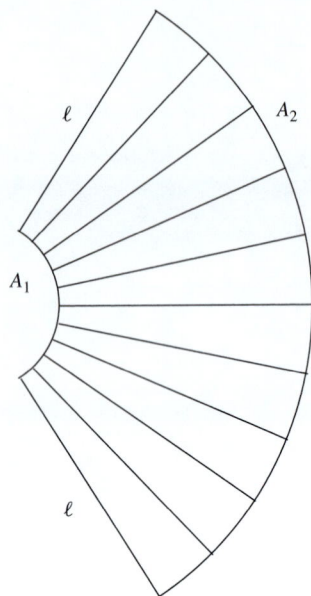

$\ell$

$A_2$

$A_1$

$\ell$

**Figure 12.62.**

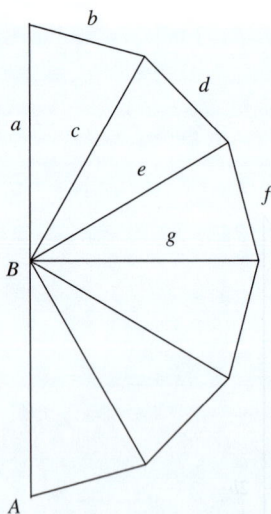

$b$

$a$    $c$    $d$

$e$

$f$

$B$    $g$

$A$

**Figure 12.63.**

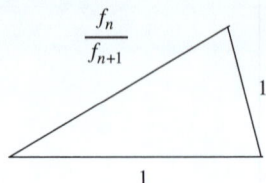

$\dfrac{f_n}{f_{n+1}}$

$1$

$1$

**Figure 12.64.**

mon center, and let a (partial) spherical slice at radius $r$ have area $A$. Since the area of a spherical slice is proportional to the radius, we have $A/r^2 = A_1/r_1^2 \implies A = A_1(r^2/r_1^2)$. Using Eq. (4.17), the total resistance between the endcaps is

$$R = \int_{r_1}^{r_2} \frac{\rho \, dr}{A} = \int_{r_1}^{r_2} \frac{\rho \, dr}{A_1(r^2/r_1^2)} = \frac{\rho r_1^2}{A_1} \int_{r_1}^{r_2} \frac{dr}{r^2} = \frac{\rho r_1^2}{A_1}\left(\frac{1}{r_1} - \frac{1}{r_2}\right)$$

$$= \frac{\rho(r_2 - r_1)}{A_1}\frac{r_1}{r_2} = \frac{\rho \ell}{A_1}\frac{\sqrt{A_1}}{\sqrt{A_2}} = \frac{\rho \ell}{\sqrt{A_1 A_2}}, \qquad (12.208)$$

where we have again used the fact that $A \propto r^2$.

Alternatively, we can consider the object to be the sum of a large number of thin objects arranged in "parallel," as shown in Fig. 12.62. All of these objects have the same length (which wasn't the case for the conical object in part (a)). And they all have a slow taper, so we can apply the result from Exercise 4.32 to each one. If you do that exercise, you will note that the resistance can be written as $R = \rho \ell / \sqrt{\alpha_1 \alpha_2}$ where $\alpha_1$ and $\alpha_2$ are the areas of the end faces. When we combine a large number $N$ of these thin objects in parallel, the resistance goes down by a factor $N$. So we have

$$R = \frac{\rho \ell}{N\sqrt{\alpha_1 \alpha_2}} = \frac{\rho \ell}{\sqrt{(N\alpha_1)(N\alpha_2)}} = \frac{\rho \ell}{\sqrt{A_1 A_2}}, \qquad (12.209)$$

as above.

4.7    *Triangles of resistors*

(a) In Fig. 12.63, resistors $a$ and $b$ are in series and can be replaced by $2R$. This combination is in parallel with $c$, so $c$ can be replaced with $(2/3)R$. This is in series with $d$, which yields $(5/3)R$. This in turn is in parallel with $e$, so $e$ can be replaced by $(5/8)R$. Continuing in this manner, $g$ can be replaced by $(13/21)R$, the next spoke by $(34/55)R$, the next by $(89/144)R$, and finally the effective resistance across $AB$ equals $(233/377)R$. The numbers in the fractions here are the familiar Fibonacci numbers, defined by the recursion relation $f_{n+1} = f_n + f_{n-1}$; 233 and 377 are the 13th and 14th Fibonacci numbers.

We can easily prove by induction that if the equivalent resistance across a given spoke (incorporating all of the resistors on its counterclockwise side) takes the form of $(f_n/f_{n+1})R$, then the equivalent resistance across the next spoke equals $(f_{n+2}/f_{n+3})R$. This is true because the effective resistance across the bottom resistor in Fig. 12.64 is (ignoring the $R$)

$$\frac{1}{\dfrac{1}{1} + \dfrac{1}{1 + \dfrac{f_n}{f_{n+1}}}} = \frac{1}{1 + \dfrac{f_{n+1}}{f_n + f_{n+1}}} = \frac{1}{1 + \dfrac{f_{n+1}}{f_{n+2}}}$$

$$= \frac{f_{n+2}}{f_{n+1} + f_{n+2}} = \frac{f_{n+2}}{f_{n+3}}, \qquad (12.210)$$

as desired. If there are $N$ triangles ($N = 6$ here), the effective resistance across the last (or first) spoke is $(f_{2N+1}/f_{2N+2})R$.

(b) We'll use the letter $r$ for $R_{\text{eff}}$ since it is quicker to write. If the resistance approaches the value $r$, then adding on another triangle must still yield an effective resistance of $r$. The effective resistance across the left spoke in Fig. 12.65 must therefore be $r$. This gives (ignoring the $R$)

$$r = \frac{1}{\dfrac{1}{1} + \dfrac{1}{r+1}} \implies r = \frac{r+1}{r+2} \implies r^2 + r - 1 = 0$$

$$\implies r = \frac{-1 + \sqrt{5}}{2} \approx 0.618. \qquad (12.211)$$

This is the limit of the ratio of consecutive Fibonacci numbers; it is also the inverse of the golden ratio. Even for fairly small Fibonacci numbers, the ratio is close to 0.618. In the above case of six triangles, $233/377$ agrees with $(-1 + \sqrt{5})/2$ to the fifth decimal place.

**4.8** *Infinite square lattice*

Let the two adjacent nodes be labeled $N_1$ and $N_2$. Consider a setup where a current of 1 A flows *into* the lattice at $N_1$ and heads out to infinity in the two-dimensional plane. If you want, you can imagine a return lead connected around a rim very far away. By symmetry, the current in each of the four resistors connected to $N_1$ is $1/4$ A (all flowing away from $N_1$). In particular, there is a current of $1/4$ A flowing from $N_1$ to $N_2$.

Consider a second setup where a current of 1 A comes in from infinity in the two-dimensional plane and flows *out of* the lattice at $N_2$. Again, you can imagine a return lead connected around a rim very far away. By symmetry, the current in each of the four resistors connected to $N_2$ is $1/4$ A (all flowing toward $N_2$). In particular, there is a current of $1/4$ A flowing from $N_1$ to $N_2$.

If we superpose the above two setups, we now have 1 A entering the lattice at $N_1$, 1 A leaving the lattice at $N_2$, and no current at infinity. This is exactly the setup we wanted to create. The total current flowing from $N_1$ to $N_2$ is $1/4 + 1/4 = 1/2$ A. This current flows across a 1 Ω resistor, so the voltage drop from $N_1$ to $N_2$ is $1/2$ V. The effective resistance is defined by $V = IR_{\text{eff}}$, where $I$ is the current that enters and leaves the circuit (which is 1 A here). So we have $1/2$ V $= (1 \text{ A})R_{\text{eff}}$, which gives $R_{\text{eff}} = 1/2 \, \Omega$. Unfortunately, this quick method works only if the nodes are adjacent.

**4.9** *Sum of the effective resistances*

(a) As stated in the problem, the sum of the effective resistances across the four resistors in the first network is 2. In the second network, three of the effective resistances (in units of $R$) are $2/3$, and two are simply 1. So the sum is 4. In the third network, all four effective resistances are $3/4$, so the sum is 3. In the fourth network, the effective resistance across the diagonal is $1/2$, and the other four resistances are $5/8$. So the sum is 3. In the fifth network, the effective resistance across each of the $n$ resistors is $1/n$, so the sum is 1. In all cases, the sum of

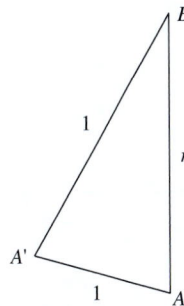

**Figure 12.65.**

the effective resistances across all of the resistors is $N - 1$ (times $R$), where $N$ is the number of points. So that's our conjecture. Now to prove it.

(b) Here's the hint: if you solved Problem 4.8, you saw that a useful technique for finding the effective resistance between two nodes, call them $A$ and $B$, is to consider the superposition of a setup where a current $I$ enters at $A$ and another setup where a current $I$ exits at $B$. But in each of these separate setups, we can't just put in (or take out) current without also taking out (or putting in) current somewhere else. (We could do so in the infinite network in Problem 4.8 because the current could just sail off to infinity.) Let's take out (and put in) the current in a symmetric manner, in the following way.

- Setup #1. Put in a current $(N - 1)I/N$ at node $A$ and take out a current $I/N$ at the other $N - 1$ nodes.
- Setup #2. Take out a current $(N - 1)I/N$ at node $B$ and put in a current $I/N$ at the other $N - 1$ nodes.

The superposition of these two setups involves a current $I$ entering at $A$ and a current $I$ exiting at $B$, and nothing happening at the other $N - 2$ nodes. That's the hint. Try to prove the conjecture without reading further.

Continuing onward... consider two nodes, $A$ and $B$, that are connected by one or more of the resistors $R$. If we put in a current $I$ at $A$, and take out a current $I$ at $B$, then the effective resistance between $A$ and $B$ is $V/I$, where $V$ is the potential difference between the two nodes. The situation where a current $I$ goes in at $A$ and comes out at $B$ can be considered as the superposition of the above two setups.

In setup #1, label the current going from $A$ to $B$, across a *particular one* of the resistors $R$ (if more than one resistor connects $A$ and $B$), as $I^A_{A \to B}$. In setup #2, label the current going from $A$ to $B$, across the same resistor, as $I^B_{A \to B}$. The superscript here denotes the node at which the current of $(N - 1)I/N$ enters or leaves. Remember that the two setups are defined independently and have nothing to do with each other.

In the combined setup, the current going from $A$ to $B$, across the given resistor, is $I^A_{A \to B} + I^B_{A \to B}$. Since this current passes along a resistor $R$, the voltage difference between $A$ and $B$ is $V = (I^A_{A \to B} + I^B_{A \to B})R$. The effective resistance between $A$ and $B$ is therefore

$$R_{AB} = \frac{V}{I} = \frac{(I^A_{A \to B} + I^B_{A \to B})R}{I}. \tag{12.212}$$

We must now add up these $R_{AB}$ contributions across all the resistors. Let the desired sum be $S$. Then

$$S = \frac{R}{I} \sum (I^A_{A \to B} + I^B_{A \to B}), \tag{12.213}$$

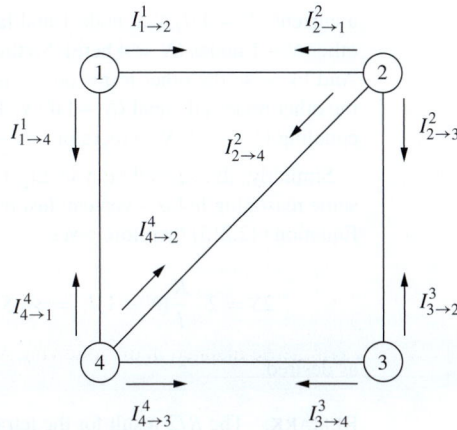

**Figure 12.66.**

where the sum runs over all the resistors (and again, *not* just over all pairs $AB$ connected by a resistor, because some pairs may be connected by more than one resistor). Let's write this sum in a more symmetric form. By reversing the roles of $A$ and $B$ (it doesn't matter which node the current goes in and which comes out), we can also write

$$S = \frac{R}{I} \sum (I_{B \to A}^B + I_{B \to A}^A). \qquad (12.214)$$

Adding the previous two equations and rearranging the grouping gives

$$2S = \frac{R}{I} \sum (I_{A \to B}^A + I_{B \to A}^B) + \frac{R}{I} \sum (I_{B \to A}^A + I_{A \to B}^B). \qquad (12.215)$$

Consider the first sum. For concreteness let's look at the fourth network discussed in part (a). The sum runs over the five resistors, and for each resistor the terms $I_{A \to B}^A$ and $I_{B \to A}^B$ are the two *influxes* of current at the ends of the resistor, as shown in Fig. 12.66. Remember that each of these $I_{C \to D}^C$ type terms equals the current that flows between nodes $C$ and $D$ when a current $(N-1)I/N$ is put in at $C$ and a current $I/N$ is taken out of the other $N-1$ nodes. Figure 12.66 suggests replacing the sum over resistors with a sum over nodes. That is, in the following equation the terms on the left-hand side are grouped by the five resistors (as in the first sum in Eq. (12.215)), while the terms on the right-hand side are grouped by the four nodes:

$$(I_{1 \to 2}^1 + I_{2 \to 1}^2) + (I_{2 \to 3}^2 + I_{3 \to 2}^3) + (I_{3 \to 4}^3 + I_{4 \to 3}^4)$$
$$+ (I_{4 \to 1}^4 + I_{1 \to 4}^1) + (I_{2 \to 4}^2 + I_{4 \to 2}^4)$$
$$= (I_{1 \to 2}^1 + I_{1 \to 4}^1) + (I_{2 \to 1}^2 + I_{2 \to 4}^2 + I_{2 \to 3}^2) + (I_{3 \to 2}^3 + I_{3 \to 4}^3)$$
$$+ (I_{4 \to 1}^4 + I_{4 \to 2}^4 + I_{4 \to 3}^4). \qquad (12.216)$$

The first term on the right-hand side, $I_{1 \to 2}^1 + I_{1 \to 4}^1$, is (by definition) the sum of the currents emanating from node 1 when we put in a current $(N-1)I/N$ at node 1 and take out a current $I/N$ from the other $N-1$ nodes ($N=4$ here). So this term must equal $(N-1)I/N$. And likewise the other terms on the right-hand side associated with the other nodes all equal $(N-1)I/N$. The sum of these $N$ terms, each equaling $(N-1)I/N$, is therefore $(N-1)I$.

Similarly, the second sum in Eq. (12.215) equals $(N-1)I$. The same reasoning holds – you can just replace "put in" with "take out." Equation (12.215) therefore gives

$$2S = 2 \cdot \frac{R}{I}(N-1)I \implies S = (N-1)R, \qquad (12.217)$$

as desired.

REMARK: The $R/2$ result for the tetrahedron in Problem 4.11 below is consistent with this result, because $N=4$ and there are six resistors. Likewise for the $7R/12$ result for the cube in Exercise 4.35(c), because $N=8$ and there are 12 resistors. The $R/2$ result for the infinite square grid in Problem 4.8 is also consistent with this result, for the following reason. If we have a very large number $N$ of nodes in the square grid, then the number of resistors is $2N$, neglecting boundary effects. (This is true because each node is connected to four resistors, but this double counts the number of resistors since each resistor has two nodes at its ends.) So the sum of the effective resistances across all these resistors is $(2N)(R/2) = NR$, which is essentially equal to $(N-1)R$ (in a multiplicative sense) for $N \to \infty$. Said in another way, if the effective resistance were anything but $R/2$, the sum would have no chance of equaling $(N-1)R$ in the $N \to \infty$ limit.

A more general result, which requires only a slight modification of the above proof, is the following. If the resistors are not necessarily identical, then we can still can say that $\sum(r_k/R_k) = N-1$, where the sum runs over all the resistors in the network, and where $r_k$ is the effective resistance across the $k$th resistor (indexed in an arbitrary manner), and $R_k$ is the actual resistance of the $k$th resistor. In our special case above where all the $R_k$ have the same value $R$, we can take the $R$ out of the sum to obtain $\sum r_k = (N-1)R$, in agreement with Eq. (12.217).

**4.10** *Voltmeter, ammeter*

**Ammeter** To measure the current across point $A$, we need to cut the wire at $A$ and insert the galvanometer (so now $A$ is represented by two points in Fig. 12.67), or ideally just unclip two leads and clip them to the ends of the galvanometer. However, if this is the only thing we do, we will certainly affect the circuit because we have just added additional resistance to it (unless $R_g$ is much smaller than the resistance of the rest of the circuit, which we can't assume in general). Furthermore, the current passing through the galvanometer might very well be larger than its limit. We can fix both of these issues by connecting a very small known resistor in parallel with the galvanometer, as shown in Fig. 12.67. Virtually all of the current in the circuit will then pass through this small resistor, called a "shunt" resistor, $R_{sh}$. This fixes the problem of affecting the circuit (because the overall resistance of the parallel combination is very small),

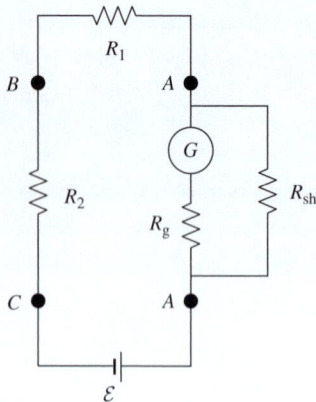

**Figure 12.67.**

and also the problem of overloading the galvanometer (because hardly any of the current passes through it).[3]

The components of an ammeter are shown in Fig. 12.68. If $R_{sh}$ is chosen to be $1/N$ times as large as $R_g$ (where $N$ is a large number, say 1000), then $N$ times as much current flows through the shunt as through the galvanometer. So if the galvanometer measures a small current $I_g$, we conclude that the total current flowing across $A$ is $I = I_g + N I_g = (N+1)I_g$. We therefore simply need to relabel all the readings on the galvanometer by a factor of $N + 1$, and we have constructed our ammeter.

**Voltmeter** To measure the voltage difference between points $B$ and $C$, we need to add the galvanometer in parallel with the given resistor $R_2$ between these points. However, if this is the only thing we do, we will certainly affect the circuit because we have just added another path for the current to take (unless $R_g$ is much larger than $R_2$, which we can't assume in general). Furthermore, the current passing through the galvanometer might very well be larger than its limit. We can fix both of these issues by connecting a very large known resistor $R_{ser}$ in series with the galvanometer, as shown in Fig. 12.69. Virtually none of the current in the circuit will then pass through the alternative galvanometer path. This fixes the problem of affecting the circuit (because the overall resistance of the parallel combination between $B$ and $C$ is essentially unchanged), and also the problem of overloading the galvanometer (because hardly any of the current passes through it).[4]

The components of a voltmeter are shown in Fig. 12.70. If $R_{ser}$ is chosen to be $N$ times as large as $R_g$ (where $N$ is a large number, say 1000), then the voltage drop across $R_{ser}$ is $N$ times as large as the voltage drop across $R_g$. So if the galvanometer measures a small current $I_g$, we conclude that the total voltage drop between $B$ and $C$ is $V = I_g R_g + I_g (N R_g) = (N+1)I_g R_g$. We therefore simply need to relabel all the readings on the galvanometer by a factor of $(N+1)R_g$ (so now the units are different), and we have constructed our voltmeter.

To summarize: An ammeter is made by combining a *small* resistor in *parallel* with a galvanometer, and the combination is then inserted in *series* with the given circuit. A voltmeter is made by combining a *large* resistor in *series* with a galvanometer, and the combination is then inserted in *parallel* with the given circuit. In both cases, there is minimal disruption to the given circuit, and the current passing through the galvanometer is very small.

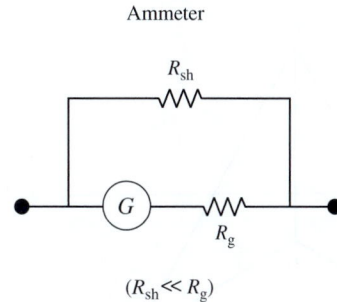

Ammeter

$(R_{sh} \ll R_g)$

**Figure 12.68.**

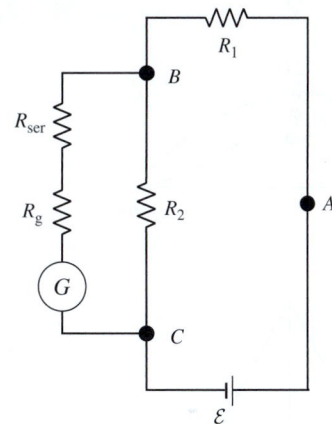

**Figure 12.69.**

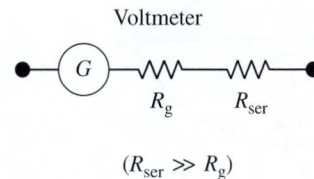

Voltmeter

$(R_{ser} \gg R_g)$

**Figure 12.70.**

---

[3] The precise conditions on $R_{sh}$ are that (1) $R_{sh}$ is much smaller than $R_g$, so that only a small fraction of the current goes through the galvanometer, and (2) $R_{sh}$ is much smaller than the resistance $R_1 + R_2$ of the original circuit. If this latter condition isn't met, then although our ammeter will give an exact measurement of the current flowing across $A$ *with* the ammeter inserted in the circuit, this current won't equal the current in the original circuit without the ammeter.

[4] The precise condition on $R_{ser}$ is that $R_{ser} + R_g$ is much larger than the resistor $R_2$ in the circuit. (Without making any assumptions about $R_g$, this means that $R_{ser}$ is much larger than $R_2$.) If this condition isn't met, then although our voltmeter will give an exact measurement of the voltage difference between $B$ and $C$ *with* the voltmeter connected in parallel, this voltage won't equal the voltage in the original circuit without the voltmeter.

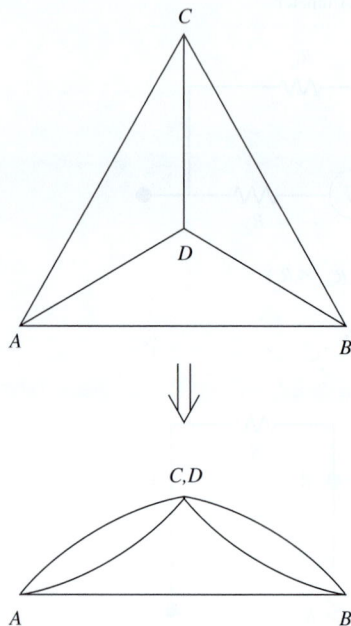

**Figure 12.71.**

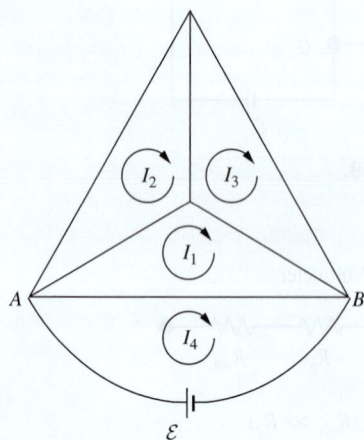

**Figure 12.72.**

**4.11    *Tetrahedron resistance***

(a) Let the two vertices be $A$ and $B$. By symmetry, the other two vertices are at the same potential, so we can collapse them to one point (because no current will flow in the resistor connecting them). We then have the equivalent network in Fig. 12.71, and the effective resistance quickly comes out to be $R/2$. Note that the sum of the effective resistances across all six resistors equals $6(R/2) = 3R$, which is consistent with the general result in Problem 4.9.

(b) The four loop equations for Fig. 12.72 are (dividing through by $R$ in all of them)

$$\mathcal{E}/R - (I_4 - I_1) = 0,$$
$$-(I_1 - I_4) - (I_1 - I_2) - (I_1 - I_3) = 0,$$
$$-I_2 - (I_2 - I_3) - (I_2 - I_1) = 0,$$
$$-I_3 - (I_3 - I_2) - (I_3 - I_1) = 0. \qquad (12.218)$$

The last two equations are symmetric in "2" and "3." Taking their difference yields $I_2 = I_3$ (as expected from the symmetry). The third equation then gives $I_1 = 2I_2$, whereupon the second equation gives $I_4 = 4I_2$, or equivalently $I_4 = 2I_1$. Finally, the first equation gives $\mathcal{E}/R - (I_4 - I_4/2) = 0 \implies \mathcal{E} = I_4(R/2)$. Since $I_4$ is the current through the battery, this implies that the effective resistance between $A$ and $B$ is $R/2$, in agreement with part (a).

**4.12    *Find the voltage difference***

With the loop currents shown in Fig. 12.73, the three clockwise loop equations are

$$0 = \mathcal{E} - I_1 R - \mathcal{E} - (I_1 - I_2)R,$$
$$0 = \mathcal{E} - I_2 R - (I_2 - I_3)R - (I_2 - I_1)R,$$
$$0 = -\mathcal{E} - (I_3 - I_2)R - I_3 R. \qquad (12.219)$$

These equations simplify to

$$0 = -2I_1 + I_2,$$
$$0 = \mathcal{E}/R - 3I_2 + I_3 + I_1,$$
$$0 = -\mathcal{E}/R - 2I_3 + I_2. \qquad (12.220)$$

The third equation plus twice the second gets rid of the $I_3$ terms: $0 = \mathcal{E}/R - 5I_2 + 2I_1$. Adding this to the first equation then gives $0 = \mathcal{E}/R - 4I_2$. So $I_2 = \mathcal{E}/4R$, from which we quickly obtain $I_1 = \mathcal{E}/8R$ and $I_3 = -3\mathcal{E}/8R$. The potential difference between points $a$ and $b$ is then

$$V_b - V_a = (I_2 - I_3)R = \left(\mathcal{E}/4R - (-3\mathcal{E}/8R)\right)R = 5\mathcal{E}/8. \qquad (12.221)$$

This is positive, so $b$ is at the higher potential. This makes sense by looking at the orientation of all the batteries.

## 4.13 Thévenin's theorem

We present two proofs. The first one is a direct "practical" type of proof. The second is rather slick. The second proof works for an arbitrary circuit $B$, but in the first proof we restrict $B$ to consist of a single emf, for simplicity. The key ingredient in both proofs is the linearity of the circuit.

**First proof** We will assume that the circuit $B$ is a single emf $\mathcal{E}$, although our reasoning can be extended to a general circuit $B$. Given the circuit $A$ and the external emf $\mathcal{E}$, imagine using Kirchhoff's rules to write down the loop equations for all the loop currents. We'll label the loop current passing through the external emf $\mathcal{E}$ as $I$. For the example shown in Fig. 12.74, the loop equations are

$$0 = \mathcal{E} + \mathcal{E}_2 - (I - I_2)R_4,$$
$$0 = -\mathcal{E}_2 + \mathcal{E}_1 - I_1 R_1 - (I_1 - I_2)R_2,$$
$$0 = \mathcal{E}_3 - (I_2 - I)R_4 - (I_2 - I_1)R_2 - I_2 R_3. \qquad (12.222)$$

All of these equations have the same form. They all involve terms that are products of $I$ and $R$ values, and also terms that are *linear* in the emfs (the emfs inside $A$, along with $\mathcal{E}$). These equations can be grouped together into one big matrix equation of the form $M\mathbf{I} = \boldsymbol{\mathcal{E}}$, where $M$ is a (symmetric) matrix whose entries are functions of the $R$ values (linear functions, although this isn't important), $\mathbf{I}$ is the vector of loop currents $(I, I_1, I_2, \ldots, I_n)$, and $\boldsymbol{\mathcal{E}}$ is a vector whose entries are linear functions of the emfs (this linearity *is* important). For the above example we have

$$\begin{pmatrix} R_4 & 0 & -R_4 \\ 0 & R_1 + R_2 & -R_2 \\ -R_4 & -R_2 & R_2 + R_3 + R_4 \end{pmatrix} \begin{pmatrix} I \\ I_1 \\ I_2 \end{pmatrix} = \begin{pmatrix} \mathcal{E} + \mathcal{E}_1 \\ \mathcal{E}_1 - \mathcal{E}_2 \\ \mathcal{E}_3 \end{pmatrix}.$$

$$(12.223)$$

The solution for all the currents is given by $\mathbf{I} = M^{-1}\boldsymbol{\mathcal{E}}$. However, we are not concerned with the exact nature of the inverse matrix $M^{-1}$. The only things we care about are that $M^{-1}$ depends only on the $R$ values, and that each loop current is linear in the emfs (because $\boldsymbol{\mathcal{E}}$ is linear in the emfs). In particular, the current $I$ passing through the external emf $\mathcal{E}$ takes the form (with $m$ being the number of internal emfs, which need not be equal to the number $n$ of loop currents),

$$I = a\mathcal{E} + a_1\mathcal{E}_1 + a_2\mathcal{E}_2 + \cdots + a_m\mathcal{E}_m, \qquad (12.224)$$

where the $a$ values are functions of *only* the $R$ values. This can be written as

$$\mathcal{E} + \mathcal{E}_{\text{eq}} = I R_{\text{eq}}, \qquad (12.225)$$

where

$$R_{\text{eq}} = \frac{1}{a} \quad \text{and} \quad \mathcal{E}_{\text{eq}} = \frac{a_1\mathcal{E}_1 + \cdots + a_m\mathcal{E}_m}{a}. \qquad (12.226)$$

But Eq. (12.225) is exactly the same equation that we would obtain if the circuit $A$ consisted of a single emf $\mathcal{E}_{\text{eq}}$ in series with a single resistor $R_{\text{eq}}$.

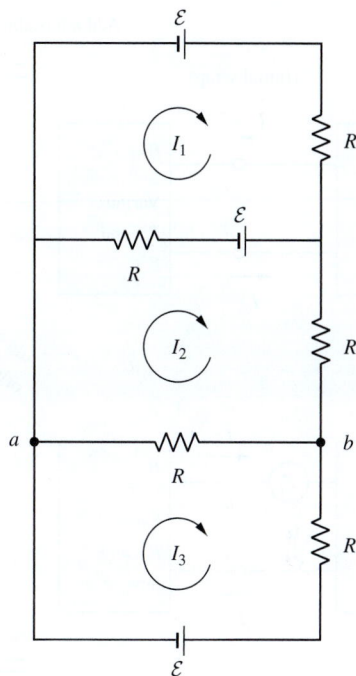

**Figure 12.73.**

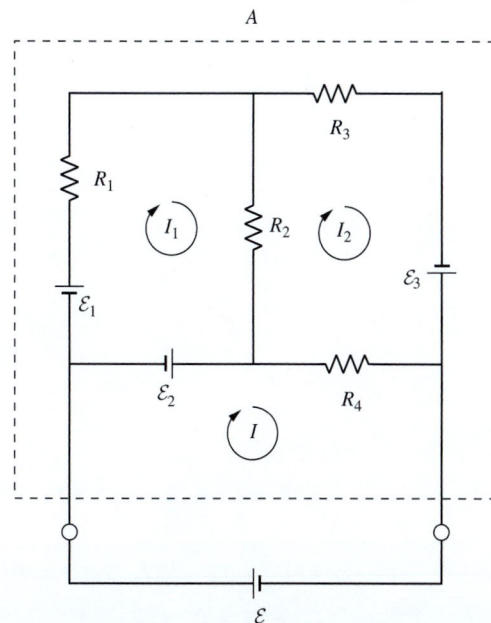

**Figure 12.74.**

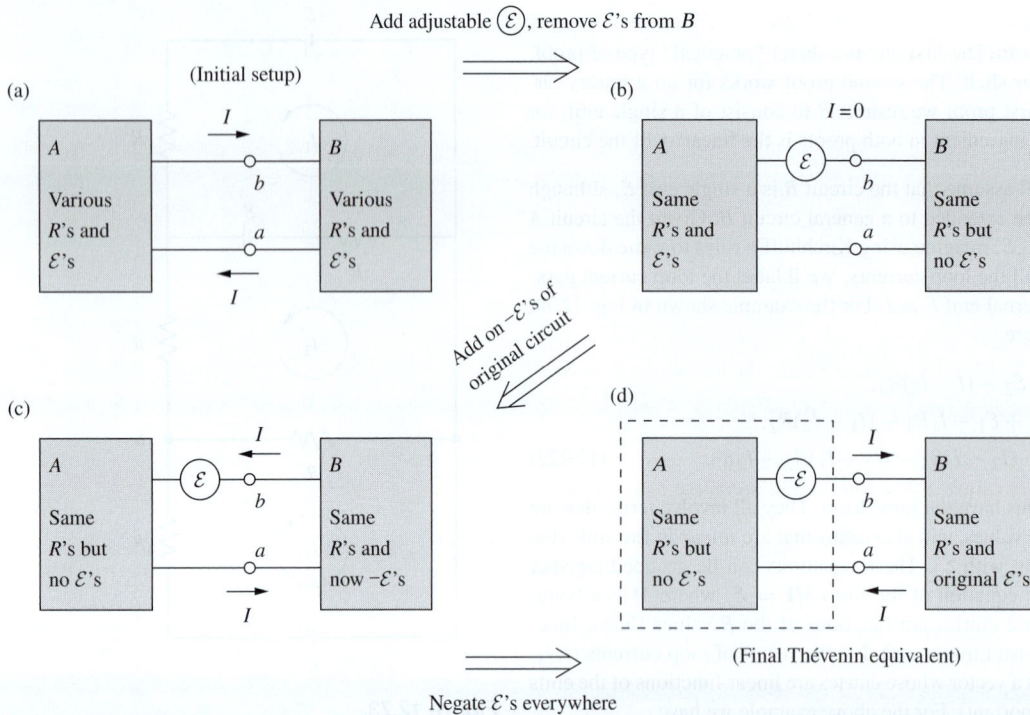

Add adjustable $\mathcal{E}$, remove $\mathcal{E}$'s from $B$

(a)

(Initial setup)

$I \rightarrow$

| $A$ | | $B$ |
|---|---|---|
| Various | $b$ | Various |
| $R$'s and | $a$ | $R$'s and |
| $\mathcal{E}$'s | | $\mathcal{E}$'s |

$\leftarrow I$

$\Longrightarrow$

(b)

$I = 0$

| $A$ | $\mathcal{E}$ | $B$ |
|---|---|---|
| Same | $b$ | Same |
| $R$'s and | $a$ | $R$'s but |
| $\mathcal{E}$'s | | no $\mathcal{E}$'s |

*Add on $-\mathcal{E}$'s of original circuit*

(c)

$\leftarrow I$

| $A$ | $\mathcal{E}$ | $B$ |
|---|---|---|
| Same | $b$ | Same |
| $R$'s but | $a$ | $R$'s and |
| no $\mathcal{E}$'s | | now $-\mathcal{E}$'s |

$I \rightarrow$

(d)

$I \rightarrow$

| $A$ | $-\mathcal{E}$ | $B$ |
|---|---|---|
| Same | $b$ | Same |
| $R$'s but | $a$ | $R$'s and |
| no $\mathcal{E}$'s | | original $\mathcal{E}$'s |

$\leftarrow I$

(Final Thévenin equivalent)

$\Longrightarrow$

Negate $\mathcal{E}$'s everywhere

**Figure 12.75.**

For any external emf $\mathcal{E}$, the current $I$ is therefore the same in the original circuit as in the simple circuit involving $\mathcal{E}_{eq}$ and $R_{eq}$. The two circuits are therefore equivalent.

How do we determine the values of $R_{eq}$ and $\mathcal{E}_{eq}$? Since $R_{eq}$ depends only on the $R$ values, and not on any of the emfs, we can determine $R_{eq}$ by picking any convenient set of values for the internal emfs. If we pick them all to be zero, we see that $R_{eq}$ is obtained by finding the resistance between the terminals when all of the internal emfs are set equal to zero. This is the claim we made, without proof, in Section 4.10.2.[5] Having found $R_{eq}$, we can determine $\mathcal{E}_{eq}$ by setting $\mathcal{E} = 0$ and calculating the current $I$ (which is the short-circuit current $I_{sc}$). Equation (12.225) then gives $\mathcal{E}_{eq} = I_{sc}R_{eq}$. Alternatively, $\mathcal{E}_{eq}$ equals the open-circuit voltage.

As an exercise, you can extend this reasoning to the case where $B$ is a general circuit instead of a single emf. *Hint:* The loop currents in circuit $A$ can be solved for in terms of the resistances and emfs solely in $A$, plus the loop current $I$ in the "connecting" loop. Likewise for $B$. You can then write down the connecting-loop equation for $I$; this equation will break up into two separate pieces depending only on $A$ and only on $B$.

**Second proof**  In the complete circuit in Fig. 12.75(a), let the current that flows in the horizontal wires be $I$. There are three steps to the proof, as

---

[5] Although it might seem obvious that this is what $R_{eq}$ must be, it was necessary to demonstrate first that $R_{eq}$ is independent of the emfs. Otherwise the value of $R_{eq}$ when all the internal emfs are set equal to zero might not be the value for another set of emfs.

outlined in the figure and discussed in detail below. Although we are often concerned only with the case where the circuit $B$ is a single emf source (possibly with a resistor in series), the proof below works for a general circuit $B$.

The first step is to change the circuit by removing all of the internal emfs from $B$ (or equivalently, by adding on the negatives of all these emfs right next to them), and also inserting an emf $\mathcal{E}$ as shown in Fig. 12.75(b) and then adjusting it until the current in the horizontal wires is zero. Since the current is now zero, and since $B$ now has no internal emfs, the voltage across the $a, b$ terminals of $A$ is (by definition) the open-circuit voltage $V_A^{\text{open}}$. Therefore, the emf $\mathcal{E}$ that we added must be equal to $-V_A^{\text{open}}$, because the current is zero.

Second, in Fig. 12.75(c) let us remove all of the internal emfs from $A$ (or equivalently, add on their negatives). Additionally, in $B$ let us add on the negatives of the original emfs. The internal emfs in $A$ are now zero, and the internal emfs in $B$ are now the negatives of what they were initially. This second step can be looked at as simply adding on the negative of all of the original circuit emfs. By linearity, this has the effect of adding on a current equal to the negative of the original current $I$. Since the current in the circuit was zero before we performed this second step, it is now $-I$ (the negative sign means counterclockwise), as shown.

The third step is to negate the emfs everywhere (that is, in $B$ and in the inserted emf; all the emfs in $A$ are zero). This yields the circuit shown in Fig. 12.75(d). The circuit $B$ and the current $I$ are now exactly what they were originally. We conclude that from $B$'s point of view, the circuit $A$ is equivalent to a voltage source $\mathcal{E}_{\text{eq}}$ that equals $-\mathcal{E} = V_A^{\text{open}}$, in series with a resistor $R_{\text{eq}}$ that is obtained by ignoring all of the voltage sources in $A$. This completes the proof.

4.14 *Thévenin $R_{\text{eq}}$ via $I_{\text{sc}}$*

If we connect $A$ and $B$ with a wire with zero resistance, then we can ignore the resistor $R_3$ in Fig. 4.24. The circuit therefore looks like the one shown in Fig. 12.76. The two loop equations are

$$0 = \mathcal{E}_1 - I_1 R_1,$$
$$0 = \mathcal{E}_2 - I_2 R_2, \qquad (12.227)$$

so the loop currents are simply $I_1 = \mathcal{E}_1/R_1$ and $I_2 = \mathcal{E}_2/R_2$. The short-circuit current between $A$ and $B$ is then

$$I_{\text{sc}} = I_1 - I_2 = \frac{\mathcal{E}_1}{R_1} - \frac{\mathcal{E}_2}{R_2} = \frac{\mathcal{E}_1 R_2 - \mathcal{E}_2 R_1}{R_1 R_2}. \qquad (12.228)$$

As mentioned in Section 4.10.2, $\mathcal{E}_{\text{eq}}$ equals the open-circuit voltage, which in the present setup is $I_3 R_3$, with $I_3$ given in Eq. (4.33). The equivalent resistance is therefore

$$R_{\text{eq}} = \frac{\mathcal{E}_{\text{eq}}}{I_{\text{sc}}} = \frac{I_3 R_3}{I_{\text{sc}}} = \frac{\mathcal{E}_1 R_2 - \mathcal{E}_2 R_1}{R_1 R_2 + R_2 R_3 + R_1 R_3} \cdot R_3 \cdot \frac{R_1 R_2}{\mathcal{E}_1 R_2 - \mathcal{E}_2 R_1}$$

$$= \frac{R_1 R_2 R_3}{R_1 R_2 + R_2 R_3 + R_1 R_3}, \qquad (12.229)$$

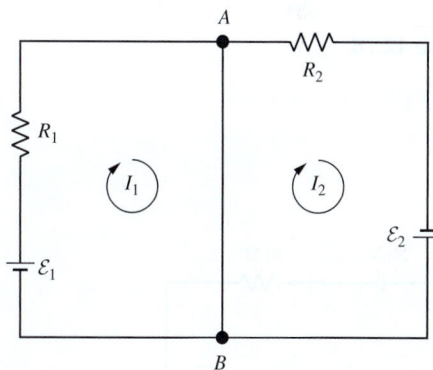

**Figure 12.76.**

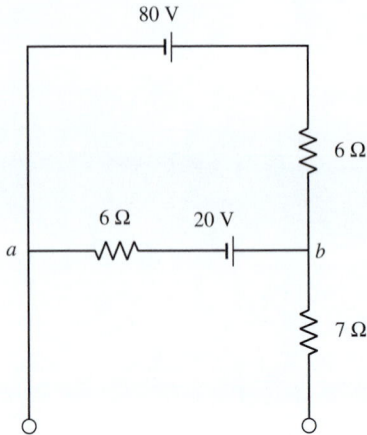

**Figure 12.77.**

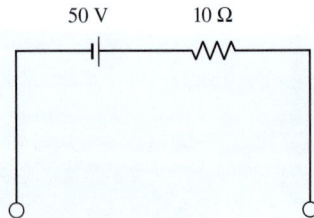

**Figure 12.78.**

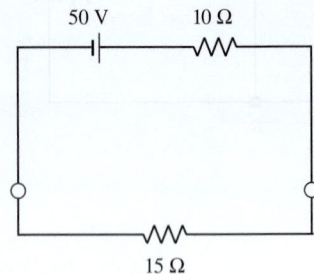

**Figure 12.79.**

in agreement with the result of the other (much quicker) method of calculating $R_{eq}$ presented right before the example in Section 4.10.2.

### 4.15   *A Thévenin equivalent*

The equivalent resistance $R_{eq}$ is quick to find. Ignoring the emfs (that is, setting them equal to zero and shorting them out) yields a circuit with a parallel combination of two $6\,\Omega$ resistors in series with a $7\,\Omega$ resistor. So $R_{eq} = 3\,\Omega + 7\,\Omega = 10\,\Omega$.

To find $\mathcal{E}_{eq}$, note that in the open circuit, no current flows through the $7\,\Omega$ resistor, so the open-circuit voltage is the same as the voltage between nodes $a$ and $b$ in Fig. 12.77. The only current flowing is around the closed loop, so Kirchhoff's rule gives (with clockwise $I$ defined to be positive)

$$80\,\text{V} - (6\,\Omega)I - 20\,\text{V} - (6\,\Omega)I = 0 \implies I = 5\,\text{A}. \tag{12.230}$$

The voltage drop from $b$ to $a$ is then $V_b - V_a = 20\,\text{V} + (6\,\Omega)(5\,\text{A}) = 50\,\text{V}$. Alternatively, the drop through the top branch (where the path from $b$ to $a$ goes against the current) is $-(6\,\Omega)(5\,\text{A}) + 80\,\text{V} = 50\,\text{V}$. So $50\,\text{V}$ is the desired $\mathcal{E}_{eq}$. We therefore have the Thévenin equivalent circuit shown in Fig. 12.78.

If we connect a $15\,\Omega$ resistor across the terminals, we can use the Thévenin equivalent circuit to find immediately from Fig. 12.79 that the current through the $15\,\Omega$ resistor is $(50\,\text{V})/(10\,\Omega + 15\,\Omega) = 2\,\text{A}$.

If you don't trust that the Thévenin circuit produces the same current through the external resistor as the original circuit does, you can directly calculate the current in the original circuit by using Kirchhoff's rules. With the two loop currents shown in Fig. 12.80, we have

$$80\,\text{V} - (6\,\Omega)I_1 - 20\,\text{V} - (6\,\Omega)(I_1 - I_2) = 0,$$
$$20\,\text{V} - (7\,\Omega)I_2 - (15\,\Omega)I_2 - (6\,\Omega)(I_2 - I_1) = 0. \tag{12.231}$$

Solving this system of equations gives $I_2 = 2\,\text{A}$, as desired. We also find $I_1 = 6\,\text{A}$; this current exists in the original circuit, but it doesn't apply to the Thévenin equivalent circuit. Note that the closed-circuit voltage across the terminals is $(15\,\Omega)(2\,\text{A}) = 30\,\text{V}$. Also, the voltage drop from $b$ to $a$ is now $44\,\text{V}$ (you should check that all three possible paths yield this drop). Neither of these voltages of $30\,\text{V}$ or $44\,\text{V}$ (which depend on our specific choice for the external resistor) equals the open-circuit voltage of $50\,\text{V}$ (which is associated with an infinite external resistor).

### 4.16   *Discharging a capacitor*

The important point to realize is that, although the field is very small outside the capacitor (except right near the edges, where it is relatively large), the line integral of the field, along *any* path starting at one plate and ending up at the other, must be equal to (plus or minus) the potential difference between the plates. This is due to the fact that the electric field is a conservative field, or equivalently that it has zero curl (for static situations). For some paths there is a large field for a small distance, and for other paths there is a small field for a large distance.

In Fig. 4.41(a), there is essentially no contribution to the line integral along the parts of the wire parallel to the plates. But the field along the short segment on the left is essentially equal to the field $E$ inside the capacitor. So the line integral equals $Es$, just as it does for a path inside the capacitor.

In Fig. 4.41(b), it turns out that essentially all of the line integral comes from the vertical parts of the wire. This can be seen in two ways. First, at points very far away from the capacitor, the capacitor looks essentially like a dipole, because the two plates can be considered to be the result of placing a large number of dipoles next to each other. And we know from Eq. (2.36) that the field from a dipole (both the radial and tangential components) falls off like $1/r^3$. So the very large semicircular part of the wire (the length of which is proportional to $r$) contributes nothing as $r \to \infty$. The full contribution must therefore come from the straight parts of the wire. And furthermore it must come from the region close to the capacitor, from the same $1/r^3$ reasoning.

A second line of reasoning is the following. If you want, you can explicitly calculate the field outside the capacitor, along the axis. (You just need to know the field from a disk; see Section 2.6.) And then you can integrate this along each of the two arms. You will obtain a result of $\sigma s/2\epsilon_0$ for each, where $\sigma$ is the surface density on each plate. (You will find that the integral is dominated by the initial part of the wire out to a few multiples of the diameter (or general length scale) of the capacitor.) With the two arms, the total line integral along the vertical part of the wire is $\sigma s/\epsilon_0$. And this equals the field inside the capacitor, $\sigma/\epsilon_0$, times the distance between the plates, $s$, as desired.

There is actually no need to calculate the field outside the capacitor explicitly in order to do the line integral. There is a much easier way. *Hint:* In each arm, the line integral is the sum of the line integrals of the fields due to the two plates. These two (oppositely signed) integrals are the same, except that one starts at zero and one starts at $s$.

4.17 *Charging a capacitor*

Let $I(t)$ be the clockwise current, and let $Q(t)$ be the charge on the left plate of the capacitor. Demanding that the total voltage drop around the loop is zero gives $\mathcal{E} - Q/C - RI = 0$. But $I = dQ/dt$, so we have

$$\mathcal{E} - \frac{Q}{C} - R\frac{dQ}{dt} = 0 \implies \frac{dQ}{dt} = -\frac{1}{RC}(Q - C\mathcal{E}). \qquad (12.232)$$

We can solve this differential equation by separating variables and integrating:

$$\int_0^Q \frac{dQ'}{Q' - C\mathcal{E}} = -\int_0^t \frac{dt'}{RC} \implies \ln\left(Q' - C\mathcal{E}\right)\Big|_0^Q = -\frac{t}{RC}$$

$$\implies \ln\left(\frac{Q - C\mathcal{E}}{-C\mathcal{E}}\right) = -\frac{t}{RC} \implies Q(t) = C\mathcal{E}\left(1 - e^{-t/RC}\right).$$

$$(12.233)$$

As a double check, we have $Q(0) = 0$, which is correct. And $Q(\infty) = C\mathcal{E}$, which is correct because eventually the voltage across the capacitor equals $\mathcal{E}$.

Alternatively, you can avoid separating variables and integrating if you note that Eq. (12.232) can be written as $d\tilde{Q}/dt = -(1/RC)\tilde{Q}$, where

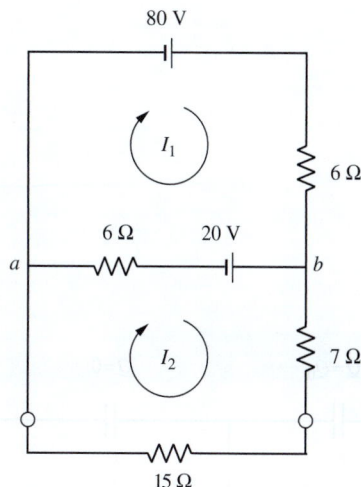

Figure 12.80.

$\tilde{Q}$ is defined by $\tilde{Q} \equiv Q - C\mathcal{E}$. This has the simple solution, $\tilde{Q} = Ae^{-t/RC}$, where the initial conditions quickly give the constant of integration as $A = -C\mathcal{E}$. Using the $\tilde{Q} \equiv Q - C\mathcal{E}$ definition, we then have $Q - C\mathcal{E} = -C\mathcal{E}e^{-t/RC} \Longrightarrow Q = C\mathcal{E}(1 - e^{-t/RC})$, as above.

The current is

$$I(t) = \frac{dQ}{dt} = \frac{\mathcal{E}}{R}e^{-t/RC}. \tag{12.234}$$

This equals $\mathcal{E}/R$ at $t = 0$ (the capacitor provides no back emf right at the start). And it equals zero at $t = \infty$, as it should.

4.18 *A discharge with two capacitors*

(a) Let $Q_1$ and $Q_2$ be the charges on the (left plates of the) left and right capacitors. And let $I_1$ and $I_2$ be the left and right loop currents, with counterclockwise positive, as shown in Fig. 12.81. Then the two loop equations are

$$\frac{Q_1}{C} - I_1R = 0, \qquad \text{and} \qquad \frac{Q_2}{C} - I_2R = 0. \tag{12.235}$$

But $I_1 = -dQ_1/dt$ and $I_2 = -dQ_2/dt$, so we have

$$\frac{Q_1}{C} + R\frac{dQ_1}{dt} = 0, \qquad \text{and} \qquad \frac{Q_2}{C} + R\frac{dQ_2}{dt} = 0. \tag{12.236}$$

These two equations are decoupled (that is, each equation involves only one of the $Q$'s), so we can solve for $Q_1$ and $Q_2$ separately. Separating variables and integrating each equation (or just realizing that we have exponential solutions), we find that both $Q_1$ and $Q_2$ are proportional to $e^{-t/RC}$. Given the initial charges of $Q_0$ and 0, we see that the charges as functions of time are $Q_1(t) = Q_0e^{-t/RC}$, and $Q_2(t) = 0$.

The left capacitor simply discharges by sending current around the left loop. The right loop effectively isn't there. Basically, the current has no desire to pass through a second resistor, when it is possible to pass through only one resistor on its journey to the other side of the left capacitor. At the junction at the bottom of the circuit, the path with resistance $R$ around the right loop has infinitely more resistance than the path with zero resistance that goes straight up the middle of the circuit.

(b) The two loop equations are now

$$\frac{Q_1}{C} - I_1R - (I_1 - I_2)R = 0 \quad \text{and} \quad \frac{Q_2}{C} - I_2R - (I_2 - I_1)R = 0. \tag{12.237}$$

These equations are coupled; they both involve both $Q_1$ and $Q_2$. If we add them, we obtain

$$\frac{(Q_1 + Q_2)}{C} - (I_1 + I_2)R = 0 \Longrightarrow \frac{(Q_1 + Q_2)}{C} + R\frac{d(Q_1 + Q_2)}{dt} = 0. \tag{12.238}$$

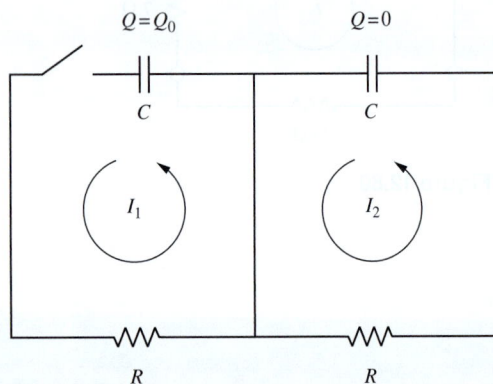

$Q = Q_0$ $\qquad$ $Q = 0$

$C$ $\qquad\qquad$ $C$

$I_1$ $\qquad\qquad$ $I_2$

$R$ $\qquad\qquad$ $R$

**Figure 12.81.**

This equation involves only the combination $Q_1 + Q_2$ of the charges. The solution is $Q_1 + Q_2 = Ae^{-t/RC}$, where $A$ is a constant, determined by the initial conditions. Similarly, if we take the difference, we obtain

$$\frac{(Q_1 - Q_2)}{C} - 3(I_1 - I_2)R = 0 \Longrightarrow \frac{(Q_1 - Q_2)}{C} + 3R\frac{d(Q_1 - Q_2)}{dt} = 0.$$
(12.239)

The solution here is $Q_1 - Q_2 = Be^{-t/3RC}$, where $B$ is another constant. Having solved for $Q_1 + Q_2$ and $Q_1 - Q_2$, we can take the sum and difference of these results to obtain

$$Q_1(t) = ae^{-t/RC} + be^{-t/3RC} \quad \text{and} \quad Q_2(t) = ae^{-t/RC} - be^{-t/3RC},$$
(12.240)

where $a \equiv A/2$ and $b \equiv B/2$. The initial condition $Q_2(0) = 0$ quickly gives $a = b$. And then the initial condition $Q_1(0) = Q_0$ gives $a = b = Q_0/2$. So the desired charges as functions of time are

$$Q_1(t) = \frac{Q_0}{2}\left(e^{-t/RC} + e^{-t/3RC}\right),$$

$$Q_2(t) = \frac{Q_0}{2}\left(e^{-t/RC} - e^{-t/3RC}\right).$$
(12.241)

Note that $Q_1(t)$ decreases monotonically with time, but $Q_2(t)$ reaches a maximum negative value at some finite time (it is zero at both $t = 0$ and $t = \infty$). Setting the derivative of $Q_2(t)$ equal to zero gives $t_{max} = RC(3/2)\ln 3 \approx (1.65)RC$. Plugging this back into $Q_2(t)$ yields a maximum negative value of $-Q_0/3\sqrt{3} \approx -(0.19)Q_0$. Plots of $Q_1(t)$ and $Q_2(t)$ are shown in Fig. 12.82(a).

With the third resistor added to the circuit, the current must pass through two resistors, no matter what path it takes to reach the other side of the left capacitor. (And the right capacitor will also have an effect, once it acquires charge.) So some of the current goes around the right loop. This puts positive charge on the right plate of the right capacitor and negative charge on its left plate (which is the plate that determines the sign of $Q_2$, by our convention). This negative charge reaches a maximum value at $t_{max} \approx (1.65)RC$, and then at this time the current $I_2(t)$ in the right loop changes sign, from being positive (counterclockwise) to being negative (clockwise). You can show that the clockwise current reaches a maximum value at a time of $2t_{max} \approx (3.3)RC$. Immediately after the switch is closed, equal currents pass through the middle and right resistors, because there is initially no voltage across the right capacitor, so these two resistors are in parallel. But once the right capacitor acquires charge, this equality is lost. Plots of $I_1(t)$ and $I_2(t)$ (the negative derivatives of $Q_1$ and $Q_2$) are shown in Fig. 12.82(b).

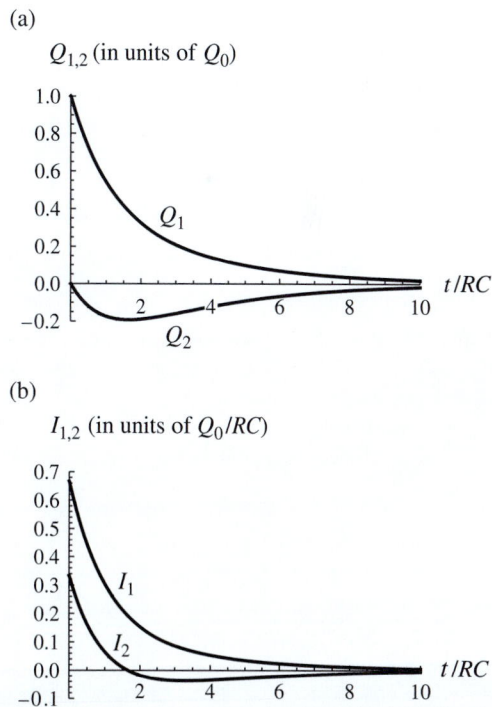

(a)

$Q_{1,2}$ (in units of $Q_0$)

(b)

$I_{1,2}$ (in units of $Q_0/RC$)

**Figure 12.82.**

## 12.5 Chapter 5

### 5.1   *Field from a filament*

(a) The magnitude of the excess charge is

$$Q = Ne = (5 \cdot 10^8)(1.6 \cdot 10^{-19}\,\text{C}) = 8 \cdot 10^{-11}\,\text{C}. \qquad (12.242)$$

The linear charge density is then

$$\lambda = \frac{Q}{\ell} = \frac{8 \cdot 10^{-11}\,\text{C}}{0.04\,\text{m}} = 2 \cdot 10^{-9}\,\text{C/m}. \qquad (12.243)$$

The electric field is therefore

$$E = \frac{\lambda}{2\pi\epsilon_0 r} = \frac{2 \cdot 10^{-9}\,\text{C/m}}{2\pi \left(8.85 \cdot 10^{-12}\,\frac{\text{s}^2\text{C}^2}{\text{kg m}^3}\right)(5 \cdot 10^{-5}\,\text{m})}$$

$$= 7.2 \cdot 10^5\,\text{V/m}. \qquad (12.244)$$

The field is directed radially toward the wire.

(b) In this frame the charge density (and hence field) is larger by a factor $\gamma = 1/\sqrt{1 - (0.9)^2} = 2.29$. So the field is $E' = \gamma E = 1.65 \cdot 10^6\,\text{V/m}$. It is still directed radially toward the wire.

### 5.2   *Maximum horizontal force*

The magnitude of the electric field is given in Eq. (5.15), with $r' = b/\sin\theta$. We are concerned with the horizontal component, so this brings in a factor of $\cos\theta$. We therefore want to maximize the function

$$E_x \propto \frac{\sin^2\theta\cos\theta}{(1 - \beta^2\sin^2\theta)^{3/2}}. \qquad (12.245)$$

Setting the derivative equal to zero and simplifying yields

$$(1 - \beta^2\sin^2\theta)(2\cos^2\theta - \sin^2\theta) + 3\beta^2\sin^2\theta\cos^2\theta = 0. \qquad (12.246)$$

Using $\cos^2\theta = 1 - \sin^2\theta$ and solving for $\sin^2\theta$ gives

$$\sin\theta = \sqrt{\frac{2}{3 - \beta^2}}. \qquad (12.247)$$

If $\beta \approx 1$ then $\theta \approx 90°$, which is reasonable. The largeness of the field near $90°$ wins out over the smallness of the $\cos\theta$ factor involved in taking the horizontal component. However, if $\theta = 90°$ exactly, then the horizontal force is zero.

If $\beta \approx 0$ then $\sin\theta \approx \sqrt{2/3} \implies \theta \approx 54.7°$ (or $125.3°$). You can quickly check from scratch that this is the correct result in the nonrelativistic case, where the horizontal component of the force is proportional to $\sin^2\theta\cos\theta$.

### 5.3   *Newton's third law*

The electric field of the stationary proton at the position of the pion takes the simple Coulomb form of $E = e/4\pi\epsilon_0 r^2$. So the force on the pion is

$$F = eE = \frac{1}{4\pi\epsilon_0}\frac{e^2}{r^2} = \left(9 \cdot 10^9\,\frac{\text{kg m}^3}{\text{s}^2\text{C}^2}\right)\frac{(1.6 \cdot 10^{-19}\,\text{C})^2}{(10^{-4}\,\text{m})^2} = 2.3 \cdot 10^{-20}\,\text{N}.$$

$$(12.248)$$

From Eq. (5.15), with $\theta' = 0$, the field of the moving pion at the position of the proton is $E = (1 - \beta^2)e/4\pi\epsilon_0 r^2$. So the force on the proton is smaller than the force on the pion by a factor of $1 - \beta^2$. Hence the force on the proton equals $(0.64)(2.3 \cdot 10^{-20}\,\text{N}) = 1.47 \cdot 10^{-20}\,\text{N}$.

We see that Newton's third law, applied to the charges, does *not* hold. Equivalently (since $F = dp/dt$), the total momentum of the proton plus pion is not conserved. However, the sacred fact that is still true is that the total momentum of the *entire system* is conserved. And the system here consists of the two charges *plus* the electromagnetic field. We will learn in Chapter 9 that there is momentum *in the field,* and the field is changing here. The total momentum (proton plus pion plus field) is indeed conserved. This is not a two-body system!

5.4 *Divergence of E*

(a) The field in Eq. (5.15) points radially, so from Eq. (F.3) in Appendix F the divergence is $(1/r^2)\partial(r^2 E_r)/\partial r$, where we have dropped the primes on the coordinates. The $1/r^2$ dependence of $E_r$ implies that $\partial(r^2 E_r)/\partial r = 0$, as desired.

Note that the $\theta$ dependence of $E_r$ is irrelevant here. If $E_r$ is the only nonzero component, then it can be an arbitrary function of $\theta$ (and $\phi$), and the divergence of **E** will still be zero. This is clear if you consider a volume with its "sides" lying along the radial direction, and with its two endcaps at constant $r$ values. The angular dependence does not change the fact that if $E_r \propto 1/r^2$, then any flux that enters through the inner endcap also exits through the outer endcap, because this property holds for each angle individually.

(b) The components in Eq. (5.13) are valid for a point in the $xz$ plane. (Again, we drop the primes on the coordinates.) But we must remember that in general the $y$ coordinate is nonzero. Since the divergence involves derivatives, and since derivatives involve values at nearby points, we need to consider points with nonzero $y$ values. The $E_y$ component is analogous to the $E_z$ component (these are the two transverse directions), so it has a $\gamma Q y$ in the numerator. And all three components now have a $[(\gamma x)^2 + y^2 + z^2]^{3/2}$ in the denominator. (When $y = 0$, this field properly reduces to the two components in Eq. (5.13), and it is properly symmetric with respect to $y$ and $z$.) For convenience, let's label the denominator as $D^{3/2}$. Then we have

$$\frac{4\pi\epsilon_0}{\gamma Q}\nabla \cdot \mathbf{E} = \frac{\partial}{\partial x}\left(\frac{x}{D^{3/2}}\right) + \frac{\partial}{\partial y}\left(\frac{y}{D^{3/2}}\right) + \frac{\partial}{\partial z}\left(\frac{z}{D^{3/2}}\right)$$

$$= \left(\frac{1}{D^{3/2}} - \frac{3\gamma^2 x^2}{D^{5/2}}\right) + \left(\frac{1}{D^{3/2}} - \frac{3y^2}{D^{5/2}}\right)$$

$$+ \left(\frac{1}{D^{3/2}} - \frac{3z^2}{D^{5/2}}\right)$$

$$= \left(\frac{3}{D^{3/2}} - \frac{3\big((\gamma x)^2 + y^2 + z^2\big)}{D^{5/2}}\right)$$

$$= \left(\frac{3}{D^{3/2}} - \frac{3D}{D^{5/2}}\right) = 0, \tag{12.249}$$

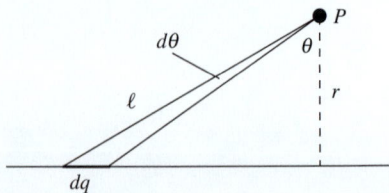

**Figure 12.83.**

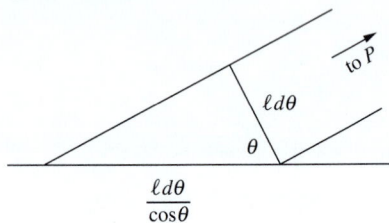

**Figure 12.84.**

as desired. Note that this wouldn't have worked out if the exponent in the denominator hadn't been $3/2$, or equivalently if the net power of length in the fraction hadn't been $-2$. This is consistent with the $E_r \propto 1/r^2$ reasoning in part (a).

5.5  *E from a line of moving charges*
To find the field at point $P$ in Fig. 12.83, consider a small interval of the charges at angle $\theta$, subtending an angle $d\theta$, as shown. This angle is the complement of the angle in Eq. (5.15), so the field at $P$ due to the small interval with charge $dq$ is

$$dE = \frac{dq}{4\pi\epsilon_0 \ell^2} \frac{1 - \beta^2}{(1 - \beta^2 \cos^2 \theta)^{3/2}}, \qquad (12.250)$$

where $\ell = r/\cos\theta$. The length of the small interval is $d(r\tan\theta) = r\,d\theta/\cos^2\theta$. (This can also be obtained from the zoomed-in view in Fig. 12.84, which shows the length to be $(\ell\,d\theta)/\cos\theta$, with $\ell = r/\cos\theta$.) So the charge is $dq = \lambda(r\,d\theta/\cos^2\theta)$. By symmetry, only the field component perpendicular to the stream will survive; this brings in a factor of $\cos\theta$. So the total field at the given point $P$ is radial and has magnitude

$$E = \int_{-\pi/2}^{\pi/2} \frac{dq}{4\pi\epsilon_0 \ell^2} \frac{1 - \beta^2}{(1 - \beta^2 \cos^2\theta)^{3/2}} \cos\theta$$

$$= \frac{1}{4\pi\epsilon_0} \int_{-\pi/2}^{\pi/2} \frac{\lambda(r\,d\theta/\cos^2\theta)}{(r/\cos\theta)^2} \frac{1 - \beta^2}{(1 - \beta^2\cos^2\theta)^{3/2}} \cos\theta$$

$$= \frac{\lambda(1 - \beta^2)}{4\pi\epsilon_0 r} \int_{-\pi/2}^{\pi/2} \frac{\cos\theta\,d\theta}{(1 - \beta^2\cos^2\theta)^{3/2}}. \qquad (12.251)$$

Using the integral table in Appendix K, this becomes

$$\frac{\lambda(1 - \beta^2)}{4\pi\epsilon_0 r} \cdot \frac{\sin\theta}{(1 - \beta^2)\sqrt{1 - \beta^2\cos^2\theta}}\bigg|_{-\pi/2}^{\pi/2} = \frac{\lambda(1 - \beta^2)}{4\pi\epsilon_0 r} \cdot \frac{2}{(1 - \beta^2)}$$

$$= \frac{\lambda}{2\pi\epsilon_0 r}, \qquad (12.252)$$

as desired.

5.6  *Maximum field from a passing charge*
In the rest frame of the proton, the antiproton flies by with a speed $\beta$ that is obtained from applying the velocity-addition formula to the lab-frame velocities: $\beta = 2\beta_{\text{lab}}/(1 + \beta_{\text{lab}}^2)$. The $\gamma$ factor associated with this speed is $\gamma = (1 + \beta_{\text{lab}}^2)/(1 - \beta_{\text{lab}}^2)$, as you can verify. With $\beta_{\text{lab}} \approx 1$, this becomes $\gamma \approx 2/(1 - \beta_{\text{lab}}^2) = 2\gamma_{\text{lab}}^2 = 2 \cdot 10^4$.

The maximum value of the field in Eq. (5.15) is achieved when $\theta = 90°$, in which case the value is $\left(Q/4\pi\epsilon_0 r^2\right)/\sqrt{1 - \beta^2} = \gamma Q/4\pi\epsilon_0 r^2$.

(We've dropped the primes on the coordinates.) The desired maximum strength of the field is therefore

$$E_{max} = \frac{1}{4\pi\epsilon_0}\frac{\gamma e}{r^2} = \left(9\cdot 10^9\,\frac{kg\,m^3}{s^2\,C^2}\right)\frac{(2\cdot 10^4)(1.6\cdot 10^{-19}\,C)}{(10^{-10}\,m)^2}$$

$$= 2.88\cdot 10^{15}\,V/m. \qquad (12.253)$$

From Eq. (5.15), the field achieves half of the maximum intensity when the $(1 - \beta^2\sin^2\theta)^{3/2}$ factor in the denominator equals twice its minimum value, which is $(1-\beta^2)^{3/2}$. (We will find that $\theta$ will be very close to $\pi/2$, so the variation in $r$ in Eq. (5.15) can be ignored.) It's a little easier to work in terms of the small angle $\alpha$ defined by $\alpha \equiv \pi/2 - \theta$. We then have $\sin^2\theta = \cos^2\alpha = 1 - \sin^2\alpha \approx 1 - \alpha^2$. The $(1 - \beta^2\sin^2\theta)^{3/2}$ factor can then be written as

$$\left(1 - \beta^2(1-\alpha^2)\right)^{3/2} = \left(1 - \beta^2 + \beta^2\alpha^2\right)^{3/2} \approx \left(\frac{1}{\gamma^2} + \alpha^2\right)^{3/2},$$

$$(12.254)$$

where we have used $1 - \beta^2 \equiv 1/\gamma^2$ and $\beta^2 \approx 1$. Therefore, the field achieves half of the maximum intensity at the $\alpha$ for which

$$\left(\frac{1}{\gamma^2} + \alpha^2\right)^{3/2} = 2\left(\frac{1}{\gamma^2}\right)^{3/2} \implies \frac{1}{\gamma^2} + \alpha^2 = 2^{2/3}\frac{1}{\gamma^2}$$

$$\implies \alpha^2 = \frac{2^{2/3} - 1}{\gamma^2}. \qquad (12.255)$$

Hence

$$\alpha \approx \frac{0.766}{\gamma} = \frac{0.766}{2\cdot 10^4} = 3.83\cdot 10^{-5}. \qquad (12.256)$$

At $r = 10^{-10}$ m, the angle $\pm\alpha$ spans a distance of $r(2\alpha) = 7.7\cdot 10^{-15}$ m, which is traversed in a time of essentially $2r\alpha/c = 2.6\cdot 10^{-23}$ s. Note that Eq. (12.256) says that the angular width of the "pancake" of field lines is on the order of $1/\gamma$.

5.7    *Electron in an oscilloscope*

**Lab frame**    Let the $x$ direction be the direction of the electron's initial velocity. In the lab frame $F$, Eq. (G.11) tells us that the momentum is $p_x = \gamma m v_0$, where $\gamma = 1/\sqrt{1 - v_0^2/c^2}$. This momentum is constant throughout the motion because there is no field and hence no force in the $x$ direction. The time spent between the plates is $t = \ell/v_0$ (ignoring the effect discussed in Exercise 5.25, since the transverse motion is nonrelativistic). The transverse force, which has the constant value of $eE$ (we'll just deal with the magnitude here), equals the rate of change of transverse momentum. So the final transverse momentum is

$$p_y = (eE)t = \frac{eE\ell}{v_0}. \qquad (12.257)$$

Since $p_y = \gamma m v_y$, the transverse velocity upon exiting is $v_y = p_y/\gamma m = eE\ell/\gamma m v_0$. (The $\gamma$ factor here involves the entire speed, which is essentially $v_0$, and *not* the transverse speed $v_y$.) Since the transverse force is constant, and since we are assuming that the transverse velocity is non-relativistic, the transverse acceleration is also constant. The average transverse velocity is therefore half of the $v_y$ we just found. That is, $\bar{v}_y = eE\ell/2\gamma m v_0$. The transverse distance traveled is then

$$y = \bar{v}_y t = \frac{eE\ell}{2\gamma m v_0}\frac{\ell}{v_0} = \frac{eE\ell^2}{2\gamma m v_0^2}. \tag{12.258}$$

It makes sense intuitively that this result grows with $e$, $E$, and $\ell$; and that it decreases with $m$ and $v_0$. Note that the angle of deflection upon exiting, which is given by $p_y/p_x$ or equivalently $v_y/v_x$, equals $eE\ell/\gamma m v_0^2$. This is twice the value of $y/x$, as is always the case for constant transverse acceleration, as you can check.

**Electron frame**   Now consider the frame $F'$ in which the electron is initially at rest. (We'll call this the electron frame, even though the electron will gradually accelerate away from it in the transverse direction.) The plates move to the left with speed $v_0$, and their length is length-contracted down to $\ell/\gamma$. So they are above and below the electron for a time $t' = (\ell/\gamma)/v_0$. The field in the electron frame $F'$ is larger than the field in the lab frame $F$ by a factor $\gamma$ so[6] $E' = \gamma E$. The transverse momentum acquired is therefore

$$p_y' = eE't' = e(\gamma E)\frac{t}{\gamma} = eEt = \frac{eE\ell}{v_0}. \tag{12.259}$$

But from Eq. (G.12) we see that the transverse momentum is unaffected by a Lorentz transformation, so $p_y$ in the lab frame also equals $eE\ell/v_0$. This agrees with the result in Eq. (12.257). In short, the transverse momenta are equal because the field $E'$ is larger than $E$ by a factor $\gamma$, but the time $t'$ is shorter than $t$ by a factor $\gamma$, so these two effects exactly cancel.

In frame $F'$ the electron is nonrelativistic, so the final $v_y'$ is simply $v_y' = p_y'/m = eE\ell/m v_0$ (note that this is larger than the final $v_y$ in the lab frame, by a factor $\gamma$). The average transverse velocity is then $\bar{v}_y' = v_y'/2 = eE\ell/2m v_0$. The total transverse distance traveled is therefore

$$y' = \bar{v}_y' t' = \frac{eE\ell}{2m v_0}\frac{\ell}{\gamma v_0} = \frac{eE\ell^2}{2\gamma m v_0^2}. \tag{12.260}$$

But from Eq. (G.2) we see that the transverse distance is also unaffected by a Lorentz transformation, so $y$ in the lab frame also equals $eE\ell^2/2\gamma m v_0^2$. (Hence the "as measured in the lab frame" qualifier in the statement of the problem wasn't actually necessary.) This agrees with the result in Eq. (12.258). In short, the transverse distances are equal because the speed

---

[6] Remember that the field is smallest in the frame of the sources; see Eq. (5.7). The relation $E' = \gamma E$ is also consistent with the fact that the force on the electron is larger in the electron frame than in any other frame; see Eq. (5.17).

$v'_y$ is larger than $v_y$ by a factor $\gamma$, but the time $t'$ is shorter than $t$ by a factor $\gamma$, so these two effects exactly cancel.

If you want to use the kinematic expression $y = a_y t^2/2$ in each frame, then you will again find that $y' = y$, because in the electron frame $F'$, the time $t'$ is shorter by a factor $\gamma$, but the acceleration $a'_y$ is larger by a factor $\gamma^2$ (because it takes a time that is $\gamma$ shorter to achieve a final velocity that is $\gamma$ larger).

You should think about how all of the following facts relate: compared with the corresponding quantities in the lab frame, $t'$ is smaller by a factor $\gamma$, $p'_y$ and $y'$ are the same, $E'$ and $v'_y$ are larger by a factor $\gamma$, and $a'_y$ is larger by a factor $\gamma^2$.

**5.8**  *Finding the magnetic field*

In the example at the end of Section 5.8, we found that the total force was $qE_2/\gamma$ and the electric force was $\gamma qE_2$. Assuming that $q$ and $\sigma$ are positive, these forces are both repulsive. If the sum of the electric and magnetic forces is to equal the total force, we need the magnetic force to be attractive with magnitude

$$\gamma qE_2 - \frac{qE_2}{\gamma} = \gamma qE_2\left(1 - \frac{1}{\gamma^2}\right) = \gamma qE_2\left(\frac{v^2}{c^2}\right) = qv\left(\frac{\gamma vE_2}{c^2}\right).$$

$$(12.261)$$

Since the magnetic force is given by $q\mathbf{v} \times \mathbf{B}$, the $qv(\gamma vE_2/c^2)$ attractive force is exactly the force that arises from a magnetic field pointing out of the page with magnitude $\gamma vE_2/c^2$, as desired.

**5.9**  *"Twice" the velocity*

(a) If the test charge sees the electrons moving backward with speed $v_0$, then the electrons see the test charge moving forward with speed $v_0$. So the test charge moves with speed $v_0$ with respect to the electrons, which in turn move (by definition) with speed $v_0$ with respect to the lab. This is exactly the situation where the velocity-addition formula applies. Relativistically adding $\beta_0$ with itself gives the $\beta$ of the test charge with respect to the lab frame as $\beta = 2\beta_0/(1 + \beta_0^2)$, as desired. You can check that relativistically subtracting $\beta_0$ from $2\beta_0/(1 + \beta_0^2)$ yields $\beta_0$. This gives another way of solving the problem.

(b) First, note that the $\gamma$ factor associated with the above value of $\beta$ is

$$\gamma = \frac{1}{\sqrt{1 - \beta^2}} = \frac{1}{\sqrt{1 - \left(\dfrac{2\beta_0}{1 + \beta_0^2}\right)^2}} = \frac{1 + \beta_0^2}{1 - \beta_0^2}.$$

$$(12.262)$$

In the test-charge frame, the test charge sees the positive ions moving backward with speed $\beta$ (because they were at rest in the lab frame). Their separation is contracted by a factor $\gamma$, so their density is increased by a factor $\gamma$ to $(1 + \beta_0^2)\lambda_0/(1 - \beta_0^2)$. The test charge sees the electrons move with the same speed $\beta_0$ that they had in the lab frame (the direction is opposite, but that doesn't matter here). So the electron charge density is still $-\lambda_0$. The net density that the test charge sees is therefore

$$\lambda' = \lambda_0 \frac{1 + \beta_0^2}{1 - \beta_0^2} - \lambda_0 = \frac{2\beta_0^2 \lambda_0}{1 - \beta_0^2}. \qquad (12.263)$$

This agrees with Eq. (5.24), because in that equation we have

$$\lambda' = \gamma \beta \beta_0 \lambda_0 = \left(\frac{1 + \beta_0^2}{1 - \beta_0^2}\right) \left(\frac{2\beta_0}{1 + \beta_0^2}\right) \beta_0 \lambda_0 = \frac{2\beta_0^2 \lambda_0}{1 - \beta_0^2}, \qquad (12.264)$$

as desired.

## 12.6 Chapter 6

6.1 *Interstellar dust grain*

The grain in Exercise 2.38 has a radius of $3 \cdot 10^{-7}$ m and is charged to a potential of $-0.15$ V. Since $\phi = q/4\pi\epsilon_0 r$, we have $q = 4\pi\epsilon_0 r\phi$, which gives

$$q = 4\pi \left(8.85 \cdot 10^{-12} \frac{\text{s}^2\,\text{C}^2}{\text{kg m}^3}\right)(3 \cdot 10^{-7}\,\text{m})(-0.15\,\text{V}) = -5 \cdot 10^{-18}\,\text{C}.$$

$$(12.265)$$

Moving through a magnetic field $B$, the grain experiences a transverse force $qvB$. If its path is a circle (which it is; see Problem 6.26 or Exercise 6.29) of radius $R$, then $F = ma$ gives $qvB = mv^2/R \Longrightarrow v/R = qB/m$. So the "cyclotron" frequency, $\omega = v/R$, equals

$$\omega = \frac{qB}{m} = \frac{(5 \cdot 10^{-18}\,\text{C})(3 \cdot 10^{-10}\,\text{T})}{10^{-16}\,\text{kg}} = 1.5 \cdot 10^{-11}\,\text{s}^{-1}. \qquad (12.266)$$

The period is then $T = 2\pi/\omega = 4.2 \cdot 10^{11}$ s $\approx 13{,}000$ years. Note that this is independent of the speed $v$ and the radius $R$ (which are proportional to each other, for given $q, B, m$).

6.2 *Field from power lines*

The power is $P = IV$, so the current equals

$$I = \frac{P}{V} = \frac{10^7\,\text{J/s}}{5 \cdot 10^4\,\text{J/C}} = 200\,\text{A}. \qquad (12.267)$$

The field due to one wire is then

$$B = \frac{\mu_0 I}{2\pi r} = \frac{(4\pi \cdot 10^{-7}\,\text{kg m/C}^2)(200\,\text{A})}{2\pi(1\,\text{m})} = 4 \cdot 10^{-5}\,\text{T}. \qquad (12.268)$$

The other wire causes an equal field (in the same direction), so the total field midway between the wires is $8 \cdot 10^{-5}$ tesla, or 0.8 gauss.

6.3 *Repelling wires*

We first need to find the location of the center of mass of $BCDE$. This can be done in various ways. Since the vertical sides are twice as long as the bottom side, the object is equivalent to a mass $(2m + 2m)$ located 15 cm below the top, plus a mass $m$ located 30 cm below the top. The center of mass is $1/(4+1)$ of the way between these two effective masses, which means 3 cm below the $4m$, or 18 cm below the top.

The total weight is $(0.75 \, \text{m})(0.08 \, \text{N/m}) = 0.06 \, \text{N}$. If $F$ is the repulsive magnetic force between $CD$ and $GH$, then balancing the torques around $BE$ gives $F(0.30 \, \text{m}) = (0.06 \, \text{N})(0.18 \, \text{m}) \sin \theta$, where $\theta$ is the angle that $BC$ makes with the vertical. Therefore,[7]

$$F = (0.06 \, \text{N}) \cdot \frac{18}{30} \cdot \frac{0.5}{30} = 6 \cdot 10^{-4} \, \text{N}. \tag{12.269}$$

We know from Eq. (6.15) that the force on a length $\ell$ of wire due to the magnetic field from an infinite wire carrying the same current is $F = \mu_0 I^2 \ell / 2\pi r$. Setting this equal to $6 \cdot 10^{-4} \, \text{N}$ gives

$$\frac{(4\pi \cdot 10^{-7} \, \text{kg m/C}^2) I^2 (0.15 \, \text{m})}{2\pi (0.005 \, \text{m})} = 6 \cdot 10^{-4} \, \text{N} \implies I = 10 \, \text{A}. \tag{12.270}$$

The equilibrium is stable, because a larger angle yields a smaller magnetic torque and a larger gravitational torque, which makes the angle decrease. And likewise a smaller angle yields a larger magnetic torque and a smaller gravitational torque, which makes the angle increase.

If you solve for $I$ symbolically, you can show that $I = \left(2\pi g \lambda_m r^2 (h + \ell)/\mu_0 h \ell\right)^{1/2}$, where $\lambda_m$ is the mass density per unit length of the wire (so $g\lambda_m$ is the weight per unit length), $h$ is the height of $BC$, $\ell$ is the length of $CD$, and $r$ is the deflection distance. We see that $I$ is proportional to $r$, and also that $I$ decreases if both $h$ and $\ell$ are scaled up by the same factor.

**6.4**    *Vector potential for a wire*
Since the unit vector $\hat{\boldsymbol{\theta}}$ equals $-\sin\theta\,\hat{\mathbf{x}} + \cos\theta\,\hat{\mathbf{y}}$, and since $\sin\theta = y/r$ and $\cos\theta = x/r$, we can write $\mathbf{B}$ in terms of Cartesian coordinates as

$$\mathbf{B} = \frac{\mu_0 I\left(-(y/r)\hat{\mathbf{x}} + (x/r)\hat{\mathbf{y}}\right)}{2\pi r} = \frac{\mu_0 I}{2\pi}\left(\frac{-y\hat{\mathbf{x}} + x\hat{\mathbf{y}}}{x^2 + y^2}\right). \tag{12.271}$$

The vector potential $\mathbf{A}(x, y, z)$ can be written as (using $\ln r^2 = 2 \ln r$)

$$\mathbf{A} = -\hat{\mathbf{z}}\frac{\mu_0 I}{4\pi}\ln r^2 = -\hat{\mathbf{z}}\frac{\mu_0 I}{4\pi}\ln(x^2 + y^2). \tag{12.272}$$

The components of $\nabla \times \mathbf{A}$ are

$$(\nabla \times \mathbf{A})_x = \frac{\partial A_z}{\partial y} - \frac{\partial A_y}{\partial z} = \frac{\mu_0 I}{2\pi}\frac{-y}{x^2 + y^2},$$

$$(\nabla \times \mathbf{A})_y = \frac{\partial A_x}{\partial z} - \frac{\partial A_z}{\partial x} = \frac{\mu_0 I}{2\pi}\frac{x}{x^2 + y^2},$$

$$(\nabla \times \mathbf{A})_z = \frac{\partial A_y}{\partial x} - \frac{\partial A_x}{\partial y} = 0, \tag{12.273}$$

which are correctly equal to the components of $\mathbf{B}$.

---

[7]  Although we've made a few (quite reasonable) small-angle approximations in this reasoning, you can show that the following result happens to be exact.

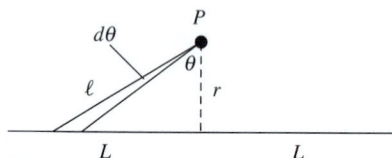

**Figure 12.85.**

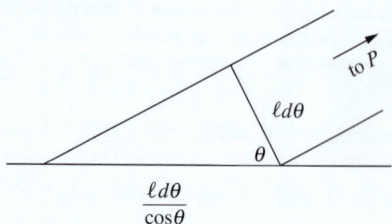

**Figure 12.86.**

6.5    *Vector potential for a finite wire*

(a) In both of Eqs. (6.34) and (12.272) we took the log of a quantity
with dimensions, either length or length squared. But it makes no
sense to take the log of a dimensionful quantity. Any function that
can be represented by a Taylor series (with more than one term) can
only be a function of a dimensionless quantity, because otherwise we
would be adding quantities with different dimensions in the Taylor
series. And a meter plus a square meter makes no sense. So the $\ln r$
in Eq. (6.34) should really be $\ln(r/a)$, where $a$ is some length. But,
having said this, you will quickly find that when taking the curl of $\mathbf{A}$
to find $\mathbf{B}$, the $a$ cancels out. It therefore doesn't matter what the value
of $a$ is, and that's why we got away with ignoring it. Equivalently,
$\ln(r/a)$ equals $\ln r - \ln a$, and adding on a constant doesn't affect the
derivative.

But this raises the following question: what does a particular value
of $a$ have to do with an infinitely long and infinitesimally thin wire?
Such a wire has no natural length scale (other than 0 and $\infty$), so there
is no way that a particular finite value of $a$ could possibly arise from a
calculation involving the wire. If a parameter doesn't exist at the start
of a calculation, then it doesn't exist at the end either. We'll answer
this question after solving part (b).

(b) Consider a small element of the current at angle $\theta$ and subtending an
angle $d\theta$, as shown in Fig. 12.85. If $\ell$ is the distance from the point
$P$ in question to the small element, then Fig. 12.86 shows that the
length of the element is $\ell\, d\theta / \cos\theta$. The expression for $\mathbf{A}$ in Eq. (6.46)
therefore gives

$$\mathbf{A} = \frac{\mu_0 I}{4\pi} \int \frac{dl}{r_{12}} = 2 \cdot \frac{\mu_0 I}{4\pi} \int_0^{\theta_0} \frac{\ell\, d\theta / \cos\theta}{\ell}\, \hat{\mathbf{x}}, \qquad (12.274)$$

where $\theta_0 = \tan^{-1}(L/r)$. The $\ell$'s cancel, so we just have $\int d\theta / \cos\theta$,
which the integral table in Appendix K gives as $\ln[(1 + \sin\theta)/\cos\theta]$.
We therefore obtain (the lower limit of the integration gives zero contribution)

$$\mathbf{A} = \frac{\mu_0 I}{2\pi} \ln\left(\frac{1 + \sin\theta_0}{\cos\theta_0}\right) \hat{\mathbf{x}} = \frac{\mu_0 I}{2\pi} \ln\left(\frac{1 + L/\sqrt{L^2 + r^2}}{r/\sqrt{L^2 + r^2}}\right) \hat{\mathbf{x}}$$

$$= \frac{\mu_0 I}{2\pi} \ln\left(\frac{\sqrt{L^2 + r^2} + L}{r}\right) \hat{\mathbf{x}}. \qquad (12.275)$$

In the $L \gg r$ limit, the $r^2$ term in the numerator is negligible,
so $\mathbf{A}$ simplifies to $\mathbf{A} = (\mu_0 I/2\pi) \ln(2L/r)\hat{\mathbf{x}}$. If we write this as
$\mathbf{A} = -(\mu_0 I/2\pi) \ln(r/2L)\hat{\mathbf{x}}$, then we see that the length $a$ referred
to in part (a) is simply the total length of the wire. Since this is
the only length scale of the wire, $a$ has no choice but to be some
multiple of $L$.

Of course, replacing the $2L$ by, say, $5L$ or by any other constant
length would still yield an expression for $\mathbf{A}$ that would produce the
correct $\mathbf{B}$ field. So it doesn't make any sense to say that $2L$ gives the

"correct" value of **A**. (We know that the same type of statement is true for the electrostatic potential $\phi$, because we can add on an arbitrary constant there, too.) But it is reassuring to see how the technically necessary parameter $a$ in $\ln(r/a)$ arises for a finite wire.

Since the final result for **B** is independent of $L$ (or rather, since it has a dependence on $L$ that goes to zero as $L \to \infty$, if we use the exact form of **A** in Eq. (12.275)), we can take the $L \to \infty$ limit and nothing changes. Note that if you calculate **B** directly by using the Biot–Savart law (see Exercise 6.45), the integral converges, so there is no need to truncate the integral by giving the wire a finite length.

6.6 *Zero divergence of A*

Let us first show that $\nabla_1(1/r_{12}) = -\nabla_2(1/r_{12})$. This follows from explicitly calculating the derivatives in Cartesian coordinates. With

$$r_{12} = \left[(x_1 - x_2)^2 + (y_1 - y_2)^2 + (z_1 - z_2)^2\right]^{1/2}, \quad (12.276)$$

the $x$ component of $\nabla_1(1/r_{12})$ is

$$\frac{\partial}{\partial x_1}\left[(x_1 - x_2)^2 + (y_1 - y_2)^2 + (z_1 - z_2)^2\right]^{-1/2} = \frac{x_2 - x_1}{r_{12}^3}. \quad (12.277)$$

Similarly, the $x$ component of $\nabla_2(1/r_{12})$ is $(x_1 - x_2)/r_{12}^3$. These are the negatives of each other, as promised. Likewise for the $y$ and $z$ components.

In calculating $\nabla \cdot \mathbf{A}$, it is understood that the $\nabla$ operator here is $\nabla_1$, because **A** is a function of the "1" coordinates. So taking the divergence of Eq. (6.44) yields (the steps are explained below)

$$\nabla_1 \cdot \mathbf{A}_1 = \frac{\mu_0}{4\pi} \int \nabla_1 \cdot \left(\frac{\mathbf{J}_2}{r_{12}}\right) dv_2$$

$$= \frac{\mu_0}{4\pi} \int \left(\frac{1}{r_{12}}\nabla_1 \cdot \mathbf{J}_2 + \mathbf{J}_2 \cdot \nabla_1\left(\frac{1}{r_{12}}\right)\right) dv_2$$

$$= \frac{\mu_0}{4\pi} \int \left(-\frac{1}{r_{12}}\nabla_2 \cdot \mathbf{J}_2 - \mathbf{J}_2 \cdot \nabla_2\left(\frac{1}{r_{12}}\right)\right) dv_2$$

$$= -\frac{\mu_0}{4\pi} \int \nabla_2 \cdot \left(\frac{\mathbf{J}_2}{r_{12}}\right) dv_2$$

$$= -\frac{\mu_0}{4\pi} \int_S \left(\frac{\mathbf{J}_2}{r_{12}}\right) \cdot d\mathbf{a}_2$$

$$= 0, \quad (12.278)$$

as desired. In the first line, we were able to bring the $\nabla_1$ operator inside the integral because the integral is with respect to the "2" coordinates. In the second line we used the given vector identity. The first terms in the second and third lines are equal because they are both zero so the minus sign doesn't matter ($\nabla_1 \cdot \mathbf{J}_2$ is zero because $\mathbf{J}_2$ doesn't depend on the "1" coordinates, and $\nabla_2 \cdot \mathbf{J}_2$ is zero because the current is assumed to be steady); and the second terms are equal due to the above result for the gradients of $1/r_{12}$. In the fourth line we used the given vector identity in reverse. In the fifth line we used the divergence theorem, with $S$ being a surface at infinity. In the sixth line we used the fact that there is zero net current flowing out of a sphere at infinity, which is the statement of global conservation of charge. (We generally assume the more strict condition that the current is actually zero at infinity, in which case the surface integral is certainly zero). Note that the main goal in the above string of equations was simply to exchange the $\nabla_1$ operator for the $\nabla_2$ operator, because this allowed us to apply the divergence theorem to the $dv_2$ integral.

6.7　*Vector potential on a spinning sphere*

(a) Consider the contribution to $\mathbf{A}$ at the point $(R, 0, 0)$ from the ring shown in Fig. 12.87; $\mathbf{A}$ is given by $(\mu_0/4\pi)\int \mathbf{J}\, dV/r$. As the ring rotates around the $z$ axis along with the sphere, all of its points have a positive velocity component $v_y$ (into the page). Points with $y < 0$ have a positive $v_x$ component, and points with $y > 0$ have a negative $v_x$ component. These $v_x$ components cancel in pairs in the above integral, so we need only worry about the $v_y$ components.

It turns out that all points on the ring have the *same* $v_y$. This can be seen by noting that the total speed $v$ is related to the distance $\sqrt{x^2 + y^2}$ from the $z$ axis by $v = \omega\sqrt{x^2 + y^2}$. But in finding the $v_y$ component of the velocity, we must multiply by $x/\sqrt{x^2 + y^2}$. So $v_y$ at all points on the ring equals $\omega x$, where $x$ is the common $x$ coordinate of the points. Therefore, for the present purposes, we can consider the ring simply to be sliding with speed $v_y = \omega x = \omega R \cos\theta$ in the $y$ direction. Hence $\mathbf{A}$ at the point $(R, 0, 0)$ has only a $y$ component.

The area of the ring is $da = (2\pi R \sin\theta)(R\, d\theta)$. If we imagine the ring to have a slight thickness $dr$ filled with volume charge density $\rho$, then the $J_y\, dV$ part of $\mathbf{A}$ can be written as $(\rho v_y)(da\, dr) = (\rho\, dr)(v_y\, da) = \sigma v_y\, da$. (We have used the fact that $J$ can be written as $\rho v$, as you should check.) The point $(R, 0, 0)$ is a distance $r = 2R \sin(\theta/2)$ from all points on the ring, so we have

$$A_y = \frac{\mu_0}{4\pi}\int \frac{J_y\, dV}{r} = \frac{\mu_0}{4\pi}\int_0^\pi \frac{\sigma(\omega R\cos\theta)\cdot(2\pi R\sin\theta)(R\, d\theta)}{2R\sin(\theta/2)}$$

$$= \frac{\mu_0 R^2 \sigma\omega}{4}\int_0^\pi \frac{\sin\theta\cos\theta\, d\theta}{\sin(\theta/2)}. \tag{12.279}$$

Writing $\sin\theta$ as $2\sin(\theta/2)\cos(\theta/2)$ and $\cos\theta$ as $1 - 2\sin^2(\theta/2)$, we obtain

$$A_y = \frac{\mu_0 R^2 \sigma\omega}{2}\int_0^\pi \cos(\theta/2)\Big(1 - 2\sin^2(\theta/2)\Big)d\theta$$

$$= \frac{\mu_0 R^2 \sigma\omega}{2}\Big(2\sin(\theta/2) - (4/3)\sin^3(\theta/2)\Big)\Big|_0^\pi$$

$$= \frac{\mu_0 R^2 \sigma\omega}{3}. \tag{12.280}$$

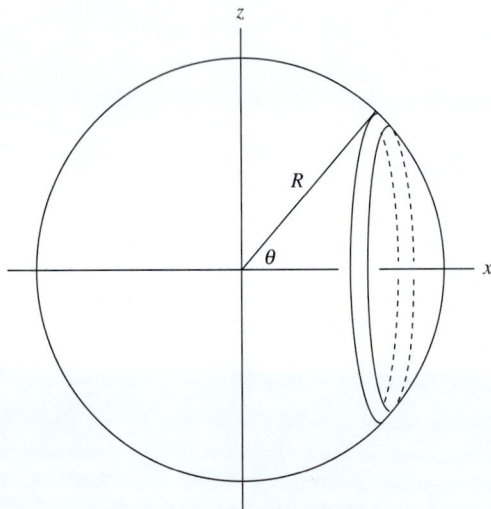

**Figure 12.87.**

(b) The rotation around the $\omega_1$ vector in Fig. 6.34 contributes nothing to **A**, because for every piece of the sphere moving one way there is another piece moving the other way, at the *same* distance $r$ from the point $(x, 0, z)$. So the contributions to the integral expression for **A** cancel in pairs.

We therefore need only worry about the $\omega_2$ rotation. But this is just the setup from part (a), with the only change being that the magnitude of $\omega$ is now $|\omega_2| = \omega \cos \beta$. **A** still has only a $y$ component (into the page), so Eq. (12.280) gives (using $R \cos \beta = x$)

$$A_y = \frac{\mu_0 R^2 \sigma (\omega \cos \beta)}{3} = \frac{\mu_0 x R \sigma \omega}{3}. \qquad (12.281)$$

(c) All points on the spherical shell with the same value of $z$ (in other words, a horizontal circle) have the same magnitude of **A**, by symmetry. Only the direction differs; **A** points tangentially around the circle. If the radius of this circle is $r$ (which was just $x$ in part (b)), then from part (b) the magnitude of **A** is $A = \mu_0 r R \sigma \omega / 3$. Figure 12.88 shows a top view, looking down along the $z$ axis. The components of **A** are $(-A \sin \alpha, A \cos \alpha, 0)$, so the desired value of **A** at an arbitrary point on the surface of the sphere is

$$\mathbf{A} = \frac{\mu_0 R \sigma \omega}{3} (-r \sin \alpha, r \cos \alpha, 0) = \frac{\mu_0 R \sigma \omega}{3} (-y, x, 0). \quad (12.282)$$

At this point, you have all the information needed to use the strategy in Problem 11.8(b) to find the magnetic field inside our rotating hollow spherical shell. But we'll save that for Chapter 11 because it builds on a strategy used in Chapter 10.

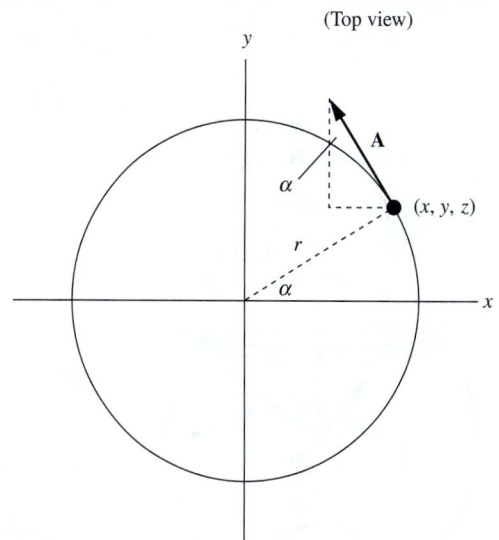

Figure 12.88.

6.8 *The field from a loopy wire*

The Biot–Savart law gives the field contribution from a piece $d\mathbf{l}$ of the wire as $d\mathbf{B} = (\mu_0/4\pi) I \, d\mathbf{l} \times \hat{\mathbf{r}}/r^2$. At a point very far from the wire, the $\hat{\mathbf{r}}$ vector and $r$ distance are essentially the same for all points in the wire. So when we integrate over the entire wire, we can take these quantities outside the integral. The field due to the wire is therefore

$$\mathbf{B} = \frac{\mu_0}{4\pi} \int \frac{I \, d\mathbf{l} \times \hat{\mathbf{r}}}{r^2} = \frac{\mu_0 I}{4\pi r^2} \left( \int d\mathbf{l} \right) \times \hat{\mathbf{r}} = \frac{\mu_0 I}{4\pi r^2} \mathbf{l} \times \hat{\mathbf{r}}, \quad (12.283)$$

where $\mathbf{l}$ is the vector from one point to the other. This result is simply the result for the field due to a straight wire between the two points, as desired.

6.9 *Scaled-up ring*

The resistance of each ring takes the form of $R = \rho L/A$, where $\rho$ is the resistivity, $L$ is the circumference, and $A$ is the cross-sectional area. The resistance of the larger ring is therefore half that of the smaller ring, because $L$ scales like length, and $A$ scales like length squared. Hence the current in the larger ring is twice that in the smaller ring, because the voltage is the same. But Eq. (6.54) gives the magnetic field at the center of a ring as $\mu_0 I/2r$. Since the larger ring has twice the current and twice the radius, the field at the center is the same.

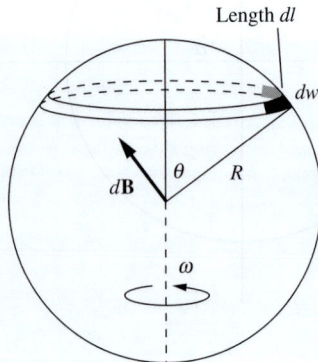

Length $dl$

$dw$

$d\mathbf{B}$

$\theta$  $R$

$\omega$

**Figure 12.89.**

6.10  *Rings with opposite currents*

On the axis, let $z = 0$ correspond to the point midway between the rings. Then from Eq. (6.53) the field on the axis at a general value of $z$ has magnitude

$$B_z = \frac{\mu_0 I a^2}{2\big(a^2 + (z - \epsilon/2)^2\big)^{3/2}} - \frac{\mu_0 I a^2}{2\big(a^2 + (z + \epsilon/2)^2\big)^{3/2}}. \quad (12.284)$$

We could use a Taylor series to calculate this difference to lowest order in $\epsilon$. But it's easier to use the fact that, by definition, the difference is simply $\epsilon$ times the (negative of the) derivative of the function $\mu_0 I a^2 / 2(a^2 + z^2)^{3/2}$. So we have

$$B_z = -\epsilon \frac{d}{dz}\left(\frac{\mu_0 I a^2}{2(a^2 + z^2)^{3/2}}\right) = \frac{3\epsilon \mu_0 I a^2}{2} \frac{z}{(a^2 + z^2)^{5/2}}. \quad (12.285)$$

We want to maximize this function of $z$. Using the quotient rule and setting the numerator of the derivative equal to zero gives

$$0 = (a^2 + z^2)^{5/2}(1) - z(5/2)(a^2 + z^2)^{3/2}(2z)$$

$$\implies 0 = (a^2 + z^2) - 5z^2 \implies z = a/2. \quad (12.286)$$

The maximum value turns out to be $24\epsilon\mu_0 I/(25\sqrt{5}a^2)$.

6.11  *Field at the center of a sphere*

In Fig. 12.89, let the axis of rotation point vertically, and consider a ring on the shell located at an angle $\theta$ down from the vertical, subtending an angle $d\theta$. The width of the ring is $dw = R\, d\theta$, and the velocity of any point on it is $v = \omega(R\sin\theta)$. The amount of charge that passes by a given point in time $dt$ is $dq = \sigma(dw)(v\, dt) = \sigma(R\, d\theta)(\omega R\sin\theta)dt$. The current produced by the ring is therefore $I = dq/dt = \sigma\omega R^2 \sin\theta\, d\theta$.

From the Biot–Savart law, a small piece of the ring with length $dl$ at the location shown in Fig. 12.89 produces a $d\mathbf{B}$ field at the origin that points up to the left as shown, with magnitude $(\mu_0/4\pi)I\, dl/R^2$. When we integrate over the whole ring, the horizontal components of the $d\mathbf{B}$ vectors cancel, and we are left with only a vertical component. This brings in a factor of $\sin\theta$. For a given ring, the $dl$ in the Biot–Savart law integrates up to the length of the ring, which is $l = 2\pi(R\sin\theta)$. The contribution to the field from a given ring at angle $\theta$, subtending an angle $d\theta$, is therefore

$$\hat{\mathbf{z}}\frac{\mu_0}{4\pi}\frac{Il}{R^2}\sin\theta = \hat{\mathbf{z}}\frac{\mu_0}{4\pi}\frac{(\sigma\omega R^2 \sin\theta\, d\theta)(2\pi R\sin\theta)}{R^2}\sin\theta$$

$$= \hat{\mathbf{z}}\frac{1}{2}\mu_0\sigma\omega R\sin^3\theta\, d\theta. \quad (12.287)$$

Integrating this from $0$ to $\pi$ gives the total field at the origin. You can either look up the integral in a table, or write $\sin^3\theta$ as $\sin\theta(1 - \cos^2\theta)$. The result is

$$\mathbf{B} = \hat{\mathbf{z}}\frac{1}{2}\mu_0\sigma\omega R \int_0^\pi \sin^3\theta\, d\theta = \hat{\mathbf{z}}\frac{1}{2}\mu_0\sigma\omega R \left(-\cos\theta + \frac{\cos^3\theta}{3}\right)\Bigg|_0^\pi$$

$$= \hat{\mathbf{z}}\frac{2}{3}\mu_0\sigma\omega R. \quad (12.288)$$

Interestingly, the magnetic field takes on this same value *everywhere* inside the sphere; see Problems 11.7 and 11.8.

The field in Eq. (12.288) happens to be 4/3 as large as the field at the center of a spinning disk with the same $R$, $\sigma$, $\omega$ values, which is $\mathbf{B}_{\text{disk}} = \hat{\mathbf{z}}\mu_0\sigma\omega R/2$; see Exercise 6.49.

6.12  *Field in the plane of a ring*

Let the current flow counterclockwise around the ring in Fig. 12.90. Consider a small piece of the ring at angle $\theta$ with respect to the line to the point $P$ in question, and subtending an angle $d\theta$. In Cartesian coordinates (with $x$ horizontal, $y$ vertical, and $z$ pointing out of the page), we have $d\mathbf{l} = (R\,d\theta)(-\sin\theta, \cos\theta, 0)$ and $\mathbf{r} = (a - R\cos\theta, -R\sin\theta, 0)$. The cross product of these two vectors is

$$d\mathbf{l} \times \mathbf{r} = (R\,d\theta) \begin{vmatrix} \hat{\mathbf{x}} & \hat{\mathbf{y}} & \hat{\mathbf{z}} \\ -\sin\theta & \cos\theta & 0 \\ a - R\cos\theta & -R\sin\theta & 0 \end{vmatrix} = (R\,d\theta)(R - a\cos\theta)\hat{\mathbf{z}}.$$

(12.289)

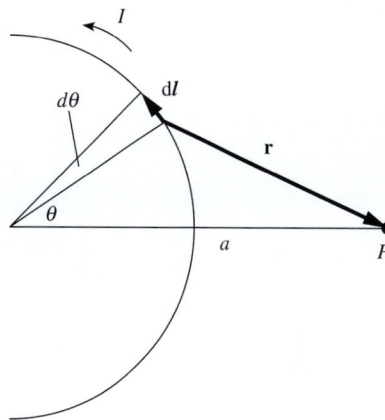

**Figure 12.90.**

Note that if $P$ is outside the ring (that is, $a > R$), then $\cos\theta = R/a$ is the cutoff angle between the field contributions pointing into or out of the page. This is correctly the angle at which the $\mathbf{r}$ vector is tangent to the ring; the small piece of current is then parallel to $\mathbf{r}$ and therefore produces no magnetic field at the point $P$. If $P$ is inside the ring (that is, $a < R$), then $d\mathbf{l} \times \mathbf{r}$ points out of the page for all $\theta$.

The law of cosines tells us that $r = (a^2 + R^2 - 2aR\cos\theta)^{1/2}$, so the Biot–Savart law gives

$$B = \frac{\mu_0 I}{4\pi} \int \frac{d\mathbf{l} \times \mathbf{r}}{r^3} = 2 \cdot \frac{\mu_0 I}{4\pi} \int_0^\pi \frac{(R - a\cos\theta)R\,d\theta}{(a^2 + R^2 - 2aR\cos\theta)^{3/2}},$$

(12.290)

with positive corresponding to pointing out of the page. The factor of 2 out front comes from the fact that we are integrating only from 0 to $\pi$. It turns out that $\mathbf{B}$ points into the page if $P$ is outside the ring (and if the current is counterclockwise, as shown in Fig. 12.90); the contributions from the closer points on the ring "win," so the net field points into the page.

In the special case where $a = 0$ we have

$$B = \frac{\mu_0 I}{2\pi} \int_0^\pi \frac{(R)R\,d\theta}{(R^2)^{3/2}} = \frac{\mu_0 I}{2\pi} \cdot \frac{\pi}{R} = \frac{\mu_0 I}{2R},$$

(12.291)

as desired.

6.13  *Magnetic dipole*

Factoring out a few powers of $a$ in the numerator and denominator in Eq. (6.94) yields

$$B = -\frac{\mu_0 I R}{2\pi a^2} \int_0^\pi \left(\cos\theta - \frac{R}{a}\right)\left(1 + \frac{R^2}{a^2} - \frac{2R}{a}\cos\theta\right)^{-3/2} d\theta.$$

(12.292)

Ignoring the very small $R^2/a^2$ term and using $(1 + \epsilon)^{-3/2} \approx 1 - 3\epsilon/2$ gives

$$B \approx -\frac{\mu_0 IR}{2\pi a^2} \int_0^\pi \left(\cos\theta - \frac{R}{a}\right)\left(1 + \frac{3}{2} \cdot \frac{2R}{a}\cos\theta\right) d\theta. \qquad (12.293)$$

The leading-order term (the one not involving a factor of $R/a$) is zero since $\int_0^\pi \cos\theta\, d\theta = 0$. We therefore have (dropping the $R^2/a^2$ term)

$$B \approx -\frac{\mu_0 IR}{2\pi a^2} \int_0^\pi \frac{R}{a}(3\cos^2\theta - 1)d\theta. \qquad (12.294)$$

You can calculate this integral, or you can just note that the average value of $\cos^2\theta$ from 0 to $\pi$ is 1/2. Replacing $\cos^2\theta$ with 1/2 yields

$$B \approx -\frac{\mu_0 IR}{2\pi a^2} \cdot \frac{R}{a} \cdot \frac{\pi}{2} = -\frac{\mu_0}{4\pi}\frac{\pi R^2 I}{a^3} \equiv -\frac{\mu_0}{4\pi}\frac{m}{a^3}, \qquad (12.295)$$

as desired. The minus sign here comes from the sign convention in Problem 6.12. But in general the direction of the field is determined by the contribution from the current in the nearer side of the ring.

6.14 *Far field from a square loop*

(a) All parts of the horizontal sides are essentially perpendicular to the radius vector to $P$, so in the Biot–Savart law we can set the $\sin\theta$ term in the cross product equal to 1. The two horizontal sides therefore give Biot–Savart contributions of $\pm(\mu_0/4\pi)Ia/(r \pm a/2)^2$. These contributions point in opposite directions, so to leading order in $a$ the net field at $P$ is

$$B = \frac{\mu_0 Ia}{4\pi r^2}\left(\frac{1}{(1 - a/2r)^2} - \frac{1}{(1 + a/2r)^2}\right)$$

$$\approx \frac{\mu_0 Ia}{4\pi r^2}\left(\frac{1}{(1 - a/r)} - \frac{1}{(1 + a/r)}\right)$$

$$\approx \frac{\mu_0 Ia}{4\pi r^2}\left(\left(1 + \frac{a}{r}\right) - \left(1 - \frac{a}{r}\right)\right) = \frac{\mu_0 Ia^2}{2\pi r^3}, \qquad (12.296)$$

as desired. The sign is determined by the edge that is closest to $P$. The field points into the page if the current is counterclockwise in Fig. 6.36.

(b) We made two approximations in the reasoning that led to the above result. One is inconsequential, the other is critical. The inconsequential approximation is that we set the $\sin\theta$ term in the Biot–Savart law equal to 1. But the horizontal edges aren't exactly perpendicular to the radius vector to $P$ at all points. However, you can show that the tiny correction is smaller than the leading-order result of order $a^2/r^3$, by a factor of order $a^2/r^2$. So the correction term is of order $a^4/r^5$, which is too small to matter. (Using the same $r$ value for all points on a given edge yields the same type of error.)

The critical approximation we made is that we neglected the field contributions from the two vertical sides. True, these sides are nearly parallel to the radius vector to $P$, so the cross product in the Biot–Savart law is very small. But it isn't small enough to neglect. The point is that the small contributions from the two vertical sides point in the *same* direction (both out of the page in Fig. 6.36), whereas the larger contributions from the two horizontal sides point in *opposite* directions. And the *sum* of the small contributions is of the same order of magnitude as the *difference* of the (nearly identical) larger contributions. Things quickly work out quantitatively as follows.

In Fig. 6.36 the fields from both of the vertical edges point out of the page with magnitude $(\mu_0/4\pi)(Ia/r^2)\big((a/2)/r\big)$, where the last factor comes from the cross product in the Biot–Savart law; the sine of the angle between a vertical edge and the radius vector to $P$ is essentially equal to $(a/2)/r$. Doubling this result because there are two vertical sides gives $\mu_0 Ia^2/4\pi r^3$. This field points out of the page, so it partially cancels the into-the-page result from part (a). The net field therefore points into the page with magnitude $\mu_0 Ia^2/4\pi r^3$, as we wanted to show.

**6.15** *Magnetic scalar "potential"*

(a) The curl in cylindrical coordinates is given in Appendix F. Since $\mathbf{B}$ has only a $\hat{\boldsymbol{\theta}}$ component, and since this component has only $r$ dependence, the only term in the curl that has a chance of being nonzero is $(1/r)\big(\partial(rB_\theta)/\partial r\big)\hat{\mathbf{z}}$. But $B_\theta \propto 1/r$, so this term is zero, as desired.

Alternatively, we can use Cartesian coordinates. With the symmetry axis pointing along the $z$ axis, the tangential $\mathbf{B}$ field lies in the $xy$ plane and is proportional to the vector $\pm(-y, x, 0)$. (This direction can be deduced from the fact that the dot product with the radial vector $(x, y, 0)$ must be zero.) We need the magnitude of $\mathbf{B}$ to be $\mu_0 I/2\pi r$, so $\mathbf{B}$ must equal $[\mu_0 I/2\pi(x^2 + y^2)](-y, x, 0)$. The only component of the curl that has a chance of being nonzero is the $z$ component, which is $\partial B_y/\partial x - \partial B_x/\partial y$. You can quickly check that this equals zero.

(b) In cylindrical coordinates, the $\hat{\boldsymbol{\theta}}$ component of the gradient of a function $\psi$ is given in Appendix F as $(1/r)(\partial\psi/\partial\theta)\hat{\boldsymbol{\theta}}$. So we want

$$\frac{1}{r}\frac{\partial\psi}{\partial\theta} = \frac{\mu_0 I}{2\pi r} \implies \frac{\partial\psi}{\partial\theta} = \frac{\mu_0 I}{2\pi} \implies \psi = \frac{\mu_0 I}{2\pi}\theta. \quad (12.297)$$

We therefore see that $\mathbf{B}$ can be written as $\nabla\psi$. However, the problem with this $\psi$ is that it is multi-valued. For example, for given $r$ and $z$, the values $\theta = 0, 2\pi, 4\pi$, etc., all correspond to the same point in space. So $\psi$ cannot be used as a potential that uniquely labels each point in space. In a limited region, however, it can be of use.

**6.16** *Copper solenoid*

Since there are two layers, the average diameter of the turns in the coil is $(8 + 2 \cdot 0.163)\,\text{cm} \approx 8.3\,\text{cm}$. The total length of the wire is

$$\left(\pi \cdot 8.3\,\frac{\text{cm}}{\text{turn}}\right)\left(4\,\frac{\text{turns}}{\text{cm}}\right)\left(32\,\frac{\text{cm}}{\text{layer}}\right)(2\text{ layers}) \approx 6700\,\text{cm} = 67\,\text{m}.$$

$$(12.298)$$

The resistance is $R = (67\,\text{m})(0.01\,\Omega/\text{m}) = 0.67\,\Omega$, so the current is $I = (50\,\text{V})/(0.67\,\Omega) = 75\,\text{A}$. The power is then $P = IV = (75\,\text{A})(50\,\text{V}) = 3750\,\text{J/s}$.

From Eq. (6.56) the field at the center of the solenoid is $B = \mu_0 nI \cos\theta$, where $n = 8$ turns/cm $= 800$ turns/m, $I = 75\,\text{A}$, and $\theta = \tan^{-1}(4/16)$. So we have

$$B = \mu_0 nI \cos\theta = \left(4\pi \cdot 10^{-7}\,\frac{\text{kg m}}{\text{C}^2}\right)(800\,\text{m}^{-1})(75\,\text{A})\cos 14°$$

$$= 0.0732\,\text{T}, \tag{12.299}$$

or 732 gauss. The field of an infinitely long solenoid would be $\mu_0 nI$, which equals 754 gauss.

6.17  *A rotating solid cylinder*

(a) Let's slice up the solid cylinder into a collection of thin shells with thickness $dr$. The effective surface charge density of such a shell is $\sigma_r = \rho\,dr$. Now, if a cylindrical shell with radius $r$ and surface charge density $\sigma_r$ spins with frequency $\omega$, the surface current density is $\mathcal{J}_r = \sigma_r v_r = \sigma_r \omega r$. This is true because the area that crosses a segment of length $\ell$ in time $dt$ is $\ell(v_r\,dt)$, so the charge per time is $\sigma_r \ell v_r$. The current per length (which is $\mathcal{J}$ by definition) is therefore $\sigma_r v_r$.

In terms of $\mathcal{J}_r$, the interior magnetic field due to the current running around the cylinder is $B_r = \mu_0 \mathcal{J}_r$. This is just the continuum limit of the $B = \mu_0 nI$ expression for a solenoid. Putting the above results together, the interior field due to one of the shells is

$$B_r = \mu_0 \mathcal{J}_r = \mu_0(\sigma_r v_r) = \mu_0(\rho\,dr)(\omega r) = \mu_0 \rho \omega r\,dr. \tag{12.300}$$

The axis is inside all of the shells, so integrating over all the shells from $r = 0$ to $r = R$ gives the total magnetic field on the axis as $B = \mu_0 \rho \omega R^2/2$.

Alternatively, we can solve this problem with Ampère's law. Consider the loop indicated by the dashed line in Fig. 12.91. To find the current enclosed, we can use the fact that $J = \rho v_r$ (which you should verify). Hence $J = \rho \omega r$. The current passing through a thin strip with width $dr$ and height $\ell$ at radius $r$ is then $dI = J\ell\,dr = \rho \omega \ell r\,dr$. Integrating from $r = 0$ to $r = R$ gives the current enclosed as $I = \rho \omega \ell R^2/2$. The field is zero outside the cylinder (because the cylinder is a superposition of solenoids), so the only contribution to the line integral in Ampère's law comes from the side of the loop lying along the axis. Therefore $B\ell = \mu_0(\rho \omega \ell R^2/2) \implies B = \mu_0 \rho \omega R^2/2$, as above.

(b) If all the charge is located on the surface, then the surface charge density is given by $\sigma 2\pi R = \rho \pi R^2$, because these are two different expressions for the charge per unit length along the cylinder. Hence $\sigma = \rho R/2$. The magnetic field on the axis (or anywhere inside) is then

$$B = \mu_0 \mathcal{J} = \mu_0(\sigma \omega R) = \mu_0(\rho R/2)\omega R = \mu_0 \rho \omega R^2/2, \tag{12.301}$$

which equals the $B$ in part (a). The reason for this is the following.

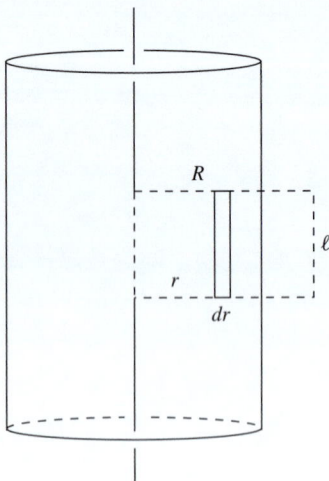

**Figure 12.91.**

Consider taking the charge on one of the many shells in the solid cylinder and moving it out to the surface. How does this affect the field on the axis? It doesn't, because the field depends only on $\mathcal{J}$, and $\mathcal{J}$ doesn't change if we move the charge outward. The same amount of charge makes one revolution in the same amount of time, independent of the radius of the shell at which it resides (because all shells rotate with the same $\omega$). We can therefore take every shell and move its charge out to the surface without affecting the field on the axis.

6.18 *Vector potential for a solenoid*

First, note that $\mathbf{A}$ must have only a $\hat{\boldsymbol{\theta}}$ component, that is, it must point in the tangential direction around the axis of the solenoid. This follows from the fact that each $d\mathbf{A}$ contribution points in the same direction as the $\mathbf{J}$ current that produces it. (See Eq. (6.44), although that equation relies on the div $\mathbf{A} = 0$ assumption in Eq. (6.40).) And every piece of current in the system points in the $\hat{\boldsymbol{\theta}}$ direction. Furthermore, $A_\theta$ can have no dependence on $\theta$ or $z$, by symmetry. So the one nonzero component, $A_\theta$, must be a function only of $r$. Our goal is therefore to find the function $A_\theta(r)$.

(a) From Problem 6.41 we know that the magnetic flux is given by $\Phi = \int_C \mathbf{A} \cdot d\mathbf{l}$. If we take the curve $C$ to be a circle with radius $r$ inside the solenoid, this relation becomes

$$B(\pi r^2) = A_\theta(2\pi r) \Longrightarrow A_\theta = \frac{(\mu_0 nI)(\pi r^2)}{2\pi r} = \frac{\mu_0 nIr}{2} \qquad \text{(inside).}$$

(12.302)

If the curve $C$ is a circle outside the solenoid, we obtain

$$B(\pi R^2) = A_\theta(2\pi r) \Longrightarrow A_\theta = \frac{(\mu_0 nI)(\pi R^2)}{2\pi r} = \frac{\mu_0 nIR^2}{2r} \qquad \text{(outside).}$$

(12.303)

(b) Since we have only one component, $A_\theta(r)$, the only nonzero term in the expression for the curl in cylindrical components given in Appendix F is $\hat{\mathbf{z}}(1/r)\partial(rA_\theta)/\partial r$. So inside the solenoid, $\mathbf{B} = \nabla \times \mathbf{A}$ becomes

$$\hat{\mathbf{z}}\mu_0 nI = \hat{\mathbf{z}}\frac{1}{r}\frac{\partial(rA_\theta)}{\partial r} \implies \frac{\partial(rA_\theta)}{\partial r} = \mu_0 nIr$$

$$\implies rA_\theta = \frac{\mu_0 nIr^2}{2} \implies A_\theta = \frac{\mu_0 nIr}{2} \qquad \text{(inside),} \quad (12.304)$$

in agreement with the result in part (a). Outside the solenoid, $\mathbf{B} = \nabla \times \mathbf{A}$ becomes

$$0 = \hat{\mathbf{z}}\frac{1}{r}\frac{\partial(rA_\theta)}{\partial r} \implies \frac{\partial(rA_\theta)}{\partial r} = 0$$

$$\implies rA_\theta = C \implies A_\theta = \frac{C}{r} \qquad \text{(outside).} \quad (12.305)$$

We see that any field proportional to $1/r$ yields zero curl outside the solenoid. The field in Eq. (12.303) is a special case of this. All fields that are proportional to $1/r$ yield the same zero curl, but not the

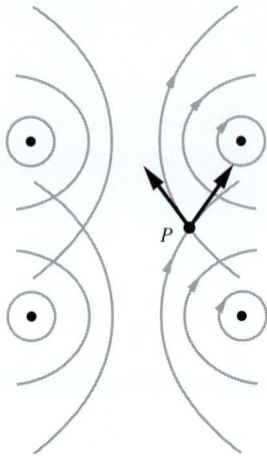

**Figure 12.92.**

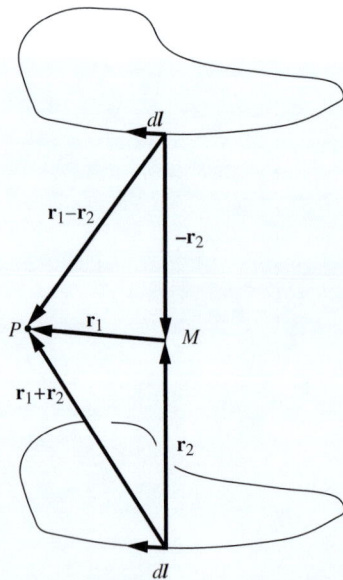

**Figure 12.93.**

same line integral around a circle of radius $r$. This is due to the fact that the curl of $1/r$ diverges at the origin; you should think about how Stokes' theorem comes into play. The issue here is similar to the one discussed in Problem 2.26.

Inside the solenoid, if we had included a constant of integration in Eq. (12.304), we would have obtained an additional term of the form $C/r$. Although this wouldn't affect the $\mathbf{B} = \mu_0 n I \hat{\mathbf{z}}$ result at points away from the origin, it would yield an infinite $\mathbf{B}$ at $r = 0$. So we must reject this term.

6.19   *Solenoid field, inside and outside*

(a) First solution   The longitudinal nature of the field follows from considering the contributions from two loops on either side of a given point $P$, equidistant from $P$. Figure 6.15 shows the field due to one ring, so the fields due to two rings are shown in Fig. 12.92. At any point $P$ on the plane midway between the rings, the magnetic field points in the longitudinal direction, because the radial components cancel, as shown. This argument holds both inside and outside the solenoid, although we will find in part (c) that the field outside is actually zero. You should convince yourself why there are cancelation effects that make it possible for the field to be zero outside, but not inside.

Second solution   We can show that the Biot–Savart contributions from corresponding small intervals of two symmetrically located circles sum to a longitudinal vector. This argument (along with all the other results in this problem) actually holds for a solenoid with a cross-section of *arbitrary* uniform shape. To see why, consider the Biot–Savart $d\boldsymbol{l} \times \mathbf{r}$ cross products involved in calculating the field at point $P$ in Fig. 12.93 due to the two $d\boldsymbol{l}$ pieces shown. The point $M$ is midway between the two pieces, and the various vectors are labeled as shown. The sum of the two $d\boldsymbol{l} \times \mathbf{r}$ contributions is

$$d\boldsymbol{l} \times (\mathbf{r}_1 + \mathbf{r}_2) + d\boldsymbol{l} \times (\mathbf{r}_1 - \mathbf{r}_2) = 2\, d\boldsymbol{l} \times \mathbf{r}_1. \tag{12.306}$$

Now, $\mathbf{r}_1$ points in a transverse direction (that is, perpendicular to the axis) because both $M$ and $P$ lie on the plane midway between the loops. And $d\boldsymbol{l}$ points in a transverse direction too, of course. Since the cross product of two vectors is perpendicular to both vectors, we see that $d\boldsymbol{l} \times \mathbf{r}_1$ points in the longitudinal direction, as desired. This reasoning holds both inside and outside the solenoid.

(b) Having shown that the field is longitudinal, we will now show that it is uniform inside (and outside) the solenoid. It is certainly uniform in the longitudinal direction, by symmetry. So the task is to show that it is uniform in the transverse direction. Consider a rectangular Amperian loop lying completely inside the solenoid, with two sides pointing in the longitudinal direction, and two sides pointing in a transverse direction, as shown in Fig. 12.94. This loop encloses zero current, so the line integral of $\mathbf{B}$ must be zero. The line integral is nonzero only along the longitudinal sides (this would effectively be true even if there existed a component of $\mathbf{B}$ in the transverse direction, because the contributions would cancel on the two transverse sides of the rectangle). So the field must have the same value on the longitudinal sides. Since the rectangle can have an arbitrary width and be positioned at an arbitrary location inside the solenoid, we conclude that the field must be uniform inside the solenoid. The same reasoning holds outside the solenoid, so the field is uniform there too (with a different value, as we will see).

We must now determine how these two uniform values (inside and outside) are related. If we draw a rectangular Amperian loop with one longitudinal side inside the solenoid and the other outside (let these sides have length $\ell$), as shown in Fig. 12.95, the loop now encloses a current of $nI\ell$. Ampère's law therefore gives (taking positive current to point into the page)

$$B_{\text{in}}\ell - B_{\text{out}}\ell = \mu_0 nI\ell \implies B_{\text{in}} = B_{\text{out}} + \mu_0 nI. \quad (12.307)$$

Note that since the transverse width of the rectangle can have any size, this argument by itself shows that the field is uniform inside and outside. We technically didn't need to include the previous paragraph.

(c) The $B_{\text{in}} = B_{\text{out}} + \mu_0 nI$ result, combined with the above results about uniformity, implies that if we can show that the field is zero at any point outside the solenoid, then we have shown that $B = 0$ everywhere outside, and $B = \mu_0 nI$ everywhere inside. And indeed, we can show that $B = 0$ at infinity, as follows.

Consider the field at a point $P$ due to a particular loop of the solenoid. Let $P$ be a large distance from the loop (large compared with the size of the loop). The field due to the loop is smaller than the field due to a straight wire segment that has the same length $b = 2\pi a$ as the loop (where $a$ is the radius of the solenoid) and that is oriented perpendicular to the vector to $P$. (This is true because the current moves in different directions around the loop, so the Biot–Savart contributions partially, or actually mostly, cancel.) Therefore, an upper bound on the field due to the solenoid is the field due to an infinite set of straight wire segments with length $b$, lined up side by side. Only the longitudinal components of the fields due to these wires will survive, but we don't need to use this fact. It turns out that we can obtain a sufficiently small upper bound on the field by adding up the *magnitudes* of the fields due to the wire segments. This certainly overestimates the net field. The magnitude of the field due to a distant wire segment is $(\mu_0/4\pi)Ib/(x^2 + r^2)$, where $r$ is the perpendicular distance from $P$ to the solenoid, and $x$ is the distance shown in Fig. 12.96. There are $n$ wire segments per unit length, so an upper bound on the field due to the solenoid is

$$B_{\text{bound}} = \frac{\mu_0 nIb}{4\pi} \int_{-\infty}^{\infty} \frac{dx}{x^2 + r^2}. \quad (12.308)$$

The integral here equals $(1/r)\tan^{-1}(x/r)\big|_{-\infty}^{\infty} = \pi/r$. So in the $r \to \infty$ limit, our upper bound on $B$ is zero; $B$ must therefore be zero at $r = \infty$, as we wanted to show. Due to the overestimates we made above, the field actually goes to zero much faster than $1/r$, but

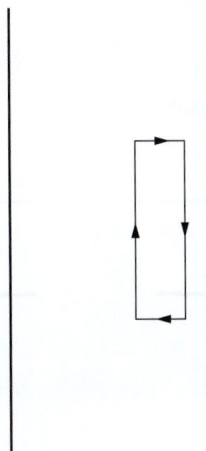

**Figure 12.94.**

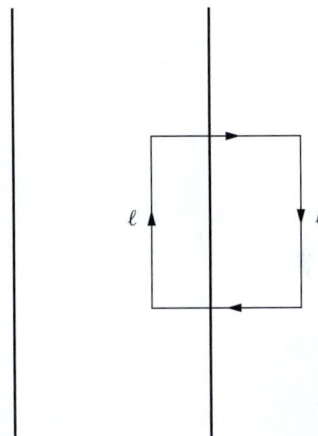

**Figure 12.95.**

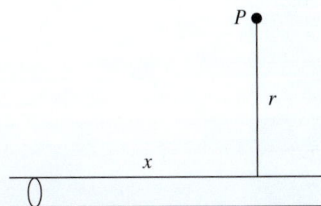

**Figure 12.96.**

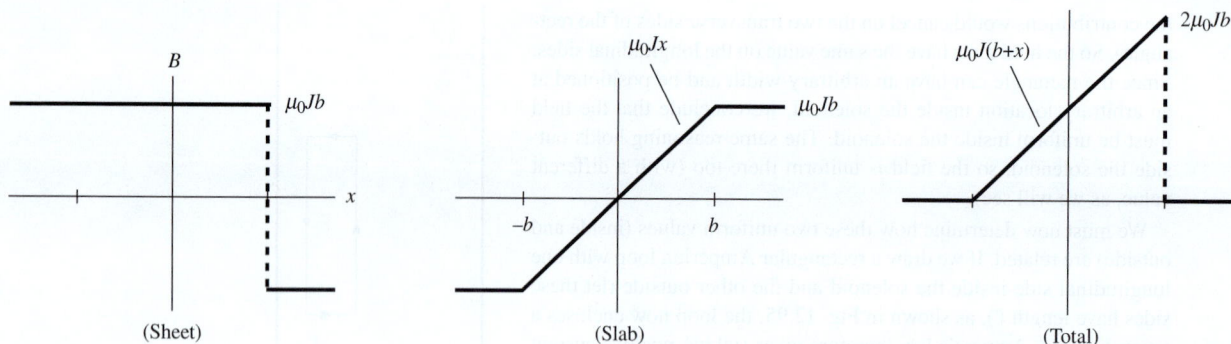

**Figure 12.97.**

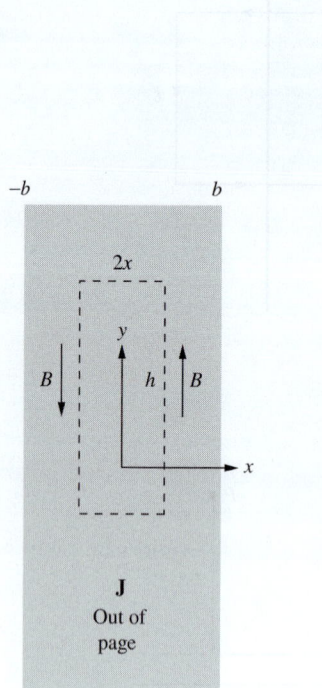

**Figure 12.98.**

our coarse estimates were good enough to get the job done. Another method is to use the fact that the field due to a ring behaves like $1/d^3$ at large distances $d$. In Eq. (6.53), we showed this for points on the axis. You can think about how to treat a general point in space; we will discuss this in Chapter 11.

This problem can also be solved by skipping part (c), and instead combining the results of parts (a) and (b) with the result in Eq. (6.57) for the field on the axis. The important point to note is that we need to calculate the actual value of $B$ at at least one point (because part (b) involves only differences in $B$ values), and the easiest options are a point on the axis or a point at infinity.

### 6.20   A slab and a sheet

(a)  The total magnetic field equals the field due to the thin sheet plus the field due to the thick slab. The field due to the thin sheet is simply $\mu_0 \mathcal{J}/2 = \mu_0(2bJ)/2 = \mu_0 Jb$. (This can be found via an Amperian loop with a side on either side of the sheet.) It points upward on the left, and downward on the right; see the step function shown in Fig. 12.97. (The direction can be found by imagining the sheet to be built up from a series of parallel wires.)

To find the magnetic field due to the thick slab, consider an Amperian loop centered in the slab, as shown in Fig. 12.98. The slab is symmetric under translations in the $y$ direction, so the field must be independent of $y$. Also, the slab is symmetric under rotations by 180° around the $z$ axis, so the $y$ component of the field must be an odd function of $x$, otherwise the field wouldn't look the same after a rotation by 180°. (Additionally, you can rule out $x$ and $z$ components by considering the slab to be built up from wires.)

The current enclosed in the Amperian loop is $I = h(2x)J$. Since only the left and right sides contribute to the line integral, we have

$$\int \mathbf{B} \cdot d\mathbf{s} = \mu_0 I \implies 2Bh = \mu_0(2xhJ) \implies B = \mu_0 Jx.$$

$$(12.309)$$

Outside the slab, the slab looks like a sheet (from the same Amperian argument that is used for an actual sheet). So on either side, the field has a constant value equal to the value at the boundary, namely $\pm\mu_0 Jb$. The slab's field is shown in Fig. 12.97. The total field, which is the sum of the sheet's field and the slab's field, is also shown. It equals zero outside the slab, and $\mu_0 J(b + x)\hat{\mathbf{y}}$ inside.

Alternatively, the interior field of the slab can be found by considering the two sub-slabs on either side of a given position. At position $x$ there is a slab with thickness $b + x$ on the left which is equivalent to a sheet with surface charge density $\mathcal{J}_{\text{left}} = J(b + x)$. And likewise there is a slab with thickness $b - x$ on the right which is equivalent to a sheet with surface charge density $\mathcal{J}_{\text{right}} = J(b - x)$. The left "sheet" produces a field $\mu_0\mathcal{J}/2 = \mu_0 J(b + x)/2$ upward, and the right "sheet" produces a field $\mu_0\mathcal{J}/2 = \mu_0 J(b - x)/2$ downward. The net interior field of the slab is therefore $\mu_0 J x$ upward (so if $x$ is negative, this points downward).

(b) Inside the slab, the curl of $\mathbf{B}$ is

$$\nabla \times \mathbf{B} = \begin{pmatrix} \hat{\mathbf{x}} & \hat{\mathbf{y}} & \hat{\mathbf{z}} \\ \partial/\partial x & \partial/\partial y & \partial/\partial z \\ 0 & \mu_0 J(b + x) & 0 \end{pmatrix} = \mu_0 J\hat{\mathbf{z}} = \mu_0\mathbf{J},$$

(12.310)

as desired. Outside the slab, $\mathbf{B}$ and $\mathbf{J}$ are both zero, so $\nabla \times \mathbf{B} = \mu_0\mathbf{J}$ is trivially true. At the boundary at $x = b$, the $B_y$ component is discontinuous, so the $\partial B_y/\partial x$ derivative in the curl is infinite. This is consistent with the fact that a nonzero $\mathcal{J}$ implies an infinite $J$.

6.21 *Maximum field in a cyclotron*

We must determine the electric field in the frame of the moving ion. In the lab frame there is no electric field, just a magnetic field $\mathbf{B}_\perp$. (The longitudinal electric field that increases the speed of the ions is relatively small.) So the Lorentz transformation in Eq. (6.76) tells us that the electric field in the frame of the moving ion is $\mathbf{E}_\perp = \gamma\mathbf{v} \times \mathbf{B}_\perp$. Since the kinetic energy equals the rest energy, the total energy equals $2mc^2$, which means that $\gamma = 2$. Hence $\beta = \sqrt{3}/2$. The condition $E_\perp < 4.5 \cdot 10^8$ V/m therefore gives

$$\gamma vB < 4.5 \cdot 10^8 \text{ V/m} \implies B < \frac{4.5 \cdot 10^8 \text{ V/m}}{(2)(\sqrt{3}/2 \cdot 3 \cdot 10^8 \text{ m/s})} = 0.866 \text{ T},$$

(12.311)

or 8660 gauss.

Alternatively, we can think in terms of forces. The force in the lab frame has magnitude $qvB$ and points in the transverse direction. Since the transverse force is always larger by a factor $\gamma$ in the frame of the particle, the force in the ion's frame is $\gamma qvB$. This must be an electric force because the ion isn't moving in its frame. Therefore, $qE = \gamma qvB \implies E = \gamma vB$, as above.

6.22 *Zero force in any frame*

Let the given frame be $F$, and consider a frame $F'$ moving with velocity $\mathbf{v}$ with respect to $F$. The only nonzero field in frame $F$ is $\mathbf{B}_\perp$, so Eq. (6.76) gives the fields in frame $F'$ as

$$\mathbf{E}'_\perp = \gamma\mathbf{v} \times \mathbf{B}_\perp \qquad \text{and} \qquad \mathbf{B}'_\perp = \gamma\mathbf{B}_\perp,$$

(12.312)

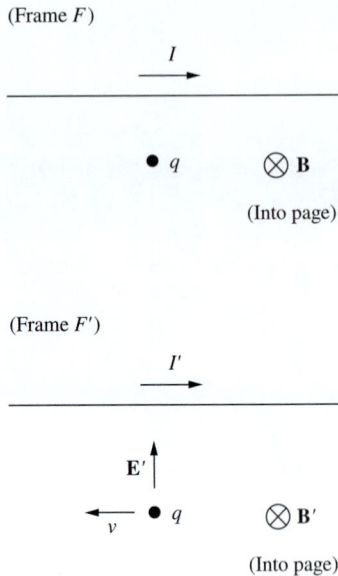

(Frame $F$)

$I$

$\bullet\ q$            $\otimes$ **B**

(Into page)

(Frame $F'$)

$I'$

**E**$'$ $\uparrow$

$\underset{v}{\longleftarrow}$ $\bullet\ q$            $\otimes$ **B**$'$

(Into page)

**Figure 12.99.**

where $\mathbf{B}_\perp = \mu_0 I/2\pi r$, although we won't need to use this explicit form. In Eq. (6.76) **v** is the velocity of $F'$ with respect to $F$, so if **v** points to the right, the situations in the two frames are shown in Fig. 12.99, with the charge $q$ moving to the left in $F'$. (If $v$ is large enough, then $I'$ may be negative, but that isn't important.) In frame $F$, $\mathbf{B}_\perp$ points into the page at the location of the charge. So in frame $F'$, $\mathbf{B}'_\perp = \gamma \mathbf{B}_\perp$ also points into the page, and $\mathbf{E}'_\perp = \gamma \mathbf{v} \times \mathbf{B}_\perp$ points toward the wire.

In frame $F'$, the charge is moving with velocity $-\mathbf{v}$, so the magnetic force is $q(-\mathbf{v}) \times \mathbf{B}'_\perp = q(-\mathbf{v}) \times (\gamma \mathbf{B}_\perp)$. And the electric force is $q\mathbf{E}'_\perp = q(\gamma \mathbf{v} \times \mathbf{B}_\perp)$. These two forces are negatives of each other, so the total force is zero, as desired. This reasoning is actually valid for any **v** in the plane of the page. You can also quickly check that the case with **v** perpendicular to the page works out.

6.23 *No magnetic shield*

The analysis in Section 5.9 showed that a test charge moving parallel to a wire containing current experiences a force which, *as observed in the rest frame of the test charge,* is due to an electric field. To understand why the introduction of a conductor, such as a metal plate, between the wire and the test charge has no effect, let us view the situation from the rest frame of the test charge. In that frame, the conducting plate, which is stationary in the lab frame, is moving. It is moving through a magnetic field *and* an electric field, and these are related precisely so as to make the total force on any charge in the plate zero (see Problem 6.22). Hence there is no redistribution whatsoever of electrons in the plate.

Problem 6.22 works things out quantitatively, but to understand the setup qualitatively, let the line $L$ on the plate be defined as the projection of the wire onto the plate. Consider the case where the test charge is moving in, say, the same direction as the current. Then you should verify that in the frame of the test charge, the electric force from the wire tends to attract the plate's charge to $L$, and the magnetic force from the wire tends to repel the plate's charge from $L$. These effects turn out to cancel exactly.

On the other hand, if we caused the plate to move along with the same velocity as the test charge, it *would* make a difference. An observer in the frame of the test charge would say that we have introduced a stationary plate into an electrostatic field (and a magnetic field, but that doesn't affect the stationary plate), with a consequent redistribution of charge on the plate and a resulting alteration of the total electric field. An observer in the lab frame, where there is no electric field from the wire, would say that the electrons in the moving plate have redistributed themselves under the influence of the $q\mathbf{v} \times \mathbf{B}$ force, and the new charge distribution itself produces an electric field. You should verify that these two effects act in the same direction. That is, they both attract charge to the line $L$, or they both repel it, depending on the relative sign of the current and the test charge's velocity.

6.24 *E and B for a point charge*

(a) In the frame of the charge, the electric field is given by Coulomb's law (exactly), and there is no magnetic field. If the charge moves with velocity **v** with respect to the lab frame, then the lab frame moves with velocity $-\mathbf{v}$ with respect to the charge frame. So the exact same reasoning that led to Eq. (6.81) now gives

$$\mathbf{B}_{\text{lab}} = -\left(\frac{-\mathbf{v}}{c^2}\right) \times \mathbf{E}_{\text{lab}}. \qquad (12.313)$$

(The rule for the signs here is that the velocity in the parentheses is the velocity of the frame whose fields appear in the equation, with respect to the frame in which $\mathbf{B}$ is zero.) Canceling the minus signs and dropping the "lab" subscript yields the desired result.

(b) The exact shape of the essentially point-like charge can't matter, so let's assume that it takes the shape of a tiny rod. If the rod has a small length $l$, then the Biot–Savart law gives the magnetic field due to the charge as $\mathbf{B} = (\mu_0/4\pi)Il \times \hat{\mathbf{r}}/r^2$. If the rod has linear charge density $\lambda$, then the current is $I = \lambda v$. The vector $Il$ therefore has length $\lambda v l$; and the direction is along $\mathbf{v}$, so the complete vector is simply $\lambda \mathbf{v} l$. But $\lambda l$ is the charge $q$, so the $Il$ vector equals $q\mathbf{v}$. The magnetic field due to the charge is therefore (using $\mu_0 = 1/\epsilon_0 c^2$)

$$\mathbf{B} = \frac{\mu_0}{4\pi} \frac{Il \times \hat{\mathbf{r}}}{r^2} = \frac{1}{4\pi \epsilon_0 c^2} \frac{q\mathbf{v} \times \hat{\mathbf{r}}}{r^2} = \frac{\mathbf{v}}{c^2} \times \frac{q\hat{\mathbf{r}}}{4\pi \epsilon_0 r^2} = \frac{\mathbf{v}}{c^2} \times \mathbf{E},$$

$$(12.314)$$

as desired.

## 6.25 *Force in three frames*

(a) Let's call the three frames (of the lab, charge $q$, and charges in the wire) $L$, $Q$, and $W$, respectively. In the given frame $L$ (see Fig. 12.100), the repulsive electric force is simply $F_E = qE = q\lambda/2\pi \epsilon_0 r$. Since $I = \lambda u$, the magnetic field is $B = \mu_0 I/2\pi r = (\lambda u)/2\pi r(\epsilon_0 c^2)$, where we have used $\mu_0 = 1/\epsilon_0 c^2$. This field points out of the page, so by the right-hand rule the magnetic force $qvB$ is also repulsive. The total force on the charge $q$ in the lab frame $L$ is therefore

$$F = F_E + F_B = \frac{q\lambda}{2\pi \epsilon_0 r}\left(1 + \frac{uv}{c^2}\right) \equiv \frac{q\lambda}{2\pi \epsilon_0 r}(1 + \beta_u \beta_v). \quad (12.315)$$

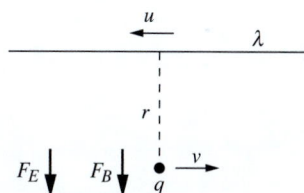

**Figure 12.100.**

(b) We'll solve this by calculating the charge density in $Q$. You should check that the Lorentz transformations give the same result. In frame $Q$ (see Fig. 12.101), the charge $q$ is at rest and the charges in the wire move to the left with speed $(u + v)/(1 + uv/c^2)$, which we label as $u \oplus v$. There is a magnetic field in this frame, but it doesn't come into play because $q$ is stationary. So we need only worry about the electric force.

To find the charge density on the wire, we can compare the densities in frames $L$ and $Q$ via frame $W$. (Remember that length contraction is applicable only if one of the frames involved is the rest frame of the object.) Due to length contraction, the density in $L$ is larger than the density in $W$ by the factor $\gamma_u$. And likewise the density in $Q$ is larger than the density in $W$ by the factor $\gamma_{u \oplus v} = \gamma_u \gamma_v (1 + \beta_u \beta_v)$. Therefore, the density $\lambda_Q$ in $Q$ is larger than the density $\lambda$ in $L$ by the

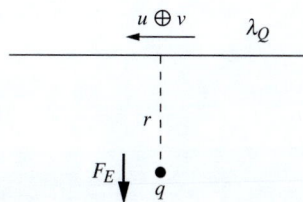

**Figure 12.101.**

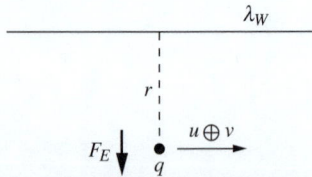

**Figure 12.102.**

factor $\gamma_v(1 + \beta_u\beta_v)$. That is, $\lambda_Q = \lambda\gamma_v(1 + \beta_u\beta_v)$. So the repulsive (electric) force on the charge $q$ in its own frame $Q$ is

$$F = F_E = \frac{q\lambda_Q}{2\pi\epsilon_0 r} = \frac{q\lambda}{2\pi\epsilon_0 r}\gamma_v(1 + \beta_u\beta_v). \qquad (12.316)$$

(c) In frame $W$ (see Fig. 12.102), the wire's charges are at rest and the charge $q$ moves to the right with speed $u \oplus v$. But this speed doesn't matter because there is no magnetic field in this frame. So we care only about the electric force. The charge density $\lambda_W$ in $W$ is smaller than the density $\lambda$ in frame $L$ by the factor $\gamma_u$. That is, $\lambda_W = \lambda/\gamma_u$. So the repulsive (electric) force on the charge $q$ in the frame $W$ of the charges on the wire is

$$F = F_E = \frac{q\lambda_W}{2\pi\epsilon_0 r} = \frac{q\lambda}{2\pi\epsilon_0 r}\frac{1}{\gamma_u}. \qquad (12.317)$$

Now let's check that these three forces relate properly. The force on a particle is largest in its rest frame. It is smaller in any other frame by the relative $\gamma$ factor. And indeed, the force in frame $Q$ in part (b) is the largest of the three forces. It is larger than the force in frame $L$ in part (a) by the factor $\gamma_v$, which is correct because $v$ is the speed of $L$ with respect to $Q$. And the force in frame $Q$ is larger than the force in frame $W$ in part (c) by the factor $\gamma_u\gamma_v(1 + \beta_u\beta_v) = \gamma_{u\oplus v}$, which is correct because $u \oplus v$ is the speed of $W$ with respect to $Q$.

**6.26** *Motion in E and B fields*

Newton's third law says that $d\mathbf{p}/dt = \mathbf{F}_B + \mathbf{F}_E$, which yields (with $p = mv$ since the velocity is nonrelativistic)

$$\frac{d(m\mathbf{v})}{dt} = q\mathbf{v} \times \mathbf{B} + q\mathbf{E} \implies \frac{d\mathbf{v}}{dt} = \frac{q}{m}\mathbf{v} \times \mathbf{B} + \frac{q}{m}\mathbf{E}. \qquad (12.318)$$

With $\mathbf{v} = (v_x, v_y, 0)$ and $\mathbf{B} = (0, 0, B)$, we have $\mathbf{v} \times \mathbf{B} = B(v_y, -v_x, 0)$. So with $\mathbf{E} = (0, E, 0)$, the $x$ and $y$ components of Eq. (12.318) can be written as

$$\frac{dv_x}{dt} = \frac{qB}{m}v_y \quad \text{and} \quad \frac{dv_y}{dt} = -\frac{qB}{m}v_x + \frac{qE}{m}. \qquad (12.319)$$

Taking the derivative of the second of these equations (which gets rid of the $E$ term), and then substituting in the value of $dv_x/dt$ from the first, gives

$$\frac{d^2v_y}{dt^2} = -\left(\frac{qB}{m}\right)^2 v_y. \qquad (12.320)$$

This is a simple-harmonic-oscillator type equation, for which the general solution takes the form

$$v_y(t) = v_0\cos(\omega t + \phi), \quad \text{where} \quad \omega = \frac{qB}{m}. \qquad (12.321)$$

The second of the equations in Eq. (12.319) then quickly gives $v_x(t) = v_0 \sin(\omega t + \phi) + E/B$. The quantities $v_0$ and $\phi$ are arbitrary constants, determined by the initial conditions, although we will see in a moment that $v_0$ has a simple interpretation.

Integrating $v_x(t)$ and $v_y(t)$ to find $x$ and $y$ gives (up to arbitrary additive constants)

$$\big(x(t), y(t)\big) = \frac{v_0}{\omega}\Big(-\cos(\omega t + \phi), \sin(\omega t + \phi), 0\Big) + \Big(\frac{Et}{B}, 0, 0\Big).$$

(12.322)

This says that, relative to the point $(Et/B, 0, 0)$, the particle moves in a circle with radius $v_0/\omega$. From the above expressions for $v_x$ and $v_y$, we see that $v_0$ is the speed of the circularly moving particle in the frame moving along with the point $(Et/B, 0, 0)$. So in terms of the momentum $p = mv$ in this frame (which is constant, unlike in the lab frame), the radius of the circle can be written as $r = v_0/\omega = (p/m)/(qB/m) = p/qB$. Back in the lab frame, we have the type of path shown in Fig. 12.103. It is a circle that drifts in the $x$ direction with speed $E/B$.

We can perform a double check on this result. Given that the particle simply moves in a circle in the frame $F'$ moving with velocity $(E/B)\hat{\mathbf{x}}$ with respect to the lab frame $F$, it must be the case that there is no electric field in $F'$. And indeed, the transformation for $\mathbf{E}'_\perp$ in Eq. (6.76) gives

$$\mathbf{E}'_\perp = \gamma(\mathbf{E}_\perp + \mathbf{v} \times \mathbf{B}_\perp) = \gamma\Big(E\hat{\mathbf{y}} + (E/B)\hat{\mathbf{x}} \times B\hat{\mathbf{z}}\Big)$$
$$= \gamma(E\hat{\mathbf{y}} - E\hat{\mathbf{y}}) = 0. \qquad (12.323)$$

Note that the drift is in the $x$ direction, even though the electric field points in the $y$ direction. This is somewhat counterintuitive, because you might think that the particle should generally head in the direction of the electric field. What happens is that as the particle speeds up in the direction of the electric field (the $y$ direction), the magnetic force $qvB$ becomes larger. At a point such as $P$ in Fig. 12.103, the large magnetic force has a component in the negative $y$ direction, and it happens to win out over the electric force and cause the particle to slow down in the $y$ direction. Eventually $v_y$ becomes zero at the top of the arc, and the particle reverses its $y$ motion and heads downward.

As exercises, you can find the net force at the top and bottom points, and also the $y$ component of the force when the particle crosses the $x$ axis. Your results should be consistent with the circular motion in the frame $F'$. (Remember that the motion is assumed to be nonrelativistic.) You can also find the radii of curvature at the top and bottom points.

From the above $v_0 \sin(\omega t + \phi) + E/B$ expression for $v_x$, we see that if the electric field is weak (more precisely, if $E/B < v_0$), then there are times when $v_x$ is negative; the path looks like the one shown in Fig. 12.103. If, on the other hand, the electric field is strong (more precisely, if $E/B > v_0$), then $v_x$ is always positive; the path looks like the one shown in Fig. 12.104.

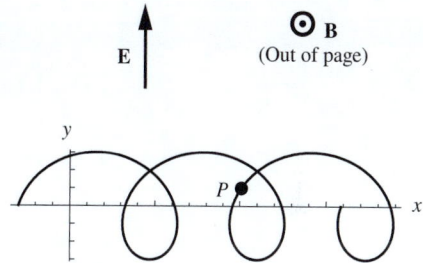

**Figure 12.103.**

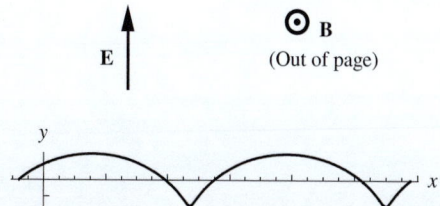

**Figure 12.104.**

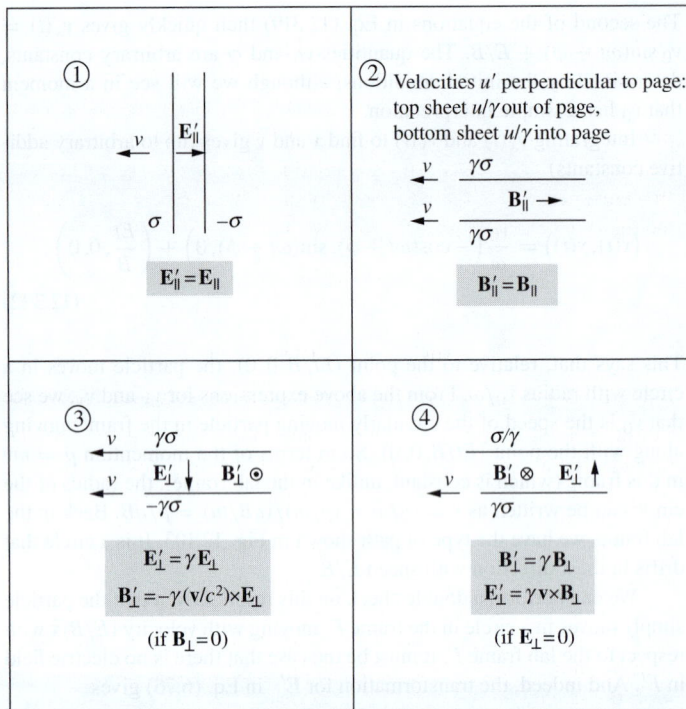

**Figure 12.105.**
Setups in frame $F'$.

6.27   *Special cases of Lorentz transformations*
Figure 12.105 shows the four setups as viewed in frame $F'$.

**Setup 1**   There is no length contraction in the longitudinal direction, so the charge densities $\sigma$ and $\sigma'$ in the two frames are equal. Since the electric fields are given by $E_\parallel = \sigma/\epsilon_0$ and $E_\parallel' = \sigma'/\epsilon_0$, the longitudinal fields are therefore equal: $E_\parallel' = E_\parallel$.

**Setup 2**   In frame $F$, the magnetic field is $B_\parallel = \mu_0 \mathcal{J}$, with $\mathcal{J} = \sigma u$, where $u$ is the speed of the sheets into and out of the page in frame $F$. (As in Section 6.6 this can be shown by drawing an Amperian rectangular loop with sides on either side of the sheet.) In frame $F'$, the longitudinal magnetic field is $B_\parallel' = \mu_0 \mathcal{J}'$, where $\mathcal{J}'$ is the current density perpendicular to the page. This current density equals $\sigma' u'$, where $u'$ is the speed perpendicular to the page. But $\sigma' u' = \sigma u \implies \mathcal{J}' = \mathcal{J}$ because the charge density is larger due to length contraction ($\sigma' = \gamma\sigma$), while the transverse speed is smaller due to the transverse velocity addition formula ($u' = u/\gamma$; see the discussion following Eq. (G.10)). Therefore $B_\parallel' = B_\parallel$.

In the case where $u$ is small, it is easy to see that the transverse speed should be smaller in $F'$, due to time dilation (this is the argument we made with the solenoid in Section 6.7). Note that each sheet also produces a magnetic field perpendicular to the page, due to the longitudinal speed $v$. But these fields cancel in the region between the sheets, so we can ignore them. In any event, they wouldn't change the above $B_\parallel' = B_\parallel$ result.

**Setup 3** First relation: due to length contraction, we have $\sigma' = \gamma\sigma$, so the field $E'_\perp = \sigma'/\epsilon_0$ is $\gamma$ times the field $E_\perp = \sigma/\epsilon_0$.

Second relation: since both sheets move to the left with speed $v$ in frame $F'$, and since they have opposite charge, the magnetic field has magnitude $B'_\perp = \mu_0 \mathcal{J}'$, where $\mathcal{J}' = \sigma'v$. It points out of the page. The direction of the $\mathbf{B}'_\perp = -\gamma(\mathbf{v}/c^2) \times \mathbf{E}_\perp$ relation therefore works out correctly (remember that $\mathbf{v}$, which is the velocity of $F'$ with respect to $F$, points to the right). Due to length contraction, we have $\sigma' = \gamma\sigma$, so

$$B'_\perp = \mu_0 \mathcal{J}' = \mu_0(\sigma' v) = \mu_0(\gamma\sigma)v$$

$$= \frac{1}{\epsilon_0 c^2}\gamma\sigma v = \gamma\frac{v}{c^2}\frac{\sigma}{\epsilon_0} = \gamma\frac{v}{c^2}E_\perp. \tag{12.324}$$

Hence the magnitude of the relation works out correctly too.

**Setup 4** First relation: in frame $F$, only the top sheet is moving. The magnetic field is directed into the page with magnitude $B_\perp = \mu_0 \mathcal{J}/2$, where $\mathcal{J} = \sigma v$. In frame $F'$, only the bottom sheet is moving (to the left). The magnetic field is again directed into the page, but now has magnitude $B'_\perp = \mu_0 \mathcal{J}'/2$, where $\mathcal{J}' = \sigma'v$. Due to length contraction, the charge density on the moving bottom sheet is $\sigma' = \gamma\sigma$. Hence $B'_\perp = \gamma B_\perp$.

Second relation: in frame $F'$, the top and bottoms sheets have densities $\sigma/\gamma$ and $\gamma\sigma$, respectively. The electric field therefore points upward with magnitude $E'_\perp = \sigma(\gamma - 1/\gamma)/2\epsilon_0$. The direction of the $\mathbf{E}'_\perp = \gamma\mathbf{v} \times \mathbf{B}_\perp$ relation therefore works out correctly (remember that $\mathbf{v}$ points to the right). And we have

$$E'_\perp = \frac{\sigma(\gamma - 1/\gamma)}{2\epsilon_0} = \frac{\gamma\sigma(1 - 1/\gamma^2)}{2\epsilon_0} = \frac{\gamma\sigma\beta^2}{2\epsilon_0}$$

$$= \frac{\gamma\sigma v^2}{2\epsilon_0 c^2} = \gamma v\frac{1}{\epsilon_0 c^2}\frac{\sigma v}{2} = \gamma v\frac{\mu_0 \mathcal{J}}{2} = \gamma v B_\perp, \tag{12.325}$$

so the magnitude of the relation works out correctly too.

6.28 *The retarded potential*

(a) When the charge crosses the $y$ axis, as shown in Fig. 12.106, the electric field at the origin in the charge's frame equals $\mathbf{E}_{charge} = -\hat{\mathbf{y}}q/4\pi\epsilon_0 r^2$. The lab frame moves with velocity $\mathbf{v} = -v\hat{\mathbf{x}}$ with respect to the charge, so the relevant Lorentz transformation in Eq. (6.76), namely $\mathbf{B}_{\perp, lab} = -\gamma(\mathbf{v}/c^2) \times \mathbf{E}_{\perp, charge}$, gives

$$\mathbf{B}_{\perp, lab} = -\gamma\frac{(-v\hat{\mathbf{x}})}{c^2} \times \left(\frac{-\hat{\mathbf{y}}q}{4\pi\epsilon_0 r^2}\right)$$

$$= -\hat{\mathbf{z}}\frac{\gamma q v}{4\pi\epsilon_0 c^2 r^2} = -\hat{\mathbf{z}}\frac{\mu_0}{4\pi}\frac{\gamma q v}{r^2}, \tag{12.326}$$

where we have used $\mu_0 = 1/\epsilon_0 c^2$. The $-\hat{\mathbf{z}}$ direction points into the page. There is no longitudinal field, so this is the entire $\mathbf{B}_{lab}$ field.

(b) Let the little segment of current have length $\ell$. Assuming $\ell$ is very small, the $\hat{\mathbf{r}}$ and $r$ in the Biot–Savart law in Eq. (6.49) are essentially constant, so they can be pulled outside the integral. The $dl$

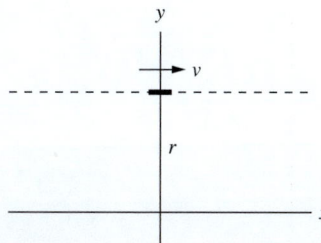

**Figure 12.106.**

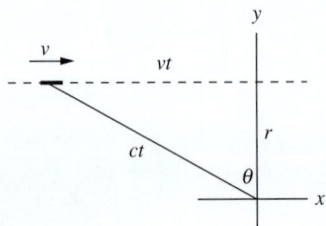

**Figure 12.107.**

integral then simply gives the $l$ for the stick, which equals $\ell \hat{\mathbf{x}}$. We therefore have

$$\mathbf{B}_{\text{lab}} = \frac{\mu_0 I}{4\pi} \frac{(\ell\hat{\mathbf{x}}) \times \hat{\mathbf{r}}}{r^2} = \frac{\mu_0 (\lambda v)}{4\pi} \frac{\ell\hat{\mathbf{x}} \times (-\hat{\mathbf{y}})}{r^2}$$

$$= -\frac{\mu_0}{4\pi} \frac{(\lambda\ell)v\hat{\mathbf{z}}}{r^2} = -\hat{\mathbf{z}} \frac{\mu_0}{4\pi} \frac{qv}{r^2}, \quad (12.327)$$

where we have used the fact that $\lambda\ell$ equals the charge $q$ on the stick. As promised, this result is a factor of $\gamma$ smaller than the result in part (a).

(c) Consider the position of the charge shown in Fig. 12.107. If we want a photon to travel from the charge to the origin in the same time $t$ that it takes the charge to reach the $y$ axis, then the two long legs of the right triangle shown must have lengths $ct$ and $vt$. The Pythagorean theorem then gives $(ct)^2 = (vt)^2 + r^2 \implies t = r/\sqrt{c^2 - v^2}$. The charge is therefore a distance $vt = rv/\sqrt{c^2 - v^2}$ from the $y$ axis. This is small if $v$ is small, and it goes to infinity as $v \to c$.

If $\ell$ is the proper length of the little stick representing the charge, then the length in the lab frame is $\ell/\gamma$. However, this won't be the length in the photograph mentioned in the statement of the problem, because if photons are released simultaneously (in the lab frame) from different points on the stick, they won't reach the origin at the same time. The photon from the front end will arrive first, because it starts closer to the origin. So in the photograph, the photon from the front end must have been emitted later (because the photograph simply records information on the set of photons that hit the camera simultaneously). Equivalently, the photon from the back end must have had a head start. How much of a head start?

In Fig. 12.108, the distances from points $B$ and $C$ to the origin are essentially equal (assuming the stick is very short). So if the back photon is emitted at the instant shown, we want the front photon to be emitted at some later time $\tau$ (when the front of the stick is at $C$) such that the distance $AB$ is $c\tau$. The back photon will then be at $B$ at the same time the front photon is emitted from the stick at $C$, which means that the two photons will arrive at the origin at the same time. Triangle $ABC$ is similar to the large right triangle in the figure, so we have

$$\frac{\ell/\gamma + v\tau}{c\tau} = \frac{ct}{vt} \implies \tau = \frac{\ell v}{\gamma(c^2 - v^2)} = \frac{\gamma \ell v}{c^2}. \quad (12.328)$$

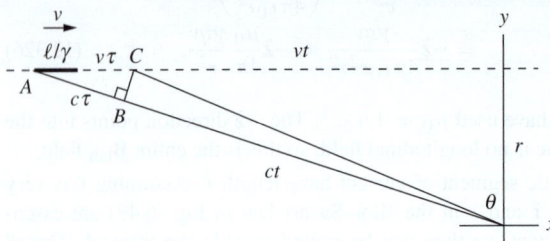

**Figure 12.108.**

The length of the stick in the photograph, which is $AC$, is therefore

$$\frac{\ell}{\gamma} + v\tau = \frac{\ell}{\gamma} + v\frac{\gamma\ell v}{c^2} = \gamma\ell\left(\frac{1}{\gamma^2} + \frac{v^2}{c^2}\right) = \gamma\ell. \qquad (12.329)$$

So the length of the stick in the photograph is $\gamma^2$ times the actual length $\ell/\gamma$ in the lab frame.

Now, the current produced by the stick is the linear density times the speed. If the density is $\lambda$ in the stick's frame, then it is $\gamma\lambda$ in the lab frame, due to the length contraction of the stick. The current is therefore $I = (\gamma\lambda)v$. In the expression for the vector potential $\mathbf{A}$ in Eq. (6.46), we need to multiply this current by its length in the photograph, which we just found to be $\gamma\ell$. We also need to divide by the distance $r_{12}$ in Eq. (6.46), which is $r/\cos\theta$ here. Since $\sin\theta = v/c$, we have $\cos\theta = \sqrt{1 - v^2/c^2} = 1/\gamma$. So the distance $r_{12}$ equals $\gamma r$. Therefore, the vector potential at the origin at the moment the charge crosses the $y$ axis is (using $\lambda\ell = q$)

$$\mathbf{A} = \frac{\mu_0 I}{4\pi}\int \frac{dl}{r_{12}} = \frac{\mu_0(\gamma\lambda v)}{4\pi}\frac{\gamma\ell\hat{\mathbf{x}}}{\gamma r} = \frac{\mu_0}{4\pi}\frac{\gamma qv}{r}\hat{\mathbf{x}}. \qquad (12.330)$$

We can now take the curl of $\mathbf{A}$ to obtain $\mathbf{B}$. Equation (F.2) in Appendix F gives the expression for the curl in cylindrical coordinates (the $x$ direction here is the axial direction). The only nonzero derivative in the lengthy expression is $\partial A_x/\partial r$, so we have

$$\mathbf{B} = \nabla \times \mathbf{A} = -\frac{\partial A_x}{\partial r}\hat{\boldsymbol{\theta}} = -\frac{\partial}{\partial r}\left(\frac{\mu_0}{4\pi}\frac{\gamma qv}{r}\right)\hat{\boldsymbol{\theta}} = \frac{\mu_0}{4\pi}\frac{\gamma qv}{r^2}\hat{\boldsymbol{\theta}},$$
$$(12.331)$$

where $\hat{\boldsymbol{\theta}}$ points into the page. This agrees with Eq. (12.326).

Note that this result is independent of the length $\ell$ of the little stick, as long as it is small. The charge can therefore take any shape, and the result will still be valid. For example, a sphere can be considered to be a collection of adjacent sticks with various lengths, all of which yield the same extra factor of $\gamma$. No matter how small and point-like we make the charge, the lengthening effect from the retarded time always exists. The vector potential in Eq. (12.330) is a special case of a more general result for moving charges, known as the *Liénard–Wiechert* vector potential. There is also a corresponding Liénard–Wiechert scalar potential.

## 12.7 Chapter 7

7.1    *Current in a bottle*

The magnitude of $\mathbf{v} \times \mathbf{B}$ is $vB = (1 \text{ m/s})(3.5 \cdot 10^{-5}\,\text{T}) = 3.5 \cdot 10^{-5}\,\text{V/m}$. So the effective electric field is $E = 3.5 \cdot 10^{-5}\,\text{V/m}$, and the current density is

$$J = \sigma E = \left(4\,(\text{ohm-m})^{-1}\right)(3.5 \cdot 10^{-5}\,\text{V/m}) = 1.4 \cdot 10^{-4}\,\text{A/m}^2.$$
$$(12.332)$$

If a bottle of seawater were carried at this speed, a current would flow only long enough to separate enough charge to establish an electric field equal and opposite to $\mathbf{v} \times \mathbf{B}$. To find roughly how long this takes, consider the charge that would pile up per m$^2$ of surface in time $t$. From the definition of $J$ (charge per area per time), this charge density is simply $Jt$. It is positive on one side, negative on the other. (This result isn't exact, because $J$ decreases continuously to zero as the charge piles up, rather than dropping suddenly to zero from a constant value. But we're just trying to make a rough estimate.) The resulting field depends somewhat on the shape of the bottle, but in order of magnitude it equals $Jt/\epsilon_0 = (\sigma E)t/\epsilon_0$. (This "$\sigma$" is conductivity, not surface charge density.) The charge stops flowing when this field equals $E$, that is, when

$$\frac{\sigma t}{\epsilon_0} = 1 \implies t = \frac{\epsilon_0}{\sigma} = \frac{8.85 \cdot 10^{-12} \, \dfrac{\text{s}^2 \, \text{C}^2}{\text{kg} \, \text{m}^3}}{4 \, (\text{ohm-m})^{-1}} = 2.2 \cdot 10^{-12} \, \text{s}.$$

$$(12.333)$$

So, except for a completely negligible time at the start, there would be no current flowing in the bottle.

7.2  *What's doing work?*

Figure 12.109 shows the situation. For simplicity, we assume that the mobile charges are positive; this doesn't affect the result. The important point to realize is that there are *two* components to a given charge's velocity $\mathbf{u}$, namely the horizontal component $u_x = v$ due to the motion of the rod, and the vertical component $u_y$ due to the current along the rod. This means that the magnetic force $\mathbf{F}_B$ points up and to the left, perpendicular to $\mathbf{u}$, as shown. Its magnitude is $F_B = quB$, and its two components have magnitudes $F_{B,x} = qu_yB$ and $F_{B,y} = qu_xB = qvB$. The latter of these is what we called $\mathbf{f}$ in Eq. (7.5). Assuming that the current is steady and the charge isn't accelerating, the total force on it equals zero. So if you are applying the force to the rod, then your force is given by $F_{\text{you}} = F_{B,x}$, and the resistive force on the charges is given by $F_R = F_{B,y}$. (All of these quantities are magnitudes, so they are defined to be positive.)

Which forces do work? As mentioned in the problem, the magnetic force does no work because $\mathbf{F}_B$ is perpendicular to $\mathbf{u}$. But if you wish, you can break this zero work into two equal and opposite pieces. The vertical component of $\mathbf{F}_B$ does work at a rate $F_{B,y}u_y = (qu_xB)u_y$. And the horizontal component does work at a rate $-F_{B,x}u_x = -(qu_yB)u_x$. These two rates are equal and opposite, as they must be. You also do work, because there is a component of $\mathbf{u}$ in the direction in which you are pulling. The rate at which you do work is $F_{\text{you}}u_x$. And due to the balancing of all the forces, this positive rate is equal and opposite to the negative rate at which $F_{B,x}$ does work. The resistive force also does work, and the rate is $-F_Ru_y$. This negative rate is equal and opposite to the positive rate at which $F_{B,y}$ does work.

We see that the magnetic force does zero net work, while the positive work you do is canceled by the negative work the resistive force does. While it is true that that a *component* of $\mathbf{F}_B$ does positive work (the vertical component, which we called $\mathbf{f}$ in Section 7.3), the other component of $\mathbf{F}_B$

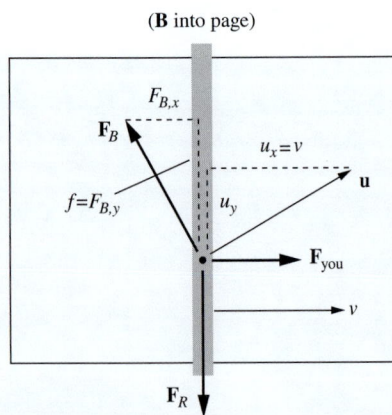

(**B** into page)

**Figure 12.109.**

does an equal and opposite amount of negative work. So it would hardly be accurate to say that *the* magnetic force does work.

This setup is essentially the same as the setup in which you push a block up a frictionless inclined plane at constant speed **u**, by applying a *horizontal* force, as shown in Fig. 12.110. This figure is simply Fig. 12.109 with the forces relabeled. The normal force replaces the magnetic force, and gravity replaces the resistive force. The vertical component of the normal force does positive work, but the horizontal component does an equal and opposite amount of negative work. You are the entity pumping energy into the system (which shows up as gravitational potential energy), just as you were the entity pumping energy into the above circuit (which showed up as heat). Although the vertical component of the normal force is the only force actually lifting the block upward, the *entire* normal force does zero net work. Conversely, you are not lifting the block upward, but you do in fact do positive work.

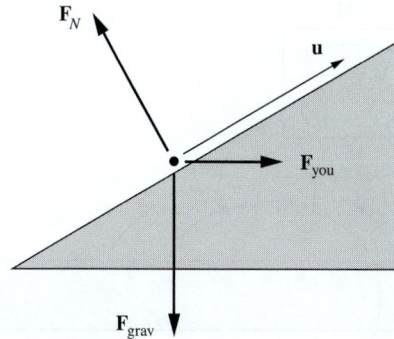

**Figure 12.110.**

7.3 *Pulling a square frame*

(a) The length of the border of the shaded region that lies inside the square is $2x$. So in a time $dt$, a thin rectangle of flux with area $(2x)(v\,dt)$ disappears from the square. Hence the magnitude of the emf is

$$\mathcal{E} = \frac{d\Phi}{dt} = \frac{B(2xv\,dt)}{dt} = 2Bxv. \tag{12.334}$$

The induced current in the square is therefore $I = 2Bxv/R$. It flows counterclockwise, to generate a $B$ field pointing out of the page to oppose the change in flux.

The force on a current-carrying wire is $F = I\ell B$. Consider the upper segment of the square with length $\ell = \sqrt{2}x$ that lies inside the shaded region. From the right-hand rule, the magnetic force on the current points up to the left. Likewise, the force on the lower segment inside the shaded region points down to the left. The vertical components cancel, so we care only about the leftward components. This brings in a factor of $\cos 45° = 1/\sqrt{2}$. The total leftward force on the square                                                                  is therefore

$$F = 2I\ell B\cos 45° = 2 \cdot \frac{2Bxv}{R} \cdot \sqrt{2}x \cdot B \cdot \frac{1}{\sqrt{2}} = \frac{4B^2x^2v}{R}. \tag{12.335}$$

If the square moves with constant velocity, your force must be rightward with the same magnitude.

(b) The work you do is

$$W = \int F\,dx = \int_0^{x_0} \frac{4B^2x^2v}{R}\,dx = \frac{4B^2x_0^3v}{3R}. \tag{12.336}$$

(Technically, the integral goes from $x_0$ to 0, but your displacement is $-dx$. So your work still comes out to be positive, as must be the case.) The total energy dissipated in the resistor is (using $dt = dx/v$)

$$\int I^2R\,dt = \int \left(\frac{2Bxv}{R}\right)^2 R\,dt = \int \frac{4B^2x^2v}{R}\,dx, \tag{12.337}$$

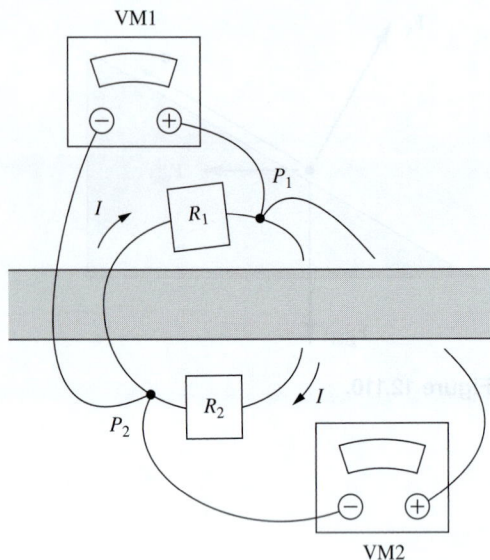

VM1

$P_1$

$I$

$R_1$

$P_2$

$R_2$

$I$

VM2

**Figure 12.111.**

which is the same as the integral in the above expression for $W$. So your work does indeed equal the energy dissipated in the resistor.

**7.4   *Loops around a solenoid***

Let VM1 and $R_1$ be the upper voltmeter and resistor, and let VM2 and $R_2$ be the lower ones in Fig. 12.111. The induced emf around a loop enclosing the solenoid is (ignoring the sign)

$$\mathcal{E} = \frac{d\Phi}{dt} = A\frac{dB}{dt} = (0.002\ \text{m}^2)(0.01\ \text{T/s}) = 2 \cdot 10^{-5}\ \text{V}, \qquad (12.338)$$

or 20 microvolts. The loop containing the resistors encloses the solenoid, so it has a current of $I = \mathcal{E}/R = (2 \cdot 10^{-5}\ \text{V})/(100\ \Omega) = 2 \cdot 10^{-7}\ \text{A}$, or 0.2 microamps. Since the flux is increasing, Lenz's law tells us that the current in the loop is clockwise.

The loop that includes VM1 and $R_1$ encloses no changing flux, so voltage differences in the loop are path independent. The voltage drop across $R_1$ is $IR = (2 \cdot 10^{-7}\ \text{A})(50\ \Omega) = 10\ \mu\text{V}$, with the more positive end connected to the $(-)$ lead on VM1, because the current flows across $R_1$ in the direction from $P_2$ to $P_1$. Therefore, VM1 will read $-10\ \mu\text{V}$.

Alternatively, the line integral of $\mathbf{E}$ around the loop from $P_1$ to $P_2$ via VM1, and then from $P_2$ back to $P_1$ via $R_1$, equals zero, because the loop encloses no changing flux. But the latter part of this integral is $+10\ \mu\text{V}$ (because the current, and hence the field, points from $P_2$ to $P_1$). So the former part must be $-10\ \mu\text{V}$, and this is, by definition, the reading on VM1.

Likewise, the loop that includes VM2 and $R_2$ encloses no changing flux. The voltage drop across $R_2$ is $10\ \mu\text{V}$, but now the more positive end is connected to the $(+)$ lead on VM2, because the current flows across $R_2$ in the direction from $P_1$ to $P_2$. Therefore, VM2 will read $+10\ \mu\text{V}$.

We can arrive at these conclusions in other ways, too. For example, consider the loop containing $R_1$ and VM2. This loop encloses changing flux. The line integral of $\mathbf{E}$ around the loop from $P_1$ to $P_2$ via VM2, and then from $P_2$ to $P_1$ via $R_1$, equals $20\ \mu\text{V}$, from Eq. (12.338). But the latter part of this integral is $+10\ \mu\text{V}$. So the former part must be $+10\ \mu\text{V}$, and this is, by definition, the reading on VM2.

The moral of this problem is that if a setup contains changing flux, it makes no sense to talk about *the* voltage difference (that is, the value of $-\int \mathbf{E} \cdot d\mathbf{s}$) between two points. It is necessary to *state the path* over which $-\int \mathbf{E} \cdot d\mathbf{s}$ is calculated. Someone looking at VM1 will give an answer of $-10\ \mu\text{V}$ for the voltage difference $V_{P_1} - V_{P_2}$, while someone looking at VM2 will give an answer of $+10\ \mu\text{V}$ for $V_{P_1} - V_{P_2}$. However, if the wire connecting VM2 to $P_1$ instead passes in *front* of the solenoid, then VM2 will read $-10\ \mu\text{V}$ just like VM1. In magnetostatics (that is, setups with constant currents), we don't have to specify the path, so we can uniquely label every point in a circuit with a definite potential. But not so if there is changing flux.

**7.5   *Total charge***

The magnetic flux through the coil is $\Phi = N\pi a^2 B$, so the induced emf equals $\mathcal{E} = d\Phi/dt = N\pi a^2(dB/dt)$. We ignore the signs for now.

The current is $I = \mathcal{E}/R = (N\pi a^2/R)(dB/dt)$, so the total charge passing through the resistor is

$$Q = \int I\, dt = \int \frac{N\pi a^2}{R}\frac{dB}{dt}\, dt = \frac{N\pi a^2}{R}\int_{B_0}^{0} dB = -\frac{N\pi a^2 B_0}{R}.$$
$$(12.339)$$

Since we haven't been keeping track of the signs, the minus sign here doesn't mean much. The correct statement to make, by Lenz's law, is that a total positive charge of $N\pi a^2 B_0/R$ flows in the direction that produces magnetic flux in the direction of the initial flux. That is, the current and the original flux are related by the right-hand rule.

We see that $Q$ depends only on the net change in $B$, and not on the rate at which this change comes about. Intuitively, if $B$ changes more slowly, then $\mathcal{E}$ (and hence $I$) is smaller, so less charge passes through the resistor during a given time interval. But the process takes longer, so this allows more charge to pass through. These two competing effects exactly cancel.

In the event that essentially all of the circuit is represented by the coil, the resistance $R$ is proportional to the number of turns $N$. The total charge $Q$ in Eq. (12.339) is then independent of $N$. Additionally, you can show that $Q$ is proportional to the volume taken up by the wire in one ring of the coil.

7.6    *Growing current in a solenoid*
The integral form of Faraday's law is $\int \mathbf{E} \cdot d\mathbf{s} = -(d/dt)\int \mathbf{B} \cdot d\mathbf{a}$. Inside the solenoid, the magnetic field is $B = \mu_0 nI$, so the magnetic flux through a circle with radius $r$ is $\Phi = B(\pi r^2) = (\mu_0 nCt)(\pi r^2)$. Faraday's law then gives the magnitude of the tangential component of $\mathbf{E}$ as (ignoring the signs)

$$E_\theta(2\pi r) = \frac{d}{dt}\left[(\mu_0 nCt)(\pi r^2)\right] \implies E_\theta = \frac{\mu_0 nCr}{2}. \qquad (12.340)$$

From Lenz's law, $E_\theta$ points in the tangential direction opposite to the current flow around the solenoid.

Outside the solenoid, the magnetic flux is $\Phi = (\mu_0 nCt)(\pi R^2)$, because the field is nonzero only inside the solenoid. So Faraday's law gives

$$E_\theta(2\pi r) = \frac{d}{dt}\left[(\mu_0 nCt)(\pi R^2)\right] \implies E_\theta = \frac{\mu_0 nCR^2}{2r}. \qquad (12.341)$$

The direction is again opposite to the current flow.

To check that the differential form of the law, namely $\nabla \times \mathbf{E} = -\partial\mathbf{B}/\partial t$, is satisfied, we can use the expression for the curl in cylindrical coordinates given in Appendix F. Since $\mathbf{E}$ has only a $\theta$ component and since this component depends only on $r$, only one term in the curl survives, and we have $\nabla \times \mathbf{E} = \hat{\mathbf{z}}(1/r)\partial(rE_\theta)/\partial r$. Now, when working with the integral form of Faraday's law, it is usually easiest to take care of the signs with Lenz's law. But when working with the differential form, we should pay attention to the actual signs of the various quantities. In Fig. 12.112, the positive $\hat{\mathbf{z}}$ direction is out of the page, and the positive

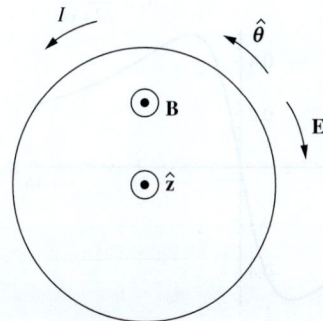

**Figure 12.112.**

$\hat{\boldsymbol{\theta}}$ direction is counterclockwise. If the current is counterclockwise, then $\mathbf{B}$ points in the $\hat{\mathbf{z}}$ direction, while the induced $\mathbf{E}$ points in the *negative* $\hat{\boldsymbol{\theta}}$ direction. So the $E_\theta$ that we want to use is the negative of the $E_\theta$ given in Eqs. (12.340) and (12.341). Therefore, inside the solenoid we have

$$\nabla \times \mathbf{E} = \hat{\mathbf{z}}\frac{1}{r}\frac{\partial}{\partial r}\left(r \cdot \frac{-\mu_0 nCr}{2}\right) = -\mu_0 nC\hat{\mathbf{z}}, \qquad (12.342)$$

which does indeed equal $-\partial \mathbf{B}/\partial t = -\partial(\mu_0 nCt\hat{\mathbf{z}})/\partial t$. And outside the solenoid we have

$$\nabla \times \mathbf{E} = \hat{\mathbf{z}}\frac{1}{r}\frac{\partial}{\partial r}\left(r \cdot \frac{-\mu_0 nCR^2}{2r}\right) = 0, \qquad (12.343)$$

which again equals $-\partial \mathbf{B}/\partial t$ because $\mathbf{B} = 0$ outside the solenoid.

The above calculations involve only the electric field that arises due to the changing magnetic field. There may very well be other fields present. We can add on any field with $\nabla \times \mathbf{E} = 0$, that is, any electrostatic field. For example, a line of charge along the axis would add on a radial field with magnitude $\lambda/2\pi\epsilon_0 r$.

7.7　*Maximum emf for a thin loop*

The setup is shown in Fig. 12.113. At a general position $x$ in the $xy$ plane, the magnitude of $B$ is $\mu_0 I/2\pi r$, where $r = \sqrt{h^2 + x^2}$. Only the $z$ component matters in the flux, and this brings in a factor of $x/r$. So

$$B_z(x) = \frac{\mu_0 I}{2\pi r}\frac{x}{r} = \frac{\mu_0 I}{2\pi}\frac{x}{h^2 + x^2}. \qquad (12.344)$$

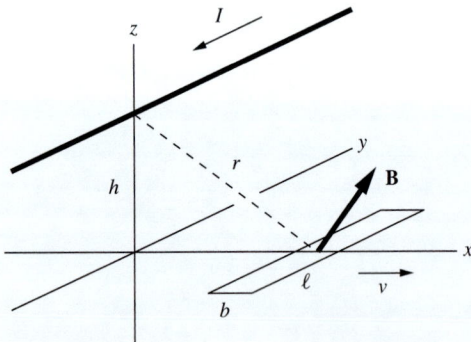

**Figure 12.113.**

A plot of $B_z(x) \propto x/(h^2 + x^2)$ is shown in Fig. 12.114. Note that $B_z$ is an increasing function of $x$, for small $x$. This is because the increase in the tilt of the $B$ vector matters more than the decrease in the magnitude of $B$ arising from the increase in $r$ (because $r$ hardly changes near $x = 0$). Conversely, $B_z$ is a decreasing function of $x$, for large $x$. This is because the decrease in the magnitude of $B$ arising from the increase in $r$ matters more than the increase in the tilt of the $B$ vector (because the tilt hardly changes for large $x$).

If we take $x$ to represent the position of the center of the thin loop, then the leading and trailing edges are at positions $x + b/2$ and $x - b/2$, respectively. So our standard argument, involving the flux gained at the leading edge and lost at the trailing edge, yields

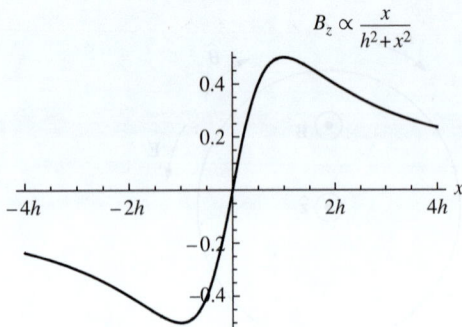

**Figure 12.114.**

$$\begin{aligned}
\mathcal{E} = \frac{d\Phi}{dt} &= \left(B_z(x + b/2) - B_z(x - b/2)\right)v\ell \\
&= \frac{\mu_0 Iv\ell}{2\pi}\left(\frac{x + b/2}{h^2 + (x + b/2)^2} - \frac{x - b/2}{h^2 + (x - b/2)^2}\right) \\
&\approx \frac{\mu_0 Iv\ell}{2\pi} \cdot b\frac{d}{dx}\left(\frac{x}{h^2 + x^2}\right) = \frac{\mu_0 Iv\ell b}{2\pi} \cdot \frac{h^2 - x^2}{(h^2 + x^2)^2},
\end{aligned}$$

$$(12.345)$$

where we have used the definition of the derivative to approximate the difference in the $B$ fields. With the signs of the various quantities indicated in Fig. 12.113, positive $\mathcal{E}$ is clockwise when viewed from above. For $x < h$, the flux is increasing, so a clockwise current is induced, which creates downward flux to oppose the change. Conversely, for $x > h$ the flux is decreasing, so a counterclockwise current is induced, which creates upward flux to oppose the change.

The plot of $\mathcal{E}(x) \propto dB_z(x)/dx \propto (h^2 - x^2)/(h^2 + x^2)^2$ is shown in Fig. 12.115. It is zero at $x = h$, consistent with the fact that $B_z$ achieves an extremum (a maximum) there. This extremum implies that the $B_z$ values at the leading and trailing edges of the loop are essentially equal, thereby yielding no change in flux.

To find where $\mathcal{E}$ achieves a local maximum or minimum, we must set the derivative equal to zero. This gives

$$\frac{d\mathcal{E}}{dx} = 0 \implies (h^2 + x^2)^2(-2x) - (h^2 - x^2)2(h^2 + x^2)(2x) = 0$$

$$\implies x = 0 \text{ or } x = \pm\sqrt{3}\,h. \tag{12.346}$$

The $x = 0$ root corresponds to the maximum clockwise emf, and the $x = \pm\sqrt{3}\,h$ roots corresponds to the maximum counterclockwise emf (which is much smaller). These three points correspond to the points where $B_z$ is changing the fastest (locally) in Fig. 12.114, because at these points the difference in the $B_z$ values at the leading and trailing edges is largest. Said in a different way, these three points are the inflection points of the $B_z(x)$ plot, where the second derivative of $B_z$ equals zero. Indeed, Eq. (12.346) is the statement that $d\mathcal{E}/dx = 0$, which is equivalent to $d^2B_z/dx^2 = 0$, because $\mathcal{E}$ is proportional to $dB_z/dx$.

7.8    *Faraday's law for a moving tilted sheet*

(a) As in the example in Section 5.5, the components of the electric field in the new frame $F'$ are $E'_\parallel = E_\parallel = E/\sqrt{2}$ and $E'_\perp = \gamma E_\perp = \gamma E/\sqrt{2}$ (where $E = \sigma/2\epsilon_0$, but we won't need this). So the magnitude of the electric field in $F'$ is $E' = (E/\sqrt{2})\sqrt{1 + \gamma^2}$. To find the component $E'_p$ parallel to the sheet, Fig. 5.12 shows that we must multiply the magnitude by $\sin(2\theta - 90°)$, where $\tan\theta = \gamma$. This trig factor can alternatively be written as $-\cos 2\theta$, which in turn can be written as $1 - 2\cos^2\theta$. So we have (using $\tan\theta = \gamma \implies \cos\theta = 1/\sqrt{1 + \gamma^2}$)

$$E'_p = E'\sin(2\theta - 90°) = \frac{E}{\sqrt{2}}\sqrt{1 + \gamma^2}\left(1 - 2\cos^2\theta\right)$$

$$= \frac{E}{\sqrt{2}}\sqrt{1 + \gamma^2}\left(1 - \frac{2}{1 + \gamma^2}\right) = \frac{E}{\sqrt{2}}\frac{\gamma^2 - 1}{\sqrt{1 + \gamma^2}}$$

$$= \frac{E}{\sqrt{2}}\frac{\gamma^2\beta^2}{\sqrt{1 + \gamma^2}}, \tag{12.347}$$

where we have used $\gamma^2 = 1/(1 - \beta^2)$. On the left side of the sheet, $E'_p$ points up along the sheet, and on the right side it points down along the sheet, as shown in Fig. 12.116 (assuming the sheet is positively charged).

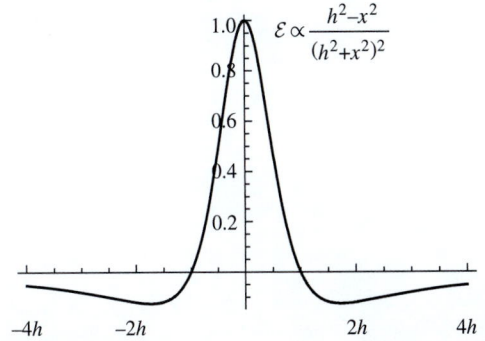

$$\mathcal{E} \propto \frac{h^2 - x^2}{(h^2 + x^2)^2}$$

**Figure 12.115.**

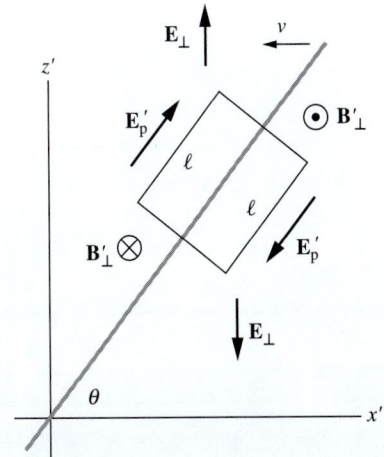

**Figure 12.116.**

(b) The Lorentz transformations in Eq. (6.76) give the transverse magnetic field as $\mathbf{B}'_\perp = -\gamma(\mathbf{v}/c^2) \times \mathbf{E}_\perp$, where the velocity $\mathbf{v}$ of $F'$ with respect to the lab frame $F$ points to the right. On the left side of the sheet, $\mathbf{E}_\perp$ points up, and on the right side it points down. So the right-hand rule tells us that $\mathbf{B}'_\perp$ points into the page on the left, and out of the page on the right. In both cases it has magnitude

$$B'_\perp = \frac{\gamma v}{c^2} E_\perp = \frac{\gamma v}{c^2} \frac{E}{\sqrt{2}}. \tag{12.348}$$

(c) Let's first check that the signs work out with Faraday's law applied to the given rectangle. As the sheet moves to the left in $F'$, the area with flux into the page decreases, and the area with flux out of the page increases. (Remember that the rectangle is fixed in $F'$.) So the flux increases out of the page. The induced emf should therefore be clockwise, to counteract the change and produce flux into the page. This is consistent with the directions of $E'_p$ we found in part (a).

Now let's see if the numbers work out. If the sides of the rectangle parallel to the sheet have length $\ell$, then $\int \mathbf{E} \cdot d\mathbf{s}$ simply equals $2E'_p\ell$. (There is zero net contribution involving $E'_n$ and the other two sides.) To find the rate of change of the magnetic flux, note that, as the sheet moves to the left, only the component of the velocity that is perpendicular to the sheet causes a change in the two areas of flux. This component is $v \sin \theta$. So in a small time $dt$, the area that is swept through is $(v \sin \theta \, dt)\ell$. Since the magnetic field goes from pointing one way to pointing the opposite way, the change in the field is $2B'_\perp$. The change in flux in a time $dt$ is therefore $d\Phi = (\ell v \sin \theta \, dt)(2B'_\perp)$. Ignoring the signs, since we already checked that they work out, we see that $\int \mathbf{E} \cdot d\mathbf{s} = -d\Phi/dt$ is true if

$$\int \mathbf{E} \cdot d\mathbf{s} = \frac{d\Phi}{dt}$$

$$\Longleftrightarrow \quad 2E'_p\ell = \frac{2\ell v \sin \theta \, B'_\perp \, dt}{dt}$$

$$\Longleftrightarrow \quad 2\left(\frac{E}{\sqrt{2}} \frac{\gamma^2 \beta^2}{\sqrt{1+\gamma^2}}\right)\ell = 2\ell v \left(\frac{\gamma}{\sqrt{1+\gamma^2}}\right)\left(\frac{\gamma v}{c^2} \frac{E}{\sqrt{2}}\right). \tag{12.349}$$

Since $\beta \equiv v/c$, you can quickly check that all the factors match up on both sides of this equation.

Note that if the sheet is horizontal in the lab frame, then in $F'$ there is no parallel $E'_p$ field. This is consistent with the fact that, although there is a nonzero $B'_\perp$ field in $F'$, the flux doesn't change because the sheet is moving parallel to itself and therefore sweeps out zero area. In the other extreme, if the sheet is vertical in the lab frame, then both $E'_p$ and $B'_\perp$ are zero, so Faraday's law is trivially satisfied.

For small speeds $v$, you might think that you could solve this problem by ignoring the various relativistic $\gamma$ factors, that is, by setting $\gamma = 1$. But you would run into difficulty, because the magnetic field in Eq. (12.348) would be $B'_\perp = (v/c^2)E/\sqrt{2}$, but the parallel electric field $E'_p$ in the second line of Eq. (12.347) would be zero. So you would conclude that there is changing flux but no emf, in violation of Faraday's law. The error is that, even for small $v$, we can't set $\gamma$ equal to 1, because this neglects effects of order $v^2/c^2$ (as is evident in the third line of Eq. (12.347)). And $v^2/c^2$ is exactly the order of the magnetic-flux term on the right-hand side of Eq. (12.349).

7.9 *Mutual inductance for two solenoids*
Let current $I_2$ flow in the outer solenoid. The field inside is approximately uniform in the region occupied by the inner solenoid, and the value is roughly that of an infinite solenoid, namely $B_2 = \mu_0 n_2 I_2 = \mu_0 (N_2/b_2) I_2$. The flux through the inner solenoid is

$$\Phi_{12} = N_1 \pi a_1^2 B_2 = N_1 \pi a_1^2 \frac{\mu_0 N_2 I_2}{b_2}. \tag{12.350}$$

The mutual inductance is obtained by erasing the $I_2$, so we have

$$M = \frac{\mu_0 \pi a_1^2 N_1 N_2}{b_2}. \tag{12.351}$$

If we want to obtain a better approximation for $M$, we can use Eq. (6.56) to find the field at the center of a finite solenoid of length $b_2$ and radius $a_2$. The correction factor is $\cos\left(\tan^{-1}(a_2/(b_2/2))\right) = b_2/\sqrt{b_2^2 + 4a_2^2}$. This will still not yield an exact result, because the inner solenoid extends over a nonzero volume in which the field varies somewhat. But for the proportions shown in the figure, the approximation will be pretty good. The flux, and hence mutual inductance, will also be smaller by the factor $b_2/\sqrt{b_2^2 + 4a_2^2}$, so the above $M$ becomes

$$M = \frac{\mu_0 \pi a_1^2 N_1 N_2}{\sqrt{b_2^2 + 4a_2^2}}. \tag{12.352}$$

In the limit $b_2 \gg a_2$ this of course reduces to the value in Eq. (12.351).

If you try to determine $M$ by the reverse method of calculating the flux through the *outer* solenoid due to the field from the *inner* solenoid, you have to be careful. The field from the inner solenoid is $B_1 = \mu_0 N_1 I_1/b_1$. Inside the inner solenoid, this field exists within a cross-sectional area of $\pi a_1^2$. So the flux through a ring of the outer solenoid, in the middle region, is $B_1(\pi a_1^2)$. If the entirety of the flux stays within the outer cylinder all the way out to the end (it actually doesn't; see below), then all $N_2$ rings have this same flux, so the total flux through the outer cylinder is $N_2 \cdot B_1(\pi a_1^2) = \mu_0 \pi a_1^2 N_1 N_2 I_1/b_1$. The mutual inductance is obtained by erasing the $I_1$, so we obtain an $M$ that does *not* agree with the $\mu_0 \pi a_1^2 N_1 N_2/b_2$ result in Eq. (12.351); the denominators differ.

The error is that the $B_1$ field doesn't stay inside the outer cylinder; it leaks out through the side. A more accurate approximation is to say that it leaks out *immediately* after it exits the inner solenoid. (The field lines diverge on a length scale of the diameter, and we're assuming that the lengths of the solenoids are fairly long compared with their diameters.) In this case, the relevant number of turns in the outer cylinder is the number that exist within the span of $b_1$, which is $b_1 n_2 = b_1 (N_2/b_2)$. Replacing $N_2$ with this value in our incorrect result turns it into the (approximately) correct result in Eq. (12.351).

The first solution above was simpler because the $B_2$ field lines diverge very little over the span of the inner cylinder, whereas the $B_1$ field lines diverge greatly over the span of the outer cylinder.

7.10   *Mutual-inductance symmetry*
The induced emfs in circuits 1 and 2 are, respectively,

$$\mathcal{E}_1 = -L_1 \frac{dI_1}{dt} - M_{12} \frac{dI_2}{dt} \qquad \text{and} \qquad \mathcal{E}_2 = -L_2 \frac{dI_2}{dt} - M_{21} \frac{dI_1}{dt}.$$

(12.353)

(A different sign convention for the currents might yield minus signs in front of the $M$ terms, but they both have the same sign in any case.) The external agency must provide opposing emfs in the two circuits to balance these emfs, so the sum of the power inputs (voltage times current) in the two circuits by the external agency is given by

$$P = \left( L_1 I_1 \frac{dI_1}{dt} + M_{12} I_1 \frac{dI_2}{dt} \right) + \left( L_2 I_2 \frac{dI_2}{dt} + M_{21} I_2 \frac{dI_1}{dt} \right). \quad (12.354)$$

The total energy input equals the time integral of $P$. The two $L$ terms are total differentials, so these terms yield final energies of $L_1 I_{1f}^2/2$ and $L_2 I_{2f}^2/2$, independent of how the changes in the currents come about. But the two $M$ terms are not total differentials, so they depend on the exact procedure.

(a) If we keep $I_2$ at zero and increase $I_1$, then since both $I_2$ and $dI_2/dt$ are zero, both of the $M$ terms are zero, so no extra work is required. In the second stage of this program, where $I_1$ is held constant at $I_{1f}$, the $M_{21}$ term is zero, but the $M_{12}$ term integrates to $M_{12} I_{1f} I_{2f}$. The total work done during this program is therefore

$$W_1 = \frac{1}{2} L_1 I_{1f}^2 + \frac{1}{2} L_2 I_{2f}^2 + M_{12} I_{1f} I_{2f}. \quad (12.355)$$

This equals the total energy in the final system, because we are assuming there are no dissipative elements (and also no radiative effects if the currents change slowly; see Appendix H).

(b) The same reasoning holds for the second program; we simply need to switch the labels 1 and 2. The final energy of the system is therefore

$$W_2 = \frac{1}{2} L_1 I_{1f}^2 + \frac{1}{2} L_2 I_{2f}^2 + M_{21} I_{2f} I_{1f}. \quad (12.356)$$

The final energy must be independent of the procedure by which the currents are brought to their final values of $I_{1f}$ and $I_{2f}$ (assuming slow changes, so that we can ignore radiative effects). So by comparing $W_1$ and $W_2$ we find $M_{12} = M_{21}$, as desired.

**7.11** *L for a solenoid*

The field in the interior of the solenoid is $B = \mu_0 n I = \mu_0 (N/\ell) I$, so the flux through the $N$ turns is

$$\Phi = N(\pi r^2) B = \frac{\mu_0 \pi r^2 N^2}{\ell} I. \qquad (12.357)$$

Since the self-inductance is defined by $L \equiv \Phi/I$, we simply need to erase the $I$ in this result. So we have $L = \mu_0 \pi r^2 N^2/\ell$.

**7.12** *Doubling a solenoid*

(a) From Problem 7.11, the self-inductance of a solenoid is $L = \mu_0 \pi r^2 N^2/\ell$. In the present scenario, both $N$ and $\ell$ are doubled, so $L$ increases by a factor of $2^2/2 = 2$.

In words: the self-inductance equals the flux $\Phi$ divided by the current $I$. When we double the length of the solenoid, the field inside stays the same (it equals $\mu_0 n I = \mu_0 (N/\ell) I$). But we now have twice as many turns, so the flux increases by a factor of 2.

(b) In this scenario, $N$ is doubled, but $\ell$ remains the same. So the self-inductance $L = \mu_0 \pi r^2 N^2/\ell$ increases by a factor of $2^2 = 4$.

In words: when we put one solenoid on top of the other, the number of turns per unit length, $n$, doubles. So the field inside doubles. And we now have twice as many turns. So we have twice the field going through twice the number of turns. The flux therefore increases by a factor of 4.

**7.13** *Adding inductors*

(a) When inductors are connected in series, the currents $I$ across them are equal, because charge can't pile up between them. The $dI/dt$ values for the two inductors are therefore also equal. The total voltage $V$ across the effective inductor is the sum of the voltages across the two inductors, so $V = V_1 + V_2$. We know that $V = L\, dI/dt$, $V_1 = L_1\, dI/dt$, and $V_2 = L_2\, dI/dt$, where the same $I$ appears everywhere here. Plugging these expressions into $V = V_1 + V_2$ gives

$$L\frac{dI}{dt} = L_1\frac{dI}{dt} + L_2\frac{dI}{dt} \implies L = L_1 + L_2. \qquad (12.358)$$

If $L_1 \to 0$ then $L \to L_2$. This makes sense because, for a given $dI/dt$, the voltage $L_1\, dI/dt$ across $L_1$ is tiny, which means that the overall voltage $L\, dI/dt$ is essentially equal to the voltage $L_2\, dI/dt$ across $L_2$. So $L \approx L_2$.

If $L_1 \to \infty$ then $L \to \infty$ also. This makes sense because, for a given $dI/dt$, the voltage $L_1\, dI/dt$ across $L_1$ is huge, which means that the overall voltage $L\, dI/dt$ is likewise huge, since it is at least as large. So $L$ must be very large.

(b) When inductors are connected in parallel, the voltages across them are equal, because the voltage drop from the left side of the overall circuit to the right side can't depend on the path. Let $V$ be this common voltage, which is of course also the voltage drop in the overall effective inductor.

The total current $I$ across the effective inductor is the sum of the currents across the two inductors, so $I = I_1 + I_2 \implies dI/dt = dI_1/dt + dI_2/dt$. But we know that $dI/dt = V/L$, $dI_1/dt = V/L_1$, and $dI_2/dt = V/L_2$, where the same $V$ appears everywhere here. Plugging these expressions into $dI/dt = dI_1/dt + dI_2/dt$ gives

$$\frac{V}{L} = \frac{V}{L_1} + \frac{V}{L_2} \implies \frac{1}{L} = \frac{1}{L_1} + \frac{1}{L_2}. \tag{12.359}$$

In short, in the series case the currents across the two inductors are the same, and the voltages add; whereas in the parallel case the voltages are the same, and the currents add.

If $L_1 \to 0$ then $L \to 0$ also. This makes sense because, for a given $V$, the $dI_1/dt$ value of $V/L_1$ is huge, which means that the overall $dI/dt$ value of $V/L$ is likewise huge, since it is at least as large. So $L$ must be very large.

If $L_1 \to \infty$ then $L \to L_2$. This makes sense because, for a given $V$, the $dI_1/dt$ value of $V/L_1$ is tiny, which means that the overall $dI/dt$ value of $V/L$ is essentially equal to the $dI_2/dt$ value of $V/L_2$. So $L \approx L_2$.

7.14 *Current in an RL circuit*
We can separate variables in Eq. (7.65) and integrate:

$$\mathcal{E}_0 - L\frac{dI}{dt} = RI \implies \int_0^I \frac{L\,dI'}{\mathcal{E}_0 - RI'} = \int_0^t dt'$$

$$\implies -\frac{L}{R}\ln(\mathcal{E}_0 - RI')\bigg|_0^I = t \implies \ln\left(\frac{\mathcal{E}_0 - RI}{\mathcal{E}_0}\right) = -(R/L)t$$

$$\implies 1 - \frac{R}{\mathcal{E}_0}I = e^{-(R/L)t} \implies I(t) = \frac{\mathcal{E}_0}{R}\left(1 - e^{-(R/L)t}\right), \tag{12.360}$$

as desired. Alternatively, we can solve Eq. (7.65) in a quick manner by shifting the $I$ variable. Equation (7.65) can be written as $dI/dt = -(R/L)(I - \mathcal{E}_0/R)$, which in turn can be written as

$$\frac{d(I - \mathcal{E}_0/R)}{dt} = -\frac{R}{L}(I - \mathcal{E}_0/R), \tag{12.361}$$

where we have used the fact that $d\mathcal{E}_0/dt = 0$ since $\mathcal{E}_0$ is a constant. This is a simple differential equation in the variable $I - \mathcal{E}_0/R$, so we can immediately write down the solution:

$$I - \mathcal{E}_0/R = De^{-(R/L)t}, \tag{12.362}$$

where $D$ is determined by the initial conditions. Since $I = 0$ at $t = 0$, we must have $D = -\mathcal{E}_0/R$. This yields the same $I(t)$ as in Eq. (12.360).

### 7.15 Energy in an RL circuit

If we integrate the equation $I^2 R = I(\mathcal{E}_0 - L\, dI/dt)$ up to a time $t$, we obtain (writing one of the $I$'s as $dQ/dt$)

$$\int_0^t I^2 R\, dt = \mathcal{E}_0 \int_0^t \frac{dQ}{dt}\, dt - L \int_0^t I \frac{dI}{dt}\, dt$$

$$\implies \int_0^t I^2 R\, dt = \mathcal{E}_0 Q - \frac{1}{2} L I^2. \qquad (12.363)$$

This is the statement that the energy dissipated in the resistor equals the energy delivered by the battery minus the energy stored in the inductor, as desired.

Of course, we can go in the other direction too, by differentiating instead of integrating. If we start with the conservation-of-energy statement, $\mathcal{E}_0 Q = L I^2/2 + \int_0^t I^2 R\, dt$, and then take the time derivative, we obtain $\mathcal{E}_0 I = L I\, dI/dt + I^2 R$. Dividing by $I$ gives $\mathcal{E}_0 = L\, dI/dt + IR$, which is just the loop equation in Eq. (7.66).

### 7.16 Energy in a superconducting solenoid

The energy density is

$$\frac{B^2}{2\mu_0} = \frac{(3\,\text{T})^2}{2(4\pi \cdot 10^{-7}\,\text{kg m/C}^2)} = 3.6 \cdot 10^6\,\text{J/m}^3. \qquad (12.364)$$

We'll work in the approximation where the $B$ field is uniform inside the solenoid and zero outside. The volume is $\pi r^2 \ell = \pi (0.45\,\text{m})^2 (2.2\,\text{m}) = 1.4\,\text{m}^3$, so the total energy in the field is $(3.6 \cdot 10^6\,\text{J/m}^3)(1.4\,\text{m}^3) = 5 \cdot 10^6\,\text{J}$. This is enough energy to raise a 2000 kg car 250 meters off the ground. But at 10 cents per kilowatt hour, the energy costs only about a penny.

A more accurate estimate could be made, if it were needed, by consulting a table of the inductance of finite solenoids, calculating the current required to produce the given central field, and then computing $LI^2/2$. The result, when that is carried out in this case, is nearly the same – only a few percent smaller.

### 7.17 Two expressions for the energy

The internal field of a solenoid is $B = \mu_0 n I = \mu_0 (N/\ell) I$, and the volume is $V = \pi r^2 \ell$. From Problem 7.11, the self-inductance of a long solenoid is $L = \mu_0 \pi r^2 N^2/\ell$. So the two expressions give the same energy if

$$\frac{1}{2} L I^2 = \frac{1}{2\mu_0} B^2 V$$

$$\iff \frac{1}{2}\left(\frac{\mu_0 \pi r^2 N^2}{\ell}\right) I^2 = \frac{1}{2\mu_0}\left(\frac{\mu_0 N I}{\ell}\right)^2 \pi r^2 \ell, \qquad (12.365)$$

which is indeed true.

7.18  *Two expressions for the energy (general)*
The vector identity is made-to-order, with no need even to change the letters. Since $\nabla \times \mathbf{A} = \mathbf{B}$ and $\nabla \times \mathbf{B} = \mu_0 \mathbf{J}$, we have

$$\nabla \cdot (\mathbf{A} \times \mathbf{B}) = \mathbf{B} \cdot \mathbf{B} - \mathbf{A} \cdot \mu_0 \mathbf{J}. \tag{12.366}$$

Integrating this over a volume and using the divergence theorem gives

$$\int_S (\mathbf{A} \times \mathbf{B}) \cdot d\mathbf{a} = \int_V B^2 \, dv - \mu_0 \int_V \mathbf{A} \cdot \mathbf{J} \, dv. \tag{12.367}$$

If the currents are contained in a finite region, and if we take the surface $S$ to infinity, then the surface integral on the left-hand side is zero. This follows from the fact that there is an $r^2$ in the denominator of the Biot–Savart law, and an $r$ in the denominator of the analogous expression for $\mathbf{A}$ in Eq. (6.44). This means that $\mathbf{A} \times \mathbf{B}$ falls off at least as fast as $1/r^3$. Since the area of the surface $S$ grows only like $r^2$, the surface integral therefore goes to zero as $r \to \infty$. So Eq. (12.367) becomes

$$\int_V B^2 \, dv = \mu_0 \int_V \mathbf{A} \cdot \mathbf{J} \, dv. \tag{12.368}$$

The volume integral $\int_V \mathbf{A} \cdot \mathbf{J}$ is zero everywhere except where there is current, so we can replace this integral with a line integral over the wire in the system. Using the fourth and fifth of the given hints, this yields (however, see the remark below)

$$\int_V \mathbf{A} \cdot \mathbf{J} \, dv = \int \mathbf{A} \cdot I \, d\mathbf{l} = I \int \mathbf{A} \cdot d\mathbf{l} = I\Phi = I(LI). \tag{12.369}$$

Equation (12.368) then becomes

$$\int_V B^2 \, dv = \mu_0 L I^2 \quad \Longrightarrow \quad \frac{1}{2\mu_0} \int_V B^2 \, dv = \frac{1}{2} L I^2, \tag{12.370}$$

as desired. The volume integral runs over all space, although there is negligible contribution from regions far away (assuming that the currents are contained in a finite region).

REMARK: We glossed over a certain issue in deriving Eq. (12.369). We assumed that $\mathbf{A}$ takes on a particular value at a given location along the wire. However, $\mathbf{A}$ varies over the wire's cross section. For example, it is proportional to $r^2$ for a straight wire, from Exercise 6.43. So although $\mathbf{A} \cdot \mathbf{J}$ is well defined, $\mathbf{A} \cdot d\mathbf{l}$ is not. However, the result in Eq. (12.369) isn't affected, because we can get around this issue by dividing the wire into a large number of very thin tubes of current (like the fibers in a fiber optic cable). Then $\mathbf{A}$ is essentially constant over the tiny cross section of each tube. ($\mathbf{A}$ will undoubtedly vary along the length of the tube, but that is perfectly fine.) Let the current in each tube be labeled as $I_n$, where the index $n$ runs over all the tubes. Then the contribution to Eq. (12.369) from a given tube is $I_n(LI)$. The second $I$ here is indeed the entire $I$ of the whole wire, because that is what appears in the $\Phi = LI$ definition of $L$. Summing over all the little tubes simply turns the $I_n$ into $I$, so we end up with the same result, $\int_V \mathbf{A} \cdot \mathbf{J} \, dv = LI^2$.

Note that there is a slight ambiguity in $\Phi$; all the little tubes aren't in exactly the same location, so there will be different fluxes through each of their circuits. This is the issue we discussed at the end of Section 7.8; it is unclear what is meant by *the* self-inductance of a circuit. But we haven't let that worry us so far, so we won't start worrying about it now.

7.19   *Critical frequency of a dynamo*

(a) The units of the three relevant quantities are

$$\sigma : \frac{C^2\,s}{kg\,m^3}, \qquad \mu_0 : \frac{kg\,m}{C^2}, \qquad d : m, \qquad (12.371)$$

where we have used the fact that the units of $\sigma$ are $(\text{ohm-m})^{-1}$. Our goal is to combine these quantities to make the units of $\omega_0$, namely inverse seconds. The coulombs and kilograms cancel out in the product $\mu_0\sigma$. The meters then cancel if we multiple by $d^2$. That leaves only seconds in the numerator, so we just need to take the inverse, yielding $1/\mu_0\sigma d^2$, up to a numerical factor.

(b) We will ignore all numerical factors in the following reasoning, so any "=" signs below should be taken with a grain of salt. The resistance of the current path in the dynamo scales like $1/\sigma d$, by dimensional analysis or equivalently because $R = \ell/\sigma A$, where $\ell \propto d$ and $A \propto d^2$. Ignoring numerical factors, we will set $R = 1/\sigma d$. To maintain a current $I$, we need an emf of $\mathcal{E} = IR$. Hence $\mathcal{E} = I/\sigma d$. But since voltage equals the line integral of the electric field, $\mathcal{E}$ is given by $Ed$, up to numerical factors. So we have[8] $Ed = I/\sigma d \Longrightarrow E = I/\sigma d^2$.

Now, the effective electric field causing the charges to move in the disk is $E = vB$ (because the force is $qvB$), where $v = \omega_0 d$. Therefore, $I/\sigma d^2 = (\omega_0 d)B$. Finally, the magnetic field produced by the current $I$ is, in order of magnitude, $B = \mu_0 I/d$ (because $B = \mu_0 I/2\pi r$ for a wire). So we arrive at

$$\frac{I}{\sigma d^2} = (\omega_0 d)\frac{\mu_0 I}{d} \implies \omega_0 = \frac{K}{\mu_0 \sigma d^2}, \qquad (12.372)$$

where we have introduced the factor $K$ to represent all the numerical factors we dropped.

For copper at room temperature, we have $\sigma \approx 6 \cdot 10^7\,(\text{ohm-m})^{-1}$. If we take $d = 1\,m$ then we obtain $\omega_0 = K(0.013)\,s^{-1}$. This seems rather slow. However, it turns out that the factor $K$ is generally much larger than 1, for various reasons. For one, the above form of $B$ for a wire tells us that we should be using $\mu_0/2\pi$ instead of $\mu_0$. This would increase $K$ by $2\pi$. But far more importantly, in the dynamo we met in Exercise 7.47, the resistance $R$ is larger than $1/\sigma d$ by something like $d^2/A$, where $A$ is the cross-sectional area of the wire of the coil. And the sliding contacts will probably have a resistance much larger than $1/\sigma d$.

---

[8] In a standard wire with a given thickness, we have $I \propto E$. The reason why $I \propto Ed^2$ here is that we are scaling the "wire" in all dimensions, so the cross-sectional area grows like $d^2$, allowing more current through.

## 12.8 Chapter 8

8.1    *Linear combinations of solutions*

If $x_1(t)$ and $x_2(t)$ are solutions to the given linear equation, then

$$A\ddot{x}_1 + B\dot{x}_1 + Cx_1 = 0,$$
$$A\ddot{x}_2 + B\dot{x}_2 + Cx_2 = 0. \tag{12.373}$$

If we add these two equations, and switch from the dot notation to the $d/dt$ notation, then we have (using the fact that the sum of the derivatives is the derivative of the sum)

$$A\frac{d^2(x_1 + x_2)}{dt^2} + B\frac{d(x_1 + x_2)}{dt} + C(x_1 + x_2) = 0. \tag{12.374}$$

But this is just the statement that $x_1 + x_2$ is a solution to our differential equation, as we wanted to show. This technique clearly works for any linear combination of $x_1$ and $x_2$, not just their sum. Note that the right-hand side of the equation needs to be zero, otherwise we wouldn't end up with the same term on the right-hand side when we add the equations. Hence the "homogeneous" qualifier in the statement of the problem.

Now consider the nonlinear equation

$$A\ddot{x} + B\dot{x}^2 + Cx = 0. \tag{12.375}$$

If $x_1$ and $x_2$ are solutions to this equation, and if we add the differential equations applied to each of them, we obtain

$$A\frac{d^2(x_1 + x_2)}{dt^2} + B\left[\left(\frac{dx_1}{dt}\right)^2 + \left(\frac{dx_2}{dt}\right)^2\right] + C(x_1 + x_2) = 0. \tag{12.376}$$

This is *not* the statement that $x_1 + x_2$ is a solution, which is instead the (false) statement that

$$A\frac{d^2(x_1 + x_2)}{dt^2} + B\left(\frac{d(x_1 + x_2)}{dt}\right)^2 + C(x_1 + x_2) = 0. \tag{12.377}$$

The preceding two equations differ by the cross term in the middle term in Eq. (12.377), namely $2B(dx_1/dt)(dx_2/dt)$. In general this term is nonzero, so Eq. (12.377) isn't valid, and we conclude that $x_1 + x_2$ isn't a solution. No matter what the order of the differential equation is, we see that these cross terms will arise if and only if the equation isn't linear.

This property of homogeneous linear differential equations – that the sum of two solutions is again a solution – is extremely useful. It means that we can build up solutions from other solutions. Systems that are governed by linear equations are *much* easier to deal with than systems that are governed by nonlinear equations. In the latter, the various solutions aren't related in an obvious way. Each one sits in isolation, in a sense. General relativity is an example of a theory that is governed by nonlinear equations, and solutions are indeed very hard to come by.

**8.2** *Solving linear differential equations*

The fundamental theorem of algebra states that any $n$th-order polynomial,

$$a_n z^n + a_{n-1} z^{n-1} + \cdots + a_1 z + a_0, \qquad (12.378)$$

can be factored into

$$a_n(z - r_1)(z - r_2) \cdots (z - r_n), \qquad (12.379)$$

where the $r_i$ roots are in general complex. This is believable, but by no means obvious. The proof is rather involved, so we'll just accept it here.

Now, because differentiation by $t$ commutes with multiplication by a constant, we can factor Eq. (8.95) just as we can factor Eq. (12.378). That is, the fundamental theorem of algebra tells us that we can rewrite Eq. (8.95) as

$$a_n \left( \frac{d}{dt} - r_1 \right) \left( \frac{d}{dt} - r_2 \right) \cdots \left( \frac{d}{dt} - r_n \right) x = 0. \qquad (12.380)$$

Furthermore, because all of these factors commute with each other, we can shuffle the order and make any one of the factors be the rightmost one. So any solution to the equation

$$\left( \frac{d}{dt} - r_i \right) x = 0 \iff \frac{dx}{dt} = r_i x \qquad (12.381)$$

is a solution to the original equation, Eq. (8.95). But the solutions to these $n$ first-order equations are simply the exponential functions, $x(t) = A_i e^{r_i t}$. (These certainly work, although if you want to show this from scratch, you can solve $dx/dt = r_i x$ by separating variables and integrating.) We have therefore found $n$ different solutions. And from Problem 8.1, we know that any linear combination of these solutions is also a solution.

There are two issues we have glossed over. First, there may be double (or triple, etc.) roots to the "characteristic equation" in Eq. (12.379). If $r$ is a double root, it turns out that, in addition to $Ae^{rt}$, another solution is $Bte^{rt}$. And a triple root would also have $Ct^2 e^{rt}$ as a solution. And so on. You can verify this by direct differentiation, but here's a general proof. If $y(t)$ is an arbitrary function, then $(d/dt - r)(ye^{rt}) = (dy/dt)e^{rt}$. So we inductively obtain $(d/dt - r)^n (ye^{rt}) = (d^n y/dt^n)e^{rt}$. This equals zero if $d^n y/dt^n = 0$, which is satisfied by any polynomial of degree $n - 1$ or less. Hence, for example, $(d/dt - r)^3 [(A + Bt + Ct^2)e^{rt}] = 0$, as we claimed above.

Second, we have found $n$ solutions, but how do we know we have found them all? Perhaps none of the factors in Eq. (12.380) alone makes the left-hand side zero, but maybe some combination of them does? We know that this doesn't happen when dealing with the normal multiplication of the factors in Eq. (12.379), but maybe something odd happens when dealing with derivatives? It turns out that this isn't the case, although it's harder to show. One way is to invoke the Fourier-analysis fact that any (reasonably well behaved) function can be written as the sum, or integral, of exponential functions. And for an exponential function, it is easy enough to show that Eq. (12.380) is zero only if one of the factors applied to it is zero.

8.3     *Underdamped motion*
Plugging an exponential solution $x(t) = Ce^{\gamma t}$ into the given differential equation gives

$$\gamma^2 C e^{\gamma t} + 2\alpha\gamma C e^{\gamma t} + \omega_0^2 C e^{\gamma t} = 0 \implies \gamma^2 + 2\alpha\gamma + \omega_0^2 = 0.$$

(12.382)

The roots of this quadratic equation are

$$\gamma_{1,2} = -\alpha \pm \sqrt{\alpha^2 - \omega_0^2},$$

(12.383)

where the subscripts 1 and 2 correspond to the + and − roots, respectively. Our two exponential solutions are therefore $x_1(t) = C_1 e^{\gamma_1 t}$ and $x_2(t) = C_2 e^{\gamma_2 t}$.

There are three cases to consider, depending on whether the quantity $\alpha^2 - \omega_0^2$ is positive, zero, or negative. We will generally be concerned with the underdamped case where $\alpha < \omega_0$, but see Problem 8.4 for a discussion of the overdamped case. If $\alpha < \omega_0$, the discriminant in Eq. (12.383) is negative, so the $\gamma$'s have an imaginary part. Let's define the real quantity $\omega$ by

$$\omega \equiv \sqrt{\omega_0^2 - \alpha^2}.$$

(12.384)

Then the $\gamma$'s in Eq. (12.383) become $\gamma_{1,2} = -\alpha \pm i\omega$, and our two exponential solutions now look like

$$x_1(t) = C_1 e^{(-\alpha+i\omega)t} \quad \text{and} \quad x_2(t) = C_2 e^{(-\alpha-i\omega)t}.$$

(12.385)

Since the given differential equation is linear, the most general solution is the sum of these two solutions, which is

$$x(t) = e^{-\alpha t}\left(C_1 e^{i\omega t} + C_2 e^{-i\omega t}\right).$$

(12.386)

Now comes the sneaky part (or the obvious part, depending on how you look at it): $x(t)$ must of course be real if it represents a physical quantity (charge, current, voltage, position, angle, etc.) The two terms in Eq. (12.386) must therefore be complex conjugates of each other, so that their imaginary parts cancel. This implies that $C_2 = C_1^*$, where the star denotes complex conjugation. If we write $C_1$ in "polar" form as $C_1 = Ce^{i\phi}$, then $C_2 = C_1^* = Ce^{-i\phi}$, so $x(t)$ becomes

$$x(t) = e^{-\alpha t}C\left(e^{i(\omega t+\phi)} + e^{-i(\omega t+\phi)}\right)$$

$$= e^{-\alpha t}C \cdot 2\cos(\omega t + \phi)$$

$$\equiv Ae^{-\alpha t}\cos(\omega t + \phi),$$

(12.387)

where $A \equiv 2C$. Applying the trig sum formula to $\cos(\omega t + \phi)$ turns it into a linear combination of $\cos \omega t$ and $\sin \omega t$ terms, which is the desired form in Eq. (8.10). In the case of the *RLC* circuit in Section 8.1, we have $\alpha = R/2L$ and $\omega_0^2 = 1/LC$. So the $\alpha$ in Eq. (12.387) agrees with the $\alpha$ in Eq. (8.8). And the frequency $\omega$ in Eq. (12.384) agrees with the frequency in Eq. (8.9).

Note that by demanding that $x$ be real (which led to the addition of the two complex conjugates in Eq. (12.387)), we are basically just taking the real part of either of the two solutions in Eq. (12.386). So in the end, you can skip much of the above reasoning and simply take the real part of an exponential solution that you obtain. This "taking the real part" strategy is discussed in detail in Section 8.3.

8.4 *Overdamped RLC circuit*
Plugging the trial solution $V(t) = Ae^{-\beta t}$ into Eq. (8.2) gives (after canceling the factor of $Ae^{-\beta t}$)

$$\beta^2 - \frac{R}{L}\beta + \frac{1}{LC} = 0. \tag{12.388}$$

The roots of this quadratic equation are

$$\beta_{1,2} = \frac{1}{2}\left(\frac{R}{L} \pm \sqrt{\frac{R^2}{L^2} - \frac{4}{LC}}\right) = \frac{R}{2L}\left(1 \pm \sqrt{1 - \frac{4L}{R^2 C}}\right), \tag{12.389}$$

where the subscripts 1 and 2 correspond to the $+$ and $-$ roots, respectively. So we have found two exponential solutions: $Ae^{-\beta_1 t}$ and $Be^{-\beta_2 t}$. Since the differential equation in Eq. (8.2) is linear, the most general solution for $V(t)$ is a linear combination of these two solutions, that is,

$$V(t) = Ae^{-\beta_1 t} + Be^{-\beta_2 t}. \tag{12.390}$$

We see from Eq. (12.389) that the roots are real if $R \geq 2\sqrt{L/C}$. In this case we have exponentially decaying motion instead of the (decaying) oscillatory motion in Eq. (8.10), relevant to the underdamped case.

If $R$ is large (more precisely, if $R^2 \gg L/C$), then in Eq. (12.389) we can use the Taylor series $\sqrt{1 - \epsilon} \approx 1 - \epsilon/2$ to write the $\beta$'s as

$$\beta_{1,2} = \frac{R}{2L}\left(1 \pm \left(1 - \frac{2L}{R^2 C}\right)\right) \implies \beta_1 \approx \frac{R}{L} \text{ and } \beta_2 \approx \frac{1}{RC}. \tag{12.391}$$

Since $R$ is large, we have $\beta_1 \gg \beta_2$. The $Ae^{-\beta_1 t}$ part of the solution therefore goes to zero much faster than the $Be^{-\beta_2 t}$ part. For large $t$ (more precisely, for $t \gg L/R$), the solution therefore looks like $V(t) \approx Be^{-t/RC}$. This is exactly the same behavior that the $RC$ circuit had in Section 4.11. So apparently we have essentially an $RC$ circuit with no $L$. This makes physical sense, because a very large $R$ means a very small current $I$, and hence a very small $dI/dt$. The voltage $L\,dI/dt$ across the inductor is therefore negligible, which means that the inductor can be ignored.

8.5 *Change in frequency*
By looking at the oscillation plot, we can make the rough estimate that the amplitude decreases by a factor of $1/e$ after about two oscillations. This means that the exponential factor $e^{-\alpha t}$ equals $e^{-1}$ after two periods. Each oscillation takes a time of $2\pi/\omega$, so $\alpha$ is given in terms of $\omega$ by

$$\alpha t = 1 \implies \alpha(2 \cdot 2\pi/\omega) = 1 \implies \alpha = \frac{\omega}{4\pi}. \tag{12.392}$$

From the expressions for $\alpha$ and $\omega$ in Eqs. (8.8) and (8.9), we then have

$$\omega^2 = \frac{1}{LC} - \frac{R^2}{4L^2} = \frac{1}{LC} - \alpha^2 = \frac{1}{LC} - \left(\frac{\omega}{4\pi}\right)^2$$

$$\implies \omega = \frac{1}{\sqrt{LC}} \cdot \frac{1}{\sqrt{1 + 1/(4\pi)^2}} \approx \frac{1}{\sqrt{LC}}\left(1 - \frac{1}{32\pi^2}\right), \quad (12.393)$$

where we have used the Taylor series, $1/\sqrt{1+\epsilon} \approx 1 - \epsilon/2$. The frequency therefore differs from the natural frequency $1/\sqrt{LC}$ by only about 0.3 percent. The moral here is that unless the oscillation *very* quickly damps out to zero, the frequency is essentially equal to the natural frequency. Even if the amplitude decreases four times as fast as in the given plot (so that $\alpha t = 1$ after half an oscillation), you can quickly show that the percentage difference in the frequency is still only about 5 percent. In any case, the frequency is always smaller than the natural frequency.

8.6　*Limits of an RLC circuit*
(a) If $R = 0$, the $\alpha$ in Eq. (8.8) equals zero, and the $\omega$ in Eq. (8.9) equals $1/\sqrt{LC} \equiv \omega_0$. So the solution in Eq. (8.4) is simply $V(t) = A\cos\omega_0 t$. The charge sloshes back and forth between the two plates of the capacitor, passing through the inductor in the process. Looking back at Eq. (8.2), we see that if $R \approx 0$ the second term is negligible compared with the first and third terms, which are the two terms relevant to an $LC$ circuit.

(b) If $L \to 0$, then $R > 2\sqrt{L/C}$, so we are in the overdamped regime. Equation (12.389) in the solution to Problem 8.4 gives the values of the $\beta$'s that appear in Eq. (8.15). For small $L$ we can use the Taylor series $\sqrt{1-\epsilon} \approx 1 - \epsilon/2$ (just as we did in Eq. (12.391)) to write the $\beta$'s in Eq. (12.389) as

$$\beta_{1,2} = \frac{R}{2L}\left(1 \pm \left(1 - \frac{2L}{R^2C}\right)\right) \implies \beta_1 \approx \frac{R}{L} \text{ and } \beta_2 \approx \frac{1}{RC}.$$
$$(12.394)$$

Since $L$ is small, $\beta_1$ is very large. So the $Ae^{-\beta_1 t}$ part of the solution goes to zero much faster than the $Be^{-\beta_2 t}$ part. The solution therefore quickly becomes essentially equal to $V(t) \approx Be^{-t/RC}$, which is the solution for an $RC$ circuit, as desired. This makes sense, because if $L \approx 0$, the voltage $L\,dI/dt$ across the inductor is zero, which means that the inductor can be ignored. Looking back at Eq. (8.2), we see that if $L \approx 0$, the first term is negligible compared with the second and third terms, which are the two terms relevant to an $RC$ circuit.

(c) If $C \to \infty$, then $R > 2\sqrt{L/C}$, so we are again in the overdamped regime. For large $C$ we can apply the same Taylor series to Eq. (12.389) as in part (b), so the $\beta$'s are again $\beta_1 \approx R/L$ and $\beta_2 \approx 1/RC$. If $C \to \infty$ then $\beta_2 \to 0$. So the $Be^{-\beta_2 t}$ part of the solution is essentially equal to the constant $B$. Hence $V(t) \approx Ae^{-(R/L)t} + B$. Ignoring the $B$ term, this is the solution for an $RL$ circuit, as desired. This makes sense, because if $C \to \infty$, the voltage $Q/C$ across the capacitor is zero (for any finite $Q$), which means that the capacitor can be ignored.

Looking back at Eq. (8.2), we see that if $C \to \infty$ the third term is negligible compared with the first and second terms, which are the two terms relevant to an $RL$ circuit.

The physical meaning of $B$ is the following. If $C \to \infty$, the capacitor is an infinite reservoir of charge, whose potential $V = Q/C$ essentially doesn't change when a finite amount of charge is added or subtracted. Imagine that there is an initial current in the inductor. This current will gradually decay according to $e^{-(R/L)t}$, and after a while it will be essentially zero. However, there will now be charge on the capacitor, and this charge (and hence the voltage, which will be very small if $C$ is large) will very slowly leak off according to $e^{-t/RC}$, which is essentially constant on any finite time scale if $C \to \infty$. In the event that the initial charge on the capacitor approaches infinity, with the voltage $Q/C$ being some finite constant (the $B$ from above), the lingering state will be one where an essentially constant current flows through the circuit, decaying negligibly according to $e^{-t/RC}$. What we found above is that the circuit will approach this state in an $e^{-(R/L)t}$ manner.

### 8.7 Magnitude and phase

If $\phi = \tan^{-1}(b/a)$, then $\phi$ is given by the triangle in Fig. 12.117. So we have $\cos\phi = a/\sqrt{a^2 + b^2}$ and $\sin\phi = b/\sqrt{a^2 + b^2}$. Using the relation $e^{i\phi} = \cos\phi + i\sin\phi$ (see Section K.5 in Appendix K), we can write $I_0 e^{i\phi}$ as

$$I_0 e^{i\phi} = I_0(\cos\phi + i\sin\phi)$$

$$= \sqrt{a^2 + b^2}\left(\frac{a}{\sqrt{a^2 + b^2}} + i\frac{b}{\sqrt{a^2 + b^2}}\right)$$

$$= a + bi, \tag{12.395}$$

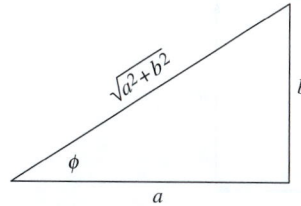

**Figure 12.117.**

as desired.

### 8.8 RLC circuit via vectors

(a) For convenience, let's repeat Eq. (8.98) here:

$$\omega L I_0 \cos(\omega t + \phi + \pi/2) + R I_0 \cos(\omega t + \phi)$$

$$+ \frac{I_0}{\omega C}\cos(\omega t + \phi - \pi/2) = \mathcal{E}_0 \cos\omega t. \tag{12.396}$$

With $I(t) = I_0 \cos(\omega t + \phi)$ we have $L\, dI/dt = -\omega L I_0 \sin(\omega t + \phi)$. And since the trig sum formula gives $\cos(\omega t + \phi + \pi/2) = -\sin(\omega t + \phi)$, the first term in Eq. (12.396) is correct. The second term is correct by definition. In the third term we have $Q = \int I\, dt$. (With our sign conventions for $I$ and $Q$, there is no minus sign here.) So $Q/C = (I_0/\omega C)\sin(\omega t + \phi)$. (As mentioned after Eq. (8.28) in the text, the constant of integration is zero.) And since the trig sum formula gives $\cos(\omega t + \phi - \pi/2) = \sin(\omega t + \phi)$, the third term Eq. (12.396) is also correct.

Equation (12.396) tells us that the voltage across the inductor, $V_L$, is 90° ahead of the voltage across the resistor, $V_R$ (which is in phase with the current), which in turn is 90° ahead of the voltage across the

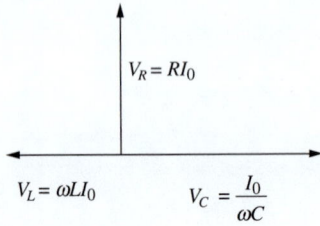

$V_R = RI_0$

$V_L = \omega L I_0$          $V_C = \dfrac{I_0}{\omega C}$

**Figure 12.118.**

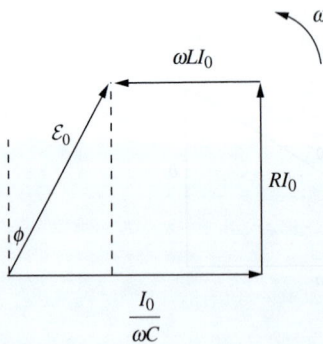

$\omega L I_0$

$\mathcal{E}_0$

$RI_0$

$\dfrac{I_0}{\omega C}$

$\phi$

$\omega$

**Figure 12.119.**

capacitor, $V_C$. The applied voltage is (in general) in phase with none of these. It is $\phi$ behind $V_R$, although we will find that $\phi$ can be positive or negative. If $\phi$ is negative, then the applied voltage is actually ahead of $V_R$, and hence also ahead of the current.

(b) Since $e^{i\theta} = \cos\theta + i\sin\theta$, the real part of a complex number $Ae^{i\theta}$ equals $A\cos\theta$. Equivalently, if a complex number is represented by a vector in the complex plane (with length $A$ and angle $\theta$ relative to the horizontal axis), then its real part equals its projection onto the horizontal axis, because the projection brings in a factor of $\cos\theta$. Since all four of the terms in Eq. (12.396) involve the cosine of some angle, they all can be considered to be the horizontal projections of four vectors with the given lengths and angles (phases).

Consider a moment in time when the phase of $V_C$ is zero (or a multiple of $2\pi$). Then the vector representing $V_C$ in the complex plane points to the right with magnitude $I_0/\omega C$, as shown in Fig. 12.118. The phase of $V_R$ is 90° larger, so $V_R$ is represented by a vector pointing upward with magnitude $RI_0$. The phase of $V_L$ is yet 90° larger, so $V_L$ is represented by a vector pointing leftward with magnitude $\omega L I_0$.

Putting these vectors tail to head, the sum is shown in Fig. 12.119. If the vector sum has length $\mathcal{E}_0$, and if its phase is $\phi$ behind the phase of $V_R$ (or ahead, if $\phi$ is negative) as shown, then the horizontal projections will add up properly and Eq. (12.396) will be satisfied. From Fig. 12.119 we see that $\phi$ is positive (or negative) if $I_0/\omega C$ is larger than (or smaller than) $\omega L I_0$. The cutoff between these cases occurs when $\omega = 1/\sqrt{LC}$.

The important point to note is that, as time goes by, the whole quadrilateral rotates around in the complex plane, due to the $\omega t$ term in all four of the phases in Eq. (12.396). The shape of the quadrilateral *doesn't change* as it rotates. The fact that the quadrilateral is closed means that the horizontal projections of the $V_C$, $V_R$, and $V_L$ vectors will always add up to the horizontal projection of the $\mathcal{E}_0$ vector. That is, Eq. (12.396) will always be satisfied if it is initially satisfied.

(c) To find the values of $I_0$ and $\phi$ (for a given $\omega$) that cause Eq. (12.396) to be satisfied, we just need to do a little geometry with our quadrilateral. If we consider the right triangle with the dashed line side in Fig. 12.119, we quickly see that

$$(RI_0)^2 + (I_0/\omega C - \omega L I_0)^2 = \mathcal{E}_0^2 \implies I_0 = \frac{\mathcal{E}_0}{\sqrt{R^2 + (1/\omega C - \omega L)^2}},$$

$$\tan\phi = \frac{I_0/\omega C - \omega L I_0}{RI_0} \implies \tan\phi = \frac{1}{R\omega C} - \frac{\omega L}{R}.$$

$$(12.397)$$

These results agree with Eqs. (8.38) and (8.39), as desired. For a given $\omega$, if $I_0$ and $\phi$ take on these values, then Eq. (12.396) will be satisfied.

The phase $\phi$ is positive if $\omega < 1/\sqrt{LC}$, which means that the current $I(t)$ (or equivalently $V_R$) is *ahead* of the applied $\mathcal{E}$; the capacitor's effect dominates the inductor's. On the other hand, $\phi$ is negative if $\omega > 1/\sqrt{LC}$, which means that the current $I(t)$ (or equivalently $V_R$)

is *behind* the applied $\mathcal{E}$; the inductors's effect dominates the capacitor's. (You should check these facts intuitively in the cases where the circuit contains only a capacitor or only an inductor.) In any case, $\phi$ lies in the range $-\pi/2 \leq \phi \leq \pi/2$, with equality achieved if $R = 0$.

For given values of $L$, $R$, $C$, and $\mathcal{E}_0$, it is instructive to look at how the frequency $\omega$ affects the shape of the quadrilateral. For this purpose, it is more convenient to consider a moment in time when the $\mathcal{E}$ vector is horizontal, as shown in Fig. 12.120. The quadrilateral can take three basic shapes, as shown. For $\omega < 1/\sqrt{LC}$, the $V_C$ side is longer; for $\omega = 1/\sqrt{LC}$, the magnitudes of $V_C$ and $V_L$ are equal; and for $\omega > 1/\sqrt{LC}$, the $V_L$ side is longer. The current amplitude $I_0$ is proportional to the length of the $V_R$ side, so we see geometrically why $I_0$ is maximum on resonance, that is, when $\omega = 1/\sqrt{LC}$.

**8.9** *Drawing the complex vectors*

**Series circuit**   For the series circuit in Fig. 8.10, the vectors are shown in Fig. 12.121. The current through all three elements (and the applied voltage source) is the same, and we have chosen the instant in time when this common complex current points along the real axis (so this is the instant when the actual current reaches its maximum value). The $\tilde{V}_R$ voltage vector also points along the real axis, because it is in phase with $\tilde{I}_R$. $\tilde{V}_L$ is 90° ahead of $\tilde{I}_L$, so $\tilde{V}_L$ points along the positive imaginary axis. And $\tilde{V}_C$ is 90° behind $\tilde{I}_C$, so $\tilde{V}_C$ points along the negative imaginary axis. From the given information about the impedances, $\tilde{V}_R$ and $\tilde{V}_L$ have the same length, and this length is twice that of $\tilde{V}_C$ (because in general $\tilde{V} = \tilde{I}\tilde{Z}$, and all the $\tilde{I}$'s are the same here).

The voltage $\tilde{V}_{\mathcal{E}}$ is the sum of the voltages across the three elements. Using what we know about the relative lengths of the voltages, we find $|\tilde{V}_{\mathcal{E}}| = (\sqrt{5}/2)|\tilde{V}_R|$. Since the current lags the applied voltage, the phase angle $\phi$ is negative. This is consistent with Eq. (8.39), because we are given that $|Z_L| > |Z_C| \Longrightarrow \omega L > 1/\omega C$. The angle $\phi$ is given by $\tan\phi = -1/2 \Longrightarrow \phi = -26.6°$.

If you wanted, you could have picked the instant when $\tilde{V}_{\mathcal{E}}$ points along the real axis, in which case all the vectors would be rotated clockwise by an angle $|\phi|$. The vectors look a little simpler the way we have drawn them, because more of them lie along the axes. The relative size of the $\tilde{I}$ and $\tilde{V}$ vectors in the figure doesn't mean much, because they have different units. But if the size of 1 volt on the page corresponds to the size of 1 amp, then, since we have drawn $\tilde{I}_R$ slightly shorter than $\tilde{V}_R$, we have apparently chosen $R$ to be slightly larger than 1 ohm.

**Parallel circuit**   For the parallel circuit in Fig. 8.20, the vectors are shown in Fig. 12.122. In this circuit the *voltage* through all three elements (and the applied voltage source) is the same. We have chosen the instant in time when this common complex voltage points along the real axis (so this is the instant when the actual voltage reaches its maximum value). The $\tilde{I}_R$ current vector also points along the real axis. $\tilde{V}_L$ is 90° ahead of $\tilde{I}_L$, so $\tilde{I}_L$ points along the *negative* imaginary axis. And $\tilde{V}_C$ is 90° behind $\tilde{I}_C$, so $\tilde{I}_C$ points along the *positive* imaginary axis. From the given information about the impedances, $\tilde{I}_R$ and $\tilde{I}_L$ have the same length, and

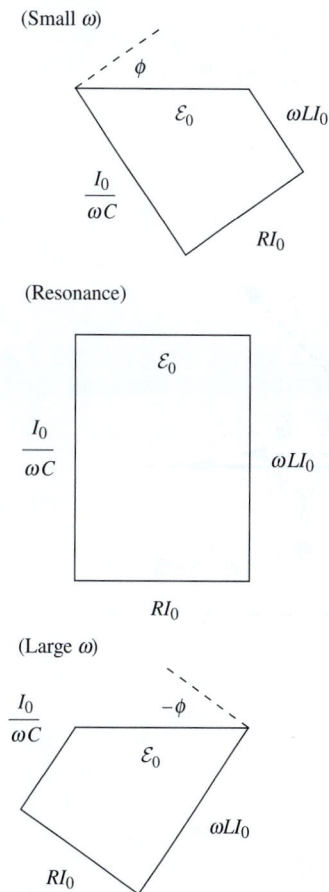

**Figure 12.120.**

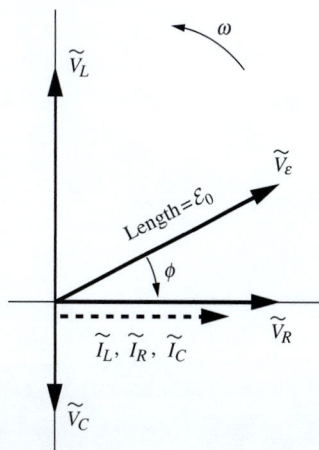

**Figure 12.121.**

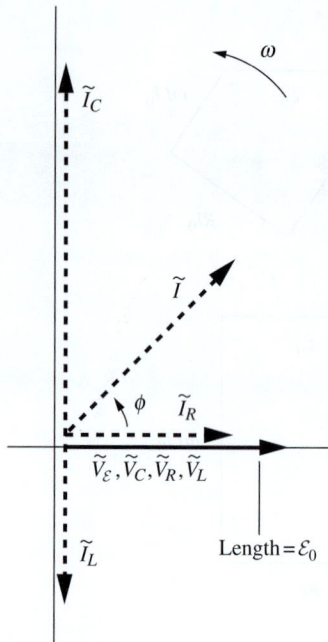

**Figure 12.122.**

this length is *half* that of $\tilde{I}_C$ (because $\tilde{V} = \tilde{I}Z$, and now all the $\tilde{V}$'s are the same).

The total current $\tilde{I}$ (which we could denote by $\tilde{I}_{\mathcal{E}}$ to parallel the above notation in the series case) is the sum of the currents across the three elements. Using what we know about the relative lengths of the currents, we find $|\tilde{I}| = \sqrt{2}|\tilde{I}_R|$. Since the total current *leads* the applied voltage, the phase angle $\phi$ is positive. This is consistent with Eq. (8.67), because as above, we are given that $\omega L > 1/\omega C$. The angle $\phi$ is given by $\tan \phi = 1 \implies \phi = 45°$.

Again, the relative size of the $\tilde{I}$ and $\tilde{V}$ vectors doesn't mean much. But if the sizes of the various units match up the natural way, we have again chosen $R$ to be slightly larger than 1 ohm. Note that, in the present case, choosing the instant when $\tilde{V}_{\mathcal{E}}$ points along the real axis leads to most of the vectors lying along the axes.

8.10  *Real impedance*

The impedances of the two branches are $R + i\omega L$ and $1/i\omega C$. Adding these in parallel gives a total impedance of

$$Z = \frac{1}{\dfrac{1}{R + i\omega L} + i\omega C}. \tag{12.398}$$

This is real if its reciprocal is real:

$$\frac{1}{Z} = \frac{1}{R + i\omega L} + i\omega C = \frac{R - i\omega L}{R^2 + \omega^2 L^2} + i\omega C. \tag{12.399}$$

Setting the imaginary part equal to zero gives

$$\frac{\omega L}{R^2 + \omega^2 L^2} = \omega C \implies \omega^2 = \frac{1}{LC} - \frac{R^2}{L^2}. \tag{12.400}$$

So the answer is "yes," provided that $R^2 < L/C$. Note that $\omega = 0$ is also a solution. In this case, the capacitor lets through no current (its impedance is infinite), and the inductor is effectively just a short-circuit wire (its impedance is zero). So we effectively have only the resistor.

8.11  *Light bulb*

The resistance of the light bulb is obtained from $P = V_{\mathrm{rms}}^2/R \implies R = (120\,\mathrm{V})^2/(40\,\mathrm{W}) = 360\,\Omega$. Since $\omega = 2\pi\nu = 2\pi(60\,\mathrm{s}^{-1}) = 377\,\mathrm{s}^{-1}$, the impedance of the capacitor is $Z_C = 1/i\omega C = -i/(377\,\mathrm{s}^{-1})(10^{-5}\,\mathrm{F}) = -265i\,\Omega$. The magnitude of the total impedance is then

$$|Z| = \sqrt{Z_R^2 + Z_C^2} = \sqrt{(360\,\Omega)^2 + (265\,\Omega)^2} = 447\,\Omega. \tag{12.401}$$

The rms current through the light bulb was originally $I_{\mathrm{rms}} = (120\,\mathrm{V})/(360\,\Omega)$, but now it is $(120\,\mathrm{V})/(447\,\Omega)$ (Ohm's law works with $|Z|$; see Eq. (8.77)). So it has decreased by a factor $360/447 = 0.81$. Since the power is proportional to the square of the current (it can be written as $I_{\mathrm{rms}}^2 R$), the brightness has therefore decreased by a factor $(0.81)^2 = 0.65$.

8.12  *Fixed voltage magnitude*

Both branches have the same voltage difference $V_0$ and the same impedance $R + 1/i\omega C$, so the complex current in both branches is given by

$$V_0 = \tilde{I}(R + 1/i\omega C) \implies \tilde{I} = \frac{V_0}{R + 1/i\omega C}. \tag{12.402}$$

The complex voltage at $A$ is $\tilde{V}_A = V_0 - \tilde{I}(1/i\omega C)$, and at $B$ it is $\tilde{V}_B = V_0 - \tilde{I}R$. Therefore (using the above value of $\tilde{I}$),

$$\tilde{V}_{AB} \equiv \tilde{V}_B - \tilde{V}_A = \tilde{I}(-R + 1/i\omega C)$$
$$= V_0 \frac{-R + 1/i\omega C}{R + 1/i\omega C} = V_0 \frac{1 - i\omega RC}{1 + i\omega RC}. \tag{12.403}$$

The numerator and denominator here have the same magnitude, so $|V_{AB}|^2 = V_0^2$, as desired. For a 90° phase difference, we need $(1 - i\omega RC)/(1 + i\omega RC) = \pm i$, since $e^{\pm i\pi/2} = \pm i$. We quickly find that if $\omega = 1/RC$, so that $\omega RC = 1$, then

$$\tilde{V}_{AB} = V_0 \left( \frac{1 - i}{1 + i} \right) = -iV_0. \tag{12.404}$$

A phase of 90° in the other direction, with $\tilde{V}_{AB} = iV_0$, would require $\omega RC = -1$, which isn't possible since all of these quantities are positive.

8.13  *Low-pass filter*

Let $I_0$ be the amplitude of the current through the resistor, which is also essentially the current through the capacitor. Then the complex $\tilde{V} = \tilde{I}Z$ statement for the voltage between the terminals at $A$ is $V_0 = \tilde{I}(R + 1/i\omega C)$. And the statement for the terminals at $B$ is $\tilde{V}_1 = \tilde{I}(1/i\omega C)$. Therefore,

$$\frac{\tilde{V}_1}{V_0} = \frac{1}{1 + i\omega RC} \implies \left| \frac{\tilde{V}_1}{V_0} \right|^2 = \frac{1}{1 + \omega^2 R^2 C^2}. \tag{12.405}$$

(It technically isn't necessary to take the magnitude of $V_0$ here, because the normal convention is to take $V_0$ to be real. But it looks a little nicer this way.) This equals 0.1 when $\omega^2 R^2 C^2 = 9$. At 5000 Hz this gives

$$RC = \frac{3}{\omega} = \frac{3}{2\pi \cdot 5000 \, \text{s}^{-1}} \approx 1 \cdot 10^{-4} \, \text{s}. \tag{12.406}$$

This is satisfied by, for example, $R = 1000 \, \Omega$ and $C = 0.1 \, \mu\text{F}$.

The power is proportional to $V^2$. If $\omega$ is large (more precisely, if $\omega RC \gg 1$), we can ignore the "1" in the denominator of Eq. (12.405). We then have $|\tilde{V}_1/V_0|^2 \propto 1/\omega^2$. Doubling $\omega$ decreases this by a factor of 4.

The physical reason why $V_1$ decreases with increasing frequency is the following. For very low frequency, the impedance of the capacitor is very large. The fixed impedance of the resistor is negligible in comparison, so essentially all of the $V_0$ voltage drop occurs across the capacitor, which is the voltage that $V_1$ registers. On the other hand, for very high frequency, the impedance of the capacitor is very small; it is essentially a short circuit.

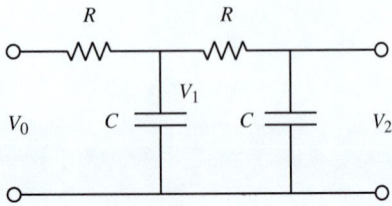

**Figure 12.123.**

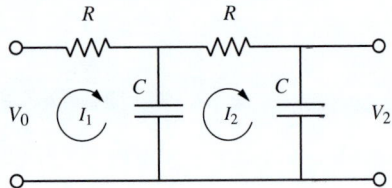

**Figure 12.124.**

Therefore, essentially all of the $V_0$ voltage drop occurs across the resistor. Very little occurs across the capacitor which, again, is the voltage that $V_1$ registers.

The circuit in Fig. 12.123 has $|V_2/V_0|^2 \propto 1/\omega^4$, for large $\omega$. (We'll drop the tildes on the $V$'s.) This is true because for large $\omega$ we have $|V_1/V_0|^2 \propto 1/\omega^2$; we are still able to invoke the above result because negligible current goes through the new circuit elements on the right (because the left capacitor has negligible impedance for large $\omega$, while the right resistor has a fixed value). And then we can use the same reasoning to say that $|V_2/V_1|^2 \propto 1/\omega^2$. So $|V_2/V_0|^2 = |V_2/V_1|^2|V_1/V_0|^2 \propto (1/\omega^2)^2 = 1/\omega^4$. Each new loop decreases the output voltage by $1/\omega^2$.

To derive this $1/\omega^4$ result from scratch, we can set up the loop currents shown in Fig. 12.124. We haven't drawn a current in the right-hand loop, because the current through the right-hand terminals is assumed to be very small. The three loop equations are (dropping the tildes)

$$V_0 - I_1 R - (I_1 - I_2)(1/i\omega C) = 0,$$
$$-I_2 R - I_2(1/i\omega C) - (I_2 - I_1)(1/i\omega C) = 0,$$
$$-V_2 + I_2(1/i\omega C) = 0. \qquad (12.407)$$

If $\omega$ is large, the first equation yields $V_0 \propto I_1$. Similarly, if $\omega$ is large, the second equation yields $I_1 \propto \omega I_2$. And the third equation yields $I_2 \propto \omega V_2$ in any case. So we have $V_0 \propto I_1$, $I_1 \propto \omega I_2$, and $I_2 \propto \omega V_2$. These three expressions yield $V_0 \propto \omega^2 V_2$. Therefore, $|V_2/V_0|^2 \propto 1/\omega^4$, as desired.

**8.14   *Series RLC power***

The current $I(t)$ and phase $\phi$ for the series *RLC* circuit are given in Eqs. (8.38) and (8.39). Since $\tan\phi = (1/\omega C - \omega L)/R$, we have

$$\cos\phi = \frac{R}{\sqrt{R^2 + (\omega L - 1/\omega C)^2}}. \qquad (12.408)$$

Equation (8.84) therefore gives the average power delivered to the circuit as

$$\overline{P} = \frac{1}{2}\mathcal{E}_0 I_0 \cos\phi$$
$$= \frac{1}{2}\mathcal{E}_0 \cdot \frac{\mathcal{E}_0}{\sqrt{R^2 + (\omega L - 1/\omega C)^2}} \cdot \frac{R}{\sqrt{R^2 + (\omega L - 1/\omega C)^2}}$$
$$= \frac{1}{2}\frac{\mathcal{E}_0^2 R}{R^2 + (\omega L - 1/\omega C)^2}. \qquad (12.409)$$

The average power dissipated in the resistor is given by Eq. (8.80), where $V_0$ is the voltage across *only* the resistor. Since this voltage is given simply by $V_0 = I_0 R$, we have

$$\overline{P}_R = \frac{1}{2}\frac{V_0^2}{R} = \frac{1}{2}\frac{(I_0 R)^2}{R} = \frac{1}{2}I_0^2 R = \frac{1}{2}\left(\frac{\mathcal{E}_0}{\sqrt{R^2 + (\omega L - 1/\omega C)^2}}\right)^2 R$$
$$= \frac{1}{2}\frac{\mathcal{E}_0^2 R}{R^2 + (\omega L - 1/\omega C)^2}, \qquad (12.410)$$

in agreement with Eq. (12.409).

8.15  *Two inductors and a resistor*

(a) The impedance of the inductor is $Z_L = i\omega L$. But since $\omega = R/L$ here, we have $Z_L = iR$. Using the standard rules for adding impedances in series and parallel, the total impedance of the circuit is

$$Z = Z_L + \frac{Z_R Z_L}{Z_R + Z_L} = iR + \frac{R(iR)}{R + iR} = R\frac{-1 + 2i}{1 + i}. \qquad (12.411)$$

This can also be written as $Z = R(1 + 3i)/2$.

(b) The complex current is

$$\tilde{I} = \frac{\mathcal{E}_0}{Z} = \frac{\mathcal{E}_0}{R}\frac{1 + i}{-1 + 2i} = \frac{\mathcal{E}_0}{R}\frac{1 - 3i}{5} = \frac{\mathcal{E}_0}{R}\frac{\sqrt{10}}{5}e^{i\phi}, \qquad (12.412)$$

where $\tan\phi = -3$. Therefore,

$$I_0 = \frac{\sqrt{10}}{5}\frac{\mathcal{E}_0}{R} \qquad \text{and} \qquad \phi = \tan^{-1}(-3) \approx -71.6°. \quad (12.413)$$

Formally,

$$I(t) = \mathrm{Re}\left[\tilde{I}e^{i\omega t}\right] = \mathrm{Re}\left[\frac{\sqrt{10}}{5}\frac{\mathcal{E}_0}{R}e^{i\phi}e^{i\omega t}\right] = \frac{\sqrt{10}}{5}\frac{\mathcal{E}_0}{R}\cos(\omega t + \phi).$$

$$(12.414)$$

(c) Since $\tan\phi = -3$ implies $\cos\phi = 1/\sqrt{10}$, Eq. (8.84) gives the average power dissipated in the circuit as

$$\frac{1}{2}\mathcal{E}_0 I_0 \cos\phi = \frac{1}{2}\mathcal{E}_0\left(\frac{\sqrt{10}}{5}\frac{\mathcal{E}_0}{R}\right)\frac{1}{\sqrt{10}} = \frac{\mathcal{E}_0^2}{10R}. \qquad (12.415)$$

Alternatively, we can find the power dissipated by finding the voltage $V_R$ across the resistor and then using $P_R = (1/2)V_R^2/R$. (The resistor is the only place where power is dissipated.) The complex voltage across the resistor equals $\mathcal{E}_0$ minus the complex voltage across the upper inductor, $V_L$. This latter voltage is

$$\tilde{V}_L = \tilde{I}Z_L = \tilde{I}(iR) = \left(\frac{\mathcal{E}_0}{R}\frac{1 - 3i}{5}\right)iR = \mathcal{E}_0\frac{3 + i}{5}. \qquad (12.416)$$

Hence

$$\tilde{V}_R = \mathcal{E}_0 - \tilde{V}_L = \mathcal{E}_0\left(1 - \frac{3 + i}{5}\right) = \mathcal{E}_0\frac{2 - i}{5}. \qquad (12.417)$$

The magnitude of this is $V_R = |\tilde{V}_R| = \mathcal{E}_0/\sqrt{5}$. Therefore,

$$P_R = \frac{1}{2}\frac{V_R^2}{R} = \frac{1}{2}\frac{\mathcal{E}_0^2/5}{R} = \frac{\mathcal{E}_0^2}{10R}, \qquad (12.418)$$

in agreement with the above result.

## 12.9 Chapter 9

9.1    *The missing term*

If we take the divergence of $\nabla \times \mathbf{B} = \mu_0 \mathbf{J} + \mathbf{W}$ and use the fact that the divergence of the curl is identically zero, we obtain $0 = \mu_0 \nabla \cdot \mathbf{J} + \nabla \cdot \mathbf{W}$. The continuity equation then gives $\nabla \cdot \mathbf{W} = \mu_0 (\partial \rho / \partial t)$. Gauss's law turns this into

$$\nabla \cdot \mathbf{W} = \mu_0 \epsilon_0 \big( \partial (\nabla \cdot \mathbf{E}) / \partial t \big) \implies \nabla \cdot \mathbf{W} = \nabla \cdot \big[ \mu_0 \epsilon_0 (\partial \mathbf{E} / \partial t) \big].$$
(12.419)

Therefore, $\mathbf{W} = \mu_0 \epsilon_0 (\partial \mathbf{E} / \partial t) + \mathbf{Z}$, where $\mathbf{Z}$ is a vector function whose divergence is identically zero. If we are allowed to work only with the given facts, then the only vector $\mathbf{Z}$ whose divergence we know is identically zero is $\mathbf{B}$. So the Maxwell equation must take the form of

$$\nabla \times \mathbf{B} = \mu_0 \mathbf{J} + \mu_0 \epsilon_0 \frac{\partial \mathbf{E}}{\partial t} + k\mathbf{B},$$
(12.420)

where $k$ is some constant. However, in the simple case of steady currents, we know that Ampère's law, $\nabla \times \mathbf{B} = \mu_0 \mathbf{J}$, is valid. Therefore $k$ must equal 0, and we arrive at the desired result.

9.2    *Spherically symmetric current*

As we showed near the end of Section 9.2, the magnetic field due to a spherically symmetric current density is zero. So the left-hand side of the given Maxwell equation is zero. Our goal is therefore to show that $\mathbf{J} = -\epsilon_0 \, \partial \mathbf{E} / \partial t$. The electric field $\mathbf{E}$ points outward (if $Q$ is positive) with magnitude $Q / 4\pi \epsilon_0 r^2$, so $\partial \mathbf{E} / \partial t = \hat{\mathbf{r}} (dQ/dt) / 4\pi \epsilon_0 r^2$. And the current density $\mathbf{J}$ points inward. The rate at which charge crosses any spherical boundary is $dQ/dt$, so the current density at radius $r$ is $\mathbf{J} = -\hat{\mathbf{r}} (dQ/dt) / 4\pi r^2$. It therefore is indeed true that $\mathbf{J} = -\epsilon_0 \, \partial \mathbf{E} / \partial t$.

9.3    *A charge and a half-infinite wire*

(a)  Consider a given point $P$ on the circle. In finding the field at $P$, there are various ways to parameterize the Biot–Savart integral. Let's work in terms of the angle $\alpha$ shown in Fig. 12.125, where $\alpha$ runs from $\alpha_0 \equiv \pi/2 - \theta$ up to $\pi/2$. The distance from a small segment of the wire to the center of the given circle is $l = b \tan \alpha$, so a short segment of the wire has length $dl = d(b \tan \alpha) = b \, d\alpha / \cos^2 \alpha$. The cross product in the Biot–Savart law brings in a factor of the sine of the angle between $dl$ and the $\hat{\mathbf{r}}$ vector to $P$, which is the same as $\cos \alpha$. So the $\mathbf{B}$ field at $P$ has magnitude

$$B = \frac{\mu_0 I}{4\pi} \int \frac{dl \cos \alpha}{r^2} = \frac{\mu_0 I}{4\pi} \int_{\alpha_0}^{\pi/2} \frac{(b \, d\alpha / \cos^2 \alpha) \cos \alpha}{(b / \cos \alpha)^2}$$

$$= \frac{\mu_0 I}{4\pi b} \int_{\alpha_0}^{\pi/2} \cos \alpha \, d\alpha = \frac{\mu_0 I}{4\pi b} (1 - \sin \alpha_0) = \frac{\mu_0 I}{4\pi b} (1 - \cos \theta).$$
(12.421)

This $\mathbf{B}$ field is tangential to the circle, so the line integral simply brings in a factor of $2\pi b$. Therefore, $\int \mathbf{B} \cdot d\mathbf{s} = (\mu_0 I / 2)(1 - \cos \theta)$. This correctly gives 0 for $\theta = 0$, and $\mu_0 I$ for $\theta = \pi$.

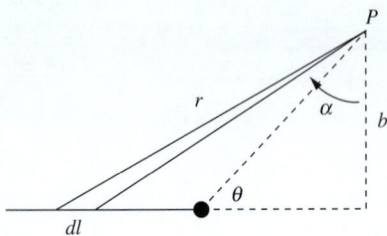

**Figure 12.125.**

(b) The $\mu_0 I$ term is in Eq. (9.59) zero in this case because the current doesn't pass through the surface. To calculate the displacement-current term, we will invoke from Problem 1.15 the result that the electric field flux through the circle is $\Phi_E = (q/2\epsilon_0)(1 - \cos\theta)$. Therefore,

$$\mu_0\epsilon_0 \int_S \frac{\partial \mathbf{E}}{\partial t} \cdot d\mathbf{a} = \mu_0\epsilon_0 \frac{\partial}{\partial t} \int_S \mathbf{E} \cdot d\mathbf{a} = \mu_0\epsilon_0 \frac{\partial \Phi_E}{\partial t}$$

$$= \mu_0\epsilon_0 \frac{dq/dt}{2\epsilon_0}(1 - \cos\theta) = \frac{\mu_0 I}{2}(1 - \cos\theta).$$

$$(12.422)$$

Since Eq. (9.59) tells us that this equals $\int \mathbf{B} \cdot d\mathbf{s}$, we obtain the same result for the line integral as in part (a). And the signs agree, given the standard right-hand convention for $d\mathbf{s}$ and $d\mathbf{a}$.

(c) We now need to include the $\mu_0 I$ term, because the flux passes through the surface. To calculate the displacement-current term, note that the union of the present surface and the surface in part (b) is a surface that completely encloses the charge $q$. So the sum of the electric field flux through both surfaces equals the total flux emanating from the charge $q$, which is $q/\epsilon_0$. The flux through the present surface is therefore $q/\epsilon_0 - (q/2\epsilon_0)(1 - \cos\theta) = (q/2\epsilon_0)(1 + \cos\theta)$. (Alternatively, the method of Problem 1.15 yields this result if $\theta$ is replaced with $\pi - \theta$.) But we must be careful about the sign; this flux pierces the surface in the direction opposite to the direction in which the current in the wire pierces it. The right-hand side of the integrated Maxwell equation is therefore

$$\mu_0 I + \mu_0\epsilon_0 \int_S \frac{\partial \mathbf{E}}{\partial t} \cdot d\mathbf{a} = \mu_0 I - \mu_0\epsilon_0 \frac{dq/dt}{2\epsilon_0}(1 + \cos\theta)$$

$$= \mu_0 I \left(1 - \frac{1}{2}(1 + \cos\theta)\right)$$

$$= \frac{\mu_0 I}{2}(1 - \cos\theta), \qquad (12.423)$$

as desired.

9.4 *B in a discharging capacitor, via conduction current*

(a) In a small time $dt$, a charge $I\,dt$ flows onto the positive plate. The fraction of this charge that ends up in the annular region between radius $r$ and the edge of the plate at radius $b$ is $\pi(b^2 - r^2)/\pi b^2 = 1 - r^2/b^2$. So this fraction of the current is what crosses a circle of radius $r$. The circumference of this circle is $2\pi r$, so the surface current density (current per length) is $\mathcal{J} = I(1 - r^2/b^2)/(2\pi r)$. If the point $P$ is close to the plate, then the plate acts essentially like an infinite plane with this surface current density.

(b) From Section 6.6, an isolated infinite sheet of current produces a field on both sides with magnitude $\mu_0 \mathcal{J}/2$. And since we have two plates with opposite currents, the sum of the two fields at point $P$ equals

$\mu_0 \mathcal{J}$. (You can verify that they add, not cancel.) So the field at $P$ due to the conduction current in the two disks is

$$B_{\text{disks}} = \mu_0 \mathcal{J} = \frac{\mu_0 I (1 - r^2/b^2)}{2\pi r} = \frac{\mu_0 I}{2\pi r} - \frac{\mu_0 I r}{2\pi b^2}. \quad (12.424)$$

This field points perpendicular to the surface current, in the tangential direction around the disks.

The field due to the (essentially continuous) wire is the standard $B_{\text{wire}} = \mu_0 I/2\pi r$. Both $B_{\text{disks}}$ and $B_{\text{wire}}$ point in the tangential direction, but you can quickly show by the right-hand rule that they point in opposite directions. So the net field is the difference of the magnitudes, which yields $B = \mu_0 I r/2\pi b^2$, as desired.

9.5   *Maxwell's equations for a moving charge*

(a) We can use the $\mathbf{B} = (1/c^2)\mathbf{v} \times \mathbf{E}$ expression for $\mathbf{B}$, along with the identity

$$\nabla \cdot (\mathbf{A} \times \mathbf{B}) = \mathbf{B} \cdot (\nabla \times \mathbf{A}) - \mathbf{A} \cdot (\nabla \times \mathbf{B}), \quad (12.425)$$

to say that

$$\nabla \cdot \mathbf{B} = \frac{1}{c^2} \nabla \cdot (\mathbf{v} \times \mathbf{E}) = \frac{1}{c^2} \mathbf{E} \cdot (\nabla \times \mathbf{v}) - \frac{1}{c^2} \mathbf{v} \cdot (\nabla \times \mathbf{E}). \quad (12.426)$$

The first term here is zero because $\mathbf{v}$ is constant. The second term is zero because $\nabla \times \mathbf{E}$ points in the $\hat{\boldsymbol{\phi}}$ direction (that is, tangential around the line of motion), so its dot product with $\mathbf{v}$ is zero. This $\hat{\boldsymbol{\phi}}$ direction follows from the expression in Eq. (F.3) for the curl in spherical coordinates ($\partial E_r/\partial\theta$ is the only nonzero derivative). Alternatively, we show in part (b) that $\nabla \times \mathbf{E}$ points in the Cartesian $\hat{\mathbf{y}}$ direction (at locations in the $xz$ plane), which is orthogonal to $\mathbf{v} \propto \hat{\mathbf{x}}$.

(b) Let's calculate $\nabla \times \mathbf{E}$ first, and then $\partial\mathbf{B}/\partial t$. Without loss of generality, we can do the calculation for a point in the $xz$ plane. Now, Eq. (5.13) is valid for points in the $xz$ plane. But since we're going to be taking derivatives (which, by definition, involve nearby points), we should be careful and include the $E_y$ component, and also the $y$ dependence of all the coordinates. (It will turn out that we could have ignored anything to do with $y$, but it is better to play it safe.) With $D \equiv (\gamma x)^2 + y^2 + z^2$, the generalization of Eq. (5.13) to any point in space is (dropping the primes on the coordinates)

$$\left( E_x, E_x, E_z \right) = \frac{\gamma Q}{4\pi\epsilon_0 D^{3/2}} (x, y, z). \quad (12.427)$$

Let's calculate the $y$ component of $\nabla \times \mathbf{E}$. It equals

$$(\nabla \times \mathbf{E})_y = \frac{\gamma Q}{4\pi\epsilon_0} \left( \frac{\partial}{\partial z}\left( \frac{x}{D^{3/2}} \right) - \frac{\partial}{\partial x}\left( \frac{z}{D^{3/2}} \right) \right)$$

$$= \frac{\gamma Q}{4\pi\epsilon_0} \left( \frac{-3xz}{D^{5/2}} + \frac{3\gamma^2 xz}{D^{5/2}} \right). \quad (12.428)$$

Using $\gamma^2 - 1 = \gamma^2 v^2/c^2$, we arrive at

$$\nabla \times \mathbf{E} = \frac{\gamma Q}{4\pi\epsilon_0} \frac{3\gamma^2 v^2 xz}{c^2 D^{5/2}} \hat{\mathbf{y}}. \qquad (12.429)$$

The $z$ component of $\nabla \times \mathbf{E}$ looks similar, with a factor of $xy$ in the numerator. But since this includes a factor of $y$, it equals zero for points in the $xz$ plane. So we can ignore it. And you can quickly show that the $x$ component of $\nabla \times \mathbf{E}$ is identically zero.

Now let's calculate $\partial \mathbf{B}/\partial t$. We have

$$\frac{\partial \mathbf{B}}{\partial t} = \frac{1}{c^2} \frac{\partial}{\partial t}(\mathbf{v} \times \mathbf{E}) = \frac{1}{c^2} \mathbf{v} \times \frac{\partial \mathbf{E}}{\partial t}. \qquad (12.430)$$

For points in the $xz$ plane, $\partial \mathbf{E}/\partial t$ has both $\hat{\mathbf{x}}$ and $\hat{\mathbf{z}}$ components, but since we're taking the cross product with $\mathbf{v} = v\hat{\mathbf{x}}$, we care only about the $\hat{\mathbf{z}}$ component. The relevant time derivatives are $dz/dt = 0$ and $dx/dt = -v$. The latter follows from the fact that $(x, y, z)$ are defined to be the coordinates relative to the charge. So, as the charge moves to the right, $x$ decreases. Equivalently, $x$ takes the form of $x = x_0 - vt$ for some $x_0$, so $dx/dt = -v$. We therefore have

$$\frac{\partial E_z}{\partial t} = \frac{\gamma Q}{4\pi\epsilon_0} \frac{\partial}{\partial t}\left(\frac{z}{D^{3/2}}\right) = \frac{\gamma Q}{4\pi\epsilon_0} z(-3/2)D^{-5/2}2\gamma^2 x(-v)$$

$$= \frac{\gamma Q}{4\pi\epsilon_0} \frac{3\gamma^2 xzv}{D^{5/2}}. \qquad (12.431)$$

Equation (12.430) then gives

$$\frac{\partial \mathbf{B}}{\partial t} = \frac{v\hat{\mathbf{x}}}{c^2} \times \frac{\gamma Q}{4\pi\epsilon_0} \frac{3\gamma^2 xzv}{D^{5/2}} \hat{\mathbf{z}} = -\frac{\gamma Q}{4\pi\epsilon_0} \frac{3\gamma^2 v^2 xz}{c^2 D^{5/2}} \hat{\mathbf{y}}. \qquad (12.432)$$

Comparing this result with Eq. (12.429) gives $\nabla \times \mathbf{E} = -\partial \mathbf{B}/\partial t$, as desired.

9.6 *Oscillating field in a solenoid*

(a) To find $E$, we can use Faraday's law (that is, the integral form of Maxwell's $\nabla \times \mathbf{E} = -\partial \mathbf{B}/\partial t$ equation) applied to a circle with radius $r$ centered on the axis. Assume that the axis of the solenoid is vertical. We define positive $B$ as pointing upward, and positive $E$ as counterclockwise when viewed from above. Then

$$\int \mathbf{E} \cdot d\mathbf{s} = -\frac{d\Phi_B}{dt} \implies 2\pi rE = -\frac{d}{dt}\left(\pi r^2 \cdot \mu_0 n I_0 \cos \omega t\right)$$

$$\implies E(r, t) = \frac{1}{2}r\mu_0 n I_0 \omega \sin \omega t. \qquad (12.433)$$

(b) Now consider a rectangular loop that has one side lying along the solenoid's axis, and one side at radius $r$; see Fig. 12.126. There is changing $E$ flux through this rectangle, due to the above $E(r, t)$. So we

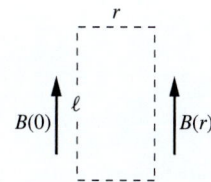

**Figure 12.126.**

want to use the integral form of Maxwell's $\nabla \times \mathbf{B} = \mu_0\epsilon_0 \partial\mathbf{E}/\partial t$ equation:

$$\int \mathbf{B} \cdot d\mathbf{s} = \mu_0\epsilon_0 \frac{d\Phi_E}{dt}. \tag{12.434}$$

If the rectangle has length $\ell$ in the direction along the axis, then the electric flux is

$$\Phi_E = \int \mathbf{E} \cdot d\mathbf{a} = \int_0^r \frac{1}{2} r' \mu_0 n I_0 \omega \sin\omega t \cdot (\ell \, dr')$$

$$= \frac{1}{4} r^2 \mu_0 n I_0 \omega \ell \sin\omega t. \tag{12.435}$$

We have defined the area vector $\mathbf{a}$ to point into the page, to match up with positive $E$ being defined to point into the page (in the right half of the solenoid). We must therefore evaluate the line integral $\int \mathbf{B} \cdot d\mathbf{s}$ in a clockwise sense, to be consistent with the right-hand rule. So the integral equals $\ell\big(B(0,t) - B(r,t)\big) \equiv \ell(-\Delta B(r,t))$. Hence Eq. (12.434) yields

$$\ell\big(B(0,t) - B(r,t)\big) = \mu_0\epsilon_0 \frac{d}{dt}\left(\frac{1}{4}r^2 \mu_0 n I_0 \omega \ell \sin\omega t\right)$$

$$\implies \Delta B(r,t) = -\mu_0\epsilon_0 \left(\frac{1}{4}r^2 \mu_0 n I_0 \omega^2 \cos\omega t\right). \tag{12.436}$$

(c) Since $B_0(t) = \mu_0 n I_0 \cos\omega t$, we have

$$\frac{\Delta B(r,t)}{B_0(t)} = -\frac{\mu_0\epsilon_0 r^2 \omega^2}{4} = -\frac{r^2\omega^2}{4c^2}, \tag{12.437}$$

where we have used $\mu_0\epsilon_0 = 1/c^2$. (The negative sign in Eq. (12.437) isn't important for the overall conclusion of this problem.) The period of the current oscillation is given by $T = 2\pi/\omega$. So $\omega = 2\pi/T$, and $\Delta B(r,t)/B_0(t)$ becomes $-r^2\pi^2/c^2 T^2$. Ignoring the numerical factor, we see that $\Delta B(r,t)/B_0(t)$ is very small if $r^2/c^2 T^2$ is very small, or equivalently if $T$ is very large compared with $r/c$. But $r/c$ is the time it takes light to travel across (half of) the solenoid, as desired.

9.7    *Traveling and standing waves*

(a) The traveling $\mathbf{B}$ fields must point in the $\pm\hat{\mathbf{y}}$ directions because they must be perpendicular to both the associated $\mathbf{E}$ field and the direction of propagation, which is $\pm\hat{\mathbf{z}}$. The magnitudes of the $\mathbf{B}$ fields are $E_0/c$. The signs are determined by the fact that $\mathbf{E} \times \mathbf{B}$ points in the direction of propagation. The two magnetic waves are therefore

$$\mathbf{B}_1 = \hat{\mathbf{y}}(E_0/c)\cos(kz - \omega t),$$
$$\mathbf{B}_2 = -\hat{\mathbf{y}}(E_0/c)\cos(kz + \omega t). \tag{12.438}$$

The sum of these waves is $\mathbf{B} = \hat{\mathbf{y}}(2E_0/c)\sin kz \sin\omega t$.

(b) We use the Maxwell equation $\nabla \times \mathbf{E} = -\partial \mathbf{B}/\partial t$ to find $\mathbf{B}$. The curl of $\mathbf{E} = 2\hat{\mathbf{x}} E_0 \cos kz \cos \omega t$ is

$$\nabla \times \mathbf{E} = \begin{vmatrix} \hat{\mathbf{x}} & \hat{\mathbf{y}} & \hat{\mathbf{z}} \\ \partial/\partial x & \partial/\partial y & \partial/\partial z \\ 2E_0 \cos kz \cos \omega t & 0 & 0 \end{vmatrix}$$

$$= -\hat{\mathbf{y}} 2kE_0 \sin kz \cos \omega t. \tag{12.439}$$

Setting this equal to $-\partial \mathbf{B}/\partial t$ gives $\mathbf{B} = \hat{\mathbf{y}}(2kE_0/\omega) \sin kz \sin \omega t$. But we know that $\omega/k = c$, because the $(kz - \omega t)$ factor in the waves can be written as $k(z - (\omega/k)t)$, and the coefficient of $t$ is the speed of the wave. So the $2kE_0/\omega$ factor in $\mathbf{B}$ equals $2E_0/c$, in agreement with the result in part (a). We have ignored the constant of integration in $\mathbf{B}$ because we are concerned only with the varying part of the field. But a constant $\mathbf{B}$ field can certainly be superposed. (It must be constant in time, but it can vary with position, as long as the rest of Maxwell's equations are satisfied.)

Alternatively, you can find $\mathbf{B}$ via the Maxwell equation $\nabla \times \mathbf{B} = \mu_0 \epsilon_0 \partial \mathbf{E}/\partial t$ (with $\mu_0 \epsilon_0 = 1/c^2$). You should check that this gives the same result.

9.8 *Sunlight*

Equation (9.37) gives the power density as $S = \overline{E^2}/(377 \, \Omega)$. So we have

$$\frac{\overline{E^2}}{377 \, \Omega} = 10^3 \, \frac{\text{J}}{\text{m}^2 \text{s}} \implies E_{\text{rms}} = 614 \, \frac{\text{V}}{\text{m}}. \tag{12.440}$$

The rms magnetic field strength is then $B_{\text{rms}} = E_{\text{rms}}/c = 2.0 \cdot 10^{-6}$ T, or 0.02 gauss. This is roughly $1/20$ of the earth's magnetic field (which varies over the surface).

9.9 *Energy flow for a standing wave*

(a) For convenience, let $k \equiv 2\pi/\lambda$, $\omega \equiv 2\pi c/\lambda$, and $A \equiv 2E_0$. Then the given standing wave can be written as

$$\mathbf{E} = \hat{\mathbf{z}} A \sin ky \cos \omega t, \qquad \mathbf{B} = -\hat{\mathbf{x}}(A/c) \cos ky \sin \omega t. \tag{12.441}$$

The energy density is (using $\mu_0 = 1/\epsilon_0 c^2$)

$$\mathcal{U} = \frac{\epsilon_0 E^2}{2} + \frac{B^2}{2\mu_0} = \frac{\epsilon_0 A^2}{2}\left(\sin^2 ky \cos^2 \omega t + \cos^2 ky \sin^2 \omega t\right). \tag{12.442}$$

At the five given times, the $\cos^2 \omega t$ and $\sin^2 \omega t$ terms take on values of 0, 1/2, or 1. So the energy densities at the five times are, respectively (in units of $\epsilon_0 A^2/2$), $\sin^2 ky$, 1/2, $\cos^2 ky$, 1/2, and finally back to $\sin^2 ky$. The plots are shown in Fig. 12.127. These plots show that the energy sloshes back and forth between regions where $ky$ is close to odd multiples of $\pi/2$ and regions where it is close to even multiples of $\pi/2$.

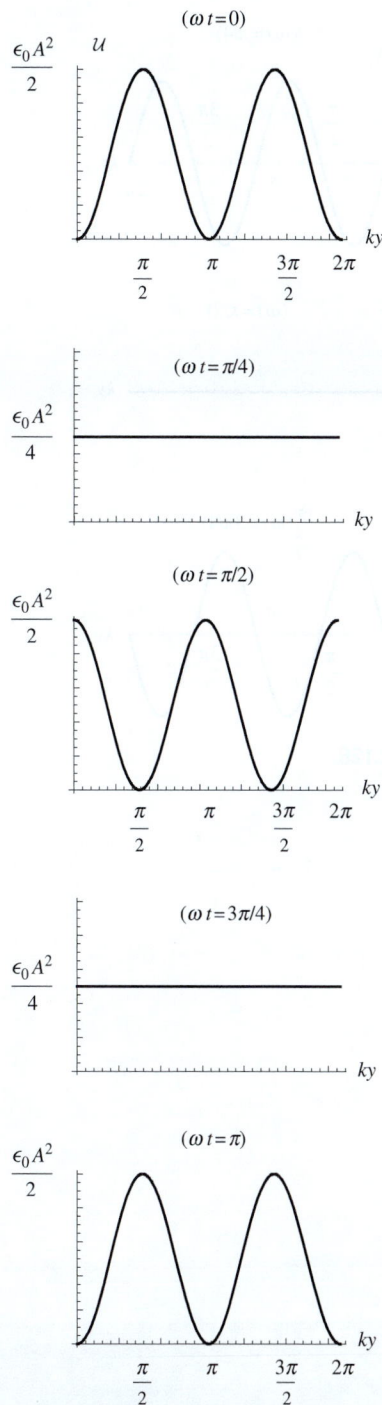

**Figure 12.127.**

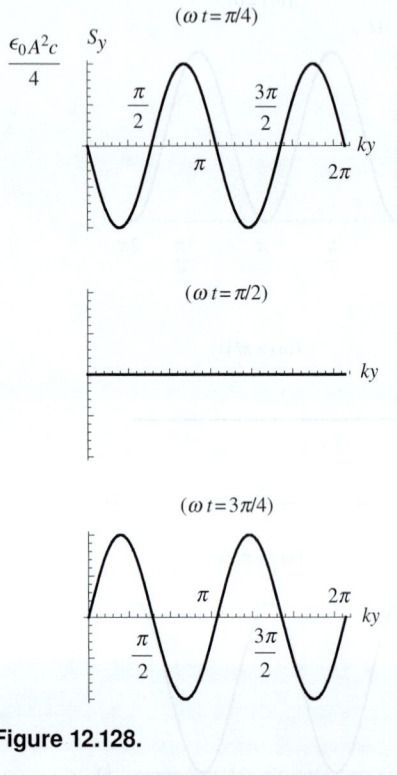

**Figure 12.128.**

(b) The Poynting vector for the standing wave is

$$\mathbf{S} = \frac{1}{\mu_0} \mathbf{E} \times \mathbf{B} = -\hat{\mathbf{y}} \frac{A^2}{\mu_0 c} \sin ky \cos ky \sin \omega t \cos \omega t$$

$$= -\hat{\mathbf{y}} \frac{\epsilon_0 A^2 c}{4} \sin 2ky \sin 2\omega t, \qquad (12.443)$$

where we have used the double-angle formula for sine, and also the relation $\mu_0 = 1/\epsilon_0 c^2$. We quickly see that the values of $S_y$ for $\omega t$ values of $\pi/4$, $\pi/2$, and $3\pi/4$ are, respectively (in units of $\epsilon_0 A^2 c/4$), $-\sin 2ky$, 0, and $\sin 2ky$. The plots are shown in Fig. 12.128. These plots show how the energy flows from one picture to the next in Fig. 12.127.

For example, consider the point $ky = \pi$ at time $\omega t = \pi/4$ in Fig. 12.128. Points to the left of this $y$ value have a positive $S_y$ (that is, a rightward flow of energy), and points to the right have a negative $S_y$. In other words, energy flows *into* the region around $ky = \pi$. This is consistent with Fig. 12.127; at $\omega t = \pi/4$, the $ky = \pi$ point is halfway through the transition between zero energy density and maximum energy density. Similarly, at $\omega t = \pi/2$ there is no flow of energy anywhere. At this time, the energy density reaches an extremum everywhere, so the flow is instantaneously at rest.

9.10   *Energy flow from a wire*

To have a nonzero Poynting vector, we need to have a nonzero magnetic field, which in our setup is the magnetic field due to the wire. Consider a thin tube of radius $b$ around the wire. The magnetic field $\mathbf{B}$ at the surface of this tube is tangential and has magnitude $\mu_0 I/2\pi b$. The electric field $\mathbf{E}$ on the surface of the tube due to the spherical shell (not the wire; see the remark at the end of the solution) is essentially parallel to the tube, with magnitude $Q/4\pi\epsilon_0 r^2$, where $r$ is the distance to the center of the shell. You can quickly verify via the right-hand rule that if the current flows toward the shell, the Poynting vector $\mathbf{S} = (\mathbf{E} \times \mathbf{B})/\mu_0$ points away from the wire. So the direction of the energy flow is correct.

Let's check that things work out quantitatively. Since $\mathbf{E}$ is perpendicular to $\mathbf{B}$, we have

$$S = \frac{EB}{\mu_0} = \frac{1}{\mu_0} \frac{Q}{4\pi\epsilon_0 r^2} \frac{\mu_0 I}{2\pi b} = \frac{QI}{(4\pi\epsilon_0 r^2)(2\pi b)}. \qquad (12.444)$$

We need to integrate this over the surface of the tube. A piece of the tube that has length $dr$ in the longitudinal direction has area $da = 2\pi b\, dr$. So we have

$$\int S\, da = \int_R^\infty \frac{QI}{(4\pi\epsilon_0 r^2)(2\pi b)} 2\pi b\, dr = \int_R^\infty \frac{QI}{4\pi\epsilon_0 r^2}\, dr$$

$$= \frac{Q(dQ/dt)}{4\pi\epsilon_0 R} = \frac{d}{dt}\left(\frac{Q^2}{8\pi\epsilon_0 R}\right). \qquad (12.445)$$

This is correct, because the energy stored in the electric field is $Q^2/8\pi\epsilon_0 R$. This can be seen in various ways. One way is to integrate $\epsilon_0 E^2/2$ over the volume external to the shell. But a quicker way is to use $U = (1/2) \int \rho\phi \, dv$. Since $\rho$ is nonzero only on the spherical shell, and since the potential takes on the constant value of $\phi = Q/4\pi\epsilon_0 R$ there, we have

$$U = \frac{1}{2}\phi \int \rho \, dv = \frac{1}{2}\frac{Q}{4\pi\epsilon_0 R}Q = \frac{Q^2}{8\pi\epsilon_0 R}, \qquad (12.446)$$

as desired. (Equivalently, you can use $U = Q^2/2C$.) The flux of $\mathbf{S}$ therefore does indeed equal the rate of change of the energy stored in the electric field. Note that the magnetic field is constant, so we don't have to worry about its energy since it doesn't change.

In the example in Section 9.6.2, we found that energy flows into the capacitor through the opening between the edges of the disks. But where does this energy flow originate? As in this problem, it originates at the wires carrying the current to (and from) the capacitor. (Of course, backing up another step, the initial energy source must be a battery or some other emf.) Problem 2.16 gives the field external to the capacitor, along the axis. As an exercise you can show that the flow works out quantitatively.

REMARK: In the above solution, we said that the electric field near the wire is directed antiparallel to the current. But doesn't the electric field *inside* the wire point *parallel* to the current (due to $\mathbf{J} = \sigma\mathbf{E}$), and doesn't this then imply that the electric field very close to the wire has a tangential component parallel to the current (because curl $\mathbf{E} = 0$), instead of antiparallel? Yes, and yes. However, as long as our wire is very thin (so that the capacitance is small) and the conductivity $\sigma$ is very high (so that the required $\mathbf{E}$ field is small), we can still consider a thin (but not too thin) tube with radius $b$ such that the field on the surface is essentially equal to the field due to the shell. This is true because, although there may be strong fields very close to the wire due to the surface charges that maintain the $\mathbf{E}$ field inside, these fields fall off quickly with distance. As mentioned at the end of Section 9.6.2, there are three types of Poynting-vector energy flow here. There is a flow *parallel* to the wire, and this flow branches off into a flow *toward* the wire to provide the resistance heating, plus another flow *away from* the wire to increase the energy density of the electric field throughout space.

9.11 *Momentum in an electromagnetic field*

The Poynting vector, $\mathbf{S} = (\mathbf{E} \times \mathbf{B})/\mu_0$, gives the amount of energy flow per area per time. Since $p = E/c$, the momentum flow per area per time is $\mathbf{S}/c = (\mathbf{E} \times \mathbf{B})/\mu_0 c$. That is, the amount of momentum that passes through a cross-sectional area $A$ during a time $t$ is $\mathbf{p} = (At)(\mathbf{E} \times \mathbf{B})/\mu_0 c$. This is the momentum that would be gained by an object that absorbed all of the wave.

We can also write the momentum $\mathbf{p}$ in terms of the momentum density, which we will label as $\tilde{\mathbf{p}}$. The volume of the wave that passes through an area $A$ during a time $t$ is $A(ct)$, because the wave travels at speed $c$. The amount of momentum that passes through the area $A$ during time $t$ is therefore $\mathbf{p} = \tilde{\mathbf{p}}(Act)$, by definition. Equating this with the above expression for $\mathbf{p}$ gives the momentum density as $\tilde{\mathbf{p}} = (\mathbf{E} \times \mathbf{B})/\mu_0 c^2$.

It is possible to show that electromagnetic waves carry momentum, without invoking relativity and by using only properties of waves. However, it's a bit tricky, although the main idea is contained in Exercise 9.21.

9.12    *Angular momentum paradox*

(a) From Faraday's law, the induced electric field at radius $r$ inside the solenoid is given by (ignoring the signs)

$$\mathcal{E} = \frac{d\Phi_B}{dt} \implies E \cdot 2\pi r = \pi r^2 \frac{dB}{dt} \implies E = \frac{r}{2}\frac{dB}{dt}. \quad (12.447)$$

By Lenz's law, since the magnetic field initially points out of the page in Fig. 9.13, and since it decreases over time, the induced $\mathbf{E}$ is counterclockwise. So the inner cylinder will rotate counterclockwise, and the outer (negative) cylinder will rotate clockwise. The force on a small piece of charge $dq$ on one of the (nonconducting) cylinders is $E\,dq$, so the torque on the piece is $rE\,dq$. The magnitude of the total torque on a cylinder is therefore $rEQ$, where $r$ equals either $a$ or $b$. Using the above form of $E$, the total torques on the two cylinders are then $(a^2 Q/2)(dB/dt)$ counterclockwise on the inner cylinder at radius $a$, and $(b^2 Q/2)(dB/dt)$ clockwise on the outer cylinder at radius $b$.

To find the total change in angular momentum of the cylinders, we must integrate the torques with respect to time. But the integral of $dB/dt$ is simply $B_0$ (or technically $-B_0$, but we've already taken care of the signs). The final angular momenta of the cylinders are therefore $a^2 Q B_0/2$ counterclockwise and $b^2 Q B_0/2$ clockwise.

(b) Taking clockwise as positive, the total final angular momentum of the cylinders is

$$L_{\text{cylinders}}^{\text{final}} = \frac{Q B_0 (b^2 - a^2)}{2}. \quad (12.448)$$

This is not zero. The initial angular momentum of the cylinders was initially zero because they were initially at rest, so it appears that angular momentum isn't conserved. However, while it is indeed true that the angular momentum of the two *cylinders* isn't conserved, the total angular momentum of the entire *system* is in fact conserved. So we must ask ourselves, what else does the system consist of?

From Problem 9.11, we know that an electromagnetic field carries momentum if $\mathbf{E} \times \mathbf{B}$ is nonzero. This means that it is also possible for the field to carry *angular* momentum. From Problem 9.11 the momentum density is $\tilde{\mathbf{p}} = (\mathbf{E} \times \mathbf{B})/\mu_0 c^2$. Both $\mathbf{E}$ and $\mathbf{B}$ must be nonzero for this to be nonzero. There is no $\mathbf{B}$ field anywhere in the final state (assuming the cylinders are rotating very slowly), so the resolution to our paradox must be that there is nonzero angular momentum in the field in the initial state. The $\mathbf{E}$ field is nonzero only between the two cylinders, where it takes on the value $E = (Q/\ell)/2\pi\epsilon_0 r$, where $\ell$ is the length (assumed to be long) of the cylinders. This field is radial, so it is perpendicular to the $\mathbf{B}$ field, which points out of the page. The momentum density therefore has magnitude

$$\tilde{p} = \frac{E B_0}{\mu_0 c^2} = \frac{Q B_0}{2\pi r \ell \epsilon_0 \mu_0 c^2} = \frac{Q B_0}{2\pi r \ell}, \quad (12.449)$$

where we have used $\mu_0 \epsilon_0 = 1/c^2$. From the right-hand rule, $\tilde{\mathbf{p}}$ points in the clockwise direction. The angular momentum is obtained by multiplying the momentum by $r$, so the angular momentum density is $r\tilde{p}$. Therefore, the total initial angular momentum contained in the field in the region between the two cylinders points in the clockwise direction and has magnitude

$$L_{\text{field}}^{\text{initial}} = \int r\tilde{p}\, dv = \int_a^b r\left(\frac{QB_0}{2\pi r\ell}\right)(2\pi r\ell\, dr)$$

$$= QB_0 \int_a^b r\, dr = \frac{QB_0(b^2 - a^2)}{2}. \qquad (12.450)$$

This is the total initial angular momentum of the system (which is all contained in the field), and it equals the final angular momentum we found above in Eq. (12.448) (which is all contained in the cylinders). So the angular momentum of the system is indeed conserved.

Here's another paradox (you should cover up the next paragraph so you can have some fun thinking about this): what if we eliminate the cylinder with radius $b$, so that we have only one cylinder with charge $Q$ and radius $a$ (along with the solenoid). The final angular momentum of the remaining cylinder will still be $-a^2 QB_0/2$ (the negative sign means counterclockwise). But the initial angular momentum of the field inside the solenoid is $QB_0(R^2 - a^2)/2$, from the above reasoning; the $E$ field now extends out to (and beyond) radius $R$, so we can replace $b$ with $R$ in Eq. (12.450). These initial and final angular momenta aren't equal, so it appears that angular momentum isn't conserved. Is this the case? Explain why or why not. (A qualitative explanation is fine.)

Angular momentum is still conserved. In the reasoning in the previous paragraph, we took into account only the angular momentum of the field *inside* the solenoid. But there is now angular momentum *outside* too, because we have a nonzero external electric field. (In the original setup, we avoided this complication by choosing two cylinders with opposite charges.) You might argue that we're still saved from having to worry about any external angular momentum because the external $B$ field is zero. However, this is not the case. No matter how long the solenoid is, the field lines still have to loop back around from one end to the other. To be sure, the external $B$ field will be small if the length of the solenoid is large. But it will still be large enough to have an effect, as we will see below.

If angular momentum is to be conserved, the initial angular momentum must contain an additional $-R^2 QB_0/2$ (counterclockwise) contribution from the external field, so that the total initial angular momentum (from both the internal and external fields) will agree with what we know the final is, namely $-a^2 QB_0/2$ from the cylinder. This is consistent with the $a \to 0$ limiting case. The external $E$ field is independent of $a$ (neglecting end effects), assuming that we keep the charge on the cylinder as $Q$. So the external angular momentum, whatever it is, is likewise independent of $a$. In the $a \to 0$ limit, the final angular momentum is certainly zero, because all of the mass is located right on the axis. The initial angular momentum of the field must therefore be zero (assuming angular momentum is conserved), which means that the external-field angular momentum must be equal and opposite to the internal-field angular momentum. And the latter equals $R^2 Q B_0/2$, from the reasoning in the original setup, with $b \to R$ and $a \to 0$ in Eq. (12.450). This argument shows that the angular momentum in the small external field can't just be ignored.

If you're still wondering where exactly in the external field this angular momentum exists, consider the following dimensional argument. (We'll need to invoke a few things from Chapter 11, although we technically know enough to derive the results here from scratch.) Let the solenoid and cylinder have length $\ell$. The $E$ field very far away ($r \gg \ell$) behaves like $1/r^2$, because the cylinder looks like a point charge. And the $B$ field behaves like $1/r^3$, because the solenoid looks effectively like a single loop, so we can invoke the general $1/r^3$ magnetic dipole behavior from Chapter 11. You can then quickly show that very large values of $r$ contribute negligibly to the angular momentum. (The $r$ factor in the angular momentum and the $r^2\,dr$ in the volume integral aren't enough to outweigh the $1/r^5$ factor from the product of the fields.)

Essentially all of the external contribution therefore comes from $r$ values within order $\ell$ from the solenoid. We claim that, in this region, $E$ and $B$ are both of order $1/\ell^2$. More precisely, for $E$, the cylinder looks roughly like a line, so $E \sim \lambda/2\pi\epsilon_0 r \sim (Q/\ell)/\epsilon_0\ell = Q/\epsilon_0\ell^2$. For $B$, we invoke the result from Exercise 11.18 in Chapter 11 that states that $B$ behaves like $B_0 R^2/\ell^2$, where $B_0$ is the internal field. The momentum density therefore behaves like $EB/\mu_0 c^2 \sim (Q/\epsilon_0\ell^2)(B_0 R^2/\ell^2)/\mu_0 c^2 = R^2 Q B_0/\ell^4$. We must multiply this by $r \sim \ell$ to obtain the angular momentum density, and then by the volume $\sim \ell^3$ to obtain the total angular momentum. All of the $\ell$ factors cancel, and we end up with a result that behaves like $R^2 Q B_0$. This is independent of $\ell$ and has the same dependence on the other parameters as the desired quantity, $-R^2 Q B_0/2$. The direction is correct too, because the "returning" external $B$ field lines point generally in the direction opposite to the internal $B$ field lines.

## 12.10 Chapter 10

10.1   *Leaky cell membrane*

(a) We are given the capacitance per area as $C/A = 1\,\mu\text{F/cm}^2 = 0.01\,\text{F/m}^2$. Since $C = \kappa\epsilon_0 A/s$, the thickness $s$ is

$$s = \frac{\kappa\epsilon_0}{C/A} = \frac{(3)\left(8.85 \cdot 10^{-12}\,\frac{\text{s}^2\,\text{C}^2}{\text{kg m}^3}\right)}{0.01\,\text{F/m}^2} = 2.7 \cdot 10^{-9}\,\text{m}. \quad (12.451)$$

(b) Since we are dealing with a dielectric, the time constant derived in Section 4.11 becomes

$$t = RC = \frac{\rho s}{A} \cdot \frac{\kappa\epsilon_0 A}{s} = \kappa\epsilon_0\rho. \quad (12.452)$$

We see that $t = RC$ is independent of $A$, because $R \propto 1/A$ and $C \propto A$. (Basically, if a given patch of the membrane leaks its charge on a given time scale, then putting a bunch of these patches together shouldn't change the time scale, because each patch doesn't care that there are others next to it.) Using the information given for 1 cm$^2$ of the membrane, we have $t = RC = (1000\,\Omega)(10^{-6}\,\text{F}) = 10^{-3}$ s.

Since $R = \rho s/A$, the resistivity is given by

$$\rho = \frac{RA}{s} = \frac{(1000\,\Omega)(10^{-4}\,\text{m}^2)}{2.7 \cdot 10^{-9}\,\text{m}} \approx 4 \cdot 10^7 \text{ ohm-m}. \qquad (12.453)$$

From Fig. 4.8, this is a little more than 100 times the resistivity of pure water.

10.2 *Force on a dielectric*

(a) The equivalent capacitance of two capacitors in parallel is simply the sum of the capacitances. (The rule is opposite to that for resistors; see Problem 3.18.) The capacitance of the part with the dielectric is $\kappa$ times what it would be if there were vacuum there. So the total capacitance is given by

$$C = C_1 + C_2 = \frac{\epsilon_0 A_1}{s} + \frac{\kappa \epsilon_0 A_2}{s}$$

$$= \frac{\epsilon_0 a(b - x)}{s} + \frac{\kappa \epsilon_0 a x}{s} = \frac{\epsilon_0 a}{s}\big[b + (\kappa - 1)x\big]. \qquad (12.454)$$

The stored energy is then

$$U = \frac{Q^2}{2C} = \frac{Q^2 s}{2\epsilon_0 a[b + (\kappa - 1)x]}. \qquad (12.455)$$

Note that as $x$ changes, the charge stays constant (by assumption), but the potential does not. So the $Q\phi/2$ and $C\phi^2/2$ forms of the energy aren't useful.

(b) The force is

$$F = -\frac{dU}{dx} = \frac{Q^2 s(\kappa - 1)}{2\epsilon_0 a[b + (\kappa - 1)x]^2}. \qquad (12.456)$$

The positive sign here means that the force points in the direction of increasing $x$. That is, the dielectric slab is pulled into the capacitor. But it's risky to trust this sign blindly. Physically, the force points in the direction of decreasing energy. And we see from the above expression for $U$ that the energy decreases as $x$ increases (because $\kappa > 1$).

The force $F$ is correctly zero if $\kappa = 1$, because in that case we don't actually have a dielectric. The $\kappa \to \infty$ limit corresponds to a conductor. In that case, both $U$ and $F$ are zero. Basically, all of the charge on the plates shifts to the overlap $x$ region, and compensating charge gathers there in the dielectric, so in the end there is no field anywhere. Note that $F$ decreases as $x$ increases. You should think about why this is the case. *Hint:* First convince yourself why the force

(a)

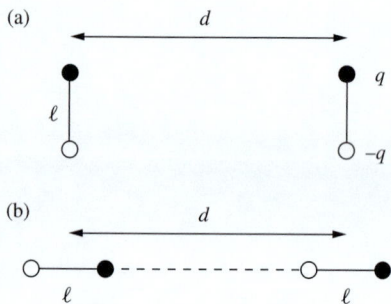

(b)

**Figure 12.129.**

should be proportional to the product of the charge *densities* (and not the total charges) on the two parts of the plates. And then look at Exercise 10.15.

**10.3** *Energy of dipoles*

The first configuration is shown in Fig. 12.129(a). There are four relevant (non-internal) pairs of charges, so the potential energy is (with $\ell \ll d$)

$$
U = \frac{1}{4\pi\epsilon_0}\left(2\cdot\frac{q^2}{d} - 2\cdot\frac{q^2}{\sqrt{d^2+\ell^2}}\right) = \frac{2q^2}{4\pi\epsilon_0 d}\left(1 - \frac{1}{\sqrt{1+\ell^2/d^2}}\right)
$$

$$
\approx \frac{2q^2}{4\pi\epsilon_0 d}\left(1 - \left(1 - \frac{\ell^2}{2d^2}\right)\right) = \frac{q^2\ell^2}{4\pi\epsilon_0 d^3} \equiv \frac{p^2}{4\pi\epsilon_0 d^3}, \tag{12.457}
$$

where we have used $1/\sqrt{1+\epsilon} \approx 1 - \epsilon/2$. The second configuration is shown in Fig. 12.129(b). The potential energy is now

$$
\frac{1}{4\pi\epsilon_0}\left(2\cdot\frac{q^2}{d} - \frac{q^2}{d-\ell} - \frac{q^2}{d+\ell}\right) = \frac{q^2}{4\pi\epsilon_0 d}\left(2 - \frac{1}{1-\ell/d} - \frac{1}{1+\ell/d}\right)
$$

$$
\approx \frac{q^2}{4\pi\epsilon_0 d}\left(2 - \left(1 + \frac{\ell}{d} + \frac{\ell^2}{d^2}\right) - \left(1 - \frac{\ell}{d} + \frac{\ell^2}{d^2}\right)\right)
$$

$$
= \frac{q^2}{4\pi\epsilon_0 d}\left(-\frac{2\ell^2}{d^2}\right) = -\frac{p^2}{2\pi\epsilon_0 d^3}, \tag{12.458}
$$

where we have used $1/(1+\epsilon) \approx 1 - \epsilon + \epsilon^2$. Note that we needed to go to second order in the Taylor expansions here. By looking at the initial expressions for $U$ for each setup, it is clear why the first $U$ is positive, but not so clear why the second $U$ is negative. However, in the limit where the dipoles nearly touch, the second $U$ is certainly negative.

**10.4** *Dipole polar components*

Remember that our convention for the angle $\theta$ is that it is measured down from the $z$ axis in Fig. 10.6. So the radial unit vector is given by $\hat{\mathbf{r}} = \sin\theta\,\hat{\mathbf{x}} + \cos\theta\,\hat{\mathbf{z}}$. The tangential unit vector, which is perpendicular to $\hat{\mathbf{r}}$, is then given by $\hat{\boldsymbol{\theta}} = \cos\theta\,\hat{\mathbf{x}} - \sin\theta\,\hat{\mathbf{z}}$; this makes the dot product of $\hat{\mathbf{r}}$ and $\hat{\boldsymbol{\theta}}$ equal to zero, and you can check that the overall sign is correct. Inverting these expressions for $\hat{\mathbf{r}}$ and $\hat{\boldsymbol{\theta}}$ gives

$$
\hat{\mathbf{x}} = \sin\theta\,\hat{\mathbf{r}} + \cos\theta\,\hat{\boldsymbol{\theta}} \quad\text{and}\quad \hat{\mathbf{z}} = \cos\theta\,\hat{\mathbf{r}} - \sin\theta\,\hat{\boldsymbol{\theta}}. \tag{12.459}
$$

Therefore,

$$
\mathbf{E} = E_x\hat{\mathbf{x}} + E_z\hat{\mathbf{z}}
$$

$$
= E_x(\sin\theta\,\hat{\mathbf{r}} + \cos\theta\,\hat{\boldsymbol{\theta}}) + E_z(\cos\theta\,\hat{\mathbf{r}} - \sin\theta\,\hat{\boldsymbol{\theta}})
$$

$$
= \hat{\mathbf{r}}(E_x\sin\theta + E_z\cos\theta) + \hat{\boldsymbol{\theta}}(E_x\cos\theta - E_z\sin\theta)
$$

$$
= \frac{p}{4\pi\epsilon_0 r^3}\left(\hat{\mathbf{r}}\Big[(3\sin\theta\cos\theta)\sin\theta + (3\cos^2\theta - 1)\cos\theta\Big]\right.
$$

$$
\left.+ \hat{\boldsymbol{\theta}}\Big[(3\sin\theta\cos\theta)\cos\theta - (3\cos^2\theta - 1)\sin\theta\Big]\right). \tag{12.460}
$$

Using $\sin^2\theta + \cos^2\theta = 1$ in the $\hat{\mathbf{r}}$ term, $\mathbf{E}$ quickly simplifies to

$$\mathbf{E} = \frac{p}{4\pi\epsilon_0 r^3}\left(2\cos\theta\,\hat{\mathbf{r}} + \sin\theta\,\hat{\boldsymbol{\theta}}\right), \qquad (12.461)$$

as desired. Alternatively, $E_r$ equals the projection of $\mathbf{E} = (E_x, E_z)$ onto $\hat{\mathbf{r}} = (\sin\theta, \cos\theta)$. Since $\hat{\mathbf{r}}$ is a unit vector, this projection equals the dot product $\mathbf{E}\cdot\hat{\mathbf{r}}$. Therefore,

$$E_r = \mathbf{E}\cdot\hat{\mathbf{r}} = (E_x, E_z)\cdot(\sin\theta, \cos\theta) = E_x\sin\theta + E_z\cos\theta, \quad (12.462)$$

in agreement with the third line in Eq. (12.460). Likewise,

$$E_\theta = \mathbf{E}\cdot\hat{\boldsymbol{\theta}} = (E_x, E_z)\cdot(\cos\theta, -\sin\theta) = E_x\cos\theta - E_z\sin\theta, \quad (12.463)$$

again in agreement with the third line in Eq. (12.460).

10.5    *Average field*

(a) From part (c) of Problem 1.28 we know that the average electric field over the volume of a sphere of radius $R$, due to a given charge $q$ at radius $r < R$, has magnitude $qr/4\pi\epsilon_0 R^3$ and points toward the center (if $q$ is positive). In vector form, this average field can be written as $-q\mathbf{r}/4\pi\epsilon_0 R^3$. If we sum this over all the charges inside the sphere, then the numerator becomes $\sum q_i\mathbf{r}_i$ (or $\int \mathbf{r}\rho\,dv$ if we have a continuous charge distribution). But this sum is, by definition, the dipole moment $\mathbf{p}$, where $\mathbf{p}$ is measured relative to the center. So the average field over the volume of the sphere is $\mathbf{E}_{\text{avg}} = -\mathbf{p}/4\pi\epsilon_0 R^3$, as desired. Note that all that matters here is the dipole moment; the monopole moment (the total charge) doesn't come into play.

(b) Since $\mathbf{E}_{\text{avg}}$ is proportional to $1/R^3$, and since volume is proportional to $R^3$, the total integral of $\mathbf{E}$ over the volume of a sphere is independent of $R$ (provided that $R$ is large enough to contain all the charges). This means that if we increase the radius by $dR$, we don't change the integral of $\mathbf{E}$. This implies that the average value of $\mathbf{E}$ over the *surface* of any sphere containing all the charges equals zero. (We actually already knew this from part (a) of Problem 1.28. Each individual charge yields zero average field over the surface.) A special case of this result is the centered point-dipole case in Exercise 10.25.

   So for the specific case shown in Fig. 10.32(a), the average value of the field over the surface of the sphere is zero. And since the dipole moment has magnitude $p = 2q\ell$ and points upward, the result from part (a) tells us that the average value over the volume of the sphere, $\mathbf{E}_{\text{avg}} = -\mathbf{p}/4\pi\epsilon_0 R^3$, has magnitude $q\ell/2\pi\epsilon_0 R^3$ and points downward.

(c) The average value of the field over the surface of the sphere in Fig. 10.32(b) is *not* zero. From part (b) of Problem 1.28, the average field due to each charge has magnitude $q/4\pi\epsilon_0\ell^2$ and points downward. So the average field over the surface, due to both charges, has magnitude $q/2\pi\epsilon_0\ell^2$ and points downward. Since this is independent of the radius of the sphere, the average field over the volume of a sphere with $R < \ell$ also has magnitude $q/2\pi\epsilon_0\ell^2$ and points downward.

The moral of all this is that "outside" the dipole, the field points in various directions and averages out over the surface of a sphere. But "inside" the dipole, the field points generally in one direction, so the average is nonzero over the surface of a sphere.

Note that *volume* average of $\mathbf{E}$ is continuous as $R$ crosses the $R = \ell$ cutoff between the two cases in parts (a) and (b); in both cases it has magnitude $q/2\pi\epsilon_0\ell^2$. If we multiply this by $\ell/\ell$ and use $p = q\ell$, we can write it as $p/2\pi\epsilon_0\ell^3$. Multiplying by the volume $4\pi\ell^3/3$ then tells us that the total volume integral of $\mathbf{E}$, over a sphere of radius $\ell$, has magnitude $2p/3\epsilon_0$ and points downward. In other words, for a fixed value of $p$, even the limit of an idealized dipole still has a nonzero value of $\int \mathbf{E}\, dv$, despite the fact that the only shells yielding nonzero contributions are infinitesimal ones.

### 10.6  *Quadrupole tensor*

Our goal is to find the potential $\phi(\mathbf{r})$ at the point $\mathbf{r} = (x_1, x_2, x_3)$. As in Section 10.2, primed coordinates will denote the position of a point in the charge distribution. The distance from $\mathbf{r}$ to a particular point $\mathbf{r}' = (x_1', x_2', x_3')$ in the distribution is

$$R = \sqrt{(x_1 - x_1')^2 + (x_2 - x_2')^2 + (x_3 - x_3')^2}$$

$$= r\sqrt{1 + \frac{r'^2}{r^2} - \frac{2\sum x_i x_i'}{r^2}} = r\sqrt{1 + \frac{r'^2}{r^2} - \frac{2\sum \hat{x}_i x_i'}{r}}, \quad (12.464)$$

where we have used $\sum x_i^2 = r^2$ and $\sum x_i'^2 = r'^2$, and where $(\hat{x}_1, \hat{x}_2, \hat{x}_3) = (x_1, x_2, x_3)/r$ is the unit vector $\hat{\mathbf{r}}$ in the $\mathbf{r}$ direction. Assuming that $r'$ is much smaller than $r$, we can use the expansion $(1 + \delta)^{-1/2} = 1 - \delta/2 + 3\delta^2/8 - \cdots$ to write (dropping terms of order $1/r^4$ and higher)

$$\frac{1}{R} = \frac{1}{r}\left[1 + \frac{\sum \hat{x}_i x_i'}{r} + \frac{3\left(\sum \hat{x}_i x_i'\right)^2}{2r^2} - \frac{r'^2}{2r^2}\right]$$

$$= \frac{1}{r}\left[1 + \frac{\sum \hat{x}_i x_i'}{r} + \frac{3\left(\sum \hat{x}_i x_i'\right)^2}{2r^2} - \frac{\left(\sum \hat{x}_i^2\right)r'^2}{2r^2}\right]. \quad (12.465)$$

In the last term here, we have multiplied by 1 in the form of the square of the length of a unit vector, for future purposes. It is easier to understand this result for $1/R$ if we write it in terms of vectors and matrices:

$$\frac{1}{R} = \frac{1}{r} + \frac{1}{r^2}(\hat{x}_1, \hat{x}_2, \hat{x}_3) \cdot \begin{pmatrix} x_1' \\ x_2' \\ x_3' \end{pmatrix} \quad (12.466)$$

$$+ \frac{1}{2r^3}(\hat{x}_1, \hat{x}_2, \hat{x}_3) \cdot \begin{pmatrix} 3x_1'^2 - r'^2 & 3x_1'x_2' & 3x_1'x_3' \\ 3x_2'x_1' & 3x_2'^2 - r'^2 & 3x_2'x_3' \\ 3x_3'x_1' & 3x_3'x_2' & 3x_3'^2 - r'^2 \end{pmatrix} \begin{pmatrix} \hat{x}_1 \\ \hat{x}_2 \\ \hat{x}_3 \end{pmatrix}.$$

You should verify that this is equivalent to Eq. (12.465). If desired, the diagonal terms of this matrix can be written in a slightly different form. Since $r'^2 = x_1'^2 + x_2'^2 + x_3'^2$, the upper left entry equals $2x_1'^2 - x_2'^2 - x_3'^2$. Likewise for the other two diagonal entries. Note that there are only five independent entries in the matrix, because it is symmetric and has trace zero.

To obtain $\phi(\mathbf{r})$, we must compute the integral,

$$\phi(\mathbf{r}) = \frac{1}{4\pi\epsilon_0} \int \frac{\rho(\mathbf{r}')dv'}{R}. \tag{12.467}$$

In other words, we must compute the volume integral of Eq. (12.466) times $\rho(\mathbf{r}')$, and then tack on a $1/4\pi\epsilon_0$. When the $1/r$ term is integrated, it simply gives $q/r$, where $q$ is the total charge in the distribution. To write the other two terms in a cleaner way, define the vector $\mathbf{p}$ to be the vector whose entries are the $\rho\,dv'$ integrals of the entries in the above $(x_1', x_2', x_3')$ vector. And likewise define the matrix $\mathbf{Q}$ to be the $\rho\,dv'$ integral of the above matrix. For example, the first component of $\mathbf{p}$ and the upper-left entry of $\mathbf{Q}$ are

$$p_1 = \int x_1'\rho(\mathbf{r}')dv' \quad \text{and} \quad Q_{11} = \int \left(3x_1'^2 - r'^2\right)\rho(\mathbf{r}')dv', \tag{12.468}$$

and so on. We can then write the result for the potential at an arbitrary point $\mathbf{r}$ in the compact form,

$$\phi(\mathbf{r}) = \frac{1}{4\pi\epsilon_0}\left[\frac{q}{r} + \frac{\hat{\mathbf{r}}\cdot\mathbf{p}}{r^2} + \frac{\hat{\mathbf{r}}\cdot\mathbf{Q}\hat{\mathbf{r}}}{2r^3}\right]. \tag{12.469}$$

The advantage of Eq. (12.469) over Eq. (10.9) in the text is the following. The latter gives the correct value of $\phi$ at points on the $z$ axis. However, if we want to find $\phi$ at another point, we must redefine $\theta$ as the angle with respect to the direction to the new point, and then recalculate all the $K_i$. The present result in Eq. (12.469) has the benefit that, although it involves calculating a larger number of quantities, it is valid for any choice of the point $\mathbf{r}$. The quantities $q$, $\mathbf{p}$, and $\mathbf{Q}$ depend *only on the distribution,* and not on the point $\mathbf{r}$ at which we want to calculate the potential. Conversely, the quantities $\hat{\mathbf{r}}$ and $r$ in Eq. (12.469) depend only on $\mathbf{r}$ and not on the distribution. So, for a given charge distribution, we can calculate (with respect to a given set of coordinate axes) $\mathbf{p}$ and $\mathbf{Q}$ once and for all. We then simply need to plug our choice of $\mathbf{r}$ into Eq. (12.469), and this correctly gives $\phi(\mathbf{r})$ up to order $1/r^3$.

In the special case where $\mathbf{r}$ lies on the $z \equiv x_3$ axis, we have $\hat{\mathbf{r}} = (0,0,1)$. Since only $\hat{x}_3$ is nonzero, only $Q_{33}$ (the lower right entry in $\mathbf{Q}$) survives in the dot product $\hat{\mathbf{r}}\cdot\mathbf{Q}\hat{\mathbf{r}}$. Furthermore, if $\theta$ is the angle of $\mathbf{r}'$ with respect to the $x_3$ axis, then we have $x_3' = r'\cos\theta$. So $Q_{33} = \int r'^2(3\cos^2\theta - 1)\rho\,dv'$. When the $1/2r^3$ factor in Eq. (12.469) is included, we correctly arrive at the result Eq. (10.9).

For a spherical shell, which we know has only a monopole moment, you can quickly verify that all of the entries in $\mathbf{Q}$ are zero. The off-diagonal entries are zero from symmetry, and the diagonal elements are zero due to the example in Section 10.2 combined with the previous paragraph. Alternatively, the average value of, say, $x_1'^2$ over the surface of a sphere equals $r'^2/3$, because it has the same average value as $x_2'^2$ and $x_3'^2$, and the sum of all three averages is $r'^2$. If you want to get some practice with $\mathbf{Q}$, Exercise 10.26 deals with the quadrupole arrangement in Fig. 10.5.

### 10.7 Force on a dipole

Let the dipole consist of a charge $-q$ at position $\mathbf{r}$ and a charge $q$ at position $\mathbf{r} + \mathbf{s}$. Then the dipole vector is $\mathbf{p} = q\mathbf{s}$. If the dipole is placed in an electric field $\mathbf{E}$, the net force on it is

$$\mathbf{F} = (-q)\mathbf{E}(\mathbf{r}) + q\mathbf{E}(\mathbf{r} + \mathbf{s}). \tag{12.470}$$

The $x$ component of this is $F_x = (-q)E_x(\mathbf{r}) + qE_x(\mathbf{r}+\mathbf{s})$. Now, the change in a function $f$ due to a small displacement $\mathbf{s}$ is $\nabla f \cdot \mathbf{s}$, by the definition of the gradient (or at least that's one way of defining it). So we can write $F_x$ as

$$F_x = q\big[E_x(\mathbf{r} + \mathbf{s}) - E_x(\mathbf{r})\big] = q\nabla E_x \cdot \mathbf{s}$$
$$= (q\mathbf{s}) \cdot \nabla E_x \equiv \mathbf{p} \cdot \nabla E_x, \tag{12.471}$$

as desired. Likewise for the other two components.

### 10.8 Force from an induced dipole

If $q$ is the charge of the ion, then the magnitude of the electric field of the ion at the location of the atom is $E = q/4\pi\epsilon_0 r^2$. If the polarizability of the atom is $\alpha$, then the induced dipole moment of the atom is $p = \alpha E = \alpha q/4\pi\epsilon_0 r^2$. This dipole moment points along the line from the ion to the atom (see Fig. 12.130), so the magnitude of the field of the induced dipole at the location of the ion is $E_{\text{dipole}} = 2p/4\pi\epsilon_0 r^3$. The magnitude of the force on the ion is therefore

$$F = qE_{\text{dipole}} = \frac{2pq}{4\pi\epsilon_0 r^3} = \frac{2(\alpha q/4\pi\epsilon_0 r^2)q}{4\pi\epsilon_0 r^3} = \frac{2\alpha q^2}{(4\pi\epsilon_0)^2 r^5}. \tag{12.472}$$

You can quickly show that the force is attractive for either sign of $q$. The potential energy relative to infinity is

$$U(r) = -\int_\infty^r F(r')dr' = -\int_\infty^r -\frac{2\alpha q^2\, dr'}{(4\pi\epsilon_0)^2 r'^5} = -\frac{\alpha q^2}{2(4\pi\epsilon_0)^2 r^4}. \tag{12.473}$$

The polarizability of sodium is given by $\alpha/4\pi\epsilon_0 = 27 \cdot 10^{-30}$ m$^3$. If the magnitude of the potential energy equals $|U| = 4 \cdot 10^{-21}$ J, then solving for $r$ and setting $q = e$ gives

$$r = \left[\frac{(\alpha/4\pi\epsilon_0)q^2}{2(4\pi\epsilon_0)|U|}\right]^{1/4} = \left[\frac{(27 \cdot 10^{-30}\ \text{m}^3)(1.6 \cdot 10^{-19}\ \text{C})^2}{2 \cdot 4\pi\left(8.85 \cdot 10^{-12}\ \frac{\text{s}^2\,\text{C}^2}{\text{kg m}^3}\right)(4 \cdot 10^{-21}\ \text{J})}\right]^{1/4}$$
$$= 9.4 \cdot 10^{-10}\ \text{m}. \tag{12.474}$$

If $r$ is larger than this, then (on average) the thermal energy is sufficient to kick the ion out to infinity.

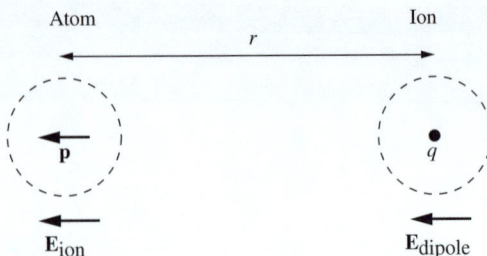

Atom　　　　　　　　　　　　Ion

$r$

$\mathbf{p}$　　　　　　　　　　　　$q$

$\mathbf{E}_{\text{ion}}$　　　　　　　　　$\mathbf{E}_{\text{dipole}}$

Figure 12.130.

10.9   *Polarized water*

We must determine the number, $n$, of molecules of water per cubic centimeter. A mole of something with molecular mass $M$ has a mass of $M$ grams. (Equivalently, since the proton mass is $1.67 \cdot 10^{-24}$ g, it takes $1/(1.67 \cdot 10^{-24}) = 6 \cdot 10^{23}$ protons to make 1 gram, and this number is essentially Avogadro's number.) Water has a molecular weight of 18, so the number of water molecules per gram ($= \text{cm}^3$) is $n = (6 \cdot 10^{23}/\text{mole})/(18 \, \text{cm}^3/\text{mole}) = 3.33 \cdot 10^{22} \, \text{cm}^{-3}$. The dipole moment of water can be written as $p = 6.13 \cdot 10^{-28}$ C-cm. Assuming the dipoles all point down, the polarization density is therefore

$$P = np = (3.33 \cdot 10^{22} \, \text{cm}^{-3})(6.13 \cdot 10^{-28} \, \text{C-cm}) = 2.04 \cdot 10^{-5} \, \text{C/cm}^2.$$
$$(12.475)$$

From the reasoning in Section 10.7, this is the surface charge density, $\sigma$. The number of electrons per square centimeter it corresponds to is $\sigma/e = (2.04 \cdot 10^{-5} \, \text{C/cm}^2)/(1.6 \cdot 10^{-19} \, \text{C}) = 1.3 \cdot 10^{14} \, \text{cm}^{-2}$. This is somewhat smaller than the number of surface molecules per square centimeter, which equals $n^{2/3} = 1.0 \cdot 10^{15} \, \text{cm}^{-2}$ because each edge of the $1 \, \text{cm}^3$ cube is (approximately) $n^{1/3}$ molecules long.

10.10   *Tangent field lines*

Consider the Gaussian surface indicated by the heavy line in Fig. 12.131. The side part of the surface is constructed to follow the field lines, so there is no flux there. Likewise, there is no flux through the top circular face, because the field is zero outside the capacitor plates. So the only flux comes from the great circle inside the sphere. From Eq. (10.53) the field inside the sphere has the uniform value of $3\mathbf{E}_0/(2+\kappa)$. So the flux out of the Gaussian surface equals $-\pi R^2 \cdot 3E_0/(2+\kappa)$, where the minus arises because the flux is inward.

The total charge enclosed in the Gaussian surface comes from two places: the negative charge in the circle on the upper capacitor plate, and the positive charge on the upper hemisphere. The former is simply $q_{\text{cap}} = (-\sigma)\pi r^2 = (-\epsilon_0 E_0)\pi r^2$, where we have used the fact that the charge densities on the capacitor plates are what cause the uniform field $E_0$; hence $E_0 = \sigma/\epsilon_0$. The latter charge is just $q_{\text{sph}} = P\pi R^2$, where $P$ is the polarization, because the top patch of the column in Fig. 10.21(a) has a charge of $P \, da$ (where $da$ is the *horizontal* cross-sectional area), independent of the tilt angle of the actual end face. And all the $da$ areas simply add up to the great-circle area, $\pi R^2$. (Or you could just integrate the $P\cos\theta$ surface density over the hemisphere.) Using the value of $P$ from Eq. (10.54), Gauss's law gives

$$\Phi = \frac{1}{\epsilon_0}\left(q_{\text{cap}} + q_{\text{sph}}\right)$$

$$\implies -\pi R^2 \frac{3E_0}{\kappa + 2} = \frac{1}{\epsilon_0}\left(-\epsilon_0 E_0 \pi r^2 + 3\frac{\kappa - 1}{\kappa + 2}\epsilon_0 E_0 \cdot \pi R^2\right)$$

$$\implies -3R^2 \frac{1}{\kappa + 2} = -r^2 + 3R^2 \frac{\kappa - 1}{\kappa + 2}$$

$$\implies r = R\sqrt{\frac{3\kappa}{\kappa + 2}}. \qquad (12.476)$$

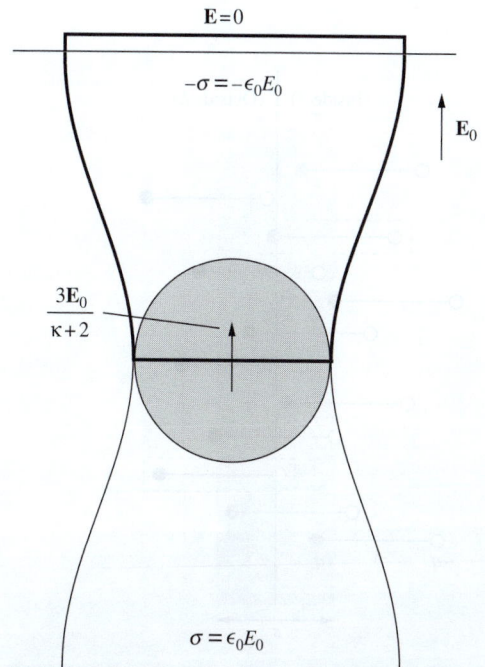

$E = 0$

$-\sigma = -\epsilon_0 E_0$

$E_0$

$\dfrac{3E_0}{\kappa + 2}$

$\sigma = \epsilon_0 E_0$

**Figure 12.131.**

As a check, we have $r = R$ when $\kappa = 1$. In this case, our dielectric is just vacuum, so the field remains $\mathbf{E}_0$ everywhere; the field lines are all straight. Also, we have $r = \sqrt{3}R$ when $\kappa \to \infty$. In this limit the sphere is a conductor. The factor of $\sqrt{3}$ isn't so obvious. Note that, in the case of a conductor, a field line can't actually be tangent to the surface, because field lines must always be perpendicular to the surface of a conductor. What happens is that the external field approaches zero along the equator (the zero vector is, in some sense, both parallel and perpendicular to the surface). But a tiny distance away from the equator, the field is nonzero, so it is meaningful to ask where that field line ends up on the distant capacitor plates.

10.11 *Bound charge and divergence of P*

If we take the volume integral of both sides of Eq. (10.61) and use the divergence theorem, we see that our goal is to show that $\int_S \mathbf{P} \cdot d\mathbf{a} = -q_{\text{bound}}$, where $q_{\text{bound}}$ is the bound charge enclosed within the surface $S$.

Assume that the polarization $\mathbf{P}$ arises from $N$ dipoles per unit volume, each with a dipole moment $\mathbf{p} = q\mathbf{s}$. Then $\mathbf{P} = N\mathbf{p} = Nq\mathbf{s}$. If the dipoles point in random directions, so that $\mathbf{P} = 0$, then there is no extra bound charge in a given volume. But if they are aligned, so that $\mathbf{P} \neq 0$, and if additionally $\mathbf{P}$ varies with position, then there may be a net bound charge in the volume. The reasoning is as follows.

Consider a collection of dipoles, as shown in Fig. 12.132. The vertical line represents a patch of the right-hand surface of $S$. How much extra negative charge is there inside $S$, that is, to the left of the line? If a given dipole lies entirely inside or outside $S$, then it contributes nothing to the net charge. But if a dipole is cut by the vertical line, then there is an extra charge of $-q$ inside $S$.

How many dipoles are cut by the line? Any dipole whose center lies within $s/2$ of the line gets cut by it. So the center must lie in a slab with thickness $s$, indicated by the shaded region in the figure. The two extreme dipole positions are indicated by the boxes. If the area of a given patch of the surface is $da$, then any dipole whose center lies in a slab of volume $s\,da$ will contribute a charge of $-q$ to $S$. Since there are $N$ dipoles per unit volume, we see that $N(s\,da)$ dipoles are cut by the line. The extra charge associated with the patch is therefore $dq_{\text{bound}} = N(s\,da)(-q)$, which can be written as $dq_{\text{bound}} = -(Nqs)da = -P\,da$.

If a dipole is tilted at an angle $\theta$ with respect to the normal to the patch, then the volume of the relevant slab is decreased by a factor of $\cos\theta$. If we tack this factor onto $P$, it simply turns $P$ into the component $P_\perp$ perpendicular to the surface. So in general the extra charge inside the volume, near a given patch with area $da$, equals $dq_{\text{bound}} = -P_\perp\,da$, which can be written as the dot product, $dq_{\text{bound}} = -\mathbf{P} \cdot d\mathbf{a}$. Integrating this over the entire surface gives the total enclosed bound charge as

$$q_{\text{bound}} = -\int \mathbf{P} \cdot d\mathbf{a}, \tag{12.477}$$

as desired.

Although we motivated this result in Section 10.11 by considering dielectrics, this problem shows (as mentioned in the text) that this result is

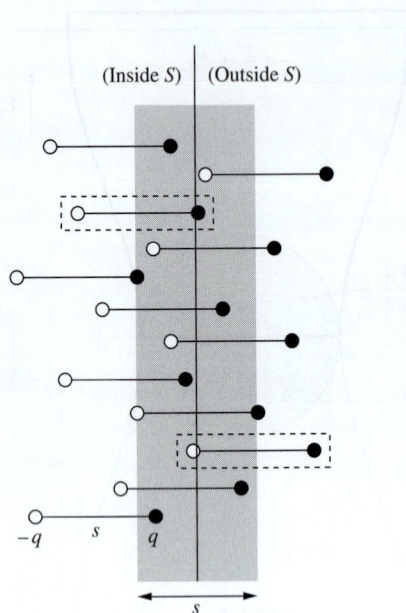

(Inside $S$)   (Outside $S$)

$-q$   $s$   $q$

$s$

**Figure 12.132.**

quite independent of dielectrics. No matter how the polarization **P** comes about, the result in Eq. (12.477) is still valid. (You can manually twist the dipoles in whatever way you want, provided that **P** changes slowly on the length scale of the dipoles, so that we can talk about smooth averages.) To emphasize what we said in the text, the logical route to Eq. (10.62) is to start with Eqs. (10.59) and (10.61), both of which are universally true, and then Eq. (10.62) immediately follows. No mention has been made of dielectrics. But if we are in fact dealing with a (linear) dielectric, then $\mathbf{P} = \chi_e \epsilon_0 \mathbf{E}$, and we can use $1 + \chi_e = \kappa$ to say that additionally

$$\epsilon_0 \mathbf{E} + \mathbf{P} = \epsilon_0 \mathbf{E} + \chi_e \epsilon_0 \mathbf{E} = \kappa \epsilon_0 \mathbf{E} \equiv \epsilon \mathbf{E}. \qquad (12.478)$$

In all cases the relation $\mathbf{D} \equiv \epsilon_0 \mathbf{E} + \mathbf{P}$ holds, but that is just a definition.

10.12 *Boundary conditions on D*

$D_\perp$ is continuous. This follows from div $\mathbf{D} = \rho_{\text{free}}$; there is no free charge in the setup, so the divergence of **D** is zero. The divergence theorem then tells us that $\int \mathbf{D} \cdot d\mathbf{a} = 0$ for any closed surface. That is, there is zero flux through any surface. So if we draw a pancake-like pillbox with one face just inside the slab and one face just outside, the inward flux through one face must equal the outward flux through the other. Hence $D_\perp^{\text{in}} A = D_\perp^{\text{out}} A \Longrightarrow D_\perp^{\text{in}} = D_\perp^{\text{out}}$. That is, $D_\perp$ is continuous across the boundary.

For $D_\parallel$, we know that $E_\parallel$ is continuous across the boundary, because all we have at the boundary is a layer of bound charge, which produces no discontinuity in $E_\parallel$. So $\mathbf{D} \equiv \epsilon_0 \mathbf{E} + \mathbf{P}$ tells us that the discontinuity in $D_\parallel$ is the same as the discontinuity in $P_\parallel$. Since $\mathbf{P} = 0$ outside, the discontinuity in $P_\parallel$ is simply $-P_\parallel^{\text{in}}$. That is, the change in $D_\parallel$ when going from inside to outside is $-P_\parallel^{\text{in}}$.

10.13 *Q for a leaky capacitor*

From Exercise 10.42, the energy density in the electric field is $\epsilon E^2 / 2$. And it is the same for the magnetic field, by plugging $B = \sqrt{\mu_0 \epsilon} E$ into $B^2 / 2\mu_0$. The total energy density is therefore $\epsilon E^2$, or $\epsilon E_0^2 \cos^2 \omega t$. But the time average of $\cos^2 \omega t$ is $1/2$, so the average energy density is $\epsilon E_0^2 / 2$.

The energy in the fields will decay due to ohmic resistance. To calculate this power dissipation, consider a tube of cross-sectional area $A$ and length $L$. The power dissipated in this tube is

$$P = I^2 R = (JA)^2 (\rho L / A) = J^2 \rho (AL)$$
$$= (\sigma E)^2 \frac{1}{\sigma} (\text{volume}) = \sigma E^2 (\text{volume}). \qquad (12.479)$$

The power dissipated per unit volume is therefore $\sigma E^2$. The time average of this is $\sigma E_0^2 / 2$. Hence

$$Q = \frac{\omega \cdot (\text{energy stored})}{\text{power loss}} = \frac{\omega (\epsilon E_0^2 / 2)}{\sigma E_0^2 / 2} = \frac{\omega \epsilon}{\sigma}, \qquad (12.480)$$

as desired. From Table 4.1, the conductivity of seawater is $\sigma = 4\,(\text{ohm-m})^{-1}$. And from Fig. 10.29, the dielectric constant $\kappa$ is still about 80 at a frequency of 1000 MHz ($10^9$ Hz). Therefore, since $\epsilon = \kappa\epsilon_0$, we have

$$Q = \frac{(2\pi \cdot 10^9\,\text{s}^{-1})\left(80 \cdot 8.85 \cdot 10^{-12}\,\frac{\text{s}^2\,\text{C}^2}{\text{kg}\,\text{m}^3}\right)}{4\,(\text{ohm-m})^{-1}} = 1.1. \tag{12.481}$$

Since $Q$ equals the number of radians of $\omega t$ required for the energy to decrease by a factor of $1/e$, we see that by the end of one cycle ($2\pi$ radians) there is practically no energy left. The wavelength corresponding to 1000 MHz is $(c/\sqrt{\kappa})/\nu = 0.033$ m. So microwave radar won't find submarines!

10.14 *Boundary conditions on E and B*
With no free charges or currents, the equations describing the system are

$$\nabla \cdot \mathbf{D} = 0, \qquad \nabla \times \mathbf{E} = -\partial\mathbf{B}/\partial t;$$
$$\nabla \cdot \mathbf{B} = 0, \qquad \nabla \times \mathbf{B} = \mu_0\,\partial\mathbf{D}/\partial t. \tag{12.482}$$

The two equations involving $\mathbf{D}$ come from Eqs. (10.64) and (10.78) with $\rho_{\text{free}}$ and $\mathbf{J}_{\text{free}}$ set equal to zero. The other two equations are two of Maxwell's equations. We can now apply the standard arguments. For the perpendicular components, we can apply the divergence theorem to the two "div" equations, with the volume chosen to be a squat pillbox, of vanishing thickness, spanning the surface. Our equations tell us that the net flux out of the volume is zero, so the perpendicular field on one side must equal the perpendicular field on the other. And for the parallel components, we can apply Stokes' theorem to the two "curl" equations, with the area chosen to be a thin rectangle, of vanishing area, spanning the surface. Our equations tell us that the line integral around the rectangle is zero, so the parallel field on one side must equal the parallel field on the other. (The finite non-zero entries on the right-hand sides of the curl equations are inconsequential, because they provide zero contribution when integrated over the area of an infinitesimally thin rectangle.) The above four equations therefore yield (with 1 and 2 labeling the two regions)

$$D_{1,\perp} = D_{2,\perp}, \qquad E_{1,\parallel} = E_{2,\parallel};$$
$$B_{1,\perp} = B_{2,\perp}, \qquad B_{1,\parallel} = B_{2,\parallel}. \tag{12.483}$$

Since $\mathbf{D} = \epsilon\mathbf{E}$ for a linear dielectric, the first of these equations gives

$$\epsilon_1 E_{1,\perp} = \epsilon_2 E_{2,\perp}. \tag{12.484}$$

So $E_\perp$ is discontinuous. But the other three components are continuous across the boundary. That is, the entire $\mathbf{B}$ field is continuous, as is the parallel component of $\mathbf{E}$.

Note that we are assuming that the materials aren't magnetic. After reading Section 11.10, you can show that in magnetic materials there is a discontinuity in $B_\parallel$.

## 12.11 Chapter 11

11.1 *Maxwell's equations with magnetic charge*

Maxwell's equations with only electric charge and electric current are given in Eq. (9.17). If magnetic charge existed, the last equation would have to be replaced, as discussed in Section 11.2, by $\nabla \cdot \mathbf{B} = b_1 \eta$, where $\eta$ is the magnetic charge density, and $b_1$ is a constant that depends on how the unit of magnetic charge is chosen. With the conventional definition of the direction of $\mathbf{B}$, a positive magnetic charge would be attracted to the north pole of the earth, so it would behave like the north pole of a compass.

Magnetic charge in motion with velocity $\mathbf{v}$ would constitute a magnetic current. Let $\mathbf{K}$ be the magnetic current density. Then $\mathbf{K} = \eta \mathbf{v}$, in analogy with $\mathbf{J} = \rho \mathbf{v}$. Conservation of magnetic charge would then be expressed by the "continuity equation," $\nabla \cdot \mathbf{K} = -\partial \eta / \partial t$, in analogy with $\nabla \cdot \mathbf{J} = -\partial \rho / \partial t$.

A magnetic current would be the source of an electric field, just as an electric current is the source of a magnetic field. So we must add to the right side of the first Maxwell equation in Eq. (9.17) a term proportional to $\mathbf{K}$. (Equivalently, if we didn't add such a term, we would end up with a contradiction, similar to the one in Section 9.1, arising from the fact that $\nabla \cdot (\nabla \times \mathbf{E}) = 0$ is identically zero.) Let this new term be $b_2 \mathbf{K}$. Then we have

$$\nabla \times \mathbf{E} = -\frac{\partial \mathbf{B}}{\partial t} + b_2 \mathbf{K}. \tag{12.485}$$

To determine the constant $b_2$, we can take the divergence of both sides of this equation. The left-hand side is identically zero because $\nabla \cdot (\nabla \times \mathbf{E}) = 0$, so we have (using the continuity equation)

$$
\begin{aligned}
0 &= -\nabla \cdot \left( \frac{\partial \mathbf{B}}{\partial t} \right) + b_2 \nabla \cdot \mathbf{K} \\
&= -\frac{\partial}{\partial t} (\nabla \cdot \mathbf{B}) + b_2 \left( -\frac{\partial \eta}{\partial t} \right) \\
&= -\frac{\partial}{\partial t} (b_1 \eta) - b_2 \frac{\partial \eta}{\partial t} \\
&= -(b_1 + b_2) \frac{\partial \eta}{\partial t}. \tag{12.486}
\end{aligned}
$$

Therefore $b_2$ must equal $-b_1$. So the generalized Maxwell's equations take the form (with $b \equiv b_1 = -b_2$),

$$\nabla \times \mathbf{E} = -\frac{\partial \mathbf{B}}{\partial t} - b\mathbf{K},$$

$$\nabla \times \mathbf{B} = \mu_0 \epsilon_0 \frac{\partial \mathbf{E}}{\partial t} + \mu_0 \mathbf{J},$$

$$\nabla \cdot \mathbf{E} = \frac{\rho}{\epsilon_0},$$

$$\nabla \cdot \mathbf{B} = b\eta. \tag{12.487}$$

The constant $b$ can be chosen arbitrarily. Two common conventions are $b = 1$ and $b = \mu_0$.

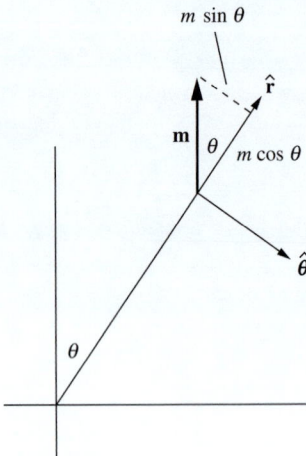

$m \sin \theta$

$\hat{\mathbf{r}}$

$\mathbf{m}$    $\theta$    $m \cos \theta$

$\hat{\boldsymbol{\theta}}$

$\theta$

$\theta$

**Figure 12.133.**

11.2   *Magnetic dipole*

If we treat the current loop like an exact dipole, then the dipole moment is $m = Ia = I\pi b^2$. Equation (11.15) gives the magnetic field at position $z$ along the axis of the dipole as $\mu_0 m/2\pi z^3$, which here equals $\mu_0(I\pi b^2)/2\pi z^3 = \mu_0 I b^2/2z^3$.

If we treat the current loop (correctly) as a loop of finite size, then Eq. (6.53) gives the field at position $z$ on the axis as $B_z = \mu_0 I b^2/2(z^2 + b^2)^{3/2}$. For $z \gg b$ we can ignore the $b^2$ term in the denominator, yielding $B_z \approx \mu_0 I b^2/2z^3$, which agrees with the above result for the idealized dipole.

The correct result is smaller than the idealized-dipole result by the factor $z^3/(z^2 + b^2)^{3/2}$. This factor approaches 1 as $z \to \infty$. It is larger than a given number $\eta$ (we are concerned with $\eta = 0.99$) if

$$\frac{z^3}{(z^2 + b^2)^{3/2}} > \eta \implies \frac{z^2}{z^2 + b^2} > \eta^{2/3} \implies z > \frac{\eta^{1/3}b}{\sqrt{1 - \eta^{2/3}}}.$$

(12.488)

For $\eta = 0.99$ this gives $z > (12.2)b$. You can show that if we want the factor to be larger than $1 - \epsilon$ (so $\epsilon = 0.01$ here), then to a good approximation (in the limit of small $\epsilon$) we need $z/b > \sqrt{3/2\epsilon}$. And indeed, $\sqrt{3/2(0.01)} = \sqrt{150} = 12.2$.

11.3   *Dipole in spherical coordinates*

Using the $\nabla \times (\mathbf{A} \times \mathbf{B})$ vector identity from Appendix K, with $\mathbf{m}$ constant, we find (ignoring the $\mu_0/4\pi$ for now)

$$\mathbf{B} \propto \nabla \times \left[\mathbf{m} \times (\hat{\mathbf{r}}/r^2)\right] = \mathbf{m}\left(\nabla \cdot (\hat{\mathbf{r}}/r^2)\right) - (\mathbf{m} \cdot \nabla)(\hat{\mathbf{r}}/r^2).$$   (12.489)

But the divergence of $\hat{\mathbf{r}}/r^2$ is zero (except at $r = 0$), because we know that the divergence of the Coulomb field is zero; alternatively we can just use the expression for the divergence in spherical coordinates. So we are left with only the second term. Therefore, using the expression for $\nabla$ in spherical coordinates,

$$\mathbf{B} \propto -\left(m_r \frac{\partial}{\partial r} + m_\theta \frac{1}{r}\frac{\partial}{\partial \theta}\right)\frac{\hat{\mathbf{r}}}{r^2}.$$   (12.490)

In the $\partial/\partial r$ term here, the vector $\hat{\mathbf{r}}$ doesn't depend on $r$, but $r^2$ does, of course, so $m_r(\partial/\partial r)(\hat{\mathbf{r}}/r^2) = -2m_r\hat{\mathbf{r}}/r^3$. In the $\partial/\partial \theta$ term, $r^2$ doesn't depend of $\theta$, but the vector $\hat{\mathbf{r}}$ *does*. If we increase $\theta$ by $d\theta$, then $\hat{\mathbf{r}}$ changes direction by the angle $d\theta$. Since $\hat{\mathbf{r}}$ has length 1, it therefore picks up a component with length $d\theta$ in the $\hat{\boldsymbol{\theta}}$ direction. See Fig. F.3 in Appendix F; that figure is relevant to the oppositely defined $\theta$ in cylindrical coordinates, but the result is the same. Hence $\partial\hat{\mathbf{r}}/\partial\theta = \hat{\boldsymbol{\theta}}$. So we have $(m_\theta/r)(\partial/\partial\theta)(\hat{\mathbf{r}}/r^2) = m_\theta\hat{\boldsymbol{\theta}}/r^3$.

Finally, in Fig. 12.133 we see that the components of the fixed vector $\mathbf{m} = m\hat{\mathbf{z}}$ relative to the local $\hat{\mathbf{r}}$-$\hat{\boldsymbol{\theta}}$ basis are $m_r = m\cos\theta$, and $m_\theta = -m\sin\theta$. The negative sign here comes from the fact that $\mathbf{m}$ points partially in the direction of decreasing $\theta$ (at least for the right half of the sphere). Putting this all together, and bringing the $\mu_0/4\pi$ back in, gives

$$\mathbf{B} = -\frac{\mu_0}{4\pi} \left( -2(m\cos\theta)\frac{\hat{\mathbf{r}}}{r^3} + (-m\sin\theta)\frac{\hat{\boldsymbol{\theta}}}{r^3} \right)$$

$$= \hat{\mathbf{r}}\frac{\mu_0 m}{2\pi r^3}\cos\theta + \hat{\boldsymbol{\theta}}\frac{\mu_0 m}{4\pi r^3}\sin\theta, \qquad (12.491)$$

in agreement with Eq. (11.15).

11.4  *Force on a dipole*

(a) The expression $(\mathbf{m} \cdot \nabla)\mathbf{B}$ is shorthand for

$$(\mathbf{m} \cdot \nabla)\mathbf{B} = \left( m_x\frac{\partial}{\partial x} + m_y\frac{\partial}{\partial y} + m_z\frac{\partial}{\partial z} \right)(B_x, B_y, B_z). \qquad (12.492)$$

The operator in parentheses is to be applied to each of the three components of $\mathbf{B}$, generating the three components of a vector. In the setup in Fig. 11.9 with the ring and diverging $\mathbf{B}$ field, $m_z$ is the only nonzero component of $\mathbf{m}$. Also, $B_x$ and $B_y$ are identically zero on the $z$ axis, so $\partial B_x/\partial z$ and $\partial B_y/\partial z$ are both zero (or negligibly small close to the $z$ axis). Therefore only one of the nine possible terms in Eq. (12.492) survives, and we have

$$(\mathbf{m} \cdot \nabla)\mathbf{B} = \left( 0, 0, m_z\frac{\partial B_z}{\partial z} \right), \qquad (12.493)$$

as desired.

(b) The expression $\nabla(\mathbf{m} \cdot \mathbf{B})$ is shorthand for

$$\nabla(\mathbf{m} \cdot \mathbf{B}) = \left( \frac{\partial}{\partial x}, \frac{\partial}{\partial y}, \frac{\partial}{\partial z} \right)(m_x B_x + m_y B_y + m_z B_z). \qquad (12.494)$$

Each derivative acts on the whole sum in the parentheses. But again, only $m_z$ is nonzero. Also, on the $z$ axis, $B_z$ doesn't depend on $x$ or $y$, to first order (because, by symmetry, $B_z$ achieves a maximum or minimum on the $z$ axis, so the slope as a function of $x$ and $y$ must be zero). Hence $\partial B_z/\partial x$ and $\partial B_z/\partial y$ are both zero (or negligibly small close to the $z$ axis). So again only one term survives and we have

$$\nabla(\mathbf{m} \cdot \mathbf{B}) = \left( 0, 0, m_z\frac{\partial B_z}{\partial z} \right), \qquad (12.495)$$

as desired.

(c) Let's first see what the two expressions yield for the force on the given square loop. Then we will calculate what the force actually is. The dipole moment $\mathbf{m}$ points out of the page with magnitude $I$(area), so we have $\mathbf{m} = \hat{\mathbf{z}}Ia^2$. Using the above expressions for $(\mathbf{m} \cdot \nabla)\mathbf{B}$ and $\nabla(\mathbf{m} \cdot \mathbf{B})$ in Eqs. (12.492) and (12.494), we obtain

$$(\mathbf{m} \cdot \nabla)\mathbf{B} = \left( 0 + 0 + (Ia^2)\frac{\partial}{\partial z} \right)(0, 0, B_0 x) = (0, 0, 0) \qquad (12.496)$$

and

$$\nabla(\mathbf{m} \cdot \mathbf{B}) = \left( \frac{\partial}{\partial x}, \frac{\partial}{\partial y}, \frac{\partial}{\partial z} \right) (0 + 0 + (Ia^2)B_0 x) = (Ia^2 B_0, 0, 0).$$
(12.497)

We see that the first expression yields zero force on the loop, while the second yields a force of $Ia^2 B_0$ in the positive $x$ direction.

Let's now explicitly calculate the force. We quickly find that the net force on the top side of the square is zero (the right half cancels the left half). Likewise for the bottom side. Alternatively, the corresponding pieces of the top and bottom sides have canceling forces. So we need only look at the left and right sides. By the right-hand rule, the force on the right side is directed to the right with magnitude $IB\ell = I(B_0 a/2)(a) = IB_0 a^2/2$. The force on the left side also points to the right (both $I$ and $\mathbf{B}$ switch sign) with the same magnitude. The total force is therefore $F = IB_0 a^2$ in the positive $x$ direction, in agreement with Eq. (12.497). So $\nabla(\mathbf{m} \cdot \mathbf{B})$ is the correct expression for the force. (Actually, all that we've done is rule out the $(\mathbf{m} \cdot \nabla)\mathbf{B}$ force. But $\nabla(\mathbf{m} \cdot \mathbf{B})$ is in fact correct in all cases.)

11.5　*Converting $\chi_m$*

Consider a setup in which the SI quantities are $M = 1$ amp/m and $B = 1$ tesla. Then $\chi_m = \mu_0 M/B = 4\pi \cdot 10^{-7}$. You can verify that the units do indeed cancel so that $\chi_m$ is dimensionless.

How would someone working with Gaussian units describe this setup? Since 1 amp/m equals $(3 \cdot 10^9 \text{ esu/s})/(100 \text{ cm})$, this would be the value of $M$ in Gaussian units if there weren't the extra factor of $c$ in the definition of $m$. This factor reduces the value of all dipole moments $m$ (and hence all magnetizations $M$) by $3 \cdot 10^{10}$ cm/s. The value of $M$ in Gaussian units is therefore

$$M = \frac{3 \cdot 10^9 \text{ esu/s}}{100 \text{ cm}} \frac{1}{3 \cdot 10^{10} \text{ cm/s}} = 10^{-3} \frac{\text{esu}}{\text{cm}^2}.$$
(12.498)

Both of the factors of 3 here are actually 2.998, so this result is exact.

The magnetic field in Gaussian units that corresponds to 1 tesla is $10^4$ gauss, so the susceptibility in Gaussian units for the given setup is

$$\chi_m = \frac{M}{B} = \frac{10^{-3} \text{ esu/cm}^2}{10^4 \text{ gauss}} = 10^{-7} \frac{\text{esu}}{\text{cm}^2 \text{ gauss}} = 10^{-7}.$$
(12.499)

The units do indeed cancel, because the expression for the Lorentz force tells us that a gauss has the units of force per charge. So the units of $\chi_m$ are $\text{esu}^2/(\text{cm}^2 \cdot \text{force})$. And these units cancel, as you can see by looking at the units in Coulomb's law. The above value of $\chi_m$ in SI units was $4\pi \cdot 10^{-7}$, which is $4\pi$ times the Gaussian value, as desired.

11.6　*Paramagnetic susceptibility of liquid oxygen*

Equation (11.20) gives the force on a magnetic moment as $F = m(\partial B_z/\partial z)$. Using the data in Table 11.1, and taking upward to be positive for all quantities, the magnetic moment of a $10^{-3}$ kg sample is

$$m = \frac{F}{\partial B_z / \partial z} = \frac{-7.5 \cdot 10^{-2} \,\text{N}}{-17 \,\text{T/m}} = 4.4 \cdot 10^{-3} \,\text{J/T}. \qquad (12.500)$$

The magnetic susceptibility is defined via $\mathbf{M} = \chi_m \mathbf{B}/\mu_0$. (The accepted $\mathbf{M} = \chi_m \mathbf{H}$ definition would give essentially the same result, because $\chi_m$ will turn out to be very small. See Exercise 11.38.) The volume of 1 gram of liquid oxygen is $V = (10^{-3} \,\text{kg})/(850 \,\text{kg/m}^3) = 1.18 \cdot 10^{-6} \,\text{m}^3$. So

$$\chi_m = \frac{M}{B/\mu_0} = \frac{(m/V)}{B/\mu_0} = \frac{m\mu_0}{BV}$$

$$= \frac{(4.4 \cdot 10^{-3} \,\text{J/T})(4\pi \cdot 10^{-7} \,\text{kg m/C}^2)}{(1.8 \,\text{T})(1.18 \cdot 10^{-6} \,\text{m}^3)} = 2.6 \cdot 10^{-3}. \qquad (12.501)$$

**11.7** *Rotating shell*

For the magnetized sphere, we know from Eq. (11.55) that near the equator the surface current density is equal to $M$, because the sphere looks essentially like a cylinder there (the surface is parallel to $\mathbf{M}$). But away from the equator, the surface is tilted with respect to $\mathbf{M}$. From the example at the end of Section 11.8, the surface current density is given by $\mathcal{J} = M_{\parallel} \implies \mathcal{J}(\theta) = M \sin\theta$, where $\theta$ is the angle down from the top of the sphere (assuming that $\mathbf{M}$ points up).

Now consider a rotating sphere with uniform surface charge density $\sigma$. The surface current density at any point is $\mathcal{J} = \sigma v$, where $v = \omega(R\sin\theta)$ is the speed due to the rotation. Hence $\mathcal{J}(\theta) = \sigma \omega R \sin\theta$. The $\mathcal{J}(\theta)$ expressions for the magnetized and rotating spheres have the same functional dependence on $\theta$, so they will be equal for all $\theta$ provided that $M = \sigma \omega R$.

**11.8** *B inside a magnetized sphere*

(a) The field in Eq. (11.15) is obtained from the field in Eq. (10.18) by letting $p \to m$ and $\epsilon_0 \to 1/\mu_0$. If we replace all the electric dipoles $p$ in a polarized sphere with magnetic dipoles $m$, then at an external point, the field from each dipole is simply multiplied by $(m/p)(\mu_0\epsilon_0)$. The integral over all the dipole fields is multiplied by this same factor, so the new magnetic field at any external point equals $(m/p)(\mu_0\epsilon_0)$ times the old electric field. We know from Section 10.9 that the old external electric field is the same as the field due to an electric dipole with strength $p_0 = (4\pi R^3/3)P$, with $P = Np$, located at the center. You can quickly check that $(m/p)(\mu_0\epsilon_0)$ times this field is the same as the magnetic field due to a magnetic dipole with strength $m_0 = (4\pi R^3/3)M$, with $M = Nm$.

(b) If $\mathbf{m_0}$ points in the $z$ direction, then from Eq. (11.12) the Cartesian components of $\mathbf{A}$ at points $(x, y, z)$ on the surface of the sphere are

$$A_x = -\frac{\mu_0}{4\pi} \frac{m_0 y}{R^3} = -\mu_0 \frac{My}{3},$$

$$A_y = \frac{\mu_0}{4\pi} \frac{m_0 x}{R^3} = \mu_0 \frac{Mx}{3},$$

$$A_z = 0. \qquad (12.502)$$

Note that the result from Problem 11.7 then tells us that the $\mathbf{A}$ on the surface of a spinning spherical shell equals $(\mu_0 \sigma \omega R/3)(-y, x, 0)$. This agrees with the $\mathbf{A}$ we found in a different manner in Problem 6.7.

Recall from Section 6.3 that $A_x$ satisfies $\nabla^2 A_x = -\mu_0 J_x$. And similarly for $A_y$. But $\mathbf{J} = 0$ inside the sphere, so both $A_x$ and $A_y$ satisfy Laplace's equation there. By the uniqueness theorem, this means that if we can find a solution to Laplace's equation inside the sphere that satisfies the boundary conditions on the surface of the sphere, then we know that we have found *the* solution. And just as with the $\phi$ for the polarized sphere in Section 10.9, the solutions for $A_x$ and $A_y$ are easy to come by. They are simply the functions given in Eq. (12.502); their second derivatives are zero, so they each satisfy Laplace's equation. The magnetic field inside the sphere is then

$$\mathbf{B} = \nabla \times \mathbf{A} = \frac{\mu_0 M}{3} \begin{vmatrix} \hat{\mathbf{x}} & \hat{\mathbf{y}} & \hat{\mathbf{z}} \\ \partial/\partial x & \partial/\partial y & \partial/\partial z \\ -y & x & 0 \end{vmatrix} = \frac{2\mu_0 M}{3} \hat{\mathbf{z}}. \quad (12.503)$$

Like the $\mathbf{E}$ inside the polarized sphere, this $\mathbf{B}$ is uniform and points vertically. But that is where the similarities end. This $\mathbf{B}$ field points upward, whereas the old $\mathbf{E}$ field pointed downward. Additionally, the numerical factor here is $2/3$, whereas it was (negative) $1/3$ in $\mathbf{E}$. The $2/3$ is exactly what is needed to make the component normal to the surface be continuous, and to make the tangential component have the proper discontinuity (see Exercise 11.31).

Equation (12.503), combined with the result from Problem 11.7, tells us that the field throughout the interior of a spinning spherical shell is uniform and has magnitude $2\mu_0 \sigma \omega R/3$. This is consistent with the result from Problem 6.11 for the field at the center of the sphere.

11.9 *B at the north pole of a solid rotating sphere*
From Problem 11.7, we know that the magnetic field due to a spinning shell with radius $r$ and uniform surface charge density $\sigma$ is the same (both inside and outside) as the field due to a sphere with uniform magnetization $M_r = \sigma \omega r$. And then from Problem 11.8 we know that the external field of a magnetized sphere is that of a dipole with strength $m = (4\pi r^3/3)M_r$ located at the center. So the (radial) field at radius $R$ outside a spinning shell with radius $r$ (at a point located above the north pole) is

$$B = \frac{\mu_0 m}{2\pi R^3} = \frac{\mu_0}{2\pi R^3} \frac{4\pi r^3 (\sigma \omega r)}{3} = \frac{2\mu_0 \sigma \omega r^4}{3R^3}. \quad (12.504)$$

We can consider the solid spinning sphere to be the superposition of many spinning shells with radii ranging from $r = 0$ to $r = R$, with uniform surface charge density $\sigma = \rho \, dr$. The north pole of the solid sphere is outside all of the shells, so we can use the above dipole form of $B$ for every shell. The total field at the north pole (that is, at radius $R$) is therefore

$$B = \int_0^R \frac{2\mu_0 (\rho \, dr)\omega r^4}{3R^3} = \frac{2\mu_0 \rho \omega R^2}{15}. \quad (12.505)$$

This field is 2/5 as large as the field at the center of the sphere; see Exercise 11.32. In terms of the total charge $Q = (4\pi R^3/3)\rho$, we can write $B$ as $B = \mu_0 \omega Q/10\pi R$.

11.10 *Surface current on a cube*

Equation (11.55) gives the surface current density as $\mathcal{J} = M$. Since the units of magnetization (J/Tm$^3$) can also be written as A/m, we have $\mathcal{J} = 4.8 \cdot 10^5$ A/m. This current density spans a ribbon that is $\ell = 0.05$ m wide, so the current is $I = \mathcal{J}\ell = (4.8 \cdot 10^5 \text{ A/m})(0.05 \text{ m}) = 24{,}000$ A.

The dipole moment of the cube is

$$m = MV = (4.8 \cdot 10^5 \text{ J T}^{-1} \text{ m}^{-3})(0.05 \text{ m})^3 = 60 \text{ J/T}. \qquad (12.506)$$

The field at a distance of 2 meters, along the axis, is given by Eq. (11.15) as

$$B = \frac{\mu_0 m}{2\pi r^3} = \frac{(4\pi \cdot 10^{-7} \text{ kg m/C}^2)(60 \text{ J/T})}{2\pi (2 \text{ m})^3} = 1.5 \cdot 10^{-6} \text{ T}, \qquad (12.507)$$

or 0.015 gauss. This is about 30 times smaller than the earth's field of $\approx 0.5$ gauss, so it wouldn't disturb a compass much.

11.11 *An iron torus*

From Fig. 11.32, a $B$ field of 1.2 tesla requires an $H$ field of about 120 A/m. Consider the line integral $\int \mathbf{H} \cdot d\mathbf{l}$ around the "middle" circle of the solenoid, with diameter 11 cm. If $I$ is the current in the wire, then $NI = 20I$ is the free current enclosed by our circular loop. Therefore,

$$\int \mathbf{H} \cdot d\mathbf{l} = I_{\text{free}} \implies (120 \text{ A/m}) \cdot \pi(0.11 \text{ m}) = 20I \implies I = 2.1 \text{ A}. \qquad (12.508)$$

# A

The table in Fig. 11.7, used at the field at the center of the square (see Figure 11.2), the magnitude of the total charge is (2.45 A·m²). Thus we can write the ... as ...

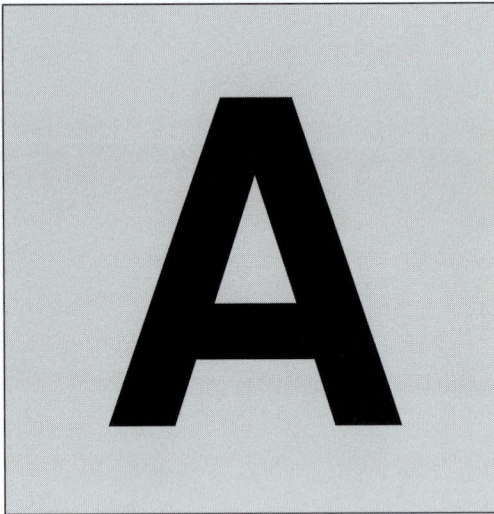

# Differences between SI and Gaussian units

In this appendix we discuss the differences between the SI and Gaussian systems of units. First, we will look at the units in each system, and then we will talk about the clear and not so clear ways in which they differ.

## A.1 SI units

Consider the SI system, which is the one we use in this book. The four main SI units that we deal with are the meter (m), kilogram (kg), second (s), and coulomb (C). The coulomb actually isn't a fundamental SI unit; it is defined in terms of the ampere (A), which is a measure of current (charge per time). The coulomb is a derived unit, defined to be 1 ampere-second.

The reason why the ampere, and not the coulomb, is the fundamental unit involving charge is one of historical practicality. It is relatively easy to measure current via a galvanometer (see Section 7.1). More crudely, a current can be determined by measuring the magnetic force that two pieces of a current-carrying wire in a circuit exert on each other (see Fig. 6.4). Once we determine the current that flows onto an object during a given time, we can then determine the charge on the object. On the other hand, although it is possible to measure charge directly via the force that two equally charged objects exert on each other (imagine two balls hanging from strings, repelling each other, as in Exercise 1.36), the setup is a bit cumbersome. Furthermore, it tells us only what the *product* of the charges is, in the event that they aren't equal. The point is that it is easy to measure current by hooking up an ammeter (the main component of which is a galvanometer) to a circuit.[1]

The exact definition of an ampere is: if two parallel wires carrying equal currents are separated by 1 meter, and if the force per meter on one wire, due to the entirety of the other wire, is $2 \cdot 10^{-7}$ newtons, then the current in each wire is 1 ampere. The power of 10 here is an arbitrary historical artifact, as is the factor of 2. This force is quite small, but by decreasing the separation the effect can be measured accurately enough with the setup shown in Fig. 6.4.

Having defined the ampere in this manner, and then having defined the coulomb as 1 ampere-second (which happens to correspond to the negative of the charge of about $6.24 \cdot 10^{18}$ electrons), a reasonable thing to do, at least in theory, is to find the force between two 1 coulomb charges located, say, 1 meter apart. Since the value of 1 coulomb has been fixed by the definition of the ampere, this force takes on a particular value. We are not free to adjust it by tweaking any definitions. It happens to be about $9 \cdot 10^9$ newtons – a seemingly arbitrary number, but in fact related to the speed of light. (It has the numerical value of $c^2/10^7$; we see why in Section 6.1.) This (rather large) number therefore appears out in front of Coulomb's law. We could label this constant with one letter, such as "$k$," but for various reasons it is labeled as $1/4\pi\epsilon_0$, with $\epsilon_0 = 8.85 \cdot 10^{-12}\,\mathrm{C^2\,s^2\,kg^{-1}\,m^{-3}}$. These units are what are needed to make the right-hand side of Coulomb's law, $F = (1/4\pi\epsilon_0)q_1 q_2/r^2$, have units of newtons (namely $\mathrm{kg\,m\,s^{-2}}$). In terms of the fundamental ampere unit, the units of $\epsilon_0$ are $\mathrm{A^2\,s^4\,kg^{-1}\,m^{-3}}$.

The upshot of all this is that because we made the choice to define current via the Lorentz force (specifically, the magnetic part of the Lorentz force) between two wires carrying current $I$, the Coulomb force between two objects of charge $q$ ends up being a number that we just have to accept. We can make the pre-factor be a nice simple number in either one of these force laws, but not both.[2] The SI system gives preference to the Lorentz force, due to the historical matters of practicality mentioned above.

It turns out that there are actually seven fundamental units in the SI system. They are listed in Table A.1. The candela isn't relevant to our study of electromagnetism, and the mole and kelvin come up only occasionally. So for our purposes the SI system consists of essentially just the first four units.

---

[1] If we know the capacitance of an object, then we *can* easily measure the charge on it by measuring the voltage with a voltmeter. But the main component of a voltmeter is again a galvanometer, so the process still reduces to measuring a current.

[2] The Biot–Savart law, which allows us to calculate the magnetic field that appears in the Lorentz force, contains what appears to be a messy pre-factor, namely $\mu_0/4\pi$. But since $\mu_0$ is defined to be exactly $4\pi \cdot 10^{-7}\,\mathrm{kg\,m/C^2}$, this pre-factor takes on the simple value of $10^{-7}\,\mathrm{kg\,m/C^2}$.

**Table A.1.**
SI base units

| Quantity | Name | Symbol |
|---|---|---|
| Length | meter | m |
| Mass | kilogram | kg |
| Time | second | s |
| Electric current | ampere | A |
| Thermodynamic temperature | kelvin | K |
| Amount of substance | mole | mol |
| Luminous intensity | candela | cd |

## A.2  Gaussian units

What do the units look like in the Gaussian system? As with the SI system, the last three of the above units (or their analogs) rarely come up, so we will ignore them. The first two units are the centimeter and gram. These differ from the SI units simply by a few powers of 10, so it is easy to convert from one system to the other. The third unit, the second, is the same in both systems.

The fourth unit, that of charge, is where the two systems fundamentally diverge. The Gaussian unit of charge is the esu (short for "electrostatic unit"), which isn't related to the coulomb by a simple power of 10. The reason for this non-simple relation is that the coulomb and esu are defined in different ways. The coulomb is a derivative unit of the ampere (which is defined via the Lorentz force) as we saw above, whereas the esu is defined via the Coulomb force. In particular, it is defined so that Coulomb's law,

$$\mathbf{F} = k\frac{q_1 q_2 \hat{\mathbf{r}}}{r^2}, \tag{A.1}$$

takes on a very simple form with $k = 1$. The price to pay for this simple form of the Coulomb force is the not as simple form of the Lorentz force between two current-carrying wires (although it isn't so bad; like the Coulomb force in SI units, it just involves a factor of $c^2$; see Eq. (6.16)). This is the opposite of the situation with the SI system, where the Lorentz force is the "nice" one. Again, in each system we are free to define things so that one, but not both, of the Lorentz force and Coulomb force takes on a nice form.

## A.3  Main differences between the systems

In Section A.2 we glossed over what turns out to be the most important difference between the SI and Gaussian systems. In the SI system, the constant in Coulomb's law,

$$k_{\mathrm{SI}} \equiv \frac{1}{4\pi\epsilon_0} = 8.988 \cdot 10^9 \, \frac{\mathrm{N\,m}^2}{\mathrm{C}^2}, \tag{A.2}$$

has nontrivial dimensions, whereas in the Gaussian system the constant

$$k_{\rm G} = 1 \qquad\qquad (\text{A.3})$$

is *dimensionless*. We aren't just being sloppy and forgetting to write the units; $k$ is simply the number 1. Although the first thing that may strike you about the two $k$ constants is the large difference in their numerical values, this difference is fairly inconsequential. It simply changes the numerical size of various quantities. The truly fundamental and critical difference is that $k_{\rm SI}$ has units whereas $k_{\rm G}$ does not. We could, of course, imagine a system of units where $k = 1$ dyne-cm$^2$/esu$^2$. This definition would parallel the units of $k_{\rm SI}$, with the only difference being the numerical value. But this is not what the Gaussian system does.

The reason why the dimensionlessness of $k_{\rm G}$ creates such a profound difference between the two systems is that it allows us to *solve for the esu in terms of other Gaussian units*. In particular, from looking at the units in Coulomb's law, we can write (using 1 dyne = 1 g · cm/s$^2$)

$$\text{dyne} = (\textit{dimensionless}) \cdot \frac{\text{esu}^2}{\text{cm}^2} \quad\Longrightarrow\quad \text{esu} = \sqrt{\frac{\text{g} \cdot \text{cm}^3}{\text{s}^2}}. \qquad (\text{A.4})$$

The esu is therefore not a fundamental unit. It can be expressed in terms of the gram, centimeter, and second. In contrast, the SI unit of charge, the coulomb, cannot be similarly expressed. Since $k_{\rm SI}$ has units of N m$^2$/C$^2$, the C's (and everything else) cancel in Coulomb's law, and we can't solve for C in terms of other units.

For our purposes, therefore, the SI system has four fundamental units (m, kg, s, A), whereas the Gaussian system has only three (cm, g, s). We will talk more about this below, but first let us summarize the three main differences between the SI and Gaussian systems. We state them in order of increasing importance.

(1) The SI system uses kilograms and meters, whereas the Gaussian system uses grams and centimeters. This is the most trivial of the three differences, because all it does is introduce some easily dealt with powers of 10.

(2) The SI unit of charge (the coulomb) is defined via the ampere, which in turn is defined in terms of the force between current-carrying wires. The Gaussian unit of charge (the esu) is defined directly in terms of Coulomb's law. This latter definition is the reason why Coulomb's law takes on a nicer form in Gaussian units. The differences between the two systems now involve more than simple powers of 10. However, although these differences can sometimes be a hassle, they aren't terribly significant. They are just numbers – no different from powers of 10, except a little messier. All of the conversions you might need to use are listed in Appendix C.

(3) In Gaussian units, the $k$ in Coulomb's law is chosen to be *dimensionless*, whereas in SI units the $k$ (which involves $\epsilon_0$) has units.[3] The result is that the esu can be expressed in terms of other Gaussian units, whereas the analogous statement is not true for the coulomb. This is the most important difference between the two systems.

## A.4 Three units versus four

Let us now discuss in more detail the issue of the number of units in each system. The Gaussian system has one fewer because the esu can be expressed in terms of other units via Eq. (A.4). This has implications with regard to checking units at the end of a calculation. In short, less information is gained when checking units in the Gaussian system, because the charge information is lost when the esu is written in terms of the other units. Consider the following example.

In SI units the electric field due to a sheet of charge is given in Eq. (1.40) as $\sigma/2\epsilon_0$. In Gaussian units the field is $2\pi\sigma$. Recalling the units of $\epsilon_0$ in Eq. (1.3), the units of the SI field are $\mathrm{kg\,m\,C^{-1}s^{-2}}$ (or $\mathrm{kg\,m\,A^{-1}s^{-3}}$ if you want to write it in terms of amperes, but we use coulombs here to show analogies with the esu). This correctly has dimensions of (force)/(charge). The units of the Gaussian $2\pi\sigma$ field are simply $\mathrm{esu/cm^2}$, but since the esu is given by Eq. (A.4), the units are $\mathrm{g^{1/2}cm^{-1/2}\,s^{-1}}$. These are the true Gaussian units of the electric field when written in terms of fundamental units.

Now let's say that two students working in the Gaussian system are given a problem where the task is to find the electric field due to a thin sheet with charge density $\sigma$, mass $m$, volume $V$, moving with a nonrelativistic speed $v$. The first student realizes that most of this information is irrelevant and solves the problem correctly, obtaining the answer of $2\pi\sigma$ (ignoring relativistic corrections). The second student royally messes things up and obtains an answer of $\sigma^3 V m^{-1} v^{-2}$. Since the fundamental Gaussian units of $\sigma$ are $\mathrm{g^{1/2}\,cm^{-1/2}\,s^{-1}}$, the units of this answer are

$$\frac{\sigma^3 V}{mv^2} \longrightarrow \frac{\left(\mathrm{g^{1/2}\,cm^{-1/2}\,s^{-1}}\right)^3 (\mathrm{cm})^3}{(\mathrm{g})(\mathrm{cm/s})^2} = \frac{\mathrm{g^{1/2}}}{\mathrm{cm^{1/2}\,s}}, \tag{A.5}$$

which are the correct Gaussian units of electric field that we found above. More generally, in view of Eq. (A.4) we see that any answer with the units of $\left(\mathrm{g^{1/2}\,cm^{-1/2}\,s^{-1}}\right)\left(\mathrm{esu\,g^{-1/2}\,cm^{-3/2}\,s}\right)^n$ has the correct units for the field. The present example has $n = 3$.

There are, of course, also many ways to obtain incorrect answers in the SI system that just happen by luck to have the correct units. Correctness of the units doesn't guarantee correctness of the answer. But the

---

[3] To draw a more accurate analogy: in SI units the defining equation for the ampere (from which the coulomb is derived) contains the *dimensionful* constant $\mu_0$ in the force between two wires.

point is that because the charge information is swept under the rug in Gaussian units, we have at our disposal the information of only three fundamental units instead of four. Compared with the SI system, there is therefore a larger class of incorrect answers in the Gaussian system that have the correct units.

## A.5 The definition of B

Another difference between the SI and Gaussian systems of units is the way in which the magnetic field is defined. In SI units the Lorentz force (or rather the magnetic part of it) is $\mathbf{F} = q\mathbf{v} \times \mathbf{B}$, whereas in Gaussian units it is $\mathbf{F} = (q/c)\mathbf{v} \times \mathbf{B}$. This means that wherever a $B$ appears in an SI expression, a $B/c$ appears in the corresponding Gaussian expression (among other possible modifications). Or equivalently, a Gaussian $B$ turns into an SI $cB$. This difference, however, is a trivial definitional one and has nothing to do with the far more important difference discussed above, where the esu can be expressed in terms of other Gaussian units.

In the Gaussian system, $E$ and $B$ have the same dimensions. In the SI system they do not; the dimensions of $E$ are velocity times the dimensions of $B$. In this sense the Gaussian definition of $B$ is more natural, because it makes sense for two quantities to have the same dimensions if they are related by a Lorentz transformation, as the $\mathbf{E}$ and $\mathbf{B}$ fields are; see Eq. (6.76) for the SI case and Eq. (6.77) for the Gaussian case. After all, the Lorentz transformation tells us that the $\mathbf{E}$ and $\mathbf{B}$ fields are simply different ways of looking at the same field, depending on the frame of reference. However, having a "$c\mathbf{B}$" instead of a "$\mathbf{B}$" in the SI Lorentz transformation can't be so bad, because $x$ and $t$ are also related by a Lorentz transformation, and they don't have the same dimensions (the direct correspondence is between $x$ and $ct$). Likewise for $p$ and $E$ (where the direct correspondence is between $pc$ and $E$). At any rate, this issue stems from the arbitrary choice of whether a factor of $c$ is included in the expression for the Lorentz force. One can easily imagine an SI-type system (where charge is a distinct unit) in which the Lorentz force takes the form $\mathbf{F} = q\mathbf{E} + (q/c)\mathbf{v} \times \mathbf{B}$, yielding the same dimensions for $E$ and $B$.

## A.6 Rationalized units

You might wonder why there are factors of $4\pi$ in the SI versions of Coulomb's law and the Biot–Savart law; see Eqs. (1.4) and (6.49). These expressions would certainly look a bit less cluttered without these factors. The reason is that the presence of $4\pi$'s in these laws leads to the absence of such factors in Maxwell's equations. And for various reasons it is deemed more important to have Maxwell's equations be the "clean" ones without the $4\pi$ factors. The procedure of inserting $4\pi$ into Coulomb's law and the Biot–Savart law, in order to keep them out of Maxwell's equations, is called "rationalizing" the units. Of course, for people concerned more with applications of Coulomb's law than with Maxwell's equations, this procedure might look like a step in the wrong direction. But since Maxwell's equations are the more fundamental equations, there is logic in this convention.

It is easy to see why the presence of $4\pi$ factors in Coulomb's law and the Biot–Savart law leads to the absence of $4\pi$ factors in Gauss's law and Ampère's law, which are equivalent to two of Maxwell's equations (or actually one and a half; Ampère's law is supplemented with another term). In the case of Gauss's law, the absence of the $4\pi$ basically boils down to the area of a sphere being $4\pi r^2$ (see the derivation in Section 1.10). In the case of Ampère's law, the absence of the $4\pi$ is a consequence of the reasoning in Sections 6.3 and 6.4, which again boils down to the area of a sphere being $4\pi r^2$ (because Eq. (6.44) was written down by analogy with Eq. (6.30)). Or more directly: the $1/4\pi$ in the Biot–Savart law turns into a $1/2\pi$ in the field from an infinite straight wire (see Eq. (6.6)), and this $2\pi$ is then canceled when we take the line integral around a circle with circumference $2\pi r$.

If there were no factors of $4\pi$ in Coulomb's law or the Biot–Savart law, then there *would* be factors of $4\pi$ in Maxwell's equations. This is exactly what happens in the Gaussian system, where the "curl **B**" and "div **E**" Maxwell equations each involve a $4\pi$; see Eq. (9.20). Note, however, that one can easily imagine a Gaussian-type system (that is, one where the pre-factor in Coulomb's law is dimensionless) that has factors of $4\pi$ in Coulomb's law and the Biot–Savart law, and none in Maxwell's equations. This is the case in a variation of Gaussian units called Heaviside–Lorentz units.

# SI units of common quantities

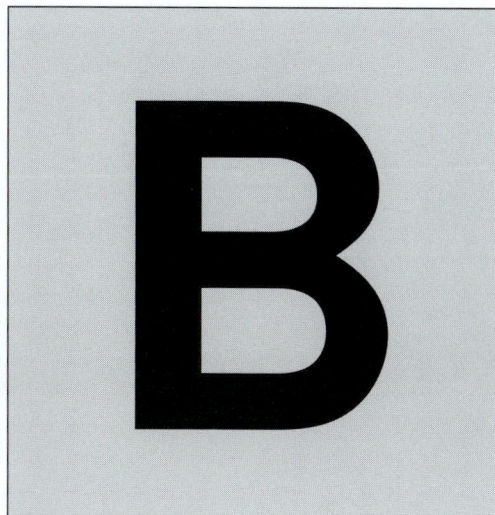

We begin this appendix with the definitions of all of the derived SI units relevant to electromagnetism (for example, the joule, ohm, etc.). We then list the units of all of the main quantities that appear in this book (basically, anything that has earned the right to be labeled with its own letter).

In SI units the ampere is the fundamental unit involving charge. The coulomb is a derived unit, being defined as one ampere-second. However, since most people find it more natural to think in terms of charge than current, we treat the coulomb as the fundamental unit in this appendix. The ampere is then defined as one coulomb/second.

For each of the main quantities listed, we give the units in terms of the fundamental units (m, kg, s, C, and occasionally K), and then also in terms of other derived units in certain forms that come up often. For example, the units of electric field are $\text{kg m C}^{-1}\text{s}^{-2}$, but they are also newtons/coulomb and volts/meter.

The various derived units are as follows:

$$\text{newton (N)} = \frac{\text{kg m}}{\text{s}^2}$$

$$\text{joule (J)} = \text{newton-meter} = \frac{\text{kg m}^2}{\text{s}^2}$$

$$\text{ampere (A)} = \frac{\text{coulomb}}{\text{second}} = \frac{\text{C}}{\text{s}}$$

$$\text{volt (V)} = \frac{\text{joule}}{\text{coulomb}} = \frac{\text{kg m}^2}{\text{C s}^2}$$

$$\text{farad (F)} = \frac{\text{coulomb}}{\text{volt}} = \frac{C^2 s^2}{kg\, m^2}$$

$$\text{ohm } (\Omega) = \frac{\text{volt}}{\text{ampere}} = \frac{kg\, m^2}{C^2 s}$$

$$\text{watt (W)} = \frac{\text{joule}}{\text{second}} = \frac{kg\, m^2}{s^3}$$

$$\text{tesla (T)} = \frac{\text{newton}}{\text{coulomb} \cdot \text{meter/second}} = \frac{kg}{C\, s}$$

$$\text{henry (H)} = \frac{\text{volt}}{\text{ampere/second}} = \frac{kg\, m^2}{C^2}$$

The main quantities are listed by chapter.

### Chapter 1

$$\text{charge } q: \ C$$

$$k \text{ in Coulomb's law:} \ \frac{kg\, m^3}{C^2 s^2} = \frac{N\, m^2}{C^2}$$

$$\epsilon_0: \ \frac{C^2 s^2}{kg\, m^3} = \frac{C^2}{N\, m^2} = \frac{C}{V\, m} = \frac{F}{m}$$

$$E \text{ field (force per charge):} \ \frac{kg\, m}{C\, s^2} = \frac{N}{C} = \frac{V}{m}$$

$$\text{flux } \Phi \ (E \text{ field times area):} \ \frac{kg\, m^3}{C\, s^2} = \frac{N\, m^2}{C} = V\, m$$

$$\text{charge density } \lambda, \sigma, \rho: \ \frac{C}{m}, \frac{C}{m^2}, \frac{C}{m^3}$$

### Chapter 2

$$\text{potential } \phi \ (\text{energy per charge):} \ \frac{kg\, m^2}{C\, s^2} = \frac{J}{C} = V$$

$$\text{dipole moment } p: \ C\, m$$

### Chapter 3

$$\text{capacitance } C \ (\text{charge per potential):} \ \frac{C^2 s^2}{kg\, m^2} = \frac{C}{V} = F$$

Chapter 4

$$\text{current } I \text{ (charge per time): } \frac{C}{s} = A$$

$$\text{current density } J \text{ (current per area): } \frac{C}{m^2\,s} = \frac{A}{m^2}$$

$$\text{conductivity } \sigma \text{ (current density per field): } \frac{C^2\,s}{kg\,m^3} = \frac{1}{\Omega\,m}$$

$$\text{resistivity } \rho \text{ (field per current density): } \frac{kg\,m^3}{C^2\,s} = \Omega\,m$$

$$\text{resistance } R \text{ (voltage per current): } \frac{kg\,m^2}{C^2\,s} = \frac{V}{A} = \Omega$$

$$\text{power } P \text{ (energy per time): } \frac{kg\,m^2}{s^3} = \frac{J}{s} = W$$

Chapter 5

$$\text{speed of light } c: \frac{m}{s}$$

Chapter 6

$$B \text{ field (force per charge-velocity): } \frac{kg}{C\,s} = T$$

$$\mu_0 : \frac{kg\,m}{C^2} = \frac{T\,m}{A}$$

$$\text{vector potential } A: \frac{kg\,m}{C\,s} = T\,m$$

$$\text{surface current density } \mathcal{J} \text{ (current per length): } \frac{C}{m\,s} = \frac{A}{m}$$

Chapter 7

$$\text{electromotive force } \mathcal{E}: \frac{kg\,m^2}{C\,s^2} = \frac{J}{C} = A\,\Omega = V$$

$$\text{flux } \Phi \text{ (}B \text{ field times area): } \frac{kg\,m^2}{C\,s} = T\,m^2$$

$$\text{inductance } M, L: \frac{kg\,m^2}{C^2} = \frac{V\,s}{A} = H$$

## Chapter 8

$$\text{frequency } \omega: \ \frac{1}{\text{s}}$$

$$\text{quality factor } Q: \ 1 \ (\textit{dimensionless})$$

$$\text{phase } \phi: \ 1 \ (\textit{dimensionless})$$

$$\text{admittance } Y \text{ (current per voltage)}: \ \frac{\text{C}^2 \, \text{s}}{\text{kg m}^2} = \frac{\text{A}}{\text{V}} = \frac{1}{\Omega}$$

$$\text{impedance } Z \text{ (voltage per current)}: \ \frac{\text{kg m}^2}{\text{C}^2 \, \text{s}} = \frac{\text{V}}{\text{A}} = \Omega$$

## Chapter 9

$$\text{power density } S \text{ (power per area)}: \ \frac{\text{kg}}{\text{s}^3} = \frac{\text{J}}{\text{m}^2 \, \text{s}} = \frac{\text{W}}{\text{m}^2}$$

## Chapter 10

$$\text{dielectric constant } \kappa: \ 1 \ (\textit{dimensionless})$$

$$\text{dipole moment } p: \ \text{C m}$$

$$\text{torque } N: \ \frac{\text{kg m}^2}{\text{s}^2} = \text{N m}$$

$$\text{atomic polarizability } \alpha/4\pi\epsilon_0: \ \text{m}^3$$

$$\text{polarization density } P: \ \frac{\text{C}}{\text{m}^2}$$

$$\text{electric susceptibility } \chi_e: \ 1 \ (\textit{dimensionless})$$

$$\text{permittivity } \epsilon: \ \frac{\text{C}^2 \, \text{s}^2}{\text{kg m}^3} = \frac{\text{C}^2}{\text{N m}^2}$$

$$\text{displacement vector } D: \ \frac{\text{C}}{\text{m}^2}$$

$$\text{temperature } T: \ \text{K}$$

$$\text{Boltzmann's constant } k: \ \frac{\text{kg m}^2}{\text{s}^2 \, \text{K}} = \frac{\text{J}}{\text{K}}$$

Chapter 11

$$\text{magnetic moment } m: \quad \frac{\text{C m}^2}{\text{s}} = \text{A m}^2 = \frac{\text{J}}{\text{T}}$$

$$\text{angular momentum } L: \quad \frac{\text{kg m}^2}{\text{s}}$$

$$\text{Planck's constant } h: \quad \frac{\text{kg m}^2}{\text{s}} = \text{J s}$$

$$\text{magnetization } M \text{ (}m\text{ per volume)}: \quad \frac{\text{C}}{\text{m s}} = \frac{\text{A}}{\text{m}} = \frac{\text{J}}{\text{T m}^3}$$

$$\text{magnetic susceptibility } \chi_m: \quad 1 \text{ (}dimensionless\text{)}$$

$$H \text{ field}: \quad \frac{\text{C}}{\text{m s}} = \frac{\text{A}}{\text{m}}$$

$$\text{permeability } \mu: \quad \frac{\text{kg m}}{\text{C}^2} = \frac{\text{T m}}{\text{A}}$$

# C

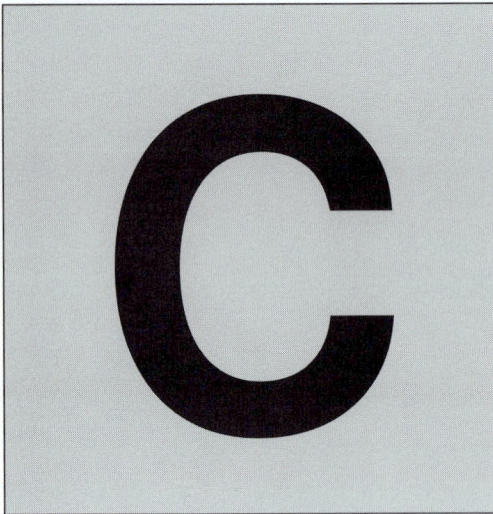

## Unit conversions

In this appendix we list, and then derive, the main unit conversions between the SI and Gaussian systems. As you will see below, many of the conversions involve simple plug-and-chug calculations involving conversions that are already known. However, a few of them (charge, $B$ field, $H$ field) require a little more thought, because the relevant quantities have different definitions in the two systems.

### C.1 Conversions

Except for the first five (nonelectrical) conversions below, we technically shouldn't be using "=" signs, because they suggest that the units in the two systems are actually the same, up to a numerical factor. This is not the case. All of the electrical relations involve charge in one way or another, and a coulomb cannot be expressed in terms of an esu. This is a consequence of the fact that the esu is defined in terms of the other Gaussian units; see Appendix A for a discussion of how the coulomb and esu differ. The proper way to express, say, the sixth relation below is "1 coulomb is *equivalent* to $3 \cdot 10^9$ esu." But we'll generally just use the "=" sign, and you'll know what we mean.

The "[3]" in the following relations stands for the "2.998" that appears in the speed of light, $c = 2.998 \cdot 10^8$ m/s. The coulomb-esu discussion below explains how this arises.

|  |  |
|---|---|
| *time:* | 1 second = 1 second |
| *length:* | 1 meter = $10^2$ centimeter |
| *mass:* | 1 kilogram = $10^3$ gram |

| | |
|---|---|
| *force:* | 1 newton $= 10^5$ dyne |
| *energy:* | 1 joule $= 10^7$ erg |
| *charge:* | 1 coulomb $= [3] \cdot 10^9$ esu |
| *E potential:* | 1 volt $= \dfrac{1}{[3] \cdot 10^2}$ statvolt |
| *E field:* | 1 volt/meter $= \dfrac{1}{[3] \cdot 10^4}$ statvolt/cm |
| *capacitance:* | 1 farad $= [3]^2 \cdot 10^{11}$ cm |
| *resistance:* | 1 ohm $= \dfrac{1}{[3]^2 \cdot 10^{11}}$ s/cm |
| *resistivity:* | 1 ohm-meter $= \dfrac{1}{[3]^2 \cdot 10^9}$ s |
| *inductance:* | 1 henry $= \dfrac{1}{[3]^2 \cdot 10^{11}}$ s$^2$/cm |
| *B field:* | 1 tesla $= 10^4$ gauss |
| *H field:* | 1 amp/meter $= 4\pi \cdot 10^{-3}$ oersted |

## C.2 Derivations
### C.2.1 Force: newton vs. dyne

$$1 \text{ newton} = 1 \frac{\text{kg m}}{\text{s}^2} = \frac{(1000\,\text{g})(100\,\text{cm})}{\text{s}^2} = 10^5 \frac{\text{g cm}}{\text{s}^2} = 10^5 \text{ dynes.}$$

$$\text{(C.1)}$$

### C.2.2 Energy: joule vs. erg

$$1 \text{ joule} = 1 \frac{\text{kg m}^2}{\text{s}^2} = \frac{(1000\,\text{g})(100\,\text{cm})^2}{\text{s}^2} = 10^7 \frac{\text{g cm}^2}{\text{s}^2} = 10^7 \text{ ergs.}$$

$$\text{(C.2)}$$

### C.2.3 Charge: coulomb vs. esu
From Eqs. (1.1) and (1.2), two charges of 1 coulomb separated by a distance of 1 m exert a force on each other equal to $8.988 \cdot 10^9$ N $\approx 9 \cdot 10^9$ N, or equivalently $9 \cdot 10^{14}$ dynes. How would someone working in Gaussian units describe this situation? In Gaussian units, Coulomb's law gives the force simply as $q^2/r^2$. The separation is 100 cm, so if 1 coulomb equals $N$ esu (with $N$ to be determined), the $9 \cdot 10^{14}$ dyne force between the charges can be expressed as

$$9 \cdot 10^{14} \text{ dyne} = \frac{(N\,\text{esu})^2}{(100\,\text{cm})^2} \implies N^2 = 9 \cdot 10^{18} \implies N = 3 \cdot 10^9.$$

$$\text{(C.3)}$$

So 1 coulomb equals $3 \cdot 10^9$ esu. If we had used the more exact value of $k$ in Eq. (1.2), the "3" in this result would have been replaced by $\sqrt{8.988} = 2.998$, which is precisely the 2.998 that appears in the speed of light, $c = 2.998 \cdot 10^8$ m/s. The reason for this is the following.

If you follow through the above derivation while keeping things in terms of $k \equiv 1/4\pi\epsilon_0$, you will see that the number $3 \cdot 10^9$ is actually $\sqrt{\{k\} \cdot 10^5 \cdot 10^4}$, where we have put the braces around $k$ to signify that it is just the number $8.988 \cdot 10^9$ without the SI units. (The factors of $10^5$ and $10^4$ come from the conversions to dynes and centimeters, respectively.) But we know from Eq. (6.8) that $\epsilon_0 = 1/\mu_0 c^2$, so we have $k = \mu_0 c^2/4\pi$. Furthermore, the numerical value of $\mu_0$ is $\{\mu_0\} = 4\pi \cdot 10^{-7}$, so the numerical value of $k$ is $\{k\} = \{c\}^2 \cdot 10^{-7}$. Therefore, the number $N$ that appears in Eq. (C.3) is really

$$N = \sqrt{\{k\} \cdot 10^9} = \sqrt{(\{c\}^2 \cdot 10^{-7})10^9} = \{c\} \cdot 10 = 2.998 \cdot 10^9 \equiv [3] \cdot 10^9.$$

(C.4)

### C.2.4 Potential: volt vs. statvolt

$$1 \text{ volt} = 1 \frac{\text{J}}{\text{C}} = \frac{10^7 \text{ erg}}{[3] \cdot 10^9 \text{ esu}} = \frac{1}{[3] \cdot 10^2} \frac{\text{erg}}{\text{esu}} = \frac{1}{[3] \cdot 10^2} \text{ statvolt.}$$

(C.5)

### C.2.5 Electric field: volt/meter vs. statvolt/centimeter

$$1 \frac{\text{volt}}{\text{meter}} = \frac{\frac{1}{[3] \cdot 10^2} \text{ statvolt}}{100 \text{ cm}} = \frac{1}{[3] \cdot 10^4} \frac{\text{statvolt}}{\text{cm}}.$$

(C.6)

### C.2.6 Capacitance: farad vs. centimeter

$$1 \text{ farad} = 1 \frac{\text{C}}{\text{V}} = \frac{[3] \cdot 10^9 \text{ esu}}{\frac{1}{[3] \cdot 10^2} \text{ statvolt}} = [3]^2 \cdot 10^{11} \frac{\text{esu}}{\text{statvolt}}.$$

(C.7)

We can alternatively write these Gaussian units as centimeters. This is true because 1 statvolt = 1 esu/cm (because the potential from a point charge is $q/r$, so 1 esu/statvolt = 1 cm. We therefore have

$$1 \text{ farad} = [3]^2 \cdot 10^{11} \text{ cm.}$$

(C.8)

### C.2.7 Resistance: ohm vs. second/centimeter

$$1 \text{ ohm} = 1 \frac{\text{V}}{\text{A}} = 1 \frac{\text{V}}{\text{C/s}} = \frac{\frac{1}{[3] \cdot 10^2} \text{ statvolt}}{[3] \cdot 10^9 \text{ esu/s}} = \frac{1}{[3]^2 \cdot 10^{11}} \frac{\text{s}}{\text{esu/statvolt}}$$

$$= \frac{1}{[3]^2 \cdot 10^{11}} \frac{\text{s}}{\text{cm}},$$

(C.9)

where we have used 1 esu/statvolt = 1 cm.

### C.2.8 Resistivity: ohm-meter vs. second

$$1 \text{ ohm-meter} = \left( \frac{1}{[3]^2 \cdot 10^{11}} \frac{\text{s}}{\text{cm}} \right) (100 \text{ cm}) = \frac{1}{[3]^2 \cdot 10^9} \text{ s}. \quad \text{(C.10)}$$

### C.2.9 Inductance: henry vs. second²/centimeter

$$1 \text{ henry} = 1 \frac{\text{V}}{\text{A/s}} = 1 \frac{\text{V}}{\text{C/s}^2} = \frac{\frac{1}{[3] \cdot 10^2} \text{ statvolt}}{[3] \cdot 10^9 \text{ esu/s}^2}$$

$$= \frac{1}{[3]^2 \cdot 10^{11}} \frac{\text{s}^2}{\text{esu/statvolt}} = \frac{1}{[3]^2 \cdot 10^{11}} \frac{\text{s}^2}{\text{cm}}, \quad \text{(C.11)}$$

where we have used 1 esu/statvolt = 1 cm.

### C.2.10 Magnetic field B: tesla vs. gauss

Consider a setup in which a charge of 1 C travels at 1 m/s in a direction perpendicular to a magnetic field with strength 1 tesla. Equation (6.1) tells us that the force on the charge is 1 newton. Let us express this fact in terms of the Gaussian force relation in Eq. (6.9), which involves a factor of $c$. We know that $1 \text{ N} = 10^5$ dyne and $1 \text{ C} = [3] \cdot 10^9$ esu. If we let 1 tesla = $N$ gauss, then the way that Eq. (6.9) describes the given situation is

$$10^5 \text{ dyne} = \frac{[3] \cdot 10^9 \text{ esu}}{[3] \cdot 10^{10} \text{ cm/s}} \left( 100 \frac{\text{cm}}{\text{s}} \right) (N \text{ gauss}). \quad \text{(C.12)}$$

Since 1 gauss equals 1 dyne/esu, all the units cancel (as they must), and we end up with $N = 10^4$, as desired. This is an exact result because the two factors of [3] cancel.

### C.2.11 Magnetic field H: ampere/meter vs. oersted

The $H$ field is defined differently in the two systems (there is a $\mu_0$ in the SI definition), so we have to be careful. Consider a $B$ field of 1 tesla in vacuum. What $H$ field does this $B$ field correspond to in each system? In the Gaussian system, $B$ is $10^4$ gauss. But in Gaussian units $H = B$ in vacuum, so $H = 10^4$ oersted, because an oersted and a gauss are equivalent units. In the SI system we have (you should verify these units)

$$H = \frac{B}{\mu_0} = \frac{1 \text{ tesla}}{4\pi \cdot 10^{-7} \text{ kg m/C}^2} = \frac{10^7}{4\pi} \frac{\text{A}}{\text{m}}. \quad \text{(C.13)}$$

Since this is equivalent to $10^4$ oersted, we arrive at 1 amp/meter $= 4\pi \cdot 10^{-3}$ oersted. Going the other way, 1 oersted equals roughly 80 amp/meter.

# D

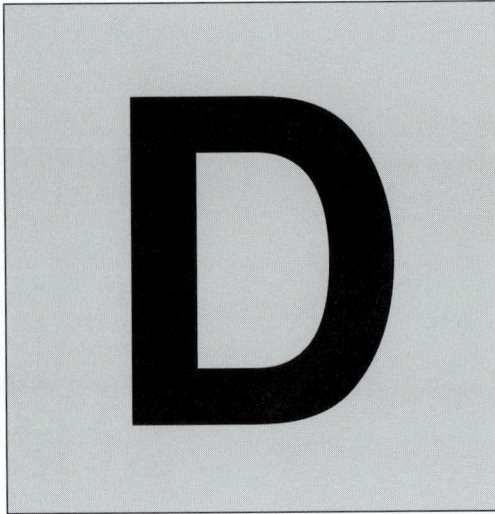

# SI and Gaussian formulas

The following pages provide a list of all the main results in this book, in both SI and Gaussian units. After looking at a few of the corresponding formulas, you will discover that transforming from SI units to Gaussian units involves one or more of the three types of conversions discussed below.

Of course, even if a formula takes exactly the same form in the two systems of units, it says two entirely different things. For example, the formula relating force and electric field is the same in both systems: $\mathbf{F} = q\mathbf{E}$. But in SI units this equation says that a charge of 1 coulomb placed in an electric field of 1 volt/meter feels a force of 1 newton, whereas in Gaussian units it says that a charge of 1 esu placed in an electric field of 1 statvolt/centimeter feels a force of 1 dyne. When we say that two formulas are the "same," we mean that they look the same on the page, even though the various letters mean different things in the two systems.

The three basic types of conversions from SI to Gaussian units are given in Sections D.1 to D.3. We then list the formulas in Section D.4 by chapter.

## D.1 Eliminating $\epsilon_0$ and $\mu_0$

Our starting point in this book was Coulomb's law in Eq. (1.4). The SI expression for this law contains the factor $1/4\pi\epsilon_0$, whereas the Gaussian expression has no factor (or rather just a 1). To convert from SI units to Gaussian units, we therefore need to set $4\pi\epsilon_0 = 1$, or equivalently $\epsilon_0 = 1/4\pi$ (along with possibly some other changes, as we will see below). That is, we need to erase all factors of $4\pi\epsilon_0$ that appear, or equivalently replace all $\epsilon_0$'s with $1/4\pi$'s. In many formulas this change is all that is needed. A few examples are: Gauss's law, Eq. (1.31) in the list in Section D.4;[1] the field due to a line or sheet, Eqs. (1.39) and (1.40); the energy in an electric field, Eq. (1.53); and the capacitance of a sphere or parallel plates, Eqs. (3.10) and (3.15).

A corollary of the $\epsilon_0 \to 1/4\pi$ rule is the $\mu_0 \to 4\pi/c^2$ rule. We introduced $\mu_0$ in Chapter 6 via the definition $\mu_0 \equiv 1/\epsilon_0 c^2$, so if we replace $\epsilon_0$ with $1/4\pi$, we must also replace $\mu_0$ with $4\pi/c^2$. An example of this $\mu_0 \to 4\pi/c^2$ rule is the force between two current-carrying wires, Eq. (6.15).

It is also possible to use these rules to convert formulas in the other direction, from Gaussian units to SI units, although the process isn't quite as simple. The conversion must (at least for conversions where only $\epsilon_0$ and $\mu_0$ are relevant) involve multiplying by some power of $4\pi\epsilon_0$ (or equivalently $4\pi/\mu_0 c^2$). And there is only one power that will make the units of the resulting SI expression correct, because $\epsilon_0$ has units, namely $C^2\,s^2\,kg^{-1}\,m^{-3}$. For example, the Gaussian expression for the field due to a sheet of charge is $2\pi\sigma$ in Eq. (1.40) in the list below, so the SI expression must take the form of $2\pi\sigma(4\pi\epsilon_0)^n$. You can quickly show that $2\pi\sigma(4\pi\epsilon_0)^{-1} = \sigma/2\epsilon_0$ has the correct units of electric field (it suffices to look at the power of any one of the four units: kg, m, s, C).

## D.2 Changing *B* to *B*/*c*

If all quantities were defined in the same way in the two systems of units (up to factors of $4\pi\epsilon_0$ and $4\pi/\mu_0 c^2$), then the above rules involving $\epsilon_0$ and $\mu_0$ would be sufficient for converting from SI units to Gaussian units. But unfortunately certain quantities are defined differently in the two systems, so we can't convert from one system to the other without knowing what these arbitrary definitions are.

The most notable example of differing definitions is the magnetic field. In SI units the Lorentz force (or rather the magnetic part of it) is $\mathbf{F} = q\mathbf{v} \times \mathbf{B}$, while in Gaussian units it is $\mathbf{F} = (q/c)\mathbf{v} \times \mathbf{B}$. To convert from an SI formula to a Gaussian formula, we therefore need to replace every $B$ with a $B/c$ (and likewise for the vector potential $A$). An example of this is the $B$ field from an infinite wire, Eq. (6.6). In SI units we have $B = \mu_0 I/2\pi r$. Applying our rules for $\mu_0$ and $B$, the Gaussian $B$ field is obtained as follows:

$$B = \frac{\mu_0 I}{2\pi r} \quad \longrightarrow \quad \left(\frac{B}{c}\right) = \left(\frac{4\pi}{c^2}\right)\frac{I}{2\pi r} \quad \Longrightarrow \quad B = \frac{2I}{rc}, \qquad \text{(D.1)}$$

which is the correct result. Other examples involving the $B \to B/c$ rule include Ampère's law, Eqs. (6.19) and (6.25); the Lorentz transformations, Eq. (6.76); and the energy in a magnetic field, Eq. (7.79).

---

[1] The "double" equations in the list in Section D.4, where the SI and Gaussian formulas are presented side by side, are labeled according to the equation number that the SI formula has in the text.

## D.3 Other definitional differences

The above two conversion procedures are sufficient for all formulas up to and including Chapter 9. However, in Chapters 10 and 11 we encounter a number of new quantities ($\chi_e$, $\mathbf{D}$, $\mathbf{H}$, etc.), and many of these quantities are defined differently in the two systems of units,[2] mainly due to historical reasons. For example, after using the $\epsilon_0 \to 1/4\pi$ rule in Eq. (10.41), we see that we need to replace $\chi_e$ by $4\pi \chi_e$ in going from SI to Gaussian units. The Gaussian expression is then given by

$$\chi_e = \frac{P}{\epsilon_0 E} \longrightarrow (4\pi \chi_e) = \left(\frac{4\pi}{1}\right)\frac{P}{E} \Longrightarrow \chi_e = \frac{P}{E}, \qquad (D.2)$$

which is correct. This $\chi_e \to 4\pi \chi_e$ rule is consistent with Eq. (10.42). Similarly, Eq. (10.63) shows that $\mathbf{D}$ is replaced by $\mathbf{D}/4\pi$.

On the magnetic side of things, a few examples are the following. Equation (11.9) shows that $\mathbf{m}$ (and hence $\mathbf{M}$) is replaced by $c\mathbf{m}$ when going from SI to Gaussian units (because $\mathbf{m} = I\mathbf{a} \to c\mathbf{m} = I\mathbf{a} \Rightarrow \mathbf{m} = I\mathbf{a}/c$, which is the correct Gaussian expression). Also, Eqs. (11.69) and (11.70) show that $\mathbf{H}$ is replaced by $(c/4\pi)\mathbf{H}$. Let's check that Eq. (11.68) is consistent with these rules. The SI expression for $\mathbf{H}$ is converted to Gaussian as follows:

$$\mathbf{H} = \frac{1}{\mu_0}\mathbf{B} - \mathbf{M} \longrightarrow \left(\frac{c}{4\pi}\mathbf{H}\right) = \left(\frac{c^2}{4\pi}\right)\left(\frac{\mathbf{B}}{c}\right) - (c\mathbf{M})$$

$$\Longrightarrow \mathbf{H} = \mathbf{B} - 4\pi\mathbf{M}, \qquad (D.3)$$

which is the correct Gaussian expression. Although it is possible to remember all the different rules and then convert things at will, there are so many differing definitions in Chapters 10 and 11 that it is probably easiest to look up each formula as you need it. But for Chapters 1–9, you can get a lot of mileage out of the first two rules above, namely (1) $\epsilon_0 \to 1/4\pi$, $\mu_0 \to 4\pi/c^2$, and (2) $B \to B/c$.

## D.4 The formulas

In the pages that follow, the SI formula is given first, followed by the Gaussian equivalent.

---

[2] The preceding case with $B$ is simply another one of these differences, but we have chosen to discuss it separately because the $B$ field appears so much more often in this book than other such quantities.

## Chapter 1

Coulomb's law (1.4): $\quad \mathbf{F} = \dfrac{1}{4\pi\epsilon_0}\dfrac{q_1 q_2 \hat{\mathbf{r}}}{r^2} \qquad \mathbf{F} = \dfrac{q_1 q_2 \hat{\mathbf{r}}}{r^2}$

potential energy (1.9): $\quad U = \dfrac{1}{4\pi\epsilon_0}\dfrac{q_1 q_2}{r} \qquad U = \dfrac{q_1 q_2}{r}$

electric field (1.20): $\quad \mathbf{E} = \dfrac{1}{4\pi\epsilon_0}\dfrac{q\hat{\mathbf{r}}}{r^2} \qquad \mathbf{E} = \dfrac{q\hat{\mathbf{r}}}{r^2}$

force and field (1.21): $\quad \mathbf{F} = q\mathbf{E} \qquad$ (same)

flux (1.26): $\quad \Phi = \displaystyle\int \mathbf{E}\cdot d\mathbf{a} \qquad$ (same)

Gauss's law (1.31): $\quad \displaystyle\int \mathbf{E}\cdot d\mathbf{a} = \dfrac{q}{\epsilon_0} \qquad \displaystyle\int \mathbf{E}\cdot d\mathbf{a} = 4\pi q$

field due to line (1.39): $\quad E_r = \dfrac{\lambda}{2\pi\epsilon_0 r} \qquad E_r = \dfrac{2\lambda}{r}$

field due to sheet (1.40): $\quad E = \dfrac{\sigma}{2\epsilon_0} \qquad E = 2\pi\sigma$

$\Delta\mathbf{E}$ across sheet (1.41): $\quad \Delta\mathbf{E} = \dfrac{\sigma}{\epsilon_0}\hat{\mathbf{n}} \qquad \Delta\mathbf{E} = 4\pi\sigma\hat{\mathbf{n}}$

field near shell (1.42): $\quad E_r = \dfrac{\sigma}{\epsilon_0} \qquad E_r = 4\pi\sigma$

$F/$(area) on sheet (1.49): $\quad \dfrac{F}{A} = \dfrac{1}{2}(E_1 + E_2)\sigma \qquad$ (same)

energy in $E$ field (1.53): $\quad U = \dfrac{\epsilon_0}{2}\displaystyle\int E^2\, dv \qquad U = \dfrac{1}{8\pi}\displaystyle\int E^2\, dv$

## Chapter 2

electric potential (2.4): $\quad \phi = -\displaystyle\int \mathbf{E}\cdot d\mathbf{s} \qquad$ (same)

field and potential (2.16): $\quad \mathbf{E} = -\nabla\phi \qquad$ (same)

potential and density (2.18): $\quad \phi = \displaystyle\int \dfrac{\rho\, dv}{4\pi\epsilon_0 r} \qquad \phi = \displaystyle\int \dfrac{\rho\, dv}{r}$

potential energy (2.32): $\quad U = \dfrac{1}{2}\displaystyle\int \rho\phi\, dv \qquad$ (same)

dipole potential (2.35): $\quad \phi = \dfrac{q\ell\cos\theta}{4\pi\epsilon_0 r^2} \qquad \phi = \dfrac{q\ell\cos\theta}{r^2}$

dipole moment (2.35): $\quad p = q\ell \qquad$ (same)

dipole field (2.36): $\qquad$ $\mathbf{E} = \dfrac{q\ell}{4\pi\epsilon_0 r^3}\left(2\cos\theta\,\hat{\mathbf{r}} + \sin\theta\,\hat{\boldsymbol{\theta}}\right)$ $\qquad$ $\mathbf{E} = \dfrac{q\ell}{r^3}\left(2\cos\theta\,\hat{\mathbf{r}} + \sin\theta\,\hat{\boldsymbol{\theta}}\right)$

divergence theorem (2.49): $\qquad$ $\displaystyle\int_S \mathbf{F}\cdot d\mathbf{a} = \int_V \operatorname{div}\mathbf{F}\,dv$ $\qquad$ (same)

$\mathbf{E}$ and $\rho$ (2.52): $\qquad$ $\operatorname{div}\mathbf{E} = \dfrac{\rho}{\epsilon_0}$ $\qquad$ $\operatorname{div}\mathbf{E} = 4\pi\rho$

$\mathbf{E}$ and $\phi$ (2.70): $\qquad$ $\operatorname{div}\mathbf{E} = -\nabla^2\phi$ $\qquad$ (same)

$\phi$ and $\rho$ (2.72): $\qquad$ $\nabla^2\phi = -\dfrac{\rho}{\epsilon_0}$ $\qquad$ $\nabla^2\phi = -4\pi\rho$

Stokes' theorem (2.83): $\qquad$ $\displaystyle\int_C \mathbf{F}\cdot d\mathbf{s} = \int_S \operatorname{curl}\mathbf{F}\cdot d\mathbf{a}$ $\qquad$ (same)

## Chapter 3

charge and capacitance (3.7): $\qquad$ $Q = C\phi$ $\qquad$ (same)

sphere $C$ (3.10): $\qquad$ $C = 4\pi\epsilon_0 a$ $\qquad$ $C = a$

parallel-plate $C$ (3.15): $\qquad$ $C = \dfrac{\epsilon_0 A}{s}$ $\qquad$ $C = \dfrac{A}{4\pi s}$

energy in capacitor (3.29): $\qquad$ $U = \dfrac{1}{2}C\phi^2$ $\qquad$ (same)

## Chapter 4

current, current density (4.7): $\qquad$ $I = \displaystyle\int \mathbf{J}\cdot d\mathbf{a}$ $\qquad$ (same)

$\mathbf{J}$ and $\rho$ (4.10): $\qquad$ $\operatorname{div}\mathbf{J} = -\dfrac{\partial\rho}{\partial t}$ $\qquad$ (same)

conductivity (4.11): $\qquad$ $\mathbf{J} = \sigma\mathbf{E}$ $\qquad$ (same)

Ohm's law (4.12): $\qquad$ $V = IR$ $\qquad$ (same)

resistivity (4.16): $\qquad$ $\mathbf{J} = \left(\dfrac{1}{\rho}\right)\mathbf{E}$ $\qquad$ (same)

resistance, resistivity (4.17): $\qquad$ $R = \dfrac{\rho L}{A}$ $\qquad$ (same)

power (4.31): $\qquad$ $P = IV = I^2 R$ $\qquad$ (same)

$R, C$ time constant (4.43): $\qquad$ $\tau = RC$ $\qquad$ (same)

## Chapter 5

Lorentz force (5.1):
$$\mathbf{F} = q\mathbf{E} + q\mathbf{v} \times \mathbf{B} \qquad\qquad \mathbf{F} = q\mathbf{E} + \frac{q}{c}\mathbf{v} \times \mathbf{B}$$

charge in a region (5.2):
$$Q = \epsilon_0 \int \mathbf{E} \cdot d\mathbf{a} \qquad\qquad Q = \frac{1}{4\pi} \int \mathbf{E} \cdot d\mathbf{a}$$

$E$ transformations (5.7):
$$E'_\parallel = E_\parallel, \;\; E'_\perp = \gamma E_\perp \qquad\qquad \text{(same)}$$

$E$ from moving $Q$ (5.15):
$$E' = \frac{Q}{4\pi\epsilon_0 r'^2}\,\frac{1-\beta^2}{(1-\beta^2\sin^2\theta')^{3/2}} \qquad E' = \frac{Q}{r'^2}\,\frac{1-\beta^2}{(1-\beta^2\sin^2\theta')^{3/2}}$$

$F$ transformations (5.17):
$$\frac{dp_\parallel}{dt} = \frac{dp'_\parallel}{dt'}, \;\; \frac{dp_\perp}{dt} = \frac{1}{\gamma}\frac{dp'_\perp}{dt'} \qquad\qquad \text{(same)}$$

$F$ from current (5.28):
$$F_y = \frac{qv_x I}{2\pi\epsilon_0 rc^2} \qquad\qquad F_y = \frac{2qv_x I}{rc^2}$$

## Chapter 6

$B$ due to wire (6.3), (6.6):
$$\mathbf{B} = \hat{\mathbf{z}}\frac{I}{2\pi\epsilon_0 rc^2} = \hat{\mathbf{z}}\frac{\mu_0 I}{2\pi r} \qquad \mathbf{B} = \hat{\mathbf{z}}\frac{2I}{rc}$$

speed of light (6.8):
$$c^2 = \frac{1}{\mu_0\epsilon_0} \qquad\qquad \text{(no analog)}$$

$F$ on a wire (6.14):
$$F = IBl \qquad\qquad F = \frac{IBl}{c}$$

$F$ between wires (6.15):
$$F = \frac{\mu_0 I_1 I_2 l}{2\pi r} \qquad\qquad F = \frac{2I_1 I_2 l}{c^2 r}$$

Ampère's law (6.19):
$$\int \mathbf{B} \cdot d\mathbf{s} = \mu_0 I \qquad\qquad \int \mathbf{B} \cdot d\mathbf{s} = \frac{4\pi}{c} I$$

(differential form) (6.25):
$$\text{curl } \mathbf{B} = \mu_0 \mathbf{J} \qquad\qquad \text{curl } \mathbf{B} = \frac{4\pi}{c}\mathbf{J}$$

vector potential (6.32):
$$\mathbf{B} = \text{curl } \mathbf{A} \qquad\qquad \text{(same)}$$

$\mathbf{A}$ and $\mathbf{J}$ (6.44):
$$\mathbf{A} = \frac{\mu_0}{4\pi}\int \frac{\mathbf{J}\,dv}{r} \qquad\qquad \mathbf{A} = \frac{1}{c}\int \frac{\mathbf{J}\,dv}{r}$$

Biot–Savart law (6.49):
$$d\mathbf{B} = \frac{\mu_0 I}{4\pi}\frac{d\mathbf{l} \times \hat{\mathbf{r}}}{r^2} \qquad\qquad d\mathbf{B} = \frac{I}{c}\frac{d\mathbf{l} \times \hat{\mathbf{r}}}{r^2}$$

$B$ in solenoid (6.57):
$$B_z = \mu_0 n I \qquad\qquad B_z = \frac{4\pi n I}{c}$$

$\Delta B$ across sheet (6.58):
$$\Delta B = \mu_0 \mathcal{J} \qquad\qquad \Delta B = \frac{4\pi \mathcal{J}}{c}$$

$F/$(area) on sheet (6.63):
$$\frac{F}{A} = \frac{(B_z^+)^2 - (B_z^-)^2}{2\mu_0} \qquad\qquad \frac{F}{A} = \frac{(B_z^+)^2 - (B_z^-)^2}{8\pi}$$

E, B transforms (6.76):     $\mathbf{E}'_\| = \mathbf{E}_\|$                                              (same)

$\mathbf{B}'_\| = \mathbf{B}_\|$                                              (same)

$\mathbf{E}'_\perp = \gamma\left(\mathbf{E}_\perp + \boldsymbol{\beta} \times c\mathbf{B}_\perp\right)$     $\mathbf{E}'_\perp = \gamma\left(\mathbf{E}_\perp + \boldsymbol{\beta} \times \mathbf{B}_\perp\right)$

$c\mathbf{B}'_\perp = \gamma\left(c\mathbf{B}_\perp - \boldsymbol{\beta} \times \mathbf{E}_\perp\right)$     $\mathbf{B}'_\perp = \gamma\left(\mathbf{B}_\perp - \boldsymbol{\beta} \times \mathbf{E}_\perp\right)$

Hall $\mathbf{E}_t$ field (6.84):     $\mathbf{E}_t = \dfrac{-\mathbf{J} \times \mathbf{B}}{nq}$     $\mathbf{E}_t = \dfrac{-\mathbf{J} \times \mathbf{B}}{nqc}$

## Chapter 7

electromotive force (7.5):     $\mathcal{E} = \dfrac{1}{q} \displaystyle\int \mathbf{f} \cdot d\mathbf{s}$     (same)

Faraday's law (7.26):     $\mathcal{E} = -\dfrac{d\Phi}{dt}$     $\mathcal{E} = -\dfrac{1}{c}\dfrac{d\Phi}{dt}$

(differential form) (7.31):     $\text{curl }\mathbf{E} = -\dfrac{\partial \mathbf{B}}{\partial t}$     $\text{curl }\mathbf{E} = -\dfrac{1}{c}\dfrac{\partial \mathbf{B}}{\partial t}$

mutual inductance (7.37), (7.38):     $\mathcal{E}_{21} = -M_{21}\dfrac{dI_1}{dt}$     (same)

self-inductance (7.57), (7.58):     $\mathcal{E}_{11} = -L_1\dfrac{dI_1}{dt}$     (same)

$L$ of toroid (7.62):     $L = \dfrac{\mu_0 N^2 h}{2\pi} \ln\left(\dfrac{b}{a}\right)$     $L = \dfrac{2N^2 h}{c^2} \ln\left(\dfrac{b}{a}\right)$

$R, L$ time constant (7.69):     $\tau = L/R$     (same)

energy in inductor (7.74):     $U = \dfrac{1}{2}LI^2$     (same)

energy in $B$ field (7.79):     $U = \dfrac{1}{2\mu_0} \displaystyle\int B^2 \, dv$     $U = \dfrac{1}{8\pi} \displaystyle\int B^2 \, dv$

## Chapter 8

$RLC$ time constant (8.8):     $\tau = \dfrac{1}{\alpha} = \dfrac{2L}{R}$     (same)

$RLC$ frequency (8.9):     $\omega = \sqrt{\dfrac{1}{LC} - \dfrac{R^2}{4L^2}}$     (same)

$Q$ factor (8.12):     $Q = \omega \cdot \dfrac{\text{energy}}{\text{power}}$     (same)

$I_0$ for series $RLC$ (8.38):     $I_0 = \dfrac{\mathcal{E}_0}{\sqrt{R^2 + (\omega L - 1/\omega C)^2}}$     (same)

$\phi$ for series $RLC$ (8.39):     $\tan\phi = \dfrac{1}{R\omega C} - \dfrac{\omega L}{R}$    (same)

resonant $\omega$ (8.41):     $\omega_0 = \dfrac{1}{\sqrt{LC}}$    (same)

width of $I$ curve (8.45):     $\dfrac{2|\Delta\omega|}{\omega_0} = \dfrac{1}{Q}$    (same)

admittance (8.61):     $\tilde{I} = Y\tilde{V}$    (same)

impedance (8.62):     $\tilde{V} = Z\tilde{I}$    (same)

impedances (Table 8.1):     $R, i\omega L, -i/\omega C$    (same)

average power in $R$ (8.81):     $\overline{P}_R = \dfrac{V^2_{\text{rms}}}{R}$    (same)

average power (general) (8.85):     $\overline{P} = V_{\text{rms}}I_{\text{rms}}\cos\phi$    (same)

## Chapter 9

displacement current (9.15):     $\mathbf{J}_{\text{d}} = \epsilon_0\dfrac{\partial\mathbf{E}}{\partial t}$       $\mathbf{J}_{\text{d}} = \dfrac{1}{4\pi}\dfrac{\partial\mathbf{E}}{\partial t}$

Maxwell's equations (9.17):     $\text{curl }\mathbf{E} = -\dfrac{\partial\mathbf{B}}{\partial t}$       $\text{curl }\mathbf{E} = -\dfrac{1}{c}\dfrac{\partial\mathbf{B}}{\partial t}$

$\text{curl }\mathbf{B} = \mu_0\epsilon_0\dfrac{\partial\mathbf{E}}{\partial t} + \mu_0\mathbf{J}$       $\text{curl }\mathbf{B} = \dfrac{1}{c}\dfrac{\partial\mathbf{E}}{\partial t} + \dfrac{4\pi}{c}\mathbf{J}$

$\text{div }\mathbf{E} = \dfrac{\rho}{\epsilon_0}$       $\text{div }\mathbf{E} = 4\pi\rho$

$\text{div }\mathbf{B} = 0$       (same)

speed of wave (9.26), (9.27):     $v = \dfrac{1}{\sqrt{\mu_0\epsilon_0}} = c$       $v = c$

$E, B$ amplitudes (9.26), (9.27):     $E_0 = \dfrac{B_0}{\sqrt{\mu_0\epsilon_0}} = cB_0$       $E_0 = B_0$

power density (9.34):     $S = \epsilon_0\overline{E^2}c$       $S = \dfrac{\overline{E^2}c}{4\pi}$

Poynting vector (9.42):     $\mathbf{S} = \dfrac{\mathbf{E}\times\mathbf{B}}{\mu_0}$       $\mathbf{S} = \dfrac{c}{4\pi}\mathbf{E}\times\mathbf{B}$

invariant 1 (9.51):     $\mathbf{E}'\cdot\mathbf{B}' = \mathbf{E}\cdot\mathbf{B}$       (same)

invariant 2 (9.51):     $E'^2 - c^2B'^2 = E^2 - c^2B^2$       $E'^2 - B'^2 = E^2 - B^2$

## Chapter 10

| | | |
|---|---|---|
| dielectric constant (10.3): | $\kappa = Q/Q_0$ | (same) |
| dipole moment (10.13): | $\mathbf{p} = \int \mathbf{r}' \rho \, dv'$ | (same) |
| dipole potential (10.14): | $\phi(\mathbf{r}) = \dfrac{\hat{\mathbf{r}} \cdot \mathbf{p}}{4\pi\epsilon_0 r^2}$ | $\phi(\mathbf{r}) = \dfrac{\hat{\mathbf{r}} \cdot \mathbf{p}}{r^2}$ |
| dipole $(E_r, E_\theta)$ (10.18): | $\dfrac{p}{4\pi\epsilon_0 r^3}(2\cos\theta, \sin\theta)$ | $\dfrac{p}{r^3}(2\cos\theta, \sin\theta)$ |
| torque on dipole (10.21): | $\mathbf{N} = \mathbf{p} \times \mathbf{E}$ | (same) |
| force on dipole (10.26): | $F_x = \mathbf{p} \cdot \text{grad } E_x$ | (same) |
| polarizability (10.29): | $\mathbf{p} = \alpha\mathbf{E}$ | (same) |
| polarization density (10.31): | $\mathbf{P} = \mathbf{p}N$ | (same) |
| $\phi$ due to column (10.34): | $\phi = \dfrac{P \, da}{4\pi\epsilon_0}\left(\dfrac{1}{r_2} - \dfrac{1}{r_1}\right)$ | $\phi = P \, da\left(\dfrac{1}{r_2} - \dfrac{1}{r_1}\right)$ |
| surface density (10.35): | $\sigma = P$ | (same) |
| average field (10.37): | $\langle \mathbf{E} \rangle = -\dfrac{\mathbf{P}}{\epsilon_0}$ | $\langle \mathbf{E} \rangle = -4\pi\mathbf{P}$ |
| susceptibility (10.41): | $\chi_e = \dfrac{P}{\epsilon_0 E}$ | $\chi_e = \dfrac{P}{E}$ |
| $\chi_e$ and $\kappa$ (10.42): | $\chi_e = \kappa - 1$ | $\chi_e = \dfrac{\kappa - 1}{4\pi}$ |
| $\mathbf{E}$ in polar sphere (10.47): | $\mathbf{E}_{\text{in}} = -\dfrac{\mathbf{P}}{3\epsilon_0}$ | $\mathbf{E}_{\text{in}} = -\dfrac{4\pi\mathbf{P}}{3}$ |
| permittivity (10.56): | $\epsilon = \kappa\epsilon_0$ | (no analog) |
| $\mathbf{P}$ divergence (10.61): | $\text{div } \mathbf{P} = -\rho_{\text{bound}}$ | (same) |
| displacement $\mathbf{D}$ (10.63): | $\mathbf{D} = \epsilon_0\mathbf{E} + \mathbf{P}$ | $\mathbf{D} = \mathbf{E} + 4\pi\mathbf{P}$ |
| $\mathbf{D}$ divergence (10.64): | $\text{div } \mathbf{D} = \rho_{\text{free}}$ | $\text{div } \mathbf{D} = 4\pi\rho_{\text{free}}$ |
| $\mathbf{D}$ for linear (10.65): | $\mathbf{D} = \epsilon\mathbf{E}$ | $\mathbf{D} = \kappa\mathbf{E}$ |
| $\chi_e$ for weak $E$ (10.73): | $\chi_e \approx \dfrac{Np^2}{\epsilon_0 kT}$ | $\chi_e \approx \dfrac{Np^2}{kT}$ |
| bound current $\mathbf{J}$ (10.74): | $\mathbf{J}_{\text{bound}} = \dfrac{\partial \mathbf{P}}{\partial t}$ | (same) |
| curl of $\mathbf{B}$ (10.78): | $\text{curl } \mathbf{B} = \mu_0 \dfrac{\partial \mathbf{D}}{\partial t} + \mu_0\mathbf{J}$ | $\text{curl } \mathbf{B} = \dfrac{1}{c}\dfrac{\partial \mathbf{D}}{\partial t} + \dfrac{4\pi}{c}\mathbf{J}$ |

speed of wave (10.83): $\qquad v = \dfrac{c}{\sqrt{\kappa}}$ (same)

$E, B$ amplitudes (10.83): $\qquad E_0 = \dfrac{cB_0}{\sqrt{\kappa}} = vB_0 \qquad E_0 = \dfrac{B_0}{\sqrt{\kappa}}$

## Chapter 11

dipole moment (11.9): $\qquad \mathbf{m} = I\mathbf{a} \qquad\qquad \mathbf{m} = \dfrac{I\mathbf{a}}{c}$

vector potential (11.10): $\qquad \mathbf{A} = \dfrac{\mu_0}{4\pi}\dfrac{\mathbf{m} \times \hat{\mathbf{r}}}{r^2} \qquad \mathbf{A} = \dfrac{\mathbf{m} \times \hat{\mathbf{r}}}{r^2}$

dipole $(B_r, B_\theta)$ (11.15): $\qquad \dfrac{\mu_0 m}{4\pi r^3}(2\cos\theta, \sin\theta) \qquad \dfrac{m}{r^3}(2\cos\theta, \sin\theta)$

force on dipole (11.23): $\qquad \mathbf{F} = \nabla(\mathbf{m} \cdot \mathbf{B})$ (same)

orbital $\mathbf{m}$ for $e$ (11.29): $\qquad \mathbf{m} = \dfrac{-e}{2m_e}\mathbf{L} \qquad\qquad \mathbf{m} = \dfrac{-e}{2m_e c}\mathbf{L}$

polarizability (11.41): $\qquad \dfrac{\Delta m}{B} = -\dfrac{e^2 r^2}{4m_e} \qquad \dfrac{\Delta m}{B} = -\dfrac{e^2 r^2}{4m_e c^2}$

torque on dipole (11.47): $\qquad \mathbf{N} = \mathbf{m} \times \mathbf{B}$ (same)

polarization density (11.51): $\qquad \mathbf{M} = \dfrac{\mathbf{m}}{\text{volume}}$ (same)

susceptibility $\chi_m$ (11.52): $\qquad \mathbf{M} = \chi_m\dfrac{\mathbf{B}}{\mu_0} \qquad\qquad \mathbf{M} = \chi_m\mathbf{B}$

$\chi_{pm}$ for weak $B$ (11.53): $\qquad \chi_{pm} \approx \dfrac{\mu_0 Nm^2}{kT} \qquad \chi_{pm} \approx \dfrac{Nm^2}{kT}$

surface density $\mathcal{J}$ (11.55): $\qquad \mathcal{J} = M \qquad\qquad \mathcal{J} = Mc$

volume density $\mathbf{J}$ (11.56): $\qquad \mathbf{J} = \operatorname{curl}\mathbf{M} \qquad \mathbf{J} = c\,\operatorname{curl}\mathbf{M}$

$\mathbf{H}$ field (11.68): $\qquad \mathbf{H} = \dfrac{\mathbf{B}}{\mu_0} - \mathbf{M} \qquad \mathbf{H} = \mathbf{B} - 4\pi\mathbf{M}$

curl of $\mathbf{H}$ (11.69): $\qquad \operatorname{curl}\mathbf{H} = \mathbf{J}_{\text{free}} \qquad \operatorname{curl}\mathbf{H} = \dfrac{4\pi}{c}\mathbf{J}_{\text{free}}$

(integrated form) (11.70): $\qquad \displaystyle\int \mathbf{H} \cdot d\mathbf{l} = I_{\text{free}} \qquad \displaystyle\int \mathbf{H} \cdot d\mathbf{l} = \dfrac{4\pi}{c}I_{\text{free}}$

$\chi_m$ (accepted def.) (11.72): $\qquad \mathbf{M} = \chi_m\mathbf{H}$ (same)

permeability (11.74): $\qquad \mu = \mu_0(1 + \chi_m) \qquad \mu = 1 + 4\pi\chi_m$

$\mathbf{B}$ and $\mathbf{H}$ (11.74): $\qquad \mathbf{B} = \mu\mathbf{H}$ (same)

Appendix H

tangential $E_\theta$ (H.3):     $E_\theta = \dfrac{qa\sin\theta}{4\pi\epsilon_0 c^2 R}$     $E_\theta = \dfrac{qa\sin\theta}{c^2 R}$

power (H.7):     $P_{\mathrm{rad}} = \dfrac{q^2 a^2}{6\pi\epsilon_0 c^3}$     $P_{\mathrm{rad}} = \dfrac{2q^2 a^2}{3c^3}$

# E

# Exact relations among SI and Gaussian units

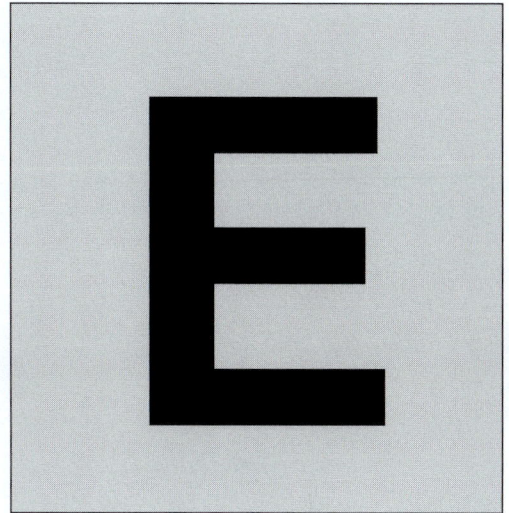

In 1983 the General Conference on Weights and Measures officially redefined the meter as the distance that light travels in vacuum during a time interval of 1/299,792,458 of a second. The second is defined in terms of a certain atomic frequency in a way that does not concern us here. The nine-digit integer was chosen to make the assigned value of $c$ agree with the most accurate measured value to well within the uncertainty in the latter. Henceforth the velocity of light is, *by definition*, 299,792,458 meters/second. An experiment in which the passage of a light pulse from point $A$ to point $B$ is timed is to be regarded as a measurement of the distance from $A$ to $B$, not a measurement of the speed of light.

While this step has no immediate practical consequences, it does bring a welcome simplification of the exact relations connecting various electromagnetic units. As we learn in Chapter 9, Maxwell's equations for the vacuum fields, formulated in SI units, have a solution in the form of a traveling wave with velocity $c = (\mu_0 \epsilon_0)^{-1/2}$. The SI constant $\mu_0$ has always been defined exactly as $4\pi \cdot 10^{-7} \, \text{kg m/C}^2$, whereas the value of $\epsilon_0$ has depended on the experimentally determined value of the speed of light, any refinement of which called for adjustment of the value of $\epsilon_0$. But now $\epsilon_0$ acquires a permanent and perfectly precise value of its own, through the requirement that

$$(\mu_0 \epsilon_0)^{-1/2} = 299{,}792{,}458 \text{ meters/second.} \qquad \text{(E.1)}$$

In the Gaussian system no such question arises. Wherever $c$ is involved, it appears in plain view, and all other quantities are defined exactly, beginning with the electrostatic unit of charge, the esu, whose definition by Coulomb's law involves no arbitrary factor.

With the adoption of Eq. (E.1) in consequence of the redefinition of the meter, the relations among the units in the systems we have been using can be stated with unlimited precision. These relations are listed at the beginning of Appendix C for the principal quantities we deal with. In the list the symbol [3] stands for the precise decimal 2.99792458.

The exact numbers are uninteresting and for our work quite unnecessary. It is sheer luck that [3] happens to be so close to 3, an accidental consequence of the length of the meter and the second. When 0.1 percent accuracy is good enough we need only remember that "300 volts is a statvolt" and "$3 \cdot 10^9$ esu is a coulomb." Much less precisely, but still within 12 percent, a capacitance of 1 cm is equivalent to 1 picofarad.

An important SI constant is $(\mu_0/\epsilon_0)^{1/2}$, which is a resistance in ohms. Since $\epsilon_0 = 1/\mu_0 c^2$, this resistance equals $\mu_0 c$. Using the exact values of $\mu_0$ and $c$, we find $(\mu_0/\epsilon_0)^{1/2} = 40\pi \cdot [3]$ ohms $\approx 376.73$ ohms. One tends to remember it, and even refer to it, as "377 ohms." It is the ratio of the electric field strength $E$, in volts/meter, in a plane wave in vacuum, to the strength, in amperes/meter, of the accompanying magnetic field $H$. For this reason the constant $(\mu_0/\epsilon_0)^{1/2}$ is sometimes denoted by $Z_0$ and called, rather cryptically, the *impedance of the vacuum.* In a plane wave in vacuum in which $E_{\text{rms}}$ is the rms electric field in volts/meter, the mean density of power transmitted, in watts/m$^2$, is $E_{\text{rms}}^2/Z_0$.

The logical relation of the SI electrical units to one another now takes on a slightly different aspect. Before the redefinition of the meter, it was customary to designate one of the electrical units as *primary,* in this sense: its precise value could, at least in principle, be established by a procedure involving the SI mechanical and metrical units only. Thus the ampere, to which this role has usually been assigned, was defined in terms of the force in newtons between parallel currents, using the relation in Eq. (6.15). This was possible because the constant $\mu_0$ in that relation has the precise value $4\pi \cdot 10^{-7}$ kg m/C$^2$. Then with the ampere as the primary electrical unit, the coulomb was defined precisely as 1 ampere-second. The coulomb itself, owing to the presence of $\epsilon_0$ in Coulomb's law, was not eligible to serve as the primary unit. Now with $\epsilon_0$ as well as $\mu_0$ assigned an exact numerical value, the system can be built up with any unit as the starting point. All quantities are in this sense on an equal footing, and the choice of a primary unit loses its significance. Never a very interesting question anyway, it can now be relegated to history.

# Curvilinear coordinates

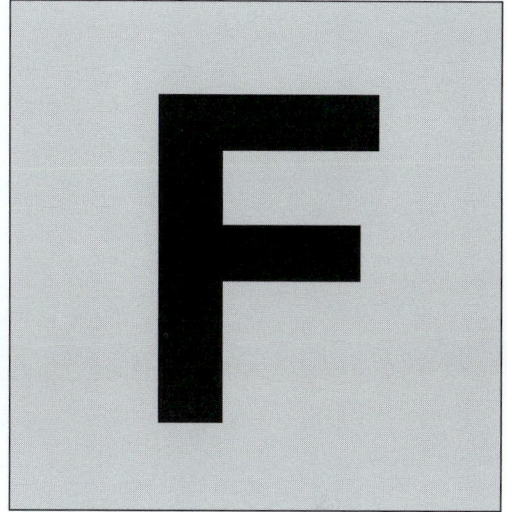

We begin this appendix by listing the main vector operators (gradient, divergence, curl, Laplacian) in Cartesian, cylindrical, and spherical coordinates. We then talk a little about each operator – define things, derive a few results, give some examples, etc. You will note that some of the expressions below are rather scary looking. However, you won't have to use their full forms in this book. In the applications that come up, invariably only one or two of the terms in the expressions are nonzero.

## F.1 Vector operators
### F.1.1 Cartesian coordinates

$$d\mathbf{s} = dx\,\hat{\mathbf{x}} + dy\,\hat{\mathbf{y}} + dz\,\hat{\mathbf{z}},$$

$$\nabla = \hat{\mathbf{x}}\frac{\partial}{\partial x} + \hat{\mathbf{y}}\frac{\partial}{\partial y} + \hat{\mathbf{z}}\frac{\partial}{\partial z},$$

$$\nabla f = \frac{\partial f}{\partial x}\hat{\mathbf{x}} + \frac{\partial f}{\partial y}\hat{\mathbf{y}} + \frac{\partial f}{\partial z}\hat{\mathbf{z}},$$

$$\nabla \cdot \mathbf{A} = \frac{\partial A_x}{\partial x} + \frac{\partial A_y}{\partial y} + \frac{\partial A_z}{\partial z},$$

$$\nabla \times \mathbf{A} = \left(\frac{\partial A_z}{\partial y} - \frac{\partial A_y}{\partial z}\right)\hat{\mathbf{x}} + \left(\frac{\partial A_x}{\partial z} - \frac{\partial A_z}{\partial x}\right)\hat{\mathbf{y}} + \left(\frac{\partial A_y}{\partial x} - \frac{\partial A_x}{\partial y}\right)\hat{\mathbf{z}},$$

$$\nabla^2 f = \frac{\partial^2 f}{\partial x^2} + \frac{\partial^2 f}{\partial y^2} + \frac{\partial^2 f}{\partial z^2}. \tag{F.1}$$

## F.1.2 Cylindrical coordinates

$$ds = dr\,\hat{\mathbf{r}} + r\,d\theta\,\hat{\boldsymbol{\theta}} + dz\,\hat{\mathbf{z}},$$

$$\nabla = \hat{\mathbf{r}}\frac{\partial}{\partial r} + \hat{\boldsymbol{\theta}}\frac{1}{r}\frac{\partial}{\partial \theta} + \hat{\mathbf{z}}\frac{\partial}{\partial z},$$

$$\nabla f = \frac{\partial f}{\partial r}\hat{\mathbf{r}} + \frac{1}{r}\frac{\partial f}{\partial \theta}\hat{\boldsymbol{\theta}} + \frac{\partial f}{\partial z}\hat{\mathbf{z}},$$

$$\nabla \cdot \mathbf{A} = \frac{1}{r}\frac{\partial (rA_r)}{\partial r} + \frac{1}{r}\frac{\partial A_\theta}{\partial \theta} + \frac{\partial A_z}{\partial z},$$

$$\nabla \times \mathbf{A} = \left(\frac{1}{r}\frac{\partial A_z}{\partial \theta} - \frac{\partial A_\theta}{\partial z}\right)\hat{\mathbf{r}} + \left(\frac{\partial A_r}{\partial z} - \frac{\partial A_z}{\partial r}\right)\hat{\boldsymbol{\theta}}$$

$$+ \frac{1}{r}\left(\frac{\partial (rA_\theta)}{\partial r} - \frac{\partial A_r}{\partial \theta}\right)\hat{\mathbf{z}},$$

$$\nabla^2 f = \frac{1}{r}\frac{\partial}{\partial r}\left(r\frac{\partial f}{\partial r}\right) + \frac{1}{r^2}\frac{\partial^2 f}{\partial \theta^2} + \frac{\partial^2 f}{\partial z^2}. \tag{F.2}$$

## F.1.3 Spherical coordinates

$$ds = dr\,\hat{\mathbf{r}} + r\,d\theta\,\hat{\boldsymbol{\theta}} + r\sin\theta\,d\phi\,\hat{\boldsymbol{\phi}},$$

$$\nabla = \hat{\mathbf{r}}\frac{\partial}{\partial r} + \hat{\boldsymbol{\theta}}\frac{1}{r}\frac{\partial}{\partial \theta} + \hat{\boldsymbol{\phi}}\frac{1}{r\sin\theta}\frac{\partial}{\partial \phi},$$

$$\nabla f = \frac{\partial f}{\partial r}\hat{\mathbf{r}} + \frac{1}{r}\frac{\partial f}{\partial \theta}\hat{\boldsymbol{\theta}} + \frac{1}{r\sin\theta}\frac{\partial f}{\partial \phi}\hat{\boldsymbol{\phi}},$$

$$\nabla \cdot \mathbf{A} = \frac{1}{r^2}\frac{\partial (r^2 A_r)}{\partial r} + \frac{1}{r\sin\theta}\frac{\partial (A_\theta \sin\theta)}{\partial \theta} + \frac{1}{r\sin\theta}\frac{\partial A_\phi}{\partial \phi},$$

$$\nabla \times \mathbf{A} = \frac{1}{r\sin\theta}\left(\frac{\partial (A_\phi \sin\theta)}{\partial \theta} - \frac{\partial A_\theta}{\partial \phi}\right)\hat{\mathbf{r}} + \frac{1}{r}\left(\frac{1}{\sin\theta}\frac{\partial A_r}{\partial \phi} - \frac{\partial (rA_\phi)}{\partial r}\right)\hat{\boldsymbol{\theta}},$$

$$+ \frac{1}{r}\left(\frac{\partial (rA_\theta)}{\partial r} - \frac{\partial A_r}{\partial \theta}\right)\hat{\boldsymbol{\phi}},$$

$$\nabla^2 f = \frac{1}{r^2}\frac{\partial}{\partial r}\left(r^2\frac{\partial f}{\partial r}\right) + \frac{1}{r^2\sin\theta}\frac{\partial}{\partial \theta}\left(\sin\theta\frac{\partial f}{\partial \theta}\right) + \frac{1}{r^2\sin^2\theta}\frac{\partial^2 f}{\partial \phi^2}. \tag{F.3}$$

# F.2 Gradient

The gradient produces a vector from a scalar. The gradient of a function $f$, written as $\nabla f$ or $\mathrm{grad}\,f$, may be defined[1] as the vector with the

---

[1] We used a different definition in Section 2.3, but we will show below that the two definitions are equivalent.

property that the change in $f$ brought about by a small change $d\mathbf{s}$ in position is

$$df = \nabla f \cdot d\mathbf{s}. \tag{F.4}$$

The vector $\nabla f$ depends on position; there is a different gradient vector associated with each point in the parameter space.

You might wonder whether a vector that satisfies Eq. (F.4) actually exists. We are claiming that if $f$ is a function of, say, three variables, then at every point in space there exists a *unique* vector, $\nabla f$, such that for *any* small displacement $d\mathbf{s}$ from a given point, the change in $f$ equals $\nabla f \cdot d\mathbf{s}$. It is not immediately obvious why a single vector gets the job done for all possible displacements $d\mathbf{s}$ from a given point. But the existence of such a vector can be demonstrated in two ways. First, we can explicitly construct $\nabla f$; we will do this below in Eq. (F.5). Second, any (well-behaved) function looks like a linear function up close, and for a linear function a vector $\nabla f$ satisfying Eq. (F.4) does indeed exist. We will explain why in what follows. However, before addressing this issue, let us note an important property of the gradient.

From the definition in Eq. (F.4), it immediately follows (as mentioned in Section 2.3) that $\nabla f$ points in the direction of steepest ascent of $f$. This is true because we can write the dot product $\nabla f \cdot d\mathbf{s}$ as $|\nabla f||d\mathbf{s}| \cos \theta$, where $\theta$ is the angle between the vector $\nabla f$ and the vector $d\mathbf{s}$. So for a given length of the vector $d\mathbf{s}$, this dot product is maximized when $\theta = 0$. We therefore want the displacement $d\mathbf{s}$ to point in the direction of $\nabla f$, if we want to produce the maximum change in $f$.

If we consider the more easily visualizable case of a function of two variables, the function can be represented by a surface above the $xy$ plane. This surface is locally planar; that is, a sufficiently small bug walking around on it would think it is a (generally tilted) flat plane. If we look at the direction of steepest ascent in the local plane, and then project this line onto the $xy$ plane, the resulting line is the direction of $\nabla f$; see Fig. 2.5. The function $f$ is constant along the direction perpendicular to $\nabla f$. The magnitude of $\nabla f$ equals the change in $f$ per unit distance in the parameter space, in the direction of $\nabla f$. Equivalently, if we restrict the parameter space to the one-dimensional line in the direction of steepest ascent, then the gradient is simply the standard single-variable derivative in that direction.

We could alternatively work "backwards" and define the gradient as the vector that points in the direction (in the parameter space) of steepest ascent, with its magnitude equal to the rate of change in that direction. It then follows that the general change in $f$, for *any* displacement $d\mathbf{s}$ in the parameter space, is given by Eq. (F.4). This is true because the dot product picks out the component of $d\mathbf{s}$ along the direction of $\nabla f$. This component causes a change in $f$, whereas the orthogonal component does not.

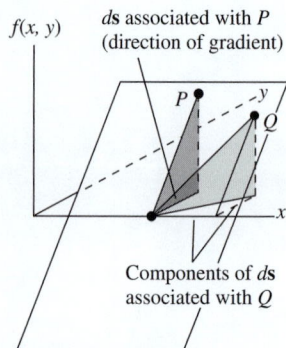

**Figure F.1.**
Only the component of $d\mathbf{s}$ in the direction of the gradient causes a change in $f$.

Figure F.1 shows how this works in the case of a function of two variables. We have assumed for simplicity that the local plane representing the surface of the function intersects the $xy$ plane along the $x$ axis. (We can always translate and rotate the coordinate system so that this is true at a given point.) The gradient then points in the $y$ direction. The point $P$ shown lies in the direction straight up the plane from the given point. The projection of this direction onto the $xy$ plane lies along the gradient. The point $Q$ is associated with a $d\mathbf{s}$ interval that doesn't lie along the gradient in the $xy$ plane. This $d\mathbf{s}$ can be broken up into an interval along the $x$ axis, which causes no change in $f$, plus an interval in the $y$ direction, or equivalently the direction of the gradient, which causes the change in $f$ up to the point $Q$.

The preceding two paragraphs explain why the vector $\nabla f$ defined by Eq. (F.4) does in fact exist; any well-behaved function is locally linear, and a unique vector $\nabla f$ at each point will get the job done in Eq. (F.4) if $f$ is linear. But as mentioned above, we can also demonstrate the existence of such a vector by simply constructing it. Let's calculate the gradient in Cartesian coordinates, and then in spherical coordinates.

### F.2.1 Cartesian gradient

In Cartesian coordinates, a general change in $f$ for small displacements can be written as $df = (\partial f/\partial x)dx + (\partial f/\partial y)dy + (\partial f/\partial z)dz$. This is just the start of the Taylor series in three variables. The interval $d\mathbf{s}$ is simply $(dx, dy, dz)$, so if we want $\nabla f \cdot d\mathbf{s}$ to be equal to $df$, we need

$$\nabla f = \left(\frac{\partial f}{\partial x}, \frac{\partial f}{\partial y}, \frac{\partial f}{\partial z}\right) \equiv \frac{\partial f}{\partial x}\,\hat{\mathbf{x}} + \frac{\partial f}{\partial y}\,\hat{\mathbf{y}} + \frac{\partial f}{\partial z}\,\hat{\mathbf{z}}, \quad \text{(F.5)}$$

in agreement with the $\nabla f$ expression in Eq. (F.1). In Section 2.3 we took Eq. (F.5) as the definition of the gradient and then discussed its other properties.

### F.2.2 Spherical gradient

In spherical coordinates, a general change in $f$ is given by $df = (\partial f/\partial r)dr + (\partial f/\partial \theta)d\theta + (\partial f/\partial \phi)d\phi$. However, the interval $d\mathbf{s}$ takes a more involved form compared with the Cartesian $d\mathbf{s}$. It is

$$d\mathbf{s} = (dr, r\,d\theta, r\sin\theta\,d\phi) \equiv dr\,\hat{\mathbf{r}} + r\,d\theta\,\hat{\boldsymbol{\theta}} + r\sin\theta\,d\phi\,\hat{\boldsymbol{\phi}}. \quad \text{(F.6)}$$

If we want $\nabla f \cdot d\mathbf{s}$ to be equal to $df$, then we need

$$\nabla f = \left(\frac{\partial f}{\partial r}, \frac{1}{r}\frac{\partial f}{\partial \theta}, \frac{1}{r\sin\theta}\frac{\partial f}{\partial \phi}\right) \equiv \frac{\partial f}{\partial r}\,\hat{\mathbf{r}} + \frac{1}{r}\frac{\partial f}{\partial \theta}\,\hat{\boldsymbol{\theta}} + \frac{1}{r\sin\theta}\frac{\partial f}{\partial \phi}\,\hat{\boldsymbol{\phi}}, \quad \text{(F.7)}$$

in agreement with Eq. (F.3).

We see that the extra factors (compared with the Cartesian case) in the denominators of the gradient come from the coefficients of the unit vectors in the expression for $d\mathbf{s}$. Similarly, the form of the gradient in cylindrical coordinates in Eq. (F.2) can be traced to the fact that the interval $d\mathbf{s}$ equals $dr\,\hat{\mathbf{r}} + r\,d\theta\,\hat{\boldsymbol{\theta}} + dz\,\hat{\mathbf{z}}$. Since the extra factors that appear in $d\mathbf{s}$ show up in the denominators of the $\nabla$-operator terms, and since the $\nabla$ operator determines all of the other vector operators, we see that every result in this appendix can be traced back to the form of $d\mathbf{s}$ in the different coordinate systems. For example, the big scary expression listed in Eq. (F.3) for the curl in spherical coordinates is a direct consequence of the $d\mathbf{s} = dr\,\hat{\mathbf{r}} + r\,d\theta\,\hat{\boldsymbol{\theta}} + r\sin\theta\,d\phi\,\hat{\boldsymbol{\phi}}$ interval.

Note that the consideration of units tells us that there must be a factor of $r$ in the denominators in the $\partial f/\partial\theta$ and $\partial f/\partial\phi$ terms in the spherical gradient, and in the $\partial f/\partial\theta$ term in the cylindrical gradient.

## F.3 Divergence

The divergence produces a scalar from a vector. The divergence of a vector function was defined in Eq. (2.47) as the net flux out of a given small volume, divided by the volume. In Section 2.10 we derived the form of the divergence in Cartesian coordinates, and it turned out to be the dot product of the $\nabla$ operator with a vector $\mathbf{A}$, that is, $\nabla \cdot \mathbf{A}$. We use the same method here to derive the form in cylindrical coordinates. We then give a second, more mechanical, derivation. A third derivation is left for Exercise F.2.

### F.3.1 Cylindrical divergence, first method

Consider the small volume that is generated by taking the region in the $r$-$\theta$ plane shown in Fig. F.2 and sweeping it through a span of $z$ values from a particular $z$ up to $z + \Delta z$ (the $\hat{\mathbf{z}}$ axis points out of the page). Let's first look at the flux of a vector field $\mathbf{A}$ through the two faces perpendicular to the $\hat{\mathbf{z}}$ direction. As in Section 2.10, only the $z$ component of $\mathbf{A}$ is relevant to the flux through these faces. In the limit of a small volume, the area of these faces is $r\,\Delta r\,\Delta\theta$. The inward flux through the bottom face equals $A_z(z)\,r\,\Delta r\,\Delta\theta$, and the outward flux through the top face equals $A_z(z+\Delta z)\,r\,\Delta r\,\Delta\theta$. We have suppressed the $r$ and $\theta$ arguments of $A_z$ for simplicity, and we have chosen points at the midpoints of the faces, as in Fig. 2.22. The net outward flux is therefore

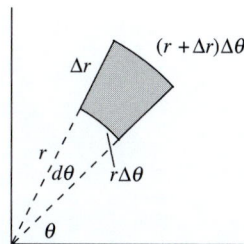

**Figure F.2.**
A small region in the $r$-$\theta$ plane.

$$\Phi_{z\text{ faces}} = A_z(z+\Delta z)\,r\,\Delta r\,\Delta\theta - A_z(z)\,r\,\Delta r\,\Delta\theta$$

$$= \left(\frac{A_z(z+\Delta z) - A_z(z)}{\Delta z}\right) r\,\Delta r\,\Delta\theta\,\Delta z$$

$$= \frac{\partial A_z}{\partial z}\, r\,\Delta r\,\Delta\theta\,\Delta z. \tag{F.8}$$

Upon dividing this net outward flux by the volume $r \, \Delta r \, \Delta \theta \, \Delta z$, we obtain $\partial A_z / \partial z$, in agreement with the third term in $\nabla \cdot \mathbf{A}$ in Eq. (F.2). This was exactly the same argument we used in Section 2.10. The $z$ coordinate in cylindrical coordinates is, after all, basically a Cartesian coordinate. However, things get more interesting with the $r$ coordinate.

Consider the flux through the two faces (represented by the curved lines in Fig. F.2) that are perpendicular to the $\hat{\mathbf{r}}$ direction. The key point to realize is that *the areas of these two faces are not equal*. The upper right one is larger. So the difference in flux through these faces depends not only on the value of $A_r$, but also on the area. The inward flux through the lower left face equals $A_r(r)\left[ r \, \Delta \theta \, \Delta z \right]$, and the outward flux through the upper right face equals $A_r(r + \Delta r)\left[ (r + \Delta r) \, \Delta \theta \, \Delta z \right]$. As above, we have suppressed the $\theta$ and $z$ arguments for simplicity, and we have chosen points at the midpoints of the faces. The net outward flux is therefore

$$\Phi_{r \text{ faces}} = (r + \Delta r)A_r(r + \Delta r) \, \Delta \theta \, \Delta z - rA_r(r) \, \Delta \theta \, \Delta z$$

$$= \left( \frac{(r + \Delta r)A_r(r + \Delta r) - rA_r(r)}{\Delta r} \right) \Delta r \, \Delta \theta \, \Delta z$$

$$= \frac{\partial (rA_r)}{\partial r} \, \Delta r \, \Delta \theta \, \Delta z. \tag{F.9}$$

Upon dividing this net outward flux by the volume $r \, \Delta r \, \Delta \theta \, \Delta z$, we have a leftover $r$ in the denominator, so we obtain $(1/r)\big(\partial(rA_r)/\partial r\big)$, in agreement with the first term in Eq. (F.2).

For the last two faces, the ones perpendicular to the $\hat{\boldsymbol{\theta}}$ direction, we don't have to worry about different areas, so we quickly obtain

$$\Phi_{\theta \text{ faces}} = A_\theta(\theta + \Delta \theta) \, \Delta r \, \Delta z - A_\theta(\theta) \, \Delta r \, \Delta z$$

$$= \left( \frac{A_\theta(\theta + \Delta \theta) - A_\theta(\theta)}{\Delta \theta} \right) \Delta r \, \Delta \theta \, \Delta z$$

$$= \frac{\partial A_\theta}{\partial \theta} \, \Delta r \, \Delta \theta \, \Delta z. \tag{F.10}$$

Upon dividing this net outward flux by the volume $r \, \Delta r \, \Delta \theta \, \Delta z$, we again have a leftover $r$ in the denominator, so we obtain $(1/r)(\partial A_\theta / \partial \theta)$, in agreement with the second term in Eq. (F.2).

If you like this sort of calculation, you can repeat this derivation for the case of spherical coordinates. However, it's actually not too hard to derive the general form of the divergence for *any* set of coordinates; see Exercise F.3. You can then check that this general formula reduces properly for spherical coordinates.

## F.3.2 Cylindrical divergence, second method

Let's determine the divergence in cylindrical coordinates by explicitly calculating the dot product,

$$\nabla \cdot \mathbf{A} = \left( \hat{\mathbf{r}} \frac{\partial}{\partial r} + \hat{\boldsymbol{\theta}} \frac{1}{r} \frac{\partial}{\partial \theta} + \hat{\mathbf{z}} \frac{\partial}{\partial z} \right) \cdot \left( \hat{\mathbf{r}} A_r + \hat{\boldsymbol{\theta}} A_\theta + \hat{\mathbf{z}} A_z \right). \quad \text{(F.11)}$$

At first glance, it appears that $\nabla \cdot \mathbf{A}$ doesn't produce the form of the divergence given in Eq. (F.2). The second two terms work out, but it seems like the first term should simply be $\partial A_r / \partial r$ instead of $(1/r)(\partial(rA_r)/\partial r)$. However, the dot product does indeed correctly yield the latter term, because we must remember that, in contrast with Cartesian coordinates, *in cylindrical coordinates the unit vectors themselves depend on position.* This means that in Eq. (F.11) the derivatives in the $\nabla$ operator also act on the unit vectors in $\mathbf{A}$. This issue doesn't come up in Cartesian coordinates because $\hat{\mathbf{x}}$, $\hat{\mathbf{y}}$, and $\hat{\mathbf{z}}$ are fixed vectors, but that is more the exception than the rule. Writing $\mathbf{A}$ in the abbreviated form $(A_r, A_\theta, A_z)$ tends to hide important information. The full expression for $\mathbf{A}$ is $\hat{\mathbf{r}} A_r + \hat{\boldsymbol{\theta}} A_\theta + \hat{\mathbf{z}} A_z$. There are six quantities here (three vectors and three components), and if any of these quantities vary with the coordinates, then these variations cause $\mathbf{A}$ to change. The derivatives of the unit vectors that are nonzero are

$$\frac{\partial \hat{\mathbf{r}}}{\partial \theta} = \hat{\boldsymbol{\theta}} \quad \text{and} \quad \frac{\partial \hat{\boldsymbol{\theta}}}{\partial \theta} = -\hat{\mathbf{r}}. \quad \text{(F.12)}$$

To demonstrate these relations, we can look at what happens to $\hat{\mathbf{r}}$ and $\hat{\boldsymbol{\theta}}$ if we rotate them through an angle $d\theta$. Since the unit vectors have length 1, we see from Fig. F.3 that $\hat{\mathbf{r}}$ picks up a component of length $d\theta$ in the $\hat{\boldsymbol{\theta}}$ direction, and $\hat{\boldsymbol{\theta}}$ picks up a component of length $d\theta$ in the $-\hat{\mathbf{r}}$ direction. The other seven of the nine possible derivatives are zero because none of the unit vectors depends on $r$ or $z$, and furthermore $\hat{\mathbf{z}}$ doesn't depend on $\theta$.

Due to the orthogonality of the unit vectors, we quickly see that, in addition to the three "corresponding" terms that survive in Eq. (F.11), one more term is nonzero:

$$\hat{\boldsymbol{\theta}} \cdot \frac{1}{r} \frac{\partial}{\partial \theta} (\hat{\mathbf{r}} A_r) = \hat{\boldsymbol{\theta}} \cdot \frac{1}{r} \left( \frac{\partial \hat{\mathbf{r}}}{\partial \theta} A_r + \hat{\mathbf{r}} \frac{\partial A_r}{\partial \theta} \right) = \hat{\boldsymbol{\theta}} \cdot \frac{1}{r} \hat{\boldsymbol{\theta}} A_r + 0 = \frac{A_r}{r}. \quad \text{(F.13)}$$

Equation (F.11) therefore becomes

$$\nabla \cdot \mathbf{A} = \frac{\partial A_r}{\partial r} + \frac{1}{r} \frac{\partial A_\theta}{\partial \theta} + \frac{\partial A_z}{\partial z} + \frac{A_r}{r}. \quad \text{(F.14)}$$

The sum of the first and last terms here can be rewritten as the first term in $\nabla \cdot \mathbf{A}$ in Eq. (F.2), as desired.

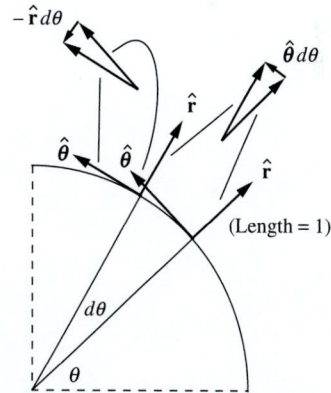

**Figure F.3.**
How the $\hat{\mathbf{r}}$ and $\hat{\boldsymbol{\theta}}$ unit vectors depend on $\theta$.

## F.4 Curl

The curl produces a vector from a vector. The curl of a vector function was defined in Eq. (2.80) as the net circulation around a given small area, divided by the area. (The three possible orientations of the area yield the three components.) In Section 2.16 we derived the form of the curl in Cartesian coordinates, and it turned out to be the cross product of the $\nabla$ operator with the vector $\mathbf{A}$, that is, $\nabla \times \mathbf{A}$. We'll use the same method here to derive the form in cylindrical coordinates, after which we derive it a second way, analogous to the above second method for the divergence. Actually, we'll calculate just the $z$ component; this should make the procedure clear. As an exercise you can calculate the other two components.

### F.4.1 Cylindrical curl, first method

The $z$ component of $\nabla \times \mathbf{A}$ is found by looking at the circulation around a small area in the $r$-$\theta$ plane (or more generally, in some plane parallel to the $r$-$\theta$ plane). Consider the upper right and lower left (curved) edges in Fig. F.2. Following the strategy in Section 2.16, the counterclockwise line integral along the upper right edge equals $A_\theta(r + \Delta r)\big[(r + \Delta r)\,\Delta\theta\big]$, and the counterclockwise line integral along the lower left edge equals $-A_\theta(r)\big[r\,\Delta\theta\big]$. We have suppressed the $\theta$ and $z$ arguments for simplicity, and we have chosen points at the midpoints of the edges. Note that we have correctly incorporated the fact that the upper right edge is longer than the lower left edge (the same issue that came up in the above calculation of the divergence). The net circulation along these two edges is

$$
\begin{aligned}
C_{\theta \text{ sides}} &= (r + \Delta r)A_\theta(r + \Delta r)\,\Delta\theta - rA_\theta(r)\,\Delta\theta \\[4pt]
&= \left(\frac{(r + \Delta r)A_\theta(r + \Delta r) - rA_\theta(r)}{\Delta r}\right)\Delta r\,\Delta\theta \\[4pt]
&= \frac{\partial(rA_\theta)}{\partial r}\Delta r\,\Delta\theta.
\end{aligned}
\tag{F.15}
$$

Upon dividing this circulation by the area $r\,\Delta r\,\Delta\theta$, we have a leftover $r$ in the denominator, so we obtain $(1/r)\big(\partial(rA_\theta)/\partial r\big)$, in agreement with the first of the two terms in the $z$ component of $\nabla \times \mathbf{A}$ in Eq. (F.2).

Now consider the upper left and lower right (straight) edges. The counterclockwise line integral along the upper left edge equals $-A_r(\theta + \Delta\theta)\,\Delta r$, and the counterclockwise line integral along the lower right edge equals $A_r(\theta)\,\Delta r$. The net circulation along these two edges is

$$
\begin{aligned}
C_{r \text{ sides}} &= -A_r(\theta + \Delta\theta)\,\Delta r + A_r(\theta)\,\Delta r \\[4pt]
&= -\left(\frac{A_r(\theta + \Delta\theta) - A_r(\theta)}{\Delta\theta}\right)\Delta r\,\Delta\theta \\[4pt]
&= -\frac{\partial A_r}{\partial\theta}\Delta r\,\Delta\theta.
\end{aligned}
\tag{F.16}
$$

Upon dividing this circulation by the area $r \, \Delta r \, \Delta \theta$, we again have a left-over $r$ in the denominator, so we obtain $-(1/r)(\partial A_r / \partial \theta)$, in agreement with Eq. (F.2).

### F.4.2 Cylindrical curl, second method

Our goal is to calculate the cross product,

$$\nabla \times \mathbf{A} = \left( \hat{\mathbf{r}} \frac{\partial}{\partial r} + \hat{\boldsymbol{\theta}} \frac{1}{r} \frac{\partial}{\partial \theta} + \hat{\mathbf{z}} \frac{\partial}{\partial z} \right) \times \left( \hat{\mathbf{r}} A_r + \hat{\boldsymbol{\theta}} A_\theta + \hat{\mathbf{z}} A_z \right), \quad \text{(F.17)}$$

while remembering that some of the unit vectors depend on the coordinates according to Eq. (F.12). As above, we'll look at just the $z$ component. This component arises from terms of the form $\hat{\mathbf{r}} \times \hat{\boldsymbol{\theta}}$ or $\hat{\boldsymbol{\theta}} \times \hat{\mathbf{r}}$. In addition to the two obvious terms of this form, we also have the one involving $\hat{\boldsymbol{\theta}} \times (\partial \hat{\boldsymbol{\theta}} / \partial \theta)$, which from Eq. (F.12) equals $\hat{\boldsymbol{\theta}} \times (-\hat{\mathbf{r}}) = \hat{\mathbf{z}}$. The complete $z$ component of the cross product is therefore

$$(\nabla \times \mathbf{A})_z = \hat{\mathbf{r}} \times \frac{\partial (\hat{\boldsymbol{\theta}} A_\theta)}{\partial r} + \hat{\boldsymbol{\theta}} \times \frac{1}{r} \frac{\partial (\hat{\mathbf{r}} A_r)}{\partial \theta} + \hat{\boldsymbol{\theta}} \times \frac{1}{r} \frac{\partial (\hat{\boldsymbol{\theta}} A_\theta)}{\partial \theta}$$

$$= \hat{\mathbf{z}} \left( \frac{\partial A_\theta}{\partial r} - \frac{1}{r} \frac{\partial A_r}{\partial \theta} + \frac{A_\theta}{r} \right). \quad \text{(F.18)}$$

The sum of the first and last terms here can be rewritten as the first term in the $z$ component of $\nabla \times \mathbf{A}$ in Eq. (F.2), as desired.

## F.5 Laplacian

The Laplacian produces a scalar from a scalar. The Laplacian of a function $f$ (written as $\nabla^2 f$ or $\nabla \cdot \nabla f$) is defined as the divergence of the gradient of $f$. Its physical significance is that it gives a measure of how the average value of $f$ over the surface of a sphere compares with the value of $f$ at the center of the sphere. Let's be quantitative about this.

Consider the average value of a function $f$ over the surface of a sphere of radius $r$. Call it $f_{\text{avg},r}$. If we choose the origin of our coordinate system to be the center of the sphere, then $f_{\text{avg},r}$ can be written as (with $A$ being the area of the sphere)

$$f_{\text{avg},r} = \frac{1}{A} \int f \, dA = \frac{1}{4\pi r^2} \int f \, r^2 d\Omega = \frac{1}{4\pi} \int f \, d\Omega, \quad \text{(F.19)}$$

where $d\Omega = \sin \theta \, d\theta \, d\phi$ is the solid-angle element. We are able to take the $r^2$ outside the integral and cancel it because $r$ is constant over the sphere. This expression for $f_{\text{avg},r}$ is no surprise, of course, because the integral of $d\Omega$ over the whole sphere is $4\pi$. But let us now take the $d/dr$ derivative of both sides of Eq. (F.19), which will allow us to invoke the divergence theorem. On the right-hand side, the integration doesn't involve $r$, so we can bring the derivative inside the integral. This yields (using $\hat{\mathbf{r}} \cdot \hat{\mathbf{r}} = 1$)

$$\frac{df_{\text{avg},r}}{dr} = \frac{1}{4\pi} \int \frac{\partial f}{\partial r} d\Omega = \frac{1}{4\pi} \int \hat{\mathbf{r}} \frac{\partial f}{\partial r} \cdot \hat{\mathbf{r}} \, d\Omega = \frac{1}{4\pi r^2} \int \hat{\mathbf{r}} \frac{\partial f}{\partial r} \cdot \hat{\mathbf{r}} \, r^2 d\Omega.$$

(F.20)

(Again, we are able to bring the $r^2$ inside the integral because $r$ is constant over the sphere.) But $\hat{\mathbf{r}} r^2 d\Omega$ is just the vector area element of the sphere, $d\mathbf{a}$. And $\hat{\mathbf{r}}(\partial f/\partial r)$ is the $\hat{\mathbf{r}}$ component of $\nabla f$ in spherical coordinates. The other components of $\nabla f$ give zero when dotted with $d\mathbf{a}$, so we can write

$$\frac{df_{\text{avg},r}}{dr} = \frac{1}{4\pi r^2} \int \nabla f \cdot d\mathbf{a}.$$ (F.21)

The divergence theorem turns this into

$$\frac{df_{\text{avg},r}}{dr} = \frac{1}{4\pi r^2} \int \nabla \cdot \nabla f \, dV \implies \boxed{\frac{df_{\text{avg},r}}{dr} = \frac{1}{4\pi r^2} \int \nabla^2 f \, dV}$$

(F.22)

There are two useful corollaries of this result. First, if $\nabla^2 f = 0$ everywhere, then $df_{\text{avg},r}/dr = 0$ for all $r$. In other words, the average value of $f$ over the surface of a sphere doesn't change as the sphere grows (while keeping the same center). So all spheres centered at a given point have the same average value of $f$. In particular, they have the same average value that an infinitesimal sphere has. But the average value over an infinitesimal sphere is simply the value at the center. Therefore, if $\nabla^2 f = 0$, then the average value of $f$ over the surface of a sphere (of *any* size) equals the value at the center:

$$\nabla^2 f = 0 \implies f_{\text{avg},r} = f_{\text{center}}.$$ (F.23)

This is the result we introduced in Section 2.12 and proved for the special case of the electrostatic potential $\phi$.

Second, we can derive an expression for how $f$ changes, for *small* values of $r$. Up to this point, all of our results have been exact. We will now work in the small-$r$ approximation. In this limit we can say that $\nabla^2 f$ is essentially constant throughout the interior of the sphere (assuming that $f$ is well-enough behaved). So its value everywhere is essentially the value at the center. The volume integral in Eq. (F.22) then equals $(4\pi r^3/3)(\nabla^2 f)_{\text{center}}$, and we have

$$\frac{df_{\text{avg},r}}{dr} = \frac{1}{4\pi r^2} \frac{4\pi r^3}{3} (\nabla^2 f)_{\text{center}} \implies \frac{df_{\text{avg},r}}{dr} = \frac{r}{3} (\nabla^2 f)_{\text{center}}. \quad \text{(F.24)}$$

Since $(\nabla^2 f)_{\text{center}}$ is a constant, we can quickly integrate both sides of this relation to obtain

$$f_{\text{avg},r} = f_{\text{center}} + \frac{r^2}{6}(\nabla^2 f)_{\text{center}} \qquad \text{(for small } r\text{)}, \qquad \text{(F.25)}$$

where the constant of integration has been chosen to give equality at $r = 0$. We see that the average value of $f$ over a (small) sphere grows quadratically, with the quadratic coefficient being $1/6$ times the value of the Laplacian at the center.

Let's check this result for the function $f(r, \theta, \phi) = r^2$, or equivalently $f(x, y, z) = x^2 + y^2 + z^2$. By using either Eq. (F.1) or Eq. (F.3) we obtain $\nabla^2 f = 6$. If our sphere is centered at the origin, then Eq. (F.25) gives $f_{\text{avg},r} = 0 + (r^2/6)(6) = r^2$, which is correct because $f$ takes on the constant value of $r^2$ over the sphere. In this simple case, the result is exact for all $r$.

### F.5.1 Cylindrical Laplacian

Let's explicitly calculate the Laplacian in cylindrical coordinates by calculating the divergence of the gradient of $f$. As we've seen in a few cases above, we must be careful to take into account the position dependence of some of the unit vectors. We have

$$\nabla \cdot \nabla f = \left( \hat{\mathbf{r}} \frac{\partial}{\partial r} + \hat{\boldsymbol{\theta}} \frac{1}{r} \frac{\partial}{\partial \theta} + \hat{\mathbf{z}} \frac{\partial}{\partial z} \right) \cdot \left( \hat{\mathbf{r}} \frac{\partial f}{\partial r} + \hat{\boldsymbol{\theta}} \frac{1}{r} \frac{\partial f}{\partial \theta} + \hat{\mathbf{z}} \frac{\partial f}{\partial z} \right). \qquad \text{(F.26)}$$

In addition to the three "corresponding" terms, we also have the term involving $\hat{\boldsymbol{\theta}} \cdot (\partial \hat{\mathbf{r}} / \partial \theta)$, which from Eq. (F.12) equals $\hat{\boldsymbol{\theta}} \cdot \hat{\boldsymbol{\theta}} = 1$. So this fourth term reduces to $(1/r)(\partial f / \partial r)$. The Laplacian is therefore

$$\nabla^2 f = \frac{\partial}{\partial r} \left( \frac{\partial f}{\partial r} \right) + \frac{1}{r} \frac{\partial}{\partial \theta} \left( \frac{1}{r} \frac{\partial f}{\partial \theta} \right) + \frac{\partial}{\partial z} \left( \frac{\partial f}{\partial z} \right) + \frac{1}{r} \frac{\partial f}{\partial r}$$

$$= \frac{\partial^2 f}{\partial r^2} + \frac{1}{r^2} \frac{\partial^2 f}{\partial \theta^2} + \frac{\partial^2 f}{\partial z^2} + \frac{1}{r} \frac{\partial f}{\partial r}. \qquad \text{(F.27)}$$

The sum of the first and last terms here can be rewritten as the first term in the $\nabla^2 f$ expression in Eq. (F.2), as desired.

## Exercises

F.1 *Divergence using two systems* **

    (a) The vector $\mathbf{A} = x \hat{\mathbf{x}} + y \hat{\mathbf{y}}$ in Cartesian coordinates equals the vector $\mathbf{A} = r \hat{\mathbf{r}}$ in cylindrical coordinates. Calculate $\nabla \cdot \mathbf{A}$ in both Cartesian and cylindrical coordinates, and verify that the results are equal.

    (b) Repeat (a) for the vector $\mathbf{A} = x \hat{\mathbf{x}} + 2y \hat{\mathbf{y}}$. You will need to find the cylindrical components of $\mathbf{A}$, which you can do by using $\hat{\mathbf{x}} = \hat{\mathbf{r}} \cos \theta - \hat{\boldsymbol{\theta}} \sin \theta$ and $\hat{\mathbf{y}} = \hat{\mathbf{r}} \sin \theta + \hat{\boldsymbol{\theta}} \cos \theta$. Alternatively, you can project $\mathbf{A}$ onto the unit vectors, $\hat{\mathbf{r}} = \hat{\mathbf{x}}\cos\theta + \hat{\mathbf{y}}\sin\theta$ and $\hat{\boldsymbol{\theta}} = -\hat{\mathbf{x}}\sin\theta + \hat{\mathbf{y}}\cos\theta$.

**F.2   *Cylindrical divergence*  ***

Calculate the divergence in cylindrical coordinates in the following way. We know that the divergence in Cartesian coordinates is $\nabla \cdot \mathbf{A} = \partial A_x/\partial x + \partial A_y/\partial y + \partial A_z/\partial z$. To rewrite this in terms of cylindrical coordinates, show that the Cartesian derivative operators can be written as (the $\partial/\partial z$ derivative stays the same)

$$\frac{\partial}{\partial x} = \cos\theta\frac{\partial}{\partial r} - \sin\theta\frac{1}{r}\frac{\partial}{\partial\theta},$$

$$\frac{\partial}{\partial y} = \sin\theta\frac{\partial}{\partial r} + \cos\theta\frac{1}{r}\frac{\partial}{\partial\theta}, \tag{F.28}$$

and that the components of $\mathbf{A}$ can be written as ($A_z$ stays the same)

$$A_x = A_r\cos\theta - A_\theta\sin\theta,$$

$$A_y = A_r\sin\theta + A_\theta\cos\theta. \tag{F.29}$$

Then explicitly calculate $\nabla \cdot \mathbf{A} = \partial A_x/\partial x + \partial A_y/\partial y + \partial A_z/\partial z$. It gets to be a big mess, but it simplifies in the end.

**F.3   *General expression for divergence*  ***

Let $\hat{\mathbf{x}}_1$, $\hat{\mathbf{x}}_2$, $\hat{\mathbf{x}}_3$ be the (not necessarily Cartesian) basis vectors of a coordinate system. For example, in spherical coordinates these vectors are $\hat{\mathbf{r}}$, $\hat{\boldsymbol{\theta}}$, $\hat{\boldsymbol{\phi}}$. Note that the $d\mathbf{s}$ line elements listed at the beginning of this appendix all take the form of

$$d\mathbf{s} = f_1\,dx_1\,\hat{\mathbf{x}}_1 + f_2\,dx_2\,\hat{\mathbf{x}}_2 + f_3\,dx_3\,\hat{\mathbf{x}}_3, \tag{F.30}$$

where the $f$ factors are (possibly trivial) functions of the coordinates. For example, in Cartesian coordinates, $f_1, f_2, f_3$ are $1, 1, 1$; in cylindrical coordinates they are $1, r, 1$; and in spherical coordinates they are $1, r, r\sin\theta$. As we saw in Section F.2, these values of $f$ determine the form of $\nabla$ (the $f$ factors simply end up in the denominators), so they determine *everything* about the various vector operators. Show, by applying the first method we used in Section F.3, that the general expression for the divergence is

$$\nabla \cdot \mathbf{A} = \frac{1}{f_1 f_2 f_3}\left[\frac{\partial(f_2 f_3 A_1)}{\partial x_1} + \frac{\partial(f_1 f_3 A_2)}{\partial x_2} + \frac{\partial(f_1 f_2 A_3)}{\partial x_3}\right]. \tag{F.31}$$

Verify that this gives the correct result in the case of spherical coordinates. (The general expression for the curl can be found in a similar way.)

F.4  *Laplacian using two systems*  **

(a) The function $f = x^2 + y^2$ in Cartesian coordinates equals the function $f = r^2$ in cylindrical coordinates. Calculate $\nabla^2 f$ in both Cartesian and cylindrical coordinates, and verify that the results are equal.

(b) Repeat (a) for the function $f = x^4 + y^4$. You will need to determine what $f$ looks like in cylindrical coordinates.

F.5  *"Sphere" averages in one and two dimensions*  **

Equation (F.25) holds for a function $f$ in 3D space, but analogous results also hold in 2D space (where the "sphere" is a circle bounding a disk) and in 1D space (where the "sphere" is two points bounding a line segment). Derive those results. Although it is possible to be a little more economical in the calculations by stripping off some dimensions at the start, derive the results in a 3D manner exactly analogous to the way we derived Eq. (F.25). For the 2D case, the relevant volume is a cylinder, with $f$ having no dependence on $z$. For the 1D case, the relevant volume is a rectangular slab, with $f$ having no dependence on $y$ or $z$. The 1D result should look familiar from the standard 1D Taylor series.

F.6  *Average over a cube*  ***

By using the second-order Taylor expansion for a function of three Cartesian coordinates, show that the average value of a function $f$ over the surface of a cube of side $2\ell$ (with edges parallel to the coordinate axes) is

$$f_{\text{avg}} = f_{\text{center}} + \frac{5\ell^2}{18}\,(\nabla^2 f)_{\text{center}}.  \tag{F.32}$$

You should convince yourself why the factor of $5/18$ here is correctly larger than the $1/6$ in Eq. (F.25) and smaller than $(\sqrt{3})^2/6$.

# G

# A short review of special relativity

## G.1 Foundations of relativity

We assume that the reader has already been introduced to special relativity. Here we shall review the principal ideas and formulas that are used in the text beginning in Chapter 5. Most essential is the concept of an inertial frame of reference for space-time events and the transformation of the coordinates of an event from one inertial frame to another.

A frame of reference is a coordinate system laid out with measuring rods and provided with clocks. Clocks are everywhere. When something happens at a certain place, the time of its occurrence is read from a clock that was at, and stays at, that place. That is, time is measured by a *local* clock that is *stationary* in the frame. The clocks belonging to the frame are all *synchronized*. One way to accomplish this (not the only way) was described by Einstein in his great paper of 1905. Light signals are used. From a point $A$, at time $t_A$, a short pulse of light is sent out toward a remote point $B$. It arrives at $B$ at the time $t_B$, as read on a clock at $B$, and is immediately reflected back toward $A$, where it arrives at $t'_A$. If $t_B = (t_A + t'_A)/2$, the clocks at $A$ and $B$ are synchronized. If not, one of them requires adjustment. In this way, all clocks in the frame can be synchronized. Note that the job of observers in this procedure is merely to record local clock readings for subsequent comparison.

An *event* is located in space and time by its coordinates $x$, $y$, $z$, $t$ in some chosen reference frame. The event might be the passage of a particle at time $t_1$, through the space point $(x_1, y_1, z_1)$. The history of the particle's motion is a sequence of such events. Suppose the sequence has the special property that $x = v_x t$, $y = v_y t$, $z = v_z t$, at every time $t$, with $v_x$, $v_y$, and $v_z$ constant. That describes motion in a straight line at constant speed with respect to this frame. An *inertial frame of reference* is a frame in which an isolated body, free from external influences, moves in this way. An inertial frame, in other words, is one in which Newton's first law is obeyed. Behind all of this, including the synchronization of clocks, are two assumptions about empty space: it is *homogeneous* (that is, all locations in space are equivalent) and it is *isotropic* (that is, all directions in space are equivalent).

Two frames, let us call them $F$ and $F'$, can differ in several ways. One can simply be displaced with respect to the other, the origin of coordinates in $F'$ being fixed at a point in $F$ that is not at the $F$ coordinate origin. Or the axes in $F'$ might not be parallel to the axes in $F$. As for the timing of events, if $F$ and $F'$ are not moving with respect to one another, a clock stationary in $F$ is stationary also in $F'$. In that case we can set all $F'$ clocks to agree with the $F$ clocks and then ignore the distinction. Differences in frame location and frame orientation only have no interesting consequences if space is homogeneous and isotropic. Suppose now that the origin of frame $F'$ is *moving* relative to the origin of frame $F$. The description of a sequence of events by coordinate values and clock times in $F$ can differ from the description of the same events by space coordinate values in $F'$ and times measured by clocks in $F'$. How must the two descriptions be related? In answering that we shall be concerned only with the case in which $F$ is an inertial frame and $F'$ is a frame that is moving relative to $F$ at constant velocity and without rotating. In that case $F'$ is also an inertial frame.

Special relativity is based on the postulate that physical phenomena observed in different inertial frames of reference appear to obey exactly the same laws. In that respect one frame is as good as another; no frame is unique. If true, this relativity postulate is enough to determine the way a description of events in one frame is related to the description in a different frame of the same events. In that relation there appears a universal speed, the same in all frames, whose value must be found by experiment. Sometimes added as a second postulate is the statement that a measurement of the velocity of light in any frame of reference gives the same result whether the light's source is stationary in that frame or not. One may regard this as a statement about the nature of light rather than an independent postulate. It asserts that electromagnetic waves in fact travel with the limiting speed implied by the relativity postulate. The deductions from the relativity postulate, expressed in the formulas of special relativity, have been precisely verified by countless experiments. Nothing in physics rests on a firmer foundation.

## G.2 Lorentz transformations

Consider two events, $A$ and $B$, observed in an inertial frame $F$. *Observed*, in this usage, is short for "whose space-time coordinates are determined with the measuring rods and clocks of frame $F$." (Remember that our observers are equipped merely with pencil and paper, and we must post an observer at the location of every event!) The displacement of one event from the other is given by the four numbers

$$x_B - x_A, \quad y_B - y_A, \quad z_B - z_A, \quad t_B - t_A. \tag{G.1}$$

The *same two events* could have been located by giving their coordinates in some other frame $F'$. Suppose $F'$ is moving with respect to $F$ in the manner indicated in Fig. G.1. The spatial axes of $F'$ remain parallel to those in $F$, while, as seen from $F$, the frame $F'$ moves with speed $v$ in the positive $x$ direction. This is a special case, obviously, but it contains most of the interesting physics.

Event $A$, as observed in $F'$, occurred at $x'_A, y'_A, z'_A, t'_A$, the last of these numbers being the reading of a clock belonging to (that is, *stationary in*) $F'$. The space-time displacement, or *interval* between events $A$ and $B$ in $F'$, is not the same as in $F$. Its components are related to those in $F$ by the *Lorentz transformation*,

$$x'_B - x'_A = \gamma(x_B - x_A) - \beta\gamma c(t_B - t_A),$$

$$y'_B - y'_A = y_B - y_A,$$

$$z'_B - z'_A = z_B - z_A,$$

$$t'_B - t'_A = \gamma(t_B - t_A) - \beta\gamma(x_B - x_A)/c. \tag{G.2}$$

In these equations $c$ is the speed of light, $\beta = v/c$, and $\gamma = 1/\sqrt{1 - \beta^2}$. The inverse transformation has a similar appearance – as it should if no frame is unique. It can be obtained from Eq. (G.2) simply by exchanging primed and unprimed symbols and reversing the sign of $\beta$, as you can verify by explicitly solving for the quantities $x_B - x_A$ and $t_B - t_A$.

Two events $A$ and $B$ are *simultaneous* in $F$ if $t_B - t_A = 0$. But that does not make $t'_B - t'_A = 0$ unless $x_B = x_A$. Thus events that are simultaneous in one inertial frame may not be so in another. Do not confuse this fundamental "relativity of simultaneity" with the obvious fact that an observer not equally distant from two simultaneous explosions will receive light flashes from them at different times. The times $t'_A$ and $t'_B$ are recorded by *local* clocks at each event, clocks stationary in $F'$ that have previously been perfectly synchronized.

Consider a rod stationary in $F'$ that is parallel to the $x'$ axis and extends from $x'_A$ to $x'_B$. Its length in $F'$ is just $x'_B - x'_A$. The rod's length as measured in frame $F$ is the distance $x_B - x_A$ between two points in the frame $F$ that its ends pass *simultaneously* according to clocks in $F$. For these two events, then, $t_B - t_A = 0$. With this condition the first of the Lorentz transformation equations above gives us at once

$$x_B - x_A = (x'_B - x'_A)/\gamma. \tag{G.3}$$

(a)

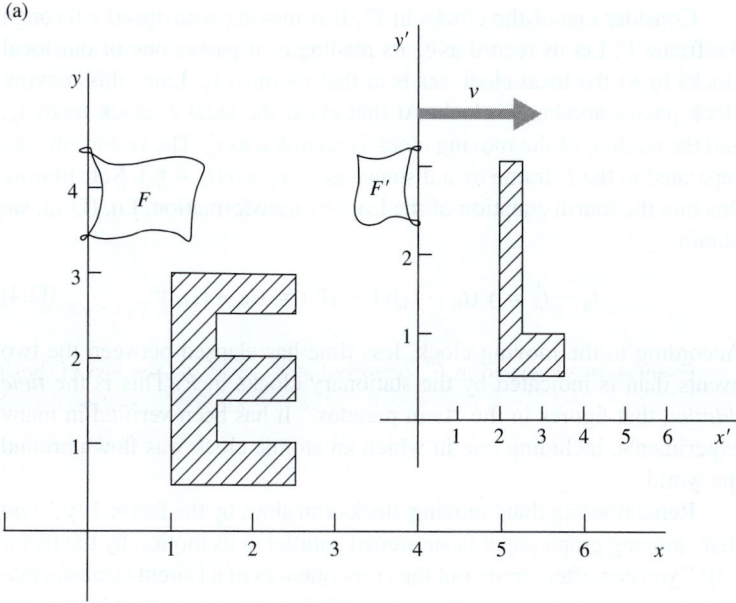

(b)

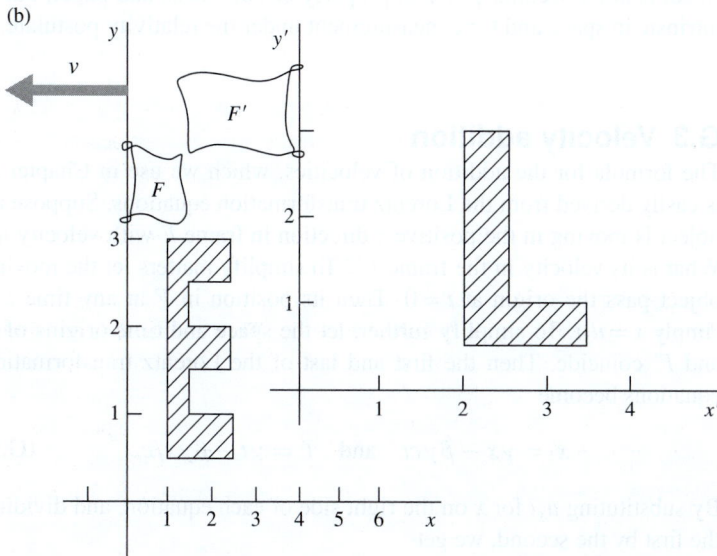

**Figure G.1.**
Two frames moving with relative speed $v$. The "E" is stationary in frame $F$. The "L" is stationary in frame $F'$. In this example $\beta = v/c = 0.866$; $\gamma = 2$. (a) Where everything was, as determined by observers in $F$ at a particular instant of time $t$ according to clocks in $F$. (b) Where everything was, as determined by observers in $F'$ at a particular instant of time $t'$ according to clocks in $F'$.

*Question:* Suppose the clocks in the two frames happened to be set so that the left edge of the E touched the left edge of the L at $t = 0$ according to a local clock in $F$, and at $t' = 0$ according to a local clock in $F'$. Let the distances be in feet and take $c$ as 1 foot/nanosecond. What is the reading $t$ of all the $F$ clocks in (a)? What is the reading $t'$ of all the $F'$ clocks in (b)?

*Answer:* $t = 4.62$ nanoseconds; $t' = 4.04$ nanoseconds. If you don't agree, study the example again.

This is the famous *Lorentz contraction.* Loosely stated, lengths between fixed points in $F'$, if parallel to the relative velocity of the frames, are judged by observers in $F$ to be shorter by the factor $1/\gamma$. This statement remains true if $F'$ and $F$ are interchanged. Lengths perpendicular to the relative velocity measure the same in the two frames.

Consider one of the clocks in $F'$. It is moving with speed $v$ through the frame $F$. Let us record as $t'_A$ its reading as it passes one of our local clocks in $F$; the local clock reads at that moment $t_A$. Later this moving clock passes another $F$ clock. At that event the local $F$ clock reads $t_B$, and the reading of the moving clock is recorded as $t'_B$. The two events are separated in the $F$ frame by a distance $x_B - x_A = v(t_B - t_A)$. Substituting this into the fourth equation of the Lorentz transformation, Eq. (G.2), we obtain

$$t'_B - t'_A = \gamma(t_B - t_A)(1 - \beta^2) = (t_B - t_A)/\gamma. \tag{G.4}$$

According to the moving clock, less time has elapsed between the two events than is indicated by the stationary clocks in $F$. This is the *time dilation* that figures in the "twin paradox." It has been verified in many experiments, including one in which an atomic clock was flown around the world.

Remembering that "moving clocks run slow, by the factor $1/\gamma$," and that "moving graph paper is shortened parallel to its motion by the factor $1/\gamma$," you can often figure out the consequences of a Lorentz transformation without writing out the equations. This behavior, it must be emphasized, is not a peculiar physical property of our clocks and paper, but is intrinsic in space and time measurement under the relativity postulate.

## G.3 Velocity addition

The formula for the addition of velocities, which we use in Chapter 5, is easily derived from the Lorentz transformation equations. Suppose an object is moving in the positive $x$ direction in frame $F$ with velocity $u_x$. What is its velocity in the frame $F'$? To simplify matters let the moving object pass the origin at $t = 0$. Then its position in $F$ at any time $t$ is simply $x = u_x t$. To simplify further, let the space and time origins of $F$ and $F'$ coincide. Then the first and last of the Lorentz transformation equations become

$$x' = \gamma x - \beta\gamma ct \quad \text{and} \quad t' = \gamma t - \beta\gamma x/c. \tag{G.5}$$

By substituting $u_x t$ for $x$ on the right side of each equation, and dividing the first by the second, we get

$$\frac{x'}{t'} = \frac{u_x - \beta c}{1 - \beta u_x/c}. \tag{G.6}$$

On the left we have the velocity of the object in the $F'$ frame, $u'_x$. The formula is usually written with $v$ instead of $\beta c$.

$$u'_x = \frac{u_x - v}{1 - u_x v/c^2}. \tag{G.7}$$

By solving Eq. (G.7) for $u_x$ you can verify that the inverse is

$$u_x = \frac{u'_x + v}{1 + u'_x v/c^2},$$ (G.8)

and that in no case will these relations lead to a velocity, either $u_x$ or $u'_x$, larger than $c$. As with the inverse Lorentz transformation, you can also obtain Eq. (G.8) from Eq. (G.7) simply by exchanging primed and unprimed symbols and reversing the sign of $v$.

A velocity component perpendicular to $v$, the relative velocity of the frames, transforms differently, of course. Analogous to Eq. (G.5), the second and last of the Lorentz transformation equations are

$$y' = y \quad \text{and} \quad t' = \gamma t - \beta \gamma x/c.$$ (G.9)

If we have $x = u_x t$ and $y = u_y t$ in frame $F$ (in general the object can be moving diagonally), then we can substitute these into Eq. (G.9) and divide the first equation by the second to obtain

$$\frac{y'}{t'} = \frac{u_y}{\gamma(1 - \beta u_x/c)} \implies u'_y = \frac{u_y}{\gamma(1 - u_x v/c^2)}.$$ (G.10)

In the special case where $u_x = 0$ (which means that the velocity points in the $y$ direction in frame $F$), we have $u'_y = u_y/\gamma$. That is, the $y$ speed is slower in the frame $F'$ where the object is flying by diagonally. In the special case where $u_x = v$ (which means that the object travels along with the $F'$ frame, as far as the $x$ direction is concerned), you can show that Eq. (G.10) reduces to $u'_y = \gamma u_y \implies u_y = u'_y/\gamma$. This makes sense; the object has $u'_x = 0$, so this result is analogous to the preceding $u'_y = u_y/\gamma$ result for the $u_x = 0$ case. In effect we have simply switched the primed and unprimed labels. These special cases can also be derived directly from time dilation.

## G.4 Energy, momentum, force

A dynamical consequence of special relativity can be stated as follows. Consider a particle moving with velocity $\mathbf{u}$ in an inertial frame $F$. We find that energy and momentum are conserved in the interactions of this particle with others if we attribute to the particle a momentum and an energy given by

$$\mathbf{p} = \gamma m_0 \mathbf{u} \quad \text{and} \quad E = \gamma m_0 c^2,$$ (G.11)

where $m_0$ is a constant characteristic of that particle. We call $m_0$ the *rest mass* (or just *the mass*) of the particle. It could have been determined in a frame in which the particle is moving so slowly that Newtonian mechanics applies – for instance, by bouncing the particle against some standard mass. The factor $\gamma$ multiplying $m_0$ is $(1 - u^2/c^2)^{-1/2}$, where $u$ is the speed of the particle as observed in our frame $F$.

Given $\mathbf{p}$ and $E$, the momentum and energy of a particle as observed in $F$, what is the momentum of that particle, and its energy, as observed in another frame $F'$? As before, we assume $F'$ is moving in the positive $x$ direction, with speed $v$, as seen from $F$. The transformation turns out to be this:

$$p'_x = \gamma p_x - \beta\gamma E/c,$$
$$p'_y = p_y,$$
$$p'_z = p_z,$$
$$E' = \gamma E - \beta\gamma c p_x. \tag{G.12}$$

Note that $\beta c$ is here the relative velocity of the two frames, as it was in Eq. (G.2), not the particle velocity.

Compare this transformation with Eq. (G.2). The resemblance would be perfect if we considered $cp$ instead of $p$ in Eq. (G.12), and $ct$ rather than $t$ in Eq. (G.2). A set of four quantities that transform in this way is called a *four-vector*.

The meaning of *force* is rate of change of momentum. The force acting on an object is simply $d\mathbf{p}/dt$, where $\mathbf{p}$ is the object's momentum in the chosen frame of reference and $t$ is measured by clocks in that frame. To find how forces transform, consider a particle of mass $m_0$ initially at rest at the origin in frame $F$ upon which a force $f$ acts for a short time $\Delta t$. We want to find the rate of change of momentum $dp'/dt'$, observed in a frame $F'$. As before, we shall let $F'$ move in the $x$ direction as seen from $F$. Consider first the effect of the force component $f_x$. In time $\Delta t$, $p_x$ will increase from zero to $f_x \Delta t$, while $x$ increases by

$$\Delta x = \frac{1}{2}\left(\frac{f_x}{m_0}\right)(\Delta t)^2, \tag{G.13}$$

and the particle's energy increases by $\Delta E = (f_x \Delta t)^2/2m_0$; this is the kinetic energy it acquires, as observed in $F$. (The particle's speed in $F$ is still so slight that Newtonian mechanics applies there.) Using the first of Eqs. (G.12) we find the change in $p'_x$:

$$\Delta p'_x = \gamma \Delta p_x - \beta\gamma \Delta E/c, \tag{G.14}$$

and using the fourth of Eqs. (G.2) gives

$$\Delta t' = \gamma \Delta t - \beta\gamma \Delta x/c. \tag{G.15}$$

Now both $\Delta E$ and $\Delta x$ are proportional to $(\Delta t)^2$, so when we take the limit $\Delta t \to 0$, the last term in each of these equations will drop out, giving

$$\frac{dp'_x}{dt'} = \lim_{\Delta t' \to 0} \frac{\Delta p'_x}{\Delta t'} = \frac{\gamma (f_x \Delta t)}{\gamma \Delta t} = f_x. \tag{G.16}$$

*Conclusion:* the force component *parallel* to the relative frame motion has the same value in the moving frame as in the rest frame of the particle.

A transverse force component behaves differently. In frame $F$, $\Delta p_y = f_y \Delta t$. But now $\Delta p'_y = \Delta p_y$, and $\Delta t' = \gamma \Delta t$, so we get

$$\frac{dp'_y}{dt'} = \frac{f_y \Delta t}{\gamma \Delta t} = \frac{f_y}{\gamma} . \tag{G.17}$$

A force component perpendicular to the relative frame motion, observed in $F'$, is *smaller* by the factor $1/\gamma$ than the value determined by observers in the rest frame of the particle.

The transformation of a force from $F'$ to some other moving frame $F''$ would be a little more complicated. We can always work it out, if we have to, by transforming to the rest frame of the particle and then back to the other moving frame.

We conclude our review with a remark about Lorentz invariance. If you square both sides of Eq. (G.12) and remember that $\gamma^2 - \beta^2 \gamma^2 = 1$, you can easily show that

$$c^2 (p'^2_x + p'^2_y + p'^2_z) - E'^2 = c^2 (p^2_x + p^2_y + p^2_z) - E^2 . \tag{G.18}$$

Evidently this quantity $c^2 p^2 - E^2$ is *not changed* by a Lorentz transformation. It is often called the *invariant four-momentum* (even though it has dimensions of energy squared). It has the same value in every frame of reference, including the particle's rest frame. In the rest frame the particle's momentum is zero and its energy $E$ is just $m_0 c^2$. The invariant four-momentum is therefore $-m_0^2 c^4$. It follows that in any other frame

$$E^2 = c^2 p^2 + m_0^2 c^4 . \tag{G.19}$$

The invariant constructed in the same way with Eq. (G.2) is

$$(x_B - x_A)^2 + (y_B - y_A)^2 + (z_B - z_A)^2 - c^2 (t_B - t_A)^2 . \tag{G.20}$$

Two events, $A$ and $B$, for which this quantity is positive are said to have a *spacelike* separation. It is always possible to find a frame in which they are simultaneous. If the invariant is negative, the events have a *timelike* separation. In that case a frame exists in which they occur at different times, but at the same place. If this "invariant interval" is zero, the two events can be connected by a flash of light.

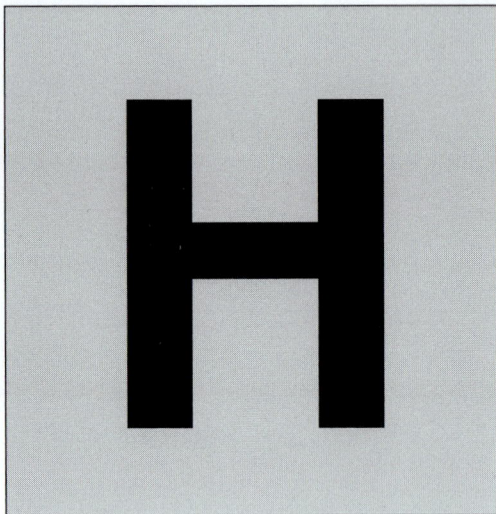

# Radiation by an accelerated charge

A particle with charge $q$ has been moving in a straight line at constant speed $v_0$ for a long time. It runs into something, let us imagine, and in a short period of constant deceleration, of duration $\tau$, the particle is brought to rest. The graph of velocity versus time in Fig. H.1 describes its motion. What must the electric field of this particle look like after that? Figure H.2 shows how to derive it.

We shall assume that $v_0$ is small compared with $c$. Let $t = 0$ be the instant the deceleration began, and let $x = 0$ be the position of the particle at that instant. By the time the particle has completely stopped it will have moved a little farther on, to $x = v_0\tau/2$. That distance, indicated in Fig. H.2, is small compared with the other distances that will be involved.

We now examine the electric field at a time $t = T \gg \tau$. Observers farther away from the origin than $R = cT$ cannot have learned that the particle was decelerated. Throughout that region, region I in Fig. H.2, the field must be that of a charge that has been moving *and is still moving* at the constant speed $v_0$. That field, as we discovered in Section 5.7, appears to emanate from the present position of the charge, which for an observer anywhere in region I is the point $x = v_0T$ on the $x$ axis. That is where the particle would be now if it hadn't been decelerated. On the other hand, for any observer whose distance from the origin is less than $c(T - \tau)$, that is to say, for any observer in region II, the field is that of a charge at rest close to the origin (actually at $x = v_0\tau/2$).

What must the field be like in the transition region, the spherical shell of thickness $c\tau$? Gauss's law provides the key. A field line such as $AB$ lies on a cone around the $x$ axis that includes a certain amount of flux from the charge $q$. If $CD$ makes the same angle $\theta$ with the axis, the cone on which it lies includes that same amount of flux. (Because $v_0$ is small, the relativistic compression of field lines visible in Fig. 5.15 and Fig. 5.19 is here negligible.) Hence $AB$ and $CD$ must be parts of the same field line, connected by a segment $BC$. This tells us the *direction* of the field $\mathbf{E}$ within the shell; it is the direction of the line segment $BC$. This field $\mathbf{E}$ within the shell has both a radial component $E_r$ and a transverse component $E_\theta$. From the geometry of the figure their ratio is easily found:

$$\frac{E_\theta}{E_r} = \frac{v_0 T \sin\theta}{c\tau}. \tag{H.1}$$

Now $E_r$ must have the same value within the shell thickness that it does in region II near $B$. (Gauss's law again!) Therefore $E_r = q/4\pi\epsilon_0 R^2 = q/4\pi\epsilon_0 c^2 T^2$, and substituting this into Eq. (H.1) we obtain

$$E_\theta = \frac{v_0 T \sin\theta}{c\tau} E_r = \frac{q v_0 \sin\theta}{4\pi\epsilon_0 c^3 T\tau}. \tag{H.2}$$

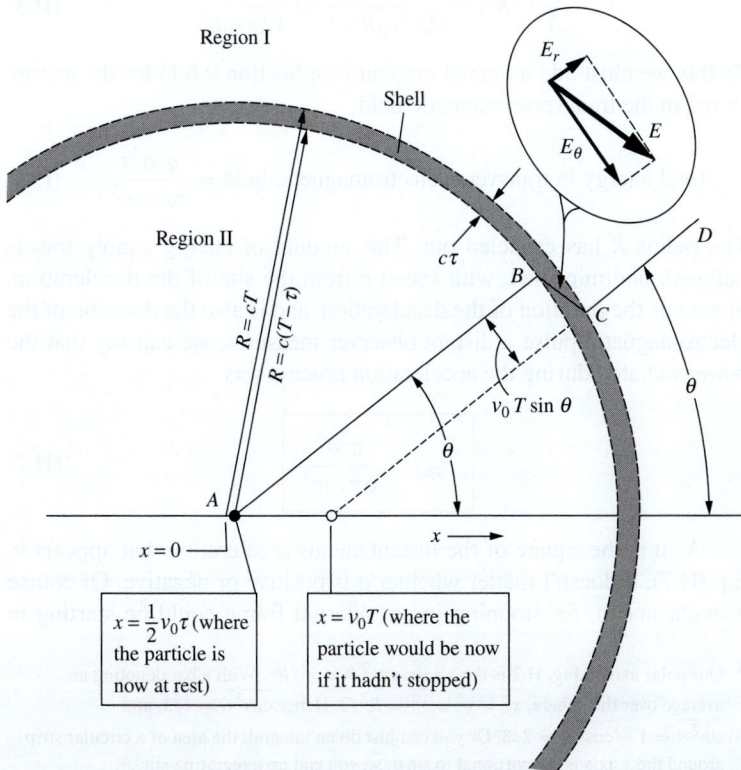

**Figure H.1.**
Velocity-time diagram for a particle that traveled at constant speed $v_0$ until $t = 0$. It then experienced a constant negative acceleration of magnitude $a = v_0/\tau$, which brought it to rest at time $t = \tau$. We assume $v_0$ is small compared with $c$.

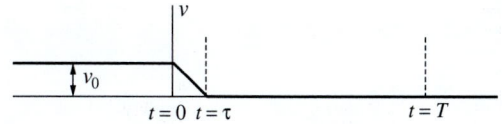

**Figure H.2.**
Space diagram for the instant $t = T \gg \tau$, a long time after the particle has stopped. For observers in region I, the field must be that of a charge located at the position $x = v_0 T$; for observers in region II, it is that of a particle at rest close to the origin. The transition region is a shell of thickness $c\tau$.

But $v_0/\tau = a$, the magnitude of the (negative) acceleration, and $cT = R$, so our result can be written as follows:

$$E_\theta = \frac{qa\sin\theta}{4\pi\epsilon_0 c^2 R} \tag{H.3}$$

A remarkable fact is here revealed: $E_\theta$ is proportional to $1/R$, *not* to $1/R^2$! As time goes on and $R$ increases, the transverse field $E_\theta$ will eventually become very much stronger than $E_r$. Accompanying this transverse (that is, perpendicular to **R**) electric field will be a magnetic field of strength $E_\theta/c$ perpendicular to both **R** and **E**. This is a general property of an electromagnetic wave, explained in Chapter 9.

Let us calculate the energy stored in the transverse electric field above, in the whole spherical shell. The energy density is

$$\frac{\epsilon_0 E_\theta^2}{2} = \frac{q^2 a^2 \sin^2\theta}{32\pi^2 \epsilon_0 R^2 c^4}. \tag{H.4}$$

The volume of the shell is $4\pi R^2 c\tau$, and the average value of $\sin^2\theta$ over a sphere[1] is $2/3$. The total energy of the transverse electric field is therefore

$$\frac{2}{3} 4\pi R^2 c\tau \frac{q^2 a^2}{32\pi^2 \epsilon_0 R^2 c^4} = \frac{q^2 a^2 \tau}{12\pi\epsilon_0 c^3}. \tag{H.5}$$

To this we must add an equal amount (see Section 9.6.1) for the energy stored in the transverse magnetic field:

$$\text{Total energy in transverse electromagnetic field} = \frac{q^2 a^2 \tau}{6\pi\epsilon_0 c^3}. \tag{H.6}$$

The radius $R$ has canceled out. This amount of energy simply travels outward, undiminished, with speed $c$ from the site of the deceleration. Since $\tau$ is the duration of the deceleration, and is also the duration of the electromagnetic pulse a distant observer measures, we can say that the *power* radiated during the acceleration process was

$$P_{\text{rad}} = \frac{q^2 a^2}{6\pi\epsilon_0 c^3} \tag{H.7}$$

As it is the square of the instantaneous acceleration that appears in Eq. (H.7), it doesn't matter whether $a$ is positive or negative. Of course it ought not to, for stopping in one inertial frame could be starting in

---

[1] Our polar axis in Fig. H.2 is the $x$ axis: $\cos^2\theta = x^2/R^2$. With a bar denoting an average over the sphere, $\overline{x^2} = \overline{y^2} = \overline{z^2} = R^2/3$. Hence $\overline{\cos^2\theta} = 1/3$, and $\overline{\sin^2\theta} = 1 - \overline{\cos^2\theta} = 2/3$. Or you can just do an integral; the area of a circular strip around the $x$ axis is proportional to $\sin\theta$, so you end up integrating $\sin^3\theta$.

another. Speaking of different frames, $P_{\mathrm{rad}}$ itself turns out to be Lorentz-invariant, which is sometimes very handy. That is because $P_{\mathrm{rad}}$ is *energy/time*, and energy transforms like time, each being the fourth component of a four-vector, as noted in Appendix G.

We have here a more general result than we might have expected. Equation (H.7) correctly gives the instantaneous rate of radiation of energy by a charged particle moving with variable acceleration – for instance, a particle vibrating in simple harmonic motion. It applies to a wide variety of radiating systems from radio antennas to atoms and nuclei.

## Exercises

H.1  *Ratio of energies*  *

An electron moving initially at constant (nonrelativistic) speed $v$ is brought to rest with uniform deceleration $a$ lasting for a time $t = v/a$. Compare the electromagnetic energy radiated during the deceleration with the electron's initial kinetic energy. Express the ratio in terms of two lengths, the distance light travels in time $t$ and the classical electron radius $r_0$, defined as $e^2/4\pi\epsilon_0 mc^2$.

H.2  *Simple harmonic motion*  **

An elastically bound electron vibrates in simple harmonic motion at frequency $\omega$ with amplitude $A$.

(a)  Find the average rate of loss of energy by radiation.

(b)  If no energy is supplied to make up the loss, how long will it take for the oscillator's energy to fall to $1/e$ of its initial value? (Answer: $6\pi\epsilon_0 mc^3/e^2\omega^2$.)

H.3  *Thompson scattering*  **

A plane electromagnetic wave with frequency $\omega$ and electric field amplitude $E_0$ is incident on an isolated electron. In the resulting sinusoidal oscillation of the electron the maximum acceleration is $E_0 e/m$ (the maximum force divided by $m$). How much power is radiated by this oscillating charge, averaged over many cycles? (Note that it is independent of the frequency $\omega$.) Divide this average radiated power by $\epsilon_0 E_0^2 c/2$, the average power density (power per unit area of wavefront) in the incident wave. This gives a constant $\sigma$ with the dimensions of area, called a *scattering cross section*. The energy radiated, or scattered, by the electron, and thus lost from the plane wave, is equivalent to that falling on an area $\sigma$. (The case here considered, involving a free electron moving nonrelativistically, is often called *Thomson scattering* after J. J. Thomson, the discoverer of the electron, who first calculated it.)

H.4  *Synchrotron radiation*  **

Our master formula, Eq. (H.7), is useful for relativistically moving particles, even though we assumed $v_0 \ll c$ in the derivation.

All we have to do is transform to an inertial frame $F'$ in which the particle in question is, at least temporarily, moving slowly, apply Eq. (H.7) in that frame, then transform back to any frame we choose. Consider a highly relativistic electron ($\gamma \gg 1$) moving perpendicular to a magnetic field **B**. It is continually accelerated perpendicular to the field, and must radiate. At what rate does it lose energy? To answer this, transform to a frame $F'$ moving momentarily along with the electron, find $E'$ in that frame, and $P'_{rad}$. Now show that, because power is (energy)/(time), $P_{rad} = P'_{rad}$. This radiation is generally called *synchrotron radiation*. (Answer: $P_{rad} = \gamma^2 e^4 B^2 / 6\pi \epsilon_0 m^2 c$.)

# Superconductivity

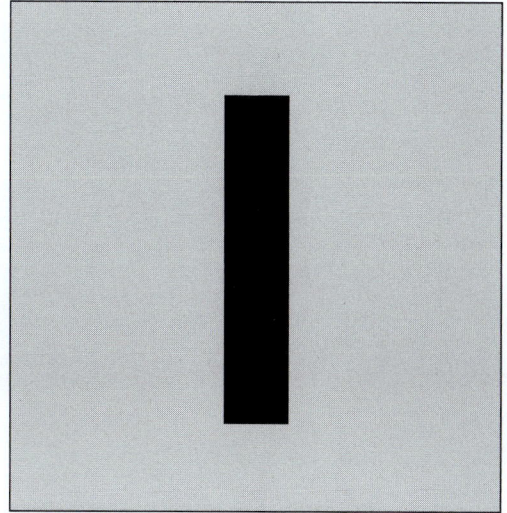

The metal lead is a moderately good conductor at room temperature. Its resistivity, like that of other pure metals, varies approximately in proportion to the absolute temperature. As a lead wire is cooled to 15 K its resistance falls to about 1/20 of its value at room temperature, and the resistance continues to decrease as the temperature is lowered further. But as the temperature 7.22 K is passed, there occurs without forewarning a startling change: the electrical resistance of the lead wire vanishes! So small does it become that a current flowing in a closed ring of lead wire colder than 7.22 K – a current that would ordinarily die out in much less than a microsecond – will flow for *years* without measurably decreasing. This phenomenon has been directly demonstrated. Other experiments indicate that such a current could persist for billions of years. One can hardly quibble with the flat statement that the resistivity is zero. Evidently something quite different from ordinary electrical conduction occurs in lead below 7.22 K. We call it *superconductivity.*

Superconductivity was discovered in 1911 by the great Dutch low-temperature experimenter Kamerlingh Onnes. He observed it first in mercury, for which the critical temperature is 4.16 K. Since then hundreds of elements, alloys, and compounds have been found to become superconductors. Their individual critical temperatures range from roughly a millikelvin up to the highest yet discovered, 138 K. Curiously, among the elements that do *not* become superconducting are some of the best normal conductors such as silver, copper, and the alkali metals.

Superconductivity is essentially a quantum-mechanical phenomenon, and a rather subtle one at that. The freely flowing electric current consists of electrons in perfectly orderly motion. Like the motion of an electron in an atom, this electron flow is immune to small disturbances – and for a similar reason: a finite amount of energy would be required to make any change in the state of motion. It is something like the situation in an insulator in which all the levels in the valence band are occupied and separated by an energy gap from the higher energy levels in the conduction band. But unlike electrons filling the valence band, which must in total give exactly zero net flow, the lowest energy state of the superconducting electrons can have a net electron velocity, hence current flow, in some direction. Why should such a strange state become possible below a certain critical temperature? We can't explain that here.[1] It involves the interaction of the conduction electrons not only with each other, but also with the whole lattice of positive ions through which they are moving. That is why different substances can have different critical temperatures, and why some substances are expected to remain normal conductors right down to absolute zero.

In the physics of superconductivity, magnetic fields are even more important than you might expect. We must state at once that the phenomena of superconductivity *in no way* violate Maxwell's equations. Thus the persistent current that can flow in a ring of superconducting wire is a direct consequence of Faraday's law of induction, given that the resistance of the ring is really zero. For if we start with a certain amount of flux $\Phi_0$ threading the ring, then because $\int \mathbf{E} \cdot d\mathbf{s}$ around the ring remains always zero (otherwise there would be infinite current due to the zero resistance), $d\Phi/dt$ must be zero. The flux cannot change; the current $I$ in the ring will automatically assume whatever value is necessary to maintain the flux at $\Phi_0$. Figure I.1 outlines a simple demonstration of this, and shows how a persistent current can be established in an isolated superconducting circuit.

Superconductors can be divided into two types. In Type 1 superconductors, the magnetic field inside the material itself (except very near the surface) is always zero. That is *not* a consequence of Maxwell's equations, but a property of the superconducting state, as fundamental, and once as baffling, a puzzle as the absence of resistance. The condition $\mathbf{B} = 0$ inside the bulk of a Type 1 superconductor is automatically maintained by currents flowing in a thin surface layer. In Type 2 superconductors, quantized magnetic flux tubes may exist for a certain range of temperature and external magnetic field. These tubes are surrounded by vortices of current (essentially little solenoids) which allow the magnetic field to be zero in the rest of the material. Outside the flux tubes the material is superconducting.

A strong magnetic field destroys superconductivity, although Type 2 superconductors generally can tolerate much larger magnetic fields than

---

[1] The abrupt emergence of a state of order at a certain critical temperature reminds us of the spontaneous alignment of electron spins that occurs in iron below its Curie temperature (mentioned in Section 11.11). Such *cooperative* phenomena always involve a large number of mutually interacting particles. A more familiar cooperative phenomenon is the freezing of water, also characterized by a well-defined critical temperature.

(a)

String

Ring of solder (lead–tin
alloy); normal conductor;
current zero; permanent
magnet causes flux $\Phi_0$
through ring.

Magnet

Liquid
helium
4.2 K

(b)

Ring cooled below its critical
temperature. (Some helium
has boiled away.) Flux through
ring unchanged. Ring is now
a superconductor.

(c)

$I$

Magnet removed. Persistent
current $I$ now flows in ring
to maintain flux at value $\Phi_0$.
Compass needle responds to
field of persistent current.

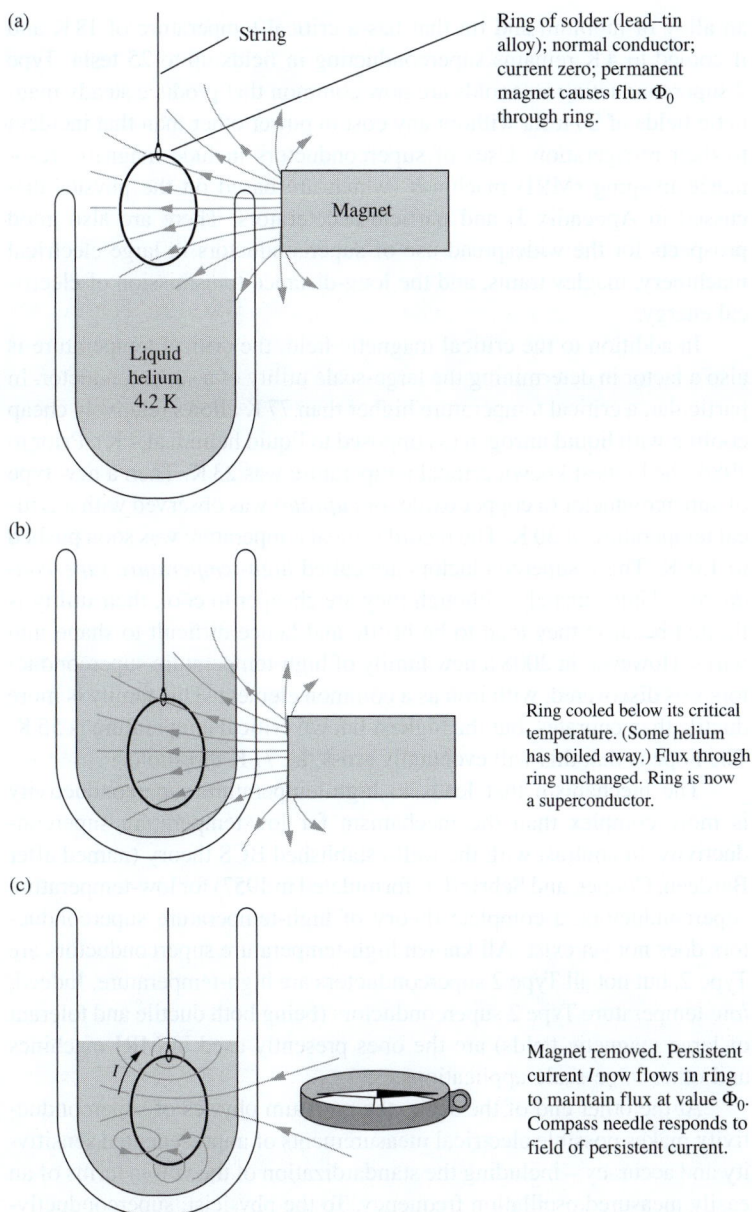

**Figure I.1.**
Establishing a persistent current in a
superconducting ring. The ring is made of
ordinary solder, a lead–tin alloy. (a) The ring, not
yet cooled, is a normal conductor with ohmic
resistance. Bringing up the permanent magnet
will induce a current in the ring that will quickly
die out, leaving the magnetic flux from the
magnet, in amount $\Phi$, passing through the ring.
(b) The helium bath is raised without altering the
relative position of the ring and the permanent
magnet. The ring, now cooled below its critical
temperature, is a superconductor with
resistance zero. (c) The magnet is removed.
The flux through the zero resistance ring cannot
change. It is maintained at the value $\Phi$ by a
current in the ring that will flow as long as the
ring remains below the critical temperature. The
magnetic field of the persistent current can be
demonstrated with the compass.

Type 1. None of the superconductors known before 1957 could stand
more than a few hundred gauss. That discouraged practical applications
of zero-resistance conductors. One could not pass a large current through
a superconducting wire because the magnetic field of the current itself
would destroy the superconducting state. But then a number of Type 2
superconductors were discovered that could preserve zero resistance in
fields up to 10 tesla or more. A widely used Type 2 superconductor is

an alloy of niobium and tin that has a critical temperature of 18 K and if cooled to 4 K remains superconducting in fields up to 25 tesla. Type 2 superconducting solenoids are now common that produce steady magnetic fields of 20 tesla without any cost in power other than that incident to their refrigeration. Uses of superconductors include magnetic resonance imaging (MRI) machines (which are based on the physics discussed in Appendix J) and particle accelerators. There are also good prospects for the widespread use of superconductors in large electrical machinery, maglev trains, and the long-distance transmission of electrical energy.

In addition to the critical magnetic field, the critical temperature is also a factor in determining the large-scale utility of a superconductor. In particular, a critical temperature higher than 77 K allows relatively cheap cooling with liquid nitrogen (as opposed to liquid helium at 4 K). Prior to 1986, the highest known critical temperature was 23 K. Then a new type of superconductor (a copper oxide, or *cuprate*) was observed with a critical temperature of 30 K. The record critical temperature was soon pushed to 138 K. These superconductors are called *high-temperature superconductors*. Unfortunately, although they are cheaper to cool, their utility is limited because they tend to be brittle and hence difficult to shape into wires. However, in 2008 a new family of high-temperature superconductors was discovered, with iron as a common element. This family is more ductile than cuprates, but the highest known critical temperature is 55 K. The hope is that this will eventually cross the 77 K threshold.

The mechanism that leads to high-temperature superconductivity is more complex than the mechanism for low-temperature superconductivity. In contrast with the well-established BCS theory (named after Bardeen, Cooper, and Schrieffer; formulated in 1957) for low-temperature superconductors, a complete theory of high-temperature superconductors does not yet exist. All known high-temperature superconductors are Type 2, but not all Type 2 superconductors are high-temperature. Indeed, *low*-temperature Type 2 superconductors (being both ductile and tolerant of large magnetic fields) are the ones presently used in MRI machines and other large-scale applications.

At the other end of the scale, the quantum physics of superconductivity makes possible electrical measurements of unprecedented sensitivity and accuracy – including the standardization of the volt in terms of an easily measured oscillation frequency. To the physicist, superconductivity is a fascinating large-scale manifestation of quantum mechanics. We can trace the permanent magnetism of the magnet in Fig. I.1 down to the intrinsic magnetic moment of a spinning electron – a kind of supercurrent in a circuit less than $10^{-10}$ m in size. The ring of solder wire with the persistent current flowing in it is, in some sense, like a gigantic atom, the motion of its associated electrons, numerous as they are, marshaled into the perfectly ordered behavior of a single quantum state.

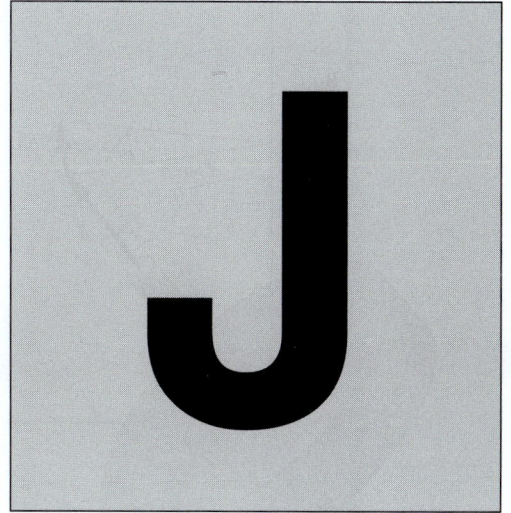

# Magnetic resonance

The electron has angular momentum of spin, **J**. Its magnitude is always the same, $h/4\pi$, or $5.273 \cdot 10^{-35}$ kg m$^2$/s. Associated with the axis of spin is a magnetic dipole moment $\mu$ of magnitude $0.9285 \cdot 10^{-23}$ joule/tesla (see Section 11.6). An electron in a magnetic field experiences a torque tending to align the magnetic dipole in the field direction. It responds like any rapidly spinning gyroscope: instead of lining up with the field, the spin axis *precesses* around the field direction. Let us see why any spinning magnet does this. In Fig. J.1 the magnetic moment $\mu$ is shown pointing opposite to the angular momentum **J**, as it would for a negatively charged body like an electron. The magnetic field **B** (the field of some solenoid or magnet not shown) causes a torque equal to $\mu \times \mathbf{B}$. This torque is a vector in the negative $\hat{x}$ direction at the time of our picture. Its magnitude is given by Eq. (11.48); it is $\mu B \sin\theta$. In a short time $\Delta t$, the torque adds to the angular momentum of our top a vector increment $\Delta\mathbf{J}$ in the direction of the torque vector and of magnitude $\mu B \sin\theta \, \Delta t$. The horizontal component of **J**, in magnitude $J \sin\theta$, is thereby rotated through a small angle $\Delta\psi$ given by

$$\Delta\psi = \frac{\Delta J}{J \sin\theta} = \frac{\mu B \, \Delta t}{J}. \qquad (J.1)$$

As this continues, the upper end of the vector **J** will simply move around the circle with constant angular velocity $\omega_p$:

$$\omega_p = \frac{\Delta\psi}{\Delta t} = \frac{\mu B}{J}. \qquad (J.2)$$

**Figure J.1.**
The precession of a magnetic top in an external field. The angular momentum of spin **J** and the magnetic dipole moment $\mu$ are oppositely directed, as they would be for a negatively charged rotor.

This is the rate of precession of the axis of spin. Note that it is the same for any angle of tip; $\sin\theta$ has canceled out.

For the electron, $\mu/J$ has the value $1.761 \cdot 10^{11} \text{ s}^{-1}\text{tesla}^{-1}$. In a field of 1 gauss ($10^{-4}$ tesla) the spin vector precesses at $1.761 \cdot 10^7$ radians/s, or $2.80 \cdot 10^6$ revolutions per second. The proton has exactly the same intrinsic spin angular momentum as the electron, $h/4\pi$, but the associated magnetic moment is smaller. That is to be expected since the mass of the proton is 1836 times the mass of the electron; as in the case of orbital angular momentum (see Eq. (11.29)), the magnetic moment of an elementary particle with spin ought to be inversely proportional to its mass, other things being equal. Actually the proton's magnetic moment is $1.411 \cdot 10^{-26}$ joule/tesla, only about 660 times smaller than the electron moment, which shows that the proton is in some way a composite particle. In a field of 1 gauss the proton spin precesses at 4258 revolutions per second. About 40 percent of the stable atomic nuclei have intrinsic angular momenta and associated magnetic dipole moments.

We can detect the precession of magnetic dipole moments through their influence on an electric circuit. Imagine a proton in a magnetic field $B$, with its spin axis perpendicular to the field, and surrounded by a small coil of wire, as in Fig. J.2. The precession of the proton causes some alternating flux through the coil, as would the end-over-end rotation of a little bar magnet. A voltage alternating at the precession frequency will be induced in the coil. As you might expect, the voltage thus induced by a single proton would be much too feeble to detect. But it is easy to provide more protons – 1 cm$^3$ of water contains about $7 \cdot 10^{22}$ protons (we're concerned with the two hydrogen atoms in each water molecule), and all of them will precess at the same frequency. Unfortunately they will not all be pointing in the same direction at the same instant. In fact, their spin axes and magnetic moments will be distributed so uniformly over all possible directions that their fields will very nearly cancel one another. But not quite, if we introduce another step. If we apply a strong magnetic field $B$ to water, for several seconds there will develop a slight excess of proton moments pointing in the direction of **B**, the direction they energetically favor. The fractional excess will be $\mu B/kT$ in order of magnitude, as in ordinary paramagnetism. It may be no more than one in a million, but these uncanceled moments, if they are now caused to precess in our coil, will induce an observable signal.

A simple method for observing nuclear spin precession in weak fields, such as the earth's field, is described in Fig. J.3. Many other schemes are used to observe the spin precession of electrons and of

**Figure J.2.**
A precessing magnetic dipole moment at the center of a coil causes a periodic change in the flux through the coil, inducing an alternating electromotive force in the coil. Note that the flux from the dipole **m** that links the coil is that which loops around outside it. See Exercise J.1.

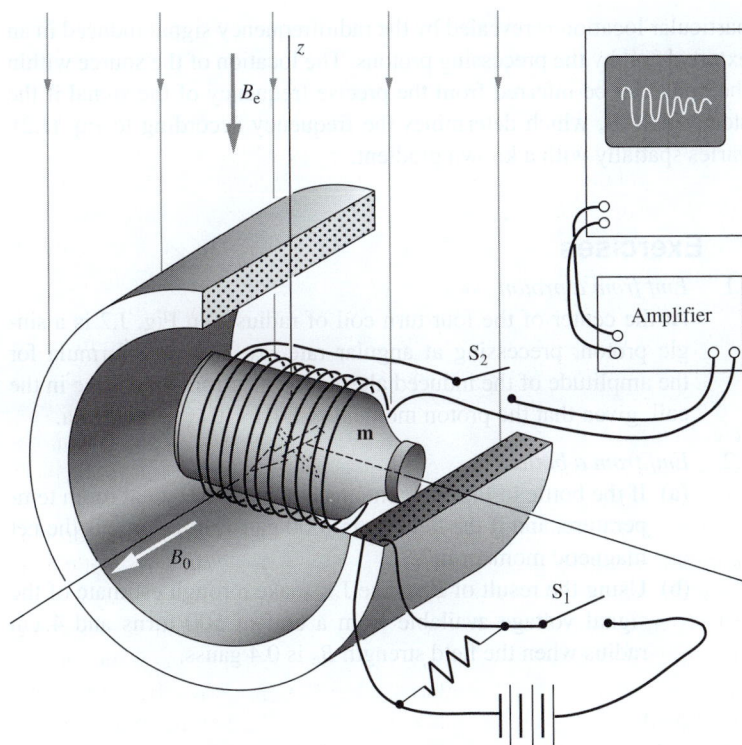

**Figure J.3.**
Apparatus for observing proton spin precession in the earth's field $B_e$. A bottle of water is surrounded by two orthogonal coils. With switch $S_2$ open and switch $S_1$ closed, the large solenoid creates a strong magnetic field $B_0$. As in ordinary paramagnetism (Section 11.6), the energy is lowered if the dipoles point in the direction of the field, but thermal agitation causes disorder. Our dipoles here are the protons (hydrogen nuclei) in the molecules of water. When thermal equilibrium has been attained, which in this case takes several seconds, the magnetization is what you would get by lining up with the magnetic field the small fraction $\mu B_0/kT$ of all the proton moments. We now switch off the strong field $B_0$ and close switch $S_2$ to connect the coil around the bottle to the amplifier. The magnetic moment **m** now precesses in the $xy$ plane around the remaining, relatively weak, magnetic field $B_e$, with precession frequency given by Eq. (J.2). The alternating $y$ component of the rotating vector **m** induces an alternating voltage in the coil which can be amplified and observed. From its frequency, $B_e$ can be very precisely determined. This signal itself will die away in a few seconds as thermal agitation destroys the magnetization the strong field $B_0$ had brought about. Magnetic resonance magnetometers of this and other types are used by geophysicists to explore the earth's field, and even by archaeologists to locate buried artifacts.

nuclei. They generally involve a combination of a steady magnetic field and oscillating magnetic fields with frequency in the neighborhood of $\omega_p$. For electron spins (*electron paramagnetic resonance,* or EPR) the frequencies are typically several thousand megahertz, while for nuclear spins (*nuclear magnetic resonance,* or NMR) they are several tens of megahertz. The exact frequency of precession, or resonance, in a given applied field can be slightly shifted by magnetic interactions within a molecule. This has made NMR, in particular, useful in chemistry. The position of a proton in a complex molecule can often be deduced from the small shift in its precession frequency.

Magnetic fields easily penetrate ordinary nonmagnetic materials, and that includes alternating magnetic fields if their frequency or the electric conductivity of the material is not too great. A steady field of 2000 gauss applied to the bottle of water in our example would cause any proton polarization to precess at a frequency of $8.516 \cdot 10^6$ revolutions per second. The field of the precessing moments would induce a signal of 8.516 MHz frequency in the coil outside the bottle. This applies as well to the human body, which, viewed as a dielectric, is simply an assembly of more or less watery objects. In *NMR imaging* (or *magnetic resonance imaging, MRI*) the interior of the body is mapped by means of nuclear magnetic resonance. The concentration of hydrogen atoms at a particular location is revealed by the radiofrequency signal induced in an external coil by the precessing protons. The location of the source within the body can be inferred from the precise frequency of the signal if the steady field $B$, which determines the frequency according to Eq. (J.2), varies spatially with a known gradient.

## Exercises

J.1 *Emf from a proton* **

At the center of the four-turn coil of radius $a$ in Fig. J.2 is a single proton, precessing at angular rate $\omega_p$. Derive a formula for the amplitude of the induced alternating electromotive force in the coil, given that the proton moment is $1.411 \cdot 10^{-26}$ joule/tesla.

J.2 *Emf from a bottle* ***

(a) If the bottle in Fig. J.3 contains 200 cm$^3$ of $H_2O$ at room temperature, and if the field $B_0$ is 1000 gauss, how large is the net magnetic moment **m**?

(b) Using the result of Exercise J.1, make a rough estimate of the signal voltage available from a coil of 500 turns and 4 cm radius when the field strength $B_e$ is 0.4 gauss.

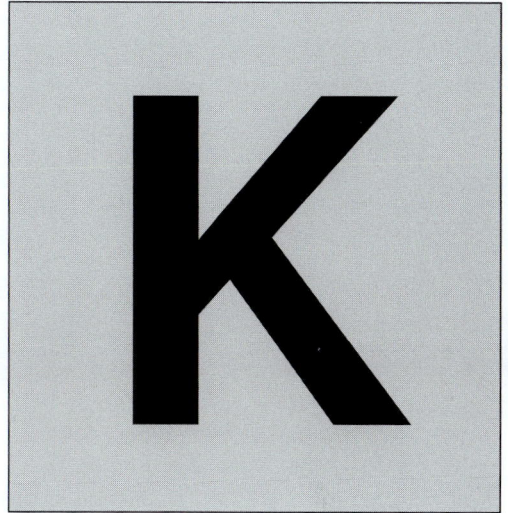

## K.1 Fundamental constants

| | | |
|---|---|---|
| speed of light | $c$ | $2.998 \cdot 10^8$ m/s |
| elementary charge | $e$ | $1.602 \cdot 10^{-19}$ C |
| | | $4.803 \cdot 10^{-10}$ esu |
| electron mass | $m_e$ | $9.109 \cdot 10^{-31}$ kg |
| proton mass | $m_p$ | $1.673 \cdot 10^{-27}$ kg |
| Avogadro's number | $N_A$ | $6.022 \cdot 10^{23}$ mole$^{-1}$ |
| Boltzmann constant | $k$ | $1.381 \cdot 10^{-23}$ J/K |
| Planck constant | $h$ | $6.626 \cdot 10^{-34}$ J s |
| gravitational constant | $G$ | $6.674 \cdot 10^{-11}$ m$^3$/(kg s$^2$) |
| electron magnetic moment | $\mu_e$ | $9.285 \cdot 10^{-24}$ J/T |
| proton magnetic moment | $\mu_p$ | $1.411 \cdot 10^{-26}$ J/T |
| permittivity of free space | $\epsilon_0$ | $8.854 \cdot 10^{-12}$ C$^2$/(N m$^2$) |
| permeability of free space | $\mu_0$ | $1.257 \cdot 10^{-6}$ T m/A |

The exact numerical value of $\mu_0$ is $4\pi \cdot 10^{-7}$ (by definition).

The exact numerical value of $\epsilon_0$ is $(4\pi \cdot [3]^2 \cdot 10^9)^{-1}$, where $[3] \equiv$ 2.99792458 (see Appendix E).

# Helpful formulas/facts

**K**

## K.2 Integral table

$$\int \frac{dx}{x^2 + r^2} = \frac{1}{r} \tan^{-1}\left(\frac{x}{r}\right) \tag{K.1}$$

$$\int \frac{dx}{\sqrt{1 - x^2}} = \sin^{-1} x \tag{K.2}$$

$$\int \frac{dx}{\sqrt{x^2 - 1}} = \ln\left(x + \sqrt{x^2 - 1}\right) \tag{K.3}$$

$$\int \frac{dx}{\sqrt{x^2 + a^2}} = \ln\left(\sqrt{x^2 + a^2} + x\right) \tag{K.4}$$

$$\int \frac{dx}{(a^2 + x^2)^{3/2}} = \frac{x}{a^2(a^2 + x^2)^{1/2}} \tag{K.5}$$

$$\int \ln x \, dx = x \ln x - x \tag{K.6}$$

$$\int x^n \ln\left(\frac{a}{x}\right) dx = \frac{x^{n+1}}{(n+1)^2} + \frac{x^{n+1}}{n+1} \ln\left(\frac{a}{x}\right) \tag{K.7}$$

$$\int x e^{-x} \, dx = -(x + 1)e^{-x} \tag{K.8}$$

$$\int x^2 e^{-x} \, dx = -(x^2 + 2x + 2)e^{-x} \tag{K.9}$$

$$\int \sin^3 x \, dx = -\cos x + \frac{\cos^3 x}{3} \tag{K.10}$$

$$\int \cos^3 x \, dx = \sin x - \frac{\sin^3 x}{3} \tag{K.11}$$

$$\int \frac{dx}{\cos x} = \ln\left(\frac{1 + \sin x}{\cos x}\right) \tag{K.12}$$

$$\int \frac{dx}{\sin x} = \ln\left(\frac{1 - \cos x}{\sin x}\right) \tag{K.13}$$

$$\int \frac{\cos x \, dx}{(1 - a^2 \cos^2 x)^{3/2}} = \frac{\sin x}{(1 - a^2)\sqrt{1 - a^2 \cos^2 x}} \tag{K.14}$$

$$\int \frac{\sin x \, dx}{(1 - a^2 \sin^2 x)^{3/2}} = \frac{-\cos x}{(1 - a^2)\sqrt{1 - a^2 \sin^2 x}} \tag{K.15}$$

$$\int \frac{\cos x \, dx}{\left(1 - b^2 \sin^2(x - a)\right)^{3/2}} = \frac{(2 - b^2)\sin x + b^2 \sin(2a - x)}{2(1 - b^2)\sqrt{1 - b^2 \sin^2(a - x)}} \tag{K.16}$$

$$\int \frac{\sin x(a \cos x - b) \, dx}{(a^2 + b^2 - 2ab \cos x)^{3/2}} = \frac{-a + b \cos x}{b^2\sqrt{a^2 + b^2 - 2ab \cos x}} \tag{K.17}$$

## K.3 Vector identities

$$\nabla \cdot (\nabla \times \mathbf{A}) = 0$$

$$\nabla \cdot (f\mathbf{A}) = f\nabla \cdot \mathbf{A} + \mathbf{A} \cdot \nabla f$$

$$\nabla \cdot (\mathbf{A} \times \mathbf{B}) = \mathbf{B} \cdot (\nabla \times \mathbf{A}) - \mathbf{A} \cdot (\nabla \times \mathbf{B})$$

$$\nabla \times (\nabla f) = 0$$

$$\nabla \times (f\mathbf{A}) = f\nabla \times \mathbf{A} + (\nabla f) \times \mathbf{A}$$

$$\nabla \times (\nabla \times \mathbf{A}) = \nabla(\nabla \cdot \mathbf{A}) - \nabla^2 \mathbf{A}$$

$$\nabla \times (\mathbf{A} \times \mathbf{B}) = \mathbf{A}(\nabla \cdot \mathbf{B}) - \mathbf{B}(\nabla \cdot \mathbf{A}) + (\mathbf{B} \cdot \nabla)\mathbf{A} - (\mathbf{A} \cdot \nabla)\mathbf{B}$$

$$\mathbf{A} \times (\mathbf{B} \times \mathbf{C}) = \mathbf{B}(\mathbf{A} \cdot \mathbf{C}) - \mathbf{C}(\mathbf{A} \cdot \mathbf{B})$$

$$\nabla(\mathbf{A} \cdot \mathbf{B}) = (\mathbf{A} \cdot \nabla)\mathbf{B} + (\mathbf{B} \cdot \nabla)\mathbf{A} + \mathbf{A} \times (\nabla \times \mathbf{B}) + \mathbf{B} \times (\nabla \times \mathbf{A})$$

## K.4 Taylor series

The general form of a Taylor series is

$$f(x_0 + x) = f(x_0) + f'(x_0)x + \frac{f''(x_0)}{2!}x^2 + \frac{f'''(x_0)}{3!}x^3 + \cdots . \quad \text{(K.18)}$$

This equality can be verified by taking successive derivatives and then setting $x = 0$. For example, taking the first derivative and then setting $x = 0$ gives $f'(x_0)$ on the left, and also $f'(x_0)$ on the right, because the first term is a constant and gives zero when differentiated, the second term gives $f'(x_0)$, and all the rest of the terms give zero once we set $x = 0$ because they all contain at least one power of $x$. Likewise, if we take the second derivative of each side and then set $x = 0$, we obtain $f''(x_0)$ on both sides. And so on for all derivatives. Therefore, since the two functions on each side of Eq. (K.18) are equal at $x = 0$ and also have their $n$th derivatives equal at $x = 0$ for all $n$, they must in fact be the same function (assuming that they are nicely behaved functions, which we generally assume in physics).

Some specific Taylor series that come up often are listed below; they are all expanded around $x_0 = 0$. We use these series countless times throughout this book when checking how expressions behave in the limit of some small quantity. The series are all derivable via Eq. (K.18), but sometimes there are quicker ways of obtaining them. For example, Eq. (K.20) is most easily obtained by taking the derivative of Eq. (K.19), which itself is simply the sum of a geometric series.

$$\frac{1}{1 - x} = 1 + x + x^2 + x^3 + \cdots \quad \text{(K.19)}$$

$$\frac{1}{(1 - x)^2} = 1 + 2x + 3x^2 + 4x^3 + \cdots \quad \text{(K.20)}$$

$$\ln(1 - x) = -x - \frac{x^2}{2} - \frac{x^3}{3} - \cdots \quad \text{(K.21)}$$

$$e^x = 1 + x + \frac{x^2}{2!} + \frac{x^3}{3!} + \cdots \tag{K.22}$$

$$\cos x = 1 - \frac{x^2}{2!} + \frac{x^4}{4!} - \cdots \tag{K.23}$$

$$\sin x = x - \frac{x^3}{3!} + \frac{x^5}{5!} - \cdots \tag{K.24}$$

$$\sqrt{1+x} = 1 + \frac{x}{2} - \frac{x^2}{8} + \cdots \tag{K.25}$$

$$\frac{1}{\sqrt{1+x}} = 1 - \frac{x}{2} + \frac{3x^2}{8} + \cdots \tag{K.26}$$

$$(1+x)^n = 1 + nx + \binom{n}{2}x^2 + \binom{n}{3}x^3 + \cdots \tag{K.27}$$

## K.5 Complex numbers

The imaginary number $i$ is defined to be the number for which $i^2 = -1$. (Of course, $-i$ also has its square equal to $-1$.) A general complex number $z$ with both real and imaginary parts can be written in the form $a + bi$, where $a$ and $b$ are real numbers. Such a number can be described by the point $(a, b)$ in the complex plane, with the $x$ and $y$ axes being the real and imaginary axes, respectively.

The most important formula involving complex numbers is

$$e^{i\theta} = \cos\theta + i\sin\theta. \tag{K.28}$$

This can quickly be proved by writing out the Taylor series for both sides. Using Eq. (K.22), the first, third, fifth, etc. terms on the left-hand side of Eq. (K.28) are real, and from Eq. (K.23) their sum is $\cos\theta$. Similarly, the second, fourth, sixth, etc. terms are imaginary, and from Eq. (K.24) their sum is $i\sin\theta$. Writing it all out, we have

$$e^{i\theta} = 1 + i\theta + \frac{(i\theta)^2}{2!} + \frac{(i\theta)^3}{3!} + \frac{(i\theta)^4}{4!} + \frac{(i\theta)^5}{5!} + \cdots$$

$$= \left(1 - \frac{\theta^2}{2!} + \frac{\theta^4}{4!} + \cdots\right) + i\left(\theta - \frac{\theta^3}{3!} + \frac{\theta^5}{5!} + \cdots\right)$$

$$= \cos\theta + i\sin\theta, \tag{K.29}$$

as desired.

Letting $\theta \to -\theta$ in Eq. (K.28) yields $e^{-i\theta} = \cos\theta - i\sin\theta$. Combining this with Eq. (K.28) allows us to solve for $\cos\theta$ and $\sin\theta$ in terms of the complex exponentials:

$$\cos\theta = \frac{e^{i\theta} + e^{-i\theta}}{2}, \qquad \sin\theta = \frac{e^{i\theta} - e^{-i\theta}}{2i}. \tag{K.30}$$

A complex number $z$ described by the Cartesian coordinates $(a, b)$ in the complex plane can also be described by the polar coordinates $(r, \theta)$. The radius $r$ and angle $\theta$ are given by the usual relation between Cartesian and polar coordinates (see Fig. K.1),

$$r = \sqrt{a^2 + b^2} \qquad \text{and} \qquad \theta = \tan^{-1}(b/a). \qquad \text{(K.31)}$$

Using Eq. (K.28), we can write $z$ in *polar form* as

$$a + bi = (r \cos \theta) + (r \sin \theta)i = r(\cos \theta + i \sin \theta) = re^{i\theta}. \qquad \text{(K.32)}$$

We see that the quantity in the exponent (excluding the $i$) equals the angle of the vector in the complex plane.

The *complex conjugate* of $z$, denoted by $z^*$ (or by $\bar{z}$), is defined to be $z^* \equiv a - bi$, or equivalently $z^* \equiv re^{-i\theta}$. It is obtained by reflecting the Cartesian point $(a, b)$ across the real axis. Note that either of these expressions for $z^*$ implies that $r$ can be written as $r = \sqrt{zz^*}$. The radius $r$ is known as the *magnitude* or *absolute value* of $z$, and is commonly denoted by $|z|$. The complex conjugate of a product is the product of the complex conjugates, that is, $(z_1 z_2)^* = z_1^* z_2^*$. You can quickly verify this by writing $z_1$ and $z_2$ in polar form. The Cartesian form works too, but that takes a little longer. The same result holds for the quotient of two complex numbers.

As an example of the use of Eq. (K.28), we can quickly derive the double-angle formulas for sine and cosine. We have

$$\cos 2\theta + i \sin 2\theta = e^{i2\theta} = \left(e^{i\theta}\right)^2 = (\cos \theta + i \sin \theta)^2$$

$$= (\cos^2 \theta - \sin^2 \theta) + i(2 \sin \theta \cos \theta). \qquad \text{(K.33)}$$

Equating the real parts of the expressions on either end of this equation gives $\cos 2\theta = \cos^2 \theta - \sin^2 \theta$. And equating the imaginary parts gives $\sin 2\theta = 2 \sin \theta \cos \theta$. This method easily generalizes to other trig sum formulas.

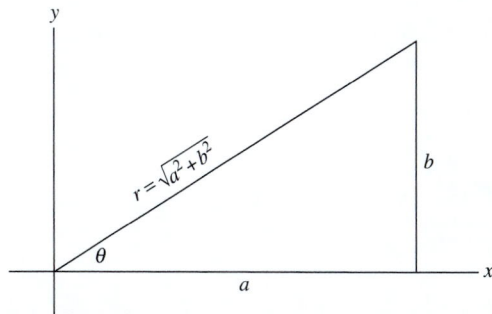

**Figure K.1.**
Cartesian and polar coordinates in the complex plane.

## K.6  Trigonometric identities

$$\sin 2\theta = 2 \sin \theta \cos \theta, \qquad \cos 2\theta = \cos^2 \theta - \sin^2 \theta \qquad \text{(K.34)}$$

$$\sin(\alpha + \beta) = \sin \alpha \cos \beta + \cos \alpha \sin \beta \qquad \text{(K.35)}$$

$$\cos(\alpha + \beta) = \cos \alpha \cos \beta - \sin \alpha \sin \beta \qquad \text{(K.36)}$$

$$\tan(\alpha + \beta) = \frac{\tan \alpha + \tan \beta}{1 - \tan \alpha \tan \beta} \qquad \text{(K.37)}$$

$$\cos\frac{\theta}{2} = \pm\sqrt{\frac{1+\cos\theta}{2}}, \qquad \sin\frac{\theta}{2} = \pm\sqrt{\frac{1-\cos\theta}{2}} \qquad \text{(K.38)}$$

$$\tan\frac{\theta}{2} = \pm\sqrt{\frac{1-\cos\theta}{1+\cos\theta}} = \frac{1-\cos\theta}{\sin\theta} = \frac{\sin\theta}{1+\cos\theta} \qquad \text{(K.39)}$$

The hyperbolic trig functions are defined by analogy with Eq. (K.30), with the $i$'s omitted:

$$\cosh x = \frac{e^x + e^{-x}}{2}, \qquad \sinh x = \frac{e^x - e^{-x}}{2} \qquad \text{(K.40)}$$

$$\cosh^2 x - \sinh^2 x = 1 \qquad \text{(K.41)}$$

$$\frac{d}{dx}\cosh x = \sinh x, \qquad \frac{d}{dx}\sinh x = \cosh x \qquad \text{(K.42)}$$

# References

Andrews, M. (1997). Equilibrium charge density on a conducting needle. *Am. J. Phys.*, **65**, 846–850.

Assis, A. K. T., Rodrigues, W. A., Jr., and Mania, A. J. (1999). The electric field outside a stationary resistive wire carrying a constant current. *Found. Phys.*, **29**, 729–753.

Auty, R. P. and Cole, R. H. (1952). Dielectric properties of ice and solid $D_2O$. *J. Chem. Phys.*, **20**, 1309–1314.

Blakemore, R. P. and Frankel, R. B. (1981). Magnetic navigation in bacteria. *Sci. Am.*, **245**, (6), 58–65.

Bloomfield, L.A. (2010). *How Things Work*, 4th edn. (New York: John Wiley & Sons).

Boos, F. L., Jr. (1984). More on the Feynman's disk paradox. *Am. J. Phys.*, **52**, 756–757.

Bose, S. K. and Scott, G. K. (1985). On the magnetic field of current through a hollow cylinder. *Am. J. Phys.*, **53**, 584–586.

Crandall, R. E. (1983). Photon mass experiment. *Am. J. Phys.*, **51**, 698–702.

Crawford, F. S. (1992). Mutual inductance $M_{12} = M_{21}$: an elementary derivation. *Am. J. Phys.*, **60**, 186.

Crosignani, B. and Di Porto, P. (1977). Energy of a charge system in an equilibrium configuration. *Am. J. Phys.*, **45**, 876.

Davis, L., Jr., Goldhaber, A. S., and Nieto, M. M. (1975). Limit on the photon mass deduced from Pioneer-10 observations of Jupiter's magnetic field. *Phys. Rev. Lett.*, **35**, 1402–1405.

Faraday, M. (1839). *Experimental Researches in Electricity* (London: R. and J. E. Taylor).

Feynman, R. P., Leighton, R. B., and Sands, M. (1977). *The Feynman Lectures on Physics*, vol. II (Reading, MA: Addision-Wesley).

Friedberg, R. (1993). The electrostatics and magnetostatics of a conducting disk. *Am. J. Phys.*, **61**, 1084–1096.

Galili, I. and Goihbarg, E. (2005). Energy transfer in electrical circuits: a qualitative account. *Am. J. Phys.*, **73**, 141–144.

Goldhaber, A. S. and Nieto, M. M. (1971). Terrestrial and extraterrestrial limits on the photon mass. *Rev. Mod. Phys.*, **43**, 277–296.

Good, R. H. (1997). Comment on "Charge density on a conducting needle," by David J. Griffiths and Ye Li [*Am. J. Phys.* **64**(6), 706–714 (1996)]. *Am. J. Phys.*, **65**, 155–156.

Griffiths, D. J. and Heald, M. A. (1991). Time-dependent generalizations of the Biot–Savart and Coulomb laws. *Am. J. Phys.*, **59**, 111–117.

Griffiths, D. J. and Li, Y. (1996). Charge density on a conducting needle. *Am. J. Phys.*, **64**, 706–714.

Hughes, V. W. (1964). In Chieu, H. Y. and Hoffman, W. F. (eds.), *Gravitation and Relativity*. (New York: W. A. Benjamin), chap. 13.

Jackson, J. D. (1996). Surface charges on circuit wires and resistors play three roles. *Am. J. Phys.*, **64**, 855–870.

Jefimenko, O. (1962). Demonstration of the electric fields of current-carrying conductors. *Am. J. Phys.*, **30**, 19–21.

King, J. G. (1960). Search for a small charge carried by molecules. *Phys. Rev. Lett.*, **5**, 562–565.

Macaulay, D. (1998). *The New Way Things Work* (Boston: Houghton Mifflin).

Marcus, A. (1941). The electric field associated with a steady current in long cylindrical conductor. *Am. J. Phys.*, **9**, 225–226.

Maxwell, J. C. (1891). *Treatise on Electricity and Magnetism*, vol. I, 3rd edn. (Oxford: Oxford University Press), chap. VII. (Reprinted New York: Dover, 1954.)

Mermin, N. D. (1984a). Relativity without light. *Am. J. Phys.*, **52**, 119–124.

Mermin, N. D. (1984b). Letter to the editor. *Am. J. Phys.*, **52**, 967.

Nan-Xian, C. (1981). Symmetry between inside and outside effects of an electrostatic shielding. *Am. J. Phys.*, **49**, 280–281.

O'Dell, S. L. and Zia, R. K. P. (1986). Classical and semiclassical diamagnetism: a critique of treatment in elementary texts. *Am. J. Phys.*, **54**, 32–35.

Page, L. (1912). A derivation of the fundamental relations of electrodynamics from those of electrostatics. *Am. J. Sci.*, **34**, 57–68.

Press, F. and Siever, R. (1978). *Earth*, 2nd edn. (New York: W. H. Freeman).

Priestly, J. (1767). *The History and Present State of Electricity*, vol. II, London.

Roberts, D. (1983). How batteries work: a gravitational analog. *Am. J. Phys.*, **51**, 829–831.

Romer, R. H. (1982). What do "voltmeters" measure? Faraday's law in a multiply connected region. *Am. J. Phys.*, **50**, 1089–1093.

Semon, M. D. and Taylor, J. R. (1996). Thoughts on the magnetic vector potential. *Am. J. Phys.*, **64**, 1361–1369.

Smyth, C. P. (1955). *Dielectric Behavior and Structure* (New York: McGraw-Hill).

Varney, R. N. and Fisher, L. H. (1980). Electromotive force: Volta's forgotten concept. *Am. J. Phys.*, **48**, 405–408.

Waage, H. M. (1964). Projection of electrostatic field lines. *Am. J. Phys.*, **32**, 388.

Whittaker, E. T. (1960). *A History of the Theories of Aether and Electricity*, vol. I (New York: Harper), p. 266.

Williams, E. R., Faller, J. E., and Hill, H. A. (1971). New experimental test of Coulomb's law: a laboratory upper limit on the photon rest mass. *Phys. Rev. Lett.*, **26**, 721–724.

# Index